Constants

UNIVERSAL GAS CONSTANT

$\cdot$ °R

'R

STANDARD ACCELERATION OF GRAVITY

$$g = \begin{cases} 9.80665 \text{ m/s}^2 \\ 32.174 \text{ ft/s}^2 \end{cases}$$

STANDARD ATMOSPHERIC PRESSURE

$$1 \text{ atm} = \begin{cases} 101.325 \text{ kPa} \\ 14.696 \text{ lbf/in.}^2 \end{cases}$$

TEMPERATURE RELATIONS

$$T(°R) = 1.8\, T(K)$$
$$T(°C) = T(K) - 273.15$$
$$T(°F) = T(°R) - 459.67$$

STEFAN-BOLTZMANN CONSTANT

$$\sigma = 5.670 \times 10^{-8} \frac{W}{m^2 \cdot K^4}$$

$$= 0.1714 \times 10^{-8} \frac{Btu}{hr \cdot ft^2 \cdot °R^4}$$

Thermal Environmental Engineering

Third Edition

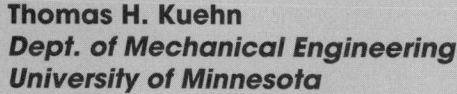

Thomas H. Kuehn
Dept. of Mechanical Engineering
University of Minnesota

James W. Ramsey
Dept. of Mechanical Engineering
University of Minnesota

James L. Threlkeld

Prentice Hall, *Upper Saddle River, New Jersey 07458*

Contents

Contents

Preface

This book is intended for use by advanced undergraduate or graduate students in mechanical engineering and as a basic reference for practicing mechanical engineers. Principally, the book covers refrigeration, psychrometrics, solar radiation, heating and cooling loads in buildings, and design of duct and piping systems. Theory and analysis are emphasized throughout.

It is intended that this book may serve as a text for a course of about two semesters in length; for shorter courses, certain chapters may be deleted without loss of continuity. The reader is expected to have a background in basic engineering thermodynamics, fluid mechanics, and heat transfer.

The authors have retained the emphasis on basic principles that was the philosophy used by Professor Threlkeld in earlier editions of this book. Current design guidelines practiced in the United States are introduced where appropriate. The reader is often given a choice of methods to apply to a particular problem.

Changes to the previous editions include the use of SI units in addition to English units in tables, figures, worked examples, and homework problems. A substantial number of new worked examples and problems have been added. The heat-transfer review has been expanded, and updated correlations have been included in Chapter 2. In Chapters 3 and 4, refrigerant HCFC-22 has replaced CFC-12 as the fluorocarbon refrigerant. Ammonia has been retained as the second refrigerant. Newer replacement refrigerants have been added and environmental issues are discussed. The section on refrigeration compressors covers all of the types commonly used in addition to the traditional reciprocating designs. In the psychrometric section, additional emphasis has been placed on psychrometric processes and applications. More recent humidity-measurement techniques have been added to the discussion of wet-bulb psychrometers, and updated calibration and standard measurement methods are covered.

In Chapter 11, the number-of-transfer-units (NTU) method has been added to the log-mean-temperature-difference (LMTD) method for heat-exchanger design. Updated cooling-tower design information and an approximate sizing method have been included in Chapter 10. Chapter 12 contains mostly new material on human thermal comfort and indoor air quality. The ASHRAE clear sky solar-radiation model has been included in Chapter 13. The discussion of winter design heat loss has been significantly enhanced to include framing correction methods, below-grade heat transfer, and infiltration estimation methods. Moisture transport through structures has been added. Chapters 15 and 16 expand upon the previous analytical approach to heat-gain and cooling-load estimation by including total-equivalent-temperature-difference (TETD) and cooling-load-temperature-difference (CLTD) methods and computer calculations of heat gain and cooling loads in buildings. Most of the material in Chapter 17 is new, with degree day, bin, and computerized energy-calculation methods covered. Chapters 18 and 19 have been added to provide the reader with background on hydronic system design, air motion in rooms, duct design, and pump and fan selection.

It is our intent that this text provide the fundamental background necessary to understand the operation of HVAC&R systems. With a good understanding of the principles, the reader, with some experience, should become a proficient engineer in this field.

Acknowledgments

The authors wish to acknowledge Dr. Hongmei Liang for generating most of the thermodynamic property tables and charts used in the appendix. We also wish to thank many colleagues and friends who have stimulated discussion on various points over the years and who have encouraged us to complete this revision. Finally, and most importantly, we thank Professor Threlkeld, whose classic earlier editions provided the framework and philosophy for this text.

Thomas H. Kuehn
James W. Ramsey

part I
Introduction

chapter 1
Introduction to Heating, Ventilating, Air Conditioning, and Refrigeration

1.1 HISTORICAL OVERVIEW

The title of this book, *Thermal Environmental Engineering,* refers to the science and practice of controlling an environmental condition through the use of thermal processes and systems. Included in the design of an environmental control system are the end use or application considered, the environmental parameters to be controlled, the energy usage rates necessary to provide this control, the processes involved, equipment, systems, and controls. The designer has latitude in many of the choices that need to be made along the way toward a final design. As in most engineering fields, considerable prior experience has been incorporated into recommended practice, standard methods, and codes.

An acronym commonly used in this field is HVAC&R, which stands for Heating, Ventilating, Air Conditioning, and Refrigeration. A brief description and background is given on each of these topics.

Heating

Space heating has been used by humans probably since the controlled use of fire to provide more comfortable conditions than in the natural environment. Fire provided not only a convenient method of cooking but space heating as well. Heating occurred from thermal radiation from the fire and from the combustion byproducts of unvented fires. Early open fires were later placed into fireplaces to better remove the emissions from the living space. In warm climates the cooking fires were sometimes located in a separate building to prevent space heating.

The Koreans invented a floor radiant heating system [1], called the *Ondol* system. Hot gases from the cooking fire were passed under large flat rocks that formed the floor to a chimney at the other end of the building. The rocks provided not only radiant heat but also thermal storage at night when the fire was not maintained. Modern versions of the Ondol system are still very popular in Korea.

Further refinements to the use of combustion as the heat source included the invention of the Franklin stove and the introduction of coal as the fuel. Later technology allowed the heat from combustion to be conveyed to the heated space by natural air convection rather than by direct radiation from the fire enclosure. Thus the space-heating furnace was born. Later forced-air circulation was developed into the forced air furnace with fuel oil or natural gas as the primary fuels. Steam boilers were also developed with single-pipe heating systems for large buildings. Steam and hot-water boilers are used today for many large space-heating applications.

Ventilation

The smoke associated with the fires used for cooking and space heating was irritating when not properly removed from the occupied space. Several means were developed to improve the introduction of outside air and the removal of these unwanted byproducts. The term *ventilation* came to be associated with the introduction of clean outside air. King Charles I of England decreed in the year 1600 that the minimum ceiling height in a house must be 10 ft (3 m) and that windows had to be higher than they were wide to assist the introduction of outdoor air and the removal of smoke [2]. Janssen [3] indicated that a Cornish mining engineer, T. Tredgold, published his minimum requirement of outside air per person of about 4 ft^3/min (2 L/s) in 1836. Both the reduction of noxious odors and the prevention of the spread of disease were factors in the development of a ventilation standard. It was not until 1893 that the American Society of Heating and Ventilating Engineers (ASHVE) adopted 30 ft^3/min (14 L/s) per person as the minimum ventilation rate. Since that time the minimum ventilation rate has varied considerably, as shown in Fig. 1.1.

Air Conditioning

The layman takes the term *air conditioning* to refer to cooling provided in a building on a hot day. This term is defined more broadly for the HVAC engineer as the provision of conditioned air to a space. The air supplied should have controlled temperature, humidity, and airborne-contaminant concentration.

The application of air conditioning to a building in the form of cooling and dehumidification was first developed and widely promoted by Willis Carrier, who is often called "the father of air conditioning" [4]. Carrier's work followed the development of mechanical vapor-compression refrigeration machines in the early twentieth century and better understanding of humidity-control processes. Carrier also laid the foundation for the present-day understanding of moist-air mixtures, the science of psychrometrics, in a paper presented at the 1911 ASME meeting [5]. However, it was not until the 1960s that central cooling systems were installed in commercial buildings on a large scale.

Today, air conditioning encompasses the control of temperature, humidity, air cleanliness, and air circulation in a space as required by the occupants or materials. A complete air conditioning system includes heating, cooling, moisture addition and removal, and air cleaning by removal of gas and particles. Outdoor air must be provided in a minimum amount as determined by the use of the space. Air distribution within the space must ensure that conditioned air reaches all critical regions in the space.

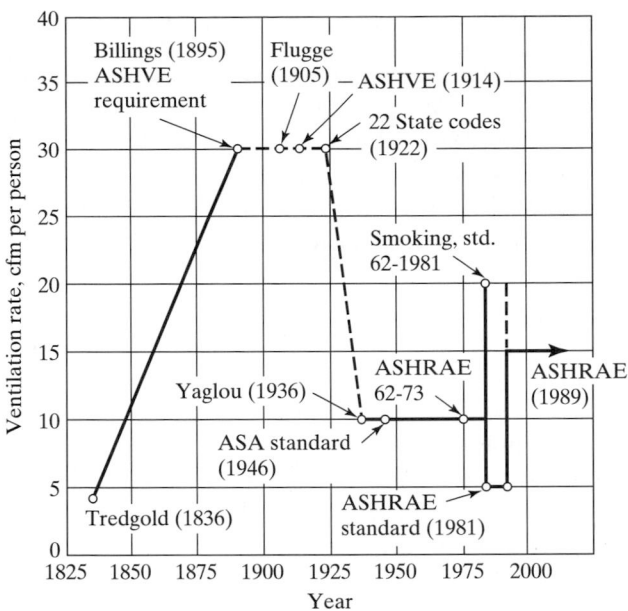

Figure 1.1 Recommended minimum ventilation rate over the past one and one-half centuries. [Reprinted by permission from *ASHRAE Journal*, 36:8 (1994), 127.]

Refrigeration

Refrigeration of food to retard spoilage has been practiced for millennia. The Chinese, Egyptians, and Indians are all known to have made ice in shallow clay vessels by leaving the water-filled containers outside overnight in the cold desert environment. The Romans recommended covering certain dishes, such as prawns, with snow, probably as a method of refrigeration. The addition of certain chemicals, such as saltpeter, to water was known to depress the solution temperature below ambient. Naturally formed ice was widely used to preserve food in ice boxes until the middle of the twentieth century. Ice harvesting and shipping was a major industry in the New England states of the United States.

Vapor-compression refrigeration was first developed by physicians who were seeking a method to cool fever patients. Ether was the first refrigerant used, followed by ammonia, using reciprocating compressors. By 1900 large steam-engine-driven ammonia refrigeration plants were manufacturing "artificial ice" that was distributed to residential and commercial ice boxes. In the 1920s the availability of electric power and the development of smaller refrigeration systems led to residential vapor-compression systems. The General Electric OC-2 unit used sulfur dioxide as the refrigerant [6].

Widespread use of small refrigeration systems did not occur until safe refrigerants were developed. This occurred in the early 1930s, when Thomas Midgley at DuPont developed CFC-12 and other halocarbon refrigerants. Other developments include the introduction of absorption machines and other types of compressors such as the rotary, screw, and scroll designs.

In the 1980s the environmental issues, ozone depletion, and greenhouse effects associated with fully halogenated refrigerants sent the industry scrambling for alternative refrigerants [7]. Newer halocarbon refrigerants, including HFC-134a, are now used instead of CFC-12 and others found to have negative environmental impacts.

Food preservation and storage were the primary motivating factors in the early development of refrigeration. That legacy has not been lost. In the United States more than 90 percent of the industrial refrigeration system capacity is associated with food processing, storage, and transport.

1.2 OVERVIEW OF HVAC SYSTEM DESIGN

Applications

The application or use of a space dictates the level of the necessary environmental variables (e.g., temperature). There are many possible applications to consider.

Human occupancy is a very important application. In the developed countries approximately 90 percent of a person's time is spent indoors. The indoor environment must provide safe, comfortable conditions for the occupants. This is usually accomplished in a building. Types of buildings include residences, schools, commercial office buildings, and industrial facilities. Special considerations are needed in applications such as hospital operating rooms and bone-marrow-transplant recovery rooms, where the patient's immune system is not capable of defeating airborne fungal spore infection. Modes of transportation are also important and include automobiles, buses, aircraft, and spacecraft.

Animals constitute another inhabitant of buildings. The widespread use of animal-confinement buildings has lead to the development of environmental conditions most suitable for animal welfare and weight gain in production applications. Zoos, laboratory animal buildings, and others also have special requirements.

Vegetation is the primary occupant of greenhouses. Wholesale and retail florist facilities require different temperature, humidity, and gas control than human- or animal-occupied buildings.

Preservation and storage facilities require special environmental conditions. Food warehouses, transport, and retail display cases comprise a large need for refrigeration. Museums require very careful control of temperature, humidity, sunlight, and airborne contaminants to prevent degradation of artwork and other historic objects.

Special industrial processes have specific needs. These are becoming more common as industrial manufacturing processes need tighter tolerances. One example is the control of the environment of a semiconductor manufacturing clean room. With computer chip feature sizes in the submicron range, small changes in temperature causing differential thermal expansion of overlaying masks can have devastating effects on the yield of a process. Humidity control is also very important, as is the control of airborne particulate matter and reactive gases.

Indoor Environmental Parameters

In all of the applications listed above, a set of basic parameters should be controlled in the conditioned space. The relative importance of each parameter, the desired setpoint, and the allowable variation about the mean are all application specific. These parameters include:

Temperature—both dry-bulb and mean radiant

Humidity

Air motion

Airborne contaminants—both gaseous and particulate

Discussions of temperature and humidity are given in Chapters 7 (on psychrometrics) and 12. Air motion in rooms is discussed in Chapter 18. Airborne-contaminant control is described in Chapter 12.

Heating and Cooling Loads

The difference in indoor and outdoor levels of temperature and humidity drives energy between the indoor and outdoor environments. Ventilation air requirements, solar radiation, and internal heat gains from occupants, lights, and equipment also contribute to energy gains or losses. Figure 1.2 is a flowchart of the information needed and the sequence used to determine the energy use of an occupied building. The building function dictates the level of the indoor environmental parameters. The location prescribes the local climate and weather information. These, together with the building envelope construction, provide the heating and cooling energy loads through the envelope. Additional loads are imposed by the internal heat and moisture gains. Requirements to condition the ventilation air add another portion of the total building heating and cooling system load. The ventilation load can be greatly reduced by using some form of heat recovery between the exhaust and makeup air streams. The procedure for calculating these loads during periods of peak demand (winter and summer design loads) is discussed in detail in Chapters 14, 15, and 16. The basic procedure is developed for forced-air heating and cooling systems in occupied buildings and should be used with care with other types of systems and applications.

Processes

Some basic air-treatment processes are required to condition the air to meet the requirements discussed above. The basic processes are:

1. Sensible heating or cooling
2. Latent heating or cooling (moisture addition or removal)
3. Gas and vapor removal
4. Particulate removal

Processes 1 and 2 are often combined, as in the case of cooling and dehumidification. Others, such as mixing, are needed in the entire system. These are psychrometric processes and are covered in Chapter 8.

Processes 3 and 4 are air-cleaning processes and are discussed in Chapter 12. Air cleaning alone is not sufficient. The cleaned air must be adequately distributed to the building occupants. This topic is covered in Chapters 12 and 18. This general subject area is termed *indoor air quality* (*IAQ*).

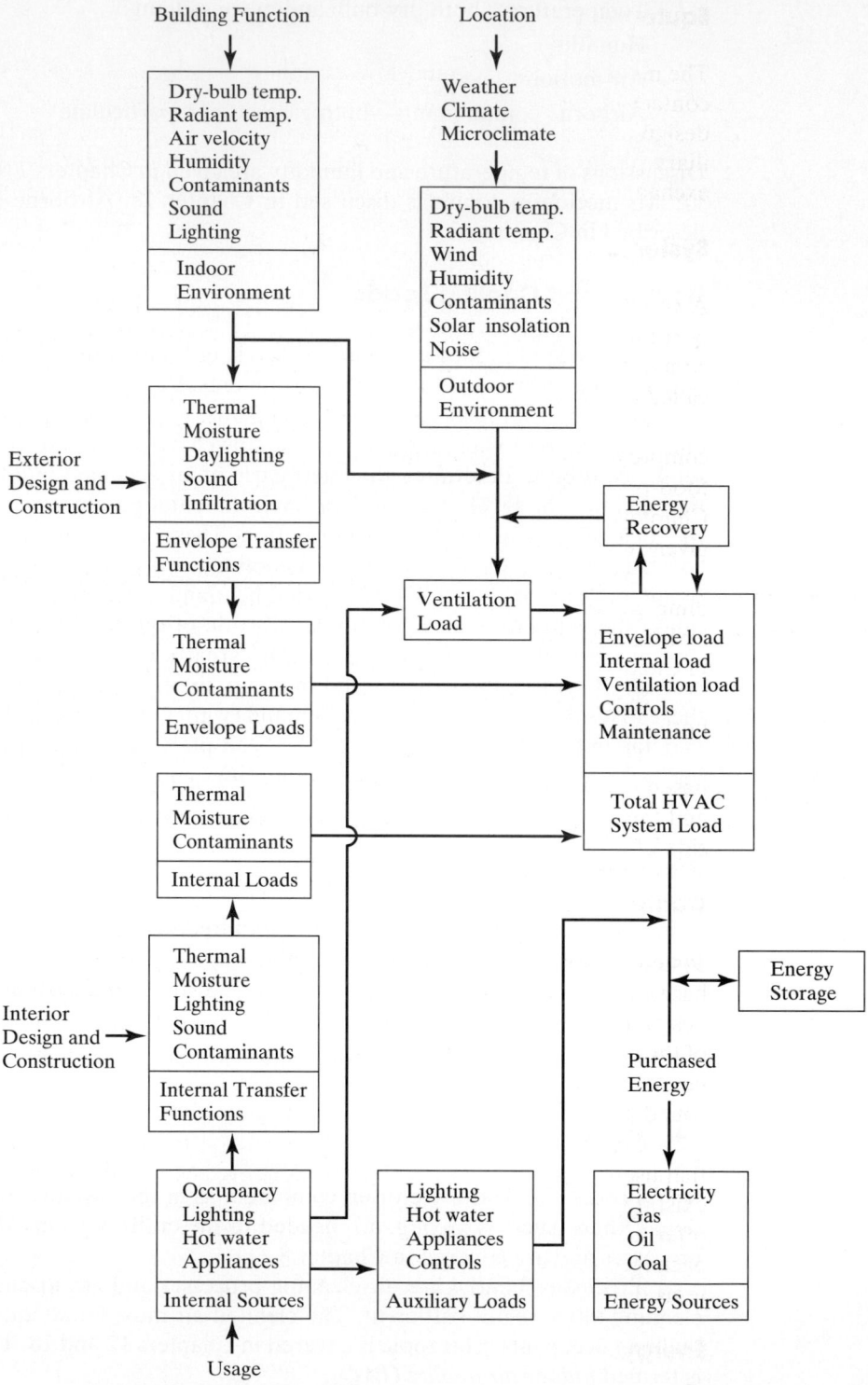

Figure 1.2 Flowchart illustrating the procedure to estimate the energy requirements of an occupied building.

Equipment

The main pieces of equipment needed for these processes include heat exchangers, direct-contact humidifiers and air washers, and gas and particulate filters. The operation and design of these devices are discussed in Chapters 11, 10, and 12, respectively. Other auxiliary pieces of equipment also needed are boilers and furnaces (special types of heat exchangers), pumps, and fans. Topical coverage is included in Chapters 11, 18, and 19.

Systems

A *system* is required to interconnect the pieces of equipment described above and transport the air, water, steam, or refrigerant to and from the components. Some systems are completely assembled at the factory except for external utility connections. These are called *packaged* or *unitary* systems. Larger systems are assembled on site.

One common system is a refrigeration system. This is usually a mechanical vapor-compression or an absorption system, although cryogenic systems are widely used in the food-processing industry. The principles of operation of these systems are given in Chapters 3, 5, and 6, with a more detailed discussion of vapor-compression systems given in Chapter 4.

Other types include hydronic systems to transport hot or chilled water and air-handling systems to condition, supply, and remove air from the controlled space. Design of hydronic systems is discussed in Chapter 19. Types of air-handling systems are described in Chapter 8, with duct design included in Chapter 18.

Some system types include both air and water transport, so that elements of both are used.

The energy requirements of the system in an occupied building are estimated based on the envelope and ventilation loads, the equipment performance information, and the system configuration. The total HVAC system load is shown on the right-hand side of Fig. 1.2.

Controls

System design including equipment sizing, pipe and ductwork sizing, and energy use, is based primarily on peak or design loads. However, the system rarely operates at the design load. Most of the time the system is at a part-load condition, when the capacities of the central equipment and fluid flowrates can be much less than the peak or design values. Adequate controls are necessary to provide controlled environmental conditions year round and to minimize the annual energy use and operating cost of the system.

Control systems have traditionally been very simple and robust. Pneumatic controls that use compressed air have been widely used and are found in a significant share of the existing building stock. Newer direct digital control systems have been developed that offer more options, such as reprogramming and offsite monitoring capabilities. Pieces of packaged equipment such as refrigeration chillers often have built-in digital control systems that can be tied into the overall building control and monitoring system.

Energy Use and Cost

The heating, ventilating, and air conditioning system with the appropriate controls will require a certain amount of purchased energy on an annual basis. The peak instantaneous demand relates to a demand charge, and the total energy use can be converted with the appropriate rate structure to an annual energy-usage charge. The lower right-hand portion of Fig. 1.2 illustrates the conversion from system energy use to purchased energy cost.

The energy-demand charge may be significantly reduced by incorporating energy storage into the building or by relaxing the indoor environmental conditions for a few hours near the time of peak demand. Some thermal storage is inherent in the building structure itself. This can shift the peak cooling load into the evening or night hours. Additional energy storage can be added in the form of chilled-water or hot-water storage. The storage system can be charged during times of low rate such as with off-peak electric power rates at night, and discharged during peak demand times during the day.

Energy estimation methods are discussed in Chapter 17. The simplest method for estimating space-heating energy use is the degree-day method, which is based strictly on the average daily outdoor air temperature. The classical bin method is also discussed, which is more appropriate for heating systems such as air-to-air heat pumps. The heating capacity of a heat pump is strongly affected by the outdoor air temperature. An introduction to more sophisticated computer simulation methods is also given.

1.3 SUMMARY

We have introduced the concepts of thermal environmental engineering and the science and engineering knowledge and skills necessary to analyze and design systems for various applications. Complete information necessary to design a particular system is not included in this text. Instead, the fundamentals are emphasized. However, representative tables and charts have been included to provide the reader with some basic tools with which to begin a system design.

Additional design information can be obtained from manuals and design guides, such as the ASHRAE handbook series [8]. Additional historical information is given in the historical articles published in the *ASHRAE Journal* and in the *ASHRAE Transactions* in the years 1994 and 1995.

ENDNOTES

1. S.-D. Park, "A Review on Ondol Heating System and the Thermal Performance," *Proc. Int. Ondol Conference,* Seoul, Korea, July 28–31, pp. 20–40, 1996.

2. J. Woods, "Air Quality," *Encyclopedia of Architecture: Design Engineering and Construction,* Vol. 1 (New York: John Wiley & Sons, 1988).

3. J. E. Janssen, "The V in ASHRAE: An Historical Perspective," *ASHRAE J.*, 36:8 (August 1994), 126–132.

4. Willis Carrier, *"Father of Air Conditioning"* (Louisville, KY: Fetter Printing Co., 1991).

5. H. J. Sauer, Jr. and R. H. Howell, *Principles of Heating, Ventilating and Air Conditioning* (Atlanta: ASHRAE, 1990).

6. W. L. Holladay, "The General Electric Monitor Top Refrigerator," *ASHRAE J.*, 36:9 (1994), 49–55.

7. M. O. McLinden and D. A. Didion, "Quest for Alternatives," *ASHRAE J.*, 29:12 (1987), 32–42.

8. ASHRAE Handbook Series (Atlanta: American Society of Heating, Refrigerating and Air Conditioning Engineers).

chapter 2
Review of Thermodynamics, Fluid Mechanics, and Heat Transfer

2.1 INTRODUCTION

This book is concerned with the control of the thermal environment within enclosed spaces and therefore involves transfers of heat, mass, and work. It is assumed throughout that the reader has completed courses in engineering thermodynamics, fluid mechanics, and heat transfer of about one year's length. This chapter reviews a few principles in these areas of thermal science pertinent to air conditioning and refrigeration.

Thermodynamic concepts are used extensively when analyzing systems, both in terms of conservation of mass and energy and in evaluating the thermodynamic properties of the working substances. We will review the necessary equations and property formulations used throughout this book.

Energy transfers in the form of heat are encountered in many areas of thermal environmental engineering. Examples include the heat transfer from heating or cooling heat exchangers, heat transfer through building envelopes, and heat transfer from occupants. We must be able to predict the rates of heat transfer for many of these applications from basic concepts and correlations. We will review the essentials of heat conduction, single-phase forced and natural convection, boiling, condensation, and thermal radiation.

2.2 THERMODYNAMIC PROPERTIES

A *thermodynamic property* is any observable, measurable, or calculable characteristic of a substance which depends only upon the state of the substance. In an air conditioning system, moist air is always a working substance, but associated systems may involve refrigerants, steam, water, and other fluids. In order to study an air conditioning or refrigeration system, we must always define the state of the working substance in terms of its thermodynamic properties.

2.3 SYSTEMS OF UNITS

Two systems of units are used throughout this book, the pound-second-foot or English system and the kilogram-second-meter system or Système Internationale (SI). In the English system the unit of mass is the pound mass, which is determined by reference to a standard quantity of material. The unit of force is the pound force, which is the force required to accelerate the pound mass at the standard acceleration of gravity of 32.17 ft/sec². Mass, force, and acceleration are related by Newton's second law of motion

$$F = kma \qquad (2.1)$$

where F = force in pounds (lbf),

        m = mass in pounds (lbm),

        a = acceleration of mass m (ft/sec²),

        k = proportionality constant (lbf · sec²/lbm · ft).

The weight of a body in pounds, which is the force of gravity upon the body at its location, is numerically equal to its mass in pounds at the earth's surface, if k in Eq. (1.1) is set equal to 1/32.17. Thus,

$$F = \frac{ma}{32.17} \qquad (2.2)$$

and the weight W of a body at any location where the acceleration of gravity is g is

$$W = \frac{mg}{32.17} \qquad (2.3)$$

Even at high altitudes g differs from 32.17 by only a fraction of 1 percent. Thus, for practical purposes, weight in pounds and mass in pounds are numerically equal.

In the SI system of units, the unit of mass is the kilogram and the unit of force is the newton, which is the force required to accelerate the kilogram mass at the standard acceleration of gravity of 9.807 m/sec². The force given by Newton's second law becomes

$$F = \frac{ma}{9.807} \qquad (2.4)$$

where F = force in newtons (N), m = mass in kilograms (kg), and a = acceleration in m/sec².

2.4 SPECIFIC VOLUME, DENSITY, SPECIFIC WEIGHT, AND SPECIFIC GRAVITY

The *specific volume* of a substance is its volume per unit mass and is expressed in ft³/lbm or m³/kg. The *density* of a substance is its mass per unit volume expressed in lbm/ft³ or kg/m³ and is the reciprocal of specific volume. The *specific weight* of a substance is its weight per unit volume measured in lbm/ft³ or N/m³.

The terms *density* and *specific weight* are often confused. Density involves mass, whereas specific weight involves force. In the pound-second-foot system of units, as discussed in Sec. 2.3, density and specific weight are numerically equal for practical purposes, provided g in Eq. (2.3) is equal or almost equal to 32.17. The distinction between density and specific weight should always be recognized.

The specific volume and density of a vapor or gas are affected by both pressure and temperature, and values must be found from tables of properties, or often, in the case of gases, calculated. The density of liquids is, to some extent, affected by pressure but is usually taken as a function of temperature only.

The specific gravity of a liquid is defined as the ratio of the weight of the given liquid to the weight of an equal volume of water at some standard temperature, usually 60 °F or 20 °C. For practical purposes, the specific gravity of a liquid is its specific weight in lbf/ft^3 divided by 62.4 or its specific weight in N/m^3 divided by 9800.

2.5 PRESSURE

The pressure of a substance is the force it exerts per unit area. In engineering, pressures may be designated as absolute, gauge, or vacuum. *Absolute pressure* is a thermodynamic property. If the absolute pressure is higher than atmospheric or barometric pressure, the *gauge pressure* is the difference between the absolute pressure and atmospheric pressure. If the absolute pressure is less than atmospheric pressure, the *vacuum pressure* is the difference between atmospheric pressure and the absolute pressure.

A variety of units are used for expressing pressure, those commonly used being pounds per square foot, pounds per square inch, inches of mercury, feet or inches of water, and pascals. Since the specific weight of liquids varies with temperature, pressure in terms of liquid height implies a standard liquid specific weight. Taking the specific weight of water as 62.4 lbf/ft^3 and the specific gravity of mercury as 13.6, we have the following conversion relations:

   1 psi = 144 psf = 2.04 in. Hg = 27.7 in. water = 6.895 kPa

   1 in. Hg = 0.491 psi = 13.6 in. water = 3.38 kPa

   1 in. water = 0.0361 psi = 0.0735 in. Hg = 0.249 kPa

   1 standard atmosphere = 14.696 psia = 29.921 in. Hg = 101.33 kPa

   1 kPa = 0.145 psi = 0.296 in. Hg = 4.02 in. water

If fluids other than mercury or water are used for measuring pressure heads, or if mercury or water is used at temperatures other than standard temperatures, a useful relation is

$$z_1 w_1 = z_2 w_2 \tag{2.5}$$

where z_1 and z_2 are heights of the two fluids for respective specific weights w_1 and w_2. Consistent units must be used for z and w.

2.6 TEMPERATURE

The temperature of a substance may be expressed in either relative or absolute units. On the Fahrenheit scale, the temperature of melting ice is arbitrarily called 32 degrees, and the temperature of boiling water at 14.696 psia is called 212 degrees. On the Celsius scale, similar markings are 0 degrees for melting ice and 100 degrees for boiling water. Conversions between scales are given by

$$t_F = \frac{9}{5} t_C + 32 \tag{2.6}$$

$$t_C = \frac{5}{9} (t_F - 32) \tag{2.7}$$

By means of the second law of thermodynamics we can prove that there is a "lowest conceivable temperature." This temperature is absolute zero, and any temperature reckoned above it is an absolute temperature. The absolute Fahrenheit scale (commonly called the Rankine scale) and the absolute Celsius scale (or Kelvin scale) are used for expressing absolute temperatures. Absolute zero occurs at -459.67 degrees on the ordinary Fahrenheit scale and at -273.15 degrees on the ordinary Celsius scale. Absolute temperature on the two scales may be found by

$$T_F = t_F + 459.67 \tag{2.8}$$

$$T_C = t_C + 273.15 \tag{2.9}$$

In approximate calculations, the constants used are 460 and 273.

2.7   HEAT AND POWER UNITS

Heat is energy transferred from one body to another because of a temperature difference. The units used to quantitatively express heat are the British thermal unit (Btu) and the kilojoule (kJ). In ordinary use, one Btu is defined as the quantity of heat required to raise the temperature of one pound of water through one degree Fahrenheit.

The *specific heat* of a substance is the heat required to raise the temperature of unit mass of the substance one degree. The units used are Btu/lbm · °F or kJ/kg · °C. The specific heat of most substances varies with temperature, but over limited temperature ranges average values may be used.

In the case of gases, two specific heats are of importance, specific heat at constant pressure, c_p, and at constant volume, c_v. In air conditioning work, values of c_p are usually needed to compute specific enthalpy changes.

Calculation of heat quantities involving specific heats is given by

$$Q = mc(t_2 - t_1) \tag{2.10}$$

where Q = heat transfer in Btu (kJ), m = mass in lbm (kg), c = specific heat in Btu/lbm · °F (kJ/kg · °C), and $(t_2 - t_1)$ = temperature change in °F (°C).

In air conditioning work, two types of heat are often considered. *Sensible heat* is heat added to a substance resulting in a temperature rise; *latent heat* is heat associated with a change of phase at constant temperature. The heat necessary to melt a solid without change of temperature is called the *latent heat of fusion,* while the heat required to vaporize a liquid without temperature change is called the *latent heat of vaporization.*

Work is energy transferred between a system and its surroundings when either of them exerts a force that acts on the other through a distance. *Power* is defined as the time rate at which work is performed. The basic units for expressing work are the ft · lbf and

N · m. Power may be expressed in terms of ft · lbf/min, horsepower, or kilowatts. The following conversion units are useful:

$$1 \text{ hp} = 33{,}000 \text{ ft} \cdot \text{lbf/min} = 2545 \text{ Btu/hr} = 0.746 \text{ kW}$$

$$1 \text{ kW} = 3413 \text{ Btu/hr} = 1.34 \text{ hp}$$

A *ton* of refrigeration is the removal of heat at a rate of 200 Btu/min, 12,000 Btu/hr, or 3.52 kW. These units are customarily used for expressing the capacity of refrigeration systems.

2.8   FIRST AND SECOND LAWS OF THERMODYNAMICS

The science of engineering thermodynamics is based upon two empirical principles called the first and second laws of thermodynamics. Neither of these principles can be proved, but since no exceptions to them have ever been observed, they are accepted as correct.

The *first law of thermodynamics* states that if any system undergoes a process during which work or heat is added to or removed from it, none of the energy added is destroyed within the system and none of the energy removed is created within the system. According to the first law, heat and work are interconvertible.

It is a matter of experience that heat will not flow spontaneously out of one system into another at higher temperature. In order to transfer heat to a higher temperature level, a refrigerating machine is needed, wherein energy from an external source is supplied. These statements embody the principles of the *second law of thermodynamics*, which, according to Clausius, states that it is impossible for a self-acting machine, unaided by any external agency, to transfer heat from one body to another at higher temperature.

2.9   INTERNAL ENERGY, ENTHALPY, AND ENTROPY

In many thermodynamic processes the amount of energy added to a system in the form of work or heat is not equal to the heat and/or work removed from it. Therefore, since the first law of thermodynamics states that it is impossible to create or destroy energy, it must be possible to store energy in a fluid. By this reasoning we obtain the concept of the property of internal energy. The *internal energy* of a substance includes all types of energy stored within its molecules in both kinetic and potential form.

The *enthalpy* of a substance is a composite energy term defined by the following equation:

$$h = u + Pv \tag{2.11}$$

where h = enthalpy, u = internal energy, P = pressure, and v = specific volume. The importance of enthalpy is due to its presence in all steady-flow energy-balance relations.

As a consequence of the second law of thermodynamics, we derive the concept of a property called entropy. *Entropy s* is defined by the following equation:

$$ds = \frac{dQ}{T} \quad \text{(reversible process)}$$

where ds = differential change in entropy, dQ = differential heat reversibly added, and T = absolute temperature. For a finite amount of heat added,

$$s_2 - s_1 = \int_1^2 \frac{dQ}{T} \tag{2.12}$$

Entropy, like enthalpy, is a mathematical property and is not evident by direct measurement. In engineering, entropy is useful in the solution of problems involving isothermal or adiabatic reversible processes. In more advanced phases of thermodynamics, entropy is used as a criterion of equilibrium.

2.10 THE PERFECT GAS

An *equation of state* expresses the relationships among pressure, specific volume, and temperature of a substance. It must be determined at least in part by experiment. Such equations are available for many fluids. In the case of a perfect gas, an important special equation of state is:

$$Pv = RT \qquad (2.13)$$

where P = absolute pressure, v = specific volume, T = absolute temperature, and R = gas constant. The gas constant R varies with different gases and is approximately

$$R = \frac{1545}{M} \frac{\text{ft} \cdot \text{lbf}}{\text{lbm} \cdot {}^\circ\text{R}} = \frac{8.314}{M} \frac{\text{kJ}}{\text{kg} \cdot \text{K}} \qquad (2.14)$$

where M is the molecular weight of the gas.

Equation (2.13) is satisfactory for real gases at relatively high temperatures and at low pressures. In air conditioning work Eq. (2.13) is a valuable tool, since moist air behaves very much like a perfect gas.

Several useful relationships involving specific heats may be derived for perfect gases. For any process, we may prove that the change of internal energy is a sole function of temperature, given by

$$u_2 - u_1 = c_v(t_2 - t_1) \qquad (2.15)$$

Likewise, the change of enthalpy of a perfect gas is, for any process,

$$h_2 - h_1 = c_p(t_2 - t_1) \qquad (2.16)$$

A useful relationship between c_p and c_v for a perfect gas is

$$c_p - c_v = R \qquad (2.17)$$

Numerous other relations for perfect gases may be derived from fundamental thermodynamic concepts. For complete discussion and derivation, reference should be made to any standard text on engineering thermodynamics.

2.11 MIXTURES OF PERFECT GASES

The air conditioning engineer is continually involved with gaseous mixtures. This section discusses some basic concepts regarding mixtures of perfect gases.

Let us first consider a given volume of a mixture of two perfect gases x and y. Each gas occupies the entire volume V, and each gas is at the same temperature T. Since we are dealing with perfect gases, there are no interactions between them, and each separately follows Eq. (2.13). The following relations will then apply:

$$m = m_x + m_y$$

$$V = V_x = V_y$$

$$T = T_x = T_y$$

$$P = P_x + P_y$$

Since each gas is assumed to behave as if the other gas were not present, by Eq. (2.13) we have

$$P_x V = m_x R_x T \tag{2.18}$$

$$P_y V = m_y R_y T \tag{2.19}$$

and for the mixture of the two gases,

$$PV = mRT \tag{2.20}$$

Thus, from Eqs. (2.18)–(2.20), we have

$$R = \frac{m_x R_x + m_y R_y}{m} \tag{2.21}$$

The gas constant of the perfect-gas mixture is thus the weighted mean of the gas constants of the components.

When gases are mixed adiabatically without work being done, the first law of thermodynamics requires that the enthalpy of the system remain constant. Thus, we may write

$$h = \frac{m_x h_x + m_y h_y}{m} \tag{2.22}$$

If Eq. (2.22) is written in differential form,

$$m\, dh = m_x\, dh_x + m_y\, dh_y$$

and since, for perfect gases,

$$dh = c_p\, dT$$

we have the following relation for the specific heat c_p of the mixture of gases:

$$c_p = \frac{m_x c_{px} + m_y c_{py}}{m} \tag{2.23}$$

2.12 DRY AIR

A clean atmosphere is composed of dry air and water (principally in vapor form). On a volume basis, dry air contains 78.084 percent nitrogen, 20.948 percent oxygen, and traces of approximately 15 other gases. The composition of dry air is nearly uniform over the earth's surface and is scarcely changed by altitude up to a height of 100 miles or more.

The molecular weight of dry air is 28.9645 and its gas-law constant R is 53.352 ft · lbf/lbm · °R (0.2870 kJ/kg · K). Precise measurements have shown that Eq. (2.13) is only approximately correct for dry air. At sufficiently low pressures and high temperatures the correlation is excellent. For other conditions the departure from perfect-gas behavior may be considerable.

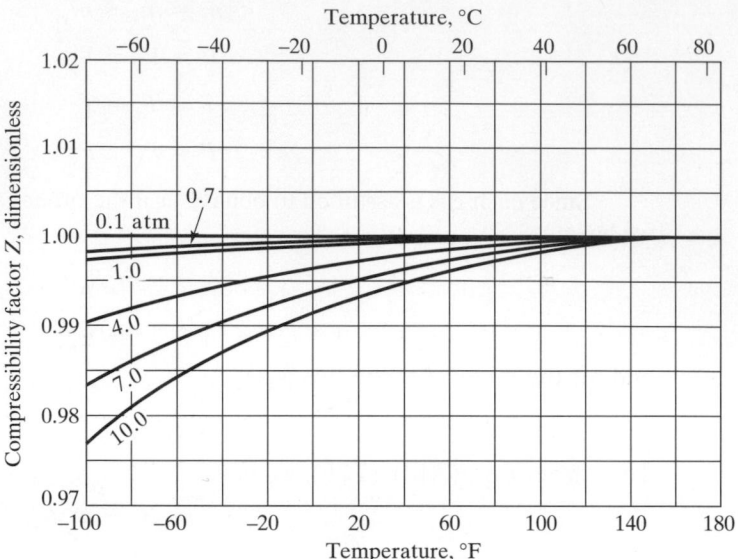

Figure 2.1 Compressibility factor for dry air.

A reliable equation of state for real gases may be derived from fundamental considerations of statistical mechanics. This equation, called a *virial equation of state*, is given by the power series

$$\frac{Pv}{RT} = 1 + A_2(T)P + A_3(T)P^2 + \cdots = Z \qquad (2.24)$$

The sum of the series Z is called the *compressibility factor*. The coefficients $A_2(T)$, $A_3(T)$, etc. are called *virial coefficients* and represent the departure from the perfect-gas law caused by reactions between molecules. These coefficients are functions of temperature.

The National Bureau of Standards publication *Tables of Thermal Properties of Gases* [1] shows accurately determined thermodynamic properties for dry air and several other gases. It was the source of information for Figs. 2.1 and 2.2. Figure 2.1 shows the compressibility factor for dry air at various pressures and temperatures. At ordinary atmospheric pressures, Z is essentially unity in the range of −100 °F to 180 °F (−70 °C to 80 °C). Thus, for this range, Eq. (2.13) is an excellent equation of state for dry air. Figure 2.2 shows the specific heat c_p for dry air as a function of pressure and temperature. At ordinary atmospheric pressures, c_p varies from about 0.240 to 0.242 Btu/lbm · °F (0.558 to 0.563 kJ/kg · °C) in the range of −100 °F to 250 °F (−70 °C to 120 °C).

2.13 PROPERTIES OF WATER AND STEAM

A thorough understanding of the properties of ice, water, and water vapor (steam) is essential for air conditioning engineers. Water vapor is a constituent of the atmosphere, and a knowledge of its characteristics is required for almost all air conditioning calculations. Steam and hot water are common media for heating buildings, while chilled water is a popular means of cooling atmospheric air in summer. The air conditioning engineer may be involved, to a lesser extent, with solid water or ice.

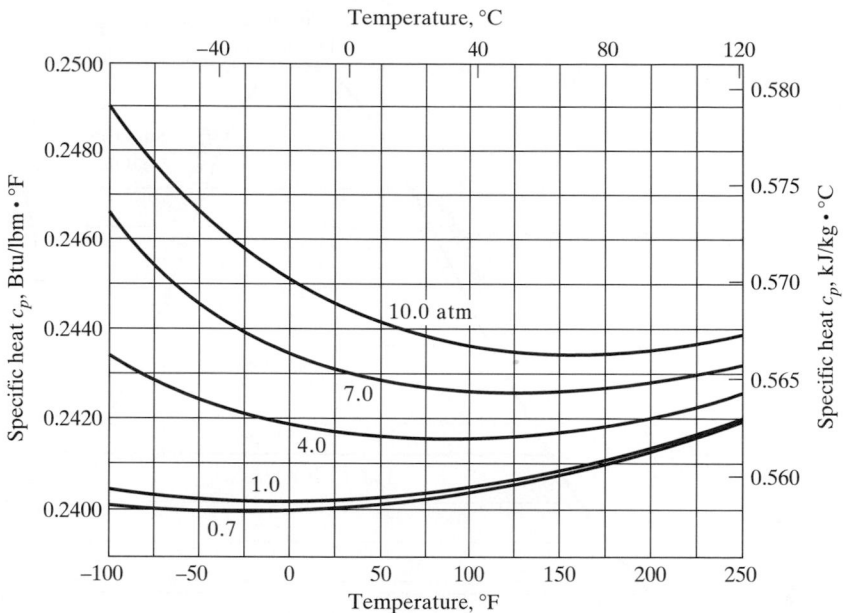

Figure 2.2 Specific heat c_p for dry air.

Thermodynamic properties of water and steam may be found from tables or charts based on experimental data or from real-gas equations of state that are useful in computer calculations. Table A.1E* shows thermodynamic properties of water and steam in English units for the saturation temperature range from −150 °F to 500 °F and Table A.1SI shows these properties in SI units from −100 °C to 250 °C. These data were computed using the algorithms from Hyland and Wexler [2] from −150 °F (−100 °C) to 32 °F (0 °C) and from Reynolds [3] from 32 °F (0 °C) to 500 °F (250 °C).

If water (or any other pure fluid) is heated at constant pressure, the temperature at which it boils is called its *saturation temperature*. Tables A.1E and A.1SI show, for various saturation temperatures, the pressure, the specific volume of saturated liquid v_f, saturated ice v_i, and saturated vapor v_g; the enthalpy of saturated liquid h_f, saturated ice h_i, and saturated vapor h_g; and the entropy of saturated liquid s_f, saturated ice s_i, and saturated vapor s_g. Enthalpy and entropy values in Tables A.1E and A.1SI are reckoned above an arbitrary zero for saturated liquid at 32 °F (0 °C).

Figure 2.3 shows properties of water and steam on temperature-entropy coordinates; it was adapted from a diagram in Keenan, Keyes, Hill, and Moore, *Steam Tables* [4]. The saturated-liquid line is a locus of points representing liquid water at the boiling temperature for various pressures. The saturated-vapor line is a locus of points showing steam at the condensation temperature for various pressures. Any state point located between the two saturation lines, such as 1 or 2, is a mixture of saturated water and saturated steam with each component at the same temperature. Any state point to the right of the saturated-vapor line, such as 3, is superheated, since its temperature is higher than the saturation temperature corresponding to its pressure. The diagram reaches a peak at the critical point of 705.4 °F (374.1 °C) and 3206.2 psia (22.09 MPa). The critical point is the maximum saturation condition at which evaporation and condensation may occur.

---

*All tables with the letter A preceding the number are in the Appendix.

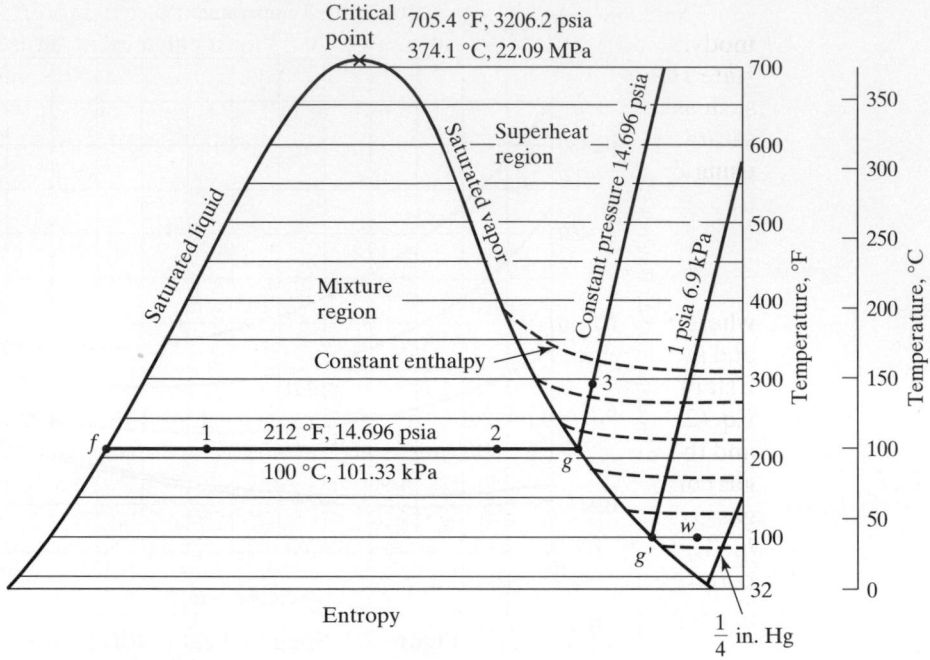

Figure 2.3 Temperature-entropy diagram for water.

The specific volume, enthalpy, and entropy of a mixture state may be found if the *quality x*, in pounds of saturated vapor per pound of mixture, is known. For each pound of mixture there are x pounds of vapor and $(1 - x)$ pounds of liquid. The following equations are then evident:

$$v = (1 - x)v_f + xv_g$$

$$h = (1 - x)h_f + xh_g$$

$$s = (1 - x)s_f + xs_g$$

The above equations may be changed into the following forms:

$$v = v_g - (1 - x)v_{fg} \tag{2.25}$$

$$h = h_g - (1 - x)h_{fg} \tag{2.26}$$

$$s = s_g - (1 - x)s_{fg} \tag{2.27}$$

or alternatively:

$$v = v_f + xv_{fg} \tag{2.28}$$

$$h = h_f + xh_{fg} \tag{2.29}$$

$$s = s_f + xs_{fg} \tag{2.30}$$

All properties with the subscript *fg* are values equal to the difference between properties with subscripts *g* and *f*. In the case of enthalpy, h_{fg} is also called the latent heat of vaporization.

In Eqs. (2.25) and (2.28), since v_f is very small compared to v_g, it is usually satisfactory to use the approximation

$$v = xv_g \tag{2.31}$$

When computer algorithms rather than tabular or graphical data are used for thermodynamic property determination, it is often convenient to use a real-gas equation of state rather than Eq. (2.13) to evaluate the P-v-T properties and additional properties such as h, u, and s for saturated and superheated vapor. The enthalpy difference between saturated liquid and saturated vapor at a specified temperature, $h_{fg}(T)$, can be evaluated using the Clapeyron equation

$$h_{fg}(T) = T v_{fg}(T) \left(\frac{dP_{\text{sat}}}{dT_{\text{sat}}} \right)_T \tag{2.32}$$

where $v_f(T)$ and $(dP_{\text{sat}}/dT_{\text{sat}})_T$ must be known independently of the real-gas correlation and are given for H_2O by Hyland and Wexler and for a variety of substances by Reynolds. Establishing a reference enthalpy such as $h_f(32\ °F) = 0$ or $h_f(0\ °C) = 0$ and using Eq. (2.32), one can evaluate the enthalpy of saturated vapor at the reference temperature and therefore set the enthalpy at any saturated- or superheated-vapor state using the real gas correlations. Values for h_f at other temperatures may also be computed using the real-gas correlations to determine h_g at the specified temperature and Eq. (2.32) to compute h_f. The entropy change between saturated liquid and saturated vapor at a specified temperature, $s_{fg}(T)$, can be computed from

$$s_{fg}(T) = \frac{h_{fg}(T)}{T} \tag{2.33}$$

Establishing a reference entropy such as $s_f(32\ °F) = 0$ or $s_f(0\ °C) = 0$ and using Eq. (2.33) and the real-gas correlations uniquely establishes values of entropy for all saturated-liquid, saturated-vapor, and superheated-vapor states.

2.14 LOW-PRESSURE WATER VAPOR

The properties of low-pressure water vapor are of particular significance in air conditioning, since water vapor existing in the atmosphere typically exerts a pressure less than one psia (6.9 kPa). At these low pressures, observations have shown that water vapor closely exhibits perfect-gas behavior.

Equation (2.24) may also be used to represent the equation of state for low-pressure water vapor. However, we do not need to analyze the virial equation of state, since extensive steam-table data and correlating equations exist.

Figure 2.3 shows several lines of constant enthalpy in the superheat region. At pressures of about one atmosphere, enthalpy lines are distinctly curved close to the saturated-vapor line. However, as the pressure is reduced, we see that the enthalpy lines progressively flatten, and at pressures less than about one psia (6.9 kPa) the enthalpy of superheated vapor is very nearly equal to the enthalpy of saturated vapor at the same temperature. Thus, for any superheat state w as shown in Fig. 2.3, where the pressure is less than about one psia or 6.9 kPa,

$$h_w = h_{g'} \tag{2.34}$$

Figure 2.4 shows a plot of the enthalpy of low-pressure water vapor as a function of temperature. The plotted values were taken from Keenan, Keyes, Hill, and Moore, *Steam*

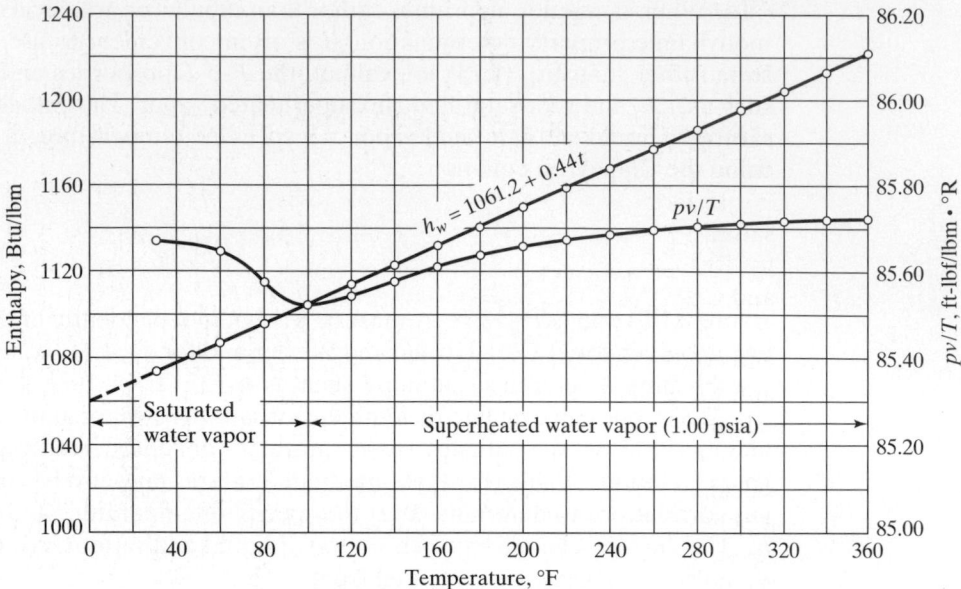

Figure 2.4 Approximate perfect-gas behavior of low-pressure water vapor.

Tables [4]. We see that a straight line fits the plotted points quite well. The equation of this line evaluated from 32 °F to 200 °F is approximately

$$h_w = 1061 + 0.44t \text{ Btu/lbm}$$
$$= h_g(0 \text{ °F}) + c_{p,w}t \tag{2.35}$$

In SI units the equation between 0 °C and 100 °C is approximately

$$h_w = 2501 + 186t \text{ kJ/kg}$$
$$= h_g(0 \text{ °C}) + c_{p,w}t \tag{2.36}$$

If low-pressure water vapor behaved as a perfect gas, its gas-law constant by Eq. (2.14) would be

$$R_w = \frac{1545}{18.015} = 85.76 \text{ ft} \cdot \text{lbf/lbm} \cdot \text{°R}$$

$$= \frac{8.314}{18.015} = 0.4615 \text{ kJ/kg} \cdot \text{K}$$

Figure 2.4 shows a plot of Pv/T for water vapor using Keenan, Keyes, Hill, and Moore values. Saturated-vapor values are shown up to 100 °F, and at higher temperatures superheat values at 1 psia are plotted. In the temperature range of 32 °F to 200 °F (0 °C to 100 °C), excellent correlation is obtained at a value of Pv/T equal to 85.6 ft · lbf/lbm · °R (0.461 kJ/kg · K). Thus, we have further proof that at pressures of less than about one psia (6 kPa), water vapor exhibits approximate perfect-gas behavior. Although use of Eq. (2.13) with water vapor is not accurate enough for many thermodynamic applications, nevertheless it is a valid and valuable method for many problems in air conditioning.

2.15 THERMODYNAMIC PROPERTIES OF REFRIGERANTS

Broadly, we may define a *refrigerant* as any substance which absorbs heat from another substance or from a space. However, we will restrict our discussion here to phase-change refrigerants. There are many available refrigerants, two of the most common being chlorodifluoromethane (Refrigerant 22) and ammonia.

Tables A.2E and A.2SI show thermodynamic properties for saturated-liquid and saturated-vapor states for ammonia, and Tables A.3E and A.3 SI show similar properties for Refrigerant 22. Figures C-1E and C-1 SI are *P-h* diagrams for ammonia. Figures C-2E and C-2SI are *P-h* diagrams for Refrigerant 22.

All phase-change refrigerants behave similarly in a qualitative sense. Nomenclature and equations given in Sec. 2.13 for water and steam apply to any phase-change refrigerant. Figure 2.3, although restricted quantitatively to water, represents qualitatively and schematically a *T-s* diagram for any phase-change refrigerant.

The most common diagram of thermodynamic properties used in refrigeration-cycle calculations is the *P-h* diagram. However, the *T-s* diagram is highly useful, particularly as a schematic tool. From the definition of entropy, it follows that the area under a reversible-process line on *T-s* coordinates represents the heat added or withdrawn during the process. Figure 2.5 shows a schematic case where a superheated vapor (state 1) is condensed to a saturated liquid (state 2).

By the general property relation

$$dh = T \, ds + v \, dP$$

it follows that the change of specific enthalpy Δh for a reversible process is equal to the area under the process line on *T-s* coordinates if the pressure is constant. Thus, assuming constant pressure for the process 1-2 in Fig. 2.5, the area under the process line is also equal to $(h_1 - h_2)$.

A frequent problem in refrigeration-cycle calculations is that of evaluating the properties of a subcooled liquid. In Fig. 2.5, suppose that the saturated liquid at state 2 is further cooled at constant pressure to some subcooled condition (state 3). We now consider

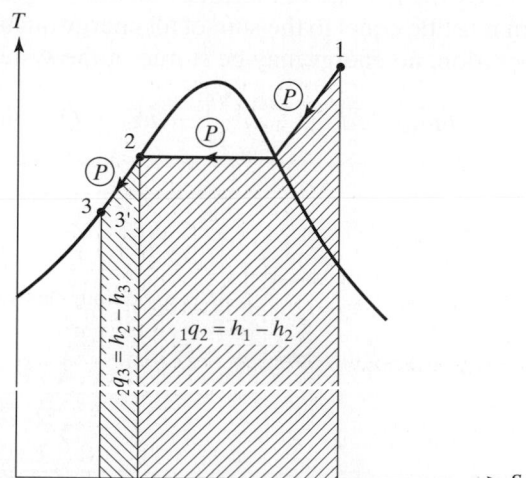

Figure 2.5 Schematic heat removal processes on *T-s* coordinates.

how the thermodynamic properties at state 3 differ from those of a saturated liquid at the same temperature (state 3'). If we assume that the liquid is incompressible and that the internal energy is a function of temperature only, from the general property relation it follows that, since

$$T\,ds = du + P\,dv$$

then $s_3 = s_{3'}$. Therefore, states 3 and 3' are coincident on the T-s diagram. The enthalpy h_3 of the subcooled state may be calculated by

$$h_3 = h_{3'} + (P_3 - P_{3'})v_{3'} \qquad (2.37)$$

However, for subcooled refrigerant liquids, the term $(P_3 - P_{3'})v_{3'}$ is usually so small compared to $h_{3'}$ that it may be neglected. Thus, for most cases, the enthalpy of a subcooled liquid may be assumed equal to that of the saturated liquid at the same temperature.

2.16 THE STEADY-FLOW ENERGY EQUATION

Most of the thermodynamic processes considered in this book are of a steady-flow nature. Usually these processes are approximately steady with respect to time. The steady-flow energy equation is a powerful tool in the solution of many problems throughout this text and, because of its importance, we will review it here.

Figure 2.6 shows a schematic thermodynamic system to illustrate the steady-flow process. Fluid at a constant rate of $\dot{m}$ crosses the boundaries of the system at section 1 in a uniform state P_1, t_1, v_1, etc. The fluid leaves the system at section 2 at the same rate and in a uniform state. Also crossing the boundaries of the system between sections 1 and 2 may be heat at a constant rate of $\dot{Q}$ (assumed inward) and shaft work at a constant rate of $\dot{W}$ (assumed outward). The thermodynamic state of the fluid at any location within the system is constant with respect to time.

For each incremental mass of fluid that crosses the boundaries of the system, certain energy quantities are convected by the fluid. These energy quantities, written as energy per unit mass, are internal energy u, flow work Pv, potential energy due to elevation z, and kinetic energy due to velocity of flow $\mathbf{V}^2/2g$.

By the first law of thermodynamics the sum of all energy quantities entering the system must be equal to the sum of all energy quantities leaving the system, since, for steady operation, no energy may be stored in the system. Thus:

$$\dot{m}u_1 + \dot{m}P_1v_1 + \dot{m}\frac{\mathbf{V}_1^2}{2g} + \dot{m}z_1 + \dot{Q} = \dot{m}u_2 + \dot{m}P_2v_2 + \dot{m}\frac{\mathbf{V}_2^2}{2g} + \dot{m}z_2 + \dot{W}$$

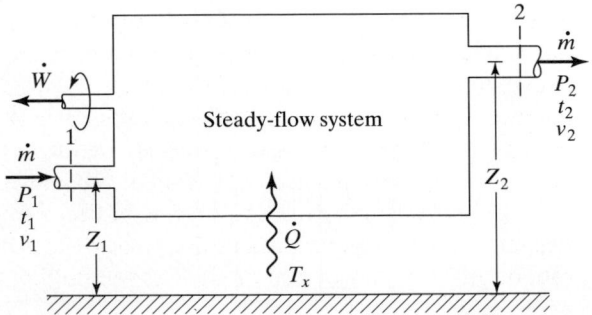

Figure 2.6 Schematic thermodynamic system for steady flow.

Since by definition, specific enthalpy $h = u + Pv$, we have

$$\dot{m}h_1 + \dot{m}\,\frac{\mathbf{V}_1^2}{2g} + \dot{m}z_1 + \dot{Q} = \dot{m}h_2 + \dot{m}\,\frac{\mathbf{V}_2^2}{2g} + \dot{m}z_2 + \dot{W} \qquad (2.38)$$

Equation (2.38) is the general steady-flow energy equation on a unit-time basis. The use of this system of units is particularly advantageous if multiple streams of fluid enter and/or leave the system. Thus, if in Fig. 2.6 one fluid stream $\dot{m}_1$ entered the system and two streams $\dot{m}_2$ and $\dot{m}_3$ left the system, we would have the mass balance and the energy balance

$$\dot{m}_1 = \dot{m}_2 + \dot{m}_3 \qquad (2.39)$$

$$\dot{m}_1\left(h_1 + \frac{\mathbf{V}_1^2}{2g} + z_1\right) + \dot{Q} = \dot{m}_2\left(h_2 + \frac{\mathbf{V}_2^2}{2g} + z_2\right) + \dot{m}_3\left(h_3 + \frac{\mathbf{V}_3^2}{2g} + z_3\right) + \dot{W} \qquad (2.40)$$

The use of units of energy per unit mass is often more convenient if only a single stream enters and leaves the system. If each term of Eq. (2.38) is divided by $\dot{m}$, we have

$$h_1 + \frac{\mathbf{V}_1^2}{2g} + z_1 + q = h_2 + \frac{\mathbf{V}_2^2}{2g} + z_2 + w \qquad (2.41)$$

where q and w are, respectively, heat added and work removed per unit mass of fluid.

The relative importance of individual terms in Eq. (2.38) or (2.41) depends on the steady-flow problem to be analyzed. In most problems some terms are usually negligible or nonexistent. Because of the wide variety of problems for which the steady-flow energy equation applies, there is no general rule which defines the negligible terms. Experience shows that in many problems the potential-energy term $(z_2 - z_1)$ is negligible. In heating devices such as steam boilers, warm-air furnaces, and duct-type coils the potential- and kinetic-energy terms are usually negligible and the work term is zero. In fluid flow through nozzles and venturi meters the work and heat terms are zero and the potential-energy term is negligible but the kinetic-energy term is significant. In many locations throughout this book the steady-flow energy equation will be used and its application to special problems will be discussed.

2.17 THE STEADY-FLOW ENTROPY EQUATION

In addition to the steady-flow mass and energy equations discussed in Sec. 2.16, the steady-flow entropy equation is useful to compute the entropy production or irreversibility of a given process. This is not necessary for system analysis but provides additional insight into the nature of the processes in the system. The process with the largest irreversibility impairs the performance of a work-producing or work-consuming device the most. This is the process for which improved design or retrofit may result in the largest improvement in system performance. Therefore the entropy equation serves as a tool to quantify losses using a second-law-of-thermodynamics approach.

In Figure 2.6, by the second law of thermodynamics, the sum of all entropy quantities leaving the system minus the entropy quantities entering the system must equal the rate of entropy production or the irreversibility rate. Thus

$$\dot{m}s_2 = \dot{m}s_1 + \frac{\dot{Q}}{T_x} + \dot{I} \qquad (2.42)$$

where T_x is the absolute temperature of the heat source or sink and $\dot{I}$ is the rate of entropy production or the irreversibility rate. If one stream $\dot{m}_1$ entered and two streams $\dot{m}_2$ and $\dot{m}_3$ left the system, and heat $\dot{Q}_A$ was added from T_A, and $\dot{Q}_B$ removed to T_B with $\dot{Q}_A - \dot{Q}_B = \dot{Q}$, we would have

$$\dot{m}_2 s_2 + \dot{m}_3 s_3 = \dot{m}_1 s_1 + \frac{\dot{Q}_A}{T_A} - \frac{\dot{Q}_B}{T_B} + \dot{I} \tag{2.43}$$

Thus Eqs. (2.39), (2.40), and (2.43) form the set of governing equations for the mass, energy, and entropy balances, respectively.

The sum of all the irreversibility rates in a system, $\dot{I}_{\text{total}}$, such as a refrigerator or heat pump, multiplied by the ambient absolute temperature, T_0, is equal to the additional rate of high-grade energy, exergy, or work required above the minimum required by a completely reversible system operating between the same temperature levels with the same capacity. Thus the actual power required by a refrigeration system can be evaluated by

$$\dot{W}_{\text{actual}} = \dot{W}_{\text{min}} + T_0 \dot{I}_{\text{total}} \tag{2.44}$$

where $\dot{W}_{\text{min}}$ is usually evaluated from a Carnot refrigerator operating between the same two fixed temperatures as the actual system with the same refrigerating capacity and $\dot{I}_{\text{total}}$ is the sum of the irreversibility rates in all system components evaluated using equations similar to Eqs. (2.42) and (2.43).

2.18 HEAT CONDUCTION

Conduction is the transfer of heat through a material by molecular diffusion due to a temperature gradient. Solutions of thermal conduction problems are based on the empirical expression

$$\dot{Q}_x = -kA \frac{dt}{dx} \tag{2.45}$$

called Fourier's law. The proportionality factor, k, is called the *thermal conductivity*. Equation (2.45) expresses the heat-transfer rate in the x direction resulting from the temperature gradient that exists in that direction. In general, the direction of heat flow is normal to a surface of constant temperature, called an *isothermal surface*. Thus, in order to calculate the heat-transfer rate in any direction within a material it is necessary to determine the temperature distribution. That is, once the temperature distribution is known, the conduction heat-transfer rate at any point in the material or on its surface may be computed from Fourier's law. The three-dimensional, transient heat-diffusion equation for calculating the temperature distribution in Cartesian coordinates is

$$\frac{\partial}{\partial x}\left(k\frac{\partial t}{\partial x}\right) + \frac{\partial}{\partial y}\left(k\frac{\partial t}{\partial y}\right) + \frac{\partial}{\partial z}\left(k\frac{\partial t}{\partial z}\right) + \dot{Q}' = \rho c_p \frac{\partial t}{\partial \tau} \tag{2.46}$$

where $\dot{Q}'$ is the internal heat-generation rate per unit volume and τ is time.

In cylindrical coordinates, r, ϕ, z, the transient diffusion equation is

$$\frac{1}{r}\frac{\partial}{\partial r}\left(kr\frac{\partial t}{\partial r}\right) + \frac{1}{r^2}\frac{\partial}{\partial \phi}\left(k\frac{\partial t}{\partial \phi}\right) + \frac{\partial}{\partial z}\left(k\frac{\partial t}{\partial z}\right) + \dot{Q}' = \rho c_p \frac{\partial t}{\partial \tau} \tag{2.47}$$

In analyzing heat exchangers and estimating heat losses through pipe insulation, a special case of Eq. (2.47) is used in which there is a temperature gradient only in the radial direction, the thermal conductivity is constant, there is no internal heat generation, and steady-state conditions exist. For this set of conditions Eq. (2.47) becomes

$$\frac{k}{r}\frac{\partial}{\partial r}\left(r\frac{\partial t}{\partial r}\right) = 0 \tag{2.48}$$

and the resulting heat-transfer rate per unit length for an annulus of inner and outer radii r_i and r_o with inner and outer surface temperatures t_i and t_o is

$$\frac{\dot{Q}}{L} = \frac{2\pi k (t_i - t_o)}{\ln (r_o/r_i)} \tag{2.49}$$

2.19 FORCED CONVECTION FOR INTERNAL FLOWS

Forced convection is the transfer of thermal energy by means of large-scale fluid motion. Most forced-convection applications that we are concerned with occur at a solid-fluid interface. Forced convection occurs in the outdoor environment by wind and water motion. However, most engineering applications achieve forced convection by using a pump or fan to move the fluid.

One very common example is flow of a fluid through a circular pipe or duct. The rate of heat transfer between the fluid and the wall is governed by the nature of the fluid motion and the temperature difference. In fully developed laminar flow, the fluid molecules do not move in the radial direction, toward or away from the wall, so heat transfer with the wall is mainly by diffusion or conduction. However, when the flow becomes turbulent, the eddies move fluid in the direction normal to the wall, which greatly increases the rate of heat transfer. When the flow is in a long tube or duct and is in the fully developed region, the nature of the flow depends on the Reynolds number:

Laminar flow: $0 \le \mathrm{Re}_D \le 2200$
Turbulent flow: $\mathrm{Re}_D \ge 10{,}000$

where the Reynolds number is $\mathrm{Re}_D = \mathbf{V}D/\nu$; $\mathbf{V}$ is the average velocity of the fluid, D is the inside diameter of a round tube or pipe or the hydraulic diameter of a noncircular duct, $D_h = 4A/p$ (A is the cross-sectional area of the duct and p is the wetted perimeter of the duct), and ν is the kinematic viscosity of the fluid. For values of Reynolds number between 2200 and 10,000 the flow is in the transition region between laminar and fully turbulent flow. In HVAC equipment we wish to have the largest heat-transfer coefficient possible, so we design flows in pipes and ducts to be turbulent to take advantage of the larger heat-transfer coefficient.

The convective heat transfer between a fluid and a solid surface can be described by

$$\dot{Q} = hA(t_1 - t_2) \tag{2.50}$$

where h is the convective heat-transfer coefficient, A is the surface area of the interface, and t_1 and t_2 are the temperatures of the fluid and solid surface. In many applications the dimensionless heat-transfer coefficient, the Nusselt number, can be related to the fluid motion through the Reynolds number and Prandtl number:

$$\mathrm{Nu} = C\,\mathrm{Re}_D^m\,\mathrm{Pr}^n \tag{2.51}$$

or

$$h\frac{D}{k} = C\left(\frac{\mathbf{V}D}{\nu}\right)^m \left(\frac{\mu c_p}{k}\right)^n \tag{2.52}$$

McAdams [5] recommended the values; $C = 0.023$, $m = 0.8$, and $n = 0.4$, which are still regarded as satisfactory for many turbulent flows in tubes and pipes. Others recommend that $n = 0.4$ when the fluid is being heated and $n = 0.3$ when the fluid is being cooled, which accounts for the effect of variable fluid properties. The fluid properties in Eq. (2.52) should be determined at the fluid bulk temperature. The necessary thermophysical properties for air, water, R-22, and ammonia can be found in Tables A.5, A.6, A.8, and A.9, respectively. Values for other fluids can be found in the *ASHRAE Handbook of Fundamentals* [6] and many heat-transfer textbooks.

2.20 FORCED CONVECTION FOR EXTERNAL FLOWS

Forced convection is important from the exterior surfaces of walls and roofs, as wind is often the driving force for the convective heat transfer. Forced convection is also employed in heat-exchanger coils to transfer energy from one fluid stream to another. The three most common configurations of external forced convection encountered in environmental engineering are the flat plate, cylinder, and sphere.

Flow over a *flat plate* begins with a laminar boundary layer that becomes turbulent when the Reynolds number reaches about 5×10^5. In most of our applications the flow is turbulent. A correlation for the average Nusselt number on a flat plate for $\mathrm{Re} > 2 \times 10^5$ is given by Ozisik [7] as

$$\mathrm{Nu}_L = \frac{h_m L}{k} = 0.036 \, \mathrm{Pr}^{0.43} \left(\mathrm{Re}_L^{0.8} - 9200\right)\left(\frac{\mu_\infty}{\mu_w}\right)^{0.25} \tag{2.53}$$

where the length parameter in the Nusselt and Reynolds numbers is the length of the plate in the direction of the flow, μ_∞ is the fluid dynamic viscosity in the free stream, and μ_w is the viscosity at the wall temperature. The remaining fluid properties are evaluated at the wall temperature. For gases, the viscosity correction term is often neglected.

Flow perpendicular to a circular cylinder is found for flow across piping and ducts, simple heat exchangers, and often is an approximation to flow over the human body. Whitacker [8] correlated the average heat-transfer coefficient for flow around a single cylinder for $40 < \mathrm{Re}_D < 10^5$ by

$$\mathrm{Nu}_D = \frac{h_m D}{k} = \left(0.4 \, \mathrm{Re}_D^{0.5} + 0.06 \, \mathrm{Re}_D^{2/3}\right) \mathrm{Pr}^{0.4}\left(\frac{\mu_\infty}{\mu_w}\right)^{0.25} \tag{2.54}$$

where the characteristic length is the cylinder diameter. The fluid property values are evaluated at the free-stream temperature, and the viscosity correction term can be neglected for gases.

Heat transfer from a sphere reduces to conduction when the fluid velocity approaches zero. Therefore correlations for forced convection from a sphere often incorporate the conduction limit; $\mathrm{Nu}_D = 2$. The correlation given by Whitacker (1972) is valid in the range $3.5 < \mathrm{Re}_D < 8 \times 10^4$ for most gases and liquids. The average Nusselt number for a sphere is obtained by adding 2.0 onto Eq. (2.54) given above. At large Reynolds numbers the addition of 2.0 does not change the Nusselt number significantly from the value for the cylinder.

2.21 NATURAL CONVECTION FOR EXTERNAL FLOWS

Natural convection is an important means of heat transfer inside buildings, where thermal buoyancy is often the main driving force for the air flow. Temperature gradients are established on cold wall and window surfaces, hot walls and roofs, adjacent to human bodies, and on most electric equipment. In some ventilation methods, natural convection is relied upon to convect airborne contaminants away from the source.

The most common configuration encountered is a *vertical flat plate*. As with forced convection on a flat plate, the boundary layer near the beginning of the plate may be laminar, but the majority of the flow in the boundary layer in our applications is turbulent. The heat transfer is governed by the Rayleigh number rather than the Reynolds number. The boundary layer is laminar when $Ra < 10^8$ and is mainly turbulent when $Ra > 10^9$. An estimate of the average heat-transfer coefficient on a vertical flat plate valid for $Ra > 10^3$ is given by

$$\text{Nu}_H = \frac{h_m H}{k} = \left[\left(0.67 \, \text{Ra}_H^{1/4} \left[1 + \left(\frac{0.559}{\text{Pr}} \right)^{3/5} \right]^{-5/12} \right)^{15} + (0.1 \, \text{Ra}_H^{1/3})^{15} \right]^{-1/15} \quad (2.55)$$

The first term in this equation proportional to $Ra^{1/4}$ represents the heat transfer by laminar flow. The second term proportional to $Ra^{1/3}$ represents the heat transfer by turbulent flow. The powers of 15 are used to smooth between these two limiting cases. When the flow is in the turbulent regime, as in most building applications, Eq. (2.55) reduces to

$$\text{Nu}_H = \frac{h_m H}{k} = 0.1 \, \text{Ra}_H^{1/3} = 0.1 \left(\frac{g\beta H^3 \Delta t}{\nu \alpha} \right)^{1/3} \quad (2.56)$$

An interesting result of this correlation is that the mean heat-transfer coefficient, h_m, is independent of the surface height, H. An estimate of the heat-transfer coefficient for an inclined surface can be obtained by replacing Ra_H in Eqs. (2.55) and (2.56) by $Ra_H \cos \theta$, where θ is the angle from the vertical. However, the angle of inclination should not exceed 60 degrees.

Another common configuration is natural convection from a horizontal cylinder. A correlation valid for $Ra_D > 10^{-2}$ is

$$\text{Nu}_D = \frac{h_m D}{k} = \frac{1}{\ln \left[1 + \dfrac{2}{\left[\left(0.518 \, \text{Ra}_D^{1/4} \left[1 + \left(\frac{0.559}{\text{Pr}} \right)^{3/5} \right]^{-5/12} \right)^{15} + (0.1 \, \text{Ra}_D^{1/3})^{15} \right]^{1/15}} \right]} \quad (2.57)$$

The term containing $Ra_D^{1/4}$ represents the laminar-flow correlation, and the term with $Ra_D^{1/3}$ is valid for turbulent flow. The logarithm accounts for the curvature in the boundary layer. For values of Rayleigh number larger than 10^8 this correlation can be approximated by

$$\text{Nu}_D = \frac{h_m D}{k} = 0.1 \, \text{Ra}_D^{1/3} \quad (2.58)$$

Natural-convection heat transfer from spheres can be correlated for $Ra_D > 10^{-2}$ by

$$\mathrm{Nu}_D = \frac{h_m D}{k} = 2 + \left[\left(0.56\, \mathrm{Ra}_D^{1/4} \left[1 + \left(\frac{0.559}{\mathrm{Pr}} \right)^{3/5} \right]^{-5/12} \right)^{15} + (0.1\, \mathrm{Ra}_D^{1/3})^{15} \right]^{-1/15} \qquad (2.59)$$

where the conduction limit of 2.0 is included.

Equation (2.56) can be used with reasonable accuracy for most geometries in the turbulent-flow regime at large values of the Rayleigh number for gases and liquids when a specific correlation is not available.

2.22 NATURAL CONVECTION FOR CAVITY FLOWS

Heat transfer by natural convection across a cavity will be equal to the value for heat conduction when the Rayleigh number based on the cavity spacing is low, usually less than 1000. On the other hand, when the Rayleigh number is large and the boundary-layer thickness is small compared to the cavity width, the boundary layers act independently, and we can use the correlations developed for external flows given above. However, there are usually two boundary layers in series in a cavity, so the heat-transfer correlation is a combination of the two boundary-layer correlations.

A common configuration is an air space such as the space between two panes of glass in a window. A heat-transfer correlation that combines the conduction limit and the two-boundary-layer limit that is valid for all values of Rayleigh number is given as

$$\mathrm{Nu}_L = \frac{h_m L}{k} = [1 + (\mathrm{Nu}_{L,conv})^{15}]^{1/15} \qquad (2.60)$$

where L is the cavity width and

$$\mathrm{Nu}_{L,conv} = \frac{1}{2} \left[\left(0.67 \left[\mathrm{Ra}_L \cos\theta\, \frac{L}{H} \right]^{1/4} \left[1 + \left(\frac{0.559}{\mathrm{Pr}} \right)^{3/5} \right]^{-5/12} \right)^{15} + (0.1\, [\mathrm{Ra}_L \cos\theta]^{1/3})^{15} \right]$$

When the cavity is vertical, such as an air space in a vertical window, the angle of inclination becomes zero and the cosines become equal to 1.0. Note that the mean heat-transfer coefficient, h_m, is proportional to $1/L$ when conduction dominates and is independent of L in the convective boundary-layer regime for both laminar and turbulent flow. Therefore one can select the optimum spacing at which the total heat transfer is a minimum by setting the Rayleigh number to 1000.

2.23 POOL BOILING

One common method of boiling a refrigerant in an evaporator is to allow the fluid to boil on a surface with low fluid velocities. This is called a *flooded evaporator*. The thermal buoyancy of the fluid or the buoyant bubbles create the fluid movement necessary to bring liquid to the surface and transport the heat away from the surface. Figure 2.7, taken from Farber and Scorah [9], illustrates the various boiling regimes for one of these evaporators. We will consider here only boiling from the exterior surface of horizontal cylinders.

Boiling from a horizontal cylinder in the free-convection regime can be described by Eq. (2.57). Numerous correlations have been presented for boiling in the nucleate regime. Some of these are reviewed in the *Handbook of Heat Transfer Fundamentals* [10]

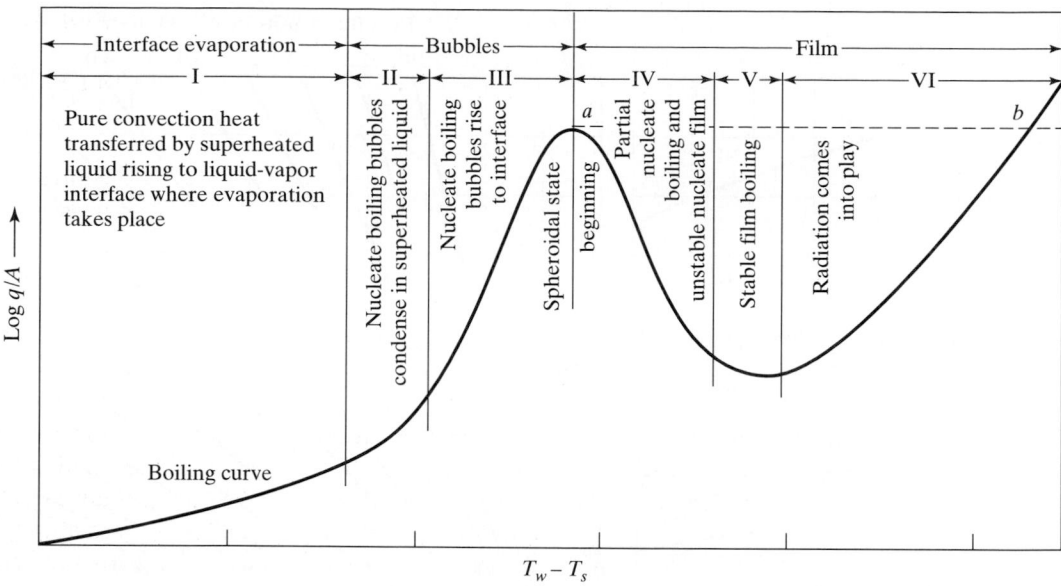

Figure 2.7 Characteristic boiling curve for pool boiling. [Reprinted by permission from *ASHRAE Fundamentals 1993* (IP & SI), p. 4.1.]

and the *ASHRAE Handbook of Fundamentals* [6]. Figure 2.8 gives nomographs that can be used to estimate nucleate-boiling heat transfer coefficients from a horizontal tube of diameter 1.18 in. (30 mm) for various fluids. The heat-transfer coefficient normally varies between 200 and 600 Btu/hr · ft^2 · °F (1000–3000 W/m^2 · °C).

2.24 FORCED-CONVECTION BOILING

Most applications of forced-convection boiling that we will consider use horizontal tubes. The most common example is the evaporation of a refrigerant in copper tubes.

The evaporation process is a very complex phenomenon. Some of the factors involved are surface roughness, tube diameter, fluid velocity, liquid-vapor ratio or quality, fluid thermophysical properties, and heat-flux rate. Many correlations for local and average heat-transfer coefficient have been proposed. Some general trends have been observed, such as that the heat-transfer coefficient increases with increasing heat-flux rate, and the coefficient increases as the tube diameter is made smaller at the same heat-flux rate. However, no single uniform correlation incorporates all governing factors.

Consider an evaporator in a refrigeration system that uses horizontal tubes. The refrigerant enters the evaporator as a liquid-vapor mixture with a quality of about 20 percent. Depending on the fluid mass velocity, the flow then passes through a bubble-and-plug flow regime or a wavy regime, where the liquid flows along the bottom of the tube and the vapor on top. The flow then changes to an annular flow, where liquid covers the walls and high-velocity vapor passes through the core. The flow progresses through a spray annular flow, a mist flow, and finally changes to a superheated-vapor flow. Each flow regime has a different heat-transfer coefficient.

One heat-transfer correlation is given here from Pierre [11]. This is an average value that was obtained using R-12 and R-22 flowing in copper tubes of 0.471 in. (12.0 mm) and

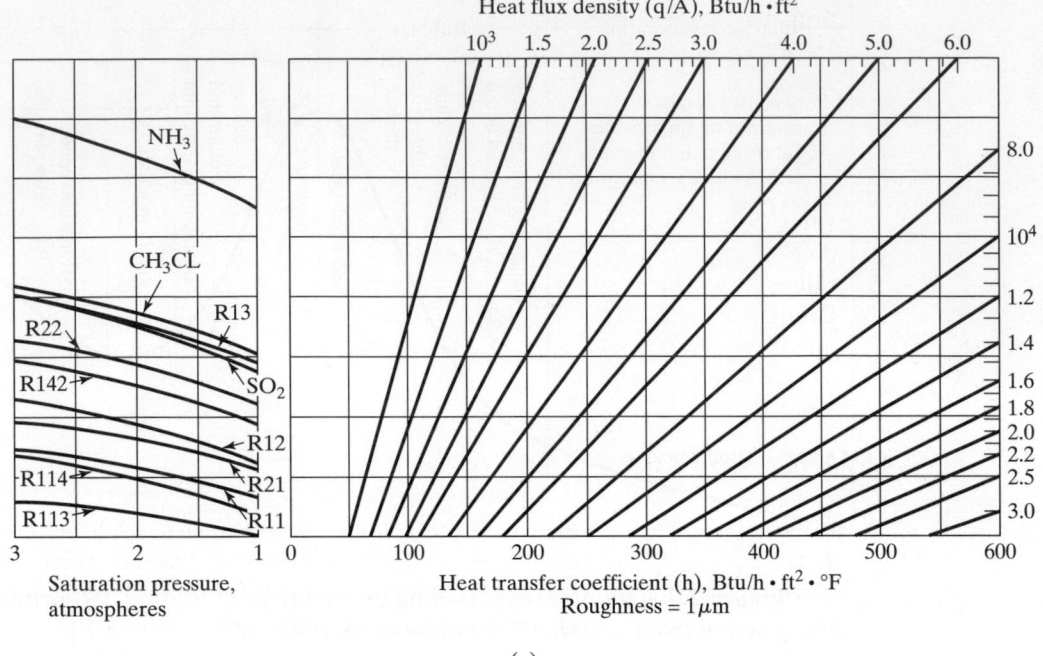

Heat flux density (q/A), Btu/h · ft²

Saturation pressure, atmospheres

Heat transfer coefficient (h), Btu/h · ft² · °F
Roughness = 1 μm

(a)

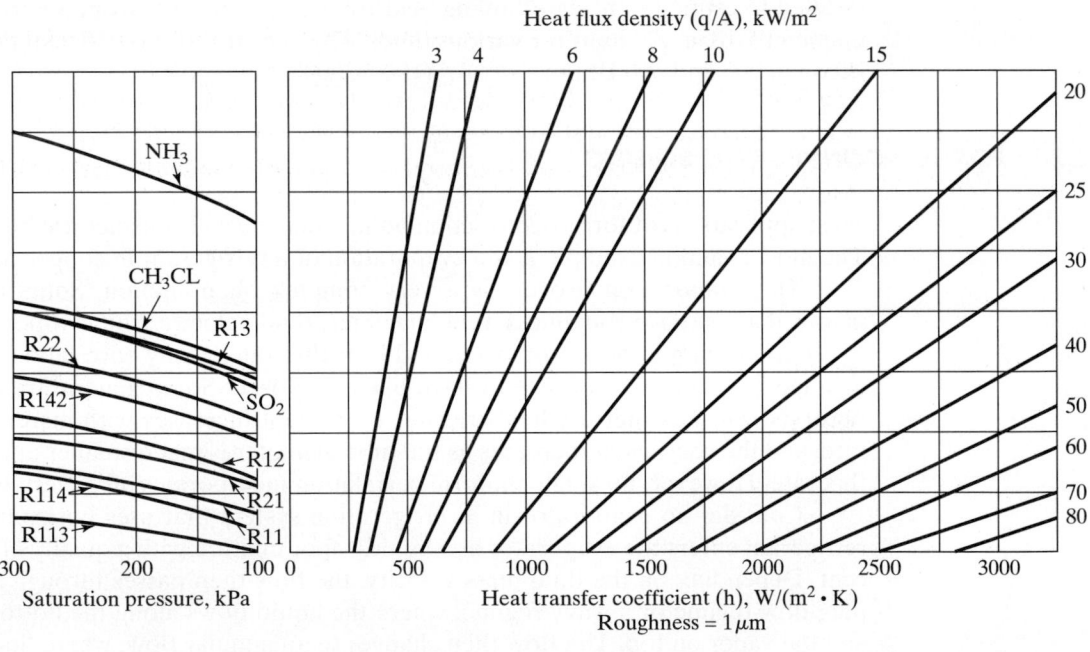

Heat flux density (q/A), kW/m²

Saturation pressure, kPa

Heat transfer coefficient (h), W/(m² · K)
Roughness = 1 μm

(b)

Figure 2.8 Heat transfer coefficient nomographs for pool boiling from a horizontal cylinder. [Reprinted by permission from *ASHRAE Fundamentals 1993* (IP & SI), p. 4.4.]

0.708 in. (18.0 mm) inside diameter, and 11.7 ft (3.6 m) to 27 ft (8.2 m) long. The saturation temperature ranged from −4 to 32 °F (−20 to 0 °C), and the refrigerant varied from 15 percent quality to 11 °F (6 °C) superheat.

$$\text{Nu}_m = \frac{h_m D_i}{k_\ell} = 0.0082(\text{Re}_\ell^2 K)^{0.4} \tag{2.61}$$

where $\text{Re}_\ell = GD_i/\mu_\ell$ and $K = C\,\Delta x h_{fg}/L$; $C = 778$ ft · lbm/Btu in English units and 0.255 m · kg/kJ in SI units. This correlation is valid for $10^9 < \text{Re}_\ell^2 K < 7 \times 10^{12}$. The usual range of heat-transfer coefficient for halocarbon refrigerants is between 100 and 800 Btu/hr · ft² · °F (500 and 4500 W/m² · °C). The values for ammonia are larger, owing mainly to the larger value of the thermal conductivity of the fluid.

2.25 CONDENSATION ON EXTERNAL SURFACES

Condensation on external surfaces occurs in either film or dropwise fashion. In *film condensation* the condensate forms a liquid film over the surface. The heat transfer must occur through this film, so the film thermal conductance plays a major role in the heat-transfer coefficient. Most film condensation is in the laminar-flow regime. In *dropwise condensation* the surface is exposed mainly to vapor, with drops covering a small portion of the surface area. Film condensation is the predominant form of condensation in HVAC equipment, so it is the mechanism discussed here.

Condensation of refrigerant often occurs on the external surface of horizontal tubes in a shell-and-tube condenser. The tubes are often aligned in the vertical direction, so that the film falling from one tube impinges on the one below. This increases the film velocity, which also reduces the film thickness and thereby increases the heat-transfer coefficient.

A correlation for N horizontal tubes of diameter D in a vertical plane given by McAdams is presented here:

$$h = 0.79F_1\left(\frac{h_{fg}}{ND\,\Delta t}\right)^{0.25} \tag{2.62}$$

where $F_1 = \left(\dfrac{k_\ell^3 \rho_\ell^2 g}{\mu_\ell}\right)^{0.25}$

with English units of

$$\left(\frac{\text{Btu}^3 \cdot \text{lbm}}{\text{hr}^4 \cdot \text{ft}^7 \cdot {}^\circ\text{F}^3}\right)^{0.25}$$

and SI units of

$$\left(\frac{\text{W}^3 \cdot \text{kg}}{\text{s} \cdot \text{m}^7 \cdot \text{K}^3}\right)^{0.25}$$

Values for F_1 are listed in Table 2.1.

TABLE 2.1 Values For Condensing-Coefficient Factors, F_1

Refrigerant	Film Temperature, °F (°C)	F_1
R-22	75 (24)	153 (80.3)
	100 (38)	144 (75.5)
	125 (68)	132 (69.2)
Ammonia	75 (24)	409 (214.5)
	100 (38)	408 (214.0)
	125 (68)	408 (214.0)

2.26 CONDENSATION IN INTERNAL FLOWS

Condensation inside horizontal tubes is the most common type of condensation found in HVAC equipment. Examples include steam heating coils, perimeter steam heaters, and various types of refrigerant condensers. The heat-transfer coefficient depends on the velocity of the vapor and liquid phases within the tube.

Correlations from Ackers and Rosson [12] are given here for values of the liquid Reynolds number, $DG_\ell/\mu_\ell \leq 5000$.

$$1000 \leq \text{Re}_C \leq 20{,}000, \qquad \text{Nu}_\ell = 13.8\, \text{Pr}_\ell^{1/3}\, M^{1/6}\, \text{Re}_C^{0.2}$$

$$20{,}000 \leq \text{Re}_C \leq 100{,}000, \qquad \text{Nu}_\ell = 0.1\, \text{Pr}_\ell^{1/3}\, M^{1/6}\, \text{Re}_C^{2/3} \qquad (2.63)$$

where

$$\text{Re}_C = \frac{DG_v}{\mu_\ell} \left(\frac{\rho_\ell}{\rho_v}\right)^{0.5}$$

$$\text{Nu}_\ell = \frac{hD}{k_\ell}, \qquad \text{Pr}_\ell = \frac{c_{p,\ell}\mu_\ell}{k_\ell}, \quad \text{and} \quad M = \frac{h_{fg}}{c_{p,\ell}\,\Delta t}$$

2.27 THERMAL RADIATION

Thermal radiation involves the transfer of heat from one body to another at lower temperature by electromagnetic waves passing through a separating medium. Thermal radiation waves have properties similar to other types of electromagnetic waves, differing only in wavelength. The approximate wavelength range for thermal radiation is 0.1 to 100 μm and includes a portion of the ultraviolet (UV) and all of the visible and infrared (IR). The thermal radiation emitted by a surface encompasses a range of wavelengths, and the magnitude of the radiation varies with wavelength; the term *spectral* is used to refer to the nature of this dependence. The spectral distribution of the radiation emitted by a surface is a function of its temperature and physical properties. Most thermal-radiation problems in HVAC applications involve infrared rays. In the case of solar radiation, a significant amount of the energy transfer occurs in the visible range of wavelengths.

A second feature of thermal radiation relates to its *directionality*. A surface may emit preferentially in certain directions, creating a directional distribution of the emitted radiation. To fully treat the spectral and directional effects of thermal radiation we use the concept of *radiation intensity*. This radiation intensity, $I_{\lambda,e}(\lambda, \theta, \phi)$, is defined by

$$I_{\lambda,e}(\lambda, \theta, \phi) \equiv \frac{dQ}{dA \cos\theta \, d\omega \, d\lambda} \tag{2.64}$$

where dQ is the radiation heat transfer coming from the surface area dA in the wavelength interval $d\lambda$ about λ and which passes through the solid angle $d\omega$. The solid angle $d\omega$ is about the (θ, ϕ) direction, where θ and ϕ are the zenith and azimuth angles, respectively, for the surface area dA. A special case is that of a *diffuse surface* defined as one whose intensity is not a function of direction. Many surfaces encountered in HVAC applications are diffuse. This is important, since it greatly simplifies the calculation of radiant-heat transfer between surfaces.

Blackbody Radiation

An important concept in thermal radiation is that of a blackbody. A *blackbody* or *black surface* is an ideal surface having the following properties.

1. A blackbody absorbs all incident radiation, regardless of wavelength and direction.
2. For a given temperature and wavelength, no surface can emit more energy than a blackbody.
3. The radiation emitted by a blackbody is independent of direction. That is, a blackbody is a diffuse emitter.

Thus, as a perfect emitter and absorber, the blackbody is the standard against which the radiation properties of real surfaces are compared. The term *black* should not be confused with the color of the same name, since a white-colored surface may absorb the same amount of infrared energy as a black-colored surface. An approximation to a blackbody is a hollow enclosure with small opening. Three important laws for the emission of blackbody radiation exist. The Stefan-Boltzmann law states that for a blackbody

$$E_b = \sigma T^4 \tag{2.65}$$

where E_b is the emissive power of the blackbody, Btu/hr $\cdot$ ft^2 (W/m^2), T is the absolute temperature, °R (K), and σ is the Stefan-Boltzmann constant.

$$\sigma = 0.1713 \times 10^{-8} \text{ Btu/hr} \cdot \text{ft}^2 \cdot \text{°R}^4$$

$$\sigma = 5.670 \times 10^{-8} \text{ W/m}^2 \cdot \text{K}^4$$

The wavelength dependence of blackbody radiation is given by Planck's distribution, which states

$$E_{b\lambda} = \frac{C_1 \lambda^{-5}}{e^{C_2/\lambda T} - 1} \tag{2.66}$$

where $E_{b\lambda}$ is the emissive power at a particular wavelength (i.e., monochromatic emissive power) of a blackbody, Btu/hr $\cdot$ ft^2 $\cdot$ μm (W/m^2 $\cdot$ μm), $C_1 = 1.187 \times 10^8$ Btu $\cdot$ μm^4/hr $\cdot$ ft^2 (3.742×10^8 W $\cdot$ μm^4/m^2), and $C_2 = 2.59 \times 10^4$ μm $\cdot$ °R (1.439×10^4 μm $\cdot$ K).

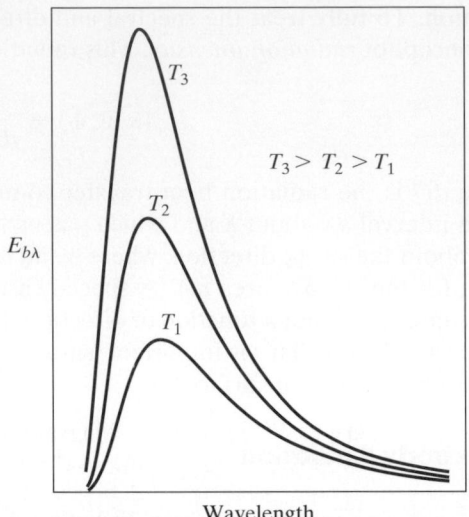

Figure 2.9 Schematic variation of monochromatic emissive power for a blackbody.

Wein's displacement law, which may be derived from Eq. (2.66), states that the product of the wavelength at which the maximum monochromatic emissive power is attained and the absolute temperature is a constant, or

$$\lambda_{max} T = C_3 \tag{2.67}$$

where $C_3 = 5216.4 \ \mu m \cdot \degree R$ ($C_3 = 2897.8 \ \mu m \cdot K$).

Figure 2.9 schematically illustrates the three blackbody radiation laws given by Eqs. (2.65)–(2.67). Points for each curve can be calculated using Eq. (2.66). As the temperature increases, the maximum monochromatic emissive power is shifted to a shorter wavelength. This is required by Eq. (2.67). The area under any curve is the total emissive power, which is equal to the value calculated by Eq. (2.65).

Radiation Properties

Most surfaces do not behave like blackbodies. To characterize the radiation properties of nonblack surfaces, the dimensionless quantities emittance, absorptance, reflectance, and transmittance are used to relate the emitting, absorbing, reflecting, and transmitting capabilities of a real surface to those of a blackbody. The radiation properties of real surfaces are functions of temperature, wavelength, and direction. The properties that describe how a surface behaves as a function of wavelength are referred to as *spectral* or *monochromatic* properties. *Directional* properties describe the distribution of radiation as related to angular direction.

Engineering calculations usually can be carried out with acceptable accuracy by a simplified approach using radiation properties averaged over direction and the wavelength range of interest. Radiation properties averaged over all directions and wavelengths are called *total properties*. The definition of total-radiation properties is illustrated in Fig. 2.10. When radiation is incident on a surface, a fraction of that radiation is absorbed, a fraction is reflected, and a fraction is transmitted. The *absorptance*, α, is the

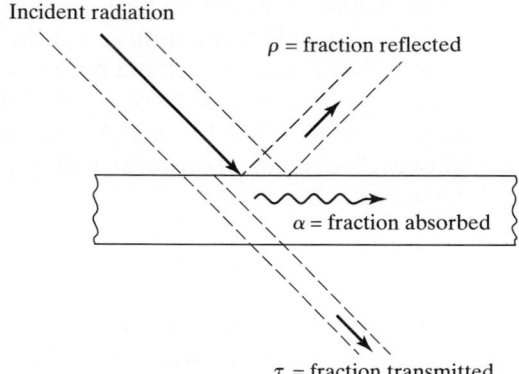

Incident radiation

ρ = fraction reflected

α = fraction absorbed

τ = fraction transmitted

Figure 2.10 Schematic diagram illustrating incident, reflected, absorbed, and transmitted radiation.

fraction absorbed; the *reflectance*, ρ, is the fraction reflected; and the transmittance, τ, is the fraction transmitted. From an energy balance one obtains

$$\alpha + \rho + \tau = 1 \tag{2.68}$$

Another important total-radiation property of real surfaces is the emittance. The *emittance* of a surface, ε, is defined as the total radiation emitted divided by the total radiation that would be emitted by a blackbody at the same temperature, or

$$\varepsilon = \frac{E(T)}{E_b(T)} = \frac{E(T)}{\sigma T^4} \tag{2.69}$$

Monochromatic Radiation Properties, Kirchoff's Law, and Gray Surfaces

Total-radiation properties can be obtained from spectral or monochromatic properties which apply at a single wavelength. The *hemispherical spectral emittance*, ε_λ, is defined in terms of the monochromatic emissive power of a real surface and the the monochromatic blackbody emissive power at the same temperature, or

$$\varepsilon_\lambda(T) = \frac{E_\lambda(T)}{E_{b\lambda}(T)} \tag{2.70}$$

where $E_\lambda(T)$ is the rate at which radiation is emitted by a real surface per unit area and $d\lambda$ centered at λ, Btu/hr $\cdot$ ft^2 $\cdot$ μm (W/m^2 $\cdot$ μm). The term *hemispherical* is used to indicate that the quantities have been integrated over the hemisphere above the surface. Similarly, the *hemispherical monochromatic absorptance*, α_λ, reflectance, ρ_λ, and transmittance, τ_λ, values are defined as the fractions of the total incident radiation at wavelength λ that are absorbed, reflected, and transmitted, respectively. An energy balance on a monochromatic basis yields

$$\alpha_\lambda + \rho_\lambda + \tau_\lambda = 1 \tag{2.71}$$

Kirchoff's law relates the monochromatic emittance and absorptance values for a surface. If the surface is diffuse or the irradiation striking the surface is diffuse, Kirchoff's law states

$$\alpha_\lambda = \varepsilon_\lambda \tag{2.72}$$

The prediction of radiation heat exchange between surfaces is greatly simplified when the surfaces involved are diffuse and their radiation properties are not a function of wavelength. The first condition is a reasonable approximation for many surfaces, particularly for electrically nonconducting materials. Surfaces meeting the second condition are called *gray surfaces*. That is, a gray surface is one in which both the emittance and absorptance are independent of wavelength ($\varepsilon_\lambda = \varepsilon$ and $\alpha_\lambda = \alpha$). Thus, for gray surfaces Eq. (2.72) becomes

$$\alpha = \varepsilon \qquad (2.73)$$

Although the assumption of a gray surface is reasonable for many practical applications, caution should be exercised in its use, particularly if the spectral regions of irradiation and emission are widely separated. A very important case in point is when the incident radiation on a surface is solar radiation while the surface is at a typical environmental temperature. For example, approximately 98 percent of the solar radiation falls in a wavelength range from 0.2 to 3.0 μm with the peak radiation near 0.5 μm, while a surface at a temperature of 100 °F (38 °C) has 98 percent of its blackbody emission in the range of 4.5 to 64 μm with a peak at 9.3 μm. Most surfaces can not be treated as gray when solar radiation is present in the energy calculations. Table 2.2 shows emittance and absorptance values for various surfaces.

TABLE 2.2 Emittance and Absorptance Values for Various Surfaces

Surface	Emittance or Absorptance		Absorptance for Solar Radiation
	50–100 °F	1000 °F	
A small hole in a large box, sphere, furnace, or enclosure	0.97 to 0.99	0.97 to 0.99	0.97 to 0.99
Black nonmetallic surfaces such as asphalt, carbon, slate, paint, paper	0.90 to 0.98	0.90 to 0.98	0.85 to 0.98
Red brick and tile, concrete and stone, rusty steel and iron, dark paints (red, brown, green, etc.)	0.85 to 0.95	0.75 to 0.90	0.65 to 0.80
Yellow and buff brick and stone, firebrick, fire clay	0.85 to 0.95	0.70 to 0.85	0.50 to 0.70
White or light-cream brick, tile, paint or paper, plaster, whitewash	0.85 to 0.95	0.60 to 0.75	0.30 to 0.50
Window glass	0.90 to 0.95	—	—
Bright aluminum paint; gilt or bronze paint	0.40 to 0.60	—	0.30 to 0.50
Dull brass, copper, or aluminum; galvanized steel; polished iron	0.20 to 0.30	0.30 to 0.50	0.40 to 0.65
Polished brass, copper, monel metal	0.02 to 0.05	0.05 to 0.15	0.30 to 0.50
Highly polished aluminum, tin plate, nickel, chromium	0.02 to 0.04	0.05 to 0.10	0.10 to 0.40

SOURCE: Abstracted by permission from *ASHRAE Handbook of Fundamentals* (Atlanta: American Society of Heating, Refrigerating and Air Conditioning Engineers, 1993) p. 3.8.

Heat Exchange Between Two Diffuse Gray Surfaces

When two surfaces within visual range of each other are separated by a medium which does not absorb radiation, a radiation energy exchange will occur between the surfaces. Let us denote these surfaces as 1 and 2 and apply the appropriate subscripts to their area, temperature, and radiation properties. Then, if the surfaces are diffuse and gray, the net rate of heat transfer from surface 1 to surface 2 is given by

$$\dot{Q}_{1-2} = \frac{A_1 \sigma (T_1^4 - T_2^4)}{\left(\dfrac{1 - \varepsilon_1}{\varepsilon_1}\right) + \dfrac{1}{F_{1-2}} + \dfrac{A_1}{A_2}\left(\dfrac{1 - \varepsilon_2}{\varepsilon_2}\right)} \tag{2.74}$$

where F_{1-2} is the fraction of radiation that leaves surface A_1 that strikes A_2. For diffuse surfaces this factor is a function only of the geometry of the surfaces and their orientation relative to each other. This factor is tabulated for a wide range of configurations in most heat-transfer texts and is commonly called a view factor or angle factor or shape factor, depending on the terminology of the particular heat-transfer book. Equation (2.74) will be a convenient starting point for several radiation heat-transfer analyses encountered in later chapters. Heat exchange by radiation for other conditions may be found in various references on heat transfer.

Radiation Heat-Transfer Coefficient

In many problems it is convenient to express the rate of heat transfer by radiation, $\dot{Q}_{rad}$, by

$$\dot{Q}_{rad} = h_r A(T_1 - T_2) \tag{2.75}$$

where h_r is a radiation heat-transfer coefficient and T_1 and T_2 are the appropriate temperatures for the heat exchange. For radiation heat exchange between two surfaces as described by Eq. (2.74), the radiation heat-transfer coefficient becomes

$$h_r = \frac{\sigma (T_1^4 - T_2^4)/(T_1 - T_2)}{\left(\dfrac{1 - \varepsilon_1}{\varepsilon_1}\right) + \dfrac{1}{F_{1-2}} + \dfrac{A_1}{A_2}\left(\dfrac{1 - \varepsilon_2}{\varepsilon_2}\right)} \tag{2.76}$$

In some instances it is desirable to estimate the radiation heat-transfer coefficient when the actual temperature differences are unknown but the temperature level is known. This can be achieved by observing that

$$\frac{T_1^4 - T_2^4}{T_1 - T_2} = 4(T_{avg})^3 \left[1 + \frac{1}{4}\left(\frac{\Delta T}{T_{avg}}\right)^2\right] \tag{2.77}$$

where $T_{avg} = \dfrac{T_1 + T_2}{2}$ and $\Delta T = T_1 - T_2$.

(Note that T_{avg} must be in °R or K.) In many HVAC applications the term $\frac{1}{4}(\Delta T/T_{avg})^2$ is small enough compared to 1.0 that it can be neglected, and a reasonable approximation is

$$\frac{T_1^4 - T_2^4}{T_1 - T_2} \approx 4(T_{avg})^3$$

and Eq. (2.76) may be approximated by

$$h_r \approx \frac{4\sigma (T_{\text{avg}})^3}{\left(\dfrac{1 - \varepsilon_1}{\varepsilon_1}\right) + \dfrac{1}{F_{1-2}} + \dfrac{A_1}{A_2}\left(\dfrac{1 - \varepsilon_2}{\varepsilon_2}\right)} \tag{2.78}$$

2.28 COMBINED MODES OF HEAT TRANSFER

Most HVAC applications in which we want to evaluate the heat transfer from a surface (e.g., an inside wall) or between two surfaces (e.g., across the air gap in a double-glazed window) involve combined convection and radiation. Thus it is conventional practice to define a combined heat-transfer coefficient. Figure 2.11 schematically depicts a surface at temperature T_w which is exchanging heat by convection with air which is at temperature T_a and by radiation with surroundings which are at temperature T_s. The total heat flux from the wall, $\dot{q}$, is the sum of the convective heat flux, $\dot{q}_c$, and radiation heat flux, $\dot{q}_r$, and these heat fluxes can be expressed in terms of their respective heat-transfer coefficients, h_c and h_r.

$$\dot{q} = \dot{q}_c + \dot{q}_r = h_c(T_w - T_a) + h_r(T_w - T_s) \tag{2.79}$$

Notice that in Eq. (2.79) only temperature differences appear. Therefore, these temperature differences can be written in terms of either absolute temperatures (°R or K) or nonabsolute temperatures (°F or °C). Thus we can write

$$\dot{q} = \dot{q}_c + \dot{q}_r = h_c(t_w - t_a) + h_r(t_w - t_s) \tag{2.79a}$$

For convenience, it is common practice to write the heat exchange in terms of a single heat-transfer coefficient, h, which includes both convection and radiation and in terms of the temperature difference $(t_w - t_a)$.

$$\dot{q} = h(t_w - t_a) \tag{2.80}$$

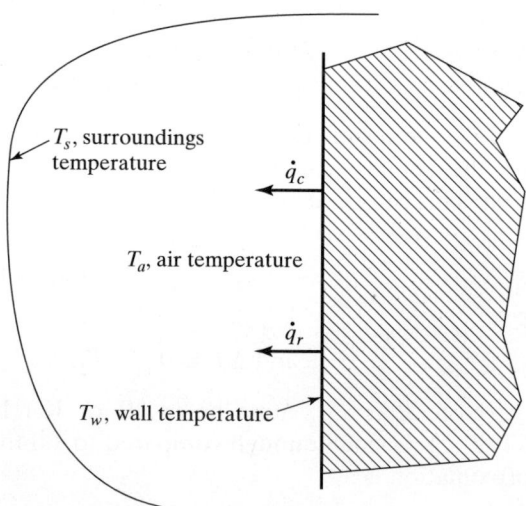

Figure 2.11 Combined convection and radiation heat transfer from a surface.

By Eqs. (2.79a) and (2.80) we have

$$h = h_c + h_r \left(\frac{t_w - t_s}{t_w - t_a} \right) \qquad (2.81)$$

In many applications the surrounding temperature for radiation heat exchange is equal to the surrounding air temperature, and the combined heat-transfer coefficient becomes the sum of the convective and radiative heat-transfer coefficients.

2.29 OVERALL TRANSFER OF HEAT

Most practical problems in heat transfer involve two or more of the heat-transfer modes. Such problems must be analyzed from a total or overall aspect. Two important determinations are the overall heat-transfer coefficient, U, and the true mean temperature difference, Δt_m.

Figure 2.12 shows schematically two fluids separated by a pipe. We will consider a short section dL, where the various temperatures have constant values. We will also assume steady-state conditions exist. Thus, the rate of heat transfer through each part of the system is the same.

The following equations may be written for the heat-transfer rate:

$$d\dot{Q} = h_i\, dA_i\, (t_1 - t_2) = \frac{2\pi k (t_2 - t_3)\, dL}{\ln(D_o/D_i)} = h_o\, dA_o (t_3 - t_4) \qquad (2.82)$$

The coefficient h_i is determined from an appropriate equation for the heat-transfer case involved (e.g., forced convection, condensing vapor, etc.). If the cold fluid shown in Fig. 2.12 is a gas such as air, h_o must be a combined coefficient as described in the previous section. The equalities of Eq. (2.82) are inconvenient, since the pipe surface temperatures t_2 and t_3 are involved. It is preferred to write the heat-transfer rate in terms of the fluid temperatures t_1 and t_4. This is accomplished by defining an overall heat-transfer coefficient, U, such that

$$d\dot{Q} = U_o\, dA_o (t_1 - t_4) \qquad (2.83)$$

The subscript on U_o denotes that the outside surface area, dA_o, is used in defining the overall heat-transfer coefficient. Whether the inner or outer surface area is selected is

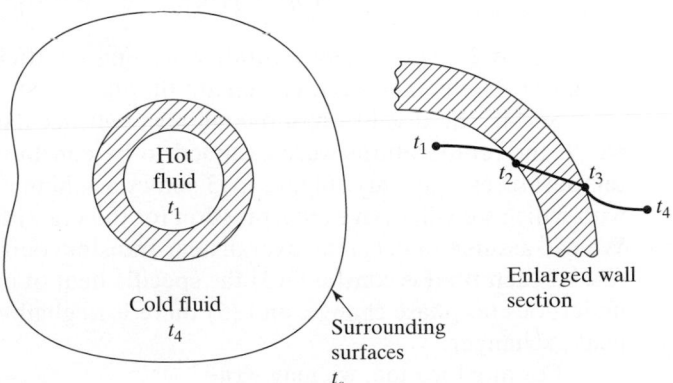

Figure 2.12 Schematic illustration of overall heat transfer.

arbitrary. The important feature is that the overall heat-transfer coefficient, or U- value, allows the heat transfer to be expressed in terms of the fluid temperatures.

By combining Eqs. (2.82) and (2.83) and noting that $dA_i = \pi D_i \, dL$ and $dA_o = \pi D_o \, dL$, we can show that

$$\frac{1}{U_o} = \frac{D_o}{D_i h_i} + \frac{D_o \ln (D_o/D_i)}{2k} + \frac{1}{h_o} \qquad (2.84)$$

Therefore, for pipes or cylinders, by Eq. (2.84) we have

$$U_o = \frac{1}{\dfrac{D_o}{D_i h_i} + \dfrac{D_o \ln (D_o/D_i)}{2k} + \dfrac{1}{h_o}} \qquad (2.85)$$

The term $1/U_o$ is called the *total thermal resistance*, R_t. Similarly, the quantities $D_o/D_i h_i$, $D_o \ln (D_o/D_i)/2k$, and $1/h_i$ are, respectively, the inside-surface resistance, R_i, the tube-wall resistance, R_w, and the outside-surface resistance, R_o. Thus

$$1/U_o = R_t = R_i + R_w + R_o = \Sigma R \qquad (2.86)$$

The resistance concept to overall heat transfer is important. Through simple addition of individual resistances, the total resistance may be found. In a multiple-layer pipe or wall, the total wall resistance is found by adding the similar terms for each layer. Furthermore, in approximate calculations, the numerical significance of each resistance term may be studied, and the relative importance of each separate coefficient analyzed.

Equation (2.85) makes no allowance for fouling, since the tube of Fig. 2.12 was assumed to be perfectly clean. This condition is true for new equipment only. After use, heat-transfer surfaces may accumulate deposits of scale, oil, or other foreign matter. Such deposits are usually thin but provide resistance to heat flow and should be accounted for in equipment design. Conventional practice is to express the resistance of the deposit in terms of a *fouling coefficient* or *deposit factor*, h_d. Values of h_d are determined experimentally. In the case where deposits exist on both the inner and outer surfaces of the tube, Eq. (2.85) would be modified to read

$$U_o = \frac{1}{\dfrac{D_o}{D_i h_i} + \dfrac{D_o}{D_i h_{d,i}} + \dfrac{D_o \ln (D_o/D_i)}{2k} + \dfrac{1}{h_{d,o}} + \dfrac{1}{h_o}} \qquad (2.87)$$

Table 2.3 shows representative fouling coefficients for ordinary heat-exchanger design problems. The values given are thermal resistances $(1/h_d)$.

So far, the discussion of overall heat transfer has been limited to a local section in which the temperatures were assumed to be constant. In a finite heat exchanger, fluid temperatures may vary. Figure 2.13 shows a schematic of counterflow heat exchanger, with which we will derive an expression for the logarithmic mean temperature difference. We will assume that: (1) the overall heat-transfer coefficient is constant, (2) the mass flow rate of each fluid is constant, (3) the specific heat of each fluid is constant, (4) each fluid undergoes no phase change, and (5) there is negligible heat loss from the outside of the heat exchanger.

For any location, we may write

$$U_o \, dA_o (t_h - t_c) = \dot{m}_c c_{p,c} \, dt_c$$

and

$$t_h = t_{h,i} - \frac{\dot{m}_c c_{p,c}}{\dot{m}_h c_{p,h}} (t_{c,o} - t_c)$$

TABLE 2.3 Recommended Deposit Coefficients $(1/h_d)$, $(hr \cdot ft^2 \cdot °F)/Btu$

Application	Nonferrous Tubes	Ferrous Tubes
Brine coolers (brine side)		
Inhibited salt brines	0.0005	0.001
Noninhibited salt brines	0.001	0.002
Solvent brines (methylene chloride, halocarbon refrigerants)	0.0	0.0
Water coolers (water side)		
Recirculated water, closed system	0.0005	0.001
Recirculated water, open system	0.001	0.002
Condensers (water side)		
City or well water (typical)	0.001	0.002
River water (typical)	0.003	0.005
Brackish water	0.002	0.002
Condensers (refrigerant side)		
Ammonia	—	0.0003
Halocarbon refrigerants (miscible with oil)	0.0	0.0002
Evaporators (refrigerant side)		
Ammonia	—	0.001
Halocarbon refrigerants (miscible with oil)	0.0	0.0002

SOURCE: Abstracted by permission from *Air Conditioning, Refrigerating Data Book Design Volume* (American Society of Refrigerating Engineers, 1957), pp. 20–14, 20–16.

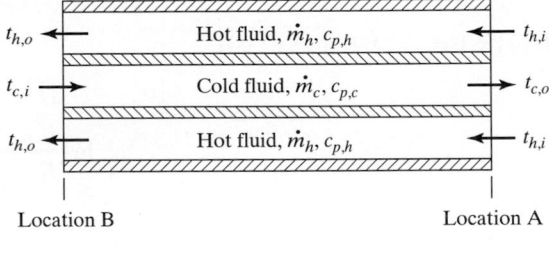

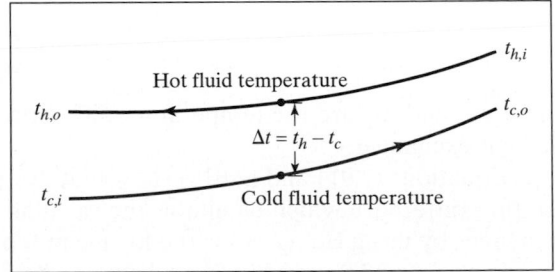

Figure 2.13 Schematic temperature changes in a counterflow heat exchanger.

Thus we obtain

$$\frac{dt_c}{K_1 - K_2 t_c} = \frac{U_o \, dA_o}{\dot{m}_c c_{p,c}}$$ (2.88)

where

$$K_1 = t_{h,i} - \left(\frac{\dot{m}_c c_{p,c}}{\dot{m}_h c_{p,h}}\right) t_{c,o}$$

$$K_2 = 1 - \frac{\dot{m}_c c_{p,c}}{\dot{m}_h c_{p,h}}$$

Integration of Eq. (2.88) gives

$$\ln\left(\frac{K_1 - K_2 t_{c,i}}{K_1 - K_2 t_{c,o}}\right) = K_2 \frac{U_o A_o}{\dot{m}_c c_{p,c}}$$

Substituting back for K_1 and K_2 and noting that

$$\frac{\dot{m}_c c_{p,c}}{\dot{m}_h c_{p,h}} = \frac{t_{h,i} - t_{h,o}}{t_{c,o} - t_{c,i}}$$

and

$$\dot{Q} = \dot{m}_c c_{p,c}(t_{c,o} - t_{c,i})$$

gives

$$\ln\left(\frac{t_{h,o} - t_{c,i}}{t_{h,i} - t_{c,o}}\right) = [(t_{h,o} - t_{c,i}) - (t_{h,i} - t_{c,o})] \frac{U_o A_o}{\dot{Q}}$$ (2.89)

Define Δt_m such that

$$\dot{Q} = U_o A_o \, \Delta t_m$$ (2.90)

Thus by Eqs. (2.89) and (2.90)

$$\Delta t_m = \frac{[(t_{h,o} - t_{c,i}) - (t_{h,i} - t_{c,o})]}{\ln\left(\dfrac{t_{h,o} - t_{c,i}}{t_{h,i} - t_{c,o}}\right)}$$ (2.91)

or

$$\Delta t_m = \frac{\Delta t_A - \Delta t_B}{\ln\left(\dfrac{\Delta t_A}{\Delta t_B}\right)}$$ (2.91a)

where Δt_A and Δt_B are the temperature differences between the fluids at the two ends of the heat exchanger.

Equations (2.91) and (2.91a) show that, for pure counterflow, the appropriate temperature difference which facilitates the calculation of the heat-transfer rate of the heat exchanger by using Eq. (2.90) is the log mean temperature difference.

Equations (2.90) and (2.91a) also apply to any heat-exchanger arrangement where one fluid temperature remains constant. For parallel flow, Eq. (2.91a) should be used. For other cases, such as cross flow, Eq. (2.91) or Eq. (2.91a) must be modified in order to provide the appropriate expression for Δt_m for application of Eq. (2.90). Heat-transfer books such as Incropera and DeWitt [13] present solutions for a number of arrangements.

In the design of a practical heat exchanger, the required capacity, $\dot{Q}$, is usually known. Other factors, such as one or more of the fluid temperatures, may also be known. In general, it is desirable to have a large magnitude for U_o, since this reduces the required A_o and/or temperature difference. High fluid velocities may give higher values of U_o, but they may also result in higher pressure drops through the heat exchanger and, therefore, an increase in the pumping power required to operate the heat exchanger. The optimum design will depend on which parameters are most critical to the specific application.

2.30 FLUID FLOW IN DUCTS AND PIPES

Consider steady flow of a fluid in a straight duct or pipe as shown in Fig. 2.14. The mass balance for the control volume bounded by the dashed lines can be written

$$\dot{m}_1 = \dot{m}_2$$

or, writing this in terms of fluid density, mean velocity, and flow cross-sectional area,

$$\rho_1 \mathbf{V}_1 A_1 = \rho_2 \mathbf{V}_2 A_2 \tag{2.92}$$

Relationships among pressure, velocity, and work can be obtained using an energy balance. We will begin with an equation for the amount of reversible work for a steady-flow, open system with one stream in and one leaving, as shown in Fig. 2.14. If the fluid is incompressible (i.e., density is constant), the equation can be written

$$w_{\mathrm{rev}_{1-2}} = \frac{(P_2 - P_1)}{\rho} + \frac{(\mathbf{V}_2^2 - \mathbf{V}_1^2)}{2} + g(z_2 - z_1) \tag{2.93}$$

However, in actual systems, the actual work is equal to the reversible work given above plus the lost work due to irreversibilities, $w_{\mathrm{actual}} = w_{\mathrm{rev}} + w_{\mathrm{loss}}$. Combining this with Eq. (2.93),

$$w_{\mathrm{actual}_{1-2}} = w_{\mathrm{rev}_{1-2}} + w_{\mathrm{loss}_{1-2}} = \frac{P_2 - P_1}{\rho} + \frac{\mathbf{V}_2^2 - \mathbf{V}_1^2}{2} + g(z_2 - z_1) + w_{\mathrm{loss}_{1-2}} \tag{2.94}$$

Other forms of Eq. (2.94) can be obtained. Multiplying Eq. (2.94) by the fluid density, ρ, gives a form written in terms of pressure:

$$\rho w_{\mathrm{actual}_{1-2}} = \Delta P_{\mathrm{tot}_{1-2}} = (P_2 - P_1) + \rho \frac{(\mathbf{V}_2^2 - \mathbf{V}_1^2)}{2} + \rho g(z_2 - z_1) + \rho w_{\mathrm{loss}_{1-2}} \tag{2.95}$$

| total pressure | static pressure | velocity pressure | elevation pressure | lost pressure |

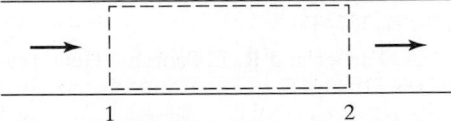

Figure 2.14 Fluid flowing steadily through a straight section of duct or pipe.

The relationships among static pressure, velocity pressure, and total pressure can be seen from this equation. For example, in a pitot tube where the change in elevation is zero and there is no lost work,

$$\Delta P_{\text{tot}_{1-2}} = \Delta P_{\text{static}_{1-2}} + \rho \, \frac{(\mathbf{V}_2^2 - \mathbf{V}_1^2)}{2} \tag{2.96}$$

Dividing Eq. (2.94) by gravitational acceleration, g, gives a form written in terms of head:

$$\underbrace{\frac{w_{\text{actual}_{1-2}}}{g} = \Delta H_{1-2}}_{\substack{\text{total} \\ \text{head}}} = \underbrace{\frac{P_2 - P_1}{g\rho}}_{\substack{\text{pressure} \\ \text{head}}} + \underbrace{\frac{\mathbf{V}_2^2 - \mathbf{V}_1^2}{2g}}_{\substack{\text{velocity} \\ \text{head}}} + \underbrace{(z_2 - z_1)}_{\substack{\text{elevation} \\ \text{head}}} + \underbrace{\frac{w_{\text{loss}_{1-2}}}{g}}_{\substack{\text{lost} \\ \text{head}}} \tag{2.97}$$

The last term in Eq. (2.97), the lost head or head loss, is represented by the symbol ℓ. For straight, round duct or pipe the head loss is usually determined using a friction factor

$$\ell = f \, \frac{L}{D} \, \frac{\mathbf{V}^2}{2g} \tag{2.98}$$

where f is the friction factor, L is the distance along the duct or pipe, D is the inner diameter, and $\mathbf{V}$ is the mean velocity. The friction factor is a function of the flow Reynolds number, $\text{Re} = \mathbf{V}D/\nu$, and the relative roughness or roughness height divided by the inner diameter, ε/D.

ENDNOTES

1. *Tables of Thermal Properties of Gases,* U.S. Department of Commerce, National Bureau of Standards Circular 564 (Washington, D.C.: Government Printing Office, 1955).

2. R. W. Hyland and A. Wexler, "Formulations for the Thermodynamic Properties of the Saturated Phases of H_2O from 173.15 K to 473.15 K," *ASHRAE Trans.,* 89:2A (1983), 500–519.

3. W. C. Reynolds, *Thermodynamic Properties in SI: graphs tables and computational equations for 40 substances* (Stanford CA: Department of Mechanical Engineering, Stanford University, 1979).

4. J. H. Keenan, F. G. Keyes, P. G. Hill, and J. G. Moore, *Steam Tables* (New York: John Wiley & Sons, Inc., 1969).

5. W. H. McAdams, *Heat Transmission*, 3d ed. (New York: McGraw-Hill, 1954).

6. *ASHRAE Handbook of Fundamentals* (Atlanta: American Society of Heating, Refrigerating and Air Conditioning Engineers, 1993).

7. M. N. Ozisik, *Heat Transfer: A Basic Approach* (New York: McGraw-Hill, 1985), 372.

8. S. Whitacker, "Forced Convection Heat Transfer Calculations for Flow in Pipes, Past Flat Plates, Single Cylinders and for Flow in Packed Beds and Tube Bundles," *AIChE J.*, 18 (1972), 361–371.

9. E. A. Farber and R. L. Scorah, "Heat Transfer to Water Boiling Under Pressure," *ASME Trans.*, 1948, 373.

10. *Handbook of Heat Transfer Fundamentals*, W. M. Rohsenow, J. P. Hartnet, and E. N. Ganic, eds. 2d ed. (New York: McGraw-Hill, 1985).

11. B. Pierre, *S. F. Review*, A. B. Svenska Flaktafabriken, Stockholm, Sweden, 2:1 (1955), 55.

12. W. W. Akers and H. F. Rosson, "Condensation Inside a Horizontal Tube," *Chemical Engineering Progress Symposium Series*, 56:30 (1960).

13. F. P. Incropera and D. P. DeWitt, *Fundamentals of Heat and Mass Transfer,* 4th ed. (New York: John Wiley & Sons, Inc., 1996), 581–619.

PROBLEMS

2.1 Convert a barometric pressure of 28.75 in. Hg at 32 °F (0 °C) to
  (a) psia,
  (b) inches of water,
  (c) meters of liquid whose specific gravity is 0.8, and
  (d) kPa.

2.2 Moist air at 70 °F and 14.32 psia total pressure contains 0.012 lbm of water vapor per lbm of dry air. Assuming perfect-gas behavior, determine
  (a) the partial pressure of the water vapor,
  (b) the density of the mixture,
  (c) gas-law constant for the mixture, and
  (d) the specific heat of the mixture unit mass of dry air.

2.3 Repeat Prob. 2.2 for moist air at 20 °C and 95 kPa total pressure with a moisture content of 0.01 kg of water vapor per kg of dry air.

2.4 Determine the percent error if Eq. (2.13) is used in calculating the density of dry air
  (a) at a temperature of −20 °F and a pressure of 10 atmospheres, and
  (b) at a temperature of −30 °C and a pressure of 1 MPa.

2.5 Calculate the compressibility factor for saturated water vapor at 100 °F.

2.6 Steam enters a radiator at 16 psia and 0.97 quality. The steam flows through the radiator, is condensed, and leaves as liquid water at 200 °F. If the heating capacity of the radiator is 5000 Btu/hr, at what rate in lbm/hr must the steam be supplied?

2.7 Ammonia enters an evaporator at −20 °C and 0.3 quality at a mass flow rate of 1 kg/s. Compute the cooling capacity of the evaporator in kilowatts if the ammonia leaves as saturated vapor at −20 °C.

2.8 Compute the irreversibility rate for the process described in Prob. 2.6 if the heat is being added to air at 70 °F.

2.9 Determine the irreversibility rate for the process described in Prob. 2.7 if the heat is added to the ammonia from air at −10 °C.

2.10 Steady heat conduction occurs through a wall of a building. The inside temperature is 70 °F and the outside temperature 0 °F. The inside surface heat-transfer coefficient is 1.5 Btu/hr · ft^2 · °F and the value on the outside surface is 6.0 Btu/hr · ft^2 · °F. The three layers of the wall are as given below:
  Layer 1: gypsum wall board, 0.5 in. thick, k = 0.1 Btu/hr · ft · °F
  Layer 2: insulation, 3.5 in. thick, k = 0.03 Btu/hr · ft · °F
  Layer 3: brick, 4 in. thick, k = 0.75 Btu/hr · ft · °F
Compute the rate of heat transfer through this wall per unit surface area (Btu/hr · ft^2).

2.11 Calculate the convective heat-transfer coefficient for water flowing in a round pipe with an inner diameter of 3.0 cm. The water flow rate is 2 L/s and the water temperature is 30 °C.

2.12 Air at 70 °F is flowing with a velocity of 10 ft/s over a heated horizontal pipe with an outside diameter of 5 in. and a surface temperature of 150 °F. Determine the convective heat-transfer rate per foot of pipe in units of Btu/hr.

2.13 Determine the convective heat-transfer rate per unit length from the pipe described in Prob. 2.12 if the forced-air flow is stopped and heat is transferred to the air by natural convection.

2.14 Compute the natural convective heat transfer across a window cavity that is 1 cm wide, 1 m high, and 1 m wide. Assume that the cavity is filled with air at atmospheric pressure and that the cold surface is at a temperature of 0 °C and the warm surface at 15 °C.

2.15 Estimate the heat-transfer coefficient for ammonia from Fig. 2.8 when pool boiling occurs from a horizontal tube at a pressure of 150 kPa and a surface heat flux of 10 kW/m^2. What is the temperature of the tube surface?

2.16 Refrigerant 22 is boiling in a 0.5-in. inner-diameter copper tube. The mass velocity is 10 lbm/min, the saturation temperature is 10 °F, and the change in quality between the inlet and outlet is 0.8.

(a) Determine the average boiling heat-transfer coefficient.

(b) Compute the length of tube needed.

2.17 Calculate the condensation heat-transfer coefficient for R-22 flowing in a horizontal tube given the following information: mass velocity = 10 lbm/min, tube inner diameter = 0.5 in., saturation temperature = 100 °F, $\Delta t = 8$ °F.

2.18 A horizontal roof exchanges heat by thermal radiation with the sky on a cold winter night. Compute the radiative heat loss per unit area of the roof if the roof surface temperature is 10 °F, the sky temperature is −40 °F, and blackbody radiation is assumed.

2.19 Compute the radiative heat transfer from the pipe described in Prob. 2.13 if the surroundings are at 70 °F and the pipe surface emissivity is 0.85.

2.20 Calculate the radiation heat transfer across the window cavity described in Prob. 2.14 if the emissivity of both surfaces is 0.9.

2.21 Determine the combined convective-radiative heat-transfer coefficient for the pipe described in Probs. 2.13 and 2.19.

2.22 Determine the combined convective-radiative heat-transfer coefficient for the window cavity described in Probs. 2.14 and 2.20.

SYMBOLS

A	Virial coefficient in Eq. (2.24); Surface area.
a	Acceleration.
C	Constant.
c	Specific heat; c_p at constant pressure; c_v at constant volume.
D	Diameter.
E	Emissive power.
F	Force; View factor.
f	Friction factor.
G	Mass velocity = $\rho \mathbf{V}$.
g	Acceleration of gravity.
H	Height of plate.
ΔH	Change in head.
h	Specific enthalpy h_f for saturated liquid; h_g for saturated vapor; $h_{fg} = h_g - h_f$; h_w for low-pressure water vapor; Heat-transfer coefficient.
I	Radiation intensity.
$\dot{I}$	Irreversibility rate.
k	F/ma or Thermal conductivity.
L	Length; Cavity width.
l	Head loss.
M	Molecular weight.
m	Mass.
$\dot{m}$	Mass flow rate.
N	Number of tubes in row.
Nu	Nusselt number.
P	Static pressure.
p	Perimeter.
ΔP	Pressure change.
Pr	Prandtl number.

Q	Quantity of heat.
$\dot{Q}$	Rate of heat transfer.
$\dot{Q}'$	Internal heat-generation rate per unit volume.
q	Heat transfer per unit mass.
$\dot{q}$	Heat transfer rate per unit surface area.
R	Pv/T for perfect gas; Thermal resistance.
r	Radius.
Ra	Rayleigh number.
Re	Reynolds number.
s	Specific entropy; s_f for saturated liquid; s_g for saturated vapor; $s_{fg} = s_g - s_f$.
T	Absolute temperature.
t	Temperature.
Δt	Temperature difference.
U	Overall heat-transfer coefficient.
u	Specific internal energy.
V	Volume.
$\mathbf{V}$	Velocity.
v	Specific volume; v_f for saturated liquid; v_g for saturated vapor; $v_{fg} = v_g - v_f$.
W	Weight.
w	Work per unit mass; Specific weight.
$\dot{W}$	Rate of work transfer (power).
x	Quality, mass of saturated vapor per unit mass of mixture.
x, y, z	Coordinates.
Z	Dimensionless compressibility factor for a gas $= Pv/RT$.
z	Height or elevation above arbitrary datum.

Greek Symbols

α	Absorptance.
β	Thermal coefficient of volumetric expansion.
ε	Emittance; Roughness height.
ϕ	Azimuthal angle.
λ	Wavelength.
μ	Dynamic viscosity.
θ	Angle of inclination from vertical.
ν	Kinematic viscosity.
ρ	Density; Reflectance.
σ	Stefan-Boltzmann constant.
τ	Time; Transmittance.
ω	Solid angle.

Subscripts

A	Location A.
a	Air; Ambient.
avg	Average.
B	Location B.
b	Blackbody.

C	Celsius.
c	Convection; Cold fluid.
conv	Convection.
D	Diameter is length scale.
d	Deposit or fouling coefficient.
F	Fahrenheit.
f	Saturated liquid.
fg	Difference between saturated liquid and saturated vapor.
g	Saturated vapor.
H	Plate height is length scale.
h	Hot fluid.
i	Saturated solid; Inner; Inner diameter; Inlet conditions.
L	Length of plate or width of cavity is length scale.
ℓ	Liquid.
m	Mean.
o	Ambient; Outer; Outlet conditions.
r	Radiation.
rad	Radiation.
rev	Reversible.
s	Surroundings.
sat	Saturation.
t	Total.
v	Vapor.
w	Water; Wall.
x, y	Gas-mixture constituents.
λ	Spectral.
∞	Free stream.

chapter 3

Mechanical Vapor-Compression Refrigeration Cycles

3.1 INTRODUCTION

A most important aspect of thermal environmental engineering is refrigeration. *Refrigeration* is the withdrawal of heat, producing in a substance or within a space a temperature lower than that of the natural surroundings. Thus any methods available for lowering temperature, in the range from ambient temperature to absolute zero, involve refrigerating processes.

Refrigeration may be produced by several means including (1) mechanical vapor-compression systems, (2) absorption systems in which the vapor is absorbed into a liquid prior to pumping, and (3) gas-compression systems involving throttling or unrestrained expansion of the compressed gas. All three methods will be studied in this text. Method 1 is used in the majority of refrigeration systems, with compression accomplished through a mechanical compressor. Method 2 has advantages when low-cost heat sources are available. Method 3 has found application in the liquefaction, storage, and separation of various gases.

Mechanical compression systems using refrigerants that change phase are predominant among refrigerating methods. Cooling is accomplished by evaporation of a liquid refrigerant under reduced pressure and temperature. The saturation temperature of the vapor is then elevated by mechanical compression, allowing the vapor to be condensed by heat rejection to ordinary cooling water or atmospheric air. The relatively high-pressure liquid is then expanded to the heat exchanger, where evaporation occurs. The expansion process is usually accomplished by throttling through a valve. A mechanical refrigeration system is a closed thermodynamic cycle.

Several arbitrary terms are used in expressing the performance of a refrigeration cycle. A common unit of capacity is the ton of refrigeration. *One ton of refrigeration* is defined as the useful withdrawal of heat at a rate of 200 Btu/min (3.52 kW). The word "ton" is an anachronism—a carryover from the days when ice was the principal means of refrigeration. If we take the heat of fusion of water as 144 Btu/lbm (335 kJ/kg), the heat

49

absorbed by one ton (2000 lbm, 907 kg) of ice in melting during a 24-hour period is 288,000 Btu (304,000 kJ). This rate of heat absorption is equivalent to an average of 200 Btu/min (3.52 kW).

The coefficient of performance expresses the energy effectiveness of a refrigeration system. It is a dimensionless ratio defined by the expression

$$\text{C.O.P.} \equiv \frac{\text{useful refrigerating effect}}{\text{net energy supplied from external sources}} \tag{3.1}$$

For a mechanical compression system, work must be supplied by an external source. Thus

$$\text{C.O.P.} = \frac{q_R}{w_{\text{net}}} \tag{3.2}$$

The refrigerating efficiency η_R expresses the approach of the cycle or system to that of an ideal reversible refrigerating cycle. By definition,

$$\eta_R \equiv \frac{\text{C.O.P.}}{(\text{C.O.P.})_{\text{rev}}} \tag{3.3}$$

The Carnot cycle usually serves as the ideal reversible cycle.

This chapter considers theoretical cycles. The single-stage cycle analysis is followed by a discussion of multistage cycles commonly found in commercial and industrial applications.

3.2 THE CARNOT REFRIGERATION CYCLE

The *Carnot cycle,* being completely reversible, is a model of perfection for a refrigeration cycle operating between two fixed temperature limits, or between two different-temperature fluids, each with an infinitely large heat capacity. Two important concepts involving reversible cycles are: (1) no refrigerating cycle may have a coefficient of performance higher than that for a reversible cycle which would be operated between the same temperature limits, and (2) all reversible cycles when operated between the same temperature limits would have the same coefficient of performance. Proof of both statements may be found in almost any book on elementary engineering thermodynamics.

Figure 3.1 shows the processes of the Carnot cycle on temperature-entropy coordinates. Heat is withdrawn at the constant temperature T_R from the region to be refriger-

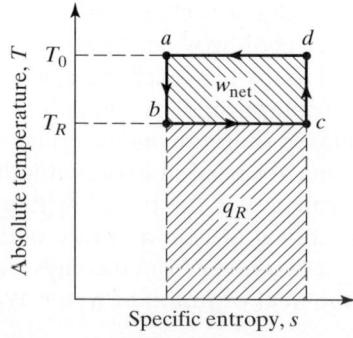

Figure 3.1 Processes of the Carnot refrigeration cycle.

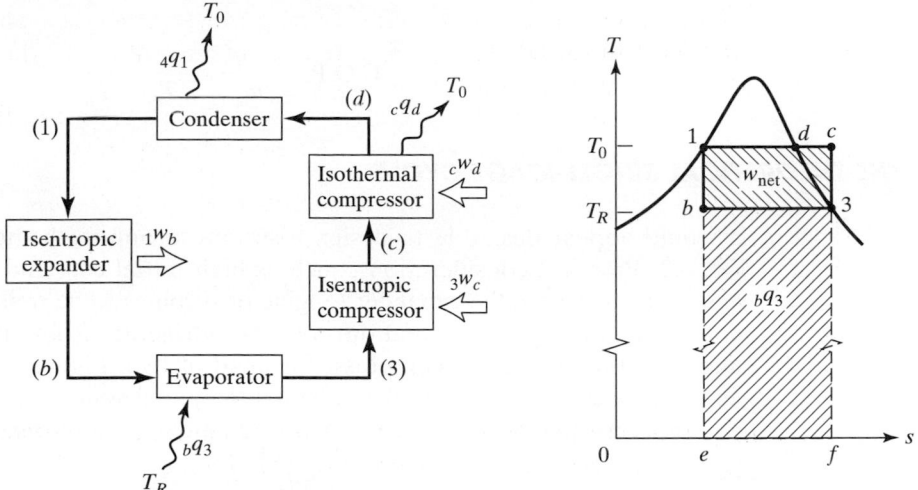

Figure 3.2 Carnot vapor-compression cycle.

ated. Heat is rejected at the constant ambient temperature T_0. The cycle is completed by an isentropic expansion and an isentropic compression. The energy transfers are given by

$$q_0 = T_0(s_d - s_a)$$
$$q_R = T_R(s_d - s_a)$$
$$w_{\text{net}} = q_0 - q_R$$

Thus, by Eq. (3.2),

$$\text{C.O.P.} = \frac{T_R}{T_0 - T_R} \qquad (3.4)$$

For refrigerants that operate in a two-phase region where the heat-addition and heat-rejection processes occur at constant pressure, it is desirable to consider the Carnot cycle as shown by Fig. 3.2. Saturated liquid at state 1 is expanded isentropically to the low temperature and pressure of the cycle at state b. Heat is added isothermally and isobarically by evaporating the liquid-phase refrigerant from state b to state 3. The cold saturated vapor at state 3 is compressed isentropically to the high temperature in the cycle at state c. However, the pressure at state c is below the saturation pressure corresponding to the high temperature in the cycle. The compression process is completed by an isothermal compression process from state c to state d. The cycle is completed by an isothermal and isobaric heat-rejection or condensing process from state d to state 1. By applying the steady-flow energy equation on a unit-mass basis, we obtain

$$_1w_b = h_1 - h_b$$
$$_3w_c = h_c - h_3$$
$$_cw_d = T_c(s_c - s_d) - (h_c - h_d)$$
$$_bq_3 = h_3 - h_b = \text{Area } bef3b$$

The net work for the cycle is

$$w_{\text{net}} = {}_3w_c + {}_cw_d - {}_1w_b = \text{Area } b3cd1b$$

and

$$\text{C.O.P.} = \frac{_bq_3}{w_{\text{net}}} = \frac{T_R}{T_0 - T_R} \tag{3.4a}$$

3.3 THE THEORETICAL SINGLE-STAGE CYCLE

It would appear desirable to design a system to approach the ideal model shown in Fig. 3.2. Practical considerations, such as high initial cost and increased maintenance requirements, cause the expander (engine or turbine) to be replaced by a simple expansion valve in most systems that throttles the refrigerant from high pressure to low pressure, and the use of one compressor instead of two. Figure 3.3 shows the theoretical single-stage cycle which is used as a model for actual systems.

By applying the steady-flow energy equation on a unit-mass basis, we obtain

$$_2q_3 = h_3 - h_2$$
$$_3w_4 = h_4 - h_3$$
$$_4q_1 = h_4 - h_1$$
$$h_1 = h_2$$

The constant-enthalpy throttling process assumes no heat transfer or change in potential or kinetic energy through the expansion valve.

The coefficient of performance is

$$\text{C.O.P.} = \frac{_2q_3}{_3w_4} = \frac{h_3 - h_2}{h_4 - h_3} \tag{3.5}$$

The theoretical compressor displacement C.D. (100 percent volumetric efficiency) is given by

$$\text{C.D.} = \dot{m}v_3 \tag{3.6}$$

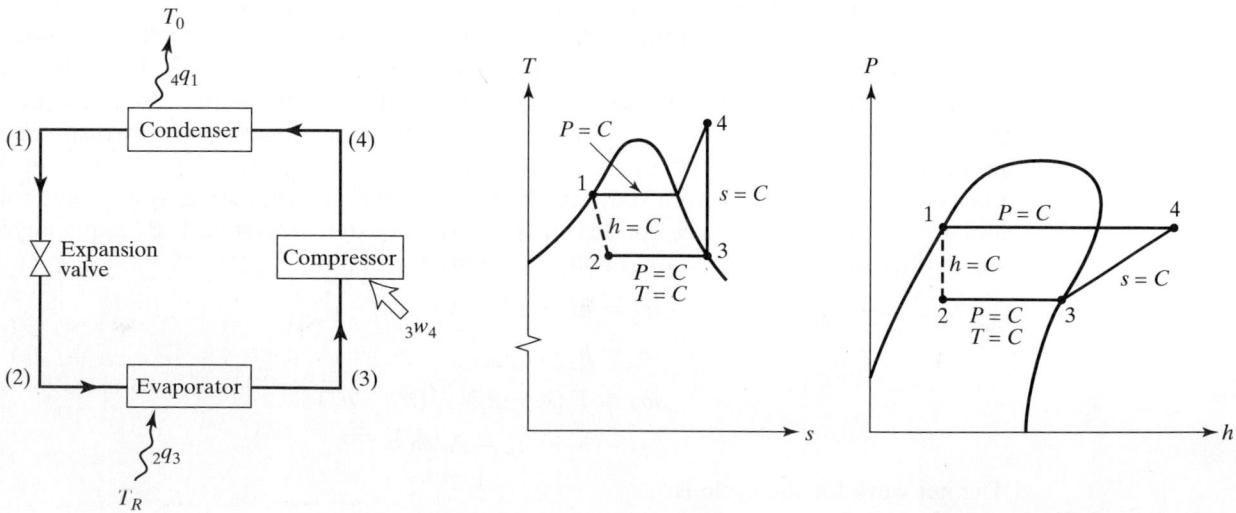

Figure 3.3 The theoretical single-stage vapor-compression cycle.

which is a measure of the physical size and speed of the compressor required to handle the prescribed refrigeration load.

EXAMPLE 3.1

An ammonia refrigerating plant following the theoretical single-stage cycle operates with a condensing temperature of 90 °F and an evaporating temperature of 0 °F. The system produces 15 tons of refrigeration. Determine (a) the coefficient of performance, (b) refrigerating efficiency, (c) rate of refrigerant flow in lbm per min, (d) theoretical horsepower input to compressor, and (e) theoretical displacement of the compressor in ft³/min.

Solution: Figure 3.4 shows a schematic *P-h* diagram for the problem with numerical property data. Saturated-liquid and saturated-vapor properties were read from Table A.2E. The enthalpy h_4 = 723.5 Btu/lbm was read from Fig. C-1E (Mollier chart for ammonia) at the intersection of s_3 = 1.3326 Btu/lbm · °R and P_4 = 180.73 psia. The property data are provided in Table 3.1.

(a) By Eq. (3.5)

$$\text{C.O.P.} = \frac{610.8 - 143.3}{723.5 - 610.8} = 4.15$$

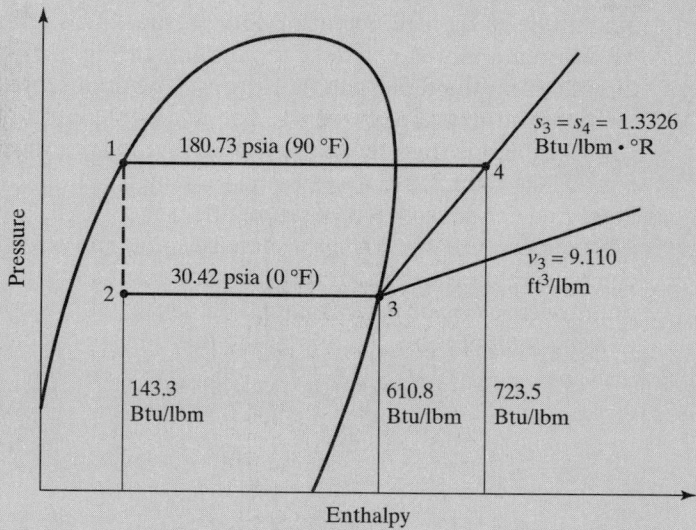

Figure 3.4 Schematic *P-h* diagram for Ex. 3.1.

TABLE 3.1 Thermodynamic Property Data for Ex. 3.1

State	T, °F	P, psia	v, ft³/lbm	h, Btu/lbm	s, Btu/lbm · °R
1	90.0	180.73	0.027	143.3	0.2951
2	0.0	30.42	1.637	143.3	0.3158
3	0.0	30.42	9.110	610.8	1.3326
4	230.9	180.73	2.284	723.5	1.3326

(b) By Eq. (3.3)

$$\eta_R = \frac{\text{C.O.P.}\,(T_1 - T_2)}{T_2} = \frac{(4.15)(90)}{459.6} = 0.81 \text{ or } 81 \text{ percent}$$

(c) The rate of refrigerant flow is obtained from an energy balance on the evaporator. Thus

$$\dot{m}(h_3 - h_2) = (\text{tons})(200)$$

and

$$\dot{m} = \frac{(15)(200)}{h_3 - h_2} = \frac{3000}{467.5} = 6.42 \text{ lbm/min}$$

(d) The theoretical horsepower may be found by

$$\text{Hp} = \frac{\dot{m}(h_4 - h_3)}{42.4} = \frac{(6.42)(112.7)}{42.4} = 17.1 \text{ hp}$$

(e) The theoretical compressor displacement (100 percent volumetric efficiency) may be found by

$$\text{C.D.} = \dot{m}v_3 = (6.42)(9.110) = 58.5 \text{ ft}^3/\text{min}$$

The saturation temperatures of the single-stage cycle have a strong influence on the magnitude of the coefficient of performance. This influence may be readily appreciated by an area analysis on the T-s diagram. In Fig. 3.5, the area representing $_2q_3$ follows directly from the definition of entropy. The area representing $_4q_1$ is the total area under the constant-pressure curve 41. The net work required, $_3w_4$, is equal to the difference, $_4q_1 - {_2q_3}$ and is therefore represented by the area shown in Fig. 3.5.

Since the C.O.P. $= {_2q_3}/{_3w_4}$, we may observe how changes in evaporating temperature and condensing temperature affect the C.O.P. We find that a decrease in evaporating temperature significantly increases $_3w_4$ and slightly decreases $_2q_3$. An increase in condensing pressure produces the same results but with lesser effect on $_3w_4$. For maximum coefficient of performance, the cycle should operate at the lowest possible condensing temperature, T_0, and at the maximum possible evaporating temperature, T_R.

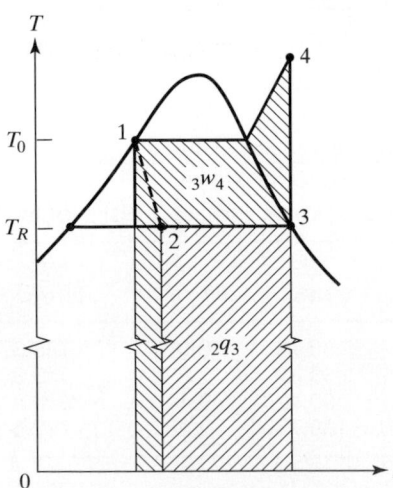

Figure 3.5 Areas on the T-s diagram representing the refrigerating effect and the work supplied for the theoretical single-stage cycle.

3.4 COMPARISON OF THE THEORETICAL SINGLE-STAGE CYCLE WITH THE CARNOT CYCLE

An interesting and informative analysis consists of comparing the processes of the single-stage cycle with those of the Carnot cycle on T-s coordinates. Figure 3.6 shows the single-stage cycle superimposed on the Carnot cycle. We assume that states 1 and 3 are common to both cycles.

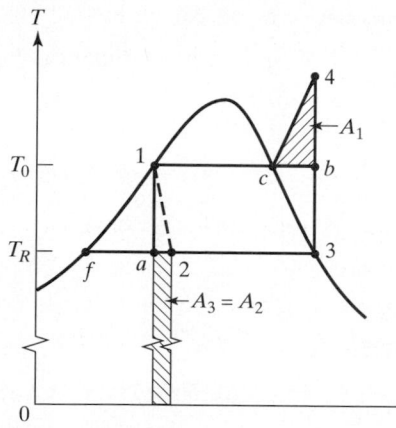

Figure 3.6 Area deviations on the T-s diagram for the theoretical single-stage cycle, with respect to the Carnot cycle.

On an area basis, the single-stage cycle deviates from the Carnot cycle in three ways. Area A_1 represents additional work required for the single-stage cycle because of the superheat horn. Area A_2 represents additional work required for the single-stage cycle because no work is recovered in the expansion process. Area A_3 represents a loss of cooling effect caused by throttling as compared to the isentropic expansion of the Carnot cycle. On an energy-per-unit-mass basis

$$A_1 = (h_4 - h_c) - T_1(s_3 - s_c) \tag{3.7}$$

$$A_2 = h_2 - h_a = h_1 - h_a = T_R(s_2 - s_1) \tag{3.8}$$

$$A_3 = h_2 - h_a = h_1 - h_a = A_2 \tag{3.9}$$

Thus throttling causes an equal loss in work recovered and in refrigerating capacity. Since

$$_3w_4 = w_{\text{Carnot}} + A_1 + A_2$$

and

$$_2q_3 = q_{\text{Carnot}} - A_2$$

we may show that for the single-stage cycle

$$\eta_R = \frac{1 - (A_2/q_{\text{Carnot}})}{1 + \dfrac{A_1 + A_2}{w_{\text{Carnot}}}} \tag{3.10}$$

EXAMPLE 3.2

A Refrigerant-22 theoretical single-stage cycle operates with a condensing temperature of 30 °C and an evaporating temperature of 0 °C. As shown in Fig. 3.6, we will assume a Carnot cycle operating between the same temperatures. Determine (a) the Carnot-cycle work of compression in kJ/kg, (b) the Carnot-cycle refrigerating effect in kJ/kg, (c) the excess work of compression in kJ/kg for the single-stage cycle caused by the superheat horn, (d) the excess work of compression in kJ/kg for the single-stage cycle caused by throttling, (e) the loss in refrigerating effect in kJ/kg for the single-stage cycle caused by throttling, and (f) the refrigerating efficiency.

Solution: Following the nomenclature of Fig. 3.6, we have

(a) $w_{Carnot} = (T_1 - T_2)(s_3 - s_1) = (30)(0.9270 - 0.3004) = 18.80 \text{ kJ/kg}.$

(b) $q_{Carnot} = T_2(s_3 - s_1) = (273.15)(0.6266) = 171.16 \text{ kJ/kg}.$

(c) By Eq. (3.7)

$$A_1 = (271.47 - 259.12) - (303.15)(0.9270 - 0.8872) = 0.28 \text{ kJ/kg}$$

(d) To evaluate A_2, we must first calculate h_a. We have

$$s_1 = s_a = s_f + x_a(s_3 - s_f)$$

$$x_a = \frac{s_1 - s_f}{s_3 - s_f} = \frac{0.3004 - 0.1751}{0.9270 - 0.1751} = 0.1666$$

$$h_a = h_f + x_a(h_3 - h_f) = 44.59 + (0.1666)(249.95 - 44.59) = 78.80 \text{ kJ/kg}$$

By Eq. (3.8),

$$A_2 = 81.26 - 78.80 = 2.46 \text{ kJ/kg}$$

(e) The loss of refrigerating effect $A_3 = A_2 = 2.46 \text{ kJ/kg}.$

(f) The refrigerating efficiency may be calculated in two ways. Thus

$$\eta_R = \frac{(h_3 - h_2)(T_1 - T_2)}{(h_4 - h_3)T_2} = \frac{(168.69)(30)}{(21.52)(273.15)} = 0.86$$

or, by Eq. (3.10),

$$\eta_R = \frac{1 - (2.46/171.16)}{1 + \dfrac{0.28 + 2.46}{18.80}} = 0.86$$

Thus for Refrigerant 22 the theoretical single-stage cycle has an extremely small superheat horn. The throttling process accounts for almost the entire deviation from the Carnot cycle.

The deviations shown in Fig. 3.6 are dependent upon the shapes of the saturation lines on the *T-s* diagram. The area A_2 would disappear if the saturated liquid line were vertical. However, this would require the liquid to have a specific heat of zero. The superheat horn (A_1) would disappear if the saturated-vapor line were vertical or if it had a positive slope.

The shapes of the saturation lines vary widely among the various refrigerants. Ammonia has symmetrically shaped saturation lines with both area deviations being of significant size. Figure 3.7 shows that Refrigerant 22 has an extremely small superheat horn but a relatively large deviation area caused by throttling. Refrigerant 11 is closely similar in these respects. Refrigerant 12 is also similar but with a relatively smaller superheat horn. Refrigerants 113 and 114 have no superheat horn.

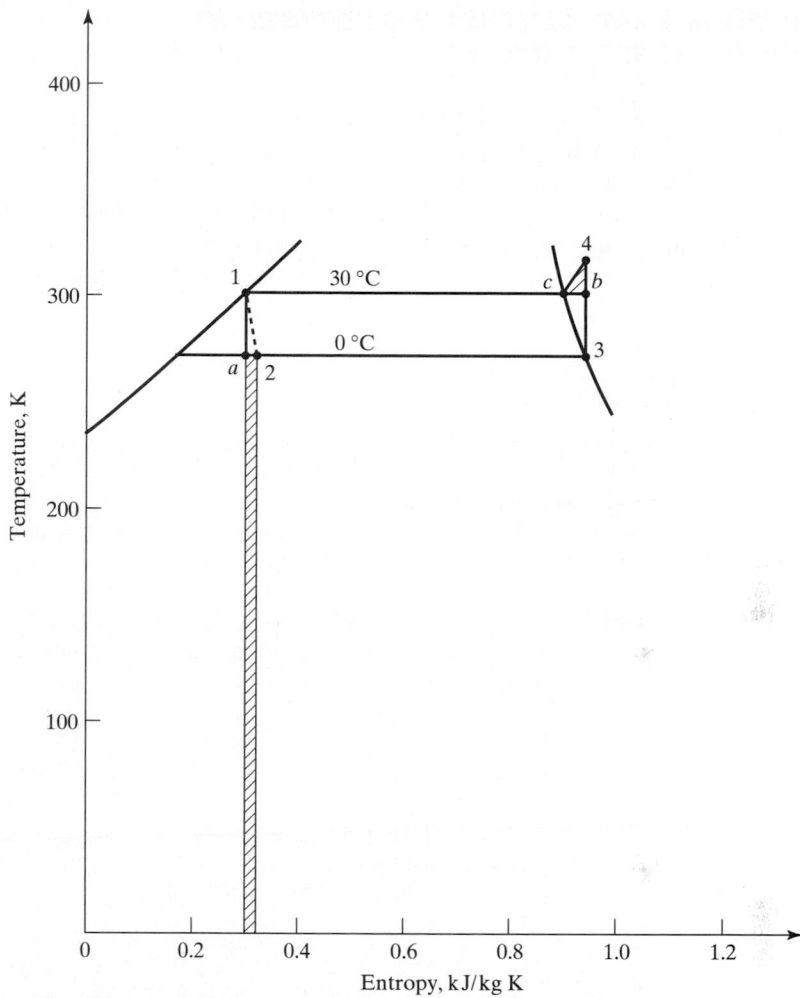

Figure 3.7 Temperature-entropy diagram for Refrigerant 22, showing area deviations of theoretical single-stage cycle with respect to the Carnot cycle.

The combined effects of the deviations from the Carnot cycle are included in the refrigerating efficiency. The refrigerating efficiency depends on the refrigerant being used in addition to the saturated evaporating and condensing temperatures.

The comparison of the theoretical single-stage cycle and the Carnot cycle on a *T-s* diagram cannot be readily extended to more complex systems such as multistage cycles or real refrigeration systems. Multistage cycles have different refrigerant mass-flow rates through various processes. Actual systems do not have internally reversible processes, so the process lines drawn on a *T-s* diagram do not form boundaries for heat-transfer or work areas. Another method of comparison, a second-law irreversibility analysis, can be used to compare any refrigeration system with a Carnot cycle. This method of analysis is straightforward to apply and is demonstrated for a real refrigeration system in Chapter 4.

3.5 SUBCOOLING AND SUPERHEATING REFRIGERANT IN THE SINGLE-STAGE CYCLE

Frequently, the liquid refrigerant entering the expansion valve may be *subcooled*, whereby its temperature is less than the saturation temperature corresponding to its pressure. Subcooling may occur within the condenser or within the liquid line by heat exchange with the ambient surroundings. The vapor may be superheated a few degrees in the evaporator and may be further superheated in the compressor suction line. However, the only useful refrigerating effect is the enthalpy change within the evaporator.

EXAMPLE 3.3

Liquid refrigerant leaves the condenser of an ammonia-vapor compression system under a pressure of 1.35 MPa and at a temperature of 30 °C. Evaporating pressure is 0.1195 MPa and the vapor leaves the evaporator at −20 °C. Otherwise, the system follows the theoretical single-stage cycle. Find (a) the enthalpy of the liquid entering the expansion valve, (b) the refrigerating effect in kJ/kg, and (c) the coefficient of performance.

Solution: Figure 3.8 shows schematic *T-s* and *P-h* diagrams for the problem. If the liquid is assumed to be incompressible, the entropy of the subcooled liquid is equal to the entropy of saturated liquid at the same temperature.

  (a) Assuming the liquid to be incompressible, we may show that

$$h_1 = h_a + (P_1 - P_a)v_a = 322.6 + (1.35 - 1.167)(1000)(0.00168)$$

$$= 322.9 \text{ kJ/kg}$$

The above calculation shows that h_1 differs from h_a by a negligible amount. In ordinary refrigeration-cycle problems it is not necessary to correct for this small difference. The

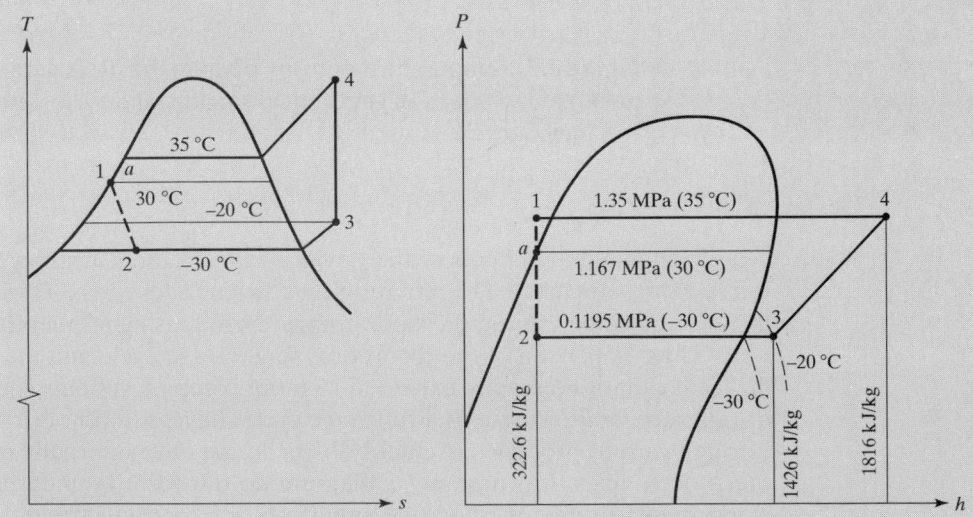

Figure 3.8 Schematic *T-s* and *P-h* diagrams for Ex. 3.3.

enthalpy of a subcooled liquid refrigerant may be taken as equal to the enthalpy of the saturated liquid at the same temperature. Thus, use

$$h_1 = 322.6 \text{ kJ/kg}$$

(b) The evaporating temperature is -30 °C and the vapor leaving the evaporator is at -20 °C. Thus, the vapor is superheated 10 °C. State 3 may be located graphically on the Mollier chart by the intersection of $P_3 = 0.1195$ MPa and $t_3 = -20$ °C. Thus $h_3 = 1426$ kJ/kg and

$$_2q_3 = h_3 - h_2 = 1426 - 322.6 = 1103.4 \text{ kJ/kg}$$

(c) State 4 for the vapor leaving the compressor may also be located graphically through use of the Mollier chart. A sufficiently accurate procedure is as follows. Observe the position of state 3 with regard to the two lines of constant entropy which include it. With a pencil trace a path upward between the two lines of constant entropy. Locate state 4 on the line representing P_4 and at the same proportional position with respect to entropy. Thus, at $P_4 = 1.35$ MPa, $h_4 = 1816$ kJ/kg. A more accurate approach is to use thermodynamic equations of state for superheated NH_3 vapor. Setting $P_4 = 1.35$ MPa and $s_4 = s_3 = 5.868$ kJ/kg · K, $h_4 = 1816$ kJ/kg. By Eq. (3.5),

$$\text{C.O.P.} = \frac{1426 - 322.6}{1816 - 1426} = 2.83$$

3.6 PERFORMANCE OF SINGLE-STAGE CYCLE AT LOW EVAPORATING TEMPERATURES

The single-stage vapor compression cycle is an efficient refrigerating method at relatively high evaporating temperatures. As the evaporating temperature is reduced, the coefficient of performance decreases, power requirements increase for a given refrigerating load, and the required compressor displacement increases. There are three principal disadvantages of single-stage operation at low evaporating temperatures. These are (1) low refrigerating efficiency, (2) large compressor displacement, and (3) high compressor discharge temperature, which may cause valve failure and lubrication problems.

In Sec. 3.4 we discussed the three area deviations on the T-s diagram for the theoretical single-stage cycle as compared to the Carnot cycle. In Eq. (3.10) the ratios A_2/q_{Carnot}, A_1/w_{Carnot}, and A_2/w_{Carnot} all increase with decrease in evaporating temperature, thus causing the refrigerating efficiency η_R to decrease.

Figure 3.9 shows the variation of coefficient of performance and refrigerating efficiency for the theoretical single-stage cycle for refrigerants 12 and 22 and for ammonia. A relatively low C.O.P. is inevitable at the lower evaporating temperatures. However, the decrease in η_R shows that the single-stage cycle deviates increasingly from the Carnot cycle as the evaporating temperature is reduced.

A sharp increase of compressor displacement for the theoretical single-stage cycle occurs as the evaporating temperature is reduced. This is caused primarily by the increase in specific volume of the refrigerant entering the compressor. Decrease of compressor volumetric efficiency imposes a practical limitation on the single-stage cycle and makes staging of compressors necessary in low-temperature systems, irrespective of other considerations.

Figure 3.10 shows that at relatively low evaporating temperatures, the compressor discharge temperature in an ammonia system becomes extremely high. Practically, in

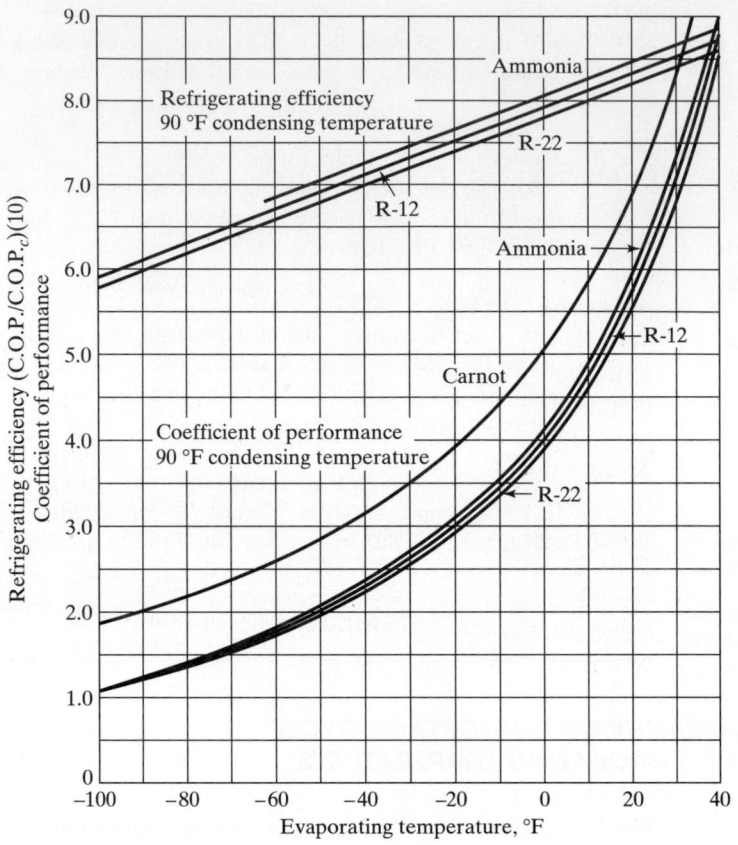

Figure 3.9 Variation of coefficient of performance and refrigerating efficiency for theoretical single-stage cycle.

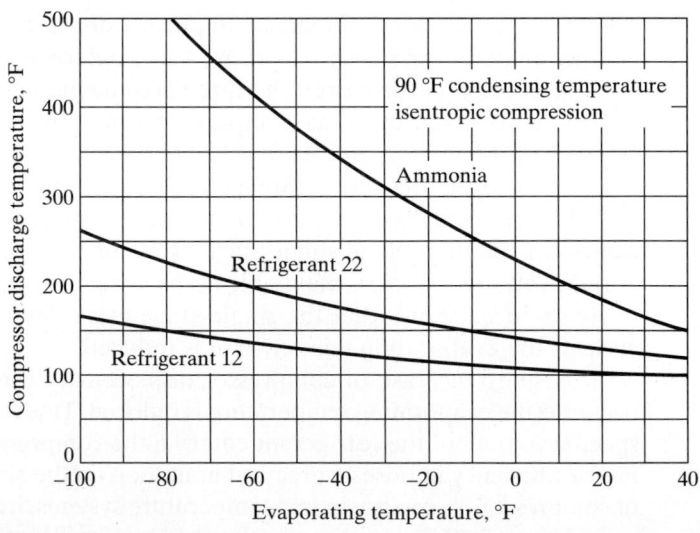

Figure 3.10 Variation of compressor discharge temperature for theoretical single-stage cycle.

Refrigerant-22 systems as well as in ammonia systems, staging of compressors and vapor intercooling becomes necessary at low evaporating temperatures in order to prevent overheating of the compressor.

3.7 THEORETICAL MULTISTAGE VAPOR-COMPRESSION CYCLES

By using multiple compressors and with intercooling of liquid and vapor refrigerant, the disadvantages of the single-stage cycle may be largely overcome. At low evaporating temperatures, staging of compressors is necessary because of reduced volumetric efficiency. With refrigerants such as ammonia, compressor staging and vapor intercooling are required to prevent excessive discharge temperature. Fortunately, the same arrangements allow a significant improvement in coefficient of performance also. In this section we will deal principally with multistage arrangements where ammonia and Refrigerant 22 are used.

Figure 3.11 shows a two-stage cycle suitable for use with ammonia. The high-pressure side of the cycle (states 8, 9, 1, 2) can be analyzed as a theoretical single-stage cycle. The only difference is that the evaporator has been replaced by the flash intercooler and the other components that comprise the low-pressure side of the cycle. The flash intercooler is a pressure vessel or tank in which a fixed liquid level of refrigerant is maintained by a float valve. The float valve also serves as expansion valve A. Saturated liquid at the cycle intermediate pressure and temperature is withdrawn from the flash intercooler (state 3) and is expanded through expansion valve B to the evaporator pressure. Feeding expansion valve B with pure liquid prevents surging due to vapor bubbles and increases the enthalpy difference across the evaporator. The liquid and vapor mixture that enters the evaporator (state 4) is changed to saturated vapor at state 5. The vapor is compressed isentropically back to the intermediate pressure in the cycle through the low-pressure compressor. The superheated vapor leaving the low-pressure compressor is first cooled by water in an intercooler. This heat exchanger may not be feasible if the temperature of

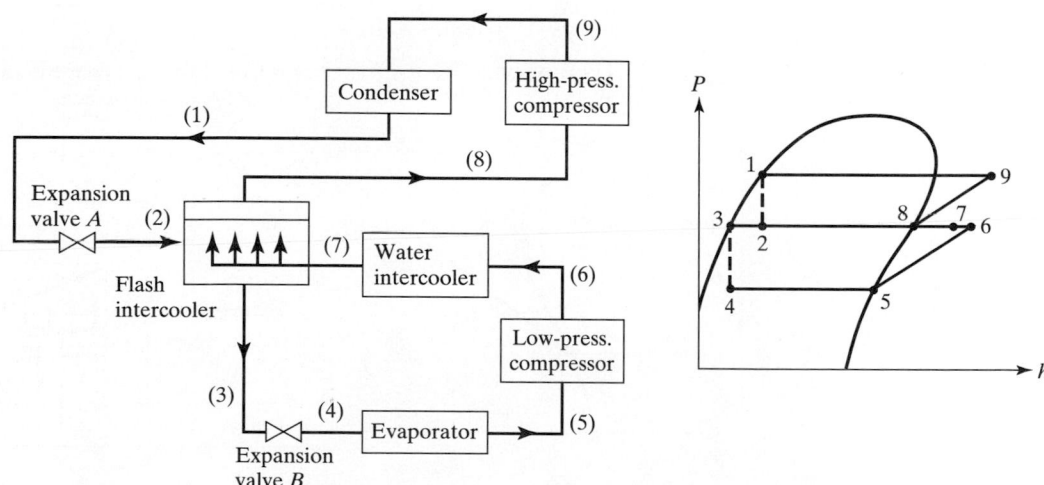

Figure 3.11 Schematic diagrams for cycle with two-stage compression, water intercooler, and flash intercooler.

the available cooling water is not much less than T_6. The vapor is further cooled to saturated-vapor conditions (state 8) by direct contact with the relatively cold liquid in the flash intercooler.

The flash intercooler of Fig. 3.11 is of particular interest. The flash vapor formed by the throttling process in the float valve (expansion valve A) is removed to the suction line of the high-pressure compressor. The superheated vapor at state 7 is admitted to the bottom of the tank through a slotted pipe or through orifices, and it bubbles up through the liquid. The vapor is de-superheated by evaporation of liquid. The extent of de-superheating depends upon the direct-contact heat-transfer rate. Complete de-superheating is assumed in Fig. 3.11, although this is impossible practically. The vapor entering the high-pressure compressor thus includes three components: (a) the flash vapor formed in the float valve, (b) the refrigerant circulated through the evaporator, and (c) the vapor formed by evaporation of liquid in the flash intercooler.

Since equipment arrangement in a multistage system may vary with the purpose of the system and with the refrigerant used, no development of special formulae will be given here. Solution of a typical problem involving the system of Fig. 3.11 is shown by the following example, which illustrates the advantages of the multistage cycle compared to a theoretical single-stage cycle operating with the same load and temperature levels.

EXAMPLE 3.4

Determine (a) the coefficient of performance, (b) the maximum cycle temperature, and (c) the theoretical compressor displacement in ft³/min · ton refrigeration for the theoretical single-stage cycle of Fig. 3.3 and the two-stage cycle of Fig. 3.11. For each cycle assume an evaporating temperature of $-50\ °F$, a condensing temperature of $100\ °F$, isentropic compression with a compressor volumetric efficiency of 100 percent, and ammonia as the refrigerant. For the two-stage cycle assume that the intermediate saturation temperature is 12 °F and that the vapor leaves the water intercooler at 100 °F.

Solution: We will first consider the single-stage cycle. Figure 3.12(a) shows a schematic *P-h* diagram. By Table A.2E and Fig. C-1E we obtain

$$h_1 = h_2 = 155.0 \text{ Btu/lbm}, \qquad h_3 = 593.1 \text{ Btu/lbm}, \qquad v_3 = 33.09 \text{ ft}^3/\text{lbm}$$

$$s_3 = 1.4479 \text{ Btu/lbm} \cdot °R, \qquad h_4 = 827.0 \text{ Btu/lbm}, \qquad t_4 = 410 \text{ °F}$$

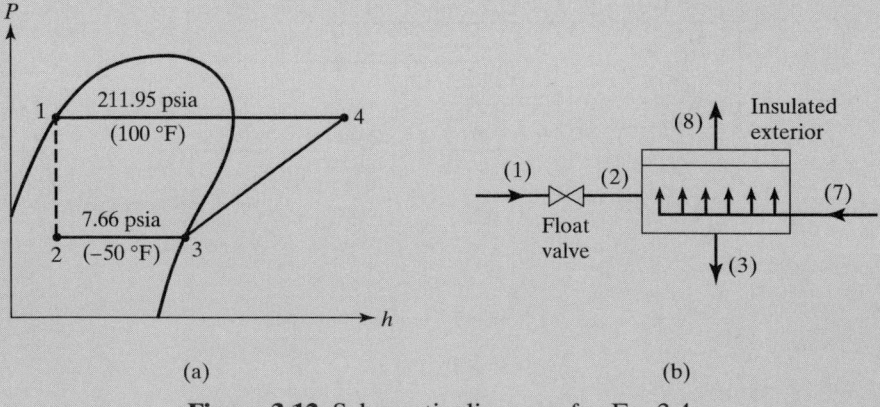

Figure 3.12 Schematic diagrams for Ex. 3.4.

(a) By Eq. (3.6)

$$\text{C.O.P.} = \frac{593.1 - 155.0}{827.0 - 593.1} = 1.87$$

(b) $t_{\max} = t_4 = 410\ °F$

(c) $\text{C.D.} = m v_3 = \dfrac{(200\ \text{Btu/min} \cdot \text{ton}) v_3}{h_3 - h_2} = \dfrac{(200\ \text{Btu/min} \cdot \text{ton}) 33.09\ \text{ft}^3/\text{lbm}}{(593.1 - 155.0)\ \text{Btu/lbm}}$

$$= 15.1\ \text{ft}^3/\text{min} \cdot \text{ton}$$

For the two-stage cycle of Fig. 3.11 we obtain the following property data:

$h_1 = h_2 = 155.0\ \text{Btu/lbm},\quad h_3 = h_4 = 55.4\ \text{Btu/lbm},\quad h_5 = 593.1\ \text{Btu/lbm}$

$v_5 = 33.09\ \text{ft}^3/\text{lbm},\quad s_5 = 1.4479\ \text{Btu/lbm} \cdot °R,\quad h_6 = 689\ \text{Btu/lbm}$

$t_6 = 145\ °F,\quad h_7 = 664\ \text{Btu/lbm},\quad h_8 = 614.5\ \text{Btu/lbm},\quad v_8 = 6.989\ \text{ft}^3/\text{lbm}$

$s_8 = 1.3092\ \text{Btu/lbm} \cdot °R,\quad h_9 = 719.4\ \text{Btu/lbm},\quad t_9 = 229\ °F$

(a) The coefficient of performance must be dimensionless. Since the mass flows through the two loops of the system are different, we may *not* express the C.O.P. as a ratio of enthalpy differences alone. For the low-pressure loop

$$\dot{m}_3 = \frac{200\ \text{Btu/min} \cdot \text{ton}}{h_5 - h_4} = \frac{200\ \text{Btu/min} \cdot \text{ton}}{537.7\ \text{Btu/lbm}} = 0.372\ \text{lbm/min} \cdot \text{ton}$$

Figure 3.12(b) shows the flash intercooler as an isolated thermodynamic system. The fundamental steady-flow energy and mass-balance equations are

$$\dot{m}_2 h_2 + \dot{m}_7 h_7 = \dot{m}_3 h_3 + \dot{m}_8 h_8$$

$$\dot{m}_2 + \dot{m}_7 = \dot{m}_3 + \dot{m}_8$$

However, for this cycle, $\dot{m}_2 = \dot{m}_8$ and $\dot{m}_3 = \dot{m}_7$. Thus

$$\dot{m}_8 (h_8 - h_2) = \dot{m}_3 (h_7 - h_3)$$

By Eq. (3.1),

$$\text{C.O.P.} = \frac{\dot{m}_3 (h_5 - h_4)}{\dot{m}_3 (h_6 - h_5) + \dot{m}_8 (h_9 - h_8)} = \frac{(h_5 - h_4)}{(h_6 - h_5) + \dfrac{\dot{m}_8}{\dot{m}_3} (h_9 - h_8)}$$

$$= \frac{(h_5 - h_4)}{(h_6 - h_5) + \dfrac{(h_7 - h_3)}{(h_8 - h_2)} (h_9 - h_8)} = \frac{537.7}{95.5 + \dfrac{(609)}{(459.5)} (105)} = 2.29$$

(b) $t_{\max} = t_9 = 229\ °F$

(c) For the low-pressure compressor

$$\text{C.D.} = \dot{m}_3 v_5 = (0.372)(33.09) = 12.3\ \text{ft}^3/\text{min} \cdot \text{ton}$$

For the high-pressure compressor

$$\dot{m}_8 = \dot{m}_3 \frac{(h_7 - h_3)}{(h_8 - h_2)} = (0.372) \frac{(609)}{(459.5)} = 0.493\ \text{lbm/min} \cdot \text{ton}$$

and

$$\text{C.D.} = \dot{m}_8 v_8 = (0.493)(6.989) = 3.45\ \text{ft}^3/\text{min} \cdot \text{ton}$$

TABLE 3.2 Calculated Results for Ex. 3.4

Cycle	C.O.P.	t_{max}, °F	C.D. ft³/min · ton		
			L.P.	H.P.	Total
Single-stage	1.87	410	—	—	15.1
Two-stage	2.29	229	12.3	3.45	15.8

The results of the example are summarized in Table 3.2. The total compressor displacements are nearly equal in this example, but the value becomes much larger for the single-stage cycle when realistic volumetric efficiencies are used. The compressor displacement for the single-stage cycle can be twice the value for the two-stage cycle in practice.

In Example 3.4, the decrease of maximum temperature with the two-stage cycle is due to the intercooling between the two stages of compression. We will further examine the reasons for the substantial increase in the coefficient of performance. The method of analysis, described in Sec. 3.4 and shown in Fig. 3.6, which was used for the single-stage cycle, is not straightforward for the two-stage cycle. Complications arise with areas on the *T-s* diagram because of different mass flowrates in the system.

For the two-stage cycle, we can easily show that the throttling loss (work recoverable by an isentropic expansion) is considerably less than it is for the single-stage cycle. This is apparent from practical reasoning as well, since the flash vapor formed in the throttling process from state 1 to state 2 is compressed in the high-pressure compressor only. Plank [1] has shown that the intercooling of vapor by evaporation of liquid in the flash intercooler results in an increase of C.O.P. except for conditions near the critical point.

A further matter of thermodynamic importance with the cycle of Fig. 3.11 is determination of the optimum interstage pressure which would maximize the C.O.P. for a given set of operating temperatures. It is not possible to determine mathematically an optimum pressure for a vapor refrigerant, as would be the case for a perfect gas. For two-stage compression of a perfect gas with complete intercooling we may show that, for minimum work, the interstage pressure P_i is given by

$$P_i = (P_A P_B)^{0.5} \tag{3.11}$$

where P_A is the suction pressure of the low-pressure compressor and P_B is the discharge pressure of the high-pressure compressor. Figure 3.13 shows that for maximum C.O.P. the optimum interstage pressure is slightly higher than that given by Eq. (3.11). Another method of determining interstage pressure is to select the value which provides equal compressor displacement in the two compressors.

The flash intercooler of Fig. 3.11 has practical disadvantages. The liquid refrigerant in the tank is at the interstage pressure and is saturated. If expansion valve *B* is above the flash intercooler, or if heat is absorbed in the liquid line ahead of it, some liquid will evaporate ahead of the expansion valve, which will cause surging. In addition, the operation of the expansion valve may be sluggish because the pressure differential is too small.

Figure 3.14 shows another two-stage cycle used in ammonia systems. The flooded-type shell-and-coil intercooler effectively subcools the liquid refrigerant and eliminates the flashing of liquid ahead of expansion valve *B*. In addition, the pressure differential across expansion valve *B* is larger, because the liquid is at the condensing pressure. How-

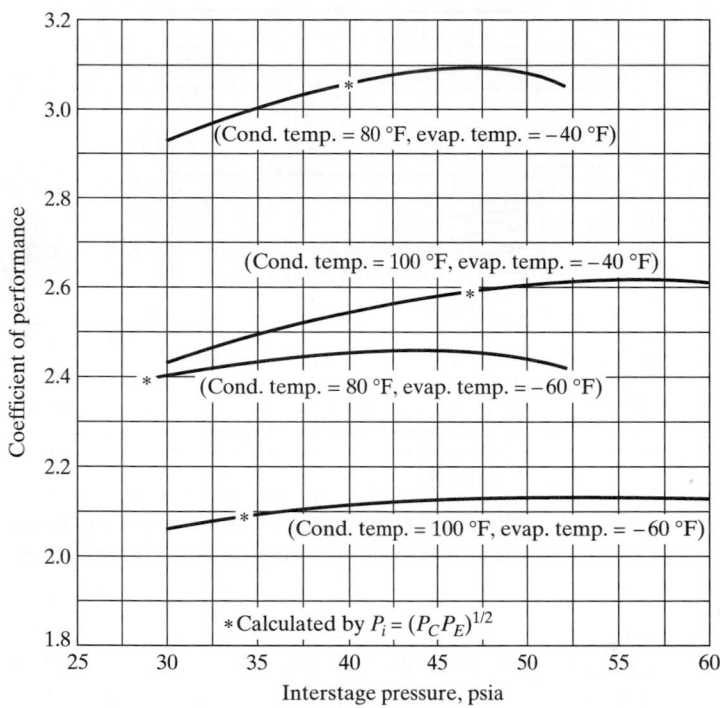

Figure 3.13 Optimum interstage pressure for cycle of Fig. 3.11.

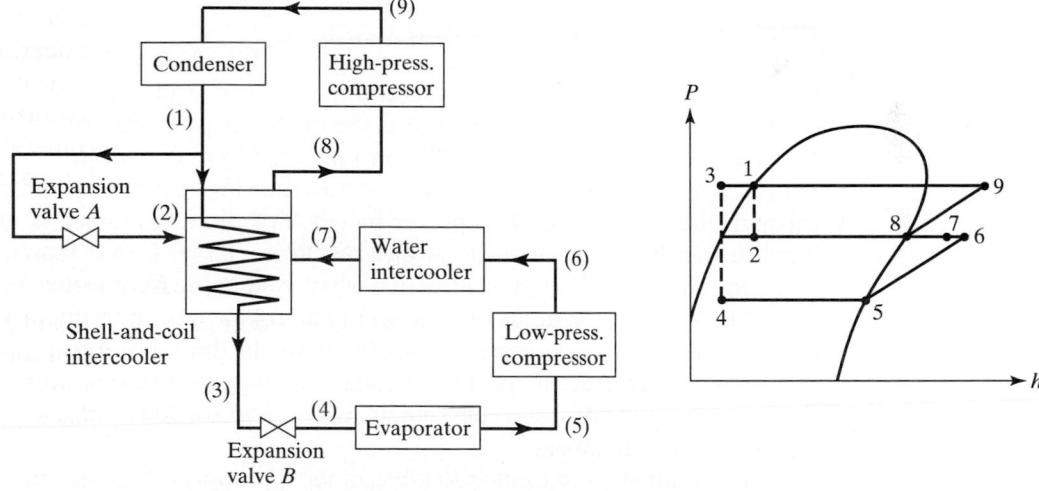

Figure 3.14 Schematic diagrams for cycle with two-stage compression, water intercooler, and flooded-type shell-and-coil intercooler.

ever, the use of the flooded shell-and-coil intercooler results in a slightly lower coefficient of performance, since it is not practically possible to intercool the liquid (state 3) as much as in the flash intercooler.

Figure 3.15 shows a two-stage cycle suitable for use with Refrigerant 22. The vapor discharged by the low-pressure compressor is not intercooled, except by mixing with

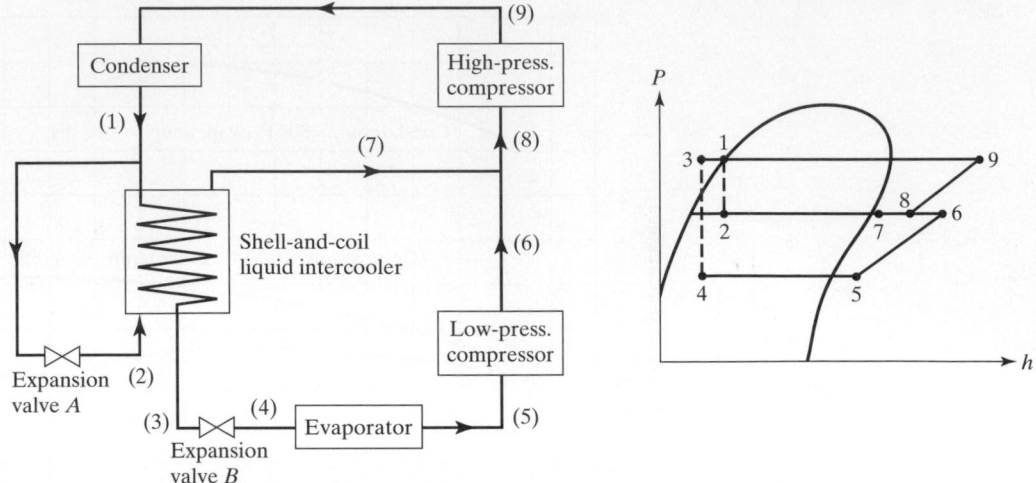

Figure 3.15 Schematic diagrams for cycle with two-stage compression and dry-type, shell-and-coil liquid intercooler.

refrigerant from the intercooler. The intercooler is typically a dry type. Instead of a float valve, as shown in Figs. 3.11 and 3.14, a thermostatic expansion valve is used for expansion valve *A*. The feeler bulb of the valve is usually placed in the suction line of the high-pressure compressor after the mixing of the two streams.

A variation of this arrangement is called an *economizer cycle*. The two compressors shown in Figure 3.15 are replaced by a single compressor, usually a screw compressor, in which the refrigerant vapor leaving the intercooler enters through a secondary suction port. Vapor at state 5 enters the compressor and is compressed to state 6. Then the gas from the intercooler mixes with this compressed gas inside the compressor to form state 8. The mixed gas is then compressed to the discharge pressure, state 9. The mass flow rate through the compressor at state 5 is not affected by the mass flow at state 7. Therefore the refrigerating capacity of the cycle increases, owing to the increased refrigerating effect through the evaporator. The power increase of the compressor is not as large as the increase in refrigerating capacity, which results in a larger C.O.P. than a single-stage cycle. The economizer cycle is most effective when system pressure ratios are larger than 2.

The two-stage systems of Figs. 3.11 and 3.14 have been frequently used with ammonia for industrial and commercial applications. In the food industry, ice cream holding rooms and sharp-freezer rooms are usually refrigerated in this way. Refrigerant 22 has been widely applied in systems similar to Fig. 3.15 for low-temperature test rooms and environmental chambers.

The multistage system is flexible in its application. One or more evaporators may be operated at the intermediate pressure, in addition to the low-temperature evaporator. Figure 3.16 shows such a system in which an evaporator operating at the intermediate pressure is added to the cycle of Fig. 3.14. Note that the locations of the state points on the pressure-enthalpy diagrams of Figs. 3.14 and 3.16 are identical.

The number of stages of compression is determined by economic as well as practical considerations. For Refrigerants 12, 22, and ammonia, single-stage operation is indicated for evaporating temperatures above about −20 °F (−30 °C). The single-stage system may be more successfully used at lower temperatures with Refrigerants 12 and 22

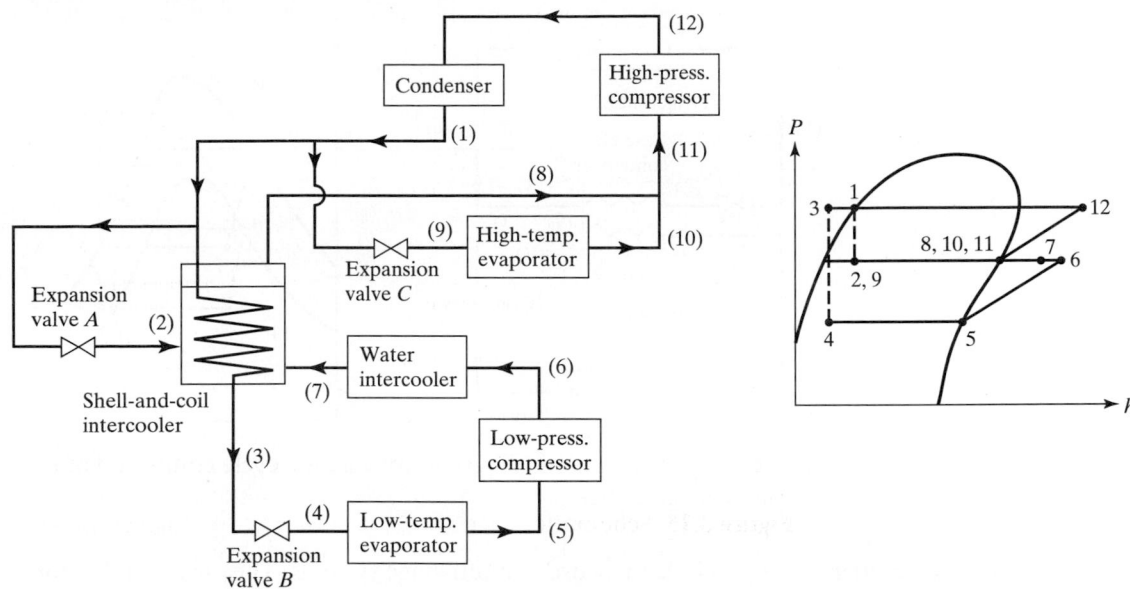

Figure 3.16 Schematic diagrams for two-stage cycle with both low-pressure and intermediate-pressure evaporators.

than with ammonia. In the approximate range of evaporating temperatures from $-75\,°F$ to $-20\,°F$ ($-60\,°C$ to $-30\,°C$), two-stage operation is usually indicated.

The multistage vapor-compression system is highly successful but has limitations. It uses only one refrigerant, so that a refrigerant with a reasonably high critical temperature is required. The minimum evaporating temperature depends primarily upon the pressure-temperature characteristics of the refrigerant.

Table 3.3 shows thermodynamic data for several vapor refrigerants. Of those shown, Refrigerants 12, 22, and ammonia are the only ones suitable for multistage systems. If, for practical purposes, we assume an evaporating pressure of one psia (0.007 MPa) as a minimum, Table 3.3 shows minimum evaporating temperatures of approximately $-105\,°F$ ($-76\,°C$) for ammonia, $-109\,°F$ ($-78\,°C$) for Refrigerant 12, and $-122\,°F$ ($-85\,°C$) for Refrigerant 22. In general, we may obtain minimum space or product temperatures of about $-100\,°F$ ($-75\,°C$) in such systems.

Table 3.3 shows that Refrigerants 13 and 14 evaporate at much lower temperatures than Refrigerants 12, 22, and ammonia for a given pressure. However, their low critical

TABLE 3.3 Pressure and Temperature Data for Various Refrigerants

Refrigerant	Saturation Temperature at 1 atm, °F (°C)	Saturation Temperature at 1 psia, °F (0.007 MPa, °C)	Freezing Temperature, °F (°C)	Critical Temperature, °F (°C)	Critical Pressure, psia (MPa)
Ammonia	−28.0 (−33.0)	−105 (−76)	−107.9 (−77.7)	271.4 (133.0)	1657 (11.417)
12	−21.6 (−29.8)	−109 (−78)	−252 (−158)	233.6 (112.0)	597 (4.113)
13	−114.6 (−81.4)	−185 (−120)	−294 (−181)	83.9 (28.8)	561 (3.865)
14	−198.3 (−127.9)	−250 (−157)	−299 (−184.9)	−50.2 (−45.7)	543 (3.741)
22	−41.4 (−40.8)	−122 (−85)	−256 (−160)	204.8 (96.0)	721.9 (4.974)

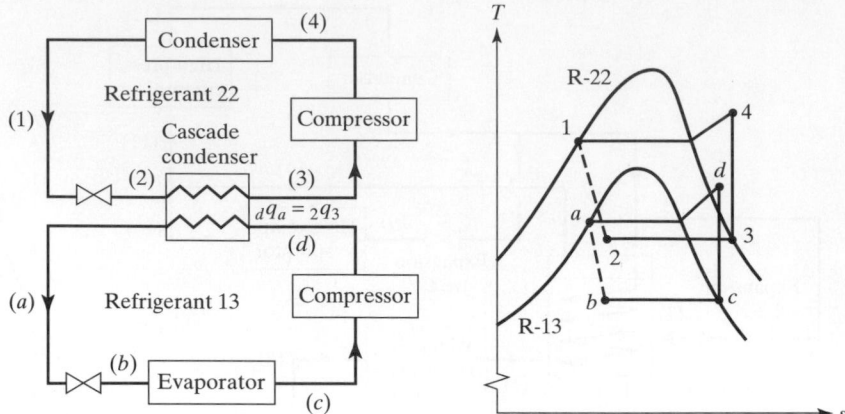

Figure 3.17 Schematic diagrams for cascade cycle composed of two single-stage cycles.

temperatures preclude their use in multi-stage systems. They are suitable for cascade systems, which are discussed in the next section.

3.8 THEORETICAL CASCADE VAPOR-COMPRESSION CYCLES

We may produce space temperatures lower than about -100 °F (-75 °C) by mechanical compression cycles through cascade or binary arrangements. Figure 3.17 shows a simple scheme using two single-stage cycles. The two cycles are independent systems. The cascade condenser is a heat exchanger where the Refrigerant 13 vapor is condensed and the Refrigerant 22 liquid is evaporated. We may, of course, multistage either or both sides of the system. The analysis of a cascade system follows directly from the analyses for single-stage and/or multistage systems.

The cascade system does not compete with multistage systems. It only provides a vapor-compression method for obtaining lower temperatures. The main thermodynamic disadvantage of a cascade system is the finite temperature difference between the two flow streams in the cascade condenser.

Cascade arrangements are applicable to various industrial processes including the liquefaction of petroleum vapors, the liquefaction of atmospheric gases, and the manufacture of dry ice.

3.9 THEORETICAL SINGLE-STAGE CYCLE USING A REFRIGERANT MIXTURE

When a binary mixture that is not azeotropic is used as the refrigerant and the evaporation and condensation processes occur at constant pressure, the temperatures change during the evaporation and condensation processes, and the theoretical single-stage cycle can be shown on T-s coordinates as in Fig. 3.18. This can be compared with Fig. 3.5, in which the cycle is shown operating with a pure simple substance as the refrigerant. An analysis of areas representing additional work and reduced refrigerating effect from the Carnot

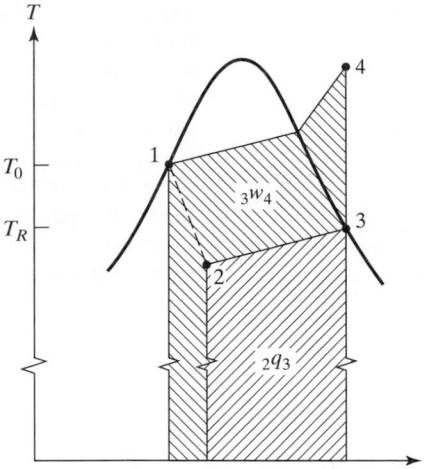

Figure 3.18 Areas on the *T-s* diagram representing the refrigerating effect and the work supplied for the theoretical single-stage cycle using a binary nonazeotropic mixture as the refrigerant.

cycle can be performed using an approach similar to that given in Sec. 3.4, where the theoretical single-stage cycle and the Carnot cycle were compared.

The Carnot cycle is not a practical cycle with which a real system can be compared. The Carnot cycle has no thermal resistance between the working refrigerant and the external fluid in both the condenser and evaporator. Also, the heat-transfer capacity rates of the two external fluids are assumed to be infinitely large through the condenser and evaporator, so the fluid temperature remains fixed at T_0 and T_R, respectively. If one includes both finite thermal resistance and finite capacity rates, it has been shown that a cycle using a refrigerant mixture has a higher coefficient of performance than a cycle using a simple pure substance as refrigerant, providing both cycles have the same thermal resistance in the heat exchangers [2]. However, the improvement in coefficient of performance is usually only a few percentage points. The use of refrigerant mixtures is attractive in heat pumps where the mixture proportion could be changed seasonally to change the density of the refrigerant entering a constant displacement compressor and thereby changing the system capacity.

ENDNOTES

1. R. Plank, "Über den Ideal-Prozess von Kältemaschinen bei Verbund-Kompression," *Zeitschrift für die gesamte kälte-Industrie*, 35 (1928), 17–24.
2. T. H. Kuehn and R. Gronseth, "The Effect of a Nonazeotropic Binary Refrigerant Mixture on the Performance of a Single-Stage Refrigeration Cycle," *Proc. Int. Inst. Refrigeration Conf.*, Purdue, August (1986), 119–127.

PROBLEMS

3.1 Consider a Carnot refrigeration cycle that uses R-22 and operates between 10 °C and 30 °C. The refrigerant leaving the condenser is saturated liquid and the refrigerant leaving the evaporator is saturated vapor. Prepare a table listing the refrigerant temperature, pressure, specific enthalpy, specific volume, and specific entropy at the beginning of each of the four processes in the cycle. Compute the cycle coefficient of performance from the enthalpy values and compare this with the coefficient of performance determined by the operating temperatures.

3.2 A theoretical single-stage refrigeration cycle using R-22 operates between 10 °C and 30 °C. Determine the extra work of compression due to the superheat horn (A_1) and the expansion process (A_2) as shown in Fig. 3.6 over a Carnot cycle operating between the same temperatures.

3.3 Solve Ex. 3.1 for R-22.

3.4 In a theoretical single-stage ammonia vapor-compression system, liquid leaves the condenser at 250 psia and 104 °F. Evaporator pressure is 18.30 psia. Vapor leaves the evaporator at 0 °F. The system produces 25 tons of refrigeration. Determine

   (a) the coefficient of performance, and

   (b) the piston displacement in ft³/min, assuming a volumetric efficiency of 100 percent.

3.5 A Refrigerant-22 system is arranged as shown in Fig. 3.19. Compression is isentropic. Assume frictionless flow. The known data are: $t_1 = 100$ °F, $t_2 = 80$ °F, and $t_4 = -10$ °F. Find the system horsepower per ton.

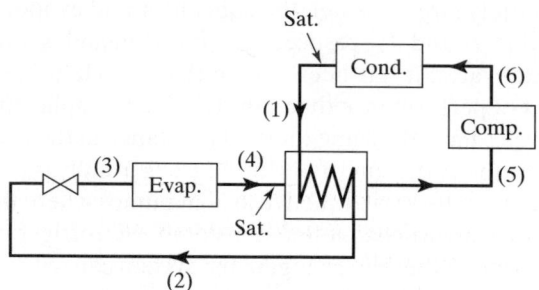

Figure 3.19 Illustration for Prob. 3.5.

3.6 Repeat Prob. 3.5 for values of $t_2 = 60$ °F, 40 °F, 20 °F, and 0 °F. Explain the trend observed in system horsepower per ton as t_2 decreases.

3.7 Solve Ex. 3.2 for ammonia.

3.8 An ammonia system is arranged as shown in Fig. 3.20. Assume isentropic compression and frictionless flow. The load on evaporator A is 10 tons at 40 °F and the load on evaporator B is 5 tons at 0 °F. The condensing temperature is 90 °F.

   (a) Determine the overall coefficient of performance if two Carnot refrigeration cycles were used to handle the loads between 40 °F and 90 °F for A, and 0 °F and 90 °F for B.

   (b) Compute the refrigerating efficiency of the cycle shown in Fig. 3.20 if both evaporator A

and evaporator B were operated at a saturation temperature of 0 °F such that states 3, 4, and 6 were equal.

   (c) Compute the refrigerating efficiency of the cycle shown in Fig. 3.20 if $t_3 = 40$ °F and $t_6 = 0$ °F.

   (d) Compute the refrigerating efficiency if the cycle shown in Fig. 3.20 were replaced with one having a separate compressor for each evaporator and $t_3 = 40$ °F and $t_6 = 0$ °F.

   (e) Comment on the differences observed in the results from parts (b), (c), and (d).

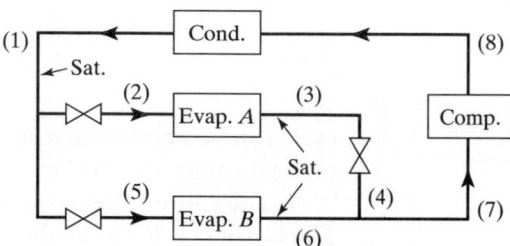

Figure 3.20 Illustration for Prob. 3.8.

3.9 An ammonia vapor-compression system is arranged as shown in Fig. 3.21. Assume isentropic compression and frictionless flow. Given data are $t_1 = 30$ °C, $t_6 = -20$ °C, and $t_8 = -30$ °C.

   (a) Compare the coefficient of performance for the cycle of Fig. 3.21 with that obtained if the liquid cooler were omitted.

   (b) What conclusion can be drawn from theoretical reasoning on the use of liquid intercoolers in single-stage systems?

   (c) Using practical reasoning, discuss the conditions under which the arrangement of Fig. 3.21 might be advantageous.

3.10 A Refrigerant-22 economizer cycle is arranged as shown in Fig. 3.22. Assume isentropic compression and frictionless flow. State 5 is compressed isentropically to state 6 which mixes with state 7 at the same pressure to form state 8. This refrigerant is then compressed isentropically to state 9. Known data are: $t_1 = 90$ °F, $t_3 = 20$ °F, and $t_5 = 0$ °F.

   (a) Calculate the coefficient of performance for this cycle.

   (b) Compute the coefficient of performance for this cycle when the mass flow rate at state 7 is reduced to zero.

   (c) Comment on the difference between the results obtained in parts (a) and (b).

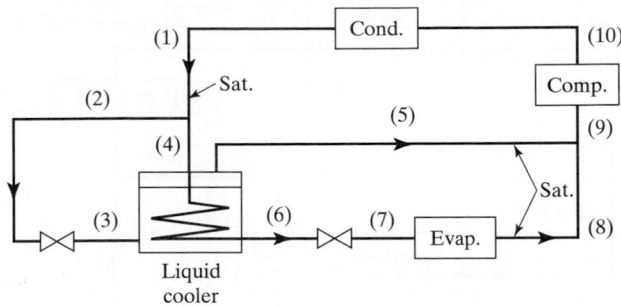

Figure 3.21 Illustration for Prob. 3.9.

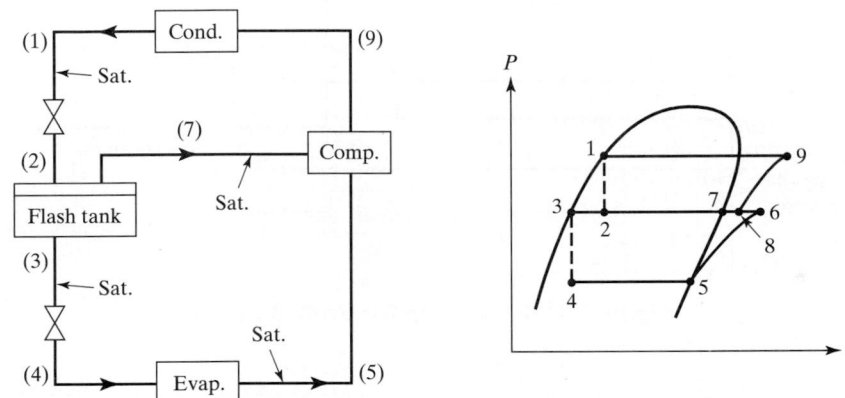

Figure 3.22 Illustration for Prob. 3.10.

3.11 A two-stage ammonia system is arranged as shown in Fig. 3.14 except that the water intercooler is omitted. Condensing temperature is 90 °F, intermediate saturation temperature is 15 °F, and evaporating temperature is −40 °F. Liquid leaves the intercooler at a temperature of 25 °F. Saturated liquid leaves the condenser. Saturated vapor leaves the evaporator. Saturated vapor leaves the intercooler. The compressor displacement of the second-stage (high-pressure) compressor is 100 ft³/min. Calculate the compressor displacement of the first-stage compressor.

3.12 A Refrigerant-22 system is arranged as shown in Fig. 3.15. Condensing pressure is 1.5 MPa, intermediate pressure is 0.5 MPa, and evaporating pressure is 0.2 MPa. The following temperatures are known: $t_1 = 35$ °C, $t_3 = 5$ °C, $t_5 = −20$ °C, and $t_7 = 5$ °C. Assume isentropic compression and frictionless flow. Calculate the coefficient of performance.

3.13 An ammonia system is arranged as shown in Fig. 3.16. Condensing temperature is 100 °F, intermediate saturation temperature is 12 °F, low-side evaporating temperature is −50 °F, $t_3 = 20$ °F, and

$t_7 = 100$ °F. Saturation states occur as shown in the $P\text{-}h$ diagram of Fig. 3.16. Assume isentropic compression and frictionless flow. Tons capacity of the high-temperature evaporator is four times that of the low-temperature evaporator. Calculate the ratio of the compressor displacement of the low-pressure compressor to that of the high-pressure compressor.

3.14 A Refrigerant-22 system is arranged as shown in Fig. 3.23. Assume isentropic compression and frictionless flow. The following data are given: condensing temperature = 90 °F; evaporator A has a capacity of 5 tons and an evaporating temperature of −20 °F; evaporator B has a capacity of 10 tons and an evaporating temperature of −70 °F; vapor leaves each evaporator in dry and saturated condition; $t_1 = 84$ °F, $t_6 = −5$ °F, $t_{11} = 20$ °F, $t_{16} = −5$ °F, and $t_{17} = −55$ °F. Draw schematic $P\text{-}h$ and $T\text{-}s$ diagrams for the cycle and determine
(a) the C.O.P. for the system,
(b) the compressor displacement, ft³/min, for each compressor, and
(c) the theoretical horsepower input for each compressor.

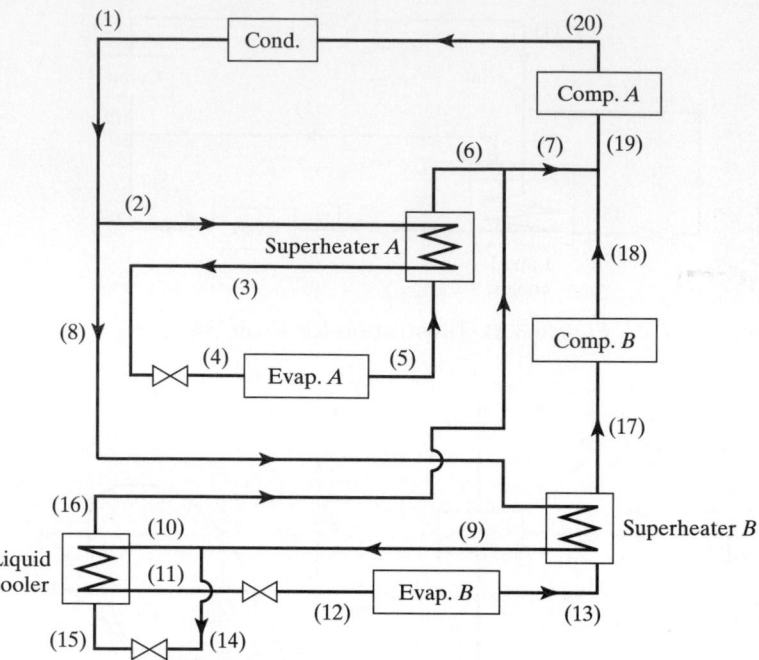

Figure 3.23 Illustration for Prob. 3.14.

SYMBOLS

A_1 Additional work compared to Carnot cycle, Btu/lbm (kJ/kg), required by theoretical single-stage cycle because of superheat horn on *T-s* diagram.

A_2 Additional work compared to Carnot cycle, Btu/lbm (kJ/kg), required by theoretical single-stage cycle because of throttling process.

A_3 Loss of refrigerating effect, Btu/lbm (kJ/kg), compared to Carnot cycle because of throttling process in theoretical single-stage cycle.

C.D.   Compressor displacement, ft³/min (m³/s).

C.O.P.   Coefficient of performance.

Hp   Horsepower.

h   Specific enthalpy, Btu/lbm (kJ/kg).

$\dot{m}$ Mass rate of flow, lbm/min (kg/s).

P   Pressure, psia (MPa).

q   Heat transfer per unit mass of refrigerant, Btu/lbm (kJ/kg).

s   Specific entropy, Btu/lbm · °R (kJ/kg K).

T   Absolute temperature, °R (K).

t   Temperature, °F (°C).

v   Specific volume, ft³/lbm (m³/kg).

w   Work per unit mass refrigerant, Btu/lbm (kJ/kg).

x   Quality of saturated vapor per lbm of mixture.

Greek Letter

η_R Refrigerating efficiency.

chapter 4
Mechanical Vapor-Compression Refrigeration Components and Systems

4.1 REFRIGERANTS

In this section we will deal with the working mediums used in compression refrigeration systems. Such a medium is called a *refrigerant*. Most of our attention will be given to refrigerants that change phase at a fixed temperature when the pressure remains at a constant value.

The design of a vapor-compression refrigeration system is greatly influenced by the properties of the refrigerant employed. The suitability of a refrigerant for a certain application is determined by its physical, thermodynamic, and chemical properties and by various practical factors.

Many substances may be used as refrigerants, including halocarbon compounds, hydrocarbon compounds, inorganic compounds, and others. The American Society of Heating, Refrigerating, and Air Conditioning Engineers [1] has adopted a specific designation system for refrigerants. Table 4.1 shows this system for a few common compounds of commercial importance in refrigeration. The identifying numbers assigned to refrigerants of the methane, ethane, propane, and cyclobutane families have a special meaning. The first digit on the right is the number of fluorine (F) atoms in the refrigerant. The second digit from the right is one more than the number of hydrogen (H) atoms present. The third digit from the right is one less than the number of carbon (C) atoms, but when this digit is zero it is omitted. The fourth number from the right equals the number of carbon-carbon bonds in the molecule, although the number is omitted when it equals zero. Thus, ethane (CH_3CH_3) is Refrigerant 170, while chlorodifluoromethane ($CHClF_2$) is Refrigerant 22. A number followed by a capital B designates that one or more chlorine atoms have been replaced by bromine atoms. The number of chlorine atoms present is determined by subtracting the sum of fluorine, bromine, and hydrogen atoms from the total number of atoms that could be connected to the carbon atoms. In some cases the number is followed by a lower-case letter such as a or b. This indicates another molecular arrangement or isomer using the same chemical compound. The more unsymmetric the molecule, the higher the subscript letter.

TABLE 4.1 ASHRAE Designations for Refrigerants

Number	Type of Compound and Chemical Name	Chemical Formula	Molecular Weight
	Halocarbons		
11	Trichloromonofluoromethane	CCl_3F	137.4
12	Dichlorodifluoromethane	CCl_2F_2	120.9
13	Monochlorotrifluoromethane	$CClF_3$	104.5
14	Carbon tetrafluoride	CF_4	88.0
21	Dichloromonofluoromethane	$CHCl_2F$	102.9
22	Chlorodifluoromethane	$CHClF_2$	86.5
113	1,1,2-Trichlorotrifluoroethane	CCl_2FCClF_2	187.4
114	1,2-Dichlorotetrafluoroethane	$CClF_2CClF_2$	170.9
114a	1,1-Dichlorotetrafluoroethane	CCl_2FCF_3	170.9
114B2	1,2-Dibromotetrafluoroethane	$CBrF_2CBrF_2$	259.8
123	2,2-Dichloro1,1,1-trifluoroethane	$CHCl_2CF_3$	153.0
143a	1,1,1,2-Tetrafluoroethane	CH_2FCF_3	102.0
	Azeotropes		
500	R-12/152a (73.8/26.2)	CCl_2F_2/CH_3CHF_2	99.3
502	R-22/115 (48.8/51.2)	$CHClF_2/CClF_2CF_3$	112.0
	Hydrocarbons		
50	Methane	CH_4	16.0
170	Ethane	CH_3CH_3	30.0
290	Propane	$CH_3CH_2CH_3$	44.0
	Inorganic Compounds		
717	Ammonia	NH_3	17.0
718	Water	H_2O	18.0
744	Carbon dioxide	CO_2	44.0

SOURCE: Abstracted by permission from *ANSI/ASHRAE Standard 34-1992.*

Refrigerant mixtures or blends are designated by the respective constituent refrigerant numbers and the mass fractions of each in the mixture. For example, a mixture of 92 percent R-502 (the azeotrope of R-22 and R-115) with 8 percent propane (R-290) would be indicated as R-290/22/115 (8/45/47).

Most mixtures of refrigerants would change their composition and saturation temperature as they change phase at a constant pressure. These are called *zeotropic* mixtures or blends. Zeotropic mixtures that have been commercialized are given a number in the 400 series.

An azeotrope is a mixture of two or more refrigerants in a proportion such that the mixture behaves as a simple, pure substance. Its concentration does not change during a phase-change process, and its saturation temperature is fixed during a phase change whenever the pressure is held constant. Azeotropes that have been commercially developed have been assigned a number in the 500 series. An example is R-502, which is a mixture of R-22 and R-115 on a 48.8 percent/51.2 percent mass-ratio basis.

Miscellaneous organic refrigerants are numbered serially in the 600 series.

The inorganic refrigerants are designated by adding 700 to the molecular weight of the compound. This identification was adopted arbitrarily. Thus, ammonia is Refrigerant 717 while water is Refrigerant 718.

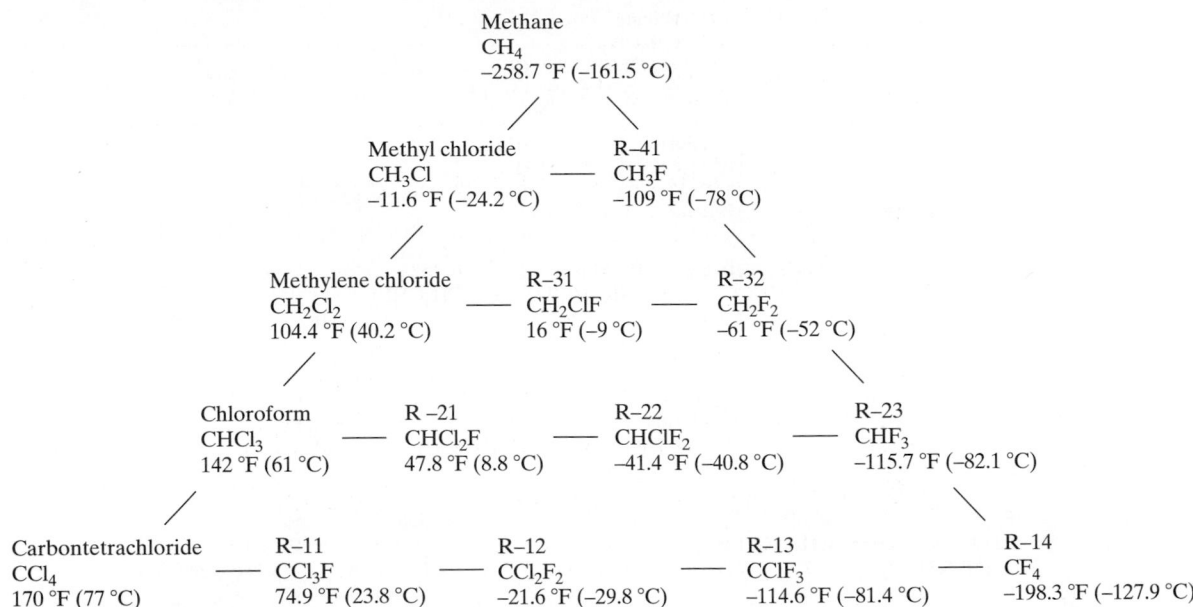

Figure 4.1 Normal boiling points in °F (°C) for the methane series of chloro-fluorocarbon refrigerants arranged according to molecular structure.

Figures 4.1 and 4.2 show the methane and ethane families of refrigerants, respectively. Compounds with the most hydrogen are near the top, those with the most chlorine atoms are near the left-hand side, and those with the most fluorine atoms are near the right-hand side. The normal boiling point at atmospheric pressure is given for each compound. The boiling points increase as one moves down on each figure as the hydrogen atoms are replaced with heavier chlorine or fluorine atoms. The boiling points also increase as one moves from right to left as the lighter fluorine atoms are replaced with heavier chlorine atoms. The compounds near the top of each figure are flammable, owing to the large number of hydrogen atoms present. Those on the bottom row are fully halogenated and are extremely stable. Those on the left-hand side with large amounts of chlorine are toxic.

Composition-designating prefixes have been adopted by ASHRAE. The refrigerant number is preceded by the letter C for carbon, preceded by B, C, or F, or a combination of these, to signify the presence of bromine, chlorine or fluorine atoms. Compounds containing hydrogen are further preceded by the letter H to signify the presence of one or more hydrogen atoms. For example, Refrigerant 22 becomes HCFC-22 and Refrigerant 502 becomes HCFC/CFC-502.

Before proceeding further in discussing definite compounds, we will examine several properties which a desirable refrigerant should possess. The ideal refrigerant should have, at least, the following characteristics:

Thermodynamic

1. *Positive evaporating gauge pressures.* Positive evaporating or low-side gauge pressures prevent possible leakage of atmospheric air into the system during operation.

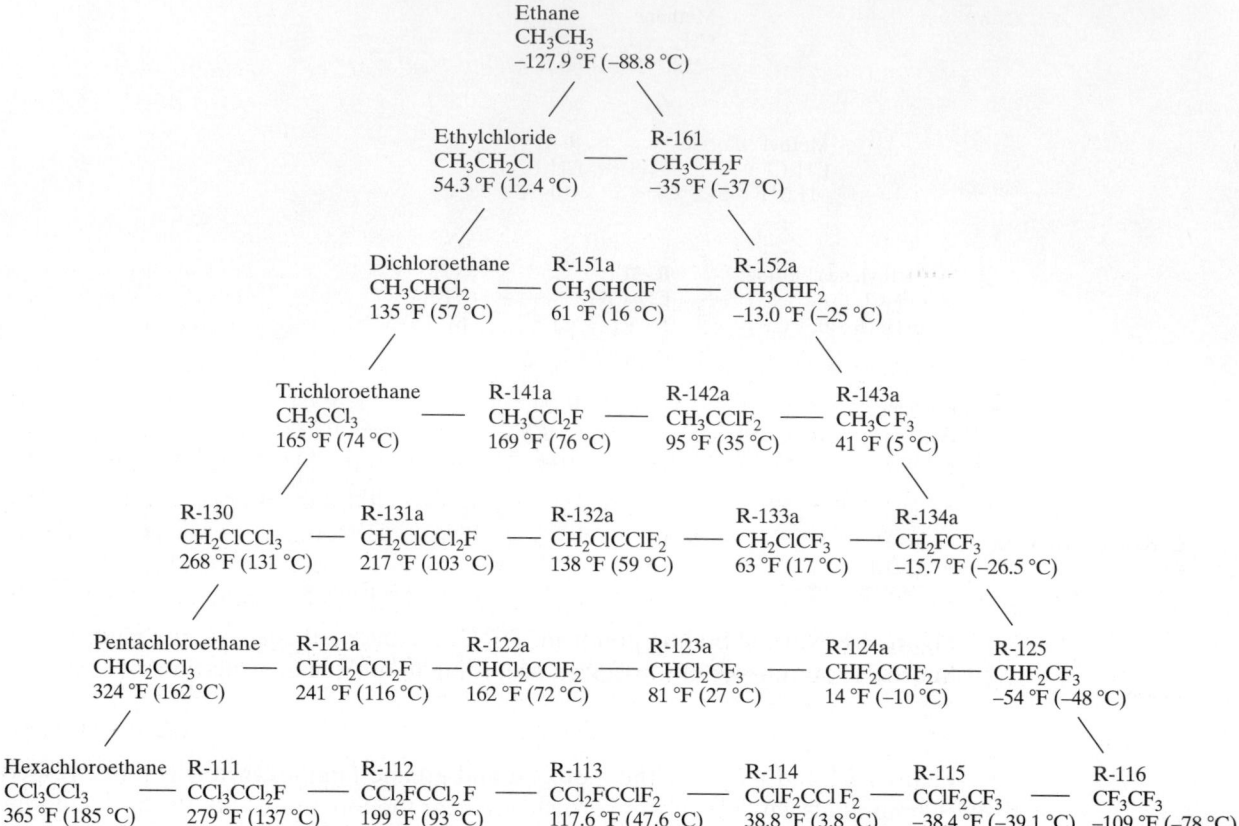

Figure 4.2 Normal boiling points in °F (°C) for the ethane series of chlorofluorocarbon refrigerants arranged according to molecular structure.

2. *Moderately low condensing pressures.* This feature permits the use of lightweight equipment and piping on the high-pressure side of the system.

3. *Relatively high critical temperature.* The critical temperature of the refrigerant should be much higher than normal operating condensing temperatures in order to prevent unduly large power requirements.

4. *High lateWnt heat of vaporization.* A high latent heat of vaporization means a high refrigerating effect per unit mass of refrigerant circulated. This feature is generally desirable, although in small-capacity systems low rates of refrigerant flow may result in control difficulties.

Transport

1. *High heat-transfer characteristics.* This is a broad statement involving such properties as density, specific heat, and thermal conductivity. High heat-transfer coefficients reduce the surface area required in heat exchangers.

2. *Low viscosity.* A low viscosity ensures minimal pressure drop through the piping, heat exchangers, and other system components.

Chemical

Inertness and stability. The refrigerant should be inert to reactions with materials of the system. It should be noncorrosive in the presence of water. It should be stable in its chemical make-up through the entire range of operating conditions.

Health, Safety, and Environmental

1. *Nontoxicity.* The refrigerant should be nonpoisonous to human beings and to food-stuffs. Besides the relative toxicity of the refrigerant vapor, important factors are (a) the concentration or the percentage of vapor in the air, and (b) the duration of exposure. ASHRAE has established a toxicity rating for refrigerants, where A is the designation for low toxicity and B is given to compounds with higher toxicity.

2. *Nonflammability.* The vapor should not burn or support combustion in any concentration with atmospheric air. ASHRAE rates the flammability of refrigerants as follows: 1—no flame propagation, 2—low flammability, 3—higher flammability.

3. *Low ozone-depletion potential.* Compounds containing chlorine are particularly effective at reducing the concentration of ozone in the earth's atmosphere. Refrigerants with no chlorine often have the lowest ozone-depletion potential as referenced to R-11, which has been assigned an ozone-depletion potential of 1.0. Refrigerants containing hydrogen tend to have weak bonds, which allows them to degrade and be removed from the atmosphere faster than fully halogenated compounds along the bottom row of Figs. 4.1 and 4.2.

4. *Low greenhouse potential.* Certain refrigerant molecules can absorb large amounts of infrared radiation. This contributes to the greenhouse effect in the earth's atmosphere. Molecules containing hydrogen are better, as they should have shorter lives in the atmosphere. Global-warming potentials are usually referenced to R-11, which has been assigned a value of 1.0.

Miscellaneous

1. *Low cost of refrigerant.* This feature is obvious, although its significance depends upon the size of the system. Cost of the refrigerant may be rather immaterial where the entire charge is very small, as in a household refrigerator.

2. *Easy leakage detection.* Leaks in refrigerant lines and equipment should be detectable by a simple and positive method.

3. *Satisfactory oil solubility.* This characteristic is discussed in Sec. 4.4.

4. *Low water solubility.* This factor is discussed in Sec. 4.4.

5. *High dielectric strength of vapor.* This characteristic is important in hermetically sealed compressor units where the refrigerant vapor may come into contact with the motor windings.

Unfortunately, no single compound is available which is absolutely satisfactory for all refrigeration systems. We find that some refrigerants are well suited for particular applications but wholly unsuitable for others. Recent environmental issues are causing a shift from traditional, very stable refrigerants such as R-11 and R-12 to others with less environmental stability. From Figs. 4.1 and 4.2 one can see that the normal boiling points of R-123 and R-11 are nearly equal, as are R-134a and R-12. Other thermodynamic, transport, and toxicity levels are similar, so that the less environmentally stable compounds R-123 and R-134a are replacing the more stable, traditional refrigerants R-11 and R-12.

4.2 THE INORGANIC REFRIGERANTS

Until about 1930, refrigerants in use were almost exclusively inorganic compounds. Ammonia, carbon dioxide, and sulfur dioxide were commonly used in vapor-compression systems. Of these three, only ammonia is of commercial importance today. Various halo-carbon refrigerants have largely superseded carbon dioxide and sulfur dioxide.

(a) Ammonia. Although it is one of the oldest refrigerants, ammonia is still widely used, particularly in the larger industrial applications. Ammonia is an excellent refrigerant from a thermodynamic standpoint. Its boiling point of −28 °F (−33 °C) at 1 atm pressure allows positive evaporating pressures in a majority of refrigeration needs. Its critical temperature of 271 °F (133 °C) is relatively high, while its freezing temperature of −108 °F (−78 °C) is sufficiently low. One of its outstanding characteristics is its high latent heat of vaporization. This feature accounts for the relatively small compressor displacement required for ammonia systems, and for the modest size of pipe lines. Ammonia has excellent heat-transfer characteristics; it is relatively cheap, too.

In the presence of water, ammonia strongly attacks copper and cuprous alloys. Since some water may exist in any refrigeration system, cuprous metals are never used with ammonia. Ferrous materials are used for equipment and piping. Ammonia is nonmiscible or insoluble with mineral lubricating oils. However, it is soluble in all proportions with water.

The primary disadvantages of ammonia are certain rather hazardous characteristics it has. It is a strong irritant and is intolerable even in small concentrations. It is mildly toxic. In concentrations in air of between 16 to 25 percent by volume it will burn feebly.

(b) Carbon Dioxide. The principal refrigeration uses of CO_2 today are as dry ice and as a coolant in food processing. Carbon dioxide is nontoxic, nonirritating, and nonflammable. The triple point of CO_2 is −69.9 °F (−56.6 °C), and at atmospheric pressure dry ice sublimes at −109 °F (−78 °C). As a refrigerant in a vapor-compression system, carbon dioxide has distinct disadvantages. Its critical temperature of 87.8 °F (31 °C) is below normal condensing temperatures attainable in most systems. Compared to most other refrigerants, operating pressures are extremely high and power costs relatively large.

(c) Water. The principal refrigeration use of water is as ice. The high freezing temperature of water limits its use in vapor-compression systems. Water is used as the refrigerant in some absorption systems and in systems with steam-jet compressors.

4.3 THE HALOCARBON REFRIGERANTS

Although the American Society of Heating, Refrigerating and Air Conditioning Engineers identifies some 78 hydrocarbon, halocarbon, zeotropic, and azeotropic compounds as refrigerants, only Refrigerants 12, 22, 134a, and 502 will be considered here. These compounds are all synthetically produced. In 1930 the first of the family, Refrigerant 12, was developed [2] and was introduced as "Freon 12." Introduction of the Freon refrigerants provided a tremendous impetus to the refrigeration industry. Their outstanding characteristics are their nonhazardous features. All of them are nontoxic, nonirritating, and nonflammable. Figure 4.3 illustrates the methane and ethane families of refrigerants and the

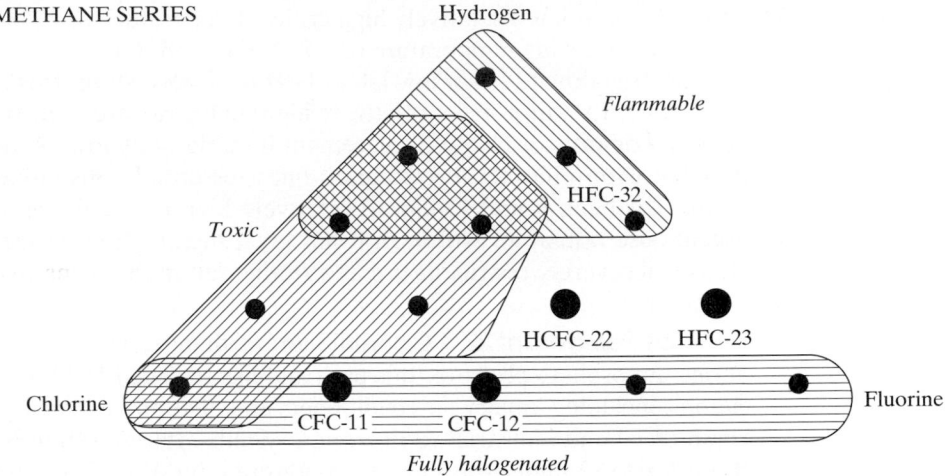

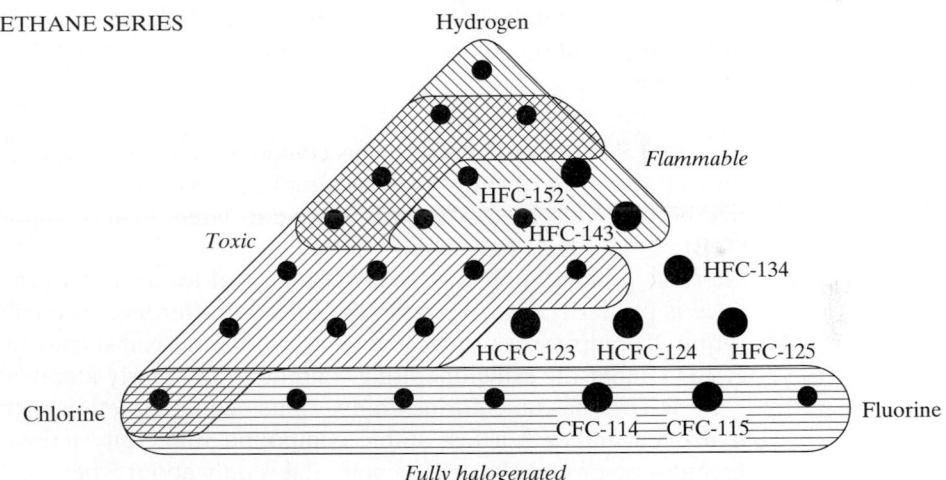

Figure 4.3 Undesirable features of halocarbon refrigerants in the methane and ethane families. Source: Allied Signal.

compounds that are flammable, toxic, and fully halogenated. The fully halogenated molecules are extremely stable and were the first successful halocarbon refrigerants developed. Alternatives to the fully halogenated compounds are shown and include R-22, R-123, and R-134 [3, 4].

(a) Refrigerant 12. Dichlorodifluoromethane (CCl_2F_2) was the first halocarbon refrigerant to be developed. It is a clear and colorless liquid with a normal boiling point of $-21.6\ °F$ ($-29.8\ °C$), thus allowing positive evaporating pressures for a wide range of

applications. It has a relatively high critical temperature of 233.6 °F (112.0 °C) and a relatively low freezing temperature of −252 °F (−158 °C).

Refrigerant 12 has a low latent heat of vaporization. It is a highly inert and stable compound. Its vapor has a relatively high dielectric strength. Because of its high vapor density, compressor piston displacement is moderately low. Refrigerant 12 is completely miscible with mineral lubricating oil but is essentially insoluble with water. This compound has been found to have a relatively high ozone-depletion potential and a high greenhouse potential. Although it has great historical importance, environmental concerns will severely restrict its use as a refrigerant in the future.

(b) Refrigerant 22. Chlorodifluoromethane ($CHClF_2$) boils at −41.4 °F (−40.8 °C) at one atmosphere pressure. It is used and has replaced R-12 in many refrigeration applications, including unitary air conditioners and heat pumps, refrigeration chillers, and both commercial and industrial refrigeration systems. Its saturation pressures are higher than those for R-12 under the same temperature operating conditions. Refrigerant 22 has a greater latent heat of vaporization than R-12, and its vapor is more dense. It is semimiscible with oil, being highly miscible at room temperature but relatively immiscible at temperatures well below 0 °F (−18 °C). This compound has a hydrogen atom that replaces one of the chlorine atoms found in R-12. Therefore it has a lower ozone-depletion potential than R-12 (approximately 10 percent of R-12's) and a lower greenhouse or global-warming potential (about 6 percent of R-12's), which makes it preferable from the environmental viewpoint.

(c) Refrigerant 134a. This compound has thermodynamic properties very similar to those of R-12 and is used as a replacement for R-12 in some applications. Its normal boiling point is −15 °F (−26 °C) and its latent heat of vaporization is approximately 90 Btu/lbm (200 kJ/kg), slightly exceeding that of R-12. The C.O.P. for a theoretical single-stage cycle using R-134a is lower than that for a similar cycle using R-12. The difference is primarily due to the larger enthalpy difference of compression for R-134a. The required compressor displacements of two theoretical single-stage cycles using R-12 and R-134a under the same operating conditions are nearly identical.

R-134a has no chlorine atoms, so its ozone-depletion potential is nearly zero. The hydrogen makes it a less stable compound than fully halogenated molecules, so its greenhouse or global warming potential is only about 5 percent of R-12's, or slightly less than R-22's.

(d) Refrigerant 502. This substance is an azeotropic mixture of R-22 and R-115 and has a normal boiling temperature of −49.8 °F (−45.4 °C). Refrigerant 502 is widely used in retail food refrigeration equipment and in both commercial and industrial refrigeration systems.

4.4 REACTION OF REFRIGERANTS WITH MOISTURE AND OIL

A refrigeration system should be charged with only the refrigerant compound intended. However, it is extremely difficult to prevent traces of moisture from existing in systems. Various refrigerants react differently in the presence of water.

The two principal effects resulting from moisture in a refrigeration system are corrosion and freeze-up of expansion devices. Almost all refrigerants will form corrosive acids or bases in the presence of water; these corrosive compounds may be highly destructive to valves, seals, and other metallic parts.

When water comes in contact with a refrigerant, it may go into solution with the refrigerant or it may remain as free water. Water is soluble in ammonia in all proportions so that free water does not occur in ammonia systems. Water is also highly soluble in carbon dioxide and sulfur dioxide. Any water in excess of the soluble amount would occur as free water. Freeze-up of expansion valves and formation of ice in evaporators result from presence of free water. Refrigerant-12 systems are particularly susceptible to this difficulty, which is one reason why Refrigerant 22 has replaced Refrigerant 12 in most applications.

An important practical problem is to keep a refrigeration system dry. Any system should be thoroughly cleaned and dehydrated before being charged with a refrigerant. The refrigerant and the compressor oil that are used should be moisture free. Every precaution should be observed to prevent possible entry of moisture, both during erection and during operation of a system. As a further protection, many systems employ permanently installed dryers. The dryer element typically uses a desiccant such as silica gel or activated alumina. In some driers a color change is used to indicate the need for replacement of the dryer.

In most refrigeration compressors the refrigerant comes into direct contact with lubricating oil. Depending upon the mutual solubility characteristics of the refrigerant and of the oil, some oil may go into solution with the refrigerant vapor and some refrigerant may be dissolved into the oil. Some oil may also be picked up physically by the vapor without going into solution. Therefore, the working fluid in the system is not pure refrigerant, but a mixture of refrigerant and oil.

Reaction of refrigerants with oils is important in several ways. Oil carried from the compressor into other parts of the system may reduce heat transfer in the condenser and evaporator. The pressure-temperature characteristic of the refrigerant-oil solution may differ from that of the pure refrigerant. Also, lubrication of the compressor may be affected, since viscosity of the oil changes by dilution with the refrigerant.

Some refrigerants are fully miscible with oil. The solution formed is homogeneous, and oil tends to stay with the refrigerant throughout the system. Miscible refrigerants include Refrigerants 11, 12, 21, and 113.

On the other hand, many refrigerants are immiscible with oil. If oil is picked up by the refrigerant vapor, a heterogeneous mixture is formed. In such a case, the oil is mechanically separable from the refrigerant. Refrigerants which are immiscible with oil include ammonia, sulfur dioxide, carbon dioxide, and Refrigerants 13 and 14.

A few refrigerants are intermediate in their action with oil. Such a refrigerant may be fully miscible with oil above some critical solution temperature but only partially miscible at lower temperatures. Examples are Refrigerants 22 and 114.

The problem of oil miscibility is primarily one of system design. A system using an immiscible refrigerant must be equipped with an efficient oil separator after the compressor. Systems using a miscible refrigerant must be designed for sufficient velocity of flow in evaporators and suction lines. The greatest problem of design occurs with refrigerants which are intermediate in miscibility. Oil may come out of solution in the cold evaporator and may remain in the evaporator unless adequate velocities are present to drag the oil back into the compressor.

The thermodynamic cycle for an actual single-stage refrigeration system may depart significantly from the theoretical cycle of Fig. 3.3. The principal departure occurs in the compressor. In the condenser, evaporator, and piping, departures from the theoretical cycle may be kept small by designing the components for small refrigerant-pressure losses and by insulating components to reduce heat exchange with ambient surroundings. In this section we will be especially concerned with thermodynamic processes occurring within a reciprocating compressor. The reciprocating compressor is a positive-displacement device and was the first type of compressor used in refrigeration systems.

Figure 4.4 is not a thermodynamic-state diagram but only shows the variation of pressure with position of the piston. The diagram is somewhat idealized with respect to compressor-valve operation. Because of valve pressure drop, the cylinder pressure during intake, $P_a = P_b$, may be less than the suction-line pressure P_3, and the cylinder pressure during discharge, $P_c = P_d$, may be greater than the discharge-line pressure P_4. Because of heat exchange between the vapor and the cylinder walls, the thermodynamic state of the vapor at position a may be different from that at position b, and the state at position c may be different from that at position d. At positions b, c, and d we will assume the thermodynamic state of the vapor in the cylinder to be uniform.

For the compression process from b to c we will assume a polytropic process, where V is cylinder volume and n is a constant.

$$PV^n = \text{a constant}$$

Since the process is nonflow, we have

$$P_b v_b^n = P_c v_c^n$$

which is written in terms of thermodynamic properties, where v is specific volume. Let us further assume that the temperature change from c to d is negligible during the discharge process. This is a good assumption, because the heat transfer from the vapor during the discharge process is very small, as the process occurs very quickly and the exposed cylin-

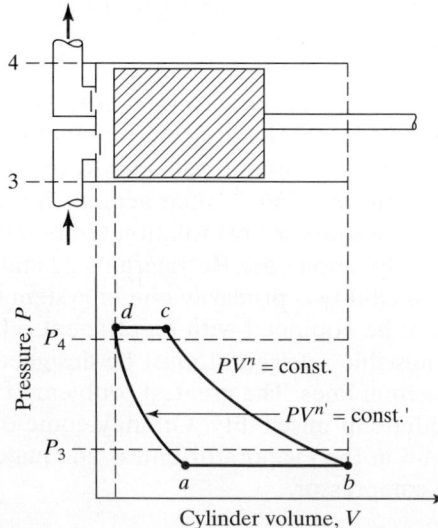

Figure 4.4 Schematic pressure-volume diagram for a reciprocating compressor.

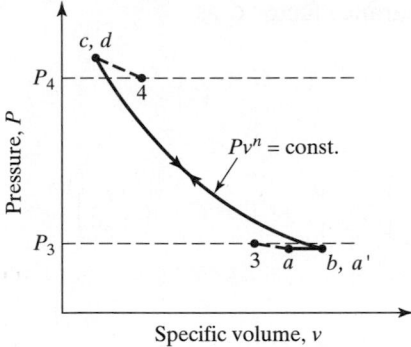

Specific volume, v

Figure 4.5 Pressure-specific volume diagram for compressor of Fig. 4.4.

der area is very small. With this approximation, state c = state d and $v_c = v_d$. For the reexpansion process

$$P_d v_d^{n'} = P_{a'} v_{a'}^{n'}$$

If we assume that $n = n'$, as the mean temperature and pressure of the two processes are equal, we have the result that $v_{a'} = v_b$. The above analysis with its several approximations establishes a P-v diagram as shown in Fig. 4.5. The compression and reexpansion curves are coincident. Thus, we may conclude that the net work required to compress the clearance vapor is zero. The clearance vapor is expanded until the pressure drops sufficiently below the suction-line pressure (P_3) to force the intake valve open. The refrigerant in the cylinder when the suction valve opens is at state a' as shown in Fig. 4.5. The vapor entering the cylinder from the suction line is throttled from state 3 to state a across the suction valve, as indicated by the dashed curve on Fig. 4.5. The refrigerant entering at state a mixes with the vapor at state a'. This mixture is then heated by the cylinder walls and reaches state b when the compression process begins. The refrigerant vapor discharged from the compressor is throttled across the discharge valve from states c and d to state 4, as indicated by the dashed curve in Fig. 4.5.

At the end of the intake stroke (position b), the total mass of vapor in the cylinder is V_b/v_b. The mass of clearance vapor is $V_a/v_{a'}$. Thus

$$\text{mass of intake vapor per cycle} = \frac{V_b}{v_b} - \frac{V_a}{v_{a'}} = \frac{V_b - V_a}{v_b} \tag{4.1}$$

The compressor volumetric efficiency η_V may be defined as the mass of vapor actually pumped by the compressor divided by the mass of vapor which the compressor could pump if it handled a volume of vapor equal to its piston displacement and if no thermodynamic state changes occurred during the intake stroke. Thus, for the compressor of Fig. 4.4 and for one cycle of events in the compressor

$$\eta_V = \frac{(V_b - V_a)v_3}{(V_b - V_d)v_b} \tag{4.2}$$

but

$$V_b - V_a = (V_b - V_d) - (V_a - V_d)$$

$$V_a = V_d\left(\frac{P_d}{P_a}\right)^{1/n} = V_d\left(\frac{P_c}{P_b}\right)^{1/n}$$

Let us define the clearance factor C as

$$C = \frac{V_d}{V_b - V_d}$$

Thus

$$\frac{V_b - V_a}{V_b - V_d} = 1 + C - C\left(\frac{P_c}{P_b}\right)^{1/n}$$

and

$$\eta_V = \left[1 + C - C\left(\frac{P_c}{P_b}\right)^{1/n}\right]\frac{v_3}{v_b} \tag{4.3}$$

Equation (4.3) accounts for the three principal factors which affect compressor volumetric efficiency: reexpansion of clearance vapor, pressure drop in suction and discharge valves, and heating of the vapor on the intake stroke. Leakage past piston rings and through valve seals is neglected. If $C = 0$, the term in brackets is unity. When $C > 0$, valve-pressure drops also contribute to decreasing the term in brackets. The quantity v_3/v_b may be less than unity because of pressure drop in the suction valve and because of cylinder-wall heating effects.

It also follows from the basic definition of volumetric efficiency that

$$\eta_V = \frac{\dot{m}v_3}{\text{C.D.}} \tag{4.4}$$

where $\dot{m}$ is the mass flow rate pumped by the compressor and C.D. is the compressor piston displacement. By Eqs. (4.3) and (4.4),

$$\dot{m} = \left[1 + C - C\left(\frac{P_c}{P_b}\right)^{1/n}\right]\frac{\text{C.D.}}{v_b} \tag{4.5}$$

Table 4.2 shows values of the isentropic exponent k for several refrigerants. This exponent may be used as an approximation to the actual exponent n where more accurate data are not known.

TABLE 4.2 Isentropic Exponent for Various Saturated Refrigerant Vapors at 50 °F (10 °C)

Refrigerant:	12	22	Ammonia
n:	1.07	1.12	1.29

EXAMPLE 4.1

Saturated R-22 vapor at -10 °C enters the compressor of a single-stage system. Discharge pressure is 1.2 MPa. Pressure drop in the compressor suction valves is 12 kPa. Pressure drop in the discharge valves is 24 kPa. Assume that the vapor is superheated 10 °C in the cylinder during the intake stroke. Compressor clearance is 5 percent. Determine (a) the volumetric efficiency, and (b) the compressor pumping capacity in kg per sec if the piston displacement is 0.025 m^3 per sec.

Part II / Refrigeration

Solution: From the given data and Table A.3SI, we have

$$C = 0.05, \qquad P_c = 1.2 + 0.024 = 1.224 \text{ MPa}$$

$$P_b = 0.354 - 0.012 = 0.342 \text{ MPa}, \qquad v_3 = 0.06536 \text{ m}^3/\text{kg}$$

By Fig. C2-SI, at $t_b = 0\,°\text{C}$ and $P_b = 0.342$ MPa we obtain

$$v_b = 0.071 \text{ m}^3/\text{kg}$$

By Table 4.2, assume that $n = 1.12$.

(a) By Eq. (4.3),

$$\eta_V = \left[1 + 0.05 - (0.05)\left(\frac{1.224}{0.342}\right)^{1/1.12} \right] \frac{0.06536}{0.071} = 0.823$$

(b) By Eq. (4.4),

$$\dot{m} = \frac{\eta_V (\text{C.D.})}{v_3} = \frac{0.823\,(0.025)}{0.06536} = 0.315 \text{ kg/sec}$$

We may derive an expression for the compressor work required, subject to the same approximations and limitations as used in the analysis for volumetric efficiency. Since the work per cycle is represented by the enclosed area of the diagram of Fig. 4.4, we may write

$$W/\text{cycle} = \int_b^c V\, dP - \int_a^d V\, dP$$

Using the relation $PV^n = $ a constant,

$$W/\text{cycle} = \frac{n}{n-1} P_b (V_b - V_a) \left[\left(\frac{P_c}{P_b}\right)^{(n-1)/n} - 1 \right] \qquad (4.6)$$

By Eqs. (4.1) and (4.6),

$$w = \frac{n}{n-1} P_b v_b \left[\left(\frac{P_c}{P_b}\right)^{(n-1)/n} - 1 \right] \qquad (4.7)$$

with w as the net work required per unit mass of refrigerant pumped.

Figure 4.6 shows schematic diagrams defining a single-stage refrigeration cycle which deviates from the theoretical cycle in several respects. Refrigerant is subcooled leaving the condenser and superheated leaving the evaporator. Pressure drop in compressor valves and heating of the vapor on the intake stroke are assumed. The compression process is assumed to be polytropic. Pressure drops in the condenser, evaporator, and piping are neglected. Heat addition in the compressor suction line and heat rejection in the compressor discharge line are assumed.

The refrigeration capacity and the power requirement are important performance data. The equations which follow are associated with thermodynamic calculations for capacity and power. State designations are given by Fig. 4.6.

$$\text{capacity} = {}_2\dot{Q}_3 = \dot{m}(h_3 - h_2) \qquad (4.8)$$

$$\dot{m} = \frac{\eta_V (\text{C.D.})}{v_4} \qquad (4.9)$$

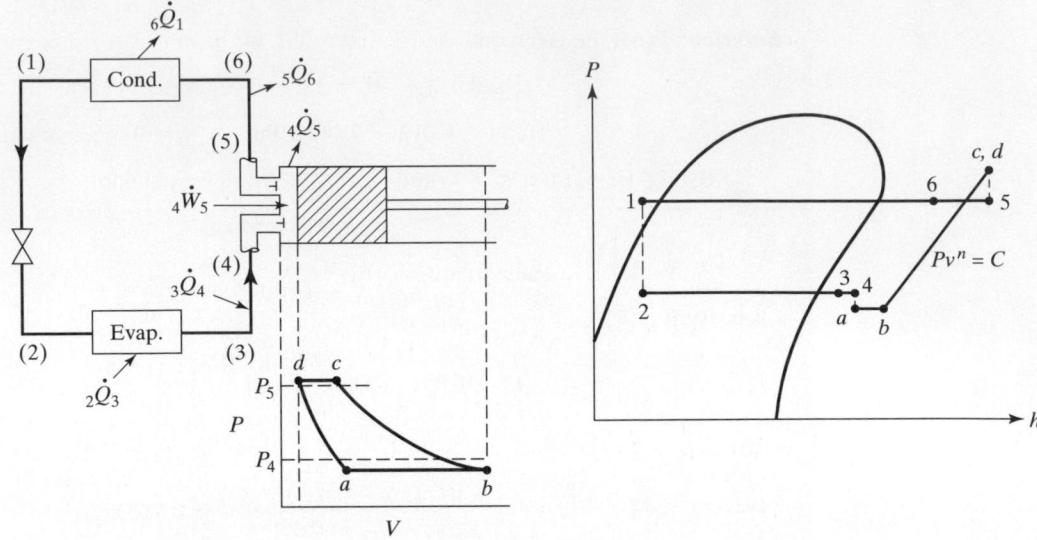

Figure 4.6 Schematic diagrams describing practical single-stage cycle with a reciprocating compressor.

$$\text{power} = {}_4\dot{W}_5 = \dot{W}_{\text{comp}} = \dot{m}_4 w_5 \tag{4.10}$$

$$_4w_5 = \frac{nP_b v_b}{n-1}\left[\left(\frac{P_c}{P_b}\right)^{(n-1)/n} - 1\right] \tag{4.11}$$

$$\frac{\text{power}}{\text{capacity}} = \frac{nP_b v_b}{(n-1)(h_3 - h_2)}\left[\left(\frac{P_c}{P_b}\right)^{(n-1)/n} - 1\right] \tag{4.12}$$

where in Eqs. (4.8)–(4.12) $\dot{m}$ = mass rate of refrigerant flow; h = specific enthalpy; η_V = compressor volumetric efficiency; C.D. = compressor piston displacement; P = pressure; $_4w_5$ = compression work per unit mass of refrigerant; n = polytropic exponent for compression path.

The power required by an actual compressor is larger than the amount computed from Eq. (4.10). The additional power is required to overcome sliding friction in bearings and between the piston and cylinder. Additional power is required to lubricate the compressor which is used to drive an oil pump in larger machines. The shaft power required by the compressor can be given as

$$\dot{W}_{\text{shaft}} = \frac{\dot{W}_{\text{comp}}}{\eta_m} \tag{4.13}$$

where the mechanical efficiency, η_m, accounts for the losses mentioned earlier.

Most compressors are driven by electric motors of either constant or variable speed. The electric-motor losses can be accounted for by an electric-motor efficiency, η_E:

$$\dot{W}_{\text{elec}} = \frac{\dot{W}_{\text{shaft}}}{\eta_E} \tag{4.14}$$

where $\dot{W}_{\text{elec}}$ is the electric power supplied to the motor driving the compressor. The electric-motor efficiency is not needed for compressors operated by power take-off, such as in automotive air conditioning units.

Combining Eqs. (4.5), (4.7), (4.13), and (4.14) allows us to estimate the electric power requirement from known or estimated compressor construction and operating information:

$$\dot{W}_{elec} = \frac{\dot{m}W_{comp}}{\eta_m \eta_E} = \frac{nP_b \text{C.D.}}{(n-1)\eta_m \eta_E}\left[1 + C - C\left(\frac{P_c}{P_b}\right)^{1/n}\right]\left[\left(\frac{P_c}{P_b}\right)^{(n-1)/n} - 1\right] \quad (4.15)$$

EXAMPLE 4.2

An ammonia refrigeration plant with a capacity of 100 tons is equipped with a two-cylinder, single-acting reciprocating compressor having 3 percent clearance and operated at 560 rpm. Evaporating temperature is -20 °F and condensing pressure is 194 psia. Liquid ammonia leaves the condenser and enters the expansion valve at 86 °F. The vapor is superheated 10 °F in the evaporator and further superheated 25 °F more in the compressor suction line. Pressure drop in the compressor suction valves is 4 psi and in the discharge valves 6 psi. The vapor is superheated 20 °F in the cylinder on the intake stroke after passing the suction valves. Compression is polytropic with $n = 1.20$. The discharge vapor is cooled 100 °F in the compressor discharge line before entering the condenser. Ignore pressure drop in condenser, evaporator, and refrigerant piping. Ignore heat transfer in the liquid line to the expansion valve and in the line between the expansion valve and the evaporator. Determine the following items: (a) volumetric efficiency of compressor, (b) required bore and stroke for compressor if a stroke-to-diameter ratio of 1.25 is assumed, (c) shaft horsepower input to compressor, assuming a mechanical efficiency of 75 percent, (d) heat rejected to compressor jacket water in Btu per min, and (e) quantity of condensing water required in gallons per minute if water temperature rise is 12 °F.

Solution: The first step is to establish the various state points. States 1, 2, 3, 4, a, and b may be located from the given data. Thus

$$P_1 = 194 \text{ psia}, \quad t_1 = 86 \text{ °F}, \quad h_1 = h_2 = 138.7 \text{ Btu/lbm}$$

$$P_2 = P_3 = P_4 = 18.3 \text{ psia}, \quad t_3 = -10 \text{ °F}, \quad h_3 = 609.6 \text{ Btu/lbm}$$

$$t_4 = 15 \text{ °F}, \quad h_4 = 623 \text{ Btu/lbm}, \quad v_4 = 16 \text{ ft}^3/\text{lbm}$$

$$P_a = P_4 - \Delta P_s = 18.3 - 4 = 14.3 \text{ psia}, \quad h_a = h_4 = 623 \text{ Btu/lbm}$$

$$t_a = 13 \text{ °F}, \quad P_b = P_a = 14.3 \text{ psia}, \quad t_b = t_a + 20 = 33 \text{ °F}$$

and $v_b = 21.4$ ft³/lbm. State c can be located by the following method:

$$P_c = P_5 + \Delta P_d = 194 + 6 = 200 \text{ psia}$$

$$P_b v_b^n = P_c v_c^n$$

$$v_c = \frac{v_b}{\left(\dfrac{P_c}{P_b}\right)^{1/n}} = \frac{21.4}{\left(\dfrac{200}{14.3}\right)^{0.833}} = 2.38 \text{ ft}^3/\text{lbm}$$

Thus

$$P_c = 200 \text{ psia}, \quad v_c = 2.38 \text{ ft}^3/\text{lbm}, \quad h_c = 776 \text{ Btu/lbm}$$

$$h_5 = h_c = 776 \text{ Btu/lbm}, \quad P_5 = 194 \text{ psia}, \quad t_5 = 323 \text{ °F}$$

$$P_6 = P_5 = 194 \text{ psia}, \quad t_6 = t_5 - 100 = 223 \text{ °F}$$

$$h_6 = 717 \text{ Btu/lbm}$$

(a) By Eq. (4.3) and using v_4 instead of v_3

$$\eta_V = \left[1.03 - 0.03 \left(\frac{200}{14.3} \right)^{0.833} \right] \frac{16}{21.4} = 0.568$$

(b) By Eq. (4.8)

$$\dot{m} = \frac{(100)(200)}{609.6 - 138.7} = 42.47 \text{ lbm/min}$$

By Eq. (4.9)

$$\text{C.D.} = \frac{(42.47)(16)}{0.568} = 1196 \text{ ft}^3/\text{min}$$

and

$$\frac{2\pi D^2 (1.25)(D)(\text{rpm})}{(4)(144)(12)} = \text{C.D.}$$

Thus

$$D = \sqrt[3]{\frac{(1196)(1728)}{\pi(1.25)(280)}} = 12.3 \text{ in.}$$

$$L = (1.25)(12.3) = 15.4 \text{ in.}$$

(c) By Eq. (3.22)

$$_4 w_5 = \frac{(1.20)(14.3)(144)(21.4)}{(0.2)(778)} \left[\left(\frac{200}{14.3} \right)^{0.1667} \right] = 187.7 \text{ Btu/lbm}$$

By Eq. (4.10)

$$_4 \dot{W}_5 = \frac{(42.47)(187.7)}{(42.4)(0.75)} = 251 \text{ hp}$$

(d) Per lbm of refrigerant, we have by the steady-flow equation

$$_4 q_5 = {}_4 w_5 - (h_5 - h_4) = 187.7 - (776 - 623) = 34.7 \text{ Btu/lbm}$$

Thus

$$_4 \dot{Q}_5 = \dot{m}_4 q_5 = (42.47)(34.7) = 1479 \text{ Btu/min}$$

(e)
$$_6 \dot{Q}_1 = \dot{m}(h_6 - h_1) = (42.47)(717 - 138.7) = 24{,}560 \text{ Btu/min}$$

and

$$\dot{\text{vol}}_w = \frac{{}_6 \dot{Q}_1}{(1)(12)(8.35)} = 245 \text{ gal/min}$$

Figure 4.7 shows variation of tons capacity, horsepower, and horsepower per ton with saturation-suction temperature and condensing temperature for a Refrigerant-12 condensing unit. Figure 4.7 was constructed from actual performance data obtained from a manufacturer. It shows that capacity decreases sharply with decrease of evaporating temperature. The horsepower required decreases and the horsepower per ton increases with the decrease of evaporating temperature.

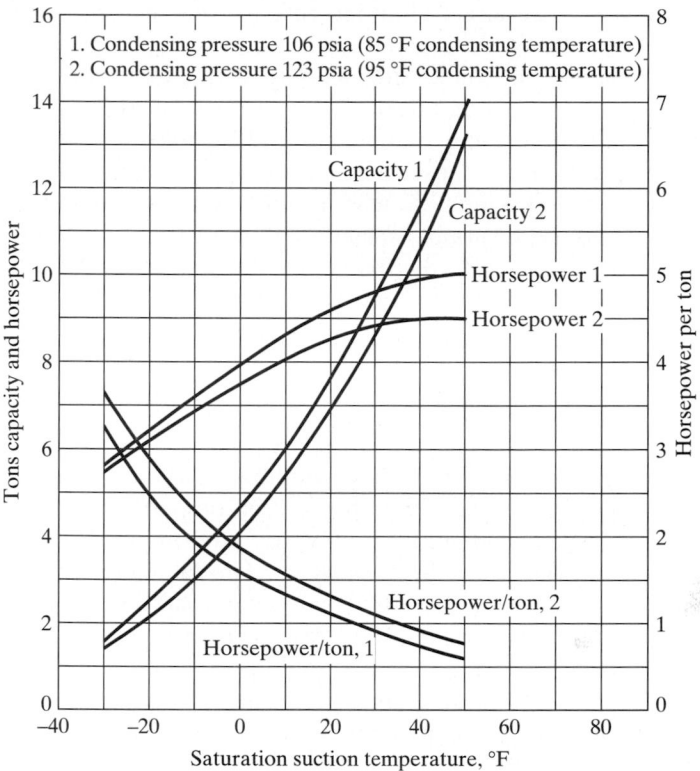

Figure 4.7 Capacity and horsepower characteristics for a typical Refrigerant-12 condensing unit equipped with a constant displacement compressor.

The trends of the curves in Fig. 4.7 may be readily explained by equations of this section. The decrease in capacity with decrease of saturation-suction temperature may be explained by examining Eqs. (4.3), (4.8), and (4.9). For this case assume a constant condensing pressure. With decrease in saturation-suction temperature, η_V will decrease, owing to the increase in the ratio P_c/P_b, while the specific volume at compressor intake, v_4, will increase. Both of these effects will cause a reduction in the mass rate of flow $\dot{m}$. The refrigerating effect $(h_3 - h_2)$ will decrease with decease in saturation-suction temperature. Thus, the reduction in tons capacity is due to three causes: (a) decrease of volumetric efficiency, (b) increase in specific volume, and (c) decrease of refrigerating effect. Of these three factors, the increase in specific volume is by far the most important.

The trend of the horsepower curve may be explained with aid of Eqs. (4.7), (4.9), and (4.10). With decrease of saturation-suction temperature, $_4w_5$ will increase, owing to increase in the pressure ratio P_c/P_b and increase in specific volume v_b. The cylinder suction pressure P_b will decrease as a compensating factor, but the net result will be an increase in $_4w_5$.

However, the mass rate of flow $\dot{m}$ decreases at a much greater rate than the increase of $_4w_5$. The net result is a decrease in horsepower with decrease in saturation-suction temperature. The trend of the horsepower-per-ton curves may be justified by a similar analysis using Eq. (4.12).

The effect of an increase in condensing pressure can be readily explained by the equations. At a given saturation-suction temperature the capacity will decrease with increase of condensing pressure, owing to decrease of volumetric efficiency and decrease of refrigerating effect. The power increases due to the increase of the work $_4w_5$. Compressor capacity may be reduced by unloading one or more cylinders of a multicylinder compressor. This is accomplished by keeping the suction valves open in the unloaded cylinders so that no mass is pumped through them. Another method of capacity control is to use a variable-speed drive or a variable-speed motor. The motor speed can be varied by changing the frequency of the alternating current supplied to it. Solid-state inverters can be used to control the output frequency and thus the motor speed.

4.6 ROTARY COMPRESSORS

Rotary compressors have found wide use in domestic refrigeration equipment because of the low noise and vibration levels compared to a reciprocating compressor [5]. Two types of rotary compressors are in use, the rolling-piston type shown in Fig. 4.8 and the rotating-vane type shown in Fig. 4.9. The rolling-piston-type compressor has a shaft and cylindrical piston that both rotate clockwise. Two gas pockets are separated by the contact point between the rolling piston and the housing and a vane held against the rolling piston by a spring. The gas in the pocket on the left-hand side of Fig. 4.8 undergoes compression and discharge, while the suction gas enters the pocket on the right-hand side. Thus the rolling-piston-type compressor is a double-acting compressor. The volume of the

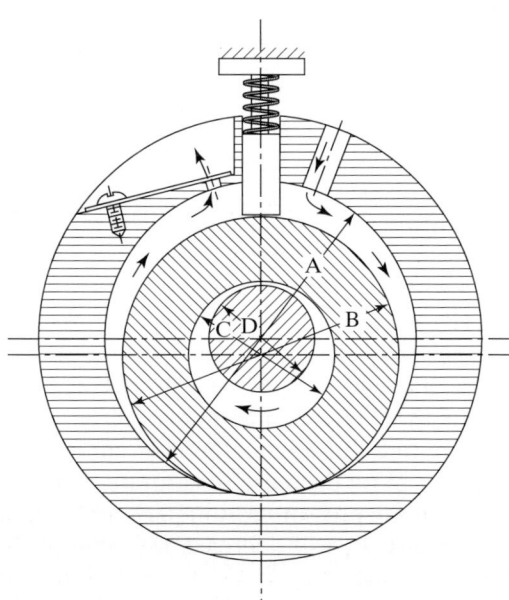

Figure 4.8 Rolling-piston-type rotary compressor. [Reprinted by permission from *ASHRAE Systems and Equipment 1996* (IP & SI), p. 34.9.]

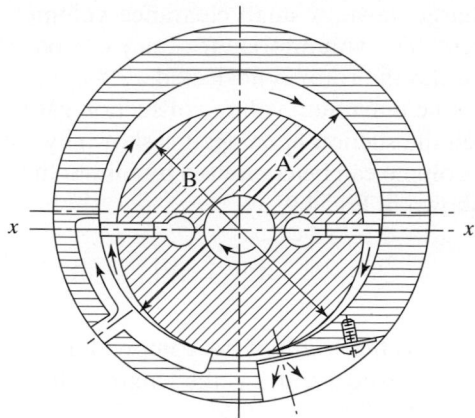

Figure 4.9 Rotating-vane-type rotary compressor. (Reprinted by permission from *ASHRAE Equipment 1988*, p. 12.11.)

suction gas at the beginning of compression is given by the difference in volume between the rolling piston of diameter B and the housing of diameter A,

$$V_s = \frac{\pi W}{4}(A^2 - B^2) \tag{4.16}$$

where W is the width of the roller and housing. This compressor has no suction valve and usually has a flexing reed-type discharge valve. As in a reciprocating compressor, the refrigerant vapor will be superheated during the intake process. The volumetric efficiency is a function of the amount of vapor remaining in the clearance volume at the end of the discharge process, the leakage around the vane, and leakage past the contact between the rolling piston and housing. If leakage is neglected, the volumetric efficiency may be computed from Eq. (4.3) and the mass flow rate from Eq. (4.5). The compressor displacement is

$$\text{C.D.} = V_s(\omega_{\text{comp}}) \tag{4.17}$$

The compression cycle is shorter than the time required for shaft rotation, owing to the difference in diameters. Since $C/D = A/B$,

$$\omega_{\text{comp}} = \omega_{\text{shaft}} / \left[\left(\frac{A}{B} - 1 \right) \frac{A}{B} \right] \tag{4.18}$$

By Eqs. (4.16), (4.17), and (4.18)

$$\text{C.D.} = \frac{\pi W}{4}(A^2 - B^2)(\omega_{\text{shaft}}) / \left[\left(\frac{A}{B} - 1 \right) \frac{A}{B} \right] \tag{4.19}$$

The rotating-vane-type rotary compressor usually has two vanes, as shown in Fig. 4.9, but may have more. The two-vane model has three gas pockets separated by the contact between the rotor and housing and the two vanes. Suction gas enters the pocket on the lower left-hand side of Fig. 4.9. The pocket above the two vanes has just ended the suction process and is beginning the compression process. The third pocket on the lower right-hand side is undergoing compression and discharge through the discharge valve. Like the rolling-piston-type compressor, this compressor has no suction valve, a reed-type discharge valve, a small clearance volume, and vapor superheating during the intake process. The volumetric efficiency can be computed from Eq. (4.3) if leakage past the vanes and the rotor is neglected.

The maximum volume of suction gas in the compressor occurs just after a vane has passed the suction opening; it is shown by the gas pocket above the two vanes in Fig. 4.9. This volume can be computed by subtracting (a) the volume in the housing under the line x-x shown in Fig. 4.9,

$$V_{\text{x-x}} = W\left[\frac{A^2}{4}\cos^{-1}\left(\frac{A-B}{A}\right) + \frac{C}{4}(B-A)\right] \tag{4.20}$$

where W is the width of the compressor, A is the diameter of the housing, B is the diameter of the rotor, and C is the length of the chord through the housing along the line x-x,

$$C = \sqrt{B(2A-B)}$$

(b) one-half the rotor volume and (c) one-half the exposed volume of the vanes from the total volume of the housing. The result becomes

$$V_s = \frac{W}{4}\left[\pi\left(A^2 - \frac{B^2}{2}\right) - A^2\cos^{-1}\left(\frac{A-B}{A}\right) + C(A-B) - 2t(C-B)\right] \tag{4.21}$$

where t is the vane thickness.

The mass flow rate pumped by the compressor can be computed from Eq. (4.5), where

$$\text{C.D.} = 2V_s(\omega_{\text{shaft}}) \tag{4.22}$$

The performances of rolling-piston-type and rotating-vane-type rotary compressors are similar. These compressors are usually hermetically sealed, so that the electric power requirement can be estimated using Eq. (4.15), where P_b equals the suction-line pressure due to the absence of an intake valve. Typical refrigerating-capacity curves and compressor electric-power-consumption curves versus evaporating temperature for Refrigerant 22 are given in Fig. 4.10. The mass flow rate through the compressor, and the corresponding refrigerating capacity, decrease sharply at low evaporating temperatures, owing primarily to the increased specific volume of the suction gas. The work required per unit mass of refrigerant increases at low evaporating temperatures, but the product of the specific-work requirement and the refrigerant mass flow rate results in the power curves shown on Fig. 4.10 that are nearly independent of evaporating temperature.

4.7 SCROLL COMPRESSORS

Like the reciprocating, rolling-piston, and rotary-vane compressors, the scroll compressor is a constant displacement device. It has very few moving parts, low noise and vibration levels, and no valves other than a check valve on the discharge port. This compressor has a fixed volume ratio and essentially no clearance volume. However, compressed gas can leak to gas pockets of lower pressure, which reduces the volumetric efficiency. Leakage can occur between faces of the scroll, which have rolling contact, and the scroll tips, which have sliding contact. The close tolerance between the two scrolls and the oil seal prevent significant leakage from occurring. The volumetric efficiency of this compressor is more difficult to quantify than that of a reciprocating compressor.

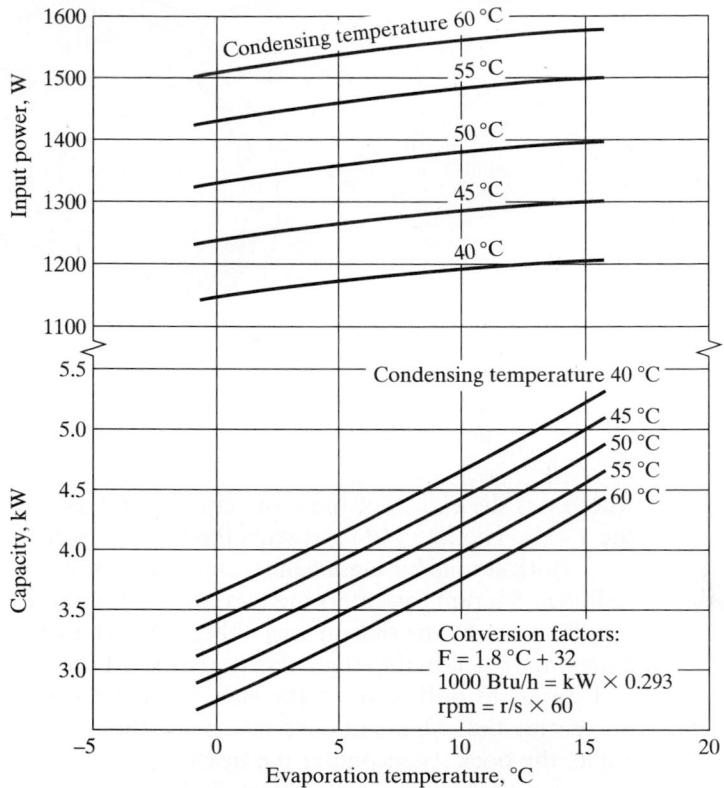

Refrigerant-22, 58 r/s speed, 33 °C ambient
temperature, 33 °C suction gas temperature,
0.25 °C liquid subcooling

Figure 4.10 Typical performance curves for a rotary compressor.
(Reprinted by permission from *ASHRAE Equipment 1988*, p. 12.12.)

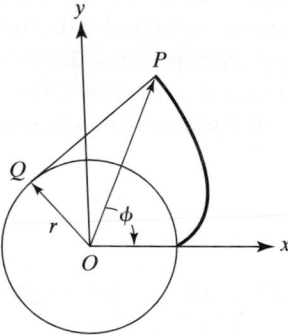

Figure 4.11 Involute spiral used in the
scroll compressor.

The basic geometry of the scroll is an involute spiral. This is the curve formed by
the end of a string as the string is unwrapped from a cylinder. A portion of an involute
spiral is shown in Fig. 4.11. The location of point P on the spiral is given by

$$x_p = r(\cos\phi + \phi\sin\phi)$$
$$y_p = r(\sin\phi - \phi\cos\phi) \tag{4.23}$$

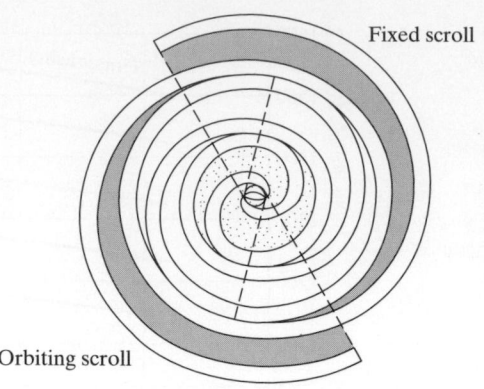

Figure 4.12 Sketch of the two scrolls showing various stages of compression.

where r is the radius of the basic circle and ϕ is the angle between the vector **OP** and the x axis. The line QP represents the imaginary string as it unwraps from the cylinder.

Both scrolls have a height h and thickness t, and the orbiting scroll has an orbiting radius R. Suction gas enters the two cavities formed between the outer portion of the two scrolls shown by the shaded area of Fig. 4.12. The volume in these suction pockets reaches a maximum just before the pockets are sealed off. As the crank angle increases, the contact points at both ends of the suction pockets move toward the center of the spiral, decreasing the volume of the pockets and thereby compressing the gas. At a given crank angle, the pockets encounter the discharge port when the discharge process begins. The volume ratio between suction and discharge is therefore fixed by the compressor design and is not a function of the operating pressures as for a reciprocating compressor. At the end of the discharge process a small amount of gas remains in the innermost sealed pocket. Several stages of compression occur simultaneously—suction, compression, and discharge—which results in low vibration and noise levels and smaller variations in torque than in a reciprocating or rotary compressor. Very little heat transfer occurs between the refrigerant and the compressor, as the refrigerant vapor passes through the scroll set in only one direction and the refrigerant and scroll temperatures tend to be equal during the various stages of the compression process. Therefore the state (a) of the vapor entering the compressor is assumed to equal the state (b) of the vapor at the beginning of the compression process. The compression process given in Fig. 4.13 shows the overcompression

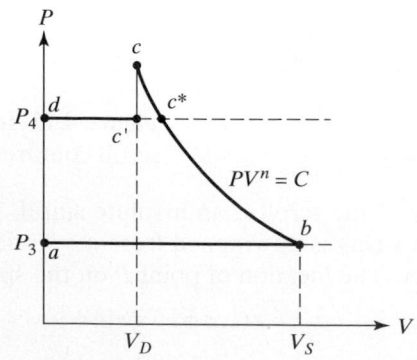

Figure 4.13 Schematic pressure-volume diagram for a scroll compressor with overcompression.

Part II / Refrigeration

that can occur when the built-in volume ratio results in a maximum pressure at state c that is larger than the discharge pressure at state d.

$$\frac{P_c}{P_b} = \frac{P_c}{P_3} = \left(\frac{V_b}{V_c}\right)^n = \left(\frac{V_s}{V_d}\right)^n = V_r^n$$

where V_r is the built-in volume ratio of the compressor.

The mass of vapor compressed per revolution of the crank (per cycle) can be expressed as

$$M/\text{cycle} = V_s/v_b \qquad (4.24)$$

The volumetric efficiency is difficult to quantify except by test, although it can be written as

$$\eta_V = (1 - L_{\text{tip}} - L_{\text{flank}})\frac{v_3}{v_b} \qquad (4.25)$$

where L_{tip} is the fraction of vapor lost through leakage past the scroll tips and L_{flank} is the loss past the rolling seal of the flank. The mass flow pumped through the compressor becomes

$$\dot{m} = \frac{\eta_V \text{C.D.}}{v_3} = (1 - L_{\text{tip}} - L_{\text{flank}})\text{C.D.}/v_b \qquad (4.26)$$

The work of compression per cycle can be found by integrating the area on the diagram of Fig. 4.13:

$$W/\text{cycle} = \int_b^c V\, dP - \int_{c'}^c V\, dP$$

Assuming polytropic compression,

$$W/\text{cycle} = \frac{n}{n-1} P_3 V_s \left[\left(\frac{P_c}{P_3}\right)^{(n-1)/n} - 1\right] - (P_c - P_4)V_d \qquad (4.27)$$

or, writing this in terms of the volume ratio V_r

$$W/\text{cycle} = \frac{n}{n-1} P_3 V_s (V_r^{n-1} - 1) - (P_3 V_r^n - P_4)\frac{V_s}{V_r} \qquad (4.28)$$

By Eqs. (4.24) and (4.28),

$$w = \frac{n}{n-1} P_3 v_b (V_r^{n-1} - 1) - (P_3 V_r^n - P_4)\frac{v_b}{V_r} \qquad (4.29)$$

The electric-motor power requirement can be determined by combining Eqs. (4.8), (4.13), (4.14), and (4.29) with $v_3 = v_b$:

$$\dot{W}_{\text{elec}} = \frac{\eta_v \text{C.D.}}{\eta_m \eta_E}\left[\frac{n}{n-1} P_3 (V_r^{n-1} - 1) - \left(\frac{P_3 V_r^n - P_4}{V_r}\right)\right] \qquad (4.30)$$

EXAMPLE 4.3

A scroll compressor operates with R-22 in a refrigeration system which has a saturated evaporating temperature of 10 °F and a saturated condensing temperature of 90 °F. The vapor enters the compressor at 20 °F and is compressed polytropically. The compressor has a volume ratio of 3.5, a displacement of 4.5 in.³/cycle, a speed of 3600 rpm, a volumetric efficiency of 0.9, a mechanical efficiency of 0.8, and an electric-motor efficiency of 0.85. Compute: (a) the mass flow rate of refrigerant through the compressor and (b) the electrical power required in kW.

Solution: The first step is to determine the specific volume at state 3

$$P_3 = P_b = 47.45 \text{ psia}, \qquad T_3 = 20 \text{ °F}, \qquad v_3 = 1.16 \text{ ft}^3/\text{lbm}$$

from Eq. (4.26).

(a) $\dot{m} = \dfrac{(0.9)(4.5)(3600)}{(1.16)(1728)} = 7.27 \text{ lbm/min}.$

(b) $P_4 = 183.05$ psia. From Table 4.2 $n = 1.12$. By Eq. (4.30)

$$\dot{W}_{\text{elec}} = \frac{(0.9)(4.5)(3600)(0.75)}{(0.8)(0.85)(12)(33,000)} \left\{ \frac{1.12}{0.12} (47.45)[(3.5)^{0.12} - 1] - \left[\frac{47.45(3.5)^{1.12} - 183.05}{3.5} \right] \right\}$$

$$\dot{W}_{\text{elec}} = 2.80 \text{ kW}$$

4.8 SCREW COMPRESSORS

The screw compressor is a positive-displacement compressor that is manufactured in two distinct types: single-screw and twin screw. The single-screw compressor is sketched in Fig. 4.14. The center helical screw drives the two star or gate rotors which form the seal in the flutes of the main rotor. The main rotor has six flutes, and each of the gate rotors has eleven teeth. No power is transferred to the gate rotors; all the input power is either transferred to the gas or is used to overcome frictional losses.

The twin-screw compressor is shown schematically in Fig. 4.15. In most designs a four-lobe rotor drives a six-fluted rotor, although other ratios such as 5/6 and 5/7 are also used. The lobed rotor drives the fluted rotor, with about 5 to 25 percent of the torque being transferred to the fluted rotor.

The operation and performance of single-screw and twin-screw compressors are similar. The discussion that follows applies to both types. The compression process begins

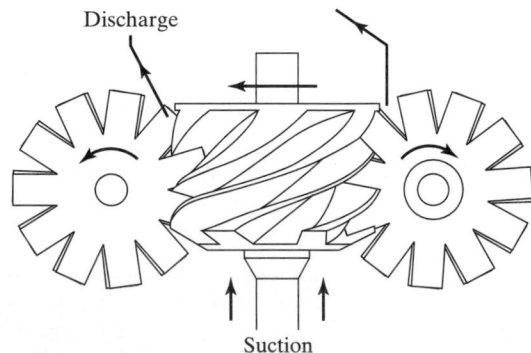

Figure 4.14 Schematic of a single-screw compressor. (Reprinted by permission from *ASHRAE Equipment 1988*, p. 12.16.)

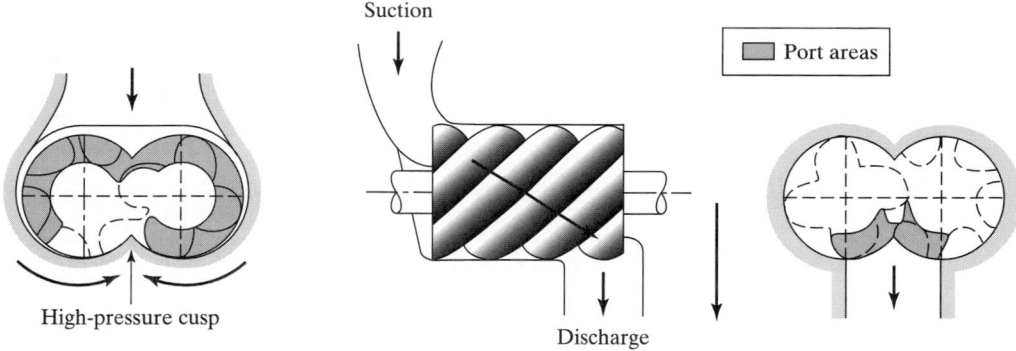

Figure 4.15 Twin-screw compressor. [Reprinted by permission from *ASHRAE Systems and Equipment 1996* (IP & SI), p. 34.17.]

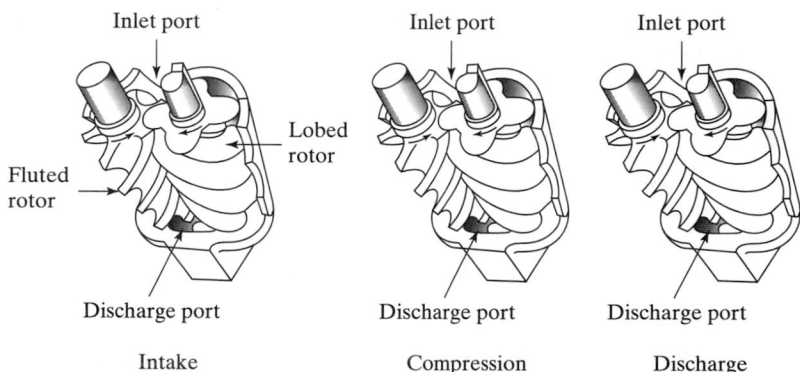

Figure 4.16 Compression process. [Reprinted by permission from *ASHRAE Systems and Equipment 1996* (IP & SI), p. 34.17.]

when the suction gas enters the space created between the flutes, lobes, or teeth and the housing. As the space leaves the inlet port, a lobe or tooth meshes with the flute, sealing off the gas pocket. Further rotation causes the lobe or tooth to progress along the flute and compress the gas. When the flute rotates sufficiently to reach the discharge port, the discharge process begins, and it continues until the compressed gas has been forced out of the flute. While this compression-discharge process is occurring, additional suction gas enters the flute behind the lobe or tooth that is causing the compression, so that the process can be repeated immediately after the discharge process. This process is shown on Fig. 4.16 for the twin-screw compressor. The screw compressor has no valves and essentially no clearance volume. Its volumetric efficiency is a function primarily of gas leakage and therefore is a function both of speed and pressure ratio. The compressor has a built-in volume ratio, V_r, which determines the pressure ratio, P_r, under given operating conditions.

$$P_r = V_r^n \tag{4.31}$$

where n is the polytropic exponent of compression. The pressure ratio of the compression may not match the pressure ratio required by the rest of the refrigeration system, which causes over- or undercompression. The overcompression process is shown in Fig. 4.13. The mismatch of pressure ratios requires additional work and reduces the efficiency of the compression process.

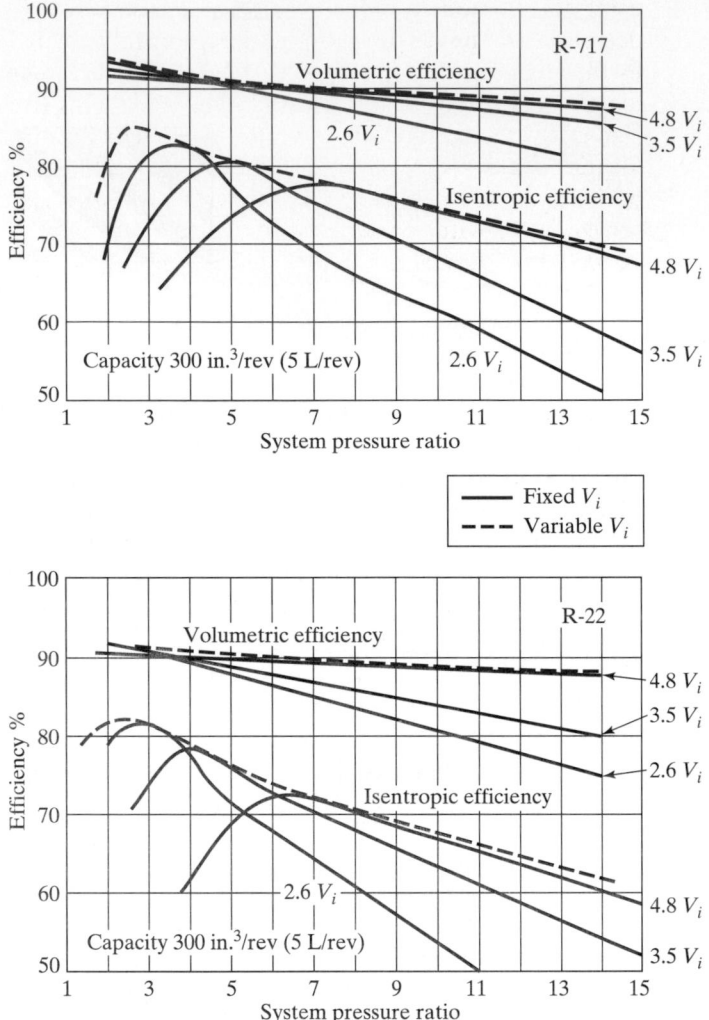

Figure 4.17 Twin-screw compressor efficiency curves. [Reprinted by permission from *ASHRAE Systems and Equipment 1996* (IP & SI), p. 34.19.]

The mass flow rate through the compressor can be computed using Eq. (4.9). Figure 4.17 shows some typical volumetric-efficiency curves for a twin-screw compressor operating with ammonia (R-717) or R-22. The curves are nearly independent of system pressure ratio when the built-in volume ratio is large or the volume ratio is variable. The compressor displacement is a function of the rotor design and speed.

The work required to compress the refrigerant may be estimated from Eq. (4.29), which includes losses due to over- or undercompression. Curves of compressor isentropic efficiency are given in Fig. 4.17 for a twin-screw compressor. Isentropic efficiency is the ratio of the isentropic enthalpy change to the actual enthalpy change of the refrigerant

$$\eta_s = \frac{(h_4 - h_3)_s}{h_d - h_3} \tag{4.32}$$

Part II / Refrigeration

using the nomenclature in Fig. 4.13. As shown in Fig. 4.17, the isentropic efficiency decreases whenever the system pressure ratio is below or above the pressure ratio given by the built-in volume ratio, Eq. (4.31). However, when a variable volume ratio is used to match the pressure ratios in the compressor and the system, the isentropic efficiency remains at the maximum value possible. This is usually accomplished by changing the location of the beginning of compression so that the volume V_S varies. The slide-valve position is varied axially along the screw by a hydraulic or electric actuator. A compressor equipped with this type of unloading device has a variable capacity that ranges from 10 to 100 percent of full load at constant speed.

4.9 CENTRIFUGAL COMPRESSORS

Large-capacity refrigeration units used in air conditioning applications may use a centrifugal compressor. This type of compressor is not a positive-displacement device like the other compressors discussed in this chapter but belongs to the family of turbomachines. These compressors may be placed in series to achieve pressure ratios up to 30, although the pressure ratio across any single stage is limited. Economizer cycles that incorporate flash cooling can be used, as suction gas can be added to the compressor at intermediate pressure levels between impellers in a multistage compressor.

A simplified analysis can be used to estimate the power required to operate a centrifugal compressor under specified operating conditions. The analysis begins under the assumption that the fluid is an ideal gas with constant specific heat. However, this restriction will be removed later.

The reversible work per unit mass of fluid during the compression process from suction pressure, P_1, to discharge pressure, P_2, can be found from the following integral:

$$w = \int_{P_1}^{P_2} v \, dP \tag{4.33}$$

In a polytropic process, $Pv^n = c$, and this specific-work requirement becomes the polytropic work,

$$w_P = \frac{n}{n-1} P_1 v_1 \left[\left(\frac{P_2}{P_1} \right)^{(n-1)/n} - 1 \right] \tag{4.34}$$

The minimum work required in an adiabatic process is the isentropic work

$$w_s = h_{2_s} - h_1 \tag{4.35}$$

For an ideal gas with constant specific heat, a reversible, adiabatic (isentropic) process can be represented by $Pv^k =$ constant, so the isentropic work can be written as

$$w_s = \frac{k}{k-1} P_1 v_1 \left[\left(\frac{P_2}{P_1} \right)^{(k-1)/k} - 1 \right] \tag{4.36}$$

The ratio of the polytropic to isentropic work requirements allows the polytropic work to be computed as follows:

$$w_P = \frac{w_P}{w_s} (h_{2_s} - h_1) = \frac{\dfrac{n}{n-1} \left[\left(\dfrac{P_2}{P_1} \right)^{(n-1)/n} - 1 \right]}{\dfrac{k}{k-1} \left[\left(\dfrac{P_2}{P_1} \right)^{(k-1)/k} - 1 \right]} (h_{2_s} - h_1) \tag{4.37}$$

However, n and k are related. Consider the ratio of polytropic to actual work required over a small section of the compression process. The polytropic efficiency becomes

$$\eta_p = \frac{\text{reversible work}}{\text{actual work}} = \frac{v\,dP}{dw} \tag{4.38}$$

For an ideal gas undergoing an adiabatic process with negligible change in fluid kinetic energy

$$dw = c_p\,dt = \frac{c_p}{R}\,d(Pv) \tag{4.39}$$

Substituting into Eq. (4.38),

$$\eta_p = \frac{Rv\,dP}{c_p\,d(Pv)} \tag{4.40}$$

Expanding the term $d(Pv)$ and rearranging,

$$n = \left(\frac{c_p\eta_p}{c_p\eta_p - R}\right)\frac{dv}{v} + \frac{dP}{P} = 0 \tag{4.41}$$

For this process to have a constant polytropic exponent, n, the term in the parentheses must equal n.

$$n = \frac{c_p\eta_p}{c_p\eta_p - R} \tag{4.42}$$

Setting $R = c_p - c_v = c_p(k-1)/k$ and solving for η_p,

$$\eta_p = \frac{n}{n-1}\frac{k-1}{k} \tag{4.43}$$

Substituting this expression into Eq. (4.37) gives

$$w_p = \eta_p \frac{\left[\left(\dfrac{P_2}{P_1}\right)^{(k-1)/\eta_p k} - 1\right]}{\left[\left(\dfrac{P_2}{P_1}\right)^{(k-1)/k} - 1\right]} (h_{2_s} - h_1) \tag{4.44}$$

Although Eq. (4.44) was derived assuming ideal-gas conditions, the polytropic efficiency, η_p, and the ratio of the two terms in the brackets are rather insensitive to the nature of the fluid. Therefore Eq. (4.44) has been found to apply to refrigerant vapors with reasonable accuracy. One should check the polytropic exponent, n, by comparing the values of pressure and specific volume at the suction and discharge ports of the compressor. The value of n must be greater than k for the polytropic efficiency given in Eq. (4.43) to be less than 1.0. The specific-heat ratio, k, can be determined for refrigerant vapors using the principle of corresponding states.

4.10 EXPANSION DEVICES

The primary functions of the expansion device are to reduce the pressure and corresponding saturation temperature of the refrigerant and to control the flow of liquid refrigerant in response to cooling load at the evaporator. Usually the condenser pressure is fixed by the heat rejected, and the suction or evaporator pressure will vary according to the load. That is, as flow increases, owing to increase in cooling load, the suction pressure will also increase (and vice versa). Therefore, it cannot be said that the expansion device will always create a fixed pressure drop. It merely throttles the pressure from the condenser pressure to the existing evaporator pressure. If no separate evaporator-pressure regulator is installed in the suction side of the evaporator line, the evaporator pressure will be an indirect result of the flow controlled by the expansion device. An exception is the automatic expansion valve, which can be adjusted to obtain a constant specific evaporator pressure. A flow will thus be obtained as an indirect result of the pressure setting.

A slight amount of superheat (5 to 10 degrees) is always desirable in the evaporator to assure that no liquid slugs are fed to the compressor. Superheat also facilitates control of flow. The disadvantages of excessive superheat include:

1. Increased compressor power consumption.
2. Decreased compressor pumping capacity, i.e., volume of vapor handled.
3. Increased discharge temperature from the compressor.

Three types of expansion devices are described here: a thermostatic expansion valve (T.E.V.), automatic expansion valve, and capillary tube. The operation of the *thermostatic expansion valve* is based on maintaining a constant superheat in the refrigerant vapor leaving the evaporator coil, a circumstance which allows the valve to keep practically the entire coil filled with evaporating refrigerant under all conditions of system loading, without fear of liquid-slugging of the compressor. This ability to make effective use of the entire evaporator surface under all load conditions makes the T.E.V. especially suitable for systems with wide and frequent fluctuations in loading. Its adaptability to any application and its high efficiency have made it today's most widely used refrigerant control.

Figure 4.18 shows the main parts of the T.E.V.: the diaphragm or pressure bellows, a needle and seat, a spring, whose tension is manually adjustable, and a fluid-charged

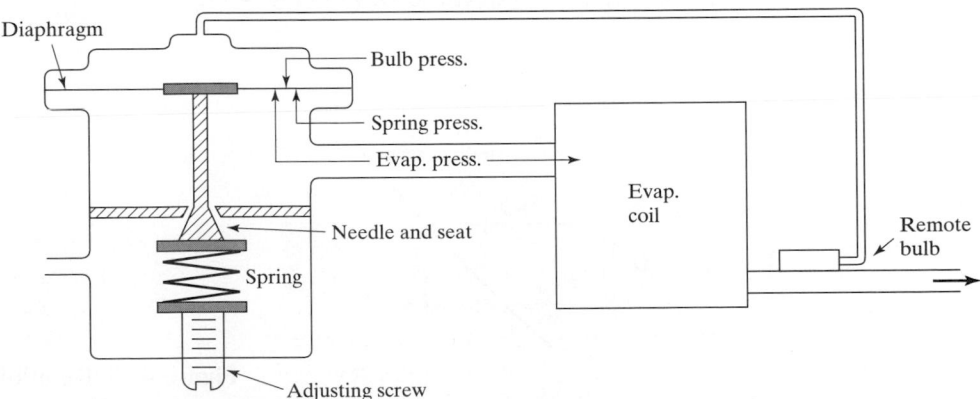

Figure 4.18 Thermostatic expansion valve (T.E.V.).

remote feeler bulb connected to the diaphragm by a capillary tube. An interaction of the spring pressure, the evaporator pressure, and the pressure exerted by the saturated liquid-vapor mixture in the sensing bulb are responsible for the valve operation.

The feeler bulb is clamped to the suction outlet of the evaporator coil and thus assumes the temperature of the leaving refrigerant vapor. Assuming that the refrigerant leaves the coil superheated and that the feeler bulb is filled with the same type of refrigerant used in the system, the bulb will exert a pressure on the valve diaphragm equal to the saturation pressure for the superheated temperature. This will be substantially greater than the evaporator pressure exerted on the opposite side of the diaphragm. For steady-state operation the spring tension must make up for this pressure differential. It thus becomes evident that the superheat can be increased or decreased by increasing or decreasing the spring tension. The state of the superheated vapor leaving the evaporator is shown as state 3 in Fig. 4.19. The pressure difference made up by the spring force is linearly proportional to the amount of superheat, ΔT, achieved. The thermostatic expansion valve has the following characteristics:

1. Accommodates a variable load.
2. Operates evaporator with some degree of superheat.
3. Can be used with pressure or temperature types of compressor motor controls.
4. Operates coil on "dry expansion."
5. Does not require "critical charge" of refrigerant in the system.
6. Is expensive.
7. Can't set specific flow of refrigerant or specific evaporating pressure. Can set only superheat.
8. Requires care in installation.
9. Capacity of a given valve will vary as the pressure range through which it operates is varied; i.e., valve must be selected with knowledge of the high- and low-side pressures to be encountered during operation.
10. Thermal bulb should be charged with same refrigerant as in the system.
11. If pressure drop through the coil shall exceed approximately 1 psi, the valve should be equipped with an external equalizer. This means that the pressure under the diaphragm equals the pressure at the evaporator exit rather than the inlet, as shown by the internally equalized valve in Fig. 4.18.

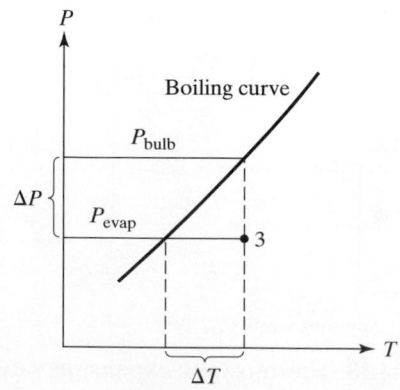

Figure 4.19 Relationship between pressure difference and evaporator superheat as controlled by a T.E.V.

The *automatic expansion valve (A.E.V.)* operates to maintain a constant pressure in the evaporator by flooding more or less of the evaporator surface in response to changes in the evaporator load. The maintaining of a constant evaporator pressure requires that the rate of vaporization in the evaporator be held constant. When the evaporator load decreases and consequently the heat-transfer capacity per unit of evaporator surface is reduced, more and more of the evaporator surface must be flooded with liquid if a constant rate of vaporization is to be maintained. The A.E.V. thus possesses the peculiar property of increasing flow as load on the evaporator decreases, and vice versa. Its application is therefore limited to small loads where the system is adequately protected by thermal compressor-motor controls against flooding.

In view of its poor efficiency under heavy load conditions, the A.E.V. is best applied only to small equipment having relatively constant loads, such as domestic refrigerators and freezers. Today, however, even in these applications the A.E.V. is seldom employed, having been replaced by more efficient and often cheaper flow controls. The only significant use might be as "condenser bypass valves."

From Fig. 4.20 it is evident that the major components of the A.E.V. are the bellows or diaphragm, the needle and seat, the spring, and the adjusting screw. The constant-pressure characteristics of the valve result from the interaction of the evaporator pressure and the spring force. Once the tension of the spring is adjusted for the desired evaporator pressure, the valve will operate automatically (hence the name) to regulate the flow of liquid refrigerant into the evaporator. The characteristic features of the automatic expansion valve are:

1. Can set specific flow or evaporator pressure.
2. Is not likely to place a high starting load on compressor motor.
3. Is easy to install.
4. Does not require "critical charge" of refrigerant in the system.
5. Cannot respond properly to a variable load.
6. Is expensive.
7. Can be used only with thermal-type compressor-motor controls (a pressure-type motor switch is useless, since the low-side pressure does not vary).

The simplest of all refrigerant flow controls, the *capillary tube*, usually consists of a small-bore copper tube of some select length positioned between the condenser and the

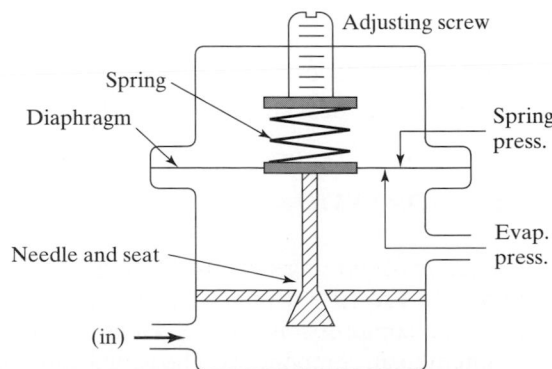

Figure 4.20 Automatic expansion valve.

evaporator in lieu of the conventional liquid line. The high frictional resistance (function of its length and small bore) serves both to meter the flow of refrigerant from condenser to evaporator and to maintain the desired pressure differential between these two components. Efficient overall system operation requires that the tube dimensions be carefully selected, so that for the desired evaporator and condenser pressures the flow through the tube will be compatible with the pumping capacity of the compressor. Too little flow results in a starved evaporator, which in turn lowers the suction pressure and raises the condensing pressure, both disadvantages to overall system efficiency. Too much flow results in overfeeding of the evaporator, a condition which soon leads to slugging of the compressor. It is evident that maximum efficiency can be maintained at only one set of operating conditions.

In addition to simple construction and low cost the capillary tube offers numerous other advantages, including the need for only a low-torque motor to drive the compressor (a result of pressure equalization during the "off" cycle). Furthermore, the small critical charge required eliminates the need for a receiver tank while at the same time reducing the quantity of refrigerant needed. These factors, all advantages from the manufacturing-cost standpoint, have led to the universal use of capillary tubes in domestic refrigeration units such as refrigerators, freezers, and room air conditioners.

Soldering the capillary tube to the suction line tends to subcool the liquid flowing through it, thus reducing the formation of flash gas in the tube. This flash gas, because of its large volume, is forced to move at a higher velocity and thus introduces increased frictional pressure drop, which can substantially reduce the flow capacity of the tube. The characteristics of the capillary tube are:

1. Cheapest of all expansion devices.
2. High- and low-side pressure equalize on "off" cycle, so there is low starting load on the compressor motor.
3. No liquid receiver is used.
4. No moving parts.
5. Eliminates the conventional liquid line between the condenser and evaporator.
6. Requires an accumulator in the evaporator to handle excess liquid refrigerant.
7. Requires that special care be taken to prevent kinking or clogging of the tube and that all moisture in the system be removed that may freeze in the tube.
8. Difficult to design except by test and, therefore, best adapted to factory assembled and charged units.
9. If used with an air-cooled condenser, it does allow some variation in the flow as the cooling load varies.
10. Because of critical charge, it should always be used with a hermitic compressor.

4.11   ACTUAL REFRIGERATION SYSTEMS

Actual systems operating steadily differ in many respects from the ideal cycles considered in Chapter 3. Pressure drops occur in the refrigerant piping everywhere in the system except in the compression process. Heat transfer occurs between the refrigerant and its environment in all components. The actual compression process differs substantially from the isentropic compression assumed in Chapter 3. The working fluid is not a pure substance but a mixture of refrigerant and oil. All of these deviations from a theoretical cycle cause irreversibilities within the system. Each irreversibility requires additional power into the compressor. It is useful to understand how these irreversibilities are distributed throughout a real system. Example 4.4 illustrates how the irreversibilities can be computed in a real system and how they require additional compressor power to overcome. The input data have been rounded off for ease of computation.

EXAMPLE 4.4

An air-cooled direct-expansion single-stage mechanical vapor-compression refrigeration system uses Refrigerant 22 and operates under steady conditions. A schematic drawing of this system is shown in Fig. 4.21. Pressure drops occur in all piping, and heat gains or losses occur as indicated. Power input includes compressor power and the power required to operate both fans. The following performance data are obtained:

Ambient air temperature, $t_0 = 90\ °F$

Refrigerated-space temperature, $t_R = 20\ °F$

Refrigeration load, $\dot{Q}_E = 2$ tons

Compressor power input, $\dot{W}_{comp} = 3.0$ hp

Condenser-fan input, $\dot{W}_{C.F.} = 0.2$ hp

Evaporator-fan input, $\dot{W}_{E.F.} = 0.15$ hp

Refrigerant pressures and temperatures are measured at the seven locations shown on Fig. 4.21. Table 4.3 lists the measured and computed thermodynamic properties of the refrigerant, neglecting the dissolved oil. A pressure-enthalpy diagram of this cycle is shown in Fig. 4.22 compared with a theoretical single-stage cycle operating between the air temperatures

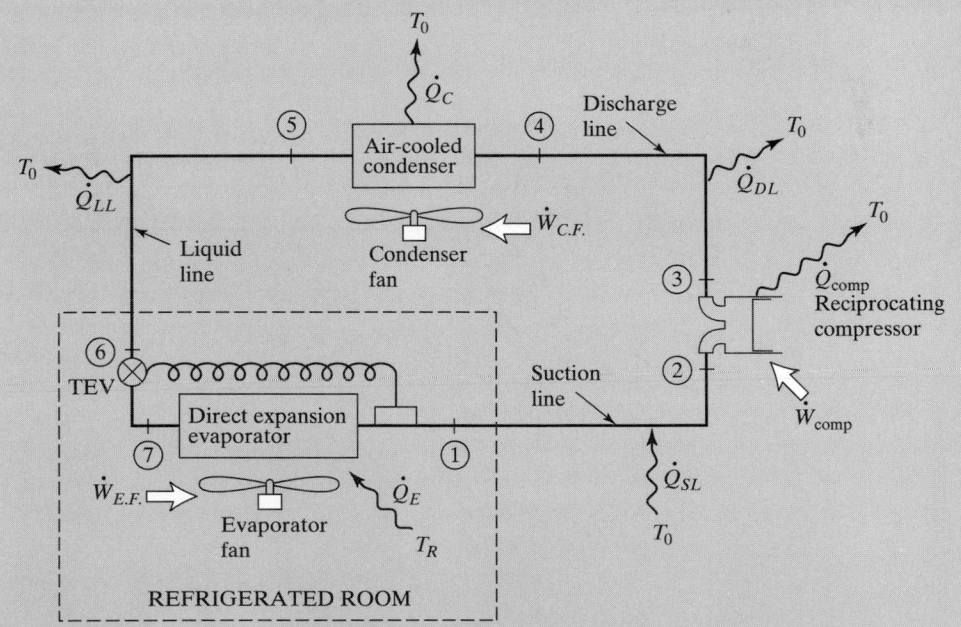

Figure 4.21 Schematic of real direct-expansion mechanical vapor-compression refrigeration system.

TABLE 4.3 Measured and Computed Thermodynamic Properties of Refrigerant 22 for Ex. 4.4.

| State | Measured | | Computed | | |
	P, psia	t, °F	h, Btu/lbm	s, Btu/lbm · °R	v, ft³/lbm
1	45.0	15.0	106.4	0.2291	1.213
2	44.0	25.0	108.1	0.2330	1.276
3	210.0	180.0	128.8	0.2374	0.331
4	208.0	160.0	124.8	0.2314	0.318
5	205.0	94.0	37.4	0.0761	0.014
6	204.0	92.0	36.8	0.0750	0.014
7	46.5	9.0	36.8	0.0800	0.308

t_R and t_0. Compute the energy transfers to the refrigerant in each component of the system and determine the second-law irreversibility in each component. Show that the total irreversibility multiplied by the ambient temperature (i.e., total exergy destruction) is equal to the difference between the actual power input and the power required by a Carnot cycle operating between t_R and t_0 with the same refrigerating load.

Solution: The mass flow rate of refrigerant is the same through all components, so it is computed only once through the evaporator. Each component in the system will be analyzed sequentially beginning with the evaporator. Note that the temperature used in the second-law analysis is the absolute temperature that the refrigerant is exchanging heat with.

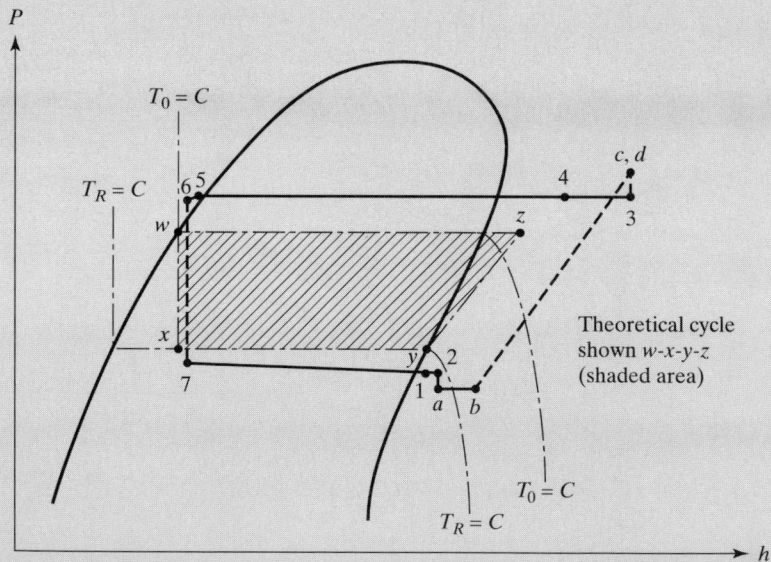

Figure 4.22 Pressure-enthalpy diagram of actual system and theoretical single-stage system operating between the same inlet air temperatures, t_R and t_0.

Evaporator

Energy balance:

$$_7\dot{Q}_1 = \dot{m}(h_1 - h_7) = 24{,}000 \text{ Btu/hr}$$

$$\dot{m} = \frac{24{,}000}{(106.4 - 36.8)} = 345 \text{ lbm/hr}$$

Second law:

$$_7\dot{I}_1 = \dot{m}(s_1 - s_7) - \frac{_7\dot{Q}_1}{T_R}$$

$$= 345(0.2291 - 0.0800) - \frac{24{,}000}{479.67}$$

$$= 1.405 \text{ Btu/hr} \cdot {}^\circ\text{R}$$

Suction Line

Energy balance:

$$_1\dot{Q}_2 = \dot{m}(h_2 - h_1)$$

$$= 345(108.1 - 106.4) = 586 \text{ Btu/hr}$$

Second law:

$$_1\dot{I}_2 = \dot{m}(s_2 - s_1) - \frac{_1\dot{Q}_2}{T_0}$$

$$= 345(0.2330 - 0.2291) - \frac{586}{549.67}$$

$$= 0.279 \text{ Btu/hr} \cdot {}^\circ\text{R}$$

Compressor

Energy balance:

$$_2\dot{Q}_3 = \dot{m}(h_3 - h_2) + {}_2\dot{W}_3$$

$$= 345(128.8 - 108.1) - 3.0(2545)$$

$$= -494 \text{ Btu/hr}$$

Second law:

$$_2\dot{I}_3 = \dot{m}(s_3 - s_2) - \frac{_2\dot{Q}_3}{T_0}$$

$$= 345(0.2374 - 0.2330) - \left(\frac{-464}{549.67}\right)$$

$$= 2.417 \text{ Btu/hr} \cdot {}^\circ\text{R}$$

Discharge Line

Energy balance:

$$_3\dot{Q}_4 = \dot{m}(h_4 - h_3)$$

$$= 345(124.8 - 128.8) = -1380 \text{ Btu/hr}$$

Second law:

$$_3\dot{I}_4 = \dot{m}(s_4 - s_3) - \frac{_3\dot{Q}_4}{T_0}$$

$$= 345(0.2314 - 0.2374) - \left(\frac{-1380}{549.67}\right)$$

$$= 0.441 \text{ Btu/hr} \cdot {}^\circ\text{R}$$

Condenser

Energy balance:

$$_4\dot{Q}_5 = \dot{m}(h_5 - h_4)$$

$$= 345(37.4 - 124.8) = -30{,}153 \text{ Btu/hr}$$

Second law:

$$_4\dot{I}_5 = \dot{m}(s_5 - s_4) - \frac{_4\dot{Q}_5}{T_0}$$

$$= 345(0.0761 - 0.2314) - \left(\frac{-30{,}153}{549.67}\right)$$

$$= 1.278 \text{ Btu/hr} \cdot {}^\circ\text{R}$$

Liquid Line

Energy balance:

$$_5\dot{Q}_6 = \dot{m}(h_6 - h_5)$$

$$= 345(36.8 - 37.4) = -207 \text{ Btu/hr}$$

Second law:

$$_5\dot{I}_6 = \dot{m}(s_6 - s_5) - \frac{_5\dot{Q}_6}{T_0}$$

$$= 345(0.0750 - 0.0761) - \left(\frac{-207}{549.67}\right) \cong 0 \text{ Btu/hr} \cdot {}^\circ\text{R}$$

Expansion Device

Energy balance:

$$_6\dot{Q}_7 = \dot{m}(h_7 - h_6) = 0$$

Second law:

$$_6\dot{I}_7 = \dot{m}(s_7 - s_6)$$

$$= 345(0.0800 - 0.0750) = 1.725 \text{ Btu/hr} \cdot {}^\circ\text{R}$$

These results are summarized in Table 4.4.
The Carnot cycle coefficient of performance is

$$\text{C.O.P.}_c = \frac{T_R}{T_0 - T_R} = \frac{479.67}{70} = 6.852$$

TABLE 4.4 Energy Transfers and Irreversibility Rates for the Refrigeration System in Ex. 4.4.

Component	$\dot{Q}$, Btu/hr	$\dot{W}$, Btu/hr	$\dot{I}$, Btu/hr · °R	$\dot{I}/\dot{I}_{total}$, percent
Evaporator	24,000	0	1.405	19
Suction line	586	0	0.279	4
Compressor	−494	7,635	2.417	32
Discharge line	−1,380	0	0.441	6
Condenser	−30,153	0	1.278	17
Liquid line	−207	0	~0	~0
Expansion device	0	0	1.725	23
Totals	−7,648	7,635	7.545	

The Carnot power requirement for the 2-ton load is

$$\dot{W}_{Carnot} = \frac{\dot{Q}_E}{C.O.P._c} = \frac{24000}{6.852} = 3502 \text{ Btu/hr}$$

The actual power requirement for the compressor is

$$\dot{W}_{comp} = \dot{W}_{Carnot} + \dot{I}_{total} T_0$$
$$= 3502 + 7.545(549.67) = 7649 \text{ Btu/hr}$$

This result is within computational error of the measured power input to the compressor of 7635 Btu/hr.

The methodology demonstrated in Ex. 4.4 can be applied to any actual vapor-compression refrigeration system. The only required information for the second-law analysis is the refrigerant thermodynamic state points and mass flow rates and the temperatures with which the system is exchanging heat. In this example, the extra compressor power required to overcome the irreversibility in each component is determined. The component with the largest loss is the compressor, owing to motor inefficiencies, frictional losses, and irreversibilities due to pressure drops, mixing, and heat transfer between the compressor and the surroundings. The unrestrained expansion in the expansion device is the next largest loss. This could be reduced by utilizing an expander rather than a throttling process. This may be economical on large machines. All the heat-transfer irreversibilities on both the refrigerant side and the air side of the condenser and evaporator are included in the analysis. The refrigerant pressure drop is also included. The only items not included are the air-side pressure-drop irreversibilities of the two heat exchangers. However, these are equal to the fan-power requirements, as all the fan power is dissipated into heat.

The results of an overall second-law analysis such as that given in Ex. 4.4 provide the system designer with some guidance as to which components are creating the most losses and therefore should be replaced or redesigned to produce the greatest improvement in the system performance. However, this type of analysis does not identify the nature of the losses. This requires a detailed second-law analysis, in which the actual processes are analyzed in terms of fluid flow and heat transfer. A detailed analysis will show that most of the irreversibilities associated with the heat exchangers are due to heat transfer, only a very small proportion are caused by pressure drop on the air side, and a negligibly small proportion are caused by refrigerant-pressure drop. This indicates that promoting refrigerant heat-transfer coefficients, at the expense of increasing the pressure drop, will usually result in improved system performance.

ENDNOTES

1. ANSI/ASHRAE Standard 34-92, *Number Designation and Safety Classification of Refrigerants* (Atlanta: American Society of Heating, Refrigerating and Air Conditioning Engineers, 1992).

2. Thomas Midgley, Jr., and Albert L. Henne, "Organic Fluorides as Refrigerants," *Industrial and Engineering Chemistry*, 22 (1930), 542.

3. UNEP, *Montreal Protocol on Substances That Deplete the Ozone Layer, Final Act* (New York: United Nations Environmental Program, 1987).

4. *ASHRAE Handbook of Fundamentals* (Atlanta: American Society of Heating, Refrigerating and Air Conditioning Engineers, 1997).

5. *ASHRAE Handbook of HVAC Systems and Equipment* (Atlanta: American Society of Heating, Refrigerating and Air Conditioning Engineers, 1996).

PROBLEMS

4.1 A Refrigerant-22 plant operating at 0 °F evaporating temperature and 70 °F condensing temperature uses a reciprocating compressor having 4 percent clearance factor. Neglect valve pressure drops and superheating during the intake stroke. Under these conditions the plant produces 20 tons of refrigeration.

(a) Calculate the capacity of the plant if the condensing temperature were to be increased to 100 °F. Assume that other conditions are the same and the compressor is operated at the same rpm. Assume a theoretical single-stage cycle but include the effect of compressor volumetric efficiency.

(b) Assume the above system is operated at 100 °F condensing temperature. What evaporating temperature would be required for refrigerating capacity to be zero?

4.2 An ammonia refrigeration plant utilizes a water-jacketed reciprocating compressor. Saturated vapor at 10 °F enters the compressor, and the vapor leaves the compressor at 180.6 psia and 170 °F. Flow through the condenser and evaporator is at constant pressure. The liquid leaving the condenser is saturated. The system produces 25 tons of refrigeration, and the power input (power delivered to refrigerant passing through compressor) is 23.2

horsepower. Find the lbm per min of cooling water which must flow through the compressor jacket, if the water temperature rise is 10 °F.

4.3 A Refrigerant-22 refrigeration system operates with a condensing temperature of 35 °C and an evaporating temperature of 0 °C. Assume frictionless flow. Saturated liquid leaves the condenser and saturated vapor leaves the evaporator. The reciprocating compressor has a suction valve pressure drop of 0.02 MPa, a discharge-valve pressure drop of 0.005 MPa, and a clearance factor of 5 percent. Assume 5 °C superheating during the intake stroke. The compressor displacement remains constant at 0.005 cu m/sec. Assume a mechanical efficiency of 0.85 and an electric-motor efficiency of 0.9. Calculate the system refrigerating capacity and the electric power required by the compressor motor.

4.4 Repeat Prob. 4.3 with clearance factors of
(a) 4 percent and
(b) 3 percent if all other parameters are unchanged.

4.5 Repeat Prob. 4.3 with valve pressure drops of
(a) 0.015 MPa, 0.03 MPa, and
(b) 0.01 MPa, 0.015 MPa for the suction and discharge valves, respectively, if all other parameters remain unchanged.

4.6 Repeat Prob. 4.3 with evaporating temperatures of
(a) 10 °C,
(b) −10 °C, and
(c) −30 °C
if all other parameters remain unchanged. Plot the system capacity, power requirement, and power-requirement-to-capacity ratio as a function of evaporating temperature similar to Fig. 4.8.

4.7 An industrial plant has available a 4-cylinder, 3-in. bore by 4-in. stroke, 800-rpm, single-acting reciprocating compressor for use with Refrigerant 22. Proposed operating conditions for the compressor are 100 °F condensing temperature and 40 °F evaporating temperature. It is estimated that the refrigerant will enter the expansion valve as a saturated liquid, vapor will leave the evaporator at a temperature of 45 °F, and vapor will enter the compressor at a temperature of 55 °F. Assume a compressor volumetric efficiency of 70 percent. Assume frictionless flow. Calculate the refrigerating capacity in tons for a system equipped with this compressor.

4.8 The following are the conditions for a single-stage ammonia refrigeration plant with a reciprocating-compressor condensing temperature 90 °F with no liquid subcooling, evaporating temperature 0 °F with saturated vapor at evaporator outlet, polytropic compression with $n = 1.24$, and compressor clearance = 5 percent. Assume no pressure drop in piping and compressor valves and no temperature changes either in piping or on intake stroke of compressor.
(a) Calculate the coefficient of performance and compare with the coefficient of performance calculated for isentropic compression.
(b) Assume that the original system is operated with one additional change: the vapor picks up heat in the suction line of the compressor so that the temperature of the vapor entering the compressor is 60 °F. Calculate the percent decrease in the coefficient of performance.
(c) Assume that the original system is operated with a single change: a 4-psi drop in the suction valves and 6-psi drop in the discharge valves of the compressor. Calculate the percent decrease in the coefficient of performance.

4.9 Work parts (b) and (c) of Prob. 4.8 but calculate percent decrease in tons capacity. Assume a constant-displacement compressor.

4.10 Assume that you are the chief engineer of a creamery. You need an ammonia compressor for an addition to your plant. You have been informed that another creamery operated by your company has a surplus compressor which you may have if it is adequate for your installation. It is estimated that the proposed installation would operate under the following conditions: capacity = 18.5 tons; condensing pressure = 195 psia; ammonia liquid leaves condenser at saturated conditions; ammonia liquid enters expansion valve at 76 °F; evaporating temperature = 10 °F; vapor leaves evaporator at saturated conditions; vapor is superheated 20 °F in compressor suction line. About the surplus compressor you know only the following: It is a 4-cylinder, vertical, reciprocating, single-acting, single-stage, water-jacketed compressor with a maximum rpm of 600. Cylinder diameter is 4 in. and the stroke is 5 in. Based on your past experience you make the following supplemental assumptions to allow further calculations: clearance = 4 percent; pressure drop in suction valves = 4 psi; vapor is superheated 15 °F in cylinder on the intake stroke after passing the suction valves; compression is polytropic with $n = 1.27$; pressure drop in discharge valves = 6 psi; compressor mechanical efficiency = 80 percent.
(a) Determine whether the surplus compressor may be used and, if so, what rpm it should operate at.
(b) Estimate the horsepower required to drive the compressor.

4.11 Compute the compressor displacement of a rolling-piston rotary compressor with a housing inner diameter of 10 cm, a piston outer diameter of 8 cm, a width of 4 cm, and a shaft speed of 60 revolutions per second.

4.12 A rotary-vane-type rotary compressor has a housing with an inner diameter of 3 in., a rotor diameter of 2.5 in., a width of 1 in., and two vanes of 0.25 in. thickness. The shaft speed is 3450 revolutions per minute. Compute the compressor displacement for this compressor.

4.13 A rotary-vane-type rotary compressor has a compressor displacement of 0.001 m³/s, a clearance factor of 1 percent, and a discharge-valve pressure drop of 0.01 MPa. Assume the condensing temperature is 50 °C, the vapor enters the compressor at 33 °C, and the product of the mechanical and electrical efficiencies is 0.54. Assuming saturated liquid leaves the condenser, compute the system refrigerating capacity and compressor electrical power requirement for evaporator temperatures of 10 °C, 5 °C, and 0 °C. Compare the results with the experimental values shown in Fig. 4.10.

4.14 A scroll compressor may operate such that the pressure at the end of the compression process, P_c, is less than the condenser pressure, P_4 (see Fig. 4.13). Refrigerant enters the compressor from the discharge line and immediately increases the pressure to the discharge pressure. The entire discharge process then occurs at constant pressure. Sketch a pressure-volume diagram for this process similar to Fig. 4.13 but with under-compression. Derive an expression similar to Eq. (4.27) for the work per cycle in this mode of operation.

4.15 A scroll compressor using Refrigerant 22 has a volume ratio of 4, a polytropic compression exponent of 1.122, and operates with suction conditions of saturated vapor at 0 °F. Compute the work per unit mass of refrigerant required for condensing temperatures of
(a) 90 °F,
(b) 80 °F, and
(c) 70 °F.
Assume frictionless flow and neglect superheating of the refrigerant during the suction process.

4.16 Repeat Prob. 4.15 with the condensing temperature fixed at 80 °F and evaporating temperature fixed at 0 °F, but with superheated vapor entering the compressor at
(a) 20 °F and
(b) 40 °F.

4.17 Determine the volume ratio required in a screw compressor so that the pressure at the end of the compression process, P_c, equals the condenser pressure, P_4. The compressor uses ammonia, the polytropic exponent for compression is 1.2, the evaporating temperature is -10 °C, and the condensing temperature is 40 °C.

4.18 Compute the isentropic efficiency of a screw compressor with a volume ratio fixed at 3 that uses Refrigerant 22 when the evaporating temperatures are **(a)** 10 °F, **(b)** 20 °F, **(c)** 30 °F. Assume that the polytropic exponent for compression equals 1.18, condensing pressure is 150 psia, and saturated vapor enters the compressor. Neglect superheating of the vapor during the intake process.

4.19 Calculate the polytropic efficiency of a centrifugal compressor that compresses ammonia from $P_1 = 48$ psia, $T_1 = 30$ °F to $P_2 = 180$ psia, $T_2 = 240$ °F.

4.20 Plot the ratio of polytropic work to isentropic work for a centrifugal compressor using ammonia over a range of pressure ratios from 1 to 100. The polytropic efficiency is 0.76 and the specific-heat ratio is 1.3.

4.21 Repeat Prob. 3.10 with the isentropic compression processes replaced by two stages of an adiabatic centrifugal compressor. Assume a polytropic efficiency of 0.8 and a specific-heat ratio of 1.15.

SYMBOLS

A	Inner diameter of rotary compressor housing, ft (m).
B	Outer diameter of rotary compressor rotor, ft (m).
C	Clearance factor.
C	Inner diameter of rolling piston, ft (m).
c_p	Constant pressure specific heat, Btu/lbm · °F (kJ/kg · °C).
c_v	Constant volume specific heat, Btu/lbm · °F (kJ/kg · °C).
C.D.	Compressor displacement, ft³/min (m³/sec).
C.O.P.	Coefficient of performance.
D	Inner diameter of compressor cylinder, ft (m).
h	Specific enthalpy, Btu/lbm (kJ/kg).
$\dot{I}$	Irreversibility rate, Btu/hr · °R (kW/K).
k	Isentropic exponent of compression.
L	Stroke of piston, ft (m); Leakage past seals of scroll compressor.
M	Mass, lbm (kg).
$\dot{m}$	Mass flow rate, lbm/min (kg/sec).
n	Polytropic exponent of compression.
P	Pressure, lbf/in.² (kPa).

P_r	Pressure ratio.
$\dot{Q}$	Heat-transfer rate, Btu/hr (kW).
q	Specific heat transfer, Btu/lbm (kJ/kg).
R	Gas constant, Btu/lbm · °R (kJ/kg · K).
r	Radius of involute spiral, ft (m).
s	Specific entropy, Btu/lbm · °R (kJ/kg · K).
t	Temperature, °F (°C); Vane thickness, ft (m).
V	Volume, ft³ (m³).
V_s	Volume of suction gas, ft³ (m³).
v	Specific volume, ft³/lbm (m³/kg).
V_r	Volume ratio.
$\dot{\text{vol}}$	Volumetric flow rate, ft³/min (m³/sec).
$\dot{W}_{\text{elec}}$	Electric power, Btu/hr (kW).
$\dot{W}_{\text{shaft}}$	Shaft power, Btu/hr (kW).
w	Specific work, Btu/lbm (kJ/kg); Width of rotary compressor, ft (m).
w_p	Polytropic work, Btu/lbm (kJ/kg).
w_s	Isentropic work, Btu/lbm (kJ/kg).
x_P	Coordinate of point on involute spiral.
y_P	Coordinate of point on involute spiral.

Greek Letters

η_E	Electric-motor efficiency.
η_M	Mechanical efficiency.
η_P	Polytropic efficiency.
η_s	Isentropic efficiency.
η_V	Volumetric efficiency.
ϕ	Angular coordinate of involute spiral.
ω	Angular frequency.

chapter 5
Absorption Refrigeration

5.1 *INTRODUCTION*

Refrigeration by a mechanical vapor-compression system may be an efficient method. However, the energy input is work, which is high-grade energy and therefore expensive. The relatively large amount of work is required because in compression the vapor undergoes a large change in specific volume.

If means were available for raising the pressure of the refrigerant without appreciably altering its volume, the amount of work required could be greatly reduced. Example 5.1 shows how this may be done by absorption of the refrigerant vapor by a liquid.

EXAMPLE 5.1

In Ex. 3.1, let the ammonia vapor leaving the evaporator be absorbed by water, producing a liquid solution with a concentration x of 0.4 lbm of ammonia per lbm of solution. Assume that the specific volume of the liquid solution is 0.0184 ft³/lbm. Compare the theoretical work required to raise the pressure of the liquid solution from 30.42 to 180.73 psia with that required for vapor compression.

Solution: We will assume that the liquid solution is incompressible. The work required per pound of ammonia is given by

$$\frac{\Delta Pv}{x} = \frac{(180.73 - 30.42)(144)(0.0184)}{(0.4)(778)} = 1.28 \text{ Btu/lbm of ammonia}$$

By Ex. 3.1, the theoretical work required for vapor compression is

$$_3w_4 = h_4 - h_3 = 112.7 \text{ Btu/lbm of ammonia}$$

Thus, the work requirement for the liquid solution is approximately 1 percent of that required for vapor compression.

114

Example 5.1 demonstrates the principal advantage of the absorption refrigeration cycle. Only a small amount of work is needed. However, we will find that a heat input many times greater than the work input of the mechanical vapor-compression cycle is required. If heat is sufficiently cheap, the absorption cycle may be attractive economically.

5.2 SIMPLE THEORETICAL ABSORPTION REFRIGERATION SYSTEM

Figure 5.1 shows a schematic arrangement of an absorption refrigeration system. This can be compared with the single-stage mechanical vapor-compression system shown in Fig. 3.3. For simplicity we assume that only pure refrigerant flows through the condenser, expansion valve A, and evaporator. These three components may be identical to those used in a mechanical vapor-compression system. However, the compressor has been replaced by the remaining components shown in Fig. 5.1. The refrigerant vapor leaving the evaporator is absorbed into a liquid solution in the absorber. The pressure of the liquid solution is then raised to the condenser pressure by the pump. The solution is preheated in the heat exchanger. By the addition of heat in the generator, refrigerant vapor is driven out of the liquid solution. The absorbant passes back through the heat exchanger and is throttled down to the evaporator pressure in expansion valve B. The generator, condenser, and heat exchanger are on the high-pressure side of the system, while the evaporator and absorber are on the low-pressure side.

An absorption refrigeration system may be properly called a *vapor-compression* system. We see that several components are required to perform the function of the compressor in a mechanical vapor-compression system. The system shown in Fig. 5.1 is not intended to represent an ideal type of absorption system. The ideal system would be completely reversible; this is not theoretically possible in Fig. 5.1. In Chapter 3 we found that

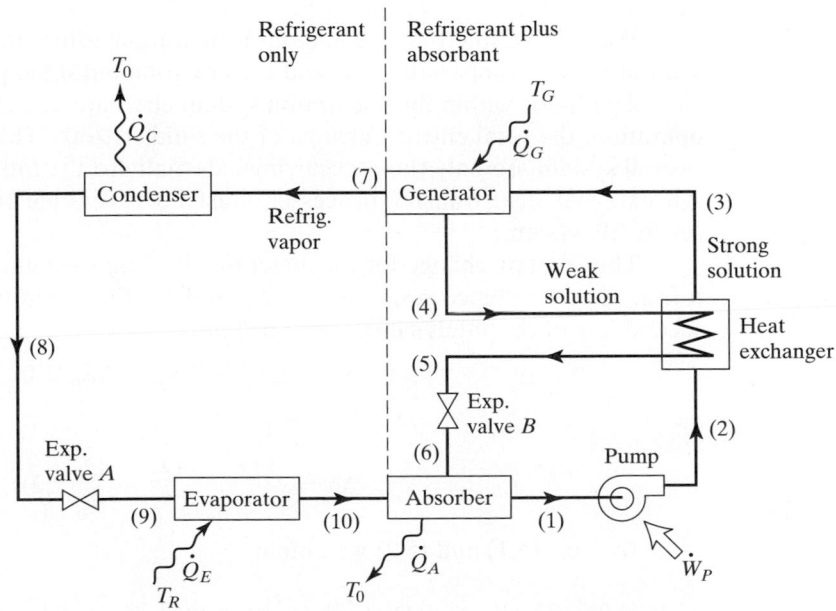

Figure 5.1 Simple absorption refrigeration system.

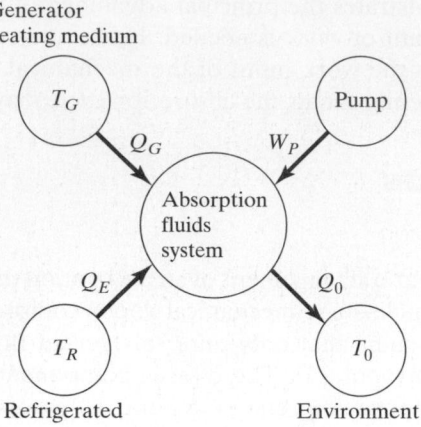

Generator
heating medium

T_G

Pump

Q_G W_P

Absorption
fluids
system

Q_E Q_0

T_R T_0

Refrigerated
space

Environment

Figure 5.2 External energy transfers for an absorption refrigeration system.

the Carnot cycle gave the maximum coefficient of performance for a mechanical vapor-compression system. We will use Bosnjakovic's [1] method to establish the maximum attainable C.O.P. for an absorption system. Figure 5.2 shows the energy transfers to and from the fluids of an absorption system. The generator heating medium adds heat Q_G to the system. The pump adds work W_P. The substance to be cooled in the evaporator adds heat Q_E to the absorption system. The system rejects heat to the environment (cooling water or atmospheric air) in the absorber (Q_A) and in the condenser (Q_C). We will combine these last quantities into the waste heat

$$Q_0 = Q_A + Q_C$$

By the first law of thermodynamics

$$Q_0 = Q_G + Q_E + W_P \tag{5.1}$$

We will assume that the generator heating-medium temperature T_G, the refrigerated-substance temperature T_R, and the environmental temperature T_0 are constants.

The fluids within the absorption system circulate in a closed cycle. For steady-state operation, the total entropy change of the fluids is zero. The changes of entropy for the overall system are only those occurring externally to the fluids of the absorption system. All external heat-transfer processes must be reversible for this to be a completely reversible system.

The entropy change for the generator heating medium is $\Delta S_G = -Q_G/T_G$, for the refrigerated substance $\Delta S_R = -Q_E/T_R$, and for the environment $\Delta S_0 = Q_0/T_0$. By the second law of thermodynamics, we must have

$$\Delta S = \Delta S_G + \Delta S_R + \Delta S_0 \geq 0$$

or

$$\Delta S = -\frac{Q_G}{T_G} - \frac{Q_E}{T_R} + \frac{Q_0}{T_0} \geq 0 \tag{5.2}$$

By Eqs. (5.1) and (5.2) we obtain

$$Q_G \left(\frac{T_G - T_0}{T_G} \right) \geq Q_E \left(\frac{T_0 - T_R}{T_R} \right) - W_P$$

If we assume that W_P is negligible, we have

$$\text{C.O.P.} = \frac{Q_E}{Q_G} \leq \frac{T_R(T_G - T_0)}{T_G(T_0 - T_R)}$$

and for a completely reversible system

$$(\text{C.O.P.})_{\text{max}} = \frac{T_R}{(T_0 - T_R)} \frac{(T_G - T_0)}{T_G} \tag{5.3}$$

Equation (5.3) shows an interesting result: the maximum attainable coefficient of performance for an absorption system is equal to the coefficient of performance for a Carnot refrigerating cycle working between the temperatures T_R and T_0 multiplied by the efficiency of a Carnot engine working between the temperatures T_G and T_0. Therefore the C.O.P. for an absorption system should not be compared directly with the C.O.P. for a mechanical vapor-compression system. Rather, the heat input to the absorption system should be compared with the heat input into the engine that produced the work to drive the compressor in the mechanical vapor-compression system. Equation (5.3) also shows that for a given environmental temperature T_0, the C.O.P. will increase with increase of the generator heating-medium temperature T_G and with increase of the temperature of the refrigerated region T_R. In practice, however, the C.O.P. is much less than that given by Eq. (5.3).

5.3 ELEMENTARY PROPERTIES OF BINARY MIXTURES

Before discussing the details of absorption-refrigeration cycles, we will first study some fundamental characteristics of binary mixtures. Our understanding of absorption refrigeration is greatly improved if we are thoroughly acquainted with the behavior of the working fluids.

Mixtures are formed by combining two or more pure substances. A mixture may be classified as either homogeneous or heterogeneous. A *homogeneous mixture* is uniform in composition, regardless of how small the particles are. The various properties such as density, pressure, and temperature are uniform throughout the mixture. A homogeneous mixture cannot be separated into its constituents by pure mechanical means such as settling or centrifuging. Practically all gaseous mixtures are homogeneous, because it is the nature of gases to diffuse into each other. An example is moist air, which is a homogeneous mixture of dry air and water vapor.

A *heterogeneous mixture* is nonuniform in composition. It can be separated by ordinary mechanical means. An example of a heterogeneous mixture is a fog or cloud, which is a mixture of liquid water and saturated air.

In general, substances may be combined to form heterogeneous or homogeneous mixtures in solid, liquid, or vapor phases. In this chapter we will be concerned only with homogeneous liquid and homogeneous vapor mixtures, and we will limit our study to binary mixtures. Bosnjakovic [2] has extensively covered thermodynamics of binary mixtures.

The thermodynamic state of a binary mixture cannot be established from two independent thermodynamic properties, such as pressure and temperature, as may be done with a pure substance. A third independent thermodynamic property is required. We will consider the quantitative composition in terms of the concentration or mass fraction x, which is the mass of one arbitrary constituent divided by the mass of the mixture. Knowledge of p, t, and x enables us to establish the thermodynamic state of the mixture.

An important characteristic of a mixture is its miscibility. A mixture is *miscible* if, through any arbitrary range of concentration values, a homogeneous mixture is formed. Thus, a nonmiscible mixture is a heterogeneous mixture. In Chapter 4 we briefly considered mixtures of oil and refrigerants. Thus, an oil-ammonia mixture is heterogeneous or nonmiscible, whereas an oil-Refrigerant 12 mixture is homogeneous or miscible. Some mixtures are miscible under certain conditions but otherwise nonmiscible. Miscibility is materially influenced by temperature. Thus at low temperatures an oil-Refrigerant 22 mixture is nonmiscible, while at higher temperatures the mixture becomes miscible.

Binary mixtures suitable for an absorption system must be completely miscible in both the liquid and vapor phases. There can be no miscibility gap or range of concentration values where a heterogeneous mixture would exist.

Two important phenomena occurring with the mixing of two liquids are the change of volume and change of temperature of the constituents during mixing. Figure 5.3(a) shows schematically a divided thermally insulated container holding x mass of liquid A and $(1 - x)$ mass of liquid B. Each liquid is at the same temperature t_1. The volume of the mass of constituents is

$$v_1 = xv_A + (1 - x)v_B$$

We assume that the dividing wall is removed and the two liquids are thoroughly mixed adiabatically. We generally observe that $v_2 \neq v_1$. If we repeated the experiment with different liquids and with various concentration values, we would find that, in some cases, contraction occurred. In other cases Δv would be positive. We would find that no general rule was followed, and that experimentation was necessary to find the result.

Our other important observation in the experiment of Fig. 5.3 would be that, in general, $t_2 \neq t_1$. In some cases we would find that a warming effect occurred; in others we might observe a decrease of temperature. The effect may be expressed in terms of a heat of solution ΔH_x. In the experiment we could measure the heat which we had to remove or add in order to keep the temperature constant. If the mixing were carried out at constant pressure, ΔH_x would be strictly related to the enthalpy of the mixture. For the original components, we have

$$h_1 = xh_A + (1 - x)h_B$$

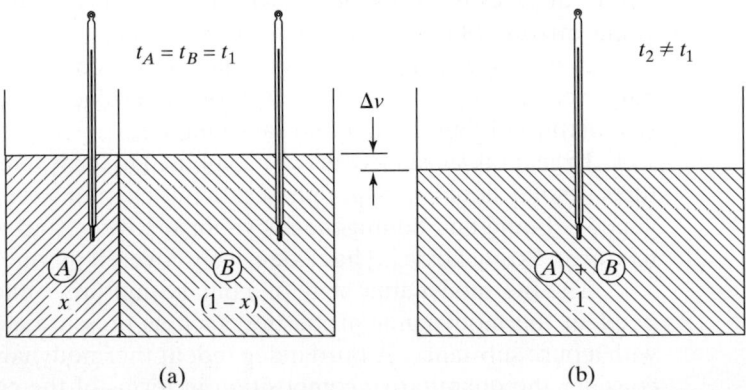

Figure 5.3 Schematic illustrations of change of volume and change of temperature in the mixing of two liquids.

and after the mixing

$$h_2 = h_1 + \Delta H_x = x h_A + (1 - x) h_B + \Delta H_x \tag{5.4}$$

Equation (5.4) allows calculation of the specific enthalpy of a mixture for a solution of known concentration at some fixed temperature and pressure, if the enthalpies of the pure components and the isothermal heat of solution are known.

The pressure-temperature relationships for a boiling-liquid binary mixture or a condensing-vapor binary mixture are of special importance in absorption refrigeration. We will discuss these relationships through another imaginary experiment, where we again restrict ourselves to using only homogeneous solutions.

Figure 5.4 schematically shows the experiment and the observed results on t-x coordinates. We start with a liquid solution (state 1) and slowly add heat to it, at all times keeping the pressure constant. (We may imagine a glass cylinder so that we may view the solution inside.) Until the temperature t_2 is reached, we observe that the solution remains entirely liquid. However, upon further addition of heat we find that the piston rises above the liquid, indicating that vaporization occurs. If we were to stop the experiment at the condition shown in Fig. 5.4(b) and chemically analyze the vapor and liquid components, we would make an interesting discovery. The concentration of the liquid and vapor would be different, and furthermore, both concentrations would be different from the original concentration x_1. However, the overall bulk concentration of both liquid and vapor phases remains at x_1 throughout the entire experiment. We would find that the liquid concentration $x_3 < x_1$, while the vapor concentration $x_4 > x_1$. If we added more heat, we would notice that the liquid would gradually disappear. We would also notice that both the liquid concentration and the vapor concentration had decreased. When we reached the state-point (5), only vapor would remain, and here we would observe that $x_5 = x_1$. Further heating would result in superheating of the vapor at constant concentration.

If we repeated the experiment with different initial concentration values, but with the same pressure, our experimental results would give the equilibrium-boiling and equilibrium-condensing lines shown in Fig. 5.4(c). If we repeated the experiments but with different pressures, we would obtain the results shown in Fig. 5.4(d). If we reversed the experiments, started with a superheated vapor, and then removed heat, we would have the results shown in Fig. 5.4(e).

Thus binary mixtures, as contrasted with pure substances, do not have a single boiling temperature or condensing temperature for a given pressure. The equilibrium or saturation temperature is also dependent on the concentration. These relationships must be determined experimentally.

The enthalpy-concentration diagram (h-x coordinates) is the most useful diagram of properties for a binary mixture. Figure 5.5 shows a schematic h-x diagram including liquid and vapor regions for a homogeneous binary mixture. Enthalpy-concentration diagrams for two actual binary mixtures are included in Appendix C. Figures C-3E and C-3SI are for ammonia-water combinations, while Figs. C-4E and C-4SI are for water-lithium bromide mixtures.

We will now examine the construction of Figs. C-3E and C-3SI in some detail. The charts are drawn for saturated or equilibrium solutions of ammonia and water. The liquid region is the lower part of the diagram. Equilibrium liquid lines (boiling lines) are shown for various pressures. Isothermal lines are also shown in the liquid region. If a liquid solution is known to be saturated, the state-point may be located by the intersection of the temperature line with the saturated-liquid line for the appropriate pressure. If the

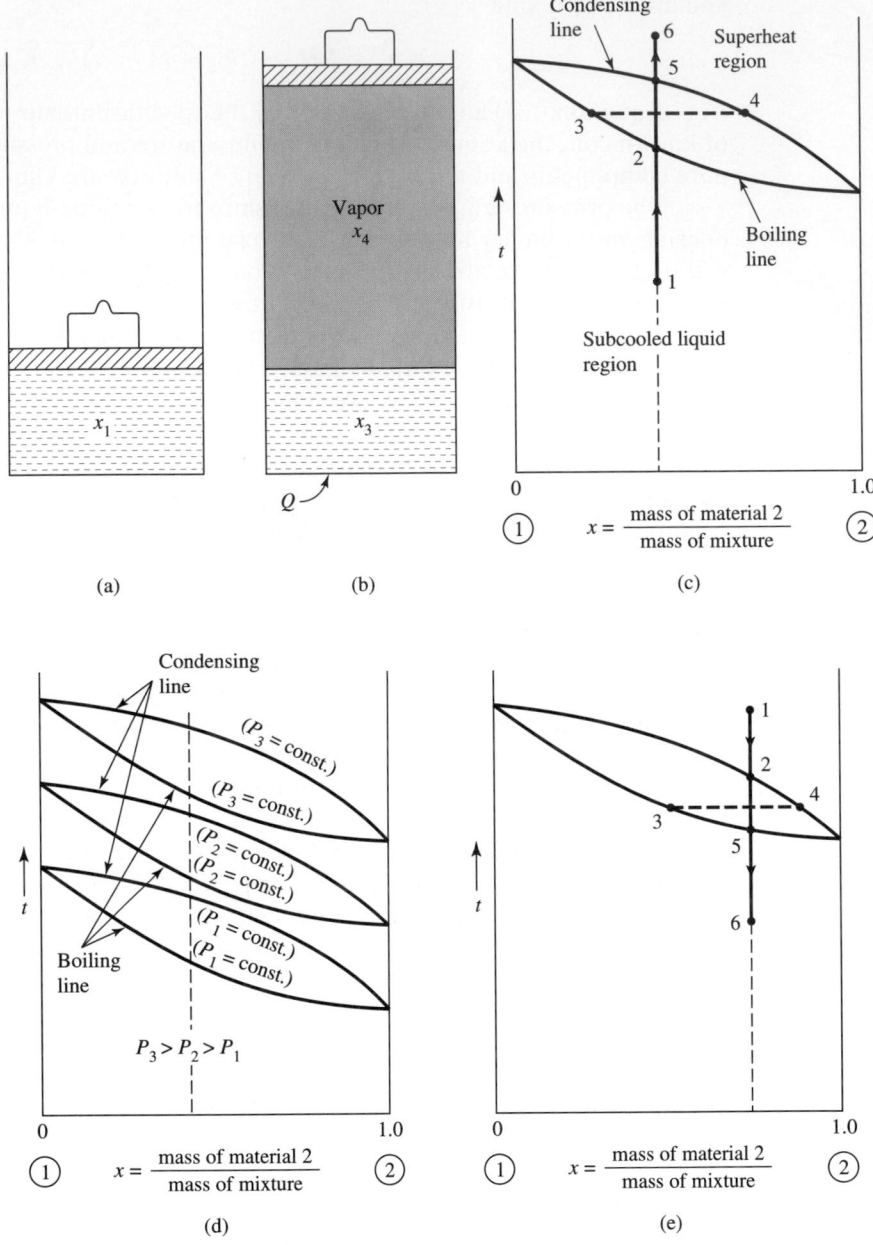

Figure 5.4 Evaporation and condensation characteristics for a homogeneous binary mixture.

liquid solution is subcooled, the state-point may be located approximately if the temperature and concentration are known.

No temperature lines are shown in the vapor region of Figs. C-3E or C-3SI. However, a saturated-vapor state in equilibrium with a known saturated-liquid state may be located through use of the equilibrium construction lines. The procedure may be illustrated by an example for saturation conditions of 100 psia and 260 °F. A vertical line is

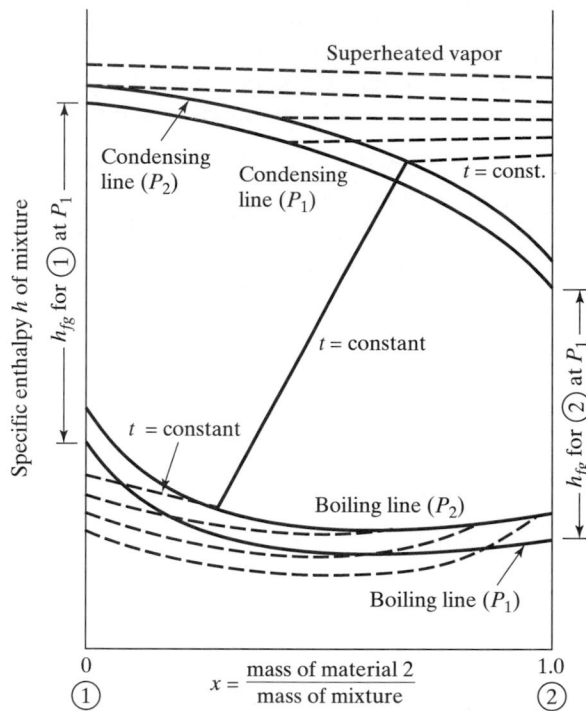

Figure 5.5 Schematic *h-x* diagram for liquid and vapor regions for a homogeneous binary mixture.

drawn upward from the saturated-liquid point to the equilibrium construction line for 100 psia. From this intersection, a horizontal line is drawn to the saturated-vapor line for 100 psia. This intersection is the vapor state-point. Thus the properties of the saturated-vapor state are known ($t = 260$ °F, $p = 100$ psia, $h = 906$ Btu per lbm, $x = 0.63$ lbm of NH_3 per lbm of mix). Furthermore, a straight line drawn from the saturated-liquid state to saturated-vapor state is an isotherm for the liquid-vapor region.

Figures C4E and C4SI contain thermodynamic property data for saturated-liquid solutions and subcooled-liquid solutions of lithium bromide and water. Saturated-liquid curves at various pressures are shown by the dashed lines, and isotherms in the saturated- and subcooled-liquid regions are indicated by the solid lines. As lithium bromide has a very low vapor pressure, saturated-liquid solutions are in equilibrium with essentially pure water vapor with a concentration of zero. These pure-water state-points are located along the left-hand side of Figs. C4E and C4SI. However, the enthalpy of water vapor is beyond the range of these figures. The enthalpy of these equilibrium water-vapor states can be found from superheated-steam data at the given solution temperature and pressure.

5.4 ELEMENTARY STEADY-FLOW PROCESSES WITH BINARY MIXTURES

In industrial systems using binary homogeneous mixtures, several types of thermodynamic processes may be involved. Some of the common processes are adiabatic mixing of fluid streams; mixing of fluid streams with heat exchange, vaporization, and condensation;

and throttling. Several specific steady-flow processes will be considered in this section. Study of these separate processes will help us later to understand more clearly how the absorption-refrigeration system functions. The examples will also illustrate how helpful the *h-x* diagram may be in graphical solution of problems.

(a) Adiabatic mixing of two streams. Figure 5.6(a) shows schematically a mixing chamber where two fluid streams are brought together adiabatically. In the following analysis, as well as in all of our subsequent processes, we will use the following symbols: specific enthalpy *h* in units of energy content per unit mass of mixture, concentration *x* in units of mass of one constituent per mass of mixture, and mass rate of flow $\dot{m}$ in units of mass of mixture per unit time.

We may write three fundamental equations for the system of Fig. 5.6. These are

$$\dot{m}_1 h_1 + \dot{m}_2 h_2 = \dot{m}_3 h_3$$

$$\dot{m}_1 + \dot{m}_2 = \dot{m}_3$$

$$\dot{m}_1 x_1 + \dot{m}_2 x_2 = \dot{m}_3 x_3$$

By elimination of $\dot{m}_3$, we obtain

$$\frac{\dot{m}_1}{\dot{m}_2} = \frac{h_2 - h_3}{h_3 - h_1} = \frac{x_2 - x_3}{x_3 - x_1} \tag{5.5}$$

Equation (5.5) defines a straight line on the *h-x* diagram as shown by Fig. 5.6(b). Thus, state 3 must lie on a straight line connecting states 1 and 2 on the *h-x* diagram. The location of state 3 on the line is determined by the mass ratios. We may also deduce that

$$x_3 = x_1 + \frac{\dot{m}_2}{\dot{m}_3}(x_2 - x_1) \tag{5.6}$$

$$h_3 = h_1 + \frac{\dot{m}_2}{\dot{m}_3}(h_2 - h_1) \tag{5.7}$$

In the adiabatic mixing of two binary-liquid mixtures, determination of the mixture temperature using the *h-x* diagram may be somewhat involved, since partial vaporization may occur during mixing.

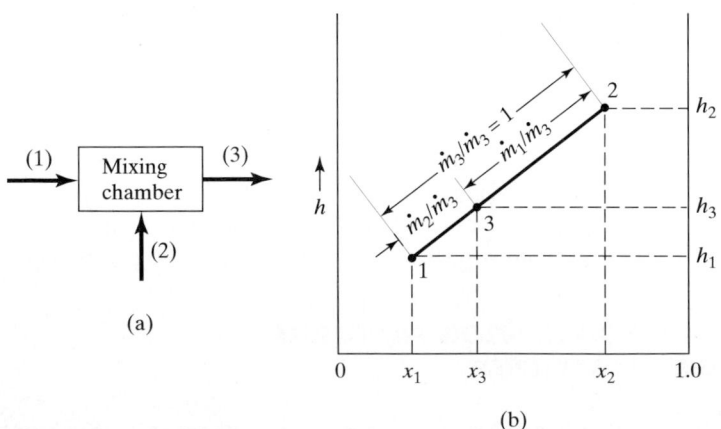

(a)

(b)

Figure 5.6 Adiabatic mixing of two binary fluid streams under steady-flow conditions.

Part II / Refrigeration

EXAMPLE 5.2

A stream of liquid aqua-ammonia ($\dot{m}$ = 10 lbm/min, x = 0.7 lbm NH$_3$/lbm mix, t = 60 °F, p = 100 psia) is adiabatically mixed with a stream of saturated liquid aqua-ammonia ($\dot{m}$ = 5 lbm/min, t = 200 °F, p = 100 psia). Assuming steady-flow conditions, determine (a) the mixture concentration, (b) the mixture specific enthalpy, (c) the equilibrium temperature of the mixture, and (d) percentage liquid and percentage vapor composition of the mixture after equilibrium has been reached.

Solution: Figure C-3E will be used for the solution, and Fig. 5.7 schematically shows the procedure.

(a) The state of the stream for the flow of 10 lbm/min we will call state 1. We find that state 1 is a subcooled condition, and we locate the point at t = 60 °F and x = 0.7. State 2, being saturated, is located at t = 200 °F and p = 100 psia. By Eq. (5.6),

$$x_3 = 0.70 + \frac{5}{15}\,(0.26 - 0.70) = 0.553 \text{ lbm NH}_3/\text{lbm mix}$$

(b) We connect states 1 and 2 by a straight line. At the intersection of x_3 with this line, the mixture state is established. From the chart, we read h_3 = 38.0 Btu/lbm mix.

(c) We know that P_3 = 100 psia. We observe that state 3 lies above the equilibrium liquid line for 100 psia. Thus, we know that the state is a mechanical mixture of saturated liquid and saturated vapor. With a straight-edge, we find by trial the line $f3g$ of Fig. 5.7. This line is the isotherm for the liquid-vapor region connecting states f and g. At f we read $t_f = t_g = t_3 = 110$ °F.

(d) Since state 3 is a mixture of liquid and vapor, the usual mixing equations apply. Thus,

$$\frac{\dot{m}_g}{\dot{m}_3} = \frac{x_3 - x_f}{x_g - x_f} = \frac{\overline{f3}}{\overline{fg}} = 0.033$$

Therefore the mixture consists of 96.7 percent liquid and 3.3 percent vapor.

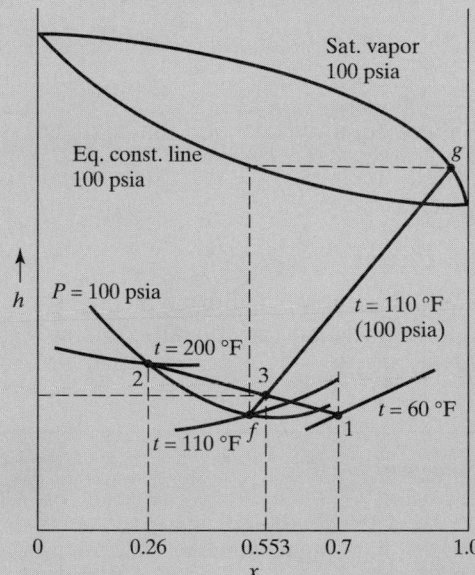

Figure 5.7 Schematic h-x diagram for Ex. 5.2.

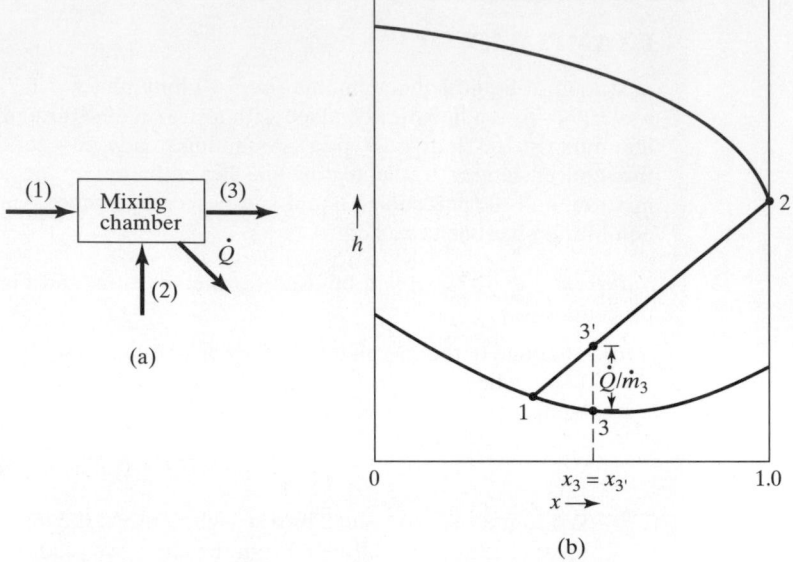

(a)

(b)

Figure 5.8 Mixing of two binary fluid streams under steady-flow conditions with heat exchange.

(b) Mixing of two streams with heat exchange.   In the mixing of binary fluids, it is often necessary to remove heat or add heat in order to obtain the desired final condition. An example process occurs in the absorber of an absorption-refrigeration system. Figure 5.8(a) shows the schematic steady-flow problem where heat removal is assumed. The fundamental mixing equations are

$$\dot{m}_1 h_1 + \dot{m}_2 h_2 = \dot{m}_3 h_3 + \dot{Q}$$

$$\dot{m}_1 + \dot{m}_2 = \dot{m}_3$$

$$\dot{m}_1 x_1 + \dot{m}_2 x_2 = \dot{m}_3 x_3$$

From these equations we find that Eq. (5.6) is still applicable, but instead of Eq. (5.7) we have

$$h_3 = h_1 + \frac{\dot{m}_2}{\dot{m}_3}(h_2 - h_1) - \frac{\dot{Q}}{\dot{m}_3} \tag{5.8}$$

With heat removal during mixing, state 3 lies at a vertical distance $\dot{Q}/\dot{m}_3$ below the state-point obtained for adiabatic mixing. Figure 5.8(b) shows a schematic h-x diagram for the process.

EXAMPLE 5.3

One lbm/min of saturated ammonia vapor ($x = 1.0$) at 30 psia is mixed with 10 lbm/min of saturated liquid aqua-ammonia at 30 psia and 100 °F. The final mixture state is saturated liquid at 30 psia. Find (a) the concentration, temperature, and enthalpy of the mixture state, and (b) the heat removal in Btu/min.

Solution: Figure 5.8(b) shows the schematic solution.

(a) By Eq. (5.6),

$$x_3 = 0.345 + \frac{1}{11}(1.0 - 0.345) = 0.405 \text{ lbm NH}_3/\text{lbm mix}$$

We draw the straight line $\overline{12}$ and locate state $3'$. We then proceed vertically downward to the saturated-liquid line for 30 psia and locate state 3. We read $t_3 = 80 \,°\text{F}$, $h_3 = -24.0$ Btu/lbm mix.

(b) By Fig. 5.8(b) we see that

$$\frac{\dot{Q}}{\dot{m}_3} = h_{3'} - h_3 = 54 - (-24) = 78 \text{ Btu/lbm mix}$$

Thus

$$\dot{Q} = (11)(78) = 858 \text{ Btu/min}$$

(c) Simple heating and cooling processes. Processes of vaporization are important in all absorption-refrigeration systems. In the case of some binary mixtures, such as aqua-ammonia, rectification or purification of the vapor is required. We will now consider two simple processes—heating and cooling. We will further see the usefulness of the *h-x* diagram in solving problems.

Figure 5.9(a) shows a simple arrangement of heat exchangers by which we may produce a high-purity refrigerant vapor from a solution with a relatively low concentration of refrigerant. In order to simplify our analysis, we assume that liquid and vapor components are adiabatically separated after each heat exchanger. For heat exchanger A, we obtain for steady-flow conditions

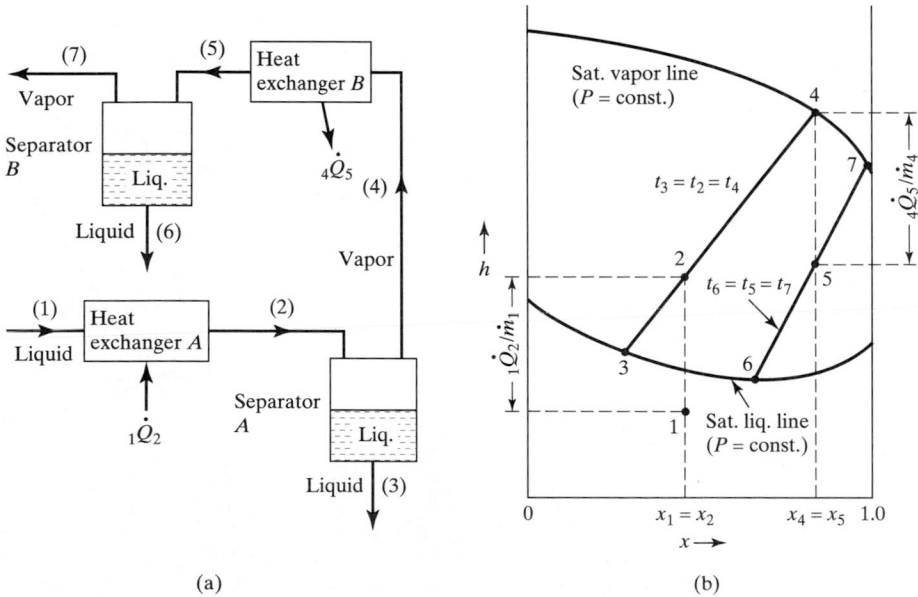

(a)   (b)

Figure 5.9 Simple heating and cooling processes for steady-flow conditions.

$$_1\dot{Q}_2 = \dot{m}_1(h_2 - h_1)$$

$$\dot{m}_1 = \dot{m}_2$$

$$x_1 = x_2$$

For separator A, we have

$$\dot{m}_2 h_2 = \dot{m}_3 h_3 + \dot{m}_4 h_4$$

$$\dot{m}_2 = \dot{m}_3 + \dot{m}_4$$

$$\dot{m}_2 x_2 = \dot{m}_3 x_3 + \dot{m}_4 x_4$$

and thus

$$\frac{\dot{m}_3}{\dot{m}_2} = \frac{x_4 - x_2}{x_4 - x_3} = \frac{h_4 - h_2}{h_4 - h_3} \tag{5.9}$$

$$\frac{\dot{m}_4}{\dot{m}_2} = \frac{x_2 - x_3}{x_4 - x_3} = \frac{h_2 - h_3}{h_4 - h_3} \tag{5.10}$$

Figure 5.9(b) shows the state-points for heat exchanger A and separator A on the h-x diagram. Equations (5.9) and (5.10) show that the fractional components for separator A may be obtained directly from the diagram. Thus $\dot{m}_3/\dot{m}_2 = \overline{24/34}$ and $\dot{m}_4/\dot{m}_2 = \overline{32/34}$. Also, the heat requirement $_1\dot{Q}_2/\dot{m}_1$ may be graphically represented as shown.

The analyses for heat exchanger B and separator B are exactly similar to those for heat exchanger A and separator A. Figure 5.9(b) shows the state-points on the h-x diagram.

(d) Throttling. The throttling process occurs in a majority of systems using binary liquids. Figure 5.10(a) shows schematically an adiabatic throttling valve. We recognize that $x_1 = x_2$ and that $h_1 = h_2$. On the h-x diagram, points 1 and 2 are identical. However,

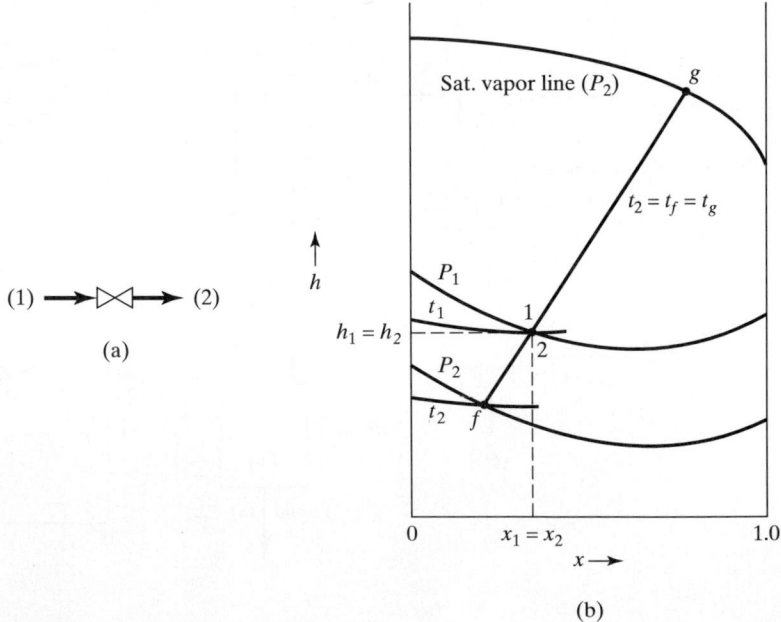

Figure 5.10 Throttling of a binary liquid mixture under steady-flow conditions.

in the throttling of a binary liquid, vaporization may occur. Figure 5.10(b) shows the procedure where we assume that the liquid and vapor components are in equilibrium. With a straight-edge we find by trial the isotherm $\overline{f2g}$. We find that t_2 may be considerably less than t_1. The fractional components of liquid and vapor may be obtained from the segment ratios of the line $\overline{f2g}$.

EXAMPLE 5.4

A saturated solution of lithium bromide and water at a pressure of 15 kPa and a temperature of 90 °C is throttled to a pressure of 6 kPa. Determine (a) the final temperature and (b) the enthalpy and concentration of the saturated lithium bromide-water solution.

Solution: Figure C-4SI will be used for the solution and Fig. 5.11 schematically shows the procedure.

(a) The final state-point 2 must lie on the straight line (isotherm) connecting the saturated-liquid-solution state (2f) and the superheated-water-vapor state (2v), both of which are at the final pressure of 6 kPa. State 2v is beyond the top of the chart, so its enthalpy will be estimated from Eq. (2.36). An initial estimate for t_2 is 90 °C, which equals the initial temperature of the solution before throttling. Denoting the initial estimates by an asterisk,

$$t_2^* = 90 \,°C \quad \text{and} \quad h_{2v}^* = 2501 + 1.86(90) = 2668 \text{ kJ/kg water}$$

The slope of the isotherm connecting states 2 and 2v* becomes

$$\frac{\Delta h^*}{\Delta x} = \frac{2668 - (-58)}{0 - 0.55} = -4950 \text{ kJ/kg LiBr}$$

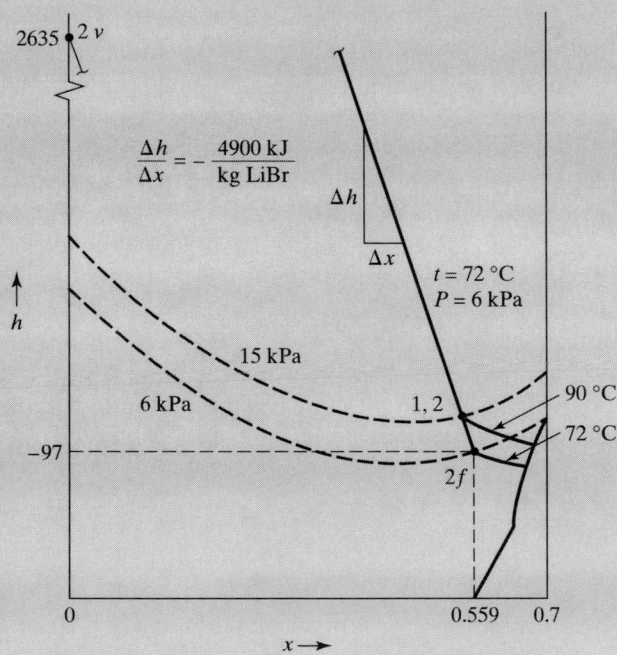

Figure 5.11 Schematic h-x diagram for Ex. 5.4.

Extending this straight line through state 2 to the intersection of the saturated-liquid curve at 6 kPa results in state $2f^*$, $t_{2f}^* = 72$ °C. Now use $t_2^{**} = 72$ °C as the next estimate for T_2.

$$h_{2v}^{**} = 2501 + 1.86(72) = 2635 \text{ kJ/kg water}$$

$$\frac{\Delta h^{**}}{\Delta x} = \frac{2635 - (-58)}{0 - 0.55} = -4900 \text{ kJ/kg LiBr}$$

Extending this straight line through state 2 results in a saturated-liquid state ($2f^{**}$) that is very nearly equal to the previous estimate for state $2f$. Therefore we can assume that state $2f$ is equal to state $2f^{**}$ so that $t_2 = 72$ °C.

(b) The enthalpy and concentration of the saturated lithium bromide-water solution at state $2f$ is obtained directly from Fig. C-4SI,

$$h_{2f} = -97 \text{ kJ/kg solution}, \qquad x_{2f} = 0.559 \text{ kg LiBr/kg solution}$$

5.5 RECTIFICATION OF A BINARY MIXTURE

In the previous section we discussed simple steady-flow processes of heating and cooling. In Fig. 5.9, heat exchanger A performs the function of a generator which generates the vapor to be purified, while heat exchanger B performs the function of a dephlegmator which produces the liquid necessary to purify the vapor. Through the combination shown, a vapor of arbitrarily high purity may be produced at the exit of the dephlegmator.

The combination of Fig. 5.9, however, is inadequate for the separation of a binary mixture. To achieve efficient purification of one of the mixture components we must introduce a rectifying column between the generator and the dephlegmator. In this section we will discuss in some detail the fundamental principles of a rectifying column. This discussion will enable us to understand better the aqua-ammonia absorption system and also to understand better in Chapter 6 how liquefied-gas mixtures are separated.

Figure 5.12(a) shows schematically a rectifying column between a generator and a dephlegmator. The rectifying column includes many capped plates or perforated plates. A vapor mixture V ascends through the column and a liquid solution L descends. The purpose of the plates is to bring the liquid and vapor solutions into intimate direct contact. The vapor rising through the tower has some liquid condensed from it, while the liquid trickling downward has some vapor evaporated from it. With regard to composition, the ascending vapor develops a progressively higher concentration, while the descending liquid solution develops a progressively lower concentration of the constituent being purified.

In Fig. 5.12(b) the dephlegmator and an arbitrary part of the column are shown. Between any two cross sections, such as a and b, for steady-state conditions we have

$$\dot{m}_{V,a} + \dot{m}_{L,b} = \dot{m}_{V,b} + \dot{m}_{L,a}$$

or

$$\dot{m}_{V,a} - \dot{m}_{L,a} = \dot{m}_{V,b} - \dot{m}_{L,b} \qquad (5.11)$$

Equation (5.11) must hold for every part of the column. Thus we have, for any cross section,

$$\dot{m}_V - \dot{m}_L = \text{a constant} = \dot{m}_3 \qquad (5.12)$$

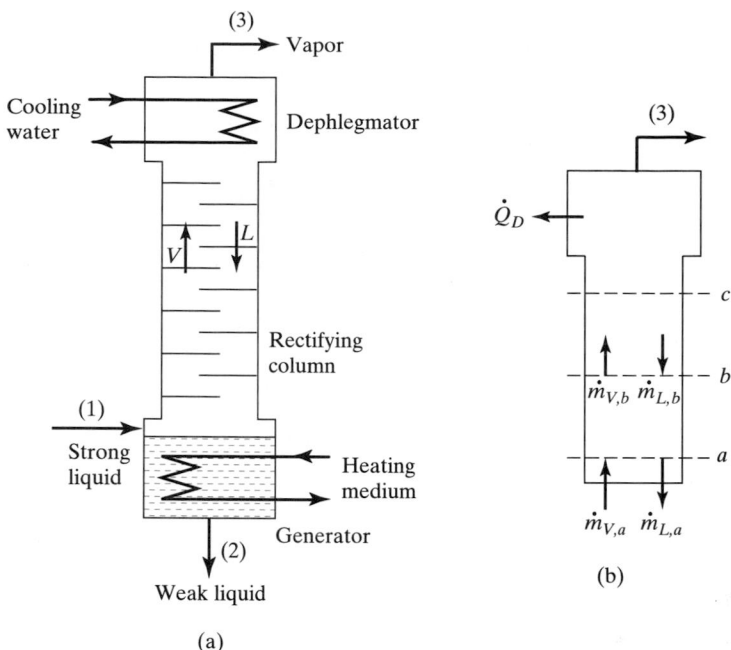

(a)

Figure 5.12 The rectifying column.

In a similar fashion we may show that

$$\dot{m}_V x_V - \dot{m}_L x_L = \dot{m}_3 x_3 \tag{5.13}$$

$$\dot{m}_V h_V - \dot{m}_L h_L = \dot{m}_3 h_3 + \dot{Q}_D \tag{5.14}$$

In Eqs. (5.12), (5.13), and (5.14) it is to be understood that the subscripts V and L always refer to the same cross section, although the location of the cross section is arbitrary. By Eqs. (5.12) and (5.13) we have

$$\frac{\dot{m}_V}{\dot{m}_3} = \frac{x_3 - x_L}{x_V - x_L} \tag{5.15}$$

By Eqs. (5.12) and (5.14) we have

$$\frac{\dot{m}_V}{\dot{m}_3}(h_V - h_L) = (h_3 - h_L) + \frac{\dot{Q}_D}{\dot{m}_3} \tag{5.16}$$

Thus, by Eqs. (5.15) and (5.16), we obtain

$$\frac{x_3 - x_L}{x_V - x_L}(h_V - h_L) = (h_3 - h_L) + \frac{\dot{Q}_D}{\dot{m}_3} \tag{5.17}$$

As shown by Fig. 5.13, Eq. (5.17) defines a straight line on the h-x diagram for steady-flow conditions. Equation (5.17) must be satisfied for the combination of dephlegmator and any connected portion of the column. This requirement leads to the graphical construction shown by Fig. 5.14. If we connect, by straight lines, the vapor and liquid state-points for the same cross section, all such lines must intersect at a common point. These straight lines are called *operating lines*. Their point of intersection P_1 is called a *pole*.

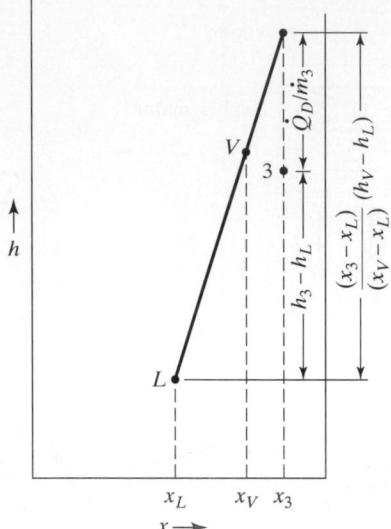

Figure 5.13 Schematic representation of Eq. (5.17) on h-x diagram.

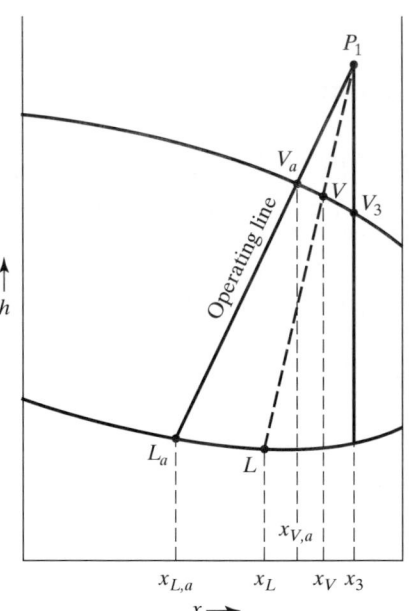

Figure 5.14 Rectification pole and operating lines.

In order to know what happens at a plate of the column, we must know the direct-contact heat-transfer rate between the liquid and the vapor. However, we may draw an important conclusion regarding rectification by making an analysis using a diagram similar to Fig. 5.14. In Fig. 5.15(a), let us assume that sections a and b represent a portion of a column containing one contact plate. In the limiting case, the vapor passing section b may be brought to the same temperature as the liquid passing section a. In a finite case, the temperature of the vapor at section b would be greater than the temperature of the liquid at section a. This would require that the vapor state V_b be located to the left of the intersection of the isotherm $t_{L,a}$ and the saturated-vapor line. Thus, for separation to occur

Part II / Refrigeration

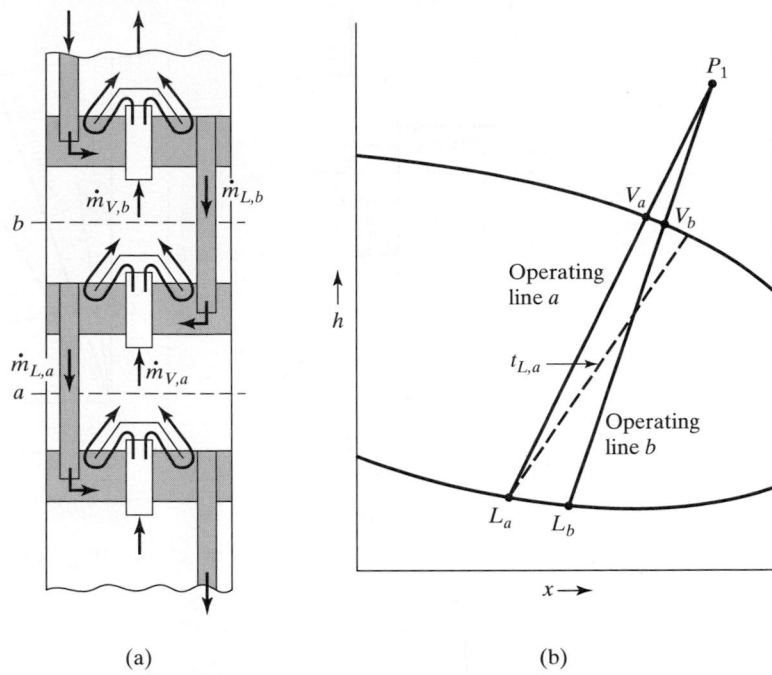

(a)   (b)

Figure 5.15 Relation of operating lines to mixture-region isotherms.

in a rectifying column, the operating line for each cross section must be steeper than the corresponding mixture-region isotherm. The pole of the operating lines must be located high enough for this requirement to be met for each and every part of the column.

The separation arrangement of Fig. 5.12(a) may be improved upon substantially if the strong liquid, state 1, is introduced part of the way up the column instead of directly into the generator. Such an arrangement combines an exhausting column with the generator. We will examine the functioning of a simple exhausting column.

Figure 5.16(a) schematically shows an exhausting column. In a procedure identical to that used for the rectifying column, we may show that for any cross section of the column

$$\dot{m}_L - \dot{m}_V = \dot{m}_2 \tag{5.18}$$

$$\dot{m}_L x_L - \dot{m}_V x_V = \dot{m}_2 x_2 \tag{5.19}$$

$$\dot{m}_L h_L - \dot{m}_V h_V = \dot{m}_2 h_2 - \dot{Q}_G \tag{5.20}$$

From Eqs. (5.18)–(5.20) we obtain

$$\frac{x_V - x_2}{x_V - x_L}(h_V - h_L) = (h_V - h_2) + \frac{\dot{Q}_G}{\dot{m}_2} \tag{5.21}$$

Equation (5.21) must be satisfied for every cross section of the column. This requirement leads to the graphical construction shown in Fig. 5.16(b). All cross-section operating lines must intersect at a common point P_2 located at a distance $\dot{Q}_G/\dot{m}_2$ below state-point 2.

The heat requirement $\dot{Q}_G/\dot{m}_2$ for the generator may be substantially decreased if the strong liquid is preheated in a heat exchanger by the weak-liquid solution. The concentration x of each solution remains constant throughout the heat exchanger. Point 1 in

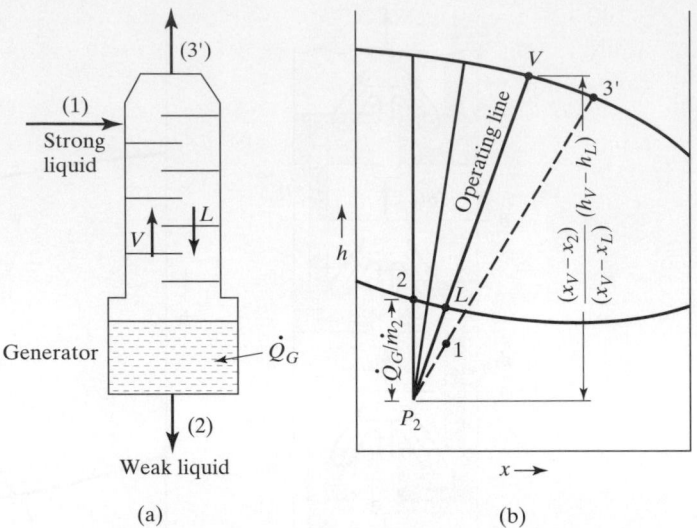

Figure 5.16 The exhausting column.

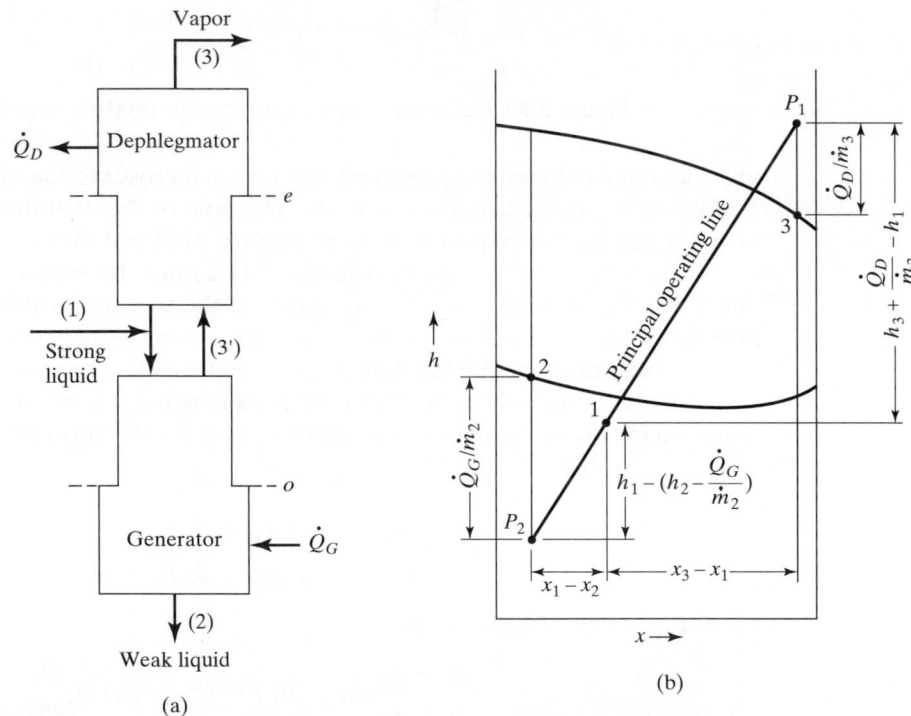

Figure 5.17 The double rectifying column.

Fig. 5.16(b) is raised, thus also raising the pole P_2 vertically and thereby decreasing the quantity $\dot{Q}_G/\dot{m}_2$.

Figure 5.17(a) schematically shows a double rectifying column. The double column is formed by coupling the exhausting column and the simple rectifying column discussed previously. Vapor boiled off in the generator rises through the exhausting column and is purified to a condition 3'. The vapor is further concentrated in the rectifying column and dephlegmator to a desired final condition 3. Strong liquid solution, state 1, is mixed with liquid leaving the upper column, and this mixture is stripped of some material in the exhausting column.

We have previously analyzed the separate columns, and will now examine the combination. We may write the following fundamental equations:

$$\dot{m}_1 = \dot{m}_2 + \dot{m}_3 \tag{5.22}$$

$$\dot{m}_1 x_1 = \dot{m}_2 x_2 + \dot{m}_3 x_3 \tag{5.23}$$

$$\dot{m}_1 h_1 + \dot{Q}_G = \dot{m}_2 h_2 + \dot{m}_3 h_3 + \dot{Q}_D \tag{5.24}$$

By Eqs. (5.22)–(5.24) we may show that

$$\frac{\left(h_3 + \dfrac{\dot{Q}_D}{\dot{m}_3}\right) - h_1}{x_3 - x_1} = \frac{h_1 - \left(h_2 + \dfrac{\dot{Q}_G}{\dot{m}_2}\right)}{x_1 - x_2} \tag{5.25}$$

Equation (5.25) requires that the two poles P_1 and P_2 and state-point 1 lie on the same straight line on the h-x diagram. This requirement is evident from Fig. 5.17(b). This straight line is called the *principal operating line*.

The location of the operating lines on the h-x diagram for the exhausting column and the rectifying column depends upon the apparatus and upon the rates of heat transfer involved. Figure 5.18 shows a possible set of circumstances. Because of agitation due to boiling in the generator, we may assume that vapor at a state V_o leaves the generator in equilibrium with the weak-liquid state-point 2. The liquid state L_o leaving the exhausting column and entering the generator would be given on the h-x diagram by the intersection of the operating line $\overline{P_2 V_o}$ with the saturated-liquid line. Bosnjakovic [3] has shown that the least number of plates are required in the two columns if the vapor state 3' leaving the exhausting column lies on the principal operating line $\overline{P_2 1 P_1}$. Thus all operating lines for the exhausting column would lie between the lines $\overline{P_2 L_o V_o}$ and $\overline{P_2 1 P_1}$.

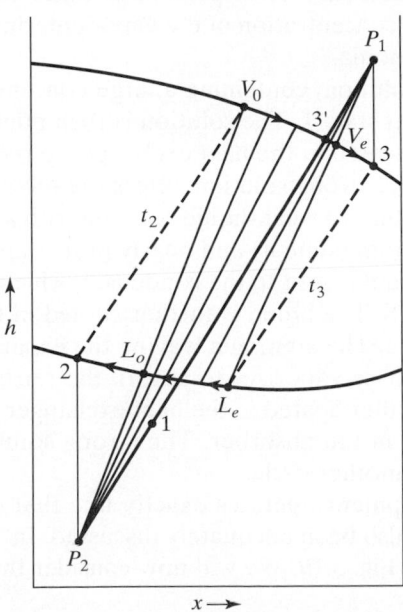

Figure 5.18 Operating lines for the double rectifying column.

In the upper column it is more difficult to estimate the state of the liquid leaving the dephlegmator and entering the rectifying column. This state-point is dependent upon the type of heat exchanger used as the dephlegmator. If we assume that liquid enters the rectifying column L_e in equilibrium with the vapor state 3, we have the circumstances shown in Fig. 5.18. All operating lines for the upper column lie between the lines $\overline{P_2 1 P_1}$ and $\overline{L_e V_e P_1}$.

The position of the principal operating line $\overline{P_2 1 P_1}$ fixes the generator heat requirement $\dot{Q}_G / \dot{m}_2$ and the dephlegmator cooling requirement $\dot{Q}_D / \dot{m}_3$ for fixed values of x_2 and x_3. Both quantities decrease as the principal operating line becomes more flat. However, the principal operating line cannot be drawn arbitrarily. The poles P_1 and P_2 must be located so that all operating lines are steeper than the mixture region isotherms for the corresponding cross sections.

5.6 THE AQUA-AMMONIA ABSORPTION REFRIGERATION SYSTEM

Having some knowledge of the characteristics of binary mixtures, we will now consider absorption-refrigeration systems. The aqua-ammonia absorption system is one of the oldest methods of refrigeration. Ammonia is the refrigerant and water is the absorbent. The system may be applicable where ammonia is a suitable refrigerant. Since the absorbent, water, is volatile, the simple system of Fig. 5.1 must be modified. If aqua-ammonia were used in the system of Fig. 5.1, the vapor leaving the generator would contain too much water and would have a high freezing temperature. Means must be provided to rectify the generator vapor and increase the ammonia concentration. Then temperatures significantly below 0 °C (32 °F) can be achieved.

Figure 5.19 schematically shows an industrial aqua-ammonia system. The principal difference between Figs. 5.1 and 5.19 is that a double rectifying column and a dephlegmator are added in Fig. 5.19. A second heat exchanger also is added. The vapor leaving the generator may contain from 5 to 10 percent of water vapor. Through the use of the rectifying equipment, the concentration of the vapor entering the condenser can be made almost equal to pure ammonia.

The strong-liquid solution, containing a large concentration of ammonia refrigerant, leaves the absorber at state 1. The solution is then pumped to the condensing pressure. The solution is preheated in the heat exchanger to reduce the heating required by the generator. This heated, strong solution enters the rectifying column partway up the column. The column produces a weak-liquid solution with a low concentration of ammonia refrigerant at the bottom (state 4) and nearly pure ammonia vapor at the top (state 7). The ammonia refrigerant is sent to the condenser, which condenses it to saturated or subcooled liquid at state 8. The liquid is further cooled in the heat exchanger before it enters the expansion valve. The ammonia leaving the expansion valve enters the evaporator, where the liquid phase vaporizes to absorb the refrigerating load on the system, $\dot{Q}_E$. The refrigerant is further heated in the heat exchanger prior to being absorbed into the weak-liquid solution in the absorber. The strong solution leaving the absorber at state 1 is ready to begin another cycle.

The rectifying equipment operates exactly like that described in Sec. 5.5. Other parts of the system have also been adequately discussed. In Sec. 5.4 we treated several of the processes involved in Fig. 5.19. We will now consider the solution of a complete system problem.

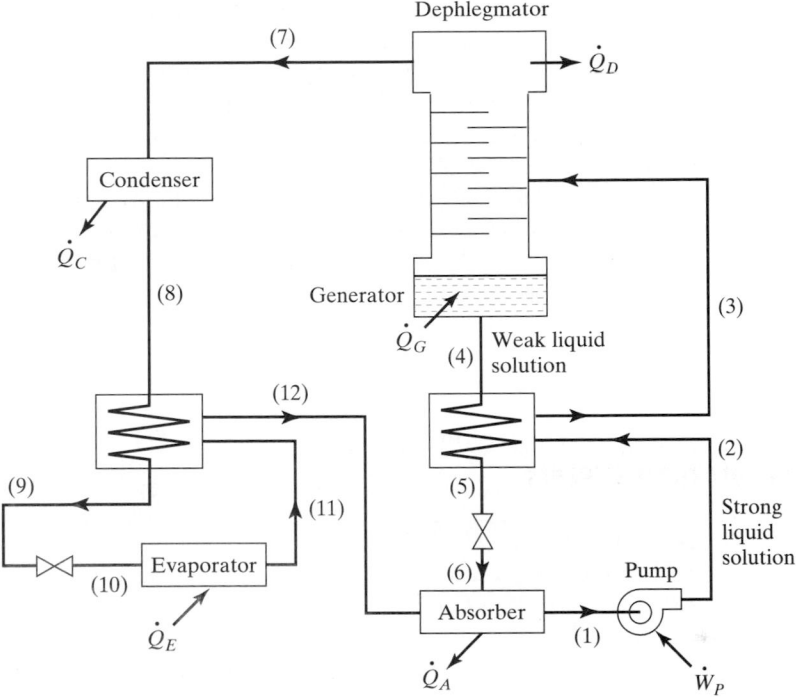

Figure 5.19 Industrial aqua-ammonia absorption system.

EXAMPLE 5.5

The following data are known for the system of Fig. 5.19: condensing pressure, 200 psia; evaporating pressure, 30 psia; generator temperature, 240 °F; temperature of vapor leaving dephlegmator, 130 °F; ambient temperature, 79 °F; and temperature of strong solution entering column, 200 °F. The temperature of liquid leaving the condenser is reduced 10 °F in the heat exchanger.

Assume equilibrium (saturated) conditions for states 1, 3, 4, 7, 8, and 12. Neglect pressure drop in components and lines. Assume that the system produces 100 tons of refrigeration. Determine (a) thermodynamic properties p, t, x, and h for all state-points of the system, (b) mass rate of flow in lbm/min for all parts of system, (c) horsepower required for the pump, if a mechanical efficiency of 75 percent is assumed, (d) system coefficient of performance, (e) system refrigerating efficiency, (f) comparison of coefficient of performance with that of a theoretical vapor-compression ammonia cycle, and (g) an energy balance for the entire system in Btu/min.

Solution:

(a) and (b) We need to work parts (a) and (b) concurrently. Table 5.1 shows thermodynamic properties and flow rates for the problem. Figure 5.20 shows the processes on a schematic h-x diagram. The procedure will now be discussed in detail.

TABLE 5.1 Thermodynamic Properties and Flow Rates for Ex. 5.5

State-Point	Pressure P, psia	Temperature t, °F	Concentration x, lbm NH3/lbm mix	Enthalpy h, Btu/lbm mix	Flow Rate, lbm mix/min
1	30	81	0.402	−24	257.1
2	200	82	0.402	−23	257.1
3	200	200	0.402	109	257.1
4	200	240	0.293	158	217.3
5	200	98	0.293	2	217.3
6	30	98	0.293	2	217.3
7	200	130	0.997	655	39.8
8	200	95	0.997	149	39.8
9	200	85	0.997	137	39.8
10	30	−2	0.997	137	39.8
11	30	46	0.997	640	39.8
12	30	61	0.997	652	39.8

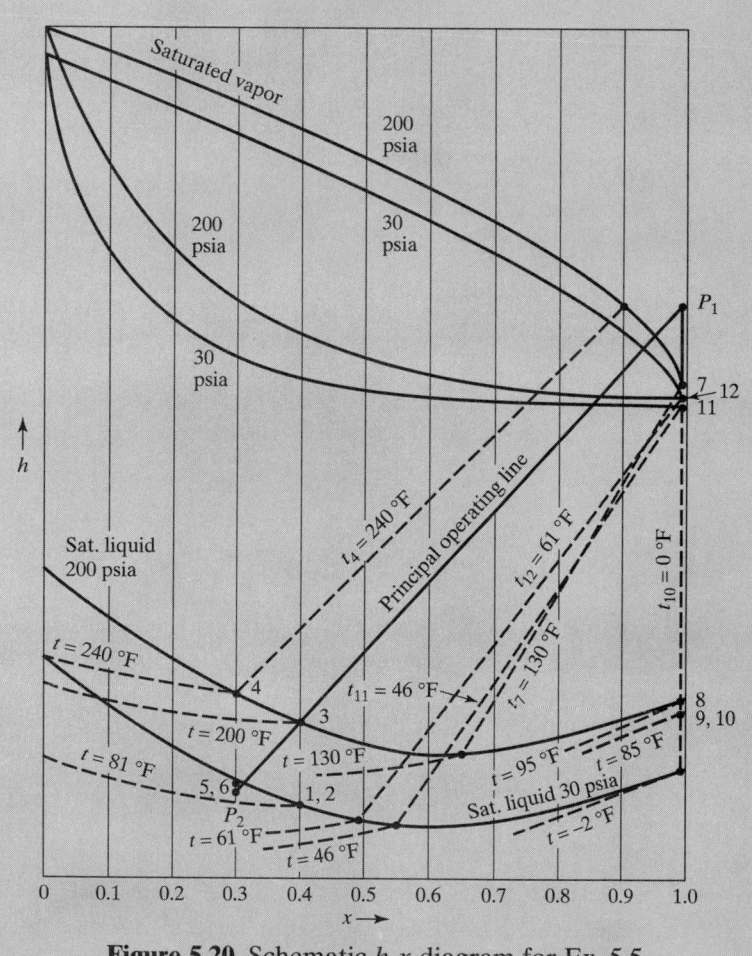

Figure 5.20 Schematic h-x diagram for Ex. 5.5.

All pressures are established from the given data. The temperatures at 3, 4, and 7 are given. Since states 3, 4, and 7 are equilibrium states, we may locate the state-points on Fig. C-3E and read concentration and enthalpy values.

We know that for points 7 through 12 the concentration is the same. We also know that $x_1 = x_2 = x_3$ and that $x_4 = x_5 = x_6$. Since states 1, 8, and 12 are equilibrium states, we may establish the points on Fig. C-3E from the known pressure and concentration values. Since $t_9 = t_8 - 10\,°F$, we may read h_9 at the known values of t and x.

For the combination of evaporator and liquid-suction heat exchanger we have

$$\dot{m}_8 h_8 + (100)(200) = \dot{m}_8 h_{12}$$

$$\dot{m}_8 = \frac{20{,}000}{h_{12} - h_8} = \frac{20{,}000}{503} = 39.8 \text{ lbm/min}$$

For the absorber, we have

$$\dot{m}_{12} + \dot{m}_6 = \dot{m}_1$$

$$\dot{m}_{12} x_{12} + \dot{m}_6 x_6 = \dot{m}_1 x_1$$

Thus

$$\dot{m}_6 = \dot{m}_{12} \frac{x_{12} - x_1}{x_1 - x_6} = (39.8) \frac{(0.595)}{(0.109)} = 217.3 \text{ lbm/min}$$

$$\dot{m}_1 = \dot{m}_{12} + \dot{m}_6 = 257.1 \text{ lbm/min}$$

Thus, all of the flow rates of the system are known. For the pump we have

$$h_2 = h_1 + (P_2 - P_1)v_1$$

The specific volume v_1 may be found by the empirical relation

$$v_1 = (1 - x_1)v_{H_2O} + (0.85)x_1 v_{NH_3}$$

Using steam and ammonia tables, we find

$$v_1 = (0.598)(0.01608) + (0.85)(0.402)\frac{1}{37.42} = 0.0187 \text{ ft}^3/\text{lbm}$$

and

$$h_2 = -24 + \frac{(200 - 30)(144)(0.0187)}{778} = -23.41 \text{ Btu/lbm}$$

For the solution heat exchanger we have

$$h_5 = h_4 - \frac{\dot{m}_2}{\dot{m}_4}(h_3 - h_2) = 158 - \frac{(257.1)(132)}{(217.3)} = 2 \text{ Btu/lbm}$$

The temperature at state 5 may be read from Fig. C-3E. Since state 6 is also a subcooled-liquid state, its temperature is the same as at state 5.

States 9 and 10 are coincidental on the h-x diagram. State 10 is a mechanical mixture of liquid and vapor. Its temperature $(-2\,°F)$ may be found by the method described for Ex. 5.2(c). The enthalpy at state 11 may be found from the energy balance for the evaporator. State 11 is also a mechanical-mixture state. We find its temperature to be 46 °F.

(c) We may find the pump horsepower by

$$w\dot{m} = \frac{\dot{m}_1(h_2 - h_1)}{42.4\eta_m} = \frac{(257.1)(0.59)}{(42.4)(0.75)} = 4.77 \text{ hp}$$

(d) Before calculating the coefficient of performance, we must first establish the generator heat requirement. The quantities $\dot{Q}_G/\dot{m}_4$ and $\dot{Q}_D/\dot{m}_7$ may be immediately found, once the principal operating line is drawn. However, additional assumptions regarding the column must be made. Referring to the discussion shown in Sec. 5.5 and Fig. 5.18, let us assume that the state of the vapor in the column lies on the principal operating line at the location where the strong solution enters the column. Let us further assume that this vapor state is saturated with a temperature 10 °F higher (210 °F) than that of the strong solution which is entering (t_3). These assumptions allow the principal operating line to be constructed. By Fig. C-3E we obtain

$$\dot{Q}_G/\dot{m}_4 = 172 \text{ Btu/lbm} \quad \text{and} \quad \dot{Q}_D/\dot{m}_7 = 129 \text{ Btu/lbm}$$

Thus, we have

$$\dot{Q}_G = (217.3)(172) = 37{,}400 \text{ Btu/min}$$
$$\dot{Q}_D = (39.8)(129) = 5130 \text{ Btu/min}$$

If we neglect the pump work,

$$\text{C.O.P.} = \frac{\dot{Q}_E}{\dot{Q}_G} = \frac{20{,}000}{37{,}400} = 0.535$$

(e) The ideal C.O.P. may be calculated by Eq. (5.3). Assuming an environmental temperature of 70 °F, then for $t_G = 240$ °F and $t_R = -2$ °F

$$\text{C.O.P.}_{\text{max}} = \frac{(458)(170)}{(72)(700)} = 1.54$$

and

$$\eta_R = \frac{\text{C.O.P.}}{\text{C.O.P.}_{\text{max}}} = \frac{0.535}{1.54} = 0.35$$

(f) For a theoretical ammonia vapor-compression cycle operating with an evaporating pressure of 30 psia and a condensing pressure of 200 psia, we have, by Eq. (3.5), that the C.O.P. = 3.82, which is about seven times the C.O.P. for the absorption system.

(g) Calculations for an energy balance are shown in Table 5.2.

TABLE 5.2 Energy-Balance Calculations for Ex. 5.5

Component	Calculations	Btu/min Gains	Btu/min Losses
Absorber	$\dot{Q}_A = \dot{m}_6 h_6 + \dot{m}_{12} h_{12} - \dot{m}_1 h_1 =$		32,300
Pump	$\dot{W}_P = \dot{m}_1(h_2 - h_1) =$	200	
Generator	$\dot{Q}_G = \dot{m}_4(\dot{Q}_G/\dot{m}_4) =$	37,400	
Dephlegmator	$\dot{Q}_D = \dot{m}_7(\dot{Q}_D/\dot{m}_7) =$		5130
Condenser	$\dot{Q}_C = \dot{m}_7(h_7 - h_8) =$		20,100
Evaporator	Given	20,000	
Totals		57,600	57,530

Other types of ammonia absorption systems have been developed. One approach is to adsorb the ammonia vapor into a solid such as silver chloride rather than a liquid. This

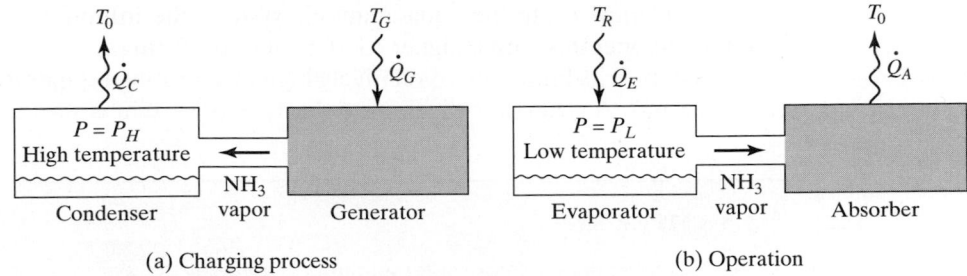

T_0 T_G T_R T_0

$\dot{Q}_C$ $\dot{Q}_G$ $\dot{Q}_E$ $\dot{Q}_A$

$P = P_H$
High temperature

Condenser   NH₃ vapor   Generator

$P = P_L$
Low temperature

Evaporator   NH₃ vapor   Absorber

(a) Charging process   (b) Operation

Figure 5.21 Schematic diagram of an intermittent absorption refrigeration system using a solid absorbent: (a) charging process, (b) operation.

simplifies the generation process, as the solid can be heated to drive off the ammonia vapor directly without resorting to a complex rectification process. As the solid cannot be easily transported throughout the system, this system must be used in an intermittent fashion. A schematic drawing of a system using this approach is given in Fig. 5.21. During the charging process, shown in Fig. 5.21(a), heat is added to generate the ammonia from the solid. The high-pressure high-temperature ammonia vapor is condensed in the other portion of the system, which acts as the condenser. In operation, the two sides of the system change roles, as shown in Fig. 5.21(b). The pressure and saturated temperature drop as ammonia vapor is readsorbed by the solid. The condenser becomes the evaporator and the generator becomes the absorber. This type of intermittent absorption system is well suited for use with solar energy, where the system can be charged during the day when solar heating is available and can provide refrigeration at a later time.

Another approach to changing the saturation temperature of ammonia refrigerant is to change its partial pressure. This is accomplished by introducing a noncondensible gas, keeping the total pressure constant, rather than changing the total pressure inside the refrigeration system. An absorption system with no moving parts has been developed to operate using ammonia as the refrigerant, water as the absorbent, and hydrogen gas as the dilutant. Vapor bubbles push liquid slugs upward in a small-bore tube above the generator, which is located near the bottom of the unit. The remaining liquid circulation is purely gravitational. Hydrogen gas enters in the evaporator to reduce the partial pressure, and thus the saturation temperature, of the ammonia refrigerant. The generator, rectifier, condenser, and absorber operate in a manner similar to the commercial system analyzed in Example 5.5. As the system pressure is uniform with the exception of a small amount of hydrostatic pressure, no pump is necessary. The only source of energy required is the heat input to the generator.

5.7 THE LITHIUM BROMIDE-WATER ABSORPTION SYSTEM

The lithium bromide-water absorption system has become prominent in refrigeration for air conditioning. Water is the refrigerant while the absorbent is lithium bromide. Pure lithium bromide is a solid, but when it is mixed with sufficient water, homogeneous liquid solutions are formed.

The outstanding feature of the system is the nonvolatility of the lithium bromide. In the generator, only water vapor is driven off. Therefore no rectifying equipment is required. Compared to the aqua-ammonia system, the lithium bromide-water system is simpler and operates with a higher coefficient of performance. The primary disadvantage of the system is its limitation to relatively high evaporating temperatures, since the refrigerant is water. A schematic diagram of this type of system is shown in Fig. 5.1.

EXAMPLE 5.6

The following data are known for a lithium bromide-water system of the type shown in Fig. 5.1: condensing temperature = 100 °F, evaporating temperature = 40 °F, temperature of strong solution leaving absorber = 100 °F, temperature of strong solution entering generator = 180 °F, generator temperature = 200 °F.

Assume saturated conditions for states 3, 4, 8, and 10. Neglect pressure drop in components and lines. Assume a system capacity of one ton of refrigeration. Determine (a) thermodynamic properties p, t, x, and h for all necessary state-points of the system, (b) mass rate of flow in lbm/min for each part of system, (c) system coefficient of performance, (d) system refrigerating efficiency, and (e) steam consumption for the generator in lbm/hr, if saturated steam at 220 °F is used.

Solution:

(a) and (b) Parts (a) and (b) will be worked concurrently. Table 5.3 shows a tabulation of thermodynamic properties and flow rates for the problem. Properties of pure water were obtained from Table A.1E, while solution properties were read from Fig. C-4E. It is important to note that Fig. C-4E expresses concentration in terms of the absorbent lithium bromide.

The two pressures are found from the given saturation temperatures at states 8 and 10. Enthalpies at states 8 and 10 are found from Table A.1E. The enthalpy at state 7 may be calculated by Eq. (2.35). Since states 3 and 4 are equilibrium states, we may locate the state-points on Fig. C-4E and read enthalpy and concentration values. Table 5.3 shows the thermodynamic properties for the problem. Figure 5.22 shows the state-points on a schematic h-x diagram. For the evaporator, we have

$$\dot{m}_{10} = \dot{m}_9 = \frac{200}{h_{10} - h_9} = \frac{200}{1011} = 0.198 \text{ lbm/min}$$

TABLE 5.3 Thermodynamic Properties and Flow Rates for Ex. 5.6

State-Point	Pressure P, mm Hg	Temperature t, °F	Concentration x, lbm LiBr/lbm mix	Enthalpy h, Btu/lbm mix	Flow Rate $\dot{m}$, lbm mix/min
1	6.3	100	0.598	−70.3	2.72
2	49.1	100	0.598	−70.3	2.72
3	49.1	180	0.598	−38.9	2.72
4	49.1	200	0.645	−32.5	2.52
5	49.1	111.3	0.645	−66.4	2.52
6	6.3	127.8	0.645	−66.4	2.52
7	49.1	200	0.000	1150	0.198
8	49.1	100	0.000	68	0.198
9	6.3	40	0.000	68	0.198
10	6.3	40	0.000	1079	0.198

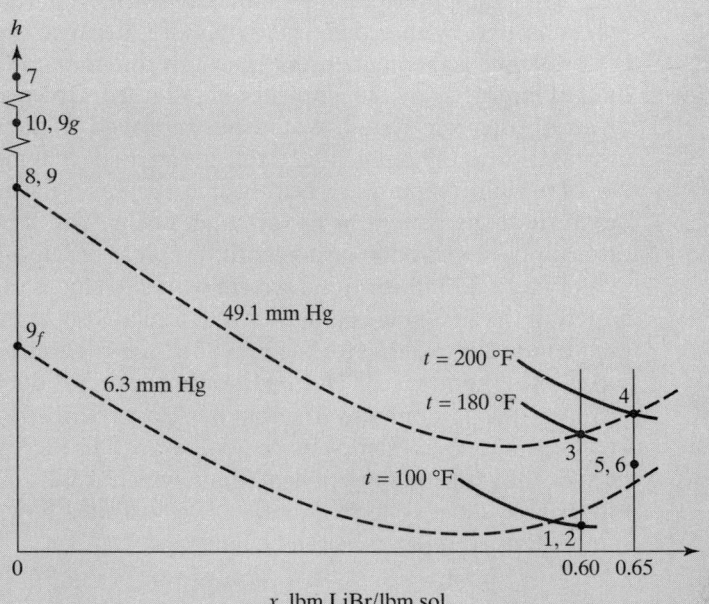

Figure 5.22 Schematic h-x diagram for Ex. 5.6.

For the absorber, we have

$$\dot{m}_6 = \frac{\dot{m}_{10}(x_1 - x_{10})}{x_6 - x_1} = \frac{(0.198)(0.598)}{(0.047)} = 2.52 \text{ lbm/min}$$

$$\dot{m}_1 = \dot{m}_6 + \dot{m}_{10} = 2.72 \text{ lbm/min}$$

(c) For the generator, we have

$$\dot{Q}_G = \dot{m}_4 h_4 + \dot{m}_7 h_7 - \dot{m}_3 h_3$$

$$= (2.52)(-32.5) + (0.198)(1150) - (2.72)(-38.9)$$

$$= 252 \text{ Btu/min}$$

Neglecting the pump work, we have

$$\text{C.O.P.} = \frac{\dot{Q}_E}{\dot{Q}_G} = \frac{200}{252} = 0.794$$

(d) For an environmental temperature of 100 °F, a generator temperature of 200 °F, and an evaporating temperature of 40 °F, we have, by Eq. (5.3),

$$\text{C.O.P.}_{\text{max}} = \frac{(500)(100)}{(60)(660)} = 1.263$$

and

$$\eta_R = \frac{\text{C.O.P.}}{\text{C.O.P.}_{\text{max}}} = \frac{0.794}{1.263} = 0.629$$

(e) We will assume that saturated water at 220 °F leaves the steam coil. Thus

$$\dot{m}_s = \frac{60\dot{Q}_G}{h_{fg,s}} = \frac{(60)(252)}{(964.9)} = 15.67 \text{ lbm/hr}$$

The temperature of the vapor leaving the generator of a single-stage absorption system is limited by the condensing pressure, because the pressure in the generator equals the pressure in the condenser. Therefore the maximum temperature of the generator is limited indirectly by the ambient temperature. However, Eq. (5.3) shows that the C.O.P. of an absorption system would be improved if the generator temperature could be increased with the ambient and refrigerated-space temperatures remaining constant. The use of a high-temperature heat source such as a natural-gas flame could raise the temperature of the generator to very high levels. The limitation on the generator temperature can be removed by using multiple stages of vapor generation.

Figure 5.23 illustrates a system with two stages of vapor generation. The vapor generated in the first-stage generator is at a relatively high pressure and temperature. This is used as the heat source for the second-stage generator, which operates at the same pressure as the condenser. The additional vapor produced in the second-stage generator increases the amount of refrigerant that passes through the evaporator for a given amount of energy added to the first-stage generator. This increases the cooling capacity of the system for the same amount of heat input, which results in a higher value for the C.O.P. than for a single-stage system. Example 5.7 illustrates how the performance of the two-stage system shown in Fig. 5.23 can be analyzed.

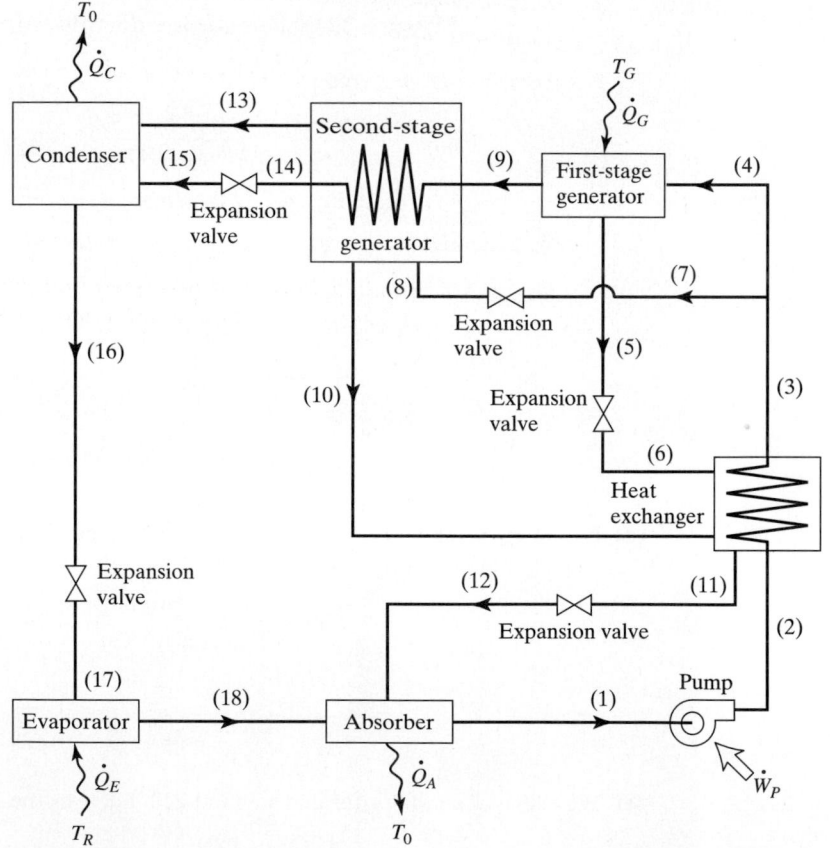

Figure 5.23 Schematic diagram of a two-stage LiBr-H$_2$O absorption refrigeration system.

EXAMPLE 5.7

The following data are known for a lithium bromide-water system as shown in Fig. 5.23: evaporating temperature = 7 °C, condensing temperature = 36 °C, first-stage generator pressure = 80 kPa, temperature of the strong solution leaving the heat exchanger = 80 °C, concentration of strong solution leaving the pump = 0.575 kg LiBr/kg mix, concentration of liquid solution leaving the first-stage generator = 0.64 kg LiBr/kg mix, concentration of liquid solution leaving second-stage generator = 0.62 kg LiBr/kg mix. Assume saturated conditions for states 1, 5, 10, 14, 16, and 18. Neglect pressure drops in components and lines except for the expansion valves. The system has a cooling capacity of 1 MW.

Determine (a) the thermodynamic properties p, t, x, and h for each of the state points indicated on Fig. 5.23, (b) mass rate of flow in kg/sec for each of the numbered locations shown on Fig. 5.23, and (c) the system coefficient of performance.

Solution:

(a) and (b) Parts (a) and (b) will be worked concurrently. Table 5.4 shows a tabulation of the thermodynamic properties and flow rates for the problem. The properties of pure water were obtained from Table A.1SI and Eq. (2.36), while solution properties were determined from Fig. C-4SI. Figure 5.24 shows a schematic h-x diagram for this example.

The condensing and evaporating pressures are found from the given saturation temperatures in the condenser and evaporator. States 1, 5, and 10 are equilibrium states and can be located on Fig. C-4SI at the intersection of the known pressure curves and the known concentration lines. Neglecting the pump work, states 1 and 2 have the same concentration and enthalpy and are at the same location on Fig. C-4SI. States 5 and 6 are also located at the same point, because neither the concentration nor the enthalpy changes during a throttling process. State 3 is located at the intersection of the known temperature isotherm and the line of constant concentration. State-points 4, 7, and 8 are located at the same point as state 3. State 9 is water vapor at 80 kPa pressure and at the

TABLE 5.4 Thermodynamic Properties and Flow Rates for Ex. 5.7

State-Point	Pressure P, kPa	Temperature t, °F	Concentration x, kg LiBr/kg mix	Enthalpy h, kJ/kg mix	Flow Rate $\dot{m}$, kg mix/sec
1	1	42	0.575	−160	4.93
2	80	42	0.575	−160	4.93
3	80	80	0.575	−84	4.93
4	80	80	0.575	−84	2.26
5	80	159	0.64	57	2.03
6	6	97	0.64	57	2.03
7	80	80	0.575	−84	2.67
8	6	76	0.575	−84	2.67
9	80	159	0	2797	0.229
10	6	86	0.62	−76	2.48
11	6	74	0.629	−99.5	4.5
12	1	55	0.629	−99.5	4.5
13	6	85	0	2659	0.194
14	80	93.5	0	392	0.229
15	6	36	0	392	0.229
16	6	36	0	150	0.423
17	1	7	0	150	0.423
18	1	7	0	2514	0.423

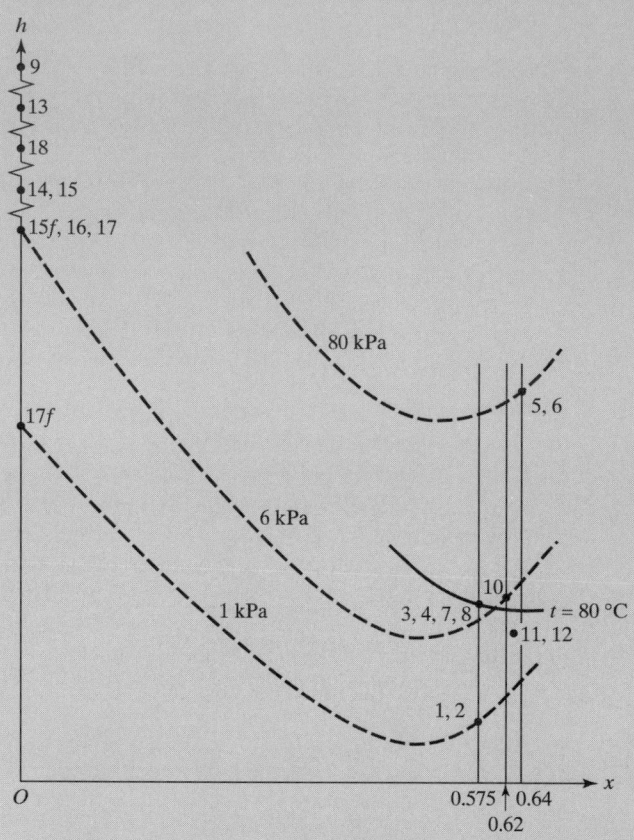

Figure 5.24 Schematic h-x diagram for Ex. 5.7.

same temperature as the solution leaving the first-stage generator (159 °C), so its enthalpy is found from Eq. (2.36). State 14 is saturated liquid water at 80 kPa pressure and is found from Table A.1SI. State 15 has the same enthalpy as state 14 but is at the condenser pressure. State 13 is water vapor at the condenser pressure and at the same temperature as the solution leaving the second-stage generator (85 °C). State 16 is saturated liquid water at the condenser temperature and is found using Table A.1SI. State 17 has the same enthalpy as state 16 but is at the evaporator pressure and temperature. State 18 is saturated water vapor at the evaporator temperature and pressure; its enthalpy is computed using Eq. (2.36). This completes all the thermodynamic-property data needed for Table 5.4 except for states 11 and 12. Determination of these state-points requires solving an energy balance on the heat exchanger, which in turn requires knowledge of the mass flow rates. Therefore the mass flow rates will be determined next.

The mass and energy balance equations for the second-stage generator can be written as follows:

$$\dot{m}_9 = \dot{m}_{14}$$
$$\dot{m}_8 = \dot{m}_{10} + \dot{m}_{13}$$
$$\dot{m}_8 x_8 = \dot{m}_{10} x_{10} + \dot{m}_{13} x_{13}$$
$$\dot{m}_9 h_9 + \dot{m}_8 h_8 = \dot{m}_{13} h_{13} + \dot{m}_{14} h_{14} + \dot{m}_{10} h_{10}$$

Realizing that x_{13} equals zero and combining these equations gives

$$\frac{\dot{m}_9}{\dot{m}_8} = \frac{(h_{13} - h_8) + \dfrac{x_8}{x_{10}}(h_{10} - h_{13})}{h_9 - h_{14}}$$

$$= \frac{[2659 - (-84)] + \dfrac{0.575}{0.62}(-76 - 2659)}{2797 - 392} = 0.0859$$

A mass balance is then performed on the first-stage generator:

$$\dot{m}_4 = \dot{m}_5 + \dot{m}_9$$

$$\dot{m}_4 x_4 = \dot{m}_5 x_5 + \dot{m}_9 x_9$$

Combining these, realizing that x_9 equals zero, and dividing by m_8, we get

$$\frac{\dot{m}_4}{\dot{m}_8} = \frac{\dfrac{\dot{m}_9}{\dot{m}_8} x_5}{x_5 - x_4} = \frac{0.0859(0.64)}{0.64 - 0.575} = 0.846$$

From the tee following the heat exchanger

$$\dot{m}_3 = \dot{m}_4 + \dot{m}_8 \quad \text{or} \quad \frac{\dot{m}_3}{\dot{m}_8} = \frac{\dot{m}_4}{\dot{m}_8} + 1.0 = 1.846$$

Then, inverting this,

$$\frac{\dot{m}_8}{\dot{m}_3} = 0.542$$

Then

$$\frac{\dot{m}_4}{\dot{m}_3} = 0.458$$

Returning to the two mass-balance equations for the first-stage generator,

$$\frac{\dot{m}_9}{\dot{m}_3} = \frac{\dot{m}_4(x_5 - x_4)}{\dot{m}_3 x_5} = \frac{0.458(0.64 - 0.575)}{0.64} = 0.0465$$

and

$$\frac{\dot{m}_5}{\dot{m}_3} = \frac{\dot{m}_4}{\dot{m}_3} - \frac{\dot{m}_9}{\dot{m}_3} = 0.458 - 0.0465 = 0.412$$

From the second and third mass-balance equations for the second-stage generator

$$\frac{\dot{m}_{10}}{\dot{m}_3} = \frac{\dot{m}_8 x_8}{\dot{m}_3 x_{10}} = 0.542\frac{0.575}{0.62} = 0.503$$

and

$$\frac{\dot{m}_{13}}{\dot{m}_3} = \frac{\dot{m}_8}{\dot{m}_3} - \frac{\dot{m}_{10}}{\dot{m}_3} = 0.542 - 0.503 = 0.039$$

A mass balance on the condenser gives $\dot{m}_{13} + \dot{m}_{15} = \dot{m}_{16}$ or

$$\frac{\dot{m}_{13}}{\dot{m}_3} + \frac{\dot{m}_9}{\dot{m}_3} = \frac{\dot{m}_{16}}{\dot{m}_3} = 0.039 + 0.0465 = 0.086$$

Now we write an energy balance on the evaporator to determine the water mass flow rate through the evaporator.

$$\dot{m}_{18} = \frac{\dot{Q}_E}{h_{18} - h_{17}} = \frac{1 \text{ MW}}{2513.8 - 150.4} = 0.423 \text{ kg/sec}$$

Combining this result with the equation above, we find $m_3 = 4.93$ kg/sec. Now that m_3 and the mass flow rate ratios are known, all the remaining mass flow rates can be determined. The results are listed in Table 5.4.

The last state-points to determine are states 11 and 12. Writing a LiBr mass balance on the absorber, we can solve for the concentration of the weak solution entering the absorber:

$$x_{12} = \frac{\dot{m}_1 x_1}{\dot{m}_{12}} = \frac{4.93(0.575)}{4.50} = 0.629 \text{ kg LiBr/kg mix}$$

An energy balance on the heat exchanger allows us to compute the enthalpy of states 11 and 12, the only remaining unknown enthalpy values.

$$h_{11} = \frac{\dot{m}_2 h_2 + \dot{m}_6 h_6 + \dot{m}_{10} h_{10} - \dot{m}_3 h_3}{\dot{m}_{11}}$$

$$= \frac{4.93(-160) + 2.03(57) + 2.48(-76) - 4.93(-84)}{4.50} = -99.5 \text{ kJ/kg}$$

(c) The coefficient of performance for this system, neglecting the pump work, is C.O.P. $= \dot{Q}_E / \dot{Q}_G$. The rate of heat input into the first-stage generator can be obtained from an energy balance on that generator:

$$\dot{Q}_G = \dot{m}_9 h_9 + \dot{m}_5 h_5 - \dot{m}_4 h_4$$

$$= 0.229(2797) + 2.03(57) - 2.26(-84) = 946 \text{ kW}$$

so the system coefficient of performance becomes

$$\text{C.O.P.} = \frac{1 \text{ MW}}{946 \text{ kW}} = 1.06$$

Examples 5.6 and 5.7 illustrate a method of obtaining a thermodynamic solution to single- and two-stage LiBr absorption systems that operate in steady state. Real systems often operate with a variety of heat-source temperatures, heat-rejection temperatures, and loads on the evaporator. Figure 5.25 illustrates how a single-stage cycle changes its capacity from the rated capacity as the heat-rejection temperature changes (condenser water, t_0) and the temperature of the load changes (chilled water, t_R).

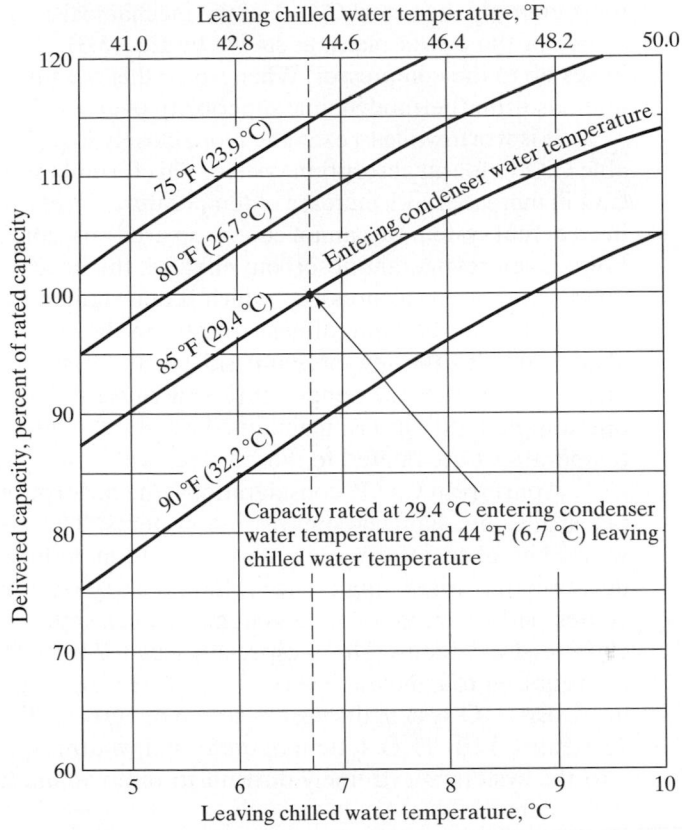

Figure 5.25 Performance characteristics of lithium-bromide-cycle water chiller. [Reprinted by permission from *ASHRAE Equipment 1988,* p. 13.6.]

5.8 COMPARISON OF ABSORPTION SYSTEMS WITH MECHANICAL VAPOR-COMPRESSION SYSTEMS

Example 5.5 showed that the coefficient of performance for an aqua-ammonia absorption system was about one-seventh of that for a theoretical single-stage ammonia system operating at approximately the same evaporating and condensing temperatures. Example 5.6 showed that the C.O.P. for a lithium bromide-water absorption system was somewhat higher than that of an aqua-ammonia system.

It is unfair to the absorption system to make a direct comparison of its C.O.P. with that of a mechanical-compression system. A C.O.P. calculated for an absorption system, as in Ex. 5.5, is more realistic than a C.O.P. calculated for a theoretical vapor-compression cycle, where isentropic compression is assumed. In order to make a comparison, we must multiply the actual C.O.P. of the mechanical compression system by the thermal efficiency of the power plant, as shown by Eq. (5.3). We must also include the transmission losses up to the compressor. When we do this, we find that a mechanical-compression system has little thermodynamic superiority over an absorption system.

It is worthwhile to examine more closely Eq. (5.3), which gives the maximum attainable C.O.P. for an absorption system. This formula shows that, for a reversible system, the C.O.P. increases with increase of temperature of the generator heating medium. However, in an actual system we cannot choose an arbitrary combination of operating temperatures. For a given refrigerant-absorbent mixture, the necessary generator temperature depends upon the condensing pressure, which is determined by the temperature of available cooling water or air and rate of heat transfer in the condenser. The generator temperature is also related to the low-pressure side of the system. In the aqua-ammonia system, an increase of generator temperature will cause a decrease in absorber pressure, if other operating conditions are maintained constant. Thus, in an absorption system, operating temperatures are limited to those values which are satisfactory in actual practice.

Apart from C.O.P. considerations, the absorption system has some practical advantages over mechanical-compression systems. The absorption system is less subject to wear. The absorption system may operate at reduced evaporating pressure with little decrease in refrigerating capacity. Liquid carry-over from the evaporator causes no difficulties. However, absorption systems are generally more bulky than mechanical vapor-compression systems. This is especially true for $LiBr$-H_2O systems, in which large volumes are required to handle the very low-density water vapor. The weak solution may solidify in a $LiBr$-H_2O system during off-design operation. This requires additional maintenance. The entire $LiBr$-H_2O system operates below atmospheric pressure, so any air that leaks into the system is extremely difficult to remove and lowers the refrigerating capacity of the system.

The choice between an absorption system and a mechanical-compression system is largely determined by economic factors. The absorption water-chilling machine used for air conditioning may be economically attractive where low-cost fuel is available, where electricity rates are high, where heating-boiler capacity is idle or partially idle during summer months, where waste steam is available, or where existing electric facilities are inadequate for installing electric motor-driven mechanical compressors. Additional discussion can be found in [4, 5].

ENDNOTES

1. Fran Bosnjakovic, *Technical Thermodynamics*, trans. Perry L. Blackshear, Jr. (New York: Holt, Rinehart & Winston, 1965), 268–270.

2. Bosnjakovic, *Technical Thermodynamics*.

3. Bosnjakovic, *Technical Thermodynamics*, p. 182.

4. M. Bogart, *Ammonia Absorption Refrigeration in Industrial Processes* (Gulf Publishing Co., 1981).

5. *ASHRAE Handbook of Equipment* (Atlanta: American Society of Heating, Refrigerating and Air Conditioning Engineers, 1980), ch. 13.

PROBLEMS

5.1 Saturated water vapor at 50 °F is mixed in a steady-flow chamber with a saturated lithium bromide-water solution having a concentration of 0.60 lbm Li Br/lbm mix. The mass of liquid solution mixed is five times the mass of the water vapor mixed. The mixing process occurs at constant pressure. Determine
  (a) the concentration of the resulting mixture, and
  (b) the heat which must be removed in Btu/lb of the final mixture, if a saturated-liquid solution is produced.

5.2 A saturated solution of LiBr-H_2O at 40 °C is heated to 100 °C. Determine
  (a) the relative mass proportions of liquid and vapor phases that exist, and
  (b) the specific enthalpy values for the liquid and vapor phases.

5.3 A saturated liquid aqua-ammonia solution at 220 °F and 200 psia is throttled to a pressure of 10 psia. Determine
  (a) the equilibrium temperature after the throttling process, and
  (b) the relative portions of liquid and vapor coexisting after throttling.

5.4 Perform the complete derivation of Eq. (5.25).

5.5 An aqua-ammonia system similar to that in Fig. 5.19 operates as follows: high-side pressure = 200 psia; t_3 = 190 °F; t_7 = 140 °F; t_4 = 210 °F; m_7 = 100 lbm/min. Assume equilibrium conditions for states 3, 4, and 7. Find (a) the lbm/min of strong solution leaving the absorber, and (b) the lbm/min of cooling water required for the dephlegmator if the water-temperature rise is 15 °F. You may make the same assumptions with regard to the rectifying column as made in part (d) of Example 5.4.

5.6 An aqua-ammonia absorption system similar to that in Fig. 5.19 operates as follows: high-side pressure = 2.0 MPa; low-side pressure = 0.2 MPa; t_3 = 100 °C; t_7 = 60 °C; t_4 = 120 °C; refrigerating capacity = 15 kW. Assume equilibrium conditions for states 1, 3, 4, 7, 8, and 12. Find
  (a) the mass flow rates at all state points indicated on Fig. 5.19,
  (b) the concentration at all state points, and
  (c) the specific enthalpy at all state points. Assume the vapor in the rectifying column where state 3 enters is saturated at a temperature 10 °C higher than t_3.

5.7 Rework Prob. 5.6 but change the condensing pressure to 1.0 MPa. What conclusions can be made regarding the choice of condensing pressure?

5.8 Rework Ex. 5.5 but omit the heat exchanger between states 8, 9, 11, and 12. Compare the results with those of Ex. 5.5. Should the heat exchanger be retained in the system?

5.9 Rework Ex. 5.5 with t_3 changed to
  (a) 180 °F and
  (b) 160 °F.
All other parameters and assumptions remain as before. Comment on the importance of the heat exchanger used to preheat the liquid that enters the rectifying column.

5.10 An aqua-ammonia system similar to that in Fig. 5.19 operates as follows: high-side pressure = 220 psia; low-side pressure = 20 psia; t_4 = 210 °F; t_8 = 80 °F; t_{12} = 40 °F; m_4 = 1000 lbm/min; m_{12} = 100 lbm/min. Assume equilibrium states at 1, 3, 4, and 12. Determine
  (a) the concentration of the strong solution leaving the heat exchanger,
  (b) the heat removed in the absorber in Btu/min, and
  (c) the tons of refrigeration produced.

5.11 A lithium bromide-water system similar to that shown in Fig. 5.1 operates with a condensing temperature of 40 °C, evaporating temperature of 5 °C, temperature of solution leaving absorber of 35 °C, temperature of solution entering generator of 80 °C, and temperature of solution leaving generator of 100 °C. Assume saturated conditions for states 3, 4, 8, and 10. Saturated steam at 0.2 MPa enters the generator and leaves as saturated water. Compute the required mass flow rate of steam in kg/hr per kW of refrigeration produced.

5.12 Rework Prob. 5.11, changing the temperature of the solution entering the generator to
  (a) 60 °C and
  (b) 40 °C.
All other parameters remain as specified in Prob. 5.11. Comment on the importance of the heat exchanger to system performance.

5.13 A lithium bromide-water system of the type shown in Fig. 5.1 operates with a condensing temperature of 110 °F, evaporating temperature of 38 °F, temperature of solution leaving absorber of 100 °F, temperature of solution entering generator of

180 °F, and temperature of solution leaving generator of 210 °F. Assume saturated conditions for states 3, 4, 8, and 10. Neglect pressure drops in components and lines. Warm water from the load returns to the machine at a temperature of 52 °F and at a rate of flow of 600 gpm. Chilled water leaves the machine at a temperature of 44 °F. Saturated steam at 25 psia enters the generator and leaves as saturated water. Calculate the required rate of flow of steam in lbm/hr.

5.14 Rework Ex. 5.6 with evaporation temperature changed to
(a) 50 °F, and
(b) 60 °F.
All other parameters remain as specified in Ex. 5.6.

5.15 Determine the coefficient of performance of a single-stage LiBr absorption-refrigeration system that uses the same condensing and evaporating pressures as the two-stage system in Ex. 5.7. Assume saturated liquid leaves the condenser, absorber, and

generator. Saturated vapor leaves the evaporator. Assume that the strong liquid that enters the generator is saturated. Compare the results from this single-stage system to the coefficient of performance for the two-stage system considered in Ex. 5.7.

5.16 Repeat Ex. 5.7 with the concentration of the two streams leaving the generators changed to $x_5 = 0.62$ and $x_{10} = 0.60$. Which set of concentrations, the values used in Ex. 5.7 or the values used here, result in the best system performance?

5.17 Compare the cost per hour per ton refrigeration for generator steam in Ex. 5.6 with cost of electricity per hour per ton refrigeration for a Refrigerant-22 vapor-compression refrigeration system. Use local costs for steam and electricity. Assume that the Refrigerant-22 system is operated at 40 °F evaporating temperature and 95 °F condensing temperature. You may obtain an estimate of the compressor power requirement from Fig. 4.8. Assume an electric-motor efficiency of 85 percent.

SYMBOLS

C.O.P.	Coefficient of performance, dimensionless.
ΔH_x	Isothermal heat of solution, Btu/lbm, kJ/kg.
h	Specific enthalpy, Btu/lbm, kJ/kg.
$h_{fg,s}$	Latent heat of condensation for steam, Btu/lbm, kJ/kg.
L	Designation for liquid state on h-x diagram.
$\dot{m}$	Mass rate of flow, $\dot{m}_L$ for liquid; $\dot{m}_V$ for vapor; $\dot{m}_s$ for steam, lbm/hr.
P	Pressure, psia, psfa, mm Hg, MPa.
P_1, P_2	Rectification poles on h-x diagram.
Q	Heat transfer, Btu, kJ.
$\dot{Q}$	Rate of heat transfer, Btu/min, kW, $\dot{Q}_C$ for condenser, $\dot{Q}_D$ for dephlegmator, $\dot{Q}_E$ for evaporator, and $\dot{Q}_G$ for generator.
S	Entropy, Btu/°R, kJ/K.
T	Absolute temperature, °R, K.
t	Temperature, °F, °C.
V	Designation for vapor state on h-x diagram.
v	Specific volume.
$\dot{W}$	Power.
w	Work per unit mass.
x	Concentration; for ammonia-water solutions, mass ammonia/mass mix; for water-lithium bromide solutions, mass lithium bromide/mass mix.

Greek Letters

η_m	Mechanical efficiency, dimensionless.
η_R	Refrigerating efficiency, dimensionless.

chapter 6
Ultralow-Temperature Refrigeration: Cryogenics

6.1 INTRODUCTION

In the previous chapters we studied a variety of refrigerating methods. With these methods, the lowest temperatures may be produced by the cascade vapor-compression arrangement. Table 3.3 shows that a cascade system using Refrigerant 14 as the low-side refrigerant should be capable of attaining a minimum temperature of about -250 °F (-157 °C). However, there remains an approximate range of 210 °F (116 °C) between this temperature and absolute zero. In this range of temperatures we must use liquefied gases to achieve refrigeration. The term *cryogenics* is used to describe such applications.

Table 6.1 shows normal boiling temperature, freezing temperature, and critical temperature and pressure for several cryogenic fluids. Figures C-5E and C-5SI give general thermodynamic properties for dry air.

Oxygen, nitrogen, argon, neon, xenon, and krypton are the main constituents of air. Each element may be obtained by liquefaction and separation of atmospheric air. Pure hydrogen may be obtained by separation from coke-oven gas and from other sources. Pure helium is commonly obtained by separation from helium-bearing natural gas.

The subject of cryogenics has become increasingly important in recent years. Many laboratories are now equipped with ultralow-temperature systems which allow realization of temperatures within a fraction of a degree of absolute zero. The most important commercial application is the separation of oxygen and nitrogen from the atmosphere. Pure oxygen and nitrogen gas have many commercial industrial uses, ranging from medical applications and steel making (oxygen) to manufacture of computer chips (nitrogen). Gas liquefaction is useful for reducing the volume for storage and is widely used for transport and storage of natural gas (methane). Liquid oxygen and liquid hydrogen are used in rocket propulsion. Liquid hydrogen and helium and more recently liquid nitrogen can be used to cool alloys to superconducting temperatures.

Cryogenics is a broad field of study, the liquefaction of gases being an important aspect. Adiabatic demagnetization is used to realize temperatures close to absolute zero.

TABLE 6.1 Pressure and Temperature Data for Various Cryogenic Fluids

Substance	Boiling Temperature at 1 atm, °F (°C)		Freezing Temperature, °F (°C)		Critical Temperature, °F (°C)		Critical Pressure, psia (MPa)
Xenon	−162.4	(−108.0)	−169.6	(−112.0)	61.9	(16.6)	852 (5.87)
Krypton	−243.8	(−153.2)	−251.0	(−157.2)	−82.8	(−63.8)	798 (5.50)
Oxygen	−297.4	(−183.0)	−361.8	(−218.8)	−181.8	(−118.8)	737 (5.08)
Argon	−302.6	(−185.9)	−308.9	(−189.3)	−188.1	(−122.3)	710 (4.90)
Air	−312.7[a]	(−191.5)	−351.2[c]	(−212.9)	−221.3	(−140.7)	547 (3.77)
	−318.1[b]	(−194.5)	−357.2[d]	(−216.2)			
Nitrogen	−320.4	(−198.8)	−346.0	(−210.0)	−232.4	(−146.9)	493 (3.40)
Neon	−411.0	(−246.1)	−415.5	(−248.6)	−379.7	(−228.7)	395 (2.72)
Hydrogen	−423.0	(−252.8)	−434.6	(−259.2)	−399.9	(−239.9)	191 (1.32)
Helium	−452.1	(−268.9)			−450.3	(−267.9)	33 (0.23)

[a] Condensing temperature of saturated vapor.

[b] Boiling temperature of saturated liquid.

[c] Freezing temperature of liquid.

[d] Melting temperature of solid.

Special problems arise in thermometry. Elaborate insulation techniques are necessary. In this chapter we will concentrate on the thermodynamics of elementary cryogenic cycles that use Joule-Thomson and expansion cooling. More complete information on systems, equipment, and applications may be obtained from cryogenic books such as *Cryogenic Engineering* by Scott [1] and *Cryocoolers* by Walker [2]

6.2 MINIMUM WORK REQUIRED TO LIQUEFY A GAS

In order to liquefy a gas, heat must be withdrawn from the gas and rejected to an environment at higher temperature. Such a process requires an expenditure of work. It is of thermodynamic importance to know the theoretical minimum work required. The minimum work input occurs when the processes involved are reversible. Figure 6.1 shows one reversible arrangement, which we will use for our analysis. The pure gas at state 1 is

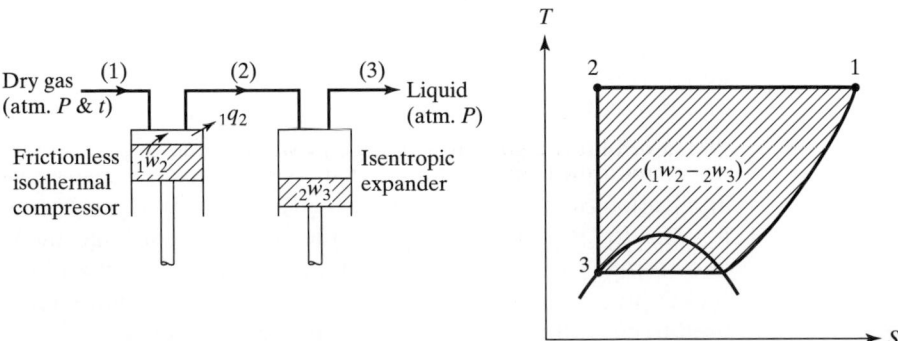

Figure 6.1 Schematic diagrams illustrating a reversible arrangement for liquefying a gas.

isothermally compressed to state 2. An isentropic expansion allows liquefaction such that at state 3 pure liquid exists. For the system shown in Fig. 6.1, for steady-flow conditions, we may write

$$_1w_2 = (h_2 - h_1) + T_1(s_1 - s_2)$$

$$_2w_3 = h_2 - h_3$$

Thus

$$w_{\text{net}} = {_1w_2} - {_2w_3} = T_1(s_1 - s_3) - (h_1 - h_3) \qquad (6.1)$$

Equation (6.1) may be generalized in the form

$$w_{Z,\text{min}} = T_0(s_0 - s_f) - (h_0 - h_f) \qquad (6.2)$$

where $w_{Z,\text{min}}$ is the minimum work required per unit mass of liquid produced. The subscript 0 refers to the gas state, and the subscript f refers to the liquid state. In practice the work required for gas liquefaction is many times more than that given by Eq. (6.2).

6.3 COOLING OF A GAS BY EXPANSION

In order to liquefy a gas, heat must be withdrawn from it. Ordinary refrigeration methods are, of course, inadequate, although they may be used for precooling the gas. Two methods of expansion may be used for reducing the gas temperature: (1) a restrained expansion, where work is performed by the gas, and (2) an unrestrained expansion or throttling of the gas.

In Fig. 6.1 an isentropic expansion was theoretically employed. The system of Fig. 6.1 would not be practical, since the pressure at state 2 would be prohibitively great. Because of the very low temperature, irreversibilities in the expander are large, so that classical cryogenic cycles do not use a restrained expansion alone. However, modern cycles use expanders that operate in the two-phase mixture region. Restrained expansion using a turbine or piston device is also an efficient means of gas cooling. Such expansion devices are often employed in gas-liquefaction apparatus.

Cooling of a gas by throttling is of major importance, and therefore we will consider this process in some detail. The Joule-Thomson coefficient,

$$\mu = \left(\frac{\partial T}{\partial P}\right)_h \qquad (6.3)$$

has particular significance. If $\mu = 0$, the temperature of the gas remains constant with throttling. If $\mu > 0$, the temperature of the gas decreases with throttling. If $\mu < 0$, the temperature of the gas increases with throttling. Thus, in cooling of a gas by throttling we require that the gas show a large positive value of μ.

The Joule-Thomson coefficient for a gas is not a constant but is a thermodynamic property of the gas that is a function of both pressure and temperature. We will now derive a functional relationship for the coefficient. We begin with the general property relation

$$T\,ds = dh - v\,dP$$

Thus

$$ds = \frac{1}{T}\,dh - \frac{v}{T}\,dP$$

Since *ds* is an exact differential, we must have

$$\left[\frac{\partial \left(\frac{1}{T} \right)}{\partial P} \right]_h = - \left[\frac{\partial \left(\frac{v}{T} \right)}{\partial h} \right]_P$$

But

$$\left[\frac{\partial \left(\frac{1}{T} \right)}{\partial P} \right]_h = - \frac{1}{T^2} \left[\frac{\partial T}{\partial P} \right]_h$$

and

$$- \left[\frac{\partial \left(\frac{v}{T} \right)}{\partial h} \right]_P = - \left[\frac{\partial \left(\frac{v}{T} \right)}{\partial T} \right]_P \left(\frac{\partial T}{\partial h} \right)_P = - \frac{\left[T \left(\frac{\partial v}{\partial T} \right)_P - v \right]}{T^2} \left(\frac{\partial T}{\partial h} \right)_P$$

Since

$$\frac{1}{c_P} = \left(\frac{\partial T}{\partial h} \right)_P$$

we have

$$\mu = \frac{1}{c_P} \left[T \left(\frac{\partial v}{\partial T} \right)_P - v \right] \tag{6.4}$$

Equation (6.4) is general and is valid for liquids as well as gases. In order to calculate the Joule-Thomson coefficient, we must know the equation of state for the substance. For a perfect gas ($Pv = RT$) we have

$$\left(\frac{\partial v}{\partial T} \right)_P = \frac{R}{P} = \frac{v}{T}$$

and the Joule-Thomson coefficient is always equal to zero.

The magnitude of the Joule-Thomson coefficient is a measure of the deviation of a gas from perfect-gas behavior. For real gases, μ may have either positive or negative values, depending upon the thermodynamic state. The temperature at which μ equals zero is called the *inversion temperature* for a given pressure. The inversion temperature of a gas may be easily determined from a diagram of properties such as Fig. C-5E or C-5SI for air. An inversion-temperature line may be drawn by connecting the peaks of the constant enthalpy lines. Figures C-5E and C-5SI show that the inversion temperature for air decreases with increase of pressure. Other gases behave similarly in this respect. Figure 6.2 shows schematically an illustration of an inversion-temperature line on *T-s* coordinates.

EXAMPLE 6.1

Estimate the value of the Joule-Thomson coefficient for air at a pressure of 10 atm (1.01 MPa) and a temperature of 135 K.

Solution: Equation (6.4) can be written in the form

$$\mu = \left(\frac{\partial T}{\partial h} \right)_P \left[T \left(\frac{\partial v}{\partial T} \right)_P - v \right]$$

The partial derivatives can be evaluated by approximating them as finite differences:

$$\mu \approx \frac{T_1(h_1, P) - T_2(h_2, P)}{h_1 - h_2} \left[T\left(\frac{v_A(T_A, P) - v_B(T_B, P)}{T_A - T_B}\right) - v \right]$$

where states (h_1, P), (h_2, P), and (T_A, P), (T_B, P) are selected arbitrarily on either side of the state point of interest (T, P).

We can now use graphical property data, tabulated property data, or thermodynamic equations of state to determine the appropriate property values. We will use graphical property data from Fig. C-5SI. Let $h_1 = 7500$ J/mole, $h_2 = 7000$ J/mole, $T_A = 155$ K, and $T_B = 115$ K. Then

$$T_1(7500 \text{ J/mole}, 10 \text{ atm}) = 142 \text{ K}$$

$$T_2(7000 \text{ J/mole}, 10 \text{ atm}) = 128 \text{ K}$$

$$v_A(155 \text{ K}, 10 \text{ atm}) = 1200 \text{ cm}^3/\text{mole}$$

$$v_B(115 \text{ K}, 10 \text{ atm}) = 750 \text{ cm}^3/\text{mole}$$

$$v(135 \text{ K}, 10 \text{ atm}) = 1000 \text{ cm}^3/\text{mole}$$

$$\mu(135 \text{ K}, 10 \text{ atm}) \approx \frac{142 \text{ K} - 128 \text{ K}}{7500 - 7000 \text{ J/mole}}$$

$$\left[135 \text{ K} \left(\frac{1200 - 750 \text{ cm}^3/\text{mole}}{155 \text{ K} - 115 \text{ K}}\right) - 1000 \text{ cm}^3/\text{mole} \right]$$

$$\mu(135 \text{ K}, 10 \text{ atm}) \approx 14.5 \text{ K cm}^3/\text{J}$$

Note that this is a positive quantity, which indicates a drop in temperature with a drop in pressure during a process of constant enthalpy. This can be seen from Figs. C-5SI and C-5E also as a negative slope to the constant-enthalpy curves in the vicinity of the state-point of interest.

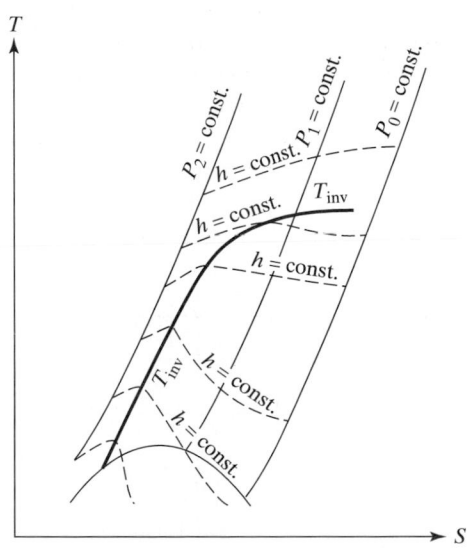

Figure 6.2 Schematic illustration of inversion-temperature line.

6.4 LINDE AIR-LIQUEFACTION CYCLE

Air liquefaction is important in the recovery of oxygen, nitrogen, and other gases from the atmosphere. In this section we will treat air-liquefaction cycles in some detail. Our fundamental analyses will be applicable to the liquefaction of other gases as well.

The most elementary air-liquefaction method is the simple Linde (or Hampson) cycle. Figure 6.3 shows the method. Equipment includes a compressor, a heat exchanger, a throttling device, and a separator. The heat exchanger and separator must be extremely well insulated. Figure 6.3 also shows a *T-s* diagram for operation after the system has adjusted itself to steady-state conditions. The compressor discharges air at a relatively high pressure (50 to 200 atm). It is necessary to cool the air as much as possible during compression to approach an isothermal compression process. In the heat exchanger the high-pressure air is cooled sufficiently so that liquid air can result from a throttling process. The system can pull itself down to operating conditions after start-up from a warm condition if the temperature of the air leaving the compressor is less than the inversion temperature. Although Fig. 6.3 shows waste gas leaving the system, in actual operation this gas stream is mixed with makeup gas at ambient conditions and the mixture is supplied to the compressor.

Two performance quantities are of particular interest in gas-liquefaction systems. These are (1) the yield Z, mass of liquid produced per unit mass of gas compressed, and (2) the specific work requirement w_Z, work per unit mass of liquid produced. We will now derive expressions for Z and w_Z for the simple Linde cycle.

Taking the combination of the separator, throttle, and heat exchanger as a thermodynamic system, we have for steady-flow conditions:

$$\dot{m}_2 h_2 = \dot{m}_5 h_5 + \dot{m}_7 h_7$$

$$\dot{m}_2 = \dot{m}_5 + \dot{m}_7$$

Eliminating $\dot{m}_7$, we have

$$Z_L = \frac{\dot{m}_5}{\dot{m}_2} = \frac{h_7 - h_2}{h_7 - h_5} \tag{6.5}$$

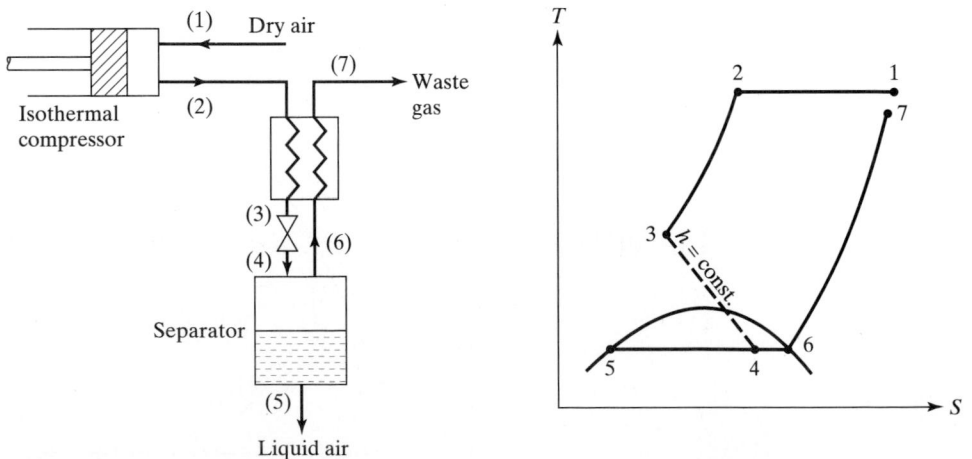

Figure 6.3 Simple Linde cycle for liquefying air.

Assuming isothermal compression, we may write the following expression for each unit mass of air passing through the compressor:

$$_1w_2 = T_1(s_1 - s_2) - (h_1 - h_2)$$

The specific-work requirement $w_{Z,L}$ becomes

$$w_{Z,L} = \frac{\dot{m}_2(_1w_2)}{\dot{m}_5} = \frac{_1w_2}{Z_L} = [T_1(s_1 - s_2) - (h_1 - h_2)]\frac{h_7 - h_5}{h_7 - h_2} \qquad (6.6)$$

It is useful to examine Eq. (6.5) closely. This equation shows that the yield for the simple Linde cycle is dependent upon states 2, 5, and 7 only. Since the denominator $(h_7 - h_5)$ is very large compared to the numerator $(h_7 - h_2)$, we see that the yield Z_L will be small. Furthermore, the numerator $(h_7 - h_2)$ is decisive, since a small change in either h_7 or h_2 may greatly change the yield Z_L. It is obvious that h_7 should be as large as possible and h_2 as small as possible. To achieve a high value of h_7, extremely effective heat exchange is necessary. In order that h_2 may be relatively low, the gas discharged by the compressor must be cooled as much as practicable.

If we examine Fig. C-5E or Fig. C-5SI, we see that for isothermal compression h_2 will decrease with increase of P_2 until the inversion pressure is reached. Beyond this pressure h_2 will increase. Thus, there is a limiting pressure beyond which the yield Z_L decreases.

EXAMPLE 6.2

Determine the temperature, pressure, specific enthalpy, and specific entropy for all state points and compute the yield, Z_L, and the specific-work requirement, $w_{Z,L}$, for the simple Linde cycle shown in Fig. 6.3, operating as follows:

$$t_1 = 70\ °F, \qquad P_1 = 14.7\ \text{psia}, \qquad P_2 = 2000\ \text{psia},$$
$$t_3 = -160\ °F \qquad t_7 = 60\ °F$$

Solution: The state-point property values are determined from Fig. C-5E and are given in Table 6.2.

$$Z_L = \frac{h_7 - h_2}{h_7 - h_5} = \frac{225 - 216}{225 - 47} = 0.051\ \text{lbm liquid/lbm gas}$$

$$w_{Z,L} = \frac{T_1(s_1 - s_2) - (h_1 - h_2)}{Z_L} = \frac{530(0.922 - 0.563) - (227 - 216)}{0.051}$$

$$w_{Z,L} = 3515\ \text{BTU/lbm liquid}$$

TABLE 6.2 Thermodynamic Property Value Data for Ex. 6.2

State	Temperature, °F	Pressure, psia	Enthalpy, BTU/lbm	Entropy, BTU/lbm · °R
1	70	14.7	227	0.922
2	70	2000	216	0.563
3	−160	2000	130	0.342
4	−314	14.7	130	0.582
5	−318	14.7	47	0.0
6	−313	14.7	134	0.610
7	60	14.7	225	0.927

The simple Linde cycle is an inefficient method for liquefying air for essentially the same basic reasons that the single-stage vapor-compression cycle is inefficient at low evaporating temperatures. Figure 6.4 shows a modified Linde cycle which requires a smaller specific-work requirement than the simple cycle. Air is liquefied at an intermediate pressure, and the flash vapor formed in the first throttling process (5 to 6) is compressed in the high-pressure compressor only. Such an arrangement considerably reduces the throttling loss of the simple Linde cycle.

Both the simple Linde cycle and the dual-pressure Linde cycle may be improved upon if the compressed air is precooled by a mechanical vapor-compression system. Figure 6.5 shows a simple Linde system equipped with a vapor-compression system evaporator for precooling. Since the vapor-compression cycle using conventional refrigerants such as ammonia or Refrigerant 22 is thermodynamically superior to the Linde system itself, the specific-work requirement may be considerably reduced. The best performance is obtained when the compressed air is precooled to a relatively low temperature. Thus a multistage or cascade vapor-compression system may be advantageous.

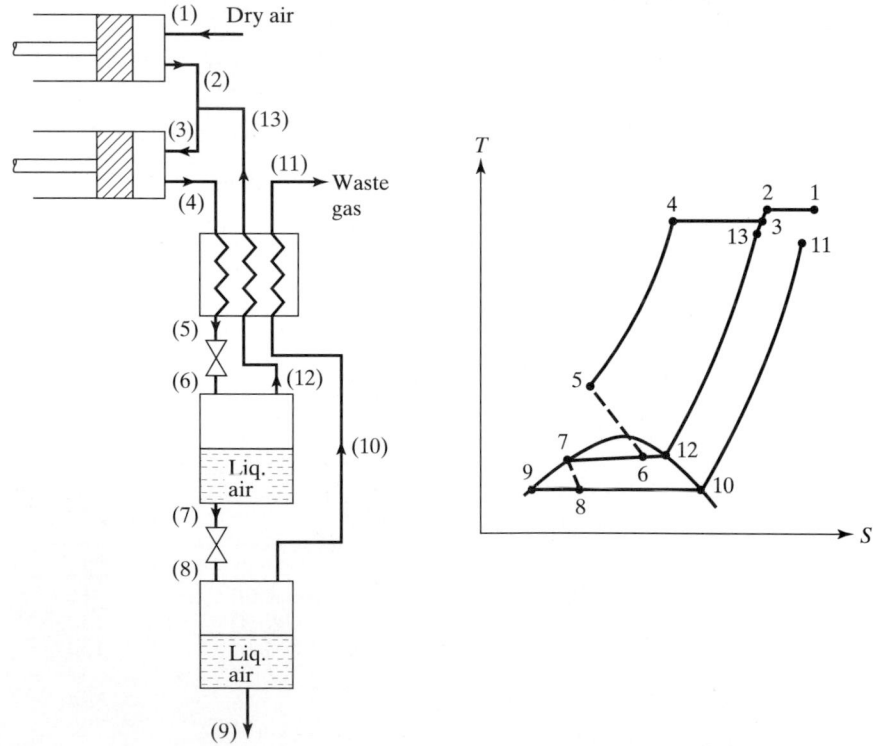

Figure 6.4 Dual-pressure Linde air-liquefaction system.

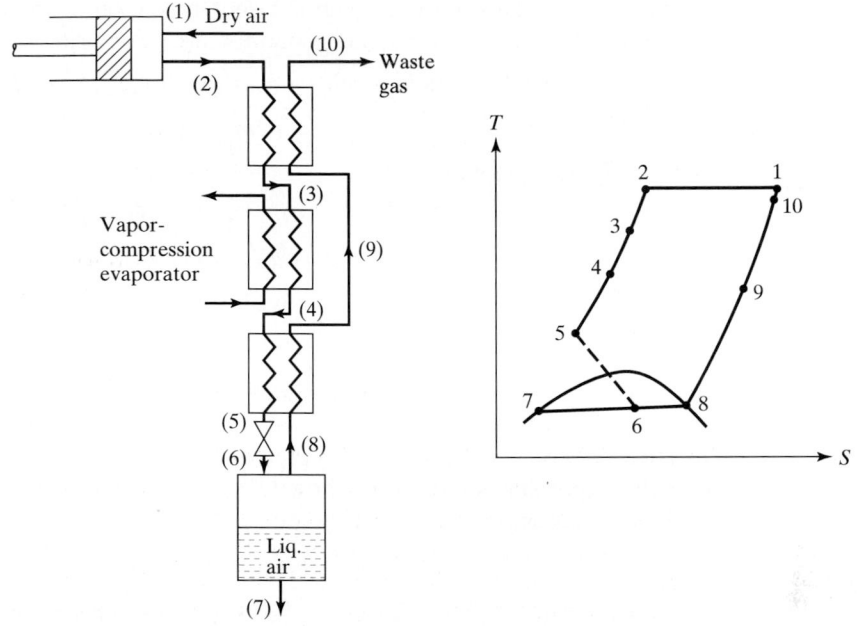

Figure 6.5 Simple Linde air-liquefaction system with precooling by mechanical-compression refrigeration system.

6.5 CLAUDE AIR-LIQUEFACTION CYCLE

The elementary Claude air-liquefaction system differs from the simple Linde cycle by the addition of an expansion engine and a second heat exchanger. Figure 6.6 shows the Claude cycle. Because of friction and heat-conduction losses, the expansion process (3 to

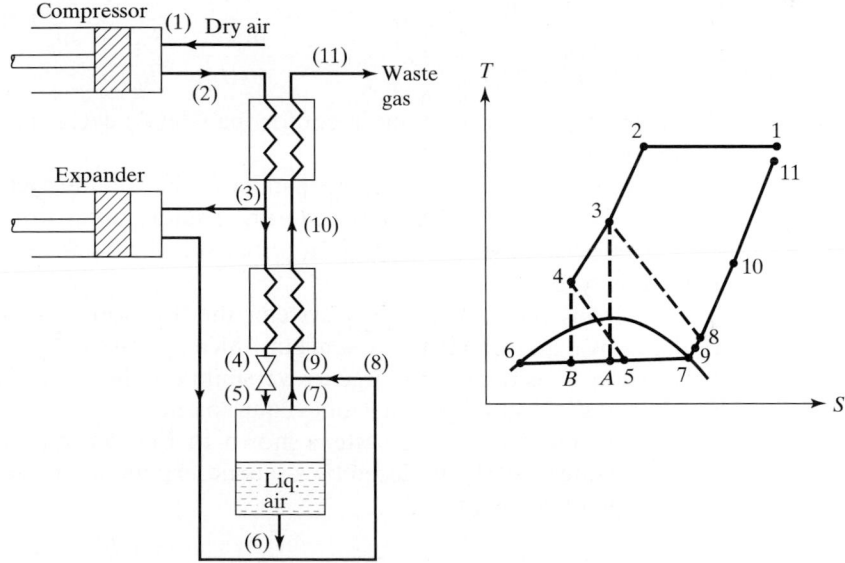

Figure 6.6 Claude air-liquefaction system.

8) deviates greatly from the theoretical isentropic case (3 to A). For the combination of two heat exchangers, throttle, and separator, we may write

$$\dot{m}_2 h_2 + \dot{m}_8 h_8 = \dot{m}_8 h_3 + \dot{m}_6 h_6 + \dot{m}_{11} h_{11}$$

$$\dot{m}_2 = \dot{m}_6 + \dot{m}_{11}$$

Eliminating $\dot{m}_{11}$, we have

$$Z_C = \frac{\dot{m}_6}{\dot{m}_2} = \frac{h_{11} - h_2}{h_{11} - h_6} + \frac{\dot{m}_8(h_3 - h_8)}{\dot{m}_2(h_{11} - h_6)} \tag{6.7}$$

The quantity $(h_{11} - h_2)/(h_{11} - h_6)$ represents the yield of the simple Linde cycle. Equation (6.7) may be written as

$$Z_C = Z_L + \frac{\dot{m}_8(h_3 - h_8)}{\dot{m}_2(h_{11} - h_6)} \tag{6.8}$$

Thus the yield of the Claude cycle is greater than that of the simple Linde cycle. Note that when the expander behaves as a throttle, $h_3 = h_8$ and $Z_C = Z_L$

Assuming isothermal compression, we have

$$_1 w_2 = T_1(s_1 - s_2) - (h_1 - h_2)$$

The specific-work requirement $w_{Z,C}$, work per unit mass of liquid air produced, is

$$w_{Z,C} = \frac{\dot{m}_2(_1 w_2) - \dot{m}_8(_3 w_8)}{\dot{m}_6} = \frac{_1 w_2}{Z_C} - \frac{\dot{m}_8}{\dot{m}_6}(_3 w_8)$$

Thus

$$w_{Z,C} = \frac{T_1(s_1 - s_2) - (h_1 - h_2)}{Z_C} - \frac{\dot{m}_8}{\dot{m}_6}(_3 w_8) \tag{6.9}$$

The quantity $[T_1(s_1 - s_2) - (h_1 - h_2)]$ is also equal to the work $w_{Z,L}$ of the simple Linde cycle. Thus,

$$w_{Z,C} = \frac{w_{Z,L}}{Z_L + \dfrac{\dot{m}_8(h_3 - h_8)}{\dot{m}_2(h_{11} - h_6)}} - \frac{\dot{m}_8}{\dot{m}_6}(_3 w_8) \tag{6.10}$$

Thus the specific-work requirement for the Claude cycle is less than that for the simple Linde cycle.

The main advantage of the expander in the Claude cycle is for cooling of the compressed air rather than for work recovery. The expander work output is often wasted.

Table 6.3 shows a comparison of the specific-work requirement in various air-liquefaction cycles.

In many modern cryogenic systems the throttling processes have been completely replaced by expansion engines, some of which operate in the two-phase liquid-vapor mixture region. This removes a large irreversibility in the system and therefore increases the yield and reduces the specific-work requirement.

Consider the Claude system shown in Fig. 6.6 but with the throttling process between states 4 and 5 replaced by a second expander. An analysis of this system shows that the yield becomes

$$Z_C^* = Z_C + \left(1 - \frac{\dot{m}_8}{\dot{m}_2}\right)\frac{h_4 - h_5}{h_{11} - h_6} \tag{6.11}$$

TABLE 6.3 Specific Work Requirement for Various Air-Liquefaction Cycles

System	Work Required, Btu/lbm Liquid Air (kJ/kg Liquid Air)	
Reversible process (calculated)	324	(754)
Simple Linde (observed)	4440	(10,300)
Simple Linde with precooling to −49 °F (−45 °C) (observed)	2390	(5560)
Dual-pressure Linde (observed)	2370	(5510)
Dual-pressure Linde with precooling to −49 °F (observed)	1370	(3190)
Claude (observed)	1370	(3190)

SOURCE: R. B. Scott, *Cryogenic Engineering* (Princeton, NJ: D. Van Nostrand Company, Inc., 1959), 17.

The second term on the right-hand side of this equation is always positive, so the yield of this modified cycle is larger than the Claude cycle. Also, if the expansion process between states 4 and 5 becomes a throttling process, $h_4 = h_5$, and the yield is again given by Eq. (6.8). The specific-work requirement can be written

$$w_{Z,C}^* = \frac{w_{Z,L}}{Z_C^*} - \frac{\dot{m}_8}{\dot{m}_6}(_3 w_8) - \frac{\dot{m}_4}{\dot{m}_6}(_4 w_5) \qquad (6.12)$$

The work of each adiabatic expander can be described in terms of its isentropic efficiency. For example, the work for the first expander can be written as $_3 w_8 = (h_3 - h_8) = \eta_{I,38}(h_3 - h_A)$. Thus the specific-work requirement when two expanders are used can be written as

$$w_{Z,C}^* = \frac{_1 w_2}{Z_C^*} - \frac{\dot{m}_8}{\dot{m}_6}\eta_{I,38}(h_3 - h_A) - \frac{\dot{m}_4}{\dot{m}_6}\eta_{I,45}(h_4 - h_B) \qquad (6.13)$$

EXAMPLE 6.3

Determine the yield and specific-work requirement for the Claude system shown in Fig. 6.6 but with the throttle replaced by a second expander. Operating parameters are: $T_1 = 300$ K, $P_1 = 1$ atm, $P_2 = 50$ atm, $T_3 = 215$ K, $T_4 = 140$ K, $T_{11} = 295$ K, $\eta_{I,38} = 0.8$, $\eta_{I,45} = 0.85$. One-half the air that is compressed is sent through the expander from 3 to 8.

Solution: The necessary specific-enthalpy values can be obtained from Fig. C-5SI. $h_1 = 12,400$ J/mole, $s_1 = 112.5$ J/mole K, $h_2 = 12,000$ J/mole, $s_2 = 79.2$ J/mole K, $h_3 = 9270$ J/mole.

$$h_A = 5400 \text{ J/mole}, \qquad h_8 = h_3 - 0.8(h_3 - h_A)$$
$$= 9270 - 0.8(9270 - 5400)$$
$$= 6170 \text{ J/mole}$$

$$h_4 = 4700 \text{ J/mole}, \qquad h_B = 3100 \text{ J/mole}$$
$$h_5 = h_4 - 0.85(h_4 - h_B)$$
$$= 4700 - 0.85(4700 - 3100)$$
$$= 3340 \text{ J/mole}$$

$$h_6 = 0, \qquad h_7 = 5900 \text{ J/mole}, \qquad h_{11} = 12,250 \text{ J/mole}$$

The yield is evaluated by using Eqs. (6.7), (6.8), and (6.11).

$$Z_L = \frac{h_{11} - h_2}{h_{11} - h_6} = \frac{12{,}250 - 12{,}100}{12{,}250 - 0} = 0.012$$

$$Z_C = Z_L + \frac{\dot{m}_8(h_3 - h_8)}{\dot{m}_2(h_{11} - h_6)} = 0.012 + 0.5\,\frac{9270 - 6170}{12{,}250 - 0} = 0.139$$

$$Z_C^* = Z_C + \left(1 - \frac{\dot{m}_8}{\dot{m}_2}\right)\frac{h_4 - h_5}{h_{11} - h_6} = 0.139 + 0.5\,\frac{4700 - 3340}{12{,}250 - 0} = 0.195$$

The units on these values for yield can be expressed either as a mole ratio or a mass ratio (kg liquid/kg gas compressed). Note the increase in yield as the Linde cycle is modified. The specific-work requirement is found using Eq. (6.12). For the separator alone we may write

$$\dot{m}_5 h_5 = \dot{m}_6 h_6 + \dot{m}_7 h_7$$

$$\dot{m}_5 = \dot{m}_6 + \dot{m}_7$$

Solving for the ratio $\dot{m}_5/\dot{m}_6 = \dot{m}_4/\dot{m}_6$,

$$\frac{\dot{m}_5}{\dot{m}_6} = \frac{h_7 - h_6}{h_7 - h_5} = \frac{5900 - 0}{5900 - 3340} = 2.30$$

$$\frac{\dot{m}_8}{\dot{m}_6} = \frac{\dot{m}_5}{\dot{m}_6} = 2.30$$

The specific work can be obtained from Eq. (6.12), assuming the work from both expanders is recovered.

$$W_{Z,C}^* = \frac{300(112.5 - 79.2) - (12{,}400 - 12{,}100)}{0.195}$$

$$-2.30(9270 - 6170) - 2.30(4700 - 3340)$$

$$W_{Z,C}^* = 49{,}700 - 7100 - 3100 = 39{,}500 \text{ J/mole liquid}$$

or

$$39{,}500 \text{ J/mole} \times \frac{\text{mole}}{0.02896 \text{ kg}} = 1360 \text{ kJ/kg liquid}$$

Note that the specific-work requirement is 49,700 J/mole liquid (1720 kJ/kg liquid) if the work of the expanders is wasted.

6.6 SEPARATION OF AIR

Air liquefaction is of great importance in the production of the pure components of dry air—oxygen, nitrogen, argon, and the rare gases neon, krypton, and xenon. The recovery of oxygen, nitrogen, and other gases from the atmosphere requires the separation of liquefied air. Such applications date from the year 1902, when Linde demonstrated that by rectification of liquid air the separate gases, oxygen and nitrogen, could be obtained.

In this section we will study in an elementary manner how dry air may be separated into its constituents. Bosnjakovic [3] has covered this topic in detail. Ruhemann [4] has also treated the subject extensively, including separation of the rare gases. We will emphasize methods for the production of oxygen and nirtogen of commercial purity.

Table 6.2 shows approximately the composition of dry air. The constituents neon, krypton, and xenon have been ignored. These gases appear in dry air only minutely, as shown by the following approximate volumetric percentages: neon, 0.0018; krypton, 0.0001; and xenon, 0.000008. For practical problems involving the commercial production of oxygen and nitrogen, it is customary to consider dry air as a binary mixture of 21 percent oxygen and 79 percent nitrogen on a volume basis.

Figure C-6E is an enthalpy-concentration diagram for nitrogen-oxygen mixtures. Coordinates are h in Btu per lb mole and mole-fraction x in moles of nitrogen per mole of mixture. Figure C-6E may be used in exactly the same manner as described in Secs. 5.2–5.4.

When we consider air as a binary mixture, rectification of liquid air follows the basic principles discussed in Sec. 5.4. Although the principles are the same, there are some important differences between rectifying liquid air and rectifying a binary mixture such as ammonia and water. The system shown in Fig. 5.14 employs external energy transfers. The heat $\dot{Q}_G$ added to the generator comes from an external source such as condensing steam. The heat $\dot{Q}_D$ rejected in the dephlegmator is absorbed by an external medium such as cooling water. These procedures are suitable to rectification of aqua-ammonia but are not permissible in rectification of liquid air. Since very low temperatures are involved, there is obviously no cheap medium available to which to reject heat in the dephlegmator. Moreover, any introduction of external heat to the generator would necessitate a vastly increased expenditure of work. In the liquefaction and rectification of air every effort should be made to minimize the addition of any external heat. Thus the rectification column must operate as nearly adiabatically as possible, and all energy transfers must be of an internal nature.

Before considering the double column used for separation of oxygen and nitrogen, we will study the more elementary single Linde column. Figure 6.7 shows an arrangement in which a simple exhausting column is used. Figure 6.8 shows a schematic h-x diagram for the exhausting column. Dry compressed air at a pressure of four to six atmospheres is cooled in the heat exchanger and is used as a heat source in the generator. Here the air

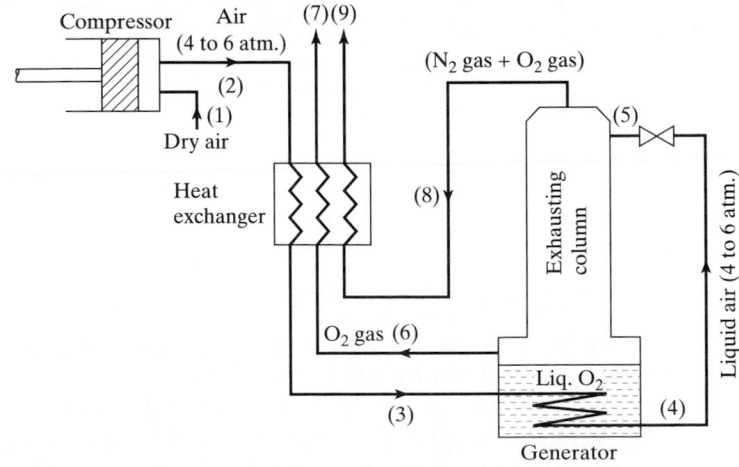

Figure 6.7 Linde single-column apparatus for separation of oxygen.

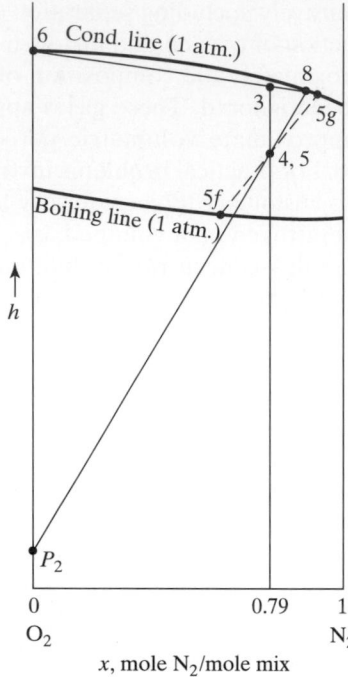

Figure 6.8 Schematic h-x diagram for exhausting column of Fig. 6.7.

is liquefied (state 4) by transfer of heat to liquid oxygen, boiling at atmospheric pressure. The liquid air is then throttled to atmospheric pressure and is admitted to the top of the exhausting column as a mechanical mixture of liquid and vapor (states $5f$ and $5g$).

The exhausting column operates similarly to that described in Sec. 5.4 for the system of Fig. 5.13. One difference is that in Figs. 6.7 and 6.8 the weak component (oxygen) is removed as a gas instead of as a liquid. All operating lines for the exhausting column in Fig. 6.8 are contained between the lines $\overline{P_2 6}$ and $\overline{P_2 58}$.

The striking disadvantage of the single Linde column is the loss of oxygen with the nitrogen gas. Figure 6.8 shows that each mole of gas withdrawn from the top of the column (state 8) will contain $(1 - x_8)$ moles of oxygen. In the limiting case the gas would contain $(1 - x_{5g})$ moles of oxygen, which would represent an oxygen content of about 9 percent. Since air contains only 21 percent oxygen by volume, the system of Fig. 6.7 could recover only somewhat less than two-thirds of the available oxygen.

In order to obtain a more complete yield of oxygen from air, a double rectifying column is needed. This method was introduced by Linde in 1910. Figure 6.9 schematically shows the Linde double column. Such an apparatus may produce both oxygen and nitrogen of arbitrary purity. Compressed air, precooled in the heat exchanger, is liquefied in the coil of generator A by heat exchange with the boiling liquid in the bottom of the lower column. The liquid air (state 2) is then throttled to the pressure P_L of the lower column and admitted to the column at an intermediate location. In a manner similar to that described in Sec. 5.5 for the system of Fig. 5.14, the vapor rising in the lower column is enriched with nitrogen, while the descending liquid acquires a greater oxygen content. Almost pure nitrogen vapor at the top of the lower column is condensed in generator B (dephlegmator for lower column). Part of this liquid falls back into the lower column, while the remainder is trapped and used as reflux for the upper column.

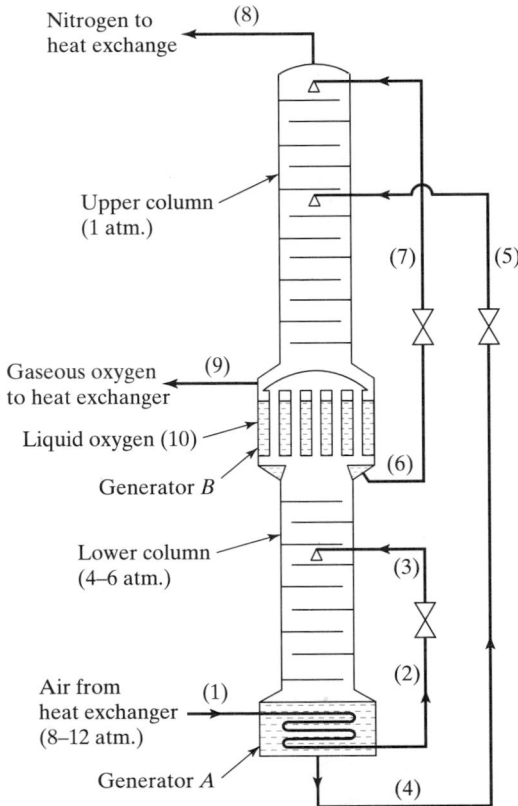

Figure 6.9 Linde double column.

Liquid (state 4) is removed from the bottom of the lower column at an intermediate concentration. This liquid is throttled to the pressure P_U of the upper column and enters the upper column as a mechanical mixture of liquid and vapor. Almost pure nitrogen liquid (state 6) is likewise throttled and admitted to the top of the upper column. Almost pure liquid oxygen (state 10) boils in the bottom of the upper column (generator B). The oxygen vapor (state 9) is removed at a location above the boiling liquid. Nitrogen vapor (state 8) is removed from the top of the upper column.

Operation of the Linde double column is better understood by consideration of the processes on the h-x diagram. Construction of Fig. 6.10 will now be explained by the following itemized statements. Many of these statements are evident from the discussion in Sec. 5.5.

1. The desired product states (8 and 9) may be located. We assume each may be of arbitrary purity.
2. The required state (1) of the inlet air may be located on the straight line $\overline{89}$ at $x = 0.79$. This statement is valid if we assume no heat leakage into the column from the surroundings. The statement is proved by the solution of the following equations:

$$\dot{m}_1 = \dot{m}_8 + \dot{m}_9$$

$$\dot{m}_1 x_1 = \dot{m}_8 x_8 + \dot{m}_9 x_9$$

$$\dot{m}_1 h_1 = \dot{m}_8 h_8 + \dot{m}_9 h_9$$

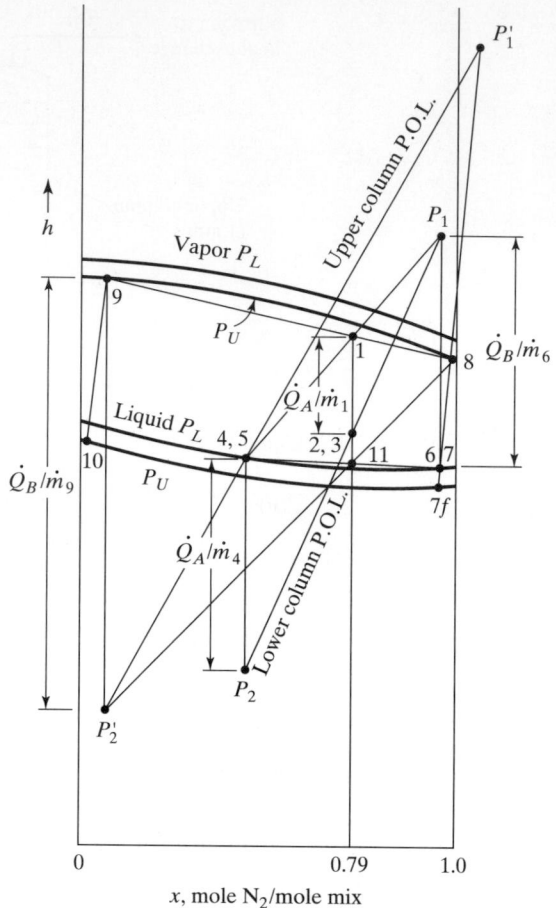

Figure 6.10 Schematic *h-x* diagram for Linde double column.

3. State-point 2 may be located on the boiling line (saturated-liquid line) for the inlet pressure at $x_2 = x_1 = 0.79$.

4. State-point 3 = state-point 2 because of the throttling process.

5. The principal operating line (P.O.L.) for the lower column may be drawn through point 3. Exact direction of the line is somewhat arbitrary. It is necessary that the P.O.L. be somewhat steeper than mixture-region isotherm t_3.

6. State-points 6, 7, and $7f$ may be located in the following manner: The operating line $\overline{7f8}$ for the top of the upper column must be somewhat steeper than the isotherm t_7 in the mixture region, since t_7 must be less than t_8. This requirement allows the saturated-liquid point $7f$, the mixture state 7, and the saturated-liquid state 6 to be located. States 6 and 7 are coincident because of the throttling process.

7. The pole P_1 may be located at the intersection of vertical line $x = x_6$ with the lower column P.O.L.

8. State-point 4 and therefore state-point 5 may be located at the intersection of the extension of line $\overline{P_1 1}$ with the P_L boiling line. This statement can be proved as follows. For the lower column we may write

$$\dot{m}_1 = \dot{m}_4 + \dot{m}_6$$

$$\dot{m}_1 x_1 = \dot{m}_4 x_4 + \dot{m}_6 x_6$$

$$\dot{m}_1 h_1 = \dot{m}_4 h_4 + \dot{m}_6 h_6 + \dot{Q}_B$$

These equations may be combined to give the relation

$$\frac{h_1 - h_4}{x_1 - x_4} = \frac{\dfrac{\dot{Q}_B}{\dot{m}_6} - (h_4 - h_6)}{x_6 - x_4}$$

which proves the original statement.

9. The pole P_2 may be located at the intersection of the vertical line $x = x_4$ with the lower column P.O.L.

10. The pole P_2' may be located at the intersection of the extension of the line $\overline{8\ 11}$ with the vertical line $x = x_9$.

11. Point 11 is a fictitious state equivalent to a state produced by a hypothetical adiabatic mixing of streams 4 and 6. Thus

$$\dot{m}_{11} = \dot{m}_4 + \dot{m}_6 = \dot{m}_1$$

$$\dot{m}_1 x_{11} = \dot{m}_4 x_4 + \dot{m}_6 x_6$$

$$\dot{m}_1 h_{11} = \dot{m}_4 h_4 + \dot{m}_6 h_6$$

For the upper column, we may write

$$\dot{m}_6 + \dot{m}_4 = \dot{m}_8 + \dot{m}_9$$

$$\dot{m}_6 x_6 + \dot{m}_4 x_4 = \dot{m}_8 x_8 + \dot{m}_9 x_9$$

$$\dot{m}_6 h_6 + \dot{m}_4 h_4 + \dot{Q}_B = \dot{m}_8 h_8 + \dot{m}_9 h_9$$

Combining the two sets of equations gives the relation

$$\frac{h_8 - h_{11}}{x_8 - x_{11}} = \frac{\dfrac{\dot{Q}_B}{\dot{m}_9} - (h_9 - h_{11})}{x_{11} - x_9}$$

which proves the original statement.

12. The upper column P.O.L. may be drawn through points P_2' and 5.

13. The pole P_1' may be located at the intersection of the upper column P.O.L. with the extension of the operating line $\overline{7f8}$.

14. State-point 10 for the liquid oxygen may be located by the assumption that $t_{10} = t_9$.

In an actual problem, construction of the h-x diagram would have to be checked to prove that all operating lines for each column were steeper than the corresponding mixture-region isotherms, or, stated in a different way, that no negative temperature differences existed. If this requirement were not satisfied, the compressed-air pressure and possibly the lower column pressure would need to be adjusted.

Figure 6.11 shows a simplified contemporary type of double column for production of gaseous oxygen. Such a column has both theoretical and practical advantages over the Linde double column of Fig. 6.9. Partially liquefied compressed air is admitted to the

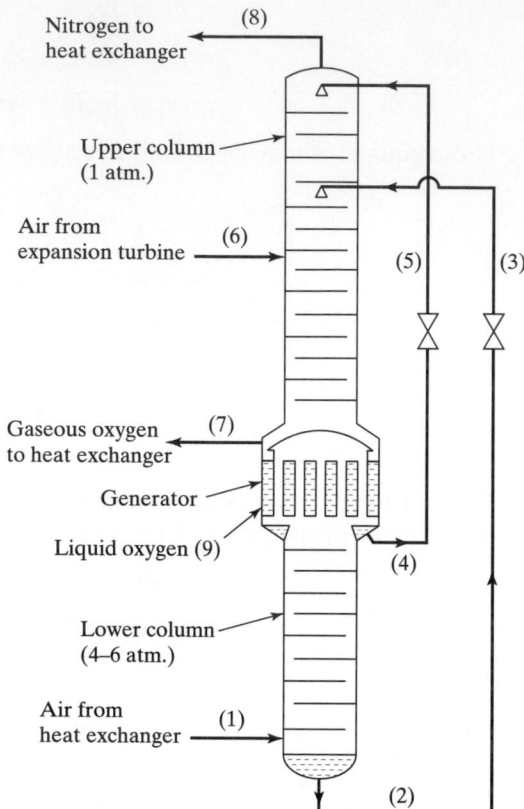

Figure 6.11 Contemporary double column.

lower column at a pressure of four to six atmospheres. In the lower column the rising vapor is enriched with nitrogen. High-purity nitrogen vapor is condensed to liquid in the generator by transfer of heat to liquid oxygen boiling at a pressure of approximately one atmosphere. Part of this liquid nitrogen is used as reflux for the upper column. Liquid (state 2) is removed from the bottom of the lower column at an intermediate concentration, throttled to the pressure of the upper column, and admitted to the upper column as a mechanical mixture of liquid and vapor. The upper column is also supplied with air vapor (state 6) from an expansion turbine. High-purity oxygen vapor (state 7) is removed at a location above the boiling liquid in the generator. High-purity nitrogen vapor (state 8) is removed from the top of the upper column.

Figure 6.12 shows a schematic h-x diagram for the column of Fig. 6.11. States 10, 11, and 12 are all fictitious and are introduced for convenience in construction of the diagram. State 10 is the adiabatic equivalent (resulting from a hypothetical adiabatic mixing) of streams 2 and 4. State 11 is the adiabatic equivalent of streams 3, 5, and 6, while state 12 is the adiabatic equivalent of streams 1 and 6, or of 7 and 8. Construction of Fig. 6.12 should be obvious from the detailed discussion given earlier for Fig. 6.10. In Fig. 6.12 we may also show that several mass flow ratios are given by the graphical construction. These include the relations

$$\frac{\dot{m}_2}{\dot{m}_4} = \frac{\overline{4\ 10}}{\overline{10\ 2}} = \frac{\overline{P_1\ 1}}{\overline{1\ 2}}$$

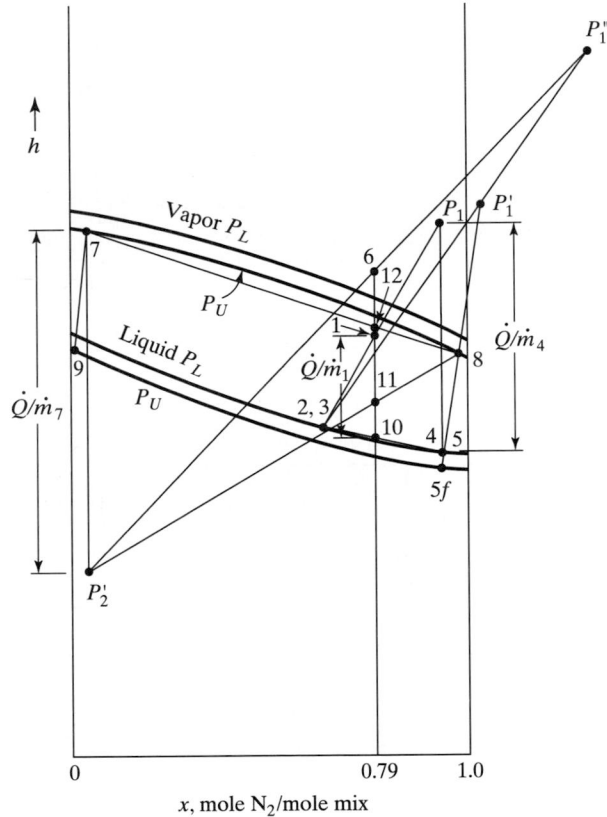

Figure 6.12 Schematic h-x diagram for column of Fig. 6.11.

$$\frac{\dot{m}_6}{\dot{m}_1} = \frac{\overline{12\ 1}}{\overline{6\ 12}} = \frac{\overline{11\ 10}}{\overline{6\ 11}}$$

$$\frac{\dot{m}_7}{\dot{m}_8} = \frac{\overline{8\ 12}}{\overline{12\ 7}} = \frac{\overline{8\ 11}}{\overline{11\ P_2'}}$$

The column of Fig. 6.11 is superior to the Linde column of Fig. 6.9. Only two pressure levels are needed instead of the three involved with the Linde double column. Thus a lower supply air pressure and therefore a lower compressor power input is needed for a system with the column of Fig. 6.11.

An analysis utilizing the second law of thermodynamics is of benefit to a study of air-separation systems. Although we will not include a second-law analysis here [5], we may observe that such a study would show that fewer irreversibilities result in a system employing the column of Fig. 6.11 than in a system using the Linde double column. This circumstance also leads to a lower power requirement for a system with the column of Fig. 6.11.

The column of Fig. 6.11 has further advantages compared to the Linde double column. Generator A is eliminated. A lesser heat-transfer rate is required in the generator of Fig. 6.11 than in generator B of Fig. 6.9.

Figures 6.13 and 6.14 show schematic diagrams of contemporary-type plants for producing oxygen. Moisture-removal and air-purification equipment and some small heat

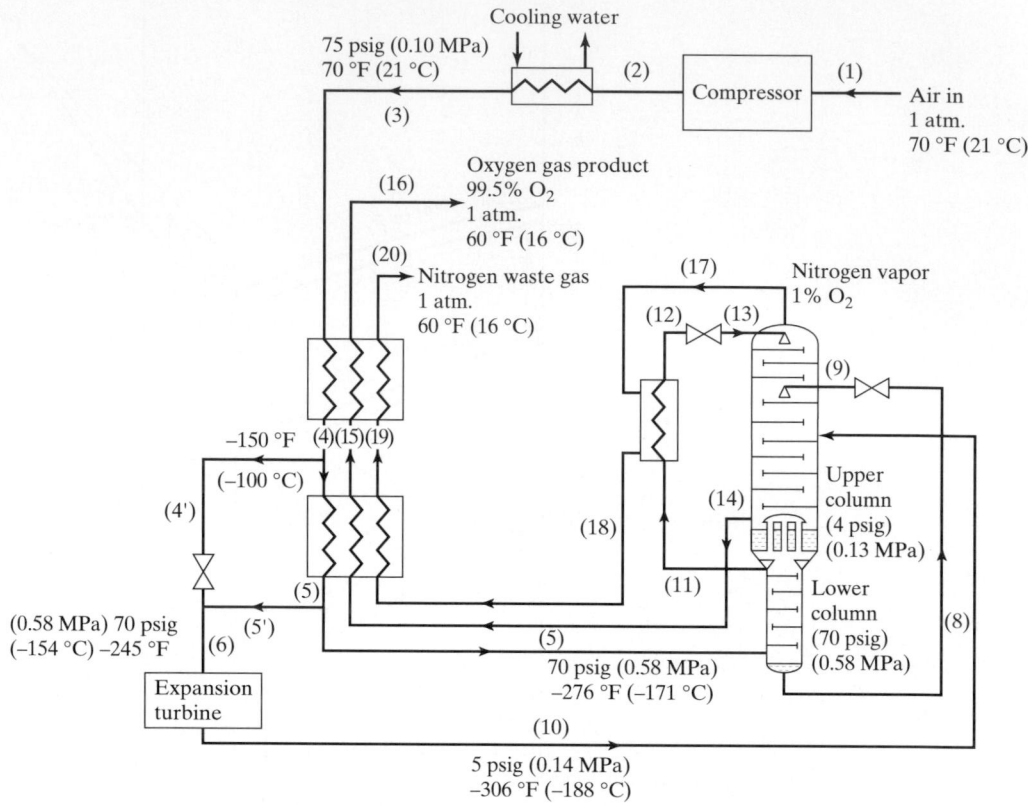

Figure 6.13 Schematic gaseous-oxygen plant.

exchangers are not shown for simplicity. The diagrams and associated operating conditions have been adapted from a paper by Zenner [6]. Regenerative heat exchangers with periodic flow reversal are often used rather than the two large continuous-flow heat exchangers shown here. The thermodynamic analysis of these systems must include a transient analysis of the regenerative heat exchangers and therefore becomes a more tedious task than the steady-flow analysis used here.

Figure 6.13 shows a schematic gaseous-oxygen plant. The rectifying column is identical to that of Fig. 6.11. Figure 6.14 shows a schematic liquid-oxygen plant. A very high compressed-air pressure is required. Zenner states that maximum yield of liquid oxygen is obtained when about 50 percent of the high-pressure air is expanded in the engine. Zenner also states that power costs for the gaseous-oxygen product are about one-third those for liquid-oxygen product.

We will only briefly consider how the rare constituents of air may be separated. Figure 6.15 shows the column of Fig. 6.11 equipped with devices which allow separation of argon, krypton and xenon, and neon and helium. The boiling point of argon is but slightly lower than that of oxygen. Argon is most concentrated at about the middle of the upper column. Vapor withdrawn at this point may contain from 10 to 15 percent of argon, while the remainder is mostly oxygen. The mixture is rectified in the argon column, where most of the oxygen is condensed and withdrawn as liquid from the bottom of the column while crude argon vapor is removed at the top. The argon vapor may be further purified by selective adsorption.

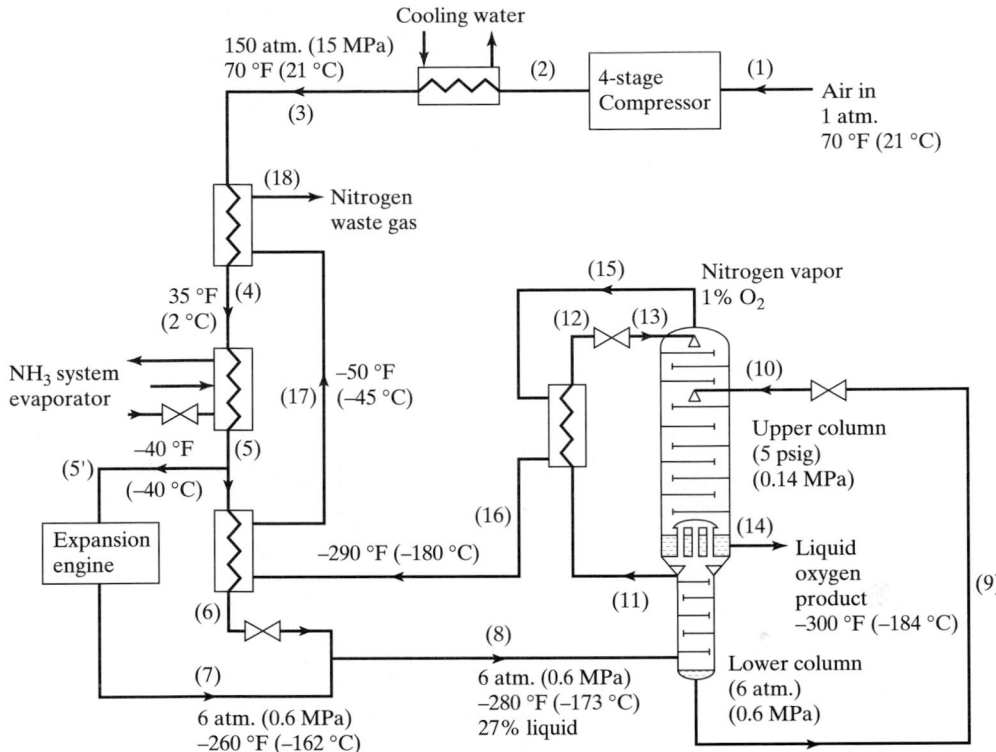

Figure 6.14 Schematic liquid-oxygen plant.

Both neon and helium appear in dry air in very small amounts. Both gases are much more volatile than nitrogen. In air separation these gases form as noncondensibles in the dome of the dephlegmator of the lower column. As shown in Fig. 6.15, gas (noncondensibles plus nitrogen, primarily) may be withdrawn from the dome of the lower column and passed through a rectifier which is refrigerated by liquid nitrogen. Most of the nitrogen vapor present with the helium and neon may be condensed and separated. Neon may be separated from the mixture of neon and helium by a selective-adsorption arrangement.

Both krypton and xenon have higher boiling temperatures than oxygen. Both remain in the liquid oxygen in the bottom of the upper column. As shown in Fig. 6.15, liquid may be withdrawn from the bottom of the upper column and admitted to the top of an auxiliary exhausting column. Here the oxygen portion may be distilled from the mixture. The crude krypton and xenon may be further purified and separated by a selective-adsorption system.

6.7 LIQUEFACTION OF HYDROGEN AND HELIUM

Hydrogen and helium are the most difficult gases to liquefy because of their extremely low liquefaction temperatures. All other substances freeze at temperatures higher than the normal boiling points (1 atm) of either element.

Figure 6.16 shows a schematic arrangement of equipment for liquefying hydrogen. Pure hydrogen gas at about 100 atmospheres pressure is precooled in a divided heat-exchanger arrangement (heat exchangers *A* and *B*) to about −300 °F (−185 °C). In heat

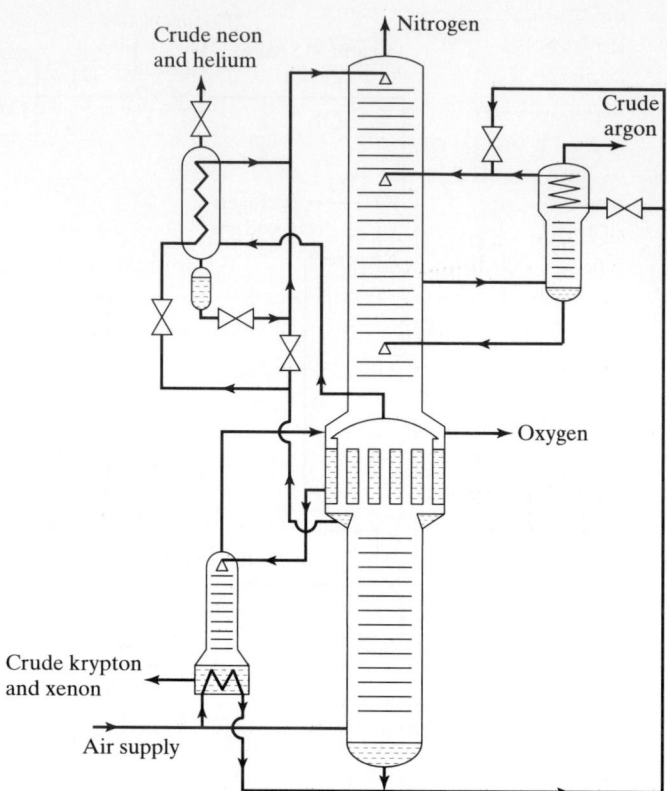

Figure 6.15 Schematic apparatus for separation of air.

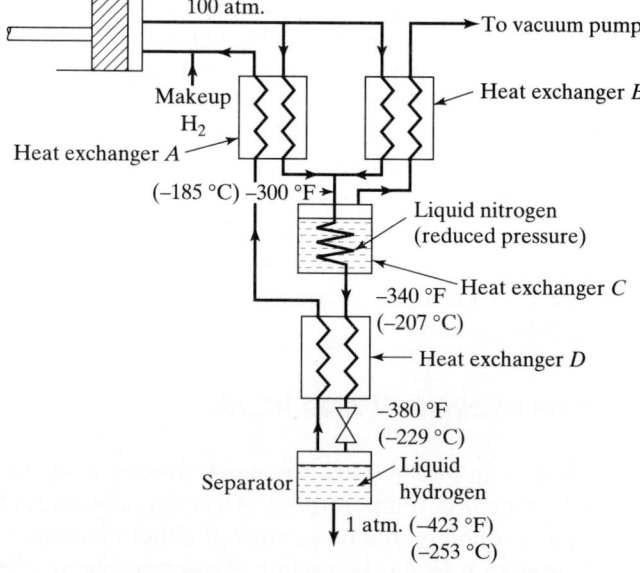

Figure 6.16 Schematic hydrogen liquefier.

exchanger C, the hydrogen vapor is further cooled to about $-340\,°F$ $(-200\,°C)$ by nitrogen boiling under reduced pressure. In heat exchanger D, the hydrogen vapor is still further cooled to about $-380\,°F$ $(-230\,°C)$ by the low-pressure hydrogen vapor returning from the separator. By throttling to atmospheric pressure, liquid hydrogen may then be produced.

Helium is the most difficult of all gases to liquefy. At 1 atmosphere pressure it boils at approximately $-452\,°F$ $(-269\,°C)$. Its maximum inversion temperature is approximately $-390\,°F$ $(-230\,°C)$. Helium was first liquefied by H. K. Onnes of the University of Leiden in 1908. Helium may be liquefied by an arrangement similar to that in Fig. 6.16, where both liquid nitrogen and liquid hydrogen are used for precooling. Disadvantages of this method include the high cost and the hazardous nature of liquid hydrogen.

Helium may also be liquefied through use of the Claude principle, where expansion engines are used for producing refrigeration. Figure 6.17 shows a system developed by Collins [7]. It illustrates the principle of the Collins Cryostat. Helium at approximately 12 atm pressure is supplied to the liquefier by a four-stage compressor. Part of the helium is precooled by liquid nitrogen. Through use of the combination of heat exchangers and expansion engines, the high-pressure helium gas may be cooled to about $-430\,°F$ $(-257\,°C)$. Liquid helium is then produced by throttling of the vapor to atmospheric pressure. Additional information is given by Collins and Cannaday [8] and Codlin [9].

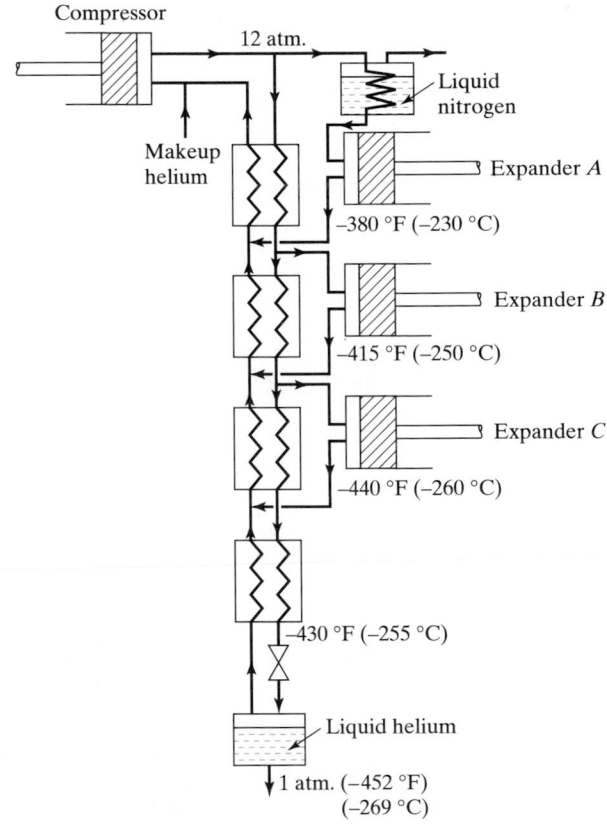

Figure 6.17 Schematic helium liquefier.

6.8 APPROACH TO ABSOLUTE ZERO BY THE ADIABATIC DEMAGNETIZATION OF A PARAMAGNETIC SALT

Through the lowering of pressure over liquid helium, it is possible to attain a temperature of about -458 °F (-272 °C). Lower temperatures may be achieved by adiabatic demagnetization of certain paramagnetic salts.

The atoms of a paramagnetic salt may be considered as tiny magnets. When the salt is not magnetized, the atoms are oriented in a random manner such that the magnetic forces are in balance. If such a substance is exposed to a strong magnetic field, the atoms attempt to align themselves in the direction of the field. The alignment of the atoms requires work. This work is converted into heat and causes a temperature rise, unless the heat is removed by some form of cooling. When the magnetic field is removed, the atoms readjust their positions to the original random arrangement. Such readjustment requires that the salt atoms perform work. In the absence of external heat exchange, the internal energy of the salt decreases. Consequently the salt must cool itself.

Figure 6.18 shows a schematic arrangement for adiabatic demagnetization of a paramagnetic salt. The inner chamber containing the salt specimen is initially filled with gaseous helium. This chamber is surrounded by a bath of liquid helium, which in turn is surrounded by a bath of liquid hydrogen. External insulation is not shown in Fig. 6.18.

The salt is first cooled to about -458 °F (-272 °C) by reducing the pressure over the liquid helium. Next, the salt is exposed to a strong magnetic field of about 25,000 gauss. Heat produced by magnetization of the salt is transferred to the liquid helium without causing an increase of the salt temperature. With the magnetic field still present, the inner chamber containing the salt is evacuated of gaseous helium. The salt is then almost completely thermally isolated. Upon release of the magnetic field, the salt temperature decreases in an almost perfectly isentropic way. Temperatures of the salt as low as approximately 0.001 K have been reported.

An interesting and important problem in adiabatic demagnetization is the determination of the very low temperatures produced. In the neighborhood of absolute zero, all

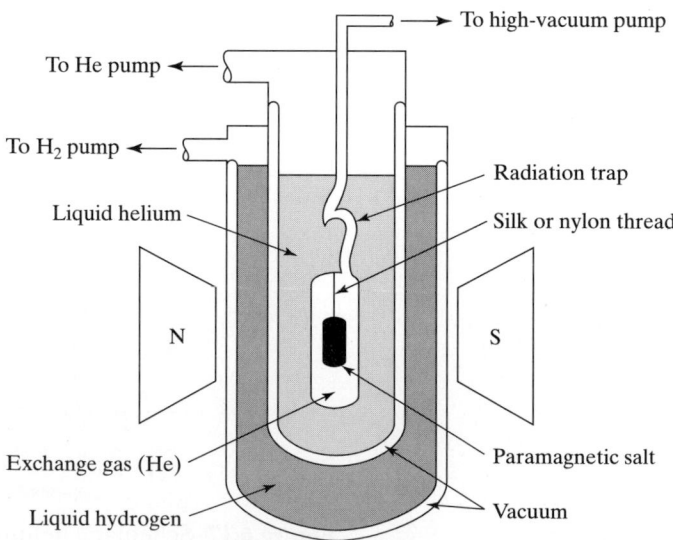

Figure 6.18 Schematic adiabatic demagnetization apparatus.

ordinary means of temperature measurement fail. Scott [10] has given a thorough review of temperature-measurement principles used in cryogenic work.

Close to absolute zero, temperature cannot be measured but must be calculated. The temperature may be calculated approximately by the Curie relation

$$x_m = \frac{C}{T}$$

where x_m is the magnetic susceptibility of the salt, T the absolute temperature, and C a constant. Through magnetic measurements the absolute temperature may be calculated.

The most accurate method of calculating the final temperature produced in a salt through adiabatic demagnetization directly utilizes the second law of thermodynamics. The demagnetization process is almost perfectly isentropic. By means of thermal measurements, the temperature may be calculated.

ENDNOTES

1. R. B. Scott, *Cryogenic Engineering* (Princeton, NJ: D. Van Nostrand Company, Inc., 1959).
2. G. Walker, *Cryocoolers*, Part 1: *Fundamentals*, Part 2: *Applications* (New York: Plenum Press, 1983).
3. Fran Bosnjakovic, *Technische Thermodynamik*, Zweiter Teil (Dresden und Leipzig: Theodor Steinkopff, 1937), pp. 142–151.
4. M. Ruhemann, *The Separation of Gases* (London: Oxford University Press, 1949).
5. H. Bliss and B. F. Dodge, "Oxygen Manufacture, Part I," *Chemical Engineering Progress,* 45: 1 (January 1949), 56–64.
6. G. H. Zenner, "Cryogenics and Mechanical Engineering," *Mechanical Engineering,* 83 (September 1961), 39–43.
7. S. C. Collins, "Helium Liquefier," *Science,* 116 (September 19, 1952), 289–294.
8. S. C. Collins and R. L. Cannaday, *Expansion Machines for Low Temperature Processes* (Oxford University Press, 1958).
9. E. M. Codlin, *Cryogenics and Refrigeration: A Bibliographical Guide—2* (London: MacDonald & Co., 1970).
10. Scott, *Cryogenic Engineering,* 109–141.

PROBLEMS

6.1 Determine the theoretical minimum work required in Btu/lbm to produce liquid air at 14.696 psia from dry air initially at 70 °F and 14.696 psia.

6.2 Determine the theoretical minimum work required in kJ/kg to produce liquid air at 1 atm pressure from dry air initially at 300 K and 1 atm pressure.

6.3 Dry air at 70 °F and 14.696 psia is to be liquefied by the simple Linde method. Assume the air is isothermally compressed at 70 °F to 2500 psia. Also assume that the waste gas leaves the system at 70 °F. Determine
  (a) the yield of liquid air in lbm per lbm of air compressed, and
  (b) the specific work requirement in Btu per lbm of liquid air.

6.4 Dry air at 300 K and 1 atm pressure is to be liquefied by the simple Linde method. Air is compressed isothermally at 300 K to 200 atm pressure. Assume that the waste gas leaves at 300 K. Determine
  (a) the yield of liquid air in kg per kg air compressed, and
  (b) the specific work requirement in kJ per kg of liquid air.

6.5 Consider the start-up of the Linde system shown in Fig. 6.3. Initially all equipment is at ambient temperature, 70 °F. In the first cycle air is compressed isothermally from 70 °F and 14.7 psia to 2500 psia. States 2 and 3 are equal, as no cooling occurs in the heat exchanger. The air is expanded into the separator and all of the gas is sent back through the heat exchanger. In the second and subsequent cycles assume that T_3 equals T_4 of the previous cycle. Determine

(a) the number of cycles required before any liquid air is produced, and

(b) the value of T_3 when the system is operated steadily.

6.6 Dry air at 300 K and 1 atm pressure is liquefied using the system shown in Fig. 6.3 with the throttling process replaced with an isentropic expander. Air is compressed isothermally at 300 K to 200 atm pressure and the waste gas leaves at 300 K. Determine

(a) the yield of liquid air in kg per kg air compressed, and

(b) the specific work requirement in kJ per kg of liquid air.

6.7 For the dual-pressure Linde cycle of Fig. 6.4, prove that the specific-work requirement $W_{Z,2L}$ in work per unit mass of liquid air is

$$W_{Z,2L} = \frac{h_{10} - h_9}{h_{10} - h_8} ({}_1W_2)$$

$$+ \left[\frac{h_{13} - h_9}{h_{13} - h_4} \right.$$

$$\left. - \frac{(h_8 - h_9)(h_{11} - h_{13})}{(h_{10} - h_8)(h_{13} - h_4)} \right] ({}_3W_4)$$

where ${}_1W_2 = T_1(s_1 - s_2) - (h_1 - h_2)$

${}_3W_4 = T_3(s_3 - s_4) - (h_3 - h_4)$

6.8 Dry air at 70 °F and 14.696 psia is to be liquefied at 14.696 psia by the dual-pressure Linde method. Intermediate pressure is 200 psia and high-side pressure is 2500 psia. Assume that air is isothermally compressed at 70 °F in both stages. Also assume that the waste gas leaves the system at 70 °F. Determine

(a) the yield of liquid air m_9/m_1 in lbm per lbm of air compressed in the low-pressure compressor, and

(b) the specific work requirement in Btu per lbm of liquid air.

6.9 Dry air at 300 K and 1 atm pressure is to be liquefied using the simple Linde air-liquefaction system with precooling as shown in Fig. 6.5. Assume the air is cooled from state 3 to state 4 using a theoretical single-stage mechanical vapor-compression refrigeration system that uses R-22 and has a saturated evaporating temperature of 240 K and a saturated condensing temperature of 300 K. The air is compressed isothermally at 300 K to a pressure of 200 atm. $T_3 = 270$ K, $T_4 = 240$ K, and $T_{10} = 300$ K. Determine

(a) the yield of the system in kg liquid air per kg air compressed, and

(b) the specific-work requirement including both the air compressor and the refrigeration compressor in kJ per kg liquid air.

6.10 Dry air at 70 °F and 14.696 psia is to be liquefied by the Claude method. Assume the air is isothermally compressed at 70 °F to 2500 psia. Assume that 80 percent of the total mass of air compressed passes through the expander. Assume that the temperature of the air entering the expander is −110 °F, while the temperature of the waste gas leaving the system is 70 °F. Assume that the enthalpy drop across the expander is 50 percent of the isentropic value. Determine

(a) the yield of liquid air in lbm per lbm of air compressed,

(b) the specific work requirement in Btu per lbm of liquid air if the enthalpy drop across the expander is fully recovered as work to help drive the compressor, and

(c) the specific-work requirement in Btu per lbm of liquid air if the expander work output is wasted.

6.11 Repeat Prob. 6.10, assuming the fraction of the compressed air that passes through the expander is changed to

(a) 60 percent, and

(b) 40 percent.

Does there appear to be a fraction of air expanded that will maximize the yield and/or minimize the specific-work requirement for this system?

6.12 For the Claude cycle of Fig. 6.6 with the throttling process replaced by an adiabatic expander, prove that the yield Z_C^* in mass of liquid air produced per unit mass of gas compressed is

$$Z_C^* = \frac{h_{11} - h_2}{h_{11} - h_6} + \frac{\dot{m}_8(h_3 - h_8)}{\dot{m}_2(h_{11} - h_6)} + \left(1 - \frac{\dot{m}_8}{\dot{m}_2}\right) \frac{h_4 - h_5}{h_{11} - h_6}$$

6.13 Gaseous oxygen (99 percent pure, molar basis) is to be produced at a rate of 350 ft³/hr in a Linde single-column plant similar to the one depicted in Fig. 6.7. The column operates at a pressure of 1.0 atm. Air leaves the compressor at a pressure of 10.0 atm. The nitrogen product is to have 91 percent purity. Dry compressed air enters the heat exchanger at 70 °F and the outgoing products leave the heat exchanger at 60 °F. You may assume that the inlet air and the gaseous products behave as ideal gases near room temperature and at atmospheric pressure. Determine

(a) the volume of air in ft³/hr which must be compressed,

(b) the oxygen yield (ratio of moles of oxygen produced to moles of oxygen available), and

(c) the temperature at which the air must leave the heat exchanger.

6.14 A Linde double-column apparatus processes 200,000 ft^3/hr of air ($T = 70$ °F, $P = 14.696$ psia). The air is compressed and cooled and passed through the boiler (generator A) at a pressure of 8.0 atm. The oxygen product has a concentration of 0.01 (molar basis) and the nitrogen product a concentration of 0.95. The upper column operates at a pressure of 1.0 atm. Determine

(a) the heat-exchange rate in Btu/hr in generator A if the air at 8.0 atm pressure leaves as a saturated liquid, and

(b) the percentage of incoming air which is exhausted as oxygen product (molar basis).

6.15 Dry air is to be separated in a Linde double-column apparatus into oxygen (99.5 percent purity, molar basis) and nitrogen (99.0 percent purity). The precooled compressed air leaves the generator as a saturated liquid at 6.0 atm pressure. The upper column operates at 1.0 atm pressure and the lower column at 4.0 atm. Assume equilibrium exists between nitrogen product and reflux at top of upper column ($t_7 = t_8$). The concentration x_4 of the oxygen-rich liquid leaving generator A is 0.59.

(a) Plot the entire h-x diagram and, in a table, show values of $p, t, h,$ and x for all ten principal locations shown in Fig. 6.9.

(b) Determine the h-x coordinates for each of the four poles.

(c) Determine the heat transferred in generator A per mole of oxygen-rich liquid removed at point 4.

(d) Determine the heat transferred in generator B per mole of oxygen product removed at point 9.

(e) Determine the oxygen yield (ratio of moles of oxygen produced to moles of oxygen available).

6.16 Analyze the schematic h-x diagram of Fig. 6.12 for the double column of Fig. 6.11. Develop a set of itemized statements on construction of the diagram similar to those shown in Sec. 6.6 for the Linde double column. Show details and fully justify each statement.

6.17 The gaseous-oxygen plant of Fig. 6.13 is to be designed to produce 10 tons/hr of oxygen product. For the compressor, you may assume polytropic compression ($Pv^{1.35} = C$) and a mechanical efficiency of 70 percent. Assume sea-level pressure and neglect pressure drop in the water-cooled heat exchanger following the compressor. Calculate the required shaft-horsepower input for the compressor.

6.18 Develop a general schematic h-x diagram analysis similar to Fig. 6.12, but for the double column of Fig. 6.14. Ignore specific property data shown in Fig. 6.14. Also ignore the heat exchanger between nitrogen vapor and reflux from the lower column.

SYMBOLS

c_p	Specific heat at constant pressure, Btu/lbm · °F (kJ/kg · °C).
h	Specific enthalpy, Btu/lbm or Btu/lb mole (kJ/kg or kJ/kg mole).
$\dot{m}$	Mass rate of flow, lbm/min or lb moles/min (kg/sec or kg moles/sec).
P	Pressure, psia, psfa, atm.
$\dot{Q}$	Rate of heat transfer, Btu/min (kW).
R	Constant in perfect-gas law, ft-lbf/lbm · °R (kJ/kg · K).
s	Specific entropy, Btu/lbm · °R (kJ/kg · K).
T	Absolute temperature, °R (K).
t	Temperature, °F (°C).
v	Specific volume, ft^3/lbm (m^3/kg).
w	Work, Btu/lbm (kJ/kg).
w_Z	Specific-work requirement, Btu/lbm of liquid produced (kJ/kg liquid produced).
x	Concentration of nitrogen in mixture of nitrogen and oxygen, moles N_2/mole mix.
Z	Yield of liquid, lbm liq/lbm of gas compressed (kg liq/kg gas compressed).

Greek Letters

η	Isentropic efficiency of expander.
μ	Joule-Thomson coefficient $(\partial T/\partial P)_h$.

chapter 7

Thermodynamic Properties of Moist Air

7.1 ATMOSPHERIC AIR

The earth's atmosphere is a mixture of gases, including nitrogen, oxygen, argon, carbon dioxide, water vapor, and traces of other gases. Atmospheric air usually contains various particulate matter. Additional vapors are often present. Dust particles and condensible vapors such as water vapor are usually concentrated in the atmosphere only within a few thousand feet of the earth's surface. Above an altitude of about 20,000 ft (6100 m), the atmosphere consists essentially of dry air.

Barometric pressure is the force per unit area due to the weight of the atmosphere. Standard sea-level pressure is 14.696 psia (101.325 kPa). As one proceeds vertically into the atmosphere, pressure decreases. Temperature also decreases until the tropopause is reached.

The U.S. Standard Atmosphere provides a reference standard with respect to barometric pressure for an air conditioning engineer. The *ASHRAE Handbook of Fundamentals* [1] has given the following information which forms the definition of the U.S. Standard Atmosphere:

1. The atmosphere consists of dry air which behaves as a perfect gas; thus
$$Pv = RT \tag{7.1}$$

2. Gravity is constant at 32.174 ft/sec^2 (9.80665 m/s^2)

3. At sea level, pressure is 29.921 in. Hg and temperature is 59 °F (101.325 kPa, 18 °C).

4. Temperature t decreases linearly with altitude z up to the lower limit of the isothermal atmosphere according to the relation
$$t = t_0 - 0.00356z \tag{7.2}$$
with t and t_0 in °F and z in ft. Or
$$t = t_0 - 0.0065z$$
with t and t_0 in °C and z in meters.

The isothermal atmosphere, $-69.7\,°F$ ($-56.5\,°C$) begins at 36,152 ft (11,000 m). Equations (7.1) and (7.2) can be combined to give the relation

$$\frac{P}{P_0} = \left(1 - \frac{Az}{T_0}\right)^{5.266} \tag{7.3}$$

with

$$A = 0.00356\,°R/ft\ (0.0065\,K/m)$$

Table 7.1 shows variation of pressure and temperature for the U.S. standard atmosphere. Although the actual atmosphere above a locality would not correspond precisely to the U.S. standard atmosphere, Table 7.1 provides a convenient means for estimating barometric pressure for a given altitude above sea level.

7.2 FUNDAMENTAL DISCUSSION OF MOIST AIR

The composition of atmospheric air is variable, particularly with regard to amounts of water vapor and particulate matter. Before we can discuss thermodynamic properties, the substance must be precisely defined. The working substance in air conditioning problems is called *moist air*. Moist air is defined as a binary mixture of dry air and water vapor. Goff [2], in a final report of the Working Subcommittee, International Joint Committee on Psychrometric Data, has defined dry air as shown in Table 7.2.

Although somewhat arbitrary, this composition is regarded as exact, by definition. The molecular weights for dry air and water vapor are 28.966 and 18.016, respectively. The respective gas constants can be obtained by dividing the universal gas constant, $\overline{R} = 1.986\ Btu/lbmole \cdot °R = 1545\ ft\text{-}lbf/lbmole \cdot °R\ (8.314\ kJ/kmole \cdot K)$, by the appropriate molecular weight.

dry air: $\quad R_a = \dfrac{\overline{R}}{28.966} = 0.0686\ Btu/lbm \cdot °R = 53.35\ ft\text{-}lbf/lbm \cdot °R = 287\ J/kg \cdot °K$

water vapor: $\quad R_v = \dfrac{\overline{R}}{18.016} = 0.110\ Btu/lbm \cdot °R = 85.78\ ft\text{-}lbf/lbm \cdot °R$
$$= 462\ J/kg \cdot °K$$

Moist air may contain variable amounts of water vapor from zero (dry air) to that of saturated moist air. Goff [3] has defined *saturation of moist air* as that condition where moist air may coexist in neutral equilibrium with associated condensed water, presenting a flat surface to it.

The humidity ratio, W, is defined as the mass of water vapor per unit mass of dry air in a moist air mixture.

$$W = \frac{m_v}{m_a}$$

Two measures of humidity relative to saturation conditions are commonly used. *Degree of saturation* is defined by the relation

$$\mu = \frac{W}{W_s} \tag{7.4}$$

TABLE 7.1 Standard Atmospheric Data for Altitudes to 60,000 ft

Altitude, ft	Temperature, °F	Pressure	
		in. Hg	psia
−1000	62.6	31.02	15.236
−500	60.8	30.47	14.966
0	59.0	29.921	14.696
500	57.2	29.38	14.430
1000	55.4	28.86	14.175
2000	51.9	27.82	13.664
3000	48.3	26.82	13.173
4000	44.7	25.82	12.682
5000	41.2	24.90	12.230
6000	37.6	23.98	11.778
7000	34.0	23.09	11.341
8000	30.5	22.22	10.914
9000	26.9	21.39	10.506
10,000	23.4	20.58	10.108
15,000	5.5	16.89	8.296
20,000	−12.3	13.76	6.758
30,000	−47.8	8.90	4.371
40,000	−69.7	5.56	2.731
50,000	−69.7	3.44	1.690
60,000	−69.7	2.14	1.051

Standard Atmospheric Data for Altitudes to 10,000 m

Altitude, m	Temperature, °C	Pressure, kPa
−500	18.2	107.478
0	15.0	101.325
500	11.8	95.461
1000	8.5	89.874
2000	2.0	79.495
3000	−4.5	70.108
4000	−11.0	61.640
5000	−17.5	54.020
6000	−24.0	47.181
7000	−30.5	41.061
8000	−37.0	35.600
9000	−43.5	30.742
10,000	−50.0	26.436

SOURCE: Abstracted by permission from *ASHRAE Handbook of Fundamentals 1993*, ch. 6.

where W_s is the humidity ratio at saturation for the same temperature and pressure as those of the actual state.

Relative humidity ϕ is defined by the relation

$$\phi = \frac{x_w}{x_{w,s}} \tag{7.5}$$

TABLE 7.2 Composition of Dry Air

Substance	Molecular Weight	Mol-fraction Composition in Dry Air	Partial Molecular Weight in Dry Air
Oxygen (O$_2$)	32.000	0.2095	6.704
Nitrogen (N$_2$)	28.016	0.7809	21.878
Argon (A)	39.944	0.0093	0.371
Carbon dioxide (CO$_2$)	44.01	0.0003	0.013
		1.0000	28.966

SOURCE: Reprinted by permission from *Trans. ASHVE 55*, 463.

where x_w is the mole fraction of the water vapor in the mixture and $x_{w,s}$ is the mole fraction of water vapor at saturation for the same temperature and pressure as those of the actual state. We may convert Eq. (7.5) to the forms

$$\phi = \frac{1 + \dfrac{0.622}{W_s}}{1 + \dfrac{0.622}{W}} = \mu \, \frac{0.622 + W_s}{0.622 + W} \qquad (7.6)$$

It is important to observe that neither μ nor ϕ are defined when the temperature of moist air exceeds the saturation temperature of pure water corresponding to the moist air pressure. For sea-level pressure, W_s approaches infinity at 212 °F (100 °C). Thus for 14.696 psia (101.325 kPa), μ and ϕ are undefined for temperatures higher than 212 °F (100 °C).

Three moist-air properties are associated with temperature. The dry-bulb temperature, t, is the true temperature of moist air at rest. The dew-point temperature, t_d, is defined as the solution $t_d(P, W)$ of the equation

$$W_s(P, t_d) = W \qquad (7.7)$$

In words, the dew-point temperature of a moist air state is the saturation temperature corresponding to the humidity ratio and pressure of the state. Another way to describe the dew-point temperature is to consider a moist air mixture defined by P, W, and t. If you slowly reduce the temperature of the mixture while holding P and W constant, then the temperature at which saturation is reached is the dew-point temperature, t_d.

The third temperature associated with moist air property is the *thermodynamic wet-bulb temperature, t^**, which is also referred to as the *adiabatic saturation temperature*. However, before defining this temperature it is necessary to introduce the enthalpy of the moist air mixture, H, which is the sum of the enthalpies of the dry air, H_a, and the water vapor, H_w.

$$H = H_a + H_w = m_a h_a + m_w h_w \qquad (7.8)$$

For convenience, the specific enthalpy of moist air, h, is defined per unit mass of dry air.

$$h \equiv \frac{H}{m_a} \qquad (7.9)$$

(Note that this form of the definition is particularly convenient for humidification and dehumidification processes where the flow rate of dry air remains constant as the moist air passes through the humidifier or dehumidifier.)

Thus

$$h = h_a + \frac{m_w}{m_a} h_w$$

or

$$h = h_a + W h_w \qquad (7.10)$$

7.3 THERMODYNAMIC WET-BULB TEMPERATURE

Adiabatic saturation temperature is that temperature at which water, by evaporating into air, can bring the air to saturation adiabatically at the same temperature. It is unnecessary to inject practical details into a discussion of adiabatic saturation, since the results of the process are given by definition. However, to help us understand it better, we may consider how such a process could be approached.

Figure 7.1 schematically shows a device which may provide adiabatic saturation. The chamber could be indefinitely long and perfectly insulated. The total quantity of water present could be large compared to that added to the air in a given length of time. We may assume no temperature gradients within the water body. Regardless of the initial temperature of the water, we would expect that after sufficient time the water would assume a constant temperature. This limiting temperature of water should be less than the entering air dry-bulb temperature, but greater than the entering air dew-point temperature.

In order to maintain steady conditions within the chamber, makeup water at temperature t_2 is supplied at a rate $\dot{m}_l$, which is equal to the rate at which vapor is added to the moist air. Thus, if the mass flow rate of dry air passing through the chamber is $\dot{m}_a$, the rate of makeup water is given by

$$\dot{m}_l = \dot{m}_a (W_{s,2} - W_1) \qquad (7.11)$$

The chamber is adiabatic, and no work is done by or on the apparatus. Therefore, the steady-flow energy equation for the process is

$$\dot{m}_a h_{a,1} + \dot{m}_a W_1 h_{w,1} + \dot{m}_l h_{f,2} = \dot{m}_a h_{a,2} + \dot{m}_a W_{s,2} h_{w,2} \qquad (7.12)$$

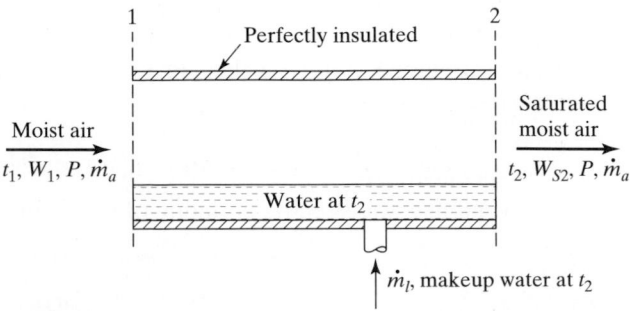

Figure 7.1 Schematic of an adiabatic saturation device.

By applying Eq. (7.10) in order to write the energy balance in terms of the moist-air enthalpy, substituting Eq. (7.11) for the rate of makeup water and noting that saturation conditions exist at the exit, the steady-flow energy equation can be written as

$$h_1 + (W_{s,2} - W_1)h_{f,2} = h_{s,2} \tag{7.13}$$

Since the leaving air is saturated, and since we assume a constant pressure P, the quantities $W_{s,2}$, $h_{s,2}$, and h_{f2}, are sole functions of temperature t_2. We may then deduce that t_2 is a function of h_1, W_1, and P, or that t_2 is a function of state 1. Therefore, t_2 is a thermodynamic property of state 1. We call this property the thermodynamic wet-bulb temperature t^*. Denoting all properties evaluated at t^* with the superscript $*$, Eq. (7.13) becomes

$$h + (W_s^* - W)h_f^* = h_s^* \tag{7.14}$$

Thus, for given values of h, W, and P (given moist air state), the thermodynamic wet-bulb temperature t^* is that value of temperature which satisfies Eq. (7.14). There are numerous practical problems where the concept of thermodynamic wet-bulb temperature is useful.

7.4 THE TABLES FOR MOIST AIR

By applying fundamental procedures of statistical mechanics, Goff and Gratch [4] calculated accurate thermodynamic properties of moist air for standard sea-level pressure. More recently, new formulations were developed by Hyland and Wexler [5]. Tables A.4E and A.4SI, extracted from the *ASHRAE Handbook, 1993 Fundamentals Volume* [6], presents properties for dry air and saturated moist air based on the new formulations. In the following list, brief explanations of the data in Tables A.4E and A.4SI are shown.

W_s = humidity ratio of saturated air, mass of water vapor per mass of dry air.

v_a = specific volume of dry air under 14.696 psia (101.325 kPa) pressure, ft^3/lbm_a (m^3/kg_a).

v_s = volume of saturated air, ft^3/lbm dry air (m^3/kg dry air).

$v_{as} = v_s - v_a$, ft^3/lbm dry air (m^3/kg dry air).

h_a = specific enthalpy of dry air, Btu/lbm_a (kJ/kg dry air). Zero enthalpy for dry air is taken at 0 °F (0 °C).

h_s = enthalpy of saturated air, Btu/lbm of dry air (kJ/kg dry air).

$h_{as} = h_s - h_a$, Btu/lbm of dry air (kJ/kg dry air).

s_a = specific entropy of dry air, $\text{Btu}/\text{lbm}_a \cdot °\text{R}$ ($\text{kJ}/\text{kg}_a\text{-K}$). Zero entropy for dry air is taken at 0 °F (0 °C).

s_s = entropy of saturated air, Btu/lbm of dry air $\cdot °\text{R}$ (kJ/kg dry air $\cdot$ K).

$s_{as} = s_s - s_a$, Btu/lbm of dry air $\cdot °\text{R}$ (kJ/kg dry air $\cdot$ K).

Calculations for volume, enthalpy, and entropy of unsaturated moist air states are closely given by the relations

$$v = v_a + \mu v_{as} \tag{7.15}$$

$$h = h_a + \mu h_{as} \tag{7.16}$$

$$s = s_a + \mu s_{as} \tag{7.17}$$

The relations for specific volume and enthalpy would be exact if both the water vapor and dry air were perfect gases. At a temperature of approximately 100 °F (38 °C) real gas effects result in maximum corrections to Eqs. (7.15) and (7.16) of 0.003 percent and 0.015 percent, respectively. As the temperature increases, the real gas effects increase. However, for the purposes of most moist air calculations the corrections to Eqs. (7.15) and (7.16) are still small enough to be neglected. For example, at approximately 190 °F (88 °C) the corrections to the two equations are 0.16 percent and 0.11 percent, respectively.

The values of entropy calculated using Eq. (7.17) closely approximate the true values up to approximately 100 °F (38 °C). Above that temperature the mixing entropy not accounted for in Eq. (7.17) is significant. For example, Eq. (7.17) is in error by 1.8 percent, 3.5 percent, and 4.2 percent at temperatures of 96 °F (36 °C), 144 °F (62 °C) and 192 °F (89 °C), respectively. Additional information regarding correction factors for Eqs. (7.15), (7.16), and (7.8), including formulae for calculating the corrections, can be found in the *ASHRAE Handbook, 1977 Fundamentals Volume* [7].

EXAMPLE 7.1

Moist air exists at 80 °F dry-bulb temperature, 60 °F dew-point temperature, and 14.696 psia pressure. Determine (a) the humidity ratio, lbm_w/lbm_a, (b) degree of saturation, (c) relative humidity, (d) enthalpy, Btu/lbm_a, and (e) the volume in ft^3/lbm_a.

Solution: **(a)** Since sea-level pressure exists, Table A.4E will be used. By Eq. (7.7), $W = W_s$ at 60 °F. Thus $W = 0.011087\ lbm_w/lbm_a$.

(b) At 80 °F, $W_s = 0.02234\ lbm_w/lbm_a$. By Eq. (7.4)

$$\mu = 0.011087/0.02234 = 0.496 \text{ or } 49.6 \text{ percent}$$

(c) By Eq. (7.6)

$$\phi = \frac{(0.496)(0.622 + 0.02234)}{0.622 + 0.01108} = 0.505 \text{ or } 50.5 \text{ percent}$$

(d) We may calculate the enthalpy by Eq. (7.16). Using Table A.4E, we have $h = 19.222 + (0.496)(24.479) = 31.36\ Btu/lbm_a$

(e) We may calculate the volume by Eq. (7.15). Using Table A.4E, we have $v = 13.602 + (0.496)(0.487) = 13.84\ ft^3/lbm_a$

EXAMPLE 7.2

Moist air exists at 66 °C dry-bulb temperature and 30 percent degree of saturation. Pressure is 101.325 kPa. Determine (a) the enthalpy, kJ/kg_a, and (b) the specific volume, m^3/kg_a.

Solution: **(a)** By Table A.4SI, $W_s = 0.21848\ kg_w/kg_a$. By Eq. (7.16),

$$h = 66.455 + (0.30)(572.116) = 238.09\ kJ/kg_a$$

(b) By Eq. (7.15),

$$v = 0.9608 + (0.30)(0.3350) = 1.061\ m^3/kg_a$$

EXAMPLE 7.3

Moist air exists at 80 °F dry-bulb temperature, 60 °F thermodynamic wet-bulb temperature, and 14.696 psia pressure. Through use of Tables A.4E and A.1E, determine (a) the degree of saturation, and (b) the enthalpy.

Solution: **(a)** By Eqs. (7.16) and (7.14),

$$h_a + \mu h_{as} = h_s^* - (W_s^* - W)h_f^* = h_s^* - W_s^* h_f^* + \mu W_s h_f^*$$

Thus

$$\mu = \frac{h_s^* - W_s^* h_f^* - h_a}{h_{as} - W_s h_f^*} = \frac{26.467 - (0.011087)(27.63) - 19.222}{24.479 - (0.02234)(27.63)}$$

$$= 0.291$$

(b) By Eq. (7.16),

$$h = 19.222 + (0.291)(24.479) = 26.34 \text{ Btu/lbm}_a$$

7.5 PERFECT-GAS RELATIONSHIPS FOR APPROXIMATE CALCULATIONS

The methods of Sec. 7.4 allow accurate calculation of moist air properties through use of Table A.4. However, Table A.4 is restricted to standard atmospheric pressure. Basic relationships shown in Secs. 7.2 and 7.3 may be applied for any existing pressure. The discussion presented in Chapter 2 regarding low-pressure water vapor showed that for water-vapor pressure below 1 psia (6.9 kPa) the vapor exhibits approximate perfect-gas behavior. Therefore, in this section we will assume that perfect-gas relations exist and formulations for this special situation will be developed for the expressions presented in Secs. 7.2 and 7.3.

A useful expression for the humidity ratio can be derived by substituting the perfect-gas equations into the definition of W:

$$W = \frac{m_w}{m_a} = \frac{P_w V/R_w T}{P_a V/R_a T} = \frac{R_a}{R_w} \frac{P_w}{P_a}$$

It is convenient to substitute for the partial pressure of dry air, P_a, by noting that the total pressure of the mixture, P, is the sum of the partial pressures. In addition, the ratio of the gas constants can be replaced by the inverse ratio of the molecular weights. These substitutions result in the equation

$$W = 0.622 \frac{P_w}{P - P_w} \tag{7.18}$$

This equation is also useful in establishing a method of evaluating the dew-point temperature, since it shows that a process in which W and P are constant corresponds to a process in which P_w is constant. Therefore, from Eqs. (7.7) and (7.18) we may deduce that the dew-point temperature, t_d, is equal to the saturation temperature corresponding to the vapor pressure P_w.

For a perfect gas the enthalpy is a function of temperature only. Thus

$$h_a = \int c_{pa} \, dT$$

$$h_w = \int c_{pw} \, dT$$

where c_{pa} and c_{pw} are the specific heats at constant pressure for dry air and water vapor, respectively. If the specific heat of dry air is assumed to be a constant and if we select the same reference states used in Tables A.4E and A.4SI, (i.e., $h_a = 0$ at zero degrees Fahrenheit or Celsius), the enthalpy of dry air can be expressed as

$$h_a = c_{pa} t \tag{7.19}$$

Similarly, if the specific heat at constant pressure for water vapor is assumed constant and we select the same reference states as those used in Tables A.1E and A.1SI, the enthalpy of the water vapor can be expressed as (see Sec. 2.14)

$$h_w = h_g = c_{pw} t + h_{g0} \tag{7.20}$$

where h_{g0} is the enthalpy of saturated water vapor at zero degrees Fahrenheit or Celsius.

The values of the quantities used in Eqs. (7.19) and (7.20) are given in Table 7.3.

The substitution of Eqs. (7.19) and (7.20) into (7.10) provides an expression for the specific enthalpy of the mixture

$$h = c_{pa} t + W h_g \tag{7.21}$$

$$h = c_{pa} t + W(c_{pw} t + h_{g0}) \tag{7.22}$$

Another form of the enthalpy equation can be written by defining a specific heat of the mixture, c_p, as

$$c_p = c_{pa} + W c_{pw} \tag{7.23}$$

which results in

$$h = c_p t + W h_{g0} \tag{7.24}$$

The specific volume of the mixture, v, is defined as the volume of the mixture per unit mass of dry air. However, since the mixture, the dry air, and the water vapor all occupy the same volume,

$$v = \frac{R_a T}{P_a}$$

TABLE 7.3 Quantities Used in Perfect-Gas Approximations for Enthalpy

t	English Units °F	S.I. Units °C
c_{pa}	0.240 Btu/lbm$_a$ · °F	1.00 kJ/kg$_a$ · °C
c_{pw}	0.444 Btu/lbm$_w$ · °F	1.86 kJ/kg$_w$ · °C
h_{g0}	1061 Btu/lbm$_w$	2501 kJ/kg$_w$

or

$$v = \frac{R_a T}{P - P_w} \tag{7.25}$$

The perfect-gas approximation for the relative humidity is obtained by observing that for a perfect gas

$$x_w = \frac{P_w}{P} \tag{7.26}$$

and thus from Eqs. (7.5) and (7.26)

$$\phi = \frac{P_w}{P_{ws}} \tag{7.27}$$

Equation (7.14) may be altered through use of the perfect-gas relations. By Eqs. (7.14), (7.23), and (7.24)

$$(W_s^* - W)h_{fg}^* = c_p(t - t^*) \tag{7.28}$$

By Eqs. (7.14) and (7.21)

$$W = \frac{W_s^* h_{fg}^* - c_{pa}(t - t^*)}{h_g - h_f^*} \tag{7.29}$$

The total entropy of a moist air mixture, S, which is at a pressure and temperature P and T, can be written as

$$S = S_{a,0} + \Delta S_{a,T,P} + \Delta S_{a,\text{mix}} + S_{w,0} + \Delta S_{w,T,P} + \Delta S_{w,\text{mix}} \tag{7.30}$$

The quantities $S_{a,0}$ and $S_{w,0}$ are the entropy values at the reference conditions for the dry air and water vapor, respectively. The terms $\Delta S_{a,T,P}$ and $\Delta S_{w,T,P}$ are the entropy changes that result in going from the reference states to the state P, T and the terms $\Delta S_{a,\text{mix}}$ and $\Delta S_{w,\text{mix}}$ are the mixing entropies for the dry air and water vapor. These latter two mixing terms account for the fact that it is the partial pressures, P_a and P_w, not the total pressure, P, that are needed in evaluating the entropies of the components.

For the case of perfect gases with constant specific heats and with P_0, T_0 as the reference state

$$\Delta S_{a,T,P} = m_a \left(c_{pa} \ln \frac{T}{T_0} - R_a \ln \frac{P}{P_0} \right) \tag{7.31}$$

$$\Delta S_{a,\text{mix}} = -m_a \left(R_a \ln \frac{P_a}{P} \right) \tag{7.32}$$

$$\Delta S_{w,T,P} = m_w \left(c_{pw} \ln \frac{T}{T_0} - R_w \ln \frac{P}{P_0} \right) \tag{7.33}$$

$$\Delta S_{w,\text{mix}} = -m_w \left(R_w \ln \frac{P_w}{P} \right) \tag{7.34}$$

The temperatures and pressures in Eqs. (7.31) thru (7.34) must be absolute values.

In the English system of units the reference condition is taken as $T_0 = 0\ °F$ and $P_0 = 14.696$ psia. It will be convenient to select $S_{a,0}$ and $S_{w,0}$ such that the values of S will correspond to tabulated values. The dry air tables are constructed with zero entropy at $0\ °F$, 14.696 psia. Therefore, $S_{a,0} = 0$. However, the steam tables are constructed with the entropy of saturated liquid equal to zero at 32 °F, and a value for $S_{w,0}$ must be calculated. Although $S_{w,0}$ is at a fictitious state, it can be evaluated by extrapolating the ideal-gas relations to the reference state at T_0, P_0. Thus

$$S_{w,0} = m_w \left[s_g(T_0, P_{w,s,T_0}) - R_w \ln \frac{P_0}{P_{w,s,T_0}} \right] \tag{7.35}$$

where P_{w,s,T_0} is the saturation pressure at T_0, the reference temperature. Substitution in to Eq. (7.35) results in

$$S_{w,0} = m_w(1.5937 \text{ Btu/lbm}_w \cdot °R) = m_w s_{w,0}$$

In the S.I. system of units the reference condition is selected as $T_0 = 0$ °C and $P_0 = 101.325$ kPa. As before, if we select values of $S_{a,0}$ and $S_{a,w}$ in order that entropy values correspond to tabulated values, we find

$$S_{a,0} = 0$$

$$S_{w,0} = m_w(6.7975 \text{ kJ/kg}_w \cdot \text{K}) = m_w s_{w,0}$$

Substituting Eqs. (7.31), (7.32), (7.33), and (7.34) into Eq. (7.30), setting $S_{a,0} = 0$, dividing through by m_a, and combining terms results in the following expression for the specific entropy for the mixture:

$$\frac{S}{m_a} = s = c_{pa} \ln \frac{T}{T_0} - R_a \ln \frac{P_a}{P_0} + W \left(s_{w,0} + c_{pw} \ln \frac{T}{T_0} - R_w \ln \frac{P_w}{P_0} \right) \tag{7.36}$$

or

$$s = (c_{pa} + c_{pw}W) \ln \frac{T}{T_0} - R_a \ln \frac{P_a}{P_0} + W \left(s_{w,0} - R_w \ln \frac{P_w}{P_0} \right) \tag{7.37}$$

As stated earlier, equations of this section are only approximate. Figure 7.2 shows percent error in calculation of humidity ratio W'_s, enthalpy h'_s, and volume v'_s by Eqs. (7.18), (7.21), and (7.25), respectively, for saturated air at 14.696 psia (101.325 kPa).

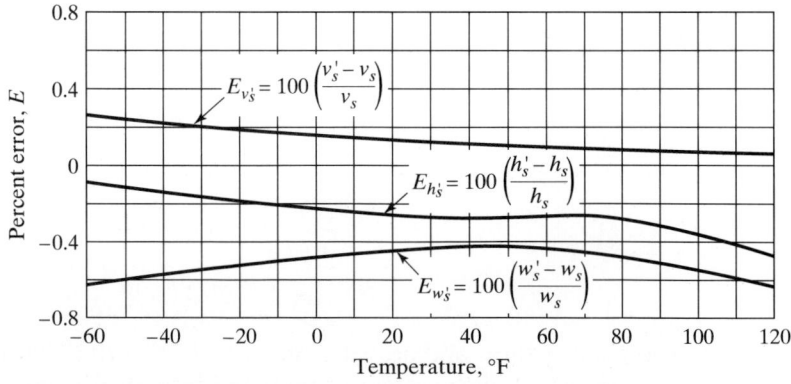

Figure 7.2 Error of perfect-gas relationships in calculation of humidity ratio, enthalpy, and volume of saturated air at 14.696 psia pressure (101.324 kPa).

Correct values W_s, h_s, and v_s were taken from Table A.4E. Figure 7.2 shows that the error in calculation of humidity ratio by Eq. (7.18) is less than about 0.6 percent in the range of $-50\,°F$ to $110\,°F$ ($-46\,°C$ to $43\,°C$). Except for temperatures above about $100\,°F$ ($38\,°C$), the error in calculation of enthalpy by Eq. (7.21) is less than about 0.4 percent. For the temperature range of $-30\,°F$ to $120\,°F$ ($-34\,°C$ to $49\,°C$), the volume may be calculated by Eq. (7.25) with an error of less than 0.2 percent.

EXAMPLE 7.4

Rework Ex. 7.1, using perfect-gas relations and steam-table data.

Solution:

(a) We may deduce from Eq. (7.18) that the dew-point temperature t_d is the saturation temperature corresponding to the partial pressure of water vapor P_w. By Table A.1E, at $60\,°F$, $P = 0.25629$ psi. By Eq. (7.18),

$$W = \frac{(0.622)(0.256)}{14.696 - 0.256} = 0.01103\ \text{lbm}_w/\text{lbm}_a$$

(b) By Table A.1E, at $80\,°F$, $P_{w,s} = 0.50734$ psia. By Eq. (7.18) for saturated air

$$W_s = \frac{(0.622)(0.507)}{14.696 - 0.507} = 0.02223\ \text{lbm}_w/\text{lbm}_a$$

By Eq. (7.4),

$$\mu = \frac{0.01103}{0.02223} = 0.496$$

(c) By Eq. (7.27),

$$\phi = \frac{0.25618}{0.50701} = 0.505$$

(d) By Table A.1E, at $80\,°F$, $h_g = 1096.20$ Btu/lbm_w. By Eq. (7.21),

$$h = (0.240)(80) + (0.01103)(1096.20) = 31.29\ \text{Btu}/\text{lbm}_a$$

(e) By Eq. (7.25),

$$v = \frac{(53.35)(540)}{(14.44)(144)} = 13.85\ \text{ft}^3/\text{lbm}_a$$

We observe that the answers of Example 7.4 differ but slightly from those of Example 7.1.

EXAMPLE 7.5

Moist air exists at $30\,°C$ dry-bulb temperature, 40 percent relative humidity, and 86.60 kPa barometric pressure. Determine (a) the dew-point temperature, (b) the specific enthalpy, and (c) the specific entropy.

Solution: **(a)** By Table A.1SI, at 30 °C, $P_{w,s}$ = 4.246 kPa. By Eq. (7.27),

$$P_w = (0.40)(4.246) = 1.698 \text{ kPa}$$

Interpolation in Table A.1SI gives t_d = 14.9 °C.

(b) By Eq. (7.18),

$$W = \frac{(0.622)(1.698)}{86.60 - 1.698} = 0.0124 \text{ kg}_w/\text{kg}_a$$

By Table A.1SI, at 30 °C, h_g = 2555.80 kJ/kg$_w$. By Eq. (7.21),

$$h = (1.0)(30) + (0.0124)(2555.80) = 61.79 \text{ kJ/kg}_a$$

(c) The specific entropy can be calculated using Eq. (7.37). First find the partial pressure of dry air:

$$P_a = P - P_w = 86.600 - 1.698 = 84.902 \text{ kPa}$$

By Eq. (7.37),

$$s = [1.0 + 1.86(0.0124)] \ln\frac{303}{273} - 0.287 \ln\frac{84.902}{101.325}$$

$$+ (0.0124)\left(6.7975 - 0.461 \ln\frac{1.698}{101.325}\right)$$

$$= 0.2651 \text{ kJ/kg}_a \cdot \text{K}$$

There is always the temptation to perform the calculations using tables and charts for standard atmospheric pressure. The pressure in this example corresponds to that which would occur at an altitude of approximately 1300 meters (see Table 7.1) and is approximately 15 percent below standard atmospheric pressure. If we had neglected that the pressure was lower and had used Table A.4SI along with Eqs. (7.6), (7.16), and (7.17), the results would have been

$$W = 0.0107 \text{ kg/kg}_a \qquad \text{(14\% lower than the correct value)}$$
$$h = 57.39 \text{ kJ/kg}_a \qquad \text{(7\% lower than the correct value)}$$
$$s = 0.1996 \text{ kJ/kg}_a \cdot \text{K} \qquad \text{(25\% lower than the correct value)}$$

Thus it is apparent that the moist air property tables can give results that are considerably different from the correct values when the pressure differs from standard atmospheric pressure.

EXAMPLE 7.6

Moist air exists at 100 °F dry-bulb temperature, 80 °F thermodynamic wet-bulb temperature, and 13.86 psia pressure. Determine (a) the humidity ratio and (b) the relative humidity.

Solution: **(a)** By Table A.1E, at 80 °F, $P_{w,s}^*$ = 0.50734 psi. By Eq. (7.18),

$$W_s^* = \frac{(0.622)(0.507)}{13.86 - 0.507} = 0.02362 \text{ lbm}_w/\text{lbm}_a$$

By Table A.1E, at 80 °F, h_{fg}^* = 1048.5 Btu/lbm$_w$, h_f^* = 47.70 Btu/lbm$_w$. At 100 °F, h_g = 1104.83 Btu/lbm$_w$. By Eq. (7.29)

$$W = \frac{(0.02362)(1048.5) - (0.240)(20)}{1104.83 - 47.70} = 0.01889 \text{ lbm}_w/\text{lbm}_a$$

(b) Equation (7.18) may be changed to the form

$$P_w = \frac{1.608PW}{1 + 1.608W}$$

Thus

$$P_w = \frac{(1.608)(13.86)(0.01889)}{1 + (1.608)(0.01889)} = 0.408 \text{ psi}$$

By Table A.1E, at 100 °F, $P_{w,s} = 0.95034$ psi. By Eq. (7.27),

$$\phi = \frac{0.408}{0.950} = 0.430 \text{ or } 43.0 \text{ percent}$$

In the United States, air conditioning engineers sometimes use an approximate procedure in performing psychrometric process calculations. The concept followed is to treat any general psychrometric process as the sum of a sensible and a latent process and to use approximate relationships to evaluate these processes.

In order to illustrate the procedure, consider the enthalpy change in a general process that goes from state 1 to state 2.

$$\Delta h = h_2 - h_1$$

Applying Eq. (7.22),

$$\Delta h = (c_{pa}t_2 + W_2 c_{pw} t_2 + W_2 h_{g0}) - (c_{pa}t_1 + W_1 c_{pw} t_1 + W_1 h_{g0}) \tag{7.38}$$

If
$$\bar{t} = (t_1 + t_2)/2$$
$$\Delta t = t_2 - t_1$$
$$\overline{W} = (W_1 + W_2)/2$$
$$\Delta W = W_2 - W_1$$

then Eq. (7.38) can be rewritten as

$$\Delta h = (c_{pa} + \overline{W}c_{pw})\,\Delta t + (c_{pw}\bar{t} + h_{g0})\Delta W \tag{7.39}$$

The terms on the right-hand side of Eq. (7.39) can be thought of as the enthalpy changes for sensible and latent processes, respectively. Thus the total enthalpy change is divided into sensible and latent components.

$$\Delta h = \Delta h_S - \Delta h_L \tag{7.40}$$

where

$$\Delta h_S = (c_{pa} + \overline{W}c_{pw})\,\Delta t: \qquad \text{sensible} \tag{7.41}$$

$$\Delta h_L = (c_{pw}\bar{t} + h_{g0})\,\Delta W: \qquad \text{latent} \tag{7.42}$$

The coefficient $(c_{pa} + \overline{W}c_{pw})$ is the average specific heat for the moist air mixture for the process from state 1 to state 2:

$$\bar{c}_p = c_{pa} + \overline{W}c_{pw} \tag{7.43}$$

The coefficient $(c_{pw}\bar{t} + h_{g0})$ can be interpreted as the average enthalpy of water vapor for the process from state 1 to state 2:

$$\bar{h}_g = c_{pw}\bar{t} + h_{g0} \tag{7.44}$$

Therefore Eqs. (7.41) and (7.42) can be written as

$$\Delta h_s = \bar{c}_p \, \Delta t \qquad (7.45)$$

$$\Delta h_L = \bar{h}_g \, \Delta W \qquad (7.46)$$

No new approximations, beyond the assumptions, of perfect gas and constant specific heat, have been used in the development of Eqs. (7.45) and (7.46). The approximate nature of the procedure is introduced in selecting the coefficients for the two equations. The numerical values of $\bar{c}_p$ and $\bar{h}_g$ are relatively insensitive to the humidity ratios and temperatures, respectively, that commonly occur in HVAC processes. Therefore, constant values are generally selected. The specific heat, $\bar{c}_p$, is commonly taken as 0.245 Btu/lbm$_a$ · °F (1.02 kJ/kg$_a$ · °C). Depending upon the reference, values selected for the average enthalpy of water vapor, $\bar{h}_g$, vary from about 1050 to 1150 Btu/lbm$_w$ (2500 to 2700 kJ/kg$_w$). The use of these constant coefficients makes the procedure easy to use but also approximate.

7.6 CONSTRUCTION OF THE PSYCHROMETRIC CHART

In previous sections of this chapter we considered various thermodynamic properties of moist air and the equations relating them. With these relations we may accurately solve problems concerning moist air. However, the calculations can be tedious and time consuming.

It is of considerable advantage to plot the relations to give a nomograph called a *psychrometric chart*. Such a chart not only allows graphical reading of the various properties but also provides for convenient graphical solutions to many process problems.

A thermodynamic state for moist air is uniquely fixed if the barometric pressure and two independent properties are known. A psychrometric chart may be constructed for some single value of barometric pressure. Traditionally, standard sea-level pressure has been used. The choice of coordinates is, of course, arbitrary. Many psychrometric charts used in the United States have employed dry-bulb temperature and humidity ratio as the basic coordinates. In 1923, Richard Mollier [8] of Dresden, Germany, introduced a chart using enthalpy and humidity ratio as the coordinates. This chart received wide acceptance in Europe.

The use of enthalpy and humidity ratio as basic coordinates presents many advantages. Thermodynamic wet-bulb temperature lines are identically straight. A majority of the common psychrometric processes appear as straight lines on h-W coordinates. In general, the Mollier type of chart allows the most fundamentally consistent treatment of air conditioning problems with a minimum of approximations.

Through use of the psychrometric relations of this chapter we may readily construct an h-W chart. Experience has shown that the best intersections result when enthalpy is used as an oblique coordinate and humidity ratio as a rectangular coordinate.

The construction method used here has been adapted from Goodman's [9] procedure. Figure 7.3 shows the basic geometry. The enthalpy lines are inclined at an angle β to the horizontal humidity ratio lines. The line $\overline{12}$ may represent any straight line. The vertical scalar distance representing $(W_2 - W_1)$ is L_W, and the horizontal scalar distance representing $(h_2 - h_1)$ is L_h.

By the law of sines, we may write

$$\frac{L_h}{\sin \alpha} = \frac{b}{\sin \beta} \qquad (7.47)$$

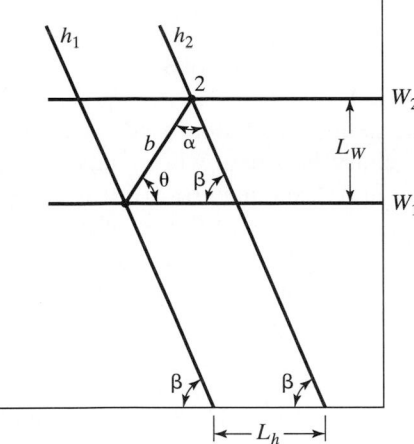

Figure 7.3 Fundamental geometry of psychrometric chart.

Also

$$\alpha = 180 - (\theta + \beta)$$

$$\sin\alpha = \sin(\theta + \beta) = \sin\theta\cos\beta + \cos\theta\sin\beta \qquad (7.48)$$

and

$$b = \frac{L_w}{\sin\theta} \qquad (7.49)$$

By Eqs. (7.47)–(7.49) we have

$$\cot\beta + \cot\theta = \frac{L_h}{L_W} \qquad (7.50)$$

We may define an enthalpy scale factor s_h by

$$s_h = \frac{h_2 - h_1}{L_h}$$

and a humidity-ratio scale factor s_W by

$$s_W = \frac{W_2 - W_1}{L_W}$$

Then

$$\frac{L_h}{L_W} = \frac{s_W(h_2 - h_1)}{s_h(W_2 - W_1)} = \frac{q'}{S} \qquad (7.51)$$

where

$$S = \frac{S_h}{S_W} = \text{chart scale factor in Btu per lbm}_w\,(\text{kJ/kg}_w)$$

and

$$q' = \frac{h_2 - h_1}{W_2 - W_1} = \text{enthalpy-moisture ratio in Btu/lbm}_w\,(\text{kJ/kg}_w)$$

By Eqs. (7.50) and (7.51),

$$\cot\theta + \cot\beta = \frac{q'}{S} \qquad (7.52)$$

Equation (7.52) is the general equation used for constructing the various straight lines on the psychrometric chart. Before any lines may be drawn, the scale factor S and the angle β must be established. Each may be arbitrarily fixed, but their choice greatly influences the appearance and usability of the resulting chart. A large scale factor S is desirable so that the angle θ may vary more uniformly with uniform changes in the enthalpy-moisture ratio q'. On the other hand, the enthalpy scale factor s_h can be made too large for accurate reading of enthalpy values.

The inclination angle β of the enthalpy lines may be fixed by choosing some property line and considering it to be vertical. Usually some one of the dry-bulb temperature lines is chosen. By Eq. (7.10), and noting that h_a and h_w are constants, it is true for a line of constant dry-bulb temperature that

$$h_2 - h_1 = (W_2 - W_1)h_w$$

or

$$q' = h_w = h_g \qquad (7.53)$$

Thus the dry-bulb temperature lines are straight but not parallel, since h_g varies with temperature. If we arbitrarily choose some dry-bulb temperature line t to be vertical, by Eqs. (7.52) and (7.53) we have

$$\tan\beta = \frac{S}{h_{g,t}} \qquad (7.54)$$

Equation (7.54) establishes the inclination of the enthalpy lines. The grid of h and W lines may then be constructed.

The remaining procedure for constructing a complete psychrometric chart will now be reviewed.

Saturation Curve. The saturation curve is a locus of points representing saturated air. For standard barometric pressure, values of h_s and W_s at various temperatures may be read from Table A.4 and the points plotted. For other barometric pressures, Eq. (7.18) and Eq. (7.21) may be used.

Dry-Bulb Temperature Lines. We have already shown that, within the accuracy of the perfect-gas approximation, these lines are straight. By Eqs. (7.52) and (7.53),

$$\cot\theta = \frac{h_g}{S} - \cot\beta$$

For various dry-bulb temperatures we may determine h_g and calculate $\cot\theta$. One point on each line may be conveniently located at $W = 0$, and knowledge of the angle θ then lets us draw the line.

Thermodynamic Wet-Bulb Temperature Lines. By Eq. (7.14),

$$q' = h_f^* = \frac{h_s^* - h}{W_s^* - W}$$

For a line of constant wet-bulb temperature, Eq. (7.52) becomes

$$\cot \theta = \frac{h_f^*}{S} - \cot \beta$$

Thus thermodynamic wet-bulb temperature lines are identically straight. For various values of t^*, we may determine θ and extend the lines through one known point. Locations are known on the saturation curve. For unsaturated air, a point may be calculated by solving Eq. (7.14) or Eq. (7.29).

Volume Lines. Lines of constant specific volume, v, are not strictly straight, but their curvature is so slight that they may be drawn as straight lines. For a line of constant volume, with the perfect-gas approximation, we may show that

$$q' = \frac{dh}{dW} = \left(\frac{c_{pw} - 1.608\, c_{pa}}{R_a} \right) \left[\frac{Pv}{(1 + 1.608W)^2} \right] - c_{pw} T_0 + h_{g0} \qquad (7.55)$$

The temperature T_0 is the reference temperature in absolute units in defining the enthalpy of the mixture, i.e., $T_0 = 460 \ °R$ (273 K). An average W may be used in Eq. (7.55) with little error. (For example, a value of $W = 0.01$ is acceptable.) This leads to the result that the slope of the constant-volume line expressed by Eq. (7.55) is only a function of the chosen v and the pressure, which is constant for the entire chart. This slope, when substituted into Eq. (7.52), yields a value for the angle θ. For the chosen v, a value of dry-bulb temperature may be found at $W = 0$. Through knowledge of θ, we may then extend the volume line.

Relative-Humidity Lines. By Eqs. (7.18) and (7.27), we have

$$W = 0.622 \frac{\phi P_{ws}}{P - \phi P_{ws}}$$

For chosen values of ϕ and t, we may calculate values of W and plot points for a line of constant relative humidity.

Enthalpy-Moisture Ratio Protractor. A convenient aid in many psychrometric-chart problems is a protractor showing the direction of straight lines for various values of the enthalpy-moisture ratio q'. Such a protractor may be directly calculated from Eq. (7.52).

Sensible-Heat-Ratio Protractor. Many psychrometric charts have a sensible-heat-ratio (SHR) protractor. For a given process from state 1 to state 2 the *sensible-heat ratio* is defined as the sensible enthalpy change divided by the total enthalpy change

$$\text{SHR} = \frac{\Delta h_s}{\Delta h} \qquad (7.56)$$

The relationship between the SHR and the enthalpy-moisture ratio can be found by combining Eqs. (7.56), (7.40), and (7.46):

$$\text{SHR} = \frac{\Delta h - \Delta h_L}{\Delta h} = 1 - \frac{\overline{h}_g \, \Delta W}{\Delta h}$$

Thus

$$\text{SHR} = 1 - \frac{\overline{h}_g}{q'} \tag{7.57}$$

or

$$q' = \frac{\overline{h}_g}{1 - \text{SHR}} \tag{7.58}$$

In order to draw a SHR protractor, a constant value of $\overline{h}_g$ is selected. Therefore, these lines are approximate.

7.7 THE ASHRAE PSYCHROMETRIC CHARTS

Eight psychrometric charts are included with this text. These are the English and S.I. versions of the three ASHRAE sea-level pressure charts (low, normal, and high temperature) and the ASHRAE charts for high elevation (5000 ft and 1500 m) at normal temperatures.

The three ASHRAE psychrometric charts are similar in format. We will limit further discussion in this section to Fig. C-8E and SI. Humidity-ratio lines are horizontal and are shown for the range from zero (dry air) to 0.03 lbm moisture/lbm dry air or to 30 g of moisture/kg dry air. Enthalpy lines are obliquely drawn across the chart in intervals of 5 Btu/lbm dry air or 10 kJ/kg dry air. All enthalpy lines are precisely parallel. Edge scales for enthalpy are shown above the saturation curve and at the bottom and right-hand margins.

Dry-bulb temperature lines are shown at intervals of 1 °F or 1 °C. The dry-bulb temperature lines are drawn straight, are inclined slightly from the vertical position, and are not strictly parallel to one another. Thermodynamic wet-bulb temperature lines are obliquely drawn across the chart in intervals of 1 °F or 1°C. Their directions differ but slightly from that of the enthalpy lines. The thermodynamic wet-bulb temperature lines are exactly straight but are not strictly parallel to each other.

Relative-humidity lines are shown in intervals of 10 percent from zero ($W = 0$) to 100 percent (saturation curve). Volume lines are obliquely drawn straight lines in intervals of 0.5 ft³/lbm dry air or 0.01 m³/kg dry air. The volume lines are not strictly parallel to one another.

A narrow region above the saturation curve has been developed for fog conditions. A *fog* is a mechanical mixture of saturated moist air and water droplets, both at the same temperature. A fog is equivalent to the condition produced by taking saturated moist air and adiabatically supersaturating it with water at the same temperature. Thus an isotherm in the fog region is an extension of a thermodynamic wet-bulb temperature line.

A protractor is shown to the left of the chart body. The protractor shows two scales—one for the ratio of enthalpy difference to humidity-ratio difference ($q' = \Delta h/\Delta W$) and one for the sensible-total heat ratio (SHR). The S.I. version includes a second sensible-heat-ratio protractor. The protractor scale is along the right-hand side of the chart, and the center is at the small circle located at 24 °C dry-bulb temperature and 50 percent relative humidity. From Eq. (7.57) and examination of the protractors, we see that values of approximately $\overline{h}_g = 1100 \text{ Btu/lbm}_w$ and $\overline{h}_g = 2500 \text{ kJ/kg}_w$ were used in the construction of the protractors in Figs. C-8E and C-8SI, respectively.

EXAMPLE 7.7

Moist air exists at a condition of 100 °F dry-bulb temperature, 65 °F thermodynamic wet-bulb temperature, and 14.696 psia pressure. Determine (a) the humidity ratio, (b) enthalpy, (c) dew-point temperature, (d) relative humidity, and (e) the volume.

Solution: The state-point may be located on Fig. C-8E at the intersection of the 100 °F dry-bulb temperature line and the 65 °F thermodynamic wet-bulb temperature line.

(a) Read $W = 0.00523$ lbm water vapor/lbm dry air.

(b) Through use of two triangles, draw a line parallel to nearest enthalpy line (30 Btu/lbm dry air) through the state-point to the nearest edge scale. Read $h = 29.80$ Btu/lbm dry air.

(c) The dew-point temperature may be read at the intersection of $W = 0.00523$ lbm water vapor per lbm dry air with the saturation curve. Thus, $t_d = 40.1$ °F.

(d) Read $\phi = 13$ percent.

(e) The volume may be accurately found by linear interpolation between the volume lines for 14.0 and 14.5 ft^3/lbm$_a$ air. Thus, $v = 14.22$ ft^3/lbm$_a$.

ENDNOTES

1. *ASHRAE Handbook, Fundamentals Volume* (Atlanta: American Society of Heating, Refrigerating and Air Conditioning Engineers, 1993), 6.1 (IP & SI Editions).

2. J. A. Goff, "Standardization of Thermodynamic Properties of Moist Air," *Trans. ASHVE*, 55 (1949), 463–464.

3. *Trans. ASHVE*, 55, 473–474.

4. J. A. Goff and S. Gratch, "Thermodynamic Properties of Moist Air," *Trans. ASHVE*, 51 (1945), 125–164.

5. R. W. Hyland and A. Wexler, "Formulations for the Thermodynamic Properties of Dry Air from 173.15 K to 473.15 K and of Saturated Moist Air from 173.15 K to 372.15 K, at Pressures to 5 MPa," *ASHRAE Trans.*, 89, pt. 2 (1983).

6. *ASHRAE Handbook, Fundamentals Volume* (Atlanta: American Society of Heating, Refrigerating and Air Conditioning Engineers, 1993).

7. *ASHRAE Handbook, Fundamentals Volume* (Atlanta: American Society of Heating, Refrigerating and Air Conditioning Engineers, 1977), 5.5–5.6.

8. Richard Mollier, "Ein neues Diagram für Dampfluftgcmischc," *ZVDI*, 67 (September 8, 1923), 869–872.

9. William Goodman, *Air Conditioning Analysis* (New York: The Macmillan Company, 1943), 271–276.

PROBLEMS

7.1 Moist air exists at 80 °F dry-bulb temperature, 0.0150 lbm$_w$/lbm$_a$ humidity ratio, and 14.696 psia pressure. Through the use of Table A.4 and fundamental relations, determine

(a) the dew-point temperature,
(b) relative humidity,
(c) volume in ft^3/lbm$_a$, and
(d) the enthalpy in Btu/lbm$_a$.

7.2 Calculate values of humidity ratio, enthalpy, and volume for saturated air at 14.696 psia pressure, using perfect-gas relations and Table A.1, for temperatures of
(a) 70 °F, and   (b) −20 °F.
Compare your results with those shown in Table A.4.

7.3 The atmosphere within a room is at 70 °F dry-bulb temperature, 50 percent degree of saturation, and 14.696 psia pressure. The inside surface temperature of the windows is 40 °F. Will moisture condense out of the air upon the window glass?

7.4 Assume that the dimensions of the room of Prob. 7.3 are 30 ft by 15 ft by 8 ft high. Calculate the number of pounds of water vapor in the room.

7.5 Calculate values of humidity ratio, enthalpy, and volume for saturated air at 101.325 kPa pressure using perfect-gas relations and Table A.1, for temperatures of
(a) 20 °C, and   (b) −25 °C.
Compare your results with those shown in Table A.4.

7.6 Moist air exists at a dry-bulb temperature of 40 °C, relative humidity of 20 percent, and 101.325 kPa pressure. Find the enthalpy in kJ/kg$_a$. Base the solution on Table A.4 and fundamental relations. Do not use perfect-gas expressions.

7.7 Moist air exists at a dry-bulb temperature of 100 °F, relative humidity of 20 percent, and 14.696 psia pressure. Find the enthalpy in Btu/lbm$_a$. Base the solution on Table A.4 and fundamental relations. Do not use perfect-gas expressions.

7.8 Moist air exists at a dew-point temperature of 65 °F, a relative humidity of 60.3 percent, and a pressure of 14.00 psia. Determine
(a) the humidity ratio in lbm$_w$/lbm$_a$, and
(b) the volume in ft^3/lbm$_a$.

7.9 Moist air exists at a dew-point temperature of 20 °C, a relative humidity of 60.3 percent, and a pressure of 96.5 kPa. Determine
(a) the humidity ratio, and
(b) the volume in m^3/kg$_a$.

7.10 The inside surface temperature of a window in a house is 5 °C. The dry-bulb temperature of air in the house is 20 °C, and the pressure is 101.325 kPa. What is the maximum relative humidity allowable in the house if no condensation is to form on the window?

7.11 Determine the relative humidity and dew-point temperature of moist air at 95 °F dry-bulb temperature, 80 °F thermodynamic wet-bulb temperature, and 13.20 psia pressure.

7.12 Determine the relative humidity and dew-point temperature of moist air at 35 °C dry-bulb temperature, 25 °C thermodynamic wet-bulb termperature, and 91.0 kPa pressure.

7.13 Calculate the dry-bulb temperature of moist air at 30 °C thermodynamic wet-bulb temperature, 0.020 kg$_w$/kg$_a$ humidity ratio, and 90.0 kPa pressure.

7.14 Calculate the enthalpy in Btu per lbm$_a$ of moist air at 70 °F thermodynamic wet-bulb temperature, 34 °F dew-point temperature, and 14.696 psia pressure.

7.15 Calculate the dry-bulb temperature of moist air at 80 °F thermodynamic wet-bulb temperature, 0.01250 lbm$_w$/lbm$_a$ humidity ratio, and 13.00 psia pressure.

7.16 Develop the complete derivation for Eq. (7.3).

7.17 Through the use of basic definitions and perfect-gas relations, derive the following equations:
(a) $v = \dfrac{R_a T}{P}(1 + 1.608W)$
(b) $\phi = 1.608\dfrac{P}{P_{ws}}\left(\dfrac{W}{1 + 1.608W}\right)$

7.18 Through the use of basic definitions and perfect-gas expressions, show that
$$v = v_a + \mu v_{as}$$
reduces to
$$v = \frac{R_a T}{P - P_w}$$

7.19 It is planned to construct an h-W diagram for sea-level pressure from Table A.4E. Scale factor S for the diagram is to be 1000 Btu/lbm water. The 80 °F dry-bulb temperature line is to be vertical. Determine the slope (tan θ) of the 60 °F thermodynamic wet-bulb temperature line.

7.20 For a line of constant volume in ft^3/lbm$_a$ on the Mollier psychrometric chart, and assuming that moist air is a perfect-gas mixture, derive Eq. (7.55). [*Hint:* A good starting place is Eqs. (7.24) and (7.23). Also, the results of Prob. 7.17(a) may be useful in substituting for t in Eq. (7.24), keeping in mind that the temperature t is in °F (°C) while T is in absolute units.]

7.21 Work Prob. 7.1 using the psychrometric chart and compare the answers to those found in Prob. 7.1.

7.22 Work Prob. 7.6 using the psychrometric chart and compare answers.

7.23 What errors would have resulted if you had neglected the deviation from standard atmospheric pressure and had used the psychrometric chart in working Prob. 7.12?

7.24 Moist air exists under conditions of 85 °F dry-bulb temperature, 40 percent relative humidity, and 14.696 psia pressure. By the psychrometric chart, determine
  (a) the dew-point temperature,
  (b) thermodynamic wet-bulb temperature,
  (c) humidity ratio,
  (d) enthalpy, and
  (e) the volume.

7.25 Moist air exists under conditions of 25 °C dry-bulb temperature, 40 percent relative humidity, and 101.325 kPa pressure. By the psychrometric chart, determine
  (a) the dew-point temperature,
  (b) thermodynamic wet-bulb temperature,
  (c) humidity ratio,
  (d) enthalpy, and
  (e) the volume.

7.26 In a psychrometric process, moist air at standard atmospheric pressure goes from state 1 of 55 °F dry-bulb temperature and 50 °F dew-point temperature to state 2 of 100 °F dry-bulb temperature and 30 percent relative humidity. Use the psychrometric chart and answer the following:
  (a) What is the enthalpy-humidity ratio, q', for the process?
  (b) What is the sensible-heat ratio, SHR, for the process?
  (c) For each pound of dry air undergoing the process, how much water vapor, lbm_w, is added?

7.27 In a psychrometric process, moist air at standard atmospheric pressure goes from state 1 of 15 °C dry-bulb temperature and 10 °C wet-bulb temperature to state 2 of 35 °C dry-bulb temperature and 10 °C dew-point temperature. By the psychrometric chart, determine
  (a) the enthalpy-humidity ratio, q', for the process, and
  (b) the sensible-heat ratio, SHR, for the process.
  Notice that the SHR scale on the right side of the chart can be read with much greater accuracy than the one on the protractor to the left of the chart.

7.28 Saturated moist air at 15 °C and standard atmospheric pressure undergoes a process which has a sensible-heat ratio of SHR = 0.70. After the process, the dry-bulb temperature is 40 °C. By the psychrometric chart, determine the following at the end of the process:
  (a) humidity ratio,
  (b) dew-point temperature,
  (c) relative humidity, and
  (d) wet-bulb temperature.

SYMBOLS

A	Coefficient in Eq. (7.3).
c_p	Specific heat of moist air at constant pressure, Btu/$\text{lbm}_a \cdot$ °F or kJ/$\text{kg}_a \cdot$ °C.
$\bar{c}_p$	Average specific heat of moist air at constant pressure for a process, Btu/$\text{lbm}_a \cdot$ °F or kJ/$\text{kg}_a \cdot$ °C.
c_{pa}	Specific heat of dry air at constant pressure, Btu/$\text{lbm}_a \cdot$ °F or kJ/$\text{kg}_a \cdot$ °C.
c_{pw}	Specific heat of water vapor at constant pressure, Btu/$\text{lbm}_a \cdot$ °F or kJ/$\text{kg}_a \cdot$ °C.
H	Enthalpy, Btu or kJ; H_a for dry air; H_w for water or water vapor.
h	Specific enthalpy of moist air, Btu/lbm_a or kJ/kg_a; h_a for dry-air; h_s for saturated moist air at t, $h_{as} = h_s - h_a$.
h_f	Specific enthalpy of liquid water, Btu/lbm_w or kJ/kg_w.
h_g	Specific enthalpy of saturated water vapor, Btu/lbm_w or kJ/kg_w.
$\bar{h}_g$	Average enthalpy of water vapor defined by Eq. (7.44), Btu/lbm_w or kJ/kg_w.
h_{fg}^*	$h_g^* - h_f^*$, Btu/lbm_w or kJ/kg_w.
h_{g0}	Enthalpy of saturated water vapor at 0 °F or 0 °C, Btu/lbm_w or kJ/kg_w.
Δh_L	Latent change of specific enthalpy of moist air for a process, Btu/lbm_a or kJ/kg_a.
Δh_S	Sensible change of specific enthalpy of moist air for a process, Btu/lbm_a or kJ/kg_a.
h_w	Enthalpy of water added to moist air, Btu/lbm_w or kJ/kg_w; $h_w = h_g$ for low-pressure water vapor.

L_h	Horizontal scalar distance on h-W chart.
L_w	Vertical scalar distance on h-W chart.
m	Mass, lbm or kg.
m_a	Mass of dry air, lbm_a or kg_a.
m_w	Mass of water or water vapor, lbm_w or kg_w.
P	Pressure of moist air, psia, psfa, in. Hg or kPa.
P_a	Partial pressure of dry air, psia, psfa, in. Hg, or kPa.
P_w	Pressure of water vapor, psia, psfa, in. Hg, or kPa.
P_0	Standard atmospheric pressure or reference pressure, psia, psfa, in. Hg, or kPa.
q'	$(h_2 - h_1)/(W_2 - W_1)$, Btu/lbm_w or kJ/kg_w.
R	Gas constant, ft-lbf/lbm · °R, Btu/lbm · °R or kJ/kg · K; R_a for dry air and R_w for water vapor.
S	Psychrometric-chart scale factor.
S	Entropy of moist air, Btu/°R or kJ/K.
$S_{a,0}$	Entropy of dry air at reference conditions, Btu/°R or kJ/K.
$S_{w,0}$	Entropy of water vapor at reference conditions Btu/°R or kJ/K.
$\Delta S_{a,T,P}$	Entropy difference between reference state and state T, P for dry air, Btu/°R or kJ/K.
$\Delta S_{a,\text{mix}}$	Entropy of mixing for dry air, Btu/°R or kJ/K.
$\Delta S_{w,T,P}$	Entropy difference between reference state and state T, P for water vapor, Btu/°R or kJ/K.
$\Delta S_{w,\text{mix}}$	Entropy of mixing for water vapor, Btu/°R or kJ/K.
s	Specific entropy of moist air, Btu/lbm_a · °R or kJ/kg_a · K; s_a for dry air; s_s for saturated moist air at t; $s_{as} = s_s - s_a$.
s_h	Enthalpy scale factor for psychrometric chart.
s_w	Humidity-ratio scale factor for psychrometric chart.
$s_{w,0}$	Specific entropy of water vapor at reference conditions, Btu/lbm_w · °R or kJ/kg_w · K.
T	Absolute dry-bulb temperature, °R or K.
t	Dry-bulb temperature, °F or °C.
t_d	Dew-point temperature, °F or °C.
t^*	Thermodynamic wet-bulb temperature, °F or °C.
$\bar{t}$	Average dry-bulb temperature for a process, °F or °C.
v	Specific volume of moist air, ft^3/lbm_a or m^3/kg_a; v_s for saturated moist air.
v_a	Specific volume of dry air, ft^3/lbm_a or m^3/kg_a.
v_{as}	$v_{as} = v_s - v_a$ ft^3/lbm_a or m^3/kg_a.
W	Humidity ratio, $\text{lbm}_w/\text{lbm}_a$ or kg_w/kg_a; W_s for air saturated at t; W_s^* for air saturated at t^*.
x_a	Mol-fraction of dry air, moles dry air per mole moist air.
x_w	Mol-fraction of water vapor, moles water vapor per mole moist air; $x_{w,s}$ for case of saturated air at t.
z	Altitude, ft or m.

Greek Letters

α	$180 - (\theta + \beta)$, deg (see Fig. 7.3).
β	Inclination angle of enthalpy lines (see Fig. 7.3).
θ	Inclination angle of any straight line on psychrometric chart (see Fig. 7.3).
μ	Degree of saturation, $\mu = W/W_s$, dimensionless.
ϕ	Relative humidity, dimensionless.

chapter 8
Psychrometric Processes and Applications

8.1 INTRODUCTION

Many of the problems in processing moist air with various apparatus result in rather complex process lines on a psychrometric chart. This chapter will address these problems by first considering a number of fundamental processes that, in various combinations, are used to construct heating, ventilating, air conditioning, and HVAC systems. The remaining sections of the chapter will describe a variety of these HVAC systems.

All of the processes considered here will be for steady-flow conditions. The total or barometric pressure will be assumed constant throughout the process. This assumption is valid in almost all psychrometric processing problems, since even in actual apparatus, such as finned coils or spray chambers, the pressure drop is typically less than 1 in. of water (less than 7 kPa).

8.2 ELEMENTARY PSYCHROMETRIC PROCESSES

The analysis of psychrometric processes is performed by considering a control volume and then applying the principles of the conservation of mass and energy to the control volume. All of the processes are steady-flow processes with negligible changes in potential and kinetic energy. In addition, generally no shaft work is involved. Thus the steady-state energy equation reduces to

$$\sum_{in} \dot{m}h + \dot{Q} = \sum_{out} \dot{m}h \tag{8.1}$$

The quantity $\dot{Q}$ is the rate of heat added to the control volume. The sign convention used is that $\dot{Q}$ will be taken as positive when heat is added to the control volume. In most cases the control volume selected will be the volume of moist air that is present in an apparatus.

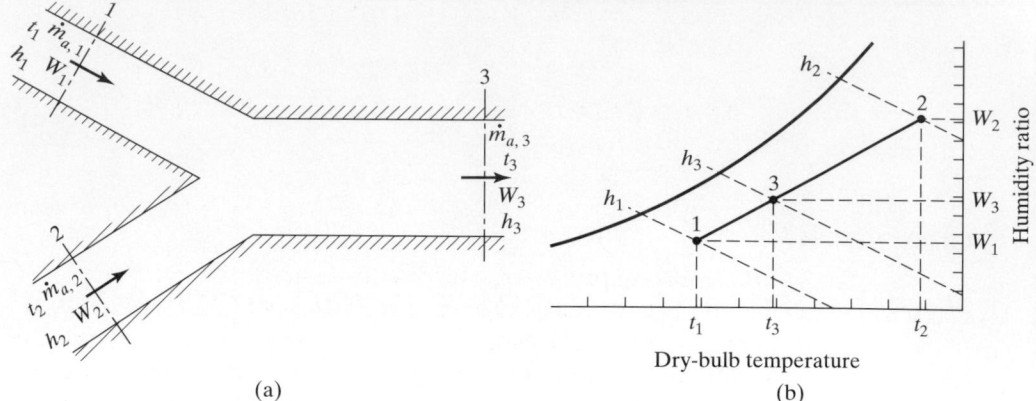

Figure 8.1 Schematic adiabatic mixing of two streams of moist air.

Adiabatic Mixing of Two Streams of Moist Air. In almost every air conditioning system the mixing of two or more air streams may occur. Usually, such mixing processes occur under essentially adiabatic conditions. Figure 8.1(a) schematically shows a mixture of two air streams. The fundamental equations applying to the process are

$$\dot{m}_{a,3} = \dot{m}_{a,1} + \dot{m}_{a,2} \tag{8.2}$$

$$W_3 = \frac{\dot{m}_{a,1}}{\dot{m}_{a,3}} W_1 + \frac{\dot{m}_{a,2}}{\dot{m}_{a,3}} W_2 \tag{8.3}$$

$$h_3 = \frac{\dot{m}_{a,1}}{\dot{m}_{a,3}} h_1 + \frac{\dot{m}_{a,2}}{\dot{m}_{a,3}} h_2 \tag{8.4}$$

By elimination of $\dot{m}_{a,3}$ we obtain

$$\frac{\dot{m}_{a,1}}{\dot{m}_{a,2}} = \frac{h_2 - h_3}{h_3 - h_1} = \frac{W_2 - W_3}{W_3 - W_1} \tag{8.5}$$

or

$$\frac{h_3 - h_2}{W_3 - W_2} = \frac{h_3 - h_1}{W_3 - W_1} \tag{8.6}$$

Equation (8.6) tells us that the process from state 2 to state 3 has the same slope, q', as the process from state 1 to state 3, and, since both processes have state 3 in common, the resulting state 3 must lie on a straight line connecting states 1 and 2 on the psychrometric chart. Furthermore, the lengths of the line segments are proportional to the mass flow rates of dry air mixed. In Fig. 8.1(b) we may write

$$\frac{\dot{m}_{a,1}}{\dot{m}_{a,2}} = \frac{\overline{32}}{\overline{13}} \quad \text{or} \quad \frac{\dot{m}_{a,1}}{\dot{m}_{a,3}} = \frac{\overline{32}}{\overline{12}} \quad \text{or} \quad \frac{\dot{m}_{a,2}}{\dot{m}_{a,3}} = \frac{\overline{13}}{\overline{12}} \tag{8.7}$$

In the United States, air conditioning engineers sometimes use an approximate equation involving the dry-bulb temperatures of the air streams to find the mixed-air condition. By Eqs. (8.4), (7.22), (8.2), and (8.3)

$$\dot{m}_{a,3}(c_{pa} + W_3 c_{pw})t_3 = \dot{m}_{a,1}(c_{pa} + W_1 c_{pw})t_1 + \dot{m}_{a,2}(c_{pa} + W_2 c_{pw})t_2 \tag{8.8}$$

or, in terms of the specific heats of the mixtures,

$$\dot{m}_{a,3} c_{p,3} t_3 = \dot{m}_{a,1} c_{p,1} t_1 + \dot{m}_{a,2} c_{p,2} t_2$$

If we make the approximation that the specific heats of the mixtures are approximately equal, the result is

$$t_3 \cong \frac{\dot{m}_{a,1}}{\dot{m}_{a,3}} t_1 + \frac{\dot{m}_{a,2}}{\dot{m}_{a,3}} t_2 \qquad (8.9)$$

For most mixing processes, the accuracy resulting from the use of Eq. (8.9) is well within the accuracy with which one can read the psychrometric chart.

There are several choices in the solution of a mixing problem. We may solve a problem algebraically using the mixing equations, or we may use a graphical solution on the psychrometric chart. The following example illustrates one convenient method of solution.

EXAMPLE 8.1

One stream of moist air (1000 ft³/min, 60 °F dry-bulb temperature, 56 °F thermodynamic wet-bulb temperature) is mixed with a second stream (400 ft³/min, 80 °F dry-bulb temperature, 67 °F thermodynamic wet-bulb temperature). Barometric pressure is 14.696 psia. Determine the dry-bulb and wet-bulb temperatures of the resulting mixture.

Solution: We may locate states 1 and 2 on Fig. C-8E and determine that $v_1 = 13.28$ ft³/lbm$_a$ and $v_2 = 13.85$ ft³/lbm$_a$. Thus

$$\dot{m}_{a,1} = \frac{1000}{13.28} = 75.3 \text{ lbm}_a/\text{min}$$

$$\dot{m}_{a,2} = \frac{400}{13.85} = 28.9 \text{ lbm}_a/\text{min}$$

By Eq. (8.2)

$$\dot{m}_{a,3} = 75.3 + 28.9 = 104.2 \text{ lbm}_a/\text{min}$$

In Fig. C-8E, connect states 1 and 2 with a straight line. Then using either Eq. (8.3) or (8.4), it is possible to locate where state 3 lies on the line. For this example problem, use Eq. (8.3). From Fig. C-8E

$$W_1 = 0.00867 \text{ lbm}_w/\text{lbm}_a$$

$$W_2 = 0.01123 \text{ lbm}_w/\text{lbm}_a$$

Thus by Eq. (8.3)

$$W_3 = \frac{75.3}{104.2} (0.00867) + \frac{28.9}{104.2} (0.01123) = 0.00938 \text{ lbm}_w/\text{lbm}_a$$

In Fig. C-8E the intersection of W_3 with the line connecting states 1 and 2 is state 3. We find that $t_3 = 65.4$ °F and $t_3^* = 59.3$ °F.

Notice that the approximation for t given by Eq. (8.9) would have resulted in

$$t_3 = \frac{75.3}{104.2} (60) + \frac{28.9}{104.2} (80) = 65.5 \text{ °F}$$

Standard Air. Before we proceed to other fundamental processes, it is useful to describe what ASHRAE defines as *standard air* [1]. Since the specific volume of moist air varies appreciably, all calculations must be made on the basis of mass, not volume, flow rates of air. However, volumetric flow rates are often required for the selection of heating or cooling coils, fans, ducts, etc. In those cases volume values based on measurements at *standard conditions* may be used for accurate results. The value specified by ASHRAE for the specific volume of *standard air*, v_{std}, is

$$v_{\text{std}} = 13.33 \text{ ft}^3/\text{lbm}_a (0.830 \text{ m}^3/\text{kg}_a)$$

or the density of standard air, ρ_{std}, is

$$\rho_{\text{std}} = 0.075 \text{ lbm}_a/\text{ft}^3 (1.204 \text{ kg}_a/\text{m}^3)$$

For standard atmospheric pressure this density corresponds to that of saturated air at about 60 °F (15 °C) and dry air at about 69 °F (20 °C). Using the definition of standard air, we can express the mass flow rate of dry air through an apparatus as

$$\dot{m}_a = \rho_{\text{std}} \dot{V}_{\text{std}} \tag{8.10}$$

where $\dot{V}_{\text{std}}$ is the volumetric flow rate of standard air.

Because air usually passes through coils, fans, ducts, etc. at a density close to standard, the accuracy desired normally requires no correction. When airflow is to be specified at a particular condition or point, such as at a coil entrance or exit, the true specific volume of the air can be read from the psychrometric chart.

EXAMPLE 8.2

Calculate the volumetric flow rate of standard air of the moist air stream at state 1 of Ex. 8.1.

Solution: In Ex. 8.1 the entering volumetric flow rate at state 1 was 1000 ft³/min at 13.28 ft³/lbm$_a$, or $\dot{m}_a$ = 75.3 lbm$_a$/min. By Eq. (8.10)

$$\dot{V}_{\text{std}} = \frac{\dot{m}_a}{\rho_{\text{std}}} = \frac{75.3}{0.075} = 1004 \text{ ft}^3/\text{min of standard air}$$

Sensible Heating or Cooling of Moist Air. If heat is added to moist air with no addition of moisture, then we speak of the process as one of *sensible heating.* Such a process may occur if moist air is passed across a heated surface, such as a bundle of finned tubes, where a medium such as hot water or steam circulates inside the tubes.

Sensible cooling is the reverse of sensible heating and may occur if air is passed across a cool surface. To restrict the process to only sensible cooling, the surface temperature must be higher than the air dew-point temperature. Figure 8.2(a) shows a schematic device for heating air, and Fig. 8.2(b) shows the process on a schematic psychrometric chart. The humidity ratio remains constant.

The steady-flow energy and material-balance equations are

$$\dot{m}_{a,1} h_1 + {}_1\dot{Q}_2 = \dot{m}_{a,2} h_2$$

$$\dot{m}_{a,1} = \dot{m}_{a,2}$$

$$\dot{m}_{a,1} W_1 = \dot{m}_{a,2} W_2$$

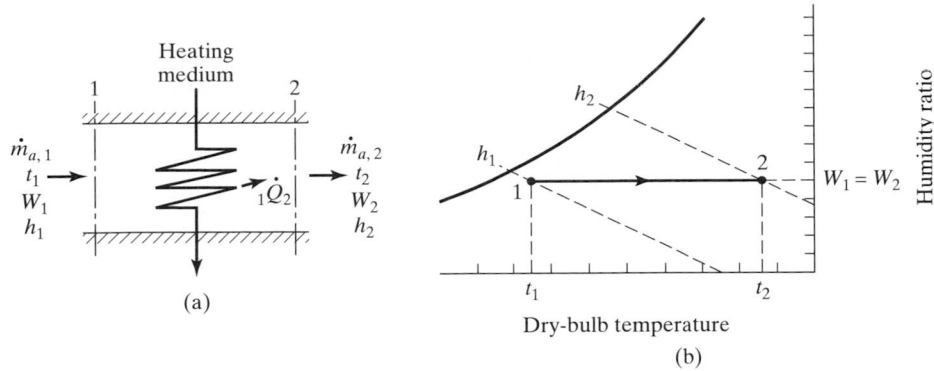

Figure 8.2 Schematic illustration of sensible heating of moist air.

Thus for sensible heating

$$_1\dot{Q}_2 = \dot{m}_a(h_2 - h_1) \tag{8.11}$$

Since the humidity ratio remains constant, it is closely true by Eqs. (7.22) and (8.11) that

$$_1\dot{Q}_2 = \dot{m}_a(c_{pa} + c_{pw}W)(t_2 - t_1) \tag{8.12}$$

In this process, as in others to follow, bulk or average properties are used at all of the inlets and outlets of the device being considered. Also, any duct walls which form the boundaries of the control volume are taken to be adiabatic.

EXAMPLE 8.3

Moist air enters a steam-heating coil at 40 °F dry-bulb temperature and 36 °F thermodynamic wet-bulb temperature at a rate of 2000 ft³/min. Barometric pressure is 14.696 psia. The air leaves the coil at a dry-bulb temperature of 140 °F. Determine the lbm/hr of saturated steam at 220 °F required, if the condensate leaves the coil at 200 °F.

Solution: By Fig. C-8E we find that $W_1 = 0.00359$ lbm$_w$/lbm$_a$, $h_1 = 13.47$ Btu/lbm$_a$, and $v_1 = 12.66$ ft³/lbm$_a$. By Fig. C-9E at $t_2 = 140$ °F and $W_2 = 0.00359$ lbm$_w$/lbm$_a$, we read $h_2 = 37.70$ Btu/lbm$_a$. Also,

$$\dot{m}_a = \frac{2000}{12.66}(60) = 9479 \text{ lbm}_a/\text{hr}$$

By Eq. (8.11),

$$_1\dot{Q}_2 = (9479)(37.70 - 13.47) = 229{,}676 \text{ Btu/hr}$$

or, by Eq. (8.12),

$$_1\dot{Q}_2 = (9479)[0.240 + (0.444)(0.00359)](140 - 40) = 229{,}000 \text{ Btu/hr}$$

For the steam, by Table A.1E, h_g (at 220 °F) = 1153.28 Btu/lbm$_w$. For the condensate, by Table A.1E, h_f (at 200 °F) = 168.34 Btu/lbm$_w$. Thus, the rate of steam flow is

$$\dot{m}_s = \frac{229{,}000}{1153.28 - 168.34} = 232.5 \text{ lbm}_w/\text{hr}$$

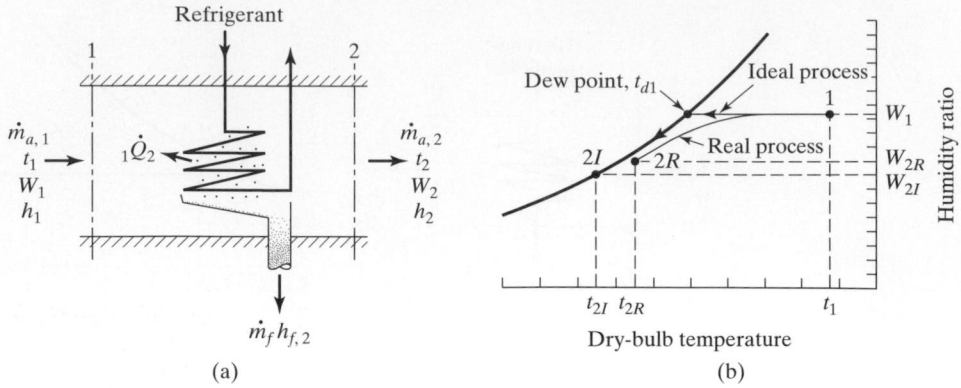

Figure 8.3 Schematic illustration of dehumidification by cooling.

Dehumidification of Moist Air by Cooling.   If moist air is cooled below its dew point, condensation of moisture will occur. Figure 8.3(a) shows a schematic cooling device, and Fig. 8.3(b) shows schematically the psychrometric-chart solution when moist air is cooled below its initial dew-point temperature. Two cases are depicted in Fig. 8.3(b), an ideal process and a real process. The ideal case corresponds to one in which the air is uniformly and perfectly contacted by the cooling coil. In this case no condensation occurs until the average or bulk temperature of the air reaches the dew-point temperature. As the temperature is further reduced, the process follows the saturation line to the final state.

In a real process the air does not come into perfect or uniform contact with the heat-exchanger surfaces. A typical heat-exchanger design consists of a series of flat, parallel, cooled metal surfaces which form passages through which the moist air flows. (A detailed description of the heat exchanger is presented in Chapter 11.) The temperature of the air flowing through the passage is nonuniform. In the entrance region of the passage the air temperature near the surfaces will drop below the dew-point temperature, while that near the center line of the passage remains above. This results in dehumidification, even though the average air temperature in the passage is above the dew point. Since process lines on the psychrometric chart represent average or bulk conditions for the air, the real process appears as the curved line shown in Fig. 8.3(b). How closely a real process approaches the ideal and whether the final state is saturated depends on the design of the cooling coil.

The condensate can leave the apparatus at various temperatures ranging from the initial dew point to the final temperature, t_2. It is customary to assume that the condensate leaves at the final temperature, t_2. Thus the steady-flow energy equation is

$$_1\dot{Q}_2 + \dot{m}_{a,1}h_1 = \dot{m}_{a,2}h_2 + \dot{m}_f h_{f,2}$$

The conservation of mass for the dry air and water results in

$$\dot{m}_{a,1} = \dot{m}_{a,2} = \dot{m}_a$$

and

$$\dot{m}_{a,1}W_1 = \dot{m}_{a,2}W_2 + \dot{m}_f$$

Thus

$$\dot{m}_f = \dot{m}_a(W_1 - W_2) \tag{8.13}$$

and

$$_1\dot{Q}_2 = \dot{m}_a[(h_2 - h_1) - (W_2 - W_1)h_{f,2}] \tag{8.14}$$

Since heat is removed from the air, the numerical value of $_1\dot{Q}_2$ will be negative. For convenience, an air conditioning equipment load or refrigeration load, $\dot{Q}_R$, is sometimes defined. This quantity is simply

$$\dot{Q}_R = |_1\dot{Q}_2| \tag{8.15}$$

Equation (8.14) represents the total rate of heat transferred from the moist air. The condensate term, $(W_2 - W_1)h_{f,2}$, is small compared to the enthalpy difference and is often neglected. Although it is becoming less common, the refrigeration load or capacity is sometimes expressed in "tons," where one ton of refrigeration is equal to 12,000 Btu/hr.

An alternate approach to analyzing the cooling/dehumidifying process uses a so-called *bypass factor*. Consider a real process shown on the psychrometric chart in Fig. 8.4. In addition to the actual process line, Fig. 8.4 contains a straight dashed line which connects states 1 and 2 and which is extended to the saturation curve at state *d*. This point represents the apparatus dew-point temperature of the cooling coil. As discussed above, in a real cooling process such as the one shown in Fig. 8.4 all of the air passing between the surfaces of the heat exchanger or cooling coil is not cooled to the surface temperature. Thus the cooling coil performs as if a portion of the air were brought to saturation at the apparatus dew point or cooling-coil temperature, t_d, while the remaining air bypassed the cooling coil. The final state, (state 2), is then thought of as the adiabatic mixture of the bypassed air, which is at state 1, and the saturated air at state *d*. Figure 8.5 schematically depicts this representation of the process.

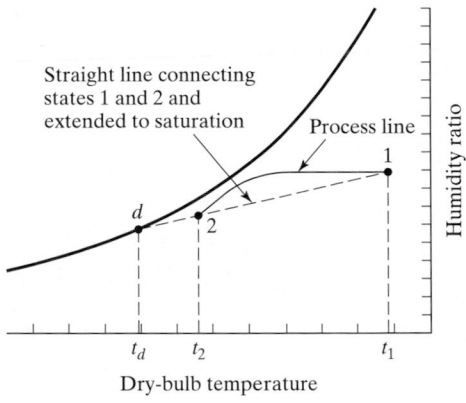

Figure 8.4 Schematic psychrometric chart of a real cooling/dehumidifying process.

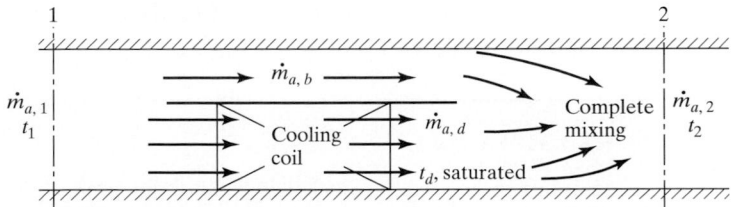

Figure 8.5 Schematic representation of the bypass-factor approach for a cooling/dehumidifying process.

The mass flow rate, $\dot{m}_{a,b}$, is thought of as bypassing the coil while $\dot{m}_{a,d}$ passes through the coil. The continuity equation for dry air gives

$$\dot{m}_{a,1} = \dot{m}_{a,2} = \dot{m}_a$$

$$\dot{m}_{a,b} + \dot{m}_{a,d} = \dot{m}_a$$

A *bypass factor*, b, is defined as

$$b = \frac{\dot{m}_{a,b}}{\dot{m}_a} \tag{8.16}$$

Applying Eq. (8.9) for the mixing process provides

$$t_2 = \frac{\dot{m}_{a,d}}{\dot{m}_a} t_d + \frac{\dot{m}_{a,b}}{\dot{m}_a} t_1$$

or

$$\frac{\dot{m}_{a,b}}{\dot{m}_a} = \frac{t_2 - t_d}{t_1 - t_d} \tag{8.17}$$

Thus from Eqs. (8.16) and (8.17)

$$b = \frac{t_2 - t_d}{t_1 - t_d} \tag{8.18}$$

This analysis approach also divides the total process into sensible and latent components $\dot{Q}_{RS}$ and $\dot{Q}_{RL}$, respectively.

$$\dot{Q}_R = \dot{Q}_{RS} + \dot{Q}_{RL} \tag{8.19}$$

By Eq. (7.45)

$$\dot{Q}_{RS} = \dot{m}_a \bar{c}_p (t_1 - t_2) \tag{8.20}$$

By Eq. (8.18)

$$t_1 - t_2 = (t_1 - t_d)(1 - b)$$

Thus an expression for the rate of sensible cooling can be written in terms of the bypass factor and apparatus dew-point temperature

$$\dot{Q}_{RS} = \dot{m}_a \bar{c}_p (t_1 - t_d)(1 - b) \tag{8.21}$$

The total cooling capacity is found through the use of the sensible-heat ratio:

$$\dot{Q}_R = \frac{\dot{Q}_{RS}}{\text{SHR}} = \frac{\dot{m}_a \bar{c}_p (t_1 - t_d)(1 - b)}{\text{SHR}} \tag{8.22}$$

The sensible-heat ratio can be read directly off the psychrometric chart for the straight line connecting states 1 and d.

EXAMPLE 8.4

Moist air enters a cooling coil at 28 °C dry-bulb temperature and 50 percent relative humidity at a flow rate of 1.5 kg_a/s. Barometric pressure is 101.325 kPa. The air leaves at 13 °C dry-bulb temperature and 90 percent relative humidity. Calculate the air conditioning load on the coil.

Solution: Using Fig. C-8SI and Table A.1SI, we find $W_1 = 0.0118$ kg$_w$/kg$_a$, $h_1 = 58.2$ kJ/kg$_a$, $W_2 = 0.0084$ kg$_w$/kg$_a$, $h_2 = 34.2$ kJ/kg$_a$, and $h_{f,2} = 53.61$ kJ/kg$_w$.

By Eq. (8.14)

$$_1\dot{Q}_2 = 1.5[(34.2 - 58.2) - (0.0084 - 0.0118)(53.61)] \text{ kJ/s} = -35.7 \text{ kJ/s}$$

or

$$\dot{Q}_R = 35.7 \text{ kW}$$

(Note that if we had neglected the condensate term, the answer would have been $\dot{Q}_R = 36.0$ kW.)

EXAMPLE 8.5

Work Ex. 8.4 using the bypass-factor approach. Find the sensible, latent, and total loads on the coil, the apparatus dew-point temperature, and the bypass factor.

Solution: Drawing a straight line through states 1 and 2 and extending it to saturation, we read the apparatus dew-point temperature as

$$t_d = 10.3 \,°\text{C}$$

By Eq. (8.18), the bypass factor is

$$b = \frac{13 - 10.3}{28 - 10.3} = 0.153$$

By Eq. (8.21), the sensible load is

$$\dot{Q}_{RS} = 1.5(1.02)(28 - 10.3)(1 - 0.153) = 22.9$$

Using the protractor on the psychrometric chart, we find the sensible heat ratio for the process to be

$$\text{SHR} = 0.63$$

By Eq. (8.22), the total load on the coil is

$$\dot{Q}_R = \frac{22.9}{0.63} = 36.3 \text{ kW}$$

By Eq. (8.19), the latent load on the coil is

$$\dot{Q}_{RL} = 36.3 - 22.9 = 13.4 \text{ kW}$$

Humidification of Moist Air. A psychrometric process employed frequently is that of adding only moisture to air passing through a chamber. No other energy is added to the air, and the moisture may be in either vapor or liquid form. (It may also be solid, although this case occurs infrequently.) Figure 8.6 shows schematically a device for humidifying moist air in steady flow. This type of device operates such that all moisture added in the chamber is retained by the air passing through. The steady-flow mass balance for water and the energy equation are

$$\dot{m}_w = \dot{m}_a(W_2 - W_1)$$
$$\dot{m}_w h_w = \dot{m}_a(h_2 - h_1)$$

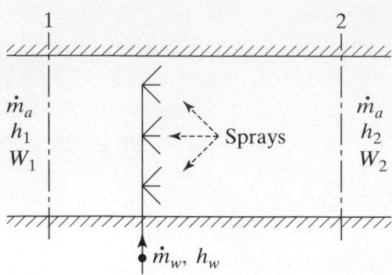

Figure 8.6 Schematic of humidification by the injection of water (liquid or vapor) into moist air.

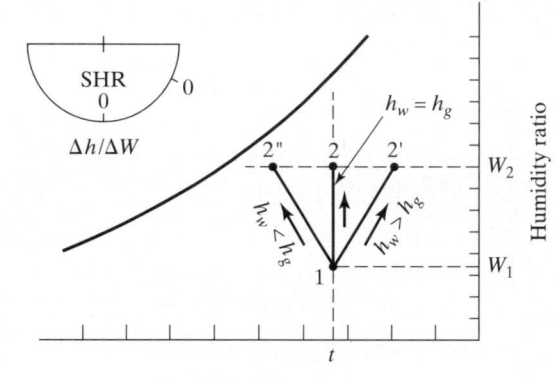

Figure 8.7 Schematic illustration of humidification processes.

Thus

$$q' = \frac{h_2 - h_1}{W_2 - W_1} = h_w \tag{8.23}$$

The direction of the condition line connecting states 1 and 2 depends on the enthalpy of the moisture added. Two unique cases will be mentioned. Each has been considered previously.

Equation (7.53) states that when air is humidified at constant dry-bulb temperature, the steam added must have a specific enthalpy equal to that of saturated steam at the air dry-bulb temperature. If water at the air thermodynamic wet-bulb temperature is added, the entering and leaving air wet-bulb temperatures must be identical.

Figure 8.7 schematically shows several humidification condition lines. The constant dry-bulb temperature line divides the processes into two categories. If $h_w > h_g$, the air will be sensibly heated as well as humidified with a moisture spray. If $h_w < h_g$, the air will be cooled during the process of humidification.

EXAMPLE 8.6

Moist air enters a humidifier at 40 °C dry-bulb temperature and 20 °C wet-bulb temperature. Wet steam at 101.3 kPa and 75 percent quality (i.e., 25 percent liquid) is injected into the moist air flow. The moist air exits the humidifier with a relative humidity of 60 percent. Determine the dry-bulb temperature of the air as it leaves the humidifier.

Solution: The solution to this example can be carried out using the psychrometric chart. Equation (8.23) shows that the slope, $\Delta h / \Delta W$ or q', is equal to the enthalpy of the steam which is injected. Therefore, through the use of two straightedges draw a line through state 1 parallel to the line on the protractor with slope $q' = h_w$. The intersection of this line with the 60 percent relative-humidity curve locates state 2.

From Table A.1SI at 101.3 kPa, $h_f = 419.51$ kJ/kg$_w$ and $h_g = 2675.60$ kJ/kg$_w$.
At a quality of 75 percent

$$h_w = 419.51 + (0.75)(2675.60 - 419.51) = 2111.58 \text{ kJ/kg}_w$$

Therefore the slope on the chart is $q' = 2.1$ kJ/gram of water.

Figure 8.8 shows the construction on the psychrometric chart. After locating state 2, we read the dry-bulb temperature as $t_2 = 34\,°C$.

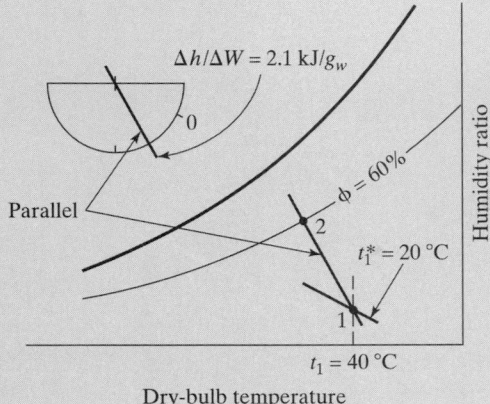

Figure 8.8 Psychrometric chart for Ex. 8.6.

EXAMPLE 8.7

Moist air is heated and humidified by passing it first over a heating coil and then adding moisture, as shown schematically in Fig. 8.9(a). The moist air enters the system at 40 °F dry-bulb and 36 °F thermodynamic wet-bulb temperature at a rate of 235 lbm$_a$/min. The humidifier injects saturated steam at 230 °F. The moist air exits the system at 90 °F dry-bulb temperature and 40 percent relative humidity. Locate state 2 on a psychrometric chart and determine the rate of heat addition by the heating coil and the rate of mass addition by the humidifier.

Solution: This example illustrates the operation of a typical heating/humidification unit, where the desired initial and final moist air states are known or specified and the problem is to size the components. State 2 can be found by the intersection of the sensible heating-process line (i.e., constant W) from state 1 and the humidifying-process line $q' = h_w$ passing through state 3. From Table A.1E at 230 °F, $h_w = 1156.93$ Btu/lbm$_w$. The psychrometric chart for this example is shown in Fig. 8.9(b). The following values are read from the chart:

$$h_1 = 13.4 \text{ Btu/lbm}_a, \qquad W_1 = 0.0036 \text{ lbm}_w/\text{lbm}_a$$

$$h_3 = 35.0 \text{ Btu/lbm}_a, \qquad W_3 = 0.0122 \text{ lbm}_w/\text{lbm}_a$$

$$W_2 = W_1 = 0.0036 \text{ lbm}_w/\text{lbm}_a, \quad t_2 = 87.6\,°F, \quad h_2 = 25.1 \text{ Btu/lbm}_a$$

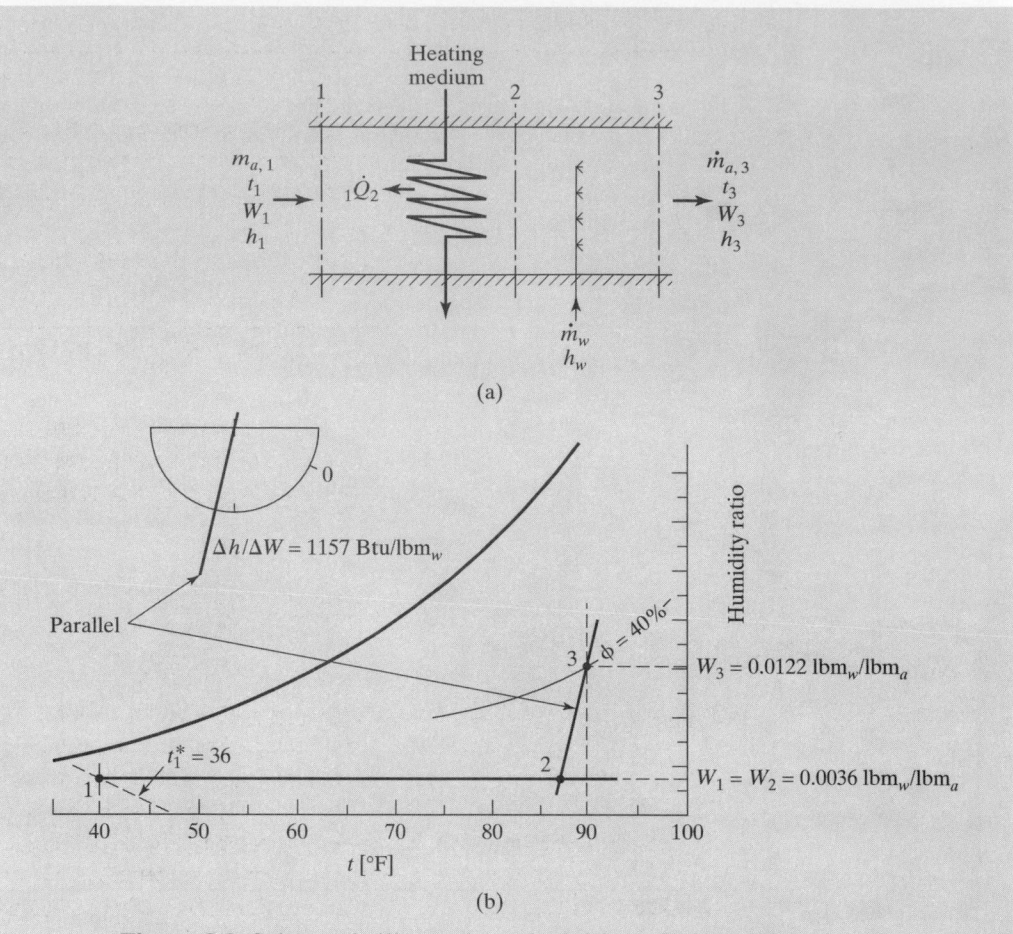

Figure 8.9 Schematic illustration of addition of heat and moisture to moist air.

The rate of heat addition is given by Eq. (8.11):

$$_1\dot{Q}_2 = 235(25.1 - 13.4) = 2750 \text{ Btu/hr}$$

The rate of water added is

$$\dot{m}_w = \dot{m}_a(W_3 - W_2) = 235(0.0122 - 0.0036) = 2.02 \text{ lbm}_w/\text{hr}$$

An alternate approach to solving the example is to solve for h_2 using Eq. (8.20) and observing from the sensible-heating process that $W_2 = W_1$.

Evaporative Cooling of Moist Air. In hot and dry climates evaporative cooling can be an effective means of reducing air temperature. Rather than pass the air through a cooling coil, we can take advantage of the low humidity to achieve cooling. This is accomplished by passing the air stream through a spray device using directly recirculated water. Such a device is called an *evaporative cooler* or *air washer*. Figure 8.10 shows a schematic diagram of such a device. Owing to the low relative humidity, part of the liquid-water stream evaporates. The energy for the evaporation process comes from the air

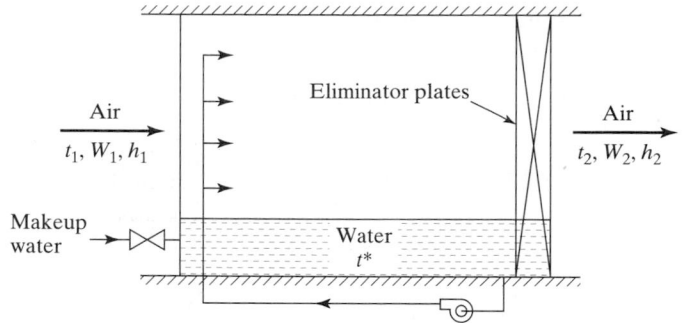

Figure 8.10 Schematic diagram of an evaporative cooler (air washer) using directly recirculated spray water.

stream. Thus, the overall effect is to cool and humidify the air. The equivalent effect may be carried out by passing the air through wetted media.

In this type of device liquid water is lifted from a sump to a distributing system from which it runs down through pads (wetted media), and the portion of the liquid not evaporated then returns to the sump to be recirculated.

A detailed analysis of the evaporative-cooling process is presented in Sec. 10.2. The analysis is carried out for the conditions that (a) the evaporation rate is much smaller than the water-recirculation rate, thus making the energy introduced by the makeup water negligible, (b) the walls of the device are adiabatic, and (c) the power of the recirculation pump is negligible. For our present purposes the key results of the analysis are:

1. The air passing through the evaporative cooler undergoes a constant wet-bulb temperature process.
2. All the liquid in the apparatus is at the wet-bulb temperature of the air stream.

Figure 8.11 shows the evaporative-cooling process on a psychrometric chart. The minimum temperature to which the air can be cooled is the wet-bulb temperature, t^*. The extent to which the leaving air temperature approaches this minimum temperature (or

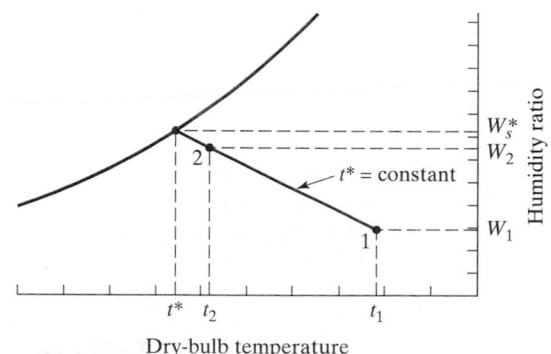

Figure 8.11 Psychrometric chart showing an evaporative-cooling process.

the extent to which complete saturation is approached) is referred to as the *saturation effectiveness*, e_c, and is defined as

$$e_c = \frac{t_1 - t_2}{t_1 - t^*} \tag{8.24}$$

Note that in Sec. 10.2 an air-washer efficiency will be defined as

$$\eta_w = \frac{W_2 - W_1}{W^* - W_1}$$

and for the assumption that the specific heat of the moist air passing through the apparatus is constant the efficiency can also be expressed as

$$\eta_w = \frac{t_1 - t_2}{t_1 - t^*}$$

However, for an evaporative cooler ASHRAE specifically points out the preferred use of the terminology *saturation effectiveness* and the definition given in Eq. (8.24) [2].

EXAMPLE 8.8

Moist air enters an evaporative cooler at 95 °F dry-bulb temperature and 10 percent relative humidity. The evaporative cooler has a saturation effectiveness of 85 percent. Determine the dry-bulb temperature and relative humidity of the air exiting the evaporative cooler.

Solution: Using Fig. C-8E at $t_1 = 95$ °F dry-bulb and $\phi_1 = 10$ percent relative humidity, we find $t^* = 60.6$ °F. From Eq. (8.24)

$$t_2 = t_1 - e_c(t_1 - t^*) = 95 - (0.85)(95 - 60.6) = 65.8 \text{ °F}$$

Thus state 2 is at $t_2 = 65.8$ °F and $t^* = 60.6$ °F, which, from Fig. C-8E, results in

$$\phi_2 = 75 \text{ percent}$$

Condition Line for a Space. A space which is to be conditioned can consist of a single room, a group of rooms, or an entire building. The space is subject to energy and moisture gains or losses (which are usually called the *loads* on the space). Typical energy gains and losses include (1) sensible-heat transfer through walls, roofs, windows, etc., (2) sensible-internal-energy gains from lighting, equipment, and occupants, (3) sensible- and latent-energy transfer due to infiltration, and (4) latent-internal-energy gains from equipment and occupants.

Typical moisture gains or losses occur due to infiltration and internal sources. Common practice is to determine the totals of the rate of sensible-energy gain or loss, the rate of latent-energy gain or loss, and the rate of moisture addition or removal for a space. The procedure for calculating the net gains or losses are presented in Chapters 14, 15, and 16.

Once the total energy and moisture loads for a space are known, the thermodynamic state of the supply air and the required air-flow rate may be determined. Figure 8.12 shows the schematic of the flow processes for an air conditioned space. The quantity $\sum \dot{Q}_S$ represents the sum of all rates of sensible-heat gain (i.e., sensible load). The quantity $\sum \dot{Q}_L = \sum \dot{m}_w h_w$ represents the sum of all rates of energy gain from added moisture (i.e., latent load). The quantity $\sum \dot{m}_w$ represents the net sum of all rates of moisture gain.

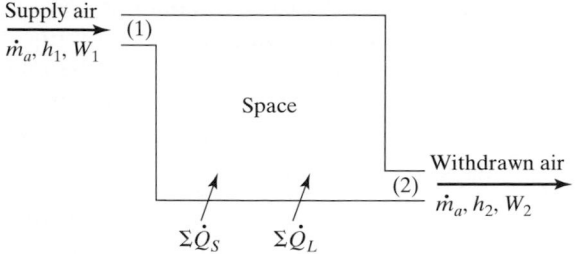

Figure 8.12 Schematic flow processes for an air conditioned space.

Assuming steady-flow conditions, we have

$$\Sigma \, \dot{Q}_S + \Sigma \, \dot{Q}_L = \dot{m}_a(h_2 - h_1) \qquad (8.25)$$

$$\Sigma \, \dot{m}_w = \dot{m}_a(W_2 - W_1) \qquad (8.26)$$

Thus the enthalpy-moisture ratio, q', is given by

$$q' = \frac{\Sigma \, \dot{Q}_S + \Sigma \, \dot{Q}_L}{\Sigma \, \dot{m}_w} = \frac{h_2 - h_1}{W_2 - W_1} \qquad (8.27)$$

Equation (8.27) reveals that for a given state of the withdrawn air, all possible psychrometric states (conditions) for the supply air must lie on the same straight line, which has a direction q' and which is drawn through state 2 on the psychrometric chart. This straight line is called the *condition line*. Figure 8.13 shows three schematic condition lines. If $q' = h_{g,2}$, the condition line coincides with the dry-bulb temperature line t_2. Once q' is known, the condition line may be drawn with the aid of the chart protractor.

According to Eq. (8.27), state 1 may lie at any location on the condition line. However, changing of state 1 requires a change in the air-flow rate $\dot{m}_a$. Practical considerations influence the location of state 1. In summer-comfort air conditioning systems, $(t_2 - t_1)$ may vary from approximately 15 to 25 °F (10 to 15 °C), depending upon the method of air distribution.

The procedure given by Eqs. (8.25)–(8.27) is consistent with basic thermodynamics. It is a realistic procedure, since all quantities involved have definite physical meaning. Further, the procedure is precisely compatible with the *h-W* type of psychrometric chart.

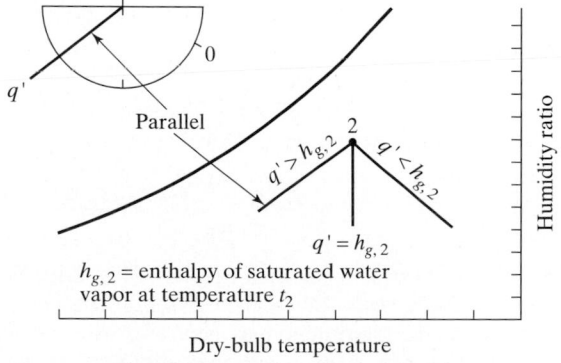

Figure 8.13 The condition line.

In the United States it is common for air conditioning engineers to use a more approximate procedure than that given by Eqs. (8.25)–(8.27). Little use is made of the property enthalpy and no use is made of the rate of moisture gain $\Sigma \dot{m}_w$. The approach followed uses the two types of heat-gain rates: sensible-heat gain $\Sigma \dot{Q}_S$ and latent-heat gain $\Sigma \dot{Q}_L$. For psychrometric calculations the approximations given by Eqs. (7.45) and (7.46) are used. Thus

$$\Sigma \dot{Q}_S = \dot{m}_a \bar{c}_p (t_2 - t_1) \qquad (8.28)$$

$$\Sigma \dot{Q}_L = \dot{m}_a \bar{h}_g (W_2 - W_1) \qquad (8.29)$$

As previously discussed, constant values are generally used for $\bar{c}_p$ and $\bar{h}_g$. The specific heat is commonly taken as 0.245 Btu/lbm$_a$ · °F (1.02 kJ/kg$_a$ · °C), and, depending upon the reference, values selected for $\bar{h}_g$ vary from about 1050 to 1150 Btu/lbm$_w$ (2500 to 2700 kJ/kg$_w$). In Sec. 7.7 it was shown that, in constructing the protractors included in the normal-temperature, sea-level psychrometric charts (Figs. C-8E and C-8SI), values of approximately $\bar{h}_g = 1100$ Btu/lbm$_w$ and $\bar{h}_g = 2500$ kJ/kg$_w$ were used. Therefore, it will be convenient to use these values for our calculations.

When using Eqs. (8.28) and (8.29) it is also customary to use the sensible-heat ratio instead of the enthalpy-moisture ratio. By definition

$$\text{SHR} = \frac{\Sigma \dot{Q}_S}{\Sigma \dot{Q}_S + \Sigma \dot{Q}_L} \qquad (8.30)$$

Thus, using this approach the space-condition line is the straight line which has a direction SHR and which is drawn through state 2 on the psychrometric chart.

EXAMPLE 8.9

The air in a restaurant is to be maintained at 75 °F dry-bulb temperature and 50 percent relative humidity. The load calculations for the restaurant estimate the sensible rate of heat gain to be $\Sigma \dot{Q}_S = 178,000$ Btu/hr. The rate of moisture gain is 95 lbm$_w$/hr with an average enthalpy of the moisture of $h_w = 1095$ Btu/lbm$_w$. Determine (a) the required dew-point temperature of the supply air, and (b) the required volume flow rate of supply air in ft^3/min. The supply air is to be at a dry-bulb temperature of 60 °F. Assume standard atmospheric pressure. The solution to the example will be given first using the enthalpy-humidity ratio and then using the procedure given by Eqs. (8.28) and (8.29).

Solution 1: (a)

$$\Sigma \dot{Q}_L = \Sigma \dot{m}_w h_w = 95(1095) = 104,000 \text{ Btu/hr}$$

By Eq. (8.27)

$$q' = \frac{\Sigma \dot{Q}_S + \Sigma \dot{Q}_L}{\Sigma \dot{m}_w} = \frac{178,000 + 104,000}{95} = 2968 \text{ Btu/lbm}_w$$

Following the notation of Fig. 8.12, we have $t_2 = 75$ °F, $\phi_2 = 50$ percent. By Fig. C-8E, we find $W_2 = 0.00928$ lbm$_w$/lbm$_a$ and $h_2 = 28.15$ Btu/lbm$_a$. As shown by Fig. 8.13, the condition line may be drawn with the aid of the chart protractor. At the intersection of $t_1 = 60$ °F with the condition line, state 1 is established. We read $t_{d,1} = 49$ °F.

(b) At state 1, $h_1 = 22.45$ Btu/lbm$_a$, $W_1 = 0.00740$ lbm$_w$/lbm$_a$, and $v_1 = 13.25$ ft^3/lbm$_a$. By Eq. (8.25)

$$\dot{m}_a = \frac{282{,}000}{28.15 - 22.45} = 49{,}474 \text{ lbm}_a/\text{hr}$$

Thus the supply volumetric flow rate is

$$\text{supply volume flow rate} = \frac{\dot{m}_a v_1}{60} = \frac{(49{,}474)(13.25)}{60} = 10{,}930 \text{ ft}^3/\text{min}$$

Solution 2: **(a)**

$$\Sigma \dot{Q}_L = \Sigma \dot{m}_w h_w = 95(1095) = 104{,}000 \text{ Btu/hr}$$

Thus the total load, $\Sigma \dot{Q}_T$, is

$$\Sigma \dot{Q}_T = \Sigma \dot{Q}_S + \Sigma \dot{Q}_L = 282{,}000 \text{ Btu/hr}$$

By Eq. (8.30)

$$\text{SHR} = \frac{178{,}000}{282{,}000} = 0.631$$

Similar to the procedure of Solution 1, but utilizing the SHR instead of q', the space-condition line may be drawn on Fig. C-8E with the aid of the chart protractor. At the intersection of $t_1 = 60$ °F with the condition line, state 1 is established. We read $t_{d,1} = 49$ °F.

(b) At state 1, $v_1 = 13.25$ ft^3/lbm$_a$. By Eq. (8.28) and with $\bar{c}_p = 0.245$ Btu/lbm$_a \cdot$ °F, we have

$$\dot{m}_a = \frac{\Sigma \dot{Q}_S}{\bar{c}_p(t_2 - t_1)} = \frac{178{,}000}{0.245(15)} = 48{,}435 \text{ lbm}_a/\text{hr}$$

Thus

$$\text{supply volume flow rate} = \frac{\dot{m}_a v_1}{60} = \frac{48{,}435(13.25)}{60} = 10{,}700 \text{ ft}^3/\text{min}$$

We may observe that the answers obtained in Ex. 8.9 using the two solution approaches are nearly equal. In later work in this chapter, the procedure given by Eqs. (8.28) and (8.29) will be emphasized.

Example 8.9 shows the actual supply volume flow rate required. As previously noted, commercial air conditioning equipment (such as fans, heating coils, etc.) is usually rated in terms of volume flow rate of standard air. Once the mass flow rate of dry air, $\dot{m}_a$, is determined, the volume flow rate of standard air is easily calculated using Eq. (8.10). (See Ex. 8.2.)

8.3 PSYCHROMETRIC SYSTEMS—SINGLE ZONE

After the heat and moisture loads of a space are calculated and the space-condition line is drawn on the psychrometric chart, a psychrometric system may be chosen to process the moist air. The equipment requirements are greatly influenced by the space conditions desired and by the magnitude of the enthalpy-moisture ratio q'. The system should be capable of maintaining acceptable space conditions at all times. Since the thermal loads of the space may be highly variable, automatic control of the processing equipment becomes an important consideration. In most large buildings the loads will vary throughout the different locations in the building. In order to maintain comfort in the entire building, it will be subdivided into zones. The term *zone* is used to identify an area of the building which has its own thermostatic control. Thus, a zone could be as small as one room or as large as the entire building. In this section we will discuss several types of psychrometric systems for a single zone or space. Our purpose will be to illustrate general principles rather than to attempt to cover all combinations. In the final section of this chapter, examples of multizone systems will be presented.

Figure 8.14(a) shows a schematic system which may produce controlled temperature and humidity conditions within a space which has both heat and moisture losses; Fig. 8.14(b) schematically shows the psychrometric processes. We will first make a few general comments about Fig. 8.14, which will apply also to all other systems discussed in this section.

Energy added to air by a fan causes some rise in temperature without change in humidity ratio. This energy input may be calculated only after the ductwork is designed and processing devices are selected. Usually the energy input is small. In our discussions we will assume that an allowance for energy input by the fan has been included in the space load calculation. Thus, in Fig. 8.14, states 1 and 8 are assumed to be the same.

For each of our systems we will schematically include filters. Although filters (pads of viscous coated fibers, electrostatic type, etc.) do not change the moist air state, they are required for control of particulate matter carried by the air.

In each system we will assume that some outdoor air is positively introduced. Such air may be required for ventilation purposes or as makeup air for exhaust fans. Introduction of outdoor air requires that an equal mass for dry air be removed from the system. We will show this removal as occurring directly from the space [location 3 in Fig. 8.14(a)]. Figure 8.14(a) shows only one fan. In large systems it is common practice to use a recirculating-air fan as well as a supply-air fan. However, this circumstance does not affect the discussions to be given here.

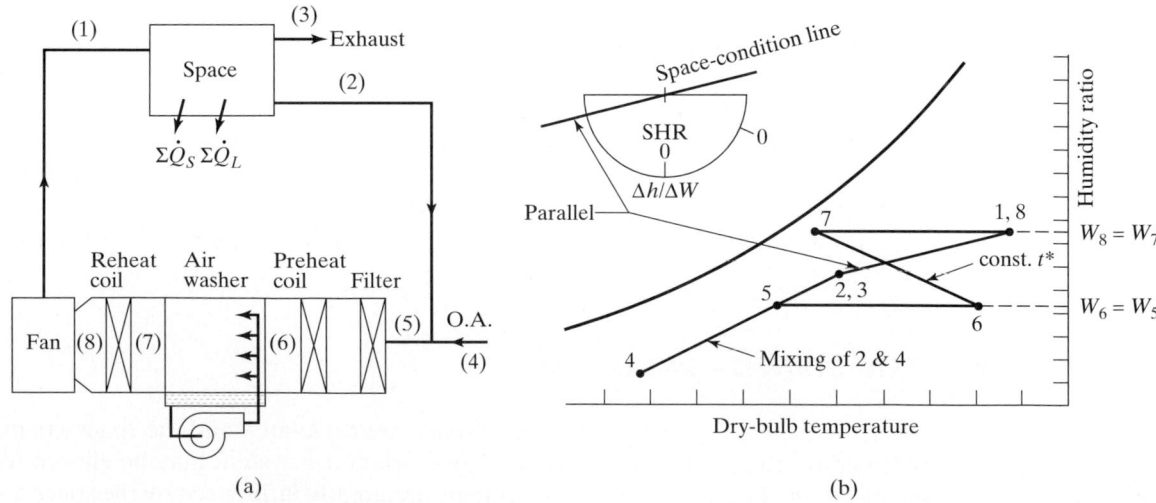

(a)   (b)

Figure 8.14 Schematic winter air conditioning system.

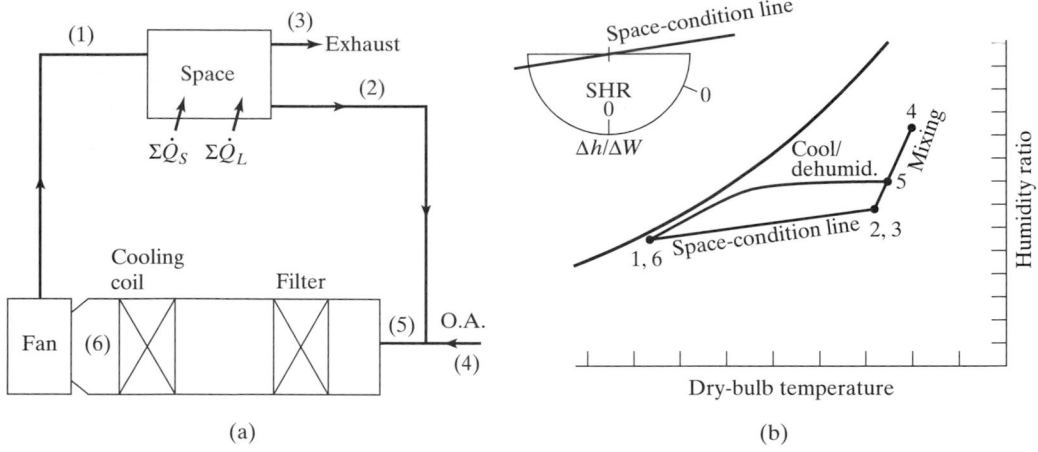

Figure 8.15 Schematic elementary summer air conditioning system.

Processing equipment in Fig. 8.14 includes a preheat coil, air washer (or evaporative cooler), and reheat coil. The heating coils would be of the finned-tube type or electric coils. The preheat coil serves two functions: it prevents water in the air washer from possible freezing, and, by regulation of its heat supply, the amount of water evaporated in the air washer may be controlled. The air washer serves as a humidifying device to offset the moisture losses of the space, and it also accomplishes a certain amount of air cleaning. Regulation of the heat supply by the reheat coil allows control of the space dry-bulb temperature.

The schematic process lines of Fig. 8.14(b) should be obvious. The line $\overline{12}$ is the space-condition line, while line $\overline{452}$ is the mixing line for the mixing of recirculated and outdoor air. The process lines for both heating coils are ones of constant humidity ratio.

Figure 8.15 shows the most elementary form of a summer-comfort air conditioning system; a cooling coil is the only processing device. This coil would be of the finned-tube type with the cooling medium being either chilled water or an evaporating refrigerant. Analysis of cooling coils is given in Chapter 11.

EXAMPLE 8.10

Assume the restaurant of Ex. 8.9 is to be provided with the simple psychrometric system of Fig. 8.15. Outside-air conditions are 92 °F dry-bulb temperature and 77 °F wet-bulb temperature. The rate of exhaust from the restaurant is 4500 standard ft³/min. As a continuation of solution 1 of Ex. 8.9, determine (a) mass flow rate of recirculated air, lbm_a/hr, (b) the thermodynamic state of the moist air entering the cooling coil, and (c) the refrigeration capacity required.

Solution: The nomenclature of Fig. 8.15 will be used. From Ex. 8.9 we have the locations of states 1 and 2 on the psychrometric chart. We also have $\dot{m}_{a,1} = 49,474$ lbm_a/hr.

(a) The rate of exhaust air is given as 4500 standard ft³/min. From Eq. (8.7)

$$\dot{m}_{a,3} = \rho_{std}\dot{V}_{std} = 0.075(4500) = 337.5 \; lbm_a/min$$

or

$$\dot{m}_{a,3} = 20{,}250 \; lbm_a/hr$$

Thus,

$$\dot{m}_{a,2} = \dot{m}_{a,1} - \dot{m}_{a,3} = 49{,}474 - 20{,}250 = 29{,}224 \text{ lbm}_a/\text{hr}$$

(b) The mass flow rate of dry air $\dot{m}_{a,4}$ introduced from outdoors is equal to the mass flow rate of exhaust air, $\dot{m}_{a,3}$. For the mixing process we can write Eq. (8.3), (8.4), or (8.6). For the purposes of this example let us use Eq. (8.3). Thus,

$$W_5 = \frac{\dot{m}_{a,2}}{\dot{m}_{a,5}} W_2 + \frac{\dot{m}_{a,4}}{\dot{m}_{a,5}} W_4$$

$$= \frac{29{,}224}{49{,}474} (0.00928) + \frac{20{,}250}{49{,}474} (0.0166)$$

$$= 0.01228 \text{ lbm}_w/\text{lbm}_a$$

State 5 is located at the intersection of W_5 and the line $\overline{24}$. We find $t_5 = 82.2$ °F, $t_5^* = 69.2$ °F, and $h_5 = 33.25$ Btu/lbm$_a$.

(c) By Eqs. (8.14) and (8.15)

$$\dot{Q}_R = \left|{}_5\dot{Q}_6\right| = \dot{m}_{a,5}[(h_5 - h_6) - (W_5 - W_6)h_{f,6}]$$

$$\dot{Q}_R = (49{,}474)[(33.25 - 22.45) - (0.01228 - 0.00740)(27.63)]$$

$$\dot{Q}_R = 528{,}000 \text{ Btu/hr}$$

As previously indicated, the capacity is occasionally expressed in units of "tons." Thus, in that terminology,

$$\dot{Q}_R = \frac{528{,}000 \text{ Btu/hr}}{12{,}000 \text{ Btu/hr} \cdot \text{ton}} = 44 \text{ tons}$$

The simple system of Fig. 8.15 has limitations. Since the cooling coil is the only processing device, only one property of the moist air can be controlled. In comfort air conditioning systems, this would always be the space dry-bulb temperature.

Figure 8.16 shows a modification to the system of Fig. 8.15 which may be useful during partial-load operation. We assume that face dampers (not shown) on the cooling coil

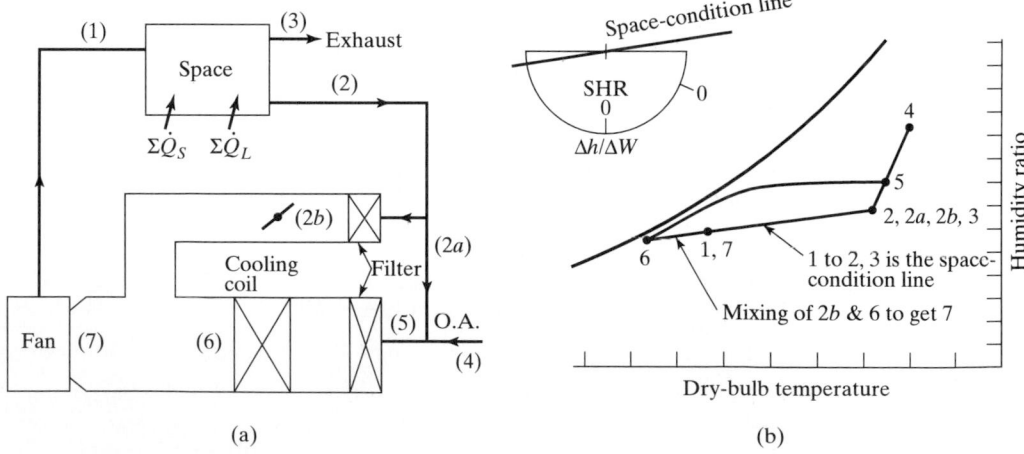

Figure 8.16 Schematic elementary summer air conditioning system with bypass of recirculated air.

and the bypass damper are controlled by a motor which positions them so as to maintain a constant space dry-bulb temperature. As the sensible-heat gain of the space decreases, more recirculated air is bypassed. However, the air which does pass across the coil may be more thoroughly dehumidified than when the full air quantity is handled. Thus, satisfactory space-humidity conditions may be maintained during some partial-load situations without the need for reheat.

Figure 8.16(b) shows the system state-points on a psychrometric chart. Although locations 2, 2a, 2b, and 3 are all at the same state, it is convenient to identify them separately to facilitate keeping track of mass flow rates in working problems. State 5 is the adiabatic mixture of return air with mass flow rate $\dot{m}_{a,2a}$ and outside air at a mass flow rate of $\dot{m}_{a,4} = \dot{m}_{a,3}$. Thus,

$$\dot{m}_{a,5} = \dot{m}_{a,2a} + \dot{m}_{a,4}$$

The supply-air condition, state 1, is assumed to be the same as that before the fan, state 7, which results from the adiabatic mixing of the air leaving the cooling coil (state 6 with a mass flow rate of $\dot{m}_{a,6} = \dot{m}_{a,5}$) and bypass air (state 2b with mass flow rate $\dot{m}_{a,2b}$). Thus state 7 is located along the line connecting states 6 and 2b. Notice that in calculating the cooling load, $_5\dot{Q}_6$, the mass flow rate through the coil is equal to $\dot{m}_{a,5}$, which is less than the supply air flow rate by an amount equal to the bypass air flow rate.

Figure 8.17 shows a summer air conditioning system with a reheat coil. Reheat is required when the space-condition line is so steep that a satisfactory intersection of it with the cooling-coil process line cannot be obtained. When a reheat coil is included, both temperature and humidity of a space may be controlled. However, the use of reheat results in more expensive system operation, since some of the heat added to the air as reheat must be removed in the cooling coil.

Figure 8.17(b) shows the system state-points on a psychrometric chart. In this type of system, the space condition (state 2) and the state of the air leaving the cooling coil (state 6) are often known. State 7, 1 then can be located at the intersection of the space-condition line drawn through state 2 and the sensible-heating process line (constant W) starting at state 6.

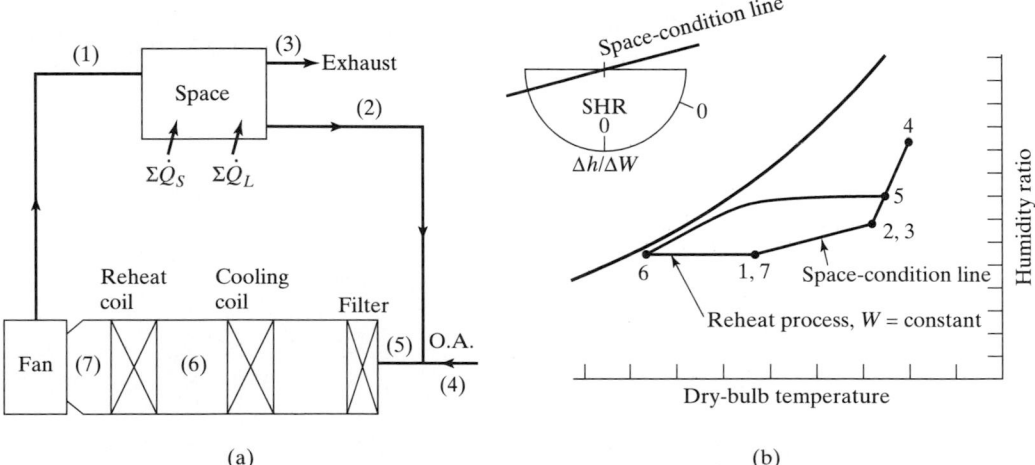

(a)   (b)

Figure 8.17 Schematic summer air conditioning system with reheat.

EXAMPLE 8.11

In order to compare the space conditions produced and the relative energy requirements of bypass and reheat systems, consider a space that has a sensible load of $\Sigma \dot{Q}_S = 205$ kW and a latent load of $\Sigma \dot{Q}_L = 88$ kW when the space is maintained at a dry-bulb temperature of 25 °C and the outdoor conditions are 35 °C dry-bulb temperature and 40 percent relative humidity. The space dry-bulb condition is to be met by using either of the systems shown in Figs. 8.16 or 8.17. Supply air is to be introduced to the space at a flow rate of 30 kg_a/s. The flow rate of exhaust air is 4.5 kg_a/s. The exit conditions of the systems cooling coils are a dry-bulb temperature of 10 °C and a relative humidity of 95 percent. For the bypass system shown in Fig. 8.16 determine (a) the relative humidity in the space and (b) the required system cooling capacity. For the reheat system shown in Fig. 8.17 determine (c) the relative humidity in the space, (d) the rate of heat required for the reheat coil, and (e) the required system cooling capacity.

Solution: The nomenclature of Fig. 8.16 will be used for parts (a) and (b). Standard atmospheric pressure is assumed and the solution carried out using psychrometric chart C-8SI. States 4 and 6 can be located from the given information. States 6, 7, and 2b fall on the mixing line connecting 6 and 2b. The slope of this line is determined by observing that it is also the space-condition line from state 1 to state 2. The sensible-heat ratio for the space-condition line is given by Eq. (8.27).

$$\text{SHR} = \frac{205}{205 + 88} = 0.70$$

(a) State 2 is located at the intersection of the line with an SHR = 0.70 drawn through state 6 and $t_2 = 25$ °C. The resulting point falls at $\phi_2 = 50$ percent. Thus using the system shown in Fig. 8.16, the space conditions will be $t_2 = 25$ °C and $\phi_2 = 50$ percent.

(b) State 5 is on the mixing line connecting states 4 and 2a. (Note that states 2, 2a, 2b, and 3 are the same.) In order to determine the division of the return air between $\dot{m}_{a,2a}$ and $\dot{m}_{a,2b}$, locate state 7, 1. By Eq. (8.28) with $\bar{c}_p = 1.02$ kJ/kg_a °C

$$(30)(1.02)(25 - t_1) = 205$$

$$t_1 = 18.3 \text{ °C}$$

By Eq. (8.9) and observing that $\dot{m}_{a,2b} = \dot{m}_{a,1} - \dot{m}_{a,6}$ and that $\dot{m}_{a,5} = \dot{m}_{a,6}$,

$$\frac{\dot{m}_{a,5}}{\dot{m}_{a,1}} = \frac{t_1 - t_2}{t_6 - t_2} = \frac{18.3 - 25}{10 - 25} = 0.45$$

Thus

$$\dot{m}_{a,5} = 0.45(30) = 13.5 \text{ } kg_a/\text{s}$$

$$\dot{m}_{a,2b} = 30 - 13.5 = 16.5 \text{ } kg_a/\text{s}$$

Also, from Fig. (8.16) it is seen that

$$\dot{m}_{a,2a} = \dot{m}_{a,1} - \dot{m}_{a,3} - \dot{m}_{a,2b} = 30 - 4.5 - 16.5 = 9.0 \text{ } kg_a/\text{s}$$

By Eq. (8.9)

$$t_5 = \frac{\dot{m}_{a,4}}{\dot{m}_{a,5}} t_4 + \frac{\dot{m}_{a,2a}}{\dot{m}_{a,5}} t_{2a} = \frac{4.5}{13.5}(35) + \frac{9.0}{13.5}(25) = 28.3 \text{ °C}$$

Thus, state 5 is located at the intersection of the line connecting 2a and 4 and a dry-bulb temperature of 28.3 °C.

By Eq. (8.14)

$$_5\dot{Q}_6 = \dot{m}_{a,5}[(h_6 - h_5) - (W_6 - W_5)h_{f,6}]$$
$$= 13.5[(28.3 - 57.8) - (0.0073 - 0.0113)(41.12)]$$
$$= -396 \text{ kW}$$

or the system cooling capacity required is

$$\dot{Q}_R = 396 \text{ kW}$$

The nomenclature of Fig. 8.17 will be used to analyze the reheat system.

(c) The dry-bulb temperature of the supply air is found in the same manner as it was in part (b) of this example. Therefore, $t_1 = 18.3$ °C. The state 1, 7 is located by following a sensible-heating process line (constant W) beginning at state 6 and ending at $t_1 = t_7 = 18.3$ °C. State 2 is located by drawing the space-condition line beginning at state 1 and ending at the prescribed space dry-bulb temperature of $t_2 = 25$ °C. In part (a) of this example the SHR for the space-condition line was calculated to be SHR = 0.70. Locating state 2 on the psychrometric chart, we read $\phi_2 = 43$ percent. Thus it is seen that use of the reheat system results in a lower relative humidity in the space than does the use of the bypass system (i.e., 43 percent vs. 50 percent).

(d) Applying Eq. (8.12), the reheat energy rate is

$$_6\dot{Q}_7 = \dot{m}_{a,1}(c_{pa} + W_6 c_{pw})(t_7 - t_6)$$
$$= 30[(1.00 + 0.0073)(1.86)](18.3 - 10)$$
$$= 252 \text{ kW}$$

(e) State 5 is the intersection of t_5 with the adiabatic mixing process line connecting states 2 and 4. Notice that in the reheat system the entire supply-airflow rate passes through the cooling coil. Therefore, $\dot{m}_{a,5} = \dot{m}_{a,1} = 30 \text{ kg}_a/\text{s}$ and

$$\dot{m}_{a,2} = \dot{m}_{a,1} - \dot{m}_{a,3} = 30 - 4.5 = 25.5 \text{ kg}_a/\text{s}$$

By Eq. (8.9)

$$t_5 = \frac{\dot{m}_{a,4}}{\dot{m}_{a,5}}t_4 + \frac{\dot{m}_{a,2a}}{\dot{m}_{a,5}}t_2 = \frac{4.5}{30}(35) + \frac{25.5}{30}(25) = 26.5 \text{ °C}$$

After locating state 5, we read $h_5 = 50.2 \text{ kJ/kg}_a$ and $W_5 = 0.0094 \text{ kg}_w/\text{kg}_a$. By Eq. (8.14)

$$_5\dot{Q}_6 = 30[(28.3 - 50.5) - (0.0094 - 0.0113)(41.12)] = -664 \text{ kW}$$

or the system cooling capacity required is

$$\dot{Q}_R = 664 \text{ kW}$$

The results of the example are summarized in Table 8.1.

TABLE 8.1 Results of Ex. 8.11

Result or Setpoint	Bypass System	Reheat System
Space dry-bulb temp.	25 °C	25 °C
Space relative humidity	50%	43%
Required cooling capacity	396 kW	664 kW
Required reheat	—	252 kW
Total energy rate required	396 kW	916 kW

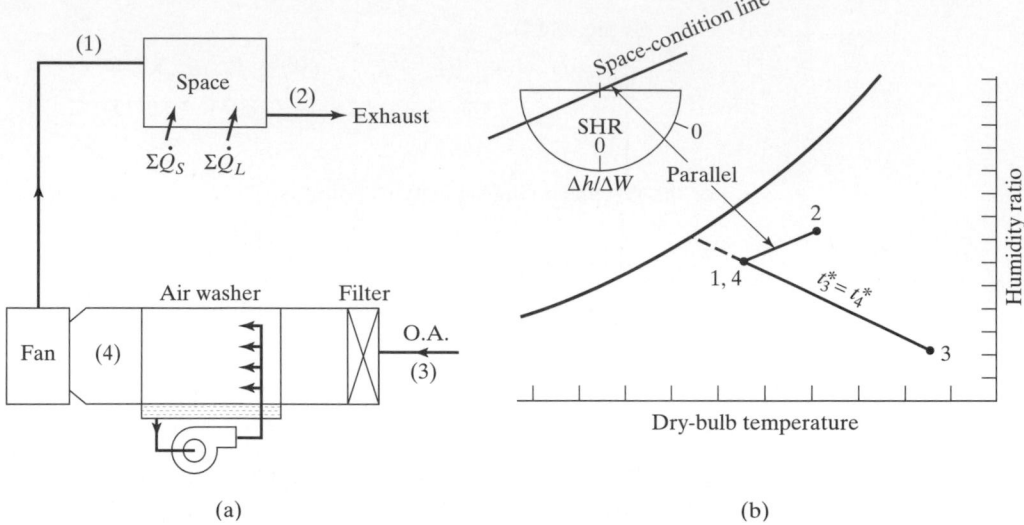

Figure 8.18 Schematic elementary evaporative-cooling system.

This example demonstrates that often a trade-off must be made between tight control of the conditions in a space and the energy needed to achieve that control.

Comfort air conditioning systems capable of maintaining optimum thermal conditions may be expensive to own and operate. Partially effective systems which involve much lesser costs may be attractive where finances preclude the installation of a completely effective system. In hot dry regions, evaporative-cooling systems may be capable of providing considerable relief in enclosed spaces. Such systems may also have application in hot industrial environments. Figure 8.18 shows an elementary evaporative-cooling system. We assume that state 2 is an acceptable space condition, although not necessarily an optimum one. State 3 of the outdoor air is assumed to be at a much higher temperature but lower humidity ratio than State 2. Through use of an air washer as the only processing device, an acceptable air conditioning system may result. Generally, a much higher flow rate of air is used with an evaporative-cooling system than with conventional systems. A high rate of air movement past a person allows the same degree of comfort but with higher effective temperatures, as compared to situations where air movement is slight.

Modifications may be made among the psychrometric systems previously discussed in order to achieve specific purposes. For example, the addition of a cooling coil between the air washer and reheater of Fig. 8.14 would provide a year-round air conditioning system.

In designing a psychrometric system, every effort should be made to utilize heat transfers internal to the system where economically feasible so as to reduce the need for external heat transfers. Figure 8.19 shows one case. We assume winter operation and a system which requires 100 percent outdoor air. The outdoor air is preheated by waste heat in the exhaust air. The same recovery scheme of Fig. 8.19 might be applied for an opposite purpose in Fig. 8.18. Here, the outdoor air might be precooled by the exhaust air, allowing state 2 to be at a lower effective temperature for the same airflow rate. Other possibilities for using internal heat transfers might include the use of refrigerant-condenser heat for reheating of moist air. When outdoor air temperatures are high, reheat-

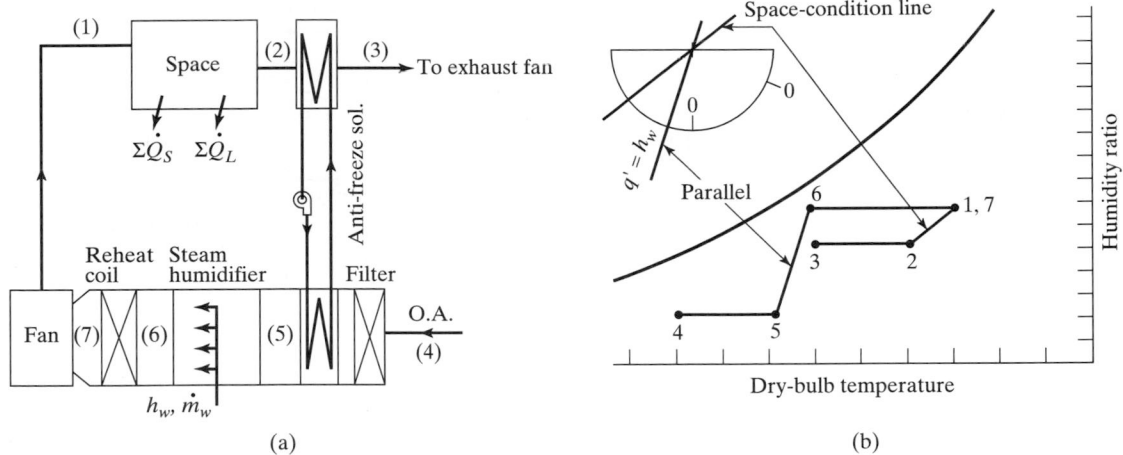

(a)   (b)

Figure 8.19 Schematic winter air conditioning system using 100 percent outdoor air with preheating by waste heat from the exhaust air.

ing may be accomplished internally by a heat-exchanger system involving the reheat coil and a coil in the outdoor-air duct.

In many applications the required refrigerating capacity chargeable to the outdoor air taken into the system is a large fraction of the total. In such cases, and in winter systems where heating is required, the use of outdoor air should be kept to a minimum. However, for spaces which may require cooling during the entire year, provision should be made for flexible use of outdoor air, so that refrigerating equipment may be shut down during cold weather.

8.4 PSYCHROMETRIC SYSTEMS—MULTIPLE ZONE

It was previously mentioned that, in most large buildings, the heating and cooling loads will vary throughout different locations in the building and that in order to maintain comfort in the entire building it is necessary to subdivide the building into zones. This section presents descriptions of three multizone systems which can be used to meet the loads in the various zones and thus provide the desired space conditions in each of them. The three systems described are simplified versions of reheat, variable-air-volume, and dual-duct systems. Again, the purpose is to illustrate general principles rather than attempt to cover all combinations and aspects of multizone systems. Multizone systems are described in greater detail in the *HVAC Systems and Applications Volume* of the *ASHRAE Handbook* series [3].

Reheat.   A multizone reheat system is a modification of the single-zone system previously presented in Fig. 8.17. As the term "reheat" implies, heat is added as a secondary process. Figure 8.20 schematically shows a simple two-zone reheat system. Generally, multizone systems consist of more than two zones. However, the principles associated with them can be illustrated considering just two zones. The system produces conditioned air from a central unit at a fixed cold-air temperature and low humidity level, state 5. The mass flow rates are generally selected to offset the maximum cooling loads

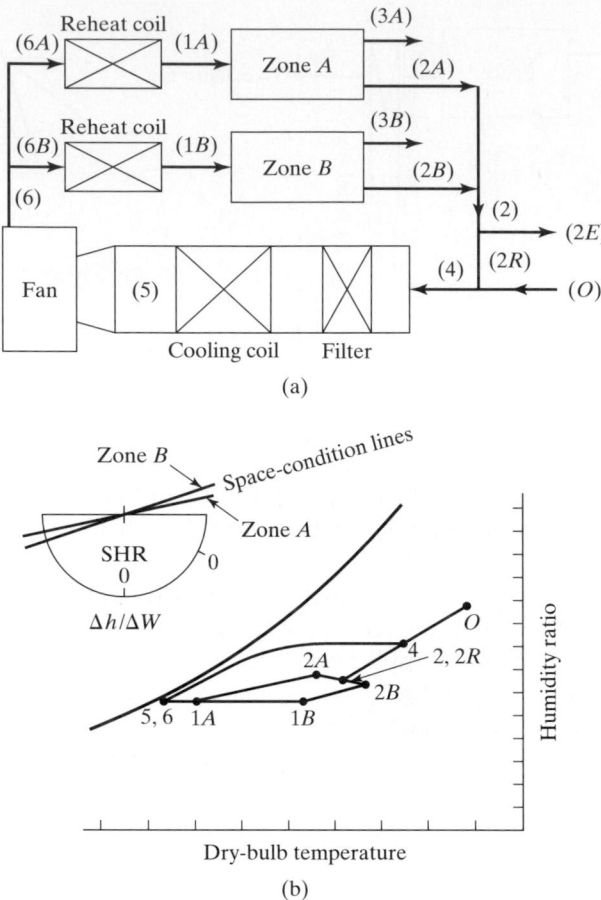

Figure 8.20 Schematic of a two-zone reheat system.

in the zones. A control thermostat in each zone calls for reheat when the cooling load in the zone drops below the design value. Return air from the zones ($\dot{m}_{a,2A}$ and $\dot{m}_{a,2B}$) mixes in the return-air duct, resulting in state 2. A portion of this air ($\dot{m}_{a,2E}$) is expelled to the outside through a main system exhaust duct, while the balance ($\dot{m}_{a,2R}$) is returned and mixed with the outside air, resulting in state 4. In addition to the main system exhaust, each zone may have its own direct exhaust flow. These are shown in Fig. 8.20 as mass flows 3A and 3B. In general, these flows are relatively small and are present to discharge localized air contaminants. The mass flow rate of outside air is equal to the sum of all exhaust flow rates.

EXAMPLE 8.12

A two-zone reheat system similar to the one shown in Fig. 8.20 is to be designed for Minneapolis summer outdoor design conditions of 32 °C dry-bulb temperature and 23 °C coincident wet-bulb temperature. In this design the only exhaust is the main system exhaust. Therefore, the mass flow rates labeled 3A and 3B in Fig. 8.20 are nonexistent. Twenty-five percent of the supply-air mass flow rate to each zone is to be fresh air (outside air). Standard atmospheric pressure is assumed to exist. The cooling coil for the system is to be selected such that the air will exit the coil saturated at 5 °C.

The design conditions for zone A are 22 °C dry-bulb temperature and 40 percent relative humidity. For these design conditions, zone A is predicted to have a sensible-heat gain of 80 kW and a latent-heat gain of 20 kW.

The design conditions for zone B are 26 °C dry-bulb temperature and 30 percent relative humidity. For these design conditions zone B is predicted to have a sensible-heat gain of 75 kW and a latent-heat gain of 25 kW.

For the above design conditions, determine the following: (a) the supply-air mass flow rate for zone A, kg_a/s, (b) the required reheat capacity for zone A, $_{6A}\dot{Q}_{1A}$ (kW), (c) the supply-air mass flow rate for zone B, kg_a/s, (d) the required reheat capacity for zone B, $_{6B}\dot{Q}_{1B}$ (kW), (e) the cooling-coil capacity, $\dot{Q}_R$ (kW).

Solution: The solution will be performed using the notation of Fig. 8.20 and using psychrometric chart Fig. C-8SI.

(a) The mass flow rate for zone A can be determined using Eq. (8.25) once state $1A$ is determined. Since only sensible processes occur between state 5 and state $1A$, $W_{1A} = W_5 = 0.0054\ kg_w/kg_a$. By Eq. (8.30) the SHR of the condition line for zone A is calculated.

$$SHR = \frac{80}{80 + 20} = 0.80$$

With the aid of the SHR protractor, the space-condition line is drawn through the given state $2A$. State $1A$ is the intersection of the condition line with humidity ratio W_{1A}. The result is $t_{1A} = 11.0$ °C. By Eq. (8.28)

$$\dot{m}_{a,1A} = \frac{\Sigma \dot{Q}_S}{c_p(t_{2A} - t_{1A})} = \frac{80}{1.02(22.0 - 11.0)} = 7.13\ kg_a/s$$

(b) By Eq. (8.12)

$$_{6A}\dot{Q}_{1A} = 7.13[1.00 + 1.86(0.0054)](11.0 - 5.0) = 43.2\ kW$$

(c) Using the same procedure as that used in part (a), we calculate the following values for Zone B.

$$SHR = 0.75$$

$$t_{1B} = 20.0\ °C$$

$$\dot{m}_{a,1B} = 12.25\ kg_a/s$$

(d) Using the same procedure as that used in part (b),

$$_{6B}\dot{Q}_{1B} = 12.25[1.00 + 1.86(0.0054)](20.0 - 5.0) = 185.6\ kW$$

(e) The refrigeration or cooling-coil capacity is calculated combining Eqs. (8.14) and (8.15):

$$\dot{Q}_R = \dot{m}_{a,5}[(h_4 - h_5) - (W_4 - W_5)h_{f,5}]$$

In order to determine state 5, we must first determine state 4, which requires knowledge of state $2R$. State $2R$ is the same as state 2, which results from the adiabatic mixing of states $2A$ and $2B$. Thus, state 2 is found by drawing a line connecting points $2A$ and $2B$ and then using any one of the mixing equations (8.3), (8.4), or (8.9) to establish the proper location on the line. By Eq. (8.9)

$$t_2 = \frac{7.13}{19.38}(22.0) + \frac{12.25}{19.38}(26.0) = 24.5\ °C$$

State 4 is a mix of 75 percent return air, state $2R$, and 25 percent outside air, state O. By Eq. (8.9)

$$t_4 = (0.75)(24.5) + (0.25)(32.0) = 26.4\,°C$$

Thus, using the mixing line connecting states $2R$ and O and the temperature $t_4 = 26.4\,°C$, state 4 is established. From the psychrometric chart

$$h_4 = 47.6\,\text{kJ/kg}_a \quad \text{and} \quad W_4 = 0.0083\,\text{kg}_w/\text{kg}_a$$

The conditions of the moist air leaving the cooling coil, state 5, were specified as part of the design. After locating state 5 on the psychrometric chart, it is found that

$$h_5 = 18.6\,\text{kJ/kg}_a \quad \text{and} \quad W_5 = 0.0054\,\text{kg}_w/\text{kg}_a$$

From Table A.6SI $h_{f,5} = 20.44\,\text{kJ/kg}_w$. Therefore, the required refrigeration capacity is

$$\dot{Q}_R = (19.38)[(47.6 - 18.6) - (0.0083 - 0.0054)(20.44)] = 560.9\,\text{kW}$$

Variable Air Volume. A variable-air-volume (VAV) system controls the dry-bulb temperature in a zone by varying the supply-airflow rate rather than the supply-air temperature. A space thermostat provides a control signal to a "VAV unit," which controls the supply airflow. Several VAV unit designs exist. Some are duct-mounted devices which control the flow by varying the settings of dampers or pressure-reducing boxes. Other VAV units control the flow at the terminal diffuser or grille. The fan system is designed to handle the largest simultaneous block load, not the sum of the peaks. At a given time, zones experiencing peak loads essentially borrow the extra air from the zones experiencing off-peak loads. Depending on the system size and complexity and the lowest expected flow rate, additional fan controls may be used to reduce fan power and to limit system noise at part-load operating conditions. Many methods are available to achieve this, including fan-speed control, variable-inlet-vane control, fan bypass, fan-discharge dampers, and variable-pitch fan control. These control methods are described in detail in the *1988 Equipment Volume* of the *ASHRAE Handbook* series [4].

Simple VAV systems typically cool only and have no requirement for simultaneous heating and cooling in various zones. In instances where there are exterior zones in a building that require heating, secondary systems are used to provide the necessary heat. Secondary systems used include baseboard heaters, radiant heaters, and independent constant-volume, variable-temperature air systems. Figure 8.21(a) schematically shows a simple two-zone VAV system. Figure 8.21(b) is a sketch of a psychrometric chart showing the state-points and process lines for the simple VAV system where the two zones are to be maintained at different temperatures, t_{2A} and t_{2B}. The locations of states $2A$ and $2B$ on a psychrometric chart can be determined by constructing the space-condition lines for the zones. Since the VAV units do not change the state of the moist air, states $1A$, $1B$, and 6 are the same. Thus, the space-condition lines can be drawn starting at state 6 and ending at the respective zone dry-bulb temperatures.

The slopes of the space-condition lines are established with the aid of the protractor, using the sensible-heat ratios for the two zones as calculated from their respective loads. Once the conditions in a zone are established, the mass flow rate for that zone can be calculated using Eq. (8.25) or Eq. (8.28). Return air from the zones ($\dot{m}_{a,2A}$ and $\dot{m}_{a,2B}$) mix in the return-air duct, resulting in state 2. A portion of this air ($\dot{m}_{a,2E}$) is expelled to

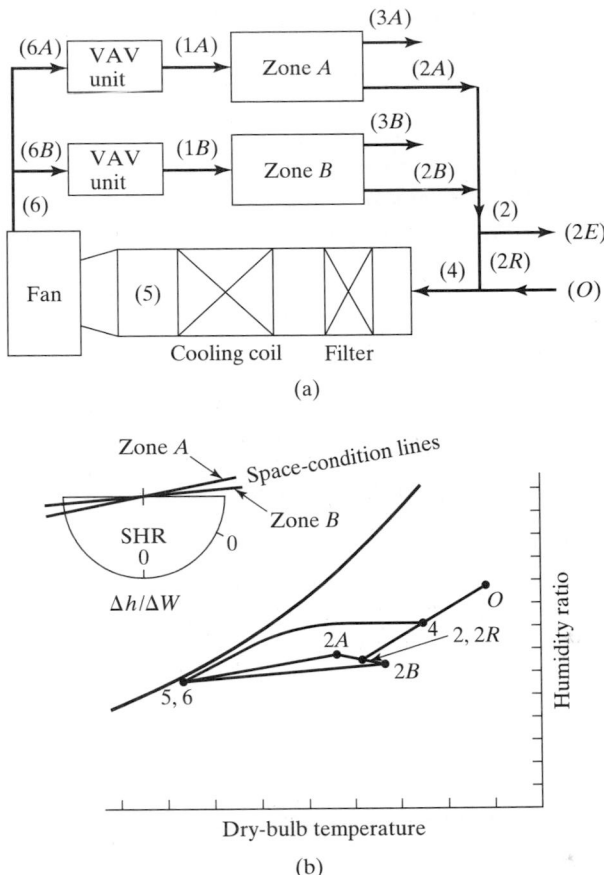

Figure 8.21 Schematic of a two-zone variable-air-volume system.

the outside through a main system exhaust duct, while the balance $(\dot{m}_{a,\,2R})$ is returned and mixed with the outside air, resulting in state 4. In addition to the main system exhaust, each zone may have its own direct exhaust flow. These are shown in Fig. 8.21 as mass flows $3A$ and $3B$. In general, these flows are relatively small and are present to discharge localized air contaminants. The mass flow rate of outside air $(\dot{m}_{a,\,O})$ is equal to the sum of all exhaust flow rates.

The *HVAC Systems and Applications Volume* of the *ASHRAE Handbook* series [3] presents an extensive list of advantages and design precautions for variable-air-volume systems. Some of the advantages listed are:

1. The VAV concept, when combined with one of the perimeter heating systems, offers inexpensive temperature control for multiple zoning.
2. Advantage may be taken of changing loads from lights, occupancy, solar, and equipment to lower the cost for fans, refrigeration, heating, and associated plant auxiliaries.
3. VAV system designs can be made virtually self-balancing.
4. It is easy and inexpensive to subdivide into new zones and to handle increased loads with new tenancy or usage if the system is designed for a potential load increase.

Some of the design precautions listed are:

1. Special care must be exercised in designing the air distribution in a zone to ensure acceptable operation at off-peak supply-airflow rates.
2. Fan controls should be considered in order to save power and for noise control. However, these controls may not be economical for systems of 10,000 ft³/min (5 m³/s) and below.
3. In zones with highly fluctuating loads, heating or reheat may be required to prevent excessive space humidity.

EXAMPLE 8.13

A two-zone variable-air-volume system similar to the one shown in Fig. 8.21 is to be designed for the same building and conditions of Ex. 8.12. Recall that the design is to be conducted for Minneapolis summer outdoor design conditions of 32 °C dry-bulb temperature and 23 °C coincident wet-bulb temperature. The only exhaust is the main system exhaust and, therefore, the mass flow rates labeled 3A and 3B in Fig. 8.21 are nonexistent. Twenty-five percent of the supply-air mass flow rate to each zone is to be fresh air (outside air). Standard atmospheric pressure is assumed to exist. The cooling coil for the system is to be selected such that the air will exit the coil saturated at 5 °C.

The design thermostat setting for zone A is 22 °C dry-bulb temperature. For the design conditions, zone A is predicted to have a sensible-heat gain of 80 kW and a latent-heat gain of 20 kW.

The design thermostat setting for zone B is 26 °C dry-bulb temperature. For the design conditions, zone B is predicted to have a sensible-heat gain of 75 kW and a latent-heat gain of 25 kW.

Notice that, unlike the case of a reheat system, it is not possible to specify both the temperature and humidity in each space when using a variable-air-volume system. Rather, the humidity levels are variables which are determined by the exit conditions of the cooling coil and the space loads.

For the above design conditions, determine the following: (a) the relative humidities in the zones, (b) the mass flow rates of supply air for each zone, $\dot{m}_{a,1A}$ and $\dot{m}_{a,1B}$, (c) the cooling-coil capacity, $\dot{Q}_R$ (kW).

Solution: The solution will be performed using the notation of Fig. 8.21 and using the psychrometric chart, Fig. C-8SI.

(a) The locations of states $2A$ and $2B$ are found by constructing the space-condition lines, using the appropriate sensible-heat ratios, starting at the supply-air state (6) and terminating at the respective zone temperatures (t_{2A} and t_{2B}). As previously determined in Ex. 8.12, the sensible-heat ratios for the two zones are:

$$(SHR)_A = 0.80 \quad \text{and} \quad (SHR)_B = 0.75$$

The construction results in:

$$W_{2A} = 0.0072 \text{ kg}_w/\text{kg}_a, \quad \phi_{2A} = 0.41, \quad h_{2A} = 40.3 \text{ kJ/kg}_a$$

and

$$W_{2B} = 0.0084 \text{ kg}_w/\text{kg}_a, \quad \phi_{2B} = 0.40, \quad h_{2B} = 47.4 \text{ kJ/kg}_a$$

(b) The mass flow rates can be calculated using either Eq. (8.25) or (8.28). Using Eq. (8.25)

$$\dot{m}_{a,1A} = \frac{80 + 20}{40.3 - 18.6} = 4.61 \text{ kg}_a/\text{s}$$

and

$$\dot{m}_{a,1B} = \frac{75 + 25}{47.4 - 18.6} = 3.47 \ \text{kg}_a/\text{s}$$

(c) In order to calculate the cooling-coil capacity, it is necessary to determine state 4, which results from the adiabatic mixing of 25 percent outside air (state O) and 75 percent of the return air (state 2). Applying Eq. (8.4),

$$h_2 = \frac{4.61}{8.08}(40.3) + \frac{3.47}{8.08}(47.4) = 43.3 \ \text{kJ/kg}_a$$

and

$$h_4 = (0.75)(43.3) + (0.25)(67.8) = 49.4 \ \text{kJ/kg}_a$$

and from the psychrometric chart

$$W_4 = 0.0091 \ \text{kg}_w/\text{kg}_a$$

Therefore the refrigeration capacity required is

$$\dot{Q}_R = (8.08)[(49.4 - 18.6) - (0.0091 - 0.0054)(20.44)] = 248.3 \ \text{kW}$$

Dual-Duct Systems. The central apparatus of a dual-duct system produces two air streams, one hot and one cold, which are carried by separate ducts to each zone. At each zone a mixing box, controlled by the zone thermostat, mixes the two air streams in the correct proportion to satisfy the existing loads and maintain the desired zone dry-bulb temperature. A simple form of a dual-duct system is shown in Fig. 8.22. The psychrometric chart shown in Fig. 8.22(b) depicts a cooling application in which the humidifier is not operational. For the purpose of illustration, the temperatures of the two zones are shown being maintained at different values, t_{2A} and t_{2B}. The return airflows from each zone mix adiabatically, resulting in state 2. A portion of the mixed air is exhausted, and the balance is mixed with outside air. The dampers that control the percentages of return and outside air that are mixed have several settings, ranging from a minimum opening for very high outdoor-air temperatures to a maximum opening when outdoor air can be used for cooling without the need for mechanical refrigeration.

When a dual-duct system is used for cooling, the temperature of the air in the cold duct typically is maintained between 50 and 60 °F (10 to 15.5 °C) and the temperature of the air in the warm duct, t_8, typically is maintained at approximately 5 °F (3 °C) higher than the temperature of the return air. Since the proper supply-air state for each zone is achieved by mixing the hot and cold air streams, the locations of the supply-air states $1A$ and $1B$ fall on the adiabatic mixing line connecting states 6 and 8. The lines connecting states $1A$ and $2A$ and states $1B$ and $2B$ are the space-condition lines for zones A and B, respectively.

The system shown in Fig 8.22(a) also provides heating capabilities for winter operation. In this case, the hot-duct temperature is automatically set progressively higher as the outdoor temperature drops. The temperature of the air in the cold duct is maintained between 50 and 60 °F (10 to 15.5 °C) either by using mechanical refrigeration or, when outdoor conditions permit, by using the proper mix of outside and return air.

The principal advantage of a dual-duct system is its ability to maintain good control of temperature and humidity in several zones which may have a relatively wide range of

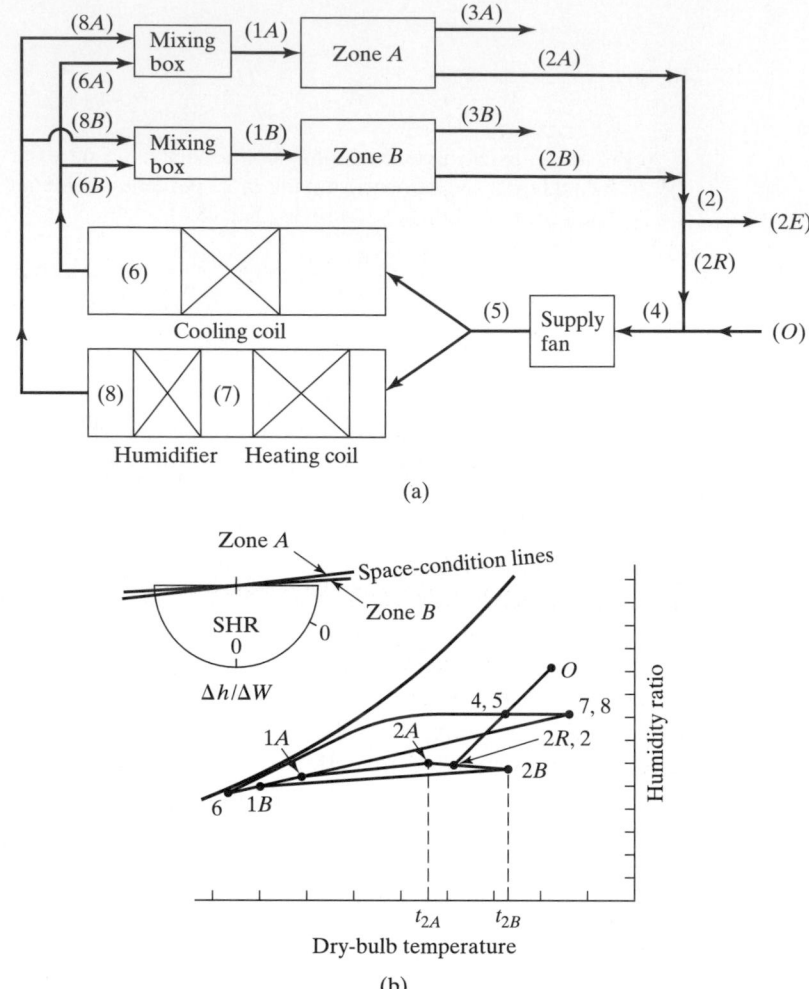

Figure 8.22 Schematic of a two-zone dual-duct system.

loads, including the condition where cooling is required in some zones while other zones require heating. On the other hand, some of the disadvantages include the relatively high energy use associated with both heating and cooling the air streams and the high initial cost associated with the installation of two air ducts.

EXAMPLE 8.14

A two-zone dual-duct system similar to the one shown in Fig. 8.22 is to be designed for the same building and conditions of Ex. 8.12 and 8.13. Recall that the design is to be conducted for Minneapolis summer outdoor design conditions of 32 °C dry-bulb temperature and 23 °C coincident wet-bulb temperature. The only exhaust is the main system exhaust, and, therefore, the mass flow rates labeled 3A and 3B in Fig. 8.22 are nonexistent. Twenty-five percent of the supply-air mass flow rate to each zone is to be fresh air (outside air). Standard atmospheric pressure is assumed to exist. The cooling coil for the system is to be selected such that the air will exit the coil saturated at 5 °C. The humidifier shown in Fig. 8.22 is not used, and the temperature in the hot duct is maintained at 28 °C.

The design thermostat setting for zone A is 22 °C dry-bulb temperature. For the design conditions, zone A is predicted to have a sensible-heat gain of 80 kW and a latent-heat gain of 20 kW. The design supply-airflow rate for zone A is 7.1 kg_a/s.

The design thermostat setting for zone B is 26 °C dry-bulb temperature. For the design conditions, zone B is predicted to have a sensible-heat gain of 75 kW and a latent-heat gain of 25 kW. The design supply-airflow rate for zone B is 12.2 kg_a/s.

This problem is posed in a manner that represents how the system would actually operate; i.e., the exiting temperature of the heating coil, the exiting temperature and humidity of the cooling coil, and thermostat settings are specified. These parameters represent control settings and the operational characteristics of the cooling coil. The humidity levels in the two zones will be determined by the system equilibrium conditions and cannot be specified.

For the above design conditions, determine the following: (a) the mass flow rates of cold and hot air that make up the supply air for zones A and B, (b) the rate of heat addition for the heating coil, $_5\dot{Q}_7$ (kW), (c) the humidity levels in the zones, and (d) the cooling-coil capacity, $\dot{Q}_R$ (kW).

Solution: The solution will be performed using the notation of Fig. 8.22 and using psychrometric chart Fig. C-8SI. Efforts to locate all the state-points on the psychrometric chart reveal that all the temperatures can be calculated but there is insufficient information to establish the humidity levels. If, on the other hand, we knew one additional humidity value, e.g., ϕ_4, we could completely solve the problem. Therefore, the approach will be to calculate as many quantities as possible, using only temperatures, and then use an iterative method to determine the balance of the state-points and thus complete the problem. The iterative method will consist of guessing a humidity level for state 4 and using this value to establish all the states including a calculated value for state 4. When the guess and calculated values for state 4 agree to within the reading accuracy of the psychrometric chart, the location of the state-points is complete. This iterative process is somewhat similar to how the system operates, in that humidity levels would continually adjust until equilibrium was achieved.

(a) First determine the supply-air temperatures for each zone using Eq. (8.28).

$$t_{1A} = t_{2A} - \frac{\Sigma \dot{Q}_S}{\bar{c}_p (\dot{m}_{a, 1A})} = 22.0 - \frac{80}{1.02(7.1)} = 11.0 \text{ °C}$$

Similarly

$$t_{1B} = 26.0 - \frac{75}{1.02(12.2)} = 20.0 \text{ °C}$$

Using Eqs. (8.9) and (8.2),

$$\frac{\dot{m}_{a, 6A}}{\dot{m}_{a, 1A}} = \frac{t_{1A} - t_8}{t_6 - t_8} = \frac{11.0 - 28.0}{5.0 - 28.0} = 0.74$$

or

$$\dot{m}_{a, 6A} = 0.74(7.1) = 5.25 \text{ kg}_a/\text{s of cold air}$$

The mass flow rate of hot air for zone A is

$$\dot{m}_{a, 8A} = 7.1 - 5.25 = 1.85 \text{ kg}_a/\text{s}$$

A similar calculation for zone B results in

$$\dot{m}_{a, 6B} = 4.27 \text{ kg}_a/\text{s of cold air}$$

and

$$\dot{m}_{a,8B} = 7.93 \text{ kg}_a/\text{s of hot air}$$

Equation (8.9) can also be used to calculate the temperatures at states 2 and 4:

$$t_2 = \frac{7.1}{19.3}(22) + \frac{12.2}{19.3}(26) = 24.5\ ^\circ\text{C}$$

and

$$t_4 = t_5 = 0.75(24.5) + 0.25(32) = 26.4\ ^\circ\text{C}$$

(b) The temperatures and mass flow rates are sufficient to calculate the heat addition of the heating coil. Since the humidifier is not in operation, states 7 and 8 are the same. Therefore:

$$\begin{aligned}
{}_5\dot{Q}_7 &= (\dot{m}_{a,8A} + \dot{m}_{a,8B})\bar{c}_p(t_8 - t_5) \\
&= (1.85 + 7.93)(1.02)(28.0 - 26.4) \\
&= 16.0 \text{ kW}
\end{aligned}$$

(c) As previously indicated, it is necessary to iterate to establish the humidity levels in the zones. A number of options exist. We could assume humidity levels in both zones, complete the calculations, and then check the validity of our assumption. Another approach is to assume (guess) a humidity level for the mixed-air condition, state 4, complete the calculations, and check our assumption. Since the latter approach requires guessing only one value, it is the one selected. When we refer to the psychrometric chart and recall that the mixed air is 75 percent return and 25 percent outside air, it appears that a reasonable first iteration would be to guess a relative humidity of $\phi_4 = 40$ percent. State 8 is found by following a line of constant humidity ratio from state 4 to $t_8 = 28\ ^\circ\text{C}$. States 1A and 1B are located along the mixing line connecting states 6 and 8 at temperatures $t_{1A} = 11.0\ ^\circ\text{C}$ and $t_{1B} = 20.0\ ^\circ\text{C}$, respectively. As previously determined in Ex. 8.12, the sensible-heat ratios for zones A and B are 0.80 and 0.75, respectively. Drawing the $(\text{SHR})_A = 0.80$ line starting at state 1A and ending at t_{2A} and drawing the $(\text{SHR})_B = 0.75$ line starting at state 1B and ending at $t_{2B} = 26.0\ ^\circ\text{C}$ results in $\phi_{2A} = 43$ percent and $\phi_{2B} = 40$ percent. State 2 is located on the mixing line connecting 2A and 2B at $t_2 = 24.5\ ^\circ\text{C}$. Thus, $\phi_2 = 41$ percent. Locating state 4 at $t_4 = 26.4\ ^\circ\text{C}$ along the line connecting states 2R and O results in $\phi_4 = 44$ percent as compared to our guess value of 40 percent. This variation is considered large enough to warrant a continuation of the solution, using a second guess value for the humidity of state 4. Recall that convergence of the iterative procedure is considered to have been achieved when the guessed and calculated values agree within the reading accuracy of the psychrometric chart. In this case, we should be able to read the relative humidity to ± 0.02.

Repeating the above calculations using a second guess value of $\phi_4 = 0.45$ satisfies the convergence criteria, and the following state-points are found.

$$t_{1A} = 11.0\ ^\circ\text{C}, \quad \phi_{1A} = 0.80 \quad\quad W_{1A} = 0.0065 \text{ kg}_w/\text{kg}_a, \quad h_{1A} = 27.4 \text{ kJ/kg}_a$$

$$t_{1B} = 20.0\ ^\circ\text{C}, \quad \phi_{1B} = 0.56 \quad\quad W_{1B} = 0.0081 \text{ kg}_w/\text{kg}_a, \quad h_{1B} = 40.6 \text{ kJ/kg}_a$$

$$t_{2A} = 22.0\ ^\circ\text{C}, \quad \phi_{2A} = 0.45 \quad\quad W_{2A} = 0.0119 \text{ kg}_w/\text{kg}_a, \quad h_{2A} = 52.2 \text{ kJ/kg}_a$$

$$t_{2B} = 26.0\ ^\circ\text{C}, \quad \phi_{2B} = 0.42 \quad\quad W_{2B} = 0.0088 \text{ kg}_w/\text{kg}_a, \quad h_{2B} = 48.4 \text{ kJ/kg}_a$$

$$t_2 = 24.5\ ^\circ\text{C}, \quad \phi_2 = 0.43 \quad\quad W_2 = 0.0082 \text{ kg}_w/\text{kg}_a, \quad h_2 = 45.4 \text{ kJ/kg}_a$$

$$t_4 = 26.4\ ^\circ\text{C}, \quad \phi_4 = 0.45 \quad\quad W_4 = 0.0097 \text{ kg}_w/\text{kg}_a, \quad h_4 = 51.1 \text{ kJ/kg}_a$$

$$t_8 = 28.0\ ^\circ\text{C}, \quad \phi_8 = 0.41 \quad\quad W_8 = 0.0097 \text{ kg}_w/\text{kg}_a, \quad h_8 = 52.8 \text{ kJ/kg}_a$$

(d) The cooling capacity can now be calculated by combining Eqs. (8.14) and (8.15) and observing that states 4 and 5 are equal.

$$\dot{Q}_R = (\dot{m}_{a,6A} + \dot{m}_{a,6B})[(h_5 - h_6) - (W_5 - W_6)h_{f,6}]$$
$$= (5.25 + 4.27)[(51.1 - 18.6) - (0.0097 - 0.0054)(20.44)]$$
$$= 308.6 \text{ kW}$$

TABLE 8.2 Comparative Results for Reheat, VAV, and Dual-Duct Systems

	Reheat	VAV	Dual Duct
Zone A temperature and humidity	22 °C, 40%	22 °C, 41%	22 °C, 45%
Zone B temperature and humidity	26 °C, 30%	26 °C, 40%	26 °C, 42%
Required reheat or heating-coil capacity, kW	228.8	N/A	16.0
Required cooling capacity, kW	560.9	248.3	308.6

Examples 8.12, 8.13, and 8.14 demonstrate the application of three different types of HVAC systems to the same building. The key results of using the different systems are tabulated in Table 8.2.

The use of either the VAV or the dual-duct system results in a higher space humidity than the use of the reheat system, but at a significant reduction in the rate of energy consumption. The supply-airflow rates for the dual-duct system in Ex. 8.14 were selected as approximately equal to those used by the reheat system. A better design may have been to use lower flow rates. This results in lower supply-air temperatures and humidities and will reduce the humidity levels in the two zones. However, the minimum flow rates allowable would be those equal to the ones determined for the VAV system, since at those flow rates the supply-air conditions correspond to the cooling-coil exit conditions (i.e., 100 percent cold air and no air from the hot deck). Thus the minimum space humidities available using the dual-duct system correspond to those of the VAV system. One feature of the dual-duct system not demonstrated in the example is its ability to simultaneously heat and cool different zones, which cannot be readily accomplished with the other two systems.

ENDNOTES

1. *ASHRAE Handbook, Fundamentals Volume* (Atlanta: American Society of Heating, Refrigerating and Air Conditioning Engineers, 1989), 26.12 (IP Edition).

2. *ASHRAE Handbook, Equipment Volume* (Atlanta: American Society of Heating, Refrigerating and Air Conditioning Engineers, 1988), 4.1.

3. *ASHRAE Handbook, HVAC Systems and Applications Volume* (Atlanta: American Society of Heating, Refrigerating and Air Conditioning Engineers, 1987).

4. *ASHRAE Handbook, Equipment Volume* (Atlanta: American Society of Heating, Refrigerating and Air Conditioning Engineers, 1988).

8.1 In a heating/humidifying system moist air first flows through a heating coil and then through an air washer. The air enters the system at 55 °F dry-bulb temperature and 40 percent relative humidity. The air exits the system at 90 °F dry-bulb temperature and 65 °F thermodynamic wet-bulb temperature. The flow rate through the system is 2200 lbm_a/hr and the process occurs at a pressure of 29.921 in. Hg. Determine:

   (a) the dry-bulb temperature of the air as it exits the heating coil,

   (b) the rate of heat addition to the air by the heating coil, and

   (c) the rate of moisture addition to the air by the adiabatic saturator.

8.2 A space to be conditioned has a sensible-heat loss of 80,000 Btu/hr and a latent-heat loss of 34,000 Btu/hr. The conditions in the space are maintained at 70 °F dry-bulb temperature and 44 °F dew-point temperature. Supply air is introduced into the space at a dry-bulb temperature of 95 °F (barometric pressure = 29.921 in. Hg). Determine

   (a) the required relative humidity of the supply air, and

   (b) the flow rate of supply air in lbm_a/hr.

8.3 Clearly show and label the following process lines and state-points on a psychrometric chart.

   (a) Outdoor air at 35 °F dry-bulb temperature and 100 percent relative humidity is heated in a furnace to 100 °F dry-bulb temperature.

   (b) Saturated steam at 200 °F is then sprayed into the air to increase its relative humidity to 40 percent.

   (c) The air is then supplied to a space with a condition line having a sensible-heat ratio of 0.7, and the air exits the space at a dry-bulb temperature of 70 °F.

8.4 Moist air enters a cooling coil at a dry-bulb temperature of 88 °F and a relative humidity of 45 percent at a volumetric flow rate of 4500 ft³/min. Barometric pressure is 12.75 psia. The air exits the coil at a dry-bulb temperature of 52 °F and a dew-point temperature of 48 °F. *Without using a psychrometric chart*, calculate the cooling capacity of the coil, Btu/hr, for these operating conditions.

8.5 Moist air enters a cooling coil at 28 °C dry-bulb temperature and 50 percent relative humidity and exits the coil at 13 °C dry-bulb temperature and 90 percent relative humidity. The flow rate through the coil is 1.50 kg_a/s and the process occurs at a pressure of 101.325 kPa. Determine:

   (a) the sensible-heat ratio for the process,

   (b) the cooling-coil capacity (heat-transfer rate),

   (c) the apparatus dew-point temperature, and

   (d) the bypass factor for the coil.

8.6 Moist air at 84 °F dry-bulb temperature and 70 °F thermodynamic wet-bulb temperature enters a perfect-contact refrigeration coil at a rate of 3500 ft³/min. The air leaves the coil at 54 °F. Assume 14.696 psia pressure. Determine the tons of refrigeration required.

8.7 Moist air enters a refrigeration coil at 89 °F dry-bulb temperature and 65 °F thermodynamic wet-bulb temperature at a rate of 1400 ft³/min. The surface temperature of the coil is 55 °F. If 3.5 tons of refrigeration are available, find the dry-bulb and wet-bulb temperatures of the air leaving the coil. Assume sea-level pressure.

8.8 Saturated steam at a pressure of 25 psia is sprayed into a stream of moist air. The initial condition of the air is 55 °F dry-bulb temperature and 35 °F dew-point temperature. The mass rate of air flow is 2000 lbm_a/min. Barometric pressure is 14.696 psia. Determine:

   (a) how much steam must be added in lbm_w/min to produce a saturated air condition, and

   (b) the resulting temperature of the saturated air.

8.9 Moist air at 70 °F dry-bulb temperature and 45 percent relative humidity is recirculated from a room and mixed with outdoor air at 97 °F dry-bulb temperature and 83 °F thermodynamic wet-bulb temperature. Determine the mixture state dry-bulb and wet-bulb temperatures if the volume of recirculated air (ft³/min) is three times the volume of outdoor air. Assume sea-level pressure.

8.10 A condition exists where it is necessary to cool and dehumidify air from 80 °F db, 67 °F wb to 60 °F db and 54 °F wb.

   (a) Discuss the feasibility of doing this in one process with a cooling coil. [*Hint:* Determine the apparatus dew-point temperature for the process.]

   (b) Describe a practical method of achieving the required process, and sketch it on a psychrometric chart.

8.11 Moist air is heated by steam condensing inside the tubes of a heating coil, as shown by Fig. 8.23. Part of the air passes through the coil and part is bypassed around the coil. Barometric pressure is 14.696 psia. Determine

   (a) the lbm_a/min which bypass the coil, and

   (b) the heat added by the coil in Btu/hr.

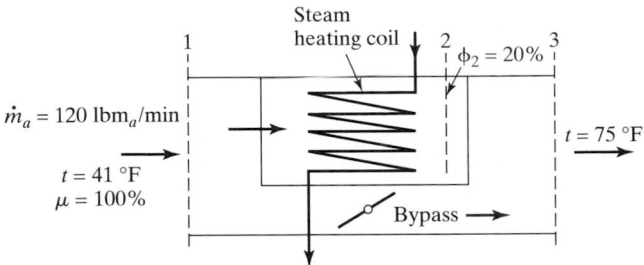

Figure 8.23 Schematic for Prob. 8.11.

8.12 Figure 8.24 schematically shows part of a winter-type air conditioning system. Barometric pressure is 14,696 psia. Determine:
(a) the temperature t_3 of the mixed air entering the heating coil, and
(b) the rate of heat addition to the air by the heating coil in Btu/hr.

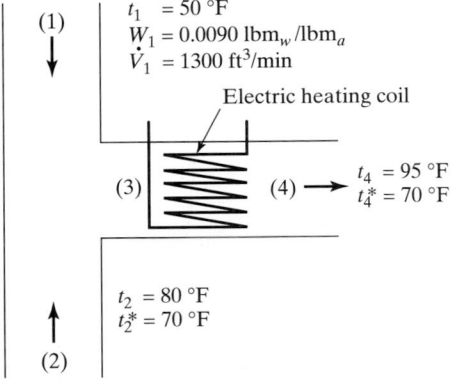

Figure 8.24 Schematic system for Prob. 8.12.

8.13 A heating/humidifying system with heat recovery is shown in Fig. 8.25. The heat-recovery unit transfers 80 percent of the heat that is removed from the exhaust air to the incoming outdoor air. The space is to be maintained at a dry-bulb temperature of 68 °F and a relative humidity of 40 percent. The space has a total load of 100,000 Btu/hr and a sensible-heat ratio 0.8. Air is supplied to the space at a dry-bulb temperature of 95 °F. The mass flow rate of air leaving the space is split such that 70 percent is recirculated and 30 percent is exhausted. Outdoor air is at a dry-bulb temperature of 40 °F and a relative humidity of 30 percent. As the exhaust air passes through the heat-recovery unit, its temperature drops to 48 °F. The heating and humidification occurs using a heating coil followed by a humidifier using saturated vapor at 240 °F.
(a) Clearly sketch and label all the points and processes on a psychrometric chart.
(b) Determine the rate at which heat is added by the heating coil, Btu/hr.
(c) Determine the rate at which moisture is added by the humidifier, lbm_w/hr.

8.14 Air at 50 °F, 40 percent relative humidity, and 13.5 psia pressure is heated in a furnace to 120 °F. The air is then supplied to a room in a building with a sensible-heat loss of 16,000 Btu/hr and a space-condition line having a sensible-heat ratio of 0.8. If the air in the room is 70 °F, determine:

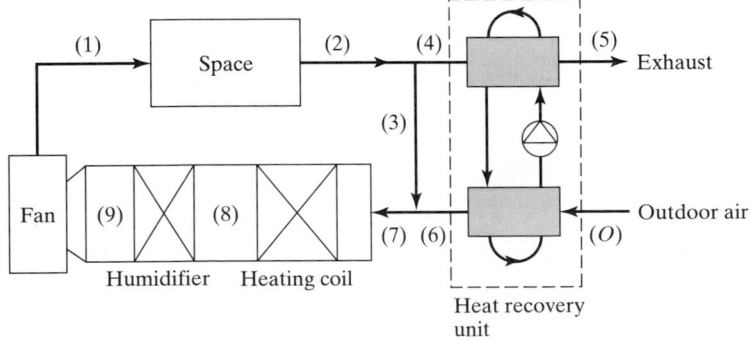

Figure 8.25 Schematic system for Prob. 8.13.

(a) the mass flow rate of dry air, and

(b) the humidity ratio, W, in the room.

Do not use the psychrometric chart to solve the problem.

8.15 Figure 8.26 shows a schematic winter-type air conditioning system. The space has a total heating load of 98,000 Btu/hr and a sensible-heat ratio of SHR = 0.6 on a day when the space temperature and relative humidity are maintained at 70 °F and 50 percent, respectively, and the outside temperature and relative humidity are 45 °F and 10 percent, respectively. The system uses 60 percent return air and 40 percent outside air (based on mass flow rates of dry air). Supply air is provided to the space at a rate of 15,000 lbm_a/hr. The saturated steam of the humidifier is at 250 °F.

(a) Carefully sketch and label all the points and process on a psychrometric chart.

(b) Calculate the rate of heat addition by the heating coil.

(c) Calculate the rate of moisture addition by the humidifier.

8.16 Air enters an air washer at a dry-bulb temperature of 100 °F and a wet-bulb temperature of 65 °F. The air exits the air washer at a dry-bulb temperature of 70 °F. Barometric pressure is 14.3 psia. *Without using a psychrometric chart*, answer the following:

(a) Calculate the relative humidity of the air *exiting* the air washer.

(b) If the mass flow rate of dry air through the air washer is 3000 lbm_a/hr, calculate the rate at which water is being added to the air stream.

8.17 An evaporative cooler (air washer) is used to cool a bus located in Phoenix. The system, similar to that shown in Fig. 8.18, takes outside air, first passes it through an evaporative cooler, then passes it through the bus, and finally exhausts it back to the outside. The system operates with a mass flow rate of 11,400 lbm_a/hr. On a very hot day the air

exits the evaporative cooler saturated at 65 °F. Atmospheric pressure is 14.7 psia.

(a) If the outdoor dry-bulb temperature is 95 °F, what is the outdoor relative humidity.

(b) If the sensible and latent heat gains of the bus are 55,200 Btu/hr and 16,800 Btu/hr, respectively, what are the dry-bulb and dew-point temperatures in the bus?

8.18 A space to be maintained at 75 °F dry-bulb temperature and 50 percent relative humidity has a rate of sensible-heat gain of 82,000 Btu/hr and a rate of moisture gain (average h_w = 1100 Btu/lbm_w) of 12.0 lbm_w/hr. Barometric pressure is 14.696 psia. Moist air is supplied to the room at 58 °F dry-bulb temperature. Determine:

(a) the dew-point temperature and thermodynamic wet-bulb temperature of the supply air, and

(b) the volume of supply air required in ft³/min of standard air.

8.19 A space is air conditioned in winter by the schematic system of Fig. 8.14. The space has a rate of sensible-heat loss of 180,000 Btu/hr and a rate of latent-heat loss of 20,200 Btu/hr. Moist air is withdrawn from the space at 70 °F dry-bulb temperature and 55 °F thermodynamic wet-bulb temperature. Barometric pressure is 14.696 psia. Moist air is supplied to the space at 100 °F dry-bulb temperature. Outdoor air is saturated at 35 °F. The dry-air flow rate of outdoor air admitted to the system is 50 percent of the dry-air flow rate of the air supplied to the space. You may assume that the air washer has an efficiency of 50 percent. Determine:

(a) the volume of air supplied to the space in ft³/min of standard air,

(b) the spray-water temperature,

(c) the lbm/hr of makeup water required for the air washer, and

(d) the rate of heat added to the air by each heating coil in Btu/hr.

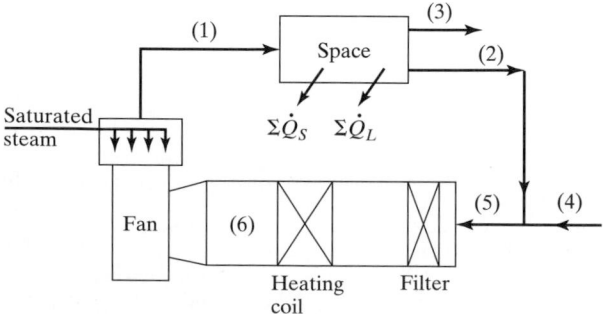

Figure 8.26 Schematic air conditioning system for Prob. 8.15.

8.20 A building is heated and humidified using the system shown in Fig. 8.14. The building has a sensible-heat loss of 260 kW and a latent-heat loss of 29 kW. The building space is maintained at 20 °C dry-bulb and 40 percent relative humidity. Outside air is at 3 °C dry-bulb and 30 percent relative humidity. Forty-five percent of the mass flow rate of dry air supplied to the space is outside air. The supply air is at a dry-bulb temperature of 35 °C. The air exiting the air washer is at 80 percent relative humidity.

(a) Sketch and label all points and processes on a psychrometric chart.

(b) Calculate the rate of heat addition to the moist air by the preheat coil.

(c) Calculate the rate of moisture addition by the air washer.

(d) Calculate the rate of heat addition to the moist air by the reheat coil.

8.21 A space to be maintained at 76 °F dry-bulb temperature and 65 °F thermodynamic wet-bulb temperature has a rate of sensible-heat gain of 84,000 Btu/hr and a rate of moisture gain (average h_w = 1120 Btu/lbm$_w$) of 20.0 lbm$_w$ per hr. Moist air enters the space at a dry-bulb temperature of 60 °F. Outdoor air at 95 °F dry-bulb temperature and 78 °F thermodynamic wet-bulb temperature is supplied for ventilation purposes at a rate of 825 ft^3/min of standard air. The space is to be conditioned by the schematic system of Fig. 8.15. Barometric pressure is 14.696 psia. Determine:

(a) the dry-bulb temperature and thermodynamic wet-bulb temperature of the air entering the cooling coil, and

(b) the tons of refrigeration required.

8.22 A space to be maintained at 80 °F dry-bulb temperature and 50 percent relative humidity has a rate of sensible-heat gain of 73,500 Btu/hr and a rate of latent-heat gain of 16,500 Btu/hr. The volume of air supplied, to the space is 7,200 ft^3/min of standard air. Outdoor air at 95 °F dry-bulb temperature and 75 °F thermodynamic wet-bulb temperature is introduced into the system at a rate of 1800 ft^3/min of standard air. The air which passes through the cooling coil is brought to 90 percent relative humidity. Barometric pressure is 14.696 psia. The space is provided with the schematic system of Fig. 8.16. Determine:

(a) the dry-bulb temperature and thermodynamic wet-bulb temperature of the air supplied to the space, and

(b) the volume of recirculated air in ft^3/min (standard air) which should bypass the cooling coil.

8.23 A space to be maintained at 27 °C dry-bulb temperature has a rate of sensible-heat gain of 13kW and a rate of latent-heat gain of 8.5 kW. The system used to condition the space is shown in Fig. 8.16. The mass flow rate of air supplied to the space is 1.1 kg$_a$/s. Outdoor air at 38 °C dry-bulb temperature and 17 °C dew-point temperature is introduced into the system at a rate of 0.28 kg$_a$/s. The air exiting the cooling coil is saturated and at a temperature of 7 °C. Standard atmospheric pressure exists.

(a) Clearly sketch and label all points and process lines on a psychrometric chart.

(b) Determine the dry-bulb temperature and relative humidity of air supplied to the space.

(c) Determine the relative humidity in the space.

(d) Determine the refrigeration load.

8.24 A space to be maintained at 75 °F dry-bulb temperature and 65 °F thermodynamic wet-bulb temperature has a rate of sensible-heat gain of 377,000 Btu/hr and a rate of moisture gain (average h_w = 1100 Btu/lbm$_w$) of 330.0 lbm$_w$ per hr. The chilled air leaves the cooling coil at 54 °F dry-bulb temperature and 90 percent relative humidity. Barometric pressure is 14.696 psia. The space is conditioned by the system shown in Fig. 8.17. Determine:

(a) the dry-bulb temperature and thermodynamic wet-bulb temperature of the air supplied to the space, and

(b) the rate of heat addition by the reheat coil in Btu/hr.

8.25 An interior space of a building is to be maintained at 72 °F during winter. The space has a rate of sensible-heat gain of 60,000 Btu/hr during working hours; moisture gain is negligible. The space is to be cooled by mixing outdoor air with recirculated air. Air is to be supplied to the space at a dry-bulb temperature of 55 °F. On a particular day, the outdoor air is saturated at −10 °F. Barometric pressure is 14.696 psia.

(a) Determine the volume of supply air required in ft^3/min of standard air.

(b) Find the percentage of the supply air (dry-air mass basis) which is outdoor air.

8.26 A space is air conditioned in winter by a system similar to that shown by Fig. 8.19. The rate of sensible-heat loss from the space is 180,000 Btu/hr and the rate of latent-heat loss from the space is 19,700 Btu/hr. Additional known data are as follows: t_1 = 100 °F, t_2 = 70 °F, t_2^* = 55 °F, t_3 = 45 °F, t_4 = 20 °F, ϕ_4 = 100 percent, and ϕ_6 = 100 percent. Barometric pressure is 14.696 psia.

(a) Locate all state-points on the psychrometric chart and read values of t_1^*, t_5^*, t_5, t_2^*, and t_6.

(b) Determine the required specific enthalpy for the steam admitted to the air by the humidifier.

8.27 Figure 8.27 schematically depicts a two-zone variable-air-volume heating/humidifying system. The following conditions exist.

Space temperatures: $t_2 = 78$ °F, $t_5 = 68$ °F

Supply air: $t_{12} = 95$ °F, $\phi_{12} = 30$ percent

Air-washer exit humidity: $\phi_{11} = 95$ percent

Outside conditions: $t_8 = 40$ °F, $\phi_8 = 50$ percent

Zone A heat losses, Btu/hr: 48,000 sensible; 32,000 latent

Zone B heat losses, Btu/hr: 24,000 sensible; 6,000 latent

Exhaust air: For both zones, the exhaust mass flow rate of dry air is 15 percent of the supply

Barometric pressure: 29.9 in. Hg

(a) Carefully sketch and label the points and processes on a psychrometric chart.

(b) Determine the rate of heat addition by the preheat coil, Btu/hr.

(c) Determine the rate of moisture addition by the air washer, lbm_w/hr.

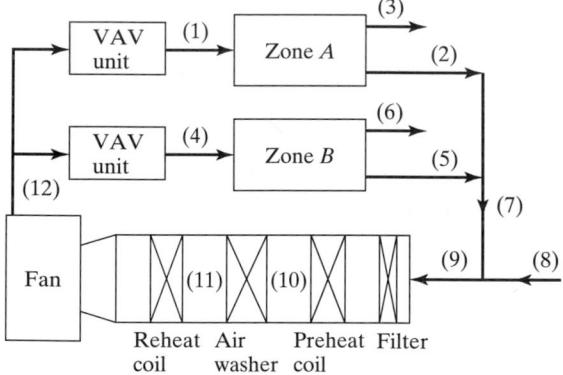

Figure 8.27 Schematic air conditioning system for Prob. 8.27.

8.28 Figure 8.28 depicts a two-zone heating/humidifying system. The following conditions exist.

Zone A is at 68 °F dry-bulb temperature and 50 percent relative humidity.

The mass flow rate of supply air into zone A is 6000 lbm_a/hr.

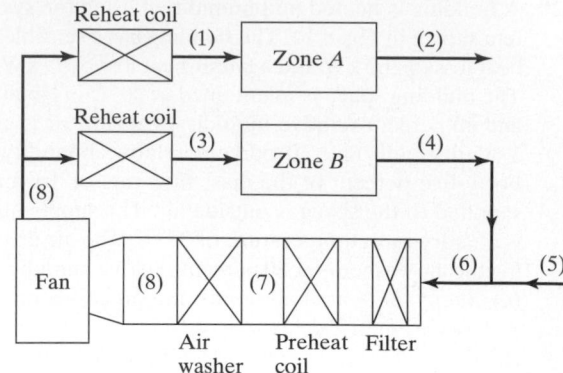

Figure 8.28 Schematic air conditioning system for Prob. 8.28.

The sensible-heat loss from zone A is 32,000 Btu/hr.

Zone B is at 72 °F dry-bulb temperature and 40 percent relative humidity.

The mass flow rate of supply air into zone B is 4000 lbm_a/hr.

The sensible- and latent-heat losses from zone B result in an SHR = 0.7.

Outside air is at 35 °F dry-bulb temperature and 20 percent relative humidity.

The preheat coil is controlled so that the air leaves the coil at a dry-bulb temperature of 80 °F.

The air exits the air washer at saturated conditions.

The system operates at standard atmospheric pressure.

Carefully sketch and label all the points and process lines on a psychrometric chart and determine:

(a) the dry-bulb temperature and relative humidity of state 6,

(b) the rate of heat input by the preheat coil,

(c) the dry-bulb and wet-bulb temperatures of state 8,

(d) the rate of moisture added to the air by the air washer,

(e) the dry-bulb temperature and relative humidity of the supply for zone A, and

(f) the rate of heat input by the reheat coil for zone B.

8.29 A dual-duct system similar to the one shown in Fig. 8.22 is used to air condition a building. On a day when the outdoor dry-bulb temperature is 60 °F and the outdoor relative humidity is 40 percent, zone A requires cooling while zone B requires heating. The following conditions exist.

Zone A	Zone B
70 °F dry-bulb temp.	72 °F dry-bulb temp.
60% relative humidity	40% relative humidity
20,000 Btu/hr sensible-heat *gain*	25,000 Btu/hr sensible-heat *loss*
16,400 Btu/hr latent-heat *gain*	Negligible latent-heat *loss*

The system operates with 50 percent exhaust and 50 percent return air flows for each zone. The air in the hot deck has a dry-bulb temperature of 95 °F and a relative humidity of 20 percent, and the air in the cold deck is saturated at 40 °F. Standard atmospheric pressure exists. Determine:

(a) the supply-airflow rates for each zone,
(b) the supply-air conditions for each zone (temperature and relative humidity), and
(c) the mass flow rate of air from the hot deck used in conditioning zone *A*.

8.30 Figure 8.22 shows a schematic of a dual-duct system. For design purposes, suppose the zones shown are two of five zones, each having identical operating design conditions. The only exhaust is the main system exhaust. The zones are to be maintained at a dry-bulb temperature of 75 °F and 50 percent relative humidity when the total heat gain of each zone is 200,000 Btu/hr and the sensible-heat ratio is 0.60. Outdoor air conditions are 95 °F dry-bulb temperature and 40 percent relative humidity. The system is designed to operate with a mass flow rate of dry air that consists of 25 percent outside air and 75 percent return air. The hot deck provides sensible heating only, and the air leaves the heating coil at a dry-bulb temperature of 105 °F. The cold deck is designed such that air leaves the cooling coil at a dry-bulb temperature of 50 °F and 90 percent relative humidity. Barometric pressure is standard atmospheric pressure.

For the above design conditions, carefully sketch and label all the points and processes on a psychrometric chart and determine:

(a) the mass flow rate, lbm_a/hr, through the heating coil,
(b) the heating-coil capacity, Btu/hr,
(c) the mass flow rate, lbm_a/hr, through the cooling coil, and
(d) the cooling-coil capacity, Btu/hr.

SYMBOLS

b	Bypass factor.
c_p	Specific heat of moist air at constant pressure, $Btu/lbm_a \cdot$ °F or $kJ/kg_a \cdot$ °C.
$\bar{c}_p$	Average specific heat of moist air at constant pressure for a process, $Btu/lbm_a \cdot$ °F or $kJ/kg_a \cdot$ °C.
c_{pa}	Specific heat of dry air at constant pressure, $Btu/lbm_a \cdot$ °F or $kJ/kg_a \cdot$ °C.
c_{pw}	Specific heat of water vapor at constant pressure, $Btu/lbm_w \cdot$ °F or $kJ/kg_w \cdot$ °C.
e_c	Saturation effectiveness for an evaporative cooler.
H	Enthalpy, Btu or kJ; H_a for dry air; H_w for water or water vapor.
h	Specific enthalpy of moist air, Btu/lbm_a or kJ/kg_a; h_a for dry air; h_s for saturated moist air at t; $h_{as} = h_s - h_a$.
h_f	Specific enthalpy of liquid water, Btu/lbm_w or kJ/kg_w.
h_g	Specific enthalpy of saturated water vapor, Btu/lbm_w or kJ/kg_w.
$\bar{h}_g$	Average enthalpy of water vapor defined by Eq. (7.44), Btu/lbm_w or kJ/kg_w.
h_{fg}^*	$h_g^* - h_f^*$, Btu/lbm_w or kJ/kg_w.
h_{g0}	Enthalpy of saturated water vapor at 0 °F or 0 °C, Btu/lbm_w or kJ/kg_w.
Δh_L	Latent change of specific enthalpy of moist air for a process, Btu/lbm_a or kJ/kg_a.
Δh_S	Sensible change of specific enthalpy of moist air for a process, Btu/lbm_a or kJ/kg_a.
h_w	Enthalpy of water added to moist air, Btu/lbm_w or kJ/kg_w; $h_w = h_g$ for low-pressure water vapor.
$\dot{m}$	Mass flow rate, lbm/hr or kg/s.

$\dot{m}_a$ Mass flow rate of dry air, lbm_a/hr or kg_a/s.

$\dot{m}_w$ Mass flow rate of water or water vapor, lbm_w/hr or kg_w/s.

$\dot{Q}$ Heat-transfer rate, Btu/hr or kW.

$_1\dot{Q}_2$ Heat-transfer rate added during the process that goes from state 1 to state 2, Btu/hr or kW.

$\dot{Q}_L$ Rate of latent-heat transfer, Btu/hr or kW.

$\dot{Q}_R$ Refrigeration load or capacity, Btu/hr or kW.

$\dot{Q}_{RL}$ Latent-refrigeration load or capacity, Btu/hr or kW.

$\dot{Q}_{RS}$ Sensible-refrigeration load or capacity, Btu/hr or kW.

$\dot{Q}_S$ Rate of sensible-heat transfer, Btu/hr or kW.

q' $(h_2 - h_1)/(W_2 - W_1)$, Btu/lbm_w or kJ/kg_w.

SHR   Sensible-heat ratio.

T Absolute dry-bulb temperature, °R or K.

t Dry-bulb temperature, °F or °C.

t_d Dew-point temperature, °F or °C.

t^* Thermodynamic wet-bulb temperature, °F or °C.

$\bar{t}$ Average dry-bulb temperature for a process, °F or °C.

$\dot{V}_{std}$ Volumetric flow rate of standard air, ft³/min, ft³/hr, or m³/s.

v Specific volume of moist air, ft^3/lbm_a or m^3/kg_a; v_s for saturated moist air.

v_a Specific volume of dry air, ft^3/lbm_a or m^3/kg_a.

v_{as} $v_{as} = v_s - v_a$ ft^3/lbm_a or m^3/kg_a.

v_{std} Specific volume of standard air, $v_{std} = 13.33$ ft^3/lbm_a (0.830 m^3/kg_a).

W Humidity ratio, $\text{lbm}_w/\text{lbm}_a$ or kg_w/kg_a; W_s for air saturated at t; W_s^* for air saturated at t^*.

Greek Letters

η_w Air-washer efficiency.

μ Degree of saturation, $\mu = W/W_s$, dimensionless.

ϕ Relative humidity, dimensionless.

ρ_{std} Density of standard air, $\rho_{std} = 0.075$ lbm_a/ft^3 (1.204 kg_a/m^3).

chapter 9
The Psychrometer and Humidity Measurement

9.1 INTRODUCTION

In Chapter 7 we discussed various thermodynamic properties of moist air and their relations to each other. In Chapter 8 we found that the psychrometric chart is a useful graphical aid in solution of moist air problems. However, to use the psychrometric chart, we must know the thermodynamic state of the air. This usually requires knowledge of the barometric pressure and two other independent properties.

Barometric pressure and dry-bulb temperature may be measured easily and precisely. Certain properties, such as specific enthalpy, specific volume, and adiabatic saturation temperature, are not directly measurable. The primary problem in psychrometric measurements is determination of some property related to the air moisture content. Accurate measurement of air humidity is difficult, and most techniques are subject to error.

The wet-bulb thermometer is one of the most convenient devices available for humidity measurement. It is a reliable instrument if properly applied. In this chapter we will study the wet-bulb thermometer in detail. We will also study briefly some of the other techniques for measurement of humidity.

One of the most intriguing problems in psychrometrics is the relationship between adiabatic saturation temperature and psychrometer wet-bulb temperature. We will closely examine that relationship in this chapter.

9.2 MASS TRANSFER AND EVAPORATION OF WATER INTO MOIST AIR

The theory of the wet-bulb thermometer involves the mechanism of evaporation of water into moist air. Some of our problems in later chapters will also deal with simultaneous heat transfer and water-vapor transfer. We will now consider in some detail the fundamental problem of evaporation from a free-water surface.

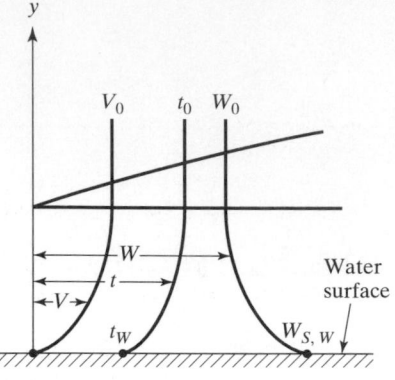

Figure 9.1 Schematic illustration of boundary-layer problem for the evaporation of water into moist air.

Figure 9.1 will serve as our schematic model. A free-water surface at temperature t_w is exposed to a moving stream of moist air. Adjacent to the water surface is a boundary layer of air whose velocity varies from zero at the water surface to the main air stream velocity V_o at its outer edge. Temperature increases within the boundary layer from t_w at the water surface to t_o of the main air-stream. Air immediately adjacent to the water surface is assumed to be saturated. Humidity ratio decreases from $W_{s,w}$ at the water surface to W_o in the main air stream. Figure 9.1 shows the thickness of all three boundary layers—velocity, temperature, and humidity ratio—to be equal. This is not exactly true, as will be shown later, but is a good approximation for air.

The two simultaneous transfer processes are heat transfer from the air to the water surface and moisture transfer from the water surface to the air stream. Heat transfer through the boundary layer is by the combined processes of convection and conduction. Vapor transfer occurs by the combined processes of convection and diffusion.

For heat transfer we may write

$$h_c(t_o - t_w) = k\left(\frac{\partial t}{\partial y}\right)_{y=0}$$

By defining a dimensionless temperature $t' = (t - t_w)/(t_o - t_w)$ and a dimensionless length $y' = y/L$, where L is a reference length, we obtain

$$\frac{h_c L}{k} = \left(\frac{\partial t'}{\partial y'}\right)_{y=0} \tag{9.1}$$

From convective heat transfer we know that the solution of Eq. (9.1) has the form

$$\frac{h_c L}{k} = f(\text{Re}_L, \text{Pr}) \tag{9.2}$$

where $h_c L/k$ is the dimensionless Nusselt number, $\text{Re}_L = L V \rho / \mu$ is the dimensionless Reynolds number, and $\text{Pr} = c_p \mu / k$ is the dimensionless Prandtl number.

The basic concept of water-vapor diffusion in air is given by Fick's law, which can be written as

$$\dot{m}_w = -D\rho_a \frac{dW}{dy}$$

where $\dot{m}_w$ = the mass flow rate of water vapor, D = the water-vapor diffusivity in air or diffusion coefficient, ρ_a = the dry-air density, mass of dry air per unit volume, W = the humidity ratio, and y = the coordinate in the direction in which the diffusion occurs.

A mass-transfer coefficient h_D for the transfer of water vapor through the boundary layer of Fig. 9.1 may be defined by the equation

$$\dot{m}_w = h_D(W_{s,w} - W_o)$$

We may then write

$$h_D(W_{s,w} - W_o) = -D\rho_a\left(\frac{\partial W}{\partial y}\right)_{y=0}$$

By use of the dimensionless quantities $W' = (W_{s,w} - W)/(W_{s,w} - W_o)$ and $y' = y/L$, we obtain

$$\frac{h_D L}{\rho_a D} = \left(\frac{\partial W'}{\partial y'}\right)_{y'=0} \tag{9.3}$$

The solution of Eq. (9.3) has the form [1]

$$\frac{h_D L}{\rho_a D} = f(\text{Re}_L, \text{Sc}) \tag{9.4}$$

where $\text{Sc} = \mu/\rho_a D$ is the dimensionless Schmidt number. The functional form of Eqs. (9.2) and (9.4) may be applied to turbulent flow of air over a variety of surfaces including flat plates, around cylinders and spheres, and through packed beds.

Equations (9.2) and (9.4) may also be expressed as

$$\frac{h_c L}{k} = a\left(\frac{L\mathbf{V}\rho}{\mu}\right)^b\left(\frac{c_p\mu}{k}\right)^c$$

$$\frac{h_D L}{\rho_a D} = a\left(\frac{L\mathbf{V}\rho}{\mu}\right)^b\left(\frac{\mu}{\rho D}\right)^c$$

We then obtain

$$\frac{h_c}{h_D} = \frac{k}{D\rho_a}\left(\frac{D}{\alpha}\right)^c \tag{9.5}$$

where $\alpha = k/\rho c_p$, the thermal diffusivity. Dividing both sides of Eq. (9.5) by c_p, we obtain

$$\frac{h_c}{h_D c_p} = \left(\frac{\alpha}{D}\right)^{1-c} \tag{9.6}$$

The dimensionless term $h_c/h_D c_p$ is called the Lewis number, Le.

Kusuda [2] made a review of correlations for calculating the Lewis number. For forced-convection air-flow he recommends the relation

$$\text{Le} = \left(\frac{\alpha}{D}\right)^{2/3} \tag{9.7}$$

For the case of natural convection, Kusuda recommends the same correlation but with the exponent of 2/3 replaced with 0.48. In the same paper, Kusuda made a study of data available on transport properties of dry and saturated moist air. Table 9.1 shows his values for α, D, and α/D for temperatures ranging from 50 to 140 °F (10 to 60 °C). In applying these values to evaporation problems, Kusuda recommends that the properties be evaluated for saturated moist air at the water surface temperature.

TABLE 9.1 Thermal and Vapor Diffusivity Data for Dry and Saturated Moist Air

Temperature, °F (°C)	Degree of Saturation	α, ft²/hr (m²/s × 10⁻⁵)	D, ft²/hr (m²/s × 10⁻⁵)	α/D
50 (10.0)	0	0.770 (1.99)	0.901 (2.33)	0.855
	1	0.769 (1.99)		0.854
60 (15.6)	0	0.799 (2.06)	0.936 (2.42)	0.854
	1	0.797 (2.06)		0.852
70 (21.1)	0	0.828 (2.14)	0.971 (2.51)	0.853
	1	0.826 (2.13)		0.850
80 (26.7)	0	0.858 (2.21)	1.007 (2.60)	0.852
	1	0.854 (2.20)		0.848
90 (32.2)	0	0.888 (2.29)	1.044 (2.70)	0.851
	1	0.883 (2.28)		0.846
100 (37.8)	0	0.919 (2.37)	1.081 (2.79)	0.850
	1	0.911 (2.35)		0.843
110 (43.3)	0	0.949 (2.45)	1.119 (2.89)	0.848
	1	0.938 (2.42)		0.838
120 (48.9)	0	0.981 (2.53)	1.157 (2.99)	0.848
	1	0.963 (2.49)		0.832
130 (54.4)	0	1.012 (2.61)	1.196 (3.09)	0.846
	1	0.985 (2.54)		0.823
140 (60.0)	0	1.044 (2.70)	1.235 (3.19)	0.845
	1	1.003 (2.59)		0.812

SOURCE: Adapted by permission from Tamami Kusuda, "Calculation of the Temperature of a Flat-Plate Wet Surface under Adiabatic Conditions with Respect to the Lewis Relation," in *Humidity and Moisture*, Vol. 1: *Principles and Methods of Measuring Humidity in Gases*, ed. Robert E. Ruskin (New York: Reinhold Publishing Corporation, 1965), 29.

9.3 THEORY OF THE PSYCHROMETER

The wet-bulb thermometer or some variation has been used for more than a century. In its simplest form, a wet-bulb thermometer consists of an ordinary thermometer whose sensing bulb is covered with a moistened cloth wick. The thermometer is ventilated by whirling it in a calm atmosphere or by exposing it to forced air motion. When both a dry-

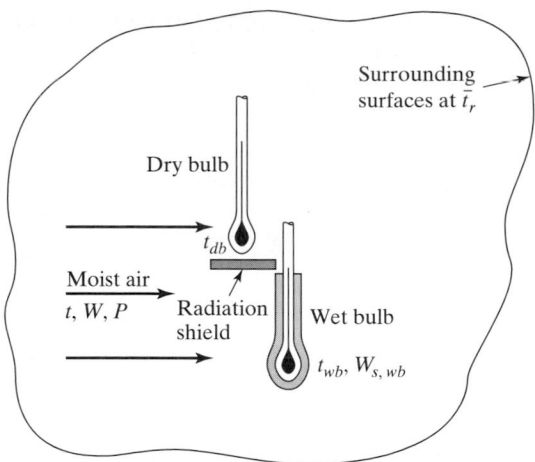

Figure 9.2 Schematic psychrometer.

bulb thermometer and a wet-bulb thermometer are included in the same instrument, the device is called a *psychrometer*.

Figure 9.2 shows a schematic psychrometer exposed to a moving stream of moist air. In the following analysis we will assume that conduction heat-transfer effects along the thermometer stems are negligible. We will also assume that air velocities are low enough so that high-velocity impact effects are negligible. If the mean radiant temperature $\bar{t}_r$ of the surrounding surfaces is different than the air dry-bulb temperature t, the dry-bulb thermometer will indicate a value t_{db} that is somewhere between $\bar{t}_r$ and t. The moistened bulb at temperature t_{wb} may receive heat energy by convection from the surrounding air stream and by radiation from surrounding surfaces. Some water may evaporate from the moistened bulb, exerting a cooling effect, which depresses the reading of the thermometer below that of the local air dry-bulb temperature.

Assuming steady state for the system of Fig. 9.2 and linearized radiation heat transfer, we may write for the dry-bulb thermometer

$$h_c(t_{\text{db}} - t) = h_r(\bar{t}_r - t_{\text{db}})$$

or

$$t = t_{\text{db}} - \left(\frac{h_r}{h_c}\right)_{\text{db}} (\bar{t}_r - t_{\text{db}}) \tag{9.8}$$

Equation (9.8) allows calculation of the air dry-bulb temperature.

For the wet-bulb thermometer of Fig. 9.2 we may write

$$h_D(W_{s,\text{wb}} - W)h_{fg,\text{wb}} = h_c(t - t_{\text{wb}}) + h_r(\bar{t}_r - t_{\text{wb}})$$

Substitution of $h_D = h_c/(\text{Le})(c_p)$ gives

$$W = W_{s,\text{wb}} - K(t - t_{\text{wb}}) \tag{9.9}$$

where

$$K = \frac{\text{Le}\, c_p}{h_{fg,\text{wb}}} \left[1 + \frac{h_r(\bar{t}_r - t_{\text{wb}})}{h_c(t - t_{\text{wb}})} \right] \tag{9.10}$$

In our procedure, we will evaluate the specific heat c_p at the arithmetic mean value of the humidity ratio. Thus by Eq. (7.23)

$$c_p = c_{pa} + c_{pw} \frac{(W + W_{s,\text{wb}})}{2} \tag{9.11}$$

By Eqs. (9.9) and (9.11) we obtain

$$K = \frac{c_{pa} + c_{pw} W_{s,\text{wb}}}{\dfrac{h_{fg,\text{wb}}}{\text{Le}\left[1 + \dfrac{h_r(\bar{t}_r - t_{\text{wb}})}{h_c(t - t_{\text{wb}})}\right]} + \dfrac{c_{pw}(t - t_{\text{wb}})}{2}} \tag{9.12}$$

Equations (9.9) and (9.12) allow calculation of the air humidity ratio W. The humidity ratio $W_{s,\text{wb}}$ of saturated moist air at the wet-bulb temperature may be calculated from Eq. (7.18) or read from Table A.4E or A.4SI for sea-level pressure. The wet-bulb coefficient K must be separately evaluated. Equation (9.12) allows calculation of K for the general case. Two special cases exist. For the common situation where the mean radiant temperature of surfaces surrounding the wet bulb may be assumed to equal the local air dry-bulb temperature ($\bar{t}_r = t$), Eq. (9.12) reduces to

$$K = \frac{c_{pa} + c_{pw} W_{s,\text{wb}}}{\dfrac{h_{fg,\text{wb}}}{\text{Le}\left(1 + \dfrac{h_{r,t}}{h_c}\right)} + \dfrac{c_{pw}(t - t_{\text{wb}})}{2}} \tag{9.13}$$

where $h_{r,t}$ means that h_r is evaluated for the condition of equal mean radiant and local air dry-bulb temperatures. If the wet bulb is completely shielded from radiation effects,

$$K = \frac{c_{pa} + c_{pw} W_{s,\text{wb}}}{\dfrac{h_{fg,\text{wb}}}{\text{Le}} + \dfrac{c_{pw}(t - t_{\text{wb}})}{2}} \tag{9.14}$$

Evaluation of the wet-bulb coefficient K is the principal problem in psychrometry. The Lewis number Le may be found from Eq. (9.7). The radiation heat-transfer coefficient h_r may be calculated from Eq. (2.76), which for the wet-bulb thermometer reduces to

$$h_r = \sigma \varepsilon_{\text{wb}} \frac{(\overline{T}_r^4 - T_{\text{wb}}^4)}{(\bar{t}_r - t_{\text{wb}})} \cong 4\sigma \varepsilon_{\text{wb}} \overline{T}_r^3 \tag{9.15}$$

where ε_{wb} is the surface emissivity of the wet bulb.

The convection coefficient h_c may be calculated from Eq. (2.54) for the flow of air normal to single wires and cylinders:

$$h_c = \frac{k}{d}\left[0.4\left(\frac{d\mathbf{V}\rho}{\mu}\right)^{0.5} + 0.06\left(\frac{d\mathbf{V}\rho}{\mu}\right)^{2/3}\right]\text{Pr}^{0.4} \qquad \text{for } 40 < \frac{d\mathbf{V}\rho}{\mu} < 10^5 \tag{9.16}$$

where d is the wet-bulb diameter and Pr and the properties k, ρ, and μ are evaluated at the local air dry-bulb temperature.

Figure 9.3 shows the ratio $h_{r,t}/h_c$ for a wet-bulb diameter of 0.3 in. (7.6 mm) calculated by Eqs. (9.15) and (9.16) for the special case where the mean radiant temperature equals the local air dry-bulb temperature. A value of $\varepsilon_{\text{wb}} = 0.9$ was used in Eq. (9.15). Figure 9.4 is similar to Fig. 9.3 except that it is for a wet-bulb diameter of 0.1 in. (2.5 mm).

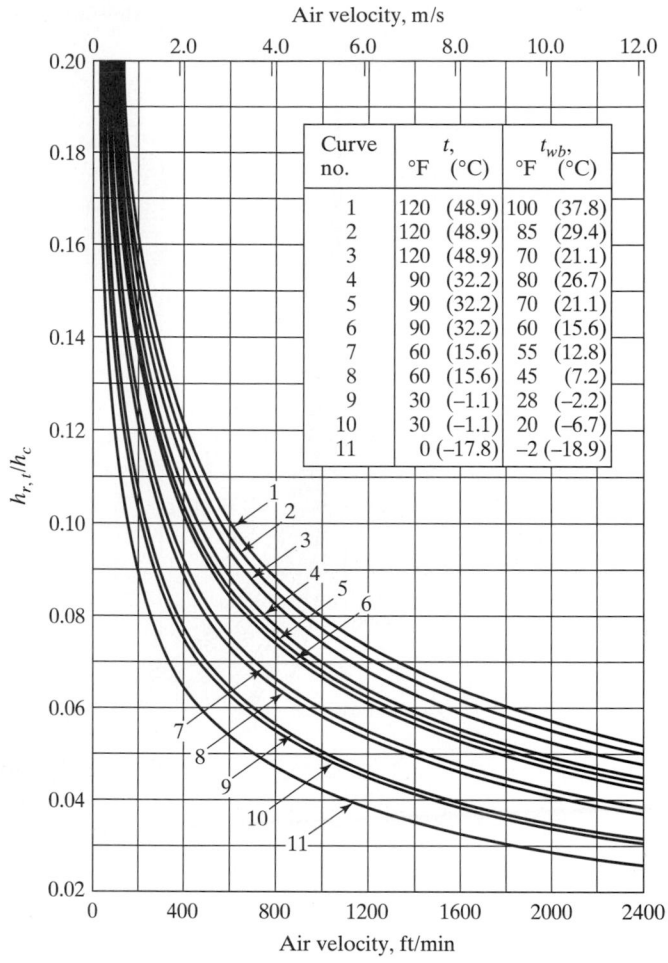

Figure 9.3 Ratio $h_{r,t}/h_c$ for a wet-bulb diameter of 0.3 in. (7.6 mm).

Figures 9.3 and 9.4 show that the ratio $h_{r,t}/h_c$ decreases with decrease of both dry-bulb temperature and wet-bulb temperature. In addition, this ratio increases rapidly with decrease of air velocities below about 500 ft/min (2.5 m/s). At velocities above about 1000 ft/min (5 m/s) the ratio $h_{r,t}/h_c$ is less dependent upon air velocity, particularly for a wet-bulb diameter of 0.1 in. (2.5 mm).

Figures 9.3 and 9.4 apply directly only to calculation of the wet-bulb coefficient by Eq. (9.13). For the general case, where the mean radiant temperature is different than the local air dry-bulb temperature, we may obtain K from the relation

$$K = \frac{\dfrac{c_{pa} + c_{pw} W_{s,\text{wb}}}{h_{fg,\text{wb}}}}{\text{Le}\left[1 + \dfrac{h_{r,t}(\overline{T}_r^4 - T_{\text{wb}}^4)}{h_c(T^4 - T_{\text{wb}}^4)}\right]} + \frac{c_{pw}(t - t_{\text{wb}})}{2} \tag{9.17}$$

where $h_{r,t}/h_c$ can be read from Fig. 9.3 or Fig. 9.4.

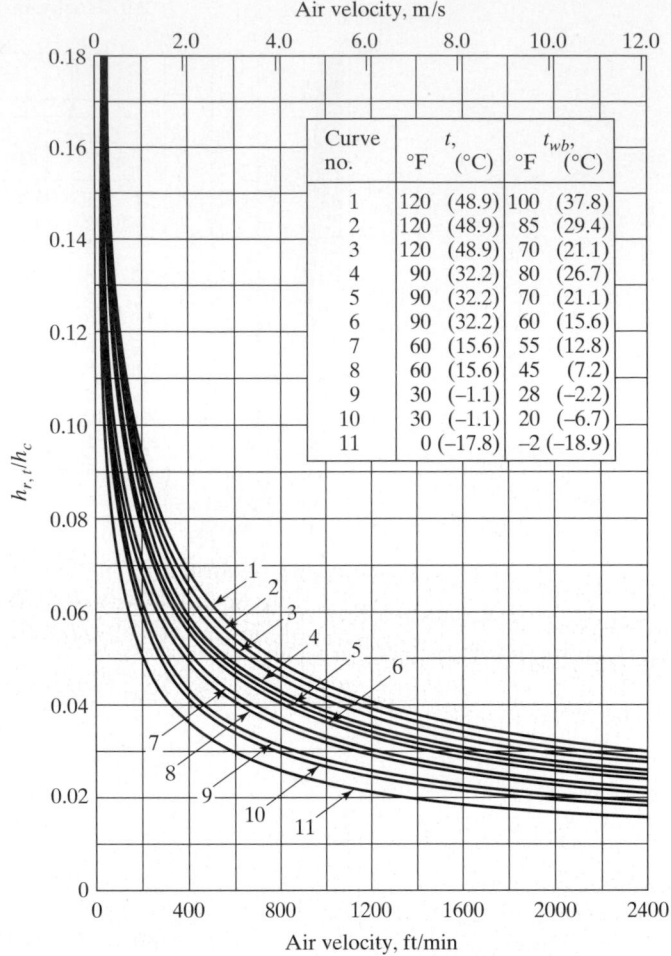

Air velocity, m/s

Curve no.	t, °F	(°C)	t_{wb}, °F	(°C)
1	120	(48.9)	100	(37.8)
2	120	(48.9)	85	(29.4)
3	120	(48.9)	70	(21.1)
4	90	(32.2)	80	(26.7)
5	90	(32.2)	70	(21.1)
6	90	(32.2)	60	(15.6)
7	60	(15.6)	55	(12.8)
8	60	(15.6)	45	(7.2)
9	30	(−1.1)	28	(−2.2)
10	30	(−1.1)	20	(−6.7)
11	0	(−17.8)	−2	(−18.9)

Air velocity, ft/min

Figure 9.4 Ratio $h_{r,t}/h_c$ for a wet-bulb diameter of 0.1 in. (2.5 mm).

EXAMPLE 9.1

Measurements with a psychrometer indicate a wet-bulb temperature reading of 50 °F and a dry-bulb temperature reading of 75 °F. Barometric pressure is 14.696 psia. Both thermometers are mercury-in-glass types with a bare-bulb diameter of 0.25 in. Air velocity is 1000 ft/min. Both thermometers are unshielded from radiation. Assume that the mean radiant temperature of the surrounding surfaces is 75 °F. Determine the humidity ratio.

Solution: Since the mean radiant temperature equals the dry-bulb temperature, we have $t = t_{db} = 75$ °F. Equation (9.13) applies for the solution of K. By Table A.1E we have for 50 °F, $h_{fg,wb} = 1065.42$ Btu/lbm. By Table 9.1, $\alpha/D = 0.854$. By Eq. (9.7)

$$\text{Le} = (0.854)^{2/3} = 0.900$$

Figure 9.3 applies for $h_{r,t}/h_c$, since the diameter of the wet-bulb wick would be about 0.3 in. We find by interpolation $h_{r,t}/h_c = 0.063$. By Table A.4E at 50 °F, $W_{s,wb} = 0.007661$ $\text{lbm}_w/\text{lbm}_a$. By Eq. (9.13) and Tables A.5E and A.6E

$$K = \frac{0.240 + (0.444)(0.007661)}{\dfrac{1065.42}{(0.90)(1.063)} + (0.225)(25)} = 0.0002175 \text{ lbm}_w/\text{lbm}_a \cdot {}^\circ\text{F}$$

By Eq. (9.9)

$$W = 0.007661 - (0.0002175)(25) = 0.00222 \text{ lbm}_w/\text{lbm}_a$$

EXAMPLE 9.2

A psychrometer with unshielded thermometers indicates a dry-bulb temperature of 70 °F and a wet-bulb temperature of 60 °F. Barometric pressure is 14.696 psia. Both thermometers are of the mercury-in-glass type with a bare-bulb diameter of 0.25 in. Air velocity is 200 ft/min. Assume that the mean radiant temperature of the surrounding surfaces is 90 °F. Determine (a) the true dry-bulb temperature, and (b) the humidity ratio.

Solution:

(a) Figure 9.3 may be used for finding $(h_r/h_c)_{\text{db}}$ if we replace t in Fig. 9.3 by $\bar{t}_r$ and t_{wb} by t_{db}. The difference in diameter, 0.05 in., should not introduce much error. By Fig. 9.3, $(h_r/h_c)_{\text{db}} = 0.146$. By Eq. (9.8)

$$t = 70 - (0.146)(90 - 70) = 67.1 \, {}^\circ\text{F}$$

(b) Equation (9.17) applies for calculation of the wet-bulb coefficient K. By identical procedures of Ex. 9.1, we find $W_{s,\text{wb}} = 0.011087 \text{ lbm}_w/\text{lbm}_a$, $h_{fg,\text{wb}} = 1059.85 \text{ Btu/lbm}_w$, $\alpha/D = 0.852$, Le $= 0.898$, and $h_{r,t}/h_c = 0.134$. By Eq. (9.17)

$$K = \frac{0.240 + (0.444)(0.011087)}{\dfrac{1059.85}{0.898 \left[1 + (0.134)\dfrac{(550^4 - 520^4)}{(527^4 - 520^4)} \right]} + 0.225(7.1)} = 0.000335 \text{ lbm}_w/\text{lbm}_a \cdot {}^\circ\text{F}$$

By Eq. (9.9)

$$W = 0.011087 - (0.000335)(67.1 - 60) = 0.00871 \text{ lbm}_w/\text{lbm}_a$$

Example 9.2 shows that radiation effects may cause considerable error in the dry-bulb thermometer reading and may also significantly effect the wet-bulb coefficient K when the air velocity is low. Since the mean radiant temperature of the surrounding surfaces is rarely known accurately, much more reliable results are obtained with a psychrometer when the air velocity is moderately high, about 1000 ft/min (5 m/s). The most reliable results, of course, are obtained when both thermometers are shielded from radiation effects. For this case, the simpler relation, Eq. (9.14), applies for the wet-bulb coefficient. Since perfect radiation shields are not possible, it is still desirable to use a moderately high velocity.

There will always be some uncertainty in the calculation of the wet-bulb coefficient K. The influence of an error in K upon W in Eq. (9.9) is most severe at low relative humidities. When W is very small compared to both $W_{s,\text{wb}}$ and $K(t - t_{\text{wb}})$, the psychrometer may yield an undesirable result. However, almost all of the commonly available techniques for measuring humidity are affected in a similar adverse manner.

9.4 PRACTICAL USE OF A PSYCHROMETER

From our discussion of the psychrometer in Sec. 9.3, we recognize that several factors may affect the readings of the thermometers. Careful application is necessary to obtain reliable results. The use of a psychrometer is covered in detail in the ASHRAE publication on psychrometry [3].

Two types of psychrometers are commonly used. Each comprises two thermometers, with the bulb of one covered by a moistened wick. It is necessary to separate the two sensing bulbs so that radiation heat exchange between them is negligible. The *sling psychrometer* is widely used for measurements involving room air or other applications where the rate of air movement is small. Air circulation past the thermometer bulbs is obtained by whirling the psychrometer. The *aspiration psychrometer* uses two stationary thermometers and is ventilated by a motor-driven blower. Unventilated psychrometers are unreliable and should not be used. Many psychrometers are mercury-in-glass thermometers. However other types of temperature sensors, such as resistance thermometers, thermocouples, and bimetallic elements, are also used.

The function of the wick is to provide a thin film of water on the wet bulb. Cotton or linen cloth of a soft mesh is satisfactory. Any factors which prevent a continuous film of water on the wet bulb may cause an erroneous reading. Thus the wick material should have no sizing or encrustations, should be clean, and should fit snugly. Wicks should be replaced frequently, and only distilled or deionized water should be used for saturating. It is desirable for the wick to extend 1 or 2 in. (2 to 5 cm) beyond the sensing bulb to help reduce the heat conduction along the stem. The wick should be maintained in a fully saturated condition, since a partially dry wick may cause an erroneous reading. The temperature of water used for saturating the wick should be close to the wet-bulb temperature.

Wile [4] has discussed application of wet-bulb psychrometers for temperatures below freezing. Here it is desirable to discard the wick and to freeze a layer of ice directly on the wet bulb. Some uncertainty may exist as to whether ice or subcooled water is in equilibrium with the wet bulb. The wet-bulb thermometer is less reliable and less convenient to use for temperatures below freezing than other methods of humidity measurement.

It is difficult to accurately measure the moisture content of air with a wet- and dry-bulb psychrometer when the dry-bulb temperature exceeds about 100 °F(40 °C). The difference between the wet- and dry-bulb readings becomes very large even for moderate humidity levels. Errors associated with stem conduction in the thermometer, radiation, and rapid water evaporation resulting in incomplete wetting of the wet bulb become very significant. The user must exercise great care in operating the psychrometer in these conditions and correcting for the errors. In most cases, other methods of humidity measurement should be used in these high-temperature applications.

9.5 CORRELATION OF PSYCHROMETER WET-BULB TEMPERATURE WITH THERMODYNAMIC WET-BULB TEMPERATURE

In Chapter 7 we discussed thermodynamic wet-bulb temperature or the temperature of adiabatic saturation. In preceding sections of this chapter we analyzed wet-bulb temperature as found from a thermometer.

Some confusion may arise because there are two types of wet-bulb temperature. We should realize that there is a distinct difference between thermodynamic wet-bulb temperature and ordinary wet-bulb temperature that may be read from a thermometer. Thermodynamic wet-bulb temperature is a hypothetical temperature which, strictly speaking, can only be approached in a limiting case and cannot be measured directly. We should emphasize that only thermodynamic wet-bulb temperature is a thermodynamic property. A wet-bulb temperature as read from a thermometer is influenced by heat- and mass-transfer rates and is therefore not a sole function of the air state to which the thermometer is exposed. Thus, in psychrometric equations and psychrometric charts where the wet-bulb temperature appears, it is always the thermodynamic wet-bulb temperature that is considered.

An interesting problem is the relationship between psychrometer wet-bulb temperature, t_{wb}, and thermodynamic wet-bulb temperature, t^*. Any air stream with the properties t, W, etc. to which we expose a wet-bulb thermometer also must have some value of t^*. Our problem is how to determine the difference between t_{wb} and t^*.

We may rewrite Eq. (7.29), obtained from an analysis of the adiabatic saturation process, as

$$W = W_s^* - K^*(t - t^*)$$
(9.18)

where

$$K^* = \frac{c_p}{h_{fg}^*}$$
(9.19)

The quantity K^* is analogous to a wet-bulb coefficient for the adiabatic saturation process. Comparison of Eqs. (9.9), (9.10), (9.18), and (9.19) shows that a shielded wet-bulb thermometer will indicate a temperature less than the thermodynamic wet-bulb temperature, since the Lewis number, Le, is less than one.

However, t_{wb} may be equal to t^*, providing that in Eq. (9.10)

$$\text{Le}\left[1 + \frac{h_r(\bar{t}_r - t_{wb})}{h_c(t - t_{wb})}\right] = 1$$

Thus radiation heat transfer to the wet bulb may compensate for the Lewis number being less than unity.

We will now derive a general relationship between the psychrometer wet-bulb temperature, t_{wb}, and thermodynamic wet-bulb temperature, t^*. By Eqs. (9.9) and (9.10),

$$(t_{wb} - t^*) + \frac{(W_{s,wb} - W_s^*)}{K^*} = \left(\frac{K}{K^*} - 1\right)(t - t_{wb})$$
(9.20)

Since we find that t_{wb} differs slightly from t^*, for a small interval of temperature, we may write

$$W_{s,wb} = A + Bt_{wb}$$

$$W_s^* = A + Bt^*$$

where A and B are constants. We obtain the difference as

$$W_{s,wb} - W_s^* = B(t_{wb} - t^*)$$
(9.21)

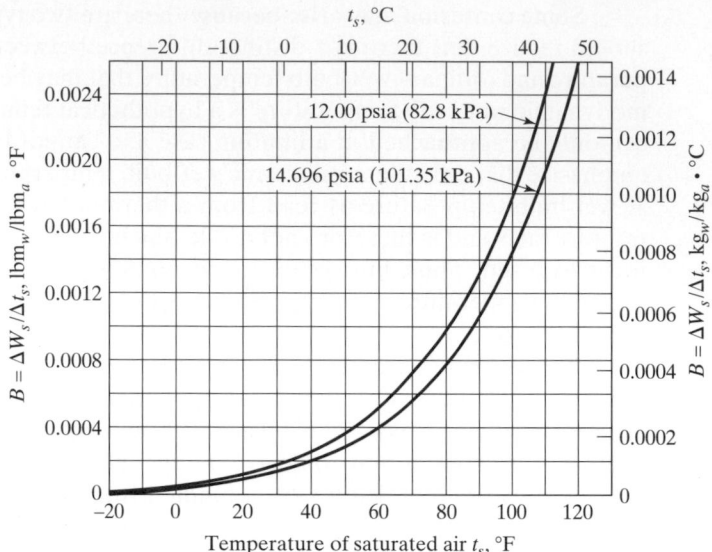

Figure 9.5 Rate of change of humidity ratio with temperature for saturated air.

By Eqs. (9.20) and (9.21),

$$\frac{t_{\mathrm{wb}} - t^*}{t - t_{\mathrm{wb}}} = \frac{(K/K^*) - 1}{1 + (B/K^*)} \tag{9.22}$$

Equation (9.22) expresses the deviation $(t_{\mathrm{wb}} - t^*)$ in terms of the wet-bulb depression $(t - t_{\mathrm{wb}})$. Evaluation of the psychrometer wet-bulb coefficient K was discussed in Sec. 9.3. Figure 9.5 shows the quantity B in Eq. (9.22) for barometric pressures of 12.00 psia (82.8 kPa) and 14.696 psia (101.325 kPa).

For the special case when the mean temperature of the surfaces surrounding the wet bulb is equal to the dry-bulb temperature $(\bar{t}_r = t)$, Eq. (9.22) reduces to

$$\frac{t_{\mathrm{wb}} - t^*}{t - t_{\mathrm{wb}}} = \frac{\mathrm{Le}\,(1 + h_{r,t}/h_c) - 1}{1 + (B/K^*)} \tag{9.23}$$

since $h_{fg,\mathrm{wb}}$ is almost equal to h_{fg}^*.

For the special case of a shielded wet-bulb thermometer, Eq. (9.22) reduces to

$$\frac{t_{\mathrm{wb}} - t^*}{t - t_{\mathrm{wb}}} = \frac{\mathrm{Le} - 1}{1 + (B/K^*)} \tag{9.24}$$

Figure 9.6 shows deviation of the psychrometer wet-bulb temperature from the thermodynamic wet-bulb temperature for a wet-bulb diameter of 0.3 in. (7.6 mm). The solid-line curves were calculated using Eq. (9.23) for an unshielded wet bulb for the case when the temperature of the surrounding surfaces equals the air dry-bulb temperature. The broken-line curves were calculated using Eq. (9.24) for the case of a shielded wet-bulb psychrometer. Three temperature combinations are shown for each case.

Figure 9.6 shows that for an unshielded wet-bulb thermometer there is some velocity that gives equality between t_{wb} and t^* for each temperature combination. The required velocity increases with an increase of temperature. Beyond a certain velocity the devia-

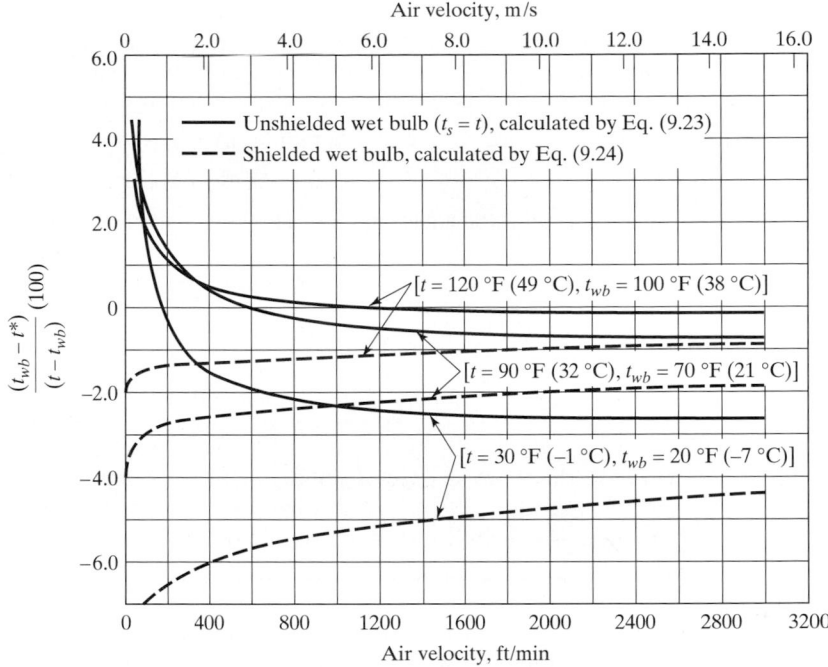

Figure 9.6 Deviation $(t_{wb} - t^*)$ in percent of the wet-bulb depression $(t - t_{wb})$ for a wet-bulb diameter of 0.3 in. (7.6 mm) and a barometric pressure of 14.696 psia (101.325 kPa).

tion is essentially constant. At velocities less than about 100 ft/min (0.6 m/s) the deviation may become large. Figure 9.6 also shows that with a shielded wet-bulb thermometer the deviation will always be negative and, in general, larger than for the unshielded wet bulb where $\bar{t}_r = t$.

Figure 9.7 shows results for conditions similar to those in Fig. 9.6 except for a wet-bulb diameter of 0.1 in. (2.5 mm). Equality of t_{wb} and t^* occurs at much lower velocities than when the wet-bulb diameter is 0.3 in. (7.6 mm).

Although Figs. 9.6 and 9.7 show results for only three temperature combinations, several general observations can be made. For atmospheric temperatures above freezing, where the wet-bulb depression does not exceed about 20 °F (10 °C), and where no unusual radiation circumstances exist, t_{wb} should differ from t^* by less than about 0.5 °F (1 °C) for an unshielded mercury-in-glass wet-bulb thermometer, as long as the air velocity exceeds about 100 ft/min (0.6 m/s). When a thermocouple is used as a wet-bulb thermometer, similar accuracy exists, except that somewhat lower air velocities are permissible.

EXAMPLE 9.3

Use Eq. (9.22) to estimate the thermodynamic wet-bulb temperature in Ex. 9.2.

Solution: Rearranging Eq. (9.22) to solve for t^*,

$$t^* = t_{wb} - (t - t_{wb}) \frac{(K/K^*) - 1}{1 + (B/K^*)}$$

We will evaluate K^* at the wet-bulb temperature of 60 °F as an estimate. Thus at 60 °F, from Table A.1E, $h_{fg}^* = 1059.85 \text{ Btu/lbm}_w$, and from Table A.4E, $W^* = 0.011087 \text{ lbm}_w/\text{lbm}_a$. From Fig. 9.5, $B = 0.00039 \text{ lbm}_w/\text{lbm}_a \cdot °F$. Therefore, from Eq. (9.19)

$$K^* = \frac{c_{pa} + c_{pw}W^*}{h_{fg}^*} = \frac{0.24 + 0.444(0.011087)}{1059.85} = 0.000231 \text{ lbm}_w/\text{lbm}_a \cdot °F$$

Substituting into the equation for t^* above,

$$t^* = 60 - (67.1 - 60)\left[\frac{(0.000335/0.000231) - 1}{1 + (0.00039/0.000231)}\right]$$

$$t^* = 58.8 °F$$

or

$$t_{wb} - t^* = 1.2 °F$$

If we use this new value for t^* and iterate these calculations, the final value for t^* becomes 58.9 °F. Therefore, the thermodynamic wet-bulb temperature is 1.1 °F lower than the psychrometer wet-bulb temperature reading for the conditions given in Ex. 9.2.

Example 9.3 illustrates the error that can occur by assuming that the wet-bulb temperature reading on the psychrometer equals the thermodynamic wet-bulb temperature of the moist air state.

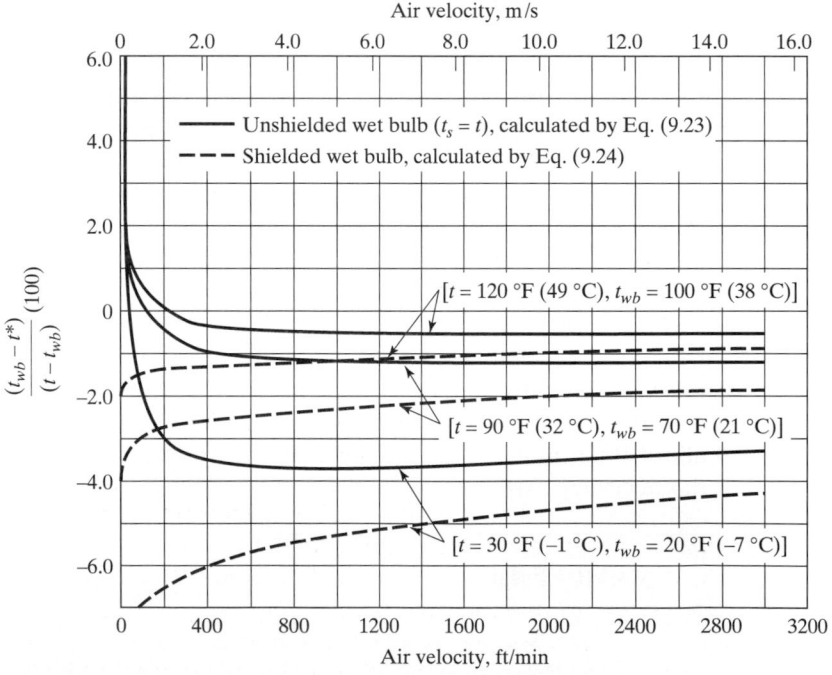

Figure 9.7 Deviation $(t_{wb} - t^*)$ in percent of the wet-bulb depression $(t - t_{wb})$ for a wet-bulb diameter of 0.1 in. (2.5 mm) and a barometric pressure of 14.696 psia (101.325 kPa).

Thus we conclude that in a majority of engineering problems, a wet-bulb temperature obtained from a properly operated unshielded psychrometer may be used directly as the thermodynamic wet-bulb temperature. The moist air state can then be read directly from a psychrometric chart or be obtained from moist air tables. When the wet-bulb depression is large but the mean temperature of the surrounding surfaces is known to differ very slightly from the air dry-bulb temperature, a more accurate procedure with an unshielded psychrometer is to find the moist air state through the use of Eqs. (9.9) and (9.10). When the mean temperature of the surrounding surfaces is believed to be substantially different from the air dry-bulb temperature, or when temperatures in direct sunshine are to be measured, both thermometers of the psychrometer should be shielded. Corrections as illustrated in Ex. 9.3 should be applied for the most accurate measurements.

9.6 OTHER METHODS FOR MEASURING AIR HUMIDITY

Besides the psychrometer, many other techniques are available for measurement of air humidity. Many of these devices are much more complicated than the psychrometer, and, unfortunately, many of them are less reliable. A brief summary of some of these devices is given below. More detailed descriptions are given in the ASHRAE publication on psychrometry [3] and in *ANSI/ASHRAE Standard 41.6-1994* [5].

Air-moisture instruments can be divided into two categories: hygrometers that measure relative humidity, and absolute-humidity instruments that measure the absolute value of the air moisture content.

Several types of hygrometers are available. The *mechanical hygrometer* uses a filament connected by a simple mechanical linkage to a pointer. Often these devices use human hair, which is hygroscopic: its length changes with relative humidity. Unfortunately, temperature also affects the elongation of the hair element. These mechanical hygrometers may be reliable to within ±3% relative humidity for ordinary room temperatures under equilibrium conditions. Since the element has a large time lag, this device is unsuitable where humidity conditions are rapidly changing.

A more recently developed hygrometer uses a thin oxide or polymer layer sandwiched between a metal substrate and a thin metal coating. Moisture diffuses through the thin coating into the middle layer, where it is adsorbed until equilibrium is attained. The change in the amount of moisture in the middle layer changes the device's electrical capacitance and resistance. Usually the capacitance is measured electronically, and the capacitance is converted into relative humidity through a calibration equation. One such device uses a one-micron-thick dielectric polymer as the middle layer. This has a response time of a few seconds. This instrument is capable of measuring air relative humidities from 0 to 100 percent with accuracies ranging from 2 to 3 percent over a temperature range of -40 to 240 °F (-40 to 115 °C). Thus this instrument covers a much wider range of temperature and relative humidity than the psychrometer and has the additional capability of being installed in a remote location with electrical connection to a data-acquisition or process-monitoring system.

The most widely used absolute-humidity measuring device is the *condensation* or *chilled-mirror dew-point instrument*. In this device a highly polished surface is cooled (either thermoelectrically, mechanically, or chemically) until dew or frost begins to form on the surface. A light source shines light at an oblique angle onto the surface. An optical detector such as a photocell is usually aligned to detect the light specularly reflected from the surface. Condensation on the surface causes more of the incident light to be scattered diffusely and less to be reflected specularly. When the reflected-light intensity at the optical detector declines, owing to the condensation on the surface, the surface has reached the dew-point temperature of the air in contact with the surface. Often the output from a photocell is sent to a feedback control system that allows the surface to warm and cool very close to the dew-point temperature. Designs that use thermoelectric refrigeration are very compact, with a response of about 4 °F (2 °C) per second. This device is very reliable and can operate over a wide range of temperature and humidity levels, usually dew-point temperatures ranging from about −95 to 200 °F (−70 to 95 °C). The condensation dew-point instrument is essentially self-calibrating and is often used as a secondary standard. Maintenance is fairly simple, as the mirror should be kept free of contamination. However, some nuclei must be present to initiate condensation on the surface.

9.7 HUMIDITY STANDARDS

As in any other measurement process, a primary standard is required for humidity measurement. To be acceptable as a primary standard for humidity measurement, a device must measure some thermodynamic property of moist air which is related to moisture content. Furthermore, the measurement must be consistent with the definition of the thermodynamic property. Humidity ratio is the only moist air property related to moisture content that is capable of being measured directly. The procedure is called the *gravimetric method* and is considered to be a primary standard. Wexler and Hyland [6] have described the operation of this method. Further discussion and error analysis has been published by the U.S. National Institute of Standards and Technology [7].

The gravimetric method uses a gravimetric hygrometer system. This system uses a humidity generator to provide a constant flow of moist air into the measurement section. The moist air passes through a drying train that adsorbs the water vapor into a solid desiccant. The drying train consists of three interchangeable glass adsorption U-tubes filled with anhydrous magnesium perchlorate and phosphorous pentoxide. The mass of water vapor adsorbed in the drying train is determined by precision weighing at the end of the run. The volume of dry gas that is used is measured by counting the number of times two calibrated cylinders are alternately evacuated and filled with the dry gas. The cylinders are immersed into a temperature-controlled bath, so that gas temperature, pressure, and volume can be measured with extreme precision. Thus the mass of the dry gas can be determined very accurately. By taking the ratio of the mass of water vapor adsorbed divided by the mass of the dry gas used, the humidity ratio of the moist air supplied to the device can be determined with precision.

The gravimetric hygrometer is a rather unwieldy instrument to use and requires up to 30 hours to obtain sufficient mass of water vapor at low humidity ratios. It is not used for normal calibration purposes. NIST has developed and calibrated a two-pressure humidity generator that is commonly used as the NIST calibration standard [7, 8] for calibrating transfer and secondary standards and for testing and evaluating hygrometers and sensors. The principle of operation is as follows. A stream of air at an elevated pressure (P_H) and controlled temperature (t) is saturated with respect to the liquid or solid phase of water. The saturated moist air is then expanded to a lower pressure (P_L) through an expansion valve. The expansion process is one of constant enthalpy. However, if both the dry air and the water vapor behave as ideal gases, the temperature remains constant also, as ideal-gas temperature and enthalpy are related by the constant-pressure specific heat. Using Dalton's law of additive pressures, the partial pressure of the water vapor is directly proportional to the total pressure. The relative humidity of the moist air stream at the low pressure can be calculated as

$$\% \ \phi_L = (P_L/P_H) \times 100 \tag{9.25}$$

As water vapor is not strictly an ideal gas, some corrections should be applied to Eq. (9.25) to achieve the most accurate results [7].

Another method used to determine relative humidity is called the *two-temperature humidity generator*. In this method, air is saturated with water at one temperature and then is heated to a higher temperature. If the total pressure remains fixed, the partial pressure of the water vapor also remains at a constant value. The relative humidity of the moist air at the higher temperature can be determined from the two temperature measurements.

A humidity-generation method that is convenient to set up is a *divided-flow humidity generator*. In this system, an air stream is divided into two parts; one part is sent through a desiccant to remove all water vapor and the other is sent through a saturator to saturate the air with water vapor. By controlling the mass ratio of the two streams that are mixed, the resulting mixture can have a relative humidity anywhere from 0 to 100 percent. Careful measurement of the volume flow rate of the two air streams and their temperature and pressure can provide accurate results.

Saturated salt solutions have been used to provide a known relative humidity for applications where the airflow is very low or in closed chambers with no airflow. A specific water/salt solution determines the equilibrium water-vapor pressure above the solution, depending on its temperature. A relative humidity from about 5 to 100 percent can be obtained using saturated salt solutions. Water vapor versus temperature data for various salt solutions have been determined experimentally [9, 10]. Table 9.2 shows relative-humidity results based on previous measurements.

TABLE 9.2 Equilibrium Relative Humidity Values over Saturated Salt Solutions

Salt	Temperature, °C										
	0	10	20	30	40	50	60	70	80	90	100
Lithium bromide	7.8	7.1	6.6	6.2	5.8	5.5	5.3	5.2	5.2	5.3	5.4
Lithium chloride	11.2	11.3	11.3	11.3	11.2	11.1	11.0	10.8	9.4	10.3	12.1
Magnesium chloride	33.7	33.5	33.1	32.4	31.6	30.5	29.3	27.8	26.1	24.1	22.0
Potassium carbonate	43.1	43.1	43.2	43.2		40.9	39.2	37.4	35.4	33.4	31.3
Magnesium nitrate	60.4	57.4	54.4	51.4	48.4	45.4					
Sodium nitrite									48.5	44.9	41.0
Sodium bromide		62.2	59.1	56.0	53.2	50.9	49.7	49.7			
Sodium nitrate									63.0	60.7	58.3
Sodium chloride	75.5	75.7	75.5	75.1	74.7	74.7	74.5	75.1	73.9	73.8	73.9
Potassium chloride	88.6	86.8	85.1	83.6	82.3	81.2	80.2	79.5			
Barium chloride									85.1	83.9	82.6
Potassium sulfate	98.8	98.2	97.6	97.0	96.4	95.8	96.6	96.3	95.8	95.2	94.5

Source: *ANSI/ASHRAE 41.6-1994.*

ENDNOTES

1. E. R. G. Eckert and R. M. Drake, *Analysis of Heat and Mass Transfer* (New York: McGraw-Hill, 1972), ch. 22.

2. T. Kusuda, "Calculation of the Temperature of a Flat Plate Wet Surface under Adiabatic Conditions with Respect to the Lewis Relation," *Humidity and Moisture*, Vol. 1: *Principles and Methods of Measuring Humidity in Gases*, R. E. Ruskin, ed. (New York: Reinhold, 1965), 16–32.

3. *Psychrometrics: Theory and Practice* (Atlanta: American Society of Heating, Refrigerating and Air Conditioning Engineers, 1996).

4. D. D. Wile, "Psychrometry in the Frost Zone," *Refrigerating Engineering*, 48 (October 1944), 291–301.

5. *ANSI/ASHRAE Standard 41.6-1994, Method for Measurement of Moist Air Properties* (Atlanta: American Society of Heating, Refrigerating and Air Conditioning Engineers, 1994).

6. A. Wexler and R. W. Hyland, "The NBS Standard Hygrometer," in *Humidity and Moisture*, Vol. 3: *Fundamentals and Standards*, ed. A. Wexler and W. A. Wildhack (New York: Reinhold Publishing Corp. 1965) 389–432.

7. NISTIR 4677, "NIST Calibration Services for Humidity Measurement" (Gaithersburg, MD: U.S. Department of Commerce, National Institute of Standards and Technology, October 1991).

8. S. Hasegawa and J. W. Little, "The NBS Two-Pressure Humidity Generator, Mark 2," *Journal of Research, National Bureau of Standards*, 81A (1977), 81.

9. L. Greenspan, "Humidity Fixed Points of Binary Saturated Aqueous Solutions," *Journal of Research of the National Bureau of Standards—A. Physics and Chemistry*, 81A: 1 (1977).

10. P. H. Huang and J. R. Whetstone, "Evaluation of Relative Humidity Values for Saturated Aqueous Salt Solutions Using Osmotic Coefficients Between 50 and 100 °C," *Humidity and Moisture*, 1985, 577–595.

PROBLEMS

9.1 A psychrometer indicates a dry-bulb temperature of 90 °F and a wet-bulb temperature of 70 °F. Barometric pressure is 13.00 psia. Air velocity is 600 ft/min. Assume that both thermometers are shielded from radiation. Determine the relative humidity in the room.

9.2 A psychrometer shielded from radiation indicates a dry-bulb temperature of 25 °C and a wet-bulb temperature of 18 °C. Barometric pressure is 105 kPa. Air velocity past the thermometers is 3 m/s. Compute the air relative humidity.

9.3 Readings from an unshielded psychrometer for an air velocity of 1200 ft/min are 120 °F dry-bulb temperature and 75 °F wet-bulb temperature. Barometer reading is 27.56 in. Hg. The wet-bulb diameter is 0.3 in. Assume that the surrounding surfaces are at 120 °F. Find the humidity ratio of the air stream.

9.4 An unshielded-type psychrometer indicates a dry-bulb temperature of 90 °F and a wet-bulb temperature of 60 °F. Air velocity past the thermometers is 500 ft/min. Diameter of the dry bulb is 0.25 in.; diameter of the wet bulb is 0.3 in. Barometric pressure is 14.696 psia. The mean temperature of the surrounding surfaces is estimated to be 120 °F. Determine the true dry-bulb temperature, humidity ratio, and thermodynamic wet-bulb temperature of the air stream.

9.5 Estimate the reading of the dry-bulb thermometer in Prob. 9.4 if the sensing bulb was tightly wrapped with a metal foil having an emissivity of 0.05.

9.6 A psychrometer with unshielded thermometers indicates a dry-bulb temperature of 250 °F and a wet-bulb temperature of 100 °F. Barometric pressure is 14.696 psia. The mean temperature of the surrounding surfaces is 250 °F. Air velocity is 900 ft/min. Diameter of the wet bulb is 0.3 in. Determine the thermodynamic wet-bulb temperature of the air stream.

9.7 A psychrometer is exposed to an air stream having true properties of 100 °F dry-bulb temperature, 70 °F thermodynamic wet-bulb temperature, and 14.00 psia barometric pressure. Surfaces surrounding the sensing bulbs are at 100 °F. Determine the required air velocity, ft/min, such that the wet-bulb thermometer reading will be equal to the thermodynamic wet-bulb temperature

(a) if the diameter of the wet-bulb is 0.1 in., and

(b) if the diameter of the wet bulb is 0.5 in.

9.8 Compressed moist air at saturation conditions under a pressure P_H is throttled to a lower pressure P_L. Assuming perfect-gas behavior, prove that the relative humidity ϕ_L is given by the relation: $\phi_L = P_L/P_H$. (*Note:* This problem demonstrates the idealized principle of the two-pressure humidity generator.)

9.9 Saturated moist air at a pressure of 100 kPa and a temperature of 15 °C is heated at constant pressure to 25 °C. Determine the resulting relative humidity at the higher temperature. (*Note:* This problem demonstrates the idealized principle of the two-temperature humidity generator.)

SYMBOLS

A	Coefficient in the equation, $W_{s,\text{wb}} = A + Bt_{\text{wb}}$, $\text{lbm}_w/\text{lbm}_a$ (kg_w/kg_a).
a	Constant, dimensionless.
B	Coefficient in the equation, $W_{s,\text{wb}} = A + Bt_{\text{wb}}$, $\text{lbm}_w/\text{lbm}_a \cdot °\text{F}$ ($\text{kg}_w/\text{kg}_a \cdot °\text{C}$).
b	Constant, dimensionless.
c	Constant, dimensionless.
c_{pa}	Specific heat of dry air at constant pressure, $\text{Btu}/\text{lbm}_a \cdot °\text{F}$ ($\text{kJ}/\text{kg}_a \cdot °\text{C}$).
c_p	Specific heat of moist air at constant pressure, $\text{Btu}/\text{lbm}_a \cdot °\text{F}$ ($\text{kJ}/\text{kg}_a \cdot °\text{C}$).
c_{pw}	Specific heat of water vapor at constant pressure, $\text{Btu}/\text{lbm}_w \cdot °\text{F}$ ($\text{kJ}/\text{kg}_w \cdot °\text{C}$).
D	Diffusivity of water vapor in air, ft^2/hr (m^2/s).
d	Diameter, ft (m).
h_c	Convective heat-transfer coefficient, $\text{Btu}/\text{hr} \cdot \text{ft}^2 \cdot °\text{F}$ ($\text{kW}/\text{m}^2 \cdot °\text{C}$).
h_D	Convective mass-transfer coefficient, $\text{lbm}_w/\text{hr} \cdot \text{ft}^2 \cdot \text{lbm}_w/\text{lbm}_a$ ($\text{kg}_w/\text{s m}^2 \cdot \text{kg}_w/\text{kg}_a$).
h_{fg}	Latent heat of vaporization for water, Btu/lbm_w (kJ/kg_w); $h_{fg,\text{wb}}$ evaluated at t_{wb}; h_{fg}^* evaluated at t^*.
h_r	Radiation heat-transfer coefficient, $\text{Btu}/\text{hr} \cdot \text{ft}^2 \cdot °\text{F}$ ($\text{kW}/\text{m}^2 \cdot °\text{C}$); $h_{r,t}$ evaluated for $\bar{t}_r = t$.
K	Psychrometer wet-bulb coefficient, $\text{lbm}_w/\text{lbm}_a \cdot °\text{F}$ ($\text{kg}_w/\text{kg}_a \cdot °\text{C}$).
K^*	Equivalent psychrometer wet-bulb coefficient for adiabatic saturation process, $\text{lbm}_w/\text{lbm}_a \cdot °\text{F}$ ($\text{kg}_w/\text{kg}_a \cdot °\text{C}$).
k	Thermal conductivity of moist air, $\text{Btu}/\text{hr} \cdot \text{ft} \cdot °\text{F}$ ($\text{kW}/\text{m} \cdot °\text{C}$); k_f evaluated at the mean film temperature, t_f.
L	Reference length, ft (m).
Le	Lewis number, $h_c/h_D c_{pa}$, dimensionless.
$\dot{m}_w$	Mass flow rate of water vapor, $\text{lbm}_w/\text{hr} \cdot \text{ft}^2$ ($\text{kg}_w/\text{s} \cdot \text{m}^2$).
P_H	High pressure, psia (kPa).
P_L	Low pressure, psia (kPa).
Pr	Prandtl number, $c_p \mu/k$, dimensionless.
Re_L	Reynolds number, $LV\rho/\mu$, dimensionless.
Sc	Schmidt number, $\mu/\rho D$, dimensionless.
T	Absolute dry-bulb temperature of the air, °R (K).

$\overline{T}_r$ Absolute mean radiant temperature surrounding psychrometer, °R (K).

T_{wb} Absolute temperature indicated by the wet-bulb thermometer, °R (K).

t Dry-bulb temperature of moist air, °F (°C).

t^* Thermodynamic wet-bulb temperature of moist air, °F (°C).

t' $(t - t_w)/(t_o - t_w)$, dimensionless temperature (see Fig. 9.1).

t_{db} Temperature indicated by dry-bulb thermometer, °F (°C).

t_f Mean film temperature, °F (°C).

t_o Temperature of bulk air stream, °F (°C).

$\overline{t}_r$ Mean radiant temperature, °F (°C).

t_w Temperature of water surface, °F (°C) (see Fig. 9.1).

t_{wb} Temperature indicated by wet-bulb thermometer, °F (°C).

$\mathbf{V}$ Air velocity, ft/hr (m/s); $\mathbf{V}_o$ for bulk stream (see Fig. 9.1).

W Humidity ratio of moist air, lbm_w/lbm_a (kg_w/kg_a); W_o for bulk air stream in Fig. 9.1.

W_s Humidity ratio of saturated moist air, lbm_w/lbm_a (kg_w/kg_a); $W_{s,w}$ evaluated at t_w of wet surface; $W_{s,wb}$ evaluated at t_{wb}; W_s^* evaluated at t^*.

W' $(W_{s,w} - W)/(W_{s,w} - W_o)$, dimensionless (see Fig. 9.1).

y Coordinate length, ft (m).

y' y/L, dimensionless length.

Greek Letters

α Thermal diffusivity, $k/\rho c_p$, ft²/hr (m²/s).

ε_{wb} Emissivity of wet bulb, dimensionless.

μ Dynamic viscosity, lbm/ft · hr (kg/m · s); μ_f evaluated at t_f.

ρ Density, lbm/ft³ (kg/m³); ρ_f evaluated at t_f.

ρ_a Density of moist air, lbm_a/ft³ (kg_a/m³).

σ Stefan-Boltzmann constant.

ϕ Relative humidity, dimensionless.

Heat- and Mass-Transfer Processes and Applications

chapter 10

Direct-Contact Transfer Processes between Moist Air and Water

10.1 INTRODUCTION

In Chapter 8 we discussed the problem of adding moisture to air. We assumed that all moisture in contact with the air was retained by the air stream. We found that the condition line on the psychrometric chart was a function solely of the enthalpy of the added moisture.

In this chapter we will consider problems involving direct contact of air and water different from those covered in Chapter 9. We will now study processes where a relatively large flow of water contacts moist air. The rate of addition (or withdrawal) of moisture to the air stream will be extremely small compared to the flow rate of the water entering the apparatus. Depending upon the moist air state and the temperature of the water, it is possible to have a variety of results. Air may be heated and humidified, cooled and humidified, or cooled and dehumidified by direct contact with water.

In this chapter we will be concerned with thermal processes only. However, in practical apparatus, atmospheric air may be partially cleansed of airborn particles and water-soluble vapors when it is washed by water. Some type of water treatment is recommended to prevent the growth of biological contaminants in the recirculated water.

10.2 DIRECTLY RECIRCULATED ISOTHERMAL SPRAY WATER: THE AIR WASHER

Spray devices using directly recirculated water have been in use for many years. Such a device is called an air washer. Figure 10.1 shows plan and elevation views of an air washer with one bank of spray nozzles. Water is withdrawn from the sump by an external pump and sprayed into the chamber in fine droplets by a bank of nozzles. At the air-outlet end, staggered metal baffles called *eliminator plates* minimize physical carry-over of water droplets with the air stream.

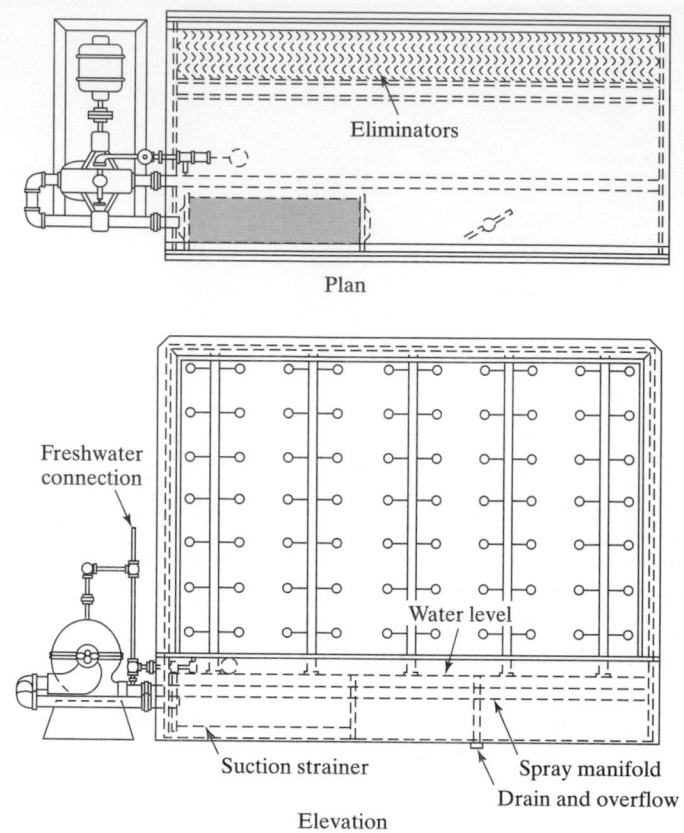

Figure 10.1 Air washer with single bank of spray nozzles. [Reprinted by permission from *ASHRAE Heating, Ventilating, Air Conditioning Guide*, Vol. 28 (1950), p. 710.]

Figure 10.2 shows a schematic diagram that we will use for our analysis. We will assume that the rate at which makeup water is added to the sump is negligibly small compared to the rate of water flow through the nozzles. We will assume that heat transfer through the walls of the chamber from the ambient surroundings may be ignored. We will further assume that the small addition of energy to the water by the pump has a negligible effect upon the water temperature. We also make the assumption that there is no temperature change of the water droplets as they pass through the washer, as the mass flow rate of water is much greater than the mass flow rate of air. Parallel-flow, cross-flow, or counterflow arrangements all provide similar results. However, we will use a parallel-flow arrangement in our analysis. Transfer processes in the chamber involve evaporation of water droplets and convection heat transfer from the air to the water. For steady-flow conditions, and for a differential volume dV, we have

$$\dot{m}_a \, dh = \dot{m}_a \, dW \, h_{f,w}$$

or

$$q' = \frac{dh}{dW} = h_{f,w} \tag{10.1}$$

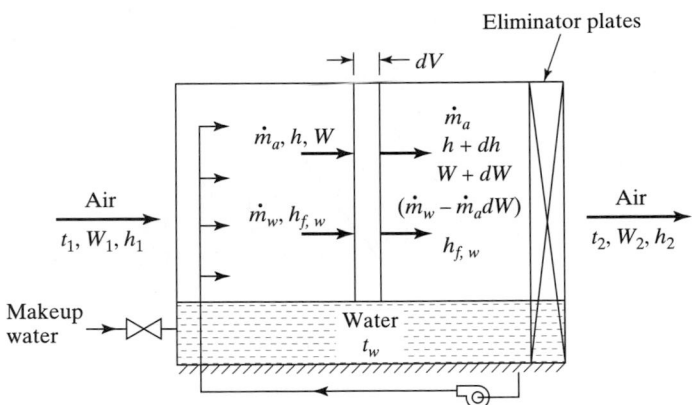

Figure 10.2 Schematic diagram of air washer using directly recirculated spray water.

Since the water temperature remains constant, the condition line on the psychrometric chart would be straight with a direction of $q' = h_{f,w}$.

By Eq. (7.21), we obtain

$$dh = c_{pa}dt + h_{g,t}\,dW \tag{10.2}$$

and by Eqs. (10.1) and (10.2)

$$\frac{dt}{dW} = \frac{-(h_{g,t} - h_{f,w})}{c_{pa}} \tag{10.3}$$

Heat transfer for evaporation of water added must come from convection cooling of the air stream. Thus

$$h_D A_V\,dV(W_{s,w} - W)h_{fg,w} = h_c A_V\,dV(t - t_w) \tag{10.4}$$

where h_D is a mass-transfer coefficient, $\text{lbm}_w/\text{hr} \cdot \text{ft}^2$ ($\text{lbm}_w/\text{lbm}_a$) or $\text{kg}_w/\text{sec} \cdot \text{m}^2$ (kg_w/kg_a); A_V is surface area of the water droplets, ft^2/ft^3 (m^2/m^3); V is contact volume, ft^3 (m^3); $W_{s,w}$ and W are, respectively, the humidity ratio of saturated moist air in equilibrium with the water and that of the air stream, $\text{lbm}_w/\text{lbm}_a$ (kg_w/kg_a); $h_{fg,w}$ is the latent heat of vaporization of the water, Btu/lbm_w (kJ/kg_w); h_c is a convection heat-transfer coefficient, $\text{Btu}/\text{hr}/\text{ft}^2 \cdot °\text{F}$ or $\text{kW}/\text{m}^2 \cdot °\text{C}$; and t and t_w are, respectively, the moist air dry-bulb temperature and the water temperature, $°\text{F}$ ($°\text{C}$).

Substitution of $\text{Le} = h_c/h_D c_{pa}$ in Eq. (10.4) gives

$$(W_{s,w} - W)h_{fg,w} = \text{Le}\,c_{pa}(t - t_w) \tag{10.5}$$

Differentiation of Eq. (10.5) with respect to W (assuming Le and t_w are constant) gives

$$\frac{dt}{dW} = \frac{-[h_{g,t} - h_{g,w} + (h_{fg,w}/\text{Le})]}{c_{pa}} \tag{10.6}$$

Since both Eqs. (10.3) and (10.6) must be satisfied, we conclude that Le must be unity for an air washer using directly recirculated spray water. This result contrasts with the development of Sec. 9.2. However, the air washer is a special problem, subject to the

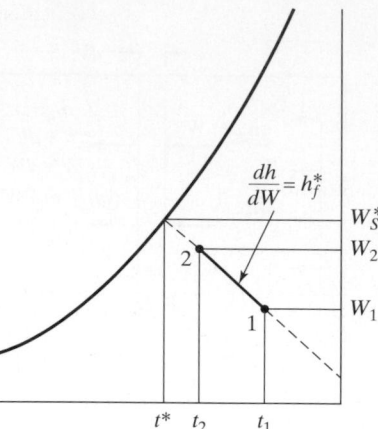

$$\frac{dh}{dW} = h_f^*$$

Figure 10.3 Schematic condition line for moist air passing through the air washer of Fig. 10.2.

thermodynamic limitations imposed by Eq. (10.3). Processes of heat transfer and mass transfer must be mutually compensating. Thus,

$$(W_{s,w} - W)h_{fg,w} = c_{pa}(t - t_w) \tag{10.7}$$

When compared to Eq. (7.28), Eq. (10.7) shows that the water temperature must be equal to the thermodynamic wet-bulb temperature of the incoming air t_1^*. Furthermore, the condition line on the psychrometric chart must coincide with a line of constant thermodynamic wet-bulb temperature. Figure 10.3 shows a schematic case. Experiments on actual air washers using directly recirculated spray water show that these conclusions are closely true.

The rate of evaporation for the volume element dV in Fig. 10.2 may be written as

$$\dot{m}_a \, dW = h_D A_V \, dV(W_s^* - W)$$

Assuming $\dot{m}_a$, $h_D A_V$, and t_w as constants, we obtain

$$\frac{W_s^* - W_2}{W_s^* - W_1} = e^{-\text{NTU}} \tag{10.8}$$

where the number of transfer units (NTU) is given by

$$\text{NTU} = \frac{h_D A_V V}{\dot{m}_a}$$

The air-washer efficiency, η_w, is defined by the relation

$$\eta_w \equiv \frac{W_2 - W_1}{W_s^* - W_1} \tag{10.9}$$

By Eqs. (10.8) and (10.9), we have

$$\eta_w = 1 - e^{-\text{NTU}} \tag{10.10}$$

If we make the approximation that c_p is constant in Eq. (7.28), we find that

$$\eta_w = \frac{t_1 - t_2}{t_1 - t^*} \tag{10.11}$$

Equation (10.9) is useful in evaluating the performance of an air washer as a humidification device. However, a dry-bulb temperature reduction occurs in the process. In hot, relatively dry climates an air washer using directly recirculated spray water (or other devices employing the same principle) may be beneficial for reducing the dry-bulb temperature of outdoor air admitted to ventilation systems. Equation (10.11) is useful in evaluating the performance of an air washer as an evaporative-cooling device.

EXAMPLE 10.1

Moist air enters an air washer similar to that of Fig. 10.2 at 90 °F dry-bulb temperature and 60 °F thermodynamic wet-bulb temperature at a rate of 5000 ft^3/min. Barometric pressure is 14.696 psia. It is desired that the humidity ratio of the air leaving the washer be 0.0080 lbm$_w$ per lbm$_a$. Face velocity of the entering air is 500 ft/min. It is estimated that the mass-transfer coefficient $h_D A_V$ has a value of 300 lbm$_w$/hr $\cdot$ ft^3 (lbm$_w$/lbm$_a$). Determine (a) the dry-bulb temperature of the air leaving the washer, (b) the air-washer efficiency, and (c) the required length of the washer.

Solution:

(a) Since $t_2^* = t_1^*$, we read t_2 = 73.4 °F from Fig. C-8E.

(b) We read W_1 = 0.00425 lbm$_w$/lbm$_a$ and W_s^* = 0.01108 lbm$_w$/lbm$_a$ from Fig. C-8E. By Eq. (10.9),

$$\eta_w = \frac{0.0080 - 0.00425}{0.01108 - 0.00425} = 0.549$$

(c) We have v_1 = 13.95 ft^3/lbm$_a$ from Fig. C-8E. Thus

$$\dot{m}_a = \frac{(5000)(60)}{13.95} = 21,505 \text{ lbm}_a/\text{hr}$$

By Eq. (10.10), $e^{-\text{NTU}}$ = 0.451 and NTU = 0.795. Thus the volume of the washer is

$$V = \frac{\dot{m}_a \, \text{NTU}}{h_D A_V} = \frac{(21,505)(0.795)}{300} = 57.0 \text{ ft}^3$$

The face area of the washer is 5000/500 = 10 ft^2. Thus, the required length is 5.70 ft.

EXAMPLE 10.2

An air washer is run under test to determine the mass-transfer coefficient of the water droplets, $h_D A_V$. Moist air enters at 101.325 kPa, 40 °C, and 20 percent relative humidity at a face velocity of 3 m/s. The washer has a face area of 4 m^2 and a length of 2 m. The water temperature in the washer is 22 °C and the exit-air dry-bulb temperature is 30 °C. Determine the value of $h_D A_V$ for this washer.

Solution: Figure C-8SI is used to obtain all the thermodynamic-property data for the moist air. W_1 = 9.25 g/kg, v_1 = 0.90 m^3/kg$_a$, W_2 = 13.4 g/kg$_a$, W_s^* = 16.7 g/kg$_a$. By Eq. (10.9)

$$\eta_w = \frac{13.4 - 9.25}{16.7 - 9.25} = 0.557$$

By Eq. (10.10), $e^{-NTU} = 0.443$ and $NTU = 0.814$.

$$h_D A_V = \frac{NTU\,\dot{m}_a}{V}$$

$$\dot{m}_a = \frac{(4\ \text{m}^2)(3\ \text{m/s})}{0.90\ \text{m}^3/\text{kg}_a} = 13.33\ \text{kg}_a/\text{s}$$

$$V = (4\ \text{m}^2)(2\ \text{m}) = 8\ \text{m}^3$$

Thus

$$h_D A_V = \frac{0.814(13.33\ \text{kg}_a/\text{s})}{8\ \text{m}^3}$$

$$= 1.36\ \text{kg}_w/\text{s} \cdot \text{m}^3\ (\text{kg}_w/\text{kg}_a)$$

10.3 COUNTERFLOW CONTACT OF MOIST AIR BY HEATED SPRAY WATER: THE COOLING TOWER

Probably the most important device utilizing direct contact between water and atmospheric air is the cooling tower. Here the objective is not the processing of the air but cooling of the spray water. Cooling towers may be used thermally to reclaim circulating water for reuse in refrigerant condensers, power-plant condensers, and other heat exchangers.

Figure 10.4(a) shows a sectional view of a counterflow cooling tower. Atmospheric air is circulated upward through the tower by a fan located below (forced-draft) or above

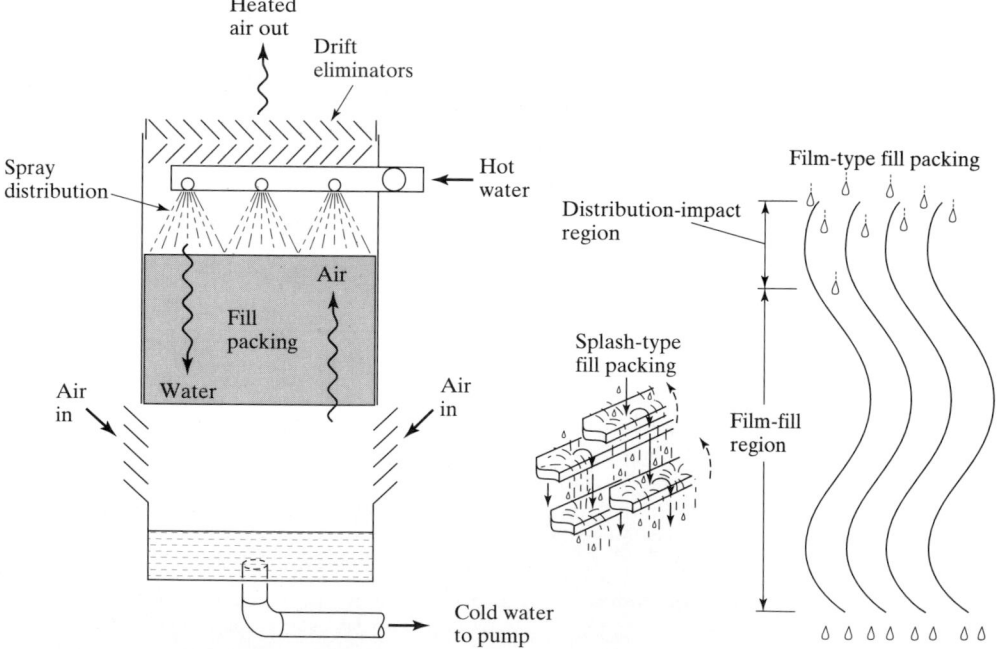

Figure 10.4 Illustration of a counterflow, induced-draft cooling tower and types of fill. [Reprinted by permission from *ASHRAE Systems and Equipment 1996* (IP & SI), pp. 36.2, 36.3.]

Part IV / Heat- and Mass-Transfer Processes and Applications

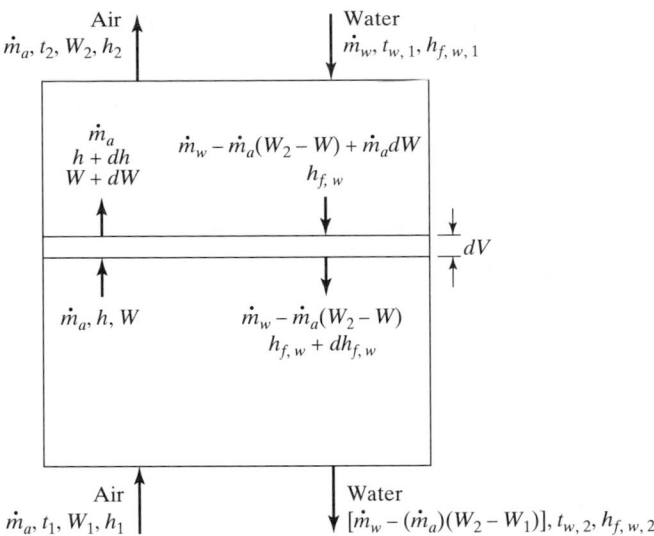

Figure 10.5 Schematic diagram of counterflow cooling tower.

(induced-draft) the tower. The warm water is admitted in the upper part of the tower and falls downward in counterflow to the air. Most towers contain a fill of some type, such as slats or latticework, as shown in Fig. 10.4(b). The fill retards the rate of water fall and increases the water surface exposed to the air. Eliminator plates at the top of the tower minimize drift or carry-over of liquid water droplets in the exhaust air. Water is lost from the tower by evaporation, drift, and blowdown. Blowdown, or wasting of some of the sump water, prevents undue concentration of solids. Total water loss depends upon the water-cooling range, but is usually 3 percent or less. More detailed practical discussion of cooling towers is given in the *ASHRAE HVAC Systems and Equipment Handbook* [1].

Figure 10.5 shows a schematic diagram of a counterflow cooling tower. We will ignore water loss by drift. We will also ignore heat transfer through the walls of the tower. For steady-flow conditions, for the differential-volume element we have the energy balance between the air and water streams

$$\dot{m}_a \, dh = -[\dot{m}_w - \dot{m}_a(W_2 - W)] \, dh_{f,w} + \dot{m}_a \, dW/h_{f,w} \tag{10.12}$$

or approximately

$$\dot{m}_a \, dh = -\dot{m}_w \, dh_{f,w} + \dot{m}_a \, dW/h_{f,w} \tag{10.13}$$

We may also write the water energy balance in terms of the heat- and mass-transfer coefficients, h_c and h_D, respectively, as

$$-\dot{m}_w \, dh_{f,w} = h_c A_V \, dV(t_w - t) + h_D A_V \, dV(W_{s,w} - W)h_{fg,w} \tag{10.14}$$

and the air-side water-vapor mass balance as

$$\dot{m}_a \, dW = h_D A_V \, dV(W_{s,w} - W) \tag{10.15}$$

By substitution of $\text{Le} = h_c/h_D c_{pa}$ in Eq. (10.14), we obtain

$$-\dot{m}_w \, dh_{f,w} = h_D A_V \, dV[\text{Le} \, c_{pa}(t_w - t) + (W_{s,w} - W)h_{fg,w}] \tag{10.16}$$

Combining Eqs. (10.13), (10.15), and (10.16),

$$\frac{dh}{dW} = \text{Le} \, c_{pa} \frac{t_w - t}{W_{s,w} - W} + h_{g,w} \tag{10.17}$$

Using the approximation of constant c_{pa} in Eq. (7.21), we have

$$h_{s,w} - h = c_{pa}(t_w - t) + h_g^0(W_{s,w} - W)$$

Equation (10.17) may then be written as

$$\frac{dh}{dW} = \text{Le} \frac{h_{s,w} - h}{W_{s,w} - W} + (h_{g,w} - h_g^0 \, \text{Le}) \tag{10.18}$$

Equation (10.18) describes the condition line on the psychrometric chart for the changes in state for moist air passing through the tower. Although data for Le for cooling towers are not widely available, it is not necessary that Le be restricted to unity as in the case of the air washer in Sec. 10.2. In our analysis we will use Eq. (10.7).

We may obtain an accurate, although approximate, solution to Eq. (10.18) graphically on the psychrometric chart by the method shown by Fig. 10.6, or we may write a program for a computer to obtain a numerical solution. We assume that the water temperatures $t_{w,1}$ and $t_{w,2}$, the water-flow rate $\dot{m}_w$, the airflow rate $\dot{m}_a$, and the entering air state are known. With this information we may solve Eq. (10.18) for the direction dh/dW of the condition line at state 1. With the chart protractor we draw a line segment with this direction through state 1. At a short distance from state 1 on this line we arbitrarily locate a new state (a). Equation (10.13) may be written as

$$-\Delta t_w = \frac{\dot{m}_a}{\dot{m}_w c_w}\left(\Delta h - \frac{W}{h_{f,w}}\right) \tag{10.19}$$

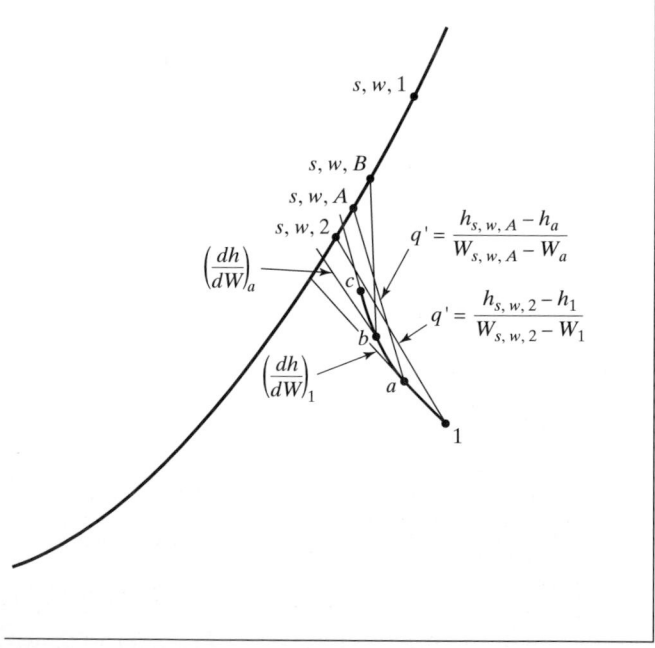

Figure 10.6 Graphical solution of Eq. (10.18) on the psychrometric chart.

With Eq. (10.19) we may calculate the water temperature $t_{w,a}$ corresponding to state a. Equation (10.18) may then be solved for $(dh/dW)_a$ and the procedure continued until the complete condition line is drawn and the final air state determined. The accuracy of the method depends upon the extent of the assumed incremental changes of air state.

For the entire cooling tower, by Fig. 10.5 we have

$$\dot{m}_a h_1 + \dot{m}_w h_{f,w,1} = \dot{m}_a h_2 + [\dot{m}_w - \dot{m}_a(W_2 - W_1)]h_{f,w,2}$$

or

$$h_2 = h_1 + \frac{\dot{m}_w c_w}{\dot{m}_a}(t_{w,1} - t_{w,2}) + (W_2 - W_1)h_{f,w,2} \tag{10.20}$$

Assuming that the complete condition line is already drawn on the psychrometric chart, the outlet air condition may be checked by Eq. (10.20).

With the condition line and the outlet-air state known, and with the average mass-transfer coefficient $h_D A_V$ known, the required tower volume may be obtained from Eq. (10.15). Thus

$$V = \frac{\dot{m}_a}{h_D A_V} \int_{W_1}^{W_2} \frac{dW}{W_{s,w} - W} \tag{10.21}$$

The integral in Eq. (10.21) may be solved by a numerical method.

A graphical method for solving integrals of the type in Eq. (10.21) is by use of the Stevens diagram. Figure 10.7 shows the diagram, where a dimensionless factor f is presented as a function of two dimensionless variables y_m/y_1 and y_m/y_2. Using the integral of Eq. (10.21) to typify a general example, we have

$$y_1 = W_{s,w,1} - W_2 \tag{10.22}$$

$$y_2 = W_{s,w,2} - W_1 \tag{10.23}$$

$$y_m = W_{s,w,m} - W_m \tag{10.24}$$

where $W_m = (W_1 + W_2)/2$ and $W_{s,w,m}$ is evaluated at the same location where W_m exists. The solution of the integral is given by

$$\int_{W_1}^{W_2} \frac{dW}{W_{s,w} - W} = \frac{W_2 - W_1}{f y_m} \tag{10.25}$$

where f is given by Fig. 10.7.

The Stevens diagram was constructed on the premise that the difference function y varies along a second-degree parabola fitted to the three known values y_1, y_m, and y_2.

EXAMPLE 10.3

A cooling tower is to be designed to cool 1500 gpm of water from 100 °F to 85 °F when the outside air is at 95 °F dry-bulb temperature and 75 °F thermodynamic wet-bulb temperature. Barometric pressure is 14.696 psia. The ratio of water flow to dry-air flow $(\dot{m}_w/\dot{m}_a)$ is 1.00. Assume a constant air-mass velocity of 1400 $\text{lbm}_a/\text{hr} \cdot \text{ft}^2$. It is estimated that the average mass-transfer coefficient $h_D A_V$ is 120 $\text{lbm}_w/\text{hr} \cdot \text{ft}^3$ ($\text{lbm}_w/\text{lbm}_a$). (a) Construct the complete condition line on the psychrometric chart; (b) determine the dry-bulb temperature, thermodynamic wet-bulb temperature, and humidity ratio of the air leaving the tower; and (c) calculate the required tower volume in ft^3.

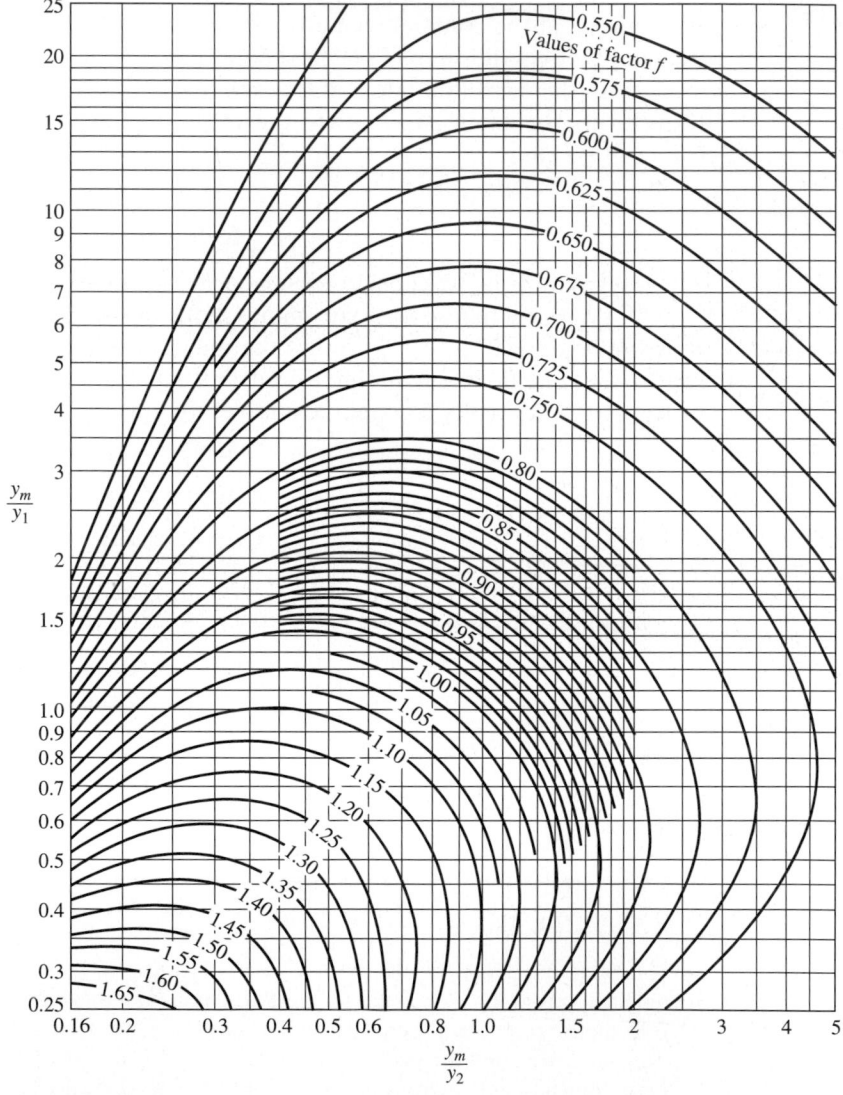

Figure 10.7 The Stevens diagram. [Reprinted from W. F. Carey and G. J. Williamson, "Gas Cooling and Humidification: Design of Packed Towers from Small-Scale Tests," *Proceedings of the Institute of Mechanical Engineers*, 163 (1950), 49.]

required. We begin the construction at the air-inlet end of the tower using Fig. C-8E. We may locate state 1 of the inlet air and state $(s, w, 2)$ of saturated air in equilibrium with the leaving water. By Table A.1E, we have $h_{f,w} = 52.7$ Btu/lbm$_w$ and $h_{g,w} = 1098$ Btu/lbm$_w$. By Eq. (9.7) and Table 9.1 at 85 °F and degree of saturation equal to 1.0, we find Le = 0.895. With the chart protractor, we obtain

$$\frac{h_{s,w,2} - h_1}{W_{s,w,2} - W_1} = 890 \text{ Btu/lbm}_w$$

By Eq. (10.18),

$$\left(\frac{dh}{dW}\right)_1 = (0.895)(890) + 1098 - (0.895)(1061) = 945 \text{ Btu/lbm}_w$$

With the chart protractor we draw a short line of direction $(dh/dW)_1$ through state 1. We arbitrarily locate state a at the intersection of $h = 40.00$ Btu/lbm$_a$ with this line. By Eq. (10.19)

$$t_{w,a} - t_{w,2} = \frac{(1)}{(1)}[(h_a - h_1) - (W_a - W_1)h_{f,w,2}]$$

We obtain $t_{w,a} = 86.52$ °F. Next we locate the state of saturated air at this temperature on the psychrometric chart. We then repeat the procedure.

Table 10.1 shows a tabulation of calculations for the complete condition line. Figure 10.8 shows a plot of the condition line on the psychrometric chart.

(b) The leaving-air state was determined as the final point on the condition line. We read $t_2 = 94.8$ °F, $t_2^* = 88.8$ °F, and $W_2 = 0.0285$ lbm$_w$/lbm$_a$. State 2 may be checked by Eq. (10.20). At $W_2 = 0.0285$ lbm$_w$/lbm$_a$, we obtain

$$h_2 = 38.39 + \frac{(1)}{(1)}(100 - 85) + (0.0285 - 0.01408)(52.7)$$

$$= 54.15 \text{ Btu/lbm}_a$$

which closely agrees with $h_2 = 54.26$ Btu/lbm$_a$ obtained by the graphical procedure.

(c) We may obtain the tower volume by graphical integration of Eq. (10.21). All necessary numerical data are known from part (a). Table 10.2 shows a tabulation of calculations for $f(W) = 1/(W_{s,w} - W)$ for various values of W. Figure 10.9 shows a plot of $f(W)$. The area under the curve is the value of the integral. We obtain

TABLE 10.1 Calculations for Part (a) of Ex. 10.3

Air state	t, °F	h, Btu/lbm$_a$	W, lbm$_w$/lbm$_a$	t_w, °F	$h_{f,w}$, Btu/lbm$_w$	Le	$\dfrac{h_{s,w} - h}{W_{s,w} - W}$, Btu/lbm$_w$	$h_{g,w}$, Btu/lbm$_w$	$\dfrac{dh}{dW}$ Btu/lbm$_w$
1	95.0	38.39	0.01408	85.00	52.7	0.895	890	1098	945
a	93.8	40.00	0.01585	86.52	54.6	0.895	940	1099	990
b	93.0	42.00	0.01782	88.41	56.4	0.895	1000	1100	1045
c	92.6	44.00	0.01975	90.30	58.3	0.895	1080	1101	1118
d	92.7	46.00	0.02154	92.20	60.2	0.895	1095	1101	1131
e	93.0	48.00	0.0233	94.09	62.1	0.895	1130	1102	1163
f	93.3	50.00	0.0250	95.97	64.0	0.895	1150	1103	1182
g	94.0	52.00	0.0267	97.85	65.9	0.895	1170	1104	1201
2	94.8	54.27	0.0285	100.00	—	—	—	—	—

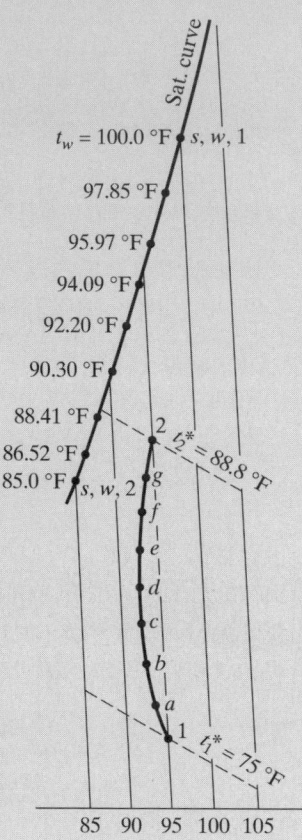

Figure 10.8 Air-process line for Ex. 10.3.

TABLE 10.2 Calculations for Part (c) of Ex. 10.3

Air State	t_w, °F	W, lbm_w/lbm_a	$W_{s,w}$, lbm_w/lbm_a	$W_{s,w} - W$, lbm_w/lbm_a	$f(W) = 1/(W_{s,w} - W)$ lbm_a/lbm_w
1	85.00	0.01408	0.02635	0.01227	81.5
a	86.52	0.01582	0.02775	0.01190	84.0
b	88.41	0.01782	0.02952	0.01170	85.5
c	90.30	0.01975	0.03157	0.01182	84.6
d	92.20	0.02154	0.0336	0.01206	82.9
e	94.09	0.0233	0.0358	0.01250	80.0
f	95.97	0.0250	0.0380	0.01300	76.9
g	97.85	0.0267	0.0404	0.01370	73.0
2	100.00	0.0285	0.0432	0.01470	68.0

$$F(W) = \int_{W_1}^{W_2} \frac{dW}{W_{s,w} - W} = 1.162$$

and

$$V = \frac{\dot{m}_a}{h_D A_V} F(W) = \frac{(746,000)(1.162)}{120} = 7220 \text{ ft}^3$$

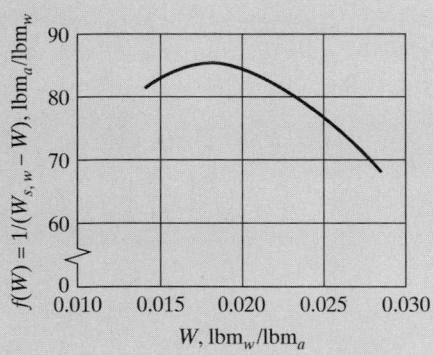

Figure 10.9 Plot of $f(W) = 1/(W_{s,w} - W)$ for Ex. 10.3.

Since air-mass velocity of 1400 lbm$_a$/hr · ft^2 was used, the tower cross-sectional area would be 746,000/1400 = 533 ft^2. The required tower height would be 7220/533 = 13.5 ft.

The Stevens diagram may be used also for solving the integral of Eq. (10.21). We have $W_m = (W_1 + W_2)/2 = 0.02129$ lbm$_w$/lbm$_a$. By Table 10.1, we may estimate $t_{w,m} = 91.83$ °F at the location where $W_m = 0.02129$ lbm$_w$/lbm$_a$ occurs. Thus, $W_{s,w,m} = 0.03305$ lbm$_w$/lbm$_a$. By Eqs. (10.22)–(10.24), we obtain $y_1 = 0.0147$ lbm$_w$/lbm$_a$, $y_2 = 0.01234$ lbm$_w$/lbm$_a$, and $y_m = 0.01176$ lbm$_w$/lbm$_a$. Thus, $y_m/y_1 = 0.80$, and $y_m/y_2 = 0.953$. By Fig. 10.7, $f = 1.045$.

By Eq. (10.25)

$$F(W) = \frac{W_2 - W_1}{fy_m} = \frac{(0.01442)}{(1.045)(0.01176)} = 1.173$$

which closely agrees with the value of 1.162 obtained before.

An interesting result, both academically and practically, is that of determining the minimum possible temperature at which water may be withdrawn from a cooling tower. By Eq. (10.13),

$$\frac{dh}{dW} = -\frac{\dot{m}_w c_w}{\dot{m}_a}\frac{dt_w}{dW} + h_{f,w}$$

The minimum water temperature would occur when $dt_w/dW = 0$, or when $dh/dW = h_{f,w}$. The transfer processes would then correspond to those for an air washer using directly recirculated spray water. For the reasons explained in Sec. 10.2, the minimum possible temperature for water leaving a cooling tower would be the thermodynamic wet-bulb temperature of the inlet air.

The methods of analysis for a cooling tower so far presented provide an accurate procedure for design purposes. Example 10.3 shows that rather lengthy calculations are required. It is necessary to have both water temperatures known in order to apply the procedure conveniently. In order to analyze tower performance at other than design conditions (given values of tower volume, $\dot{m}_w$, $\dot{m}_a$, $t_{w,1}$, and inlet-air state, but unknown values of $t_{w,2}$ and outlet-air state), a tedious trial-and-error procedure is necessary.

It is possible to simplify the analysis if certain approximations are made. Noting that $\dot{m}_w\,dh_{f,w} = \dot{m}_w c_w\,dt_w$, by Eqs. (10.13) and (10.18), we have

$$-\frac{dt_w}{h_{s,w} - h} = \frac{\dot{m}_a}{\dot{m}_w c_w}\left[\mathrm{Le} + \frac{h_{fg,w} - 10h_g^0\,\mathrm{Le}}{(h_{s,w} - h)/(W_{s,w} - W)}\right]\frac{dW}{(W_{s,w} - W)} \qquad (10.26)$$

On the right side of Eq. (10.26), the second term of the group in brackets is small compared to the Lewis number. Example (10.3) shows that Le was constant for a single problem. Thus, we may assign a mean value to the bracketed group with only a small error. By Eqs. (10.21) and (10.26),

$$\int_{t_{w,2}}^{t_{w,1}} \frac{dt_w}{h_{s,w} - h} = \frac{h_D A_v V}{\dot{m}_w c_w} \left[\text{Le} + \frac{(h_{fg,w} - h_g^0 \text{Le})}{(h_{s,w} - h)/(W_{s,w} - W)} \right]_m \qquad (10.27)$$

By Eq. (10.13),

$$\frac{dh}{dt_w} = -\frac{\dot{m}_w c_w}{\dot{m}_a} + \frac{dW}{dt_w} h_{f,w}$$

or

$$-\frac{dh}{dt_w} = \frac{\dot{m}_w c_w / \dot{m}_a}{1 - h_{f,w}/(dh/dW)} \qquad (10.28)$$

Table 10.1 shows that for Ex. 10.3, the term $h_{f,w}/(dh/dW)$ was small compared to unity and that it was also essentially constant. We may assign a mean value to this term and thus assume that approximately h varies linearly with t_w.

Solution of Eqs. (10.27) and (10.28) is simplified by use of the Stevens diagram and the analysis shown by Fig. 10.10. Figure 10.10 shows a plot of the air enthalpy h and the enthalpy of saturated air $h_{s,w}$ as a function of the water temperature. The operating line showing the variation of h with t_w is drawn by Eq. (10.28). Through use of the Stevens diagram, we have

$$\int_{t_{w,2}}^{t_{w,1}} \frac{dt_w}{h_{s,w} - h} = \frac{t_{w,1} - t_{w,2}}{f y_m} \qquad (10.29)$$

The quantities y_1, y_2, and y_m for use with Fig. 10.7 are shown in Fig. 10.10.

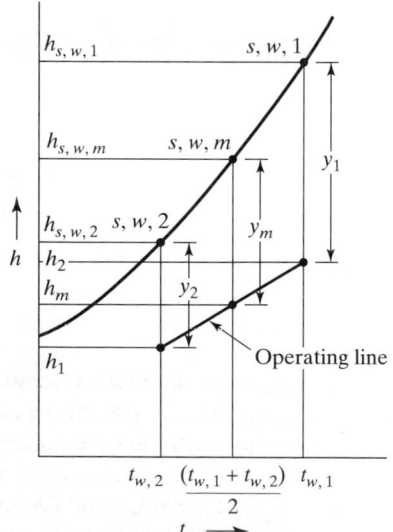

Figure 10.10 Schematic variation of $(h_{s,w} - h)$ for a cooling tower.

EXAMPLE 10.4

Determine the required tower volume for Ex. 10.3 by the approximate method given by Eqs. (10.27)–(10.29).

Solution: By Table 10.1, using air state d as an average value,

$$\left[\frac{h_{f,w}}{dh/dW}\right]_m = \frac{60.2}{1131} = 0.053$$

By Eq. (10.28),

$$h_2 = h_1 - \left(\frac{dh}{dt_w}\right)(t_{w,1} - t_{w,2}) = 38.39 + 15/0.947 = 54.23 \text{ Btu/lbm}_a$$

$$h_m = h_1 - \left(\frac{dh}{dt_w}\right)\left(\frac{t_{w,1} - t_{w,2}}{2}\right) = 38.39 + \frac{7.5}{0.947} = 46.31 \text{ Btu/lbm}_a$$

At $t_{w,1} = 100 \,^\circ\text{F}$, $h_{s,w,1} = 71.73 \text{ Btu/lbm}_a$; at $t_{w,m} = 92.5 \,^\circ\text{F}$, $h_{s,w,m} = 59.52 \text{ Btu/lbm}_a$; at $t_{w,2} = 85 \,^\circ\text{F}$, $h_{s,w,2} = 49.43 \text{ Btu/lbm}_a$. Thus

$$y_1 = h_{s,w,1} - h_2 = 71.73 - 54.23 = 17.50 \text{ Btu/lbm}_a$$

$$y_2 = h_{s,w,2} - h_1 = 49.43 - 38.39 = 11.04 \text{ Btu/lbm}_a$$

$$y_m = h_{s,w,m} - h_m = 59.52 - 46.31 = 13.21 \text{ Btu/lbm}_a$$

and $y_m/y_1 = 0.755$, $y_m/y_2 = 1.20$. By Fig. 10.7, $f = 1.01$. By Table 10.1, using air state d,

$$\left[\text{Le} + \frac{h_{fg,w} - 1061 \text{ Le}}{(h_{s,w} - h)/(W_{s,w} - W)}\right]_m = 0.895 + \frac{1041 - (1061)(0.895)}{1095}$$

$$= 0.978$$

By Eqs. (10.27) and (10.29),

$$V = \frac{\dot{m}_w c_w}{h_D A_V} \frac{\dfrac{t_{w,1} - t_{w,2}}{fy_m}}{\left[\text{Le} + \dfrac{h_{fg,w} - 1061 \text{ Le}}{(h_{s,w} - h)/(W_{s,w} - W)}\right]_m}$$

$$= \frac{(746,000)(1)(15)}{(120)(0.978)(1.01)(13.21)}$$

$$= 7140 \text{ ft}^3$$

which is 1.1 percent lower than the answer obtained for Ex. 10.3.

Our solution of Ex. 10.4 was facilitated by known calculations in Ex. 10.3. Generally, with the approximate procedure, such calculations would not be available. However, the approximations in Eqs. (10.27) and (10.29) are not critical, and satisfactory estimates may be made. For example, if we base the solution upon air state 1 in Table 10.1 (all data known from given conditions), we find $V = 7020 \text{ ft}^3$.

A simple analysis of a cooling tower has been developed which uses the concept of the *number of transfer units* or *NTU method*. This simple analysis requires no iteration and uses conventional NTU charts or equations as in conventional heat-exchanger analysis. The basic approach can be described as follows.

The evaporation of water into the air stream is neglected, so that the water mass flow rate remains constant and only sensible cooling occurs. Equation (10.13) becomes

$$\dot{m}_a \, dh = -\dot{m}_w \, dh_{f,w} = -\dot{m}_w c_{pw} \, dt_w \tag{10.30}$$

Substituting Eq. (10.30) into Eq. (10.16), assuming the Lewis number equals 1.0, and dropping the last term, as we are neglecting evaporation, we have

$$\dot{m}_a \, dh = h_D A_V (h_{s,w} - h) \, dV \tag{10.31}$$

We must relate the water surface temperature to a change in saturated moist air enthalpy, as this provides the driving force for the energy transfer. Assuming that the water surface temperature equals the bulk water temperature and that the enthalpy of saturated moist air is linearly proportional to temperature over a limited range, we may write

$$b_s = \frac{dh_{s,w}}{dt_w} \tag{10.32}$$

where b_s is the saturation specific heat. This can be determined from moist air tables, from reading values from a psychrometric chart, or reading the sensible-heat-ratio protractor of a psychrometric chart as

$$b_s = \frac{h_{s,w,1} - h_{s,w,2}}{t_{w,1} - t_{w,2}} = \frac{c_{pa}}{\text{SHR}_s} \tag{10.33}$$

The sensible-heat ratio is evaluated by drawing a line tangent to the saturated moist air curve midway between the end states and reading the value of SHR from the protractor on the psychrometric chart as explained in Chapter 7.

Substituting dt_w from Eq. (10.32) into Eq. (10.30) and taking the absolute values of both terms,

$$\dot{m}_a \, dh = \frac{\dot{m}_w c_{pw} \, dh_{s,w}}{b_s} \tag{10.34}$$

and rearranging

$$dh_{s,w} = \frac{\dot{m}_a b_s \, dh}{\dot{m}_w c_{pw}} \tag{10.35}$$

Setting $d(h_{s,w} - h) = dh_{s,w} - dh$ and substituting Eq. (10.35) for $dh_{s,w}$

$$d(h_{s,w} - h) = \frac{\dot{m}_a b_s \, dh}{\dot{m}_w c_{pw}} - dh$$

Dividing by $(h_{s,w} - h)$ from Eq. (10.31),

$$\frac{d(h_{s,w} - h)}{h_{s,w} - h} = h_D A_V \, dV \left[\frac{b_s}{\dot{m}_w c_{pw}} - \frac{1}{\dot{m}_a} \right] \tag{10.36}$$

To make Eq. (10.36) look more like a heat-transfer equation we can make the following substitution:

$$\frac{b_s}{\dot{m}_w c_{pw}} = \frac{1}{\dot{m}'_w} \tag{10.37}$$

where $\dot{m}'_w$ becomes the water-capacity rate. The air-capacity rate is $\dot{m}_a$, and the ratio of the capacity rates can be defined as

$$C \equiv \frac{\dot{m}_{\min}}{\dot{m}_{\max}} \tag{10.38}$$

where either $\dot{m}_{\min} = \dot{m}_a$ and $\dot{m}_{\max} = \dot{m}'_w$ or vice versa. Substituting Eqs. (10.37) and (10.38) into Eq. (10.36) gives

$$\frac{d(h_{s,w} - h)}{h_{s,w} - h} = \frac{h_D A_V V}{\dot{m}_{\min}} (1 - C) \tag{10.39}$$

and the number of transfer units becomes

$$\text{NTU} = \frac{h_D A_V V}{\dot{m}_{\min}} \tag{10.40}$$

The cooling-tower effectiveness, ε, can be defined in the usual manner as the ratio of the actual energy transferred to the maximum possible energy transfer for the fluid with the minimum capacity rate. When the air is the fluid with the minimum capacity rate, $\dot{m}_{\min} = \dot{m}_a$, and

$$\varepsilon = \frac{h_2 - h_1}{h_{s,w,1} - h_1} \tag{10.41}$$

The cooling-tower effectiveness, ε, number of transfer units, NTU, and fluid-capacity-rate ratio, C, are related. This relationship can be determined by integrating Eq. (10.39) between the inlet and exit air states. Introducing the definition of the effectiveness results in the following relation:

$$\varepsilon = \frac{1 - \exp\left[-\text{NTU}\,(1 - C)\right]}{1 - C \exp\left[-NTU\,(1 - C)\right]} \tag{10.42}$$

or, solving for NTU,

$$\text{NTU} = -\ln\left(\frac{1 - C}{1 - \varepsilon C}\right) \Big/ (1 - \varepsilon) \tag{10.43}$$

This is identical to the expression for a conventional-counterflow heat exchanger.

Although additional assumptions and simplifications have been made in developing the NTU analysis method, the results are reasonably correct and provide a first approximation in a design analysis. The following example illustrates the use of the NTU method and shows the accuracy that can be expected.

EXAMPLE 10.5

Determine the required cooling-tower volume for Ex. 10.3 by the NTU method given by Eqs. (10.37)–(10.43).

Solution: We need to solve for the tower volume from Eq. (10.40), so we must compute the value for NTU from known values of C and ε, using Eq. (10.43). We first must determine which fluid has the minimum capacity rate. From Eq. (10.37),

$$\dot{m}'_w = \frac{\dot{m}_w c_{pw}}{b_s} = \frac{\dot{m}_a c_{pw}}{b_s}$$

$$b_s = \frac{h_{s,w,1} - h_{s,w,2}}{t_{w,1} - t_{w,2}}$$

From Table A.4E,

$$b_s = \frac{71.761 - 49.445}{100 - 85} = 1.49$$

Then

$$\dot{m}'_w = \dot{m}_a \frac{1.0}{1.49} = 0.67 \dot{m}_a$$

and

$$\dot{m}_{\min} = \dot{m}'_w, \qquad C = \frac{\dot{m}'_w}{\dot{m}_a} = 0.67$$

Since

$$\dot{m}_{\min} = \dot{m}'_w, \varepsilon = \frac{t_{w,1} - t_{w,2}}{t_{w,1} - t_{wb,1}}$$

$$\varepsilon = \frac{100 - 85}{100 - 75} = 0.60$$

Now NTU can be determined from Eq. (10.43):

$$\text{NTU} = -\ln\left(\frac{1 - 0.67}{1 - 0.6(0.67)}\right) / (1 - 0.6) = 1.49$$

and the tower volume can be found from Eq. (10.40):

$$V = \frac{\text{NTU}\,\dot{m}'_w}{h_D A_V} = \frac{(1.49)(0.67)(746{,}000)}{120} = 6190 \text{ ft}^3$$

This volume is 14 percent less than the value computed in Ex. 10.3 and 13 percent less than the value from Ex. 10.4. Jaber and Webb [2] have shown that this one-step NTU method underpredicts the value of the tower volume and NTU in agreement with the present results.

The error between the NTU method and the more exact analysis can be reduced by subdividing the tower into increments of equal volume. The error can be reduced more than a factor of two by using two increments rather than one. As more increments are considered, the assumption of a linear relationship between saturated moist air enthalpy and temperature becomes more exact, and the detailed analysis and NTU methods converge.

Another advantage of the one-step NTU method is that flow configurations other than counterflow can be easily handled. The only difference is that the counterflow equations, Eqs. (10.42) and (10.43), must be replaced by relations for the geometry being considered. These relations are identical to those for conventional heat exchangers for both fluids unmixed.

10.4 COUNTERFLOW CONTACT OF MOIST AIR BY CHILLED SPRAY WATER: THE SPRAY DEHUMIDIFIER

It is possible to dehumidify, as well as to cool, moist air by direct contact with cold water. For effective heat exchange, counterflow must be employed.

From a schematic standpoint, Fig. 10.5 is applicable for an analysis of a spray dehumidifier. We may imagine that relatively cold well or tap water, or else refrigerated water, is supplied to the chamber. Air which leaves the contact chamber may be circulated to spaces which are to be conditioned.

For steady-flow conditions, we have for a differential volume:

$$-\dot{m}_a\, dh = \dot{m}_w\, dh_{f,w} - \dot{m}_a\, dW\, h_{f,w} \tag{10.44}$$

$$\dot{m}_w\, dh_{f,w} = \dot{m}_w c_w\, dt_w$$
$$= h_c A_V\, dV(t - t_w) + h_D A_V\, dV(W - W_{s,w})(h_{g,t} - h_{f,w}) \tag{10.45}$$

$$-\dot{m}_a\, dW = h_D A_V\, dV(W - W_{s,w}) \tag{10.46}$$

By procedures completely analogous to those for the cooling tower, we obtain

$$\frac{dh}{dW} = \text{Le}\,\frac{h - h_{s,w}}{W - W_{s,w}} + (h_{g,t} - h_g^0\,\text{Le}) \tag{10.47}$$

$$h_2 = h_1 - \frac{\dot{m}_w c_w}{\dot{m}_a}(t_{w,2} - t_{w,1}) - (W_1 - W_2)h_{f,w,2} \tag{10.48}$$

$$V = \frac{\dot{m}_a}{h_D A_V}\int_{W_2}^{W_1}\frac{dW}{W - W_{s,w}} \tag{10.49}$$

Comparison of Eqs. (10.44)–(10.49) with Eqs. (10.12)–(10.21) shows that the analysis of a counterflow spray dehumidifier is the same as that for the cooling tower. Using the psychrometric chart, we may construct the condition line and determine the outlet-air state. The required chamber volume may be found from Eq. (10.49).

Figure 10.11 shows a schematic representation of Eq. (10.47) on the psychrometric chart for the air-inlet end of the chamber. Because of the magnitude of the term

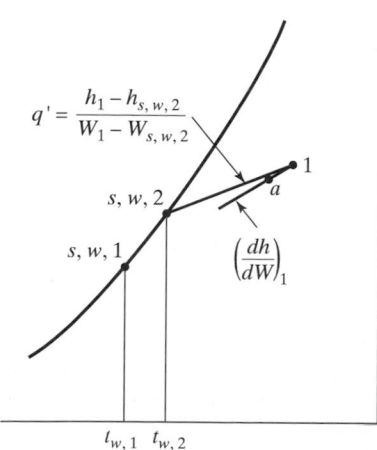

Figure 10.11 Schematic solution of Eq. (10.47) on a psychrometric chart.

$(h - h_{s,w})/(W - W_{s,w})$, the condition line of direction dh/dW may be steeper than the line of direction $q' = (h - h_{s,w})/(W - W_{s,w})$.

Spray dehumidifiers have rather infrequent practical application for cooling and dehumidifying atmospheric air compared to heat exchangers using extended surfaces. In Chapter 11 we will study in detail the cooling and dehumidifying of moist air through the use of finned-tube coils.

EXAMPLE 10.6

Compute the wet-bulb temperature of moist air that has been cooled and dehumidified from 25 °C and 80 percent relative humidity to 23 °C in a counterflow spray dehumidifier. The water leaves at 15 °C.

Solution: Since the temperature change is small, consider this to be a one-step process. Therefore we are to find the wet-bulb temperature at state a as shown in Fig. 10.11 with state 2 at 15 °C and state 1 at 25 °C, 80 percent RH. We must find the quantities needed in Eq. (10.47).

From Fig. C-7SI,

$$h_1 = 66.0 \text{ kJ/kg}_a, \qquad W_1 = 0.016 \text{ kg}_w/\text{kg}_a$$

From Table A.4SI,

$$h_{s,w,2} = 42.11 \text{ kJ/kg}_w, \qquad W_{s,w,2} = 0.0107 \text{ kg}_w/\text{kg}_a$$

From Table A.1SI,

$$h_{g,1} = 2546.73 \text{ kJ/kg}_w, \qquad h_g^0 = 2500.89 \text{ kJ/kg}_w$$

From Table 9.1 and Eq. (9.7) at 15 °C,

$$\text{Le} = 0.90$$

Substituting into Eq. (10.47),

$$\frac{dh}{dW} = \frac{0.9(66.0 - 42.11)}{(0.016 - 0.0107)} + 2546.73 - (2500.89)(0.9)$$

$$= 4057 + 2546.73 - 2251$$

$$= 4353 \text{ kJ/kg}_w = 4.35 \text{ kJ/g}_w$$

Using the protractor on Fig. C-7SI, we find the slope of the line with $\Delta h/\Delta W = 4.35$ and draw a parallel line through state 1. The intersection of this line and $t = 23$ °C locates state a. Therefore we read $t_{a,\text{wb}} = 20.95$ °C.

10.5 MASS-TRANSFER COEFFICIENTS FOR WATER-AIR DIRECT-CONTACT DEVICES

In design calculations on air washers, cooling towers, and spray dehumidifiers, we need to know the mass-transfer coefficient h_D and the surface-area-to-volume ratio A_V. Conventional procedure has been to combine the two factors into one coefficient $h_D A_V$, since it is difficult to separate them.

A large amount of performance information has been published on laboratory cooling towers. Less information is available on commercial cooling towers. Very few mass-transfer data are available on air washers or spray dehumidifiers.

Caution must be observed in applying mass-transfer-coefficient information given in the literature on cooling towers. Because of wide variation in types of packing or fill, one type of design may show a very different performance from another type. Most reported information has come from small laboratory towers, where the ratio of wall surface to volume is much larger than in commercial towers.

Most information available in the literature on the coefficient $h_D A_V$ was obtained from gross or overall measurements for the entire tower. Based upon measured inlet and outlet water temperatures, inlet and outlet air states, airflow rate, and water-flow rate, values of $h_D A_V$ were calculated according to some analytical procedure. Unfortunately, many investigators have used highly approximate methods of analysis compared to the basic analysis given in Sec. 10.3.

Cooling-tower performance data are usually correlated as

$$\frac{h_D A_V V}{\dot{m}_w} = c \left(\frac{\dot{m}_w}{\dot{m}_a} \right)^{-n} \tag{10.50}$$

where c and $-n$ are empirical constants specific for a particular tower-fill design. The value for c ranges from 1.0 to 1.7, and n varies from about 0.25 to 1.0. The smaller values of n are associated with splash-type fills and the larger values with film-type fills. In the absence of reliable data, ASHRAE [3] recommends a value of n of 0.6. With this value of n, a reasonable estimate for c becomes 1.3, using the results tabulated by Braun et al. [4]. Therefore the recommended equation in the absence of experimental data becomes

$$\frac{h_D A_V V}{\dot{m}_w} = 1.3 \left(\frac{\dot{m}_w}{\dot{m}_a} \right)^{-0.6} \tag{10.51}$$

When $\dot{m}_w$ is less than $\dot{m}_a$,

$$\frac{h_D A_V V}{\dot{m}_w} = \text{NTU} = 1.3 \left(\frac{\dot{m}_w}{\dot{m}_a} \right)^{-0.6} \tag{10.52}$$

and when $\dot{m}_a$ is less than or equal to $\dot{m}_w$,

$$\frac{h_D A_V V}{\dot{m}_a} = \text{NTU} = 1.3 \left(\frac{\dot{m}_w}{\dot{m}_a} \right)^{1-0.6} \tag{10.53}$$

Cooling-tower performance data are often plotted as shown in Fig. 10.12. The figure uses log-log coordinates so that the straight line shown between points E and F is a plot of Eq. (10.50), once the values of c and $-n$ are determined. The slope of this line is $-n$. Each figure is valid for only one inlet-air wet-bulb temperature and only one value of the range or change in water temperature between the inlet and outlet. However, various approach temperatures are considered. The approach temperature is defined as the difference between the leaving-water temperature and the entering-air wet-bulb temperature. The values for NTU and mass flow ratio, $\dot{m}_w/\dot{m}_a$, are determined by the intersection of the line E-F and the curve corresponding to the approach temperature.

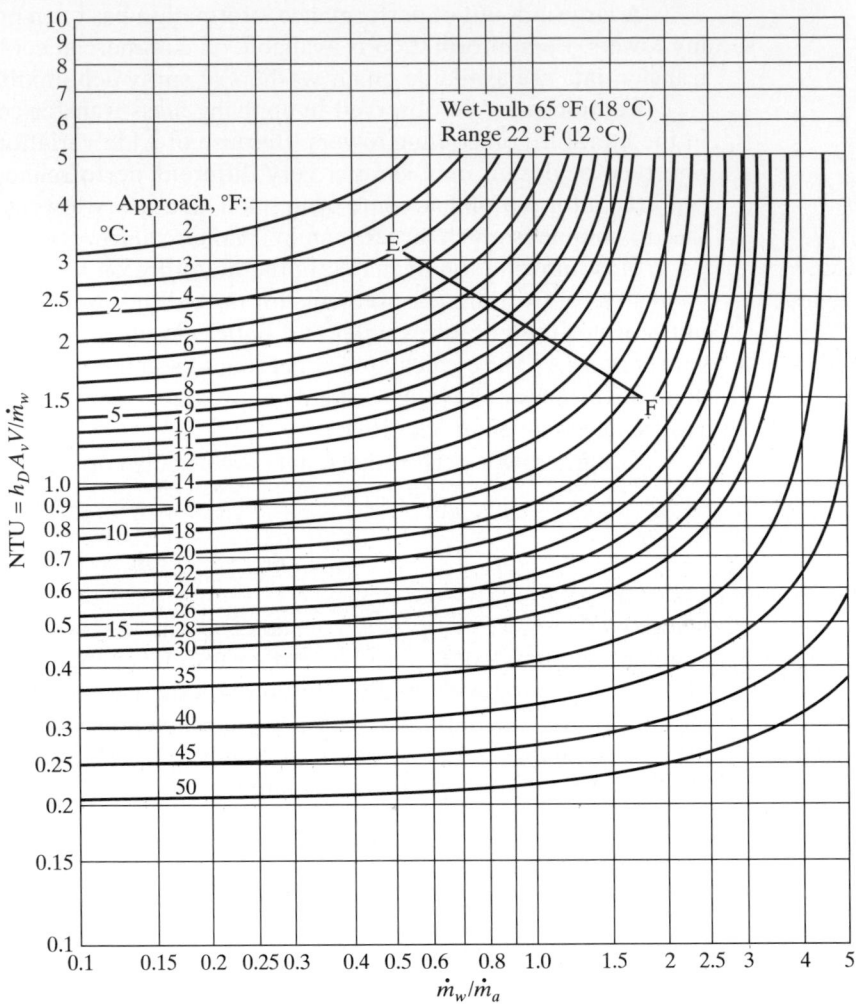

Figure 10.12 Evaporative cooling-tower design curves. [Reprinted from Warren E. Rohsenow, James P. Hartnett, and Ejup N. Ganic, *Handbook of Heat Transfer Applications*, 2d ed. (New York: McGraw Hill, 1985), p. 10-9.]

ENDNOTES

1. *ASHRAE HVAC Systems and Equipment Handbook* (Atlanta: American Society of Heating, Refrigerating and Air Conditioning Engineers, 1996), ch. 36.

2. H. Jaber and R. L. Webb, "Design of Cooling Towers by the Effectiveness-NTU Method," *ASME J. Heat Transfer*, III, (1989), 837–843.

3. *ASHRAE HVAC Systems and Equipment Handbook* (Atlanta: American Society of Heating, Refrigerating and Air Conditioning Engineers, 1996), 36.18–36.19.

4. J. E. Braun, S. A. Klein, and J. W. Mitchell, "Effectiveness Models for Cooling Towers and Cooling Coils," *ASHRAE Trans.*, 95 (1989), pt. 2, 164–174.

PROBLEMS

10.1 An air washer using directly recirculated spray water operates steadily with air entering at 30 °C dry-bulb temperature and 50 percent relative humidity at sea-level barometric pressure. The face area of the washer is 2 m² and the face velocity is 1.5 m/s. It is estimated that NTU = 0.8 for this washer. Compute:
  (a) the washer efficiency,
  (b) the leaving-air dry-bulb temperature and humidity ratio, and
  (c) the amount of water evaporated in the washer.

10.2 An air washer using directly recirculated spray water operates under steady-state conditions. Air enters the washer at 100 °F dry-bulb temperature, 65 °F thermodynamic wet-bulb temperature, and 14.696 psia barometric pressure with a velocity of 400 ft/min. The efficiency of the washer is 75 percent. Determine the dry-bulb and thermodynamic wet-bulb temperatures of the air leaving the washer.

10.3 A system including a preheating coil and an air washer (directly recirculated spray water) is to be designed to process 10,000 ft³/min of outdoor air. Barometric pressure is 14.696 psia. Outdoor air saturated at 30 °F enters the preheater, where heat is supplied by steam condensing inside the tubes of the coil. The air then passes through the air washer. The air state at the exit of the washer is 70 °F dry-bulb temperature and 60 percent degree-of-saturation. Average face velocity of the air through the washer is 500 ft/min. Use $h_D A_V$ equal to 400 $lbm_w/hr \cdot ft^3$ (lbm_w/lbm_a). Determine:

  (a) the dry-bulb and thermodynamic wet-bulb temperatures of the air entering the washer,
  (b) the required capacity of the preheating coil in Btu/hr,
  (c) the quantity of makeup water required for the air washer in gpm, and
  (d) the necessary contact volume of the air washer in ft³.

10.4 A counterflow cooling tower operates steadily with 100 L/s of water that is cooled from 30 °C to 25 °C. The ambient air is at sea-level pressure, 25 °C dry-bulb temperature and 20 °C wet-bulb temperature. The ratio of water mass flow rate to air mass flow rate is 1.0. Determine:
  (a) the final enthalpy of the leaving air and
  (b) the tower volume if $h_D A_V = 0.5$ $kg_w/s \cdot m^3$ (kg_w/kg_a), using the approximate NTU method [Eqs. (10.37)–(10.43)].

10.5 Solve Prob. 10.4 using the approximate analysis method given by Eqs. (10.27)–(10.29).

10.6 Solve Prob. 10.4 using the detailed analysis method as illustrated in Ex. 10.3. Compare the answers to the approximate NTU analysis used in Prob. 10.4.

10.7 Rework Ex. 10.5 but divide the tower into two equal-volume sections. Perform the NTU analysis for both sections and compare the total volume required with the results from Ex. 10.3 and 10.5.

10.8 A cooling tower operates under steady-state conditions as shown by Fig. 10.13. Assume that 1 percent of the inlet water is lost from the tower as drift (liquid carry-over with the exhaust air). Assume that

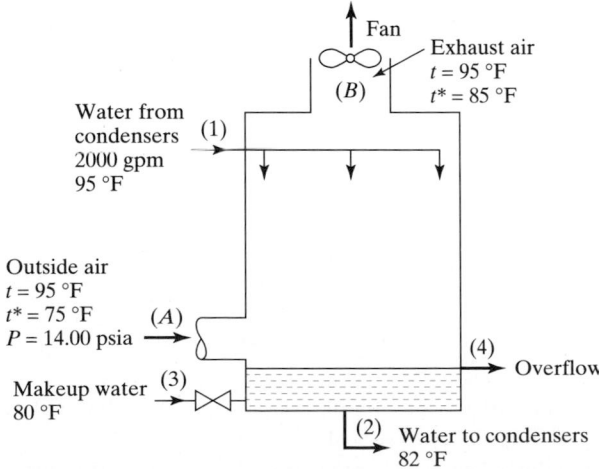

Figure 10.13 Schematic cooling tower for Prob. 10.8.

rate of overflow from the sump is 20 percent of the flow rate of makeup water. Determine:

(a) the volume of moist air entering the fan in ft^3/min, and

(b) the gpm of makeup water required.

10.9 Determine the required tower volume for Ex. 10.3 if water leaves the tower at

(a) 90 °F and

(b) 80 °F.

You may use the approximate method given by Eqs. (10.27)–(10.29).

10.10 Determine the required tower volume for Prob. 10.5 if water leaves the tower at

(a) 27 °C and

(b) 23 °C.

You may use the approximate method given by Eqs. (10.27)–(10.29).

10.11 Write a general computer program to analyze counterflow cooling towers using the detailed analysis illustrated by Ex. 10.3. Integrate Eq. (10.21) numerically by dividing the entire cooling tower into ten equal-volume sections. Assume the following functional relationships are known:

moist air enthalpy:	$h = f_1(t, W)$
saturated moist air enthalpy:	$h_s = f_2(t)$
saturated moist air humidity ratio:	$W_s = f_3(t)$
saturated liquid-water enthalpy:	$h_w = f_4(t)$
saturated water-vapor enthalpy:	$h_g = f_5(t)$

Use either English or SI units and provide:

(a) a flowchart of your program and

(b) a program listing.

10.12 A counterflow spray dehumidifier is to process moist air having an inlet condition of 85 °F dry-bulb temperature and 70 °F thermodynamic wet-bulb temperature. Barometric pressure is 14.696 psia. The rate of air flow entering the dehumidifier is 5000 ft^3/min. Water at 45 °F enters the dehumidifier at a rate of 50 gpm. The temperature of the leaving water is 55 °F. Assume an average air velocity of 500 ft/min. Construct the complete condition line and determine the dry-bulb temperature and thermodynamic wet-bulb temperature of the air leaving the chamber.

10.13 By use of Fig. 10.7 determine the contact volume in ft^3 required for Prob. 10.12. Assume a mass-transfer coefficient $h_D A_V$ of 600 lbm$_w$/hr · ft^3 (lbm$_w$ per lbm$_a$).

10.14 A counterflow spray dehumidifier to process moist air is to be designed with inlet air at 30 °C dry-bulb temperature and 20 °C wet-bulb temperature at sea-level pressure. Water enters at 10 °C at 3×10^{-3} m^3/s and leaves at 15 °C. The airflow rate is 2.5 m^3/s with an average velocity of 1 m/s. Plot the complete condition line on a psychrometric chart and determine the dry-bulb temperature and wet-bulb temperature of the air leaving the dehumidifier.

10.15 Determine the contact volume required in m^3 for Prob. 10.14 if $h_D A_V$ equals 2.5 kg$_w$/s · m^3 (kg$_w$/kg$_a$).

SYMBOLS

A_V	Surface area of water droplets, ft^2/ft^3 (m^2/m^3).
b_s	Total specific heat of saturated moist air, Btu/lbm$_a$ · °F (kJ/kg$_a$ · °C).
C	Capacity-rate ratio, dimensionless.
c_{pa}	Specific heat at constant pressure of moist air, Btu/lbm$_a$ · °F (kJ/kg$_a$ · °C).
c_w	Specific heat of water, Btu/lbm$_w$ · °F, (kJ/kg$_w$ · °C).
G_a	Mass velocity of dry air, lbm$_a$/min · ft^2 (kg/s · m^2).
h	Enthalpy of moist air, Btu/lbm$_a$ (kJ/kg$_a$).
h_c	Convection heat-transfer coefficient, Btu/hr · ft^2 · °F (kW/m^2 · °C).
h_D	Convection mass-transfer coefficient, lbm$_w$/hr · ft^2 (lbm$_w$/lbm$_a$) [kg$_w$/s · m^2 (kg$_w$/kg$_a$)].
h_f	Specific enthalpy of saturated liquid water, Btu/lbm$_w$; $h_{f,w}$ evaluated at t_w; h_f^* evaluated at t^*.
$h_{fg,w}$	$h_{g,w} - h_{f,w}$, Btu/lbm$_w$ (kJ/kg$_w$).
h_g	Specific enthalpy of saturated water vapor, Btu/lbm$_w$ (kJ/kg$_w$); $h_{g,t}$ evaluated at t; $h_{g,w}$ evaluated at t_w, h_g^0 evaluated at 0 °F or 0 °C.
$h_{s,w}$	Enthalpy of saturated moist air evaluated at t_w, Btu/lbm$_a$ (kJ/kg$_a$).
Le	Lewis number = $h_c/h_D c_{pa}$, dimensionless.

$\dot{m}_a$ Mass rate of flow of dry air, lbm_a/hr (kg_a/s).

$\dot{m}_w$ Mass rate of flow of water, lbm_w/hr (kg_w/s).

NTU $h_D A_V V/\dot{m}$, number of transfer units, dimensionless.

q' dh/dW, Btu/lbm_w (kJ/kg_w).

t Dry-bulb temperature of moist air, °F (°C).

t^* Thermodynamic wet-bulb temperature of moist air, °F (°C).

t_w Temperature of water, °F (°C).

V Volume, ft^3 (m^3).

v Specific volume of moist air, ft^3/lbm_a (m^3/kg_a).

W Humidity ratio of moist air, lbm_w/lbm_a (kg_w/kg_a).

$W_{s,w}$ Humidity ratio of saturated moist air evaluated at t_w; W_s^* evaluated at t^*; lbm_w/lbm_a (kg_w/kg_a).

Greek Letters

η_w Air washer efficiency, dimensionless.

ε Effectiveness, dimensionless.

chapter 11
Heating and Cooling of Moist Air by Extended-Surface Coils

11.1 GENERAL REMARKS

Almost every thermal environmental system involves cooling of atmospheric air. During the winter heating is a major function, while during the summer heating is often required in air conditioning systems where air humidity is controlled. In summer air conditioning systems, cooling and dehumidification are prime functions. In this chapter we will be concerned primarily with heat-transfer problems where forced-convection, turbulent air flow occurs.

Atmospheric air may be heated or cooled in *duct coils*, which are banks of bare tubes or banks of tubes which have finned or extended surfaces. Practically all modern coils are of the extended-surface type. Figure 11.1 shows two types of finned tubing. Figure 11.1(a) employs spiral fins, while Fig. 11.1(b) has continuous flat-plate fins. The heating or cooling medium passes through the tubes while moist air flows across the tubes and through the fins. The tubes are commonly made of copper or aluminum; the secondary surface is made of aluminum or copper. The fins are usually mechanically bonded to the tubes.

In contrast to bare-pipe coils of the same capacity, finned coils are much more compact, have a much smaller weight, and usually are less expensive. The secondary surface area of a finned coil may be 10 to 30 times or more that of the bare tubes. These finned coils are called *compact heat exchangers*.

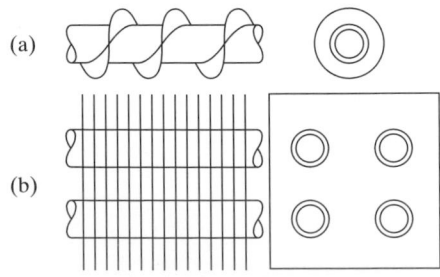

(a)

(b)

Figure 11.1 Schematic illustrations of finned tubing.

Heat transfer in finned coils is complicated. The simple expression for the overall heat-transfer coefficient for a bare pipe derived in Sec. 2.29 must be modified. In addition, the fluids are typically in some type of cross-flow arrangement, and the logarithmic mean temperature difference (Sec. 2.29) may also not be applicable.

The analysis of cooling coils is more involved than for heating coils, since mass transfer (dehumidification) may occur simultaneously with heat transfer. In this chapter we will first analyze heat transfer where the fins are dry. We will find that some of our results may be extended to the case of wet coils as well.

11.2 TRUE MEAN TEMPERATURE DIFFERENCE FOR HEAT EXCHANGERS OF THE CROSS-FLOW, FINNED-TUBE TYPE

For equal surface areas and for the same value of overall heat-transfer coefficient, a counterflow heat exchanger such as shown in Fig. 11.2 provides the maximum rate of heat transfer between two fluids. Such an arrangement gives the highest mean temperature difference Δt_m between the fluids. In Sec. 2.29 it was shown that for pure counterflow

$$\Delta t_{m,cf} = \frac{(t_{h,o} - t_{c,i}) - (t_{h,i} - t_{c,o})}{\ln\left(\dfrac{t_{h,o} - t_{c,i}}{t_{h,i} - t_{c,o}}\right)} = \frac{\Delta t_1 - \Delta t_2}{\ln\left(\dfrac{\Delta t_1}{\Delta t_2}\right)} \tag{11.1}$$

which is the *logarithmic mean temperature difference* (LMTD).

In a more general study of heat exchangers, it is conventional to express Δt_m in terms of the dimensionless quantities

$$R = \frac{t_{h,i} - t_{h,o}}{t_{c,o} - t_{c,i}} = \frac{\Delta t \text{ hot fluid}}{\Delta t \text{ cold fluid}} = \frac{\dot{m}_c c_{p,c}}{\dot{m}_h c_{p,h}} = \frac{c_c}{c_h} \tag{11.2}$$

$$P = \frac{t_{c,o} - t_{c,i}}{t_{h,i} - t_{c,i}} = \frac{\text{actual } \Delta t \text{ cold fluid}}{\text{maximum available } \Delta t} \tag{11.3}$$

By Eqs. (11.1)–(11.3), we obtain for pure counterflow

$$\Delta t_{m,cf} = \frac{(t_{c,o} - t_{c,i})(R - 1)}{\ln\left(\dfrac{1 - P}{1 - RP}\right)} \tag{11.4}$$

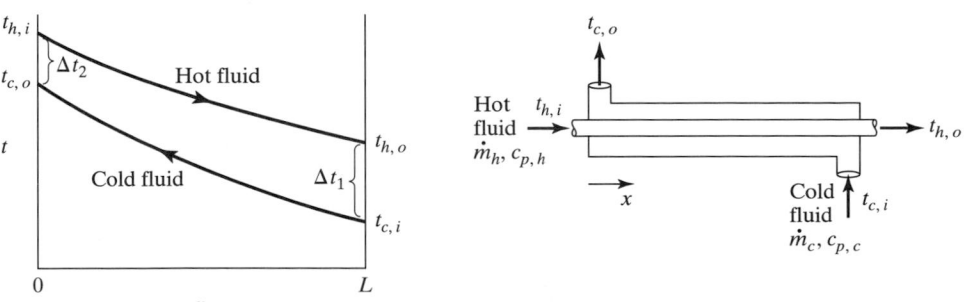

Figure 11.2 Schematic counterflow heat exchanger.

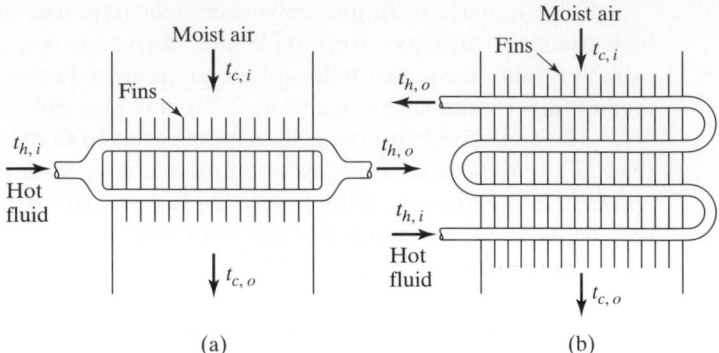

Figure 11.3 Schematic cross-flow arrangements.

In the case where air is one of the fluids, pure counterflow is generally not practicable. The most economical heat exchanger is usually the finned-tube type employing some form of *cross-flow* arrangement. Figure 11.3 shows two schematic arrangements.

Figure 11.3(a) shows a *pure cross-flow* heat exchanger with two rows of tubes. This type with one or two rows of tubes is commonly used in steam coils for heating air. Figure 11.3(b) shows a *countercross-flow* arrangement with four tube passes. This type with two or more tube passes is commonly used where hot water or chilled water passes through the tubes.

As will be shown later, the logarithmic mean temperature difference is valid for cross-flow heat exchangers only when one fluid temperature remains constant (condensing steam, evaporating refrigerant, etc.). Otherwise, the conditions required for the derivation of the logarithmic mean temperature difference do not exist when cross-flow is employed.

It is convenient to express the mean temperature difference Δt_m for a cross-flow heat exchanger as

$$\Delta t_m = F \, \Delta t_{m,cf}, \qquad 0 \le F \le 1.0 \tag{11.5}$$

where F is a correction factor and $\Delta t_{m,cf}$ is the logarithmic mean temperature difference calculated for *pure counterflow*. Solutions for Eq. (11.5) have been published for a limited number of cross-flow arrangements [1]. A derivation for F will now be given for the simple case of pure cross-flow with one row of tubes as shown in Fig. 11.4. We will assume

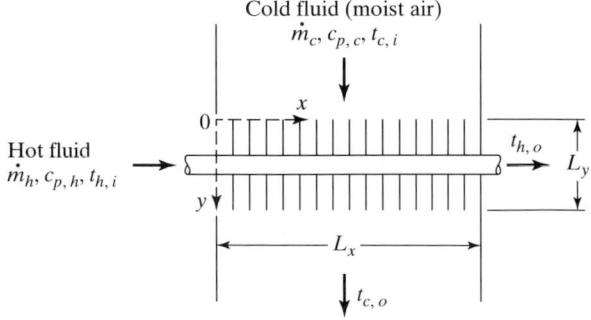

Figure 11.4 Schematic illustration of pure cross flow with one row of tubes.

that (1) the overall heat transfer coefficient U_o is constant, (2) the mass-flow rate of each fluid is constant, (3) the specific heat of each fluid is constant, (4) neither fluid undergoes a phase change, and (5) external heat losses are negligible.

The hot fluid passing through the tube may be assumed to be thoroughly mixed at any plane normal to its flow direction. Thus, the temperature of the hot fluid t_h varies only in the x direction. However, the temperature of the cold fluid t_c varies in both the x and y directions, since the baffles (fins) prevent mixing in a plane normal to flow. The final cold-fluid temperature $t_{c,o}$ is the result of the mixture of the many separate streams beyond the heat-transfer surface.

We will first consider how the cold-fluid temperature changes along a plane in the y direction for an element of heat-transfer surface. We may write that in the y direction

$$U_o \, dA_o (t_h - t_c) = \dot{m}_c \frac{dx}{L_x} c_{p,c} \, dt_c$$

but $dA = A \, (dx/L_x)(dy/L_y)$. Thus

$$\int_{t_{c,i}}^{t_c'} \frac{dt_c}{t_h - t_c} = \frac{U_o A_o}{\dot{m}_c c_{p,c} L_y} \int_0^{L_y} dy \tag{11.6}$$

The solution of Eq. (11.6) for the cold-fluid temperature t_c' leaving the heat exchanger for a strip of width dx is

$$t_c' = t_{c,i} + (t_h - t_{c,i})(1 - e^{-K_1}) \tag{11.7}$$

where

$$K_1 = \frac{U_o A_o}{\dot{m}_c c_{p,c}} \tag{11.8}$$

Let us now consider how the fluid temperatures change in the x direction. We may write

$$\dot{m}_h c_{p,h} \, dt_h = \dot{m}_c \frac{dx}{L_x} c_{p,c} (t_c' - t_{c,i})$$

With Eq. (11.7),

$$\int_{t_{h,i}}^{t_{h,o}} \frac{dt_h}{t_h - t_{c,i}} = -\frac{\dot{m}_c c_{p,c}}{\dot{m}_h c_{p,h}} \frac{1 - e^{-K_1}}{L_x} \int_0^{L_y} dx \tag{11.9}$$

The solution of Eq. (11.9) for the hot-fluid temperature $t_{h,o}$ at the exit of the heat exchanger is

$$t_{h,o} = t_{h,i} - (t_{h,i} - t_{c,i})(1 - e^{-K_2}) \tag{11.10}$$

where

$$K_2 = \frac{\dot{m}_c c_{p,c}}{\dot{m}_h c_{p,h}} (1 - e^{-K_1}) \tag{11.11}$$

Since

$$\dot{m}_c c_{p,c} (t_{c,o} - t_{c,i}) = \dot{m}_h c_{p,h} (t_{h,i} - t_{h,o})$$

we have

$$t_{c,o} = t_{c,i} + \frac{\dot{m}_h c_{p,h}}{\dot{m}_c c_{p,c}} (t_{h,i} - t_{c,i})(1 - e^{-K_2}) \qquad (11.12)$$

Thus, Eqs. (11.10) and (11.12) allow calculation of the final temperatures of the fluids.

We will now develop an expression for the mean temperature difference Δt_m between the fluids. By Eqs. (11.2), (11.3), (11.10), and (11.11),

$$e^{-K_2} = 1 - RP$$

and

$$K_2 = R(1 - e^{-K_1})$$

Thus

$$e^{K_1} = \frac{R}{R + \ln(1 - RP)} \qquad (11.13)$$

But

$$K_1 = \frac{U_o A_o}{\dot{m}_c c_{p,c}} = \frac{t_{c,o} - t_{c,i}}{\Delta t_m} \qquad (11.14)$$

By Eqs. (11.13) and (11.14)

$$\Delta t_m = \frac{t_{c,o} - t_{c,i}}{\ln\left[\dfrac{R}{R + \ln(1 - RP)}\right]} \qquad (11.15)$$

and by Eqs. (11.4), (11.5), and (11.15)

$$F = \frac{\Delta t_m}{\Delta t_{m,cf}} = \frac{\ln\left[\dfrac{1 - P}{1 - RP}\right]}{(R - 1)\ln\left[\dfrac{R}{R + \ln(1 - RP)}\right]} \qquad (11.16)$$

Equation (11.16) shows that the correction factor F is a function of only the parameters R and P. This is true for other heat-exchanger configurations also.

Three solutions for F are shown in Figs. 11.5–11.7. Figure 11.5 shows the solution of Eq. (11.16) for the single-pass, cross-flow heat exchanger of Fig. 11.4. Figure 11.6 shows the correction factor for a single-pass, cross-flow heat exchanger where both fluids are unmixed, as is typical in an air-to-air heat exchanger. Figure 11.6 applies to cases similar to Fig. 11.3(a) having many rows of tubes. Figure 11.7 shows the correction factor for a countercross-flow heat exchanger with two tube passes, with one fluid mixed and the other unmixed except between passes. Figure 11.7 applies approximately to a heat exchanger of the type shown in Fig. 11.3(b), having two tube passes. In Fig. 11.3(b) the air is unmixed between tube passes, and the correction factor F would be slightly higher than that given by Fig. 11.7. For countercross-flow heat exchangers having more than two tube passes one may usually estimate F as being between unity and the value given by Fig. 11.7.

We will now show that if one fluid temperature remains constant, the logarithmic mean temperature difference applies (i.e., $F = 1.0$) regardless of the flow arrangement. We will again analyze the heat exchanger of Fig. 11.4 but assuming the hot-fluid temper-

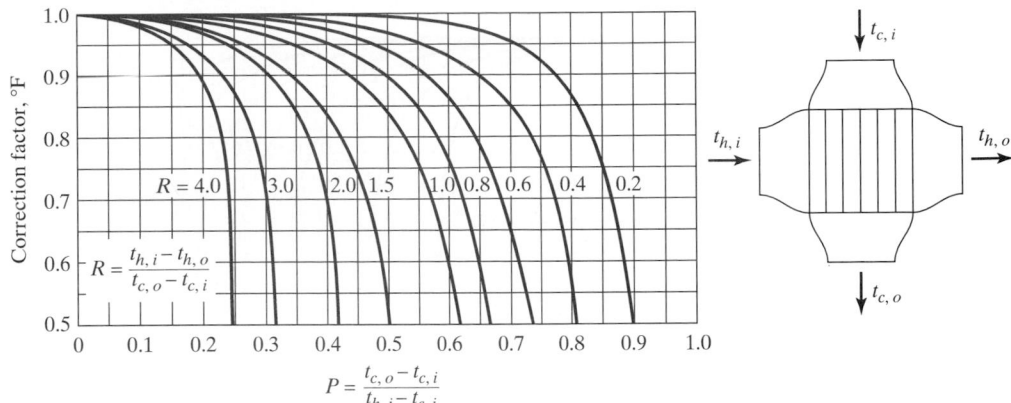

Figure 11.5 Correction factor $F = \Delta t_m / \Delta t_{m,cf}$ for a single-pass, cross-flow heat exchanger, one fluid (t_h) mixed, other fluid (t_c) unmixed. [Reprinted from R. A. Bowman, A. C. Mueller, and W. M. Nagle, "Mean Temperature Difference in Design," *ASME Trans.*, 62 (1940), 289, with permission of the publisher, Amer. Soc. Mech. Engrs.]

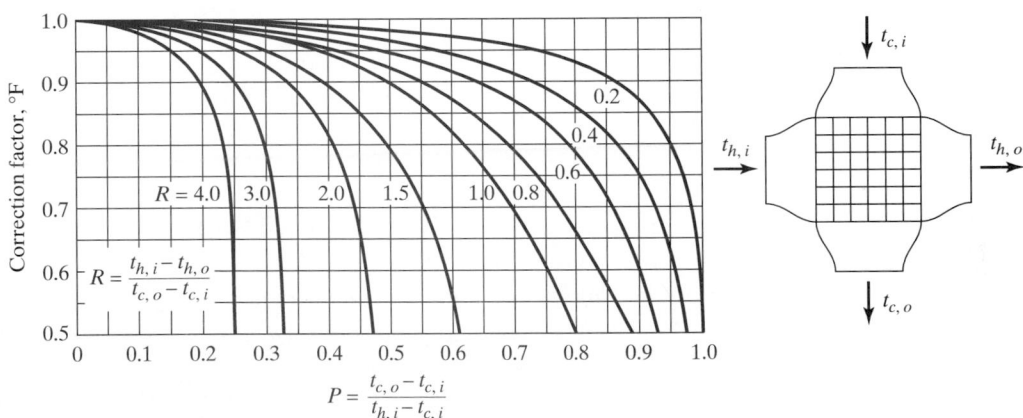

Figure 11.6 Correction factor $F = \Delta t_m / \Delta t_{m,cf}$ for a single-pass cross-flow heat exchanger, both fluids unmixed. [Reprinted from R. A. Bowman, A. C. Mueller, and W. M. Nagle, "Mean Temperature Difference in Design," *ASME Trans.*, 62 (1940), 288, with permission of the publisher, Amer. Soc. Mech. Engrs.]

ature t_h to be constant. The cold-fluid temperature t_c would vary only in the y direction. The solution of Eq. (11.6) would be

$$t_{c,o} = t_{c,i} + (t_h - t_{c,i})(1 - e^{-K_1}) \tag{11.17}$$

If there were more than one tube pass, we might easily show that Eq. (11.17) would still apply by replacing K_1 for the single pass by NK_1, where N is the number of tube passes. Furthermore, it would be immaterial how the tube passes were arranged (counter, crossflow, etc.) if the fluid temperature in the tubes remained constant.

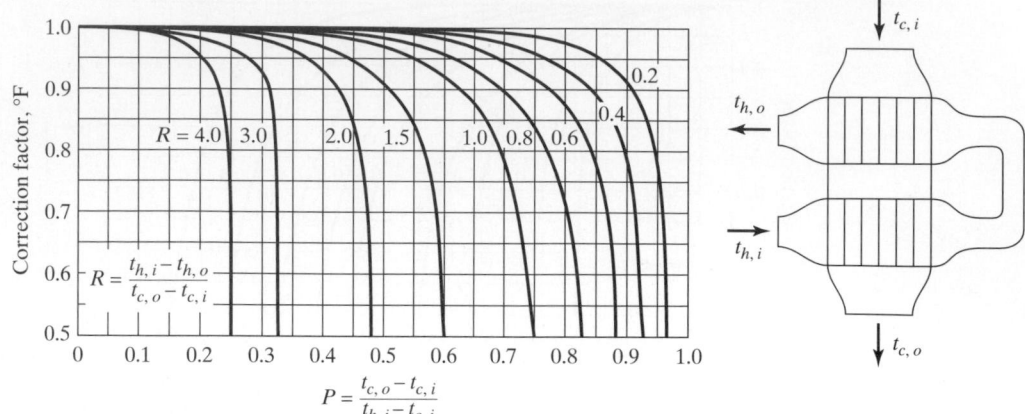

Figure 11.7 Correction factor $F = \Delta t_m / \Delta t_{m,cf}$ for a two-pass, countercross-flow heat exchanger, one fluid (t_h) mixed, other fluid (t_c) unmixed except between passes. [Reprinted from R. A. Bowman, A. C. Mueller, and W. M. Nagle, "Mean Temperature Difference in Design," *ASME Trans.*, 62 (1940), 289, with permission of the publisher, Amer. Soc. Mech. Engrs.]

By Eq. (11.17),

$$e^{K_1} = \frac{t_h - t_{c,i}}{t_h - t_{c,o}}$$

and

$$K_1 = \frac{U_o A_o}{\dot{m}_c c_{p,c}} = \frac{t_{c,o} - t_{c,i}}{\Delta t_m} = \ln\left(\frac{t_h - t_{c,i}}{t_h - t_{c,o}}\right)$$

Thus, when the *fluid temperature t_h remains constant*,

$$\Delta t_m = \frac{t_{c,o} - t_{c,i}}{\ln\left(\dfrac{t_h - t_{c,i}}{t_h - t_{c,o}}\right)} = \frac{(t_h - t_{c,i}) - (t_h - t_{c,o})}{\ln\left(\dfrac{t_h - t_{c,i}}{t_h - t_{c,o}}\right)} \tag{11.18}$$

This is identical to the log mean temperature difference for a counterflow heat exchanger. A similar result would be obtained if the cold fluid remained at constant temperature. Therefore, from Eq. (11.5), $F = 1.0$ for any heat exchanger in which one of the two fluids remains at constant temperature.

One can also obtain this result by observing the trends in the curves plotted in Figs. 11.5, 11.6, and 11.7. If the cold fluid remains isothermal, $c_{p,c} = \infty$, $\Delta t_c = 0$, and $P = 0$. All the curves approach $F = 1.0$ as P approaches zero. Similarly, if the hot fluid remains isothermal, $c_{p,h} = \infty$, $\Delta t_h = 0$, and $R = 0$. The curves approach $F = 1.0$ as R approaches zero also.

EXAMPLE 11.1

Moist air is to be heated from 70 °F to 150 °F by hot water whose temperature changes from 200 °F to 168 °F. Determine the true mean temperature difference if the heat exchanger is of the following types: (a) pure counterflow, (b) pure cross-flow with one row of tubes (Fig. 11.4), (c) pure cross-flow with four rows of tubes, and (d) countercross-flow with two tube passes.

Solution: For parts **(a)–(d)** we have

$$t_{h,i} = 200\ °F, \qquad t_{h,o} = 168\ °F, \qquad t_{c,i} = 70\ °F, \qquad t_{c,o} = 150\ °F$$

$$R = \frac{t_{h,i} - t_{h,o}}{t_{c,o} - t_{c,i}} = 0.40, \qquad P = \frac{t_{c,o} - t_{c,i}}{t_{h,i} - t_{c,i}} = 0.615$$

The solution of the entire problem is shown in Table 11.1.

TABLE 11.1 Solution for Ex. 11.1

Case	Method of Solution	F	Δt_m, °F
(a) Pure counterflow	Eq. (11.1)	1.00	71.3
(b) Pure cross-flow, one tube pass	Fig. 11.5	0.93	66.3
(c) Pure cross-flow, four tube passes	Fig. 11.6	0.94	67.1
(d) Countercross-flow, two tube passes	Fig. 11.7	0.99	70.6

11.3 EVALUATING HEAT-EXCHANGER PERFORMANCE USING THE NUMBER-OF-TRANSFER-UNITS METHOD

Another method of determining heat-exchanger performance is the number-of-transfer-units (NTU) method. This method has the advantage of being able to determine the size of heat exchanger required directly if the temperatures are known at only the two fluid inlets. The log mean temperature method requires an iterative solution if only the two inlet temperatures are known.

The purpose of the NTU method is to determine the heat-exchanger effectiveness, ε. Once the effectiveness is known, the outlet temperature of one of the two fluids can be determined. The outlet temperature of the other fluid can then be obtained from an energy balance on the entire heat exchanger. Two possibilities exist: (a) $\dot{m}_c c_{p,c} < \dot{m}_h c_{p,h}$, $c_c = c_{\min} < c_h$ or (b) $\dot{m}_h c_{p,h} \leq \dot{m}_c c_{p,c}$, $c_h = c_{\min} \leq c_c$ where the fluid capacity rates (c_c and c_h) are defined as the product of the mass flow rate and the specific heat of each fluid.

In case (a), the cold fluid will change in temperature more than the hot fluid, so the effectiveness becomes

$$\varepsilon = \frac{(t_{c,o} - t_{c,i})}{(t_{h,i} - t_{c,i})} = P = \frac{\text{actual } \Delta t \text{ cold fluid}}{\text{maximum possible } \Delta t \text{ of cold fluid}} \qquad (11.19)$$

where $0 \leq \varepsilon \leq 1.0$; the number of transfer units becomes

$$\text{NTU} = \frac{U_o A_o}{c_c} \qquad (11.20)$$

where $\text{NTU} \geq 0$; and the fluid capacity rate ratio becomes

$$c_r = \frac{c_c}{c_h} \qquad (11.21)$$

TABLE 11.2 Effectiveness Expressions for Heat Exchangers of Various Flow Configurations

Counterflow	$\varepsilon = \dfrac{1 - \exp\left[-\mathrm{NTU}(1 - c_r)\right]}{1 - c_r \exp\left[-\mathrm{NTU}(1 - c_r)\right]}$
Parallel flow	$\varepsilon = \dfrac{1 - \exp\left[-\mathrm{NTU}(1 + c_r)\right]}{1 + c_r}$
Cross-flow $c_{\min}$ fluid is unmixed $c_{\max}$ fluid is mixed	$\varepsilon = \dfrac{1}{c_r}\left[1 - \exp\left(-c_r[1 - \exp(-\mathrm{NTU})]\right)\right]$
Cross-flow $c_{\min}$ fluid is mixed $c_{\max}$ fluid is unmixed	$\varepsilon = 1 - \exp\left[-\dfrac{1}{c_r}\left(1 - \exp\left[-\mathrm{NTU}(c_r)\right]\right)\right]$

where $0 < c_r \leq 1.0$. The effectiveness is a function of (1) the heat-exchanger geometry, (2) the value for NTU, and (3) the value of c_r. Some algebraic relations for the effectiveness are given in Table 11.2. Plots for various heat-exchanger configurations are given in the *Handbook of Heat Transfer* [1].

In case (b), the hot fluid will change temperature more than the cold fluid, so

$$\varepsilon = \frac{t_{h,o} - t_{h,i}}{t_{h,i} - t_{c,i}} = RP = \frac{\text{actual } \Delta t \text{ hot fluid}}{\text{maximum possible } \Delta t \text{ of hot fluid}} \tag{11.22}$$

$$\mathrm{NTU} = \frac{U_o A_o}{c_h} \tag{11.23}$$

$$c_r = \frac{c_h}{c_c} \tag{11.24}$$

If either fluid remains isothermal, $c_r = 0$ and ε becomes

$$\varepsilon = 1 - \exp(-\mathrm{NTU}) \tag{11.25}$$

which is independent of the heat-exchanger configuration.

The use of the NTU method is illustrated in the following example.

EXAMPLE 11.2

Saturated moist air at 10 °C dry-bulb temperature is heated by hot water in a cross-flow heat exchanger. Assume the water is mixed and the air is unmixed. The water enters at 60 °C at a flowrate of 20 L/s. The air volumetric flow rate is 30 m³/s. Determine the leaving air and water temperatures if the heat exchanger has $U_o A_o = 100$ kW/°C.

Solution: The fluid-capacity rates are computed first to determine which fluid has the smaller value. The saturated-air specific-volume and humidity-ratio data are obtained from Table A.4SI, and the specific heat data are from Table 7.3.

$$\dot{m}_a c_{pa} = \frac{\dot{V}_a (c_{pa} + c_{pw} W)}{v_s} = \frac{(30)[1.00 + 1.86(0.007661)]}{0.8116} = 37.49 \text{ kW/°C}$$

The water specific volume is obtained from Table A.1SI

$$\dot{m}_w c_{pw,f} = \frac{\dot{V}_w c_{pw,f}}{v_w} = \frac{(20)(4.187)}{(1000)(0.001017)} = 82.34 \text{ kW/°C}$$

The air has the lower fluid-capacity rate, so

$$c_{\min} = c_a = 37.49 \text{ kW/°C}, \qquad c_{\max} = c_w = 82.34 \text{ kW/°C},$$

$$c_r = \frac{c_a}{c_w} = \frac{37.49}{82.34} = 0.455$$

The number of transfer units becomes

$$\text{NTU} = \frac{U_o A_o}{c_a} = \frac{100 \text{ kW/°C}}{37.49 \text{ kW/°C}} = 2.67$$

The effectiveness is computed using the expression in Table 11.2 for a cross-flow heat exchanger with $c_{\min}$ unmixed and $c_{\max}$ mixed,

$$\varepsilon = \frac{1}{0.455}[1 - \exp(-0.455[1 - \exp(-2.67)])] = 0.759$$

Using Eq. (11.19), the air exit temperature is determined:

$$t_{c,o} = (t_{h,i} - t_{c,i})\varepsilon + t_{c,i} = (60 - 10)(0.759) + 10 = 48.0 \text{ °C}$$

The exit water temperature is computed from an energy balance on the heat exchanger:

$$c_w(t_{h,o} - t_{h,i}) = c_a(t_{c,i} - t_{c,o})$$

Rearranging and solving for $t_{h,o}$,

$$t_{h,o} = c_r(t_{c,i} - t_{c,o}) + t_{h,i} = 0.455(10 - 48) + 60 = 42.7 \text{ °C}$$

11.4 THE EFFICIENCY OF VARIOUS EXTENDED SURFACES

Heat exchangers used for heating or cooling moist air may have fins of various types on the surface contacted by the air. Figure 11.1 shows two schematic arrangements with rectangular-plate fins. Circular-plate fins, bar fins, and various types of spines may also be employed.

The addition of fins to the tubes greatly increases the outer surface area but at the expense of decreasing the mean temperature difference between the surface and the air stream. Whereas the thermal resistance of the bare tube may be negligible, the thermal resistance of the extended surface may be considerable.

A significant quantity in evaluating the thermal effectiveness of fins is the *fin efficiency*, ϕ, defined as

$$\phi = \frac{t_{F,m} - t}{t_{F,B} - t} = \frac{\Delta t_{F,m}}{\Delta t_{F,B}} \tag{11.26}$$

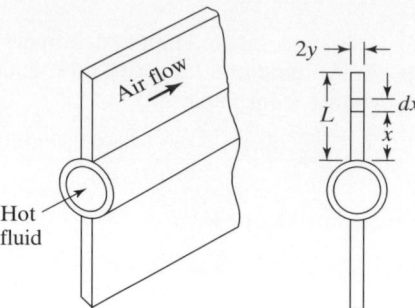

Figure 11.8 Schematic illustration of bar fin.

where $t_{F,m}$ is the mean temperature of the fin, $t_{F,B}$ is the temperature at the base of the fin, and t is the air dry-bulb temperature.

In this section we will study efficiencies of several types of finned surfaces. We will first derive the efficiency for a bar fin, which is mathematically the most elementary type. Figure 11.8 schematically shows a bar fin attached to a tube. We will assume (1) steady-state heat transfer, (2) constant thermal conductivity for the fin, (3) a constant temperature at the base of the fin, (4) one-dimensional heat conduction in the fin, (5) negligible heat transfer from the outer edge of the fin, (6) a uniform temperature of the air stream, and (7) constant outside-surface convection coefficient $h_{c,o}$.

At any cross section of unit length in Fig. 11.8, we have

$$q_F = -kA \frac{dt_F}{dx} = -2ky \frac{dt_F}{dx}$$

or

$$dq_F = -2ky \frac{d^2 t_F}{dx^2} dx = -2ky \frac{d^2 \Delta t_F}{dx^2} dx$$

But

$$dq_F = -2h_{c,o} dx (t_F - t) = -2h_{c,o} dx \Delta t_F$$

Thus

$$\frac{d^2 \Delta t_F}{dx^2} = \frac{h_{c,o}}{ky} \Delta t_F \tag{11.27}$$

Let us solve Eq. (11.27) for the following conditions: at $x = 0$, $\Delta t_F = \Delta t_{F,B}$, and at $x = L$, $d\Delta t_F/dx = 0$. We obtain:

$$\Delta t_F = \Delta t_{F,B} \left[\frac{e^{p(L-x)} + e^{-p(L-x)}}{e^{pL} + e^{-pL}} \right] \tag{11.28}$$

where $p = \sqrt{h_{c,o}/ky}$ and $\Delta t_{F,B} = t_{F,B} - t$.

We may find the total rate of heat transfer for a unit length of the fin by

$$q_F = 2h_{c,o} \int_0^L \Delta t_F dx$$

With Eq. (11.28), we have

$$q_F = \frac{2h_{c,o} \Delta t_{F,B}}{p} (\tanh pL) \tag{11.29}$$

Part IV / Heat- and Mass-Transfer Processes and Applications

By definition of the mean fin temperature $t_{F,m}$

$$q_F = h_{c,o}A_F(t_{F,m} - t) = 2h_{c,o}L\,\Delta t_{F,m} \qquad (11.30)$$

Thus, by Eqs. (11.26), (11.29), and (11.30), the efficiency of the bar fin is given by

$$\phi = \frac{\tanh pL}{pL} \qquad (11.31)$$

Figure 11.9 shows the efficiency of a bar fin as calculated by Eq. (11.31) for values of pL up to 5.0.

    Circular-plate fins are more commonly applied to heat exchangers than bar fins. Figure 11.10 shows two schematic circular-plate fins. Fin (a) has a uniform thickness, while fin (b) has a constant cross-sectional area. Plots of the fin efficiency for these two fin geometries are given in Figs. 11.11 and 11.12. Other fin configurations are considered in the *Handbook of Heat Transfer Fundamentals* [2].

    The rectangular-plate fin of uniform thickness is commonly used in finned coils for heating or cooling air. It is not possible to obtain an exact mathematical solution for the

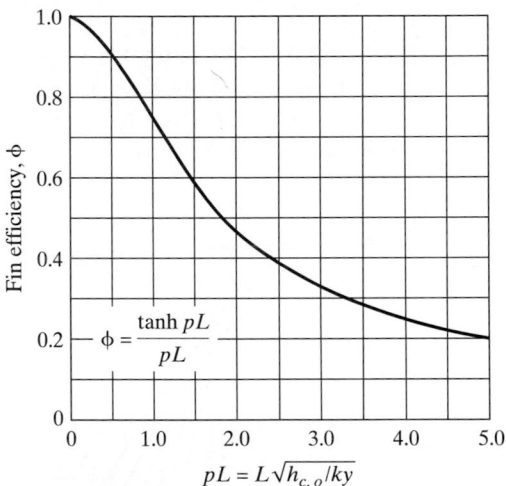

Figure 11.9 Efficiency of a bar fin.

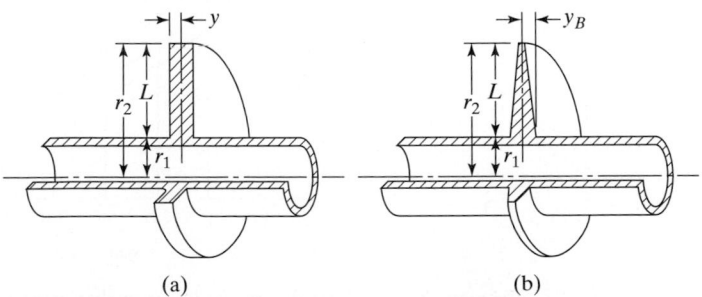

Figure 11.10 Schematic illustrations of circular-plate fins (a) having a uniform thickness, and (b) having a constant cross-sectional area.

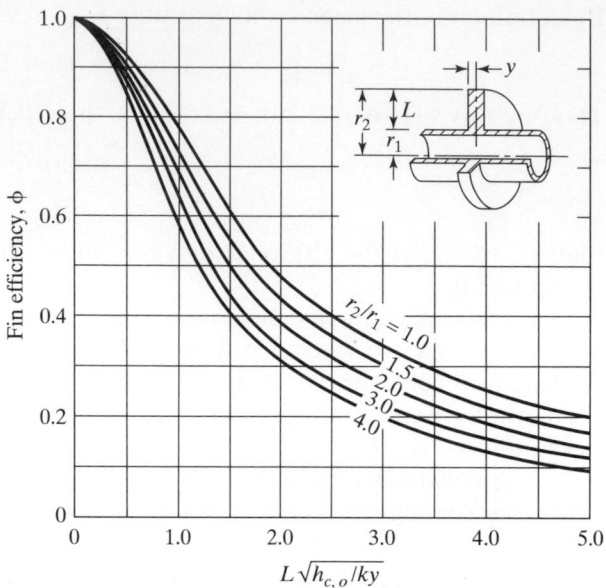

Figure 11.11 Efficiency for a circular-plate fin of uniform thickness. [Reprinted from K. A. Gardner, "Efficiency of Extended Surfaces," *ASME Trans.*, 67 (1945), 625, with permission of the publisher, Amer. Soc. Mech. Engrs.]

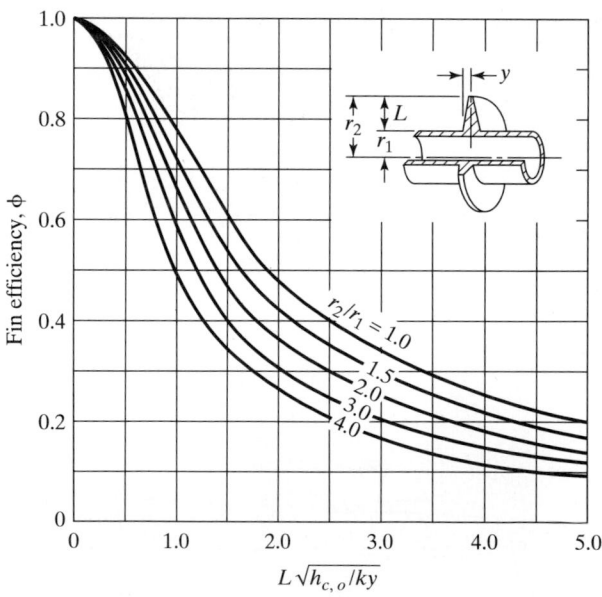

Figure 11.12 Efficiency for a circular-plate fin having a constant cross-sectional area. [Reprinted from K. A. Gardner, "Efficiency of Extended Surfaces," *ASME Trans.*, 67 (1945), 625, with permission of the publisher, Amer. Soc. Mech. Engrs.]

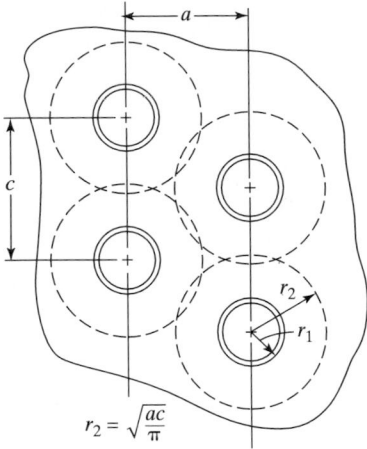

Figure 11.13 Approximation method for treating a rectangular-plate fin of uniform thickness in terms of a flat circular-plate fin of equal area.

efficiency of such a fin. An adequate approximation is to assume that the fin area served by each tube is equivalent in performance to a flat circular-plate fin of equal area. Figure 11.13 shows the method where the equivalent outer radius of the circular fin is determined as

$$r_2 = \sqrt{\frac{ac}{\pi}} \tag{11.32}$$

After determination of the equivalent outer radius, the fin efficiency may be found from Fig. 11.11.

11.5 OVERALL HEAT-TRANSFER COEFFICIENT FOR A DRY FINNED-TUBE HEAT EXCHANGER

Design problems with finned-tube heat exchangers involve solution of the equation

$$\dot{Q} = U_o A_o \, \Delta t_m \tag{11.33}$$

In Sec. 11.3 we studied calculation of the mean temperature difference Δt_m. We will now investigate calculation of the *overall heat-transfer coefficient U_o*, where we assume the fin surfaces to be dry.

Figure 11.14 shows a schematic local section of a finned-tube heat exchanger. We will assume (1) steady-state heat transfer, and (2) negligible contact resistance between

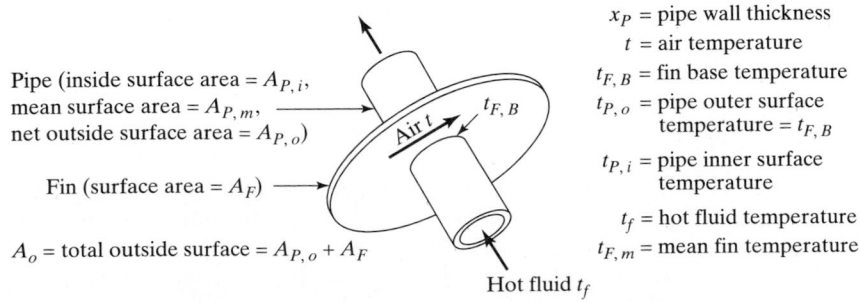

Pipe (inside surface area = $A_{P,i}$, mean surface area = $A_{P,m}$, net outside surface area = $A_{P,o}$)

Fin (surface area = A_F)

A_o = total outside surface = $A_{P,o} + A_F$

x_P = pipe wall thickness
t = air temperature
$t_{F,B}$ = fin base temperature
$t_{P,o}$ = pipe outer surface temperature = $t_{F,B}$
$t_{P,i}$ = pipe inner surface temperature
t_f = hot fluid temperature
$t_{F,m}$ = mean fin temperature

Air t $t_{F,B}$

Hot fluid t_f

Figure 11.14 Schematic illustration of a finned-tube heat exchanger.

the base of the fin and the pipe ($t_{F,B} = t_{P,o}$). We may write the following equations for the rate of heat transfer from the hot fluid to the pipe:

$$\dot{Q} = h_i A_{P,i} (t_f - t_{P,i}) \tag{11.34}$$

through the pipe:

$$\dot{Q} = \frac{k_P A_{P,m}(t_{P,i} - t_{P,o})}{x_P} \tag{11.35}$$

from the pipe and fin to the air:

$$\dot{Q} = h_{c,o,P} A_{P,o}(t_{P,o} - t) + h_{c,o,F} A_F(t_{F,m} - t) \tag{11.36}$$

and for the entire process:

$$\dot{Q} = U_o A_o(t_f - t) \tag{11.37}$$

By Eqs. (11.26) and (11.36) and assuming $h_{c,o,P} = h_{c,o,F} = h_{c,o}$, we have

$$\dot{Q} = h_{c,o}(A_{P,o} + \phi A_F)(t_{P,o} - t) \tag{11.38}$$

By Eqs. (11.34), (11.35), (11.37), and (11.38), we obtain

$$U_o = \frac{1}{\dfrac{A_o}{A_{P,i} h_i} + \dfrac{A_o x_P}{A_{p,m} k_P} + \dfrac{1 - \phi}{h_{c,o}(A_{P,o}/A_F + \phi)} + \dfrac{1}{h_{c,o}}} \tag{11.39}$$

Equation (11.39) shows that, for a heat exchanger of known dimensions, the overall coefficient U_o may be calculated if the heat-transfer coefficients h_i and $h_{c,o}$, the pipe thermal conductivity k_P, and the fin efficiency ϕ are known.

It is of considerable interest to study the influences upon U_o of the various quantities in Eq. (11.39). Some of these are shown in Example 11.3.

EXAMPLE 11.3

An air-heating coil is constructed of rectangular-plate fins bonded to copper tubes. The tubes are arranged in line in adjacent rows. Physical data are as follows: $a = 0.141$ ft, $c = 0.125$ ft, $x_P = 0.00233$ ft, $r_1 = 0.0208$ ft, fin thickness = 0.000833 ft, fin spacing = 8 per in., $A_{P,i} = 0.116$ ft^2/lineal ft of tube, $A_{P,o} = 0.120$ ft^2/lineal ft of tube, $A_{P,m} = 0.123$ ft^2/lineal ft of tube, $A_F = 3.11$ ft^2/lineal ft of tube, and $A_o = 3.23$ ft^2/lineal ft of tube. (a) Determine the individual thermal resistances and the overall heat-transfer coefficient U_o if $h_{c,o} = 10.0$ Btu/hr $\cdot$ ft^2 $\cdot$ °F, $h_i = 600$ Btu/hr $\cdot$ ft^2 $\cdot$ °F, and the fin material is aluminum. (b) Rework part (a) but with copper fins. (c) Rework part (a) but with $h_{c,o} = 5.0$ Btu/hr $\cdot$ ft^2 $\cdot$ °F. (d) Rework part (a) but with $h_i = 1200$ Btu/hr $\cdot$ ft^2 $\cdot$ °F.

Solution: All calculations are shown in Tables 11.3 and 11.4. Table 11.3 shows determination of fin efficiency ϕ and fin thermal resistance R_F. Table 11.4 shows the individual thermal resistances R_i, R_P, R_F, and R_o and their percent of the total resistance R_t, as well as the overall coefficient U_o.

Table 11.3 shows that the fin thermal resistance R_F is affected strongly by the fin thermal conductivity k but only in a small way by the outside-surface coefficient $h_{c,o}$. Table 11.3 shows that all of the thermal resistances are important *except the pipe-wall resistance R_p*, but that the outside-surface resistance R_o is dominant. Part (b) shows that use of copper fins instead of aluminum fins increased U_o by 9.3 percent. Part (c) shows that a 50 percent reduction in $h_{c,o}$ (10.0 to 5.0) decreased U_o by 36 percent. Part (d) shows that a 100 percent increase in h_i (600 to 1200) increased U_o by 14.7 percent.

TABLE 11.3 Determination of Fin Efficiency ϕ and Fin Thermal Resistance R_F for Ex. 11.3

Component or Calculation	Part (a)	Part (b)	Part (c)	Part (d)
r_1, ft	0.0208	0.0208	0.0208	0.0208
$r_2\sqrt{ac/\pi}$, ft	0.0749	0.0749	0.0749	0.0749
$L = (r_1 - r_2)$, ft	0.0541	0.0541	0.0541	0.0541
$h_{c,o}$, Btu/hr · ft² · °F	10.0	10.0	5.0	10.0
k, Btu/hr · ft · °F/ft	120	223	120	120
y, ft	0.000417	0.000417	0.000417	0.000417
r_2/r_1	3.60	3.60	3.60	3.60
$L\sqrt{h_{c,o}/ky}$	0.765	0.560	0.541	0.765
ϕ (Fig. 11.11)	0.73	0.83	0.84	0.73
$A_{P,o}/A_F$	0.0386	0.0386	0.0386	0.0386
$R_F = \dfrac{1}{h_{c,o}}\left(\dfrac{1-\phi}{\phi + A_{P,o}/A_F}\right)$	0.0351	0.0196	0.0364	0.0351

TABLE 11.4 Determination of Thermal Resistances and U_o for Ex. 11.3

Component or Calculation	Part (a)	Part (b)	Part (c)	Part (d)
Inside surface resistance, $R_i = A_o/A_{P,i}h_i$	0.0464	0.0464	0.0464	0.0232
Pipe wall resistance, $R_P = A_o x_P/A_{P,m}k_P)$	0.00027	0.00027	0.00027	0.00027
Fin resistance, R_F (Table 11.3)	0.0351	0.0196	0.0364	0.0351
Outside surface resistance, $R_o = 1/h_{c,o}$	0.1000	0.1000	0.2000	0.1000
Total resistance, $R_t = R_i + R_P + R_F + R_o$	0.18177	0.16627	0.28307	0.15857
R_i in percent of R_t	25.5	27.9	16.4	14.6
R_P in percent of R_t	0.2	0.2	0.1	0.2
R_F in percent of R_t	19.3	11.8	12.9	22.1
R_o in percent of R_t	55.0	60.1	70.6	63.1
$U_o = 1/R_t$, Btu/hr · ft² · °F	5.50	6.01	3.53	6.31

11.6 OVERALL HEAT-TRANSFER PROBLEMS INVOLVING DRY FINNED SURFACES

As stated in Sec. 11.5, a practical heat-exchanger design problem involves solution of Eq. (11.33). For a practical problem, Eq. (11.39) should be modified to allow for deposit coefficients. A minor deposit on the outside surface of a finned coil generally has little effect upon U_o because of the usually large magnitude of $1/h_{c,o}$. Sometimes an allowance is made for imperfect bonding of the fins to the tubes, but this effect is difficult to evaluate and, with good construction, it should be small. It is more important to include a deposit coefficient for the inside surface of the tubes. Example 11.3 shows that the thermal resistance R_P of the tube wall may be neglected with little error. Thus, for most cases

$$U_o = \cfrac{1}{\cfrac{A_o}{A_{P,i}h_i} + \cfrac{A_o}{A_{P,i}h_{d,i}} + \cfrac{1-\phi}{h_{c,o}(A_{P,o}/A_F + \phi)} + \cfrac{1}{h_{c,o}}} \qquad (11.40)$$

Information is given in Chapter 2 for evaluating the individual coefficients h_i and $h_{d,i}$. Equations shown in Chapter 2 for forced-convection coefficients are usually not applicable for $h_{c,o}$ for finned-tube coils. The type of fins, fin spacing, and other factors affect the value of the coefficient. It is usually necessary to have experimental information for the particular fin arrangement. A limited amount of basic information is available in published literature.

Table 11.5 shows dimensional data for two plate-fin-and-tube arrangements. The surface consists of aluminum fins bonded to copper tubes. Figure 11.15 shows a schematic sketch of the surface and also presents a heat-transfer correlation for the external surface. The information in both Table 11.4 and Fig. 11.15 was taken from *Compact Heat Exchangers* by Kays and London [3]. Similar information for many other types of extended surfaces is also given in this reference.

Figure 11.15 provides an estimation for the external-surface convection coefficient $h_{c,o}$. The dimensionless grouping $(h_{c,o}/Gc_p)(c_p\mu/k)^{2/3}$ is shown as a function of the dimensionless Reynolds number $(4r_h G/\mu)$. The mass velocity G is based upon the minimum free-flow area A_c. The Reynolds number is based upon a hydraulic diameter $4r_h$ defined by the relation $4r_h/L = 4A_c/A_o$, where L is the flow length of the heat exchanger and A_o is the total external surface area. For a given type of surface, r_h is a constant inde-

TABLE 11.5 Dimensional Data for Two Plate-Fin-and-Tube Surface Arrangements

Data	Surface I		Surface II	
Dimensions (see sketch with Fig. 11.15)				
A, tube outside diameter, in. (cm)	0.402	(1.02)	0.676	(1.72)
B, tube spacing across face, in. (cm)	1.00	(2.54)	1.50	(3.81)
C, tube spacing between rows, in. (cm)	0.866	(2.20)	1.75	(4.45)
D, spacing of fins, center to center, in. (cm)	0.125	(0.318)	0.129	(0.328)
E, thickness of aluminum fins, in. (cm)	0.013	(0.033)	0.016	(0.041)
Flow passage hydraulic diameter, $4r_h$, ft (cm)	0.01192	(0.030)	0.01268	(0.032)
Area Ratio Data				
$A_{o,1}$, ft^2 (m^2) external surface/ ft^2 (m^2) face area/row	12.92		22.86	
$A_o/A_{P,i}$, ft^2 (m^2) external surface/ ft^2 (m^2) internal surface	12.27		19.31	
A_c, minimum ft^2 (m^2) net flow area/ ft^2 (m^2) face area	0.534		0.497	
A_F/A_o, ft^2 (m^2) fin surface/ ft^2 (m^2) external surface	0.839		0.905	

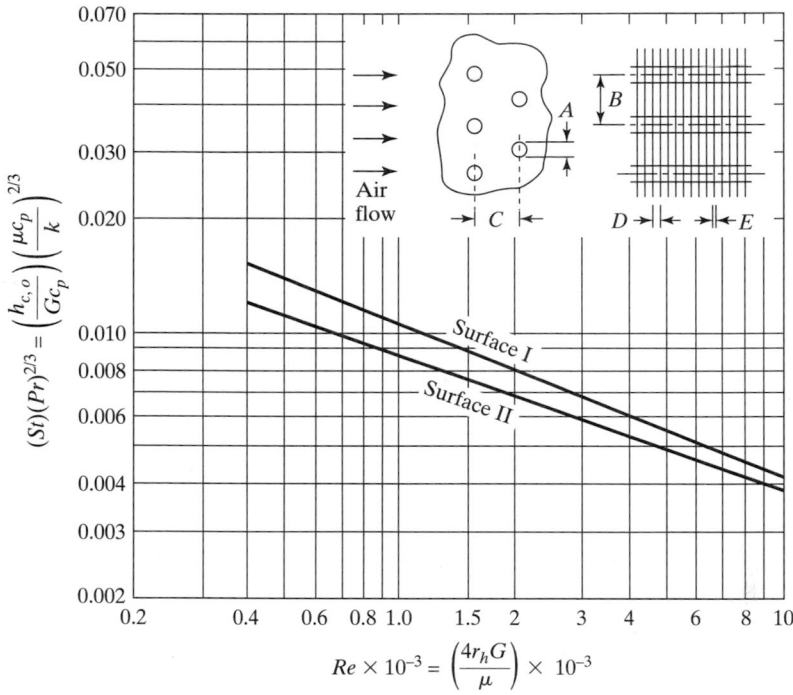

The chart shows axes labeled:

Vertical axis: $(St)(Pr)^{2/3} = \left(\dfrac{h_{c,o}}{Gc_p}\right)\left(\dfrac{\mu c_p}{k}\right)^{2/3}$

Horizontal axis: $Re \times 10^{-3} = \left(\dfrac{4r_h G}{\mu}\right) \times 10^{-3}$

Curves: Surface I, Surface II

Air flow

Figure 11.15 Correlated-external surface heat-transfer data for surfaces of Table 11.5. [Adapted by permission from W. M. Kays and A. L. London, *Compact Heat Exchangers* (The National Press, 1955), 177.]

pendent of the depth of the coil. Kays and London recommend that the fluid properties c_p, μ, and k be evaluated at the mean mixed fluid temperature.

EXAMPLE 11.4

A heat exchanger is to be sized to process 12,000 ft³/min of saturated air at 20 °F to a final temperature of 150 °F. Heating medium is dry-and-saturated steam condensing at 5 psig. Barometric pressure is 14.00 psia. Type II surface shown in Table 11.5 is to be used with copper tubes and aluminum fins. Inlet face velocity of the air is to be 500 ft/min. Determine (a) the required coil face area in ft², (b) the required total outside-surface area in ft², (c) the number of rows of tubes required, and (d) the lbm/hr of steam required if the steam condensate leaves the coil as saturated liquid.

Solution:

(a) From the given data we may calculate the coil face area as 12,000/500 = 24 ft².

(b) We must solve for A_o from Eq. (11.33). We will first solve for Δt_m, which is given by the logarithmic mean temperature difference, since the steam temperature is constant at 225.2 °F (sat. temp. at 19 psia). By Eq. (11.18), $F = 1.0$ and

$$\Delta t_m = \frac{150 - 20}{\ln\left(\dfrac{225.2 - 20}{225.2 - 150}\right)} = 129\ °F$$

To find U_o we must first evaluate the individual quantities in Eq. (11.40). By Table 11.5, $A_o/A_{P,i} = 19.31$ and $A_F/A_o = 0.905$. Thus $A_{P,o}/A_F = (A_o - A_F)/A_F = A_o/A_F - 1 = 1/0.905 - 1 = 0.105$.

Based on the discussion in Sec. 2.26, we will assume $h_i = 1200$ Btu/hr $\cdot$ ft^2 $\cdot$ °F. We will assume that the steam is free of oil, and since we have copper tubes, we will take $1/h_{d,i} = 0.0005$. We will estimate $h_{c,o}$ from Fig. 11.15. Fluid properties will be taken at the mean temperature

$$t_m = t_{\text{steam}} - \Delta t_m = 225 - 129 = 96 \text{ °F}$$

By Table A.5E, $\mu = 0.0457$ lbm/ft $\cdot$ hr and $c_p\mu/k = 0.706$. We may use $c_{pa} = 0.241$ Btu/lbm $\cdot$ °F. The mass velocity G may be found from the entering air conditions and the percent free area for the coil. By Eq. (7.18) and Table A.1E, we find $W_1 = 0.00225$ lbm$_w$/lbm$_a$. By Eq. (7.25), we find $v_1 = 12.73$ ft^3/lbm$_a$. Thus

$$\rho_1 = \frac{1 + W_1}{v_1} = \frac{1.002}{12.73} = 0.0787 \text{ lbm}_a/\text{ft}^3$$

and

$$\dot{m}_a = (12{,}000)(0.0787)(60) = 56{,}700 \text{ lbm}_a/\text{hr}$$

By Table 11.5, the fraction of the minimum face area that is between the fins (A_c) is 49.7 percent. Thus,

$$G = \frac{56{,}700}{(0.497)(24)} = 4750 \text{ lbm}_a/\text{hr} \cdot \text{ft}^2$$

By Table 11.5, $4r_h = 0.01268$ ft. Thus

$$\text{Re} = \frac{4r_h G}{\mu} = \frac{(0.01268)(4750)}{0.0457} = (1.32)(10^3)$$

By Fig. 11.15 (surface II)

$$\frac{h_{c,o}(c_p\mu/k)^{2/3}}{Gc_p} = 0.0078$$

Thus

$$h_{c,o} = \frac{(0.0078)Gc_p}{(c_p\mu/k)^{2/3}} = \frac{(0.0078)(4750)(0.241)}{(0.706)^{2/3}} = 11.3 \text{ Btu/hr} \cdot \text{ft}^2 \cdot \text{°F}$$

We may next evaluate the fin efficiency. By Eq. (11.32),

$$r_2 = \sqrt{\frac{(1.75)(1.5)}{\pi}} = 0.914 \text{ in.}$$

and

$$\frac{r_2}{r_1} = \frac{r_2}{A/2} = \frac{0.914}{0.338} = 2.70$$

For aluminum fins, $k = 120$ Btu/hr $\cdot$ ft^2 $\cdot$ °F, $L = r_2 - r_1 = 0.576$ in.,

$$L\sqrt{\frac{h_{c,o}}{ky}} = \frac{0.576}{12}\sqrt{\frac{(11.3)(2)(12)}{(120)(0.016)}} = 0.570$$

By Fig. 11.11, $\phi = 0.86$.

We may calculate U_o by Eq. (11.40). Thus,

$$U_o = \frac{1}{\dfrac{19.31}{1200} + (19.31)(0.0005) + \dfrac{0.14}{(11.3)(0.97)} + \dfrac{1}{11.3}} = 7.87 \text{ Btu/hr} \cdot \text{ft}^2 \cdot \text{°F}$$

All the heat transfer is sensible heating, which can be determined from an energy balance on the air:

$$\dot{Q} = \dot{m}_a c_{pa}(t_{c,o} - t_{c,i}) = (56,700)(0.241)(150 - 20) = 1,776,400 \text{ Btu/hr}$$

Thus, by Eq. (11.33),

$$A_o = \frac{\dot{Q}}{U_o \, \Delta t_m} = \frac{1776400}{(7.87)(129)} = 1740 \text{ ft}^2$$

(c) By Table 11.5, we find that surface II has an outside-surface area of 22.86 ft^2/ft^2 of face area · row. Thus,

$$\text{rows required} = \frac{1740}{(22.86)(24)} = 3.17 \approx 3 \text{ rows}$$

(d) We may calculate the required steam flow by performing an energy balance on the steam:

$$\dot{m}_s = \frac{\dot{Q}}{h_{fg,s}} = \frac{1,776,400}{961.9} = 1840 \text{ lbm/hr}$$

EXAMPLE 11.5

Assume the same problem as Ex. 11.4 but take the heating medium to be hot water. Assume that water enters the coil at 200 °F and leaves at 180 °F. Assume a countercross-flow arrangement similar to Fig. 11.3(b). As the face area is 24 ft^2, we will assume the coil has a height of 4 ft and a length of 6 ft. Find (a) the gpm of water required, (b) outside surface area required in ft^2, and (c) the number of rows of tubes required.

Solution:

(a) The total rate of heat transfer is the same as for Ex. 11.4. We may calculate the rate of water flow by an energy balance on the water

$$\dot{m}_w = \frac{\dot{Q}}{c_{pw} \, \Delta t_w} = \frac{1,776,400}{(1)(20)} = 88,820 \text{ lbm}_w/\text{hr}$$

From Table A.1E, at 190 °F, $\rho_w = 1/v_f = 60.4$ lbm/ft^3. One gallon of water at 190 °F has $(231)(60.4)/1728 = 8.08$ lbm. Thus, based upon the average water density,

$$\text{gpm} = \frac{\dot{m}_w}{(8.08)(60)} = \frac{88,820}{(8.08)(60)} = 183$$

(b) We will use the NTU method and either Eq. (11.20) or Eq. (11.23) to determine the total external surface area required. We must first determine which fluid has the minimum capacity rate. For the air

$$c_a = \dot{m}_a c_{pa} = (56,700)(0.241) = 13,665 \text{ Btu/hr} \cdot °\text{F}$$

and for the water

$$c_w = \dot{m}_w c_{pw} = (88,820)(1) = 88,820 \text{ Btu/hr} \cdot °\text{F}$$

Therefore the air is the minimum-capacity-rate fluid and we should use Eqs. (11.19), (11.20), and (11.21). From Eq. (11.21)

$$c_r = \frac{c_a}{c_w} = \frac{13,665}{88,820} = 0.154$$

From Eq. (11.19)

$$\varepsilon = \frac{150 - 20}{200 - 20} = 0.722$$

Solve for NTU from the relation in Table 11.2 for a counterflow heat exchanger, as at least 3 rows will be needed (see Ex. 11.4):

$$NTU = \frac{\ln\left(\frac{1 - c_r\varepsilon}{1 - \varepsilon}\right)}{1 - c_r}$$

$$= \frac{\ln\left(\frac{1 - (0.154)(0.722)}{1 - 0.722}\right)}{1 - 0.154}$$

$$= 1.374$$

The overall heat-transfer coefficient, U_o, must be modified, as hot water has replaced the steam inside the tubes. The water velocity can be computed by knowing the water volumetric flowrate and the total flow cross-sectional area. The water flowrate is

$$\dot{V}_w = \dot{m}_w v_f = (88,820)(0.01657) = 1472 \text{ ft}^3/\text{hr}$$

Assuming the water flows in parallel through all the tubes in each row, the number of tubes per row is equal to the coil height divided by the tube spacing, B, or

$$\# \text{ tubes per row} = \frac{(4)(12)}{1.5} = 32 \text{ tubes}$$

The inner diameter of the tubes is computed from the data in Table 11.5 as

$$\frac{A_{p,i}}{A_{\text{face}}} = \frac{A_o/A_{\text{face}}}{A_o/A_{p,i}} = \frac{A_{o,1}}{A_o/A_{p,i}} = \frac{\pi d_i L (\# \text{ tubes})}{A_{\text{face}}}$$

where L is the length of the heat exchanger.

Solving for the inside diameter,

$$d_i = \frac{A_{o,1} A_{\text{face}}}{(A_o/A_{pi})\pi L (\# \text{ tubes})} = \frac{(22.86)(24)(12)}{(19.31)\pi(6)(32)} = 0.565 \text{ in.}$$

The cross-sectional flow area per row becomes

$$A_w = \frac{\pi d_i^2 (\# \text{ tubes/row})}{4} = \frac{\pi(0.565)^2(32)}{4(144)} = 0.0557 \text{ ft}^2/\text{row}$$

The average water velocity is

$$\mathbf{V}_w = \dot{V}_w/A_w = \frac{1472}{(0.0557)(3600)} = 7.34 \text{ ft/sec}$$

From Table A.6E we may find the viscosity of liquid water at 190 °F to be 0.756 lbm/ft · hr, the thermal conductivity to be 0.390 Btu/hr · ft · °F and the specific heat 1.004 Btu/lbm · °F. The Reynolds number is

$$Re_w = \rho_w \mathbf{V}_w d_i/\mu_w = \frac{(60.4)(7.34)(0.565)(3600)}{(0.756)(12)} = 99,400$$

Thus the flow is well into the turbulent regime and we may use Eq. (2.51) to determine the convective heat-transfer coefficient for the water inside the tubes:

$$h_{c,i} = \frac{0.023k}{d_i} \, \text{Re}^{0.8} \, \text{Pr}^{0.3}$$

$$\text{Pr} = c_p \mu / k = \frac{(1.004)(0.756)}{0.390} = 1.95$$

$$h_{c,i} = \frac{(0.023)(0.390)(99{,}400)^{0.8}(1.95)^{0.3}(12)}{(0.565)}$$

$$= 2317 \, \text{Btu/hr} \cdot \text{ft}^2 \cdot {}^\circ\text{F}$$

The mean mixed-air temperature is approximately

$$t_m = 190 - 95 = 95 \, {}^\circ\text{F}$$

Thus $h_{c,o}$ would be the same as for Ex. 11.4. Furthermore, the fin efficiency would be the same as for Ex. 11.4. If we use a deposit coefficient $1/h_d = 0.0005$, by Eq. (11.40) we have

$$U_o = \frac{1}{\dfrac{19.31}{2317} + (19.31)(0.0005) + \dfrac{0.14}{(11.3)(0.97)} + \dfrac{1}{11.3}} = 8.38 \, \text{Btu/hr} \cdot \text{ft}^2 \cdot {}^\circ\text{F}$$

Thus, by Eq. (11.20),

$$A_o = \frac{(1.374)(13{,}665)}{8.38} = 2240 \, \text{ft}^2$$

(c) Rows of tubes required $= \dfrac{2240}{(22.86)(24)} = 4.08 \approx 4 \text{ rows}$

11.7 INTRODUCTION TO HEAT TRANSFER IN WET-SURFACE COOLING COILS

Fin-and-tube surfaces are widely used in applications for cooling atmospheric air. If no moisture is separated from the air (sensible cooling only), we may use the procedures developed in earlier sections of this chapter. However, in cooling applications it is more common that dehumidification of the air also occurs. With dehumidification, the air-side surface is wetted (liquid water or frost). Besides transfer of sensible heat, there is a transfer of latent heat because of condensation. Since water-vapor transfer is not dependent upon temperature difference alone, it follows that the analyses presented earlier in this chapter do not suffice.

Figure 11.16 shows schematically a cold surface in contact with a moving stream of moist air. A moving film of water is formed on the surface by condensation of moisture from the air stream. There is a boundary layer of air next to the water surface. In this layer, we assume that air temperature, air humidity ratio, and air velocity vary in a plane perpendicular to bulk motion of the air. Immediately next to the water film, we assume that the air is saturated at the water surface temperature t_w. The transfer processes

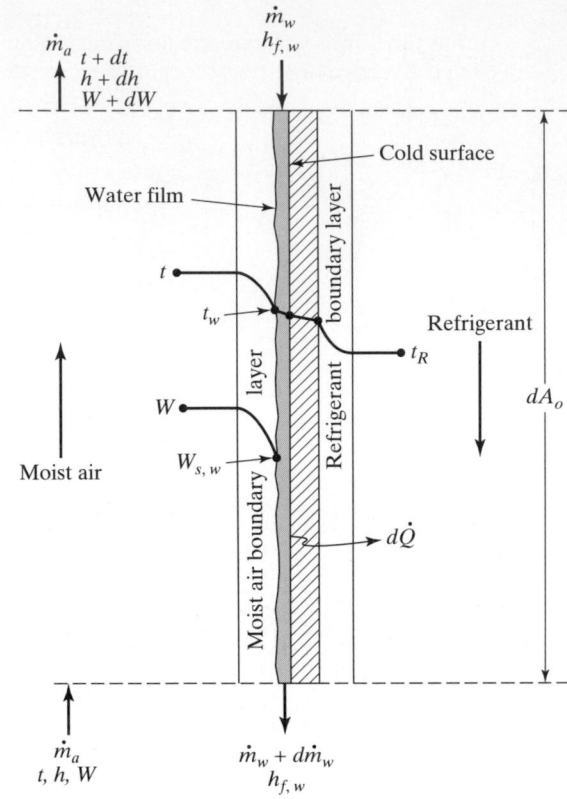

Figure 11.16 Schematic cooling and dehumidifying of moist air.

between the air stream and the water surface are similar to those described in Sec. 10.4 for the spray dehumidifier. For the differential surface area in Fig. 11.16, we have

$$-\dot{m}_a \, dh = d\dot{Q} - \dot{m}_a \, dW \, h_{f,w} \qquad (11.41)$$

$$d\dot{Q} = h_{c,o} \, dA_o (t - t_w) + h_{D,o} \, dA_o (W - W_{s,w})(h_{g,t} - h_{f,w}) \qquad (11.42)$$

$$-\dot{m}_a \, dW = h_{D,o} \, dA_o (W - W_{s,w}) \qquad (11.43)$$

Using the relation $\text{Le} = h_{c,o}/h_{D,o}c_{pa}$, Eq. (11.42) may be written as

$$d\dot{Q} = \frac{h_{c,o} \, dA_o}{c_{pa}} \left[c_{pa}(t - t_w) + \frac{(W - W_{s,w})(h_{g,t} - h_{f,w})}{\text{Le}} \right]$$

or

$$d\dot{Q} = \frac{h_{c,o} \, dA_o}{c_{pa}} \left[(h - h_{s,w}) + \frac{(W - W_{s,w})(h_{g,t} - h_{f,w} - h_g^0 \, \text{Le})}{\text{Le}} \right] \qquad (11.44)$$

By Eqs. (11.41), (11.43), and (11.44), we may show that

$$\frac{dh}{dW} = \text{Le} \, \frac{h - h_{s,w}}{W - W_{s,w}} + (h_{g,t} - h_g^0 \, \text{Le}) \qquad (11.45)$$

Equation (11.45) describes the process line on the psychrometric chart for the cooling and dehumidifying of moist air by a cold surface. Equation (11.45) is identical to Eq. (10.47) for the spray dehumidifier.

In Eq. (11.44), the latter grouping in the brackets is typically small compared to the term $(h - h_{s,w})$. For example, if the air state were 85 °F (30 °C) dry-bulb temperature, 70 °F (21 °C) thermodynamic wet-bulb temperature, and 14.696 psia (101.325 kPa) barometric pressure, and the water-film temperature were 60 °F (16 °C), we would have

$$h - h_{s,w} = 7.51 \text{ Btu/lbm}_a \text{ (17.5 kJ/kg}_a)$$

$$\frac{(W - W_{s,w})(h_{g,t} - h_{f,w} - h_g^0 \text{ Le})}{\text{Le}} = 0.142 \text{ Btu/lbm}_a \text{ (0.330 kJ/kg}_a)$$

Thus, approximately

$$d\dot{Q} = \frac{h_{c,o} \, dA_o}{c_{pa}} (h - h_{s,w}) \qquad (11.46)$$

We will find Eq. (11.46) to be of much importance in our studies on cooling coils. Note that the form of Eq. (11.46) is similar to Eq. (11.36) except the heat-transfer driving force is a difference in moist air enthalpy rather than a difference in air dry-bulb temperature. Although approximate, Eq. (11.46) allows a much easier analysis of wet cooling coils than does Eq. (11.44). Besides Eq. (11.46), another relation will be repeatedly used in subsequent sections of this chapter. We will assume that over a small range of temperature, the enthalpy of saturated air h_s, Btu per lbm$_a$ (kJ/kg$_a$), may be represented as

$$h_s = a + bt_s \qquad (11.47)$$

Figure 11.17 shows that over a narrow range of temperature, such as about 10 °F (5 °C), h may be closely given by Eq. (11.47) if the coefficients a and b are average values.

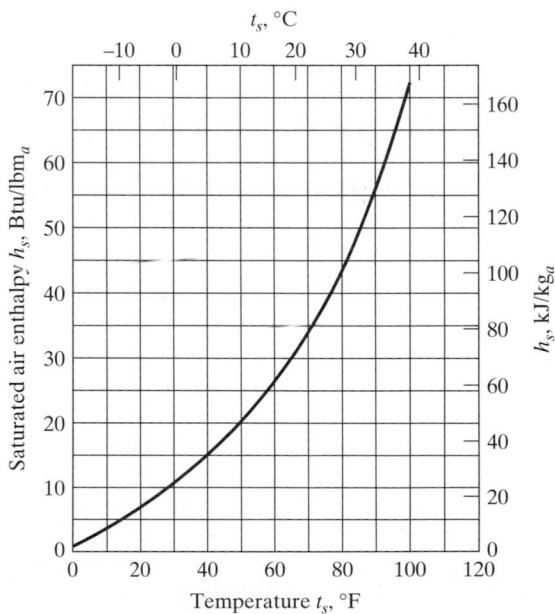

Figure 11.17 Enthalpy of saturated air as a function of temperature for standard atmospheric pressure.

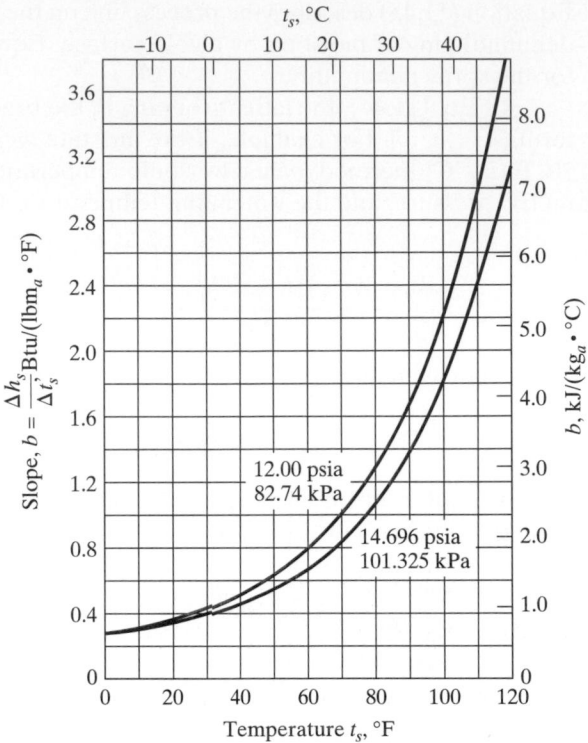

Figure 11.18 Slope $\Delta h_s / \Delta t_s$ for saturated air.

Figure 11.18 shows variation of the coefficient b in Eq. (11.47) for barometric pressures of 12.00 and 14.696 psia (82.74 and 101.325 kPa).

11.8 THE EFFICIENCY OF VARIOUS EXTENDED SURFACES WHEN BOTH COOLING AND DEHUMIDIFYING OCCUR

The efficiency for a dry bar fin was developed in Sec. 11.4. We will now study the performance of a bar fin when condensation occurs on its surfaces; Fig. 11.19 schematically shows the problem. We will make the same assumptions given in Sec. 11.4. We will also assume that heat conduction through the water film occurs in only the y direction. For a unit length of the fin, we have

$$q_F = 2k_F y \frac{dt_F}{dx} \tag{11.48}$$

where the subscript F refers to the metal fin. Also,

$$dq_F = -2 \frac{k_w}{y_w} (t_w - t_F) \, dx \tag{11.49}$$

where k_w and y_w are, respectively, the thermal conductivity and thickness of the water film. By Eqs. (11.47) and (11.49),

$$dq_F = -\frac{2k_w}{b_w y_w} (h_{s,w} - a_w - b_w t_F) \, dx$$

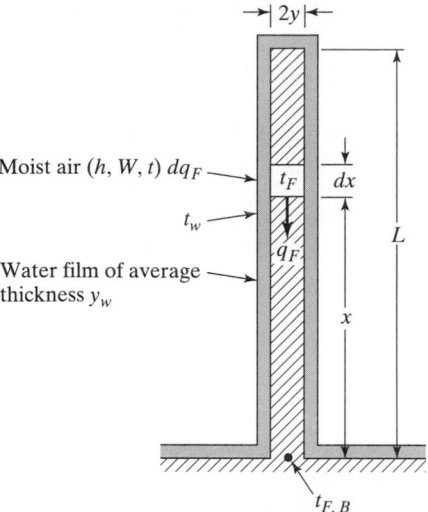

Moist air (h, W, t) dq_F

t_w

Water film of average thickness y_w

t_F

dx

q_F

L

x

$2y$

$t_{F,B}$

Figure 11.19 Schematic illustration of a bar fin wetted with moisture condensed from moist air.

But the quantity $(a_w + b_w t_F)$ has the dimensions of *moist air enthalpy*. Let us define a *fictitious air enthalpy* h_F as

$$h_F \equiv a_w + b_w t_F \qquad (11.50)$$

where the quantities a_w and b_w are evaluated at the water-air surface temperature t_w. Thus

$$dq_F = -\frac{2k_w}{b_w y_w}(h_{s,w} - h_F)\,dx \qquad (11.51)$$

By Eq. (11.46)

$$dq_F = -\frac{2h_{c,o}}{c_{pa}}(h - h_{s,w})\,dx \qquad (11.52)$$

By Eqs. (11.51) and (11.52), we obtain

$$dq_F = -\frac{2h_{o,w}}{b_w}(h - h_F)\,dx = -\frac{2h_{o,w}}{b_w}\Delta h_F\,dx \qquad (11.53)$$

where $\Delta h_F = (h - h_F)$ and

$$h_{o,w} = \frac{1}{c_{pa}/(b_w h_{c,o}) + y_w/k_w} \qquad (11.54)$$

By Eqs. (11.48) and (11.50), we have

$$q_F = \frac{2k_F y}{b_w}\frac{dh_F}{dx} = -2\frac{k_F y}{b_w}\frac{d\,\Delta h_F}{dx}$$

and thus

$$dq_F = -2\frac{k_F y}{b_w}\frac{d^2 \Delta h_F}{dx^2}\,dx \qquad (11.55)$$

By Eqs. (11.53) and (11.55)

$$\frac{d^2 \Delta h_F}{dx^2} = \frac{h_{o,w}}{k_F y}\Delta h_F \qquad (11.56)$$

The boundary conditions for Eq. (11.56) are $\Delta h_F = \Delta h_{F,B}$ at $x = 0$, and $d\,\Delta h_F / dx = 0$ at $x = L$.

Equation (11.56) and its boundary conditions are completely analogous to Eq. (11.27) and its boundary conditions. Thus, the solution of Eq. (11.56) has the same form as Eq. (11.27). Furthermore, if we define the efficiency of the wet fin as

$$\phi_w \equiv \frac{h - h_{F,m}}{h - h_{F,B}} = \frac{\Delta h_{F,m}}{\Delta h_{F,B}} \tag{11.57}$$

we find that

$$\phi_w = \frac{\tanh pL}{pL}$$

where $\quad p = \sqrt{h_{o,w}/(k_F y)}$

Thus, the solution for efficiency of the wet bar fin is of identical form to that of the dry bar fin. If we analyzed other types of fins and made similar substitutions, we would obtain analogous results. Thus we have the important conclusion that *solutions for efficiency of dry fins also apply for efficiency of wet fins (Eq. 11.57) if we substitute $h_{o,w}$ (Eq. 11.54) for the wet fin in place of $h_{c,o}$ for the dry fin.*

11.9 OVERALL HEAT-TRANSFER COEFFICIENT FOR A WET FINNED-TUBE HEAT EXCHANGER

In Sec. 11.5 we developed an expression for the overall coefficient U_o where the fin surfaces were *dry*. In this section we will develop an expression for the overall coefficient where the fins are wetted by moisture condensed from the air passing over the outside surface.

We may use Fig. 11.14 but imagine a refrigerant at temperature t_R in the tube instead of the hot fluid. We will assume that the thermal resistance of the tube wall is negligible and that the tube has a uniform temperature t_P. We will also assume that the fin and tube are covered by a thin film of water having an average thickness y_w. The air passing over the surface has an enthalpy h. We may write for the local rate of heat transfer inside the tube

$$\dot{Q} = h_i A_{P,i} (t_P - t_R) \tag{11.58}$$

By definition, let

$$b_R' = \frac{h_{s,P} - h_{s,R}}{t_p - t_R} \tag{11.59}$$

where $h_{s,P}$ and $h_{s,R}$ are fictitious enthalpies of saturated moist air evaluated at the respective temperatures t_P and t_R. By Eqs. (11.58) and (11.59), we obtain

$$\dot{Q} = \frac{h_i A_{P,i}}{b_R'} (h_{s,P} - h_{s,R}) \tag{11.60}$$

Based upon our development in Sec. 11.8 with $h_{o,w}$ given by Eq. (11.54), we have

$$\dot{Q} = \frac{h_{o,w}}{b_{w,P}} A_{P,o} (h - h_{s,P}) + \frac{h_{o,w}}{b_{w,m}} A_F (h - h_{F,m})$$

where $b_{w,P}$ is evaluated from Fig. 11.18 at the temperature of the surface of the water film on the tube and $b_{w,m}$ is evaluated at the mean surface temperature of the water film on the fin. Making the approximation $b_{w,P} = b_{w,m}$ and $h_{s,P} = h_{F,B}$, with Eq. (11.57) we have

$$\dot{Q} = \frac{h_{o,w}}{b_{w,m}} (A_{P,o} + \phi_w A_F)(h - h_{s,P}) \tag{11.61}$$

By definition of $U_{o,w}$, we may write

$$\dot{Q} = U_{o,w} A_o (h - h_{s,R}) \tag{11.62}$$

We may show by Eqs. (11.60)–(11.62) that

$$U_{o,w} = \cfrac{1}{\cfrac{b'_R A_o}{A_{P,i} h_i} + \cfrac{b_{w,m}(1 - \phi_w)}{h_{o,w}(A_{P,o}/A_F + \phi_w)} + \cfrac{b_{w,m}}{h_{o,w}}} \tag{11.63}$$

Equation (11.63) is of similar form to Eq. (11.39). Units of $U_{o,w}$ are Btu/hr · ft^2 of outside surface (Btu/lbm dry air enthalpy difference), kW per (m^2 of outside surface) (kJ/kg dry air enthalpy difference). In order to calculate $U_{o,w}$ by Eq. (11.63), we must first assume values of the mean water film surface temperature $t_{w,m}$ and of the pipe temperature t_P. These assumptions allow initial approximations to be made for $b_{w,m}$ and b'_R, respectively. After calculation of $U_{o,w}$, we must check the assumptions. We will now derive equations for these procedures. By Eqs. (11.58) and (11.62), we have for the pipe temperature

$$t_P = t_R + \frac{U_{o,w} A_o (h - h_{s,R})}{h_i A_{P,i}} \tag{11.64}$$

To establish a procedure for checking $t_{w,m}$, we begin by writing the relation

$$h - h_{F,m} = \phi_w (h - h_{s,P}) = \frac{b_{w,m} h_{c,o}}{h_{o,w} c_{pa}} (h - h_{s,w,m})$$

By Eqs. (11.60) and (11.62)

$$h - h_{s,P} = \left(1 - \frac{b'_R U_{o,w} A_o}{h_i A_{P,i}}\right)(h - h_{s,R})$$

Thus, we obtain

$$h_{s,w,m} = h - \frac{c_{pa} h_{o,w} \phi_w}{b_{w,m} h_{c,o}} \left(1 - \frac{b'_R U_{o,w} A_o}{h_i A_{P,i}}\right)(h - h_{s,R}) \tag{11.65}$$

Equation (11.65) allows determination of $t_{w,m}$ through calculation of the enthalpy of saturated air, $h_{s,w,m}$ at the same temperature.

We will now study the influences upon $U_{o,w}$ of the various quantities in Eq. (11.63). Some of these are shown in Ex. 11.6, which is a similar problem to Ex. 11.3.

EXAMPLE 11.6

The same heat-transfer surface as in Ex. 11.3 is used for cooling and dehumidifying moist air. Air conditions are 80 °F dry-bulb temperature, 68 °F thermodynamic wet-bulb temperature, and 14.696 psia barometric pressure. Refrigerant temperature is 45 °F. Assume that a film of water of 0.005 in. average thickness covers all of the outside surface. (a) Determine the individual thermal resistances and the overall heat-transfer coefficient $U_{o,w}$ if $h_{c,o}$ = 10.0 Btu/hr · ft² · °F, h_i = 600 Btu/hr · ft² · °F, and the fin material is aluminum, k_F = 120 Btu/hr · ft · °F; (b) rework part (a) but with copper fins, k_F = 223 Btu/hr · ft · °F; (c) rework part (a) but with $h_{c,o}$ = 5.0 Btu/hr · ft² · °F; and (d) rework part (a) but with h_i = 1200 Btu/hr · ft² · °F.

Solution: All dimensional factors are the same as for Ex. 11.3. We will first show the entire solution for part (a). In order to evaluate the inside-surface resistance, we will assume t_P = 55 °F. By Eq. (11.59) and with Table A.4E, we find

$$b'_R = \frac{23.229 - 17.653}{55 - 45} = 0.558 \text{ Btu/lbm}_a \cdot °F$$

Thus

$$R_{i,w} = \frac{b'_R A_o}{h_i A_{p,i}} = \frac{(0.558)(27.8)}{600} = 0.0259 \text{ (Btu/lbm}_a) \text{ hr} \cdot \text{ft}^2/\text{Btu}$$

In order to calculate $R_{F,w}$, we must first assume a mean water-film temperature. Assume $t_{w,m}$ = 60 °F. By Fig. 11.18, $b_{w,m}$ = 0.68 Btu/lbm$_a$ · °F. Also, it is satisfactory to assume that W is near 0.01 so we can write

$$c_{pa} = 0.24 + 0.444W = 0.24 + (0.444)(0.01) = 0.245 \text{ Btu/lbm}_a \cdot °F$$

For water from Table A.6E, k_w = 0.34 Btu/hr · ft · °F. By Eq. (11.54),

$$h_{o,w} = \cfrac{1}{\cfrac{0.245}{(0.68)(10)} + \cfrac{0.005}{(12)(0.34)}} = \frac{1}{0.0360 + 0.0012} = 26.9 \text{ Btu/hr} \cdot \text{ft}^2 \cdot °F$$

Thus

$$L\sqrt{\frac{h_{o,w}}{(k_F y)}} = 0.0541 \sqrt{\frac{26.9}{(120)(0.000417)}}$$

$$= 1.25$$

and we know that r_2/r_1 = 3.60. By Fig. 11.11, ϕ_w = 0.49. Thus

$$R_{F,w} = \frac{b_{w,m}(1 - \phi_w)}{h_{o,w}(A_{p,o}/A_F + \phi_w)} = \frac{(0.68)(0.51)}{(26.9)(0.529)} = 0.0245 \text{ (Btu/lbm}_a) \text{ hr} \cdot \text{ft}^2/\text{Btu}$$

$$R_{o,w} = \frac{b_{w,m}}{h_{o,w}} = \frac{0.68}{26.9} = 0.0253 \text{ (Btu/lbm}_a) \text{ hr} \cdot \text{ft}^2/\text{Btu}$$

$$R_{t,w} = R_{i,w} + R_{F,w} + R_{o,w} = 0.0757 \text{ (Btu/lbm}_a) \text{ hr} \cdot \text{ft}^2/\text{Btu}$$

$$U_{o,w} = \frac{1}{R_{t,w}} = 13.21 \text{ Btu/hr} \cdot \text{ft}^2 \text{ (Btu/lbm}_a)$$

We may now check our assumed values of t_P and $t_{w,m}$. By the psychrometric chart C-7E at the inlet air state, h = 32.30 Btu/lbm$_a$. By Eq. (11.64),

$$t_P = 45 + \frac{(13.21)(27.8)(32.30 - 17.65)}{600} = 54.0 °F$$

TABLE 11.6 Determination of Thermal Resistances and $U_{o,w}$ for Ex. 11.6

Component or Calculation	Part (a)	Part (b)	Part (c)	Part (d)
Inside-surface resistance, $R_{i,w} = b_R' A_o / A_{p,i} h_i$	0.0259	0.0259	0.0259	0.0129
Fin resistance, $R_{F,w} = \dfrac{b_{w,m}(1 - \phi_w)}{h_{o,w}(A_{p,o}/A_F + \phi_w)}$	0.0245	0.0104	0.0222	0.0245
Outside-surface resistance, $R_{o,w} = b_{w,m}/h_{o,w}$	0.0253	0.0253	0.0498	0.0253
Total resistance, $R_{t,w} = R_{i,w} + R_{F,w} + R_{o,w}$	0.0757	0.0616	0.0979	0.0627
$R_{i,w}$ in percent of $R_{t,w}$	34.2	42.0	26.4	20.6
$R_{F,w}$ in percent of $R_{t,w}$	32.4	16.9	22.7	39.1
$R_{o,w}$ in percent of $R_{t,w}$	33.4	41.1	50.8	40.4
$U_{o,w} = 1/R_{t,w}$, Btu/hr · ft^2 · Btu/lbm$_a$	13.2	16.2	10.2	15.9

which is close to our assumed value of 55 °F. We may next use Eq. (11.65) to check our assumed value for $t_{w,m}$. We find

$$h_{s,w,m} = 32.30 - \frac{(0.245)(26.9)(0.49)}{(0.68)(10.0)} \left[1 - \frac{(0.558)(13.21)(27.8)}{600} \right] \qquad (14.65)$$

$$= 27.7 \text{ Btu/lbm}_a$$

and we find that $t_{w,m}$ is 61.7 °F. Thus, our assumption of 60 °F should be close enough. Table 11.6 shows results for the complete problem.

Compared to Ex. 11.3 (Table 11.4), Table 11.6 shows that with wet fins, the inside-surface resistance $R_{i,w}$ and the fin resistance $R_{F,w}$ are more significant than when the fins are dry. Compared to aluminum fins, Table 11.6, part (b), shows that with copper fins, $U_{o,w}$ is increased by about 23 percent. Part (c) shows that reducing $h_{c,o}$ by a factor of 2 from 10.0 to 5.0 decreases $U_{o,w}$ by about 23 percent. Part (d) shows that increasing h_i by a factor of 2 from 600 to 1200 increases $U_{o,w}$ by about 21 percent.

11.10 MEAN AIR ENTHALPY DIFFERENCE FOR WET FINNED-TUBE HEAT EXCHANGERS

Equation (11.62) shows that when simultaneous cooling and dehumidifying occur, the *overall heat-transfer coefficient*, $U_{o,w}$, is based upon an *air enthalpy difference*. Furthermore, the enthalpy h is the true air enthalpy, but the quantity $h_{s,r}$ is a *fictitious enthalpy of saturated air* calculated at the refrigerant temperature. We now need to develop an expression for the mean enthalpy difference for a cooling and dehumidifying coil. The mean enthalpy difference Δh_m is defined by the equation

$$\dot{Q} = U_{o,w} A_o \, \Delta h_m \qquad (11.66)$$

We may recall that for dry fins where only sensible heat transfer occurs, Δt_m is given by the logarithmic mean temperature difference if the fluid temperature within the tubes remains constant. Furthermore, in Sec. 11.2 we observed that for a countercross-flow heat exchanger with more than two tube passes, and where the tube-side fluid temperature changed, it was generally acceptable to calculate Δt_m for pure counterflow.

Two common cases exist with cooling coils. One occurs where the coil serves as the evaporator of a direct-expansion refrigeration system. Here the refrigerant temperature remains essentially constant, and we would expect the logarithmic mean enthalpy difference to apply. The other case occurs where the refrigerant (chilled water, brine, etc.) temperature changes. However, counterflow is always desirable, and in almost all cases more than two tube passes are used. Thus, we would expect that in such cases the logarithmic mean enthalpy difference calculated for pure counterflow would be a sufficiently accurate approximation for Δh_m.

We can show that, with certain approximations, for pure counterflow the mean air enthalpy difference is given by

$$\Delta h_m = \frac{(h_1 - h_{s,R,2}) - (h_2 - h_{s,R,1})}{\ln\left(\dfrac{h_1 - h_{s,R,2}}{h_2 - h_{s,R,1}}\right)} \tag{11.67}$$

where h_1 and h_2 are, respectively, the true enthalpies, Btu/lbm$_a$ (kJ/kg$_a$), of the entering and leaving air stream, and $h_{s,R,1}$ and $h_{s,R,2}$ are, respectively, fictitious enthalpies, Btu/lbm$_a$ (kJ/kg$_a$), of saturated air calculated at the entering and leaving refrigerant temperatures. Equation (11.67) is restricted to cases where the refrigerant temperature change is small, since in the derivation it is necessary to assume the quantities a_R and b_R as constants in the relation $h_{s,R} = a_R + b_R t_R$. We must also ignore the term $\dot{m}_a\, dW h_{f,w}$ in Eq. (11.41).

11.11 OVERALL HEAT-TRANSFER PROBLEMS INVOLVING WET FINNED SURFACES

Practical cooling-coil design problems require solution of Eq. (11.66). As discussed in Sec. 11.6, it is usually necessary to include a deposit coefficient for the inside surface of the tubes in the calculation of the overall heat-transfer coefficient. Thus, for most cases, we would use

$$U_{o,w} = \frac{1}{\dfrac{b'_R A_o}{A_{P,i} h_i} + \dfrac{b'_R A_o}{A_{P,i} h_{d,i}} + \dfrac{b_{w,m}(1 - \phi_w)}{h_{o,w}(A_{p,o}/A_F + \phi_w)} + \dfrac{b_{w,m}}{h_{o,w}}} \tag{11.68}$$

Estimation of h_i in Eq. (11.68) generally poses little difficulty except in the case of evaporating refrigerants. As discussed in Sec. 2.24, correlations for h_i for boiling liquids are available, or we may resort to experiments. The wet-surface coefficient $h_{o,w}$ must be calculated by Eq. (11.54). In Eq. (11.54) the term y_w/k_w is usually small, so that an estimate of the water film thickness is not critical. However, in case of frost formation, the term y_w/k_w may be more important.

The convection heat-transfer coefficient $h_{c,o}$ in Eq. (11.54) is usually the controlling factor for $h_{o,w}$. As discussed in Sec. 11.6 for dry fins, direct experimental data are needed for an accurate estimate. Little information is available in published literature for $h_{c,o}$ for a wet cooling coil.

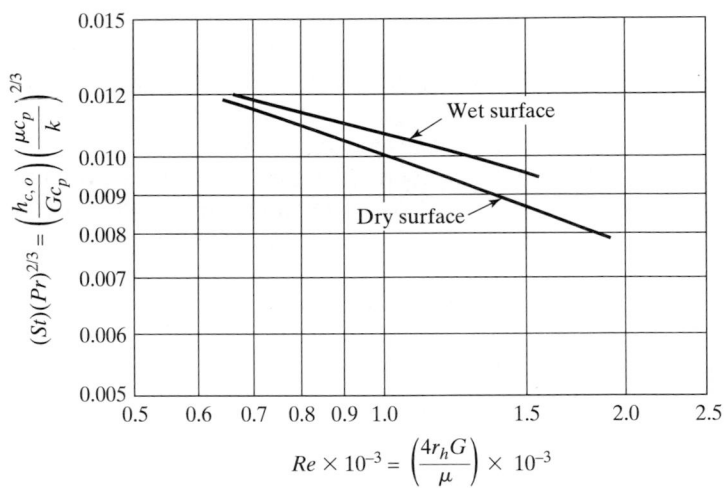

Figure 11.20 Dry-surface and wet-surface heat-transfer correlations for the external surface of a finned-tube cooling coil.

A comprehensive experimental study comparing the coefficient $h_{c,o}$ for a wet-surface cooling coil to that for the same coil operated without dehumidification was obtained by Meyers [4]. The experimental coil was similar to surface I of Table 11.5 and Fig. 11.15. Figure 11.20 shows the correlations for heat transfer to the external surface both for dry-surface operation (cooling without dehumidification) and for wet-surface operation (cooling and dehumidification). In all wet-surface calculations, a mean water film thickness of 0.004 in (0.1 mm) was assumed. Because of the presence of the water film, the core air velocity is higher for the wet coil than for the dry coil for a given face velocity.

Other experimental data for wet surfaces show similar trends. The water often condenses in dropwise form on what are called "nonwetting" surfaces. These water droplets tend to increase the roughness of the surface and increase the heat-transfer coefficient and the friction factor. The change is most pronounced at high air velocities (large Re), because the boundary layers become thinner as the air velocity increases. Other portions of the surface are wetted as a water film flows downward from regions above. The presence of grease or dirt can also change the behavior of the surface. Therefore it is very difficult to accurately predict the external heat-transfer coefficient for wet coils. An approximate method to estimate the wet-surface heat-transfer coefficient when the dry value is known for the same surface is to use the following empirical relation for the effective Reynolds number for the wet surface:

$$\text{Re}_{\text{wet}} = 4.6 \, \text{Re}^{3/4} \tag{11.69}$$

Then the data for the dry coil, such as those given in Fig. 11.15, can be used to estimate the performance of the same coil under wet conditions.

In calculation of $U_{o,w}$ by Eq. (11.68), attention must be given to the term $b_{w,m}$. It is expressed in units of Btu/lbm$_a$ · °F (kJ/kg$_a$ · °C) and represents the slope of a curve expressing the enthalpy of saturated air as a function of temperature. In our analyses, we assumed a linear relation between h_s and t_s which is permissible over a small range of t_s such as about 10 °F (5 °C) or less. The quantity $b_{w,m}$ should be evaluated at the mean water-film surface temperature. Where the mean water-film surface temperature change exceeds about 10 °F (5 °C) for the entire coil, it is more accurate to separate the coil depth into two or more parts and to treat each part separately. It is particularly interesting to analyze the influence of $b_{w,m}$ in Eqs. (11.63), (11.65), and (11.68). If the quantity y_w/k_w in Eq. (11.54) is of minor importance, $b_{w,m}$ mainly affects ϕ_w, since it is essentially cancelled in the term $b_{w,m}/h_{o,w}$. This circumstance is fortunate, and it is usually not necessary to make a precise evaluation of $b_{w,m}$.

Depending upon the circumstances of the problem, a cooling coil may operate with one or more rows of its initial external surface dry (no dehumidification) and with the remainder of its surface wet. If Eq. (11.65), when applied to inlet conditions, indicates a value of $t_{w,m}$ higher than the inlet-air dew-point temperature, then the initial surface will be dry. The procedures of Secs. 11.2–11.6 should be applied to this part of the coil.

For the dry section of a coil, we may show by procedures analogous to those used in obtaining Eq. (11.65) that

$$t_{F,m} = t - \phi\left[1 - \frac{U_o A_o}{h_i A_{p,i}}(t - t_R)\right] \tag{11.70}$$

When $t_{F,m}$ is equal to the inlet-air dew-point temperature $t_{d,1}$, we have

$$t = \frac{t_{d,1} - \phi(1 - U_o A_o/h_i A_{p,i})t_R}{1 - \phi(1 - U_o A_o/h_i A_{p,i})} \tag{11.71}$$

Equation (11.70) allows calculation of the mean fin temperature $t_{F,m}$ for a dry section for given values of air dry-bulb temperature t and refrigerant temperature t_R. Equation (11.71) allows calculation of air temperature t for the location where condensation just begins. It follows that the analysis for a wet coil would begin at this location.

An illustration of the fundamental calculation procedures for a fin-and-tube cooling coil is shown by the following example.

EXAMPLE 11.7

A direct-expansion, fin-and-tube cooling coil using Refrigerant 22 is to be sized to process 12,000 ft³/min of moist air from an initial condition of 82 °F dry-bulb temperature and 70 °F thermodynamic wet-bulb temperature, to a final condition of 54 °F dry-bulb temperature and 53 °F thermodynamic wet-bulb temperature. Barometric pressure is 14.696 psia. Refrigerant temperature is 40 °F. Type II surface (Table 11.5) is to be used. Inlet face velocity of the air is to be 500 ft/min. Experiments indicate that h_i may be estimated as 500 Btu/hr · ft² · °F. Determine (a) the required coil face area in ft², (b) the required total outside-surface area in ft², and (c) the number of rows of tubes required.

Solution:

(a) From the given data we may calculate the coil face area as 12,000/500 = 24 ft².

(b) The outside-surface area A_o must be found by solution of Eq. (11.66). We will assume that all of the coil surface is wet. The mean air enthalpy difference is given by Eq. (11.67). By Fig. C-8E, $h_1 = 33.98$ Btu/lbm$_a$ and $h_2 = 22.01$ Btu/lbm$_a$. By Table A.4E, $h_{s,R} = 15.233$ Btu/lbm$_a$. Thus

$$\Delta h_m = \frac{h_1 - h_2}{\ln\left(\dfrac{h_1 - h_{s,R}}{h_2 - h_{s,R}}\right)} = \frac{11.97}{\ln\left(\dfrac{18.75}{6.78}\right)} = 11.77 \text{ Btu/lbm}_a$$

The mean air enthalpy is then $h_{s,R} + \Delta h_m = 27.00$ Btu/lbm$_a$.

In order to find $U_{o,w}$ we must evaluate the individual quantities in Eq. (11.68). We will assume a mean pipe temperature $t_{p,m}$ of 50 °F and a mean water-film temperature $t_{w,m}$ of 54 °F as first approximations. By Eq. (11.59)

$$b'_R = \frac{20.306 - 15.233}{50 - 40} = 0.507 \text{ Btu/lbm}_a \cdot \text{°F}$$

By Fig. 11.18, $b_{w,m} = 0.60$ Btu/lbm$_a \cdot$ °F. An adequate estimate of c_{pa} is 0.245 Btu/lbm$_a \cdot$ °F. We will first determine a value of $h_{c,o}$ based upon dry-surface operation. We will use a mean air temperature of 68 °F. By Table A.5E, we obtain $\mu = 0.0439$ lbm/ft $\cdot$ hr and $c_p \mu/k = 0.709$. By Figure C-8E, $W_1 = 0.0130$ lbm$_w$/lbm$_a$ and $v_1 = 13.94$ ft^3/lbm$_a$. Thus

$$\dot{m}_a = \frac{(12,000)(60)}{13.94} = 51,600 \text{ lbm}_a/\text{hr}$$

By Table 11.5, the net area between the fins is 49.7 percent of the face area. Thus

$$G_a = \frac{51,600}{(0.497)(24)} = 4330 \text{ lbm}_a/\text{hr} \cdot \text{ft}^2$$

and

$$G = (1 + W_1) G_a = (1.013)(4330) = 4390 \text{ lbm/hr} \cdot \text{ft}^2$$

Thus the actual Reynolds number for the coil is

$$\text{Re} = \frac{(0.01268)(4390)}{0.0439} = 1268$$

By Eq. (11.69) we can determine the effective Reynolds number for the wet coil.

$$\text{Re}_{\text{wet}} = 4.6(1268)^{3/4} = 977$$

By Fig. 11.15 (surface II) we can estimate the outside-surface heat-transfer coefficient for wet operation:

$$\frac{h_{c,o}(c_p \mu/k)^{2/3}}{Gc_p} = 0.0088$$

Thus, for wet operation

$$h_{c,o} = \frac{(0.0088) G_a c_{pa}}{(c_p \mu/k)^{2/3}} = \frac{(0.0088)(4330)(0.245)}{(0.709)^{2/3}} = 11.7 \text{ Btu/hr} \cdot \text{ft}^2 \cdot \text{°F}$$

By Eq. (11.54) and assuming a water-film thickness of 0.005 in., we have

$$h_{o,w} = \frac{1}{\dfrac{0.245}{(0.60)(11.7)} + \dfrac{0.005}{(12)(0.33)}} = 27.7 \text{ Btu/hr} \cdot \text{ft}^2 \cdot \text{°F}$$

Thus (see Ex. 11.4)

$$L \sqrt{\frac{h_{o,w}}{k_F y}} = \frac{0.575}{12} \sqrt{\frac{(27.7)(2)(12)}{(120)(0.016)}} = 0.89$$

$$\frac{r_2}{r_1} = 2.70$$

By Fig. 11.11, $\phi_w = 0.68$. From Example 11.4, we know for the type II surface, $A_o/A_{p,i} = 19.31$, $A_{p,o}/A_F = 0.105$. By Table 2.3, $1/h_{d,i} = 0$. By Eq. (11.68),

$$U_{o,w} = \cfrac{1}{\cfrac{(0.507)(19.31)}{500} + \cfrac{(0.60)(0.32)}{(27.7)(0.785)} + \cfrac{0.60}{27.7}} = 20.0 \text{ Btu/hr} \cdot \text{ft}^2 \text{ (Btu/lbm}_a)$$

We may now check the assumed values of $t_{p,m}$ and $t_{w,m}$. By Eq. (11.64),

$$t_{p,m} = 40 + \frac{(20.0)(19.31)(11.77)}{500} = 49.1 \text{ °F}$$

which is sufficiently close to our assumed value of 50 °F. By Eq. (11.65),

$$h_{s,w,m} = 27.00 - \frac{(0.245)(27.7)(0.68)}{(0.60)(11.7)}\left[1 - \frac{(0.507)(20.0)(19.31)}{500}\right](27.00 - 15.23)$$

$$= 22.3 \text{ Btu/lbm}_a$$

and $t_{w,m}$ is about 53.6 °F which is very close to our assumed value of 54 °F. For the inlet conditions of the coil, by Eq. (11.65), we find that $t_{w,m}$ is about 60 °F. Since $t_{d,1} = 64.4$ °F, the coil surface is wet throughout.

The required total heat-transfer rate for the coil is

$$\dot{Q} = \dot{m}_a(h_1 - h_2) = (51,600)(33.98 - 22.01) = 618,000 \text{ Btu/hr}$$

By Eq. (11.66),

$$A_o = \frac{618,000}{(20.0)(11.77)} = 2625 \text{ ft}^2$$

(c) Table 11.5 shows that type II surface has an outside-surface area of 22.86 ft^2/(ft^2 of face area · row). Thus

$$\text{rows} = \frac{2625}{(22.86)(24)} = 4.78 \text{ or 5 rows}$$

Before leaving the solution of Ex. 11.7, we should check the mean water-film surface temperature through the coil to ascertain whether its variation exceeded about 10 °F. For the outlet conditions, we find by Eq. (11.65) that $h_{s,w,m} = 18.87$ Btu/lbm$_a$ and $t_{w,m} = 47$ °F approximately. Thus, the variation of $t_{w,m}$ is about 11 °F.

The solution of five rows of tubes for Ex. 11.7 is probably adequate. As a check to the solution, we could divide the coil into increments, such as two rows, and evaluate more accurate values of $b_{w,m}$ and $U_{o,w}$ for each increment. We may easily show that for Ex. 11.7

$$h_2 = h_{s,R} + (h_1 - h_{s,R})e^{-U_{o,w}A_o/\dot{m}_a}$$

where h_1 and h_2 are the respective enthalpies of the air entering and leaving the incremental coil and A_o is the surface area of the incremental coil. In this way, we may calculate h_2 for each increment except for the last increment, where the required *incremental* A_o would be calculated.

11.12 CALCULATION OF COOLING-COIL PERFORMANCE AT OTHER THAN DESIGN CONDITIONS

The calculation procedures given in Secs. 11.9–11.10 are adequate for calculating necessary surface area and the number of rows of tubes where operating conditions are known. Example 11.7 was such a problem, where, among other data, the final air state was specified. We may, however, encounter a problem where the coil surface area is known and we need to determine the final air state for various operating conditions. The fundamental procedure previously given would provide for determination of the final air enthalpy h_2 only. In this section we will study how the air humidity ratio W also may be determined throughout a cooling coil. Unfortunately, no convenient mathematical solution is available for W as was the case for h. However, Eq. (11.45) allows us to construct the process line on the h-W psychrometric chart for moist air passing through a cooling-and-dehumidifying coil.

We will assume that the inlet-air state, inlet-refrigerant state, the various heat-transfer coefficients, and the coil-surface data are known. If the refrigerant is a liquid which does not change phase, by alteration of Eq. (11.67) we may obtain that the enthalpy of the air leaving the coil is given by

$$h_2 = \frac{h_{s,R,1}(1 - e^{-(1-c_1)c_2}) + h_1(1 - c_1)e^{-(1-c_1)c_2}}{1 - c_1 e^{-(1-c_1)c_2}} \tag{11.72}$$

where $c_1 = \dot{m}_a b_R / \dot{m}_R c_{P,R}$ and $c_2 = U_{o,w} A_o / \dot{m}_a$, and $\dot{m}_a$ is mass-flow rate (dry basis) of the air, $\dot{m}_R$ is the mass-flow rate of the refrigerant, $c_{p,R}$ is the specific heat of the refrigerant, and other quantities are the same as those defined in Secs. 11.9 and 11.10. If the refrigerant temperature is constant, Eq. (11.72) reduces to

$$h_2 = h_{s,R} + (h_1 - h_{s,R})e^{-U_{o,w}A_o/\dot{m}_a} \tag{11.73}$$

The relationship between the mean water-film surface temperature $t_{w,m}$ and the air enthalpy h may be found from Eq. (11.65).

Equations (11.45), (11.65), and (11.72)–(11.73) in conjunction with the psychrometric chart provide a method for determining the change of air state through a cooling and dehumidifying coil. An illustration of the method is provided by Ex. 11.8.

EXAMPLE 11.8

A cooling coil having face dimensions of 3 ft by 6 ft is constructed of type II surface (Table 11.5). Refrigerant 22 evaporates in the tubes at a constant temperature of 40 °F. Air enters the coil at 82 °F dry-bulb temperature, 70 °F thermodynamic wet-bulb temperature, and 14.696 psia barometric pressure at a rate of 9000 ft³/min. For purposes of this example, it will be sufficient to assume the same fin efficiency and heat-transfer coefficients which were used for Ex. 11.7. Determine (a) the complete moist air condition line on the psychrometric chart for a coil of infinite depth, (b) the final air state for a four-row coil, and (c) the final air state if the coil was eight rows deep.

Solution:

(a) We will determine a general process line which would be applicable, regardless of the number of rows of tubes. From the given data, we obtain from the psychrometric chart,

---

$h_1 = 33.98$ Btu/lbm$_a$ and $h_{s,R} = 15.23$ Btu/lbm$_a$. Also, $\dot{m}_a = (9000)(60)/13.94 = 38,700$ lbm$_a$/hr. With the known data from Ex. 11.7, Eq. (11.65) becomes

$$h_{s,w,m} = 0.608h + 5.97$$

Since we may arbitrarily choose Δh, we may write

$$\Delta h_{s,w,m} = 0.608\,\Delta h$$

The process line may be obtained by graphical solution of Eq. (11.45). Corresponding to state 1 of the air, we have $h_{s,w,m} = (0.608)(33.98) + 5.97 = 26.63$ Btu/lbm$_a$ and $t_{w,m,1} = 60.2$ °F. Since $t_{d,1} = 64.4$ °F, the coil is wet throughout. We locate states 1 and $s,w,m,1$ on the psychrometric chart. With the chart protractor, we obtain $[(h - h_{s,w,m})/(W - W_{s,w,m})]_1 = 3800$ Btu/lbm$_w$. At 82 °F, $h_{g,t} = 1097$ Btu/lbm$_w$. By Eq. (9.7), we find Le $= 0.90$. By Eq. (11.45)

$$\left(\frac{dh}{dW}\right)_1 = (0.90)(3800) + 1097 - (1061)(0.90) = 3560 \text{ Btu/lbm}_w$$

With the aid of the chart protractor, we draw a short line segment of direction $(dh/dW)_1$ through state 1. At an arbitrary value of $h = 32.00$ Btu/lbm$_a$ we locate a new state (a) on this line with $t_a = 76.5$ °F. We then repeat the procedure until the condition line meets with the state of saturated air at the refrigerant temperature of 40 °F. Figure 11.21 shows the complete condition line.

(b) We know that the face area of the coil is 18 ft^2. By Table 11.5, the outside-surface area is 22.86 ft^2/ft^2 of face area · row. For four rows, $A_o = 1646$ ft^2, and $U_{o,w}A_o/\dot{m}_a = 0.872$. By Eq. (11.73),

$$h_2 = 15.23 + (18.75)e^{-0.872} = 23.08 \text{ Btu/lbm}_a$$

We locate state 2 at the intersection of $h_2 = 23.08$ Btu/lbm$_a$ with the process line. We obtain $t_2 = 56.6$ °F, $t_2^* = 54.9$ °F, and $W_2 = 0.00887$ lbm$_w$/lbm$_a$.

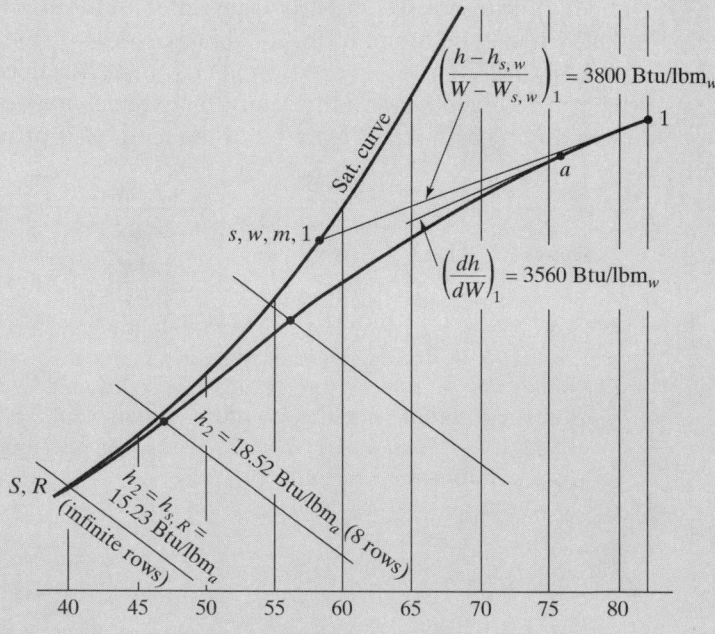

Figure 11.21 Solution of Ex. 11.8.

Example 11.8 shows how the psychrometric chart may be utilized in determining the performance of a given coil. This hand-calculation procedure is lengthy and tedious. However, the method, as well as all other procedures of this chapter, may be programmed for use with a small computer.

11.13 PRACTICAL CONSIDERATIONS INVOLVING FINNED-TUBE HEATING AND COOLING COILS

In previous sections of this chapter, procedures were presented for calculating rates of heat transfer in finned-tube coils used for heating and cooling moist air. These fundamental methods allow analysis of many of the individual variables affecting the heat-transfer performance of a coil. In the analyses, certain assumptions were necessary. In addition, the lack of information on the internal heat-transfer coefficient h_i for evaporating or condensing flows and the uncertainty of the external-surface heat-transfer coefficient $h_{c,o}$, make experimentation necessary in the design of a new type of coil surface.

In this chapter, no discussion was given on optimizing the construction of a finned-tube heat exchanger. However, in a practical design problem, attention must be given to determining the usefulness of inserts, correct fin shape and dimensions, optimum fin spacing, and the optimum air mass velocity.

Often the thermal performance of a particular heat exchanger can be improved significantly by adding an insert within the tubes which promotes turbulence and thereby increases the inner heat-transfer coefficient, h_i. The inner pressure drop is also increased, but the additional pumping power is usually offset by the reduction in fan power because a smaller coil can be used with the same capacity or heat-transfer duty. Inserts are produced in a variety of forms which add a twist or swirl to the flow to promote heat transfer. Often these are inserted after the core of the coil has been fabricated but before the headers are attached. In most cases the inserts can be removed for cleaning the tube interior surfaces.

When a fin is made larger, its efficiency decreases, so that beyond a certain size, an increase in dimensions does not appreciably increase the rate of heat transfer. For natural-convection coils where the coefficient $h_{c,o}$ is relatively small, fins may typically be larger than for forced-convection coils.

Decreasing the fin spacing may result in an increased rate of heat transfer per unit tube length provided that the boundary layer developed on one fin surface does not interfere with that of the adjacent fin surface. Thus, the length of the fins in the direction of air flow, and the air mass velocity, are important factors in determining the optimum fin spacing.

A particularly important variable is the air mass velocity G. The external-surface coefficient $h_{c,o}$ is primarily a function of G. An increase of air velocity is usually beneficial to heat transfer but is simultaneously detrimental because of the increase in pressure drop of the air flowing through the coil. The air mass velocity cannot be made arbitrarily large because the fan-power costs may become excessive. As is true with any heat exchanger, the optimum design must be a compromise which results in the highest heat-transfer capacity per unit of amortized costs. Costs must include operating expenditures as well as initial costs.

Cooling and dehumidifying coils create a good location for fungus and mold growth, as the temperature and humidity conditions are nearly optimum for some organisms to propagate. Careful cleaning of the wet coils and drain-pan surfaces and periodic use of fungicide is recommended to prevent the growth of organisms which can cause health problems for the occupants of buildings being supplied by air from the coils. Filtration of the air just before supplying it to the occupied zone is recommended to ensure safe, healthy conditions within a building where viable organism transport is a concern.

ENDNOTES

1. *Handbook of Heat Transfer Applications*, 2d ed., W. M. Rohsenow, J. P. Hartnett, and E. N. Ganié, eds. (New York: McGraw-Hill, 1985), ch. 4.
2. *Handbook of Heat Transfer Fundamentals*, 2d ed., W. M. Rohsenow, J. P. Hartnett, and E. N. Ganié, eds., (New York: McGraw-Hill, 1985).
3. W. M. Kays and A. L. London, *Compact Heat Exchangers* (Palo Alto, CA: The National Press, 1955).
4. Raymond J. Myers, "The Effect of Dehumidification on the Air Side Heat Transfer Coefficient for a Finned-tube Coil" (Master's thesis, University of Minnesota, 1967).

PROBLEMS

11.1 Two thousand cubic feet per minute of saturated moist air at 50 °F and a pressure of 14.696 psia is heated to 120 °F in a single-pass, cross-flow heat exchanger. Hot water enters at 180 °F at a volumetric flow rate of 30 gpm. The hot water is mixed, the air is unmixed. Determine:
(a) the leaving-water temperature,
(b) the value of the correction factor F,
(c) the value of LMTD,
(d) the heat-exchanger effectiveness,
(e) the fluid capacity-rate ratio,
(f) the value of NTU, and
(g) the overall UA product.

11.2 Two m³/s of moist air at 21 °C dry-bulb temperature, 13 °C wet-bulb temperature, at a pressure of 101.325 kPa is heated to 40 °C by hot water that enters at 70 °C and exits at 60 °C. The water is mixed, the air is unmixed. Determine:
(a) the mass flow rate of water required (kg/s),
(b) the value of NTU,
(c) the overall conductance of the heat exchanger (UA),

(d) the correction factor F,
(e) the logarithmic mean temperature difference (LMTD), and
(f) the heating capacity of the heat exchanger (kW).

11.3 Repeat Prob. 11.1 if the heating medium is condensing steam at a pressure of 10 psig that enters as saturated vapor and leaves as saturated liquid.

11.4 Repeat Prob. 11.2 if the heating medium is condensing refrigerant 22 at a pressure of 1.942 MPa that enters as saturated vapor and leaves as saturated liquid.

11.5 Compute the thickness ($2y$) of circular-plate fins needed to achieve a fin efficiency of 0.8 with an outer radius of the tube of 0.125 in., a fin length of 1.0 in., and an exterior-surface heat-transfer coefficient $h_{c,o} = 10$ Btu/hr · ft² · °F if the fin material is
(a) copper and
(b) aluminum.

11.6 Determine the tube row spacing (a) required to achieve a rectangular-plate fin efficiency of 0.7 when the tube outer radius is 4 mm, the tube spacing across the face of the coil is 2.0 cm, the external heat-transfer coefficient is $h_{c,o} = 50$ W/m$^2 \cdot$ °C, and the fin material is

(a) copper and

(b) aluminum.

11.7 An air heating coil utilizes circular-plate fins of uniform thickness. The aluminum fins are 0.010 in. thick and have a diameter of 2.0 in. There are 84 fins/ft. The copper tubing has an outside diameter of 0.530 in. and an inside diameter of 0.480 in. You may assume heat-transfer coefficients of 9.0 and 1000 Btu/hr $\cdot$ ft$^2 \cdot$ °F, respectively for the outside surface and inside surface. Calculate the overall heat-transfer coefficient U_o if deposit coefficients are neglected.

11.8 A heating coil utilizes aluminum rectangular-plate fins that are 0.1 mm thick with 4 fins/cm of tube length. The copper tubing outer diameter is 1.0 cm with a wall thickness of 0.5 mm. The tube center-to-center spacing is 2.5 cm, and the row center-to-center spacing is 2.0 cm. Hot water flows through the tubes with a mean velocity of 1 m/s and a mean temperature of 60 °C. It is estimated that $h_{c,o} = 50$ W/m$^2 \cdot$ °C. Compute the overall heat-transfer coefficient U_o for this coil if deposit coefficients are neglected.

11.9 Repeat Prob. 11.8 with the following exclusive modifications:

(a) change the fin spacing to 2 fins/cm and increase $h_{c,o}$ to 70 W/m$^2 \cdot$ °C,

(b) change the water mean velocity to 2 m/s, and

(c) change the tube row spacing from 2.0 cm to 3.0 cm.

11.10 A steam heating coil is constructed with rectangular-plate aluminum fins having a thickness of 0.01 in. There are eight fins per inch of tube length. The copper tubes have an outer diameter of 0.50 in. and an inner diameter of 0.44 in. Tube spacing is 1.5 in. across the face and 2.0 in. between rows. Based upon laboratory tests, it is estimated that $h_i = 1200$ Btu/hr $\cdot$ ft$^2 \cdot$ °F and that for an inlet face velocity of 500 ft/min, $h_{c,o} = 10.0$ Btu/hr $\cdot$ ft$^2 \cdot$ °F. Based upon saturated steam at 20 psia and saturated air entering the coil at 20 °F, calculate the final air temperature for a coil one row deep, two rows deep, and four rows deep. You may assume a deposit coefficient $1/h_{d,i}$ of 0.001 hr $\cdot$ ft$^2 \cdot$ °F/Btu on the steam side and a barometric pressure of 14.696 psia.

11.11 A heating coil is to be designed to heat 6000 ft^3/min of saturated moist air from 60 °F to 110 °F. Heating medium is water available at 180 °F. Type II surface (Table 11.5) is to be used. Barometric pressure

is 14.696 psia. The design is to be based upon an average water velocity of approximately 4 ft/sec, water temperature drop of 10 °F, and an inlet face velocity of the air of 600 ft/min. The tubes are to be connected to give a countercross-flow arrangement. Determine

(a) the face dimensions of the coil, and

(b) the number of rows of tubes required.

11.12 An uninsulated copper hot-water pipe with an outer diameter of 2.0 cm and an inner diameter of 1.8 cm contains water at 60 °C that flows with a velocity of 1 m/s. The exterior emissivity of the pipe is $\varepsilon = 0.2$. If the exterior convective heat-transfer coefficient is $h_{c,o} = 5$ W/m$^2 \cdot$ °C, and the ambient-air and surrounding-surface temperatures are 20 °C, determine the temperature drop of water per 100 m of pipe. Neglect deposit coefficients.

11.13 A steam pipe passes through a space where the ambient temperature is 80 °F. Temperature of the saturated steam inside the pipe is 250 °F. The wrought-iron pipe (2.375 in. O.D., 2.059 in. I.D.) is covered by insulation (2-in. thickness, $k = 0.042$ Btu/hr $\cdot$ ft $\cdot$ °F). The insulation has a highly reflective outer covering (ε negligibly small). Barometric pressure is 14.696 psia. The natural-convection coefficient $h_{c,o}$, Btu/hr $\cdot$ ft$^2 \cdot$ °F, for the outside surface of the insulation is given by the equation

$$h_{c,o} = 0.27 \left(\frac{\Delta t}{D_o}\right)^{0.25}$$

where Δt is the temperature difference between the outer surface and the surrounding air, °F, and D_o is the diameter, ft. Calculate the rate of heat loss from the pipe in Btu/hr $\cdot$ ft.

11.14 A single-row finned-tube coil having face dimensions of 18 in. by 24 in. was tested in the laboratory. The following test data were obtained.

t_1, °F	t_1^*, °F	t_2, °F	Inlet Air Flow, ft^3/min	Steam Pressure, psig
80	60	171.0	900	4.50
80	60	159.0	1200	4.50
80	60	151.0	1500	4.50
80	60	145.0	1800	4.50
80	60	140.0	2100	4.50
80	60	136.0	2400	4.50

The barometer reading for each test was 27.45 in. Hg.

(a) Plot a curve expressing $U_o A_o$ in Btu/hr · °F · ft² of face area as a function of face mass velocity in lbm/hr · ft².

(b) Consider the performance of a similar type coil but two rows deep. Assume saturated air entering at 20 °F, a face velocity of 500 ft/min, steam pressure of 10.0 psig, steam quality of 95 percent, and sea-level pressure. Calculate the temperature of the air leaving the coil.

11.15 Rework Prob. 11.5 if the fins are wet, the water film thickness can be neglected (dropwise condensation), and the mean fin temperature is 40 °F.

11.16 Rework Prob. 11.6 if the fins are wet, the water film thickness is estimated to be 1 mm, and the mean water film temperature is 10 °C.

11.17 Rework Prob. 11.6 if the fins are covered with a layer of frost with a thickness of 1 mm and a thermal conductivity of 1.0 W/m · °C. The mean temperature of the frost is −10 °C.

11.18 A cooling coil has the same basic fin-and-tube construction as in Prob. 11.10, except that it has four fins per inch of tube length. You may assume $h_i = 500$ Btu/hr · ft² · °F and $h_{c,o} = 8.0$ Btu/hr · ft² · °F. Air conditions are 0 °F, 100 percent relative humidity, and 14.696 psia barometric pressure. Refrigerant temperature is −12 °F. Determine the individual thermal resistances, their percent of the total resistance, and the overall heat-transfer coefficient $U_{o,w}$ if

(a) a uniform layer of frost ($k = 0.27$ Btu/hr · ft · °F) of 0.05 in. thickness covers all outside surfaces, and

(b) if the frost layer is of negligible thickness.

11.19 Rework Prob. 11.8 if the coil is used for cooling and dehumidifying moist air. Chilled water flows through the tubes with a mean temperature of 6 °C and a mean velocity of 1 m/s. The mean air conditions are 15 °C, saturated at a pressure of 101.325 kPa.

11.20 Derive Eq. (11.67).

11.21 Derive Eq. (11.72).

11.22 Rework Ex. 11.7 with chilled water as the refrigerant and a countercross-flow arrangement being used. Assume that water enters the coil at 40 °F and leaves at 48 °F. Assume a water velocity of approximately 7.0 ft/sec. Determine

(a) the required face dimensions of the coil, and

(b) the number of rows of tubes required.

11.23 A refrigerated storage room is maintained at an average uniform temperature of 0 °F by a bare-pipe direct-expansion coil mounted at ceiling level. Relative humidity in the room may be assumed at 100 percent. Barometric pressure is 14.696 psia. The pipes are arranged in a single horizontal bank. Total length of piping is 1000 ft. The wrought-iron pipes have an outside diameter of 2.375 in. and an inside diameter of 2.059 in. The pipes are covered with frost ($k = 0.27$ Btu/hr · ft · °F) having an average thickness of 1.0 in. Refrigerant evaporating temperature is −20 °F. It is estimated that $h_i = 500$ Btu/hr · ft² of inside pipe surface · °F and that the combined (radiation and convection) coefficient $h_o = 3.0$ Btu/hr · ft² of outside surface · °F.

(a) Determine the refrigerating capacity of the pipe coil in Btu/hr.

(b) Determine the capacity if all frost were scraped from the pipes. Assume other conditions remain the same.

11.24 A direct-expansion, fin-and-tube cooling coil is constructed of type II surface. The coil has a face area of 20 ft². It is used to process 10,000 ft³/min of moist air which enters the coil at 100 °F dry-bulb temperature and 70 °F thermodynamic wet-bulb temperature. Barometric pressure is 14.696 psia. Refrigerant temperature is 44 °F. You may assume that $h_{c,o}$ and h_i are 10.0 and 500 Btu/hr · ft² · °F, respectively.

(a) Construct the complete moist-air condition line on the psychrometric chart.

(b) Determine the dry-bulb temperature and thermodynamic wet-bulb temperature of the air leaving the coil if the coil depth is two rows, four rows, and eight rows.

SYMBOLS

A Surface area, ft² (m²); A_F refers to fin; A_i refers to inside of pipe; A_o refers to total outside surface.

A_c Net cross-sectional area, ft² (m²).

A_P Surface area of pipe, ft² (m²); $A_{P,i}$ refers to inside surface; $A_{P,m}$ refers to mean surface; $A_{P,o}$ refers to outside surface.

A_x Surface area per unit length in x direction, ft^2/ft (m^2/m).

a Row spacing, ft (m); Coefficient in Eq. (11.47), Btu/lbm$_a$ (kJ/kg$_a$); a_R evaluated at refrigerant temperature t_R; a_w evaluated at water-film temperature t_w.

b Coefficient in Eq. (11.47), Btu/lbm$_a$ · °F (kJ/kg$_a$ · °C); b_R evaluated at refrigerant temperature t_R; b_w evaluated at water-film temperature t_w; $b_{w,m}$ evaluated at mean water-film temperature $t_{w,m}$.

c_c Cold-fluid capacity rate, Btu/hr · °F (kW/°C).

c_1 $\dot{m}_a b_R / \dot{m}_h c_{p,h}$, dimensionless.

c_2 $\dot{U}_{o,w} A_o / \dot{m}_a$, dimensionless.

c_h Hot-fluid capacity rate, Btu/hr · °F (kW/°C).

c_{max} Maximum capacity rate, Btu/hr · °F (kW/°C).

c_{min} Minimum capacity rate, Btu/hr · °F (kW/°C).

$c_{p,c}$ Specific heat at constant pressure of cold fluid (usually air), Btu/lbm · °F (kJ/kg · °C).

c_{pa} Specific heat at constant pressure of moist air per unit mass of dry air, Btu/lbm$_a$ · °F (kJ/kg$_a$ · °C).

$c_{p,h}$ Specific heat at constant pressure of hot fluid, Btu/lbm · °F (kJ/kg · °C).

c_{pw} Specific heat at constant pressure of water, Btu/lbm · °F (kJ/kg · °C), $c_{p,w,f}$ refers to saturated liquid water.

c_r Capacity rate ratio, dimensionless.

F $\Delta t_m / \Delta t_{m,cf}$, dimensionless.

G Mass velocity of moist air, lbm/hr · ft^2 (kg/s · m^2).

G_a Mass velocity of dry air, lbm$_a$/hr · ft^2 (kg$_a$/s · m^2).

h Enthalpy of moist air, Btu/lbm$_a$ (kJ/kg$_a$); h_1 refers to entering air; h_2 refers to leaving air.

$h_{c,o}$ Convection heat-transfer coefficient for outside surface, Btu/hr · ft^2 · °F (kW/m^2 · °C).

$h_{D,o}$ Mass-transfer coefficient for outside surface, lbm$_w$/hr · ft^2 · lbm$_w$/lbm$_a$ (kg$_w$/s · m^2 · kg$_w$/kg$_a$).

$h_{d,i}$ Deposit coefficient for inside surface (see Table 2.3), Btu/hr · ft^2 · °F (kW/m^2 · °C).

h_F Fictitious enthalpy of moist air defined by Eq. (11.50), Btu/lbm$_a$ (kJ/kg$_a$); $h_{F,m}$ evaluated at mean fin temperature $t_{F,m}$ $h_{F,B}$ evaluated at temperature $t_{F,B}$ of fin base; $\Delta h_F = h - h_F$, etc.

$h_{f,w}$ Enthalpy of saturated liquid water at temperature t_w, Btu/lbm$_w$ (kJ/kg$_w$).

$h_{g,t}$ Enthalpy of saturated water vapor at air dry-bulb temperature, Btu/lbm$_w$ (kJ/kg$_w$).

h_g^0 Enthalpy of saturated water vapor at 0 °F (0 °C), Btu/lbm$_w$ (kJ/kg$_w$).

h_i Heat-transfer coefficient for inside surface, Btu/hr · ft^2 · °F (kW/m^2 · °C).

$h_{o,w}$ Heat-transfer coefficient, Btu/hr · ft^2 · °F (kW/m^2 · °C), defined by Eq. (11.54).

h_s Enthalpy of saturated moist air, Btu/lbm$_a$ (kJ/kg$_a$); $h_{s,P}$ evaluated at pipe temperature t_P; $h_{s,R}$ evaluated at refrigerant temperature t_R; $h_{s,w}$ evaluated at water-film temperature t_w; $h_{s,w,m}$ evaluated at mean water-film temperature $t_{w,m}$.

Δh_m Mean enthalpy difference for moist air, Btu/lbm$_a$ (kJ/kg$_a$).

K_1 $\dot{U}_o A_o / \dot{m}_c c_{pc}$, dimensionless.

K_2 $(\dot{m}_c c_{pc} / \dot{m}_h c_{p,h})(1 - e^{-K_1})$, dimensionless.

k Thermal conductivity, Btu/hr · ft^2 · °F/ft (kW/m^2 · °C/m); k_F refers to fin material; k_P refers to pipe material; k_w refers to water.

L Length, ft (m).

Le $h_{c,o} / h_{D,o} c_{pa}$, dimensionless.

L_x Total length in x direction, ft (m).

L_y Total length in y direction, ft (m).

$\dot{m}_h$ Mass rate of flow of hot fluid, lbm/hr (kg/s).

$\dot{m}_c$ Mass rate of flow of cold fluid, lbm/hr (kg/s).

$\dot{m}_a$ Mass rate of flow of dry air, lbm_a/hr (kg_a/s).

N Number of rows of tubes, dimensionless.

NTU   Number of Transfer Units, dimensionless.

P $(t_{c,o} - t_{c,i})/(t_{h,i} - t_{c,i})$, dimensionless.

p $\sqrt{h_{c,o}/ky}$ for dry fins, ft^{-1} (m^{-1}); $\sqrt{h_{o,w}/k_F y}$ for wet fins, ft^{-1} (m^{-1}).

$\dot{Q}$ Total heat-transfer rate, Btu/hr (kW).

q_F Rate of heat transfer in fins, Btu/hr (kW).

R $(t_{h,i} - t_{h,o})/(t_{c,o} - t_{c,i})$, dimensionless.

R Thermal-resistance term used with heat exchangers with dry fins, hr $\cdot$ ft^2 $\cdot$ °F/Btu ($\text{m}^2 \cdot$ °C/kW); R_i refers to inside surface; R_F refers to fins; R_P refers to pipe wall; R_o refers to outside surface; R_t is total value.

R_w Thermal-resistance term used with heat exchangers with wet fins, hr $\cdot$ ft^2 $\cdot$ Btu/lbm_a/Btu (m^2/kJ $\cdot$ kg_a/kW); $R_{i,w}$ refers to inside surface; $R_{F,w}$ refers to fins; $R_{o,w}$ refers to outside surface; $R_{t,w}$ refers to total value.

Re $4r_h G/\mu$ for fluid outside of tubes, dimensionless; $D_i G/\mu$ for fluid inside of tubes, dimensionless.

r_1 Outside radius of tube, ft (m).

r_2 Outside radius of circular fin, ft (m); $\sqrt{ac/\pi}$ for rectangular fin, ft (m) (see Fig. 11.13).

r_h LA_c/A_o, ft (m).

t Temperature, °F (°C); usually dry bulb value for air.

t_c Temperature of cold fluid, °F (°C); $t_{c,i}$ refers to entering fluid; $t_{c,o}$ refers to leaving fluid.

t^* Thermodynamic wet-bulb temperature of moist air, °F (°C).

t_F Temperature of fin, °F (°C); $t_{F,B}$ refers to base of fin; $t_{F,m}$ refers to mean fin temperature; $\Delta t_F = t_F - t$, etc.

t_f Temperature of fluid inside of tubes, °F (°C).

t_h Temperature of hot fluid, °F (°C); $t_{h,i}$ refers to entering fluid; $t_{h,o}$ refers to leaving fluid.

t_P Temperature of pipe or tube, °F (°C); $t_{P,i}$ refers to inside surface; $t_{P,o}$ refers to outside surface.

t_R Refrigerant temperature, °F (°C).

t_s Temperature of saturated moist air, °F (°C).

t_w Temperature of water film, °F (°C); $t_{w,m}$ refers to mean value.

Δt_m Mean temperature difference between fluids, °F (°C); $\Delta t_{m,cf}$ refers to pure counterflow.

Δt Temperature difference between entering and leaving fluids in a heat exchanger, °F (°C).

U_o Overall heat-transfer coefficient, Btu/hr $\cdot$ ft^2 $\cdot$ °F ($\text{kW/m}^2 \cdot$ °C).

$U_{o,w}$ Overall heat-transfer coefficient for heat exchangers with wet fins, Btu/hr $\cdot$ ft^2 $\cdot$ Btu/lbm_a ($\text{kW/m}^2 \cdot$ kJ/kg_a).

$\dot{V}$ Volumetric flow rate, ft^3/hr (m^3/s).

$\mathbf{V}_w$ Velocity of water, ft/sec (m/s).

v_a Specific volume of moist air, ft^3/lbm_a (m^3/kg_a).

v_w Specific volume of liquid water, ft^3/lbm_w (m^3/kg_w).

W Humidity ratio of moist air, $\text{lbm}_w/\text{lbm}_a$ (kg_w/kg_a); $W_{s,w}$ refers to saturated air at water-film temperature t_w.

x Distance along heat exchanger, ft (m).

x_P Thickness of pipe wall, ft (m).

| y | One-half of fin thickness, ft (m). |
| y_w | Thickness of water film, ft (m). |

Greek Letters

ε	Emissivity, dimensionless; effectiveness, dimensionless.
ρ_w	Density of water, lbm/ft^3 (kg/m^3).
μ	Fluid viscosity, lbm/ft · hr (kg/m · s).
ϕ	Fin efficiency for dry fins defined by Eq. (11.26), dimensionless.
ϕ_w	Fin efficiency for wet fins defined by Eq. (11.57), dimensionless.

chapter 12

Human Thermal Comfort and Indoor Air Quality

12.1 INTRODUCTION

A topic of fundamental importance in our study of thermal environmental engineering is the effect of the thermal environment upon people. From an engineering point of view, the human body may be likened to a heat engine. The body functions to convert the chemical energy of its food into work and heat. As is true with an engine, the harder we exercise or work, the more heat we reject. With the human body, heat rejection occurs primarily from the body surface. Through blood circulation, heat is transported to the skin, from which it is transferred to the environment. The body must continually reject heat, in summer as well as in winter. Temperature and humidity of the environment may profoundly influence the body's skin and interior temperature. Control of the thermal environment is necessary if comfort conditions are to be maintained or if physiological hazards are to be avoided in hot industries.

People are also affected by airborne contaminants. Gases, vapors, particles, and biological materials can cause adverse reactions or illness due to low-level exposure over long periods of time or higher levels of exposure over short periods. "Indoor air quality" has become the phrase that describes the control of indoor airborne contaminants. We will study recognized indoor air contaminants, documented human health implications, and various means of controlling indoor concentrations of these contaminants.

12.2 ENERGY BALANCE ON THE HUMAN BODY

A steady-state energy balance on the human body may be developed using environmental factors surrounding the body. Following the model developed by Fanger [1], we can set the rate of heat produced in the body equal to the total amount of heat rejected by the body to its environment.

$$\dot{Q}_M = \pm \dot{Q}_{\text{skin}} \pm \dot{Q}_{\text{respiration}}$$

$$= (\pm \dot{Q}_C \pm \dot{Q}_R \pm \dot{Q}_E)_{\text{skin}} + (\pm \dot{Q}_C \pm \dot{Q}_E)_{\text{respiration}} \tag{12.1}$$

where $\dot{Q}_M$ is the net rate of heat produced through metabolism, $\dot{Q}_{\text{skin}}$ is the heat loss (+) or gain (−) through the skin, and $\dot{Q}_{\text{respiration}}$ is the net heat loss (+) or gain (−) due to respiration. The subscript C designates convection, R radiation, and E evaporation. Note that there is no radiative term in the respiration heat loss or gain, as the lungs and air passages behave like a blackbody cavity for infrared radiation. The signs are significant, as the heat transfers may be to or from the body. In a hot, arid environment, for example, the skin may be receiving heat by convection and radiation, which must be dissipated in addition to the metabolized heat by evaporative cooling.

Equation (12.1) may be written for an unclothed body in the absence of solar radiation using usual heat- and mass-transfer approaches as

$$\dot{Q}_M = [h_C A_{\text{sk}}(t_{\text{sk}} - t) + h_R A_R(t_{\text{sk}} - \bar{t}_r) + h_E^v A_w(W_{s,\text{sk}} - W)h_{fg,\text{sk}}]_{\text{sk}}$$

$$+ [\dot{m}_{\text{res}} c_{pa}(t_{\text{ex}} - t) + \dot{m}_{\text{res}}(W_{\text{ex}} - W)h_{fg,b}]_{\text{res}} \tag{12.2}$$

The term h_C represents the average convective heat-transfer coefficient on the surface of the body, A_{sk} is the exposed skin surface area, t_{sk} is the average skin temperature, and t is the local air dry-bulb temperature. The radiative term uses a linearized radiative heat-transfer coefficient, h_R, and the net radiative surface area, A_R, which is less than the total exposed skin area, A_{sk}. The driving force for radiative heat transfer is the difference between the skin surface temperature and the local mean radiant temperature. The term for evaporative heat loss from the skin includes the evaporative mass-transfer coefficient, h_E^v, the wetted surface area of the body, A_w, the difference between the saturated humidity ratio evaluated at the skin temperature, $W_{s,\text{sk}}$, the humidity ratio of the surrounding moist air, W, and the latent heat of evaporation evaluated at the skin temperature, $h_{fg,\text{sk}}$. The sensible-heat transfer due to respiration is a function of the time-averaged dry air mass flow rate into the lungs, the moist air specific-heat capacity, and the difference between the inhaled-air temperature that equals the local dry-bulb temperature and the exhaled-air dry-bulb temperature. The latent-heat transfer due to respiration equals the time-averaged dry air mass flow rate into the lungs multiplied by the latent heat of evaporation evaluated at the body temperature, which is also multiplied by the difference between the exhaled-air humidity ratio and the inhaled-air humidity ratio.

Equation (12.2) is valid for an unclothed body. When clothing is considered, the heat transfer due to respiration remains unchanged but the heat transfer with the skin is modified by the clothing. The clothing adds an additional thermal resistance that is in series with the convective and radiative resistances from the surface of the clothing. The clothing also adds an additional resistance to the water transport that is in series with the exterior convective resistance. If we assume that the sensible- and latent-heat transfers are independent, and that the convective and radiative surface areas of the clothing are equal, we can set up the thermal network shown in Fig. 12.1. The outer surface area of the clothing, A_{cl}, is not necessarily equal to the skin surface area, A_b, nor is the exterior wetted area of clothing, $A_{\text{cl},w}$, necessarily equal to the wetted skin area, A_w. The clothing has a thermal resistance per unit inner surface area, R_{cl}/A_b, that is in series with the outer convective and radiative resistances. The total sensible-heat transfer can be determined by adding the convective and radiative components:

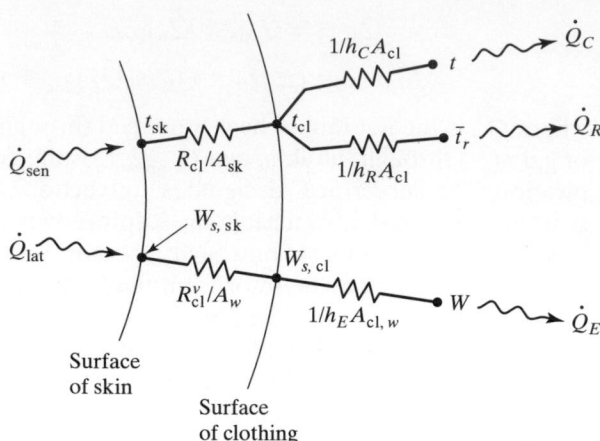

Figure 12.1 Thermal network near the skin of a clothed body.

$$\dot{Q}_{\text{sen}} = \dot{Q}_C + \dot{Q}_R$$

$$= [h_C(t_{\text{cl}} - t) + h_R(t_{\text{cl}} - \bar{t}_r)]A_{\text{cl}} \tag{12.3}$$

We can combine the convective coefficients into a single heat-transfer coefficient

$$h = h_C + h_R \tag{12.4}$$

and define the *operative temperature*, t_o, as

$$t_o \equiv \frac{h_C t + h_R \bar{t}_r}{h} \tag{12.5}$$

In some cases the operative temperature is set equal to the arithmetic average of the dry-bulb and mean radiant temperatures. This implies that the radiative and convective heat-transfer coefficients are equal. We can rewrite Eq. (12.3) as

$$\dot{Q}_{\text{sen}} = h(t_{\text{cl}} - t_o)A_{\text{cl}} \tag{12.6}$$

Placing the outer thermal resistance in series with the clothing resistance allows us to eliminate the clothing surface temperature, t_{cl}, from the heat-transfer equation:

$$\dot{Q}_{\text{sen}} = \frac{A_{\text{sk}}(t_{\text{sk}} - t_o)}{R_{\text{cl}} + \dfrac{A_{\text{sk}}}{hA_{\text{cl}}}} \tag{12.7}$$

In a similar manner, we can combine the clothing moisture resistance with the outer convective resistance to write an equation for the total latent-heat transfer from the skin:

$$\dot{Q}_{\text{lat}} = \frac{A_w(W_{s,\text{sk}} - W)h_{fg,\text{sk}}}{R_{\text{cl}}^v + \dfrac{A_w}{h_E^v A_{\text{cl},w}}} \tag{12.8}$$

Equations (12.7) and (12.8) can be combined with the respiration portion of Eq. (12.2) to yield the following heat-balance equation for a clothed body:

$$\dot{Q}_M = \frac{A_{sk}(t_{sk} - t_o)}{R_{cl} + \dfrac{A_{sk}}{hA_{cl}}}$$

$$+ \frac{A_w(W_{s,sk} - W)h_{fg,sk}}{R_{cl}^v + \dfrac{A_w}{h_E^v A_{cl,w}}}$$

$$+ \dot{m}_{res}[c_{pa}(t_{ex} - t) + h_{fg,b}(W_{ex} - W)] \tag{12.9}$$

The values for the terms included in Eq. (12.9) are given in the succeeding paragraphs. The total amount of metabolic heat that must be dissipated is given in Table 12.1 for a variety of human activities. The values in the table are given as the metabolic heat-generation rate divided by the skin surface area of an average adult male. This area is termed the *DuBois surface area* and is equal to 19.6 ft^2 (1.8 m^2). The metabolic heat-generation rate is often normalized by the metabolic rate of a seated adult. This normalized metabolic rate is given the unit of a "met," which is equal to 18.4 Btu/hr · ft^2 (58.2 W/m^2).

The time-averaged normal skin temperature, t_{sk}, in a neutral comfort state has been found to be 92.7 °F (33.7 °C) and the temperature of the body, t_b, to be 98.2 °F (36.8 °C). This gives the following values: $W_{s,sk} = 0.03406$ lbm$_w$/lbm$_a$ (kg$_w$/kg$_a$), $h_{fg,sk} = 1041.16$ Btu/lbm$_w$ (2421.79 kJ/kg$_w$), and $h_{fg,b} = 1038.0$ Btu/lbm$_w$ (2414.34 kJ/kg$_w$).

The linearized radiative heat-transfer coefficient, h_R, is given the value 0.83 Btu/hr · ft^2 · °F (4.7 W/m^2 · °C) by ASHRAE [2]. This value should suffice for normal nonmetallic clothing for which the emissivity in the infrared spectrum is near 1.0. The average convective coefficient over a body is a function of the position of the person and the relative air-velocity direction and magnitude. The correlation given by Mitchell [3] for a seated person is

$$h_C = 0.55 \text{ Btu/hr} \cdot \text{ft}^2 \cdot °F \quad (0 \leq \mathbf{V} \leq 40)$$

$$= 0.061 \mathbf{V}^{0.6} \text{ Btu/hr} \cdot \text{ft}^2 \cdot °F \quad (40 \leq \mathbf{V} \leq 800) \tag{12.10a}$$

where $\mathbf{V}$ is the relative air velocity in units of ft/min. In SI units this becomes

$$h_C = 3.1 \text{ W/m}^2 \cdot °C \quad (0 \leq \mathbf{V} \leq 0.2)$$

$$= 8.3 \, V^{0.6} \text{ W/m}^2 \cdot °C \quad (0.2 \leq \mathbf{V} \leq 4.0) \tag{12.10b}$$

where $\mathbf{V}$ is in units of m/s. A correlation developed for active people in still air by Gagge et al. [4] is

$$h_C = (M - 0.85)^{0.39} \text{ Btu/hr} \cdot \text{ft}^2 \cdot °F \quad (1.1 \leq M \leq 3.0) \tag{12.11a}$$

or in SI units

$$h_C = 5.7(M - 0.85)^{0.39} \text{ W/m}^2 \cdot °C \quad (1.1 \leq M \leq 3.0) \tag{12.11b}$$

where M is the normalized metabolic rate in units of met. Evaluating h_R and h_C is important not only because of their influence on the energy balance of a body but also because they are needed to determine the operative temperature, t_o.

TABLE 12.1 Typical Metabolic Heat Generation for Various Activities

	$Btu/hr \cdot ft^2$	W/m^2	met[a]
Resting			
Sleeping	13	40	0.7
Reclining	15	45	0.8
Seated, quiet	18	60	1.0
Standing, relaxed	22	70	1.2
Walking(on level surface)			
2.9 ft/s (0.88 m/s)	37	115	2.0
4.4 ft/s (1.3 m/s)	48	150	2.6
5.9 ft/s (1.8 m/s)	70	220	3.8
Office activities			
Reading, seated	18	55	1.0
Writing	18	60	1.0
Typing	20	65	1.1
Filing, seated	22	70	1.2
Filing, standing	26	80	1.4
Walking about	31	100	1.7
Lifting, packing	39	120	2.1
Driving/flying			
Car driving	18–37	60–115	1.0–2.0
Aircraft, routine	22	70	1.2
Aircraft, instrument landing	33	105	1.8
Aircraft, combat	44	140	2.4
Heavy vehicle	59	185	3.2
Miscellaneous occupational activities			
Cooking	29–37	95–115	1.6–2.0
Housecleaning	37–63	115–200	2.0–3.4
Seated, heavy limb movement	41	130	2.2
Machine work			
sawing (table saw)	33	105	1.8
light (electrical industry)	37–44	115–140	2.0–2.4
heavy	74	235	4.0
Handling 110-lb (50-kg) bags	74	235	4.0
Pick-and-shovel work	74–88	235–280	4.0–4.8
Miscellaneous leisure activities			
Social dancing	44–81	140–255	2.4–4.4
Calisthenics/exercise	55–74	175–235	3.0–4.0
Tennis, singles	66–74	210–270	3.6–4.0
Basketball	90–140	290–440	5.0–7.6
Wrestling, competitive	130–160	410–505	7.0–8.7

SOURCE: Adapted from *ASHRAE Fundamentals 1993*, p. 8.7.

[a] 1 met = 18.43 Btu/hr $\cdot$ ft² (58.2 W/m²).

EXAMPLE 12.1

Determine the radiative and convective heat-transfer coefficients for a person walking 2 mph (2.9 ft/s) in a building. Evaluate the operative temperature if the dry-bulb temperature is 75 °F and the mean radiant temperature is 80 °F.

Solution: We will use the standard radiative heat-transfer coefficient of 0.83 Btu/hr $\cdot$ ft² $\cdot$ °F. The convective heat-transfer coefficient will be determined using Eq. (12.11a), assuming the air in the building is nearly motionless. The value of M from Table 12.1 is 2.0 met. Therefore, the value of the convective heat-transfer coefficient becomes

$$h_C = (2 - 0.85)^{0.39} = 1.32 \text{ Btu/hr} \cdot \text{ft}^2 \cdot {}^\circ\text{F}$$

The operative temperature from Eqs. (12.4) and (12.5) becomes

$$t_o = \frac{1.32 \times 75 \,{}^\circ\text{F} + 0.83 \times 80 \,{}^\circ\text{F}}{1.32 + 0.83} = 76.9 \,{}^\circ\text{F}$$

The mass-transfer coefficient for water vapor, h_E^v, can be evaluated from the convective heat-transfer coefficient by using the Lewis relation (see Chapter 9):

$$h_E^v = h_C / \text{Le } c_{pa} \qquad (12.12)$$

In most applications the Lewis number can be set equal to 0.895.

The wetted skin surface area is usually given as a fraction of the total skin area, $w = A_w / A_{sk}$. The value for w varies from about 0.06 for normal skin moisture loss to about 0.5 for comfortable conditions.

EXAMPLE 12.2

Determine the heat transfers with the skin of an unclothed body for the following conditions: the person is seated, the air velocity is 0.3 m/s, the dry-bulb temperature is 30 °C, the mean radiant temperature is 39 °C, the dew-point temperature is 20 °C, and the skin wettedness fraction is 0.3.

Solution: Using Eq. (12.2), neglecting the respiratory heat transfer, and setting the convective and radiative skin surface areas equal,

$$\dot{Q}_{sk} = A_{sk}[h_C(t_{sk} - t) + h_R(t_{sk} - \bar{t}_r) + h_E^v w(W_{s,sk} - W)h_{fg,sk}]$$

We will use the DuBois skin surface area, $A_{sk} = 1.8 \text{ m}^2$. The convective coefficient is obtained from Eq. (12.10b):

$$h_C = 8.3(0.3)^{0.6} = 4.03 \text{ W/m}^2 \cdot {}^\circ\text{C}$$

The radiative heat-transfer coefficient will be set equal to the usual value, 4.7 W/m² · °C. The evaporative heat-transfer coefficient from Eq. (12.12) becomes

$$h_E^v = \frac{4.03 \text{ W/m}^2 \cdot {}^\circ\text{C}}{0.895 \times 1.007 \text{ kJ/kg} \cdot {}^\circ\text{C}} = 4.47 \text{ W kg}_a/\text{m}^2 \cdot \text{kJ}$$

From Tables A.4SI and A.1SI, respectively, assuming a skin temperature of 33.7 °C,

$$W_{s,sk} = 0.03406 \text{ kg}_w/\text{kg}_a, \qquad W = W_{s,20{}^\circ\text{C}} = 0.014758 \text{ kg}_w/\text{kg}_a$$

$$h_{fg,sk} = 2421.79 \text{ kJ/kg}_w$$

The convective heat transfer is

$$\dot{Q}_C = (1.8 \text{ m}^2)(4.03 \text{ W/m}^2 \cdot {}^\circ\text{C})(33.7 - 30) \,{}^\circ\text{C} = 26.84 \text{ W}$$

The radiative heat transfer is

$$\dot{Q}_R = (1.8 \text{ m}^2)(4.7 \text{ W/m}^2 \cdot {}^\circ\text{C})(33.7 - 39) \,{}^\circ\text{C} = -44.84 \text{ W}$$

The evaporative heat transfer is

$$\dot{Q}_E = (1.8 \text{ m}^2)(4.47 \text{ W kg}_a/\text{m}^2 \cdot \text{kJ})0.3(0.03406 - 0.014758)\text{kg}_w/\text{kg}_a$$
$$\times\ 2421.79 \text{ kJ/kg}_w = 112.83 \text{ W}$$

In the preceding example, the high mean radiant temperature provided a heat gain to the skin surface. However, this was more than offset by the convective and evaporative heat losses. The importance of the evaporative heat loss under hot conditions is illustrated. Note that as the ambient dew-point temperature increases, the potential to remove heat by evaporative cooling decreases significantly.

The addition of clothing introduces several factors: clothing thermal resistance, moisture resistance, and a change in wet and dry surface areas for convection, radiation, and moisture transport. Table 12.2 gives values for the thermal resistance of various clothing ensembles. The units are those of clo, where 1 clo = 0.88 hr · ft^2 · °F/Btu (0.155 m^2 · °C/W). The values for the clothing alone are given the symbol I_{cl}, and those for the total thermal resistance between the skin and the environment are given the symbol I_T. The outside surface area of the clothing should be used rather than the skin area when

TABLE 12.2 Typical Insulation and Permeability Values for Clothing Ensembles

Ensemble Description	I_{cl} (clo)	I_T (clo)	A_{cl}/A_{sk}	i_{cl}	i_T
Walking shorts, short-sleeve shirt	0.36	1.02	1.10	0.34	0.42
Trousers, short-sleeve shirt	0.57	1.20	1.15	0.36	0.43
Trousers, long-sleeve shirt	0.61	1.21	1.20	0.41	0.45
Same as above plus suit jacket	0.96	1.54	1.23		
Same as above plus vest and t-shirt	1.14	1.69	1.32	0.32	0.37
Trousers, long-sleeve shirt, long sleeve sweater, t-shirt	1.01	1.56	1.28		
Same as above plus suit jacket and long underwear bottoms	1.30	1.83	1.33		
Sweat pants, sweat shirt	0.74	1.35	1.19	0.41	0.45
Long-sleeve pajama top, long pajama trousers, short 3/4 sleeve robe, slippers, no socks	0.96	1.50	1.32	0.37	0.41
Knee-length skirt, short-sleeve shirt, pantyhose, sandals	0.54	1.10	1.26		
Knee-length skirt, long-sleeve shirt, full slip, pantyhose	0.67	1.22	1.29		
Knee-length skirt, long-sleeve shirt, half slip, panty hose, long-sleeve sweater	1.10	1.59	1.46		
Same as above, replace sweater with suit jacket	1.04	1.60	1.30	0.35	0.40
Ankle-length skirt, long-sleeve shirt, suit jacket, pantyhose	1.10	1.59	1.46		
Long-sleeve coveralls, t-shirt	0.72	1.30	1.23		
Overalls, long-sleeve shirt, t-shirt	0.89	1.46	1.27	0.35	0.40
Insulated coveralls, long-sleeve thermal underwear, long underwear bottoms	1.37	1.94	1.26	0.35	0.39

SOURCE: Adapted from *ASHRAE Fundamentals 1993*, p. 8.9.

computing the heat transfer. The surface-area ratio between the total external surface area of the clothing and the skin area, A_{cl}/A_{sk}, is also given in Table 12.2. The ratio of the external wet area to the skin wetted area is difficult to determine, owing to wicking of sweat along clothing fibers. However, a conservative estimate is to set the external wetted area ratio of the clothing equal to w for the skin alone. The last two columns in Table 12.2 give the dimensionless ratio of the clothing moisture resistance to the thermal resistance. The moisture resistance of the clothing can be calculated by

$$R_{cl}^v = 0.21 \, R_{cl}/i_{cl} \text{ hr} \cdot \text{ft}^2/\text{lbm}_a \qquad (12.13a)$$

where R_{cl} has units of hr · ft^2 · °F/Btu, or in SI units

$$R_{cl}^v = 0.90 \, R_{cl}/i_{cl} \text{ s} \cdot \text{m}^2/\text{kg}_a \qquad (12.13b)$$

where R_{cl} has units of m^2 · °C/W. The total moisture resistance between the skin and the ambient can be obtained by replacing i_{cl} by i_T in Eq. (12.13). The values in Table 12.2 are valid when the mean radiant temperature equals the dry-bulb temperature and the air velocity is less than 40 ft/min (0.2 m/s). Values for a nude body are given as $I_T = 0.72$ clo and $i_T = 0.48$.

EXAMPLE 12.3

Compute the sensible- and latent-heat loss from the skin of a person engaged in typical office work with a surrounding operative temperature of 75 °F and a relative humidity of 50 percent. The person is wearing a knee-length skirt, long-sleeve shirt, full slip, and pantyhose and has a wetted skin fraction of 0.1.

Solution: Equation (12.7) will be rewritten as

$$\dot{Q}_{sen} = \frac{A_{cl}(t_{sk} - t_o)}{R_T}$$

The skin surface area and temperature will be assumed to equal the usual values of 19.6 ft^2 and 92.7 °F, respectively. The total thermal resistance value can be computed using Table 12.2:

$$R_T = I_T \times 0.88 \text{ hr} \cdot \text{ft}^2 \cdot °F/\text{Btu}$$

$$R_T = 1.22 \times 0.88 = 1.07 \text{ hr} \cdot \text{ft}^2 \cdot °F/\text{Btu}$$

We use the total exterior clothing area as the surface area, so from Table 12.2

$$A_{cl} = A_{sk}\left(\frac{A_{cl}}{A_{sk}}\right)$$

$$= (19.6 \text{ ft}^2)(1.29) = 25.3 \text{ ft}^2$$

The total sensible-heat transfer area can be determined:

$$\dot{Q}_{sen} = (25.3 \text{ ft}^2)(92.7 - 75) °F/1.07 \text{ hr} \cdot \text{ft}^2 \cdot °F/\text{Btu} = 419 \text{ Btu/hr}$$

The total latent-heat transfer is determined by rewriting Eq. (12.8):

$$\dot{Q}_{lat} = A_{cl, w}(W_{s, sk} - W)h_{fg, sk}/R_T^v$$

The wetted clothing area is assumed to equal the wetted skin area:

$$A_{cl, w} = A_{cl}(A_w/A_{cl}) = 25.3 \text{ ft}^2 \times 0.1 = 2.53 \text{ ft}^2$$

The saturated humidity ratio at the skin temperature from Table A.4E is $W_{s,sk} = 0.03410$ lbm$_v$/lbm$_a$. The ambient humidity ratio is read from the psychrometric chart C-8E as $W = 0.0092$ lbm$_v$/lbm$_a$. The value for $h_{fg,sk}$ at the skin surface temperature of 92.7 °F from Table A1.E is 1041.8 Btu/lbm$_v$. The total moisture resistance between the skin and the ambient by modifying Eq. (12.13a) is

$$R_T^v = \frac{0.21 R_T}{i_T}$$

A value for i_T is not given in Table 12.2. ASHRAE suggests using the value $i_T = 0.4$ when more exact values are not available, so we will assume $i_T = 0.4$ here. The total moisture resistance then becomes

$$R_T^v = \frac{0.21(1.07)}{0.4} = 0.56 \text{ hr} \cdot \text{ft}^2/\text{lbm}_a$$

Substituting into the equation for latent-heat loss,

$$\dot{Q}_{lat} = (2.53 \text{ ft}^2)(0.03410 - 0.0092)\text{lbm}_v/\text{lbm}_a(1041.8 \text{ Btu/lbm}_v/0.56 \text{ hr} \cdot \text{ft}^2/\text{lbm}_a$$

$$= 117 \text{ Btu/hr}$$

The total heat loss from the skin of this person is estimated to be the sum of the sensible- and latent-heat losses or $\dot{Q}_{sk} = \dot{Q}_{sen} + \dot{Q}_{lat} = 419 + 117 = 536$ Btu/hr. A metabolic rate of 536 Btu/hr/19.6 ft^2 = 27.3 Btu/hr $\cdot$ ft^2 or 27.3 Btu/hr $\cdot$ ft^2/(18.43 Btu/hr $\cdot$ ft^2/met) = 1.5 met would be required to provide this amount of energy.

Heat transfer between the body and its environment by respiration is also an important factor. A common example is the panting of dogs to remove additional heat after strenuous activity. In humans, the heat transfer by respiration can be calculated using the last portion of Eq. (12.2). The parameters that must be known are the respiration mass flow rate, $\dot{m}_{res}$, the exhaled-air dry-bulb temperature, t_{ex}, and the exhaled-air humidity ratio, W_{ex}.

The respiration mass flow rate is a function primarily of the metabolic rate for a normal, healthy adult. This mass flow rate, or pulmonary ventilation rate, can be estimated by

$$\dot{m}_{res} = KM \tag{12.14}$$

where $\dot{m}_{res}$ has units of lbm$_a$/hr (kg$_a$/s), the proportionality constant K equals 1.19 lbm$_a$/hr $\cdot$ met (1.5×10^{-4} kg$_a$/s $\cdot$ met), and M is the metabolic rate in units of met (1 met = 18.43 Btu/hr $\cdot$ ft^2 = 58.2 W/m^2).

For typical indoor activities and environments, the respired air leaving the body is nearly saturated and near the body temperature. Empirical relations for t_{ex} and W_{ex} developed by Fanger [1] can be written as

$$t_{ex} = 88.6 + 0.066t + 57.6W \tag{12.15a}$$

$$W_{ex} = 0.0265 + 3.6 \times 10^{-5}t + 0.2W \tag{12.16a}$$

in English units and

$$t_{ex} = 32.6 + 0.066t + 32W \tag{12.15b}$$

$$W_{ex} = 0.0277 + 6.5 \times 10^{-5}t + 0.2W \tag{12.16b}$$

in SI units.

EXAMPLE 12.4

Determine the sensible- and latent-heat loss by respiration from the person and environmental conditions given in Ex. 12.3. Assume the metabolic activity rate for this person is 2 met.

Solution: From Eq. (12.2) the sensible- and latent-heat losses due to respiration can be written as

$$\dot{Q}_{sen} = \dot{m}_{res} c_{pa} (t_{ex} - t)$$

$$\dot{Q}_{lat} = \dot{m}_{res} (W_{ex} - W) h_{fg,b}$$

The respiration mass flow rate is determined from Eq. (12.14):

$$\dot{m}_{res} = (1.19 \text{ lbm}_a/\text{hr} \cdot \text{met})(2 \text{ met}) = 2.38 \text{ lbm}_a/\text{hr}$$

From Eq. (12.15a)

$$t_{ex} = 88.6 + 0.066(75) + 57.6(0.0092) = 94.1 \text{ °F}$$

From Eq. (12.16a)

$$W_{ex} = 0.0265 + 3.6 \times 10^{-5}(94.1) + 0.2(0.0092) = 0.0317 \text{ lbm}_v/\text{lbm}_a$$

The value for c_{pa} using Eq. (7.23) for the average between the inhaled and exhaled moist air is

$$c_{pa} = 0.240 + 0.444(0.0092 + 0.0317)/2 = 0.249 \text{ Btu/lbm}_a \cdot \text{°F}$$

The value for h_{fg} evaluated at the body temperature of 98.2 °F from Table A1.E is

$$h_{fg,b} = 1038.2 \text{ Btu/lbm}_v$$

From Ex. 12.3 we will assume the dry-bulb temperature equals the operative temperature, so $t = 75$ °F and $W = 0.0092 \text{ lbm}_v/\text{lbm}_a$. Substituting these values into the equations for sensible- and latent-heat loss

$$\dot{Q}_{sen} = (2.38 \text{ lbm}_a/\text{hr})(0.249 \text{ Btu/lbm}_a \cdot \text{°F})(94.1 - 75) \text{ °F} = 11.3 \text{ Btu/hr}$$

$$\dot{Q}_{lat} = (2.38 \text{ lbm}_a/\text{hr})(0.0317 - 0.0092) \text{ lbm}_v/\text{lbm}_a (1038.2 \text{ Btu/lbm}_v) = 55.6 \text{ Btu/hr}$$

Combining the results from Ex. 12.3 and 12.4, we find that under these conditions the total respiration heat loss (11.3 + 55.6 = 66.9 Btu/hr) is only 11 percent of the total heat loss due to both skin and respiratory losses (603 Btu/hr). However, this fraction will change significantly under different environmental conditions and with different metabolic rates.

12.3 ENVIRONMENTAL PARAMETERS

Fanger [1] identifies six environmental parameters that effect human thermal comfort:

1. Air dry-bulb temperature
2. Air velocity
3. Air humidity content

4. Mean radiant temperature

5. Activity level

6. Clothing

All these parameters have been used in the heat-balance equations we developed in Sec. 12.2. For example, the total heat transfer from the skin depends on the surrounding air dry-bulb temperature, air velocity (that effects the convective heat-transfer coefficient), air humidity content (that effects moisture evaporation rate), mean radiant temperature (that effects sensible-heat transfer), and clothing (that effects the thermal and moisture resistances between the skin and the environment).

The first three parameters—dry-bulb temperature, air velocity, and humidity content—are all environmental parameters that can be readily measured. Activity level and clothing were described in detail in Sec. 12.2. Mean radiant temperature is the environmental parameter that is most difficult to determine. Its definition is "the uniform temperature of an imaginary enclosure in which the radiant heat transfer from the human body equals the radiant heat transfer in the actual environment." The enclosure of uniform temperature is assumed to be a blackbody.

The mean radiant temperature can be computed when the surface temperatures of all surrounding surfaces are known and the angle factor between the person and each surface is known. The surface temperatures can be measured using conventional temperature-measuring instruments or can be computed using heat-transfer principles. The angle factors between a person and surrounding vertical or horizontal rectangular surfaces were determined by Fanger [5]. Figure 12.2 shows plots of the angle factor between a seated person and vertical or horizontal surfaces. These angle factors between a person, P, and a rectangular surface, j, can be computed using the following correlation:

$$F_{P-j} \approx \frac{1}{4\pi} \left[\frac{X}{\sqrt{1 + X^2}} \tan^{-1}\left(\frac{Y}{\sqrt{1 + X^2}}\right) + \frac{Y}{\sqrt{1 + Y^2}} \tan^{-1}\left(\frac{X}{\sqrt{1 + Y^2}}\right) \right] \quad (12.17)$$

where $X = a/1.8c$, $Y = b/1.8c$, and a, b, and c are dimensions shown in Fig. 12.2. The mean radiant temperature can then be computed from

$$\overline{T}_r = \left(\sum_{j=1}^{N} F_{P-j} T_j^4 \right)^{1/4} \quad (12.18)$$

where T_j is the absolute temperature of surface j and $\overline{T}_r$ is the absolute mean radiant temperature. Usually the mean radiant temperature is given in °F (°C), so that 459.67 °R (273.15 K) is subtracted from the value computed in Eq. (12.18).

EXAMPLE 12.5

Compute the mean radiant temperature 0.5 m above the floor in the center of an empty room. The room dimensions are 12 m long, 8 m wide, and 3 m high. Along one of the long walls, the bottom 1 m of the wall contains a heater with a surface temperature of 50 °C, and the top 2 m of the wall consists of a cold glass window at 15 °C. All remaining surfaces are at 20 °C.

Solution: The surfaces of the room can be split into three parts; those at 50 °C, those at 15 °C, and those at 20 °C. Equation (12.18) can be written

$$\overline{T}_r = (F_{P-50} T_{50}^4 + F_{P-15} T_{15}^4 + F_{P-20} T_{20}^4)^{1/4}$$

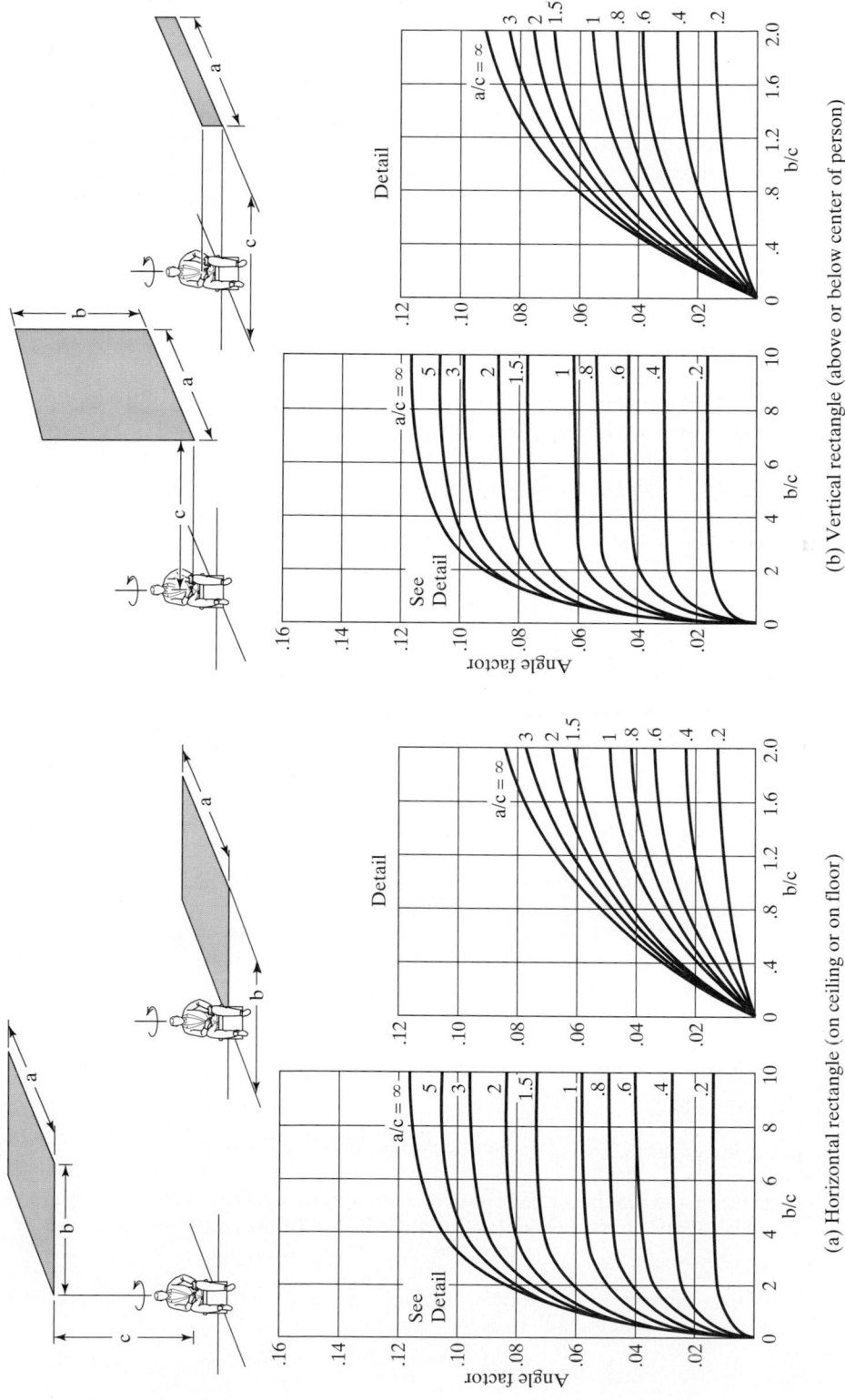

(a) Horizontal rectangle (on ceiling or on floor)

(b) Vertical rectangle (above or below center of person)

Figure 12.2 Mean value of angle factor between seated person and horizontal or vertical rectangle when person is rotated about the vertical axis. [Reprinted by permission from *ASHRAE Fundamentals 1993* (IP & SI), p. 8.12.]

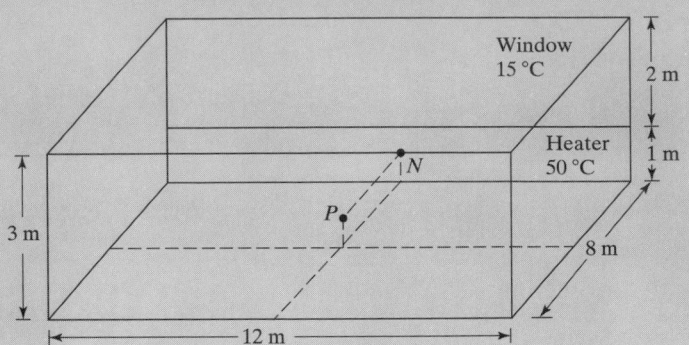

Figure 12.3 Schematic drawing of the room used in Ex. 12.5.

First the angle factors will be computed. The angle factor from the person to the rectangular heater can be determined by drawing a line perpendicular to the heater that intersects point P. This line passes through the heater 0.5 m above the floor and 6 m from the end wall and is shown in Fig. 12.3 as point N. Thus the heater is divided into four rectangular areas, which in this example are all equal in area. Each has a corner at the point N. The value for c equals the length of the line that passes between P and N, $c = 4$ m. The value for a equals one-half the room length, $a = 6$ m, and b equals the height of each rectangular element, $b = 0.5$ m. The angle factor between the person at point P and one of the four rectangular surface areas is found from Eq. (12.17):

$$X = 6 \text{ m}/(1.8 \times 4\text{m}) = 0.83$$

$$Y = 0.5 \text{ m}/(1.8 \times 4\text{m}) = 0.069$$

$$F_{P-1} = \frac{1}{4\pi} \left[\frac{0.83}{\sqrt{1 + 0.83^2}} \tan^{-1} \left(\frac{0.069}{\sqrt{1 + 0.83^2}} \right) + \frac{0.069}{\sqrt{1 + 0.069^2}} \tan^{-1} \left(\frac{0.83}{\sqrt{1 + 0.069^2}} \right) \right]$$

$$= 0.0065$$

The total angle factor to all four equal-sized areas of the heater (a, b, and c are equal for each quadrant of the heater) is four times this value or $F_{P-50} = 4 \times 0.0065 = 0.026$.

The angle factor for one-half the window must be determined by calculating the entire angle factor above and to the right (or left) of point N and subtracting the angle factor to the heater portion of this area, which was computed previously. For the combined window/heater area above and to the right of point N, $c = 4$ m, $b = 2.5$ m, and $a = 6$ m. Using Eq. (12.17) again,

$$X = 6 \text{ m}/(1.8 \times 4\text{m}) = 0.83$$

$$Y = 2.5 \text{ m}/(1.8 \times 4\text{m}) = 0.35$$

$$F_{P-1\&2} = \frac{1}{4\pi} \left[\frac{0.83}{\sqrt{1 + 0.83^2}} \tan \left(\frac{0.35}{\sqrt{1 + 0.83^2}} \right) + \frac{0.35}{\sqrt{1 + 0.35^2}} \tan \left(\frac{0.83}{\sqrt{1 + 0.35^2}} \right) \right]$$

$$= 0.151$$

Subtracting the heater portion of this and multiplying by 2 to account for both the right and left sides of the window, we obtain the angle factor for the entire window:

$$F_{P-15} = (0.151 - 0.0065)2 = 0.29$$

The angle factor for all remaining surfaces can be determined by subtracting the angle factors for the heater and the window from 1.0.

$$F_{P-20} = 1.0 - 0.29 - 0.026 = 0.68$$

Now the mean radiant temperature can be computed from Eq. (12.18):

$$\overline{T}_r = [0.026(323.15)^4 + 0.29(288.15)^4 + 0.68(293.15)^4]^{1/4}$$

$$\overline{T}_r = 292\text{ K} \quad \text{or} \quad \overline{t}_r = 19\,°\text{C}$$

This example illustrates the importance of having a warm surface to counteract a large, cold surface such as a window. The resulting mean radiant temperature is nearly equal to the temperature of the other surfaces (floor, ceiling, and three interior walls). Also note that the angle factor to the heater is quite small, 0.026, and that the angle factor to the surfaces at 20 °C is approximately 2/3. Although this calculation can be readily undertaken, nonisothermal surfaces and objects such as furniture and other occupants makes this a much more tedious procedure in real buildings than is indicated by the previous example.

Mean radiant temperature can also be determined experimentally by measuring both the air dry-bulb temperature and the temperature of a sensor in which radiation heat transfer is significant. A general approach is to consider two temperature sensors in the same thermal environment but with different absorptivities and emissivities. Figure 12.4 shows two spherical temperature sensors with surface emissivities ε_A and ε_B. Both sensors are exposed to the same air dry-bulb temperature, t, the same air velocity, $\mathbf{V}$, and the same mean radiant temperature, $\overline{t}_r$. If we assume that the only heat transfers with each sphere are convection with the surrounding air and infrared radiation with its surroundings, the following two energy-balance equations can be written.

$$h_C \pi D^2 (t_A - t) + \varepsilon_A \sigma \pi D^2 (T_A^4 - \overline{T}_r^4) = 0 \tag{12.19a}$$

$$h_C \pi D^2 (t_B - t) + \varepsilon_B \sigma \pi D^2 (T_B^4 - \overline{T}_r^4) = 0 \tag{12.19b}$$

Combining these two equations, eliminating t, and solving for the absolute mean radiant temperature, we have

$$\overline{T}_r = \left[\frac{(\varepsilon_A T_A^4 - \varepsilon_B T_B^4) + h_C(t_A - t_B)/\sigma}{\varepsilon_A - \varepsilon_B}\right]^{1/4} \tag{12.20}$$

Substituting this into Eq. (12.19a) and solving for the dry-bulb temperature, t:

$$t = \frac{\varepsilon_A \sigma}{h_C}\left[T_A^4 - \left(\frac{\varepsilon_A T_A^4 - \varepsilon_B T_B^4 + h_C(t_A - t_B)/\sigma}{\varepsilon_A - \varepsilon_B}\right)\right] + t_A \tag{12.21}$$

For the special case when $\varepsilon_B = 0$, $t_B = t$ and Eq. (12.20) reduces to

$$\overline{T}_r = \left[T_A^4 + \frac{h_C(t_A - t)}{\varepsilon_A \sigma}\right]^{1/4} \tag{12.22}$$

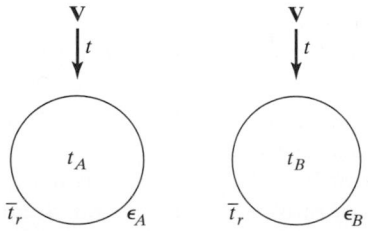

Figure 12.4 Schematic diagram of two spheres with different emissivities in the same thermal environment.

This equation can be used when the dry-bulb temperature is measured independently and only one sphere is used to determine the mean radiant temperature. This sphere should be given a coating that has a high emissivity in the infrared spectrum, usually flat black paint, to attain the most sensitivity. The hollow sphere is rather large, typically about 6 inches (15 cm) in diameter, so that the convective heat-transfer coefficient becomes small for a given air velocity. The sphere contains a temperature sensor to measure t_A. This instrument is called a "black globe thermometer." The temperature it measures is usually referred to as the "black globe temperature" and is given the symbol t_g.

12.4 ENVIRONMENTAL COMFORT INDICES

Several comfort indices have been developed that attempt to correlate human thermal comfort or heat-transfer rates with environmental conditions. A few of these indices are discussed in this section.

An interesting concept is the trade-off between temperature and humidity for the same total heat transfer. If we make some assumptions, we can find a rather simple solution. The first assumption is that the heat loss from the skin is much larger than the heat loss through respiration. This is valid at low metabolic rates. The second assumption is that the mean radiant, dry-bulb, and operative temperatures are all equal. Therefore, we can use the operative temperature as our sole temperature parameter. The third assumption is that the wetted skin area does not change for a given rate of heat loss. Incorporating these assumptions into Eq. (12.9) gives

$$\dot{Q}_{\text{sk}} = C_{\text{sen}}(t_{\text{sk}} - t_o) + C_{\text{lat}}(W_{s,\text{sk}} - W) \tag{12.23}$$

where C_{sen} includes the constant parameters in the sensible-heat-loss term and C_{lat} includes the constant parameters in the latent-heat-loss term. If we select an arbitrary operative temperature, t_o, then at 50 percent relative humidity at this temperature the heat loss using Eq. (12.23) becomes

$$\dot{Q}_{\text{sk}} = C_{\text{sen}}(t_{\text{sk}} - t_o) + C_{\text{lat}}(W_{s,\text{sk}} - W_{s,t_o}/2) \tag{12.24}$$

Once the heat loss is determined from Eq. (12.24), Eq. (12.23) can be used to determine other values of operative temperature and humidity ratio that will provide the same heat loss. All other values of t_o and W that provide the same heat loss are considered to have the same "effective temperature," or ET*, as the original conditions at t_o and 50 percent relative humidity. Note from Eq. (12.23) that for an increase in t_o, W must decrease to compensate. Note also that for every degree change in t_o, W must change by the ratio $C_{\text{lat}}/C_{\text{sen}}$. This ratio depends on the skin wettedness, clothing thermal resistance, and clothing permeability in addition to convective coefficients. Recall that the psychrometric chart has a linear scale for humidity ratio and a nearly linear scale for temperature. If we assume linear scales for both W and t_o, then the points that indicate constant values of effective temperature will fall on a straight line. Figure 12.5 illustrates a line of constant effective temperature, ET*, drawn on a psychrometric chart with linear scales for both W and t_o.

At low temperatures where the skin wetted area is small, the effect of changing the humidity ratio is negligible, as the evaporative heat loss is small. Therefore at low temperatures, lines of constant ET* are nearly vertical. At high temperatures, however, the skin wetted area and evaporative heat loss are much larger in magnitude, so that a small

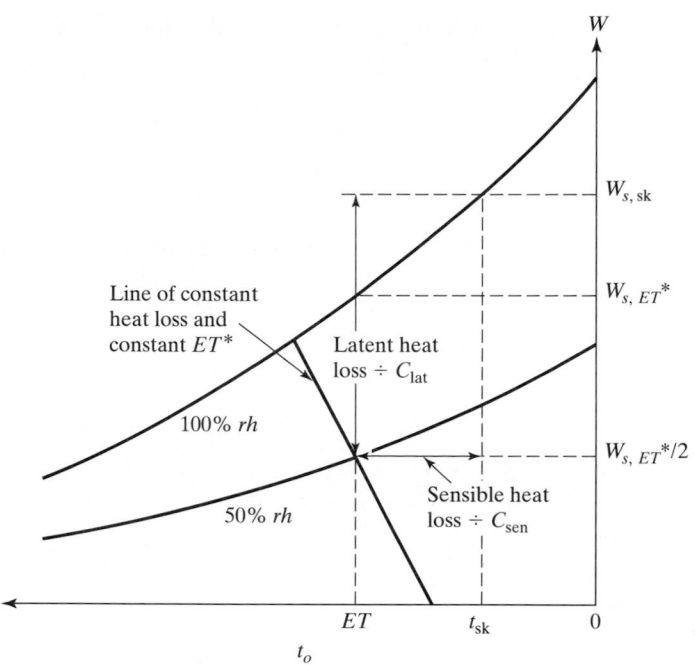

Figure 12.5 Line of constant heat loss illustrating line of constant effective temperature, ET*, on a psychrometric chart.

change in humidity ratio can result in a substantial change in operative temperature for the same value of ET*. This results in lines of constant ET* that are much less steep in regions of high temperature.

Another approach used to evaluate the combined effects of temperature and humidity is the Heat Stress Index, or H.S.I. This concept was introduced by Belding and Hatch [6]. Two important physiological criteria in the heat stress index concept are: (1) the skin temperature should not exceed 95 °F (35 °C) in order to limit the rise in body temperature and (2) the sweat rate should not exceed one quart per hour (one liter per hour) to limit the loss of body fluids. The heat stress index is defined as

$$\text{H.S.I.} = (100)\dot{Q}_E/\dot{Q}_{E,\max} \tag{12.25}$$

where $\dot{Q}_E$ is the actual evaporative heat loss and $\dot{Q}_{E,\max}$ is the maximum evaporative heat loss with the skin temperature at 95 °F (35 °C). $\dot{Q}_E$ is the evaporative cooling required to balance the body metabolic heat production and convection and radiation gains under the given environmental conditions while maintaining the skin temperature at 95 °F (35 °C). Belding and Hatch gave the following correlations to use for evaluating the two evaporative heat transfers:

$$\dot{Q}_E = \dot{Q}_M + 2\mathbf{V}^{0.5}(t - 95) + 22(\bar{t}_r - 95) \tag{12.26a}$$

$$\dot{Q}_{E,\max} = 11{,}700\mathbf{V}^{0.4}(0.03676 - W) \tag{12.27a}$$

which are valid near sea-level pressure with the metabolic rate $\dot{Q}_M$ in units of Btu/hr, $\mathbf{V}$ the air velocity in units of ft/min, t the dry-bulb temperature in units of °F, $\bar{t}_r$ the mean radiant temperature in °F, and W the humidity ratio in units of $\text{lbm}_v/\text{lbm}_a$. In SI units these equations become

$$\dot{Q}_E = \dot{Q}_M + 0.015\mathbf{V}^{0.5}(t - 35) + 0.012(\bar{t}_r - 35) \qquad (12.26\text{b})$$

$$\dot{Q}_{E,\,\text{max}} = 28.4\mathbf{V}^{0.4}(0.03676 - W) \qquad (12.27\text{b})$$

where $\dot{Q}_M$ is in kW, $\mathbf{V}$ is in m/s, t and $\bar{t}_r$ are in °C, and W is in kg_v/kg_a.

The physiological criteria and definition of the H.S.I. are based upon a so-called standard young adult male in good physical condition and acclimatized to heat. Belding and Hatch postulated that such a person could safely engage in simple physical work in a certain thermal environment over an eight-hour day provided the average skin temperature did not exceed 95 °F (35 °C) and the average sweat rate did not exceed one quart (liter) per hour. Table 12.3 shows an interpretation of the H.S.I. in regard to its physiological and hygienic implications as given by Belding and Hatch. They believed that no more than 10 to 20 percent of the working population could be expected to work successfully at an H.S.I. of 100.

The H.S.I. concept provides a seemingly sound basis for dealing with thermal problems in industrial environmental control. One problem in its application is evaluation of the effect of clothing. Determination of endurable values of H.S.I. for a particular situation is a medical problem, but the problem of altering or controlling the thermal environment is an engineering one.

TABLE 12.3 Evaluation of Heat Stress Index

Index of Heat Stress	Physiological and Hygienic Implications of 8-hr Exposures to Various Heat Stresses
−20 −10	Mild cold strain. This condition frequently exists in areas where men recover from exposure to heat.
0	No thermal strain.
+10 20 30	Mild to moderate heat strain. Where a job involves higher intellectual functions, dexterity, or alertness, subtle to substantial decrements in performance may be expected. In performance of heavy physical work, little decrement expected unless ability of individuals to perform such work under no thermal stress is marginal.
40 50 60	Severe heat strain, involving a threat to health unless men are physically fit. Break-in period required for men not previously acclimatized. Some decrement in performance of physical work is to be expected. Medical selection of personnel desirable because these conditions are unsuitable for those with cardiovascular or respiratory impairment or with chronic dermatitis. These working conditions are also unsuitable for activities requiring sustained mental effort.
70 80 90	Very severe heat strain. Only a small percentage of the population may be expected to qualify for this work. Personnel should be selected (a) by medical examination and (b) by trial on the job (after acclimatization). Special measures are needed to assure adequate water and salt intake. Amelioration of working conditions by any feasible means is highly desirable, and may be expected to decrease the health hazard while increasing efficiency on the job. Slight "indisposition" which in most jobs would be insufficient to affect performance may render workers unfit for this exposure.
100	The maximum strain tolerated daily by fit, acclimatized young men.

Source: Reprinted by permission from *ASHAE Trans.*, 62, 226.

EXAMPLE 12.6

A young man performs moderate work, $\dot{Q}_M = 900$ Btu/hr, in an environment where the air dry-bulb temperature is 100 °F, the wet-bulb temperature is 70 °F, the mean radiant temperature is 90 °F, and the local air velocity is 75 ft/min. Determine the heat stress index for this person, assuming sea-level pressure.

Solution: We have all the information necessary to evaluate Eqs. (12.25–12.27) except for the humidity ratio, W. From Fig. C-8E we find $W = 0.00883$ $\text{lbm}_v/\text{lbm}_a$. Substituting into Eq. (12.27a),

$$\dot{Q}_{E,\text{max}} = 11,700(75)^{0.4}(0.03676 - 0.00883) = 1838 \text{ Btu/hr}$$

From Eq. (12.26a)

$$\dot{Q}_E = 900 + 2(75)^{0.5}(100-95) + 22(90 - 95) = 877 \text{ Btu/hr}$$

From Eq. (12.25)

$$\text{H.S.I.} = (100)(877/1838) = 48$$

From Table 12.3 we see that this value of H.S.I. indicates conditions of severe heat strain requiring healthy, fit workers.

In studying an industrial environmental problem where a lower H.S.I. is required, it is particularly useful to analyze Eqs. (12.26) and (12.27) in order to recognize a feasible control measure. The H.S.I. may be reduced by decrease of ambient air temperature. This may be possible through general ventilation of the space with outdoor air, through spot cooling with refrigerated air, or by other means. Another effective control measure may be increase of air velocity over the worker.

Thermal radiation is often the mechanism of heat exchange which causes a high H.S.I. The mean radiant temperature may be reduced through insulating of hot surfaces or by the use of radiation shields. The coefficient 22 (or 0.012) was determined on a basis of the worker's surface being nonreflective. Through the use of aluminized clothing this coefficient may be reduced and the H.S.I. reduced, providing that the evaporation rate is not diminished by the metallic covering.

It should be realized that throughout this chapter, heat-exchange equations between the body and its environment have made no allowance for exposure to solar radiation. A high value of the H.S.I. may occur due to such exposure.

The wind-chill index and equivalent wind-chill temperature are concepts that are often used to determine the combined effects of temperature and air velocity during cold conditions, usually in winter. The wind-chill index (W.C.I.) was first developed by Siple and Passel [7], who performed measurements of heat loss from warm cylinders under different air temperature and velocity conditions. The wind-chill index is a means of correlating the heat-loss rate from an exposed surface near a skin temperature of 91.4 °F (33 °C). For example, the heat-loss rate with a low air velocity and a cold temperature can be the same as at a higher air velocity but a warmer temperature. Equation (12.7) applied to exposed skin becomes

$$\dot{Q}_{\text{sen}} = A_{\text{sk}} h(t_{\text{sk}} - t_o) \tag{12.28}$$

Here the trade-off between the heat-transfer coefficient, h, which is a function of air velocity, and the operative temperature, t_o, can be seen. The equivalent wind-chill temperature is the temperature at zero wind speed that provides the same sensible heat-loss rate as in the actual environmental conditions. Table 12.4 shows equivalent wind-chill

TABLE 12.4 Equivalent Wind-Chill Temperatures of Cold Environments

Wind Speed, mph	Actual Thermometer Reading, °F											
	50	40	30	20	10	0	−10	−20	−30	−40	−50	−60
	Equivalent Chill Temperature, °F											
0	50	40	30	20	10	0	−10	−20	−30	−40	−50	−60
5	48	37	27	16	6	−5	−15	−26	−36	−47	−57	−68
10	40	28	16	3	−9	−21	−34	−46	−58	−71	−83	−95
15	36	22	9	−5	−18	−32	−45	−59	−72	−86	−99	−113
20	32	18	4	−11	−25	−39	−53	−68	−82	−96	−110	−125
25	30	15	0	−15	−30	−44	−59	−74	−89	−104	−119	−134
30	28	13	−3	−18	−33	−48	−64	−79	−94	−110	−125	−140
35	27	11	−4	−20	−36	−51	−67	−83	−98	−114	−129	−145
40	26	10	−6	−22	−38	−53	−69	−85	−101	−117	−133	−148

Little danger: In less than 5 h, with dry skin. Maximum danger from false sense of security. (WCI less than 1400)	Increasing danger: Danger of freezing exposed flesh within one minute. (WCI between 1400 and 2000)	Great danger: Flesh may freeze within 30 seconds. (WCI greater than 2000)

Wind Speed, km/h	Actual Thermometer Reading, °C											
	10	5	0	−5	−10	−15	−20	−25	−30	−35	−40	−45	−50
	Equivalent Chill Temperature, °C											
Calm	10	5	0	−5	−10	−15	−20	−25	−30	−35	−40	−45	−50
10	8	2	−3	−9	−14	−20	−25	−31	−37	−42	−48	−53	−59
20	3	−3	−10	−16	−23	−29	−35	−42	−48	−55	−61	−68	−74
30	1	−6	−13	−20	−27	−34	−42	−49	−56	−63	−70	−77	−84
40	−1	−8	−16	−23	−31	−38	−46	−53	−60	−68	−75	−83	−90
50	−2	−10	−18	−25	−33	−41	−48	−56	−64	−71	−79	−87	−94
60	−3	−11	−19	−27	−35	−42	−50	−58	−66	−74	−82	−90	−97
70	−4	−12	−20	−28	−35	−43	−51	−59	−67	−75	−83	−91	−99

Little danger: In less than 5 h, with dry skin. Maximum danger from false sense of security. (WCI less than 1400)	Increasing danger: Danger of freezing exposed flesh within one minute. (WCI between 1400 and 2000)	Great danger: Flesh may freeze within 30 seconds. (WCI greater than 2000)

SOURCE: Reprinted by permission from *ASHRAE Fundamentals 1993*, p. 8.15.

Part IV / Heat- and Mass-Transfer Processes and Applications

temperatures for a variety of typical winter conditions in a cold climate. As an example, the heat-loss rate from exposed skin at an air dry-bulb temperature of $-20\,°C$ and a wind speed of 60 km/hr is the same at an air dry-bulb temperature of $-50\,°C$ and calm wind, as both conditions have an equivalent wind-chill temperature of $-50\,°C$ as indicated in Table 12.4.

12.5 BODY REGULATORY PROCESSES AGAINST HEAT OR COLD

A person may knowingly assist the body in maintaining proper heat balance under exposure conditions by measures such as providing suitable clothing. However, the human body has involuntarily initiated regulation means for adjusting itself to either heat or cold exposures. The basic purpose of these adjustments is to prevent abnormal change of interior body temperature, which would impair the vital organs.

For a person exercising with some constant and moderate rate of exertion, there is a narrow range of atmospheric conditions in which the body needs to take no particular action to maintain its proper heat balance. This range of conditions may be called the *neutral zone*, since the body is neutral to feelings of hot or cold. The person's degree of activity and amount of clothing affect the level of the neutral zone. For a resting nude person, the neutral zone occurs within a range of approximately 81 to 86 °F (27 to 30 °C) air dry-bulb temperature.

For atmospheric conditions colder than those of the neutral zone, the body may be able to preserve a proper temperature in its deep tissues by decreasing the blood flow at the body surface and by additional heat production. For mild exposure to cold there is a range of atmospheric conditions immediately below the neutral zone called the *zone of vasomotor regulation against cold*. Within this zone, blood vessels adjacent to the skin constrict, preventing flow of blood and transport of heat to the immediate outer surface. The outer skin tissues essentially become an insulating layer. The body surface temperature decreases somewhat, but cooling of deep tissues may be avoided. For still lower environmental temperatures, restriction of blood flow does not provide adequate protection. This range of conditions may be called the *zone of metabolic regulation against cold*. Through spontaneous increase of activity and by shivering, body heat production is increased in an effort to prevent a decrease in stored heat and to prevent the body surface temperature from further decreasing. Beyond this range of conditions the body is unable to combat cooling of its tissues and disastrous results may occur. This final range of conditions is known as the *zone of inevitable body cooling*.

On the warm side of the neutral zone there exists first a *zone of vasomotor regulation against heat*. Here the surface blood vessels dilate and allow blood flow as close as possible to the outer surface. The skin temperature increases, providing a greater temperature difference for loss of heat by convection and radiation and allowing $W_{s, sk}$ in Eq. (12.2) to become larger. For still warmer conditions there is a *zone of evaporative regulation against heat*, where the body reacts in a powerful manner to prevent further rise in skin temperature. The sweat glands become highly active, drenching the body surface with perspiration. If air humidity, velocity, and clothing permit sufficiently rapid evaporation, further rise in body temperature may be prevented. This is the last line of defense, however, and if the stored heat increases, the body enters the *zone of inevitable body heating*.

When an individual encounters the zone of inevitable body heating, the interior body temperature will rise. Several physiological hazards exist, the severity of which depends upon the extent and time duration of the body temperature rise.

Heat exhaustion is due to failure of normal blood circulation. Symptoms of heat exhaustion include fatigue, headache, dizziness, vomiting, and abnormal mental reactions such as irritability. Severe heat exhaustion may cause fainting. Heat exhaustion usually causes no permanent injury, and recovery is usually rapid when the subject is moved to a cool location.

Heat cramps result from loss of salt due to an excessive rate of body perspiration. They are painful muscle spasms that may be largely avoided by proper ingestion of salt.

Heat stroke is the most serious hazard. When a person is exposed to excessive heat, body temperature may climb rapidly to 105 °F (41 °C) or higher. At such elevated temperatures sweating ceases and the subject may enter a coma, with death imminent. People experiencing heat stroke may have permanent damage to the brain.

Figure 12.6 shows the relationship between effective temperature, ET*, thermal sensation, the human physiology involved, and the resultant health effects. The values for

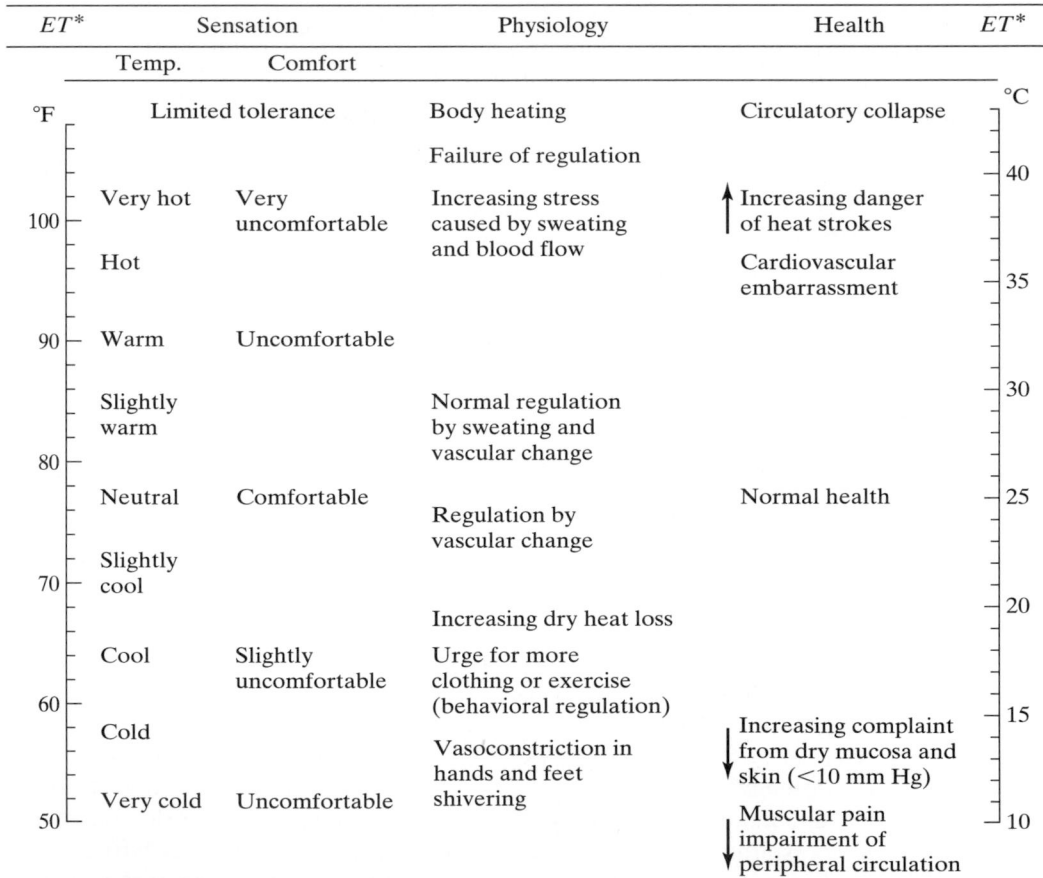

Figure 12.6 Related human sensory, physiological, and health responses for prolonged exposure. [Reprinted by permission from *ASHRAE Fundamentals 1993* (IP & SI), p. 37.8.]

effective temperature were based on the assumption of an activity level of 1 met, 0.6 clo of clothing thermal resistance, and an air velocity less than 30 ft/min (0.15 m/s).

12.6 PREDICTION OF HUMAN THERMAL COMFORT

Human thermal comfort is influenced by psychological as well as physiological factors. There is no method of stating what thermal environmental condition will effect a comfort feeling in a human being. A person's thermal comfort is influenced not only by ambient air temperature and humidity, but by the rate of air movement, body activity, and amount of clothing. It is difficult to specify a single physical quantity for evaluating human comfort. For a single individual, mean skin temperature would probably most closely suffice.

Practical application of the concept of effective temperature described in Sec. 12.4 is presented by the ASHRAE comfort chart reproduced in Fig. 12.7. This chart is applicable to reasonably still air conditions [velocity less than 30 ft/min (0.15 m/s)], to situations where occupants are seated at rest or doing light work, and to spaces whose enclosing surfaces are at a mean temperature equal to the air dry-bulb temperature. If the surrounding surfaces are below the air dry-bulb temperature, comfort would occur at a higher

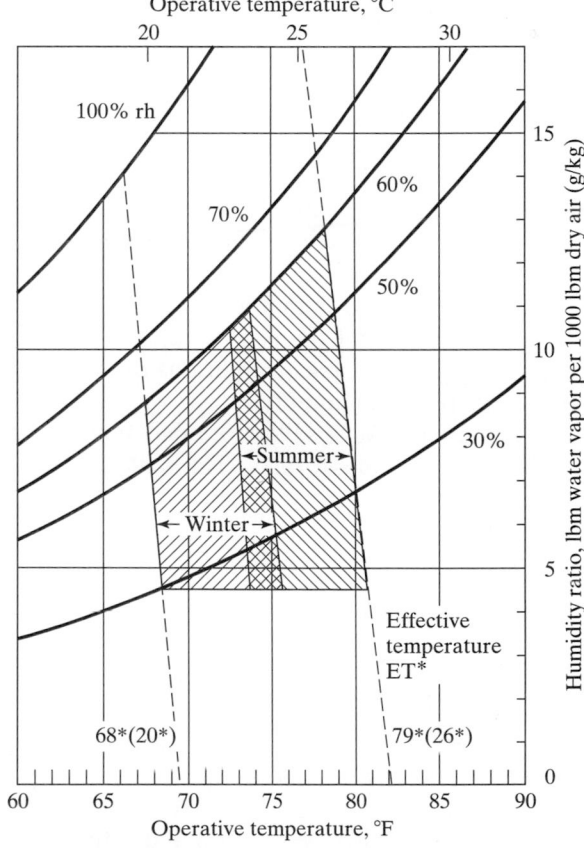

Figure 12.7 Standard effective temperature and ASHRAE comfort zones. [Reprinted by permission from *ASHRAE Fundamentals 1993* (IP & SI), p. 8.13.]

effective temperature than indicated by Fig. 12.7. An example could be a person seated adjacent to a large window expanse on a winter day. On the other hand, with panel-heated rooms, somewhat lower effective temperatures are customary. Counterradiation between occupants in densely populated spaces requires somewhat lower effective temperatures than in more sparsely occupied rooms. The difference between winter and summer ranges is primarily due to the differences in clothing worn in the two seasons. In winter, the increased clothing results in additional thermal insulation, so people prefer cooler environments. In this case, the total body heat loss remains about the same as in the summer with less clothing insulation.

Studies have indicated that women of all age groups prefer an effective temperature about one degree higher than that preferred by men, while both men and women over 40 years of age prefer an effective temperature about one degree higher than desired by younger people.

Activity of the occupants in an air conditioned space and the duration of occupancy are additional factors. Figure 12.7 applies to people doing sedentary work. People working more actively will be comfortable at lower effective temperatures. Particularly in summer, duration of occupancy must be considered. In spaces where the normal occupant stays only a short time, a higher effective temperature should be maintained than that shown by Fig. 12.7, which considers occupancy of three hours or more [8].

Figure 12.7 indicates that in winter an effective temperature of 70 to 72 °F (21 to 22 °C) is optimum for normally clothed people. Such a comfort condition may supposedly be attained by various combinations of temperature and humidity. In northern climates, space relative humidity during cold weather cannot much exceed 25 to 30 percent without condensation of moisture on inner surfaces of double-glazed windows. However, if no artificial humidification is provided, space relative humidity may fall to less than 10 percent. Such dry conditions may cause overdehydration of skin and membrane surfaces in people and excessive drying out of floors, veneered furniture, books, and other hygroscopic materials.

Figure 12.7 shows that in summer an effective temperature of about 76 °F (25 °C) should be optimum. Common practice is to design comfort air conditioning systems for summer to provide space conditions of about 75 °F (25 °C) dry-bulb temperature and 50 percent relative humidity.

Rohles and Nevins [9] developed a thermal comfort sensation scale that is called the *Predicted Mean Vote* or *PMV*. This scale is:

+3 hot

+2 warm

+1 slightly warm

 0 neutral

−1 slightly cool

−2 cool

−3 cold

Test results have been correlated with air dry-bulb temperature, humidity level, sex, and length of exposure. The basic equation used to compute the Predicted Mean Vote is

$$PMV = a^*t + b^*P_v + c^* \tag{12.29}$$

TABLE 12.5 Coefficients a*, b*, and c* Used to Calculate Predicted Mean Vote in Eq. (12.29)^a

Exposure Period, hr	Sex	English Units, t (°F), P_v (psia)			SI Units, t (°C), P_v (kPa)		
		a*	b*	c*	a*	b*	c*
1.0	Male	0.122	1.61	−9.584	0.220	0.233	−5.673
1.0	Female	0.151	1.71	−12.080	0.272	0.248	−7.245
1.0	Combined	0.136	1.71	−10.880	0.245	0.248	−6.475
3.0	Male	0.118	2.02	−9.718	0.212	0.293	−5.949
3.0	Female	0.153	1.76	−13.511	0.275	0.255	−8.622
3.0	Combined	0.135	1.92	−11.122	0.243	0.278	−6.802

SOURCE: Adapted from *ASHRAE Fundamentals 1993*, p. 8.16.

^{a} For young adult subjects with sedentary activity and wearing clothing with a thermal resistance of approximately 0.5 clo, $t_r = t$, air velocities ≤ 40 ft/min (0.2 m/s).

The coefficients a*, b*, and c* are listed in Table 12.5. Recall that P_v and W are related by

$$W = 0.6219 P_v / (P - P_v)$$

where P is the total atmospheric pressure.

EXAMPLE 12.7

Determine the difference between the Predicted Mean Vote of men and women occupants when the air dry-bulb temperature is 25 °C and the dew-point temperature is 20 °C, 3 hours after entry into the space.

Solution: We will use Eq. (12.29) with the coefficients determined from Table 12.5. The vapor pressure can be determined from Table A.1SI to be 2.339 kPa. For the men, 3 hours in the space,

$$PMV = 0.212(25) + 0.293(2.339) - 5.949 = 0.04$$

For the women, 3 hours in the space,

$$PMV = 0.275(25) + 0.255(2.339) - 8.622 = -1.15$$

The men are predicted to be thermally neutral while the women should be slightly cool.

This example illustrates the difference in thermal comfort sensation between men and women in the same environment. The women feel slightly cool, although the men are in the neutral comfort zone. The temperature necessary for the women to feel neutrally comfortable would be

$$0.275(t - 25) = 1.15$$

or t = 29.2 °C, which is about 4 °C higher than the current air temperature of 25 °C.

Not everyone will feel the same degree of comfort under the same environmental conditions. Fanger [5] developed a relationship between the Predicted Mean Vote and

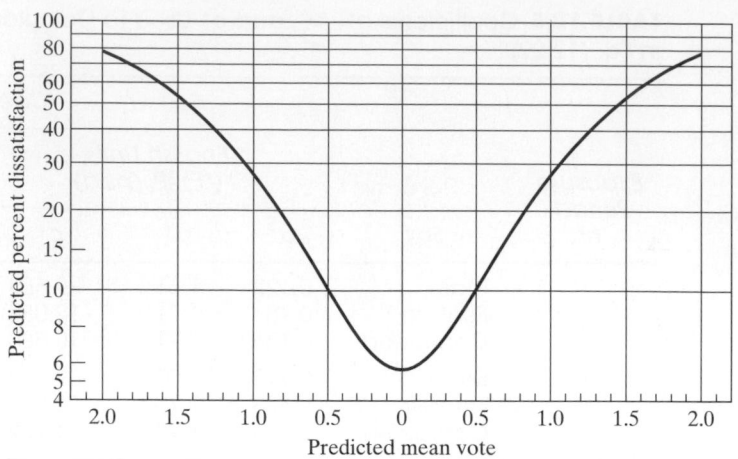

Figure 12.8 Predicted percentage of dissatisfied (PPD) as a function of predicted mean vote (PMV). [Reprinted by permission from *ASHRAE Fundamentals 1993* (IP & SI), p. 8.18.]

the Predicted Percent Dissatisfied, PPD. The dissatisfied occupants are defined as those who do not vote either $+1$, 0, or -1 on the PMV scale. Figure 12.8 shows the Predicted Percent Dissatisfied vs. the Predicted Mean Vote. Note that even when PMV = 0, 5 percent of the occupants are dissatisfied. A value of 10 percent dissatisfied is recommended by ASHRAE as an acceptable criterion for design. This results in PMV $\leq \pm 0.5$, according to the correlation shown in Fig. 12.8.

12.7 AIRBORNE CONTAMINANTS

The air inside buildings contains more than the gases considered to constitute clean, dry air (i.e., nitrogen, oxygen, argon, carbon dioxide, etc.). Psychrometrics, the study of dry air-water vapor mixtures, is important for determining energy loads on air-handling equipment. Moisture levels should be maintained between prescribed levels indoors to satisfy human comfort needs and to prevent the growth of fungi, molds, and other organisms.

Most other constituents of indoor air are considered to be contaminants. There are no lower limits to the concentrations of these contaminants, but upper levels have been developed for many of them. These contaminants are categorized as follows:

> *Particles*
> > Nonviable
> > > Dusts, fumes, smokes
> > > Mists and fogs
> > Viable
> > > Allergens, molds, bacteria
> > > Disease transmitting
> *Gases and vapors*
> > Inorganic
> > Organic

Particles can originate indoors or outdoors, by human-generated processes or naturally. *Dusts* are solid particles less than 100 microns in diameter, formed usually by mechanical action such as rubbing, scraping, or grinding. Dusts include particles from the earth's crust due to wind, volcanic action, and activities such as construction. Other forms of dust are produced by human activities and include pieces of animal fibers, grain, and wood. *Fumes* are solid particles that are formed by solidification of liquid droplets or by solidification of vapors. Examples include welding and foundry fumes. *Smokes* are produced by incomplete combustion of a fuel. Smoke particles can be in solid or liquid form. Sources include the combustion of petroleum fuels, wood, coal, and cigarettes.

Mists and fogs are liquid particles. Generally mists are considered to be larger in size than fog particles. The size of these particles can change with time as the conditions for equilibrium change. These particles can also contain other dissolved gases or vapors. Smog and acidified fog droplets in clouds are known to contain significant amounts of sulfuric acid caused by a combination of absorption and photochemical reactions.

Viable particles are of biologic origin and can propagate, given an adequate environment. Examples include fungal spores, bacteria, and viruses. Some of these particles cause reactions in humans, such as allergies. Others, such as cold viruses, cause illness and can be debilitating to the building occupants. Usually these particles can be treated as nonviable. However, conditions suitable for growth must be controlled to prevent unwanted growth and subsequent emissions of particles.

Some viable particles can cause disease in humans. Examples of such diseases are chickenpox, measles, and tuberculosis. Many of these particles are bacteria. Virus particles themselves are very small, less than 0.1 micron in diameter, but often they are attached to larger particles. Control of these particles is especially critical in hospitals, where a patient can act as a source of disease to other patients and staff in the building.

Figure 12.9 shows the size of some common particles, their settling velocity in still air, and their diffusion coefficient in standard air.

Gases and vapors can be categorized as inorganic or organic. Other factors of importance are the normal boiling point, molecular weight, and polarity. Hundreds of contaminant gases and vapors are known to exist. Table 12.6 lists some of the more common gaseous air pollutants. Radon is a radioactive gas that decays into daughter particles which may be inhaled and deposited in the lungs. Some gases are chemically reactive such as ozone. Many organic gases are irritants or may be perceived as unpleasant odors. We must know which gaseous pollutants are present before adequate control strategies can be developed.

12.8 ACCEPTABLE INDOOR LEVELS OF AIRBORNE CONTAMINANTS

Indoor pollutants can cause a variety of problems with the occupants. Some pollutants are irritants, some are toxic, some produce disease, and some produce unpleasant odors. A pollutant may be a mild irritant at low concentrations but become toxic at higher concentrations. Human reaction varies from person to person. In some people who have been sensitized to a particular contaminant, a very low level will cause severe symptoms. These factors make the establishment of acceptable levels rather difficult.

Three reference levels are important when considering indoor pollutants. One of these is the *recommended outdoor-air level*. Some countries have set standards for out-

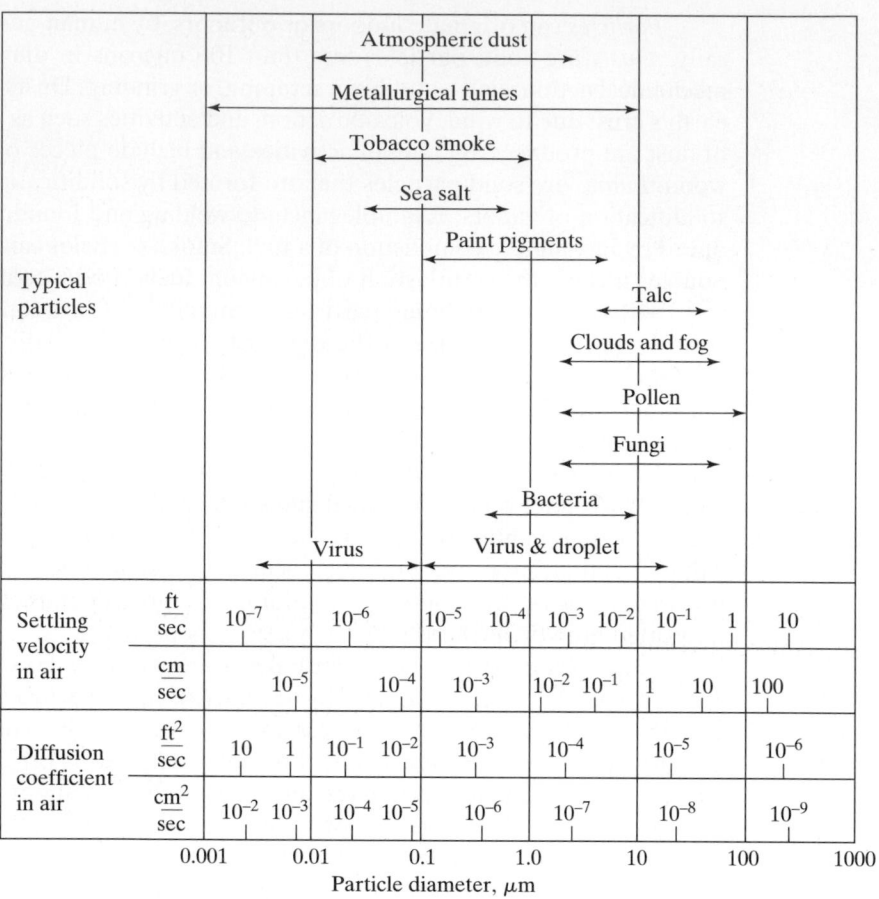

Figure 12.9 Typical airborne particle sizes, settling velocities, and diffusion coefficients. (Adapted from Fig. 1, *ASHRAE Fundamentals* 1993, p. 11.3.)

door-air concentrations, such as the U.S. National Ambient Air Quality Primary Standard [10]. This standard gives annual mean maximum values for several gaseous pollutants and particulates. Another reference is the levels that have been established for the workplace. The American Conference of Governmental Industrial Hygienists [11] maintains Threshold Limit Values (TLVs) for a large number of gaseous pollutants and particulates. The concentration limits are divided into Time Weighted Average values (TWA) and Short Term Exposure Limit values (STEL). These values are intended to prevent harmful effects to workers who may be exposed for 8 to 10 hours a day. However, the levels may not prevent unpleasant odors or mild irritation.

In many cases, a number of contaminants are present simultaneously. The following formula applies to a mixture whose constituents have similar toxicological effects:

$$C_1/\text{TLV}_1 + C_2/\text{TLV}_2 + C_3/\text{TLV}_3 + \cdots \leq 1 \qquad (12.30)$$

where C_i is the concentration of substance i and TLV_i is the threshold limit value for that substance. When this equation is satisfied, the mixture has acceptable concentrations with respect to the individual constituent threshold limit values. When the sum of the terms on

TABLE 12.6 Major Chemical Families of Gaseous Air Pollutants (with Examples)

Inorganic
1. Single-Element Molecules
    chlorine
    radon
    mercury
2. Oxidants
    ozone
    nitrogen dioxide
    nitrous oxide
    nitric oxide
3. Reducing Agents
    carbon monoxide
4. Acid Gases
    sulfuric acid
    hydrochloric acid
    nitric acid
5. Nitrogen Compounds
    ammonia
6. Sulfur Compounds
    hydrogen sulfide
7. Miscellaneous
    arsine

Organic
8. *n*-Alkanes
    methane
    n-butane
    n-hexane
    n-octane
    n-hexadecane
9. Branched Alkanes
    2-methyl pentane
    2-methyl hexane
10. Alkenes and Cyclohexanes
    1-octene
    1-decene
    cyclohexane
11. Chlorofluorocarbons
    R-11 (trichlorofluoromethane)
    1,1,1 trichloroethane
    R-114 (dichlorotetrafluoroethane)
12. Halide Compounds
    carbon tetrachloride
    chloroform
    methyl bromide
    methyl iodide
    phosgene
    carbonyl sulfide
13. Alcohols
    methanol
    ethanol
    2-propanol
    isopropanol
    phenol

cresol
diethylene glycol
14. Ethers
    vinyl ether
    methoxyvinyl ether
    n-butoxyethanol
15. Aldehydes
    formaldehyde
    acetaldehyde
    acrolein
    benzaldehyde
16. Ketones
    2-butanone (MEK)
    2-propanone
    acetone
    methyl isobutyl ketone
    chloroacetophenone
17. Esters
    ethyl acetate
    n-butyl acetate
    di-ethylhexyl phthalate (DOP)
    di-*n*-butyl phthalate
    butyl formate
    methyl formate
18. Nitrogen Compounds, and Other Than Amines
    nitromethane
    acetonitrile
    acrylonitrile
    pyrrole
    pyridine
    hydrogen cyanide
    peroxyacetal nitrate
19. Aromatic Hydrocarbons
    benzene
    toluene
    ethyl benzene
    naphthalene
    p-xylene
    benz-alpha-pyrene
20. Terpenes
    2-pinene
    limonene
21. Heterocylics
    furan
    tetrahydrofuran
    methyl furfural
    nicotine
    1,4-dioxane
    caffeine
22. Organophosphates
    malathion
    tabun
    sarin
    soman

(continued)

TABLE 12.6 (*cont.*)

23. Amines	25. Mercaptans and Other Sulfur Compounds
methylamine	*bis*-2-chloroethyl sulfide (mustard gas)
diethylamine	ethyl mercaptan
n-nitroso-dimethyamine	methyl mercaptan
24. Monomers	carbon disulfide
vinyl chloride	26. Miscellaneous
methyl formate	ethylene oxide
ethylene	

SOURCE: Reprinted by permission *ASHRAE Applications 1995*, p. 41.2.

the left-hand side is greater than 1.0, the mixture exceeds the TLV requirements and is not acceptable for the work environment.

Indoor concentration limits for nonindustrial buildings such as offices, schools, and residences are not as well established. Fewer standards and guidelines exist than for industrial environments. In some cases the recommended indoor concentrations are based on outdoor limits. Another method is to assume that the indoor concentration should not exceed 1/10 the TLV given for workplaces. Equation (12.30) can be applied for mixtures if the right-hand side is changed from 1.0 to 1/10. *ASHRAE Standard 62-1989* [12] lists concentrations of concern for some of the most common indoor-air pollutants. Table 12.7 lists several gaseous air pollutants, allowable concentrations, odor threshold levels, and some physical properties.

The units used to describe airborne concentrations of gases are given either on a number (volume) ratio basis or on a mass-per-unit-volume basis. Using the number ratio, typical units are parts per million (ppm—number of gas molecules per million molecules of air) or parts per billion (ppb—number of gas molecules per billion molecules of air). Using mass per unit volume, typical units are milligrams of contaminant per cubic meter of air (mg/m^3) and micrograms of contaminant per cubic meter of air ($\mu g/m^3$). A conversion between these two sets of units is

$$mg/m^3 = 0.01605(\text{ppm})\, MP/(273.15 + t) \tag{12.31}$$

where M is the contaminant molecular weight, P is the total pressure of the mixture in mm Hg, and t is the temperature in degrees Celsius.

For particles, the long-term ambient-air concentration allowed in the United States is 75 $\mu g/m^3$ (annual geometric mean) (National Primary & Secondary Ambient Air Quality Standards). The short-term limit is 260 $\mu g/m^3$ for 24 hours. ACGIH gives a TWA threshold limit value for nuisance dust particles of 10 mg/m^3. Particles composed of toxic chemicals are treated differently.

Particle size is an important factor when discussing contaminants. Inhaled particles of large size are deposited in the upper respiratory tract. Smaller particles tend to be deposited in the gas-exchange region of the lung. Therefore the particle size influences the deposition location and thus the resultant harmful effects. ACGIH has identified three particle-size selective threshold limit values:

1. *Inhalable particulate mass* for those materials that are hazardous when deposited anywhere in the respiratory tract.

2. *Thoracic particulate mass* for those materials that are hazardous when deposited anywhere within the lung airways and the gas-exchange region.

TABLE 12.7 Characteristics of Selected Gaseous Air Pollutants

Pollutant	Allowable Concentration, mg/m³		Odor, Threshold, mg/m³	Family	Chemical and Physical Properties	
	STEL	TWA			BP, °C	M
Acetaldehyde		360	1.2	15	21	44
Acetone	3200	2400	47	16	56	58
Acetonitrile		70	>0	18	82	41
Acrolein	0.75	0.25	0.35	15	52	56
Acrylonitrile		45	50	18	77	53
Allyl chloride	9	3	1.4	12	44	77
Ammonia	35	38	33	5	−33	17
Benzene	25	5	15	19	80	78
Benzyl chloride		5	0.2	12	179	127
2-Butanone (MEK)		590	30	16	79	72
Carbon dioxide	54000	9000	00	4	−78	44
Carbon monoxide	220	55	00	3	−192	28
Carbon disulfide	90	60	0.6	6	46	76
Carbon tetrachloride	150	60	130	12	77	154
Chlorine	1.5	3	0.007	1	−34	71
Chloroform	9.6	240	1.5	12	124	119
Chloroprene	3.6	90		12	120	89
p-Cresol		22	0.056	13	305	108
Dichlorodifluromethane		4950	5400	11	−30	121
Dioxane		360	304	21	100	68
Ethylene dibromide	233	155		12	131	188
Ethylene dichloride	410	205	25	12	84	99
Ethylene oxide	135	90	196	21	10	44
Formaldehyde	6	4	1.2	15	97	30
n-Heptane		2000	2.4	8	98	100
Hydrogen chloride	7	7	12	4	−121	37
Hydrogen cyanide		11	1	18	26	27
Hydrogen fluoride	5	2	2.7	4	19	20
Hydrogen sulfide	28	30	0.007	6	−60	34
Mercury		0.1	00	1	357	201
Methane				8	−164	16
Methanol		260	130	13	64	32
Methyl chloride	1783	1189	595	12	74	133
Methylene chloride	3480	1740	750	12	40	85
Nitric acid		5		4	84	63
Nitric oxide		30	>0	2	−152	30
Nitrogen dioxide	1.8	9	51	2	2146	2
Ozone		2	0.2	2	−112	48
Phenol	60	19	0.18	13	182	94
Phosgene	0.8	0.4	4	7	8	90
Propane			1800	8	−42	44
Sulfur dioxide		13	1.2	6	−10	64
Sulfuric acid		1	1	4	270	98
Tetrachloroethane		35	24	12	146	108
Tetrachloroethylene	1372	686	140	12	121	166
o-Toluidene		22	24	23	199	107
Toluene	1140	760	8	17	111	92
Toluene diisocyanate	0.14	0.14	15	18	251	174
1,1,1-Trichloroethane		45	1.1	11	113	133
Trichloroethylene	1080	541	120	12	87	131
Vinyl chloride monomer	0.014	0.003	1400	24	−14	63
Xylene	870	435	2	19	137	106

SOURCE: Reprinted by permission *ASHRAE Applications 1995*, p. 41.4.

TABLE 12.8 Respirable-Particulate-Mass Collection Efficiency

Particle Aerodynamic Diameter (micron)	Respirable Particulate Mass (percent)
0	100
1	97
2	91
3	74
4	50
5	30
6	17
7	9
8	5
10	1

3. *Respirable particulate mass* for those particles that are hazardous when deposited in the gas-exchange region of the lung.

The mass of particles that fall into the preceding three categories is determined using a size-selective collection efficiency. For example, the respirable particulate mass consists of those particles that are captured according to the following collection efficiency:

$$\text{SR}(d) = 0.5[1 + \exp(-0.06d)](1 - F(x)) \quad (12.32)$$

where d is the particle diameter in microns ($0 \leq d \leq 100 \ \mu\text{m}$) and F is the cumulative probability function of a standardized normal variable, $x = \ln(d/4.25 \ \mu\text{m})/\ln(1.5)$. Table 12.8 shows the respirable particulate mass versus particle size given by Eq. (12.32).

12.9 CONTROL OF INDOOR AIRBORNE CONTAMINANTS

The control of indoor airborne contaminants usually occurs using one or more of the following methods: (1) controlling the rate at which the contaminant is added to the indoor air (source control), (2) removing of the contaminant from the indoor air (removal control), and (3) reducing the concentration by dilution with cleaner outdoor air (dilution control). Figure 12.10 shows a schematic of a typical ventilation system for a single-zone commercial building. The majority of the sources are usually located within the occupied space. However, some sources can be located within the air-handling system, such as in the ductwork or in the air conditioning unit. Airborne contaminants can be removed by air cleaners. Several possible locations for air cleaners are shown on the figure. Dilution occurs naturally through infiltration/exfiltration in addition to the forced ventilation provided by the air-handling unit.

Source Control

Source control is generally believed to be the best method to control indoor levels of airborne contaminants. A variety of techniques are used. An obvious solution is to eliminate the process, product, or material that generates the contaminant. An example is to ban smoking from indoor areas. We can also choose materials such as carpeting and composite wood-based products that release a minimal amount of harmful gases into the air. The choice of paints and cleaning agents is also important. Some materials containing hazardous chemicals can be coated or packaged to reduce the rate at which the contaminants

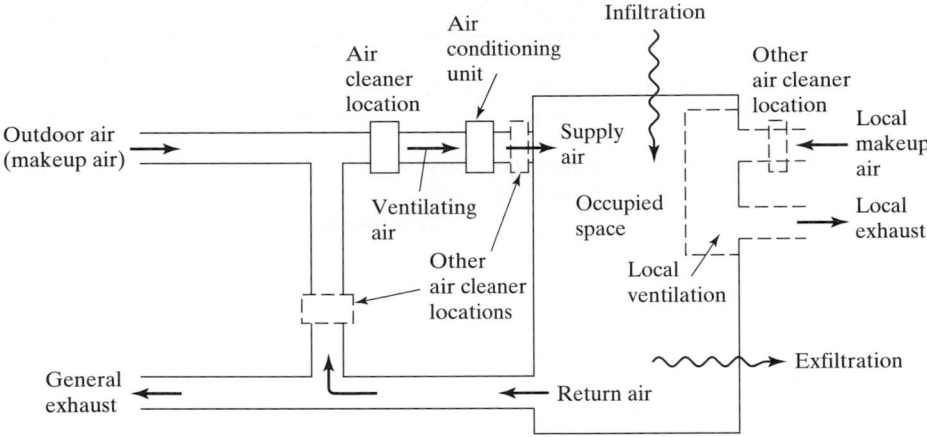

Figure 12.10 Schematic of single-zone ventilation system for controlling indoor airborne contaminants. (Reprinted by permission from *ASHRAE Standard 62-1989,* p. 4.)

are released. It may be possible to schedule processes that release contaminants into the air so that the maximum indoor concentration at any given instant in time does not exceed allowable limits.

Another very effective control option is to use local exhaust. If the contaminated air near the source can be exhausted before entering the occupied zone, the occupants will be protected. Common examples include laboratory fume hoods, kitchen exhaust hoods, and vented clothes dryers. Adequate makeup air should be available to prevent excessive depressurization of the area. Naturally vented combustion equipment, such as furnaces, fireplaces, and water heaters, may backdraft if the indoor pressure becomes too low. When backdrafting occurs, the flue gases reverse direction, so that the products of combustion enter the building rather than proceed up the flue. This condition can be lethal if the combustion gases contain sufficient amounts of carbon monoxide.

In residences, large amounts of moisture are generated by a variety of normal activities. Table 12.9 lists typical amounts. If these sources are not controlled, excessive indoor levels will exist, which can lead to mold growth and structural failure in extreme cases. Typical control measures are the use of kitchen range exhaust hoods, bathroom exhaust fans, and vented clothes dryers.

Removal Control

Removal of airborne contaminants can be placed into two categories; gas removal and particle removal.

Gas removal is most commonly accomplished using an adsorbent or a chemisorber. *Adsorption* occurs when the gas or vapor in the air stream has a higher concentration than the equilibrium concentration given by the amount of pollutant already adsorbed on the surface. *Desorption* occurs when the opposite conditions are present. Therefore, contaminant gases may be removed by adsorption or emitted by desorption, depending on the present concentration in the air stream and the previous history of the adsorbent.

Molecules must come into close contact with the adsorbing material before adsorption can occur. This requires the air stream to pass close to the surface of the adsorbent.

TABLE 12.9 Household Moisture Sources

Moisture Source by Type	Estimated Moisture Amount, lbm_w (kg_w)
Bathing:	
tub (excludes towels and spillage)	0.13 (0.06) standard-size bath
shower (excludes towels and spillage)	0.54 (0.25) 5-minute shower
Clothes washing (automatic, standpipe discharge)	Usually negligible
Clothes drying:	
vented outdoors	Usually negligible
not vented outdoors	4.88 (2.21) to 6.44 (2.92)/load
Cooking:	
breakfast (family of 4, average)	0.36 (0.17) plus 0.6 (0.27) if gas range
lunch (family of 4, average)	0.55 (0.25) plus 0.71 (0.32) if gas range
dinner (family of 4, average)	1.27 (0.58) plus 1.65 (0.75) if gas range
Dishwashing (by hand):	
breakfast (family of 4, average)	0.22 (0.10)
lunch (family of 4, average)	0.17 (0.08)
dinner (family of 4, average)	0.71 (0.32)
Floor mopping	$0.03/ft^2$ ($0.001/m^2$)
House plants (5 to 7 plants)	0.9 (0.4) to 1.0 (0.45)/day
Respiration and perspiration (family of 4, average)	0.46 (0.21)/hr

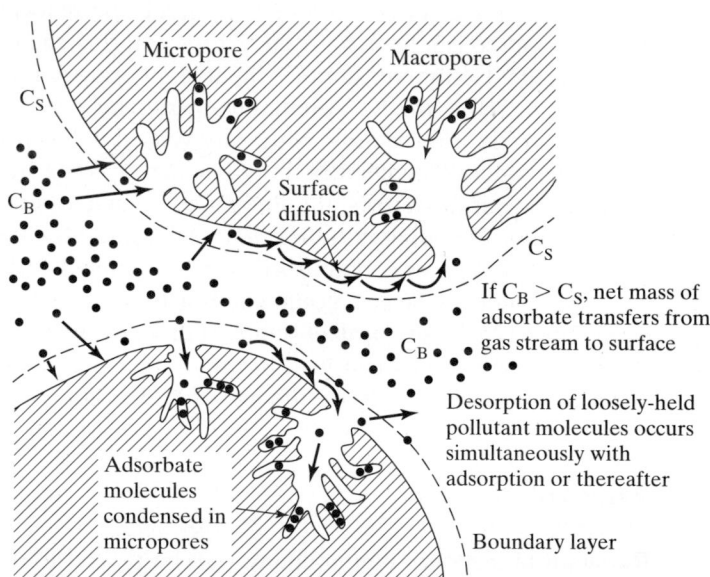

Figure 12.11 Steps in gaseous-pollutant adsorption. [Reprinted by permission from *ASHRAE Applications 1995* (IP & SI), p. 41.9.]

The molecules must penetrate the concentration boundary layer that separates the bulk air flow and the surface of the adsorbent. Then the contaminant must diffuse into the pores of the material. Finally, the molecules must adhere to the surface of the adsorbing material. Figure 12.11 shows how this process occurs in a granular structure. Large surface areas are required to provide a significant amount of contaminant retention. Typical

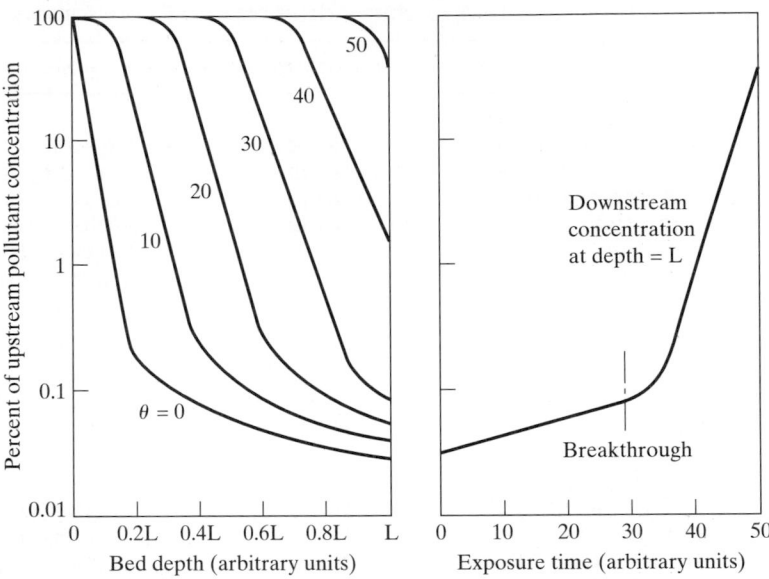

Figure 12.12 Dependence of adsorbate concentrations on bed depth and exposure time. [Reprinted by permission from *ASHRAE Applications 1995* (IP & SI), p. 41.1.]

activated alumina has surface areas of 10^6 ft²/lbm (200 m²/g), and activated carbon has from 5×10^6 to 7.5×10^6 ft²/lbm (1000 to 1500 m²/g) of adsorbing material. Temperature and the presence of other molecules, primarily water vapor, also influence the amount of contaminant that can be adsorbed.

When a new adsorbent bed is exposed to a steady air stream with a gaseous contaminant that can be adsorbed onto the surface of the material, the concentration in the air decreases as the air passes through the bed. The front portion of the bed becomes saturated with the contaminant, so that it no longer is effective in removing the molecules. The effective depth of the bed decreases with time, so that the concentration leaving the bed increases. This process is shown schematically in Fig. 12.12. Eventually the majority of the bed becomes saturated, so that the concentration leaving the bed rapidly approaches the entering concentration. This occurs near $\theta = 30$ on Fig. 12.12. The time at which this occurs is called the *breakthrough time.*

When adsorbing materials are used in building ventilation systems, the bed must operate for extended periods. Also the challenge concentration entering the bed is usually very low. The entering concentration is also often variable in time as the sources change. The bed can therefore become an adsorber during times of higher inlet concentrations and a desorber during times of low inlet concentrations. This phenomenon can lead to higher levels of contaminant downstream than would occur if the bed were not present during the times of low inlet concentration. This can be remedied using a bypass, where the cleaner air is directed around the bed into the building and the air passing through the bed is used to desorb some of the adsorbed contaminant and discharged outside. This same transient phenomenon occurs with other materials in the building also, not just with the air-cleaning beds.

Chemisorbers are similar to adsorbers except that the contaminant molecules react with the material through electron exchange. The contaminant molecules, once bonded

TABLE 12.10 Low-Temperature Adsorbers and Chemisorbers

Material	Impregnant	Typical Vapors or Gases Captured
Physical Adsorbers		
Activated carbon	None	Organic vapors, ozone, acidic gas
Activated alumina	None	Polar organic compounds[a]
Activated bauxite	None	Polar organic compounds
Silica gel	None	Water, polar organic compounds
Molecular sieves	None	Carbon dioxide, iodine
Porous polymers	None	Various organic vapors
Chemisorbers		
Activated alumina	$KMnO_4$	Hydrogen sulfide, sulphur dioxide
Activated carbon	I_2, Ag, S	Mercury vapor
Activated carbon[b]	I_2, KI_3, amines	Radioactive iodine and organic iodides
Activated carbon	$NaHCO_3$	Nitrogen dioxide
LiO_3, NaO_3, KO_3	None	Carbon dioxide
LiO_2, NaO_2, KO_2, $Ca(O_2)_2$	None	Carbon dioxide
Li_2O_2, Na_2, O_2	None	Carbon dioxide
LiOH	None	Carbon dioxide
$NaOH + Ca(OH)_2$	None	Acidic gases
Activated carbon	KI, I_2	Mercury vapor

SOURCE: Adapted from *ASHRAE Fundamentals 1995*, p. 41.11.
[a] Polar organics = alcohols, phenols, aliphatic and aromatic amines, etc.
[b] Mechanism may be isotopic exchange as well as chemisorption.

to the material, usually are not removed. Only specific chemicals will react with each chemisorber material. Many chemisorbers are formed from activated alumina or carbon, coated or impregnated with the chemisorption material. The performance of these beds is very similar to the time-dependent behavior of adsorbers as shown in Fig. 12.12. Table 12.10 lists some common materials used for adsorbers and chemisorbers and the typical gases and vapors that are removed from the air stream.

The performance of various materials and the influence of bed design are usually determined by experiment. However, many tests are run at much higher contaminant concentrations than are normally found in occupied buildings, so that extrapolation to actual operating conditions is not always reliable.

Particle removal can occur with media filters, electrostatic air cleaners, and by natural settling and deposition onto surfaces within the building. *Media filters* are the most common type of particulate air cleaner. Traditionally these air cleaners were installed to protect the HVAC equipment from becoming loaded with dirt and to protect bearings. These filters were effective at collecting the large particles that could foul equipment, but they were not very effective at capturing the small respirable particles that are of more concern to the human occupants.

The performance of particulate air cleaners is primarily a function of the pressure drop across the unit and the particle-collection efficiency. The pressure drop is important, as it requires a portion of the fan power needed to circulate air through the unit. The pressure-drop increase is proportional to the square of the velocity through the unit. The collection efficiency is traditionally defined as the ratio of the particulate mass downstream of the unit divided by the mass upstream. The traditional test procedure uses a total weight ratio between upstream and downstream. More recent test procedures utilize a concept termed "fractional efficiency," in which the ratio of the number of particles upstream versus downsteam is used to define the efficiency rather than the total mass ratio, which is dependent on the particle-size distribution in the upstream challenge aerosol.

All media filters collect particles using three mechanisms; interception, impaction, and diffusion. *Interception* occurs primarily with large particles which physically strike the filter media. As the particle size diminishes, interception becomes less likely and the collection efficiency is reduced. *Impaction* is similar in that the large particles cannot follow the air streamlines around the filter material and impact on the media. *Diffusion* occurs primarily with small particles which behave more like gas molecules. These small particles diffuse across concentration boundary layers formed around the filter media and deposit on the media. Large particles have much lower diffusivity and therefore do not deposit by diffusion nearly as well. By summing the particle losses by the three mechanisms described above, the filter media will pass more particles of intermediate size than those of very large or very small diameter. Figure 12.13 illustrates the three deposition mechanisms and the resulting filter fractional-efficiency curve versus particle diameter for a typical media filter. The particle size at which the filter efficiency is a minimum is termed the *most penetrating particle size*. These particles are usually between 0.1 and 0.5 microns in diameter.

As media filters become loaded with captured particles, the pressure drop increases at a given face velocity. The open volume in the filter decreases as it loads, requiring the air velocity to increase within the filter media itself. The filter collection efficiency also increases as the filter loads. The captured particles offer more surface area to collect additional particles. This loading is sometimes referred to as a "dust cake." When the loading becomes excessive, the filter must be cleaned or replaced. In most ventilation applications, the filters are replaced. In some industrial applications, the filters are cleaned by shaking the collected dust off or by reversing the flow momentarily using compressed air. Replacement is often determined by monitoring the pressure drop through the filter. When the pressure drop reaches a certain level, replacement is required.

Media filters are found in a variety of configurations. The most common filter used in residential and small commercial building applications is the replaceable panel filter. The filter media is loosely held between the support structure on the upstream and downstream sides of the filter. The filter is designed to be easily removed and replaced when it has become sufficiently loaded. Bag filters are used in larger air-handling units. The fil-

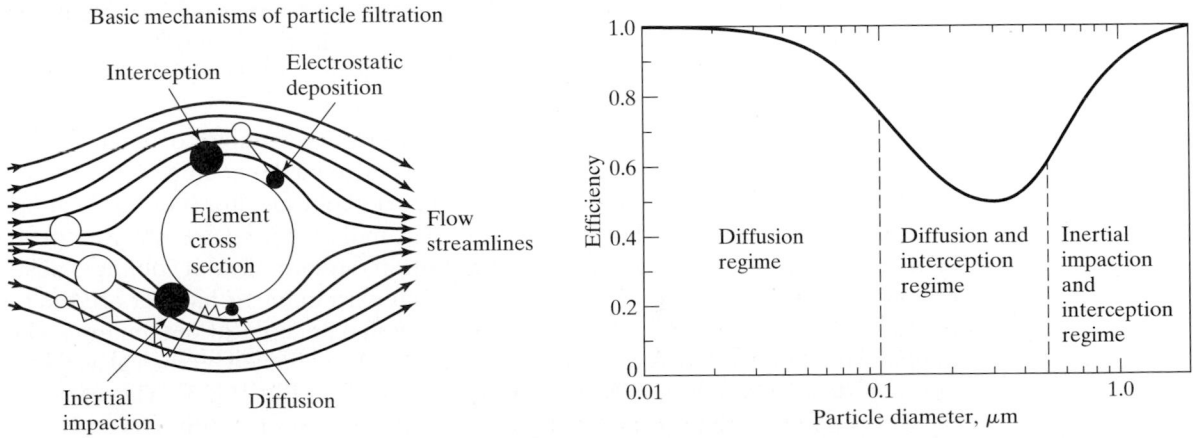

Figure 12.13 Particle collection on fibrous media filters.

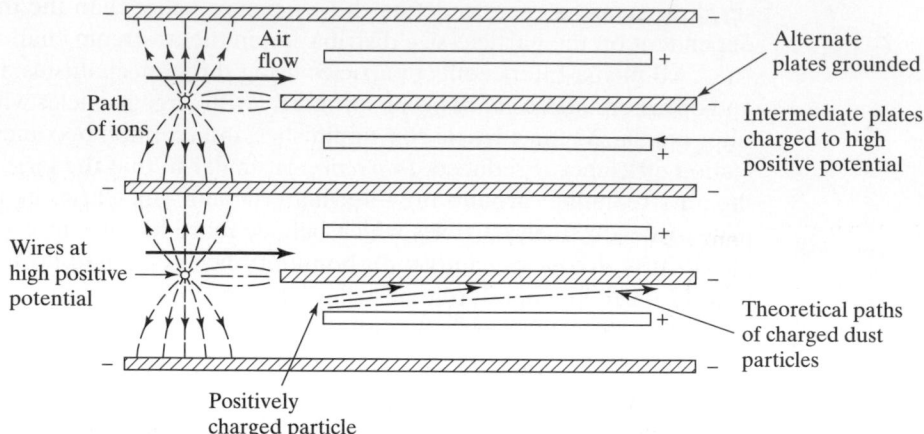

Figure 12.14 Cross section of ionizing electronic air cleaner. [Reprinted by permission from *ASHRAE Systems and Equipment 1996* (IP & SI), p. 24.8.]

ter media is in the form of woven bags, which have a much higher surface area to flow ratio than panel filters. Thus the face velocity through the bag material is much lower than for the panel filter. Renewable media filters supply clean filter media on one side of the air stream and remove loaded media on the other. These are usually motorized and have a control system to rotate the supply and takeup spools when required.

Electrostatic precipitators collect particles by charging them in an ion field and providing collection plates for the particles to deposit on. Figure 12.14 shows a schematic diagram of this type of air cleaner. Particles passing near the positively charged wire are given a positive charge. These charged particles are then attracted to the collection plates, which are negatively charged. The movement of a particle toward the collection plate depends on the particle's mobility or the ratio of the number of elementary charges to its mass. These units often contain a media prefilter that removes the very large particles and prolongs the life of the unit. The collection plates should be cleaned periodically to remove the collected dirt. This can be done by washing the plates. Cleaning is necessary when the loading becomes large enough to cause a significant current to flow between the wires and the collection plates. A simple ammeter can be used to determine when the current draw is too large and cleaning is necessary.

All media and electrostatic air cleaners offer the possibility of fungal or bacterial growth. Fungal spores and bacteria particles are collected depending on their size. These units also collect a large amount of dirt over a period of time; much of this is organic and provides nutrients to assist biological growth. The moisture level necessary for growth can be provided by rain, snow, or cooling-tower drift in outdoor air, wet cooling-coil surfaces, humidifiers, or drain pans. Air temperatures that exist in an air handler are the result of providing thermal comfort conditions in the occupied portions of the building. Often the temperature at the filters is conducive to growth of the collected biological aerosol. Anecdotal evidence and laboratory tests confirm growth on filters [13]. Therefore, maintenance is important to prevent the filter from acting as a source of bioaerosols within the building and a source of odors resulting from bacteria growth.

Dilution Control Using Ventilation

The concentration of indoor airborne contaminants can be reduced by dilution with air that has lower levels of contaminant. This is usually accomplished using outdoor air. The assumption is that the outdoor air has lower levels of the contaminant or contaminants and that it does not introduce additional hazardous contaminants into the space. This is not always a valid assumption, particularly in industrialized urban settings and in ambient excursions, such as during dust storms or volcanic activity. If the outdoor-air quality is not satisfactory, the air must be cleaned prior to introducing it into the indoor space. In the discussion that follows, we will assume that the outdoor air is acceptable or has been cleaned to an acceptable level of contaminants. We define *ventilation air* as the outdoor air (which may have been cleaned) that is used to dilute indoor air in the occupied zone of a space.

One approach to controlling indoor contaminant concentrations is to provide a specified amount of outdoor-air volumetric flow rate for each person in the conditioned space. If the occupancy is unknown, the airflow rate can be determined on a unit-floor-area basis. *ASHRAE Standard 62* uses this prescriptive method. Table 12.11 shows some of the recommended outdoor-air ventilation rates for common commercial buildings from *ASHRAE Standard 62-1989*. The values have been chosen to control indoor levels of carbon dioxide from occupant respiration and other airborne contaminants normally encountered in buildings. An assumption inherent in this approach is that the ventilation air is adequately mixed with the indoor air in the occupied zone, so the dilution that is desired actually occurs. In many cases much of the ventilation air passes through the conditioned space without actually mixing with the air in the occupied zone. An example is a situation where the air is supplied to a room through a ceiling diffuser and leaves the room through a ceiling return grille. If the supply-air jet does not penetrate down into the occupied zone because of poor diffuser selection or adjustment, or if significant internal obstructions prevent the air from reaching the occupants, the ventilation air will "short-circuit" between the supply and return under the ceiling and not reach the occupants to provide the desired dilution.

Various measures have been defined to quantify the mixing of the ventilation air with the air in the occupied zone. One of the first concepts presented was the *age-distribution* idea developed by Sandberg [14]. The *age* of air at any point in a room, τ_p, is defined as the time that has elapsed since the fluid element that passes the point entered the room. The *residual lifetime* is the time required for a fluid element at a point in the room to leave the room. The *residence time* is defined as the age of the air at a point located at the exhaust of the room and is equal to the sum of the local mean age and the residual lifetime for any point in the room. Figure 12.15 illustrates these three age definitions. The nominal time constant of a ventilated room can be determined by

$$\tau_n = \frac{\text{Vol}}{\dot{V}} \tag{12.33}$$

where Vol is the room volume and $\dot{V}$ is the volumetric airflow rate through the room. The local ventilation efficiency at point p, ε_p, is the ratio of the nominal time constant to the age of air at the point

$$\varepsilon_p = \frac{\tau_n}{\tau_p} \tag{12.34}$$

TABLE 12.11 Outdoor Air Requirements for Ventilation, Commercial Facilities (offices, stores, shops, hotels, sports facilities)

Application	Estimated Maximum Occupancy P/1000 ft² or 100 m²	Outdoor-Air Requirements				Comments
		ft³/min·person	L/s·person	ft³/min·ft²	L/s·m²	
Dry Cleaners, Laundries						Dry-cleaning processes may require more air.
Commercial laundry	10	25	13			
Commercial dry cleaner	30	30	15			
Storage, pick up	30	35	18			
Coin-operated laundries	20	15	8			
Coin-operated dry cleaner	20	15	8			
Food and Beverage Service						Supplementary smoke-removal equipment may be required.
Dining rooms	70	20	10			
Cafeteria, fast food	100	20	10			
Bars, cocktail lounges	100	30	15			
Kitchens (cooking)	20	15	8			Makeup air for hood exhaust may require more ventilating air. The sum of the outdoor air and transfer air of acceptable quality from adjacent spaces shall be sufficient to provide an exhaust rate of not less than 1.5 ft³/min·ft² (7.5 L/s·m²).
Garages, Repair, Service Stations						Distribution among people must consider worker location and concentration of running engines; stands where engines are run must incorporate systems for positive engine-exhaust withdrawal. Contaminant sensors may be used to control ventilation.
Enclosed parking garage				1.50	7.5	
Auto repair rooms				1.50	7.5	

	Outdoor Air Requirements ft³/min·person	L/s·person	ft³/min·room	L/s·room	
Hotels, Motels, Resorts, Dormitories					
Bedrooms			30	15	Independent of room size.
Living rooms			30	15	
Bath			35	18	Installed capacity for intermittent use.
Lobbies	15	8			
Conference rooms	20	10			
Assembly rooms	15	8			
Dormitory sleeping areas	15	8			See also food and beverage services, merchandising, barber and beauty shops, garages.
Gambling casinos	30	15			Supplementary smoke-removal equipment may be required.
Offices					
Office space	20	10			Some office equipment may require local exhaust.
Reception areas	15	8			
Telecommunication centers and data entry areas	20	10			
Conference rooms	20	10			Supplementary smoke-removal equipment may be required.

	ft³/min·ft²	L/s·m²	
Public Spaces			
Corridors and utilities	0.05	0.25	
Public restrooms, ft³/min·wc or urinal	50	25	Mechanical exhaust with no recirculation is recommended.
Lockers and dressing rooms	0.5	2.5	
Smoking lounge	60	30	Normally supplied by transfer air, local mechanical exhaust, with no recirculation recommended.
Elevators	1.00	5.0	Normally supplied by transfer air.

Note: Estimated maximum occupancy (persons/1000 ft²): Lobbies 30, Conference rooms 50, Assembly rooms 120, Dormitory sleeping areas 20, Gambling casinos 120, Office space 7, Reception areas 60, Telecommunication centers and data entry areas 60, Conference rooms 50, Smoking lounge 70.

Source: Adapted from *ASHRAE Standard 62-1989*, p. 8.

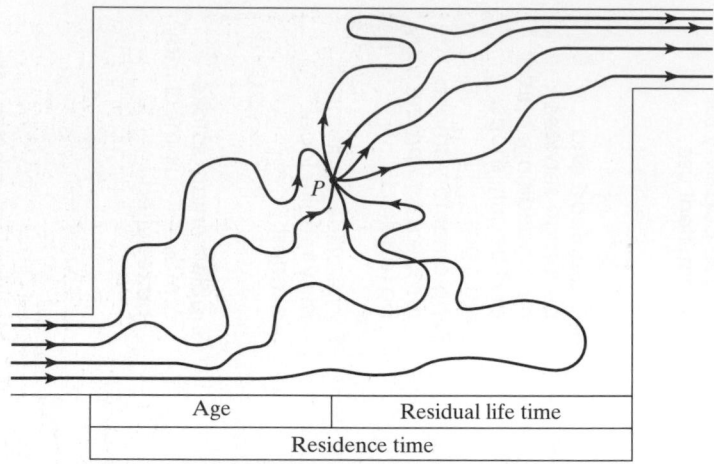

Figure 12.15 Concept of age of air. [Reprinted by permission from M. Sandberg and M. Sjoberg, "The Use of Moments for Assessing Air Quality," *Building and Environment*, 18:4, 184 (1983).]

The average ventilation efficiency for the entire room is given as ε_{ave}. These ventilation efficiencies can be determined experimentally by tagging the air molecules. This is usually done with a tracer gas, such as sulfur hexafluoride. The local ventilation efficiency becomes

$$\varepsilon_p = \frac{C_e - C_s}{C_p - C_s} \quad (12.35)$$

and the average ventilation efficiency becomes

$$\varepsilon_{ave} = \frac{C_e - C_s}{C_{ave} - C_s} \quad (12.36)$$

where C_s is the concentration in the supply air, C_p is the concentration at point p, C_{ave} is the average concentration in the room, and C_e is the concentration in the exhaust from the room.

A common experimental technique is the "step-up method." The concentration of tracer gas is initially zero in the room. A constant value of tracer-gas concentration, C_s, is then maintained at the supply, beginning at some initial time. The concentrations are monitored at the supply, at point p, and at the exhaust, C_s, C_p, and C_e, and the local ventilation efficiency is determined using Eq. (12.35). Repeated testing at different points within the room allows one to determine the average ventilation efficiency. The ventilation efficiency for a perfectly mixed room is 0.5. The value for displacement ventilation, such as in a unidirectional-flow clean room, lies between 0.5 and 1.0. The value for a room with short circuiting is less than 0.5.

EXAMPLE 12.8

Determine the amount of ventilation air required for a 120-m² dining room using the recommended values given in Table 12.11. If outdoor air is used to provide the ventilation, what fraction of the supply volumetric flow rate is outdoor air if the total amount of air supplied to the dining room is 5000 m³/hr?

Solution: The number of persons is not known in this case, so the occupancy must be determined using the floor area. From Table 12.11, for dining rooms, the estimated number of persons per 100 m^2 is 70 persons, and the amount of ventilation air required per person is 10 L/s per person. The total amount of ventilation air required is

$$(70 \text{ persons}/100 \text{ m}^2)(120 \text{ m}^2)(10 \text{ L/s per person}) = 840 \text{ L/s}$$

Converting this into units of m^3/hr,

$$(840 \text{ L/s})(\text{m}^3/10^3 \text{ L})(3600 \text{ s/hr}) = 3024 \text{ m}^3/\text{hr}$$

The fraction of the total amount of air supplied to the space is

$$(3024 \text{ m}^3/\text{hr})/(5000 \text{ m}^3/\text{hr}) = 0.6 \text{ or } 60 \text{ percent}$$

12.10 MODELING OF INDOOR CONTAMINANT CONCENTRATION

A simplified model is discussed in this section that can be used to predict concentrations of indoor airborne contaminants. Such a model can be used to determine the effectiveness of alternate control strategies and can be used to replace the prescriptive method of determining outdoor-air ventilation requirements in some cases. A basic assumption incorporated into this model is that all airflow rates, contaminant sources, and concentrations are invariant with time. Another assumption is that a uniform concentration exists within the space, i.e., the space is well mixed. It is also assumes that the volumetric flow rate is not changed by conditioning the air. The model must be applied to each contaminant of interest independently.

We will consider the case of a space with 100 percent outdoor air first. This type of system is shown schematically in Fig. 12.16. The outdoor air enters the system with a volumetric flow rate of $\dot{V}_o$ at a concentration of the contaminant of interest of C_o. The air then enters an air cleaner, which removes contaminant from the air at a rate $\dot{R}$. The air that leaves the air cleaner is the supply air to the space. Contaminant is generated at a rate of $\dot{G}$ within the space, which could be from a variety of sources. The well-mixed space concentration is C_s. This is also the concentration of air leaving the room in the exhaust. We will write an equation that balances the number of contaminant molecules or particles in this system:

$$\dot{P}_o - \dot{R} + \dot{G} = \dot{P}_e \tag{12.37}$$

where the number of contaminant molecules or particles entering the system from the outdoor air per unit time is $\dot{P}_o$, and the number exhausted from the system is $\dot{P}_e$. If we divide the terms in this equation by the outdoor-air ventilation flow rate, we have

$$\frac{\dot{P}_e}{\dot{V}_o} = \frac{\dot{P}_o}{\dot{V}_o} + \frac{\dot{G} - \dot{R}}{\dot{V}_o} \tag{12.38}$$

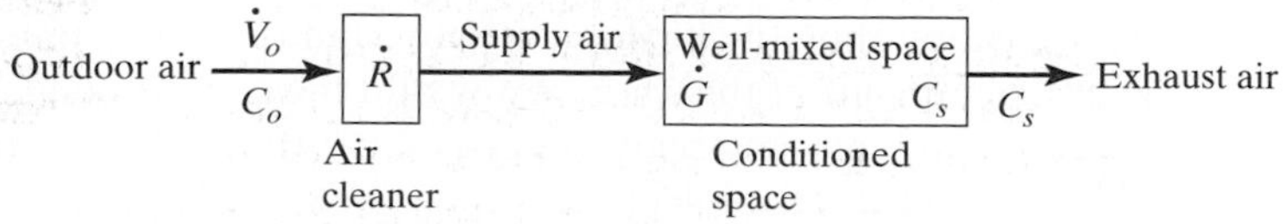

Figure 12.16 Schematic of ventilation system with 100 percent outdoor air.

Each term in this equation has units of contaminant number per unit volume, which is units of concentration. The term on the left-hand side is the indoor-space concentration of the contaminant, and the first term on the righthand side is the outdoor-air concentration. This equation can be written more compactly as

$$C_s = C_o + \frac{\dot{G} - \dot{R}}{\dot{V}_o} \tag{12.39}$$

Note the value of the indoor-air concentration under the following three cases:

a. With no source and a positive value for the removal rate, the last term on the right-hand side of the equation is negative, so that the indoor concentration is less than the outdoor concentration.

b. With a high-efficiency air cleaner such that the removal of contaminant in the air cleaner is larger than the amount generated in the space, the last term on the right-hand side is again negative, which makes the indoor concentration less than the outdoor value.

c. As the ventilation airflow rate increases, the indoor and outdoor concentrations become equal regardless of the amount of contaminant generated in the space or removed by the air cleaner.

EXAMPLE 12.9

Consider a building in which 100 percent outdoor air is supplied to the space. A media filter that has the filtration efficiency versus particle size shown in Fig. 12.13 is installed. Compute the indoor concentration of 0.1-micron particles if the outdoor concentration is $10^6/\text{ft}^3$ and the volumetric flow rate of outdoor air is 10,000 ft^3/min when (a) the indoor generation rate of 0.1-micron size particles is $10^8/\text{min}$, (b) the indoor generation rate of 0.1-micron size particles is $5 \times 10^7/\text{min}$, and (c) the generation rate is the same as in case (a), but the outdoor airflow rate is doubled and the filter efficiency is unchanged.

Solution: We will use Eq. (12.39) to estimate the indoor concentration of 0.1-micron size particles, assuming that the indoor air is well mixed. The particle removal rate by the filter can be written as $\dot{R} = \varepsilon_{ac} \dot{V}_o C_o$. The filter efficiency for 0.1-micron size particles from Fig. 12.13 is 0.75, so

$$\dot{R} = (0.75)(10,000\ \text{ft}^3/\text{min})(10^6/\text{ft}^3) = 7.5 \times 10^9/\text{min}$$

(a) $C_s = 10^6/\text{ft}^3 + (10^8/\text{min} - 7.5 \times 10^9/\text{min})/10,000\ \text{ft}^3/\text{min} = 2.6 \times 10^5/\text{ft}^3$

(b) $C_s = 10^6/\text{ft}^3 + (5 \times 10^7/\text{min} - 7.5 \times 10^9/\text{min})/10,000\ \text{ft}^3/\text{min} = 2.55 \times 10^5/\text{ft}^3$

(c) The particle removal rate now doubles to $1.5 \times 10^{10}/\text{min}$:

$$C_s = 10^6/\text{ft}^3 + (10^8/\text{min} - 1.5 \times 10^{10}/\text{min})/20,000\ \text{ft}^3/\text{min} = 2.55 \times 10^5/\text{ft}^3$$

Note that reducing the indoor generation rate by a factor of two provides the same effect as doubling the airflow rate through the filter. However, the amount of reduction is not significant in this case. We would have to provide a filter with a much higher collection efficiency to significantly reduce the indoor concentration levels below those obtained in this example.

Let us consider the more typical situation where the building has recirculated air with a small amount exhausted and a small amount of outdoor ventilation air. A

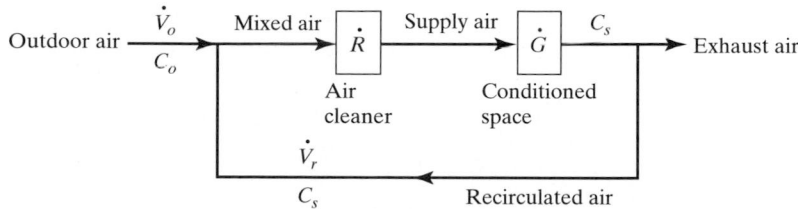

Figure 12.17 Schematic of recirculating-air ventilating system.

schematic of this type of recirculating air system is shown in Fig. 12.17. We can perform a number balance of contaminant on this system by placing a control volume that cuts through the outdoor ventilation air stream, the air leaving the space, and the recirculated air. The contaminant conservation equation becomes

$$C_o \dot{V}_o + C_s \dot{V}_r - \dot{R} + \dot{G} = (\dot{V}_o + \dot{V}_r) C_s \tag{12.40}$$

We can define the air-cleaner efficiency as

$$\varepsilon_{ac} = \frac{\dot{R}}{(\dot{V}_o + \dot{V}_r) C_m} \tag{12.41}$$

where the concentration of the mixed air entering the air cleaner, C_m, is determined by

$$C_m = \frac{C_o \dot{V}_o + C_s \dot{V}_r}{\dot{V}_o + \dot{V}_r} \tag{12.42}$$

Combining Eqs. (12.40) through (12.42), we obtain

$$C_s = \frac{C_o \dot{V}_o (1 - \varepsilon_{ac}) + \dot{G}}{\dot{V}_o + \dot{V}_r \varepsilon_{ac}} \tag{12.43}$$

for the contaminant concentration in the space.

Similar models can be developed for other air-handling system configurations. For example, one can incorporate hood-capture efficiency, so that a portion of the source is exhausted locally by an exhaust hood while the remaining contaminant is mixed with the air in the space. The main difficulty with implementing such models is the lack of data concerning the contaminant generation rates. Some values exist, such as the ones for moisture generation given in Table 12.9, but the rates for most processes and materials are unknown.

EXAMPLE 12.10

Repeat Ex. 12.9, using Eq. (12.43) for a recirculating air-handling system, and let the recirculated air volumetric flow rate equal the outdoor air flow rate.

Solution:

(a) From Eq. (12.43) we have

$$C_s = [(10^6/\text{ft}^3)(10,000 \text{ ft}^3/\text{min})(1 - 0.75) + 10^8/\text{min}]$$
$$/ (10,000 \text{ ft}^3/\text{min} + 10,000 \text{ ft}^3/\text{min} \times 0.75)$$
$$C_s = 1.49 \times 10^5/\text{ft}^3$$

(b)

$$C_s = [(10^6/\text{ft}^3)(10{,}000\ \text{ft}^3/\text{min})(1 - 0.75) + 5 \times 10^7/\text{min}]$$
$$/ (10{,}000\ \text{ft}^3/\text{min} + 10{,}000\ \text{ft}^3/\text{min} \times 0.75)$$
$$C_s = 1.46 \times 10^5/\text{ft}^3$$

(c)

$$C_s = [(10^6/\text{ft}^3)(20{,}000\ \text{ft}^3/\text{min})(1 - 0.75) + 10^8/\text{min}]$$
$$/ (20{,}000\ \text{ft}^3/\text{min} + 20{,}000\ \text{ft}^3/\text{min} \times 0.75)$$
$$C_s = 1.46 \times 10^5/\text{ft}^3$$

These models should be used for each contaminant of interest. In a typical situation, one of the contaminants will determine the minimum air-cleaner efficiency required and the necessary amount of outdoor air. Note that the air-cleaner efficiency changes when applied to different contaminants and is usually a function of airflow rate.

ENDNOTES

1. Fanger, P. O., *Thermal Comfort Analysis and Applications in Environmental Engineering* (New York: McGraw-Hill, 1970).

2. *ASHRAE Fundamentals Handbook 1993* (Atlanta: American Society of Heating, Refrigerating and Air Conditioning Engineers), 8.9.

3. Mitchell, D., *Convective Heat Transfer in Man and Other Animals, Heat Loss from Animals and Man* (London: Butterworth Publishing Inc., 1974), 59.

4. Gagge, A. P., Nishi, Y., and Nevins, R. G., "The Role of Clothing in Meeting FEA Energy Conservation Guidelines," *ASHRAE Trans.*, 82: 2 (1976), 234.

5. Fanger, P. O., *Thermal Comfort* (Malabar, FL: Robert E. Kreiger Publishing Co., 1982).

6. Belding, H. S., and Hatch, T. F., "Index for Evaluating Heat Stress in Terms of Resulting Physiological Strains," *Heating, Piping and Air Conditioning*, 207 (1955), 239.

7. Siple, P. A., and Passel, C. F., "Measurement of Dry Atmospheric Cooling in Subfreezing Temperatures," *Proceedings of the American Philosophical Society*, 89 (1945), 177.

8. *ASHRAE Standard 55-1992, Thermal Environmental Conditions for Human Occupancy* (Atlanta: American Society of Heating, Refrigerating and Air Conditioning Engineers, 1992).

9. Rohles, F. H., and Nevins, R. G., "The Nature of Thermal Comfort for Sedentary Man," *ASHRAE Trans.*, 77: 1 (1971), 239.

10. *National Primary and Secondary Ambient Air Quality Standards, Code of Federal Regulations*, Title 40, Part 50 (40 CFR50), U.S. Environmental Protection Agency.

11. *ACGIH Threshold Limit Values for Chemical Substances and Physical Agents and Biological Exposure Indices* (Cincinnati, OH: American Conference of Governmental, Industrial Hygienists, 1993).

12. *ASHRAE Standard 62-1989, Ventilation for Acceptable Indoor Air Quality* (Atlanta: American Society of Heating, Refrigerating and Air Conditioning Engineers, 1989).

13. Kuehn, T. H., Pui, D. Y. H., Vesley, D., Kemp, S. J., Streifel, A., Marx, J. and Alfred, A., *Matching Filtration to Health Requirements*, Final Report ASHRAE RP-625 Phase II (Atlanta: American Society of Heating, Refrigerating and Air Conditioning Engineers, 1994).

14. Sandberg, M., "What is Ventilation Efficiency?" *Building and Environment*, 16 (1981), 123–135.

PROBLEMS

12.1 Using fundamental heat- and mass-transfer considerations, discuss reasons for a person feeling more comfortable on a warm summer day if seated in front of an electric fan.

12.2 The mean body surface temperature of a certain unclothed person is 85 °F. The mean temperature of surrounding surfaces is 70 °F. The air within the room is at 90 °F dry-bulb temperature, 90 percent relative humidity, and 14.696 psia pressure. By what methods will this person be able to lose body heat?

12.3 Derive Eq. (12.7), starting with Eq. (12.3).

12.4 In order to maintain body thermal comfort, what change (increase or decrease) should be made in ambient air temperature to compensate for
  (a) an increase in activity,
  (b) a decrease in air velocity,
  (c) the wearing of heavier clothing,
  (d) a decrease in temperature of the surrounding surfaces, and
  (e) a decrease in air humidity ratio?

12.5 Compute the operative temperature for a seated person in an environment where the dry-bulb temperature is 25 °C, the mean radiant temperature is 35 °C, and the air velocity is 0.5 m/s.

12.6 Rework Ex. 12.2 with the dry-bulb temperature changed to 25 °C and the air velocity changed to 0.6 m/s. Assume all other parameters remain unchanged.

12.7 Rework Ex. 12.3 if the clothing is changed to walking shorts and a short-sleeve shirt.

12.8 A suggestion has been made to wrap plastic bags over a person's socks in a cold environment to trap as much insulating air as possible, which will reduce the heat loss from the feet. How would you respond to this suggestion?

12.9 The person considered in Ex. 12.4 wears a respirator. The respirator acts as a regenerative heat exchanger that preheats and prehumidifies the inhaled air. Determine the sensible and latent respiratory heat loss from this person if the sensible efficiency of the respirator is 50 percent and the latent efficiency is 30 percent.

12.10 Compute the mean radiant temperature using the room described in Ex. 12.5 if the point of interest is located 1 m above the floor, 2 m from an end wall, and 2 m from the wall that contains the windows.

12.11 Compute the mean radiant temperature distribution at a height of 1 m above the floor for the room described in Ex. 12.5. Perform enough computations so that you can plot contours of equal mean radiant temperatures for this horizontal plane.

12.12 Determine the mean radiant temperature in the center of your classroom. Assume the room is empty and that surface temperatures are uniform over each surface.

12.13 Consider two spherical thermocouples, each with a diameter of 1 mm. Compute the air dry-bulb temperature and the mean radiant temperature under the following conditions [see Eqs. (12.20) and (12.21)]: $t_A = 21$ °C, $t_B = 20$ °C, $\varepsilon_A = 0.9$, $\varepsilon_B = 0.15$, $h = 5$ W/m$^2 \cdot$ °C.

12.14 Compute the mean radiant temperature from the following information: a globe thermometer of 6 in. diameter has an emissivity of 0.95 and is at a temperature of 68 °F, the local dry-bulb temperature is 72 °F, and the air velocity is 25 ft/min.

12.15 Plot a line of constant effective temperature on a psychrometric chart that passes through 25 °C and 50 percent relative humidity when the ratio of the heat-transfer conductances, $C_{\text{sen}}/C_{\text{lat}} = 0.005$ kg$_v$/kg$_a \cdot$ °C.

12.16 Explain how the following conditions would change the slope of the lines of constant effective temperature on a psychrometric chart:
  (a) increased clothing thermal resistance,
  (b) increased clothing moisture resistance,
  (c) increased air velocity, and
  (d) increased metabolic rate.

12.17 In an industrial heat-exposure problem, labor and management have agreed that the heat stress index should not exceed 30. Measurements in the working space indicate the following conditions:

  Air dry-bulb temperature = 92 °F

  Temperature of the surrounding surfaces = 90 °F (estimate)

  Air relative humidity = 40 percent

It is estimated that the average body heat production rate is 700 Btu/hr. Determine the minimum allowable air velocity over each worker's body.

12.18 Determine the heat-stress index for the following conditions: $t = 30$ °C, $\bar{t}_r = 27$ °C, $t_{\text{wb}} = 24$ °C, air velocity = 0.5 m/s, and the metabolic rate is 3.0 met.

12.19 You have been asked to investigate a heat-exposure problem in a local company. The problem concerns 25 men doing light work requiring little mental effort. The men are selected by medical examination. The labor union believes that the working conditions are too hot and that the men are being subjected to severe heat strain. In your survey, you observe the following conditions:

> Air dry-bulb temperature = 90 °F
>
> Air wet-bulb temperature = 75 °F (assume $t^* = 75$ °F)
>
> Temperature of surrounding surfaces = 90 °F (estimate)
>
> Barometric pressure = 14.15 psia
>
> Air velocity over worker's torso = 150 ft/min
>
> Activity of workers: standing, light work at machines, mostly arm movements

What is your conclusion?

12.20 Determine whether the following conditions should be comfortable for typical people doing light work and exposed to air velocities of 15 to 25 ft/min.
   (a) Summer: $t = 75$ °F, $t^* = 65$ °F; $t = 80$ °F, $t^* = 70$ °F; $t = 95$ °F, $t^* = 75$ °F.
   (b) Winter: $t = 70$ °F, $\phi = 20$ percent; $t = 75$ °F, $\phi = 30$ percent; $t = 80$ °F, $\phi = 40$ percent.

12.21 It has been suggested to raise the humidity level in a home and reduce the indoor dry-bulb temperature in winter to save energy while maintaining the same thermal comfort conditions. If the initial indoor conditions are 20 °C and 30 percent relative humidity, determine the percentage change in energy requirements to temper outdoor ventilation air if the indoor conditions are changed to 50 percent relative humidity and the effective temperature remains the same. Outdoor conditions are 0 °C dry-bulb temperature and 100 percent relative humidity.

12.22 During an examination in a summer session class, the instructor made the following measurements in the classroom:

> Air dry-bulb temperature = 90 °F (assume $\bar{t}_r = t$)
>
> Air relative humidity = 53 percent
>
> Average air velocity = 40 ft/min

The instructor later gave these data to the class and asked them to quantitatively determine whether the conditions were favorable to good mental effort. He suggested that a body heat production rate of 500 Btu/hr should apply. What should have been the conclusion?

12.23 Determine the Predicted Mean Vote for a group of men one hour after entry into a space that has an air temperature of 19 °C and a dew-point temperature of 10 °C.

12.24 A group of mixed men and women work in an office area for at least 3 hours each day. Determine the relative humidity that should be maintained in this room to obtain a PMV of zero when

   (a) $t = 20$ °C,
   (b) $t = 25$ °C.

12.25 Estimate the number of persons dissatisfied in terms of thermal comfort out of a group of 100 men and women who spend 8 hours a day in a building in which the dry-bulb temperature is 74 °F and the dew-point temperature is 62 °F.

12.26 Estimate the amount of water vapor added to the air in a residence over a one-day period due to the following: four occupants, two showers, lunch and dinner cooked on an electric range, six house plants, and one load of laundry with the electric drier vented outdoors.

12.27 An engineer is designing the air-handling system for an office building with 10,000 ft^2 of office space. Determine the minimum amount of ventilation air required (ft^3/min) for this space according to the requirements given in Table 12.11.

12.28 Determine the total amount of ventilation air required for a building with the following areas using Table 12.11: 500 m^2 office space, 50 m^2 corridors, five restrooms with a total of 20 wc's, 30 m^2 reception area, and 50 m^2 conference rooms.

12.29 Consider a situation where the outdoor-air concentration of sulfur dioxide is 0.3 mg/m^3 and there are no sources in the building. What is the minimum allowable air-cleaner efficiency in a 100 percent outdoor-air ventilation system that will maintain the indoor concentration at 0.1 mg/m^3?

12.30 Repeat Prob. 12.29 with a recirculating air system in which 15 percent of the total amount of air supplied to the space is outdoor air and the remainder is recirculated air.

12.31 Show the complete derivation of Eq. (12.43) using Eqs. (12.40)–(12.42).

12.32 Derive an equation similar to Eq. (12.43) for a situation in which two sources exist within the occupied space but each has an exhaust hood that removes a portion of the contaminant. Assume the exhaust-hood efficiencies are $\varepsilon_{h,1}$ and $\varepsilon_{h,2}$ for the two exhaust hoods, where the efficiency is defined as the fraction of the contaminant generated that is captured and removed from the space, the remaining contaminant being assumed to mix with the air in the space. Also assume that the airflow rate through each exhaust hood is equal to one-half of the total outdoor air (makeup air) that is supplied to the space.

12.33 Repeat Ex. 12.9 with an air-cleaner efficiency of
   (i) 0.9,
   (ii) 0.99, and
   (iii) 0.999.

SYMBOLS

A	Surface area, ft^2 (m^2).
a	Length of surface in Fig. 12.2, ft (m).
a^*	Coefficient in Eq. (12.29), °F (°C).
b	Width of surface in Fig. 12.2, ft (m).
b^*	Coefficient in Eq. (12.29), psi^{-1} (kPa^{-1}).
C	Coefficients in Eqs. (12.23)–(12.24), concentration, no./vol.
c	Distance to surface in Fig. 12.2, ft (m).
c^*	Coefficient in Eq. (12.29).
c_{pa}	Specific heat of moist air, $Btu/lbm_a \cdot °F$ ($kJ/kg_a °C$).
D	Diameter of sphere, ft (m).
d	Diameter of particle, micron.
ET^*	Effective temperature, °F (°C).
F_{P-j}	Radiation angle factor between a person and surface j.
$\dot{G}$	Rate of contaminant generation, no./time.
HSI	Heat Stress Index.
h	Heat-transfer coefficient, $Btu/hr \cdot ft^2 \cdot °F$ ($W/m^2 \cdot °C$).
h^v	Mass-transfer coefficient, $lbm_a/hr \cdot ft^2$ ($kg_a/s \cdot m^2$).
h_{fg}	Latent heat of evaporation of water, Btu/lbm_w (kJ/kg_w).
I_{cl}	Thermal resistance of clothing ensemble, clo.
I_T	Total thermal resistance with clothing ensemble, clo.
i_{cl}	Ratio of clothing moisture resistance to thermal resistance, $Btu/lbm_a \cdot °F$ ($J/kg_a \cdot °C$).
i_T	Ratio of total moisture resistance to total thermal resistance, $Btu/lbm_a \cdot °F$ ($J/kg_a \cdot °C$).
K	Proportionality constant in Eq. (12.14).
Le	Lewis number.
M	Metabolic rate, met.
$\dot{m}_{res}$	Respiration mass flow rate, lbm_a/hr (kg_a/s).
P	Atmospheric pressure, psi (kPa).
$\dot{P}$	Contaminant flow rate, no./time.
P_v	Vapor pressure, psi (kPa).
PMV	Predicted Mean Vote.
$\dot{Q}$	Rate of heat transfer, Btu/hr (W).
R	Thermal resistance, $hr \cdot ft^2 \cdot °F/Btu$ ($m^2 \cdot °C/W$).
R^v	Vapor resistance, $hr \cdot ft^2/lbm_a$ ($s \cdot m^2/kg_a$).
$\dot{R}$	Rate of contaminant removal, no./time.
SR	Respirable particulate mass, Eq. (12.32).
T	Absolute temperature, °R (K).
$\overline{T}_r$	Absolute mean radiant temperature, °R (K).
TLV	Threshold Limit Value of concentration.
TWA	Time Weighted Average concentration.
t	Dry-bulb temperature, °F (°C).
t_g	Black-globe temperature, °F (°C).
$\bar{t}_r$	Mean radiant temperature, °F (°C).
t_o	Operative temperature, °F (°C).

$\mathbf{V}$	Air velocity, ft/min (m/s).
$\dot{V}$	Volumetric flow rate, ft^3/min (m^3/s).
Vol	Volume, ft^3 (m^3).
W	Humidity ratio, lbm$_w$/lbm$_a$ (kg$_w$/kg$_a$).
w	Fraction of skin area that is wetted.
X	Dimensionless geometry ratio $= a/1.8c$.
Y	Dimensionless geometry ratio $= b/1.8c$.

Subscripts

A	Sphere A.
ave	Average.
B	Sphere B.
b	Body temperature.
C	Convection.
cl	Clothing.
E	Evaporation.
e	Exhaust.
ex	Exhaled.
j	Surface j.
lat	Latent.
M	Metabolism.
m	Mixed air.
max	Maximum.
o	Outdoor.
P	Point P.
R	Radiation.
res	Respiration.
r	Recirculated.
sen	Sensible.
sk	Skin.
s	Saturated, supply, surface.
T	Total.
w	Wetted.

Greek Letters

θ	Dimensionless time.
ε	Emissivity.
ε_P	Local ventilation efficiency at point P.
ε_{ac}	Air-cleaner efficiency.
σ	Stefan-Boltzmann constant.
τ_n	Nominal time constant, min(s).
τ_P	Time for supply to reach point P, min(s).

chapter 13
Solar Radiation

13.1 INTRODUCTION

We now begin a new phase in our study of thermal environmental engineering—solar radiation. The external thermal environment of a locality results from the combined influences of solar radiation and meteorological effects. Physical influences such as topography and ocean currents may also be of great importance.

In the final analysis, the sun is the source of most energy on the earth and is a primary factor in determining a locality's thermal environment. It is important for engineers to have a working knowledge of the earth's relationship to the sun. They should be able to make estimates of solar-radiation intensity and know how to make simple solar-radiation measurements. They should also understand the thermal effects of solar radiation and know how to control or utilize them.

This chapter emphasizes how solar radiation affects building design. The amount of solar radiation entering a zone of a building, or "solar gain," is a major factor in determining the zone cooling loads. The designer must be able to estimate these solar gains. To do so it is necessary to predict the intensity of the solar radiation and the direction at which it strikes building surfaces. In addition, since much of the solar radiation enters the building through windows, it is important that one understands the solar transmission characteristics of glazing materials and how window design, including shading techniques such as window setback and exterior overhangs, can affect the solar gain of a space.

13.2 THE EARTH

The planet Earth is nearly spherical with a diameter of about 7900 miles (12.7×10^3 km). It makes one rotation about its axis every 24 hours and completes a revolution about the sun in a period of $365\frac{1}{4}$ days approximately. The earth's mean density is about 5.52 times that of water.

The earth revolves about the sun in an approximately circular path, with the sun located slightly off the circle's center. The earth's mean distance to the sun is about 9.3×10^7 miles (1.5×10^8 km). About January 1 the earth is closest to the sun, while on about July 1 it is most remote, being about 3.3 percent farther away. Since the intensity of solar radiation incident upon the top of the atmosphere varies inversely with the square of the earth-sun distance, the earth receives about 7 percent more radiation in January than in July. The earth's axis of rotation is tilted 23.5 degrees with respect to its orbit about the sun. The earth's tilted position is of profound significance, for, together with the earth's daily rotation and yearly revolution, it accounts for the distribution of solar radiation over the earth's surface, the changing length of hours of daylight and darkness, and the changing of the seasons.

Figure 13.1 schematically shows the effect of the earth's tilted axis at various times of the year. Figure 13.2 shows the position of the earth relative to the sun's rays at the time of winter solstice. At the winter solstice (December 22 approximately), the North Pole is inclined 23.5 degrees away from the sun. All points on the earth's surface north of 66.5 degrees north latitude are in total darkness, while all regions within 23.5 degrees of

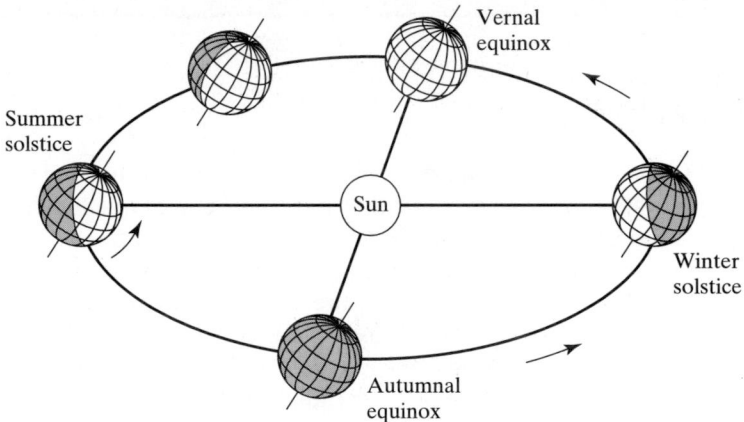

Figure 13.1 The earth's revolution about the sun.

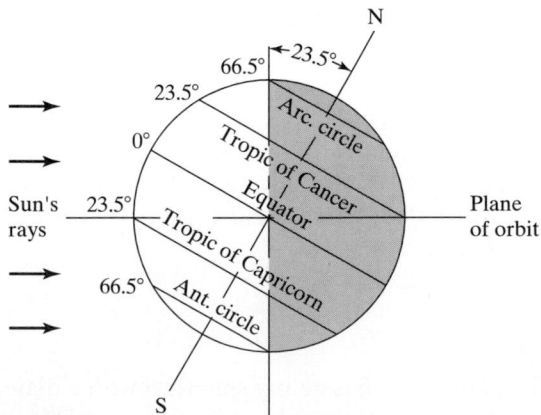

Figure 13.2 Position of earth in relation to sun's rays at the time of the winter solstice.

Part IV / Heat- and Mass-Transfer Processes and Applications

the South Pole receive continuous sunlight. At the time of the summer solstice (June 22 approximately), the situation is reversed. At the times of the two equinoxes (March 22 and September 22 approximately), both poles are equidistant from the sun and all points on the earth's surface have 12 hours of daylight and 12 hours of darkness.

Because of its tilted axis, the earth's surface has been divided into five zones. The Torrid Zone includes all locations where the sun is at the zenith (vertically overhead) at least once yearly. The Torrid Zone extends 23.5 degrees on either side of the equator. The Temperate Zones include all locations where the sun appears above the horizon each day but never at the zenith. The Temperate Zones extend from latitudes 23.5 degrees to 66.5 degrees (north and south). The Frigid Zones include all locations where the sun is below the horizon (and above) for at least one full day yearly. The Frigid Zones extend 23.5 degrees from the poles.

13.3 BASIC EARTH-SUN ANGLES

The position of a point P on the earth's surface with respect to the sun's rays is known at any instant if the latitude l and hour angle h for the point, and the sun's declination d, are known. These fundamental angles are shown by Fig. 13.3. Point P represents a location on the Northern Hemisphere. The latitude l is the angular distance of the point P north (or south) of the equator. It is the angle between the line $\overline{OP}$ and the projection of $\overline{OP}$ on the equatorial plane. Point O represents the center of the earth. The calculation of the various solar angles, performed later in this chapter, can be simplified by the adoption of a consistent sign convention. As part of this sign convention, take north latitudes as positive and south latitudes as negative.

The hour angle h is the angle measured in the earth's equatorial plane between the projection of $\overline{OP}$ and the projection of a line from the center of the sun to the center of the earth. At solar noon, the hour angle is zero. The hour angle expresses the time of day with respect to solar noon. One hour of time is represented by 360/24 or 15 degrees of hour angle. As part of the convention, take the hour angle negative before solar noon and positive after solar noon.

The sun's declination d is the angular distance of the sun's rays north (or south) of the equator. It is the angle between a line extending from the center of the sun to the center of the earth and the projection of this line upon the earth's equatorial plane. (Take

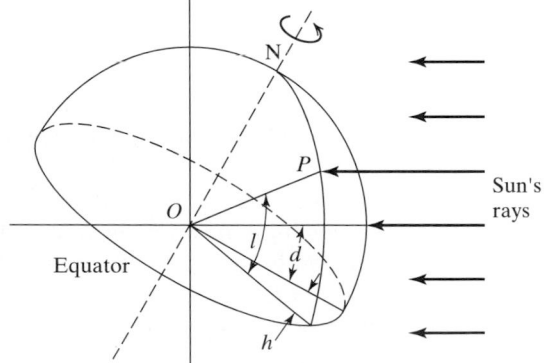

Figure 13.3 Latitude, hour angle, and sun's declination.

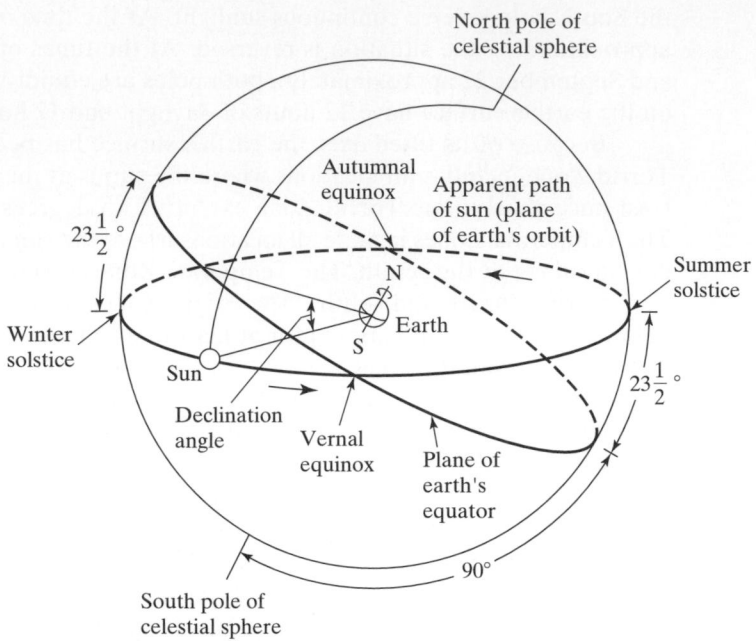

Figure 13.4 Schematic celestial sphere showing apparent path of sun and sun's declination angle.

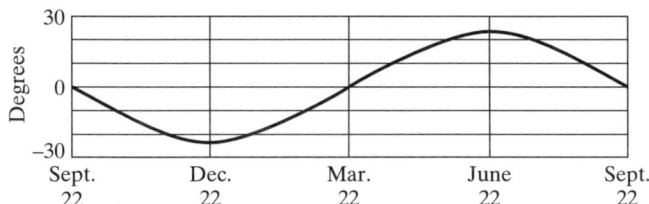

Figure 13.5 Variation of sun's declination.

the declination positive when the sun's rays are north of the equator and negative when they are south of the equator.)

Figure 13.4 shows a schematic celestial sphere where the earth is taken as the center of the universe. The sun would appear to move in the plane of the earth's orbit. Figure 13.4 shows the sun's angle of declination. At the time of the winter solstice, the sun's rays would be 23.5 degrees south of the earth's equator ($d = -23.5°$). At the time of the summer solstice, the sun's rays would be 23.5 degrees north of the earth's equator ($d = 23.5°$). At the equinoxes, the sun's declination would be zero. Figure 13.5 shows approximately the variation of the sun's declination throughout the year. Because the period of the earth's complete revolution about the sun does not coincide exactly with a calendar year, the declination varies slightly on the same day from year to year. Precise values may be obtained for a particular year from an ephemeris such as that in the *American Ephemeris and Nautical Almanac* [1]. For ordinary calculations it is sufficiently accurate to use the approximation presented by Cooper [2]:

$$d = 23.45 \sin\left(360 \frac{284 + n}{365}\right) \qquad (13.1)$$

TABLE 13.1 Variation in *n* throughout the Year for Eq. (13.1)

Month	n for the Day of the Month, D	Month	n for the Day of the Month, D
January	D	July	181 + D
February	31 + D	August	212 + D
March	59 + D	September	243 + D
April	90 + D	October	273 + D
May	120 + D	November	304 + D
June	151 + D	December	334 + D

where *n* is the day of the year. The value of *n* for any day of the month *D* can be determined easily with the aid of Table 13.1. The values of the declination angle calculated using the approximation of Eq. (13.1) are presented for the 7th, 14th, 21st, and 28th day of each month in Table 13.2.

13.4 RELATIONSHIP BETWEEN CLOCK TIME AND SOLAR TIME

Solar-radiation calculations must be made in terms of solar time. In a discussion of time, numerous designations may be used. We will consider here only a brief description to allow us to convert local clock time to solar time for engineering calculations.

Time reckoned from midnight at the Greenwich meridian (zero longitude) is known as Greenwich Civil Time or Universal Time. Such time is expressed on an hour scale from zero to 24. Thus, midnight is 0^h and noon is 12^h. Local Civil Time is reckoned from the precise longitude of the observer. On any particular meridian, Local Civil Time is more advanced at the same instant than on any meridian further west and less advanced than on any meridian further east. The difference amounts to 1/15 hour (4 minutes) of time for each degree difference in longitude.

At a given locality, clock time generally differs from civil time. Clocks are usually set for the same reading throughout an entire zone, covering about 15 degrees of longitude. The United States is divided into four time zones. The time kept in each zone is the Local Civil Time of a selected meridian near the center of the zone. Such time is called Standard Time. The four standard meridians in the United States are at west longitudes of 75 degrees (Eastern Standard Time, EST), 90 degrees (Central Standard Time, CST), 105 degrees (Mountain Standard Time, MST), and 120 degrees (Pacific Standard Time, PST). In many localities, clocks are advanced one hour beyond Standard Time in summer. In the United States such time is called Daylight Saving Time.

Time as measured by the apparent diurnal motion of the sun is called Apparent Solar Time, Local Solar Time, or Solar Time. Whereas a civil day is precisely 24 hours, a solar day is slightly different, owing to irregularities of the earth's rotation, obliquity of the earth's orbit, and other factors.

The difference between Local Solar Time (LST) and Local Civil Time (LCT) is called the Equation of Time, *E*.

The factors described above can be included in a single equation which relates solar time and clock time:

$$\text{LST} = \text{CT} + \left(\frac{1}{15}\right)(L_{\text{std}} - L_{\text{loc}}) + E - DT \tag{13.2}$$

TABLE 13.2 The Sun's Declination and Equation of Time, Calculated

	Day							
	7		14		21		28	
Month	Declination, Degrees	Eq. of Time, Hours	Declination, Degrees	Eq. of Time, Hours	Declination, Degrees	Eq. of Time, Hours	Declination, Degrees	Eq. of Time, Hours
January	−22.4	−0.10	−21.4	−0.15	−20.1	−0.19	−18.5	−0.22
February	−15.8	−0.24	−13.6	−0.24	−11.2	−0.24	−8.7	−0.22
March	−6.0	−0.20	−3.2	−0.17	−0.4	−0.13	2.4	−0.09
April	6.4	−0.04	9.0	−0.01	11.6	0.02	13.9	0.04
May	16.7	0.06	18.5	0.06	20.1	0.06	21.4	0.05
June	22.7	0.02	23.3	0.00	23.45	−0.03	23.3	−0.05
July	22.6	−0.08	21.7	−0.09	20.4	−0.10	18.9	−0.10
August	16.3	−0.09	14.1	−0.07	11.8	−0.04	9.2	−0.01
September	5.4	0.05	2.6	0.09	−0.2	0.13	−3.0	0.17
October	−6.6	0.22	−9.2	0.25	−11.8	0.27	−14.1	0.27
November	−17.1	0.27	−18.9	0.25	−20.4	0.22	−21.7	0.18
December	−22.8	0.12	−23.3	0.07	−23.45	0.02	−23.3	−0.04

where

LST = Local Solar Time, hr

CT = Clock Time, hr

L_{std} = standard meridian for the local time zone, degrees west

L_{loc} = longitude of actual location, degrees west

E = Equation of Time, hr

DT = Daylight Savings Time Correction ($DT = 0$ if not on Daylight Savings Time, otherwise DT is equal to the number of hours that the time is advanced for Daylight Savings time)

In using Eq. (13.2), all of the times should first be converted to decimal format from zero to 24 (e.g., a clock time of 3:45 p.m. should be expressed as $CT = 15.75$).

Values of the Equation of Time, E, can be calculated by the approximation given by Eq. (13.3) [3]:

$$E = 0.165 \sin 2B - 0.126 \cos B - 0.025 \sin B, \quad \text{hr} \tag{13.3}$$

where $B = \dfrac{360(n - 81)}{364}$ and n is the day of the year.

Values of E given by Eq. (13.3) are included in Table 13.2.

EXAMPLE 13.1

Determine the local solar time in Minneapolis, MN ($L_{loc} = 93\ °W$) at 2:25 p.m. Central Daylight Savings Time on July 21.

Solution: At 2:25 p.m. CT = 14 + 25/60 = 14.42 hr. For the Central Time Zone, $L_{std} = 90\ °W$. From Table 13.2 on July 21, $E = -0.10$ hr. There is a one-hour time shift for Daylight Savings Time, ∴ DT = 1. Applying Eq. (13.2),

$$\text{LST} = 14.2 + \left(\frac{1}{15}\right)(90 - 93) - 0.10 - 1 = 13.12\ \text{hr}$$

Once Local Solar Time is established, the solar hour angle, h, can be calculated. By recalling that the hour angle varies at the rate of 15 degrees per hour, that $h = 0$ at solar noon, and that our sign convention is that $h < 0$ before solar noon, the equation for the hour angle can be determined as:

$$h = 15(\text{LST} - 12) \text{ degrees} \tag{13.4}$$

Equation (13.4) provides the correct sign for the hour angle, provided that LST is expressed in a 0-to-24 hour format with LST = 12.0 at solar noon.

13.5 DERIVED SOLAR ANGLES

Besides the three basic angles—latitude, hour angle, and sun's declination—several other angles are useful in solar-radiation calculations. Such angles include the sun's zenith angle θ_H, altitude angle β, and azimuth angle ϕ. For a particular surface orientation the sun's incidence angle θ and surface-solar azimuth angle γ may be defined. All may be expressed in terms of the three basic angles.

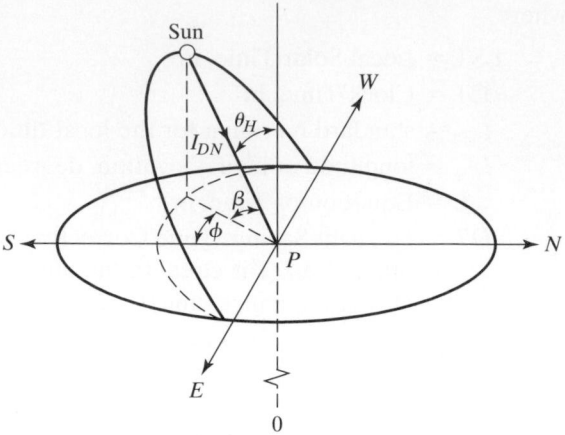

Figure 13.6 Definition of sun's zenith, altitude, and azimuth angles.

To an observer on the earth, the sun appears to move across the sky following the path of a circular arc from horizon to horizon. Figure 13.6 schematically shows one apparent solar path and defines the sun's zenith, altitude, and azimuth angles. Point P represents the position of the observer, point O is the center of the earth, and I_{DN} is a vector representing the sun's rays. The zenith angle θ_H is the angle between the sun's rays and local vertical, i.e., a line perpendicular to the horizontal plane at P (extension of $\overline{OP}$). The altitude angle β is the angle in a vertical plane between the sun's rays and the projection of the sun's rays on the horizontal plane. It follows that $\beta + \theta_H = \pi/2$. The azimuth angle ϕ is the angle in the horizontal plane measured from south to the horizontal projection of the sun's rays.

Figure 13.7 shows a coordinate system with the z axis coincident with the earth's axis. The xy plane coincides with the earth's equatorial plane. The vector I_{DN} representing the sun's rays lies in the xz plane (coinciding with a line drawn from the center of the sun to the center of the earth). The line $\overline{PS}$ pointing south from point P is perpendicular to $\overline{OP}$ and lies in the plane containing $\overline{OP}$ and the z axis.

In Fig. 13.7 let a_1, b_1, and c_1 be the direction cosines of $\overline{OP}$ with respect to the x, y, and z axes. Also let a_2, b_2, and c_2 be the corresponding direction cosines of I_{DN}. Thus,

$$a_1 = \cos l \cos h, \qquad b_1 = \cos l \sin h, \qquad c_1 = \sin l$$

$$a_2 = \cos d, \qquad b_2 = 0, \qquad c_2 = \sin d$$

The sun's zenith angle θ_H is the angle between $\overline{OP}$ and I_{DN}. By a common equation from analytic geometry, we have

$$\cos \theta_H = a_1 a_2 + b_1 b_2 + c_1 c_2$$

Thus

$$\cos \theta_H = \cos l \cos h \cos d + \sin l \sin d \qquad (13.5)$$

Since $\beta = \pi/2 - \theta_H$, we may write

$$\sin \beta = \cos l \cos h \cos d + \sin l \sin d \qquad (13.6)$$

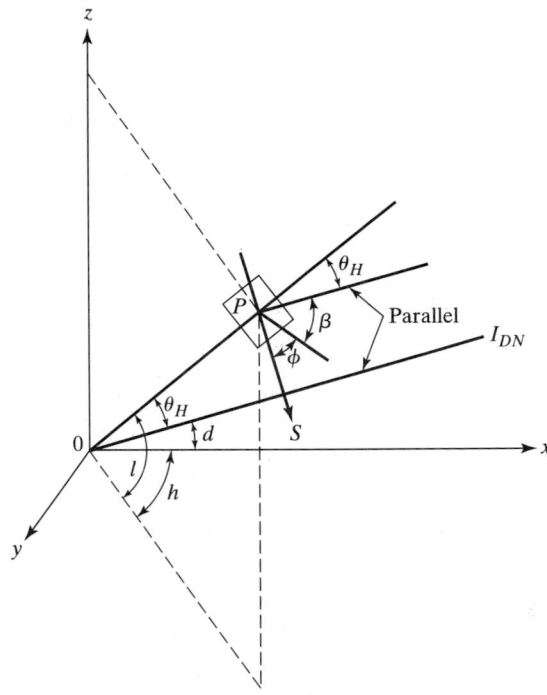

Figure 13.7 Relation of a point on the earth's surface to sun's rays.

By similar methods we may show that the sun's azimuth φ in Fig. 13.7 is given by the relation

$$\cos\phi = \frac{1}{\cos\beta}\,(\cos d \sin l \cos h - \sin d \cos l) \tag{13.7}$$

The sign convention used for the azimuth angle is to take φ negative east of south and positive west of south. Notice that this sign convention results in the hour angle, h, and the sun's azimuth angle, φ, always having the same sign. Since the cosine is an even function, calculating the right-hand side of Eq. (13.7) and taking the inverse cosine will not provide the information needed for the sign convention. The user must assign the appropriate sign.

Figure 13.8 shows the lines of Fig. 13.7 for the case of solar noon. At solar noon, $h = 0$ and $\phi = 0$ if $l > d$, or $h = 0$ and $\phi = 180°$ if $l < d$. For the case of $l = d$ the sun appears directly overhead and the azimuth φ is undefined for $h = 0$. From Fig. 13.8, we may deduce that

$$\beta_{\text{noon}} = 90° - |l - d| \tag{13.8}$$

where $|(l - d)|$ is the absolute value of $(l - d)$. Equation (13.8) allows rapid determination of the daily maximum altitude of the sun for a given location.

Equations (13.5)–(13.7) allow calculation of the sun's zenith, altitude, and azimuth angles if the declination, hour angle, and latitude are known. In applying these equations, attention must be given to correct signs. A summary of the sign convention is:

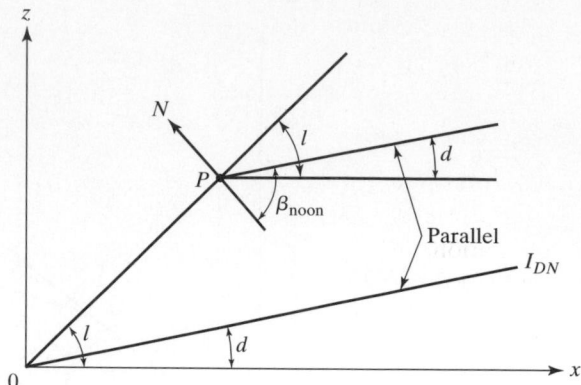

Figure 13.8 Relation of a point on the earth's surface to noon sun's rays.

l: north latitudes are positive, south latitudes are negative.

d: the declination is positive when the sun's rays are north of the equator, i.e., for the summer period in the northern hemisphere, March 22 to September 22 approximately, and negative when the sun's rays are south of the equator.

h: the hour angle is negative before solar noon and positive after solar noon.

ϕ: the sun's azimuth angle is negative east of south and positive west of south.

In calculations involving other than horizontal surfaces, it is convenient to express the sun's position relative to the surface in terms of the incidence angle θ. The sun's *angle of incidence* is the angle between the solar rays and the surface normal. (Notice that for a horizontal surface, the surface normal is local vertical and the incidence angle is equal to the zenith angle θ_H.) In order to evaluate the angle of incidence we need to specify the direction of the surface normal. This is done in terms of the surface tilt angle, Σ, and the surface azimuth angle, Ψ. These angles are defined in Fig. 13.9. The *surface tilt angle* is the

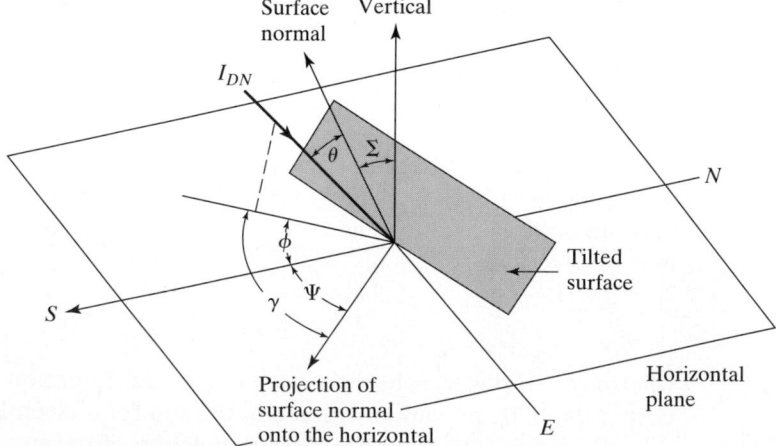

Figure 13.9 Definitions of surface azimuth, surface tilt, and surface-solar azimuth angles and the relation of sun's rays to a tilted surface.

angle between the surface normal and vertical. The *surface azimuth angle* is the angle between south and the horizontal projection of the surface normal. The same sign convention is used for the surface azimuth angle as for the solar azimuth angle; i.e., Ψ is negative for a surface that faces east of south and positive for a surface that faces west of south. The azimuth angle for a horizontal surface is undefined.

It is convenient to define one additional angle, the surface-solar azimuth angle, γ. As shown in Fig. 13.9, the *surface-solar azimuth angle* is defined as the angle between the horizontal projection of the solar rays and the horizontal projection of the surface normal. Examination of Fig. 13.9 reveals that as long as one adheres to the sign conventions for the azimuth angles ϕ and Ψ, the surface-solar azimuth angle is given by the simple relation

$$\gamma = |(\phi - \Psi)| \tag{13.9}$$

In Fig. 13.9, for the tilted surface we may derive

$$\cos\theta = \cos\beta\,\cos\gamma\,\sin\Sigma + \sin\beta\,\cos\Sigma \tag{13.10}$$

If the surface is vertical ($\Sigma = 90°$), then

$$\cos\theta = \cos\beta\,\cos\gamma \tag{13.11}$$

If the surface is horizontal ($\Sigma = 0$), then

$$\cos\theta = \sin\beta = \cos\theta_H \tag{13.12}$$

Thus, as previously described, it is seen that the incidence angle for a horizontal surface is equal to the zenith angle.

EXAMPLE 13.2

Calculate the sun's altitude and azimuth angles at 7:30 a.m. solar time on August 7 for a location at 40 degrees north latitude.

Solution: From the given data $l = 40°$, LST = 7.5. By Table 13.2, $d = 16.3°$. By Eq. (13.4), $h = 15(7.5 - 12) = -67.50°$. By Eq. (13.6),

$$\sin\beta = \cos(40)\cos(-67.5)\cos(16.3) + \sin(40)\sin(16.3) = 0.462$$

and $\beta = 27.5°$. By Eq. (13.7),

$$\cos\phi = \left[\frac{1}{\cos(27.5)}\right][\cos(16.3)\sin(40)\cos(-67.5) - \sin(16.3)\cos(40)] = 0.035$$

and $|\phi| = 88.0°$. Since the sun is east of south prior to solar noon, $\phi = -88.0°$.

EXAMPLE 13.3

Determine the solar time and azimuth angle for sunrise in Ex. 13.2.

Solution: At sunrise (or sunset), $\beta = 0$. If we let h_{sr} be the hour angle for sunrise, by Eq. (13.6) we have

$$\cos l \cos h_{sr} \cos d + \sin l \sin d = 0$$

and

$$\cos h_{sr} = -\tan l \tan d = -\tan(40)\tan(16.3) = -0.245$$

Thus, $h_{sr} = -104.2°$. By Eq. (13.4) at sunrise

$$(LST)_{sr} = \frac{h_{sr}}{15} + 12 = \frac{-104.2°}{15} + 12 = 5.05 \text{ hours}$$

or 5:03 a.m. local solar time.

Similarly, if we let h_{ss} be the hour angle at sunset, the result is $h_{ss} = +104.2°$ and $(LST)_{ss} = 18.95$ hours or 6:57 p.m. This latter result could have been determined by observing that sunrise and sunset occur symmetrically about solar noon. Thus, since sunrise occurred $(12 - 5.05) = 6.95$ hours prior to solar noon, sunset would occur 6.95 hours after solar noon or at LST = 18.95.

The azimuth angle at sunrise is found by Eq. (13.7):

$$\cos \phi = \left[\frac{1}{\cos(0)}\right][\cos(16.3)\sin(40)\cos(-104.2) - \sin(16.3)\cos(40)] = -0.366$$

and $|\phi| = 111.5°$. Thus, the azimuth angle at sunrise is $-111.5°$ and the azimuth angle at sunset is $+111.5°$. Notice this says that on August 7, for a location at 40 degrees north latitude, the sun rises and sets 21.5° north of due east and west, respectively.

EXAMPLE 13.4

Calculate the sun's incidence angle for a surface that faces 25 degrees east of south and has a tilt angle of 60 degrees at 3:00 p.m. solar time on June 7 for a location at 36 degrees north latitude.

Solution: From the given data, $l = 36°$, LST = 15.0 hours, $\Psi = -25°$, and $\Sigma = 60°$. By Table 13.2, $d = 22.7°$. By procedures identical to those in Ex. 13.2 we find $h = 45°$, $\beta = 49.0°$, and $\phi = 83.8°$. By Eq. (13.9),

$$\gamma = |83.8 - (-25)| = 108.8°$$

By Eq. (13.10),

$$\cos \theta = \cos(49.0)\cos(108.8)\sin(60) + \sin(49.0)\cos(60) = 0.194$$

and $\theta = 78.8$ degrees.

13.6 SHADING OF SURFACES FROM DIRECT SOLAR RADIATION

An obvious, but important, problem in solar-radiation calculations is determining whether a surface is sunlit. A window may be partially shaded because of setback from the external plane of a building; a flat roof may be partially shaded by parapet walls. Any external building surface may be partially or wholly shaded by nearby buildings.

A second class of problems arises when architectural projections are to be designed to prevent or control the irradiation of a surface (for example shading of windows by overhangs and awnings). Typically, the shading device is designed to exclude the sun's rays in summer but to admit them in winter.

Part IV / Heat- and Mass-Transfer Processes and Applications

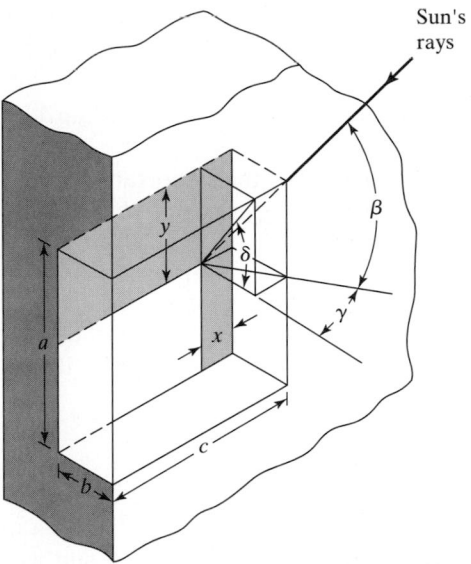

Figure 13.10 Shading of a window set back from the plane of a building.

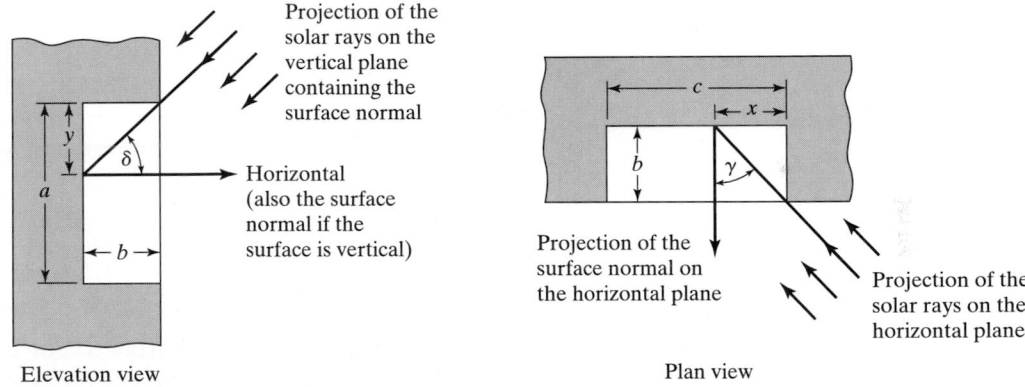

Elevation view   Plan view

Figure 13.11 Elevation and plan views of a window set back from the plane of a building.

Although each shading problem must be analyzed individually, the approach to all such problems is similar. In most cases it is desirable to make either an isometric sketch or plan and elevation sketches, showing the relations of the sun's rays to the surfaces involved.

Figure 13.10 shows a common shading problem. A window set back from the outside plane of a building will be partially shaded.

Figure 13.11 shows elevation and plan views of the window. From the elevation view it is seen that the shadow height y is given by

$$y = b \tan \delta$$

The angle δ is useful in many shading calculations. It is known as the *profile angle* or *projected altitude angle*. It is defined as: δ = profile angle = angle between horizontal and the projection of the solar rays onto the vertical plane containing the surface normal.

The profile angle δ is related to the sun's altitude angle β and the surface-solar azimuth angle γ by

$$\tan\delta = \frac{\tan\beta}{\cos\gamma} \tag{13.13}$$

In the plan view, the important angle is seen to be the surface-solar azimuth angle γ. The shadow width, x, is seen to be

$$x = b\tan\gamma$$

Another parameter of interest for partially shaded windows is the sunlit fraction, F_s, which is the area of the window exposed to direct solar radiation divided by the total area of the window. For the window of Fig. 13.10, F_s is given by

$$F_s = \frac{(a-y)(c-x)}{a \cdot c} = \frac{(a - b\tan\delta)(c - b\tan\gamma)}{a \cdot b}$$

Another typical configuration is shown in Fig. 13.12. It consists of a window with setback plus a solid overhang positioned a distance e above the window. The overhang extends a length f out from the wall and width g beyond each side of the window. One might wish to know the minimum dimensions, f and g, which will result in the window's being totally shaded at a particular time. From the plan view it is seen that the distance x must be less than or equal to the shadow width that would be cast by the setback:

$$x = (b + f)\tan\gamma - g \leq b\tan\gamma$$

or

$$g \geq f\tan\gamma$$

and from the elevation view it is seen that for total shading

$$y = (f + b)\tan\delta - e \geq a$$

or

$$f \geq (a + e)\cot\delta - b$$

Many shading applications involve south-facing surfaces. For this special case the daily variation of the profile angle follows an interesting pattern. On the equinoxes the

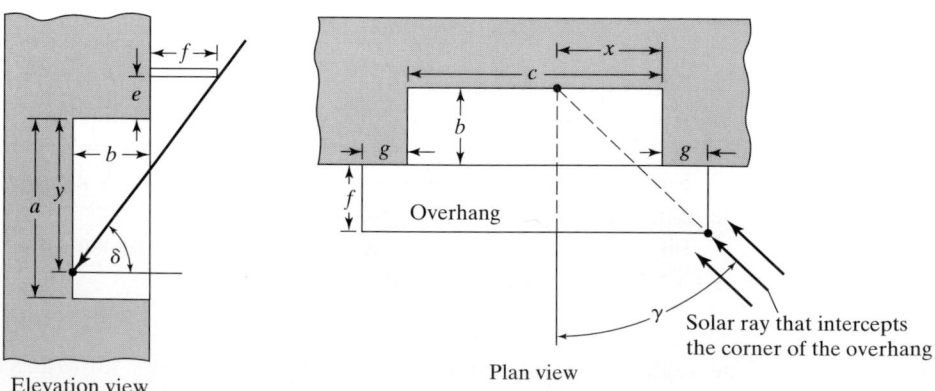

Figure 13.12 Shading of a window with both setback and overhang.

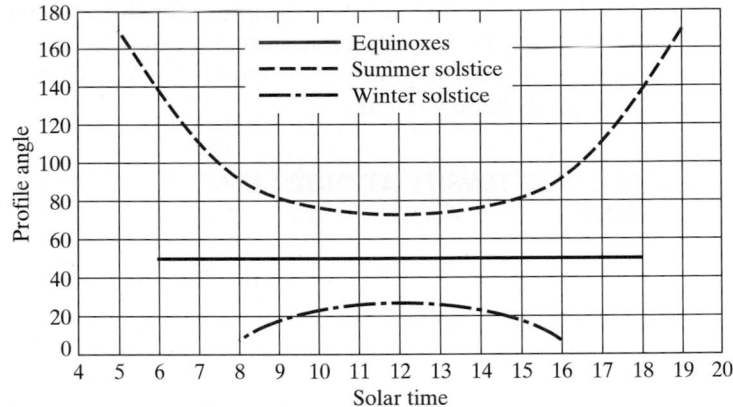

Figure 13.13 Profile angles for a south-facing surface at 40 deg north latitude.

profile angle remains constant from sunrise to sunset and is equal to $90° - l$. During the winter the profile angle is a maximum at solar noon, and during the summer it reaches its minimum at solar noon.

Figure 13.13 shows the daily variation of the profile angle for a south-facing surface at 40°N latitude. Curves are included for the winter and summer solstices and the equinoxes. The time range for the curves is from sunrise to sunset. Throughout the year the profile angle falls within the envelope defined by the curves for the winter and summer solstices. Notice that during the summer there are times when the profile angle exceeds 90°. This occurs when the sun is north of east or west (i.e., $|\phi| > 90°$). For vertical south-facing surfaces these times are not important, since they coincide with times when the direct solar radiation is behind the surface. However, these time periods can be important for nonvertical surfaces. (A quick check to see if the incidence angle is less than 90° will determine if the direct radiation is striking the surface.)

Figure 13.14 shows a flat roof partially shaded by two parapet walls. We may easily show that the shaded dimensions of the roof are

$$x_1 = \frac{a}{\tan \delta_1} \quad \text{and} \quad x_2 = \frac{a}{\tan \delta_2}$$

where δ_1 and δ_2 are the profile angles for walls 1 and 2, respectively.

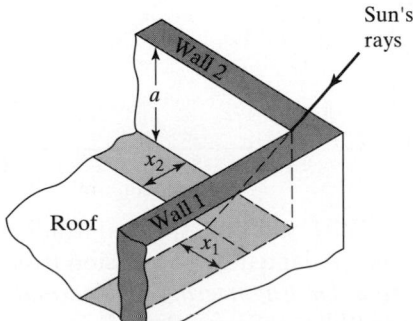

Figure 13.14 Shading of a flat roof by parapet walls.

Notice that the general procedure for the shading calculations has been to sketch elevation and plan views (or an isometric) of the geometry, locate the applicable surface-solar azimuth and profile angles, and apply the necessary trigonometric equations to find the desired shadow lengths.

13.7 SOLAR-RADIATION INTENSITY AT OUTER LIMIT OF ATMOSPHERE

When the earth is at its mean distance from the sun, the solar-radiation intensity incident upon a surface normal to the sun's rays and at the outer limit of the atmosphere is known as the *solar constant*. Early estimates of the solar constant were made from ground-based measurements of solar radiation. These estimates and later measurements made from rockets were summarized by Johnson in 1954 [4]. More recently, measurements have been made from very high altitude aircraft, balloons, and spacecraft. The World Radiation Center (WRC) has adopted a value of the solar constant, $I_{N,0}$, of 1367 W/m^2 (432 Btu/hr · ft^2). The estimated uncertainty is of the order of 1 percent [5].

Figure 13.15 shows spectral distribution of solar-radiation intensity at the outer limit of the atmosphere. Ultraviolet radiation includes the wavelength range of about 0.2 to 0.38 microns; visible radiation is contained between about 0.38 to 0.78 microns; and infrared radiation occurs at the longer wavelengths. Maximum intensity occurs within the visible range. The area under the entire curve is the solar constant.

The intensity of solar radiation $I_{N,0}$ normal to the sun's rays at the outer limit of the atmosphere varies with the earth-sun distance. Figure 13.16, calculated from data in the Smithsonian Physical Tables [6], shows the factor to be multiplied by the solar constant to give $I_{N,0}$.

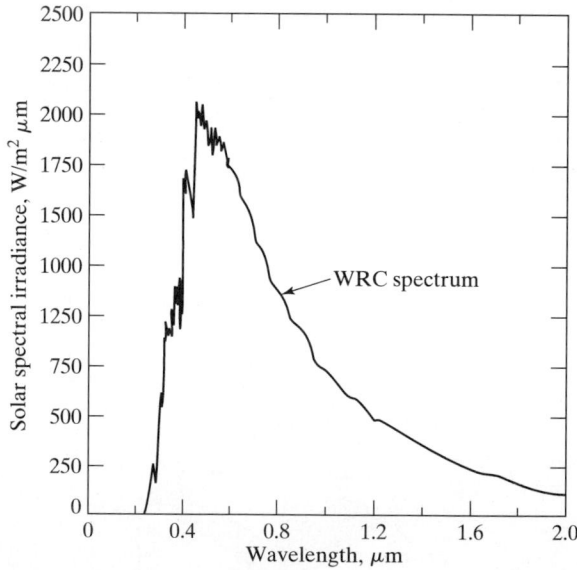

Figure 13.15 The WRC standard spectral irradiance curve at mean earth-sun distance. [Adapted by permission from J. A. Duffie and W. A. Beckman, *Solar Engineering of Thermal Processes*, 2d ed. (New York: John Wiley & Sons, 1980), 7.]

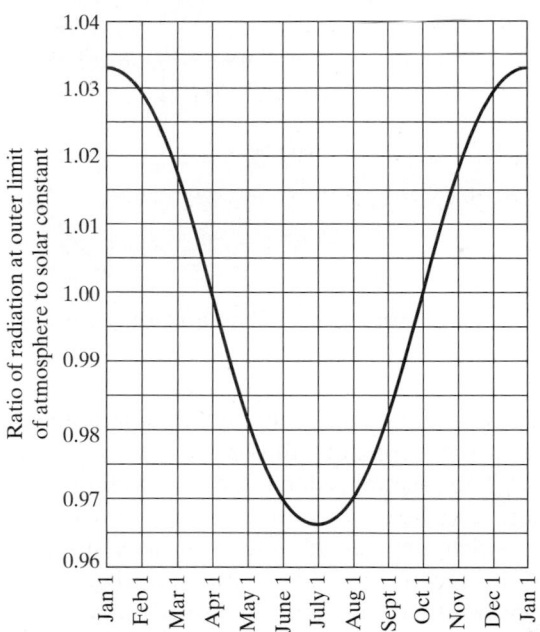

Figure 13.16 Ratio of solar-irradiation intensity at outer limit of atmosphere to the solar constant. [Reprinted by permission from *ASHRAE Trans.*, 64 (1958), 58.]

13.8 DEPLETION OF DIRECT SOLAR RADIATION BY EARTH'S ATMOSPHERE

The effects of the earth's atmosphere upon solar radiation have been studied by scientists for many years. This research has shown that when solar radiation passes through the atmosphere, part of it is intercepted by constituents such as dry-air molecules, water molecules, and dust particles, resulting in a scattering of radiation in practically all directions. Part of the radiation is absorbed, particularly by ozone in the upper atmosphere and by water vapor nearer the earth's surface. The remaining portion of the original direct radiation reaches the earth's surface unchanged in wavelength.

Some of the radiation intercepted by the atmosphere and turned aside from the direct beam reaches the earth's surface. This radiation of diffuse nature comes from the entire sky vault. Thus a surface on the earth receives solar energy of two forms—direct radiation and diffuse radiation.

The depletion of solar radiation by the atmosphere is large even during clear days, while with heavy cloudiness almost complete extinction may occur. Most environmental control problems at the earth's surface occur during clear days, when the heating effect of the sun's rays is a maximum. Fortunately, the results of depletion of solar radiation by a cloudless atmosphere are rather well established.

To understand how a clear atmosphere depletes solar radiation, we must give attention to the composition and structure of the atmosphere and consider the spectral nature of solar radiation. Experimental observations have shown that, for monochromatic radiation, the quantity of radiation depleted by absorption and scattering increases arith-

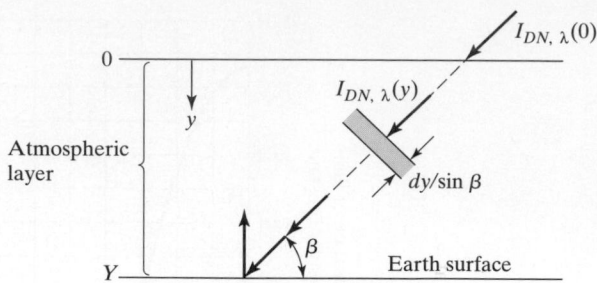

Figure 13.17 Schematic for the evaluation of solar radiation through the atmosphere.

metically with the intensity of the radiation and geometrically with the quantity of material passed through by the sun's rays.

The spectral reduction of the beam solar radiation as it passes through the atmosphere can be expressed as a function of the *monochromatic extinction coefficient*, K_λ, which includes the effects of both absorption and scattering. An equation describing the spectral reduction can be derived with the aid of Fig. 13.17. In this figure, the monochromatic direct normal solar radiation, $I_{DN,\lambda}$, is shown passing through the earth's atmosphere of depth Y at an altitude angle of β. If we focus our attention on what happens as the solar radiation moves through the atmosphere contained in the differential path length $dy/\sin\beta$, the change in the intensity, $dI_{DN,\lambda}(y)$, may be expressed as

$$dI_{DN,\lambda}(y) = -K_\lambda(y) I_{DN,\lambda}(y) \frac{dy}{\sin\beta} \tag{13.14}$$

Separating variables and integrating over the entire layer, we obtain

$$\int_{I_{DN,\lambda}(0)}^{I_{DN,\lambda}} \frac{dI_{DN,\lambda}(y)}{I_{DN,\lambda}(y)} = -\int_0^Y \frac{K_\lambda(y)\,dy}{\sin\beta}$$

or

$$I_{DN,\lambda} = I_{DN,\lambda}(0) \exp\left[-\frac{1}{\sin\beta}\int_0^Y K_\lambda(y)\,dy\right] \tag{13.15}$$

where $I_{DN,\lambda}$ is the monochromatic direct normal solar flux at the surface of the earth. This result can also be interpreted in terms of an effective monochromatic transmittance of the atmosphere. If this transmittance, τ_λ, is defined as

$$\tau_\lambda = \frac{I_{DN,\lambda}}{I_{DN,\lambda}(0)} \tag{13.16}$$

then

$$\tau_\lambda = \exp\left[-\frac{1}{\sin\beta}\int_0^Y K_\lambda(y)\,dy\right] \tag{13.17}$$

In solar-radiation calculations, unit depth of the atmosphere is taken as the depth when the sun is at the zenith. The air mass m is the ratio of the length of path of the sun's rays through the atmosphere to the length of path if the sun were at the zenith. Except for very low solar altitude angles, the air mass is equal to the cosecant of the altitude angle.

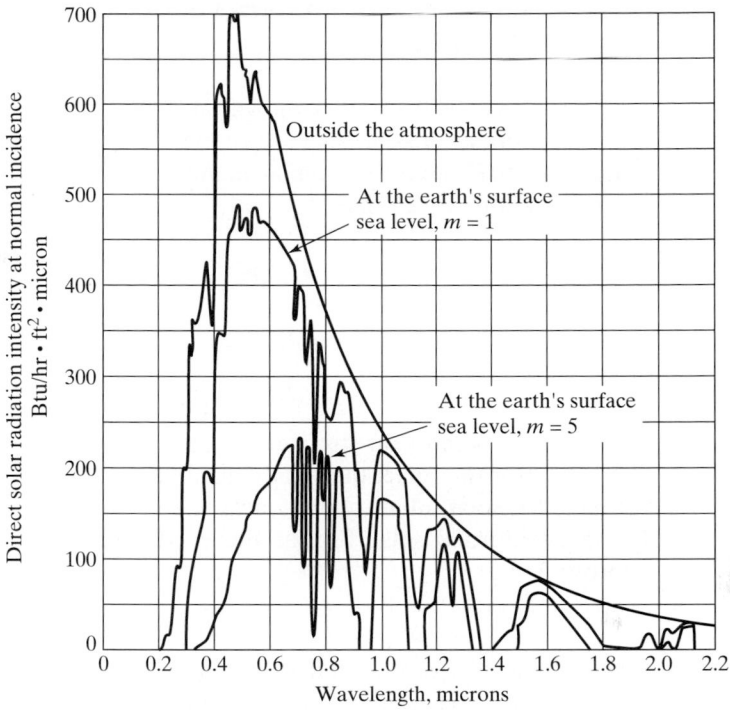

Figure 13.18 Spectral distribution of direct solar-radiation intensity at normal incidence for the upper limit of the atmosphere and at the earth's surface during clear days. [Adapted by permission from *ASHRAE Trans.*, 64 (1958), 50.]

Moon [7] correlated the work of several investigators and calculated overall transmittance for monochromatic radiation including the effects of scattering by dry-air molecules, water-vapor molecules, and dust particles, and the effects of absorption by ozone and water vapor. At any single wavelength, the overall transmittance multiplied by the intensity at the outer limit of the atmosphere gives the intensity at the earth's surface. Such calculations carried out for the entire solar spectrum allow energy-distribution curves to be drawn for a location at the earth's surface or for any point within the atmosphere.

Figure 13.18 shows typical spectral distribution curves for direct solar radiation for three sets of conditions. The upper curve applies at the outer limit of the atmosphere and is similar to the curve shown in Fig. 13.15. The two lower curves apply to the earth's surface during clear days, for a sea-level location, for a precipitable water depth of 30 mm, and for a dust scale of 400 (moderately dusty atmosphere). Conditions for the two lower curves differ only in value of the air mass. The middle curve is for $m = 1.0$ ($\beta = 90°$), and the lower curve is for $m = 5.0$ ($\beta = 11.5°$). The area under either of the lower curves divided by the area under the upper curve (solar constant) is the *atmospheric transmission factor*. For $m = 1.0$ the transmission factor is 0.63, and for $m = 5.0$ the transmission factor is 0.28. Thus the length of path of the sun's rays through the atmosphere is of extreme importance in affecting reduction of solar intensity. Atmospheric factors are available for a wide range of conditions [8].

13.9 ESTIMATION OF INTENSITY OF SOLAR RADIATION DURING AVERAGE CLEAR DAYS

In the previous section it was shown that the monochromatic intensity of the direct normal solar flux at the earth's surface can be expressed in terms of the monochromatic extinction coefficient. In a similar manner, if K is the total extinction coefficient for the solar flux defined such that

$$dI_{DN}(y) = -K(y)I_{DN}(y)\frac{dy}{\sin\beta} \tag{13.18}$$

an expression for the direct normal solar flux at the earth's surface is

$$I_{DN} = I_{DN}(0)e^{-(1/\sin\beta)\int_0^Y K(y)\,dy} \tag{13.19}$$

where $I_{DN}(0)$ is the solar flux outside the earth's atmosphere.

In order to predict the solar contributions to building cooling loads it is desirable to estimate the solar intensity on typical or average clear days. Using Eq. (13.19) as a guide, the direct normal solar flux at the earth's surface for "average clear days" can be expressed as

$$I_{DN} = Ae^{-B/\sin\beta} \tag{13.20}$$

where the coefficients A and B are empirically determined from measurements of I_{DN} made on typical clear days. Comparison of Eqs. (13.19) and (13.20) shows that the coefficients can be interpreted as

A = apparent direct normal solar flux at the outer edge of he earth's atmosphere

B = apparent atmospheric extinction coefficient

The numerical values of A and B will vary throughout the year because of seasonal changes in the dust and water-vapor content of the atmosphere and because of the changing earth-sun distance. The *ASHRAE Handbook of Fundamentals* [9] lists recommended values for the coefficients A and B for the twenty-first day of each month. These values are presented in Table 13.3. Also included in the table are the values of the declination and equation of time that ASHRAE lists in conjunction with the clear-day coefficients. These values are for the base year 1964. A comparison of the declinations listed for 1964 and those predicted by Eq. (13.1) (and listed in Table 13.2) reveals that, with the exception of October, they agree to 0.6 degree or better. The difference in the October value is only 1.3 degrees. Similarly, the calculated values for the Equation of Time agree to within 0.01 hour with those for the 1964 base year. Thus, for the purpose of making HVAC calculations the approximations of Eqs. (13.1) and (13.3) provide acceptable values for the declination angle and Equation of Time.

The use of Eq. (13.20) with the coefficients presented in Table 13.3 is commonly referred to as the "ASHRAE Clear Day Solar Flux Model." The model also approximates the average clear-day diffuse solar flux from the sky that strikes a horizontal surface, I_{dH}, by the relation

$$I_{dH} = CI_{DN} \tag{13.21}$$

The recommended values of the dimensionless coefficient C are listed in Table 13.3. The physical basis for Eq. (13.21) is not clear; in fact there may not be one. However, since the

TABLE 13.3 Coefficients for Average Clear Day Solar Radiation Calculations for the Twenty-First Day of Each Month, Base Year 1964

| | A | | B | C | | |
	$\dfrac{\text{Btu}}{\text{hr} \cdot \text{ft}^2}$	$\dfrac{\text{W}}{\text{m}^2}$	Dimensionless Ratios		Declination, deg	Equation of Time, hr
January	390	1230	0.142	0.058	−20.0	−0.19
February	385	1215	0.144	0.060	−10.8	−0.23
March	376	1186	0.156	0.071	0.0	−0.13
April	360	1136	0.180	0.097	11.6	0.02
May	350	1104	0.196	0.121	20.0	0.06
June	345	1088	0.205	0.134	23.45	−0.02
July	344	1085	0.207	0.136	20.6	−0.10
August	351	1107	0.201	0.122	12.3	−0.04
September	365	1151	0.177	0.092	0	0.13
October	378	1192	0.160	0.073	−10.5	0.26
November	387	1221	0.149	0.063	−19.8	0.23
December	391	1233	0.142	0.057	−23.45	0.03

SOURCE: Adapted by permission from *ASHRAE Handbook, Fundamentals Edition, 1993.*

diffuse solar radiation from the sky on clear days is relatively small, taking it as some fraction of the direct normal solar flux provides a sufficiently accurate estimate of its value.

The ASHRAE Clear Day Model does not give the maximum values for I_{DN} that can occur during the month but rather is representative of conditions on average, cloudless days. For very clear atmospheres, ASHRAE points out that values of I_{DN} can be 15 percent higher than those calculated using Eq. (13.20) and Table 13.3. ASHRAE also states that a "Clearness Number" modifier may be applied to adjust for local climatalogical variations. However, the data used for determining the ASHRAE proposed Clearness Number Modifier are so old that it is doubtful that they still apply. Therefore, it is the author's opinion that the application of these modifiers to the Clear Day Model is unwarranted until new data are available.

13.10 SOLAR RADIATION STRIKING A SURFACE

The solar radiation striking a surface generally consists of three components: direct, diffuse, and reflected. The direct, or beam, solar radiation is that received from the sun without having been scattered by the atmosphere. The *direct solar flux* (energy/area-time) striking a surface will be denoted by I_D. If the surface is perpendicular to the solar rays, the incident solar flux is equal to the Direct Normal flux, I_{DN}. From Fig. 13.9 it can be seen that if the solar flux strikes a surface at an angle of incidence θ, the direct solar flux striking the surface is given by

$$I_D = I_{DN} \cos\theta \qquad (13.22)$$

In the case of a horizontal surface, an additional subscript, H, will be used. Thus, I_{DH} is the direct solar flux striking the horizontal. Notice that for a horizontal surface the incidence angle is equal to the zenith angle, θ_H, and therefore

$$I_{DH} = I_{DN} \cos\theta_H = I_{DN} \sin\beta \qquad (13.23)$$

The *diffuse solar radiation* is that received from the sun after its direction has been changed by scattering by the atmosphere. The diffuse solar flux striking a surface will be denoted by I_d for the general case and by I_{dH} and I_{dV} for the special cases of horizontal and vertical surfaces, respectively. Diffuse radiation is typically of rather short wavelength, since short-wavelength radiation is scattered more by the atmosphere. Although diffuse solar radiation on clear days is usually small compared to direct radiation, it cannot be ignored in engineering calculations. During extremely cloudy days, only diffuse solar radiation may reach the ground.

Because of its nondirectional nature, diffuse solar radiation is more difficult to analyze than direct solar radiation; consequently, less is known about it. A common approximation is that the sky is a uniform radiator of diffuse radiation. If so, intensity of diffuse radiation incident upon vertical surfaces of all orientations would be the same and would be exactly one-half the intensity incident upon a horizontal surface. Experimental evidence disputes this assumption. Figure 13.19 shows results of one clear day's measurements at Minneapolis. The results show that for clear days the sky is distinctly a nonuniform radiator of diffuse radiation.

Figure 13.20 shows the ratio of diffuse sky radiation incident upon a vertical surface to that incident upon a horizontal surface as a function of the cosine of the sun's angle of incidence. These data are based upon experimental measurements for clear days at Minneapolis.

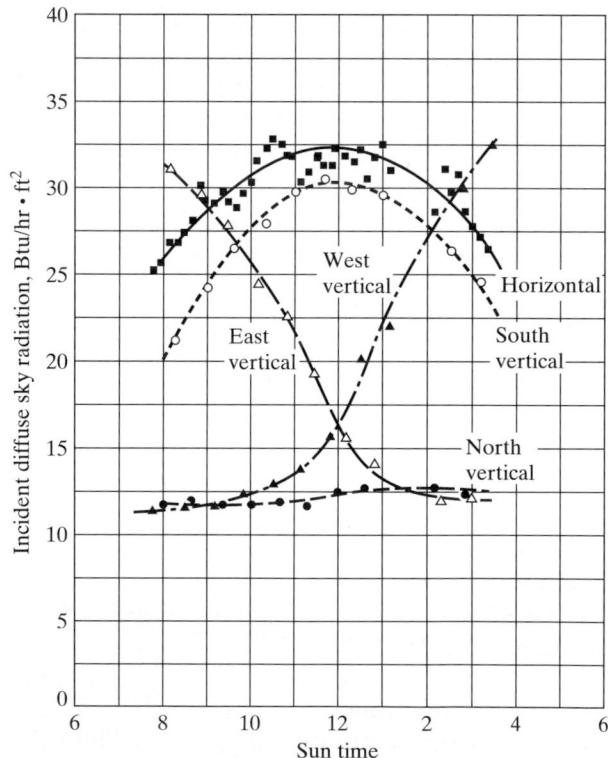

Figure 13.19 Variation of diffuse solar radiation from a clear sky incident upon various surfaces. [Reprinted by permission from *ASHRAE Trans.*, 69 (1963), 29.]

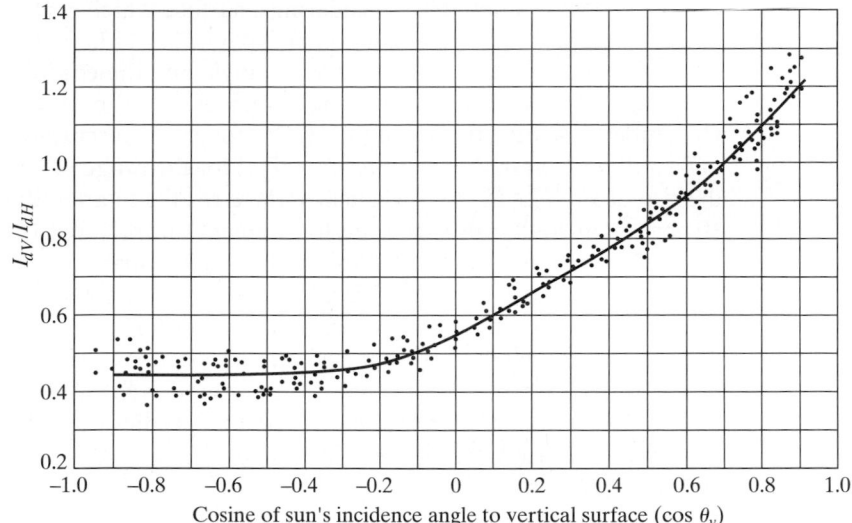

Figure 13.20 Ratio of diffuse sky radiation incident upon a vertical surface to that incident upon a horizontal surface during clear days. [Reprinted by permission from *ASHRAE Trans.*, 69 (1963), 29.]

The ASHRAE Clear Day Model, as outlined in the *Handbook of Fundamentals*, calculates the diffuse solar flux striking a surface by using a curve fit of the data in Fig. 13.20 if the surface is vertical and by using the approximation that the sky is a uniform radiator of diffuse solar radiation if the surface is not vertical. The curve fit used is

$$\frac{I_{dV}}{I_{dH}} = 0.45 \qquad\qquad \text{for } \cos\theta \le -0.2 \qquad (13.24a)$$

$$\frac{I_{dV}}{I_{dH}} = 0.55 + 0.437\cos\theta + 0.313\cos^2\theta \qquad \text{for } \cos\theta > -0.2 \qquad (13.24b)$$

where the value of I_{dH} is calculated using Eq. (13.21).

For nonvertical surfaces and assuming the sky to be a diffuse source, the ratio of the diffuse solar flux striking the surface to that striking the horizontal is given by

$$\frac{I_d}{I_{dH}} = \frac{1 + \cos\Sigma}{2} \qquad (13.25)$$

Equation (13.25) can be derived by observing that the diffuse solar radiation per unit time striking the surface from a diffuse sky source is given by the product of the flux leaving the source, the area of the source, and the radiation shape factor from the sky to the surface. The reciprocity relationship for shape factors is applied, thus replacing the product of the sky area and sky-to-surface shape factor with the product of the surface area and surface-to-sky shape factor. The ratio I_d/I_{dH} is then seen to be just the corresponding ratio of shape factors from the respective surfaces to the sky. For a surface with tilt angle Σ, the shape factor is given by $(1 + \cos\Sigma)/2$. (Note that the shape factor is unity, as expected, for a horizontal surface.) Therefore, we obtain the expression given by Eq. (13.25).

The switch from Eq. (13.25) to Eq. (13.24) as the tilt angle goes to 90 degrees results in a discontinuity in the calculated solar flux striking a surface. The magnitude of this change is best evaluated by considering a surface which is receiving the sum of direct and diffuse solar radiation. Examination of Eqs. (13.21), (13.22), (13.23), and (13.24) shows that $(I_D + I_{dV})$ is a function of the incidence angle and the coefficient C. From Table 13.3 it is seen that the coefficient C is a maximum of 0.136 in July and a minimum of 0.057 in December. For July, the use of Eq. (13.25) for a vertical surface leads to an underprediction of 8 to 10 percent in the quantity $(I_D + I_{dV})$ for the range of $0 \le \cos\theta \le 1.0$. For December the use of Eq. (13.25) for a vertical surface results in an underprediction in $(I_D + I_{dV})$ of 10 percent at $\cos\theta = 0$. This drops to 4 percent for the range $0.1 \le \cos\theta \le 1.0$.

The *reflected solar radiation* is that which strikes a surface after the radiation is reflected from surrounding surfaces. In general, the solar radiation reflected upon a surface depends on the particular location, orientation, and solar-reflectance characteristics of the surrounding surfaces. One commonly occurring situation is where the solar radiation is reflected from the ground. If the ground is horizontal and if the reflection is diffuse, a derivation similar to the one used to predict the diffuse sky radiation striking a surface can be used. If I_R is the reflected solar flux striking a surface of area A for this special case, then

$$AI_R = \rho_g I_H A_g F_{gA} \tag{13.26}$$

where

$\quad A_g$ = is the area of the ground

$\quad F_{gA}$ = shape factor from the ground to the area A (i.e., the fraction of radiation that leaves the ground that strikes the area A)

$\quad \rho_g$ = solar reflectance of the ground

$\quad I_H$ = total solar flux striking the horizontal ground.

From reciprocity

$$AF_{Ag} = A_g F_{gA} \tag{13.27}$$

where F_{Ag} is the shape factor from the surface area A to the ground. This leads to

$$I_R = \rho_g I_H F_{Ag} \tag{13.28}$$

For a diffusely reflecting surface with a tilt angle of Σ, the shape factor between it and the horizontal is given by

$$F_{Ag} = \frac{1 - \cos\Sigma}{2} \tag{13.29}$$

and

$$I_R = \frac{\rho_g I_H (1 - \cos\Sigma)}{2} \tag{13.30}$$

The reflectance of the ground varies with type of ground cover. The reflectance of browned grass is about 0.2, while that of bare soil is about 0.1. The reflectance of fresh snow cover may be as high as 0.87, with the value decreasing to less than 0.5 as the snow becomes dirty. Figure 13.21 shows experimentally determined values for several common types of ground surfaces.

The total solar flux striking a surface at any instant is the sum of the three components

$$I = I_D + I_d + I_R \tag{13.31}$$

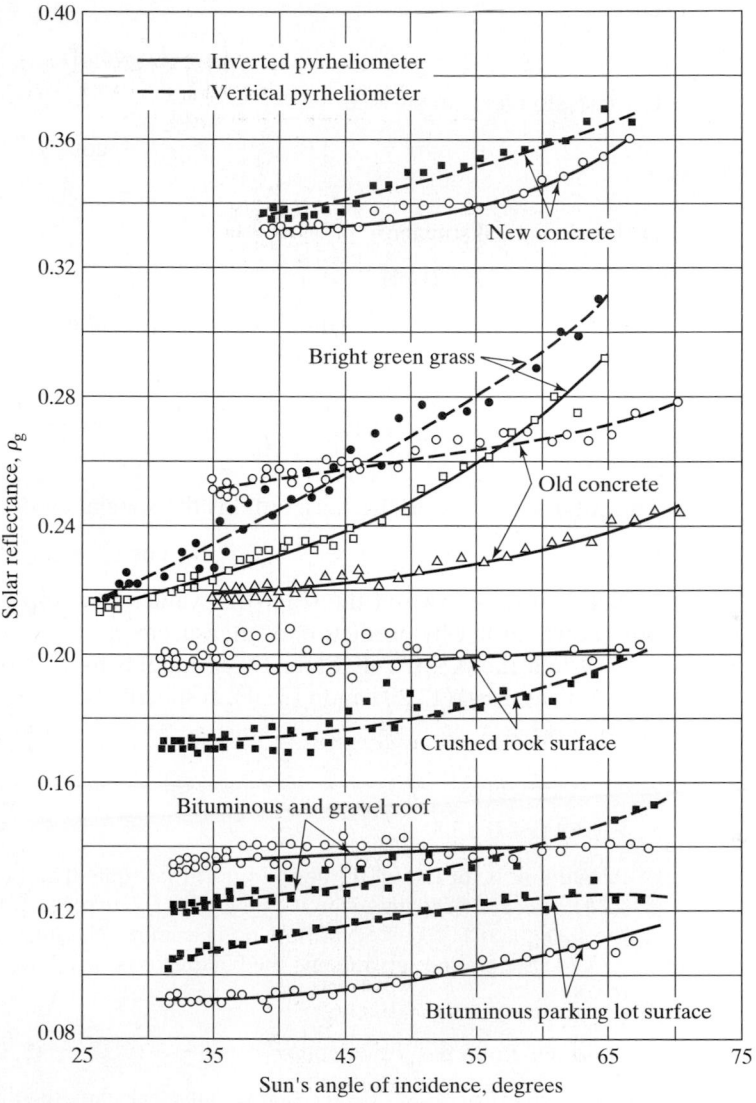

Figure 13.21 Solar reflectance for various ground surfaces. [Reprinted by permission from *ASHRAE Trans.*, 69 (1963), 31.]

In Chapter 15, calculations will be made of the rate of solar energy that enters a conditioned space through various types of fenestrations. In many instances the window design will include exterior shading such as that described in Sec. 13.6. In these instances the usual approximation is that only the direct solar radiation is affected, and the total solar flux incident on the window is approximated by

$$I = F_S I_D + I_d + I_R \tag{13.31a}$$

where F_s is the previously discussed unshaded fraction of the window.

Equation (13.31) or (13.31a), along with the previously presented equations, repeated below for convenience, permits the calculation of the solar flux striking a surface.

Direct:

$$I_D = I_{DN} \cos\theta \qquad (13.22)$$

Diffuse, nonvertical surface:

$$I_d = \frac{I_{dH}(1 + \cos\Sigma)}{2} \qquad (13.25)$$

Diffuse, vertical surface:

$$I_{dV} = I_{dH}(0.45) \qquad \text{for } \cos\theta \leq -0.2 \qquad (13.24a)$$

$$I_{dV} = I_{dH}(0.55 + 0.437\cos\theta + 0.313\cos^2\theta) \qquad \text{for } \cos\theta > -0.2 \qquad (13.24b)$$

Reflected:

$$I_R = \frac{\rho_g I_H(1 - \cos\Sigma)}{2} \qquad (13.30)$$

The total solar flux on the horizontal is the special case where $\Sigma = 0$; thus

$$I_H = I_{DN}\cos\theta_H + I_{dH} \qquad (13.32)$$

Notice that if any two of the solar-flux values that appear in the above set of equations are known, either by prediction or measurement, the rest of the quantities can be evaluated. When the ASHRAE Clear Day Model is used, the quantities I_{DN} and I_{dH} are predicted using Eqs. (13.20) and (13.21), respectively, along with the coefficients presented in Table 13.3.

EXAMPLE 13.5

A building is located at 45 degrees north latitude. The building has a skylight that faces 15 degrees east of south and is at a tilt angle of 60 degrees. Calculate the solar flux incident on the skylight at 1:30 p.m. solar time on December 21 using the ASHRAE Clear Day Model. Assume the ground surrounding the building is covered with snow with a ground reflectance of 0.8.

Solution: From the given data $l = 45°$, $\phi = -15°$, $\Sigma = 60°$, LST = 13.5 hr, and $\rho_g = 0.8$.

From Eq. (13.31) we see that we must calculate the direct, diffuse, and reflected components of the solar radiation. In addition, from Eqs. (13.20) and (13.22) we see that altitude and incidence angles are also required. By procedures identical to those in Ex. 13.2 and 13.4, and using $d = -23.45$ from Table 13.3, we find $\beta = 18.5°$, $\phi = 21.7°$, $\gamma = 36.7°$, and $\theta = 35.2°$.

From Table 13.3, $A = 391$ Btu/hr · ft^2, $B = 0.142$, and $C = 0.057$. The direct normal solar flux is calculated using Eq. (13.20):

$$I_{DN} = 391 \exp\left[-\frac{0.142}{\sin(18.5)}\right] = 250.2 \text{ Btu/hr} \cdot \text{ft}^2$$

and the diffuse solar flux striking the horizontal is calculated using Eq. (13.21):

$$I_{dH} = 0.057(250.2) = 14.3 \text{ Btu/hr} \cdot \text{ft}^2$$

The direct solar flux striking the skylight is calculated using Eq. (13.22):

$$I_D = 250.2 \cos(35.2) = 204.4 \text{ Btu/hr} \cdot \text{ft}^2$$

(The student should be aware that the incident solar flux calculated using the ASHRAE Clear Day Model is an estimate that does not justify the final result to one decimal place. However, to minimize round-off error the recommended procedure is to carry the decimal place for each of the components and then round off the final answer to the nearest whole number.)

Since the surface is not vertical, Eq. (13.25) is used to find the diffuse solar flux striking the surface:

$$I_d = \frac{14.3[1 + \cos(60)]}{2} = 10.7 \text{ Btu/hr} \cdot \text{ft}^2$$

The reflected solar flux striking the surface is estimated using Eqs. (13.32) and (13.30):

$$I_H = (250.2)\cos(71.5) + 14.3 = 94.1 \text{ Btu/hr} \cdot \text{ft}^2$$

and

$$I_R = \frac{0.8(94.1)[1 - \cos(60)]}{2} = 18.8 \text{ Btu/hr} \cdot \text{ft}^2$$

Finally, the total solar flux striking the skylight is the sum of the direct, diffuse, and reflected fluxes [Eq. (13.31)]:

$$I = 204.4 + 10.7 + 18.8 = 233.9 \text{ Btu/hr} - \text{ft}^2$$

or, rounding to the nearest whole number,

$$I = 234 \text{ Btu/hr} - \text{ft}^2$$

13.11 SOLAR-RADIATION MEASUREMENT

Experimental determination of the energy transferred to a surface by solar radiation requires instruments which will measure the heating effect of direct solar radiation and diffuse solar radiation.

There are two general classes of solar-radiation measuring devices. The instrument used to measure direct normal or beam radiation is referred to as a *pyrheliometer*. The other instrument, called a *pyranometer*, is able to measure total radiation within its hemispherical field of view. If the sensing element is shaded from the sun's direct rays, a pyranometer can also be used to measure diffuse radiation alone.

For almost a century, various solar scientists have attempted to perfect an instrument for measuring the thermal energy of the sun's rays. Just prior to 1900, Ångstrom in Sweden developed the first reliable pyrheliometer [10]. The Ångstrom pyrheliometer is based on the principle of electrical compensation. Solar energy is absorbed by a thin, blackened metallic strip at the base of a cylindrical tube. A similar strip, shaded from the sun's rays, is heated electrically, such that, at the time of a reading, both strips are at the same temperature. Response of the Ångstrom pyrheliometer defines the Ångstrom Scale of Solar Radiation.

Most development work in the United States on normal-incidence-type pyrheliometers has been done by the Smithsonian Institution. In its early work, a mercury-type pyrheliometer was used. Response of this instrument defined the Smithsonian Original Scale of Solar Radiation. Later, the Smithsonian Institution adopted the first version of Abbot's water-flow-type pyrheliometer [11]. Response of this instrument defined the Smithsonian Scale of 1913. In 1932 the Smithsonian Institution adopted Abbot's improved version of the water-flow pyrheliometer as its standard [12]. Response of this instrument defined the Smithsonian Scale of 1932.

Abbot's 1913 pyrheliometer was basically a calorimeter. Solar energy trapped by a well-insulated, cylindrical blackbody chamber was absorbed by water circulating through a coil. Measurement of temperature rise of the water and its flow rate made the calculation of the incident solar radiation possible. Abbot's 1932 pyrheliometer included modifications to the 1913 type. The device embodied two identical cylindrical chambers whose walls were cooled by water. The water flow was divided equally between the two chambers. While one chamber was open to entry of solar radiation, the other was closed, and its receiving surface was heated electrically. The electrical input was varied such that the water left each chamber at the same temperature. The heat gained by the exposed tube was equated to the electrical input of the closed tube.

Over the years many direct comparisons have been made between the Ångstrom and Smithsonian pyrheliometers. These comparisons have shown that the Smithsonian Scale of 1913 is about 3.5 percent higher than the Ångstrom Scale and about 2.4 percent higher than the Smithsonian Scale of 1932. In 1956, at the International Radiation Conference held at Davos in Switzerland, a new solar radiation scale called the International Pyrheliometric Scale of 1956 was recommended for worldwide adoption [13]. No new pyrheliometer was involved. The new scale was defined as the Ångstrom Scale plus 1.5 percent or the Smithsonian Scale of 1913 minus 2.0 percent.

Beginning with the 1956 International Radiation Conference, new comparisons have been made at approximately 5-year intervals under the auspices of the World Meteorological Organization. Comparisons made in 1975 resulted in the establishment of a new pyrheliometric scale, the World Radiometric Reference, also referred to as the Solar Constant Reference Scale. The World Radiometric Reference scale is 2.2 percent higher than the International Pyrheliometric Scale of 1956 [14].

When using pyrheliometers, close attention must be given to the scale against which the instruments were calibrated. Prior to 1957, practically all secondary-type pyrheliometers made in the United States were calibrated in terms of the Smithsonian Scale of 1913. Since July 1, 1957, most commercially available pyrheliometers have been calibrated in terms of the International Scale. Likewise, one must be careful in analyses using published radiation data. Prior to July 1, 1957, the U.S. Weather Bureau reported measurements using the Smithsonian Scale of 1913, but subsequently the Bureau has used the International Scale.

The Ångstrom pyrheliometer is the only standard type which is convenient for everyday use. The Smithsonian water-flow pyrheliometer requires auxiliary equipment and is not convenient for field measurements. For this reason, several secondary-type pyrheliometers have been developed.

The most widely accepted secondary pyrheliometer for measuring solar radiation at normal incidence is the Smithsonian silver-disk pyrheliometer [15]. This instrument, also designed by Dr. Abbot, consists of a blackened silver disk placed at the bottom of a brass tube through which the solar rays are admitted. The rate of increase of temperature of a pool of mercury under the disk is used as an indication of solar intensity. Modern adaptations of the Smithsonian silver-disk pyrheliometer use a multijunction thermopile for the detector. The field of view of a common pyrheliometer is about 5.7°, and it is typically mounted on a tracking mechanism so that a continuous record of the direct normal solar flux can be obtained.

Instruments for measuring total solar radiation (direct plus diffuse) are called *pyranometers*. Most of the available solar-radiation data have been obtained using pyranometers. A common pyranometer detector consists of a multijunction thermopile measuring the temperature difference between two circumferentially symmetric areas. Special coatings are applied to the two areas, such that one has a high solar absorptance (the high-temperature side of the thermopile) and the other has a very low solar absorptance (the low-temperature side of the thermopile). The detector is covered by one or more glass domes to eliminate wind-induced heat-transfer effects. The domes are carefully designed and manufactured so they do not introduce distortions in the pyranometer's angular (cosine) response.

ENDNOTES

1. U. S. Nautical Almanac Office, *The American Ephemeris and Nautical Almanac* (Washington, D.C.: U.S. Naval Observatory, Annual).

2. P. I. Cooper, "The Absorption of Solar Radiation in Solar Stills," *Solar Energy*, 12: 3 (1969).

3. J. A. Duffie and W. A. Beckman, *Solar Engineering of Thermal Processes* (New York: John Wiley & Sons, 1980), 9–10.

4. F. S. Johnson, "The Solar Constant," *Journal of Meteorology*, 11 (December 1954), 431–439.

5. A more detailed discussion of measurements of the solar constant is presented in J. A. Duffie and W. A. Beckman, *Solar Energy of Thermal Processes*, 2d ed. (New York: John Wiley & Sons, 1991), 5–10.

6. *Smithsonian Physical Tables*, 6th ed. (Washington, D.C.: Smithsonian Institution, 1914), Table 181.

7. P. Moon, "Proposed Standard Radiation Curves for Engineering Use," *Journal of the Franklin Institute*, 230 (November 1940), 583–617.

8. J. L. Threlkeld and R. C. Jordan, "Direct Solar Radiation Available on Clear Days," *ASHRAE Trans.,* 64 (1958), 45–56.

9. *ASHRAE Handbook, Fundamentals Volume* (Atlanta: *ASHRAE, 1993*).

10. K. Ångstrom, "The Absolute Determination of the Radiation of Heat with the Electric Compensation Pyrheliometer," *Astrophysical Journal*, 9 (1899), 332–340.

11. C. G. Abbot, F. E Fowle, and L. B. Aldrich, "Improvements and Tests of Solar-Constant Methods and Apparatus," *Annals of the Astrophysical Observatory*, 3 (1913), 39–72.

12. L. B. Aldrich and W. B. Hoover, "Pyrheliometry," *Annals of the Astrophysical Observatory*, 7 (1954), 99–104.

13. A. J. Drummond and H. W Greer, "Fundamental Pyrheliometry," *The Sun at Work*, 3 (June 1958), 3–5, 11.

14. J. A. Duffie and W. A. Beckman, "Solar Energy of Thermal Processes," 2d ed. (New York: John Wiley & Sons, 1991), 49.

15. C. G. Abbot, "The Silver Disk Pyrheliometer," *Smithsonian Miscellaneous Collections*, 56: 19 (1911), 1–10.

PROBLEMS

13.1 Calculate the sun's altitude and azimuth angles at 9:00 a.m. solar time on September 1 at 42 deg north latitude.

13.2 Determine the solar time and azimuth angle for sunrise at 50 deg north latitude on
(a) June 21 and
(b) December 21.

13.3 Determine the altitude angle of the sun at a time on July 14 at 45 deg north latitude when the horizontal projection of the sun's rays is normal to a west-facing vertical surface.

13.4 What is the maximum altitude angle of the sun for a location at
(a) 45 deg north latitude,
(b) 23.5 deg north latitude, and
(c) the equator?

13.5 Prove that at the times of the equinoxes, the sun rises due east ($\phi = -90$ deg) for all locations on the earth.

13.6 For a location at 85 deg west longitude and 43 deg north latitude on July 7, determine
(a) the incidence angle of the sun for a horizontal surface at 4:00 p.m. Central Daylight Saving Time and
(b) the time of sunset in Central Daylight Saving Time.

13.7 Calculate the angle of incidence at 11:00 a.m. EST on July 14 for a location at 36 deg north latitude and 80 deg west longitude for
(a) a horizontal surface,
(b) a south-facing vertical surface, and
(c) an inclined surface tilted 65 deg from the vertical ($\Sigma = 25°$) and facing 30 deg south of due east.

13.8 Consider a flat plate located at the equator. The plate faces due north and is tilted toward north by 30 deg from the equatorial plane. Determine the number of hours that the *south-facing* side would be sunlit on June 22.

13.9 Assuming the earth's solar constant to be 1367 W/m^2, calculate the equivalent surface temperature of the sun, if the sun is assumed to be a blackbody radiator.

13.10 Calculate the daily total solar radiation in $Btu/hr \cdot ft^2$ incident upon a horizontal surface at the outer limit of the atmosphere on March 1 at 48 deg north latitude.

13.11 A solar collector device is located at the outer limit of the atmosphere at 45 deg north latitude. The collector faces due south with its flat receiving surface placed in a vertical position. The collector is capable of absorbing 50 percent of the incident solar radiation. Calculate the total energy absorbed by the collector per sq ft of surface during one full day on June 15.

13.12 Calculate the intensity of direct solar radiation incident upon a south-facing vertical surface at solar noon for a location at 42 deg north latitude for
(a) June 21, and
(b) December 21.

13.13 Based upon average clear-day conditions, calculate the incidence of total solar radiation at 12:00 noon CST on July 21 upon a flat roof of a building in
(a) Minneapolis ($l = 45°$ N, $L_{loc} = 93°$ W), and
(b) Dallas ($l = 32.9°$ N, $L_{loc} = 96.8°$ W).

13.14 **(a)** At what time (Eastern Standard Time) will the sun set February 7 in Atlanta, Georgia (33.8° N latitude and 84.4° W longitude)?
(b) Calculate the solar angle of incidence immediately before sunset for a west-facing vertical surface.

13.15 Chicago is at 42 deg north latitude, 88 deg west longitude, and is in the Central Time Zone. For November 21 in Chicago determine the following:
(a) What time (Central Standard Time) is sunrise?
(b) What is the azimuth angle at sunrise?
(c) What is the altitude angle at 10:30 a.m. *solar time*?
(d) If a surface faces 20 degrees west of south and has a tilt angle of 65 degrees, what is the angle of incidence at 10:30 a.m. *solar time*?
(e) What is the profile angle at 10:30 a.m. *solar time* for the surface defined in part (d)?

13.16 Students are spending their spring break in Fort Lauderdale, Florida (27° north latitude and 83° west longitude). Their room has a skylight that faces southeast and has an angle of tilt of 55°. March 26 they decide to get out of bed at 1:30 p.m. Eastern Standard Time. What is the angle of incidence that the direct solar rays make with the skylight at the time they get out of bed?

13.17 Tucson, Arizona, is at 32 deg north latitude, 111 deg west longitude, and is in the Mountain Time Zone. Some students may be spending their quarter break in Tucson. Consider December 15th in Tucson and determine the following:
(a) If the students want to wake up at solar noon, for what time should they set their alarm clock?

(b) If the windows of their room face 45 degrees west of south, what is the angle of incidence on the windows at 10:00 a.m. *Local Solar Time?*

(c) At what time of day (Local Solar Time) do the direct solar rays just begin to strike the windows of their room?

13.18 A building in Phoenix, Arizona ($l = 33°$ N) has a south-facing window-wall which is at a slope angle of 75° from the horizontal. How many hours of the day on June 21 does *direct* sunlight enter the building through the window-wall?

13.19 A building has an inclined section of its roof which faces 20 degrees north of west and is sloped at an angle of 30 degrees from the horizontal. Calculate the *total* solar flux incident on the roof section at 1:40 p.m. Central Daylight Saving Time on June 15 if the building is located at 33° N latitude and 100° W longitude. Ground reflectance is 0.3.

13.20 A window panel 8 ft high in a south-facing vertical wall is located at 45 deg north latitude. The glass has no setback. It is desired to design a solid overhang such that the window will just be completely shaded at solar noon on the day of the summer solstice, but will just be completely sunlit at solar noon on the day of the winter solstice. Determine the length the overhang extends out from the wall and the distance above the top of the window at which the overhang should be located.

13.21 A window panel 2 meters high in a south-facing vertical wall is located at 48 degrees north latitude. The glass has no setback. It is desired to design a solid overhang such that the window will just be completely shaded at 2 p.m. solar time on May 1 but also will just be completely sunlit at solar noon on November 1. Determine the length that the overhang must extend out from the surface of the window panel and the distance above the top of the window at which the overhang should be located.

13.22 Windows on the west side of the University of Minnesota Mechanical Engineering Building (45° N latitude and 93° W longitude) are 100 cm wide by 2.5 m high and are set back 36 cm from the outside surface of the building. What fraction of the area of a window is not shaded from direct sunlight at 3:00 p.m. Central Daylight Saving Time on June 7?

13.23 Two downtown buildings are arranged as schematically shown in Fig. 13.22. Location is at 42 deg north latitude. Find the dimensions of the south side of building *A* shaded from direct sunlight by building *B* at 11:00 a.m. solar time on September 1.

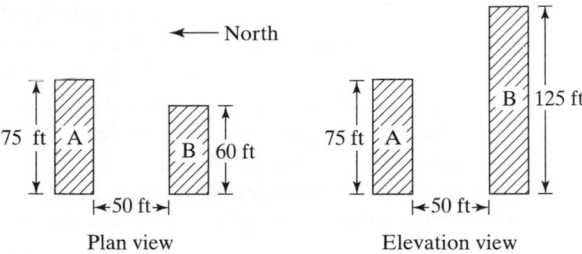

Figure 13.22 Schematic buildings for Prob. 13.23.

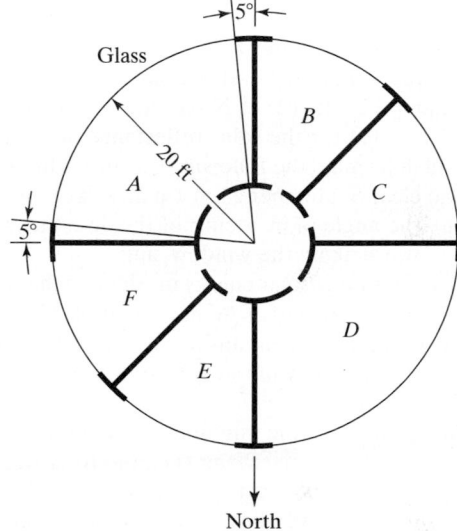

Figure 13.23 Schematic building for Prob. 13.24.

13.24 Figure 13.23 shows a schematic plan view of one floor of a cylindrically shaped office building. Location is at 45 deg north latitude and 93 deg west longitude. The glass panels are 9 ft high. Even with the top of each glass panel is a continuous solid overhang whose outer edge describes an arc of a circle of radius 22 ft. Find the surface area of the glass wall of office *A* shaded from direct sunlight on a clear August 1 day at 11:00 a.m. Central Standard Time. Describe the approximate shape of the shaded area.

13.25 The windows of a skylight face 22 deg east of south and are at a tilt angle of $\Sigma = 35$ deg. The surface surrounding the skylight has a reflectance of $\rho_g = 0.2$ and is horizontal. The building containing the skylight is located in Las Vegas, Nevada, which is at a latitude of 36 degrees north. Calculate the average clear-day direct, diffuse, reflected, and total solar flux striking the skylight on July 21 at 2:00 p.m. Local Solar Time.

13.26 A window of a building is located in a vertical wall that faces 25 degrees east of south. The window is 6 ft tall and 4 ft wide and has a setback of 0.75 ft. The building is located at a latitude of 40° N. Consider a time of 1:00 p.m. solar time on January 21.
  (a) Calculate the angle of incidence.
  (b) Calculate the unshaded fraction.
  (c) Calculate the direct, diffuse, and reflected solar flux striking the unshaded part of the window. Take ground reflectance equal to 0.7 and assume average clear-day conditions.

13.27 A window 7 ft high and 4 ft wide is built into a wall that is 1 ft thick, and the glass is flush with the inside surface of the wall. The wall is vertical, faces a direction 35 deg east of south, and is located in Miami, Florida (25.8° N latitude). Assume a value of $\rho_g = 0.4$ for the solar reflectance of the ground and determine the following quantities for an average clear September 21 at 1 p.m. solar time:
  (a) the angle of incidence of the direct rays of the sun striking the window, and
  (b) the rate of solar energy incident on the window.

13.28 A building located at 36° north latitude has a south-facing wall that contains a window that is 4 ft wide by 6 ft tall. The window is flush with the outer surface of the wall.
  (a) Determine the minimum dimensions and location for an overhang (i.e., the total width, the length it must extend beyond the wall, and the height above the top of the window), such that the window will be totally shaded at 10 a.m. solar time on June 1 and totally sunlit at solar noon on December 1.
  (b) With the overhang you specified in part (a) in place, calculate the total solar radiation per unit time incident on the window at solar noon on an average clear February 21.

13.29 A wall of a house is 60 ft long and faces 30 deg east of south. A window is located midway along the length of the wall. The window is 6.5 ft high by 4 ft wide. The house is constructed such that the roof extends 1.0 ft out beyond the exterior surface of the walls, and the bottom of this overhang is 1.5 ft above the top of the window. Calculate the rate of solar energy incident on the window on an average clear July 7 at 1:00 p.m. Local Solar Time if the house is located at 42° N latitude and 110° W longitude. Assume a value of $\rho_g = 0.3$ for the solar reflectance of the ground.

13.30 A window 2 m high by 1 m wide is in a south-facing vertical wall and is flush with the outside surface of the wall. A solid overhang is placed 15 cm above the window. The overhang is 1.5 m wide and extends 46 cm out from the wall. What is the total rate of solar radiation incident on the window at solar noon, July 1, and 42°N latitude if the solar reflectance of the ground is $\rho_g = 0.2$?

13.31 A building is located at 45° north latitude and has windows which are tilted from the horizontal at an angle of 45 deg and face directly south. There is a horizontal overhang above the window, as shown in Fig. 13.24. Calculate the rate of solar energy incident on the window per unit window area at solar noon on a standard clear day in October. Take a value of $\rho_g = 0.3$ for the solar reflectance of the ground.

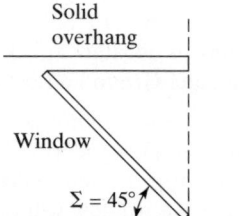

Figure 13.24 Schematic for Prob. 13.31.

13.32 A daylighting scheme for an earth-sheltered building in Minneapolis (45° N latitude and 93° W longitude) is shown in Fig. 13.25. The system is to be designed such that at 9:00 a.m. Central Standard Time (CST) on December 15 the direct solar rays reflected off the mirror will go straight down into the building. The mirror is perfectly flat and smooth, so that the angle of incidence is equal to the angle of reflection.

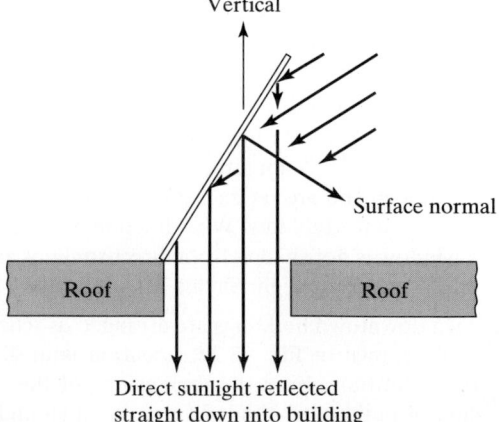

Figure 13.25 Schematic for Prob. 13.32.

(a) Calculate the azimuth, φ, and tilt, Σ, angles at which the mirror surface should be built.

(b) If the mirror dimensions are 1.5 m by 10 m, the mirror is a perfect reflector (reflectance = 1), and the window is a perfect transmitter, (transmittance = 1), calculate the rate at which direct solar radiation is reflected into the building (watts) at 9:00 a.m. CST on December 15. Assume average clear-day conditions.

SYMBOLS

A	Apparent direct normal solar flux at the outer edge of the earth's atmosphere (used in the ASHRAE Clear Day Solar Radiation Model).
a	Window height or parapet wall height.
B	Coefficient used in Eq. (13.3); Apparent atmospheric extinction coefficient used in the ASHRAE Clear Day Solar Radiation Model.
b	Setback dimension used in shading calculations.
C	Coefficient used in the ASHRAE Clear Day Solar Radiation Model.
c	Window width.
CT	Clock Time.
D	Day of the month.
d	Declination angle.
DT	Correction for Daylight Saving Time when using Eq. (13.2) for calculating LST.
E	Equation of Time.
e	Distance of an overhang above a window.
f	Length that a window overhang extends out from the wall.
g	Width that a window overhang extends beyond the edge of the window.
h	Hour angle.
h_{sr}	Sunrise hour angle.
h_{ss}	Sunset hour angle.
I	Total solar flux striking a surface.
I_D	Direct solar flux striking a surface.
I_d	Diffuse solar flux striking a surface.
I_{dH}	Diffuse solar flux striking a horizontal surface.
I_{DN}	Direct normal solar flux.
$I_{DN,\lambda}$	Monochromatic direct normal solar flux.
I_{dV}	Diffuse solar flux striking a vertical surface.
I_R	Reflected solar flux striking a surface (the reflection is from a horizontal surface).
K	Extinction coefficient.
K_λ	Monochromatic extinction coefficient.
l	Latitude angle.
LCT	Local Civil Time.
L_{loc}	Longitude of the location considered.
LST	Local Solar Time.
L_{std}	Standard meridian for the local time zone.
m	Air mass.
n	Day of the year.
x	Linear dimension.

| Y | Depth of earth's atmosphere. |
| y | Linear dimension. |

Greek Letters

β	Altitude angle.
δ	Profile angle.
ϕ	Azimuth angle.
γ	Surface-solar azimuth angle.
θ	Incidence angle.
θ_H	Zenith angle.
λ	Wavelength.
Σ	Surface tilt angle.
τ_λ	Effective monochromatic transmittance of the atmosphere.
Ψ	Surface azimuth angle.

part V
Heating- and Cooling-Load Calculations in Buildings

chapter 14
Winter Design Heat Loss

14.1 INTRODUCTION

The design of a building heating system requires the estimate of the design heat loss for each room or zone of the building. Two types of heat losses are included: (1) the heat transmitted through the building envelope and (2) the heat required to warm outdoor air entering the space. These losses are transient, since variations occur in outdoor temperature, indoor-temperature set point, solar radiation, occupancy level, internal heat sources, and energy storage. However, the goal in performing the load calculations is to provide the necessary information to select the proper equipment so that acceptable conditions can be maintained at nearly all times.

A conservative approach is to exclude heat gains which may not be present when needed. As a result, solar and internal gains are not included in the calculations. Further, since the maximum loads may occur during prolonged cold periods, energy-storage effects are usually neglected. This leads to a steady-state approach for calculating the heating loads.

Reasonable environmental conditions must be selected for the design. If one bases the design on the minimum expected temperature and maximum wind speed, the system will be oversized for all but perhaps one or two hours of the entire heating season. On the other hand, selecting too high an outdoor-air design temperature leads to lack of comfort for an unacceptable number of hours.

In this chapter we describe how to perform the various steady-state heat-transfer calculations that make up the design load. Weather data are presented along with recommendations on what design temperatures to use. At the end of the chapter we consider moisture transport in building materials. Although it is not part of the heat-loss estimating procedure, it is critical that the moisture transport through the building envelope, and the problems that occur when condensation occurs in the structure, be understood and dealt with in the design.

The steady-state heat transfer through sections of building envelopes (walls, floors, windows, roofs, etc.) can be calculated based on the thermal resistances (*R*-values) of the components which make up the section. The *R*-values may be calculated from the thermal properties of the building materials or by direct measurements using laboratory equipment such as the guarded hot box (*ASTM Standard C 236*) or the calibrated hot box (*ASTM Standard C 976*). In addition, thermal resistances due to air films and air cavities play a significant role in estimating the heat transfer through building envelopes.

This section of the chapter will present thermal-transport properties and heat-transfer-coefficient data abstracted from the *ASHRAE Handbook of Fundamentals (1993)* and will discuss methods of applying these data to estimate heat-transfer rates through building envelopes. In presenting the data, ASHRAE cautions the user that the values presented were developed under ideal conditions and that in practice the thermal performance can be reduced significantly by such factors as improper installation and shrinkage, settling, or compression of insulation materials. It is pointed out that the performance of materials fabricated in the field is especially subject to the quality of workmanship during construction and installation, and, therefore, some engineers include additional insulation or other safety factors in their design, based on experience.

Surface Conductances and Resistances for Air

Table 14.1 presents the heat-transfer coefficients and corresponding *R*-values for air films adjacent to exposed surfaces, such as interior and exterior walls. These are given as functions of wall position, direction of heat flow, and air velocity. It is important to note that these include the combined effects of convection and radiation. Thus, they are also functions of the surface emittance.

Gas-Filled Cavities

A number of applications include heat transfer across a gas-filled cavity. In the case of a wall cavity the gas is air, while in the case of a multiglazed window the gas between the sheets of glass may be either air or argon. Figure 14.1 depicts an example of heat transfer between two plane surfaces separated by a gas space. The total heat flux is the sum of the convection and radiation heat fluxes:

$$\dot{q} = \dot{q}_c + \dot{q}_r = (h_c + h_r)(t_1 - t_2) \tag{14.1}$$

The convective heat-transfer mode is natural convection as described in Sec. 2.22. Therefore, the convective heat-transfer coefficient is a function of the position of the gas space, the direction of heat flow, the temperature difference across the space, the width of the gas space, and the thermal conductivity of the gas. The thermal conductivity of the gas is evaluated at the mean air temperature in the space which is taken as the average of the surface temperatures.

The radiation heat-transfer coefficient is given by Eq. (2.76) and approximated by Eq. (2.78). For many applications the gap spacing is small compared to the dimensions of

TABLE 14.1 Surface Conductances and Resistances for Air

		Surface Emittances										
		ε = 0.9				ε = 0.2				ε = 0.05			
		h		R		h		R		h		R	
Position of Surface	Direction of Heat Flow	Btu hr·ft²·°F	W m²·C	°F·ft²·hr Btu	m²·C W	Btu hr·ft²·°F	W m²·C	°F·ft²·hr Btu	m²·C W	Btu hr·ft²·°F	W m²·C	°F·ft²·hr Btu	m²·C W
Still Air													
Horizontal	Upward	1.63	9.26	0.61	0.11	0.91	5.17	1.10	0.19	0.76	4.32	1.32	0.23
Sloping—45°	Upward	1.60	9.09	0.62	0.11	0.88	5.00	1.14	0.20	0.73	4.15	1.37	0.24
Vertical	Horizontal	1.46	8.29	0.68	0.12	0.74	4.20	1.35	0.24	0.59	3.35	1.70	0.30
Sloping—45°	Downward	1.32	7.50	0.76	0.13	0.60	3.41	1.67	0.29	0.45	2.56	2.22	0.39
Horizontal	Downward	1.08	6.13	0.92	0.16	0.37	2.10	2.70	0.48	0.22	1.25	4.55	0.80

		h		R	
		Btu hr·ft²·°F	W m²·C	°F·ft²·hr Btu	m²·C W
Moving Air (Any Position)					
Wind (for winter) 15 mph (6.7 m/s, 24 km/h)	Any	6.00	34.0	0.17	0.030
Wind (for summer) 7.5 mph (3.4 m/s, 12 km/h)	Any	4.00	22.7	0.25	0.044

Source: Adapted by permission from *ASHRAE Fundamentals 1993*.

Conductances are for surfaces of the stated emittance facing virtual blackbody surroundings at the same temperature as the ambient air. Values are based on a surface-air temperature difference of 10 °F (5.5 °C) and for surface temperatures of 70 °F (21 °C).

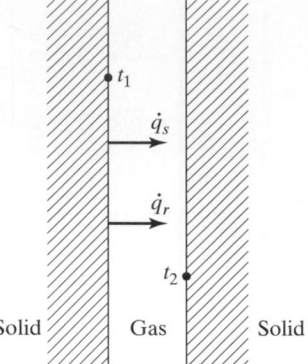

Solid — Gas — Solid **Figure 14.1** Schematic of heat transfer across a gas-filled gap.

the surrounding surfaces (therefore the view factor is approximately unity; i.e., $F_{1-2} \approx 1$) and the surface areas are approximately equal. For these conditions Eq. (2.78) becomes

$$h_r \approx \frac{4\sigma(T_{\text{avg}})^3}{\dfrac{1}{\varepsilon_1} + \dfrac{1}{\varepsilon_2} - 1} \approx 4\varepsilon_{\text{eff}}\sigma(T_{\text{avg}})^3 \qquad (14.2)$$

where ε_{eff} is called the *effective emittance* of the air space and is defined as

$$\varepsilon_{\text{eff}} \equiv \frac{1}{\dfrac{1}{\varepsilon_1} + \dfrac{1}{\varepsilon_2} - 1}. \qquad (14.3)$$

From Eq. (14.2) it is seen that the radiation heat-transfer coefficient is a function of the average of the surface temperatures (which also corresponds to the mean air temperature) and the effective emittance as defined in Eq. (14.3). Table 14.2 presents the thermal resistances of plane air spaces for selected combinations of air-space positions, heat-flow directions, mean air temperatures, temperature differences, gap spacings, and effective emittances. Table 14.3 includes example values of surface emittances and effective emittances of air spaces.

Common Building and Insulating Materials

Table 14.4 contains thermal transport properties for a representative sample of common building and insulating materials. These materials fall into two general categories. In the one case, where the material is relatively homogeneous, the thermal conductivity and its inverse, the resistance per unit thickness, are reported. The other category includes materials of specific shapes or thicknesses, e.g., concrete blocks, plaster board, built-up roofing, etc. For these materials, the thermal properties listed are the thermal conductance, C, and thermal resistance, R. In general, the *thermal conductance* for any building material is defined by writing the heat flux, $\dot{q}$, through a material as

$$\dot{q} = C\,\Delta t \qquad (14.4)$$

where Δt is the temperature difference between the two surfaces of the material.

The thermal conductance accounts for all the modes of heat transfer that exist within a material. For example, in a concrete block with hollow openings, conduction, convection, and radiation all occur at various locations in the block. Where conductance values are reported directly, these have generally been experimentally determined by

TABLE 14.2IP Thermal Resistance of Plane Air Spaces, IP Units

Units of Thermal Resistance^a: °F · ft² · hr/Btu

Position of Air Space	Direction of Heat Flow	Air Space Mean Temp., °F	Air Space Temp. Diff., °F	0.5-in. Air Space Effective Emittance ε_{eff} 0.03	0.05	0.2	0.5	0.82	3.5-in. Air Space Effective Emittance ε_{eff} 0.03	0.05	0.2	0.5	0.82
Horiz.	Up ←	90	10	2.13	2.03	1.51	0.99	0.73	2.84	2.66	1.83	1.13	0.80
		50	30	1.62	1.57	1.29	0.96	0.75	2.09	2.01	1.58	1.10	0.84
		50	10	2.13	2.05	1.60	1.11	0.84	2.80	2.66	1.95	1.28	0.93
		0	20	1.73	1.70	1.45	1.12	0.91	2.25	2.18	1.79	1.32	1.03
		0	10	2.10	2.04	1.70	1.27	1.00	2.71	2.62	2.07	1.47	1.12
45° Slope	Up ↗	90	10	2.44	2.31	1.65	1.06	0.76	3.18	2.96	1.97	1.18	0.82
		50	30	2.06	1.98	1.56	1.10	0.83	2.26	2.17	1.67	1.15	0.86
		50	10	2.55	2.44	1.83	1.22	0.90	3.12	2.95	2.10	1.34	0.96
		0	20	2.20	2.14	1.76	1.30	1.02	2.42	2.35	1.90	1.38	1.06
		0	10	2.63	2.54	2.03	1.44	1.10	2.98	2.87	2.23	1.54	1.16
Vertical	Horiz. →	90	10	2.47	2.34	1.67	1.06	0.77	3.69	3.40	2.15	1.24	0.85
		50	30	2.57	2.46	1.84	1.23	0.90	2.67	2.55	1.89	1.25	0.91
		50	10	2.66	2.54	1.88	1.24	0.91	3.63	3.40	2.32	1.42	1.01
		0	20	2.82	2.72	2.14	1.50	1.13	2.88	2.78	2.17	1.51	1.14
		0	10	2.93	2.82	2.20	1.53	1.15	3.49	3.33	2.50	1.67	1.23
45° Slope	Down ↗	90	10	2.48	2.34	1.67	1.06	0.77	4.81	4.33	2.49	1.34	0.90
		50	30	2.64	2.52	1.87	1.24	0.91	3.51	3.30	2.28	1.40	1.00
		50	10	2.67	2.55	1.89	1.25	0.92	4.74	4.36	2.73	1.57	1.08
		0	20	2.91	2.80	2.19	1.52	1.15	3.81	3.63	2.66	1.74	1.27
		0	10	2.94	2.83	2.21	1.53	1.15	4.59	4.32	3.02	1.88	1.34
Horiz.	Down →	90	10	2.48	2.34	1.67	1.06	0.77	10.07	8.19	3.41	1.57	1.00
		50	30	2.66	2.54	1.88	1.24	0.91	9.60	8.17	3.86	1.88	1.22
		50	10	2.67	2.55	1.89	1.25	0.92	11.15	9.27	4.09	1.93	1.24
		0	20	2.94	2.83	2.20	1.53	1.15	10.90	9.52	4.87	2.47	1.62
		0	10	2.96	2.85	2.22	1.53	1.16	11.97	10.32	5.08	2.52	1.64

Source: Adapted by permission from *ASHRAE Fundamentals 1993.*
^{a} Effective emittance of the space, ε_{eff}, is given by Eq. (14.3).

TABLE 14.2SI Thermal Resistance of Plane Air Spaces, SI Units

Position of Air Space	Direction of Heat Flow	Air Space Mean Temp., °C	Air Space Temp. Diff., °C	13-mm Air Space — Effective Emittance ε_{eff} 0.03	0.05	0.2	0.5	0.82	90-mm Air Space — Effective Emittance ε_{eff} 0.03	0.05	0.2	0.5	0.82
Horiz.	Up ←	32.2	5.6	0.37	0.36	0.27	0.17	0.13	0.50	0.47	0.32	0.20	0.14
		10.0	16.7	0.29	0.28	0.23	0.17	0.13	0.27	0.35	0.28	0.19	0.15
		10.0	5.6	0.37	0.36	0.28	0.20	0.15	0.49	0.47	0.34	0.23	0.16
		−17.8	11.1	0.30	0.30	0.26	0.20	0.16	0.40	0.38	0.32	0.23	0.18
		−17.8	5.6	0.37	0.36	0.30	0.22	0.18	0.48	0.46	0.36	0.26	0.20
45° Slope	Up ↗	32.2	5.6	0.43	0.41	0.29	0.19	0.13	0.56	0.52	0.35	0.21	0.14
		10.0	16.7	0.36	0.35	0.27	0.19	0.15	0.40	0.38	0.29	0.20	0.15
		10.0	5.6	0.45	0.43	0.32	0.21	0.16	0.55	0.52	0.37	0.24	0.17
		−17.8	11.1	0.39	0.38	0.31	0.23	0.18	0.43	0.41	0.33	0.24	0.19
		−17.8	5.6	0.46	0.45	0.36	0.25	0.19	0.52	0.51	0.39	0.27	0.20
Vertical	Horiz. →	32.2	5.6	0.43	0.41	0.29	0.19	0.14	0.65	0.60	0.38	0.22	0.15
		10.0	16.7	0.45	0.43	0.32	0.22	0.16	0.47	0.45	0.33	0.22	0.16
		10.0	5.6	0.47	0.45	0.33	0.22	0.16	0.64	0.60	0.41	0.25	0.18
		−17.8	11.1	0.50	0.48	0.38	0.26	0.20	0.51	0.49	0.38	0.27	0.20
		−17.8	5.6	0.52	0.50	0.39	0.27	0.20	0.61	0.59	0.44	0.29	0.22
45° Slope	Down ↗	32.2	5.6	0.44	0.41	0.29	0.19	0.14	0.85	0.76	0.44	0.24	0.16
		10.0	16.7	0.46	0.44	0.33	0.22	0.16	0.62	0.58	0.40	0.25	0.18
		10.0	5.6	0.47	0.45	0.33	0.22	0.16	0.83	0.77	0.48	0.28	0.19
		−17.8	11.1	0.51	0.49	0.39	0.27	0.20	0.67	0.64	0.47	0.31	0.22
		−17.8	5.6	0.52	0.50	0.39	0.27	0.20	0.81	0.76	0.53	0.33	0.24
Horiz.	Down →	32.2	5.6	0.44	0.41	0.29	0.19	0.14	1.77	1.44	0.60	0.28	0.18
		10.0	16.7	0.47	0.45	0.33	0.22	0.16	1.69	1.44	0.68	0.33	0.21
		10.0	5.6	0.47	0.45	0.33	0.22	0.16	1.96	1.63	0.72	0.34	0.22
		−17.8	11.1	0.52	0.50	0.39	0.27	0.20	1.92	1.68	0.86	0.43	0.29
		−17.8	5.6	0.52	0.50	0.39	0.27	0.20	2.11	1.82	0.89	0.44	0.29

Units of Thermal Resistance[a]: $K \cdot m^2/W$

Source: Adapted by permission from *ASHRAE Fundamentals 1993*.
[a] Effective emittance of the space, ε_{eff}, is given by Eq. (14.3).

TABLE 14.3 Emittance Values of Various Surfaces and Effective Emittances of Air Spaces[a]

| Surface | Average Emittance, ε | Effective Emittance ε_{eff} of Air Space | |
		One Surface Emittance, ε; Other, 0.9	Both Surfaces Emittance, ε
Aluminum foil, bright	0.05	0.05	0.03
Aluminum foil, with condensate just visible (> 0.7 gr/ft²)	0.30	0.29	—
Aluminum foil, with condensate clearly visible (> 2.9 gr/ft²)	0.70	0.65	—
Aluminum sheet	0.12	0.12	0.06
Aluminum-coated paper, polished	0.20	0.20	0.11
Steel, galvanized, bright	0.25	0.24	0.15
Aluminum paint	0.50	0.47	0.35
Building materials: wood, paper, masonry, nonmetallic paints	0.90	0.82	0.82
Regular glass	0.84	0.77	0.72

SOURCE: Adapted by permission from *ASHRAE Fundamentals 1993*.
[a] These values apply in the 4 to 40 μm range of the electromagnetic spectrum.

using a guarded or calibrated hot box. These apparatuses are designed to establish a one-dimensional steady-state heat transfer through the material while maintaining uniform surface temperatures. Measurements are made of the surface temperatures and heat flux, and Eq. (14.4) is applied to evaluate the conductance.

The concept of a thermal conductance is useful for uniform materials and for film coefficients. For a material of thickness L and thermal conductivity k the conductance is $C = k/L$, and for a surface with heat-transfer coefficient h the conductance is $C = h$. In all cases, including heat transfer across gaps, the relation between the thermal conductance and thermal resistance is simply $R = 1/C$.

Heat Transfer through Multilayered Structures

Most structures forming building envelopes consist of multiple layers of materials. Figure 14.2 (see pg. 428) depicts such a wall made from three materials having resistance values R_1, R_2, and R_3. The indoor and outdoor air temperatures are t_i and t_o, respectively. The heat-transfer coefficients between the indoor and outdoor air and the adjacent surfaces are h_i and h_o. If the heat transfer is assumed to be one dimensional and steady, the heat flux through the wall, $\dot{q}$, is

$$\dot{q} = \frac{t_i - t_o}{R_i + R_1 + R_2 + R_3 + R_o} = \frac{t_i - t_o}{R_t} = U(t_i - t_o) \tag{14.5}$$

where $R_i = 1/h_i$ and $R_o = 1/h_o$. The quantity R_t is the total resistance for the wall, including the resistances due to the air films at the inside and outside surfaces. It is important

TABLE 14.4 IP Typical Thermal Properties of Common Building and Insulating Materials—Design Values

	Density, lbm/ft³	Conductivity (k), Btu·in. / hr·ft²·°F	Conductance (C), Btu / hr·ft²·°F	Resistance (R) Per Inch Thickness (1/k), °F·ft²·hr / Btu·in.	Resistance (R) For Thickness Listed (1/C), °F·ft²·hr / Btu	Specific Heat, Btu / lbm·°F
BUILDING BOARD						
Asbestos-cement board	120	4.0	—	0.25	—	0.24
Gypsum or plaster board 0.375 in.	50	—	3.10	—	0.32	0.26
Gypsum or plaster board 0.5 in.	50	—	2.22	—	0.45	—
Plywood (Douglas fir)	34	0.80	—	1.25	—	0.29
Plywood or wood panels 0.75 in.	34	—	1.07	—	0.93	0.29
Vegetable fiber board						
Sheathing, regular density 0.5 in.	18	—	0.76	—	1.32	0.31
0.78125 in.	18	—	0.49	—	2.06	—
Sheathing, intermed. density 0.5 in.	22	—	0.92	—	1.09	0.31
Hardboard						
Medium density	50	0.73	—	1.37	—	0.31
High density, standard-tempered grade	63	1.0	—	1.00	—	0.32
Particleboard, medium density	50	0.94	—	1.06	—	0.31
Wood subfloor 0.75 in.	—	—	1.06	—	0.94	0.33
BUILDING MEMBRANE						
Vapor—permeable felt	—	—	16.70	—	0.06	—
Vapor—seal, 2 layers of mopped 15-lb felt	—	—	8.35	—	0.12	—
Vapor—seal, plastic film	—	—	—	—	Negl.	—
FINISH FLOORING MATERIALS						
Carpet and rubber pad	—	—	0.81	—	1.23	0.33
Tile—asphalt, linoleum, vinyl, rubber	—	—	20.	—	0.05	0.30
Wood, hardwood finish 0.75 in.	—	—	1.47	—	0.68	—
INSULATING MATERIALS						
Blanket and batt						
Mineral fiber, fibrous form processed from rock, slag, or glass						
approx. 3.5 in.	0.4–2.0	—	0.077	—	13	—
approx. 5.5–6.5 in.	0.4–2.0	—	0.053	—	19	—
approx. 8.25–10 in.	0.4–2.0	—	0.033	—	30	—
approx. 10–13 in.	0.4–2.0	—	0.026	—	38	—
Board and slabs						
Cellular glass	8.0	0.33	—	3.03	—	0.18

Glass fiber, organic bonded		4.0–9.0	0.25	—	4.00	—	0.23
Expanded polystyrene extruded (smooth skin surface, HCFC-142b exp.)		1.8–3.5	0.20	—	5.00	—	0.29
Expanded polystyrene, molded beads		1.0	0.26	—	3.85	—	—
Loose fill							
Cellulosic insulation (milled paper or wood pulp)		2.3–3.2	0.27–0.32	—	3.70–3.13	—	0.33
Mineral fiber (rock, slag, or glass)							
approx. 6.5–8.75 in.		0.6–2.0	—	—	—	19.0	0.17
approx. 7.5–10 in.		0.6–2.0	—	—	—	22.0	—
approx. 10.25–13.75 in.		0.6–2.0	—	—	—	30.0	—
Mineral fiber (rock, slag, or glass) approx. 3.5 in. (closed sidewall application)		2.0–3.5	—	—	—	12.0–14.0	—
Vermiculite, exfoliated		7.0–8.2	0.47	—	2.13	—	0.32
Spray applied							
Polyurethane foam		1.5–2.5	0.16–0.18	—	6.25–5.56	—	—
Cellulosic fiber		3.5–6.0	0.29–0.34	—	3.45–2.94	—	—
Glass fiber		3.5–4.5	0.26–0.27	—	3.85–3.70	—	—
ROOFING							
Asbestos-cement shingles		120	—	4.76	—	0.21	0.24
Asphalt roll roofing		70	—	6.50	—	0.15	0.36
Asphalt shingles		70	—	2.27	—	0.44	0.30
Built-up roofing	0.375 in.	70	—	3.00	—	0.33	0.35
Slate	0.5 in.	—	—	20.00	—	0.05	0.30
Wood shingles, plain and plastic film faced		—	—	1.06	—	0.94	0.31
PLASTERING MATERIALS							
Cement plaster, sand aggregate		116	5.0	—	0.20	—	0.20
Gypsum plaster:							
Lightweight aggregate	0.5 in.	45	—	3.12	—	0.32	—
Lightweight aggregate	0.625 in.	45	—	2.67	—	0.39	—
Lightweight aggregate on metal lath	0.75 in.	—	—	2.13	—	0.47	—
MASONRY MATERIALS							
Masonry units							
Brick, fired clay		120	5.6–6.8	—	0.18–0.15	—	0.19
Concrete blocks							
Normal-weight aggregate (sand and gravel)							
8 in., 33–36 lb, 126–136 lb/ft³ concrete, 2 or 3 cores		—	—	0.90–1.03	—	1.11–0.97	0.22
Same with vermiculite-filled cores		—	—	0.52–0.73	—	1.92–1.37	—
12 in., 50 lb, 125 lb/ft³ concrete, 2 cores		—	—	0.81	—	1.23	0.22

(continued)

TABLE 14.4 IP (cont.)

	Density, lbm/ft³	Conductivity (k), Btu·in./hr·ft²·°F	Conductance (C), Btu/hr·ft²·°F	Resistance (R) Per Inch Thickness (1/k), °F·ft²·hr/Btu·in.	Resistance (R) For Thickness Listed (1/C), °F·ft²·hr/Btu	Specific Heat, Btu/lbm·°F
Lightweight aggregate (expanded shale, clay, slate, or slag, pumice)						
6 in., 16–17 lb, 85–87 lb/ft³ concrete,						
2 or 3 cores	—	—	0.52–0.61	—	1.93–1.65	—
Same with vermiculite-filled cores	—	—	0.33	—	3.0	—
8 in., 19–22 lb, 72–86 lb/ft³ concrete,						
2 or 3 cores	—	—	0.32–0.54	—	3.2–1.90	0.21
Same with vermiculite-filled cores	—	—	0.19–0.26	—	5.3–3.9	—
12 in., 32–36 lb, 80–90 lb/ft³ concrete,						
2 or 3 cores	—	—	0.38–0.44	—	2.6–2.3	—
Same with vermiculite-filled cores	—	—	0.17	—	5.8	—
Concretes						
Sand and gravel or stone aggregate concretes	140	9.0–18.0	—	0.11–0.06	—	0.19–0.24
Lightweight aggregate concrete; expanded shale, clay or slate; expanded slags; cinders; pumice	80	3.3–4.1	—	0.30–0.24	—	0.20
SIDING MATERIALS (on flat surface)						
Hardboard siding, 0.4375 in.	—	—	1.49	—	0.67	0.28
Wood, drop, 1 by 8 in.	—	—	1.27	—	0.79	0.28
Wood, plywood, 0.375 in., lapped	—	—	1.59	—	0.59	0.29
Aluminum or steel, over sheathing						
Hollow-backed	—	—	1.61	—	0.61	0.29
Insulating-board backed						
nominal 0.375 in.	—	—	0.55	—	1.82	0.32
nominal 0.375 in., foil faced	—	—	0.34	—	2.96	—
Architectural (soda-lime float) glass	158	6.9	—	—	—	0.21
WOODS						
Maple, oak, and similar hardwoods	40–47	1.1	—	0.9	—	0.39
Fir, pine, spruce, hem., and similar softwoods	25–31	0.8	—	1.25	—	0.39
METALS						
Aluminum	171	1536	—	0.00065	—	0.214
Copper	556	2724	—	0.00037	—	0.092
Steel, mild	489	314	—	0.00318	—	0.12

SOURCE: Adapted by permission from *ASHRAE Fundamentals 1993.*

TABLE 14.4 SI Typical Thermal Properties of Common Building and Insulating Materials—Design Values

	Density, kg/m³	Conductivity (k), W/m·K	Conductance (C), W/m²·K	Resistance (R) $(1/k)$, m·K/W	Resistance (R) For Thickness Listed $(1/C)$, m²·K/W	Specific Heat kJ/kg·K
BUILDING BOARD						
Asbestos-cement board	1900	0.58	—	1.73	—	1.00
Gypsum or plaster board 9.5 mm	800	—	17.6	—	0.056	1.09
Gypsum or plaster board 12.7 mm	800	—	12.6	—	0.079	—
Plywood (Douglas fir)	540	0.12	—	8.66	—	1.21
Plywood or wood panels 19.0 mm	540	—	6.1	—	0.16	1.21
Vegetable fiber board						
Sheathing, regular density 12.7 mm	290	—	4.3	—	0.23	1.30
19.8 mm	290	—	2.8	—	0.36	—
Sheathing, intermed. density 12.7 mm	350	—	5.2	—	0.19	1.30
Hardboard						
Medium density	800	0.105	—	9.50	—	1.30
High density, standard-tempered grade	1010	0.144	—	6.93	—	1.34
Particleboard, medium density	800	0.135	—	7.35	—	1.30
Wood subfloor 19.0 mm	—	—	6.0	—	0.17	1.38
BUILDING MEMBRANE						
Vapor—permeable felt	—	—	94.9	—	0.011	—
Vapor—seal, 2 layers of mopped 0.73 kg/m² felt	—	—	47.4	—	0.21	—
Vapor—seal, plastic film	—	—	—	—	Negl.	—
FINISH FLOORING MATERIALS						
Carpet and rubber pad	—	—	4.60	—	0.22	1.38
Tile—asphalt, linoleum, vinyl, rubber	—	—	113.6	—	0.009	1.26
Wood hardwood finish 19.0 mm	—	—	8.35	—	0.12	—
INSULATING MATERIALS						
Blanket and batt						
Mineral fiber, fibrous form processed from rock, slag, or glass						
approx. 90 mm	6.4-3.2	—	0.44	—	2.29	—
approx. 140-165 mm	6.4-3.2	—	0.30	—	3.32	—
approx. 210-250 mm	6.4-3.2	—	0.19	—	5.34	—
approx. 250-300 mm	6.4-3.2	—	0.15	—	6.77	—

(continued)

TABLE 14.4 SI (cont.)

	Density, kg/m³	Conductivity (k), W/m·K	Conductance (C), W/m²·K	Resistance (R) (1/k), m·K/W	Resistance (R) For Thickness Listed (1/C), m²·K/W	Specific Heat kJ/kg·K
Board and slabs						
Cellular glass	136	0.050	—	19.8	—	0.75
Glass fiber, organic bonded	64–140	0.36	—	27.7	—	0.96
Expanded polystyrene extruded (smooth skin surface, HCFC-142b exp.)	29–56	0.029	—	34.7	—	1.21
Expanded polystyrene, molded beads	16	0.037	—	26.7	—	—
Loose fill						
Cellulosic insulation (milled paper or wood pulp)	37–51	0.039–0.046	—	25.6–21.7	—	1.38
Mineral fiber (rock, slag, or glass)						
approx. 170–220 mm	9.6–32	—	—	—	3.35	0.71
approx. 190–250 mm	9.6–32	—	—	—	3.87	—
approx. 260–350 mm	9.6–32	—	—	—	5.28	—
Mineral fiber (rock, slag, or glass) approx. 90 mm (closed sidewall application)	32–56	—	—	—	2.1–2.5	—
Vermiculite, exfoliated	110–130	0.068	—	14.8	—	1.34
Spray applied						
Polyurethane foam	24–40	0.023–0.026	—	43.3–38.5	—	—
Cellulosic fiber	56–96	0.042–0.049	—	23.9–20.4	—	—
Glass fiber	56–72	0.038–0.039	—	26.7–25.6	—	—
ROOFING						
Asbestos-cement shingles	1900	—	27.0	—	0.037	1.00
Asphalt roll roofing	1100	—	36.9	—	0.026	1.51
Asphalt shingles	1100	—	12.9	—	0.077	1.26
Built-up roofing 10 mm	1100	—	17.0	—	0.058	1.46
Slate 13 mm	—	—	114	—	0.009	1.26
Wood shingles, plain and plastic film faced	—	—	6.0	—	0.166	1.30
PLASTERING MATERIALS						
Cement plaster, sand aggregate	1860	0.72	—	1.39	—	0.84
Gypsum plaster:						
Low-density aggregate 12.7 mm	720	—	17.7	—	0.056	—
Low-density aggregate 16 mm	720	—	15.2	—	0.066	—
Lightweight aggregate on metal lath 19 mm	—	—	12.1	—	0.083	—
MASONRY MATERIALS						
Masonry units						
Brick, fired clay	1920	0.81–0.98	—	1.24–1.02	—	0.79

Material						
Concrete blocks						
Normal-mass aggregate (sand and gravel)						
200 mm, 15–16 kg, 2020–2180 kg/m³						
concrete, 2 or 3 cores	—	—	—	5.1–5.8	0.20–0.17	0.92
Same with vermiculite-filled cores	—	—	—	3.0–4.1	0.34–0.24	—
300 mm, 22.7 kg, 2000 kg/m³ concrete, 2 cores	—	—	—	4.60	0.217	0.92
Low-mass aggregate (expanded shale, clay, slate, or slag, pumice)						
150 mm, 7.3–7.7 kg, 1360–1390 kg/m³						
concrete, 2 or 3 cores	—	—	—	3.0–3.5	0.34–0.29	—
Same with vermiculite-filled cores	—	—	—	1.87	0.53	—
200 mm, 8.6–10 kg, 1150–1380 kg/m³						
concrete, 2 or 3 cores	—	—	—	1.8–3.1	0.56–0.33	0.88
Same with vermiculite-filled cores	—	—	—	1.1–1.5	0.93–0.69	—
300 mm, 14.5–16.3 kg, 1280–1440 kg/m³						
concrete, 2 or 3 cores	—	—	—	2.2–2.5	0.46–0.40	—
Same with vermiculite-filled cores	—	—	—	0.97	1.0	—
Concretes						
Sand and gravel or stone aggregate concretes	2240	1.3–2.6	0.77–0.39	—	—	0.8–1.0
Low-density aggregate concrete, expanded shale, clay, or slate; expanded slags; cinders; pumice	1280	0.48–0.59	2.10–1.69	—	—	0.84
SIDING MATERIALS (on flat surface)						
Hardboard siding, 11 mm	—	—	—	8.46	0.12	1.17
Wood, drop, 20 by 200 mm	—	—	—	7.21	0.14	1.17
Wood, plywood, 9.5 mm., lapped	—	—	—	9.03	0.10	1.22
Aluminum or steel, over sheathing						
Hollow-backed	—	—	—	9.14	0.11	1.22
Insulating-board backed						
9.5 mm nominal	—	—	—	3.12	0.32	1.34
9.5 mm nominal, foil backed	—	—	—	1.93	0.52	—
Architectural (soda-lime float) glass	—	—	—	56.8	0.018	0.84
WOODS						
Maple, oak, and similar hardwoods	640–750	0.17	6.0	—	—	1.63
Fir, pine, spruce, hem., and similar softwoods	400–500	0.14	7.0	—	—	1.63
METALS						
Aluminum	2740	221	0.00453	—	—	0.90
Copper	8910	393	0.00255	—	—	0.39
Steel, mild	7830	45.3	0.0221	—	—	0.50

Source: Adapted by permission from *ASHRAE Fundamentals 1993*.

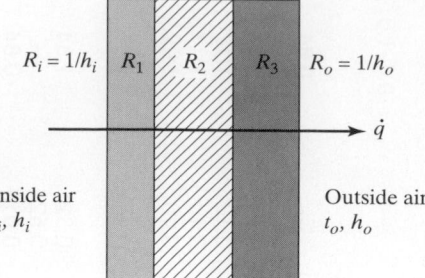

Figure 14.2 Schematic of one-dimensional heat transfer through a multilayer wall.

to keep in mind that when the air temperatures are used, the resistances of the air films must be included. The total resistance is then

$$R_t = R_i + R_1 + R_2 + R_3 + R_o = \Sigma R_j \qquad (14.6)$$

where j includes the resistances of all the materials and the film resistances.

The quantity U in Eq. (14.5) is called the "U-value" for the wall. The U-value is always defined so that the heat-transfer rate is calculated based on the temperature difference between the fluid on one side of a structure and the fluid on the other side. The U-value is given by

$$U = \frac{1}{R_t} = \frac{1}{\Sigma R_j} \qquad (14.7)$$

Most walls or roofs do not have uniform properties in the lateral direction. Examples include wood studs in frame walls or roofs, concrete webs in concrete blocks, and metal ties or other elements in insulated wall panels. A detailed calculation of the heat transfer through such a structure often requires a two- or three-dimensional numerical analysis. However, reasonable estimates can usually be obtained using approximations to the two-dimensional case. Two such approximations are referred to as the *parallel-path method* and the *isothermal-plane method*. These methods can be described by applying them to the wood-frame wall section shown in Fig. 14.3. Let A_s be the area perpendicular to one-dimensional heat flow through the studs and A_b be the area perpendicular to one-dimensional heat flow between the studs. Define the area ratios

$$a_s = \frac{A_s}{A_s + A_b} \qquad (14.8)$$

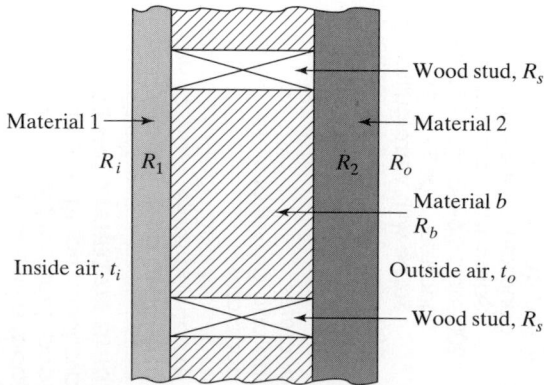

Figure 14.3 Schematic of a wood-frame wall section.

and

$$a_b = \frac{A_b}{A_s + A_b} \qquad (14.9)$$

In typical wood-frame walls, in addition to studs there are wooden plates, sills, and headers. These areas would be included in calculating the fraction a_s. The value of a_s is often referred to as the "framing factor." Framing factors are dependent on the specific type of construction and may vary for the same type of construction based on local construction practices. When detailed construction information is available, it should be used to calculate the framing factors. However, in many instances that level of detail is not available and approximations need to be used.

According to the *ASHRAE Handbook of Fundamentals (1993)* the framing factor is approximately 0.25 for stud walls 16 in. (40.6 cm) on center and approximately 0.22 for stud walls 24 in. (61.0 cm) on center. In the case of unfinished walls, i.e., when the framing is visible from the inside, a good approximation is to neglect the framing and assume the entire area has a U-value equal to that between the studs. Both methods use the same basic relationship between the total heat-transfer rate through the wall, $\dot{Q}_t$, the total average thermal resistance, $R_{t(\text{av})}$, the total surface area, A_t, and the inside and outside air temperatures, t_i and t_o, respectively:

$$\dot{Q}_t = \frac{A_t(t_i - t_o)}{R_{t(\text{av})}} \qquad (14.10)$$

Alternatively, the heat-transfer rate is expressed in terms of the average U-value,

$$\dot{Q}_t = U_{\text{av}} A_t (t_i - t_o) \qquad (14.11)$$

where

$$U_{\text{av}} = \frac{1}{R_{t(\text{av})}} \qquad (14.12)$$

The difference between the two approaches is how the average thermal resistance, $R_{t(\text{av})}$, is estimated.

Parallel-Path Method

The parallel-path method of estimating the heat flow through a section such as that shown in Fig. 14.3 assumes there is no lateral heat transfer, and you calculate the heat flows through the areas A_s and A_b independently and add them. The equivalent circuit diagram for this is shown in Fig. 14.4.

The total heat-transfer rate through the wall is given by

$$\dot{Q}_t = \dot{Q}_s + \dot{Q}_b \qquad (14.13)$$

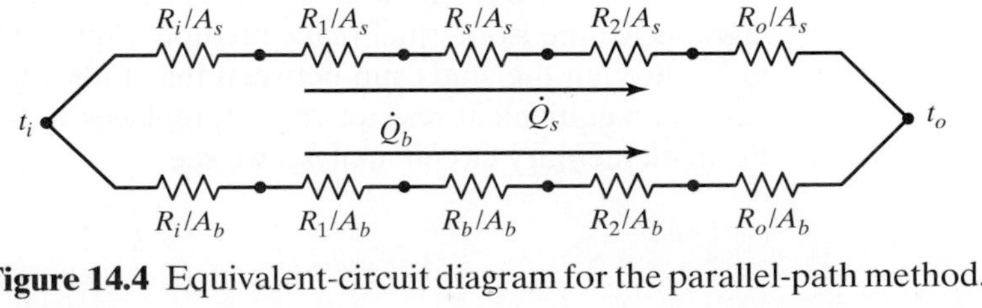

Figure 14.4 Equivalent-circuit diagram for the parallel-path method.

The individual heat-transfer rates through the areas A_s and A_b can be written in terms of the total thermal resistances for each path:

$$\dot{Q}_s = \frac{A_s(t_i - t_o)}{R_{t,s}} \qquad (14.14)$$

and

$$\dot{Q}_b = \frac{A_b(t_i - t_o)}{R_{t,b}} \qquad (14.15)$$

where

$$R_{t,s} = R_i + R_1 + R_s + R_2 + R_o = \sum_{\text{path } s} R \qquad (14.16)$$

and

$$R_{t,b} = R_i + R_1 + R_b + R_2 + R_o = \sum_{\text{path } b} R \qquad (14.17)$$

Based on the total resistances and area ratios, we see that the total average thermal resistance is given by

$$R_{t(\text{av})} = \frac{1}{\dfrac{a_s}{R_{t,s}} + \dfrac{a_b}{R_{t,b}}} \qquad (14.18)$$

The parallel-path method is easily expanded to several paths. For a wall with M parallel paths the value of the average total resistance is quickly deduced from Eq. (14.18):

$$R_{t(\text{av})} = \frac{1}{\displaystyle\sum_{j=1}^{M} \dfrac{a_j}{R_{t,j}}} \qquad (14.19)$$

Combining Eqs. (14.12) and (14.19) and recognizing that $1/R_{t,j}$ is the U-value, U_j, for path j, we derive another useful expression:

$$U_{\text{av}}A_t = \sum_{j=1}^{M} U_j A_j \qquad (14.20)$$

Isothermal-Plane Method

The isothermal-plane method assumes excellent lateral heat transfer in the materials so that the temperature is constant along each lateral interface plane. Figure 14.5 shows circuit diagrams for the isothermal-plane method. Figure 14.5 (a) depicts the separate heat-flow paths through the studs and between the studs, and the corresponding resistors. In Fig. 14.5 (b) an equivalent resistor $R_{s,b}/A_t$ replaces those of the parallel paths.

From elementary circuit analysis we see

$$\frac{A_t}{R_{s,b}} = \frac{A_s}{R_s} + \frac{A_b}{R_b} \qquad (14.21)$$

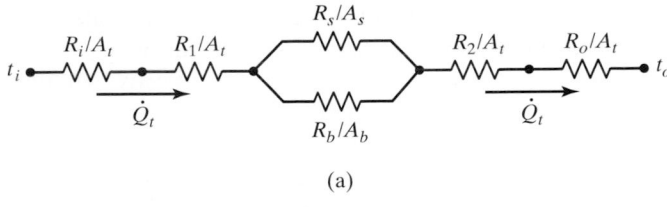

(a)

(b)

Figure 14.5 Equivalent-circuit diagrams for the isothermal-plane method.

or

$$R_{s,b} = \left(\frac{a_s}{R_s} + \frac{a_b}{R_b} \right)^{-1} \tag{14.22}$$

Thus the total average resistance becomes

$$R_{t(av)} = R_i + R_1 + R_{s,b} + R_2 + R_o \tag{14.23}$$

This is then applied to Eqs. (14.10), (14.11), and (14.12).

The value of the total average thermal resistance, $R_{t(av)}$, calculated using the parallel-path method is usually higher than that calculated using the isothermal-plane method. The actual resistance generally is some value between the two calculated values (*ASHRAE Handbook of Fundamentals 1993* [1] and Burch, D. M., et al. [2]). In the absence of test values, examination of the construction usually reveals which method gives the better approximation. Generally, if the construction contains a layer in which lateral conduction is high compared to the transmission through the construction, the isothermal-plane method is preferred. If the construction has no layer of high lateral conductance, the parallel-path method is preferred.

Windows

In the absence of sunlight, air infiltration, and moisture condensation, the rate of heat transfer, $\dot{Q}_o$, through a window system can be expressed in terms of an overall U-value for the window:

$$\dot{Q}_o = U_o A_o (t_i - t_o) \tag{14.24}$$

where U_o = overall heat-transfer coefficient (U-value)

A_o = combined glazing plus frame area projected to a plane parallel to glass as viewed from the outside

The total rate of heat transfer through a fenestration system can be calculated knowing the separate heat-transfer contributions of the glazing and the frame. In the case where there is more than one glazing lite (i.e., multiple sheets of glass or other material), the glazing heat transfer includes a one-dimensional center-of-glass contribution and a two-dimensional edge-of-glass contribution resulting from the spacers separating the glazing lites. The frame contribution is primarily two dimensional. The overall U-value is estimated using area-weighted U-values for each contribution by:

$$U_o = \frac{U_{cg}A_{cg} + U_{eg}A_{eg} + U_fA_f}{A_{cg} + A_{eg} + A_f} \tag{14.25}$$

where subscripts cg, eg, and f refer to center-of-glass, edge-of-glass, and frame, respectively.

Center-of-glass U-values can be calculated using the one-dimensional method described for multilayered structures [Eqs. (14.6) and (14.7)]. Insulating glass units usually have continuous members around the glass perimeter to separate the glazing lites and provide an edge seal. These spacers have a hollow cross section whose walls are often made of thin aluminum. In addition, the wall of the section facing the gap between the lites is often perforated and the interior of the cross section filled with a desiccant. This provides a method of maintaining a very low dew point in the gas in the gap. Since the spacers are often constructed of aluminum, the heat-transfer path near the edges of the window has a significantly lower thermal resistance than the center-of-glass. Laboratory measurements by Peterson [3] showed the region influenced by the spacer to be limited to a band 2.5 in. (6.4 cm) wide around the perimeter of the glazing unit.

Estimation of the rate of heat transfer through the framing elements of a fenestration system is complicated by the wide variety of window configurations and materials used. Available windows are usually framed in wood, aluminum, or vinyl or combinations of these materials. Some aluminum-framed units have thermal breaks to reduce the conductive losses through the framing element.

The "Fenestration" chapter of the *1993 ASHRAE Handbook of Fundamentals* contains overall U-values for nearly 2000 fenestration products. A representative sample, extracted from the *1989 ASHRAE Handbook of Fundamentals* [4], is presented in Table 14.5. The values in Table 14.5 are for commercial-size windows in which the panes of glazing are approximately 24 ft^2 (2.2 m^2) in area. The areas of the units are assumed to be 15 percent frame, 15 percent edge, and 70 percent center-of-glass for the aluminum-framed units. For wood and vinyl the fractions are taken as 18 percent frame, 15 percent edge, and 67 percent center-of-glass. The U-values, U_f, for the various frame types are indicated in parentheses in the table header.

To estimate the overall U-value of a fenestration product that differs significantly from the assumptions in Table 14.5, first determine the percentage of area that is frame, edge-of-glass (based on a 2.5-in. band around the perimeter of each glazing unit), and center-of-glass. Next, determine the appropriate component U-values. These can be either from the standard values listed in Table 14.5 or from some other source such as test data or computed values. Finally, substitute the values into Eq. (14.25) to calculate U_o. For example, smaller residential windows usually have framing and edge effects that are larger than those of commercial windows, and ASHRAE recommends that the respective percentages of framing, edge-of-glass, and center-of-glass be taken as 25, 27, and 48 percent for aluminum-framed windows and 30, 26, and 44 percent for wood- or vinyl-framed windows.

Part A of Table 14.5 is for vertical installation and outdoor wind speed of 15 mph (6.7 m/s). Part B lists approximate U-values for simple single- and double-glazed units installed at 45° or horizontally. After determining the U-value for a vertical installation from Part A, enter Part B on the 90° slope line (vertical) and find that value. Then read

TABLE 14.5 Overall Coefficients of Heat Transmission of Various Fenestration Products

Part A: U-Values for Vertical Installation^a, Btu/hr · ft^2 · °F

Glazing Type^c	Glass Only		Aluminum Frame, No Thermal Break (U_f = 1.9)	Aluminum Frame, Thermal Break (U_f = 1.0)	Wood or Vinyl Frame (U_f = 0.4)
	Center of Glass	Edge^b of Glass			
Single glazing glass	1.11	n/a	1.23	1.10	0.98
1/8 in. acrylic	1.03	n/a	1.16	1.03	0.92
Double glass					
1/4 in. air space	0.57	0.66	0.78	0.65	0.55
3/8 in. air space	0.52	0.62	0.74	0.60	0.51
1/2 in. and greater air space	0.49	0.59	0.72	0.59	0.49
Double glass, ε = 0.40 on surface 2 or 3					
1/4 in. air space	0.50	0.60	0.73	0.59	0.50
3/8 in. air space	0.43	0.55	0.67	0.54	0.45
1/2 in. and greater air space	0.41	0.54	0.65	0.52	0.42
Double glass, ε = 0.15 on surface 2 or 3					
1/4 in. air space	0.45	0.56	0.68	0.55	0.46
3/8 in. air space	0.36	0.51	0.62	0.48	0.39
1/2 in. and greater air space	0.34	0.50	0.60	0.46	0.37
Double glass					
1/4 in. argon space	0.52	0.62	0.74	0.61	0.51
3/8 in. argon space	0.48	0.59	0.71	0.57	0.48
1/2 in. and greater argon space	0.46	0.57	0.69	0.56	0.47
Double glass, ε = 0.40 on surface 2 or 3					
1/4 in. argon space	0.43	0.55	0.67	0.54	0.45
3/8 in. argon space	0.38	0.52	0.63	0.49	0.40
1/2 in. and greater argon space	0.36	0.51	0.62	0.48	0.39
Double glass, ε = 0.15 on surface 2 or 3					
1/4 in. argon space	0.36	0.51	0.62	0.48	0.39
3/8 in. argon space	0.30	0.48	0.57	0.43	0.34
1/2 in. and greater argon space	0.28	0.47	0.55	0.42	0.33
Double glazing, 1/8 in. acrylic or polycarbonate					
1/4 in. air space	0.52	0.62	0.74	0.61	0.51
3/8 in. air space	0.48	0.59	0.71	0.57	0.48
1/2 in. and greater air space	0.46	0.57	0.69	0.56	0.47
Double glazing, 1/4 in. acrylic or polycarbonate					
1/4 in. air space	0.48	0.59	0.71	0.57	0.48
3/8 in. air space	0.44	0.56	0.68	0.54	0.45
1/2 in. and greater air space	0.42	0.54	0.66	0.53	0.43
Triple glass					
1/4 in. air space	0.38	0.52	0.64	0.50	0.41
3/8 in. air space	0.34	0.50	0.60	0.46	0.38
1/2 in. and greater air space	0.32	0.49	0.58	0.45	0.36

(continued)

TABLE 14.5 (cont.)

Part A: U-Values for Vertical Installation[a], Btu/hr · ft² · °F

Glazing Type[c]	Glass Only		Aluminum Frame, No Thermal Break ($U_f = 1.9$)	Aluminum Frame, Thermal Break ($U_f = 1.0$)	Wood or Vinyl Frame ($U_f = 0.4$)
	Center of Glass	Edge[b] of Glass			
Triple glass, $\varepsilon = 0.40$ on surface 2, 3, 4, or 5					
1/4 in. air spaces	0.35	0.50	0.61	0.48	0.39
3/8 in. air spaces	0.30	0.48	0.57	0.44	0.35
1/2 in. and greater air spaces	0.28	0.47	0.55	0.41	0.33
Triple glass or double glass with polyester film suspended in between, $\varepsilon = 0.15$ on surface 2, 3, 4, or 5					
1/4 in. air spaces	0.33	0.49	0.59	0.45	0.37
3/8 in. air spaces	0.27	0.46	0.54	0.41	0.32
1/2 in. and greater air spaces	0.24	0.45	0.52	0.39	0.30
Triple glass or double glass with polyester film suspended in between, $\varepsilon = 0.15$ on surfaces 2 or 3 and 4 or 5					
1/4 in. air spaces	0.28	0.47	0.55	0.42	0.33
3/8 in. air spaces	0.22	0.45	0.51	0.37	0.29
1/2 in. and greater air spaces	0.19	0.44	0.48	0.35	0.26
Triple glass					
1/4 in. argon spaces	0.34	0.50	0.60	0.46	0.38
3/8 in. argon spaces	0.31	0.48	0.57	0.44	0.35
1/2 in. and greater argon spaces	0.29	0.47	0.56	0.42	0.34
Triple glass, $\varepsilon = 0.40$ on surface 2, 3, 4, or 5					
1/4 in. argon spaces	0.30	0.48	0.57	0.44	0.35
3/8 in. argon spaces	0.26	0.46	0.54	0.41	0.32
1/2 in. and greater argon spaces	0.25	0.46	0.53	0.39	0.31
Triple glass or double glass with polyester film suspended in between, $\varepsilon = 0.15$ on surface 2, 3, 4, or 5					
1/4 in. argon spaces	0.27	0.46	0.54	0.41	0.32
3/8 in. argon spaces	0.22	0.45	0.51	0.37	0.29
1/2 in. and greater argon spaces	0.20	0.44	0.50	0.36	0.28
Triple glass or double glass with polyester film suspended in between, $\varepsilon = 0.15$ on surfaces 2 or 3 and 4 or 5					
1/4 in. argon spaces	0.22	0.45	0.51	0.37	0.29
3/8 in. argon spaces	0.17	0.43	0.47	0.34	0.25
1/2 in. and greater argon spaces	0.15	0.43	0.46	0.32	0.24

Part B: U-Value Conversion Table for Sloped and Horizontal Glazing for Upward Heat Flow

Slope	U-Value, Btu/hr · ft² · °F												
90° (vertical)	0.10	0.20	0.30	0.40	0.50	0.60	0.70	0.80	0.90	1.00	1.10	1.20	1.30
45°	0.14	0.25	0.36	0.47	0.57	0.68	0.79	0.90	1.00	1.11	1.22	1.33	1.44
0 (horizontal)	0.19	0.29	0.40	0.51	0.61	0.72	0.82	0.93	1.04	1.14	1.25	1.35	1.46

Source: Adapted by permission from *ASHRAE Fundamentals 1989.*

Note: To convert U-values to W/m² · °C multiply by 5.68.

[a] All U-factors are based on standard ASHRAE winter conditions of 70 °F indoor and 0 °F outdoor-air temperature, with 15 mph outdoor-air velocity and zero solar flux. The outside-surface coefficient at these conditions is approximately 5.1 Btu/hr · ft² · °F, depending on the glass surface temperature. With the exception of single glazing, small changes in the interior and exterior temperatures do not significantly affect overall U-factors.

[b] Based on aluminum spacers data. Edge of glass effect assumed to extend over the 2.5-in. band around perimeter of each glazing unit.

[c] Glazing layer surfaces are numbered from the outside to the inside. Double and triple refer to the number of glazing lites. All data are based on 1/8-in. glass unless otherwise noted. Thermal conductivities are: 0.53 Btu/hr · ft · °F for glass, and 0.11 Btu/hr · ft · °F for acrylic and polycarbonate.

434

TABLE 14.6 Glazing *U*-Factor
for Various Wind Speeds

Wind Speed, mph		
15	7.5	0
U-Factor, Btu/hr · ft² · °F		
0.10	0.10	0.10
0.20	0.20	0.19
0.30	0.29	0.28
0.40	0.38	0.37
0.50	0.47	0.45
0.60	0.56	0.53
0.70	0.65	0.61
0.80	0.74	0.69
0.90	0.83	0.78
1.00	0.92	0.86
1.10	1.01	0.94
1.20	1.10	1.02
1.30	1.19	1.10

SOURCE: Adapted by permission from
ASHRAE Fundamentals 1989.

down to find the *U*-value for the desired slope. For slopes other than those in Part B, the recommended procedure given in the *ASHRAE Handbook of Fundamentals 1989* is to use the 90° value (vertical) for products installed 60° to 90° above horizontal, the 45° value for installations 30° to 60° above horizontal, and the zero-degree value (horizontal) for installations within 30° of horizontal. Table 14.6 can be used to adjust the values to outdoor wind speeds of 7.5 mph (3.4 m/s) or zero.

Doors

The *U*-values for doors can be estimated using the basic principles described above in this chapter. In addition, values for several specific wood and steel doors are presented in Table 14.7. The values for wood doors were calculated, and those for steel doors were taken from hot-box tests or from manufacturers' test reports [1]. The indoor heat-transfer coefficient used in establishing the values in Table 14.7 is that for still air on a vertical surface with an emittance of 0.9, and the outdoor heat-transfer coefficient is that for a wind speed of 15 mph (6.7 m/s). (These h_i and h_o values are contained in Table 14.1.) All values in Table 14.7 are for exterior doors without glazing. If an exterior door contains glazing, the glazing should be analyzed as a window, using the procedure described in the preceding section.

14.3 BELOW-GRADE HEAT TRANSFER IN BUILDINGS

Heat transfer from basements is particularly important for residential applications. The basement interior is considered conditioned space if a minimum of 50 °F (10 °C) is maintained over the heating season. The heat transmission from below-grade portions of the basement wall to the ambient temperature cannot be estimated by simple, one-dimensional heat conduction. In fact, a complete solution must include a transient two- or three-dimensional coupled heat- and mass-transfer analysis. Coupled heat and mass transfer is

TABLE 14.7 Transmission Coefficients U for Wood and Steel Doors, Btu/hr · ft² · °F

Nominal Door Thickness, in.	Description	No Storm Door	Wood Storm Door[c]	Metal Storm Door[d]
Wood Doors[a,b]				
1-3/8	Panel door with 7/16-in. panels[e]	0.57	0.33	0.37
1-3/8	Hollow-core flush door	0.47	0.30	0.32
1-3/8	Solid-core flush door	0.39	0.26	0.28
1-3/4	Panel door with 7/16-in. panels[e]	0.54	0.32	0.36
1-3/4	Hollow-core flush door	0.46	0.29	0.32
1-3/4	Panel door with 1-1/8-in. panels[e]	0.39	0.26	0.28
1-3/4	Solid-core flush door	0.40	—	0.26
2-1/4	Solid-core flush door	0.27	0.20	0.21
Steel Doors[b]				
1-3/4	Fiberglass or mineral wool core with steel stiffeners, no thermal break[f]	0.60	—	—
1-3/4	Paper honeycomb core without thermal break[f]	0.56	—	—
1-3/4	Solid urethane foam core without thermal break[a]	0.40	—	—
1-3/4	Solid fire-rated mineral fiberboard core without thermal break[f]	0.38	—	—
1-3/4	Polystyrene core without thermal break (18 gage commercial steel)[f]	0.35	—	—
1-3/4	Polyurethane core without thermal break (18 gage commercial steel)[f]	0.29	—	—
1-3/4	Polyurethane core without thermal break (24 gage residential steel)[f]	0.29	—	—
1-3/4	Polyurethane core with thermal break and wood perimeter (24 gage residential steel)[f]	0.20	—	—
1-3/4	Solid urethane foam core with thermal break[a]	0.20	—	0.16

SOURCE: *Adapted by permission from ASHRAE Fundamentals 1993.*

Note: All *U*-factors for exterior doors in this table are for doors with no glazing, except for the storm doors which are in addition to the main exterior door. Any glazing area in exterior doors should be included with the appropriate glass type and analyzed as a window. Interpolation and moderate extrapolation are permitted for door thicknesses other than those specified.

To convert *U*-values to W/m² · °C multiply by 5.68.

[a] Values are based on a nominal 32 by 80 in. door size with no glazing.

[b] Outside-air conditions: 15 mph wind speed, 0 °F air temperature; inside-air conditions: natural convection, 70 °F air temperature.

[c] Values for wood storm door are for approximately 50 percent glass area.

[d] Values for metal storm door are for any percent glass area.

[e] 55 percent panel area.

[f] ASTM C 236 hotbox data on a nominal 3 by 7 ft door size with no glazing.

needed because the thermal conductivity of the soil depends on moisture content, and the moisture distribution is related to the temperature distribution in the soil. However, for heating-load calculations simple two-dimensional analyses are commonly used for basement walls and floors. Comparisons by Szydlowski and Kuehn [5] of these methods with more detailed computations demonstrate that the simplified analysis provides good estimates for design heat losses. A simple one-dimensional analyses is commonly used for floor slabs. These procedures are described below.

Through Basement Walls

The method used to estimate the heat loss from below-grade walls is based on the research of Latta and Boileau [6], which showed that the isotherms near the wall approximate radial lines centered at the intersection of the grade line and the wall. Therefore, the heat-flow paths approximately follow a set of concentric circular patterns similar to those shown in Fig. 14.6. The total heat-transfer resistance along each path is calculated, including the convective/radiative resistance at the inside surface, the resistance of the wall structure, any added insulation, and the resistance through the soil. The temperature difference driving the heat transfer along each path is taken as the difference between the indoor-air temperature and a representative ground-surface temperature. This ground-surface temperature subsequently will be discussed in greater detail. (Note that since the ground-surface temperature is used, the thermal resistance between the ground surface and air is not included in the total resistance along each path.) The heat-transfer analysis is performed with the aid of Fig. 14.7, a schematic of a partially insulated wall which includes the nomenclature.

Two heat-transfer paths are shown in Fig. 14.7, one through the insulated section of the wall and one through the uninsulated section. The total resistance for path 1 is $R_{t,I}$, where the presence of the subscript "I" indicates that insulation is present. The value of $R_{t,I}$ is given by

$$R_{t,I} = R_i + R_w + R_{INS} + \frac{\pi y}{2k_s} \tag{14.26}$$

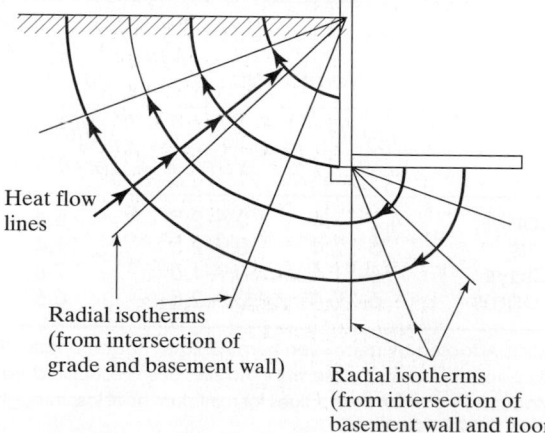

Figure 14.6 Heat flow from basement.

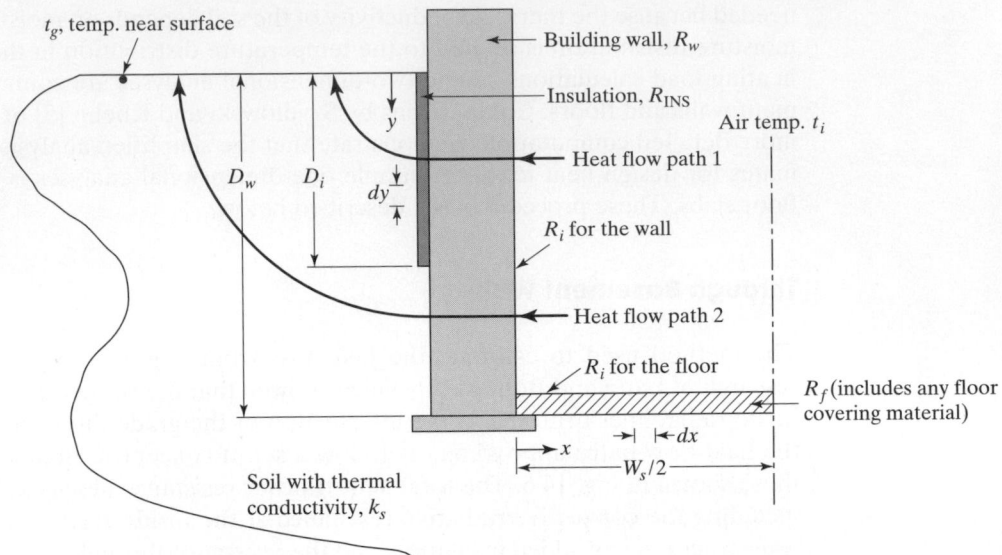

Figure 14.7 Geometry and nomenclature for heat-loss estimates from a basement wall.

In this equation the value of the thermal conductivity of the soil, k_s, is assumed constant, and the path length through the soil is calculated neglecting the thickness of the insulation. Examples of apparent thermal conductivities for soils are presented in Table 14.8.

The total resistance of the uninsulated path, path 2, is R_t.

$$R_t = R_i + R_w + \frac{\pi y}{2k_s} \tag{14.27}$$

The total heat-transfer rate from a wall with horizontal length, L, is then given by

$$\dot{Q}_w = L(t_i - t_g)\left[\int_0^{D_i} \frac{dy}{R_{t,I}} + \int_{D_i}^{D_w} \frac{dy}{R_t}\right] \tag{14.28}$$

TABLE 14.8 Typical Apparent Thermal Conductivity Values for Soils

| | Normal Range | | Recommended Values for Design[a] | | | |
| | | | Low[b] | | High[c] | |
	$\dfrac{Btu \cdot in.}{hr \cdot ft^2 \cdot °F}$	$W/m \cdot K$	$\dfrac{Btu \cdot in.}{hr \cdot ft^2 \cdot °F}$	$W/m \cdot K$	$\dfrac{Btu \cdot in.}{hr \cdot ft^2 \cdot °F}$	$W/m \cdot K$
Sands	4.2–17.4	0.6–2.5	5.4	0.78	15.6	2.25
Silts	6–17.4	0.9–2.5	11.4	1.64	15.6	2.25
Clays	6–11.4	0.9–1.6	7.8	1.12	10.8	1.56
Loams	6–17.4	0.9–2.5	6.6	0.95	15.6	2.25

SOURCE: Adapted by permission from *ASHRAE Fundamentals 1993*.

[a] Reasonable values for use when no site- or soil-specific data are available.

[b] Moderately conservative values for minimum heat loss through soil (e.g., use in soil heat exchanger or earth-contact cooling calculations).

[c] Moderately conservative values for maximum heat loss through soils (e.g., use in peak winter heat-loss calculations).

Equation (14.28) can also be written in terms of an average U-value for the wall, U_w, as

$$\dot{Q}_w = U_w (L \cdot D_w)(t_i - t_g) \tag{14.29}$$

Notice that this is a special form of a U-value, since one of the temperatures is actually the ground temperature rather than the outdoor-air temperature. The value for U_w is found by substituting Eqs. (14.26) and (14.27) into Eq. (14.28), integrating, and comparing the results with Eq. (14.29). The resulting value is

$$U_w = \frac{2k_s}{\pi D_w} \left\{ \ln\left[1 + \frac{\pi D_i}{2k_s (R_i + R_w + R_{INS})} \right] + \ln\left[1 + \frac{\pi D_w}{2k_s (R_i + R_w)} \right] - \ln\left[1 + \frac{\pi D_i}{2k_s (R_i + R_w)} \right] \right\} \tag{14.30}$$

EXAMPLE 14.1

A wall extends 8 ft below grade. The wall is constructed of 8-in. concrete block and is surrounded by a sandy soil. What is the effectiveness of insulating the wall by adding a 2-in. layer of expanded polystyrene insulation on the exterior of the wall?

Solution: To evaluate the effect of adding insulation, we will calculate the U-value for various depths of insulation using Eq. (14.30). The value of R_i is found in Table 14.1. Most common building materials and/or coatings for the inside surface of the wall will have an emittance of 0.9 and, therefore, the value of R_i is 0.68 °F · ft² · hr/Btu. Based on the data in Table 14.4IP, the R-values for the concrete block and the 2-in.-thick insulation are taken as $R_w = 1.11$ °F · ft² · hr/Btu and $R_{INS} = 10$ °F · ft² · hr/Btu. The recommended design value of the soil thermal conductivity is selected from Table 14.8. Thus, $k_s = 15.6$ Btu · in./hr · ft² · °F = 1.30 Btu/hr · ft · °F. The results of the calculations are shown in Fig. 14.8. The average U-value for the wall varies from 0.192 Btu/hr · ft² · °F with no insulation to 0.062 Btu/hr · ft² · °F when insulated along the full depth. Figure 14.8 also demonstrates how the value of adding insulation diminishes with the depth of insulation.

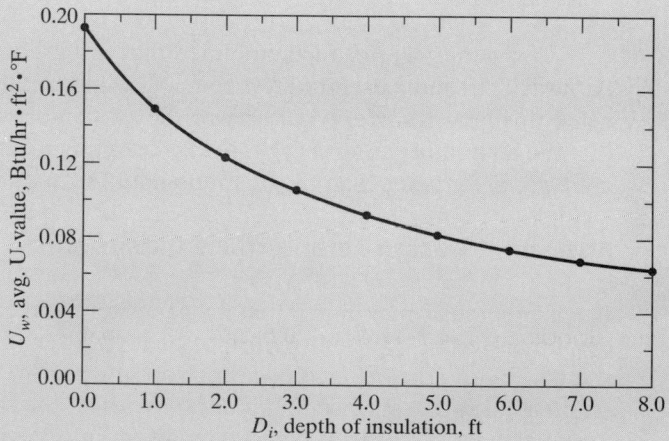

Figure 14.8 Average U-value for an 8-ft-deep below-grade wall as a function of insulation depth. (Calculated for $R_{INS} = 10$ °F · ft² · hr/Btu, $k_s = 15.6$ Btu · in./hr · ft² · °F, and $R_w = 1.11$ °F · ft² · hr/Btu.)

Through Basement Floors

The same steady-state design approach used for the basement wall can be applied to the basement floor. However, the lengths of the heat-flow paths are much longer, as seen in Fig. 14.6, and so the heat flow through the basement floor is much smaller than that from the walls. A two-dimensional steady-state analysis based on the nomenclature shown in Fig. 14.7 gives an estimate of the heat loss from the floor. The width of the basement for the analysis is taken as the shortest width of the house, W_s. At a floor location a distance x from the intersection of the basement wall and the floor the total thermal resistance along a heat-flow path to the surface of the ground $R_{t,f}$ is approximately

$$R_{t,f} = R_i + R_f + \frac{\pi}{k_s}\left(\frac{D_w}{2} + x\right) \tag{14.31}$$

Therefore, the heat-transfer rate from the floor with area, A_f, is given by

$$\dot{Q}_f = 2\left(\frac{A_f}{W_s}\right)(t_i - t_g)\int_0^{W_s/2} \frac{dx}{R_i + R_f + \frac{\pi}{k_s}\left(\frac{D_w}{2} + x\right)} \tag{14.32}$$

The equation for the heat-transfer rate from the floor can also be written in terms of an average U-value for the floor, U_f.

$$\dot{Q}_f = U_f A_f(t_i - t_g) \tag{14.33}$$

As in the case of heat loss from the basement wall, this is a special form of a U-value, since one of the temperatures is actually the ground temperature rather than the outdoor-air temperature. From Eqs. (14.32) and (14.33) we get

$$U_f = \frac{2k_s}{\pi W_s}\ln\left[\frac{R_i + R_f + \frac{\pi D_w}{2k_s} + \frac{\pi W_s}{2k_s}}{R_i + R_f + \frac{\pi D_w}{2k_s}}\right] \tag{14.34}$$

It is generally not recommended that insulation be placed below a basement floor. Although this reduces the winter heat loss, it also reduces the heat loss during the summer, when this heat transfer would lower the air conditioning load. Kuehn [7] demonstrated that the air conditioning energy savings resulting from an uninsulated floor in the summer are typically greater than the heating-energy savings achieved by insulating the floor.

Basement Design Temperature Differences

It was previously mentioned that heat transfer from earth-contact surfaces is transient. To demonstrate the transient nature of heat transfer in soils, consider the temperatures in the soil below the surface of a horizontal open area. Assume the ground water is far enough below the surface that it does not affect the soil-temperature profiles. Figure 14.9 shows the results of heat transfer in a semiinfinite solid with periodic boundary conditions. This can be applied to soil, providing the surface-temperature changes in a sinusoidal manner over a 12-month period, which is a reasonable representation of actual weather conditions, and the soil thermal conductivity is constant. On the figure, Δt is the temperature deviation from the deep-ground temperature. This deep-ground tempera-

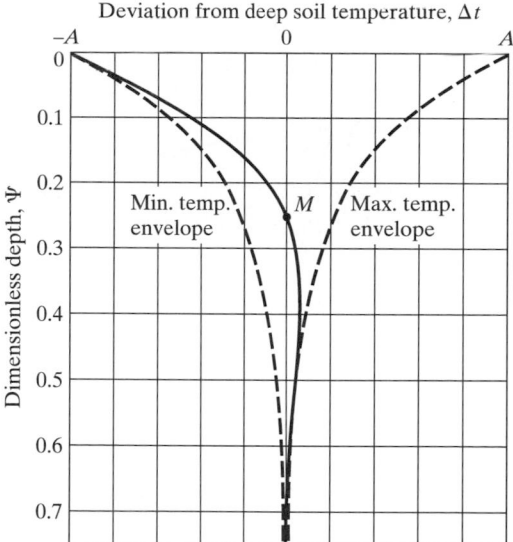

Deviation from deep soil temperature, Δt

Min. temp. envelope M Max. temp. envelope

Figure 14.9 Soil temperature vs. depth in an open field.

ture is often equal to the temperature of well water. The solution can be cast in dimensionless form. If y is the distance below grade, the dimensionless depth, Ψ, becomes

$$\Psi = \frac{y}{2\sqrt{\pi \alpha \tau_0}} \qquad (14.34a)$$

where α is the soil thermal diffusivity and τ_0 is the period (12 months in this case). The dashed lines in Fig. 14.9 show the maximum temperature swing versus depth. The temperature amplitude continuously diminishes with depth. The solid curve represents the temperature profile at one extreme of the surface fluctuation. We can interpret this as the winter design temperature profile. Note that the temperature is a minimum at the surface, is equal to the deep-ground temperature at a dimensionless depth of $\Psi = 0.25$, and is a local maximum at $\Psi = 0.4$.

On an annual basis the cross-over point, labeled as M in the figure, is well below the depth of most basements, as shown in the following example. With a typical soil thermal diffusivity of $\alpha = 0.04$ ft²/hr (1.0×10^{-6} m²/s) and a period of 12 months the depth of point M calculated from Eq. (14.34a) is approximately 17 ft (5.2 m). Therefore, it is not appropriate to use deep-ground soil temperatures for most earth-contact structure heat-loss calculations.

Daily temperature fluctuations are superimposed on the profiles shown in Fig. 14.9. However, these fluctuations are damped within a few inches (centimeters) of the surface. Agronomists have compiled a significant quantity of soil-temperature data measured at a depth of about 4 in. (10 cm). A contour plot showing lines of constant temperature amplitude, A, at the 4-in. (10-cm) soil depth is presented in Fig. 14.10. We can use this data to estimate an appropriate minimum sink temperature of the soil for winter design heat-loss calculations. The minimum temperature at the 4-in. soil depth for an open field is equal to the deep-ground temperature minus the amplitude, A. However, tests and computer simulations have shown that the use of this minimum soil temperature underestimates the winter design heat loss from earth-contact walls and floors. Better agreement has been

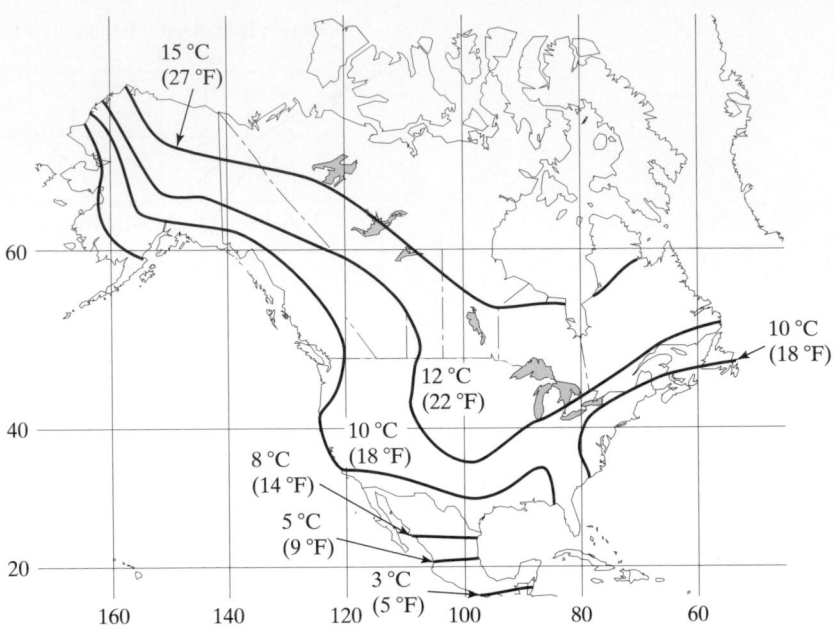

Figure 14.10 Lines of constant temperature amplitude. [Reprinted by permission from *ASHRAE Fundamentals 1993* (IP & SI), p. 25.12.]

found when the design soil temperature, t_g, is computed from Eq. (14.35), where the deep-ground temperature has been replaced by the average winter air temperature, $\bar{t}_a$.

$$t_g = \bar{t}_a - A \qquad (14.35)$$

Values of the monthly average air temperatures are given for several locations in the design data of Table B.2. The average winter air temperature, $\bar{t}_a$, is taken as the average of the values for the months of October through April.

Heat Loss from Floor Slabs

Concrete slabs on or near grade level are either unheated, relying on heat from above for warmth, or heated by air ducts around the perimeter of the slab. In both cases most of the heat loss is from the edge of the slab. Although the amount of heat loss from the slab may be small compared to other losses from the house, it is essential to insulate around the perimeter of the slab to maintain a warm, comfortable floor in cold climates. Figure 14.11 shows four typical floor-slab-on-grade configurations. Since the total heat loss is more nearly proportional to the length of the perimeter than to the area of the floor, it can be estimated by the following equation for either heated or unheated slab floors:

$$\dot{Q}_s = F_s P(t_i - t_o) \qquad (14.36)$$

where $\dot{Q}_s$ = rate of heat loss from the slab, Btu/hr or W

F_s = heat-loss coefficient per unit length of perimeter, Btu/hr · ft · °F or W/m · C

P = length of the perimeter of the slab, ft or m

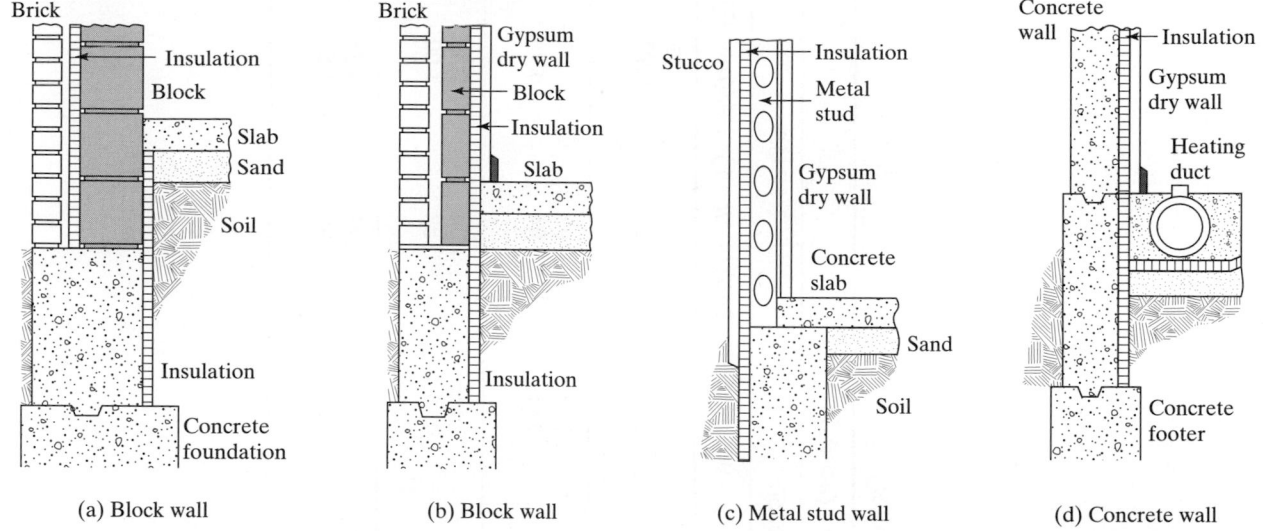

| (a) Block wall | (b) Block wall | (c) Metal stud wall | (d) Concrete wall |

Figure 14.11 Example slab-on-grade configuration. [Reprinted by permission from *ASHRAE Fundamentals 1993* (IP & SI), p. 25.13.]

t_i = indoor-air temperature (for a heated slab, t_i is weighted average heating-duct or pipe temperature)

t_o = outdoor-air temperature

Examples of perimeter heat-loss coefficients, F_s, for configurations similar to those shown in Fig. 14.11 are given in Table 14.9.

14.4 INFILTRATION

Infiltration is uncontrolled air flow through cracks and other unintentional openings in buildings. Infiltration and exfiltration air flows are driven by outdoor-to-indoor pressure differences which result from wind, indoor-outdoor temperature differences, and appliance operation. Infiltration can be of particular importance in calculating heating and cooling loads for residences and small commercial buildings in which air exchange is due primarily to envelope or shell infiltration. For larger buildings the air exchange is primarily controlled mechanical ventilation, and infiltration plays a much smaller role in the load calculation. However, the physical principles discussed in this section apply to both large and small buildings.

The net exchange of air by infiltration and exfiltration leads to both heat and moisture loads for a space. In writing the energy- and mass-balance equations for the space, the air flow is assumed to enter the space at outdoor conditions and exit at indoor conditions. Therefore, the winter heating load that results from infiltration is

$$\dot{Q} = \dot{m}_a(h_i - h_o) \tag{14.37}$$

where $\dot{m}_a$ = mass flow rate of infiltration

h_i = enthalpy of the indoor air

h_o = enthalpy of outdoor air

TABLE 14.9 Heat-Loss Coefficient F_s of Slab-Floor Construction per Unit Length of Perimeter

		Degree Days (Base 65 °F)[a]					
		2950		5350		7433	
Construction	Insulation	Btu/(hr·ft·°F)	W/(m·°C)	Btu/(hr·ft·°F)	W/(m·°C)	Btu/(hr·ft·°F)	W/(m·°C)
8-in. (200-mm) block wall, brick face	Uninsulated R-5.4 (R-0.95) slab to footing	0.62 0.48	1.07 0.83	0.68 0.50	1.18 0.87	0.72 0.56	1.25 0.97
4-in. (100-mm) block wall, brick face	Uninsulated R-5.4 (R-0.95) slab to footing	0.80 0.47	1.39 0.81	0.84 0.49	1.45 0.85	0.93 0.54	1.61 0.94
Metal stud wall, stucco	Uninsulated R-5.4 (R-0.95) slab to footing	1.15 0.51	1.99 0.88	1.20 0.53	2.08 0.92	1.34 0.58	2.32 1.00
Poured concrete wall, with perimeter ducts[b]	Uninsulated R-5.4 (R-0.95) slab to footing	1.84 0.64	3.19 1.11	2.12 0.72	3.67 1.25	2.73 0.90	4.73 1.56

Source: Adapted by permission from *ASHRAE Fundamentals 1993.*
[a] Degree-day data are given in Table B.2.
[b] Weighted average temperature of the heating duct was assumed at 110 °F (43 °C) during the heating season.
Note: R-value units in °F · ft² · hr/Btu (°C · m²/W).

This heating load can be divided into sensible and latent components by applying Eqs. (7.45) and (7.46), respectively, resulting in the following:

Sensible infiltration load: $\dot{Q}_S = \dot{m}_a \bar{c}_p (t_i - t_o)$ (14.37a)

Latent infiltration load: $\dot{Q}_L = \dot{m}_a \bar{h}_g (W_i - W_o)$ (14.37b)

Recall from the discussion in Chapter 7 that the quantity $\bar{c}_p$ is the average specific heat for the mixture for the process, i.e.,

$$\bar{c}_p = c_{pa} + \overline{W} c_{pw}$$ (7.43)

where $\overline{W}$ is the average humidity ratio for the process.

If indoor conditions are 70 °F (21 °C) and 30 percent relative humidity, the value of $\bar{c}_p$ varies from 0.241 Btu/lbm$_a$ · °F (1.00 kJ/kg$_a$ · °C) for cold, dry outdoor conditions to 0.242 Btu/lbm$_a$ · °F (1.01 kJ/kg$_a$ · °C) for 40 °F (4.4 °C) and 70 percent relative humidity outdoor conditions. Therefore, a reasonable approach is to use the dry air value of the specific heat 0.240 Btu/lbm$_a$ · °F (1.00 kJ/kg$_a$ · °C) in applying Eq. (14.37a).

The rate of moisture loss from the space that results from the infiltration is

$$\dot{m}_w = \dot{m}_a (W_i - W_o)$$ (14.38)

These equations are deceptively simple, since obtaining good estimates of the mass flow rate of dry air due to infiltration, $\dot{m}_a$, is difficult.

Techniques for estimating infiltration rates have improved in recent years. Typically, the estimates are made in terms of the volumetric flow rate of air entering the space, $\dot{V}$, and, therefore,

$$\dot{m}_a = \rho \dot{V}$$ (14.39)

where ρ is the density of the air. The convention is to use the standard air density of 0.075 lbm$_a$/ft^3 (1.2 kg$_a$/m^3) in Eq. (14.39).

Two of the earlier approaches used to estimate the infiltration rate were the "air-change method" and the "crack method." In applying the *air-change method*, one simply makes an assumption of the *number of air changes per hour (ACH)* that a building will experience based on an appraisal of the building type, construction, and use. Experience and judgment are required to obtain satisfactory results with this method. Using this approach, the infiltration rate is expressed as

$$\dot{V} = ACH \cdot V$$ (14.40)

where V is the gross volume of the space.

The *crack method* applies the principle that the infiltration rate is a function of the outside-to-inside pressure difference, ΔP. In addition, the flow rate depends on the type and total area of the cracks that exist around doors, windows, lighting fixtures, and joints between walls and floor. In applying the crack method, the flow rate is represented by

$$\dot{V} = C \Delta P^n$$ (14.41)

where C = flow coefficient, which depends on the type of crack, the nature of the flow in the crack and the area of the crack

      n = exponent that depends on the nature of the flow in the crack

A detailed description of the crack method is presented by F. C. McQuiston and J. D. Spitler [8]. More recent methods of estimating the infiltration rate are described in the *ASHRAE Handbook of Fundamentals* [1]. Before discussing these methods, it is important to look in greater detail at the inside-outside pressure difference. This pressure difference results from three effects:

$$\Delta P = \Delta P_w + \Delta P_s + \Delta P_p \tag{14.42}$$

where ΔP_w = pressure difference due to wind
ΔP_s = pressure difference due to stack effects
ΔP_p = pressure difference due to building pressurization

Pressure Difference Due to Wind

An idealized or theoretical value of the pressure difference due to wind can be calculated for the side of the building that faces directly into the wind by setting the pressure inside the building equal to the static pressure in the exterior air stream and assuming that the velocity of the air goes to zero when striking the outer surface. This theoretical pressure difference, ΔP_{wt}, will be equal to the difference between the total and static pressures in the air stream, or

$$\Delta P_{wt} = 0.5\rho V_w^2 \tag{14.43}$$

where ρ is the air density and V_w is the wind speed. However, the wind speed does not go to zero as the air moves around the building, and most surfaces do not face directly into the wind. To account for this, the actual wind-pressure difference, ΔP_w, is related to the theoretical wind-pressure difference by applying a wind-pressure coefficient, C_p.

$$\Delta P_w = C_p \, \Delta P_{wt} = 0.5 C_p \rho V_w^2 \tag{14.44}$$

Example values of pressure coefficients for walls are given for a low-rise building in Fig. 14.12 and for a tall building in Fig. 14.13. Average roof-pressure coefficients for a tall building are shown in Fig. 14.14.

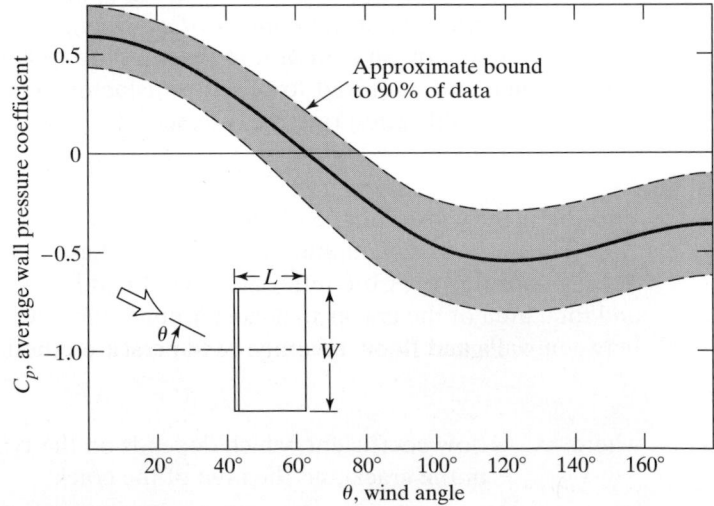

Figure 14.12 Variation of wall-averaged pressure coefficients for low-rise building. [Reprinted by permission from *ASHRAE Cooling and Heating Load Calculation Manual,* 2d ed. (1992), p. 6.5.]

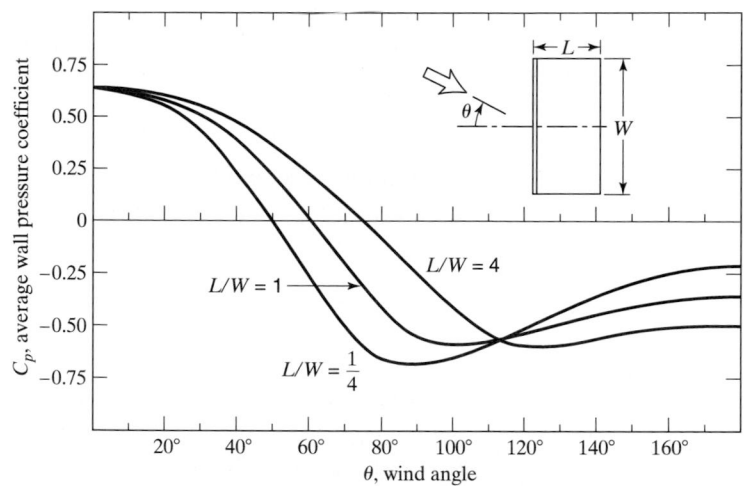

Figure 14.13 Variation of wall-averaged pressure coefficients for tall building. [Reprinted by permission from *ASHRAE Cooling and Heating Load Calculation Manual,* 2d ed. (1992), p. 6.5.]

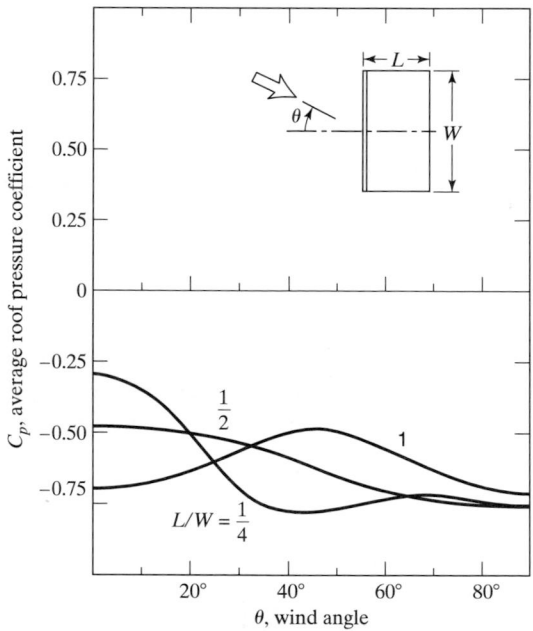

Figure 14.14 Average roof-pressure coefficients for tall building. [Reprinted by permission from *ASHRAE Cooling and Heating Load Calculation Manual,* 2d ed. (1992), p. 6.5.]

Pressure Difference Due to Stack Effect

The pressure at a given elevation results from the mass of the air column above that elevation, which is a function of the density of the air. When the indoor and outdoor temperatures are different, the air densities, and thus the column masses and pressures, will be different. This pressure difference is referred to as the *stack effect*. The idealized or theoretical local hydrostatic outside-to-inside pressure difference due to the air-density differences, ΔP_{st}, can be written as

$$\Delta P_{st} = (h - y)g(\rho_o - \rho_i) \tag{14.45}$$

where
g = gravitational acceleration
h = height of the neutral pressure level (described below)
y = height above ground level
ρ_o, ρ_i = outside and inside densities, respectively

Under the influence of stack effect, there will be a vertical location in the building where the inside and outside pressures are equal. This is defined as the *neutral pressure level* and appears as h in Eq. 14.45. If cracks and other openings are uniformly distributed vertically, the neutral pressure level will be at approximately mid-height. For nonuniformly distributed openings, the neutral pressure level shifts in the direction of the larger openings.

The densities that appear in Eq. (14.45) can be replaced by substituting the appropriate expressions using the ideal-gas equation of state:

$$\Delta P_{st} = (h - y)g\left(\frac{P_o}{R_o T_o} - \frac{P_i}{R_i T_i}\right) \tag{14.46}$$

The gas constants R_o and R_i are for the outside and inside air mixtures, respectively. Since the moisture levels will have only a minor effect on their value, they can be replaced by the gas constant for dry air, R_a. In addition, recall that the outside and inside pressures, P_o and P_i, are absolute pressures, and, since their difference is small compared to their absolute value, a good approximation is that they are both equal to the atmospheric pressure, P. Applying these substitutions to Eq. (14.46) results in

$$\Delta P_{st} = \frac{(h - y)gP}{R_a}\left(\frac{1}{T_o} - \frac{1}{T_i}\right) \tag{14.47}$$

Figure 14.15 shows the stack-effect pressure distribution for cold outside air. At heights where the outside pressure is greater than the inside pressure, as it is below the neutral height in Fig. 14.15, the air flow is into the building. For heights where the out-

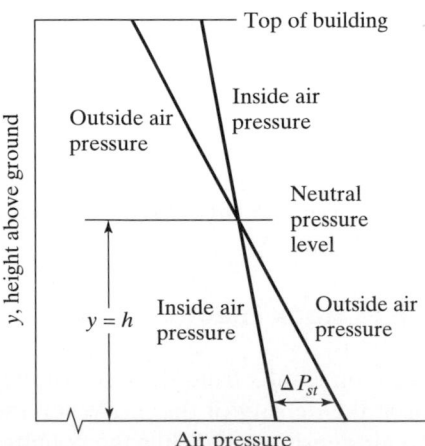

Figure 14.15 Winter stack effect showing theoretical pressure difference vs. height.

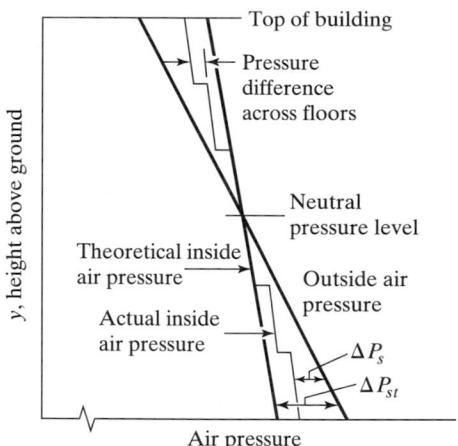

Figure 14.16 Winter stack effect showing actual pressure difference vs. height for six-story building.

side pressure is lower than the indoor pressure, the air flow will be out of the building. If the air temperature inside the building is colder than the outside temperature, the relative positions of the pressure lines in Fig. 14.15 are reversed, with air entering in the upper regions of the building and exiting in the lower regions.

The idealized pressure difference, ΔP_{st}, is valid only for buildings with no vertical separations (e.g., no floors) or for an atrium, auditorium, or stair tower. Doors and stairways between floors in a conventional building offer resistances to the vertical flow of air that results from the stack effect. If these resistances are approximately equal from floor to floor, a single correction factor, called the *thermal draft coefficient, C_d,* can be used to relate the actual pressure difference due to stack effect, ΔP_s, to the theoretical value, ΔP_{st}.

$$\Delta P_s = C_d \, \Delta P_{st} = \frac{C_d (h - y) g P}{R_a} \left(\frac{1}{T_o} - \frac{1}{T_i} \right) \qquad (14.48)$$

Figure 14.16 shows the effect of the pressure differences between floors for winter conditions. The upward air flow through the building causes the pressure to decrease at each floor. Therefore, $\Delta P_s \leq \Delta P_{st}$ and $C_d \leq 1$. Within each floor the slope of the pressure curve is equal to the slope of the theoretical curve. Discussions of the values of thermal draft coefficients are contained in McQuiston and Spitler [8] and the *ASHRAE Handbook of Fundamentals* [1]. The value of C_d depends on the resistance to the vertical air flow. The tighter the stair doors, etc., the smaller the value of C_d. Stairwells with no doors have values of C_d approaching 1.0. Values of C_d determined experimentally for a few modern office buildings ranged from 0.63 to 0.82.

EXAMPLE 14.2

A building is 500 ft tall and has openings which are approximately uniformly distributed along its height. Estimate the theoretical outside-to-inside pressure difference due to stack effect at heights of 125 and 350 ft when the outdoor temperature is 10 °F, the indoor temperature is 70 °F, and barometric pressure is 14.7 psia. Determine the pressure differences in both psia and equivalent inches of water.

Solution: The answer is obtained applying Eq. (14.47). Since the openings are approximately uniform over the height of the building, the height of the neutral pressure location is assumed to be the mid-height, $h = 250$ ft. The temperatures that appear in Eq. (14.47) must be absolute temperatures. Therefore, $T_o = 470$ °R and $T_i = 530$ °R. The gas constant for dry air is $R_a = 53.35$ (ft · lbf)/(lbm · °R), and the acceleration of gravity is 32.2 ft/s². Substituting these values into Eq. (14.47), we obtain

$$\Delta P_{st} = 0.267 \ (\text{lbm} \cdot \text{ft/s}^2) \cdot (1/\text{in.}^2) \ \text{at} \ y = 125 \ \text{ft}$$

and

$$\Delta P_{st} = -0.214 \ (\text{lbm} \cdot \text{ft/s}^2) \cdot (1/\text{in.}^2) \ \text{at} \ y = 350 \ \text{ft}$$

To convert these values to units of psia we note that 1 lbm · ft/s² = (1/32.2) lbf. Therefore, dividing the above values by 32.2 we get

$$\Delta P_{st} = 0.0083 \ \text{lbf/in}^2 \ \text{at} \ y = 125 \ \text{ft}$$

and

$$\Delta P_{st} = -0.0066 \ \text{lbf/in.}^2 \ \text{at} \ y = 350 \ \text{ft}$$

Many of the tables available that relate infiltration rates to pressure differences use the equivalent column height of water, usually in inches, as the units for ΔP. If h_w is the column height of water and ρ_w is the density of water, the hydrostatic pressure, P, for a column of water of height h_w is

$$P = \rho_w g h_w \quad \text{or} \quad h_w = P/\rho_w g$$

Applying this to our results and taking the density of water as $\rho_w = 62.4$ lbm/ft³, we get

$$\Delta P_{st} = 0.267 \left(\frac{\text{lbm} \cdot \text{ft}}{\text{s}^2 \cdot \text{in.}^2} \right) \left[\frac{1728 \ \text{in.}^3/\text{ft}^3}{(62.4 \ \text{lbm/ft}^3)(32.2 \ \text{ft/s}^2)} \right] = 0.23 \ \text{in. water}$$

and

$$\Delta P_{st} = -0.214 \left(\frac{\text{lbm} \cdot \text{ft}}{\text{s}^2 \cdot \text{in.}^2} \right) \left[\frac{1728 \ \text{in.}^3/\text{ft}^3}{(62.4 \ \text{lbm/ft}^3)(32.2 \ \text{ft/s}^2)} \right] = -0.18 \ \text{in. water}$$

at the heights of 125 and 350 ft, respectively. Notice that the pressure difference is positive at the position below the neutral pressure height—i.e., outside pressure is greater than inside pressure—and negative at the position above the neutral pressure height. Therefore, the air flow will be into the building at the 125-ft elevation and out of the building at the 350-ft elevation.

Pressure Difference Due to Building Pressurization

If the mechanical ventilation intake and exhaust are not equal, a pressure difference between outside and inside, ΔP_p, will be established which provides the necessary driving force so that the infiltration or exfiltration rate will equal the difference between the system intake and exhaust. A small amount of building pressurization is often designed into commercial buildings. However, careful design is required in order to prevent overpressurization problems, such as doors that do not close properly due to the pressure difference, or underpressurization problems, such as drawing air into the building at regions where outdoor contaminant levels are high.

Combination of Pressure-Difference Effects

The total outside-to-inside pressure difference at any building location is a combination of the effects described above, as seen from Eq. (14.42). As an example, consider winter conditions for a building which is slightly pressurized and has approximately uniform openings along its height. Figure 14.17 schematically shows the pressure distributions inside and outside the building and how they combine to give the pressure difference across the windward and leeward sides of the building. Figure 14.17(a) depicts the stack effect without and with pressurization. Without pressurization the neutral height is at the mid-height of the building. When the inside pressure increases due to pressurization, the neutral pressure level is seen to shift to a lower elevation.

The effect of wind alone on the pressure distribution is shown in Fig. 14.17(b). The wind effect produces a positive outside-to-inside pressure difference, ΔP_w, on the windward side and a negative ΔP_w on the leeward side. For the purposes of illustration, the magnitudes of ΔP_w on the two surfaces are taken as approximately equal. In practice, the relative magnitudes are proportional to the wind-pressure coefficients, C_p, which, as seen in Figs. 14.12 and 14.13, are functions of the building configuration.

Figure 14.17(c) depicts the combined effects of wind, stack, and pressurization. In this example, on the windward side of the building there is inward flow—infiltration—along approximately the lower 60 percent of the surface and outward flow—exfiltration—on the upper 40 percent, while on the leeward side infiltration occurs on approximately the lower 15 percent of the surface and exfiltration on the upper 85 percent. Based on this example, it is clear that the regions of a building's envelope that experience infiltration or exfiltration, and the corresponding flow magnitudes, vary widely with wind speed and direction, indoor and outdoor temperature, and control of building pressurization.

As previously mentioned, techniques for calculating building air-exchange rates have improved in recent years. *The ASHRAE Handbook of Fundamentals 1993* [1] describes several calculation procedures. One of them was developed at the Lawrence Berkeley Laboratory and is, therefore, referred to as the LBL model. This model has been widely used for residential applications and will be presented here. It is a single-zone approach to calculating air infiltration which uses test results to characterize house air

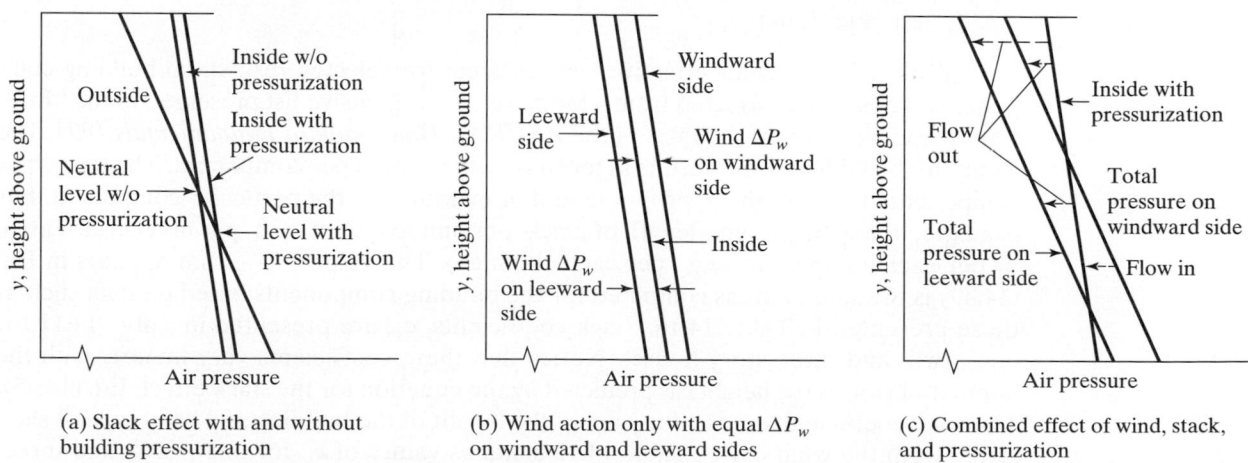

Figure 14.17 Distribution of inside and outside pressures over the height of a building for winter conditions.

leakage. In general, there are two test procedures used to measure air leakage in a building. One of these uses the introduction of a *tracer gas*, often SF_6, into the space to be evaluated. Either transient or steady-state tracer-gas methods are used. Transient tests consist of introducing a given amount of the tracer gas and measuring the concentration decay rate of the tracer as dilution occurs due to infiltration. Thus the infiltration rate is related to the decay rate of tracer gas. In performing steady-state measurements, the tracer gas is introduced at the rate required to maintain constant concentration—i.e., the rate necessary to offset the dilution that would result due to infiltration. Consequently, for this procedure the infiltration rate is related to the rate of introduction of tracer gas.

The other test procedure is referred to as the *pressurization method* or *blower-door method*. In this approach the space is slightly pressurized using a blower, and the flow rate needed to maintain the inside-to-outside pressure difference is measured. The airflow rate is typically measured for a number of pressure differences ranging from about 0.04 to 0.30 in. of water (10 to 75 Pa). The airflow rate needed to maintain a pressure difference is a measure of the tightness of the building. The LBL model for estimating infiltration uses pressurization-test results, taken at an indoor-to-outdoor pressure difference of 0.016 in. of water (4 Pa), to determine what are referred to as "effective leakage areas" for doors, windows, walls, joints, etc. It combines these leakage areas with the indoor-to-outdoor temperature difference and the wind speed to estimate the infiltration rate using Eq. (14.49).

$$\dot{V} = A_l \, (a_s \, \Delta t + a_w V_w^2)^{0.5} \qquad (14.49)$$

where

$\dot{V}$ = airflow rate, ft^3/min (L/s)

A_l = effective leakage area, in^2 (cm^2)

a_s = stack coefficient, $(ft^3/min)^2 \cdot in.^{-4} \cdot {}^\circ F^{-1}$ $[(L/s)^2 \cdot cm^{-4} \cdot {}^\circ C^{-1}]$

Δt = average indoor-outdoor temperature difference for time interval of calculation, $^\circ F$ ($^\circ C$)

a_w = wind coefficient, $(ft^3/min) \cdot in^{-4} \cdot mph^{-2}$ $[(L/s)^2 \cdot cm^{-4} \cdot (m/s)^{-2}]$

V_w = average wind speed measured at local weather station for time interval of interest, mph (m/s)

Table 14.10 presents effective leakage areas for selected residential building components. These are abstracted from a far more comprehensive list presented in the "Infiltration and Ventilation" chapter of the *ASHRAE Handbook of Fundamentals 1993*. The values in the table present results in terms of leakage area per component. The term "per component" refers to the appropriate unit of measure for the particular component. For example, it may be per unit length of crack, per unit length of sash, per unit surface area, or per each component (e.g., per each chimney). The value of A_w that appears in Eq. (14.49) is the sum of areas computed for the building components based on data such as those presented in Table 14.10. Stack coefficients, a_s, are presented in Table 14.11 for one-, two- and three-story houses. Notice that these coefficients vary linearly with the number of stories (or height) as predicted by the equation for the stack effect, Eq. (14.45). The wind coefficient, a_w, is a function of the height of the building and what type of sheltering from the wind exists. Table 14.12 contains values of a_w for one-, two-, and three-story houses for five different classes of wind shielding along with the descriptions of the wind-shielding classes.

TABLE 14.10 Effective Leakage Areas (Low-Rise Residential Applications)

	IP Units				SI Units			
	Units (see note)	Best Estimate	Mini-mum	Maxi-mum	Units (see note)	Best Estimate	Mini-mum	Maxi-mum
Chimney	in.^{2}/ea	4.5	3.3	5.6	cm^2/ea	29	21	36
Door frame								
General	in.^{2}/ea	1.9	0.37	3.9	cm^2/ea	12	2.4	25
Wood wall, uncaulked	in.^{2}/ft^2	0.024	0.009	0.024	cm^2/m^2	1.7	0.6	1.7
Wood wall, caulked	in.^{2}/ft^2	0.004	0.001	0.004	cm^2/m^2	0.3	0.1	0.3
Doors								
Attic fold down, weatherstripped	in.^{2}/ea	3.4	2.2	6.7	cm^2/ea	22	14	43
Storm (subtract from door value)	in.^{2}/ea	0.9	0.46	0.96	cm^2/ea	6	3	6.2
Single, not weatherstripped	in.^{2}/ea	3.3	1.9	8.2	cm^2/ea	21	12	53
Single, weatherstripped	in.^{2}/ea	1.9	0.6	4.2	cm^2/ea	12	4	27
Fireplace								
With damper closed	in.^{2}/ft^2	0.62	0.14	1.3	cm^2/m^2	43	10	92
With glass doors	in.^{2}/ft^2	0.58	0.06	0.58	cm^2/m^2	40	4	40
Gas water heater	in.^{2}/ea	3.1	2.3	3.9	cm^2/ea	20	15	25
Joints								
Ceiling-wall	in.^{2}/lftc	0.071	0.008	0.12	cm^2/lmc	1.5	0.16	2.5
Sole plate, floor/wall, uncaulked	in.^{2}/lftc	0.19	0.018	0.27	cm^2/lmc	4	0.38	5.6
Top plate, band joist	in.^{2}/lftc	0.005	0.004	0.018	cm^2/lmc	0.1	0.075	0.38
Vents								
Bathroom, with damper closed	in.^{2}/ea	1.6	0.39	3.1	cm^2/ea	10	2.5	20
Dryer with damper	in.^{2}/ea	0.46	0.45	1.1	cm^2/ea	3	2.9	7
Kitchen with damper closed	in.^{2}/ea	0.8	0.16	1.1	cm^2/ea	5	1	7
Window framing								
Framing, wood wall, uncaulked	in.^{2}/ft^2	0.025	0.022	0.039	cm^2/m^2	1.7	1.5	2.7
Framing, wood wall, caulked	in.^{2}/ft^2	0.004	0.004	0.007	cm^2/m^2	0.3	0.3	0.5
Windows								
Casement with weatherstripping	in.^{2}/lftc	0.011	0.005	0.14	cm^2/lmc	0.24	0.1	3
Double horizontal slider, wood with weatherstripping	in.^{2}/lftc	0.026	0.007	0.081	cm^2/lmc	0.55	0.15	1.72
Double hung with weatherstripping, with storm	in.^{2}/lftc	0.037	0.021	0.057	cm^2/lmc	0.79	0.44	1

SOURCE: Adapted by permission from *ASHRAE Fundamentals 1993*.

Abbreviations: ft^2 = gross area in square feet
             m^2 = gross area in square meters
             ea = each
             lftc = linear feet of crack
             lmc = linear meters of crack

TABLE 14.11 Stack Coefficient, a_s

	IP Units			SI Units		
	House Height (Stories)			House Height (Stories)		
	One	Two	Three	One	Two	Three
Stack coefficient	0.0156	0.0313	0.0471	0.000145	0.000290	0.000435

SOURCE: Adapted by permission from *ASHRAE Fundamentals 1993*.
IP units for a_s: $(ft^3/min)^2 \cdot in.^{-4} \cdot {}^\circ F^{-1}$
SI units for a_s: $(L/s)^2 \cdot cm^{-4} \cdot ({}^\circ C)^{-1}$

TABLE 14.12 Wind Coefficients and Shielding-Class Descriptions

Shielding Class[a]	Wind Coefficient, a_w					
	IP Units			SI Units		
	House Height (Stories)			House Height (Stories)		
	One	Two	Three	One	Two	Three
1	0.0119	0.0157	0.0184	0.000319	0.000420	0.000494
2	0.0092	0.0121	0.0143	0.000246	0.000325	0.000382
3	0.0065	0.0086	0.0101	0.000174	0.000231	0.000271
4	0.0039	0.0051	0.0060	0.000104	0.000137	0.000161
5	0.0012	0.0016	0.0018	0.000032	0.000042	0.000049

SOURCE: Adapted by permission from *ASHRAE Fundamentals 1993*.
IP units for a_w: $(ft^3/min)^2 \cdot in^{-4} \cdot mph^{-2}$.
SI units for a_w: $(L/s)^2 \cdot cm^{-4} \cdot (m/s)^{-2}$.
[a]Descriptions of shielding classes:
 1 No obstructions or local shielding.
 2 Light local shielding; few obstructions, few trees, or small shed.
 3 Moderate local shielding; some obstructions within two house heights, thick hedge, solid fence, or one neighboring house.
 4 Heavy shielding; obstructions around most of perimeter, buildings or trees within 30 ft (10 m) in most directions; typical suburban shielding.
 5 Very heavy shielding; large obstructions surrounding perimeter within two house heights; typical downtown shielding.

EXAMPLE 14.3

A wood-construction wall of a single-story residence contains four casement windows each 2.5 ft by 6 ft and a single 3-ft by 7-ft weather-stripped door with a storm door. The window and door frames are caulked. The length of the wall is 25 ft. The sole plate of the wall is uncaulked. The house is in a typical suburban location. Use the LBL model to estimate the infiltration rate for the wall if the indoor temperature is 70 °F, the outdoor temperature is −12 °F, and the wind speed is 15 mph.

Solution: Taking numbers from Table 14.10, the total leakage area for each of the components, and adding, we obtain

Component	Component Area or Perimeter (Given)	Leakage Area per Area or Perimeter	Component Area
Windows, total			
frames	60 ft²	0.004 in.²/ft²	0.24 in.²
windows	68 lftc	0.011 in.²/lftc	0.75 in.²
Door			
frame	21 ft²	0.004 in.²/ft²	0.084 in.²
door	1 door	1.9 in.²/ea	1.9 in.²
storm door	1 door	−0.9 in.²/ea	−0.9 in.²
Wall joints			
ceiling wall	25 ft	0.071 in.²/lftc	1.78 in.²
sole plate	25 ft	0.19 in.²/lftc	4.75 in.²
		Total infiltration area for the wall	8.6 in.²

From Table 14.11 the stack coefficient for a single-story house is $a_s = 0.0156$ (ft³/min) · in.$^{-4}$ · °F^{-1}. The wind coefficient from Table 14.12 for a single story house and shielding class 4 (typical suburban shielding) is $a_w = 0.0039$ (ft³/min) · in.$^{-4}$ · mph^{-2}. Substituting the values into Eq. (14.49) results in $\dot{V} = 12.6$ ft³/min.

14.5 UNHEATED-SPACE TEMPERATURES AND HEAT LOSSES

Many buildings have unheated enclosed spaces, such as attics, crawl spaces, and porches, adjacent to heated spaces. There are two reasons one may need to evaluate the temperature in these spaces: (1) to calculate the heat transfer across the structure separating the heated and unheated regions, and (2) to establish whether items in the unheated space may undergo damage due to low temperatures. For winter conditions, the approach used to calculate the temperature in the unheated space is:

1. Assume a uniform air temperature, t_u, in the unheated space.
2. Establish a control volume around the space and write the appropriate equations for the heat transfer across the surfaces of the control volume. These equations should be written in terms of the temperature t_u.
3. Write the steady-state energy balance for the space using the equations from step 2. Include any energy transport due to air flows, such as infiltration or forced ventilation. If any internal heat sources are always present in the space, include them in the energy balance.
4. Solve the energy-balance equation for the temperature t_u.

EXAMPLE 14.4

As an example of the procedure, calculate the temperature in an unheated attic of a house which is 25 ft wide by 50 ft long. The interior of the house is maintained at a temperature $t_i = 70$ °F and the ceiling is insulated such that the R-value between the inside of the house and the attic is $R_c = 30$ (hr · ft² · °F)/Btu. The height of the attic is 4.5 ft at the peak of the roof, and the peak is midway along the 25-ft width of the house. The ridge of the roof runs the 50-ft length of the house. Therefore, the vertical gables at the ends of the attic are isosceles triangles with 25-ft bases and 4.5-ft heights. The gables have resistance values of $R_e = 1.5$ (hr $\cdot$ ft^2 $\cdot$ °F)/Btu. The roof resistance is $R_r = 6$ (hr $\cdot$ ft^2 $\cdot$ °F)/Btu. The attic is vented to the outside, and the airflow rate of outside air entering and passing through the attic is 20 ft^3/min. The outdoor-air temperature is −12 °F.

Solution: The control-volume boundary for the attic consists of the two gables, the two slanted roof surfaces, and the ceiling. The heat-transfer rates into the attic across the boundaries are

$$\text{Ends: } \dot{Q}_e = \frac{A_e(t_o - t_u)}{R_e} \text{ and } A_e = 112.5 \text{ ft}^2, \text{ the total area of the gables}$$

$$\text{Roof: } \dot{Q}_r = \frac{A_r(t_o - t_u)}{R_r} \text{ and } A_r = 1329 \text{ ft}^2, \text{ the total roof surface area.}$$

$$\text{Ceiling: } \dot{Q}_c = \frac{A_c(t_i - t_u)}{R_c} \text{ and } A_c = 1250 \text{ ft}^2, \text{ the ceiling area}$$

We can probably assume that the ceiling is constructed so that very little moisture enters the attic from the house. Therefore, the air will undergo a sensible heating process, entering at the outside temperature and exiting at the unheated-space temperature. The energy-balance equation is

$$\frac{A_e(t_o - t_u)}{R_e} + \frac{A_r(t_o - t_u)}{R_r} + \frac{A_c(t_i - t_u)}{R_c} = \dot{m}_a c_p (t_u - t_o)$$

For a temperature as low as −12 °F the humidity ratio is very low, even for saturated conditions. Therefore, it is reasonable to take the specific heat as $c_p = 0.240$ Btu/lbm$_a$ $\cdot$ °F, the value for dry air. The specific volume of air at one atmosphere and −12 °F is approximately 11.3 ft^3/lbm$_a$ (this value can be found on the low-temperature psychrometric chart, C-7E). This gives a mass flow rate of

$$\dot{m}_a = \frac{\left(20 \dfrac{\text{ft}^3}{\text{min}}\right)\left(60 \dfrac{\text{min}}{\text{hr}}\right)}{11.3 \dfrac{\text{ft}^3}{\text{lbm}_a}} = 106 \frac{\text{lbm}_a}{\text{hr}}$$

Substituting the known quantities into the energy-balance equation and solving results in an unheated-space temperature of $t_u = -2.6$ °F.

The heat transfer across the ceiling into the attic contributes to the heating load of the house. This heat-transfer rate is calculated from the above equation for $\dot{Q}_c$ using the value of t_u just established.

$$\dot{Q}_c = \frac{(1250 \text{ ft}^2)(70 + 2.6) \text{ °F}}{30 \dfrac{\text{hr} \cdot \text{ft}^2 \cdot \text{°F}}{\text{Btu}}} = 3025 \frac{\text{Btu}}{\text{hr}}$$

14.6 WINTER DESIGN CONDITIONS

The primary purpose of a heating system is to maintain comfortable temperature and humidity conditions in the space while maintaining good indoor-air quality. Human thermal comfort was addressed in Chapter 12. It is influenced by psychological as well as physiological factors. There is no method of stating what thermal environmental condition will effect a comfort feeling in a human being. A person's thermal comfort is influenced not only by ambient air temperature and humidity, but also by the rate of air movement, body activity, and amount of clothing. Figure 12.7 indicates that in winter an effective temperature of 70–72 °F (21–22 °C) is optimum for normally clothed people. Such a comfort condition may supposedly be attained by various combinations of temperature and humidity. In northern climates, space relative humidity during cold weather cannot exceed much above 25–30 percent without condensation of moisture on inner surfaces of double-glazed windows. *ASHRAE/IES Standard 90.1*, "Energy Efficient Design of New Buildings," recommends that indoor design temperature and humidity be selected in accordance with the criteria in *ANSI/ASHRAE Standard 55*, which gives considerable latitude in selecting design conditions. For heating-load calculations, a design dry-bulb temperature of 70 °F (21 °C) and a relative humidity less than or equal to 30 percent are widely used.

Outdoor weather data for selected stations are presented in Table B.1. These data have been statistically analyzed for a recent 15-year period of record and are tabulated to the nearest degree Fahrenheit. The cities included in Table B.1 are extracted from an extensive list presented in the *ASHRAE Handbook of Fundamentals 1993*. The winter portion of the table lists dry-bulb temperature data for two frequencies. The 97.5 percent value is the temperature equaled or exceeded 97.5 percent of the time in the months of December, January, and February (a total of 2160 hours) for stations in the Northern Hemisphere or for the months of June, July, and August (a total of 2208 hours) for stations in the Southern Hemisphere. The 99 percent value is the temperature equaled or exceeded 99 percent of the hours in the same time period. The two design temperatures for Canadian cities are based only on data for the month of January.

The outdoor design temperature commonly recommended is the 97.5 percent value and the ASHRAE standard for energy-efficient design stipulates that the design temperature shall not be less than the 99 percent value. However, the designer and building owner need to be aware that under extreme conditions the outdoor temperature has been observed to remain below the 99 percent or 97.5 percent value for 3 to 5 days [1]. During such severe conditions, the indoor temperature may drop below the indoor design temperature. Therefore, in special applications such as nursing homes, where maintaining indoor temperature above a set level may be critical, abnormal local conditions should be considered. It is always good engineering practice to attempt to include local weather data and owner expectations in the design process.

14.7 WINTER DESIGN HEAT-LOSS SUMMARY

The design heating load for a building is performed by first calculating the design load for each room or zone of the building and summing. In this way not only is the calculation broken into manageable segments, it also provides the zone by zone information needed to design the heating system. For air handling systems zone information is required to layout the ducts, select proper air diffusers and specify supply air flow rates and conditions. Design of hydronic systems require zone information for pipe layout and sizing and equipment selection. The general procedure for calculating the sensible heating load for a given zone or room is summarized below. Typically, energy from internal sources, such as lights or office equipment, or from solar radiation are not included in the heating load calculation since one can not be certain that they will be present when needed.

Sensible-Heating Load-Calculation Summary for a Zone

1. Determine the appropriate indoor design temperature and humidity for the zone, t_i and ϕ_i. (Sec. 14.6).

2. Determine the appropriate outdoor design conditions, t_o, ϕ_o, and wind speed (Sec. 14.6).

3. For elements of the zone envelope that directly separate the indoor and outdoor conditions:
   (a) Calculate the average U-values for multilayer walls or roof sections using either the parallel-path or isothermal-plane method [Eqs. (14.12), (14.18), and (14.23)].
   (b) Determine the U-values for windows by applying Eq. (14.25) and/or tables such as Table 14.5.
   (c) Determine the U-values for doors using the same procedures used for multi-layer structures, or select values from tables such as Table 14.7.
   (d) Calculate the area for each element from the plans.
   (e) Calculate the heat-loss rate for each component using $\dot{Q} = UA(t_i - t_o)$.

4. For elements of the zone envelope adjacent to unheated spaces:
   (a) Determine the temperatures in the unheated spaces, t_u (Sec. 14.5).
   (b) Determine the U-values and areas of the elements separating the zone and unheated spaces.
   (c) Calculate the heat-loss rate for each element using $\dot{Q} = UA(t_i - t_u)$.

5. For below-grade walls and floors (Sec. 14.3):
   (a) Determine the appropriate ground temperature, t_g, for the calculation [Eq. (14.35)].
   (b) Calculate the U-value for the wall or floor [Eq. (14.30) or (14.34)].
   (c) Calculate the wall and floor areas from the plans.
   (d) Calculate the heat-loss rate for the walls and floor using Eq. (14.29), $\dot{Q} = UA(t_i - t_g)$.

6. For slab-on-grade floors (Sec. 14.3):
   (a) Determine the heat-loss coefficient per unit length of perimeter, F_s, using tables such as Table 14.9.
   (b) Calculate the floor perimeter from the plans.
   (c) Calculate the heat-loss rate for the floor using Eq. (14.36), $\dot{Q}_s = F_s P(t_i - t_o)$.

7. Infiltration (Sec. 14.4):
   (a) Determine the volumetric flow rate for infiltration, $\dot{V}$. For residential or simple buildings the LBL model can be used and the infiltration rate estimated using Eq. (14.49). The infiltration heating load for large commercial buildings is often neglected, since it is small compared to other loads.
   (b) Calculate the sensible-heating load due to infiltration using Eq. (14.37a), $\dot{Q}_s = \dot{m}_a \bar{c}_p (t_i - t_o)$, and Eq. (14.39), $\dot{m}_a = \rho \dot{V}$. Also, it is customary to use $\bar{c}_p = 0.24$ Btu/lbm$_a \cdot$ °F (1.0 kJ/kg$_a \cdot$ °C) and $\rho = 0.075$ lbm$_a$/ft^3 (1.2 kg$_a$/m^3) in performing the infiltration calculations.

8. *Total:* The total heating load for the zone is the sum of the loads calculated in items 3 through 7.

14.8 MOISTURE TRANSPORT IN BUILDING STRUCTURES

Moisture control in buildings is necessary to avoid problems related to energy performance, building durability, and human health. Moisture problems can remain hidden in a building structure until severe damage has occurred. By the time visual evidence of a moisture problem appears, the growth of mold or mildew, the decay of wood-based materials, the damage to materials due to freeze-thaw cycles, and the hydration of plastic materials very likely have taken place.

While mold and mildew problems will develop at high relative humidity, structural problems usually begin when condensation of water vapor occurs in the building envelope. Water vapor migrates in building materials by convection of moist air and by diffusion. Openings around electrical outlets, recessed light fixtures, poorly sealed window and door frames, etc. provide access for convection through walls and ceilings. Although it is difficult to evaluate air flows through a structure, it is known that even small air flows can carry large amounts of water vapor when compared to vapor diffusion. Therefore, special attention must be paid to inhibit hot moist airflow into the envelope structure. A variety of sealants and airflow retarders are designed to inhibit air from entering a structure by convection.

The diffusion of water vapor through building materials can more easily be predicted. Consequently, we can design a wall or ceiling structure to eliminate condensation problems caused by moisture diffusion. The general approach is to determine the temperature profile in the structure and, thereby, the corresponding saturation vapor pressure. The design must ensure that the actual vapor pressure everywhere in the structure is below the saturation value. In an insulated structure the major temperature drop and, therefore, the major saturation pressure drop occurs across the insulation. On the other hand, most insulation products offer a low resistance to vapor diffusion. Consequently, some means of reducing the vapor flow through the insulation must be provided. Typically this is accomplished by including a vapor retarder, often a relatively thin sheet of plastic, in the structure. The vapor retarder must be located on the warm side of the insulation.

The analysis of vapor diffusion presented in this section is based on a simple one-dimensional steady-state model. The equation used to calculate the water-vapor diffusion flux through materials is based on a form of Fick's law:

$$\frac{\dot{m}_w}{A} = -\mu \left(\frac{dP_w}{dx} \right) \tag{14.50}$$

where A = area of vapor transport
  $\dot{m}_w$ = mass flow rate of water vapor
  P_w = vapor pressure
  x = length in the direction of transport
  μ = permeability of the material

This equation simply states that the water-vapor flow by diffusion is proportional to the water-vapor pressure gradient. It is similar in form to the diffusion (i.e., conduction) equation for heat transfer, which says the heat-transfer rate is proportional to the temperature gradient. The actual diffusion through a material is complex, since the permeability for many materials is a function of the relative humidity and temperature that exist in the material. Both of these will vary along the flow path. In addition, vapor storage within the material and two-dimensional effects further complicate the problem. For our simplified model we will define a moisture transport conductance, or permeance, M such that the mass flow rate equation is written as

$$\frac{\dot{m}_w}{A} = M \, \Delta P_w \tag{14.51}$$

where ΔP_w is the vapor-pressure difference across the material.

Notice that the permeance in moisture flows plays a similar role to the conductance, C, in heat transfer. Examination of Eqs. (14.50) and (14.51) reveals that, if the permeability of a material is constant, or an average value is specified, the relationship between M and μ for that material is

$$M = \frac{\mu}{L}$$

where L is the thickness of the material.

Generally we think of the permeability as applying to homogeneous materials such as batt or loose-fill fiberglass insulation. For nonhomogeneous building material, such as plywood or concrete blocks, the concept of an overall water-vapor transmittance, or permeance, applies.

The analogy between heat and moisture diffusion is carried one step further by defining a vapor-flow resistance, Z, as the inverse of the permeance, M.

$$Z = \frac{1}{M}$$

and, therefore, Eq. (14.51) becomes

$$\frac{\dot{m}_w}{A} = \frac{\Delta P_w}{Z} \tag{14.52}$$

Values of the permeability or permeance of a selection of building materials are presented in Table 14.13. The unit of permeance in the table is called "Perm." The thickness of materials for which the permeability is given is taken in inches (in.). Therefore, the unit of permeability is the "Perm-in." The vapor-flow resistance, Z, or resistance per inch, $Z/L = 1/\mu$, is also included in the table. The corresponding units are called reciprocal Perm, "Rep," and reciprocal Perm/inch, "Rep/in." Using these units for the material properties and expressing the vapor-pressure difference in inches of mercury, in.-Hg, results in a mass flow rate of water vapor per unit area of grains of water vapor/hr · ft^2, "gr/hr · ft^2" (1 lbm = 7000 grains). Although these units are unusual, they serve the purpose for our needs, since the working equations for determining what vapor retarder, if any, is needed to prevent condensation will involve only vapor pressures and vapor resistances. And as long as we express all vapor pressures in the same units and all vapor resistances in the same units, no problems are encountered.

Moisture problems can occur during the winter months in cold climates when warm, relatively humid conditions are maintained inside a building. A substantial temperature drop occurs across the building envelope, and, unless controlled, the movement of moisture through the wall results in regions where the vapor pressure equals the saturation value and condensation begins. If this persists for any prolonged period, significant damage will occur. In cold-climate locations moisture problems seldom occur for occupied buildings during the summer months. Interior temperatures are maintained in the 70 to

TABLE 14.13 Typical Water-Vapor Permeance and Permeability Values for Common Building Materials

Material	Thickness, in.	Permeance, Perm	Resistance, Rep	Permeability, Perm-in.	Resistance/in., Rep/in.
Construction Materials					
Concrete (1:2:4 mix)				3.2	0.31
Brick masonry	4	0.8	1.3		
Concrete block (cored, limestone aggregate)	8	2.4	0.4		
Plaster on metal lath	0.75	15	0.067		
Gypsum wallboard, (plain)	0.375	50	0.02		
Structural insulating board (sheathing quality)				20-50	0.050-0.020
Built-up roofing (hot mopped)		0.0			
Wood (sugar pine)				0.4-5.4	2.5-0.19
Plywood (Douglas fir, exterior glue)	0.25	0.7	1.4		
Plywood (Douglas fir, interior glue)	0.25	1.9	0.53		
Thermal Insulations					
Air (still)				120	0.0083
Mineral wool or glass fiber (unprotected)				116	0.0086
Expanded polystyrene—extruded				1.2	0.83
Expanded polystyrene—bead				2.0-5.8	0.50-0.17
Plastic and Metal Foils and Films					
Aluminum foil	0.001	0.0			
Polyethylene	0.002	0.16	6.3		3100
Polyethylene	0.004	0.08	12.5		3100
Polyethylene	0.006	0.06	17		3100
Polyethylene	0.008	0.04	25		3100
Liquid-Applied Coating Materials					
Commercial latex paints (dry film thickness)					
Vapor retarder paint	0.0031	0.45	2.22		
Primer-sealer	0.0012	6.28	0.16		
Semigloss vinyl-acrylic enamel	0.0024	6.61	0.15		
Exterior acrylic house and trim	0.0017	5.47	0.18		
Paint—2 coats					
Various primers plus 1 coat flat oil paint on plaster		1.6-3.0	0.63-0.33		
Flat paint on interior insulation board		4	0.25		

SOURCE: Adapted by permission from *ASHRAE Fundamentals 1993.*

Note: This table permits comparisons of materials; but in the selection of vapor-retarder materials, exact values for permeance or permeability should be obtained from the manufacturer or from laboratory tests. The values shown indicate variations among mean values for materials that are similar but of different density, orientation, lot, or source. The values should not be used as specification data.

Units:

Permeance	Perm = gr/hr · ft² · in. Hg
Resistance	Rep = in. Hg · ft² · hr/gr
Permeability	Perm · in. = gr/hr · ft² · in. Hg/in.
Resistance/unit thickness	Rep/in. = in. Hg · ft² · hr/in.

80 °F (21 to 26 °C) range, which are typically above the outdoor dew-point temperature. Therefore, vapor pressures in the wall seldom, if ever, reach saturation levels.

Vapor control is provided by adding a vapor retarder, typically a relatively thin sheet of plastic with a high vapor resistance, to the wall construction. The vapor retarder must be located so that it is on the warm side of the insulation during the winter months—i.e., between the inside and the insulation.

In air conditioned buildings located in hot, humid climates where the indoor temperatures remain below the outdoor dew-point for prolonged periods, the moisture flow from the outside to inside must be reduced. Again, the vapor retarder is placed on the warm side of the insulation. But in a hot, humid climate that means somewhere between the outside of the wall and the insulation. Special applications such as cold-storage rooms also must be designed to avoid condensation in their envelopes. Typically these will require a vapor retarder located on warm side of the wall insulation. However, in some instances the inner wall of the cold room is nearly impermeable to vapor flow, and it may be very difficult to provide enough vapor-flow resistance on the outside to alleviate problems. In this case, venting the wall with low-humidity air or, if the cold room is within a building, reducing the outer dew-point below the inside dew-point may be required.

The resistance of the vapor retarder needed to avoid condensation will be evaluated with the aid of the notation given in Fig. 14.18, which schematically depicts a composite wall with many sections. Let us focus our attention on the plane of the wall identified as location-x. Since many of the thermal and vapor resistances are for complete component thicknesses, the planes considered are at interfaces between materials. Warm, humid air at temperature t_w and vapor pressure $P_{w,w}$ is present to the left of the wall. Cooler air at temperature t_c and vapor pressure $P_{w,c}$ is to the right of the wall. If we assume one-dimensional steady-state heat transfer through the wall, we can solve for the temperature at location-x.

$$t_x = t_w - \frac{R_{w \to x}}{R_t}(t_w - t_c)$$ (14.53)

where $R_{w \to x}$ = the sum of the thermal resistances from the warm air up to location-x, including the warm-side air-film resistance

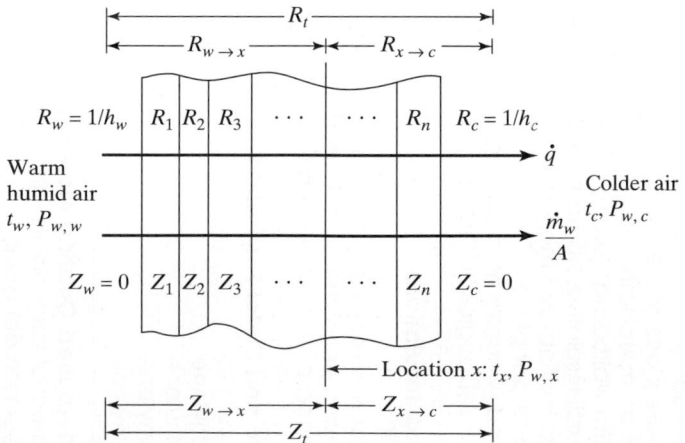

Figure 14.18 Schematic of one-dimensional heat and moisture transfer through a multilayer wall.

R_t = the total thermal resistance of the wall, including the warm- and cold-side air-film resistances, similar to that defined in Eq. (14.6)

If the vapor flux, $\dot{m}_w/A$, is continuous through the wall, it can be written in a number of ways:

$$\frac{\dot{m}_w}{A} = \frac{P_{w,w} - P_{w,c}}{Z_t} \tag{14.54a}$$

and

$$\frac{\dot{m}_w}{A} = \frac{P_{w,w} - P_{w,x}}{Z_{w \to x}} \tag{14.54b}$$

and

$$\frac{\dot{m}_w}{A} = \frac{P_{w,x} - P_{w,c}}{Z_{x \to c}} \tag{14.54c}$$

where Z_t = sum of vapor resistances from warm air to cool air
$Z_{w \to x}$ = sum of vapor resistances from warm air to location x
$Z_{x \to c}$ = sum of vapor resistances from location x to cool air

Although the above definitions of the various summations of vapor resistances formally include the resistances between the air on either or both the warm and cool sides of the structure and their respective adjacent surfaces, Z_w and Z_c, those vapor resistances are negligible; i.e.,

$$Z_w \cong 0 \quad \text{and} \quad Z_c \cong 0$$

Combining Eqs. (14.54a) and (14.54b), we obtain an equation for the vapor pressure at location-x, whose form is similar to that of Eq. (14.53):

$$P_{w,x} = P_{w,w} - \frac{Z_{w \to x}}{Z_t} (P_{w,w} - P_{w,c}) \tag{14.55}$$

By calculating the temperatures at all locations within the wall and determining the saturation vapor pressures, $P_{w,s}$, at all locations, we find the maximum allowable vapor pressure. A comparison of those values is then made with the vapor pressures calculated using Eq. (14.55). For all locations $P_{w,x} < P_{w,s,x}$ is a necessary condition to avoid condensation in the wall. If condensation does occur, Eq. (14.55) is not valid, since it was derived for vapor-flow continuity. Any plane where the vapor pressure predicted by Eq. (14.55) exceeds the saturation value is, thus, identified as a problem location, and condensation will occur in the vicinity of that plane unless the wall design is changed. Typically the problems occur on the cold side of any insulation present in the structure or, in the case of un-insulated structures, on the warm side of any high-vapor-resistance material which is in the colder regions of the structure.

The remedial action is to locate a vapor retarder as close to the warm surface of the structure as possible. For a frame wall in a cold climate the vapor retarder often consists of a thin sheet of polyethylene located just behind the wallboard, or other finish material, used as the inside layer of the wall. Vapor-retarder paints can also be applied as the inner surface finish. Typically, the thermal resistance of the vapor retarder is negligibly small, i.e., $R_{vr} \cong 0$. The minimum vapor resistance needed to avoid condensation at location-x can be evaluated by equating the right-hand sides of any two of Eqs. (14.54a),

(14.54b), and (14.54c) and replacing the vapor pressure at location-x with the corresponding saturation value. If the vapor retarder is positioned to the warm side of location-x, the most convenient pair will be Eqs. (14.54a) and (14.54c), since the vapor-retarder resistance, Z_{vr}, will be included in the total resistance, Z_t, but not in $Z_{x \to c}$.

$$Z_{vr} > Z_{x \to c} \left(\frac{P_{w,w} - P_{w,c}}{P_{w,s,x} - P_{w,c}} \right) - Z_{t-vr} \qquad (14.56)$$

where Z_{t-vr} is the sum of all the vapor flow resistances not including that of the vapor retarder.

The inequality in Eq. (14.56) replaced the equal sign to indicate that one should select a value greater than the minimum needed to avoid condensation. Many structures will have multiple locations where condensation may occur. In such instances, the vapor-retarder resistance needed to avoid condensation should be determined for all potential problem planes, and the largest installed in the design. Once a vapor retarder and its position have been selected, a check of the predicted vapor pressures vs. saturation vapor pressures can be made to verify that all condensation problems have been eliminated.

EXAMPLE 14.5

An exterior wall is constructed, from inside to outside, of $\frac{1}{4}$-in. plywood, a vapor retarder which has negligible thermal resistance, a 5.5-in. insulating batt of mineral wool, 8-in. concrete blocks (three-oval core—sand and gravel aggregate), and 4-in. thick brick. The indoor temperature is 72 °F, indoor relative humidity is 30 percent, outdoor temperature is 0 °F, outdoor relative humidity is 75 percent, and the outdoor wind speed is 15 mph. What is the minimum vapor-retarder resistance, Z_{vr}, needed to avoid condensation in the wall?

Solution: Let's number the interface planes from 1, starting at the inside surface. Potential moisture problems are expected to occur in the regions of planes 4 and 5, the insulation/block and block/brick interfaces, respectively. Therefore, we will calculate the vapor-retarder resistances needed to ensure no condensation at these locations and select the larger. The thermal and vapor resistances are shown below. These are taken from Tables 14.1, 14.4IP, and 14.13. For materials where the resistances are calculated based on the thermal conductivity or permeability, the values used are indicated. For materials where a range of properties is presented, the value was arbitrarily selected.

Material	Thermal Resistance, $(hr \cdot ft^2 \cdot °F)/Btu$	Vapor Resistance, Rep
Inside air	0.68	0
Plywood ($k = 1.25$)	0.20	0.53
Vapor retarder	0	Z_{vr}
Insulation ($\mu = 116$)	19	0.47
Block	1.11	0.4
Brick ($k = 5.6$)	0.71	1.3
Outside air	0.17	0
Totals	21.87	$2.70 + Z_{vr}$

The vapor pressures in the warm (inside) and cool (outside) are calculated using the relationship between relative humidity, ϕ, and vapor pressures, Eq. (7.27):

$$\phi = \frac{P_w}{P_{w,s}}$$

The saturation vapor pressures are found in Table A.1E. At $t_w = 72$ °F and $t_c = 0$ °F we find $P_{w,s,w} = 0.38878$ psia and $P_{w,s,c} = 0.01850$ psia. Thus

$$P_{w,w} = 0.3(0.38878) = 0.11663 \text{ psia}$$

and

$$P_{w,c} = 0.75(0.01850) = 0.01388 \text{ psia}$$

From Eq. (14.53) we obtain the temperatures at planes 4 and 5:

$$t_4 = 72 - \frac{19.88}{21.87}(72 - 0) = 6.55 \text{ °F}$$

and

$$t_5 = 72 - \frac{20.99}{21.87}(72 - 0) = 2.90 \text{ °F}$$

The saturation vapor pressures at these temperatures are obtained using linear interpolation in Table A.1E, $P_{w,s,4} = 0.02596$ psia and $P_{w,s,5} = 0.02152$ psia.

The minimum vapor-retarder values are calculated with Eq. (14.56).

$$\text{At plane 4,} \quad (Z_{vr})_4 > 1.7\left(\frac{0.11663 - 0.01388}{0.02596 - 0.01388}\right) - 2.70 = 11.8 \text{ Rep}$$

and

$$\text{At plane 5,} \quad (Z_{vr})_5 > 1.3\left(\frac{0.11663 - 0.01388}{0.02152 - 0.01388}\right) - 2.70 = 14.8 \text{ Rep}$$

Therefore we would select a vapor retarder with a resistance greater than 14.8 Rep. The 0.006-in. thick polyethylene film listed in Table 14.13, with its vapor resistance of 17 Rep, would be a good choice for this structure and these design conditions. Keep in mind that this calculation predicts only what is necessary to eliminate moisture problems due to vapor diffusion. Airflows through the structure from the warm side to the cold side may transport significant amounts of moisture into the cavity by convection.

ENDNOTES

1. *ASHRAE Handbook of Fundamentals* (Atlanta: American Society of Heating, Refrigerating and Air Conditioning Engineers, 1993).

2. D. M. Burch, B. A. Licitra, D. F. Ebberts, and R. R. Zarr, "Thermal Resistance Measurements and Calculations of an Insulated Concrete Block Wall," *ASHRAE Trans.,* 95, pt. 1, 398–404.

3. C. O. Peterson, Jr., "How Is Low-E Performance Criteria Determined?" *Glass Digest*, 1 (1987), 70–76.

4. *ASHRAE Handbook of Fundamentals* (Atlanta: American Society of Heating, Refrigerating and Air Conditioning Engineers, 1989).

5. R. F. Szydlowski and T. H. Kuehn, "Analysis of Transient Heat Loss in Earth-Sheltered Structures," *Underground Space*, 5 (1981), 237–246.

6. J. K. Latta and G. G. Boileau, "Heat Losses from House Basements," *Canadian Building*, 19: 10 (1969), 39–42.

7. T. H. Kuehn, "Temperature and Heat Flow Measurements from an Insulated Concrete Bermed Wall and Adjacent Floor," *Journal of Solar Energy*, 104 (1982), 15–22.

8. F. C. McQuiston and J. D. Spitler, *Cooling and Heating Load Calculation Manual*, 2d ed. (Atlanta: American Society of Heating, Refrigerating, and Air Conditioning Engineers, Inc., 1992), 6.1–6.10.

PROBLEMS

14.1 An exterior wall is constructed, from inside to outside, of $\frac{3}{8}$-inch plywood, a 5.5-in. insulating batt of mineral wool, 8-in. concrete blocks (three-oval core—sand and gravel aggregate), and 4-in.-thick brick. The indoor temperature is 72 °F, outdoor temperature is 0 °F, and the outdoor wind speed is 15 mph. Determine the heat-transfer rate per unit area through the wall.

14.2 A building has a flat roof consisting, from inside to outside, of $\frac{3}{8}$-in. gypsum wallboard, 2 × 8 studs located on 12-in. centers, $\frac{3}{4}$-in. plywood, and $\frac{3}{8}$-in. built-up roofing. (Note that the actual dimensions of a 2 × 8 stud are 1.5 in. × 7.5 in. and it is made out of softwood.) The 7.5-in.-high space between the studs is filled with loose cellulose insulation, which has a thermal conductivity of 0.30 Btu · in./hr · ft^2 · °F. If the outdoor temperature is 95 °F, outdoor wind speed is 7.5 mph, and indoor-air temperature is 78 °F:
(a) Calculate the heat-loss rate per unit area using the parallel-path method.
(b) Calculate the heat loss rate per unit area using the isothermal-plane method.

14.3 An exterior wall of a building is constructed using 2 × 6 framing (actual dimensions of a 2 × 6 are 1.5 in. × 5.5 in.). The inside of the wall is $\frac{3}{8}$-inch gypsum wallboard. The space between the framing is filled with fiberglass insulation. On the outside of the framing, the wall is built with a layer of $\frac{25}{32}$-inch regular-density sheathing, a 1.5-in.-thick layer of expanded polystyrene, and 4-in. brick. The framing makes up 20 percent of the area of the wall. The inside air is at 68 °F dry-bulb temperature and the outside air is at 5 °F dry-bulb temperature. Wind speed is 7.5 mph. Calculate the average heat-transfer rate per unit area for the wall, using:
(a) The parallel-path method.
(b) The isothermal-plane method.

14.4 An exterior wall is constructed, from inside to outside, of 10.0-mm plywood, framing made of "2 × 6 studs," 200-mm concrete blocks (three-oval core—sand and gravel aggregate), and 10-mm-thick brick. (Note that the actual dimensions of a "2 × 6 stud" are 38 mm × 140 mm and it is made out of softwood.) The space between the studs is filled with an insulating batt of fiberglass. The area of the framing is 25 percent of the total wall area. If the outdoor wind speed is 6.7 m/s, calculate the average U-value for the wall, using:
(a) The parallel-path method.
(b) The isothermal-plane method.

14.5 A portion of an exterior wall is constructed, from inside to outside, of $\frac{3}{8}$-in. plywood, 2 × 4 studs located on 16-in. centers, 8-in. concrete blocks (three-oval core—sand and gravel aggregate), and 4-in.-thick brick. (Note that the actual dimensions of a 2 × 4 stud are 1.5 in. × 3.5 in. and it is made out of softwood.) If the indoor-air temperature is 70 °F, the outdoor-air temperature is 10 °F, and the outdoor wind speed is 7.5 mph, calculate the average heat-transfer rate per unit area for the wall, using:
(a) The parallel-path method.
(b) The isothermal-plane method.
(c) The parallel-path method if the space between the studs is filled with an insulating batt of fiberglass.

14.6 It is desired to maintain winter indoor conditions of a building at 68 °F and 35 percent relative humidity. The windows in the building are double glass with a 0.5-in. air space between the sheets of glass. If the outdoor wind speed is 15 mph, how low can the outdoor temperature get before condensation begins to form on the inside surface of the windows? Be sure to state any assumptions made and the source of any data used in making your calculations.

14.7 A window is double glazed (i.e., double glass) with a 9.5-mm argon space between the sheets of glass. The indoor dry-bulb temperature is 23 °C. The outdoor dry-bulb temperature is the 97.5 percent design temperature for Minneapolis. What is the maximum indoor relative humidity that can be maintained and still not have condensation form on the inside of the windows? Is your answer for the "center of glass" or "edge of glass," or does it matter?

14.8 A skylight for a building is triple glazed (i.e., three sheets of glass) with 0.5-in. air gaps between the sheets of glass. The inside and outside sheets of glass are standard glass with an emittance of 0.9 on all surfaces. The center sheet of glass has special surface coatings such that it has emittance values of 0.02 on both of its surfaces. Each sheet of glass is 0.190 in. thick. The thermal conductivity of glass is 6.4 Btu · in./hr · ft^2 · °F. The skylight is mounted at a slope of 45 degrees.
   (a) Without using Table 14.5, calculate the U-value of the skylight for winter conditions if the outdoor wind speed is 15 mph.
   (b) If we wish to maintain indoor conditions of 68 °F dry-bulb temperature and 36 percent relative humidity, how cold can it get outside before any condensation will form on the inside surface of the skylight?

14.9 The windows in a building are double glazed with a 13-mm air space between the two sheets of glass. They are made of standard glass which has an infrared emittance of 0.9, and the windows are mounted at a slope of 45 degrees. Consider the winter condition when the inside temperature is approximately 18 °C, the outdoor temperature is approximately 2 °C, and outdoor wind speed is 6.7 m/s.
   (a) Without the aid of Table 14.5 calculate the center-of-glass U-value of the windows.
   (b) Compare the answer in part (a) to the value presented in Table 14.5 for a similar window design.

14.10 A basement wall extends 6 ft below grade. The wall is made of 8-in.-thick poured concrete. The complete height of the inside of the wall is finished with 0.375-in.-thick plywood paneling. A 2-in. layer of extruded polystyrene insulation (molded bead type) is located between the plywood and concrete. The total horizontal length of basement wall is 300 ft. The soil is sandy.
   (a) Calculate the average U-value for the basement walls.
   (b) Calculate the design heat-loss rate for the basement walls if the building is located in Madison, Wisconsin.

14.11 The dimensions of the basement floor in Prob. 10 are 100 ft × 50 ft. The floor is made of 6 in. of concrete. It is completely carpeted, and the carpet has a thermal resistance of approximately 3 hr · ft^2 · °F/Btu. Estimate the heat-loss rate from the floor.

14.12 A building in Denver, Colorado, is constructed with a slab-on-grade floor. The construction type is an uninsulated 100-mm block wall with a brick face. The perimeter of the building is 350 meters. Calculate the design heat-loss rate from the floor.

14.13 A two-story wood-frame construction suburban house is 20 m wide by 12 m deep. There are 52 double-hung, weatherstripped windows with storms. Each window is 1 m wide by 2 m high. There are three single-size weatherstripped doors, each 1 m wide by 2.1 m tall; each has a storm door. All the window and door framing is caulked. The house has a gas hot-water heater and furnace connected to a common chimney, a fireplace with glass doors, three bathroom vents, a dryer vent, and a kitchen vent. All the vents have dampers.
   (a) Estimate the infiltration rate for winter design temperatures for Madison, Wisconsin, and a wind speed of 4 m/s.
   (b) Estimate the sensible component of the infiltration heating load.

14.14 A building has an attached porch which is 10 ft long × 10 ft wide × 9 ft high and has a flat roof. The wall separating the porch from the rest of the building is constructed of (from inside to outside) $\frac{3}{8}$-in. plasterboard, 3.5 inches of insulation, and $\frac{3}{8}$-in. plasterboard (neglect the framing). The area of the outside walls of the porch is 60 percent window with a U-value of 0.9 Btu/hr · ft^2 · °F and 40 percent frame construction with a U-value of 0.3 Btu/hr · ft^2 · °F. The roof of the porch has a U-value of 0.5 Btu/hr · ft^2 · °F. There is negligible heat loss from the floor of the porch. The windows of the porch are not very tight and the infiltration rate through the porch is 3 air changes per hour. (The airflow rate between the building and porch is negligible.) On a night when the outside temperature is −13 °F ($v_a = 11.2$ ft^3/lbm$_a$) and the building is maintained at 65 °F, estimate:
   (a) The air temperature in the porch.
   (b) The heat-loss rate, Btu/hr, from the building into the porch.

14.15 A house has one of its lower-level walls that is completely above grade. The wall is constructed using $\frac{3}{8}$-in. gypsum wallboard on the inside, followed by a 3.5-in. layer of mineral wool insulation and hollow concrete block (8-in., three-oval core) on the outside. Environmental design conditions are: inside dry-bulb temperature 72 °F, inside relative humidity 35 percent, outdoor dry-bulb temperature −5 °F, outdoor relative humidity 60 percent, and outdoor wind speed 15 mph.

(a) For the design conditions, does the wall require a vapor retarder to prevent condensation from forming in the wall? (Provide appropriate calculations to back up your answer.)

(b) If a vapor retarder is needed, where should it be placed?

14.16 An exterior wall is constructed, from inside to outside, of $\frac{1}{2}$-in. plywood, a 3.5-in.-thick insulating batt of mineral wool, 8-in. concrete blocks, and 4-in.-thick brick. The indoor air temperature is 72 °F, indoor relative humidity is 30 percent, the outdoor air temperature is 5 °F, outdoor relative humidity is 50 percent, and the outdoor wind speed is 7.5 mph.

(a) What is the heat flux through the wall?

(b) What locations in the wall may have moisture-condensation problems?

(c) Calculate the minimum vapor-retarder vapor resistance, Z_{vr}, needed to ensure that no condensation occurs at one of the locations that you identified in part (b), and specify where it should be located.

(d) If the vapor retarder you recommended in part (c) is incorporated in the design, will condensation be prevented at all locations in the wall?

14.17 It is proposed that the exterior wall of a building be constructed, from inside to outside, of a layer of vinyl wallpaper ($R = 0$, $Z = 4.5$ Rep), $\frac{3}{8}$-in. gypsum wallboard, 3.5-in.-thick fiberglass insulation, and 8-in. concrete block. The indoor conditions are 72 °F dry-bulb temperature and 40 percent relative humidity. The building is located in Galveston, Texas, which has summer design conditions of 89 °F dry-bulb temperature and 79 °F wet-bulb temperature. (*Note:* These result in an outdoor dew-point temperature of 76 °F.) Wind speed is 15 mph. Determine if there are moisture problems associated with the proposed design; if there are, suggest a design modification to eliminate them. Back up all of your conclusions and recommendations with calculations.

14.18 An exterior wall of a building is constructed using 2 × 6 framing (actual dimensions of a 2 × 6 stud are 1.5 in. × 5.5 in.). The inside of the wall is $\frac{3}{8}$-in. gypsum wallboard. A vapor retarder is placed on the

back side of the wallboard. The space between the framing is filled with fiberglass insulation. On the outside of the framing, the wall is built with a layer of $\frac{25}{32}$-inch regular-density sheathing ($\mu = 35$ Perm · in) a 1.5-in.-thick layer of expanded polystyrene (extruded) and 4-in.-thick brick. The framing makes up 15 percent of the area of the wall. The inside air is at 68 °F dry-bulb temperature and 30 percent relative humidity. The outside air is at 5 °F dry-bulb temperature and 100 percent relative humidity. Wind speed is 7.5 mph. Calculate the average heat-transfer rate per unit area for the wall, using:

(a) The parallel-path method.

(b) The isothermal-plane method.

(c) Assume a one-dimensional vapor-flow path through the wall in the areas containing the fiberglass insulation and determine the minimum vapor-flow resistance of the vapor retarder, Z_{vr}, needed to ensure no condensation will form in these areas of the wall.

14.19 A house has one of its lower-level walls that is completely above grade. The wall is constructed using $\frac{1}{4}$-in. plywood paneling on the inside (made with interior-grade glue), followed by 1.5 in. of expanded polystyrene (extruded type) and 6 in. of concrete (sand and gravel aggregate). Design conditions are: inside dry-bulb temperature 68 °F, inside relative humidity 35 percent, outdoor dry-bulb temperature −8 °F, outdoor relative humidity 70 percent, and outdoor wind speed 15 mph.

(a) What location in the wall may have a moisture-condensation problem, and where should a vapor retarder be located to ensure condensation does not occur in the wall?

(b) Calculate the minimum vapor-retarder vapor resistance, Z_{vr}, needed to ensure that no condensation occurs in the wall.

14.20 A small single-story office building has 10-ft-high wood-frame walls built using 2 × 6 studs (actual dimensions 1.5 in. × 5.5 in.). The framing factor for the wall is 0.25. The inside of the wall is $\frac{3}{8}$-in. gypsum wallboard. The space between the studs is filled with fiberglass batt insulation. On the outside of the studs, the wall is built with a layer of $\frac{25}{32}$-inch regular-density fiberboard sheathing, a 1.5-in.-thick layer of expanded polystyrene, and 4-in. brick.

Windows in the building are wood-framed triple glass with $\frac{3}{4}$-in. air gaps between the sheets of glass. The inside and outside sheets of glass are standard glass with an emittance of 0.9 on all surfaces. The center sheet of glass has a special coating on its inner surface such that the surface has an emittance value of $\varepsilon = 0.4$.

The building has a "built-up" flat roof which has a total thermal resistance of $R_t = 9.0$ hr · ft² · °F/Btu. The exterior doors of the building are steel, 1.75-in., solid urethane foam core with thermal break. No storm doors are used. Each side of the building is 50 ft long by 10 ft high. The north side has no windows and one door, which is 6.75 ft high by 3 ft wide. The south side has 200 ft² of window area and one 6.75-ft × 3-ft door. The east and west sides each have 250 ft² of window area and no doors. The doors and windows are weatherstripped and caulked.

The building has a basement whose walls are made of 8-in. concrete block (sand and gravel aggregate). The basement walls are 8 ft high with 2 ft above grade and 6 ft below grade. The sole plate is uncaulked. The basement floor is 6-in.-thick concrete and unfinished. The soil around the building is sand.

Using an indoor design temperature of 70 °F and the 97.5 percent outdoor design temperature for Chicago, calculate the following:

(a) The average U-value for the total frame wall and roof area of the building (including doors and windows) and the total heat loss for this part of the building.

(b) The rate of heat loss for the above-grade portion of the basement walls.

(c) The rate of heat loss for the below-grade portion of the basement walls.

(d) The rate of heat loss for the basement floor.

(e) The infiltration rate of the building (based on the LBL model) and the infiltration design heating load. Assume a typical downtown location for the building.

(d) The total design heating load for the building.

SYMBOLS

A	Area; temperature amplitude.
a	Area ratio, Eqs. (14.8) and (14.9); infiltration coefficients in Eq. (14.49).
ACH	Air changes per hour.
C	Thermal conductance; flow coefficient.
C_d	Thermal draft coefficient.
C_p	Wind-pressure coefficient.
$\bar{c}_p$	Average specific heat of moist air for a process.
D	Depth below grade.
F_s	Heat-loss coefficient per unit length of perimeter.
F_{1-2}	View factor from surface 1 to surface 2.
g	Gravitational acceleration.
h	Heat-transfer coefficient; Enthalpy; Height of the neutral pressure level; Column height.
$\bar{h}_g$	Average enthalpy of water vapor for a process.
k	Thermal conductivity.
L	Thickness of a material; Length of wall.
M	Permeance (moisture transport conductance).
$\dot{m}$	Mass flow rate.
n	Exponent in Eq. (14.41).
P	Length of the perimeter of a slab; Pressure.
ΔP	Pressure difference.
$\dot{Q}$	Heat-transfer rate.
$\dot{q}$	Heat-transfer rate per unit area.

R	Thermal resistance; Gas constant.
$R_{w \to x}$	Sum of the thermal resistances from the warm air up to location-x in a structure.
T	Absolute temperature.
t	Temperature.
$\bar{t}_a$	Average winter air temperature.
Δt	Temperature difference.
U	Overall heat-transfer coefficient.
V	Volume; Wind speed.
$\dot{V}$	Volumetric flow rate.
W	Humidity ratio; Width of basement.
$\overline{W}$	Average humidity ratio for a process.
x, y, z	Coordinates.
Z	Vapor-flow resistance.
$Z_{w \to x}$	Sum of vapor resistances from warm air to location-x.
$Z_{x \to c}$	Sum of vapor resistances from location-x to cool air.
Z_{t-vr}	Sum of all the vapor-flow resistances of a structure, not including that of the vapor retarder.

Greek Letters

α	Soil thermal diffusivity.
ε	Surface emittance.
ε_{eff}	Effective emittance, Eq. (14.3).
μ	Water-vapor permeability of a material.
ϕ	Relative humidity.
Ψ	Dimensionless depth, Eq. (14.34).
ρ	Density.
τ_0	Time period.

Subscripts

a	Air; Dry air.
av	Average.
avg	Average.
b	Between the studs.
c	Ceiling; Cold side of a structure.
cg	Center of glass.
e	Ends.
eff	Effective.
eg	Edge of glass.
f	Frame; Floor.
g	Ground.
i	Inside; Indoor.
I	Indicates presence of insulation in Eq. (14.26).
INS	Insulation.
l	Leakage.
o	Outside, outdoor.
p	Building pressurization.
r	Roof.

s	Sensible; Studs; Soil; Shortest width; Stack effect; Saturated conditions.
s, b	Indicates equivalent thermal resistance of R_s and R_b in parallel.
t	Total; Theoretical.
u	Unheated.
vr	Vapor retarder.
w	Wall; Water; Water vapor; Wind effect; Warm side of a structure.

chapter 15

Instantaneous Heat Gain

15.1 INTRODUCTION

In Chapter 14 we considered the steady heat loss from a structure for winter design conditions to estimate the required capacity of the heating system. We must develop an analogous method to estimate the building heat gains. The heat gains will be used to determine the required capacity of the cooling equipment.

The instantaneous-heat-gain estimate requires a more sophisticated approach than the heat-loss estimate. The main reasons for this are outlined below:

1. Residences and small commercial buildings have heat gains that are dominated by the building envelope (walls, roof, windows, and doors). However, internal gains from occupants, lights, equipment, and appliances play an important role and often dominate in large buildings. Therefore, accurate estimates of the internal loads are important.

2. Heat gains through the building envelope are driven primarily by external absorption of solar radiation. Outdoor-to-indoor temperature difference is much less important.

3. Heat transfer through the building walls and roofs is transient in nature, as the outdoor solar radiation and air dry-bulb temperature are time dependent. Thermal storage and time lags are important in all types of wall and roof construction.

4. Outdoor-air humidity levels are significant at most locations during the cooling season. The outdoor air used for ventilation usually must be cooled and dehumidified. The moisture gains within the conditioned space must also be removed by the cooling system. Thus, the latent loads must be considered in addition to the sensible loads.

The following sections in this chapter discuss these issues in detail and describe current engineering practice used to estimate the instantaneous heat gains in occupied buildings.

15.2 SUMMER DESIGN CONDITIONS

The maximum heat gain ever likely to be achieved in a given building depends on the outdoor weather conditions, the indoor temperature and humidity levels, and the usage of the space. Indoor conditions are discussed in detail in Chapter 12. Space occupancy and other internal gains are discussed in a later section in this chapter. We will focus on the outdoor weather conditions that can be expected.

The maximum heat gain will occur on a very hot, sunny day. Therefore, the solar radiation is often treated as clear-sky radiation for the entire day. Solar angles and insolation levels for clear-sky conditions are given in Chapter 13.

Table 15.1 lists some summer design temperatures for selected cities around the world. These temperatures are measured at airports and other weather stations. The data in Table 15.1 have been statistically analyzed for a 15-year period for which hourly weather data exist.

Column 2 presents the design dry-bulb temperatures in degrees Fahrenheit followed by the mean coincident wet-bulb temperature in degrees Fahrenheit after the slash. The mean coincident wet-bulb data are not available at some locations. The dry-bulb temperature data represent values that have been equaled or exceeded 1 percent, 2.5 percent, and 5 percent of the time in the summer months of June through September in the Northern Hemisphere and December through March in the Southern Hemisphere. The data from Canada consider only the month of July. The number of hours that represent the

TABLE 15.1 Summer Design Climatic Conditions

Col. 1 Location	Col. 2 Design Dry-Bulb/Mean Coincident Wet-Bulb Temperature (°F)			Col. 3 Mean Daily Range (°F)	Col. 4 Design Wet-Bulb (°F)			Col. 5 Prevailing Wind Direction
	1%	2.5%	5%		1%	2.5%	5%	
Sydney, Australia	89/	84/	80/	13	74	73	72	NE
Rio de Janeiro, Brazil	94/	92/	90/	11	80	79	78	S
Edmonton, Canada	85/66	82/65	79/63	23	68	66	65	SE
Quebec, Canada	87/72	84/70	81/68	20	74	72	70	
Paris, France	89/	86/	83/	21	70	68	67	E
Bombay, India	96/	94/	92/	13	82	81	81	NW
Tokyo, Japan	91/	89/	87/	14	81	80	79	S
Mexico City, Mexico	83/	81/	79/	25	61	60	59	N
Cape Town, South Africa	93/	90/	86/	20	72	71	70	
Kiev, Ukraine	87/	84/	81/	22	69	68	67	
Moscow, Russia	84/	81/	78/	21	69	67	65	S
Stockholm, Sweden	78/	74/	72/	15	64	62	60	S
Tunis, Tunisia	102/	99/	96/	22	77	76	74	E
Denver, USA	93/59	91/59	89/59	28	64	63	62	SE
Los Angeles, USA	83/68	80/68	77/67	15	70	69	68	WSW
Miami, USA	91/77	90/77	89/77	15	79	79	78	SE
Minneapolis, USA	92/75	89/73	86/71	22	77	75	73	S
New York, USA	92/74	89/73	87/72	17	76	75	74	SW

Source: Adapted by permission from *ASHRAE Fundamentals 1993*, pp. 24.4–24.22.

TABLE 15.2 Percentage of the Daily Range vs. Local Solar Time*

Time, hr	%	Time, hr	%	Time, hr	%
1	87	9	71	17	10
2	92	10	56	18	21
3	96	11	39	19	34
4	99	12	23	20	47
5	100	13	11	21	58
6	98	14	3	22	68
7	93	15	0	23	76
8	84	16	3	24	82

SOURCE: Reprinted by permission from *ASHRAE Fundamentals 1993*, p. 26.6.

1 percent value are 29. Therefore, in a typical summer, 29 hours are at or above a dry-bulb temperature of 89 °F in Sydney, Australia.

The data in column 3 represent the mean daily range or variation of the dry-bulb temperature in degrees Fahrenheit. An estimate of hourly outdoor dry-bulb temperatures for a clear summer day can be obtained by combining the design dry-bulb temperature data in column 2 of Table 15.1 with the mean daily range data in column 3 and the hourly fluctuation estimate given in Table 15.2. For example, the maximum dry-bulb temperature that would occur on a 1 percent summer design day in Sydney, Australia, would be 89 °F, and this maximum would occur at 1500 hours local solar time. The minimum dry-bulb temperature would be $89 - 13 = 76$ °F, and this would occur at 0500 hours local solar time. The dry-bulb temperatures at other hours would lie between these two extremes, with the variation with time shown in Table 15.2.

The data in column 4 of Table 15.1 present the design wet-bulb temperatures in degrees Fahrenheit. These data do not necessarily correspond to the dry-bulb temperature data given in Column 2.

The prevailing wind direction in summer is given in column 5 of Table 15.1. The wind velocity is usually taken to be 7.5 mph (0.45 m/s). Thus the exterior-surface convective and radiative film heat-transfer coefficient becomes approximately 3 Btu/hr · ft^2 · °F (17 W/m^2 · °C).

15.3 SOL-AIR TEMPERATURE

Solar radiation absorbed by the exterior surface of a wall or roof often provides the majority of the energy added to the surface in summer conditions. One must include absorbed solar radiation, infrared radiation exchange, and convective heat gain or loss on an exterior surface to obtain an accurate energy balance. This can be done by treating each energy component separately. However, an alternative method has been derived which combines the three modes of energy transport to or from an exterior surface. This method utilizes the sol-air temperature. Its definition is: "The sol-air temperature is a fictitious outdoor dry-bulb temperature such that in the absence of solar radiation, the surface will exchange the same net amount of energy to air at the sol-air temperature as is exchanged in the actual environment." The following discussion illustrates the basis of the sol-air temperature and outlines the main assumptions inherent in the analysis.

Consider an exterior surface of a building which is not transparent to solar or infrared radiation. This surface absorbs a portion of the incident direct (I_D), diffuse (I_d), and reflected (I_R) solar radiation. The total amount absorbed becomes

$$q_s = (I_D + I_d + I_R)\alpha_s = I_T\alpha_s \tag{15.1}$$

provided the solar absorptivity is equal for all three components. This surface also absorbs heat by convection from the ambient air based on the temperature difference between the surface (t_w) and the air dry-bulb temperature (t_o),

$$q_c = h_{o,c}(t_o - t_w) \tag{15.2}$$

The infrared radiation emitted by the surface becomes

$$q_{\text{IR,out}} = \varepsilon_w \sigma T_w^4 \tag{15.3}$$

The infrared radiation absorbed by the surface originates from a variety of other surfaces and the sky:

$$q_{\text{IR,in}} = \sum_{i=1}^{n} (\varepsilon_i F_{w,i}\sigma T_i^4) \tag{15.4}$$

The net energy flux to the surface becomes

$$q_{\text{net}} = q_s + q_c + q_{\text{IR,in}} - q_{\text{IR,out}}$$
$$= I_T\alpha_s + h_{o,c}(t_o - t_w) + \sum_{i=1}^{n} (\varepsilon_i F_{w,i}\sigma T_i^4) - \varepsilon_w \sigma T_w^4 \tag{15.5}$$

This expression can be simplified if some assumptions are made:

1. All infrared emissivity values are equal, $\varepsilon_i = \varepsilon_w = \varepsilon$. This is a good assumption for nonmetal surfaces where $0.9 \leq \varepsilon \leq 0.95$. The net infrared radiation to the surface then becomes

$$q_{\text{IR,net}} = \varepsilon\sigma \left[\left(\sum_{i=1}^{n} F_{w,i} T_i^4 \right) - T_w^4 \right] \tag{15.6}$$

2. The infrared radiation is linearized as the temperature differences are small. This eliminates the need for absolute temperatures to the fourth power and allows us to introduce a radiative heat-transfer coefficient.

3. The temperatures of all surfaces except the one in question are at the local air dry-bulb temperature, t_o. This is a reasonable approximation, although the sky temperature is usually lower than the local dry-bulb temperature and the surrounding surfaces receiving solar radiation have a higher temperature. A correction factor is introduced to account for the error introduced with this assumption.

The net energy transfer to the surface now can be written

$$q_{\text{net}} \cong I_T\alpha_s + h_{o,c}(t_o - t_w) + h_{o,R}(t_o - t_w) - \varepsilon \Delta R \tag{15.7}$$

where $h_{o,R}$ is the linearized infrared radiative heat-transfer coefficient and $\varepsilon \Delta R$ is the correction factor mentioned before. As shown in Chapter 14, it is convenient to add the convective and linearized radiative heat-transfer coefficients into an exterior film coefficient, $h_{o,c} + h_{o,R} = h_o$. Equation (15.7) then becomes

$$q_{\text{net}} \cong I_T\alpha_s + h_o(t_o - t_w) - \varepsilon \Delta R \tag{15.8}$$

From the definition of the sol-air temperature, t_e, we can write the following equation for the same net energy transfer to the surface:

$$q_{\text{net}} = h_o(t_e - t_w) \tag{15.9}$$

Combining Eqs. (15.8) and (15.9) gives the following expression for the sol-air temperature:

$$t_e = t_o + \frac{I_T \alpha_s}{h_o} - \frac{\varepsilon \, \Delta R}{h_o} \tag{15.10}$$

ASHRAE recommends that the correction factor ($\varepsilon \, \Delta R / h_o$) be given a value of 7 °F (4 °C) for horizontal surfaces facing up. Thus the sol-air temperature is 7 °F (4 °C) cooler due to the reduced infrared radiation coming from the sky. The correction factor is specified to be zero for vertical surfaces, as the warmer sunlit surfaces compensate for the cooler sky temperature. An estimate of the correction factor for other tilt angles based upon radiation shape-factor geometry is

$$\frac{\varepsilon \, \Delta R}{h_o} = 7 \text{ °F}(4 \text{ °C}) \cos\Sigma \tag{15.11}$$

where Σ is the surface tilt angle measured between the surface normal and vertical.

The ratio α_s / h_o in the second term of Eq. (15.10) is a function of the color of the surface. Dark colors have solar absorptance values near 0.9, and lighter colors have values near 0.45. This ratio then becomes

$$\frac{0.9}{3} = 0.3 \text{ hr} \cdot \text{ft}^2 \cdot \text{°F/Btu} \ (0.052 \text{ m}^2 \cdot \text{°C/W})$$

for dark-colored surfaces and

$$\frac{0.45}{3} = 0.15 \text{ hr} \cdot \text{ft}^2 \cdot \text{°F/Btu} \ (0.026 \text{ m}^2 \cdot \text{°C/W})$$

for light-colored surfaces.

The value for I_T in Eq. (15.10) can be estimated using the procedures given in Chapter 13, and values for t_o can be estimated from data such as those given in Tables 15.1 and 15.2.

Tabulated values for sol-air temperature are valid only for specific locations and times of the year, owing to their dependence on both dry-bulb temperature and solar radiation. Table 15.3 presents hourly dry-bulb temperatures and hourly sol-air temperatures for dark- and light-colored surfaces with different orientations. These values were computed for a clear day on July 21, 40° north latitude, with a maximum dry-bulb temperature of 95 °F (35 °C) and a daily range of 21 °F (11.7 °C).

The variations shown by Table 15.3 may be assumed to be repetitive for successive 24-hr cycles. Any of the values may be mathematically expressed in terms of a Fourier series. Thus, if $t_e = f(\theta)$, where θ is the number of hours measured from midnight solar time, we have

$$t_e = t_{e,m} + M_1 \cos\omega_1\theta + N_1 \sin\omega_1\theta + M_2 \cos\omega_2\theta + N_2 \sin\omega_2\theta + \cdots \tag{15.12}$$

where the coefficients $t_{e,m}$, M_n, and N_n are given by

$$t_{e,m} = \frac{1}{24} \int_0^{24} t_e \, d\theta \approx \frac{1}{24} \sum_{\theta=0}^{23} t_e$$

TABLE 15.3 Sol-Air Temperatures, t_e for July 21, 40° N Latitude (°F)/(°C)

Time	Air Dry-Bulb Temp	N	NE	E	SE	S	SW	W	NW	HOR
					$\alpha/h_o = 0.15/0.026$ (light color)					
1	76/24.4	76/24.4	76/24.4	76/24.4	76/24.4	76/24.4	76/24.4	76/24.4	76/24.4	69/20.5
2	76/24.4	76/24.4	76/24.4	76/24.4	76/24.4	76/24.4	76/24.4	76/24.4	76/24.4	69/20.5
3	75/23.8	75/23.8	75/23.8	75/23.8	75/23.8	75/23.8	75/23.8	75/23.8	75/23.8	68/20.0
4	74/23.3	74/23.3	74/23.3	74/23.3	74/23.3	74/23.3	74/23.3	74/23.3	74/23.3	67/17.4
5	74/23.3	74/23.3	74/23.3	74/23.3	74/23.3	74/23.3	74/23.3	74/23.3	74/23.3	67/19.4
6	74/23.3	82/27.7	95/35.0	97/36.1	86/30.0	75/23.8	75/23.8	75/23.8	75/23.8	74/23.3
7	75/23.8	82/27.7	103/39.4	109/42.7	97/36.1	78/25.5	78/25.5	78/25.5	78/25.5	85/29.4
8	77/25.0	82/27.7	103/39.4	114/45.5	105/40.5	83/28.3	81/27.2	81/27.2	81/27.2	96/35.5
9	80/26.6	85/29.4	101/38.3	114/45.5	110/43.3	92/33.3	85/29.4	85/29.4	85/29.4	106/41.1
10	83/28.3	89/31.6	96/35.5	110/43.3	112/44.4	100/37.7	89/31.6	89/31.6	89/31.6	115/46.1
11	87/30.5	93/33.8	94/34.4	104/40.0	111/43.8	108/42.2	96/35.5	93/33.8	93/33.8	123/50.5
12	90/32.2	96/35.5	96/35.5	97/36.1	107/41.6	112/44.4	107/41.6	97/36.1	96/35.5	127/52.7
13	93/33.8	99/37.2	99/37.2	99/37.2	102/38.8	114/45.5	117/47.2	110/43.3	100/37.7	129/53.8
14	94/34.4	100/37.7	100/37.7	100/37.7	100/37.7	111/43.8	123/50.5	121/49.4	107/41.6	126/52.2
15	95/35.0	100/37.7	100/37.7	100/37.7	100/37.7	107/41.6	125/51.6	129/53.8	116/46.6	121/49.4
16	94/34.4	99/37.2	98/36.6	98/36.6	98/36.6	100/37.7	122/50.0	131/55.0	120/48.8	113/45.0
17	93/33.8	100/37.7	96/35.5	96/35.5	96/35.5	96/35.5	115/46.1	127/52.7	121/49.4	103/39.4
18	91/32.7	99/37.2	92/33.3	92/33.3	92/33.3	92/33.3	103/39.4	114/45.5	112/44.4	91/32.7
19	87/30.5	87/30.5	87/30.5	87/30.5	87/30.5	87/30.5	87/30.5	87/30.5	87/30.5	80/26.6
20	85/29.4	85/29.4	85/29.4	85/29.4	85/29.4	85/29.4	85/29.4	85/29.4	85/29.4	78/25.5
21	83/28.3	83/28.3	83/28.3	83/28.3	83/28.3	83/28.3	83/28.3	83/28.3	83/28.3	76/24.4
22	81/27.2	81/27.2	81/27.2	81/27.2	81/27.2	81/27.2	81/27.2	81/27.2	81/27.2	74/23.3
23	79/26.1	79/26.1	79/26.1	79/26.1	79/26.1	79/26.1	79/26.1	79/26.1	79/26.1	72/22.2
24	77/25.0	77/25.0	77/25.0	77/25.0	77/25.0	77/25.0	77/25.0	77/25.0	77/25.0	70/21.1
Avg.	83/28.3	86/30.0	89/31.6	91/32.7	90/32.2	89/31.6	90/32.2	91/32.7	89/31.6	91/32.7
					$\alpha/h_o = 0.30/0.052$ (dark color)					
1	76/24.4	76/24.4	76/24.4	76/24.4	76/24.4	76/24.4	76/24.4	76/24.4	76/24.4	69/20.5
2	76/24.4	76/24.4	76/24.4	76/24.4	76/24.4	76/24.4	76/24.4	76/24.4	76/24.4	69/20.5
3	75/23.8	75/23.8	75/23.8	75/23.8	75/23.8	75/23.8	75/23.8	75/23.8	75/23.8	68/20.0
4	74/23.3	74/23.3	74/23.3	74/23.3	74/23.3	74/23.3	74/23.3	74/23.3	74/23.3	67/19.4
5	74/23.3	74/23.3	74/23.3	74/23.3	74/23.3	74/23.3	74/23.3	74/23.3	74/23.3	67/19.4
6	74/23.3	90/32.2	117/47.2	121/49.4	99/37.2	77/25.0	77/25.0	77/25.0	77/25.0	81/27.2

(continued)

TABLE 15.3 (cont.)

$\alpha / h_o = 0.30/0.052$ (dark color)

Time	Air Dry-Bulb Temp	N	NE	E	SE	S	SW	W	NW	HOR
7	75/23.8	90/32.2	131/55.0	144/62.2	120/48.8	82/27.7	82/27.7	82/27.7	82/27.7	102/38.8
8	77/25.0	87/30.5	130/54.4	151/66.1	134/56.6	89/31.6	86/30.0	86/30.0	86/30.0	122/50.0
9	80/26.6	91/32.7	122/50.0	148/64.4	141/60.5	105/40.5	91/32.7	91/32.7	91/32.7	140/60.0
10	83/28.3	95/35.0	109/42.7	137/58.3	141/60.5	118/47.7	96/35.5	95/35.0	95/35.0	155/68.3
11	87/30.5	100/37.7	101/38.3	122/50.0	136/57.7	129/53.8	105/40.5	100/37.7	100/37.7	166/74.4
12	90/32.2	103/39.4	103/39.4	104/40.0	125/51.6	134/56.6	125/51.6	104/40.0	103/39.4	172/77.7
13	93/33.8	106/41.1	106/41.1	106/41.1	111/43.8	135/57.2	142/61.1	128/53.3	107/41.6	172/77.7
14	94/34.4	106/41.1	106/41.1	106/41.1	107/41.6	129/53.8	152/66.6	148/64.4	120/48.8	166/74.4
15	95/35.0	106/41.1	106/41.1	106/41.1	106/41.1	120/48.8	156/68.8	163/72.7	137/58.3	155/68.3
16	94/34.4	104/40.0	103/39.4	103/39.4	103/39.4	106/41.1	151/66.1	168/75.5	147/63.8	139/59.4
17	93/33.8	108/42.2	100/37.7	100/37.7	100/37.7	100/37.7	138/58.8	162/72.2	149/65.0	120/48.8
18	91/32.7	107/41.6	94/34.4	94/34.4	94/34.4	94/34.4	116/46.6	138/58.8	134/56.6	98/36.6
19	87/30.5	87/30.5	87/30.5	87/30.5	87/30.5	87/30.5	87/30.5	87/30.5	87/30.5	80/26.6
20	85/29.4	85/29.4	85/29.4	85/29.4	85/29.4	85/29.4	85/29.4	85/29.4	85/29.4	78/25.5
21	83/28.3	83/28.3	83/28.3	83/28.3	83/28.3	83/28.3	83/28.3	83/28.3	83/28.3	76/24.4
22	81/27.2	81/27.2	81/27.2	81/27.2	81/27.2	81/27.2	81/27.2	81/27.2	81/27.2	74/23.3
23	79/26.1	79/26.1	79/26.1	79/26.1	79/26.1	79/26.1	79/26.1	79/26.1	79/26.1	72/22.2
24	77/25.0	77/25.0	77/25.0	77/25.0	77/25.0	77/25.0	77/25.0	77/25.0	77/25.0	70/21.1
Avg.	83/28.3	89/31.6	95/35.0	100/37.7	99/37.2	95/35.0	99/37.2	100/37.7	95/35.0	107/41.6

SOURCE: Reprinted by permission from *ASHRAE Fundamentals 1993*, p. 26.6.

$$M_n = \frac{1}{12} \int_0^{24} t_e \cos \omega_n \theta \; d\theta \approx \frac{1}{12} \sum_{\theta=0}^{23} t_e \cos \omega_n \theta$$

$$N_n = \frac{1}{12} \int_0^{24} t_e \sin \omega_n \theta \; d\theta \approx \frac{1}{12} \sum_{\theta=0}^{23} t_e \sin \omega_n \theta$$

and $\omega_1 = \pi/12$ rad/hr or 15 deg/hr and $\omega_n = n\omega_1$. Alternatively, Eq. (15.12) may be written as

$$t_e = t_{e,m} + \sqrt{M_1^2 + N_1^2} \cos(\omega_1 \theta - \psi_1) + \sqrt{M_2^2 + N_2^2} \cos(\omega_2 \theta - \psi_2) + \cdots$$

or

$$t_e = t_{e,m} + t_{e,1} \cos(\omega_1 \theta - \psi_1) + t_{e,2} \cos(\omega_2 \theta - \psi_2) + \cdots \qquad (15.13)$$

where

$$\tan \psi_n = \frac{N_n}{M_n} \qquad (15.14)$$

In Eq. (15.14), the quadrant in which ψ_n lies is determined by the requirement that $\sin \psi_n$ must have the sign of N_n and that $\cos \psi_n$ must have the sign of M_n.

As an illustration of the method, Ex. 15.1 shows a harmonic analysis for the sol-air temperature given by the horizontal light-colored surface in Table 15.3. The procedures follow those given by Alford, Ryan, and Urban [1].

EXAMPLE 15.1

Make a harmonic analysis for sol-air temperature given by the light-colored horizontal surface of Table 15.3. Include two harmonics.

Solution: The solution is shown in Table 15.4. Column 2 shows values of t_e read from Table 15.3. Column 3 shows values of $w_1 \theta$ in degrees where $w_1 = 360/24 = 15$ deg/hr. Zero time is taken as midnight. Columns 1 and 4–7 are self-explanatory. The mean value $t_{e,m}$ is obtained by summing the values of column 2 and dividing by 24. The coefficient M_1 is obtained by adding the values of column 5 and dividing by 12. The coefficient N_1 is obtained by adding the values of Column 7 and dividing by 12. Columns 8–12 are a repetition of the procedure for the second harmonic, where $w_2 = 2w_1 = 30$ deg/hr. The expression for t_e including the second harmonic is by Eq. (15.13):

$$t_e = 91.6 + 30.3 \cos(15\theta - 196) + 8.3 \cos(30\theta - 13)$$

Column 13 shows the results of this equation. We see that column 13 rather closely fits the given variation of sol-air temperature (column 2) with a standard deviation of 1.6 °F or about 1 °C. The accuracy of the computed value continues to improve as more harmonics are added to the solution.

The methods described in this section may be used to calculate sol-air temperature variations for clear days at any time of year at any location. Representative diurnal outdoor-air temperature variations may be found from local weather data. Equations 13.22–13.31 may be used for solar-radiation intensities. The entire procedure of Ex. 15.1

TABLE 15.4 Harmonic Analysis for Sol-Air Temperature Given in Table 15.3 of a Light-Colored Horizontal Surface

1	2	3	4	5	6	7	8	9	10	11	12	13
Solar Time	t_e, °F	$\omega_1\theta$, deg	$\cos\omega_1\theta$	$t_e\cos\omega_1\theta$, °F	$\sin\omega_1\theta$	$t_e\sin\omega_1\theta$, °F	$\omega_2\theta$, deg	$\cos\omega_2\theta$	$t_e\cos\omega_2\theta$, °F	$\sin\omega_2\theta$	$t_e\sin\omega_2\theta$, °F	t_e calc., °F
0	70	0	1.000	70.0	0.000	0.0	0	1.000	70.0	0.000	0.0	70.6
1	69	15	0.966	66.7	0.259	17.9	30	0.866	59.8	0.500	34.5	69.2
2	69	30	0.866	59.8	0.500	34.5	60	0.500	34.5	0.866	59.8	67.9
3	68	45	0.707	48.1	0.707	48.1	90	0.000	0.0	1.000	68.0	67.0
4	67	60	0.500	33.5	0.866	58.0	120	-0.500	-33.5	0.866	58.0	67.4
5	67	75	0.259	17.9	0.966	64.7	150	-0.866	-58.0	0.500	33.5	69.9
6	74	90	0.000	0.0	1.000	74.0	180	-1.000	-74.0	0.000	0.0	75.2
7	85	105	-0.259	-22.0	0.966	82.1	210	-0.866	-73.6	-0.500	-42.5	83.1
8	96	120	-0.500	-48.0	0.866	83.1	240	-0.500	-48.0	-0.866	-83.1	93.3
9	106	135	-0.707	-74.9	0.707	74.9	270	0.000	0.0	-1.000	-106.0	104.4
10	115	150	-0.866	-99.6	0.500	57.5	300	0.500	57.5	-0.866	-99.6	115.1
11	123	165	-0.966	-118.8	0.259	31.9	330	0.866	106.5	-0.500	-61.5	123.6
12	127	180	-1.000	-127.0	0.000	0.0	360	1.000	127.0	0.000	0.0	128.8
13	129	195	-0.966	-124.6	-0.259	-33.4	30	0.866	111.7	0.500	64.5	129.8
14	126	210	-0.866	-109.1	-0.500	-63.0	60	0.500	63.0	0.866	109.1	126.7
15	121	225	-0.707	-85.5	-0.707	-85.5	90	0.000	0.0	1.000	121.0	120.0
16	113	240	-0.500	-56.5	-0.866	-97.9	120	-0.500	-56.5	0.866	97.9	111.0
17	103	255	-0.259	-26.7	-0.966	-99.5	150	-0.866	-89.2	0.500	51.5	101.1
18	91	270	0.000	0.0	-1.000	-91.0	180	-1.000	-91.0	0.000	0.0	91.9
19	80	285	0.259	20.7	-0.966	-77.3	210	-0.866	-69.3	-0.500	-40.0	84.2
20	78	300	0.500	39.0	-0.866	-67.5	240	-0.500	-39.0	-0.866	-67.5	78.6
21	76	315	0.707	53.7	-0.707	-53.7	270	0.000	0.0	-1.000	-76.0	75.0
22	74	330	0.866	69.1	-0.500	-37.0	300	0.500	37.0	-0.866	-64.1	73.0
23	72	345	0.966	69.6	-0.259	-18.6	330	0.866	62.4	-0.500	-36.0	71.7
Summations	2199			-350.1		-97.7			97.3		21.5	

$t_{e,m}=91.6$

$M_1=-29.2$ $N_1=-8.1$ $M_2=8.1$ $N_2=1.8$

$t_{e,1}=\sqrt{M_1^2+N_1^2}=30.3$

$\psi_1=\tan^{-1}\dfrac{N_1}{M_1}=196$ deg

$t_{e,2}=\sqrt{M_2^2+N_2^2}=8.3$

$\psi_2=\tan^{-1}\dfrac{N_2}{M_2}=13$ deg

TABLE 15.5 Constants in Eq. (15.13) for a Horizontal Surface for Clear Days at Minneapolis

Time of Year	$t_{e,m}$ °F/°C	$t_{e,1}$ °F/°C	ψ_1 deg	$t_{e,2}$ °F/°C	ψ_2 deg
Average January day	7.9/−13.4	14.4/8.0	202.2	8.5/4.7	11.9
Winter design day	−8.6/−22.6	13.6/7.6	204.0	8.2/4.6	11.9
Average April day	66.0/18.9	36.2/20.1	196.0	11.9/6.6	0.0
Average July day	95.7/35.4	39.2/21.8	193.0	9.3/5.2	−1.8
Summer design day	105.5/40.8	40.3/22.4	191.0	9.2/5.1	−1.8
Average October day	63.8/17.7	27.0/15.0	199.0	12.0/6.7	10.2

can be programmed on a small computer or calculator for calculation through any desired number of harmonics.

Example 15.1 shows that sol-air temperature variation for a horizontal surface can be adequately represented by a Fourier series with two harmonics. However, variation of t_e for vertical surfaces differs much more from pure sine-wave behavior than for a horizontal surface. In the case of north, east, and west vertical surfaces, six or more harmonics may be needed.

Generally, periodic heat transfer through a sunlit vertical wall is small compared to heat transfer through a sunlit flat roof of the same area. Thus, usually it is more important to have accurate sol-air temperature information for a horizontal surface than for vertical surfaces.

Following the same procedures described here, James C. Dunn developed a computer program and calculated diurnal sol-air temperature variations for horizontal and vertical surfaces for clear days at Minneapolis for six different times of year [2]. Table 15.5 shows his results for a horizontal surface. Constants in Eq. (15.13) are shown through two harmonics. Constants for the winter design day were determined for clear December–February days whose minimum temperatures were approximately −20 °F (−29 °C). Constants for the summer design day were determined for clear June–August days whose maximum temperatures were approximately 95 °F (35 °C).

15.4 PERIODIC HEAT GAIN THROUGH WALLS AND ROOFS

Heat transfer through walls and roofs can be treated as diurnal for purposes of design calculations. Therefore the sol-air temperature becomes the external thermal boundary condition and the indoor temperature becomes the indoor thermal boundary condition. These boundary conditions are coupled to the solid material through the external and internal convective-radiative surface film coefficients. Thus the problem becomes one in which the solid has convective coefficients on both sides with the fluid temperatures changing in a diurnal manner. The most important assumptions to be made on the solid wall or roof are: (1) internal heat transfer is by conduction only (or pseudoconduction if natural convection and infrared radiation are present in air cavities or in porous insulation), (2) contact resistances between layers of material are neglected, and (3) air infiltration/exfiltration through the wall or roof construction is negligible.

Nearly all methods that are used to compute the diurnal heat transfer through walls and roofs assume that the heat flows only in the direction perpendicular to the exposed surfaces. Thus the heat transfer is one-dimensional. If we assume that the coordinate direction perpendicular to the wall surface is x, the governing equation within each layer, j, becomes

$$\frac{\partial t}{\partial \theta} = \alpha_j \frac{\partial^2 t}{\partial x^2} \tag{15.15}$$

The temperature and heat flux must be equal in both materials at an interface between them

$$k_j \frac{\partial t}{\partial x}\bigg|_j = k_{j+1} \frac{\partial t}{\partial x}\bigg|_{j+1}, \qquad t_j = t_{j+1} \tag{15.16}$$

at the same value of x. The boundary condition on the inside surface is

$$q_i = h_i (t_{w,i} - t_i) \tag{15.17}$$

and on the outside surface is

$$q_o = h_o (t_e - t_{w,o}) \tag{15.18}$$

where the heat flux is positive when the energy is moving into the building.

Equation (15.15) with boundary conditions Eqs. (15.17) and (15.18) and the matching condition Eq. (15.16) can be solved using a variety of techniques. Analytical, numerical, and approximate hand-calculation methods will be considered in the remainder of this section.

We will now consider the periodic transfer of heat through a wall formed by a single homogeneous material. Figure 15.1 shows the schematic problem. We will assume that (1) the wall is of infinite height and length, and heat transfer occurs only in the x direction, (2) the wall is homogeneous with constant material properties, (3) the surface coefficients h_i and h_o are constants, (4) the solar absorptivity of the outside surface is independent of angle of incidence and is constant, (5) the variations of t_o and I are periodic (identical with time on consecutive days), and (6) the internal thermal environment is constant.

At any location within the wall, we may write

$$\frac{\partial t}{\partial \theta} = \alpha \frac{\partial^2 t}{\partial x^2} \tag{15.19}$$

where $\alpha = k/\rho c$ is the thermal diffusivity of the wall.

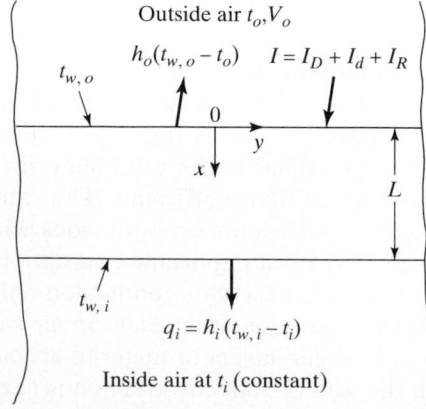

Figure 15.1 Schematic diagram for periodic heat-transfer analysis.

Equation (15.19) must be solved subject to two boundary conditions. At the inside surface we must have

$$q_i = -k \left(\frac{\partial t}{\partial x} \right)_{x=L} = h_i (t_{w,i} - t_i) \qquad (15.20)$$

At the outside surface, by Eq. (15.9), we must have

$$q_o = -k \left(\frac{\partial t}{\partial x} \right)_{x=0} = h_o (t_e - t_{w,o}) \qquad (15.21)$$

where t_e is given by Eq. (15.13).

The analytical solution of Eqs. (15.19)–(15.21) has the form

$$t = A + Bx + \sum_1^{\infty} (C_n \cos p_n mx + D_n \sin p_n mx) e^{-m^2 \omega_n \theta}$$

where A, B, C_n, D_n, and p_n are constants and $m = \sqrt[4]{-1}$. The coefficients A, B, C_n, and D_n may be either real or complex numbers, but the solution involves only the real parts.

The complete solution of Eqs. (15.19)–(15.21) has been given by Alford, Ryan, and Urban [3]. The temperature of the inside-wall surface $t_{w,i}$ may be written as

$$t_{w,i} = t_i + \frac{1}{h_i} [U(t_{e,m} - t_i) + V_1 t_{e,1} \cos(\omega_1 \theta - \psi_1 - \Phi_1)$$

$$+ V_2 t_{e,2} \cos(\omega_2 \theta - \psi_2 - \Phi_2) + \cdots] \qquad (15.22)$$

where

$$U = \frac{1}{\dfrac{1}{h_i} + \dfrac{L}{k} + \dfrac{1}{h_o}} \qquad (15.23)$$

$$V_n = \frac{h_o h_i}{\sigma_n k \sqrt{Y_n^2 + Z_n^2}} \qquad (15.24)$$

$$\sigma_n = \sqrt{\frac{\omega_n}{2\alpha}} \qquad (15.25)$$

$$Y_n = \left(\frac{h_o h_i}{2\sigma_n^2 k^2} + 1 \right) \cos \sigma_n L \sinh \sigma_n L + \left(\frac{h_o h_i}{2\sigma_n^2 k^2} - 1 \right) \sin \sigma_n L \cosh \sigma_n L$$

$$+ \frac{(h_o + h_i)}{\sigma_n k} \cos \sigma_n L \cosh \sigma_n L \qquad (15.26)$$

$$Z_n = \left(\frac{h_o h_i}{2\sigma_n^2 k^2} + 1 \right) \sin \sigma_n L \cosh \sigma_n L - \left(\frac{h_o h_i}{2\sigma_n^2 k^2} - 1 \right) \cos \sigma_n L \sinh \sigma_n L$$

$$+ \frac{(h_o + h_i)}{\sigma_n k} \sin \sigma_n L \sinh \sigma_n L \qquad (15.27)$$

$$\Phi_n = \tan^{-1} \frac{Z_n}{Y_n} \qquad (15.28)$$

In Eq. (15.28), $\sin \Phi_n$ has the sign of Z_n and $\cos \Phi_n$ has the sign of Y_n.

The rate of heat transfer to the interior is termed the "instantaneous heat gain," or IHG, and is given by

$$q_i = IHG = h_i\,(t_{w,i} - t_i) \tag{15.29}$$

By Eqs. (15.22) and (15.29),

$$q_i = U\{[t_{e,m} + \lambda_1 t_{e,1} \cos(\omega_1 \theta - \psi_1 - \Phi_1) + \lambda_2 t_{e,2} \cos(\omega_2 \theta - \psi_2 - \Phi_2) + \cdots] - t_i\} \tag{15.30}$$

where

$$\lambda_n = \frac{V_n}{U} \tag{15.31}$$

The quantity λ_n in Eq. (15.30) is called the *decrement factor*. The angle Φ_n is the angular displacement or lag between a harmonic of the sol-air temperature and the same harmonic of the inside-surface temperature and heat flux.

EXAMPLE 15.2

The light-colored roof of a building is sunlit throughout a clear day on July 21. Its location is 40 deg north latitude. Variation of sol-air temperature throughout the day is given in Table 15.4. The roof consists of 6 in. of concrete, covered by a thin layer of roofing which may be neglected in the solution. The temperature of the inside air and interior surrounding surfaces is 78 °F. Determine (a) the rate of heat transmission through the roof to the room below over the full day and (b) the rate of heat transmission into the room if heat-storage effects of the concrete slab are neglected.

Solution:

(a) Since we are using Table 15.4; we have $h_o = 3.00$ Btu/hr $\cdot$ ft^2 $\cdot$ °F. By Table 14.1, $h_i = 1.08$ Btu/hr $\cdot$ ft^2 $\cdot$ °F. By Table 14.4 IP, $\rho = 140$ lbm/ft^3, $k = 12/12 = 1.0$ Btu/hr $\cdot$ ft $\cdot$ °F, $c = 0.21$ Btu/lbm $\cdot$ °F. Thus, $\alpha = 1.0/(140)(0.21) = 0.034$ ft^2/hr. By Eq. (15.23),

$$U = \frac{1}{\dfrac{1}{1.08} + \dfrac{0.5}{1.0} + \dfrac{1}{3.00}} = 0.57 \text{ Btu/hr} \cdot \text{ft}^2 \cdot \text{°F}$$

The fundamental angular velocity $\omega_1 = 2\pi/24 = 0.2618$ rad/hr $= 15$ deg/hr. By Eq. (15.25),

$$\sigma_1 = \sqrt{\frac{0.2618}{(2)(0.034)}} = 1.96 \text{ ft}^{-1}$$

and $\sigma_2 = \sqrt{2}\,\sigma = 2.77$ ft^{-1}. Thus $\sigma_1 L = 0.98$ and $\sigma_2 L = 1.385$. By Eq. (15.26),

$$Y_1 = \left[\frac{(3.0)(1.08)}{2(1.96)^2(1.0)^2} + 1\right](0.5570)(1.145)$$

$$+ \left[\frac{(3.0)(1.08)}{2(1.96)^2(1.0)^2} - 1\right](0.8305)(1.520) + \left[\frac{3.0 + 1.08}{(1.96)(1.0)}\right](0.5570)(1.520)$$

$$= 1.939$$

By Eq. (15.27),

$$Z_1 = \left[\frac{(3.0)(1.08)}{2(1.96)^2(1.0)^2} + 1\right](0.8305)(1.520)$$

$$-\left[\frac{(3.0)(1.08)}{2(1.96)^2(1.0)^2} - 1\right](0.5570)(1.145) + \left[\frac{3.0 + 1.08}{(1.96)(1.0)}\right](0.8305)(1.145)$$

$$= 3.405$$

Likewise, we find $Y_2 = -0.649$ and $Z_2 = 9.584$. By Eq. (15.28), $\Phi_1 = \tan^{-1} 3.405/1.939 = 60$ deg and $\Phi_2 = \tan^{-1} 9.584/-0.649 = 86$ deg. By Eqs. (15.24) and (15.31),

$$\lambda_1 = \frac{h_o h_i}{U\sigma_1 k \sqrt{Y_1^2 + Z_1^2}} = \frac{(3.0)(1.08)}{(0.57)(1.96)(1.0)\sqrt{(1.939)^2 + (3.405)^2}} = 0.740$$

Likewise, we find $\lambda_2 = 214$. By Ex. 15.1, we know that

$$t_e = 91.6 + 30.3\cos(15\theta - 196) + 8.3\cos(30\theta - 13) + \cdots$$

We may now write the expression for $q_{i,2}$, including the second harmonic, by Eq. (15.30). We have

$$q_{i,2} = 0.57[13.6 + 22.4\cos(15\theta - 256) + 1.8\cos(30\theta - 99)]$$

Curve A of Fig. 15.2 shows the variation of q_i throughout the day.

(b) If heat-storage effects are neglected (no harmonics used), $\lambda_n = 1.00$ and $\Phi_n = 0$. The expression for $q_{i,0}$ is in phase with the sol-air temperature t_e. Equation (15.30) reduces to

$$q_{i,0} = 0.57[13.6 + 30.3\cos(15\theta - 196) + 8.3\cos(30\theta - 13)]$$

Curve B of Fig. 15.2 shows the fictitious variation of $q_{i,0}$, where we neglect heat-storage effects. Curve A lags curve B by approximately 3.9 hr, which closely compares with the fundamental time lag $\Phi_1/15 = 4.0$ hr. Figure 15.2 also shows that the maximum value of q_i (curve A) is about 70 percent of the maximum value which would occur if heat storage were absent, which closely compares with the fundamental decrement factor $\lambda_1 = 0.74$.

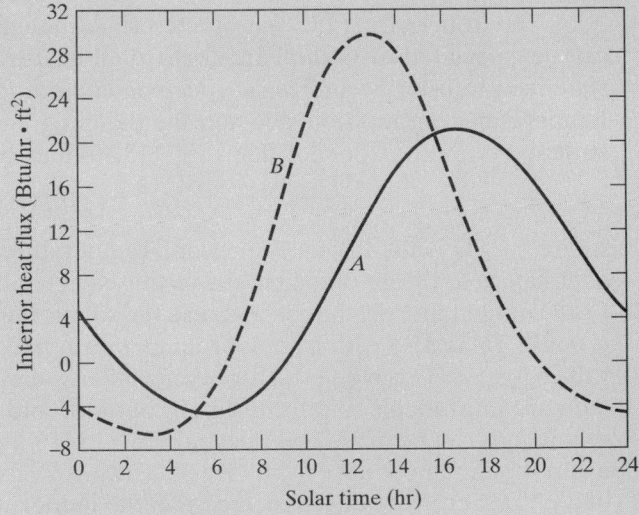

Figure 15.2 Results of Ex. 15.2.

The development of this section on periodic heat transfer is valid only for a homogeneous wall. Mackey and Wright [4] have extended a similar analysis to cover composite walls. Stewart [5] has applied the procedures of Mackey and Wright to several practical wall constructions.

The general form of the solution to Eqs. (15.19)–(15.21) for composite walls is identical to Eq. (15.30). However, the formulations for λ_n and Φ_n become progressively more complicated as the number of layers is increased. Bullock [6] developed solutions for various composite-wall constructions. He programed his solutions on a computer and calculated U, λ_n, and Φ_n values for a large number of wall and roof constructions. Table 15.6 shows Bullock's values for these constants through two harmonics for various flat-roof constructions. Only sol-air temperature data are further required to solve Eq. (15.30) for the roof constructions of Table 15.6.

EXAMPLE 15.3

Calculate the instantaneous rate of heat gain, W/m^2, through a flat-roof construction similar to No. 2 of Table 15.6 at 1600 hrs solar time for an average clear October day at Minneapolis. Assume an interior temperature of 23 °C and summer weather conditions.

Solution: By Table 15.5, we have $t_{e,m} = 17.7$ °C, $t_{e,1} = \sqrt{M_1^2 + N_1^2} = 15.0$ °C, $\psi_1 = 199.0$ deg, $t_{e,2} = \sqrt{M_2^2 + N_2^2} = 6.7$ °C, and $\psi_2 = 10.2$ deg. By Table 15.6, we have $U = 0.710$ $W/m^2 \cdot$ °C, $\lambda_1 = 0.241$, $\Phi_1 = 88.5$ deg, $\lambda_2 = 0.121$, and $\Phi_2 = 108.0$ deg. By Eq. (15.30),

$$q_i = 0.710\{[17.7 + (0.241)(15.0)\cos(240 - 199.0 - 88.5)$$

$$+ (0.121)(6.7)\cos(480 - 10.2 - 108.0)] - 23\}$$

$$= 0.710(17.7 + 2.44 + 0.81 - 23)$$

$$= -1.46 \text{ W/m}^2$$

The form of Eq. (15.30) is of interest. Although q_i may be continuously changing, it can be calculated by multiplying the overall heat-transmission coefficient U for steady-state heat transfer by an *equivalent temperature difference* which accounts for the sol-air temperature variation outside and the damping and time lag associated with the construction of the wall or roof. Thus Eq. (15.30) may be written

$$q_i = U(\text{TETD}) \tag{15.32}$$

where TETD is the "Total Equivalent Temperature Difference." Once the time-dependent values for q_i are obtained, the results can be tabulated in the form of TETD values. Conventional practice has been to use the sol-air temperatures for dark-colored surfaces given in Table 15.3 with an indoor temperature fixed at 78 °F (25.6 °C). These tabulated values for TETD can be used for other similar walls and roofs for which detailed thermal data are unavailable and for different outdoor and indoor temperatures. Temperature corrections can be made by inspecting Eq. (15.30). For every degree increase in the daily mean sol-air temperature $t_{e,m}$ above 85 °F (29.4 °C), add one degree to the tabulated value of TETD. For every degree decrease in the indoor temperature t_i below 78 °F (25.6 °C), add one degree to the tabulated value of TETD. Corrections can also be made for changes in absorbed solar radiation to account for surface color, month of the year, and latitude.

TABLE 15.6 Constants for Eq. (15.30) for Various Flat-Roof Constructions. All include built-up roofing. Nos. 1–3 have an air space below layer B and above a metal lath and plaster ceiling. Nos. 6–8 have a 0.5 in. (1.3 cm) gypsum plaster ceiling under layer B.

No.	Description	U Btu/hr·ft²·°F (W/m²·°C) Winter	Summer	λ_1 Winter	Summer	Φ_1, deg Winter	Summer	λ_2 Winter	Summer	Φ_2, deg Winter	Summer
1	A: 2 in. (5 cm) cellular glass insulation B: 2 in. (5 cm) concrete	0.137 (0.778)	0.128 (0.727)	0.506	0.447	66.0	70.0	0.280	0.241	86.6	89.2
2	A: 2 in. (5 cm) cellular glass insulation B: 4 in. (10 cm) concrete	0.134 (0.761)	0.125 (0.710)	0.279	0.241	86.2	88.5	0.142	0.121	106.5	108.0
3	A: 4 in. (10 cm) concrete B: 2 in. (5 cm) cellular glass insulation	0.134 (0.761)	0.125 (0.710)	0.566	0.519	68.5	72.2	0.320	0.286	97.3	100.2
4	Same as No. 1 except no air space and no ceiling	0.159 (0.903)	0.150 (0.852)	0.788	0.674	44.5	54.1	0.536	0.414	70.3	78.5
5	Same as No. 2 except no air space and no ceiling	0.155 (0.880)	0.146 (0.829)	0.518	0.405	70.8	78.4	0.286	0.213	97.3	102.2
6	A: 4 in. (10 cm) expanded polystyrene insulation B: 4 in. (10 cm) concrete	0.0653 (0.371)	0.0637 (0.362)	0.444	0.348	76.2	82.7	0.240	0.018	101.8	105.8
7	A: 4 in. (10 cm) cellular glass insulation B: 4 in. (10 cm) concrete	0.0773 (0.439)	0.0838 (0.476)	0.444	0.345	89.5	93.8	0.222	0.171	127.3	127.5
8	A: 8 in. (20 cm) corkboard B: 4 in. (10 cm) concrete	0.0321 (0.182)	0.0317 (0.180)	0.164	0.125	204.3	211.0	0.035	0.026	297.5	301.9

An approximate method of obtaining TETD values is to assume that the variation in q_i lags the sol-air variation by an amount of time δ hours and that the magnitude is damped by an amount λ_m. This can be seen between the curves A and B on Fig. 15.2, where curve A is the actual heat gain and curve B is the heat gain with no time lag or damping. The approximate expression for TETD can be written

$$\text{TETD}_\theta \approx (t_{e_{ave}} - t_i) + \lambda_m(t_{e_\delta} - t_{e_{ave}}) \tag{15.33}$$

where $t_{e_{ave}}$ is the daily average sol-air temperature. The time lag δ ranges from about 1.5 hr for lightweight construction to 15 hr for heavy construction. The decrement or damping factor λ_m ranges from nearly 1.0 for lightweight construction to 0.03 for heavy construction. For the flat roof considered in Ex. 15.2, the time lag is approximately 4 hr and the decrement factor is about 0.7. The TETD method has been replaced by the CLTD (Cooling Load Temperature Difference) method in many hand calculations. The CLTD method is described in detail in a later section.

EXAMPLE 15.4

Use the results of Ex. 15.2 to compute the Total Equivalent Temperature Difference (TETD) values for the roof used in that example. Also, estimate the TETD values using a time lag, δ, of 4 hr, a damping factor, λ_m, of 0.7, and a constant indoor temperature of 78 °F.

Solution: The TETD values from Ex. 15.2 can be determined by dividing the hourly interior-surface heat-gain values by the roof overall U-value, 0.57 Btu/hr · ft² · °F. The heat-gain values and corresponding TETD values are listed in the second and third columns of Table 15.7.

TABLE 15.7 Comparison of TETD Values

Solar Time, hr	Interior Heat Gain, Ex. 15.2, Btu/hr · ft²	TETD, Example 15.2, °F	Estimated TETD, Eq. (15.33), °F
0	4.49	7.88	4.5
1	1.91	3.35	2.0
2	−0.34	−0.60	0.6
3	−2.20	−3.86	−0.3
4	−3.59	−6.30	−1.1
5	−4.40	−7.72	−2.1
6	−4.51	−7.91	−3.0
7	−3.81	−6.68	−3.6
8	−2.26	−3.96	−3.3
9	0.14	0.25	−1.6
10	3.26	5.72	2.1
11	6.88	12.1	7.7
12 noon	10.7	18.8	14.8
13	14.3	25.1	22.6
14	17.4	30.5	30.1
15	19.7	34.6	36.0
16	21.0	36.8	39.6
17	21.2	37.2	40.3
18	20.3	35.6	38.2
19	18.6	32.6	33.5
20	16.2	28.4	27.2
21	13.3	23.3	20.3
22	10.3	18.1	13.8
23	7.3	12.8	8.4

The estimated TETD values are computed using Eq. (15.33) and the values for t_e from column 13 of Table 15.4. The first three hourly computations are shown below with the results shown in the fourth column of Table 15.7.

$$t_{e_{ave}} = 91.6\ °F, \qquad t_i = 78\ °F$$

0 hr: $\text{TETD}_0 \approx (91.6 - 78) + 0.7(78.6 - 91.6)$

$$= 13.6 - 9.1 = 4.5\ °F$$

1 hr: $\text{TETD}_1 = 13.6 + 0.7(75.0 - 91.6) = 2.0\ °F$

2 hr: $\text{TETD}_2 = 13.6 + 0.7(73.0 - 91.6) = 0.6\ °F$

Comparing the third and fourth columns in Table 15.7 shows that the use of Eq. (15.33) provides a reasonable estimate for the TETD values and is easy to use. The maximum and minimum values are predicted to within 6 °F, and these occur within 1 hr of the more exact distribution given in column 3.

Another method of computing heat gain through walls and roofs is the *transfer-function method* developed by Stephenson and Mitalas [7]. The basic features of this method are outlined here. The transient heat conduction through a multilayered slab with linear boundary conditions [Eqs. (15.15), (15.17), (15.18)] can be solved using the Laplace-transform (*s*-transform) method. The solution takes the following matrix form:

$$\begin{bmatrix} \theta_e \\ \phi_o \end{bmatrix} = \begin{bmatrix} A, & B \\ C, & D \end{bmatrix} \cdot \begin{bmatrix} \theta_i \\ \phi_i \end{bmatrix} \tag{15.34}$$

where θ_e and θ_i are the Laplace transforms of the sol-air temperature t_e and t_i, respectively, ϕ_o and ϕ_i are the Laplace transforms of the outside- and inside-surface heat fluxes, and the matrix in the center becomes the transmission matrix $[H]$, which is solely a function of the wall or roof construction. The transmission matrix is the product of the individual transmission matrices for the various layers in the structure in the proper sequence

$$[H] = [H_1] \cdot [H_2] \cdot \cdots \cdot [H_M] \tag{15.35}$$

The individual layers can be either a pure resistance with no thermal storage, such as an air film, or a material with prescribed thickness and thermal diffusivity. An air film becomes

$$[H_{\text{film}}] = \begin{bmatrix} 1 & R_{\text{film}} \\ 0 & 1 \end{bmatrix} \tag{15.36}$$

where the film thermal resistance is the only parameter. A solid layer becomes

$$[H_{\text{layer}\,j}] = \begin{bmatrix} \cosh\sqrt{\tau_j\,s}, & \dfrac{R_j \sinh\sqrt{\tau_j\,s}}{\sqrt{\tau_j\,s}} \\ \dfrac{\sqrt{\tau_j\,s}\,\sinh\sqrt{\tau_j\,s}}{R_j}, & \cosh\sqrt{\tau_j\,s} \end{bmatrix} \tag{15.37}$$

where

$$\tau_j = \frac{L_j^2}{\alpha_j} \tag{15.38}$$

and L_j is the material thickness, α_j is the thermal diffusivity, and R_j is the thermal resistance of the layer. A property of each individual transmission matrix and the composite matrix $[H]$ is that the determinant must always equal 1.0. Therefore, Eq. (15.34) can be rearranged into the following form:

$$\begin{bmatrix} \phi_o \\ \phi_i \end{bmatrix} = \frac{1}{B} \begin{bmatrix} D, & -1 \\ 1, & -A \end{bmatrix} \cdot \begin{bmatrix} \theta_e \\ \theta_i \end{bmatrix} \tag{15.39}$$

in which the Laplace transforms of the two unknown surface heat-flux values, q_o and q_i, appear in the left-hand matrix. Usually we are concerned only with the internal-surface heat flux or heat gain, so that Eq. (15.39) can be reduced to

$$\phi_i = \frac{\theta_e}{B} - \frac{A\theta_i}{B} \tag{15.40}$$

The input functions, θ_e and θ_i, are obtained using the *Z-transform* method. Consider a continuous function of time $f(\tau)$ that is sampled periodically with time interval $\Delta\tau$. The resulting series of output sample pulses is transformed using the Laplace-transform technique, which results in the following transformed signal:

$$f(0) + f(\Delta\tau)e^{-s\Delta\tau} + f(2\,\Delta\tau)e^{-s2\Delta\tau} + \cdots \tag{15.41}$$

If we substitute Z for $e^{s\Delta\tau}$, Eq. (15.41) becomes

$$f(0)Z^0 + f(\Delta\tau)Z^{-1} + f(2\,\Delta\tau)Z^{-2} + \cdots \tag{15.42}$$

which is a polynomial that is called the "Z-transform" of the original function $f(\tau)$. If both the input and the output function, IN and OUT, of a system are expressed in terms of their Z-transforms, the ratio of these functions becomes the system "transfer function." This transfer function is solely a property of the system and not the boundary conditions or forcing functions. It can be written

$$K(Z) = \frac{\text{OUT } (Z)}{\text{IN } (Z)} = \frac{a_0 + a_1 Z^{-1} + a_2 Z^{-2} + \cdots}{b_0 + b_1 Z^{-1} + b_2 Z^{-2} + \cdots} \tag{15.43}$$

where a and b are the coefficients of the output and input Z-transforms. The Z-transform of the output, $O(Z)$, from any input function, $I(Z)$, can be written

$$O(Z) = K(Z) \cdot I(Z)$$

or

$$O(Z) \cdot \text{IN}(Z) = I(Z) \cdot \text{OUT}(Z) \tag{15.44}$$

Both sides of this equation are polynomials. Therefore the coefficients of like powers of Z must be equal. For example, for term m (i.e., $\tau = m\,\Delta\tau$),

$$O_m b_0 + O_{m-1} b_1 + O_{m-2} b_2 + \cdots + O_{m-p} b_p = I_m a_0 + I_{m-1} a_1 + I_{m-2} a_2 + \cdots + I_{m-j} a_j$$

or

$$
\begin{aligned}
O_m b_o = {} & I_m a_o + I_{m-1} a_1 + I_{m-2} a_2 + \cdots + I_{m-j} a_j \\
& - (O_{m-1} b_1 + O_{m-2} b_2 + \cdots + O_{m-p} b_p) \quad (15.45)
\end{aligned}
$$

It is useful to examine the form of Eq. (15.45). The output at time $m\,\Delta\tau$ equals the input at time $m\,\Delta\tau$, $I_m a_o$, inputs at previous times, $I_{m-1} a_1 + \cdots$, and outputs at previous times, $O_{m-1} b_1 + \cdots$. We can determine the outputs sequentially from previously determined inputs and outputs.

Writing Eq. (15.40) in the form of Eq. (15.45), we have an equation for instantaneous heat gain at time τ as a function of the current sol-air and indoor temperatures, previous sol-air and indoor temperatures, and previous values for instantaneous heat gain:

$$\text{IHG}_\tau = \sum_{n=0} (b_n t_{e,\tau-n\Delta\tau} - c_n t_{i,\tau-n\Delta\tau}) - \sum_{n=1} d_n \text{IHG}_{\tau-n\Delta\tau} \tag{15.46}$$

Usual practice is to assume a constant indoor temperature, so Eq. (15.46) is written

$$\text{IHG}_\tau = \sum_{n=0} b_n t_{e,\tau-n\Delta\tau} - \sum_{n=1} d_n \text{IHG}_{\tau-n\Delta\tau} - t_i \sum_{n=0} c_n \tag{15.47}$$

The coefficients that appear in Eq. (15.47), b_n, d_n, and $\sum_{n=0} c_n$, have been determined for a large number of typical wall and roof constructions with time steps of 1 hr and have been tabulated in the *ASHRAE Handbook of Fundamentals* [8]. Some representative values are listed in Tables 15.8 for walls and Table 15.9 for horizontal roofs. Detailed description of these wall and roof construction code letters are given in Table 15.10.

The overall heat-transmission coefficient, or U-value, is also listed for each wall and roof construction. If the actual construction is similar to one of the tabulated constructions but the overall heat-transmission coefficient differs, the b_n and $\sum c_n$ coefficients should be corrected by multiplying the tabulated values by the ratio of the actual U-value divided by the tabulated U-value. The d_n coefficients do not change, as they are always used to multiply previous heat-gain results. The last two columns in Tables 15.8 and 15.9 give the time lag, δ, and decrement factor, λ_m, for each construction. These can be used in Eq. (15.33) to compute approximate values for TETD, which can then be used in Eq. (15.32) to estimate the instantaneous heat gain. The actual U-value should always be used when known rather than the value from the tables.

A situation often encountered is that of a building with more than one exterior wall orientation, where each orientation has the same construction. We can combine all these walls into a single calculation if the sol-air temperature used in Eq. (15.47) is replaced by the area weighted average sol-air temperature for all of these walls.

EXAMPLE 15.5

Use the transfer-function method to compute the hourly instantaneous heat gain for a 24 hr period through a flat roof of a warehouse. The roof has the construction given by #R4 in Table 15.9. Use the sol-air temperature distribution given for July 21, 40° N latitude, in Table 15.3 for dark-colored surfaces and assume the indoor temperature remains constant at 25 °C.

Solution: As the construction is identical to #R4 in Table 15.9, the coefficients will be used unchanged. Sol-air temperature and heat-gain results are needed for the 3 hr prior to the hour being computed. The sol-air temperatures are assumed to be periodic, so the temperature 1 hr before 0100 equals the value at 2400, the value 2 hr before 0100 equals the value at 2300, and so on. When the calculations first begin, there are no known previous values for the heat

TABLE 15.8 Wall-Conduction Transfer-Function Coefficients, Time Lag and Decrement Factor

Construction		Coefficients b_n and d_n						Σc_n	U	δ	λ_m
		$n = 0$	1	2	3	4	5				
#W1 Wood Frame Wall	b	0.00089	0.03097	0.05456	0.01224	0.00029		0.098947	0.1345	3.21	0.91
E0 E1 B14 A1 A0	d		−0.93389	0.27396	−0.02561	0.00014					
#W2 Metal Curtain Wall	b	0.04361	0.19862	0.04083	0.00032			0.283372	0.3724	1.30	0.98
E0 A3 B1 B13 A3 A0	d		−0.24072	0.00168							
#W3 Insulated Concrete Block	b	0.00002	0.00349	0.01641	0.01038	0.00105	0.00001	0.031356	0.61945	7.1	0.37
E0 E1 C8 B6 A1 A0	d		−1.52480	0.67146	−0.09844	0.00239					
#W4 203.2 mm (8 in.) Concrete	b	0.00009	0.01125	0.04635	0.02654	0.00249	0.00003	0.086751	1.9235	7.3	0.33
E0 E1 B1 C10 A1 A0	d		−1.51660	0.64261	−0.08382	0.00289	−0.00001				
#W5 Face Brick, Insulation, Concrete	b	0.00015	0.01152	0.03411	0.01326	0.00074		0.059779	1.4277	7.2	0.28
E0 A2 C5 B19 A6 A0	d		−1.41350	0.48697	−0.03218	0.00057					

SOURCE: Abstracted from *ASHRAE Fundamentals 1993*, pp. 26.26–26.27.
Units on U, b_n, and c_n are W/m$^2 \cdot$ °C; multiply by 0.1761 to convert to Btu/hr $\cdot$ ft$^2 \cdot$ °F. Values of d_n are dimensionless.
Time interval is 1.0 hr; time lag, δ, is in hours.

TABLE 15.9 Roof-Conduction Transfer-Function Coefficients, Time Lag and Decrement Factor

Construction		Coefficients b_n and d_n						$\Sigma\, c_n$	U	δ	λ_m
		$n = 0$	1	2	3	4	5				
#R1 152.4 mm (6 in.) Concrete, 3.8 mm (0.15 in.) Insulation	b	0.00559	0.11007	0.11826	0.01243	0.00008		0.246431	2.40935	5.5	0.47
E0 B16 C13 E3 E2 A0	d		-1.10230	0.20750	-0.00287						
#R2 101.6 mm (4 in.) Concrete, 7.6 mm (0.3 in.) Insulation	b	0.01647	0.17849	0.12003	0.00682	0.00001		0.321826	2.10793	4.6	0.6
E0 C5 B17 E3 E2 A0	d		-0.97905	0.13444	-0.00272						
#R3 101.6 mm (4 in.) Concrete, Insulation & Ceiling	b	0.00006	0.00376	0.00924	0.00277	0.00011		0.015936	0.50920	7.2	0.16
E0 E5 E4 C5 B6 E3 E2 A0	d		-1.2435	0.28741	-0.01274	0.00009					
#R4 Steel Deck and 85 mm (3.3 in.) Insulation	b	0.02766	0.19724	0.07752	0.00203			0.304451	0.45569	1.6	0.97
E0 A3 B25 E3 E2 A0	d		-0.35451	0.02267	-0.00005						
#R5 Attic Roof with 152 mm (6 in.) Insulation	b	0.00002	0.00371	0.01923	0.01361	0.00164	0.00003	0.038233	0.24311	4.9	0.82
E0 E1 B15 E4 B7 A0	d		-1.34660	0.59384	-0.09295	0.00296	-0.00001				

Source: Abstracted from *ASHRAE Fundamentals 1993*, pp. 26.21–26.22.
Units on U, b_n, and c_n are W/m$^2 \cdot$ °C; multiply by 0.1761 to convert to Btu/hr $\cdot$ ft$^2 \cdot$ °F. Values of d_n are dimensionless.
Time interval is 1.0 hr; time lag, δ, is in hours.

TABLE 15.10 Thermal Properties and Code Numbers of Wall and Roof Materials Used in Tables 15.8 and 15.9

L, mm	k, W/m·°C	ρ, kg/m³	c, kJ/kg·°C	R, m²·°C/W	m, kg/m²	Description	Code Number	L, in.	k, Btu/hr·ft·°F	ρ, lbm/ft³	c, Btu/lbm·°F	R, hr·ft²·°F/Btu	m, lbm/ft²
				0.059		Outside air film	A0					0.333	
25	0.692	1858	0.84	0.037	47.3	Stucco, wood siding, etc.	A1	1.0	0.4	116	0.20	0.21	9.7
100	1.333	2002	0.92	0.076	203.5	Face brick	A2	4.0	0.77	125	0.22	0.43	41.7
2	45	7689	0.42	0.000	11.7	Steel roof deck	A3	0.06	26.0	480	0.10	0.0	2.40
13	0.415	1249	1.09	0.031	16.1	Finish	A6	0.5	0.24	78	0.26	0.17	3.3
				0.160		Vertical air space	B1					0.91	
51	0.043	91	0.84	1.173	4.9	Insulation	B6	2.0	0.025	5.7	0.2	6.67	1.0
25	0.121	593	2.51	1.760	15.1	Wood	B7	1.0	0.07	37	0.6	10	3.1
100	0.043	91	0.84	2.347	9.3	Insulation	B13	4.0	0.025	5.7	0.2	13.3	1.90
125	0.043	91	0.84	2.933	11.7	Insulation	B14	5.0	0.025	5.7	0.2	16.67	2.4
100	1.731	2243	0.84	0.059	227.9	Heavyweight concrete	C5	4.0	1.00	140.0	0.2	0.333	46.7
200	1.038	977	0.84	0.196	198.6	H.W. concrete block	C8	8.0	0.6	61.0	0.2	1.11	40.7
200	1.731	2243	0.84	0.117	455.8	Heavyweight concrete	C10	8.0	1.00	140.0	0.2	0.67	93.4
150	1.731	2243	0.84	0.088	341.6	Heavyweight concrete	C13	6.0	1.0	140.0	0.2	0.5	70.0
				0.121		Inside air film	E0					0.69	
20	0.727	1602	0.84	0.026	30.7	Plaster or gypsum	E1	0.75	0.42	100	0.2	0.15	6.3
12	1.436	881	1.67	0.009	11.2	Slag or stone	E2	0.5	0.83	55	0.40	0.05	2.3
10	0.190	1121	1.67	0.050	10.7	Felt and membrane	E3	0.375	0.11	70	0.40	0.29	2.2
				0.176		Ceiling air space	E4					1.0	
19	0.061	481	0.84	0.314	9.3	Acoustic tile	E5	0.75	0.035	30	0.20	1.79	1.9
150	0.043	91	0.84	3.52	14.2	Insulation	B15	6.0	0.025	5.7	0.20	20.0	2.9
4	0.043	91	0.84	0.088	0.5	Insulation	B16	0.15	0.025	5.7	0.20	0.5	0.1
8	0.043	91	0.84	0.176	0.5	Insulation	B17	0.3	0.025	5.7	0.20	1.0	0.1
15	0.043	91	0.84	0.352	1.5	Insulation	B19	0.6	0.025	5.7	0.20	2.0	0.3
85	0.043	91	0.84	1.936	7.81	Insulation	B25	3.3	0.025	5.7	0.20	11.0	1.6

SOURCE: Abstracted from *ASHRAE Fundamentals 1993*, p. 26.19.

gain, so all previous values are assumed to be zero. This lack of information results in incorrect values for the heat gain near the beginning of the calculation. However, as the calculations proceed hour by hour, this error is gradually reduced. Eventually after several successive 24-hr simulations, the results from a given time of day will agree with the results 24 hr earlier to some degree of tolerance. At this point the calculations can be terminated and the results for the last 24-hr period retained as the final solution.

The first hour of calculation, 0100, becomes

$$IHG_{0100} = \begin{bmatrix} b_0 t_{e,0100} + \\ b_1 t_{e,2400} + \\ b_2 t_{e,2300} + \\ b_3 t_{e,2200} \end{bmatrix} - \begin{bmatrix} d_1 IHG_{2400} + \\ d_2 IHG_{2300} + \\ d_3 IHG_{2200} \end{bmatrix} - t_i \sum_{n=0} c_n$$

Substituting the values into this equation gives

$$IHG_{0100} = \begin{bmatrix} 0.02766 \times 20.5 + \\ 0.19724 \times 21.1 + \\ 0.07752 \times 22.2 + \\ 0.00203 \times 23.3 \end{bmatrix} - \begin{bmatrix} -0.35451 \times 0 + \\ 0.02267 \times 0 + \\ -0.00005 \times 0 \end{bmatrix} - 25 \times 0.304451$$

$$= -1.114 \, W/m^2$$

The calculation for the second hour becomes

$$IHG_{0200} = \begin{bmatrix} 0.02766 \times 20.5 + \\ 0.19724 \times 20.5 + \\ 0.07752 \times 21.1 + \\ 0.00203 \times 22.2 \end{bmatrix} - \begin{bmatrix} -0.35451 \times -1.114 + \\ 0.02267 \times \quad 0 \quad + \\ -0.00005 \times \quad 0 \end{bmatrix} - 25 \times 0.304451$$

$$= -1.715 \, W/m^2$$

Table 15.11 shows the results through three successive 24-hr periods. This roof construction does not have much thermal mass (the decrement factor is 0.97 and the time lag is only 1.6 hr). Therefore, only 6 hr of computation are required before the poor initial values of IHG no longer affect the solution. From hour 0600 onward, the solution is periodic to three decimal places. Constructions with much more thermal mass may require several days before the solution becomes periodic. The total instantaneous heat gain from the roof into the building can be obtained by multiplying the values listed for either day 2 or day 3 in Table 15.11 by the total roof area of 10,000 m². It is best to perform these calculations on a computer, as any error in hand calculations at a particular time will affect the results at later times also.

TABLE 15.11 Results of Ex. 15.5

b-Values	d-Values	i-Value	Sum of c_n	Room Temp.
0.02766	1	0	0.304451	25
0.19724	−0.35451	1		
0.07752	0.02267	2		
0.00203	−0.00005	3		
0	0	4		
0	0	5		
0	0	6		

(continued)

TABLE 15.11 (cont.)

		Heat Gain				
Time	Sol-Air Temp. t_e	IHG_τ Day 1	Time	IHG_τ Day 2	Time	IHG_τ Day 3
19	26.6	0		8.671		8.671
20	25.5	0		4.043		4.043
21	24.4	0		1.467		1.467
22	23.3	0		0.305		0.305
23	22.2	0		−0.383		−0.383
24	21.1	0		−0.936		−0.936
1	20.5	−1.114	25	−1.437	49	−1.437
2	20.5	−1.715	26	−1.808	50	−1.808
3	20.0	−1.965	27	−1.991	51	−1.991
4	19.4	−2.157	28	−2.164	52	−2.164
5	19.4	−2.376	29	−2.378	53	−2.378
6	27.2	−2.282	30	−2.282	54	−2.282
7	38.8	−0.385	31	−0.385	55	−0.385
8	50.0	3.488	32	3.488	56	3.488
9	60.0	8.218	33	8.218	57	8.218
10	68.3	12.901	34	12.901	58	12.901
11	74.4	17.058	35	17.058	59	17.058
12	77.7	20.384	36	20.384	60	20.384
13	77.7	22.610	37	22.610	61	22.610
14	74.4	23.501	38	23.501	62	23.501
15	68.3	22.953	39	22.953	63	22.953
16	59.4	21.034	40	21.034	64	21.034
17	48.8	17.838	41	17.838	65	17.838
18	36.6	13.618	42	13.618	66	13.618
19	26.6	8.671	43	8.671	67	8.671
20	25.5	4.043	44	4.043	68	4.043
21	24.4	1.467	45	1.467	69	1.467
22	23.3	0.305	46	0.305	70	0.305
23	22.2	−0.383	47	−0.383	71	−0.383
24	21.1	−0.936	48	−0.936	72	−0.936

The transfer-function method for estimating instantaneous heat gain can also be used to predict the instantaneous cooling load. This is discussed in Chapter 16. The transfer-function method is also used in building simulation programs. Actual weather data can be used to provide the necessary temperature inputs rather than clear-sky conditions. After the first few hours (or days) of simulation, the results will provide accurate values of IHG, although they will not be periodic as in Ex. 15.5.

Although a large number of wall and roof constructions have been analyzed and the resulting transfer-function coefficients have been tabulated, one can encounter different constructions which have not been previously analyzed, or one may wish to have more thermal information about the structure than the interior-surface heat flux—for example, the exterior-surface temperature or the temperature distribution throughout the structure. For these reasons, numerical solutions are employed to analyze the transient thermal response of a wall or roof construction. One method that can be used to simulate the wall or roof is shown in Fig. 15.3. This is a one-dimensional approximation. The thermal

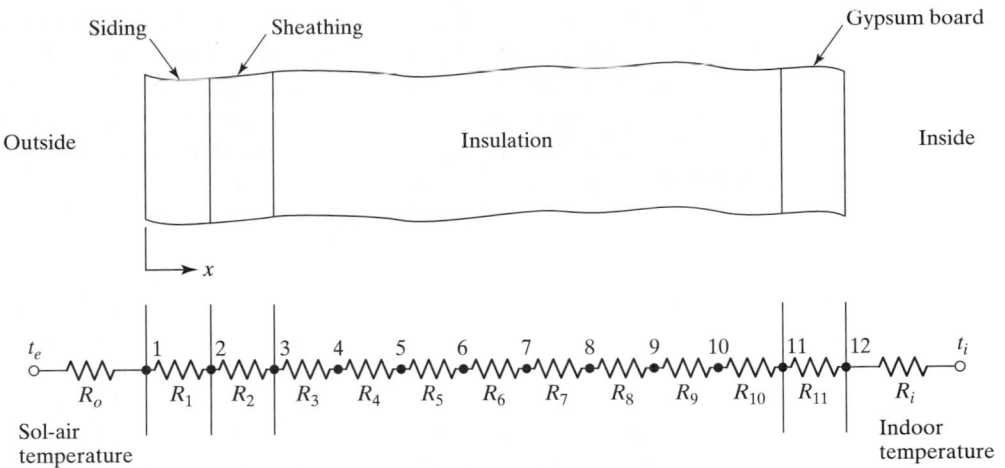

Figure 15.3 One-dimensional thermal network for transient thermal simulation of a multilayer wall.

resistances R_i and R_0 are the reciprocals of the respective surface convective-radiative film coefficients. The interior resistances are computed as

$$R_j = \frac{\Delta x_j}{k_j} \tag{15.48}$$

where Δx_j is the spacing between adjacent temperature nodes and k_j is the thermal conductivity of the material between the nodes. This method assumes that all heat is transferred internally by conduction, so any convective or radiative transport must be approximated by conduction through a material with an appropriate equivalent thermal conductivity. Each temperature node has a thermal capacitance associated with it due to the mass and heat capacity of the material surrounding the node:

$$C_j = (\Delta x_j \rho_j c_j + \Delta x_{j-1} \rho_{j-1} c_{j-1})/2 \tag{15.49}$$

Surface air films and interior air spaces are assumed to have negligible heat capacity.

An energy balance can be set up for each interior node:

$$\underbrace{\frac{t_{j+1}^k - t_j^k}{R_j}}_{\substack{\text{conduction} \\ \text{from right}}} - \underbrace{\frac{t_j^k - t_{j-1}^k}{R_{j-1}}}_{\substack{\text{conduction} \\ \text{to left}}} = \underbrace{\frac{(t_j^k - t_j^{k-1})C_j}{\Delta \tau}}_{\substack{\text{net change of} \\ \text{stored energy}}} \tag{15.50}$$

The superscript k indicates the value at the present time, $k - 1$ indicates the value at the previous time, and $\Delta \tau$ indicates the time difference between the transient solutions, or the time step used. Equation (15.50) is written for each interior temperature node in the wall or roof. The bounding temperatures, indoor and sol-air temperatures, are presumed known as functions of time. The temperature distribution through the wall or roof is given some initial distribution as a starting point. Then all the temperatures at the interior nodes are solved at each time step. A good method for solution is to use a tridiagonal matrix technique on a computer, as every interior node is affected only by the two neighboring nodes. This solution technique is not dependent on the value of the time step chosen.

Previous experience has shown that 24 hr of simulation may be necessary before a fairly repeatable diurnal response is obtained. Thick, massive walls may require more time. Thin layers such as papers, films, and paint layers should be neglected or combined with other material layers. Once a repeatable, diurnal solution is obtained, the temperatures at the nodes may be plotted as a function of time. The interior-surface heat flux, or instantaneous heat gain, may be determined as

$$\text{IHG}^k = h_i (t_{w,i}^k - t_i^k) \tag{15.51}$$

More accurate results are obtained when smaller time steps are used. Accuracy is also improved by dividing the structure into smaller segments. The time step and number of nodes is usually a compromise between accuracy and computer run time, as a smaller time step and an increase in the number of nodes both increase the computer run time. Whenever possible it is recommended to obtain two solutions, each using a different time step and node distribution, to obtain an estimate of the numerical error.

Another solution method is to rewrite Eq. (15.50) as

$$\frac{t_{j+1}^{k-1} - t_j^k}{R_j} - \frac{t_j^k - t_{j-1}^{k-1}}{R_{j-1}} = \frac{(t_j^k - t_j^{k-1})C_i}{\Delta\tau} \tag{15.52}$$

Here, only one temperature, t_j^k, can be computed directly in terms of the previous value at that node and previous values at the adjacent nodes. One can then update the temperature of each node sequentially in a node-by-node solution scheme. However, the most recent value should be used whenever possible. For example, when sweeping from left to right in Fig. 15.3, the node to the left should always contain the most recent computed temperature, so the second term in Eq. (15.52) is actually $(t_j^k - t_{j-1}^k)/R_{j-1}$. This method is amenable to hand calculation and simple computer programming. However, the solution becomes unstable if the time step becomes too large. The following example illustrates the use of Eq. (15.52).

EXAMPLE 15.6

Use an explicit finite-difference method to simulate the periodic heat gain through the roof used in Ex. 15.5. Use a time step of 1 hr and English units.

Solution: The roof is divided into seven control volumes with the nodes spaced as shown in Fig. 15.4. Using the roof construction information given in Table 15.9 and the material property values given in Table 15.10, the following resistance and capacitance values can be determined:

$$R_0 = 0.333 \; \text{hr} \cdot \text{ft}^2 \cdot \text{°F/Btu}$$

$$R_{12} = L_{E2}/k_{E2} = (0.5 \; \text{in.}/0.83 \; \text{Btu/hr} \cdot \text{ft} \cdot \text{°F})/12 \; \text{in./ft}$$

$$= 0.05 \; \text{hr} \cdot \text{ft}^2 \cdot \text{°F/Btu}$$

$$R_{23} = L_{E3}/k_{E3} = (0.375 \; \text{in.}/0.11 \; \text{Btu/hr} \cdot \text{ft}^2 \cdot \text{°F})/12 \; \text{in./ft}$$

$$= 0.284 \; \text{hr} \cdot \text{ft}^2 \cdot \text{°F/Btu}$$

$$R_{34} = R_{45} = R_{56} = R_{67} = (L_{B25}/4)/k_{B25}$$

$$= [(3.3 \; \text{in.}/4)/0.025 \; \text{Btu/hr} \cdot \text{ft} \cdot \text{°F}]/12 \; \text{in./ft}$$

$$= 2.75 \; \text{hr} \cdot \text{ft}^2 \cdot \text{°F/Btu}$$

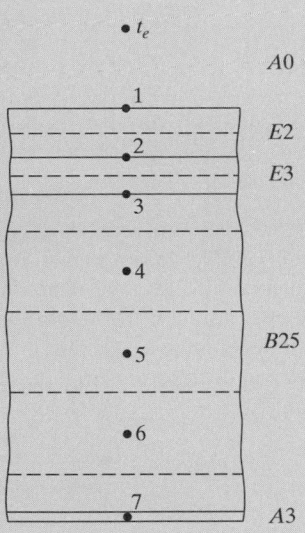

*A*0

*E*2

*E*3

*B*25

*A*3

*E*0

Figure 15.4 Nodal and control-volume distribution for the roof in Ex. 15.6.

The resistance of the steel deck is neglected. It could be included in the inside air film resistance if desired.

$$R_i = 0.69 \text{ hr} \cdot \text{ft}^2 \cdot {}^\circ\text{F/Btu}$$

$$C_1 = (L_{E2}/2)(\rho_{E2})(c_{E2})$$

$$= (0.5 \text{ in.}/2)(55 \text{ lbm/ft}^3)(0.40 \text{ Btu/lbm} \cdot {}^\circ\text{F})/(12 \text{ in./ft})$$

$$= 0.46 \text{ Btu/ft}^2 \cdot {}^\circ\text{F}$$

$$C_2 = (L_{E2}/2)(\rho_{E2})(c_{E2}) + (L_{E3}/2)(\rho_{E3})(c_{E3})$$

$$= 0.46 + (0.375 \text{ in.}/2)(70 \text{ lbm/ft}^3)(0.40 \text{ Btu/lbm} \cdot {}^\circ\text{F})/(12 \text{ in./ft})$$

$$= 0.90 \text{ Btu/ft}^2 \cdot {}^\circ\text{F}$$

$$C_3 = (L_{E3}/2)(\rho_{E3})(c_{E3}) + (L_{B25}/8)(\rho_{B25})(c_{B25})$$

$$= 0.44 + (3.3 \text{ in.}/8)(5.7 \text{ lbm/ft}^3)(0.2 \text{ Btu/lbm} \cdot {}^\circ\text{F})/(12 \text{ in./ft})$$

$$= 0.48 \text{ Btu/ft}^2 \cdot {}^\circ\text{F}$$

$$C_4 = C_5 = C_6 = (L_{B25}/4)(\rho_{B25})(c_{B25})$$

$$= (3.3 \text{ in.}/4)(5.7 \text{ lbm/ft}^3)(0.2 \text{ Btu/lbm} \cdot {}^\circ\text{F})/(12 \text{ in./ft})$$

$$= 0.078 \text{ Btu/ft}^2 \cdot {}^\circ\text{F}$$

The thermal mass of the steel deck is included in the thermal mass of node 7.

$$C_7 = (L_{B25}/8)(\rho_{B25})(c_{B25}) + (m_{A3})(c_{A3})$$

$$= (3.3 \text{ in.}/8)(5.7 \text{ lbm/ft}^3)(0.2 \text{ Btu/lbm} \cdot {}^\circ\text{F})/(12 \text{ in./ft})$$

$$+ (2.40 \text{ lb/ft}^2)(0.10 \text{ Btu/lbm} \cdot {}^\circ\text{F})$$

$$= 0.28 \text{ Btu/ft}^2 \cdot {}^\circ\text{F}$$

The sol-air temperature for a horizontal dark-colored surface from Table 15.3 is used as the outdoor boundary condition. The indoor temperature is fixed at 78 °F. All nodes in the roof must be given an initial temperature. A good approximation is the arithmetic mean between the daily average sol-air temperature and the indoor temperature, (107 °F + 78 °F)/2 = 92.5 °F.

The calculations will be initiated at 0000 hr, or $k = 0$. The results at the end of the first hour, $k = 1$, are shown here in detail using Eq. (15.51).

Node #1:
$$\frac{t_2^0 - t_1^1}{R_{12}} - \frac{t_1^1 - t_e^1}{R_0} = \frac{(t_1^1 - t_1^0)C_1}{\Delta\tau}$$

Solving for t_1^1,

$$t_1^1 = \frac{t_2^0/R_{12} + t_e^1/R_0 + t_1^0 C_1/\Delta\tau}{1/R_{12} + 1/R_0 + C_1/\Delta\tau}$$

$$= \frac{92.5/0.05 + 69/0.333 + (92.5)(0.46)/1.0}{1/0.05 + 1/0.333 + 0.46/1.0} = 89.5 \text{ °F}$$

Node #2:
$$\frac{t_3^0 - t_2^1}{R_{23}} - \frac{t_2^1 - t_1^1}{R_{12}} = \frac{(t_2^1 - t_2^0)C_2}{\Delta\tau}$$

$$t_2^1 = \frac{t_3^0/R_{23} + t_1^1/R_{12} + t_2^0 C_2/\Delta\tau}{1/R_{23} + 1/R_{12} + C_2/\Delta\tau}$$

$$= \frac{92.5/0.284 + 89.5/0.05 + (92.5)(0.90)/1.0}{1/0.284 + 1/0.05 + 0.90/1.0} = 90.0 \text{ °F}$$

Likewise for Nodes #3, 4, 5, 6, and 7:

$$t_3^1 = 90.5 \text{ °F}, \quad t_4^1 = 91.6 \text{ °F}, \quad t_5^1 = 92.1 \text{ °F}, \quad t_6^1 = 92.3 \text{ °F}, \quad t_7^1 = 82.4 \text{ °F}$$

The solution now repeats beginning with node 1 at the next time step, 0200 hr or $k = 2$.

Node #1:
$$t_1^2 = \frac{t_2^1/R_{12} + t_e^2/R_0 + t_1^1 C_1/\Delta\tau}{1/R_{12} + 1/R_0 + C_1/\Delta\tau}$$

$$= \frac{90.0/0.05 + 69/0.333 + (89.5)(0.46)/1.0}{1/0.05 + 1/0.333 + (0.46)/1.0} = 87.3 \text{ °F}$$

The solution continues in this manner for several 24-hr time periods. The results from the first two days are given in Table 15.12. By the end of the second day, the results are approximately equal to the results at the end of the first day, which indicates that only 2 or 3 days of simulation are required to achieve a periodic solution. This is approximately the same amount of simulation time required using the transfer-function method in Ex. 15.5. Table 15.13 shows the periodic inside-surface temperature and instantaneous heat gain.

TABLE 15.12 Temperature Distributions through the Roof of Ex. 15.6 for the First 48 Hours of Simulation

RESULTS FROM DAY 1

NODE #	INITIAL TEMP	TEMP HR 1	TEMP HR 2	TEMP HR 3	TEMP HR 4	TEMP HR 5	TEMP HR 6	TEMP HR 7	TEMP HR 8	TEMP HR 9	TEMP HR 10	TEMP HR 11	TEMP HR 12
1	92.5	89.5	87.3	85.3	83.5	81.8	82.1	84.8	89.2	94.9	101.5	108.3	114.9
2	92.5	90.0	87.9	85.9	84.0	82.3	82.3	84.4	88.3	93.6	99.8	106.5	113.1
3	92.5	90.5	88.5	86.5	84.7	82.9	82.5	84.1	87.4	92.2	97.9	104.2	110.6
4	92.5	91.6	90.4	89.1	86.9	84.7	83.4	83.4	84.7	87.1	90.4	94.5	98.8
5	92.5	92.1	91.4	88.6	86.1	84.1	82.7	82.2	82.5	83.7	85.5	87.9	90.7
6	92.5	92.3	87.5	84.7	83.0	81.7	80.8	80.4	80.5	81.0	82.0	83.2	84.7
7	92.5	82.4	80.2	79.5	79.1	78.8	78.6	78.5	78.5	78.6	78.8	79.0	79.3

TEMP HR 13	TEMP HR 14	TEMP HR 15	TEMP HR 16	TEMP HR 17	TEMP HR 18	TEMP HR 19	TEMP HR 20	TEMP HR 21	TEMP HR 22	TEMP HR 23	TEMP HR 24
120.7	125.0	127.6	128.1	126.5	122.8	117.7	113.4	109.4	105.7	102.1	98.7
118.5	123.5	126.5	127.5	126.5	123.3	118.6	114.3	110.3	106.5	103.0	99.5
116.3	121.0	124.2	125.7	125.2	122.6	118.5	114.5	110.6	106.9	103.4	100.0
103.1	106.9	109.9	111.9	112.7	112.0	110.2	107.8	105.2	102.6	100.0	97.4
93.6	96.3	98.6	100.3	101.3	101.4	100.6	99.4	97.8	96.1	94.3	92.6
86.3	87.8	89.1	90.2	90.8	91.0	90.6	90.0	89.2	88.3	87.3	86.3
79.6	79.9	80.2	80.4	80.5	80.6	80.5	80.4	80.3	80.1	79.9	79.7

(continued)

TABLE 15.12 (cont.)

RESULTS FROM DAY 2

NODE #	FROM DAY 1	TEMP HR 25	TEMP HR 26	TEMP HR 27	TEMP HR 28	TEMP HR 29	TEMP HR 30	TEMP HR 31	TEMP HR 32	TEMP HR 33	TEMP HR 34	TEMP HR 35	TEMP HR 36
1	98.7	95.6	92.9	90.3	87.9	85.8	85.7	88.0	92.0	97.4	103.7	110.3	116.6
2	98.5	96.4	93.6	91.0	88.6	86.4	85.9	87.6	91.2	96.1	102.1	108.5	114.7
3	100.0	96.9	94.1	91.5	89.1	86.9	86.1	87.3	90.3	94.7	100.2	106.2	112.2
4	97.4	95.0	92.7	90.6	88.6	86.8	85.6	85.6	86.8	89.1	92.3	96.1	100.3
5	92.6	90.9	89.2	87.7	86.2	84.9	83.9	83.6	83.9	85.1	86.8	89.2	91.8
6	86.3	85.4	84.5	83.6	82.8	82.0	81.4	81.2	81.3	81.8	82.7	83.9	85.3
7	79.7	79.5	79.3	79.2	79.0	78.8	78.7	78.6	78.7	78.7	78.9	79.1	79.4

TEMP HR 37	TEMP HR 38	TEMP HR 39	TEMP HR 40	TEMP HR 41	TEMP HR 42	TEMP HR 43	TEMP HR 44	TEMP HR 45	TEMP HR 46	TEMP HR 47	TEMP HR 48
122.1	126.2	128.6	128.9	127.2	123.4	118.2	113.8	109.8	106.0	102.4	99.0
120.3	124.7	127.4	128.3	127.1	123.8	119.1	114.8	110.7	106.9	103.3	99.8
117.8	122.2	125.2	126.5	125.8	123.2	119.0	114.9	110.9	107.2	103.7	100.3
104.3	108.0	110.9	112.7	113.3	112.6	110.6	108.2	105.6	102.9	100.3	97.7
94.5	97.1	99.3	101.0	101.8	101.8	101.0	99.7	98.1	96.3	94.5	92.8
86.8	88.3	89.5	90.5	91.1	91.2	90.9	90.2	89.4	88.4	87.4	86.4
79.7	80.0	80.3	80.5	80.6	80.6	80.6	80.5	80.3	80.1	79.9	79.7

TABLE 15.13 Periodic Interior Surface Temperatures and Heat Gains from Ex. 15.6

Solar Time, hr	Internal Surface Temperature, °F	Instantaneous Heat Gain, Btu/hr · ft²
0100	79.5	1.04
0200	79.3	0.90
0300	79.2	0.83
0400	79.0	0.69
0500	78.8	0.55
0600	78.7	0.48
0700	78.6	0.48
0800	78.7	0.48
0900	78.7	0.55
1000	78.9	0.62
1100	79.1	0.83
1200	79.4	0.97
1300	79.7	1.17
1400	80.0	1.38
1500	80.3	1.59
1600	80.5	1.73
1700	80.6	1.79
1800	80.6	1.79
1900	80.6	1.79
2000	80.5	1.73
2100	80.3	1.59
2200	80.1	1.45
2300	79.9	1.31
2400	79.7	1.17

The results of Ex. 15.6 show that the minimum heat gain is 0.48 Btu/hr · ft² (1.51 W/m²) which occurs near 0700 hrs and the maximum heat gain is 1.79 Btu/hr · ft² (5.64 W/m²) which occurs near 1800 hrs. These magnitudes and times differ from those of Ex. 15.5. The main reason is the choice of the 1-hr time step. Use of a smaller time step such as 5 or 10 min usually results in more accurate simulations. Adding more nodes with thinner control volumes also improves the accuracy of these solutions. The finite-difference method does provide the complete internal temperature distribution and both surface temperatures, which may be useful in determining interior heat transfer and thermal degradation.

15.5 INSTANTANEOUS HEAT GAIN THROUGH FENESTRATIONS

A *fenestration* is defined as an aperture in a building which is covered by a glazing material that transmits a portion of the incident visible radiation. Examples include windows, skylights, and glass doors. Unlike walls and roofs, fenestration materials are thin and have very little thermal mass. Therefore, the materials respond quickly to changing environmental conditions. A steady-state energy balance is adequate for most applications in which the thermal mass is neglected.

Several parameters affect the amount of energy that passes through a fenestration:

1. Incident solar radiation, $I_T = I_D + I_d + I_R$
2. Properties of the glazing material, including films and coatings
3. Interior shading devices
4. Outdoor and indoor temperatures
5. Outside and inside surface film coefficients

We will begin by considering the properties of glazing materials in the solar spectrum that must be known before an energy balance on the glazing can be developed.

The most important example of a glazing material in building construction is glass. We will study in some detail the disposal of solar radiation incident upon glass, much of our discussion being based upon the treatise by Parmelee [9].

Figure 15.5 shows the disposal of a quantity of monochromatic direct solar radiation I_λ incident upon a single sheet of glass of thickness L. Part of the incident radiation is reflected from the front surface and part is absorbed by the glass material. Because of successive internal reflections, the reflected, absorbed, and transmitted radiation quantities are given by the sums of infinite series. Let r be the fraction of each component reflected, and a be the fraction of each component available after absorption. The total monochromatic transmissivity τ_λ is given by

$$\tau_\lambda = (1 - r)^2 a + r^2(1 - r)^2 a^3 + r^4(1 - r)^2 a^5 + \cdots$$

Since this is a convergent geometric series, we have

$$\tau_\lambda = \frac{(1 - r)^2 a}{1 - r^2 a^2} \tag{15.53}$$

In a similar way, we obtain the total monochromatic reflectivity ρ_λ as

$$\rho_\lambda = r + \frac{r(1 - r)^2 a^2}{1 - r^2 a^2} \tag{15.54}$$

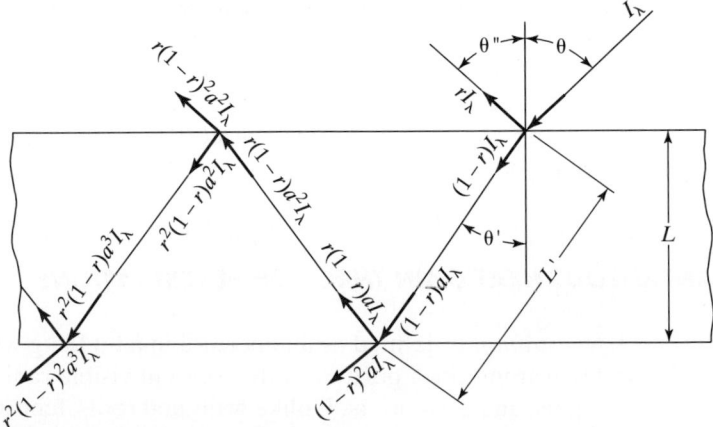

Figure 15.5 Multiple reflections of direct solar radiation by a single sheet of glass.

TABLE 15.14 Extinction Coefficient K for Various Types of Glass for Solar Radiation

Type of Glass	K, in.$^{-1}$	K, m^{-1}
1. Water white	0.10	4
2. Double-strength, A-quality	0.5	20
3. Heat-absorbing	3.3	130

and since $\alpha_\lambda = 1 - \tau_\lambda - \rho_\lambda$, we have

$$\alpha_\lambda = 1 - r - \frac{(1 - r)^2 a}{1 - ra} \tag{15.55}$$

To evaluate Eqs. (15.53) to (15.55) we must know the quantities r and a. We will first consider the absorption coefficient a. We assume that the absorbed radiation is proportional to the intensity of incident radiation and to the length of path of the refracted beam. Thus

$$-dI_\lambda = KI_\lambda dL'$$

and

$$a = \frac{I_{\lambda,2}}{I_{\lambda,1}} = e^{-KL'} \tag{15.56}$$

where K is called the *extinction coefficient*. Table 15.14 shows values of K for several types of window glass. Since $L' = L/\cos\theta'$, using Snell's law we have

$$L' = \frac{L}{\sqrt{1 - \dfrac{\sin^2\theta}{n^2}}} \tag{15.57}$$

where L is the glass thickness, θ is the angle of incidence of the sun's rays, and n is the *index of refraction* for the glass. For almost all types of window glass, $n = 1.526$.

The component reflectivity r in Eqs. (15.53)–(15.55) may be found from the Fresnel relations. Natural or unpolarized light may be assumed to consist of two vibrating components—one vibrating in a plane normal to the plane of the glass and the other vibrating in a plane parallel to the plane of the glass. If the components are of equal intensity,

$$r = \frac{1}{2}\left[\frac{\sin^2(\theta - \theta')}{\sin^2(\theta + \theta')} + \frac{\tan^2(\theta - \theta')}{\tan^2(\theta + \theta')}\right] \tag{15.58}$$

At normal incidence, $\theta = 0$, and Eq. (15.58) becomes

$$r = \left(\frac{n - 1}{n + 1}\right)^2 \tag{15.59}$$

Substituting $n = 1.526$ into Eq. (15.59) gives $r = 0.0434$ for solar radiation at normal incidence.

Figure 15.6 shows the solution of Eq. (15.58) for glass having an index of refraction 1.526.

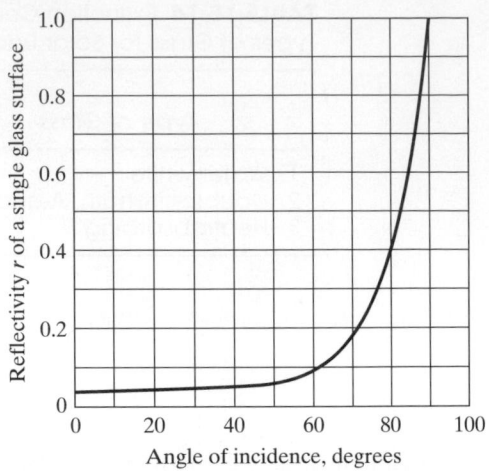

Figure 15.6 Reflection of direct solar radiation from a single surface of glass having an index of refraction of 1.526.

EXAMPLE 15.7

Compute the solar transmissivity, absorptivity, and reflectivity (τ, α, ρ) for a single sheet of 0.125-in. (3.175-mm) double-strength A-quality glass (DSA) at normal incidence.

Solution: The wavelength-averaged properties are used rather than monochromatic properties. By Eq. (15.59), $r = 0.0434$. Using Eq. (15.56) and Table 15.14,

$$a = \exp\left(-0.5 \text{ in.}^{-1}\, 0.125 \text{ in.}\right) = 0.939$$

By Eq. (15.53),

$$\tau = \frac{(1 - 0.0434)^2\, 0.939}{1 - (0.0434)^2 (0.939)^2} = 0.861$$

By Eq. (15.55),

$$\alpha = 1 - 0.0434 - \frac{(1 - 0.0434)^2\, 0.939}{1 - (0.0434)(0.939)} = 0.0608$$

Using the relation $\rho = 1 - \tau - \alpha$,

$$\rho = 1 - 0.861 - 0.0608 = 0.0782$$

Example 15.7 shows that for normal incidence, the solar properties of a single sheet of double-strength A-quality (DSA) glass are approximately: $\tau = 0.86$, $\alpha = 0.06$, and $\rho = 0.08$. This type of glass with these property values is used as the reference glass for many of the calculations later in this chapter.

An alternate method for computing the transmission and absorption coefficients is given by ASHRAE. Figure 15.7 shows that the transmittance of regular sheet glass is nearly independent of wavelength over the solar spectrum. Therefore, the wavelength dependence can be neglected to a first approximation. The equations that can be used to calculate the transmittance and absorptance for DSA glass versus incidence angle are

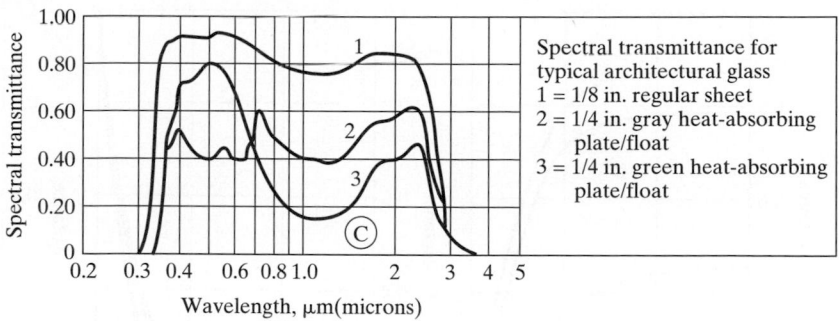

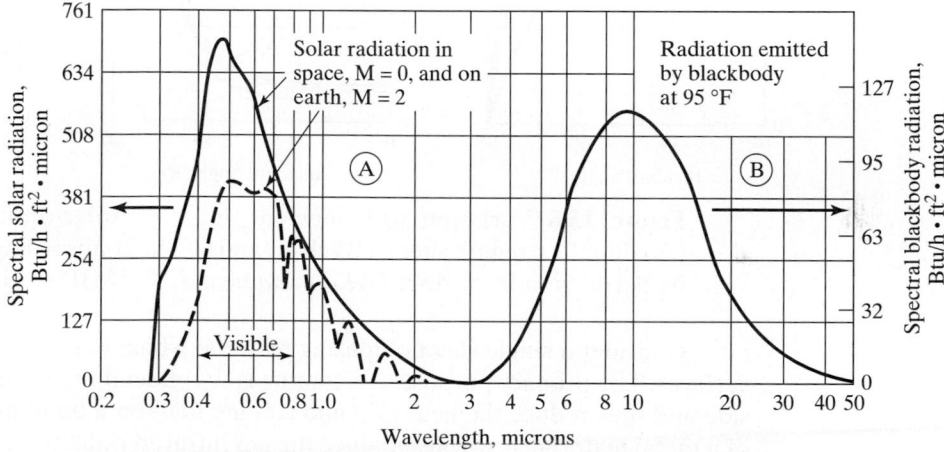

Figure 15.7 Spectral transmittance and solar radiation. [Reprinted by permission from *ASHRAE Fundamentals 1989* (IP & SI), p. 27.21.]

$$\tau = \sum_{j=0}^{5} t_j \, (\cos\theta)^j \tag{15.60}$$

$$\alpha = \sum_{j=0}^{5} a_j \, (\cos\theta)^j \tag{15.61}$$

The coefficients t_j and a_j are listed in Table 15.15. The results of using Eqs. (15.60) and (15.61) and Table 15.15 are plotted as curve A in Fig. 15.8.

TABLE 15.15 Coefficients for DSA Glass for Calculation of Transmittance and Absorptance

i	a_j	t_j
0	0.01154	−0.00885
1	0.77674	2.71235
2	−3.94657	−0.62062
3	8.57881	−7.07329
4	−8.38135	9.75995
5	3.01188	−3.89922

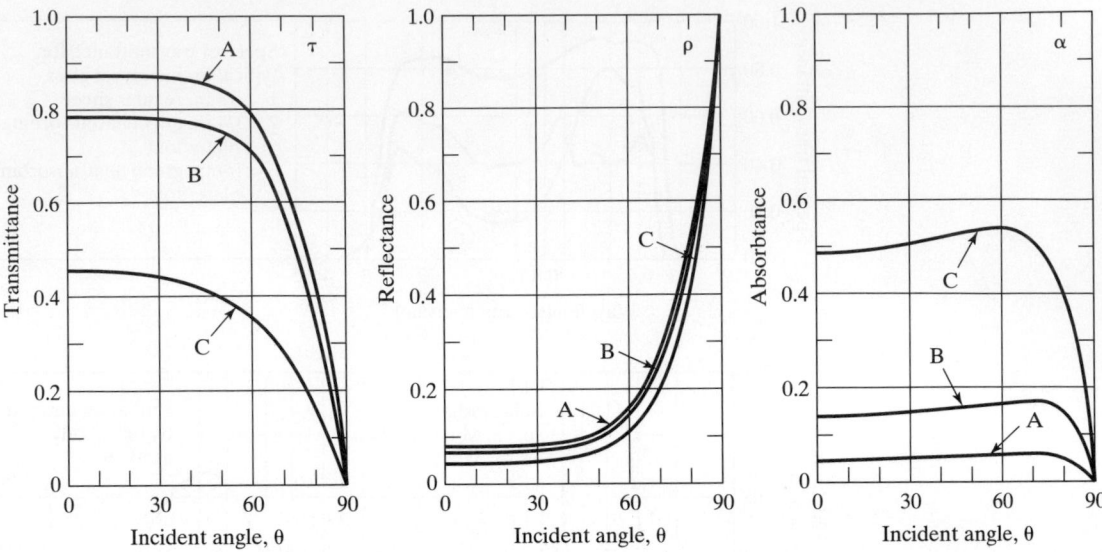

Figure 15.8 Variation with incident angle of solar-optical properties for (A) double-strength sheet, (B) clear, and (C) heat-absorbing glass. [Reprinted by permission from *ASHRAE Fundamentals 1989* (IP & SI), p. 27.21.]

Consider a single sheet of glazing material which has a thin coating on one or both surfaces. The primary purpose of a coating is to reflect more solar energy toward the outside and thus reduce the heat gain and cooling load on a building. If the coating consists of a metal material, it will also reduce the net infrared radiation exchange with the coated side, as the surface emittance is reduced. Let the front surface have a component reflectivity r_1 and the rear surface a component reflectivity r_2. The total monochromatic transmissivity now becomes

$$\tau_\lambda = (1 - r_1)(1 - r_2)a + r_1 r_2 (1 - r_1)(1 - r_2)a^3 + (r_1 r_2)^2 (1 - r_1)(1 - r_2)a^5 \cdots$$

This geometric series can be written

$$\tau_\lambda = \frac{(1 - r_1)(1 - r_2)a}{1 - r_1 r_2 a^2} \tag{15.62}$$

Note that this equation is the same regardless of which side the solar radiation strikes the glazing material. We can also sum the reflected energy to obtain

$$\rho_{\lambda, 1} = r_1 + \frac{r_2 (1 - r_1)^2 a^2}{1 - r_1 r_2 a^2} \tag{15.63}$$

and, using $\alpha_{\lambda, 1} = 1 - \tau_\lambda - \rho_{\lambda, 1}$,

$$\alpha_{\lambda, 1} = \frac{(1 - r_1)(1 - a)(1 + ar_2)}{1 - r_1 r_2 a^2} \tag{15.64}$$

Similar relations can be obtained if the radiation is incident on surface 2 rather than surface 1.

Consider two panes of glazing material in series as shown in Fig. 15.9. We will assume the radiation properties are independent of wavelength over the solar spectrum.

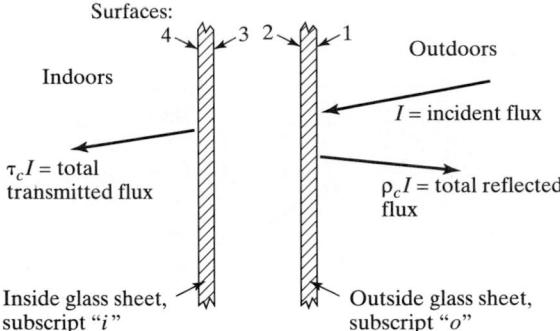

Figure 15.9 Nomenclature for solar-radiation transmission, reflection, and absorption on two layers of glazing.

The reflectivities of surfaces 1, 2, 3, and 4 could be different due to surface coatings, and the absorptivities of the two glazings could be different because of different materials or thickness. The total solar radiation incident on surface 3 due to multiple reflections between the two glazings is

$$I_3 = I\tau_o[1 + \rho_2\rho_3 + (\rho_2\rho_3)^2 + (\rho_2\rho_3)^3 + \cdots]$$

This infinite series can be written as

$$I_3 = \frac{I\tau_o}{1 - \rho_2\rho_3} \tag{15.65}$$

The total solar radiation incident on surface 3 multiplied by the transmittance of the inner sheet, τ_i, equals the total incident solar radiation times the effective transmittance of the two-glazing combination:

$$\tau_i I_3 = \tau_c I \tag{15.66}$$

Combining Eqs. (15.65) and (15.66) results in an equation for the overall transmittance of the two-layer combination:

$$\tau_c = \frac{\tau_i\tau_o}{1 - \rho_2\rho_3} \tag{15.67}$$

Similarly for the overall reflectance,

$$\rho_c I = \rho_1 I + \tau_o I_2 \tag{15.68}$$

where I_2 is the incident flux arriving at surface 2 that was reflected from the inner glazing. Equation (15.68) can be written

$$\rho_c I = I\{\rho_1 + \tau_o^2\rho_3[1 + \rho_2\rho_3 + (\rho_2\rho_3)^2 + (\rho_2\rho_3)^3 + \cdots]\}$$

or, solving for ρ_c,

$$\rho_c = \rho_1 + \frac{\tau_o^2\rho_3}{1 - \rho_2\rho_3} \tag{15.69}$$

The absorption in the outside sheet is

$$\alpha_{o,c} I = \alpha_1 I + \alpha_2 I_2$$

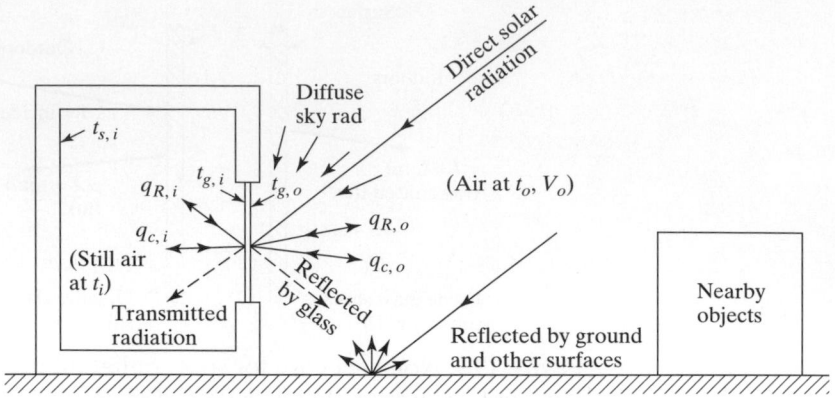

Figure 15.10 Energy exchanges through a glass window.

which can be written

$$\alpha_{o,c} = \alpha_1 + \frac{\alpha_2 \tau_o \rho_3}{1 - \rho_2 \rho_3} \tag{15.70}$$

Similarly for the absorption in the inside sheet,

$$\alpha_{i,c} I = \alpha_3 I_3 + \alpha_4 I_4$$

Assuming I_4 to be zero due to the effective absorption of solar radiation inside the building,

$$\alpha_{i,c} = \frac{\alpha_3 \tau_o}{1 - \rho_2 \rho_3} \tag{15.71}$$

Energy transmission through a single pane of window glass may result from solar-radiation effects and from a temperature difference between the internal and external thermal environments. Figure 15.10 shows the schematic problem. We assume no external shading except by setback, and no internal shades such as drapes or blinds. Direct solar radiation, diffuse sky radiation, and reflected solar radiation may be incident upon the outer surface of the window. Part of this radiation may be directly transmitted through the glass, part may be reflected, and part may be absorbed. Energy exchange by convection may occur between the glass outer surface and the outside air. Also, the glass outer surface may transfer heat by long-wave radiation exchange with the sky and surrounding objects. Similar convection and radiation exchanges may occur between the glass inner surface and the interior of the structure. In general, we may state that q_i, the rate of heat gain (or loss) by the interior of the structure through the glass, is given by

$$q_i = F_s \tau_D I_D + \tau_d I_d + \tau_R I_R + h_i (t_{g,i} - t_i) \tag{15.72}$$

Since the angles of incidence for direct radiation, diffuse sky radiation, and reflected radiation may differ, separate calculations for the products τI are required. Equations (15.53) and (15.60) or charts such as Fig. 15.8 allow determination of each transmissivity. The angle of incidence θ for direct solar radiation is given by Eqs. (13.6), (13.7), (13.9), and (13.10). Parmelee has given a graphical-integration method for determining the transmissivity, reflectivity, and absorptivity of window glass for diffuse sky radiation [10]. For ordinary calculations, he states that for a vertical window, a *mean angle of incidence of 60 deg* may be used. Similarly to diffuse sky radiation, diffusely reflected radiation may strike the window from a variety of angles. In the usual problem, it is satisfactory to use a mean angle of incidence of 60 deg. The sunlit fraction of the window surface F_s in Eq. (15.72) is discussed in Sec. 13.6. Shading of the window from diffuse sky radiation and reflected radiation is assumed to be negligible. The inside-surface coefficient h_i is given in Table 14.1.

We may also write an energy balance for the glass sheet itself. We have

$$F_s I_D \alpha_D + I_d \alpha_d + I_R \alpha_R = h_i (t_{g,i} - t_i) + h_o (t_{g,o} - t_o) \pm q_{g,s} \qquad (15.73)$$

The left side of Eq. (15.73) represents gain of heat by the glass because of absorption of solar radiation. The right side represents heat dissipation, where the surface coefficients h_i and h_o are given in Table 14.1. The term $q_{g,s}$ is the rate of change of heat stored in the glass sheet.

In many problems, Eqs. (15.72) and (15.73) may be simplified. Since $t_{g,i}$ may differ only slightly from $t_{g,o}$, an average glass temperature t_g may be used for each. For glass having a low solar absorptivity, the term $q_{g,s}$ is small and may be neglected. However, for thick heat-absorbing glasses $q_{g,s}$ may be substantial. With these approximations, we may show by Eqs. (15.72) and (15.73) for a window glass with low absorptivity that

$$q_i = (F_s \tau_D I_D + \tau_d I_d + \tau_R I_R) + \frac{F_s \alpha_D I_D + \alpha_d I_d + \alpha_R I_R}{1 + (h_o/h_i)} + U(t_o - t_i) \qquad (15.74)$$

where

$$U = \frac{1}{(1/h_i) + (1/h_o)} \qquad (15.75)$$

Further comments should be made concerning the coefficients h_i and h_o in Eqs. (15.72)–(15.75). Radiation requires that the average interior-surfaces temperature $t_{s,i}$ must be known. In most cases, we may assume $t_{s,i} = t_i$. The average temperature of outdoor radiating objects $t_{s,o}$ is difficult to estimate. Where the window glass views principally the ground and nearby buildings, it is usually satisfactory to assume $t_{s,o} = t_o$. This approximation is generally valid also where the window glass views the sky during daylight hours. However, at night the effective sky temperature for radiation exchange may be substantially less than the outdoor-air temperature.

Figure 15.10 and Eqs. (15.72)–(15.74) make no allowance for internal shading of the window from direct solar radiation. Internal shading devices such as Venetian blinds, roller shades, and drapes are more difficult to analyze than external shades. Section 13.6 covers methods for analyzing external shading devices.

EXAMPLE 15.8

Calculate the instantaneous rate of heat gain through a single-glazed window of common window glass with a thickness of 0.125 in. and an extinction coefficient of 0.5 cm^{-1}. Assume the window is vertical and faces west with no internal or external shading. Use the ASHRAE clear-sky solar-radiation model for 1600 hr local solar time on July 21, 40° N latitude, with a diffuse ground reflectivity of 0.2. The outdoor and indoor temperatures are 95 °F and 78 °F, respectively, and the exterior- and interior-surface heat-transfer coefficients are 4.0 and 1.46 Btu/hr · ft^2 · °F, respectively.

Solution: The solar angles are calculated first to obtain the direct component of the solar radiation. From Eq. (13.4), $h = 15(16 - 12) = 60$ degrees.

The declination angle is obtained from Eq. (13.1) and Table 13.1:

$$n = 181 + 21 = 202$$

$$d = 23.45 \sin\left(360\,\frac{284 + 202}{365}\right) = 20.44°$$

The solar altitude and azimuth angles are calculated using Eqs. (13.6) and (13.7):

$$\sin\beta = \cos 40° \cos 60° \cos 20.44° + \sin 40° \sin 20.44°$$

$$\beta = 35.7°$$

$$\cos\phi = \frac{1}{\cos 35.7°}\,(\cos 20.44° \sin 40° \cos 60° - \sin 20.44° \cos 40°)$$

$$\phi = 87.6°$$

The surface-solar azimuth angle, γ, from Eq. (13.9) is

$$\gamma = \left|(87.6° - 90°)\right| = 2.4°$$

and the incidence angle, θ, from Eq. (13.11) is

$$\cos\theta = \cos 35.7° \cos 2.4°, \qquad \theta = 35.8°$$

Using Eqs. (13.20)–(13.29) and Table 13.3,

$$I_{DN} = (344 \text{ Btu/hr} \cdot \text{ft}^2) \exp(-0.207/\sin 35.7°)$$

$$I_{DN} = 241 \text{ Btu/hr} \cdot \text{ft}^2$$

$$I_D = 241 \cos 35.8° = 195 \text{ Btu/hr} \cdot \text{ft}^2$$

$$I_{dH} = (0.136)241 \text{ Btu/hr} \cdot \text{ft}^2 = 33 \text{ Btu/hr} \cdot \text{ft}^2$$

$$I_d = (0.55 + 0.437 \cos 35.8° + 0.313 \cos^2 35.8°)(33 \text{ Btu/hr} \cdot \text{ft}^2)$$

$$I_d = 37 \text{ Btu/hr} \cdot \text{ft}^2$$

$$I_H = I_{DN} \sin\beta + I_{dH}$$

$$= (241 \text{ Btu/hr} \cdot \text{ft}^2) \sin 35.7° + 33 \text{ Btu/hr} \cdot \text{ft}^2$$

$$I_H = 174 \text{ Btu/hr} \cdot \text{ft}^2$$

$$I_R = (0.2)(174 \text{ Btu/hr} \cdot \text{ft}^2)(0.5) = 17 \text{ Btu/hr} \cdot \text{ft}^2$$

By Fig. 15.6, $r_D = 0.06$. By Eq. (15.57),

$$L' = \frac{0.125 \text{ in.}}{\sqrt{1 - \dfrac{\sin^2 35.8°}{1.526^2}}} = 0.135 \text{ in.}$$

By Eq. (15.56)

$$a_D = \exp[(-0.5 \text{ in.}^{-1})(0.135 \text{ in.})] = 0.935$$

By Eq. (15.53),

$$\tau_D = \frac{(1 - 0.06)^2 0.935}{1 - (0.06^2)(0.935^2)} = 0.83$$

$$\alpha_D = 1.0 - 0.06 - \frac{(1 - 0.06)^2 0.935}{1 - (0.06)(0.935)} = 0.06$$

The mean incidence angles for both the sky-diffuse and ground-reflected radiation are assumed to be 60°. Using procedures similar to those for the direct radiation,

$$\tau_d = \tau_R = 0.77, \qquad \alpha_d = \alpha_R = 0.07$$

By Eq. (15.75),

$$U = \frac{1}{(1/1.46) + (1/4.0)} = 1.07 \text{ Btu/hr} \cdot \text{ft}^2 \cdot {}^\circ\text{F}$$

Then, from Eq. (15.74),

$$q_i = (1.0)(0.83)(195) + (0.77)(37) + (0.77)(17)$$

$$+ \frac{(1.0)(0.06)(195) + (0.07)(37) + (0.07)(17)}{1 + 4.0/1.46} + 1.07(95 - 78)$$

$$q_i = 203 + 4 + 18 = 225 \text{ Btu/hr} \cdot \text{ft}^2$$

Inspection of Eq. (15.74) shows that for an unshaded glazing this equation can be represented by

$$q_i = \tau_{\text{ave}} I + N_i \alpha_{\text{ave}} I + U(t_o - t_i)$$

where τ_{ave} and α_{ave} are the solar-intensity weighted-average transmission and absorption coefficients, respectively, and $N_i = h_i/(h_i + h_o)$ is the fraction of the absorbed radiation that becomes a heat gain on the inside. For hand calculations, it is convenient to tabulate the sum of the first two terms for DSA window glass. This sum is defined as the *Solar Heat-Gain Factor (SHGF):*

$$\text{SHGF} \equiv (\tau_{\text{ave}} + N_i \alpha_{\text{ave}})I \tag{15.76}$$

The parameters upon which this is based are:

1. ASHRAE Clear Day solar radiation model (Sec. 13.9) with a clearness number of 1.0.
2. Ground reflectance of 0.2.
3. No external or internal shading.
4. Solar optical properties of DSA glass from Eqs. (15.60) and (15.61) and Table 15.15; θ for diffuse and reflected radiation is 60°.
5. Interior and exterior heat-transfer film coefficients of 1.46 Btu/hr · ft^2 · °F and 4.0 Btu/hr · ft^2 · °F (8.3 W/m^2 · °C and 22.7 W/m^2 · °C), respectively, which give a value of $N_i = 0.267$.

ASHRAE uses a slightly different method to compute the transmissivity and absorptivity for the diffuse-sky and diffuse-ground-reflected radiation. Rather than use $\theta = 60$ deg in Eqs. (15.60) and (15.61), the following equations are used:

$$\tau_d = \tau_R = 2 \sum_{j=0}^{5} t_j/(j + 2) \tag{15.77}$$

$$\alpha_d = \alpha_R = 2 \sum_{j=0}^{5} a_j/(j + 2) \tag{15.78}$$

with the t_j and a_j coefficients given in Table 15.15. However, the two methods provide nearly identical results.

Solar heat-gain factors computed using the ASHRAE method are tabulated in Table 15.16 for vertical and horizontal surfaces at 40° N latitude. The direct normal solar radiation is also given for reference. Note that only half-day values are given, as the values after noon can be obtained from the before-noon values. For example, a value for SHGF for 0900 hrs local solar time for an east-facing surface (top and left-hand headings) is equal to the value for 1500 hrs for a west-facing surface (bottom and right-hand headings). North, south, and horizontal surface orientations are completely symmetric from morning to afternoon. Half-day totals are also listed, so that estimates of cumulative heat gain can be obtained easily in addition to instantaneous values.

EXAMPLE 15.9

Determine the instantaneous heat gain for the conditions given in Ex. 15.8 using the SHGF values in Table 15.16.

Solution: The conditions specified for Ex. 15.8 match those used in the definition of SHGF. Therefore, Eqs. (15.74) and (15.76) can be combined as

$$q_i = \text{SHGF} + U(t_o - t_i)$$

The value for SHGF can be found in Table 15.16, as the latitude is 40° N. The value listed for a west-facing vertical surface, 1600 hrs local solar time on July 21, is

$$\text{SHGF} = 216 \ \text{Btu/hr} \cdot \text{ft}^2$$

The total instantaneous heat gain becomes

$$q_i = 216 \ \text{Btu/hr} \cdot \text{ft}^2 \cdot {}^\circ\text{F} + 1.07 \ \text{Btu/hr} \cdot \text{ft}^2 \cdot {}^\circ\text{F} (95 - 78) \ {}^\circ\text{F}$$
$$q_i = 234 \ \text{Btu/hr} \cdot \text{ft}^2$$

The difference between the results of Exs. 15.8 and 15.9 is due primarily to the use of different methods to calculate the solar transmissivity and diffusivity values; a more fundamental approach was used in Ex. 15.8, while the ASHRAE polynomial expressions were used to generate the SHGF value used in Ex. 15.9.

The solar heat-gain factor is a convenient parameter. However, it is limited to a single layer of DSA glass and the specific environmental parameters listed earlier. The concept can be generalized to other conditions using the following equation:

$$q_i = (\text{SC})(\text{SHGF}) + U(t_o - t_i) \tag{15.79}$$

where the *shading coefficient (SC)* accounts for all deviations from the single sheet of DSA reference glass. The definition of the shading coefficient is

$$\text{SC} \equiv \frac{\text{solar heat gain of fenestration}}{\text{solar heat gain of DSA glass}} \tag{15.80}$$

The shading coefficient accounts for one or more of the following variations:

1. Multiple glazings
2. Type of glazing material (extinction coefficient)

TABLE 15.16a English Units—Solar Intensity and Solar Heat-Gain Factors for 40° North Latitude

Date	Solar Time	Direct Normal, Btu/hr·ft^2	N	NE	E	SE	S	SW	W	NW	HOR	Solar Time
Jan. 21	0800	142	5	17	111	133	75	6	5	5	14	1600
	0900	239	12	13	154	224	160	13	12	12	55	1500
	1000	274	16	16	124	241	213	51	16	16	96	1400
	1100	289	19	19	61	222	244	118	19	19	124	1300
	1200	294	20	20	21	179	254	179	21	20	133	1200
HALF-DAY TOTALS			61	73	452	904	813	273	62	61	354	
Feb. 21	0700	55	2	23	51	47	14	2	2	2	4	1700
	0800	219	10	50	183	199	94	10	10	10	43	1600
	0900	271	16	22	186	245	157	17	16	16	98	1500
	1000	294	21	21	143	246	203	38	21	21	143	1400
	1100	304	23	23	71	219	231	103	23	23	171	1300
	1200	307	24	24	25	170	241	170	25	24	180	1200
HALF-DAY TOTALS			84	152	648	1049	821	250	85	84	548	
Mar. 21	0700	171	9	93	163	135	22	8	8	8	26	1700
	0800	250	16	91	218	211	74	16	16	16	85	1600
	0900	282	21	47	203	236	128	22	21	21	143	1500
	1000	297	25	27	153	229	171	29	25	25	186	1400
	1100	305	28	28	78	198	197	77	28	28	213	1300
	1200	307	29	29	31	145	206	145	31	29	223	1200
HALF-DAY TOTALS			114	302	832	1087	694	220	114	113	764	
Apr. 21	0600	89	11	72	88	52	5	5	5	5	11	1800
	0700	206	16	140	201	143	16	14	14	14	61	1700
	0800	252	22	128	224	188	41	21	21	21	123	1600
	0900	274	27	80	202	203	83	27	27	27	177	1500
	1000	286	31	37	152	193	121	32	31	31	217	1400
	1100	292	33	34	81	160	146	52	33	33	243	1300
	1200	293	34	34	36	108	154	108	36	34	252	1200
HALF-DAY TOTALS			154	501	957	994	488	199	148	147	957	
May 21	0500	1	0	1	1	0	0	0	0	0	0	1900
	0600	144	36	128	141	71	10	10	10	10	31	1800
	0700	216	28	165	209	131	20	19	19	19	87	1700
	0800	250	27	149	220	164	29	25	25	25	146	1600
	0900	267	31	105	197	175	53	30	30	30	195	1500
	1000	277	34	54	148	163	83	35	34	34	234	1400
	1100	283	36	38	81	130	105	42	36	36	257	1300
	1200	284	37	37	40	82	113	82	40	37	265	1200
HALF-DAY TOTALS			215	666	1024	881	358	200	176	174	1083	
Jun. 21	0500	22	10	21	20	6	1	1	1	1	3	1900
	0600	155	48	143	151	70	13	13	13	13	40	1800
	0700	216	37	172	207	122	22	21	21	21	97	1700
	0800	246	30	156	216	152	29	27	27	27	153	1600
	0900	263	33	114	192	161	45	32	32	32	201	1500
	1000	272	35	63	145	148	69	36	35	35	238	1400
	1100	277	38	40	81	116	88	41	38	38	260	1300
	1200	279	38	38	41	72	95	72	41	38	267	1200
HALF-DAY TOTALS			253	734	1038	818	315	204	188	186	1126	
			N	NW	W	SW	S	SE	E	NE	HOR	PM

(continued)

TABLE 15.16a (cont.)

Date	Solar Time	Direct Normal, Btu/hr·ft²	N	NE	E	SE	S	SW	W	NW	HOR	Solar Time
Jul. 21	0500	2	1	2	2	1	0	0	0	0	0	1900
	0600	138	37	125	137	68	11	11	11	11	32	1800
	0700	208	30	163	204	127	21	20	20	20	88	1700
	0800	241	28	148	216	160	30	26	26	26	145	1600
	0900	259	32	106	193	170	52	31	31	31	194	1500
	1000	269	35	56	146	159	81	36	35	35	231	1400
	1100	275	37	40	81	127	102	43	37	37	254	1300
	1200	276	38	38	41	80	109	80	41	38	262	1200
HALF-DAY TOTALS			223	666	1008	858	352	204	181	180	1076	
Aug. 21	0600	81	12	68	82	48	6	5	5	5	12	1800
	0700	191	17	135	191	135	17	16	16	16	62	1700
	0800	237	24	126	216	180	41	23	23	23	122	1600
	0900	260	28	82	197	196	80	28	28	28	174	1500
	1000	272	32	40	150	187	116	34	32	32	214	1400
	1100	278	35	36	81	156	141	52	35	35	239	1300
	1200	280	35	35	38	106	149	106	38	35	247	1200
HALF-DAY TOTALS			164	498	928	956	474	205	157	156	946	
Sep. 21	0700	149	9	84	146	121	21	9	9	9	25	1700
	0800	230	17	87	205	199	71	17	17	17	82	1600
	0900	263	22	47	194	226	124	23	22	22	138	1500
	1000	280	27	28	148	221	165	30	27	27	180	1400
	1100	287	29	29	78	192	191	77	29	29	206	1300
	1200	290	30	30	32	142	200	142	32	30	215	1200
HALF-DAY TOTALS			119	291	787	1033	672	222	119	118	738	
Oct. 21	0700	48	2	20	45	42	12	2	2	2	4	1700
	0800	204	11	49	173	188	89	11	11	11	43	1600
	0900	257	17	23	180	235	151	18	17	17	97	1500
	1000	280	21	22	139	238	196	38	21	21	140	1400
	1100	291	24	24	71	212	224	101	24	24	168	1300
	1200	294	25	25	27	165	234	165	27	25	177	1200
HALF-DAY TOTALS			88	152	623	1006	791	247	89	88	540	
Nov. 21	0800	136	5	18	108	129	72	6	5	5	14	1600
	0900	232	12	13	151	219	156	13	12	12	55	1500
	1000	268	16	16	122	237	209	50	16	16	96	1400
	1100	283	19	19	61	218	240	116	19	19	123	1300
	1200	288	20	20	21	176	250	176	21	20	132	1200
HALF-DAY TOTALS			63	75	445	887	798	269	63	63	354	
Dec. 21	0800	89	3	8	67	84	50	3	3	3	6	1600
	0900	217	10	11	135	205	151	13	10	10	39	1500
	1000	261	14	14	113	232	210	55	14	14	77	1400
	1100	280	17	17	56	217	242	120	17	17	104	1300
	1200	285	18	18	19	178	253	178	19	18	113	1200
HALF-DAY TOTALS			52	56	374	822	775	276	53	52	282	
			N	NW	W	SW	S	SE	E	NE	HOR	PM

SOURCE: Reprinted by permission from *ASHRAE Fundamentals 1993* (IP), p. 27.23.

Date	Solar Time	Direct Normal, W/m^2	N	NE	E	SE	S	SW	W	NW	HOR	Solar Time
Jan. 21	8	446	17	55	350	420	236	17	17	17	44	4
	9	753	37	41	485	706	504	42	37	37	172	3
	10	865	51	51	390	761	671	161	51	51	303	2
	11	912	59	59	193	699	769	372	59	59	390	1
	12	926	62	62	66	563	802	563	66	62	419	12
HALF-DAY TOTALS			194	231	1425	2851	2565	860	196	194	1117	
Feb. 21	7	175	6	71	160	150	43	6	6	6	11	5
	8	691	33	158	576	628	296	33	33	33	136	4
	9	856	52	70	587	773	496	55	52	52	52	3
	10	926	65	67	449	777	640	120	65	65	450	2
	11	957	73	73	224	690	729	325	73	73	538	1
	12	967	76	76	80	536	759	536	80	76	567	12
HALF-DAY TOTALS			266	478	2042	3307	2590	789	269	266	1729	
Mar. 21	7	540	27	295	514	425	69	26	26	26	83	5
	8	789	50	288	686	664	232	50	50	50	267	4
	9	888	67	147	639	743	404	69	67	67	450	3
	10	937	80	85	482	722	538	91	80	80	587	2
	11	960	88	88	247	623	622	244	88	88	672	1
	12	967	91	91	97	457	650	457	97	91	702	12
HALF-DAY TOTALS			358	953	2624	3425	2189	693	359	357	2409	
Apr. 21	6	282	36	228	279	164	16	15	15	15	34	6
	7	651	50	442	633	451	51	45	45	45	193	5
	8	794	69	402	705	593	130	67	67	67	389	4
	9	864	84	253	637	640	260	84	84	84	557	3
	10	901	96	117	480	608	380	101	96	96	684	2
	11	919	104	107	255	506	459	163	104	104	766	1
	12	925	106	106	114	341	486	341	114	106	793	12
HALF-DAY TOTALS			486	1579	3018	3133	1538	628	467	463	3018	
May. 21	5	3	1	3	3	1	0	0	0	0	0	7
	6	452	113	403	445	223	33	33	33	33	96	6
	7	681	89	520	658	412	63	59	59	59	275	5
	8	787	86	470	693	518	92	80	80	80	460	4
	9	843	99	330	620	550	168	96	96	96	616	3
	10	874	107	171	467	512	262	111	107	107	736	2
	11	890	114	121	255	409	330	133	114	114	811	1
	12	895	117	117	126	257	355	257	126	117	835	12
HALF-DAY TOTALS			679	2100	3228	2775	1128	629	553	549	3414	
Jun. 21	5	68	32	68	63	20	4	4	4	4	8	7
	6	488	150	450	478	222	39	39	39	39	126	6
	7	681	118	543	654	385	68	65	65	65	306	5
	8	776	94	492	680	480	93	85	85	85	483	4
	9	829	105	358	606	507	142	100	100	100	633	3
	10	858	112	197	457	468	218	115	112	112	749	2
	11	874	119	128	254	367	279	130	119	119	821	1
	12	879	121	121	131	227	301	227	131	121	844	12
HALF-DAY TOTALS			799	2314	3274	2579	995	642	592	587	3550	
			N	NW	W	SW	S	SE	E	NE	HOR	PM

(continued)

TABLE 15.16b (*cont.*)

Date	Solar Time	Direct Normal, W/m^2	N	NE	E	SE	S	SW	W	NW	HOR	Solar Time
Jul. 21	5	7	3	7	6	2	0	0	0	0	1	7
	6	434	116	395	433	216	34	34	34	34	100	6
	7	656	95	513	643	400	66	62	62	62	278	5
	8	762	90	468	680	505	94	83	83	83	459	4
	9	818	102	333	610	537	165	99	99	99	611	3
	10	850	110	177	462	501	254	114	110	110	729	2
	11	866	117	125	256	400	321	135	117	117	802	1
	12	871	120	120	130	253	344	253	130	120	826	12
HALF-DAY TOTALS			705	2102	3180	2706	1110	643	572	567	3395	
Aug. 21	6	255	38	214	259	151	18	17	17	17	38	6
	7	603	55	426	602	426	55	49	49	49	196	5
	8	747	75	397	681	569	128	72	72	72	386	4
	9	819	89	259	621	618	251	89	89	89	549	3
	10	857	102	126	472	589	367	107	102	102	674	2
	11	876	109	113	257	492	443	165	109	109	753	1
	12	882	112	112	120	333	469	333	120	112	779	12
HALF-DAY TOTALS			518	1571	2928	3014	1495	647	496	492	2982	
Sep. 21	7	471	28	265	460	381	66	27	27	27	80	5
	8	725	52	275	646	626	224	52	52	52	258	4
	9	830	71	148	613	712	390	73	71	71	434	3
	10	881	84	89	468	696	521	95	84	84	567	2
	11	906	92	92	245	604	603	242	92	92	650	1
	12	914	95	95	101	446	631	446	101	95	679	12
HALF-DAY TOTALS			374	917	2483	3256	2119	699	376	373	2329	
Oct. 21	7	152	6	64	142	132	38	6	6	6	12	5
	8	643	35	155	545	592	279	35	35	35	136	4
	9	811	54	73	567	742	476	57	54	54	305	3
	10	884	67	70	439	752	619	119	67	67	443	2
	11	917	76	76	223	670	707	317	76	76	529	1
	12	927	78	78	84	521	737	521	84	78	557	12
HALF-DAY TOTALS			277	479	1964	3172	2493	780	279	277	1704	
Nov. 21	8	429	17	55	339	406	228	18	17	17	44	4
	9	732	38	42	476	690	492	42	38	38	173	3
	10	846	52	52	385	748	658	159	52	52	303	2
	11	894	60	60	192	688	756	367	60	60	388	1
	12	907	63	63	67	555	789	555	67	63	417	12
HALF-DAY TOTALS			197	235	1402	2795	2515	849	199	197	1115	
Dec. 21	8	279	9	25	212	264	157	10	9	9	20	4
	9	684	31	33	427	645	477	40	31	31	124	3
	10	824	45	45	357	732	661	173	45	45	243	2
	11	881	53	53	177	685	764	379	53	53	327	1
	12	897	56	56	60	560	798	560	60	56	356	12
HALF-DAY TOTALS			165	178	1180	2593	2445	869	167	165	891	
			N	NW	W	SW	S	SE	E	NE	HOR	PM

SOURCE: Reprinted by permission from *ASHRAE Fundamentals 1993* (SI), p. 27.23.

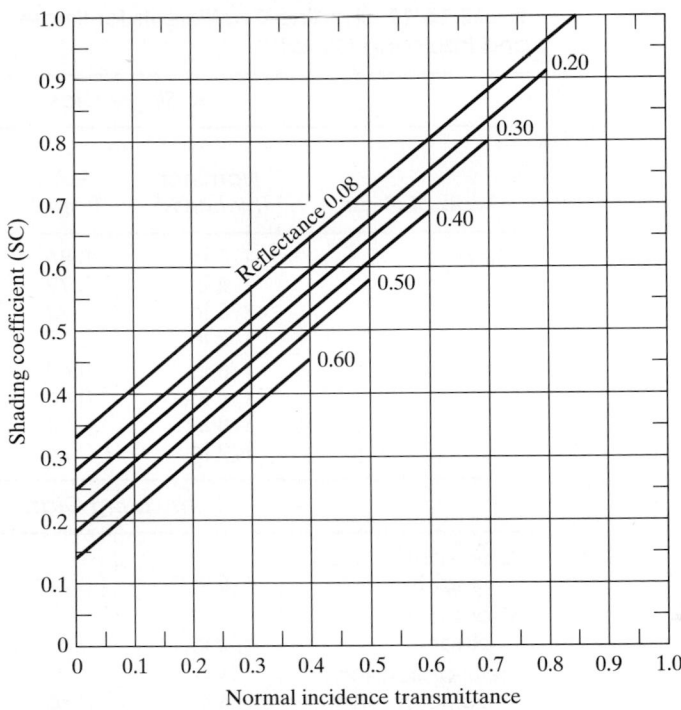

Figure 15.11 Approximate shading coefficient vs. transmittance for coated single glass. [Reprinted by permission from *ASHRAE Fundamentals 1989* (IP & SI), p. 27.29.]

3. Surface coatings
4. Indoor shading devices

Figure 15.11 and Tables 15.17 and 15.18 provide some representative values for the shading coefficient.

The solar heat-gain factor can be corrected for ground reflectivities other than 0.2. This can be done by subtracting the nominal ground-reflected component and adding the ground-reflected component using the actual ground reflectivity. Figure 13.21 shows typical values for ground solar reflectance. The ground-reflected component from Eq. (15.74) is

$$q_{i,R} = (\tau_R + N_i \alpha_R) I_R \tag{15.81}$$

The ground-reflected solar radiation can be approximated as the solar heat-gain factor for a sunlit horizontal surface divided by the sum $(\tau_H + N_i \alpha_H)$ for direct solar radiation on the horizontal surface multiplied by the ground reflectivity and ground-surface angle factor

$$I_R \approx \frac{(\text{SHGF}_H)(\rho_g)(F_{gA})}{\tau_H + N_i \alpha_H} \tag{15.82}$$

Combining Eqs. (15.81) and (15.82),

$$q_{i,R} \approx \frac{(\tau_R + N_i \alpha_R)(\text{SHGF}_H)(\rho_g)(F_{gA})}{\tau_H + N_i \alpha_H} \tag{15.83}$$

TABLE 15.17 Shading Coefficients for Single Glass and Insulating Glass[a]

A. Single Glass

Type of Glass	Nominal Thickness[b]	Solar Trans.[b]	Shading Coefficient	
			$h_o = 4.0$	$h_o = 3.0$
Clear	1/8 in.	0.86	1.00	1.00
	1/4 in.	0.78	0.94	0.95
	3/8 in.	0.72	0.90	0.92
	1/2 in.	0.67	0.87	0.88
Heat-Absorbing	1/8 in.	0.64	0.83	0.85
	1/4 in.	0.46	0.69	0.73
	3/8 in.	0.33	0.60	0.64
	1/2 in.	0.24	0.53	0.58

B. Insulating Glass

Clear out, clear in	1/8 in.[c]	0.71[e]	0.88	0.88
Clear out, clear in	1/4 in.	0.61	0.81	0.82
Heat-absorbing[d] out, clear in	1/4 in.	0.36	0.55	0.58

SOURCE: Reprinted by permission from *ASHRAE Fundamentals 1993*, p. 27.19.

[a] Refers to factory-fabricated units with 3/16-, 1/4-, or 1/2-in. air space or to prime windows plus storm sash.

[b] Refer to manufacturer's literature for values.

[c] Thickness of each pane of glass, not thickness of assembled unit.

[d] Refers to gray-, bronze-, and green-tinted heat-absorbing float glass.

[e] Combined transmittance for assembled unit.

TABLE 15.18 Shading Coefficients for Domed Horizontal Skylights

Dome	Light Diffuser (Translucent)	Curb		Shading Coefficient
		Height, in./cm	Width-to-Height Ratio	
Clear $\tau = 0.86$	Yes $\tau = 0.58$	0/0	∞	0.61
		9/23	5	0.58
		18/46	2.5	0.50
Clear $\tau = 0.86$	None	0/0	∞	0.99
		9/23	5	0.88
		18/46	2.5	0.80
Translucent $\tau = 0.52$	None	0/0	∞	0.57
		18/46	2.5	0.46
Translucent $\tau = 0.27$	None	0/0	∞	0.34
		9/23	5	0.30
		18/46	2.5	0.28

SOURCE: Reprinted by permission from *ASHRAE Fundamentals 1993*, p. 27.30.

If we assume that for most conditions the incidence angles are similar,

$$\tau_R + N_i \alpha_R \approx \tau_H + N_i \alpha_H$$

then Eq. (15.83) reduces to

$$q_{i,R} \approx (\text{SHGF}_H)(\rho_g)(F_{gA}) \tag{15.84}$$

The correction to the SHGF on a vertical surface for a ground reflectance other than 0.2 becomes

$$\text{SHGF}_{\text{corr}} \approx \text{SHGF}_{\text{table}} + 0.5(\text{SHGF}_H)(\rho_g - 0.2) \tag{15.85}$$

The solar heat-gain factor should also be corrected to account for any external shading present. The average solar heat-gain factor for a partially shaded window is

$$\text{SHGF}_{\text{ave}} = (F_s)(\text{SHGF}_D) + (\text{SHGF}_d + \text{SHGF}_R) \tag{15.86}$$

This relation assumes that the shading affects only the beam radiation and not the diffuse-sky or ground-reflected radiation. The shaded area can be computed using the procedures given in Sec. 13.6. The value for $(\text{SHGF}_d + \text{SHGF}_R)$ is nearly equal to the value of SHGF on a surface that has an incidence angle for beam radiation of equal to or larger than 90 deg. This surface has no incident beam radiation and has approximately the same diffuse-sky and ground-reflected radiation as the shaded area of the surface being considered.

EXAMPLE 15.10

Compute the instantaneous heat gain through a vertical south-facing window located at 40° N lat, on a clear day, at local solar noon on December 21. The window is 1 m tall × 2 m wide, double glazed, DSA glass on both sides, with a 1.25-mm air space. The window is set back into the wall a distance of 10 cm. The ground is covered with fresh snow with a solar reflectivity of 0.5. Indoor temperature is 21 °C; outdoor temperature is −10 °C with a wind velocity of 24 km/hr.

Solution: The solar-heat-gain-factor method will be used.

$$q_i = (\text{SC})[(F_s)(\text{SHGF}_D) + (\text{SHGF}_d + \text{SHGF}_{R,\text{corr}})] + U(t_o - t_i)$$

Only the ground-reflected component of the radiation must be corrected for the change in ground reflectivity. The shading coefficient can be obtained from Table 15.17, SC = 0.88, as the wind velocity nearly equals the value used in the solar-heat-gain-factor calculations (4 Btu/hr · ft² · °F or 22.7 W/m² · °C).

The solar heat-gain factor that contains only the sky-diffuse and ground-reflected radiation can be chosen as the east- or west-facing surface orientations at solar noon, December 21, in Table 15.16b, as both have an incidence angle of 90 deg. Therefore $(\text{SHGF}_d + \text{SHGF}_R) = 60 \text{ W/m}^2$. The value for a horizontal surface at the same location and time from Table 15.16b is $\text{SHGF}_H = 356 \text{ W/m}^2$. Using Eq. (15.85), the corrected diffuse and reflected components are

$$\text{SHGF}_d + \text{SHGF}_{R,\text{corr}} = 60 \text{ W/m}^2 + 0.5(356 \text{ W/m}^2)(0.5 - 0.2)$$

$$= 113 \text{ W/m}^2$$

The direct component of solar radiation becomes

$$\text{SHGF}_D = \text{SHGF} - (\text{SHGF}_d + \text{SHGF}_R)$$

$$= 798 \text{ W/m}^2 - 60 \text{ W/m}^2$$

$$= 738 \text{ W/m}^2$$

Next the sunlit fraction of the window will be determined. Using the method detailed in Sec. 13.10, $a = 1$ m, $c = 2$ m, $b = 0.1$ m, $\gamma = 0$ (south-facing surface at solar noon), and we must find δ, the profile angle. For a south-facing surface, the profile angle equals the altitude angle, $\delta = \beta$, so from Eq. (13.6) and Table 13.2,

$$\sin\beta = \cos 40° \cos 0 \cos(-23°27') + \sin 40° \sin(-23°27') = 0.45 \text{ or } \beta = \delta = 26.6°$$

Using the relation for F_s developed for the window shown in Fig. 13.10 with $\tan\gamma = 0$,

$$F_s = \frac{(a - b\tan\delta)}{a}$$

$$= \frac{1 \text{ m} - 0.1 \text{ m} \tan 26.6°}{1 \text{ m}} = 0.95$$

The overall heat-transfer coefficient for this window is $U = 2.78$ W/m$^2 \cdot$ °C from Table 14.5 (center of glass, 1/2 in. air space). The total instantaneous heat gain becomes

$$q_i = 0.88[(0.95)(738) + 113] + 2.78(-10 - 21) \text{ W/m}^2$$

$$q_i = 716 - 86 = 630 \text{ W/m}^2$$

Note that the combined sky-diffuse and ground-reflected radiation nearly doubled in magnitude over the standard value, owing to the high solar reflectivity of the snow. The small solar altitude angle resulted in very little shading of the window. However, the heat gain by sky-diffuse and ground-reflected radiation alone more than offsets the heat loss due to the cold outdoor temperature.

15.6   HEAT GAIN FROM INTERNAL SOURCES

In addition to heat gains from the envelope of a building (i.e., walls, roof, and windows), energy is also added to the interior by the occupants, lights, appliances, and any other heat-producing object. These heat sources are termed *internal heat gains*. Some of these internal sources contribute both energy and moisture to the space. Therefore they add energy in the form of both sensible heat and latent heat. Some of the most common forms of internal heat gain are discussed in this section.

People

Human metabolism produces both sensible heat and moisture from respiration and from the surface of the skin. The rate of metabolism is governed by the physical activity level and the amount of stress. A common assumption is that an adult woman generates 85 percent as much and a child 75 percent as much heat gain as an adult male. For design purposes a mix of adult men and women is assumed, based on the application. Children are included in some cases. Table 15.19 lists typical heat-gain data for a variety of applications. The values in the table are given per person. The table gives the total heat gain from an adult male, the total adjusted heat gain for a mix of men, women, and children, and the amount of the male heat gain that is sensible and the amount that is latent. Similar data are available for animals and plants for the design of buildings, such as animal-confinement facilities and greenhouses.

TABLE 15.19 Rates of Heat Gain from Occupants of Conditioned Spaces[a]

Degree of Activity	Typical Application	Total Heat Adults, Male, Btu/hr(W)	Total Heat Adjusted, Btu/hr(W)	Sensible Heat, Btu/hr(W)	Latent Heat, Btu/hr(W)
Moderately active office work	Offices, hotels, apartments	475 (140)	450 (130)	250 (75)	200 (55)
Standing, light work; walking	Department store, retail store	550 (160)	450 (130)	250 (75)	200 (55)
Walking; standing	Drug store, bank	550 (160)	500 (145)	250 (75)	250 (70)
Light bench work	Factory	800 (235)	750 (220)	275 (80)	475 (140)
Walking 3 mph; light machine work	Factory	1000 (295)	1000 (295)	375 (110)	625 (185)
Heavy work	Factory	1500 (440)	1450 (425)	580 (170)	870 (255)
Athletics	Gymnasium	2000 (585)	1800 (525)	710 (210)	1090 (315)

SOURCE: Adapted by permission from *ASHRAE Fundamentals 1993*, p. 26.8.

[a] Tabulated values are based on 75 °F (24 °C) room dry-bulb temperature. For 80 °F (27 °C) room dry-bulb, the total heat remains the same, but the sensible-heat values should be decreased by approximately 20 percent, and the latent-heat values increased accordingly.

Lighting

Heat gains from lighting are all sensible-energy gains. The amount of heat gain depends on the number of light fixtures installed, the electric power dissipated in each fixture when operated, and the number of fixtures that are energized at a given time. Fluorescent light fixtures have heat gains due to the ballasts in addition to the bulbs. The ballast typically adds 20 percent more heat gain than the lamp itself. High-intensity lamps such as sodium lamps may also have additional heat gains of the order of 20 percent.

In some designs the light fixture is located in the ceiling and replaces the return-air grille for the room. In this case, some of the heat generated by the fixture is convected by the return air into the return-air plenum. This heat becomes a load on the cooling equipment but should not be included as a heat gain into the conditioned space. The amount of heat that can be convected to the return air ranges from about 40 percent to 60 percent.

Appliances

Cooking equipment, office equipment, and other miscellaneous heat-producing items should be included in the space heat gain. Equipment that has an effective exhaust hood will produce much less heat gain in the conditioned space than unvented equipment. An effective exhaust captures nearly all the sensible convective heat and all the latent gain. The only significant heat gain to the space is by infrared radiation from hot surfaces. Table 15.20 lists heat gains from some typical commercial cooking equipment. Table 15.21 gives the sensible-heat gain from various items found in offices.

Infiltration

Warm, moist air that enters the building by natural air leakage also constitutes a heat gain on the conditioned space. This infiltrating air must be cooled to the indoor dry-bulb temperature and must be dehumidified to the indoor humidity level. The sensible-heat gain is

TABLE 15.20 Recommended Rate of Heat Gain from Selected Restaurant Equipment

Appliance	Size	Input Rating, Btu/hr (W)	Recommended Rate of Heat Gain, Btu/hr (W)			
			Without Hood			With Hood, Sensible
			Sensible	Latent	Total	
Electric, No Hood Required						
Coffee brewer	12 cups/2 burners	5660 (1660)	3750 (1100)	1910 (560)	5660 (1660)	1810 (530)
Dishwasher (conveyor-type water-sanitizing), per 100 dishes/hr	5000–9000 dishes/hr	1160 (380)	150 (56)	370 (123)	520 (179)	170 (56)
Display case (refrigerated), per ft³ (m³) of interior	6–67 ft³ (0.2–2 m³)	154 (1590)	62 (640)	0	62 (640)	0
Griddle/grill (large), per ft² (m²) of cooking surface	4.6–11.8 ft² (0.43–1.1 m²)	9200 (29,000)	620 (1940)	340 (1080)	960 (3020)	340 (1080)
Electric, Exhaust Hood Required						
Fryer (deep fat), per lbm (kg) of fat capacity	15–70 lbm (7–32 kg)	1270 (820)				14 (9)
Range (burners), per 2-burner section	2–10 burners	7170 (2100)				2660 (780)
Gas, No Hood Required						
Dishwasher (conveyor-type water-sanitizing), per 100 dishes/hr	5000–9000 dishes/hr	1370 (400)	370 (97)	80 (21)	450 (118)	140 (38)
Griddle/grill (large), per ft² (m²) of cooking surface	4.6–11.8 ft² (0.43–1.1 m²)	17,000 (53,600)	1140 (3600)	610 (1930)	1750 (5530)	460 (1450)
Gas, Exhaust Hood Required						
Fryer (deep fat), per lbm (kg) of fat capacity	11–70 lbm (5–32 kg)	2270 (1470)				160 (100)
Range (burners), per 2-burner section	2–10 burners	33,600 (9840)				6590 (1930)

SOURCE: Abstracted by permission from *ASHRAE Fundamentals 1993*, pp. 26.12–26.13.

TABLE 15.21 Recommended Rate of Heat Gain from Selected Office Equipment

Appliance	Size	Maximum Input		Standby Input		Recommended Rate of Heat Gain	
		Watts	Btu/hr	Watts	Btu/hr	Watts	Btu/hr
Microcomputer/ wordprocessor	16–640 kbytes[a]	100–600	340–2050	90–530	300–1800	90–530	300–1800
Printer (laser)	8 pages/min	870	3000	180	600	300	1000
Copiers	6–30[a] copies/min	1570–5800	460–1700	1570–5800	300–900	1000–3100	460–1700
Cold food/beverage		1150–1920	3900–6600			575–960	1960–3280
Coffee maker	10-cup	1500	5120	Sensible:		1050	3580
				Latent:		450	1540

SOURCE: Abstracted by permission from *ASHRAE Fundamentals 1993*, p. 26.14.
[a]Input is not proportional to capacity.

$$\dot{Q}_s = (\dot{V})(\rho)(c_p)(t_o - t_i) \tag{15.87}$$

and the latent heat gain is

$$\dot{Q}_l = (\dot{V})(\rho)(h_{fg})(W_o - W_i) \tag{15.88}$$

The moist air property values and methods to estimate the infiltration flow rate are given in Chapters 7 and 14, respectively.

ENDNOTES

1. J. S. Alford, J. E. Ryan, and F. O. Urban, "Effect of Heat Storage and Variation in Outdoor Temperature and Solar Intensity on Heat Transfer through Walls," *ASHVE Trans.*, 45 (1939), 393–395.

2. James C. Dunn, "Sol-Air Temperature Data at Minneapolis, Minnesota," Plan B Master's Degree paper, University of Minnesota, 1965.

3. *ASHVE Trans.*, 45, 387–392.

4. C. O. Mackey and L. T. Wright, Jr., "Periodic Heat Flow—Composite Walls or Roofs," *ASHVE Trans.*, 52 (1946), 283–296.

5. J. P. Stewart, "Solar Heat Gain through Walls and Roofs for Cooling Load Calculations," *ASHVE Trans.*, 54 (1948), 361–388.

6. Charles E. Bullock, "Periodic Heat Transfer in Walls and Roofs," Plan B Master's Degree paper, University of Minnesota, 1961.

7. D. G. Stephenson and G. P. Mitalas, "Calculation of Heat Conduction Transfer Functions for Multilayer Slabs," *ASHRAE Trans.*, 77, pt. 2 (1971), 117.

8. *ASHRAE Handbook of Fundamentals* (Atlanta: American Society of Heating, Refrigerating and Air Conditioning Engineers, 1989).

9. G. V. Parmelee, "Transmission of Solar Radiation through Flat Glass," *ASHVE Trans.*, 51 (1945), 317–350.

10. *ASHVE Trans.*, 51, 322–324, 334–335.

PROBLEMS

15.1 Compute the sol-air temperature in degrees Celsius at solar noon on a dark-colored horizontal surface at 35° N lat on July 21. Use the ASHRAE clear-sky solar-radiation model.

15.2 Estimate the surface temperature of a dark-colored horizontal roof that has the sol-air temperature computed in Prob. 15.1. Assume the roof is well insulated, so the amount of heat conducted into the roof from the outer surface is negligible.

15.3 Calculate the sol-air temperature in degrees Fahrenheit at solar noon for a vertical, south-facing, dark-colored wall at 45° N lat on December 21. Use the ASHRAE clear-sky solar-radiation model with a snow-covered-ground reflectivity of 0.5.

15.4 Determine the numerical coefficients for the equation for the sol-air temperature t_e of the form given by Eq. (15.13) for the dark-colored south-facing vertical surface described in Prob. 15.3. Include two harmonics.

15.5 Based upon the sol-air temperature variation determined in Prob. 15.4, calculate the rate of heat transmission q_i to the interior through each square foot of a 12-in. common brick wall at
 (a) 10:00 a.m. solar time, and
 (b) 8:00 p.m. solar time.
 Use the procedure demonstrated in Ex. 15.2.

15.6 Calculate the instantaneous rate of heat gain, Btu/hr · ft^2, through a sunlit flat-roof construction

similar to No. 5 of Table 15.6 at 3:00 p.m. solar time on an average clear day at Minneapolis for
(a) January,
(b) April,
(c) July, and
(d) October.
Assume a constant interior temperature of 75 °F.

15.7 Roofs 2 and 3 of Table 15.6 differ only in order of placement of layers A and B. Compare the maximum rate of heat gain and corresponding solar time for these two roof constructions for a summer design day at Minneapolis. Assume an interior temperature of 75 °F.

15.8 Compute the instantaneous heat gain each hour for a 24-hr period for wall #W4 in Table 15.8. The wall has a light-colored exterior surface and faces west. Use the sol-air temperature data from Table 15.3 and assume the indoor temperature is 26 °C. Use the TETD method [Eqs. (15.32) and (15.33)] with the decrement and time lag from Table 15.8.

15.9 Determine the hourly instantaneous heat gain through the wall given in Prob. 15.8, but use the transfer-function method [Eq. (15.47)] and the coefficients given in Table 15.8.

15.10 Simulate the transient heat transfer through the wall given in Prob. 15.8 using an explicit finite-difference method [Eq. (15.52)]. Space the temperature nodes 1.77 cm apart and use a 15-min time step. Plot the following periodic results for one 24-hr cycle:
(a) the outside-surface temperature,
(b) the inside-surface temperature, and
(c) the inside-surface heat flux.

15.11 Use the transfer-function method to compute the hourly instantaneous heat gain through roof #R2 given in Table 15.9. Use the sol-air temperature data for a dark-colored horizontal surface from Table 15.3 and assume an indoor temperature of 78 °F.

15.12 Compute the hourly instantaneous heat gain through the roof of Prob. 15.11 on an average January day in Minneapolis using the transfer-function method. Assume a light color and an indoor temperature of 70 °F. Use the sol-air coefficients listed in Table 15.5.

15.13 Determine the instantaneous heat gain at 10:00 p.m. solar time through the walls of a building with wall construction #W1 given in Table 15.8. The north and south walls have areas of 30 m^2 each and the east and west walls have areas of 20 m^2 each. Use the dark-colored sol-air temperature values from Table 15.3 and assume an indoor temperature of 75 °F.

15.14 Repeat Prob. 15.13 for the case when all the walls are shaded all day by thick vegetation.

15.15 Measurements on a clear day with a pyrheliometer indicate that a certain glass sample ($n = 1.526$), 0.25 in. thick, has a solar transmissivity of 0.82 when the sun's angle of incidence is zero. Through use of monochromatic formulae, calculate the solar transmissivity for an angle of incidence of 75 deg.

15.16 A window faces due east. The glass ($n = 1.526$, $K = 3.30$ in.$^{-1}$) is 0.25 in. thick. Calculate the monochromatic absorptivity of the glass on July 1 at 9:00 a.m. solar time for a location at 42 deg north latitude.

15.17 Figure 15.12 shows a schematic cross section of a greenhouse. Roof A faces due east and has an unobstructed vision of the sky. Location is at 42 deg north latitude. Interior air temperature is maintained constant at 60 °F. Transmissivity of the glass is given by Fig. 15.8 (curve A). Absorption of solar radiation by the glass may be neglected. You may also neglect ground-reflected radiation. During daylight hours the outside-surface heat-transfer coefficient h_o may be taken as 6.0 Btu/hr · ft^2 · °F, at night as 8.0 Btu/hr · ft^2 · °F. Calculate the rate of energy transmission through each square foot of glass of roof A on a clear January 1 day at

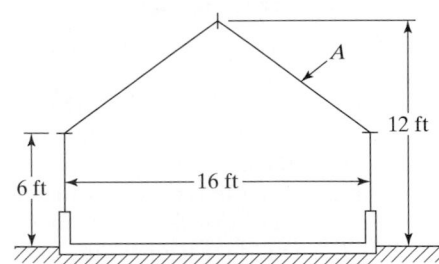

Figure 15.12 Schematic cross section of a greenhouse.

(a) 10:00 a.m. solar time when the outdoor-air temperature is 0 °F, and
(b) 10:00 p.m. when the outdoor-air temperature is −10 °F.

15.18 Assume that the building of Fig. 15.13 is located at 36 deg north latitude. The glass panels are 9 ft high. Figure 15.8 (curve C) shows variation of glass transmissivity. Calculate the rate in Btu/hr at which direct solar radiation is transmitted through the glass wall of office A on a clear August 1 day at 10:00 a.m. solar time.

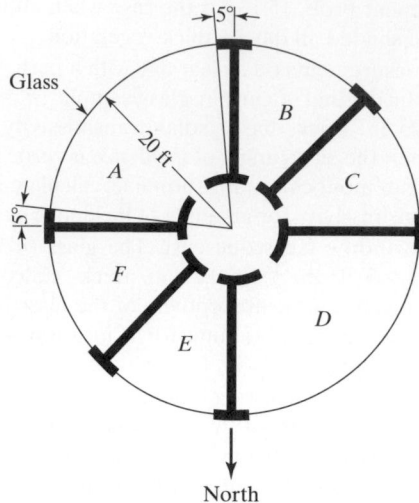

Figure 15.13 Schematic building for Prob. 15.18.

15.19 Compute the overall transmittance at an incidence angle of 60 deg of two sheets of DSA glass. Surface #2 (see Fig. 15.9) has a reflective coating such that $\rho_2 = 0.3$. The other surfaces are uncoated.

15.20 Derive an expression for the overall transmittance of three layers of glazing similar to Eq. (15.67) that was developed for two layers.

15.21 Determine the instantaneous heat gain through a 1 m × 2 m west-facing window at 6 p.m. solar time, on a clear day, July 21, at 40 deg north latitude. The window has two sheets of glass with a 1.7-cm air space between them. The outer layer is gray heat-absorbing glass; the inner layer is standard DSA glass. Assume an exterior film coefficient of 8 W/m² · °C, an outdoor temperature of 33 °C, and an indoor temperature of 27 °C.

15.22 Compute the instantaneous heat gain through a vertical south-facing window at 10:00 a.m. solar time, on a clear day, July 21, at 40 deg north lati-

tude. The window has two layers of 1/8-in. DSA glass with no coatings and a 1/2-in. air space between the glazings. The window is 4 ft high and 3 ft wide with a long 2-ft-wide overhang that is located 1 ft above the top of the window. Assume an outdoor temperature of 90 °F and an indoor temperature of 75 °F.

15.23 Work Prob. 15.22 without the overhang but with a coating on the glass. Determine the shading coefficient needed to provide the same total heat gain as with the overhang.

15.24 Work Prob. 15.22 without the overhang but with the window set back into the wall 6 inches.

15.25 A small commercial building is to be constructed in Denver, Colorado (40 deg north latitude). Part of the horizontal flat roof will be dark color with the construction given by #R4 in Table 15.9. Part of the roof will have translucent skylights, $\tau = 0.52$, with no curb. Compute the daily total heat gain

(a) per square meter of roof and
(b) per square meter of skylight for a clear, summer design day.
(c) Compare the answers from (a) and (b) and make some recommendations to the architect to reduce the daily heat gain into the building.

15.26 An office will contain 150 people, 3500 W of fluorescent lighting, 4 copy machines, 2 coffee makers, and 100 computer terminals. Determine the instantaneous sensible- and latent-heat gain in this office space due to the workers, lights, and equipment.

15.27 A commercial kitchen is to be designed with three deep-fat fryers, each with a capacity of 10 kg of fat.

(a) Compare the operating-energy cost for a 16-hr day between electric and gas units if electricity costs 8 cents per kilowatthour and gas costs 10 cents per thousand Btu.
(b) Compare the sensible-heat gain from the electric and gas fryers.
(c) Make a recommendation to purchase either the electric or gas fryers if all other costs are equal.

SYMBOLS

a	Radiation absorption coefficient, dimensionless.
a_j	Coefficients in Eq. (15.61).
a_n	Transfer-function coefficients, Btu/hr · ft² · °F (W/m² · °C).
A	Constant, °F (°C).
A, B, C, D	Constants in Laplace-transform matrix, Eq. (15.34).
b_n	Transfer-function coefficients, Btu/hr · ft² · °F (W/m² · °C).
B	Constant, °F (°C).

c_p	Specific heat capacity of moist air, Btu/lbm · °F (kJ/kg · °C).
c	Specific heat capacity, Btu/lbm · °F (kJ/kg · °C); c_j refers to layer j.
c_n	Transfer-function coefficients, Btu/hr · ft^2 · °F (W/m^2 · °C).
C_n	Constants, °F (°C).
C_j	Heat capacity of node j, Btu/ft^2 · °F (kJ/m^2 · °C).
d	Declination angle.
d_n	Transfer-function coefficients, dimensionless.
D_n	Constants, °F (°C).
f	Arbitrary function of time, $f(\tau)$.
F_{gA}	Shape factor between ground and aperture, dimensionless.
$F_{w,i}$	Shape factor between wall and surface i, dimensionless.
F_s	Sunlit fraction, dimensionless.
H	Transmission matrix defined by Eqs. (15.36) and (15.37).
h_i	Combined convective and radiative heat-transfer coefficient for inside surface, Btu/hr · ft^2 · °F (W/m^2 · °C).
h_{fg}	Latent heat of vaporization of water at room temperature, Btu/lbm (kJ/kg).
h_o	Combined convection and radiation heat-transfer coefficient for outside surface, Btu/hr · ft^2 · °F (W/m^2 · °C).
$h_{o,c}$	Convection heat-transfer coefficient for outside surface, Btu/hr · ft^2 · °F (W/m^2 · °C).
$h_{o,R}$	Infrared radiative heat-transfer coefficient for outside surface, Btu/hr · ft^2 · °F (W/m^2 · °C).
I	Incident solar radiation, Btu/hr · ft^2 (W/m^2); I_d for diffuse, I_D for direct; I_R for reflected; I_T for total; I_γ for monochromatic radiation; I_{DN} for direct normal; I_{dH} for diffuse horizontal; I_H for total horizontal.
IHG	Instantaneous heat gain, Btu/hr · ft^2 (W/m^2); IHG$_\tau$ refers to time τ.
K	Extinction coefficient, in.$^{-1}$ (m^{-1}).
$K(Z)$	Transfer function.
k	Thermal conductivity, Btu/hr · ft · °F (W/m · °C); k_j refers to layer j.
L	Thickness of material, in. (m).
M_n	Constants in Eqs. (15.12)–(15.14), °F (°C).
n	Index of refraction, dimensionless.
N_n	Constants in Eqs. (15.12)–(15.14), °F (°C).
N_i	Fraction of absorbed radiation that becomes an internal heat gain, dimensionless.
p_n	Constants, ft^{-1} (m^{-1}).
$\dot{Q}_l$	Latent heat gain, Btu/hr (W).
$\dot{Q}_s$	Sensible heat gain, Btu/hr (W).
q	Surface heat flux, Btu/hr · ft^2 (W/m^2); q_c for convective heat flux; q_{IR} for infrared radiation; q_{net} for net heat flux; q_s for absorbed solar energy; q_i for inside surface; q_o for outside surface.
r	Fraction of radiation reflected, dimensionless; r_n refers to surface n.
R	Thermal resistance, hr · ft^2 · °F/Btu (m^2 · °C/W); R_{film} refers to surface film; R_j refers to layer j.
SC	Shading coefficient defined by Eq. (15.80), dimensionless.
SHGF	Solar heat-gain factor defined by Eq. (15.76), Btu/hr · ft^2 (W/m^2).
t	Temperature, °F (°C); t_j refers to layer j.

t_e	Sol-air temperature given by Eq. (15.10), °F (°C); $t_{e,m}$ refers to 24-hr mean value, $t_{e,n}$ refers to harmonic coefficient.
t_g	Glazing temperature, °F (°C).
t_i	Inside dry-bulb temperature, °F (°C).
t_j	Coefficients in Eq. (15.60).
t_o	Outside dry-bulb temperature, °F (°C).
t_w	Wall surface temperature, °F (°C); $t_{w,i}$ refers to inside surface; $t_{w,o}$ refers to outside surface.
T	Absolute temperature, °R (K); T_i for surface i; T_w for wall surface.
TETD	Total equivalent temperature difference, °F (°C).
U	Heat-transmission coefficient, Btu/hr · ft^2 · °F (W/m^2 · °C).
V_n	Factor given by Eq. (15.24), Btu/hr · ft^2 · °F (W/m^2 · °C).
$\dot{V}$	Volumetric flow rate of moist air, ft^3/min (m^3/s).
W	Humidity ratio of moist air, lbm$_w$/lbm$_a$ (kg$_w$/kg$_a$).
x	Coordinate perpendicular to wall surface, ft (m).
Y_n	Factor defined by Eq. (15.26), dimensionless.
Z	Z-transform variable that equals $e^{s\Delta\tau}$.
Z_n	Factor defined by Eq. (15.27), dimensionless.

Greek Letters

α	Thermal diffusivity $k/\rho c$, ft^2/hr (m^2/s); α_j refers to layer j.
α	Solar absorptivity, dimensionless; α_γ for monochromatic absorptivity; α_d for diffuse; α_D for direct; α_R for reflected radiation.
α_s	Absorptivity for solar radiation, dimensionless.
β	Solar altitude angle.
γ	Surface-solar azimuth angle.
Δx	Spacing between adjacent nodes, ft (m).
ΔR	Infrared radiation correction factor given by Eq. (15.11), Btu/hr · ft^2 (W/m^2).
$\Delta\tau$	Time interval, hr (s).
ϵ	Surface infrared emissivity, dimensionless; ε_i for surface i; ε_w for wall surface.
θ	Time, hr.
θ	Laplace transform of temperature; θ_i refers to interior temperature; θ_e refers to sol-air temperature.
θ	Solar incidence angle.
λ_n	Decrement factors given by Eq. (15.31), dimensionless.
ρ	Density, lbm/ft^3 (kg/m^3); ρ_j refers to material j.
ρ	Reflectivity, dimensionless, ρ_λ for monochromatic reflectivity; ρ_g for ground reflectivity.
σ	Stefan-Boltzmann constant, Btu/hr · ft^2 · °R^4 (W/m^2 · K^4).
σ_n	Factor given by Eq. (15.25), ft^{-1} (m^{-1}).
Σ	Surface tilt angle measured between the surface normal and vertical, dimensionless.
τ	Time, hr (s).
τ	Solar transmissivity, dimensionless; τ_λ for monochromatic transmissity; τ_d for diffuse; τ_D for direct; τ_R for reflected.
τ_j	Time constant for layer j, hr (s).
Φ_n	Lag angle defined by Eq. (15.28).

ϕ	Laplace transform of surface heat flux; ϕ_i refers to interior surface; ϕ_o refers to exterior surface.
ϕ	Solar azimuth angle.
ψ_n	Angle defined by Eq. (15.14).
ω_n	Angular velocity, rad/hr.

Subscripts

c	For combination of glazing layers.
corr	Corrected.
d	Diffuse.
D	Direct.
i	Inside.
IN	Into the surface.
j	Node j.
$j - 1$	Node to left of node j.
$j + 1$	Node to right of node j.
m	Mean.
n	Order of harmonic in Fourier series.
o	Outside.
OUT	Away from the surface.
R	Reflected.
δ	Time lag, hr.
θ	Hourly value.

Superscripts

k	Time step or current time.
$k - 1$	Evaluated at previous time step.
$'$	Refers to refracted beam.

chapter 16

Instantaneous Cooling Load

16.1 INTRODUCTION

The instantaneous heat gains discussed in Chapter 15 must be modified to determine the instantaneous cooling load experienced by a forced-air cooling coil. The instantaneous cooling load is usually not equal to the instantaneous heat gain. The main reason for the difference is that the sensible, or energy, heat gains in the conditioned space are partially convective and partly radiative. The energy convected into the air becomes a nearly instantaneous load on a forced-air cooling coil. However, the infrared-radiation portion of the heat gain is primarily absorbed and stored by the construction and furnishing materials within the conditioned space. This stored energy gradually raises the temperature of these materials until they, too, become convective heat sources within the space, but at a later time.

Some of the latent, or moisture, heat gains are also stored in the interior materials to become a latent cooling load at a later time. However, the time constants associated with moisture storage are usually so large that this storage mechanism is often neglected in hourly transient analyses. Contaminant gases are also stored and behave much like stored water vapor.

In this chapter we will discuss various methods that can be used to convert the instantaneous heat gains of a building into instantaneous cooling loads for a forced-air cooling system. The primary purpose is to determine the instantaneous cooling requirement on the cooling coil. These methods are used primarily for nonresidential buildings. Simplified methods for residential construction are given in the *ASHRAE Handbook of Fundamentals* [1] and the Cooling and Heating Load Calculation Manual [2].

16.2 INSTANTANEOUS COOLING LOAD FOR FORCED-AIR SYSTEMS

Figure 16.1 illustrates the typical heat gains that occur in a small, single-zone office building. This building is cooled with a forced-air cooling system that includes a cooling coil to remove sensible and latent energy from the air passing through it. Only energy and mois-

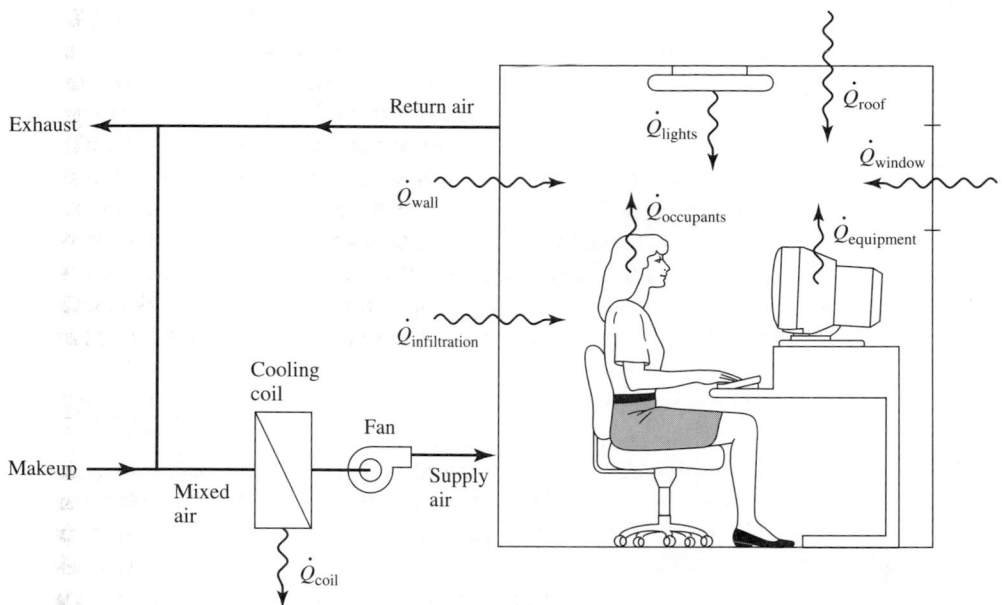

Figure 16.1 Schematic of heat gains in a single zone with forced-air cooling.

ture convected into the air in the building will be transported out of the room in the return-air stream. Some of this return air and the makeup air enters the coil. The coil must remove energy and moisture from the mixed-air stream to achieve the desired supply-air conditions. The operation of the air-handling system was discussed in Chapter 8 and the performance of a cooling coil in Chapter 11. In this chapter we will focus on converting the sensible instantaneous heat gains in the building to sensible instantaneous cooling load.

The *sensible instantaneous cooling load* is the rate at which energy is added to the air in the conditioned space by convection. The usual assumption made is that the space temperature is maintained at a constant value. The sensible cooling load is also equal to the mass flow rate of dry air supplied to the space multiplied by the change in dry-bulb temperature between the supply and return and by the air specific-heat capacity:

$$\dot{Q}_{\text{sen}} = \dot{m}_a c_{pa} (t_{\text{ret}} - t_{\text{sup}}) = \Sigma \dot{Q}_{\text{sen, conv}}$$

For a detailed analysis of the convective heat gains we must know the convective heat-transfer coefficients and the surface temperatures:

$$\dot{Q}_{\text{sen, conv}} = h_{\text{conv}} A (t_{\text{sur}} - t_a)$$

The convective heat-transfer coefficients are difficult to quantify, as they depend on a variety of factors including the air velocity, surface temperature, and shape and orientation of the surface. The surface temperature is difficult to determine precisely. For materials it is a result of coupling the surface convective heat transfer, absorbed infrared radiation, and the thermal storage within the material. Absorbed solar radiation may also occur. We can perform a detailed energy balance within a room only if we know all the parameters discussed here. Computer programs can be written to perform the necessary computations. However, surface properties and convection coefficients are often estimates of the actual values, which inherently results in some error in the calculations.

Approximate methods have been developed which are not as rigorous as a detailed energy-balance method. These approximate methods are simpler to use and are of sufficient accuracy for computerized or manual design cooling-load calculations and for estimating energy use over a specified time period. Two of these approximate methods, the transfer-function method, and the ASHRAE cooling-load temperature-difference method, are described in this chapter. In each of these methods the individual heat gains and cooling loads are assumed to be independent. Thus the total cooling load can be calculated by summing the individual components, such as the roof, walls, fenestrations, etc. However, the heat gains are not strictly independent. For example, some of the radiant-heat gain from the roof will be absorbed on the interior surfaces of the walls, which will affect the wall heat gain and cooling load. Usually these interactions are neglected, so that a simple sum of the individual cooling-load components yields the total cooling load for the conditioned space.

16.3 TRANSFER-FUNCTION METHOD

The use of the *transfer-function method* to compute instantaneous heat gains through walls and roofs was described in Sec. 15.4. The method used to compute cooling loads is very similar. The basic equation is

$$\mathrm{ICL}_\tau/A = (v_0 \mathrm{IHG}_\tau + v_1 \mathrm{IHG}_{\tau - \Delta\tau} + v_2 \mathrm{IHG}_{\tau - 2\Delta\tau} + \cdots)$$
$$- (w_1 \mathrm{ICL}_{\tau - \Delta\tau} + w_2 \mathrm{ICL}_{\tau - 2\Delta\tau} + \cdots) \tag{16.1}$$

Equation (16.1) is applied to each of the heat gains on a space separately (walls, roof, fenestrations, lights, occupants, and equipment). The individual cooling loads are summed. The coefficients in Eq. (16.1) depend on the ratio of convective and radiative heat gain from each component and on the total heat capacity or thermal mass of the structure. The thermal mass often can be well represented by the thermal mass of the floor.

Equation (16.1) can be approximated by a truncated series which retains only the first two v coefficients and the first w coefficient:

$$\mathrm{ICL}_\tau/A = (v_0 \mathrm{IHG}_\tau + v_1 \mathrm{IHG}_{\tau - \Delta\tau}) - w_1 \mathrm{ICL}_{\tau - \Delta\tau} \tag{16.2}$$

Tables 16.1 and 16.2 list coefficients for Eq. (16.2) for hourly instantaneous cooling load. Example 16.1 illustrates the use of Eq. (16.2) to determine instantaneous cooling load.

EXAMPLE 16.1

Use the transfer-function method to compute the hourly instantaneous cooling load per unit area for the roof of Ex. 15.5. Assume a medium construction, an uncovered concrete floor, and a very high air-circulation rate.

Solution: We will use Eq. (16.2) with the hourly instantaneous-heat-gain results from Ex. 15.5 and the v and w coefficients from Tables 16.1 and 16.2. The w_1 coefficient is -0.94 because of the very high air-circulation rate, the thickness of the concrete floor, and the specification that the floor is uncovered. The v_0 and v_1 coefficients become 0.681 and -0.621 from

---

TABLE 16.1 w_1 Coefficients for Cooling-Load Transfer Function

Room Air Circulation[a]	Room Envelope Construction[b]				
	2-in. (50 mm) Wood Floor	3-in. (75 mm) Concrete Floor	6-in. (150 mm) Concrete Floor	8-in. (200 mm) Concrete Floor	12-in. (300 mm) Concrete Floor
	Specific Mass per Unit Floor Area, lbm/ft² (kg/m²)				
	10 (50)	40 (200)	75 (370)	120 (590)	160 (780)
Low	−0.88	−0.92	−0.95	−0.97	−0.98
Medium	−0.84	−0.90	−0.94	−0.96	−0.97
High	−0.81	−0.88	−0.93	−0.95	−0.97
Very High	−0.77	−0.85	−0.92	−0.95	−0.97
	−0.73	−0.83	−0.91	−0.94	−0.96

SOURCE: Reprinted by permission from *ASHRAE Fundamentals 1993*, p. 26.30.

[a] *Circulation rate:*
   Low: Minimum required to handle cooling load from lights and occupants in interior zone. Supply through floor, wall, or ceiling diffuser; ceiling space not used for return. h_i = 0.4 Btu/hr · ft² · °F (2.3 W/m² · °C) (h_i = inside surface convection coefficient).
   Medium: Supply through floor, wall, or ceiling diffuser; ceiling space not used for return. h_i = 0.6 Btu/hr · ft² · °F (3.4 W/m² · °C).
   High: Room air circulation induced by primary air of induction unit or by room fan and coil unit. Ceiling used for return. h_i = 0.8 Btu/hr · ft² · °F (4.5 W/m² · °C).
   Very High: Air circulation used to minimize temperature gradients in room; ceiling space used for return. h_i = 0.8 Btu/hr · ft² · °F (4.5 W/m² · °C).

[b] Floor covered with carpet and pad. For bare floor take next value of w_1 down the column.

Table 16.2 for this roof with a medium building construction. The results for the first 2 hr of computation are:

$$ICL_1/A = 0.681(-1.437) - 0.621(-0.936) + 0.94(0) = -0.397 \text{ W/m}^2$$

$$ICL_2/A = 0.681(-1.808) - 0.621(-1.437) + 0.94(-0.397) = -0.712 \text{ W/m}^2$$

Table 16.3 summarizes the results. Note that seven days of calculations are required to reach a periodic solution to three decimal places. The long simulation time required is due to the large amount of thermal mass present in the building. The actual thermal response of this building is also slow for the same reason. The magnitude of the maximum instantaneous cooling load is about 18.5 W/m², whereas the maximum instantaneous heat gain is 23.5 W/m², and these maximums occur at the same time. This shows that the thermal storage in this example building reduces the peak cooling load below the instantaneous heat gain. Also note that the heat gain is negative at certain times of the day, whereas the cooling load remains positive. The cooling load is more uniform than the heat gain.

EXAMPLE 16.2

Compute the hourly cooling load per square meter of floor area due to lighting for the building used in Ex. 15.5 and 16.1. Assume 5 W/m² of unvented lighting is used from 0800 hrs in the morning until 1700 hrs in the afternoon each day. Begin the calculations on a Monday, assuming that the lights have not been used for several days so there is no residual heat in the building from previous use of the lights. Compute the hourly instantaneous cooling load for the next five days of the week ending at midnight on the following Friday.

TABLE 16.2 v Coefficients for Cooling-Load Transfer Function[a]

Heat-Gain Component	Room Envelope Construction[b]	v_0	v_1 Dimensionless
Solar heat gain through glass[c] with no interior shade; radiant heat from equipment and people	Light	0.224	$1 + w_1 - v_0$
	Medium	0.197	$1 + w_1 - v_0$
	Heavy	0.187	$1 + w_1 - v_0$
Conduction heat gain through exterior walls, roofs, partitions, doors, windows w/blinds or drapes	Light	0.703	$1 + w_1 - v_0$
	Medium	0.681	$1 + w_1 - v_0$
	Heavy	0.676	$1 + w_1 - v_0$
Convective heat generated by equipment and people, and from ventilation and infiltration air	Light	1.000	0.0
	Medium	1.000	0.0
	Heavy	1.000	0.0

Heat Gain From Lights[d]

Furnishings	Air Supply and Return	Type of Light Fixture	v_0	v_1
Heavyweight simple furnishings, no carpet	Low rate; supply and return below ceiling $(V \le 0.5(25))^e$	Recessed, not vented	0.450	$1 + w_1 - v_0$
Ordinary furnishings, no carpet	Medium to high rate, supply and return below or ceiling $(V \ge 0.5(25))^e$	Recessed, not vented	0.550	$1 + w_1 - v_0$
Ordinary furnishings, with or without carpet on floor	Medium to high rate, or induction unit or fan and coil, supply and return below, or through ceiling, return-air plenum $(V \ge 0.5(25))^e$	Vented	0.650	$1 + w_1 - v_0$
Any type furniture, with or without carpet	Ducted returns through light fixtures	Vented or free-hanging in air-stream with ducted returns	0.750	$1 + w_1 - v_0$

SOURCE: Reprinted by permission from *ASHRAE Fundamentals 1993*, p. 26.30.

[a] The transfer functions in this table were calculated by procedures outlined in G. P. Mitalas and D. G. Stephenson, "Room Thermal Response Factors," *ASHRAE Trans.*, 73(2): III.2.1. and are acceptable for cases where all heat-gain energy eventually appears as cooling load. The computer program used was developed at the National Research Council of Canada, Division of Building Research.

[b] The construction designations denote the following:
Light construction: such as frame exterior wall, 2-in. (50-mm) concrete floor slab, approximately 30 lbm of material/ft^2 (150 kg/m^2) of floor area.
Medium construction: such as 4-in. (100-mm) concrete exterior wall, 4-in. concrete floor slab approximately 70 lbm of building material/ft^2 (340 kg/m^2) of floor area.
Heavy construction: such as 6-in. concrete exterior wall, 6-in. (150-mm) concrete floor slab, approximately 130 lbm of building material/ft^2 (630 kg/m^2) of floor area.

[c] The coefficients of the transfer function that relate room cooling load to solar heat gain through glass depend on where the solar energy is absorbed. If the window is shaded by an inside blind or curtain, most of the solar energy is absorbed by the shade and is transferred to the room by convection and long-wave radiation in about the same proportion as the heat gain through walls and roofs; thus the same transfer coefficients apply. Radiant portion is 70 percent of heat gain from equipment and people.

[d] If room supply air is exhausted through the space above the ceiling and lights are recessed, such air removes some heat from the lights that would otherwise have entered the room. This removed light heat is still a load on the cooling plant if the air is recirculated, even though it is not a part of the room heat gain as such. The percent of heat gain appearing in the room depends on the type of lighting fixture, its mounting, and the exhaust airflow. Lighting heat gain is assumed to be 59 percent radiant.

[e] V is room air-supply rate in ft^3/min/ft^2 (L/s/m^2) of floor area.

TABLE 16.3 Results of Ex. 16.1

v-Values	w-Values	i-Value
0.681	1	0
−0.621	−0.94	1
0	0	2

		Cooling Load, W/m²						
Time	Heat Gain	ICL Day 1	ICL Day 2	ICL Day 3	ICL Day 4	ICL Day 5	ICL Day 6	ICL Day 7
	0.305	0	2.797	3.220	3.315	3.337	3.342	3.343
	−0.383	0	2.179	2.576	2.666	2.687	2.691	2.692
	−0.936	0	1.649	2.022	2.107	2.126	2.130	2.131
1	−1.437	−0.397	1.152	1.503	1.583	1.601	1.605	1.606
2	−1.808	−0.712	0.744	1.074	1.149	1.166	1.170	1.171
3	−1.991	−0.903	0.467	0.777	0.847	0.863	0.867	0.867
4	−2.164	−1.086	0.201	0.493	0.559	0.574	0.577	0.578
5	−2.378	−1.296	−0.086	0.188	0.250	0.264	0.267	0.268
6	−2.282	−1.296	−0.158	0.099	0.158	0.171	0.174	0.174
7	−0.385	−0.063	1.006	1.248	1.303	1.315	1.318	1.319
8	3.488	2.555	3.560	3.788	3.839	3.851	3.854	3.854
9	8.218	5.832	6.777	6.991	7.039	7.050	7.053	7.053
10	12.901	9.164	10.052	10.254	10.299	10.309	10.312	10.312
11	17.058	12.220	13.054	13.243	13.286	13.296	13.298	13.299
12	20.384	14.775	15.560	15.737	15.777	15.787	15.789	15.789
13	22.61	16.627	17.365	17.532	17.570	17.578	17.580	17.581
14	23.501	17.593	18.286	18.443	18.479	18.487	18.489	18.489
15	22.953	17.574	18.226	18.374	18.407	18.415	18.416	18.417
16	21.034	16.590	17.203	17.342	17.373	17.380	17.382	17.382
17	17.838	14.680	15.256	15.387	15.416	15.423	15.424	15.425
18	13.618	11.996	12.537	12.660	12.688	12.694	12.695	12.696
19	8.671	8.724	9.233	9.348	9.375	9.381	9.382	9.382
20	4.043	5.570	6.048	6.156	6.181	6.186	6.188	6.188
21	1.467	3.724	4.173	4.275	4.298	4.303	4.305	4.305
22	0.305	2.797	3.220	3.315	3.337	3.342	3.343	3.343
23	−0.383	2.179	2.576	2.666	2.687	2.691	2.692	2.692
24	−0.936	1.649	2.022	2.107	2.126	2.130	2.131	2.131
		Day 1	Day 2	Day 3	Day 4	Day 5	Day 6	Day 7

Solution: The value for the coefficient w_1 remains the same as in Ex. 16.1, because it depends only on the thermal mass of the building and the air-circulation rate. The values for the v coefficients are obtained from Table 16.2. Warehouse lighting is usually freely suspended in the air and is unvented. The closest description in Table 16.2 is that of recessed, unvented lighting with high air-circulation rate. The remaining coefficients become

$$v_0 = 0.550$$

$$v_1 = 1 + w_1 - v_0 = 1 - 0.94 - 0.55 = -0.49$$

The lighting heat gain is zero when the lights are not used, and it is 5 W/m² when the lights are on. Table 16.4 shows the results of the transfer-function calculations using the coefficients and hourly heat gains described previously. Note that the cooling load is zero until

TABLE 16.4 Results of Ex. 16.2

v-Values	w-Values	i-Value
0.55	1	0
−0.49	−0.94	1
0	0	2

		Cooling Load, W/m²						
Time	Heat Gain	ICL Day 1	ICL Day 2	ICL Day 3	ICL Day 4	ICL Day 5	ICL Day 6	ICL Day 7
	0	0	0.705	0.865	0.901	0.909	0.911	0.911
	0	0	0.663	0.813	0.847	0.855	0.856	0.857
	0	0	0.623	0.764	0.796	0.803	0.805	0.805
1	0	0.000	0.586	0.718	0.748	0.755	0.757	0.757
2	0	0.000	0.551	0.675	0.703	0.710	0.711	0.712
3	0	0.000	0.517	0.635	0.661	0.667	0.669	0.669
4	0	0.000	0.486	0.597	0.622	0.627	0.628	0.629
5	0	0.000	0.457	0.561	0.584	0.590	0.591	0.591
6	0	0.000	0.430	0.527	0.549	0.554	0.555	0.556
7	0	0.000	0.404	0.496	0.516	0.521	0.522	0.522
8	5	2.750	3.130	3.216	3.235	3.240	3.241	3.241
9	5	2.885	3.242	3.323	3.341	3.345	3.346	3.346
10	5	3.012	3.347	3.423	3.441	3.445	3.445	3.446
11	5	3.131	3.447	3.518	3.534	3.538	3.539	3.539
12	5	3.243	3.540	3.607	3.622	3.626	3.626	3.627
13	5	3.349	3.627	3.691	3.705	3.708	3.709	3.709
14	5	3.448	3.710	3.769	3.783	3.786	3.786	3.786
15	5	3.541	3.787	3.843	3.856	3.858	3.859	3.859
16	5	3.628	3.860	3.912	3.924	3.927	3.928	3.928
17	0	0.961	1.178	1.228	1.239	1.241	1.242	1.242
18	0	0.903	1.108	1.154	1.164	1.167	1.167	1.168
19	0	0.849	1.041	1.085	1.095	1.097	1.097	1.097
20	0	0.798	0.979	1.020	1.029	1.031	1.032	1.032
21	0	0.750	0.920	0.958	0.967	0.969	0.970	0.970
22	0	0.705	0.865	0.901	0.909	0.911	0.911	0.912
23	0	0.663	0.813	0.847	0.855	0.856	0.857	0.857
24	0	0.623	0.764	0.796	0.803	0.805	0.805	0.805
		Day 1	Day 2	Day 3	Day 4	Day 5	Day 6	Day 7

the lights are first turned on at 0800 hrs on the first day. The cooling load increases during the day and then diminishes after the lights are turned off at 1700 hrs. The cooling load for the second day is higher than for the first day because of the heat stored in the building from the first day. The cooling load gradually increases until it reaches a periodic condition. Successive days should be nearly the same after day 5.

Example 16.2 shows that the maximum cooling load occurs after several days of periodic heat gain—in this case, five days. The reason is that the residual cooling load from the previous day is nonzero at 0800 hrs in the morning when the lights are turned on again. The residual cooling from the previous day becomes approximately 0.5 W/m², which is about 10 percent of the lighting instantaneous heat gain. For design purposes, one should use the maximum values for the cooling load which occur when the heat gain is periodic.

The coefficients listed in Table 16.2 assume that the heat gain from people and equipment is 70 percent radiative and 30 percent convective and that the heat gain from lighting is 59 percent radiative and 41 percent convective. If the split between convective and radiative heat gain is known to differ substantially from these assumed values, the coefficients should be corrected. ASHRAE recommends that the v coefficients be changed to

$$v_0' = \left(\frac{r'}{r}\right)v_0 + \left(1 - \frac{r'}{r}\right) \tag{16.3}$$

$$v_1' = \left(\frac{r'}{r}\right)v_1 + \left(1 - \frac{r'}{r}\right)w_1 \tag{16.4}$$

where the unprimed quantities are the assumed or tabulated values and the primed quantities are the new or corrected values. The percentage of the heat gain that is radiant is designated by r and r'. The w coefficients do not change, as the previous cooling load depends only on the thermal storage of the building, not on the convective-radiative heat-gain split.

Once the cooling loads have been determined for all the heat gains associated with a conditioned space, they are summed to obtain the total space cooling load. It is the maximum of this sum that determines the maximum cooling load and the time at which it occurs. Note that the maximum total cooling load usually does not equal the sum of the maximums of the components. The reason is that the various components reach their maximum values at different times of the day. In Ex. 16.1, the roof cooling load reached a maximum at 1400 hrs. In Ex. 16.2, the cooling load due to the lights reached a maximum at 1600 hrs. The maximum total cooling load of these two components will occur between 1400 and 1600 hrs. Figure 16.2 illustrates qualitatively what the cooling-load distributions are for three components, each with a different time of peak and each with a different magnitude of maximum cooling load. The component with the largest peak cooling load tends to dominate the total, both in magnitude and time of maximum load. If one component dominates all others, then only that component needs to be considered in detail to obtain a reasonable estimate of the time and magnitude of the space total maximum cooling load.

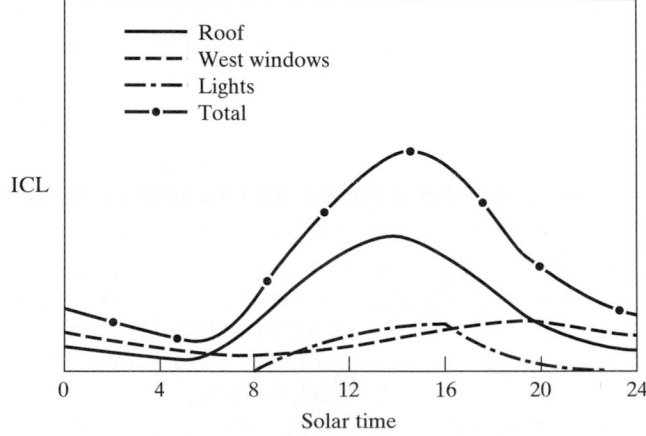

Figure 16.2 Three cooling-load components and their sum vs. time.

The previous discussion pertains to estimating the peak cooling load for proper cooling-equipment sizing when the indoor-air temperature is assumed to remain fixed. The transfer-function approach can be extended to predict conditioned-space air temperatures and heat-removal rates when the space temperature is allowed to float between prescribed limits. This often occurs when the cooling equipment is turned off during unoccupied periods. Details of this procedure are not covered here. The reader is referred to the *ASHRAE Handbook of Fundamentals* [1] for further discussion.

16.4 HAND-CALCULATION (CLTD) METHOD

The transfer-function method is convenient to use on a computer and can be used with any prescribed weather data. Often we wish to know only the peak cooling load during the summer to size the cooling system. If the building does not have very many separately controlled zones, a hand-calculation method is sufficient. The *ASHRAE Cooling-Load Temperature-Difference (CLTD) method* is described here as an approximate hand-calculation method for determining summer design cooling load. As the procedures for walls and roofs, fenestrations, internal loads, and infiltration are slightly different, each will be considered separately.

Walls and Roofs

One approximate method to compute cooling loads from structures such as walls and roofs is to assume that a fraction of the sensible-heat gain is convective, which gives an instantaneous cooling load on the cooling coil, and the remainder is radiative. The radiative portion is averaged over a specified time period to estimate the cooling load from the stored energy within the building. The instantaneous cooling load at time τ, ICL_τ, can be written

$$\text{ICL}_\tau = A \left[q_{\tau,c} + \frac{\displaystyle\sum_{n=0}^{N} q_{\tau - n\Delta\tau}}{N} \right] \tag{16.5}$$

where the total number of hours over which the radiant portion of the heat gain is averaged is given by N. The instantaneous heat gain may be computed by using any of the methods described previously. For hand calculations it is convenient to compute the instantaneous heat gain using the *Total Equivalent Temperature Difference (TETD)*. Combining Eqs. (15.32) and (16.5),

$$\text{ICL}_\tau = \text{UA} \left[(\% \text{ conv.})\text{TETD}_\tau + \frac{\% \text{ rad.}}{N} \sum_{n=0}^{N} \text{TETD}_{\tau - n\Delta\tau} \right] \tag{16.6}$$

ASHRAE recommends that the values for the convective and radiative percentages be 0.4 and 0.6, respectively, as approximately 60 percent of the interior surface film coefficient is infrared radiation and 40 percent is convection. ASHRAE also recommends that the value of N be 2 to 3 hr for lightweight construction such as wood-frame walls and floors and 6 to 8 hours for heavy construction such as concrete and brick material.

The term in the brackets in Eq. (16.6) is only a function of time when the wall or roof construction, geographic location, orientation, and month of year are determined. It is more convenient to tabulate the entire term in the square brackets than to tabulate hourly values of TETD. The units of this term are those of temperature. Thus this term can be defined as the cooling-load temperature difference at time τ, or $CLTD_\tau$. Equation (16.6) then can be written

$$ICL_\tau = UA(CLTD_\tau) \qquad (16.7)$$

Values for CLTD are given in Table 16.5 for walls and in Table 16.6 for horizontal roofs. The wall and roof designations are given in Tables 15.8 and 15.9, respectively, with the material properties listed in Table 15.10. The walls have been grouped into categories, depending on the amount of mass and the amount of insulation in the wall. Walls with similar thermal time constants can be assigned similar values for CLTD. Note that the CLTD values are based on clear-sky weather conditions for July 21, 40 deg north latitude. Corrections for other indoor and outdoor conditions are given in the footnotes for each table. Tables for other latitudes and other wall or roof orientations can be generated in a similar manner. Table 16.7 contains correction factors to be used for different surface orientations and different months of the year. The following examples illustrate the use of the CLTD method for walls and roofs.

EXAMPLE 16.3

Determine the peak summer design cooling load for a west-facing wall of construction #W3 as given in Table 15.8. The wall is located at 40 deg north latitude and has a light-colored exterior. The indoor temperature of the building is maintained at 75 °F.

Solution: The uncorrected value for the peak CLTD is obtained from Table 16.5, CLTD = 49 °F, which occurs at 2000 hrs local solar time. This tabulated value should be corrected for the exterior color of the wall and for the change in indoor temperature from the assumed values. Therefore the corrected CLTD is

$$CLTD_{corr} = (49\ °F + 0)0.65 + (78\ °F - 75\ °F) = 35\ °F$$

The instantaneous cooling load per square foot of wall-surface area is

$$\frac{ICL}{A} = U(CLTD_{corr})$$

$$= 0.61945\ W/m^2 \cdot °C \left(\frac{0.1761\ Btu/hr \cdot ft^2 \cdot °F}{W/m^2 \cdot °C} \right) 35\ °F$$

$$= 3.82\ Btu/hr\ ft^2$$

The thermal conductance (*U*-value) was obtained from Table 15.8. The change from a dark exterior color to a light color reduces the CLTD value by a factor of 0.65. This indicates that only about 65 percent of the incident solar radiation is absorbed as compared to a dark-colored wall. The decrease in interior temperature from 78 to 75 °F directly adds 3 °F to the tabulated value.

EXAMPLE 16.4

Compute the hourly cooling load from a 1000-ft^2 horizontal roof similar to #R4 with an *R*-value of 20 hr $\cdot$ ft^2 $\cdot$ °F/Btu. The calculations are for a clear day in October with a daily average outdoor temperature of 60 °F and an indoor temperature of 70 °F.

TABLE 16.5 Cooling-Load Temperature Differences (CLTD) for Calculating Cooling Load from Sunlit Walls Described in Table 15.8

N Latitude Wall Facing	0100	0200	0300	0400	0500	0600	0700	0800	0900	1000	1100	1200	1300	1400	1500	1600	1700	1800	1900	2000	2100	2200	2300	2400
Walls #W1 and #W2																								
N (°F)	3	2	1	0	-1	2	7	8	9	12	15	18	21	23	24	24	25	26	22	15	11	9	7	5
N (°C)	2	1	0	0	0	1	4	5	5	7	8	10	12	13	13	14	14	15	12	8	6	5	4	3
E (°F)	4	2	1	0	-1	11	31	47	54	55	50	40	33	31	30	29	27	24	19	15	12	10	8	6
E (°C)	2	1	1	0	0	6	17	26	30	31	28	22	19	17	17	16	15	13	11	8	7	5	4	3
S (°F)	4	2	1	0	-1	0	1	5	12	22	31	39	45	46	43	37	31	25	20	15	12	10	8	5
S (°C)	2	1	0	0	0	0	1	3	7	12	17	22	25	26	24	21	17	14	11	8	7	5	4	3
W (°F)	6	5	3	2	1	1	2	5	8	11	15	19	27	41	56	67	72	67	48	29	20	15	11	8
W (°C)	4	3	2	1	1	1	1	3	5	6	8	10	15	23	31	37	40	37	27	16	11	8	6	5
Wall #W3																								
N (°F)	12	10	8	7	5	4	3	4	5	6	7	9	11	13	15	17	19	20	21	23	20	18	16	14
N (°C)	7	6	5	4	3	2	2	2	3	3	4	5	6	7	8	10	10	11	12	12	11	10	9	8
E (°F)	14	12	10	8	6	5	6	11	18	26	33	36	38	37	36	34	33	32	30	28	25	22	20	17
E (°C)	8	7	6	5	4	3	3	6	10	15	18	20	21	21	20	19	18	18	17	15	14	12	11	9
S (°F)	15	12	10	8	7	5	4	3	4	5	5	7	9	11	14	18	19	18	17	16	14	13	11	10
S (°C)	8	7	6	5	4	3	2	2	2	3	3	4	5	6	8	10	11	10	9	9	8	7	7	7
W (°F)	25	21	17	14	11	9	7	6	6	6	7	9	11	14	20	27	36	43	49	49	45	40	34	29
W (°C)	14	12	10	8	6	5	4	3	3	4	4	5	6	8	11	15	20	24	27	27	25	22	19	16
Walls #W4 and #W5																								
N (°F)	15	14	13	12	11	10	9	8	8	7	7	8	8	9	10	12	13	14	15	16	17	17	17	16
N (°C)	9	8	7	7	6	5	5	4	4	4	4	4	5	5	6	6	7	8	9	9	9	10	9	9
E (°F)	22	21	19	17	15	14	12	12	14	16	19	22	25	27	29	29	30	30	30	29	28	27	26	24
E (°C)	13	12	11	10	9	8	7	7	8	9	11	13	14	15	16	16	17	17	16	16	16	15	14	13
S (°F)	21	19	18	16	15	13	12	10	9	9	9	10	11	14	17	20	22	24	25	26	25	25	24	22
S (°C)	12	11	10	9	8	7	6	6	5	5	5	5	6	8	9	11	12	13	14	14	14	14	13	12
W (°F)	31	29	27	25	22	20	18	16	14	13	12	12	12	13	14	16	20	24	29	32	35	35	35	33
W (°C)	17	16	15	14	12	11	10	9	8	7	7	7	7	7	8	9	11	13	16	18	19	20	19	18

(Column group header: Solar Time, hr)

SOURCE: Adapted by permission from *ASHRAE Fundamentals 1989*, p. 26.37.

(1) *Direct Application of the Table without Adjustments:*
Values in this table were calculated using the same conditions outlined for walls as outlined for the roof CLTD table, Table 16.6. These values may be used for all normal air conditioning estimates, usually without correction (except where noted below) when the load is calculated for the hottest weather. For totally shaded walls use the North orientation values.

(2) *Adjustments to Table Values:*
The following equation makes adjustments for conditions other than those listed in note (1) above.

$$CLTD_{corr} = (CLTD + LM) K + (78 \text{ °F} (25.5 \text{ °C}) - t_i) + (t_o - 85 \text{ °F} (29.4 \text{ °C}))$$

where   CLTD is from Table 16.5 at the wall orientation.
LM is the latitude-month correction from Table 16.7.
K is a color-adjustment factor applied after first making the month-latitude adjustment.

K = 1.0 if the surface is dark-colored or light-colored in an industrial area.
K = 0.83 if the surface is permanently medium-colored in a nonindustrial area.
K = 0.65 if the surface is permanently light-colored in a nonindustrial area.
   Light color: cream
   Medium color: medium blue, medium green, bright red, light brown, unpainted wood, concrete
   Dark color: dark blue, red, brown, green
($78 \text{ °F} - t_i$) or ($25.5 \text{ °C} - t_i$) is a correction for an indoor design temperature not equal to 78 °F (25.5 °C).
($t_o - 85 \text{ °F}$) or ($t_o - 29.4 \text{ °C}$) is a correction for a daily average outdoor temperature not equal to 85 °F (29.4 °C).

(3) *Wall Construction Not Given in Table:*
For walls that have similar thermal resistance, thermal mass, and heat capacity as those listed in the table, use the CLTD values from the table corrected by note (2) above.

TABLE 16.6 Cooling-Load Temperature Differences (CLTD) for Calculating Cooling Load from Sunlit Horizontal Roofs Described in Table 15.9

Description	Weight, lbm/ft² kg/m²		Solar Time 1	2	3	4	5	6	7	8	9	10	11	12	13	14	15	16	17	18	19	20	21	22	23	24	Hour of Max. CLTD	Min. CLTD	Max. CLTD	CLTD Range
R1	24	°F	22	17	13	9	6	3	1	1	3	7	15	23	33	43	51	58	62	64	62	57	50	42	35	28	18	1	64	63
	117	°C	12	10	7	5	3	2	1	0	2	4	8	13	18	24	29	33	35	36	35	32	28	24	19	16	18	0	36	36
R2	52	°F	25	22	18	15	12	9	8	8	10	14	20	26	33	40	46	50	53	53	52	48	43	38	34	30	18	8	53	45
	254	°C	14	12	10	8	7	5	4	4	6	8	11	15	18	22	25	28	29	30	29	27	24	21	19	16	18	4	30	26
R3	54	°F	30	29	27	26	24	22	21	20	20	21	22	24	27	29	32	34	36	38	38	38	37	36	34	33	19	20	38	18
	264	°C	17	16	15	14	13	13	12	11	11	11	12	13	15	16	18	19	20	21	21	21	21	20	19	18	19	11	21	10
R4	8	°F	1	−2	−3	−3	−5	−3	6	19	34	49	61	71	78	79	77	70	59	45	30	18	12	8	5	3	14	−5	79	84
	39	°C	0	−1	−2	−2	−3	−2	3	11	19	27	34	40	43	44	43	39	33	25	17	10	7	5	3	1	14	−3	44	47
R5	9	°F	3	0	−3	−4	−5	−7	−6	−3	5	16	27	39	49	57	63	64	62	57	48	37	26	18	11	7	16	−7	64	71
	44	°C	2	0	−2	−3	−4	−4	−4	−2	3	9	15	22	27	32	35	36	35	32	27	20	14	10	6	3	16	−4	36	40

SOURCE: Adapted by permission from *ASHRAE Fundamentals 1989*, p. 26.35.

(1) *Direct Application of Table without Adjustments:*

Values were calculated using the following conditions:

- Dark flat surface roof ("dark" for solar radiation absorption)
- Indoor temperature of 78 °F (25.5 °C)
- Outdoor maximum temperature of 95 °F (35 °C) with outdoor mean temperature of 85 °F (29.4 °C) and an outdoor daily range of 21 °F (11.6 °C)
- Solar radiation typical of 40 deg north latitude on July 21
- Outside surface resistance, $R_o = 0.333$ ft² · °F · hr/Btu (0.059 m² · °C/W)
- Without and with suspended ceiling, but no attic fans or return-air ducts in suspended ceiling space
- Inside surface resistance, $R_i = 0.685$ ft² · °F · hr/Btu (0.121 m² · °C/W)

(2) *Adjustments to Table Values:*
The following equation makes adjustments for deviations of design and solar conditions from those listed in note (1) above.

$$CLTD_{corr} = ((CLTD + LM)K + (78\ °F\ (25.5\ °C) - t_i) + (t_o - 85\ °F\ (29.4\ °C)))\ f$$

where CLTD is from this table.

(a) LM is latitude-month correction from Table 16.7 for a horizontal surface.
(b) *K* is a color-adjustment factor applied after first making month-latitude adjustments. Credit should not be taken for a light-colored roof except where permanence of light color is established by experience, as in rural areas or where there is little smoke.

$K = 1.0$ if dark colored or light in an industrial area
$K = 0.5$ if permanently light-colored (rural area)

(c) (78 °F − t_i) or (25.5 °C − t_i) is indoor design temperature correction.
(d) (t_o − 85 °F) or (t_o − 29.4 °C) is outdoor design temperature correction, where t_o is the average outside temperature on design day.
(e) *f* is a factor for attic fan and or ducts above ceiling applied after all other adjustments have been made.

$f = 1.0$ no attic or ducts
$f = 0.75$ positive ventilation

Values in Table 16.6 were calculated without and with suspended ceiling, but make no allowances for positive ventilation or return ducts through the space. If ceilling is insulated and fan is used between ceiling and roof, CLTD may be reduced 25 percent ($f = 0.75$). Analyze use of the suspended ceiling space for a return-air plenum or with return-air ducts separately.

(3) *Roof Constructions Not Listed in Table:*
The *U*-values listed are only guides. The actual value of *U* as obtained from tables or as calculated for the actual roof construction should be used.

An actual roof construction not in this table would be thermally similar to a roof in the table, if it has similar mass and similar heat capacity. In this case, use the CLTD from this table as corrected by note (2) above.

(4) *Additional Insulation:*
For each 7 hr · ft² · °F/Btu (1.23 m² · °C/W) increase in *R*-value from insulation added to the roof structure, use a CLTD for a roof whose weight and heat capacity are approximately the same, but whose CLTD has a maximum value 2 hr later. If this is not possible, because a roof with longest time lag has already been selected, use an effective CLTD in cooling-load calculation equal to 29 °F (16 °C).

Month	N	NNE, NNW	NE, NW	ENE, WNW	E, W	ESE, WSW	SE, SW	SSE, SSW	S	Hor.
				Fahrenheit						
Dec.	−6	−8	−10	−13	−10	−7	0	7	10	−21
Jan./Nov.	−5	−7	−10	−12	−9	−6	1	8	11	−19
Feb./Oct.	−5	−7	−8	−9	−6	−3	3	8	12	−14
Mar./Sept.	−4	−5	−5	−6	−3	−1	4	7	10	−8
Apr./Aug.	−2	−3	−2	−2	0	0	2	3	4	−3
May/Jul.	0	0	0	0	0	0	0	0	1	1
Jun.	1	1	1	0	1	0	0	−1	−1	2
				Celsius						
Dec.	−3.3	−4.4	−5.5	−7.2	−5.5	−3.8	0	3.8	5.5	−11.6
Jan./Nov.	−2.7	−3.8	−5.5	−6.6	−5.0	−3.3	0.5	4.4	6.1	−10.5
Feb./Oct.	−2.7	−3.8	−4.4	−5.0	−3.3	−1.6	1.6	4.4	6.6	−7.7
Mar./Sept.	−2.2	−2.7	−2.7	−3.3	−1.6	0.5	2.2	3.8	5.5	−4.4
Apr./Aug.	−1.1	−1.6	−1.1	−1.1	0	0	1.1	1.6	2.2	1.6
May/Jul.	0	0	0	0	0	0	0	0	0.5	0.5
Jun.	0.5	0.5	0.5	0	0.5	0	0	−0.5	−0.5	1.1

SOURCE: Abstracted by permission from *ASHRAE Fundamentals 1989*, p. 26.37.

Solution: The actual R-value for this roof will be compared to the tabulated value to determine whether the CLTD values for roof #R4 should be used rather than other values.

$$R_{actual} = 20 \text{ hr} \cdot \text{ft}^2 \cdot {}^\circ\text{F/Btu}$$

$$R_{table} = \frac{1}{U_{table}} = \frac{1}{0.45569} \text{ W/m}^2 \cdot {}^\circ\text{C}$$

$$= 2.19 \text{ m}^2 \cdot {}^\circ\text{C/W} \left(\frac{1 \text{hr} \cdot \text{ft}^2 \cdot {}^\circ\text{F/Btu}}{0.1761 \text{ m}^2 \cdot {}^\circ\text{C/W}} \right) = 12.5 \text{ hr} \cdot \text{ft}^2 \cdot {}^\circ\text{F/Btu}$$

The difference in R-values is approximately 7 hr $\cdot$ ft^2 $\cdot$ °F/Btu, so an entry in Table 16.6 with a maximum CLTD 2 hr later than roof #R4 should be used. This is to account for the increased damping and time lag associated with the additional insulation. The roof entry that has its maximum CLTD 2 hr later is roof #R5, so these values will be used. The CLTD values for roof #R5 must be corrected to account for the change in month(incident solar radiation) and the changes in both indoor and outdoor daily average temperatures:

$$\text{CLTD}_{corr} = (\text{CLTD}_{table} - 14 \,{}^\circ\text{F}) + (78 \,{}^\circ\text{F} - 70 \,{}^\circ\text{F}) + (60 \,{}^\circ\text{F} - 85 \,{}^\circ\text{F})$$

$$= \text{CLTD}_{table} - 31 \,{}^\circ\text{F}$$

The corrected values for this roof are given in Table 16.8. The instantaneous cooling load is calculated by

$$\text{ICL}_\tau = U(A)\text{CLTD}_{corr, \tau}$$

At 0100 hrs the instantaneous cooling load is

$$\text{ICL}_{0100} = \frac{1}{20} \frac{\text{Btu}}{\text{hr} \cdot \text{ft}^2 \cdot {}^\circ\text{F}} 1000 \text{ ft}^2 (- 28 \,{}^\circ\text{F}) = -1400 \text{ Btu/hr}$$

Table 16.8 lists the hourly values for ICL_τ. Owing to the combination of reduced incident solar radiation in October compared to July and the low average outdoor-air temperature, the only positive cooling loads occur between noon and 2000 hrs. The cooling load is negative for the remaining hours, which indicates a heat loss rather than a heat gain. The daily average cooling load is nearly zero. The absorbed solar radiation during the day approximately compensates for the heat loss at night.

Fenestrations

Equation (15.79) describes a method to compute the instantaneous heat gain through a fenestration per unit surface area:

$$q_i = (\mathrm{SC})(\mathrm{SHGF}) + U(t_o - t_i) \qquad (15.79)$$

The first term in this equation accounts for the solar radiation that is transmitted through or absorbed by the glazing material and added to the interior. The second term adds the energy transmitted through the material by conduction, convection, and infrared radiation.

We will consider the second term in Eq. (15.79) first. The thermal mass of a fenestration material is very small compared to that of other architectural elements of a building. Therefore, we have ignored the heat-gain time delay through the fenestration and treated it as steady heat transmission.

The cooling load is delayed and reduced in magnitude, owing to the storage of absorbed infrared radiation inside the building. Computer calculations using the transfer-function method have shown that the fenestration-transmission cooling load for most buildings can be estimated using the cooling-load temperature differences listed in Table 16.9. As doors also have very little thermal mass, these same CLTD values can be used for doors that do not receive direct solar radiation. The second term in Eq. (15.79) now can be written as

$$\mathrm{ICL}_{\tau,\,\mathrm{cond}} = (U)(A)\mathrm{CLTD}_\tau \qquad (16.8)$$

The first term in Eq. (15.79) must also be corrected for the thermal storage in the building. The transfer-function method has been used to provide cooling-load results from the solar radiant-heat gain. These cooling-load results are primarily a function of the thermal mass in the building. Several hand-calculation methods have been devised that rely on tables to provide the hourly conversion from heat gain to cooling load. One of these methods uses the following equation:

$$\mathrm{ICL}_{\tau,\,\mathrm{rad}} = A_{\mathrm{glass}}\mathrm{SHGF}_\tau\,(\mathrm{SC})\mathrm{CLF}_\tau \qquad (16.9)$$

This equation requires knowledge of the hourly solar-heat-gain factor, SHGF, and the hourly cooling-load factor, CLF. Tables of both of these quantities need to be provided for hand calculations. Only the net glazing area is used here. The total area, glazing plus framing, is used in Eq. (16.8) with the average U-value for the entire window.

A second method is to base the cooling-load factors on the maximum solar-heat-gain factor for the day rather than on instantaneous values. Thus Eq. (16.9) can be modified to

$$\mathrm{ICL}_{\tau,\,\mathrm{rad}} = A_{\mathrm{glass}}\mathrm{SHGF}_{\mathrm{max}}(\mathrm{SC})\mathrm{CLF}'_\tau \qquad (16.10)$$

This reduces the number of tabulated values that are necesary, as only one value for SHGF is needed for each surface orientation at each latitude for each month.

TABLE 16.8 Results of Ex. 16.4

												Local Solar Time												
	1	2	3	4	5	6	7	8	9	10	11	12	13	14	15	16	17	18	19	20	21	22	23	24
Corrected CLTD (°F)	-28	-31	-33	-35	-36	-38	-37	-34	-26	-15	-4	8	18	26	32	33	31	26	17	6	-5	-13	-20	-24
ICL (Btu/hr)	-1400	-1550	-1650	-1750	-1800	-1900	-1850	-1700	-1300	-750	-200	400	900	1300	1600	1650	1550	1300	850	300	-250	-650	-1000	-1200

TABLE 16.9 Cooling-Load Temperature Differences (CLTD) for Conduction through Glass and Doors

												Solar Time, hr												
	1	2	3	4	5	6	7	8	9	10	11	12	13	14	15	16	17	18	19	20	21	22	23	24
CLTD, °F	1	0	−1	−2	−2	−2	−2	0	2	4	7	9	12	13	14	14	13	12	10	8	6	4	3	2
CLTD, °C	1	0	−1	−1	−1	−1	−1	0	1	2	4	5	7	7	8	8	7	7	6	4	3	2	2	1

Source: Reprinted by permission from *ASHRAE Fundamentals 1993*, p. 26.49.

Corrections: The values in the table were calculated for an inside temperature of 78 °F (25.5 °C) and an outdoor maximum temperature of 95 °F (35 °C) with an outdoor daily range of 21 °F (12 °C). The table remains approximately correct for other outdoor maximums (93 to 102 °F) (34 to 39 °C) and other outdoor daily ranges (16 to 34 °F) (9 to 19 °C), provided the outdoor daily average temperature remains approximately 85 °F (29.4 °C). If the room-air temperature is different from 78 °F (25.5 °C) and/or the outdoor daily average temperature is different from 85 °F (29.4 °C), see note 2, Table 16.5.

TABLE 16.10 Zone Types for Use with SCL and CLF Tables, Single-Story Building

No. Walls	Floor Covering	Partition Type	Inside Shade	Glass Solar	People and Equipment	Lights
	Zone Parameters				Zone Type	
1 or 2	Carpet	Gypsum	—	A	B	B
1 or 2	Carpet	Concrete block	—	B	C	C
1 or 2	Vinyl	Gypsum	Full	B	C	C
1 or 2	Vinyl	Gypsum	Half to None	C	C	C
1 or 2	Vinyl	Concrete block	Full	C	D	D
1 or 2	Vinyl	Concrete block	Half to None	D	D	D
3	Carpet	Gypsum	—	A	B	B
3	Carpet	Concrete block	Full	A	B	B
3	Carpet	Concrete block	Half to None	B	B	B
3	Vinyl	Gypsum	Full	B	C	C
3	Vinyl	Gypsum	Half to None	C	C	C
3	Vinyl	Concrete block	Full	B	C	C
3	Vinyl	Concrete block	Half to None	C	C	C
4	Carpet	Gypsum	—	A	B	B
4	Vinyl	Gypsum	Full	B	C	C
4	Vinyl	Gypsum	Half to None	C	C	C

SOURCE: Adapted by permission from *ASHRAE Fundamentals 1993*, p. 26.49.

A third method is to combine the product of $SHGF_{max}$ and CLF', which further reduces the number of tables required. This product is given the term *Solar Cooling Load Factor (SCL)*. This factor is primarily a function of building thermal mass, latitude, time of year, and surface orientation. The equation for instantaneous cooling load now becomes

$$ICL_\tau = A_{glass}(SC)SCL_\tau + U(A)CLTD_\tau \qquad (16.11)$$

ASHRAE has identified four types of zones, into which the SCL values have been categorized. Table 16.10 lists the zone type to use for the cooling loads due to glass, people and equipment, and lights as a function of number of exterior walls, floor covering, partition, and inside shade for a single-story building. This table was obtained by correlating many building cooling-load simulations using the transfer function method. Values for the solar cooling load (SCL) are given in Table 16.11 for 40 deg north latitude in July. These values are to be used for fenestrations that are not externally shaded. The shading coefficient, SC, accounts for interior shading such as blinds and drapes. ASHRAE recommends that north-facing values be used for shaded areas, as this includes sky-diffuse and ground-reflected radiation but not direct solar radiation.

The amount of shaded area varies with time of day, as the solar angles are constantly changing. The instantaneous cooling load depends not only on the current shaded area but also on the shaded area at previous times. This makes a hand calculation of the cooling load for a partially shaded window very cumbersome. An approach that provides reasonable results is to use the current shading angles and areas when computing the instantaneous cooling load, as heat gains at previous times are not as important. However, one should use caution when computing a cooling load shortly after a sudden change in shading, such as when a window becomes shaded by a surrounding building. If the win-

TABLE 16.11 July Solar Cooling Load for Sunlit Glass 40° North Latitude

Zone Type A

Glass Face	\ Solar Time Hour 1	2	3	4	5	6	7	8	9	10	11	12	13	14	15	16	17	18	19	20	21	22	23	24
N	0	0	0	0	1	25	27	28	32	35	38	40	40	39	36	31	31	36	12	6	3	1	1	0
NE	0	0	0	0	2	85	129	134	112	75	55	48	44	40	37	32	26	18	7	3	2	1	0	0
E	0	0	0	0	2	93	157	185	183	154	106	67	53	45	39	33	26	18	7	3	2	1	0	0
SE	0	0	0	0	1	47	95	131	150	150	131	97	63	49	41	34	27	18	7	3	2	1	0	0
S	0	0	0	0	0	9	17	25	41	64	85	97	96	84	63	42	31	20	8	4	2	1	1	0
SW	0	0	0	0	0	9	17	24	30	35	39	64	101	133	151	152	133	93	35	17	8	4	2	1
W	1	0	0	0	0	9	17	24	30	35	38	40	65	114	158	187	192	156	57	27	13	6	3	2
NW	1	0	0	0	0	9	17	24	30	35	38	40	40	50	84	121	143	130	46	22	11	5	3	1
Hor.	0	0	0	0	0	24	69	120	169	211	241	257	259	245	217	176	125	70	29	14	7	3	2	1

Zone Type B

Glass Face	\ Solar Time Hour 1	2	3	4	5	6	7	8	9	10	11	12	13	14	15	16	17	18	19	20	21	22	23	24
N	2	2	1	1	1	22	23	24	28	32	35	37	38	37	35	32	31	35	16	10	7	5	4	3
NE	2	1	1	1	2	73	109	116	101	73	58	52	48	45	41	36	30	23	13	9	6	5	3	3
E	2	2	1	1	2	80	133	159	162	143	105	74	63	55	48	41	34	25	15	10	7	5	4	3
SE	2	2	1	1	1	40	81	112	131	134	122	96	69	58	49	42	35	26	15	10	8	6	4	3
S	2	2	1	1	1	8	15	21	36	56	74	86	87	79	63	46	37	27	16	11	8	6	4	3
SW	6	5	4	3	2	9	16	22	27	31	36	58	89	117	135	138	126	94	46	31	21	15	11	8
W	8	6	5	4	3	9	16	22	27	31	35	37	59	101	139	166	173	147	66	43	30	21	15	11
NW	6	5	4	3	2	9	16	22	27	31	34	37	37	46	76	108	128	119	51	33	22	16	11	8
Hor.	8	6	5	4	3	22	60	104	147	185	214	233	239	232	212	180	137	90	53	37	27	19	14	11

Zone Type C

Glass Face	\ Solar Time Hour 1	2	3	4	5	6	7	8	9	10	11	12	13	14	15	16	17	18	19	20	21	22	23	24
N	5	5	4	4	4	24	23	24	27	30	33	34	35	34	32	29	29	34	14	10	8	7	6	6
NE	7	6	6	5	6	75	106	107	88	61	49	47	45	43	40	36	31	25	16	13	11	10	9	8
E	9	8	8	7	8	83	130	148	145	124	89	62	56	52	47	43	37	30	20	16	15	13	12	11
SE	9	8	7	6	6	45	82	107	121	121	107	82	59	51	47	42	36	29	19	13	14	13	11	10
S	7	7	6	5	5	12	18	23	36	54	70	79	79	70	54	40	33	26	16	13	12	10	9	8
SW	14	12	11	10	9	15	21	26	29	33	36	57	86	98	124	125	111	80	37	28	23	20	17	15
W	17	15	13	12	11	17	22	27	31	34	36	37	59	101	132	153	156	128	50	35	28	24	21	19
NW	12	11	10	9	8	14	20	25	29	32	34	36	36	44	73	102	118	107	39	26	21	17	15	13
Hor.	24	21	19	17	16	34	68	107	144	175	199	212	215	207	189	160	123	83	53	44	38	34	30	27

(continued)

TABLE 16.11 (cont.)

Zone Type D

Glass Face	Hour 1	2	3	4	5	6	7	8	9	10	Solar Time 11	12	13	14	15	16	17	18	19	20	21	22	23	24
N	8	7	6	6	6	21	21	21	24	27	29	31	32	31	30	28	29	32	17	14	12	11	10	9
NE	11	10	9	8	9	63	87	90	77	58	49	48	46	44	42	39	35	29	22	19	17	15	14	12
E	15	13	12	11	11	70	107	123	124	110	85	65	60	57	53	48	43	37	29	25	22	20	18	16
SE	14	13	11	10	10	39	68	90	102	104	95	78	60	55	51	47	42	35	27	24	21	19	17	16
S	11	10	9	8	7	12	17	21	32	46	59	67	69	63	52	41	36	30	22	19	17	15	14	12
SW	21	19	17	15	14	18	22	25	28	31	34	51	74	94	106	109	100	78	45	37	33	29	26	23
W	25	23	20	18	17	21	24	28	30	33	34	35	53	84	112	130	135	116	57	46	39	35	31	28
NW	18	16	15	13	12	17	21	24	27	30	32	33	34	41	64	87	101	94	42	33	29	25	22	20
Hor.	37	33	30	27	24	38	64	95	124	150	171	185	191	188	176	156	128	96	72	63	56	50	45	41

Source: Reprinted by permission from *ASHRAE Fundamentals 1993*, p. 26.50.

Notes:
1. Values are in Btu/hr · ft². Multiply by 3.154 to convert to W/m².
2. Apply data directly to standard double-strength glass with no inside shade.
3. Data applies to 21st day of July.
4. For other types of glass and internal shade, use shading coefficients as multiplier. See text. For externally shaded glass, use north orientation. See text.

dow had been sunlit during several preceding hours, it should still be considered to be sunlit in order to ensure a conservative estimate of the cooling load.

EXAMPLE 16.5

Determine the instantaneous cooling load at 1200 solar time on July 21, 40 deg north latitude, for an east-facing double-glazed window with a 0.5-in. air space and a wood frame. The window dimensions are 3 ft wide by 4 ft high with a 2-in. frame around the perimeter. The interior temperature is 75 °F and the daily average outdoor temperature is 87 °F. The room being considered is in a single-story building and has wood-frame walls, two of which are on the exterior. The floor is carpet over concrete. The window has no internal or external shading.

Solution: We will use Eq. (16.11) to compute the cooling load. The net glazing area is

$$A_{glass} = (3 \text{ ft} - 4 \text{ in.})(4 \text{ ft} - 4 \text{ in.}) = 9.8 \text{ ft}^2$$

The shading coefficient from Table 15.17 is 0.88. The zone type to use for this building is type A from Table 16.10. The solar cooling load at 1200 hrs for this window is 67 Btu/hr · ft² from Table 16.11. The conduction heat transmission coefficient from Table 14.5 is 0.49 Btu/hr · ft² · °F. The cooling-load temperature difference from Table 16.9 is 9 °F. The corrected CLTD is

$$\text{CLTD}_{corr} = 9 \text{ °F} + (78 \text{ °F} - 75 \text{ °F}) + (87 \text{ °F} - 85 \text{ °F}) = 14 \text{ °F}$$

Substituting these values into Eq. (16.11),

$$\text{ICL}_{1200} = 9.8 \text{ ft}^2 (0.88)(67 \text{ Btu/hr} \cdot \text{ft}^2) + (0.49 \text{ Btu/hr} \cdot \text{ft}^2 \cdot \text{°F})(12 \text{ ft}^2)(14 \text{ °F})$$

$$\text{ICL}_{1200} = 578 + 82 = 660 \text{ Btu/hr}$$

This example shows that the cooling load is significant at solar noon, even though the window is no longer receiving direct solar radiation. The first term in Eq. (16.11) dominates for this window with 88 percent of the total cooling load.

Lights, People, and Equipment

The instantaneous sensible cooling load caused by lights, people, and equipment is governed primarily by the magnitude of the heat gain, the convective and infrared radiation split, and the thermal mass of the building. The magnitude of these heat gains was covered in Chapter 15. Here we will focus on the conversion of these heat gains to cooling loads. As for walls, roofs, and fenestrations, the cooling load lags the heat gain because of the amount of infrared radiation energy that is stored in the building. The sensible-heat gain can be converted into sensible instantaneous cooling load using the following equation:

$$\text{ICL}_\tau = \text{IHG}_\tau \text{CLF} \tag{16.12}$$

The cooling-load factor, CLF, is used for the conversion. Values for CLF have been deduced from many computer simulations using the transfer-function method. The results have been separated by zone type, as discussed earlier in the section on fenestrations. The recommended zone type to use is given in Table 16.10. The cooling-load factors are listed versus zone type, hour of day, and the amount of time the heat gain is present each day in Tables 16.12, 16.13, and 16.14. These tables assume the same heat-gain distribution

TABLE 16.12 Cooling-Load Factors for Lights*

Lights On for (Hours)	\multicolumn{24}{Number of Hours after Lights Turned On}

Lights On for (Hours)	1	2	3	4	5	6	7	8	9	10	11	12	13	14	15	16	17	18	19	20	21	22	23	24
Zone Type A																								
8	0.85	0.92	0.95	0.96	0.97	0.97	0.97	0.98	0.13	0.06	0.04	0.03	0.02	0.02	0.02	0.01	0.01	0.01	0.01	0.01	0.01	0.01	0.01	0.01
10	0.85	0.93	0.95	0.97	0.97	0.97	0.98	0.98	0.98	0.98	0.14	0.07	0.04	0.03	0.02	0.02	0.02	0.02	0.02	0.02	0.01	0.01	0.01	0.01
12	0.86	0.93	0.96	0.97	0.97	0.98	0.98	0.98	0.98	0.98	0.98	0.98	0.14	0.07	0.04	0.03	0.03	0.02	0.02	0.02	0.02	0.02	0.02	0.02
14	0.86	0.93	0.96	0.97	0.98	0.98	0.98	0.98	0.98	0.98	0.99	0.99	0.99	0.99	0.15	0.07	0.05	0.03	0.03	0.03	0.02	0.02	0.02	0.02
16	0.87	0.94	0.96	0.97	0.98	0.98	0.98	0.99	0.99	0.99	0.99	0.99	0.99	0.99	0.99	0.99	0.15	0.08	0.05	0.04	0.03	0.03	0.03	0.02
Zone Type B																								
8	0.75	0.85	0.90	0.93	0.94	0.95	0.95	0.96	0.23	0.12	0.08	0.05	0.04	0.04	0.03	0.03	0.03	0.02	0.02	0.02	0.02	0.02	0.02	0.01
10	0.75	0.86	0.91	0.93	0.94	0.95	0.95	0.96	0.96	0.97	0.24	0.13	0.08	0.06	0.05	0.04	0.04	0.03	0.03	0.03	0.03	0.02	0.02	0.02
12	0.76	0.86	0.91	0.93	0.95	0.95	0.96	0.96	0.97	0.97	0.97	0.97	0.24	0.14	0.09	0.07	0.05	0.05	0.04	0.04	0.03	0.03	0.03	0.02
14	0.76	0.87	0.92	0.94	0.95	0.96	0.96	0.97	0.97	0.97	0.97	0.98	0.98	0.98	0.25	0.14	0.09	0.07	0.06	0.05	0.05	0.04	0.04	0.03
16	0.77	0.88	0.92	0.95	0.96	0.96	0.97	0.97	0.97	0.98	0.98	0.98	0.98	0.98	0.98	0.99	0.25	0.15	0.10	0.07	0.06	0.05	0.05	0.04
Zone Type C																								
8	0.72	0.80	0.84	0.87	0.88	0.89	0.90	0.91	0.23	0.15	0.11	0.09	0.08	0.07	0.07	0.06	0.05	0.05	0.05	0.04	0.04	0.03	0.03	0.03
10	0.73	0.81	0.85	0.87	0.89	0.90	0.91	0.92	0.92	0.93	0.25	0.16	0.13	0.11	0.09	0.08	0.08	0.07	0.06	0.06	0.05	0.05	0.04	0.04
12	0.74	0.82	0.86	0.88	0.90	0.91	0.92	0.92	0.93	0.94	0.94	0.95	0.26	0.18	0.14	0.12	0.10	0.09	0.08	0.08	0.07	0.06	0.06	0.05
14	0.75	0.84	0.87	0.89	0.91	0.92	0.92	0.93	0.94	0.94	0.95	0.95	0.96	0.96	0.27	0.19	0.15	0.13	0.11	0.10	0.09	0.08	0.08	0.07
16	0.77	0.85	0.89	0.91	0.92	0.93	0.93	0.94	0.95	0.95	0.95	0.96	0.96	0.97	0.97	0.97	0.28	0.20	0.16	0.13	0.12	0.11	0.10	0.09
Zone Type D																								
8	0.66	0.72	0.76	0.79	0.81	0.83	0.85	0.86	0.25	0.20	0.17	0.15	0.13	0.12	0.11	0.10	0.09	0.08	0.07	0.06	0.06	0.05	0.04	0.04
10	0.68	0.74	0.77	0.80	0.82	0.84	0.86	0.87	0.88	0.90	0.28	0.23	0.19	0.17	0.15	0.14	0.12	0.11	0.10	0.09	0.08	0.07	0.06	0.06
12	0.70	0.75	0.79	0.81	0.83	0.85	0.87	0.88	0.89	0.90	0.91	0.92	0.30	0.25	0.21	0.19	0.17	0.15	0.13	0.12	0.11	0.10	0.09	0.08
14	0.72	0.77	0.81	0.83	0.85	0.86	0.88	0.89	0.90	0.91	0.92	0.93	0.94	0.94	0.32	0.26	0.23	0.20	0.18	0.16	0.14	0.13	0.12	0.10
16	0.75	0.80	0.83	0.85	0.87	0.88	0.89	0.90	0.91	0.92	0.93	0.94	0.94	0.95	0.96	0.96	0.34	0.28	0.24	0.21	0.19	0.17	0.15	0.14

SOURCE: Reprinted by permission from *ASHRAE Fundamentals* 1993, p. 26.52.
* See Table 16.10 for zone types. Data based on a radiative/convective fraction of 0.59/0.41.

TABLE 16.13 Cooling-Load Factors for People and Unhooded Equipment*

Hours in Space	Number of Hours after Entry into Space or Equipment Turned On																							
	1	2	3	4	5	6	7	8	9	10	11	12	13	14	15	16	17	18	19	20	21	22	23	24
Zone Type A																								
2	0.75	0.88	0.18	0.08	0.04	0.02	0.01	0.01	0.01	0.01	0.00	0.00	0.00	0.00	0.00	0.00	0.00	0.00	0.00	0.00	0.00	0.00	0.00	0.00
4	0.75	0.88	0.93	0.95	0.22	0.10	0.05	0.03	0.02	0.02	0.01	0.01	0.01	0.01	0.00	0.01	0.00	0.00	0.00	0.00	0.00	0.00	0.00	0.00
6	0.75	0.88	0.93	0.95	0.97	0.97	0.33	0.11	0.06	0.04	0.03	0.02	0.02	0.01	0.01	0.01	0.01	0.00	0.00	0.00	0.00	0.00	0.00	0.00
8	0.75	0.88	0.93	0.95	0.97	0.97	0.98	0.98	0.24	0.11	0.06	0.04	0.03	0.02	0.02	0.01	0.01	0.01	0.01	0.01	0.01	0.01	0.00	0.00
10	0.75	0.88	0.93	0.95	0.97	0.97	0.98	0.98	0.99	0.99	0.24	0.12	0.07	0.04	0.03	0.02	0.02	0.01	0.01	0.01	0.01	0.01	0.01	0.00
12	0.75	0.88	0.93	0.96	0.97	0.98	0.98	0.99	0.99	0.99	0.99	0.99	0.25	0.12	0.07	0.04	0.03	0.02	0.02	0.02	0.02	0.01	0.01	0.01
14	0.76	0.88	0.93	0.96	0.97	0.98	0.98	0.99	0.99	0.99	0.99	0.99	1.00	1.00	0.25	0.12	0.07	0.05	0.03	0.03	0.03	0.02	0.01	0.01
16	0.76	0.89	0.94	0.96	0.97	0.98	0.98	0.99	0.99	0.99	0.99	0.99	1.00	1.00	1.00	1.00	0.25	0.12	0.07	0.05	0.03	0.03	0.02	0.02
18	0.77	0.89	0.94	0.96	0.97	0.98	0.98	0.99	0.99	0.99	0.99	1.00	1.00	1.00	1.00	1.00	1.00	1.00	0.25	0.12	0.07	0.05	0.03	0.03
Zone Type B																								
2	0.65	0.74	0.16	0.11	0.08	0.06	0.05	0.04	0.03	0.02	0.02	0.01	0.01	0.01	0.01	0.00	0.00	0.00	0.00	0.00	0.00	0.00	0.00	0.00
4	0.65	0.75	0.81	0.85	0.24	0.17	0.13	0.10	0.07	0.06	0.04	0.03	0.03	0.02	0.02	0.01	0.01	0.01	0.00	0.00	0.00	0.00	0.00	0.00
6	0.65	0.75	0.81	0.85	0.89	0.91	0.29	0.20	0.15	0.12	0.09	0.07	0.05	0.04	0.03	0.02	0.02	0.01	0.01	0.01	0.01	0.01	0.00	0.00
8	0.65	0.75	0.81	0.85	0.89	0.91	0.93	0.95	0.31	0.22	0.17	0.13	0.10	0.08	0.06	0.05	0.04	0.03	0.02	0.02	0.01	0.01	0.01	0.01
10	0.65	0.75	0.81	0.86	0.89	0.91	0.93	0.95	0.96	0.97	0.33	0.24	0.18	0.14	0.11	0.08	0.06	0.05	0.04	0.03	0.02	0.02	0.01	0.01
12	0.66	0.76	0.81	0.86	0.89	0.92	0.94	0.95	0.96	0.97	0.98	0.98	0.34	0.24	0.19	0.14	0.11	0.08	0.06	0.05	0.04	0.03	0.02	0.02
14	0.67	0.76	0.82	0.86	0.89	0.92	0.94	0.95	0.96	0.97	0.98	0.98	0.99	0.99	0.35	0.25	0.19	0.15	0.11	0.09	0.07	0.05	0.04	0.03
16	0.69	0.78	0.83	0.87	0.90	0.92	0.94	0.95	0.96	0.97	0.98	0.98	0.99	0.99	0.99	0.99	0.35	0.25	0.19	0.15	0.11	0.09	0.07	0.05
18	0.71	0.80	0.85	0.88	0.91	0.93	0.95	0.96	0.97	0.98	0.99	0.99	0.99	0.99	0.99	0.99	1.00	1.00	0.35	0.25	0.19	0.15	0.11	0.09
Zone Type C																								
2	0.60	0.68	0.14	0.11	0.09	0.07	0.06	0.05	0.04	0.03	0.03	0.02	0.02	0.01	0.01	0.01	0.01	0.01	0.01	0.00	0.00	0.00	0.00	0.00
4	0.60	0.68	0.74	0.79	0.23	0.18	0.14	0.12	0.10	0.08	0.06	0.05	0.04	0.04	0.03	0.02	0.02	0.02	0.01	0.01	0.01	0.01	0.01	0.01
6	0.61	0.69	0.74	0.79	0.83	0.86	0.28	0.22	0.18	0.15	0.12	0.10	0.08	0.07	0.06	0.05	0.04	0.03	0.03	0.02	0.02	0.01	0.01	0.01
8	0.61	0.69	0.75	0.79	0.83	0.86	0.89	0.91	0.32	0.26	0.21	0.17	0.14	0.11	0.09	0.08	0.06	0.05	0.04	0.04	0.03	0.02	0.02	0.02
10	0.62	0.70	0.75	0.80	0.83	0.86	0.89	0.91	0.92	0.94	0.35	0.28	0.23	0.18	0.15	0.12	0.10	0.08	0.07	0.06	0.05	0.04	0.03	0.03
12	0.63	0.71	0.76	0.81	0.84	0.87	0.89	0.91	0.93	0.94	0.95	0.96	0.37	0.29	0.24	0.19	0.16	0.13	0.11	0.09	0.07	0.06	0.05	0.04
14	0.65	0.72	0.77	0.82	0.85	0.88	0.90	0.92	0.93	0.94	0.95	0.96	0.97	0.97	0.38	0.30	0.25	0.20	0.17	0.14	0.11	0.09	0.08	0.06
16	0.68	0.74	0.79	0.83	0.86	0.89	0.91	0.92	0.94	0.95	0.96	0.96	0.97	0.98	0.98	0.98	0.39	0.31	0.25	0.21	0.17	0.14	0.11	0.09
18	0.72	0.78	0.82	0.85	0.88	0.90	0.92	0.93	0.94	0.95	0.96	0.97	0.97	0.98	0.98	0.99	0.99	0.99	0.39	0.31	0.26	0.21	0.17	0.14
Zone Type D																								
2	0.59	0.67	0.13	0.09	0.08	0.06	0.05	0.05	0.04	0.04	0.03	0.03	0.02	0.02	0.02	0.01	0.01	0.01	0.01	0.01	0.01	0.01	0.01	0.00
4	0.60	0.67	0.72	0.76	0.20	0.16	0.13	0.11	0.10	0.08	0.07	0.06	0.05	0.05	0.04	0.03	0.03	0.03	0.02	0.02	0.02	0.01	0.01	0.01
6	0.61	0.68	0.73	0.77	0.80	0.83	0.26	0.20	0.17	0.15	0.13	0.11	0.09	0.08	0.07	0.06	0.05	0.05	0.04	0.03	0.03	0.03	0.02	0.02
8	0.62	0.69	0.74	0.77	0.80	0.83	0.85	0.86	0.30	0.24	0.20	0.17	0.14	0.13	0.11	0.10	0.08	0.07	0.06	0.05	0.05	0.04	0.04	0.03
10	0.63	0.70	0.75	0.78	0.81	0.84	0.86	0.88	0.89	0.91	0.33	0.27	0.22	0.19	0.17	0.14	0.12	0.11	0.09	0.08	0.07	0.06	0.05	0.05
12	0.65	0.71	0.76	0.79	0.82	0.84	0.87	0.88	0.90	0.91	0.92	0.93	0.35	0.29	0.24	0.21	0.18	0.16	0.13	0.12	0.10	0.09	0.08	0.07
14	0.67	0.73	0.78	0.81	0.83	0.86	0.88	0.89	0.91	0.92	0.93	0.94	0.95	0.95	0.37	0.30	0.25	0.22	0.19	0.16	0.14	0.12	0.11	0.09
16	0.70	0.76	0.80	0.83	0.83	0.87	0.89	0.90	0.92	0.93	0.94	0.95	0.95	0.96	0.96	0.97	0.38	0.31	0.26	0.23	0.20	0.17	0.15	0.13
18	0.74	0.80	0.83	0.85	0.87	0.89	0.91	0.92	0.93	0.94	0.95	0.95	0.96	0.97	0.97	0.97	0.98	0.98	0.39	0.32	0.27	0.23	0.20	0.17

*See Table 6.10 for zone type. Data based on a radiative/convective fraction of 0.70/0.30.

TABLE 16.14 Cooling-Load Factors for Hooded Equipment*

Number of Hours after Equipment Turned On

Hours in Operation	1	2	3	4	5	6	7	8	9	10	11	12	13	14	15	16	17	18	19	20	21	22	23	24
Zone Type A																								
2	0.64	0.83	0.26	0.11	0.06	0.03	0.01	0.01	0.01	0.01	0.00	0.00	0.00	0.00	0.00	0.00	0.00	0.00	0.00	0.00	0.00	0.00	0.00	0.00
4	0.64	0.83	0.90	0.93	0.31	0.14	0.07	0.04	0.03	0.03	0.01	0.01	0.01	0.01	0.00	0.00	0.00	0.00	0.00	0.00	0.00	0.00	0.00	0.00
6	0.64	0.83	0.90	0.93	0.96	0.96	0.33	0.16	0.09	0.06	0.04	0.03	0.03	0.01	0.01	0.01	0.01	0.00	0.00	0.00	0.00	0.00	0.00	0.00
8	0.64	0.83	0.90	0.93	0.96	0.96	0.97	0.97	0.34	0.16	0.09	0.06	0.04	0.03	0.03	0.03	0.03	0.01	0.01	0.01	0.01	0.00	0.00	0.00
10	0.64	0.83	0.90	0.93	0.96	0.96	0.97	0.97	0.99	0.99	0.34	0.17	0.10	0.06	0.10	0.06	0.04	0.03	0.03	0.01	0.01	0.01	0.01	0.00
12	0.64	0.83	0.90	0.94	0.96	0.97	0.97	0.99	0.99	0.99	0.99	0.99	0.36	0.17	0.36	0.17	0.10	0.07	0.04	0.04	0.03	0.03	0.03	0.01
14	0.66	0.83	0.90	0.94	0.96	0.97	0.97	0.99	0.99	0.99	0.99	0.99	1.00	1.00	1.00	1.00	0.36	0.17	0.10	0.07	0.04	0.04	0.04	0.03
16	0.66	0.84	0.91	0.94	0.96	0.97	0.97	0.99	0.99	0.99	0.99	0.99	1.00	1.00	1.00	1.00	1.00	1.00	0.36	0.17	0.10	0.08	0.07	0.10
18	0.67	0.84	0.91	0.94	0.96	0.97	0.97	0.99	0.99	0.99	0.99	1.00	1.00	1.00	1.00	1.00	1.00	1.00	0.36	0.17	0.10	0.08	0.07	0.04
Zone Type B																								
2	0.50	0.63	0.23	0.16	0.11	0.09	0.07	0.06	0.04	0.03	0.03	0.01	0.01	0.01	0.01	0.00	0.00	0.00	0.00	0.00	0.00	0.00	0.00	0.00
4	0.50	0.64	0.73	0.79	0.34	0.24	0.19	0.14	0.10	0.09	0.06	0.04	0.04	0.03	0.03	0.01	0.01	0.01	0.01	0.00	0.00	0.00	0.00	0.00
6	0.50	0.64	0.73	0.79	0.84	0.87	0.41	0.29	0.21	0.17	0.13	0.10	0.07	0.06	0.04	0.03	0.03	0.01	0.01	0.01	0.01	0.01	0.01	0.01
8	0.50	0.64	0.73	0.79	0.84	0.87	0.90	0.93	0.44	0.31	0.24	0.19	0.14	0.11	0.09	0.07	0.06	0.04	0.03	0.03	0.03	0.03	0.03	0.03
10	0.50	0.64	0.73	0.79	0.84	0.87	0.90	0.93	0.94	0.96	0.47	0.34	0.26	0.20	0.16	0.11	0.09	0.07	0.04	0.04	0.03	0.03	0.03	0.03
12	0.51	0.66	0.73	0.80	0.84	0.89	0.91	0.93	0.94	0.96	0.97	0.97	0.49	0.34	0.27	0.20	0.16	0.11	0.09	0.07	0.06	0.05	0.04	0.03
14	0.53	0.66	0.74	0.80	0.84	0.89	0.91	0.93	0.94	0.96	0.97	0.97	0.99	0.99	0.50	0.36	0.27	0.21	0.16	0.13	0.10	0.08	0.07	0.06
16	0.56	0.69	0.76	0.81	0.86	0.89	0.91	0.93	0.94	0.96	0.97	0.97	0.99	0.99	0.99	0.99	0.50	0.36	0.27	0.21	0.16	0.14	0.13	0.10
18	0.59	0.71	0.79	0.83	0.87	0.90	0.93	0.94	0.96	0.97	0.97	0.97	0.99	0.99	0.99	0.99	1.00	1.00	0.50	0.36	0.27	0.23	0.21	0.16
Zone Type C																								
2	0.43	0.54	0.20	0.16	0.13	0.10	0.09	0.07	0.06	0.04	0.04	0.03	0.03	0.01	0.01	0.01	0.01	0.01	0.01	0.00	0.00	0.00	0.00	0.00
4	0.43	0.54	0.63	0.70	0.33	0.26	0.20	0.17	0.14	0.11	0.09	0.07	0.06	0.06	0.04	0.03	0.03	0.03	0.01	0.01	0.01	0.01	0.01	0.01
6	0.44	0.56	0.63	0.70	0.76	0.80	0.40	0.31	0.26	0.21	0.17	0.14	0.11	0.10	0.09	0.07	0.06	0.04	0.04	0.03	0.03	0.02	0.01	0.01
8	0.44	0.56	0.64	0.70	0.76	0.80	0.84	0.87	0.46	0.37	0.30	0.24	0.20	0.16	0.13	0.11	0.09	0.07	0.06	0.06	0.04	0.03	0.03	0.03
10	0.46	0.57	0.64	0.71	0.76	0.80	0.84	0.87	0.89	0.91	0.50	0.40	0.33	0.26	0.21	0.17	0.14	0.11	0.10	0.09	0.07	0.06	0.06	0.04
12	0.47	0.59	0.66	0.73	0.77	0.81	0.84	0.87	0.90	0.91	0.93	0.94	0.53	0.41	0.34	0.27	0.23	0.19	0.16	0.13	0.10	0.09	0.09	0.07
14	0.50	0.60	0.67	0.74	0.79	0.83	0.86	0.89	0.90	0.91	0.93	0.94	0.96	0.96	0.54	0.43	0.36	0.29	0.24	0.20	0.16	0.14	0.13	0.11
16	0.54	0.63	0.70	0.76	0.80	0.84	0.87	0.89	0.91	0.93	0.94	0.94	0.96	0.97	0.97	0.97	0.56	0.44	0.36	0.30	0.24	0.22	0.20	0.16
18	0.60	0.69	0.74	0.79	0.83	0.86	0.89	0.90	0.91	0.93	0.94	0.96	0.96	0.97	0.97	0.99	0.99	0.99	0.56	0.44	0.37	0.33	0.30	0.24
Zone Type D																								
2	0.41	0.53	0.19	0.13	0.11	0.09	0.07	0.07	0.06	0.06	0.04	0.04	0.03	0.03	0.03	0.01	0.01	0.01	0.01	0.01	0.01	0.01	0.01	0.01
4	0.43	0.53	0.60	0.66	0.29	0.23	0.19	0.16	0.14	0.11	0.10	0.09	0.07	0.07	0.06	0.04	0.04	0.04	0.03	0.03	0.03	0.02	0.01	0.01
6	0.44	0.54	0.61	0.67	0.71	0.76	0.37	0.29	0.24	0.21	0.19	0.16	0.13	0.11	0.10	0.09	0.07	0.07	0.06	0.04	0.04	0.04	0.04	0.03
8	0.46	0.56	0.63	0.67	0.71	0.76	0.79	0.81	0.43	0.34	0.29	0.24	0.21	0.19	0.16	0.14	0.11	0.10	0.09	0.07	0.07	0.06	0.06	0.06
10	0.47	0.57	0.64	0.69	0.73	0.77	0.80	0.83	0.84	0.87	0.47	0.39	0.31	0.27	0.24	0.20	0.17	0.16	0.13	0.11	0.10	0.09	0.09	0.07
12	0.50	0.59	0.66	0.70	0.74	0.77	0.81	0.83	0.86	0.87	0.89	0.90	0.50	0.41	0.34	0.30	0.26	0.23	0.19	0.17	0.14	0.13	0.13	0.11
14	0.53	0.61	0.69	0.73	0.76	0.80	0.83	0.84	0.87	0.89	0.90	0.91	0.93	0.93	0.53	0.43	0.36	0.31	0.27	0.23	0.20	0.18	0.17	0.16
16	0.57	0.66	0.71	0.76	0.79	0.81	0.84	0.86	0.89	0.90	0.91	0.93	0.93	0.94	0.94	0.96	0.54	0.44	0.37	0.33	0.29	0.26	0.24	0.21
18	0.63	0.71	0.76	0.79	0.81	0.84	0.87	0.89	0.90	0.91	0.93	0.93	0.94	0.96	0.96	0.96	0.97	0.97	0.56	0.46	0.39	0.35	0.33	0.29

Source: Reprinted by permission from *ASHRAE Fundamentals 1997*, p. 28.53.
* See Table 16.10 for zone types. Data based on a radiative/convective fraction of 100/0.

each day and that the heat gain is continuous and constant during the time indicated. If the heat gain is known to fluctuate with time, it can be divided into several portions of constant magnitude with the superposition of these portions equal to the time-varying heat gain. The cooling load can be determined for each of these portions separately and then the results summed. The cooling-load factors given in Tables 16.12, 16.13, and 16.14 are based on specified fractions of convective/radiative heat gains. Correction for other fractions is the same as that given for the v coefficients given in Eqs. (16.3) and (16.4).

The instantaneous cooling load due to latent-heat gains from people and equipment is usually assumed to equal the instantaneous heat gain. This neglects moisture storage in the building and corresponding lag in the cooling load. Moisture storage should be included in applications where large surface areas of materials of large moisture diffusivity are present. In most commercial and institutional buildings the moisture effect is small. Moisture storage is more significant in residences, where gypsum wallboard surfaces, carpeting, and fabrics may store a significant amount of moisture in short time periods.

EXAMPLE 16.6

Determine the total instantaneous cooling load for 100 people in a commercial office building 8 hr after entry. The occupants are doing moderately active office work 8 hr per day. The building has concrete floors covered with carpeting, gypsum board partition walls, and no interior shading on the windows. Each occupied space in the building has one or two exterior walls.

Solution: The instantaneous cooling load consists of sensible and latent portions. The sensible portion is considered first. The total instantaneous heat gain from the occupants is

$$(75 \text{ W/person})(100 \text{ people}) = 7500 \text{ W}$$

from Table 15.19. The zone type from Table 16.10 is type B. The cooling-load factor from Table 16.13 is 0.95 (zone type B, 8 hr in space, 8 hr after entry). The total sensible instantaneous cooling load is then

$$ICL_{sen} = (7500 \text{ W})(0.95) = 7125 \text{ W}$$

The latent cooling load equals the instantaneous latent heat gain, as no moisture storage is considered. The latent-heat gain from Table 15.19 is 55 W per person, so the latent instantaneous cooling load is

$$ICL_{lat} = (55 \text{ W/person})(100 \text{ people}) = 5500 \text{ W}$$

The total cooling load due to these occupants is

$$ICL_t = ICL_{sen} + ICL_{lat} = 7125 \text{ W} + 5500 \text{ W} = 12{,}625 \text{ W}$$

ENDNOTES

1. *ASHRAE Handbook of Fundamentals* (Atlanta: American Society of Heating, Refrigerating and Air Conditioning Engineers, 1993).
2. F. C. McQuiston and J. D. Spitler, *Cooling and Heating Load Calculation Manual*, 2d ed. (Atlanta: ASHRAE, 1992).

16.1 Use the transfer-function method to compute the instantaneous cooling load for the wall in Prob. 15.9 with medium construction, 370 kg/m^2, and a high air-circulation rate.

16.2 Compute the instantaneous cooling load through the roof of Prob. 15.11, assuming heavy construction with an 8-in. concrete floor and low air-circulation rate.

16.3 Work Ex. 16.1 with the air-circulation rate changed from very high to low. How much does this change affect the cooling load from this roof?

16.4 Determine the hourly cooling load for the window of Prob. 15.22 using the transfer-function method. Assume the air-circulation rate is high and the building has medium construction with 6-in. concrete floors.

16.5 Use the transfer-function method to compute the hourly instantaneous cooling load from 20 kW of lights that are used 10 hr a day. The building has typical office furnishings with no carpet, a 6-in. floor, and a medium air-circulation rate through vented light fixtures.

16.6 Compute the design cooling load in July per square foot of wall area at 6:00 p.m. solar time for a west-facing wall of construction #W4 as given in Table 16.5. Use the CLTD method and assume the maximum outdoor-air temperature is 94 °F with a daily range of 18 °F, the indoor temperature is 73 °F, and the exterior-surface color is a dark brown.

16.7 Repeat Prob. 16.6 for the case where the outside wall-surface color is changed to:
(a) a light color and
(b) a medium color.
Comment on the effect of exterior color on this type of wall.

16.8 Use the CLTD method to determine the summer design cooling load for the walls of a building in August, 40 deg north latitude. The four walls face N-S-E-W and are of construction #W2 in Table 16.5. Each wall has a surface area of 100 m^2 and is of medium color. The outdoor daily average temperature is 27 °C and the indoor air is maintained at 23 °C.
(a) Plot the cooling load for each of these four walls as a function of time for one 24-hr period beginning at midnight solar time.
(b) Determine the magnitude and time of day for the peak cooling load for each of the four walls.

16.9 Repeat Prob. 16.8 for walls that have construction #W5 in Table 16.5.

16.10 Determine the hourly cooling load per square meter of wall-surface area for a south-facing wall of construction #W3 in Table 16.5 in January, 40 deg north latitude. Assume a dark color, average daily outdoor temperature of 20 °F, and an indoor temperature of 68 °F. What can you suggest to increase the cooling load (or negative heating load) for this wall?

16.11 Compute the summer design hourly cooling load through 10,000 m^2 of roof #R3 in Table 16.6, assuming standard conditions.

16.12 Determine the time and magnitude of the summer design cooling load for a 10,000 ft^2 roof of construction #R5 in Table 16.6. The roof has an additional R-7 hr $\cdot$ ft^2 $\cdot$ °F/Btu of insulation. The average daily outdoor temperature is 87 °F and the indoor temperature is 78 °F.

16.13 Use the CLTD method to compute the daily maximum cooling load from a 10,000-ft^2 flat roof of construction #R1 in Table 16.6 in October at 40 deg north latitude. Assume a daily maximum outdoor temperature of 65 °F, a daily range of 21 °F, and an indoor temperature of 70 °F.

16.14 Plot the hourly summer design cooling loads per unit area through roofs #R1, R3, and R5 given in Table 16.6. Assume standard conditions. Explain why the magnitude and the time of the peak cooling load differ for these three roof constructions.

16.15 Calculate the maximum summer design cooling load through an unshaded west-facing window with a total area of 100 ft^2 and a framing area of 10 ft^2. The window is double glazed and is located in a room with two exposed walls, no carpet, and no interior shade. Assume standard conditions.

16.16 Compare the peak summer design cooling load per square meter of the following windows. The windows face west, are in a type C zone, and experience standard temperature conditions:
(a) double-glazed sunlit window,
(b) double-glazed window with external shading until 3:00 p.m. solar time, sunlit after that time, and
(c) double glazed window with light-colored Venetian blinds installed inside.

16.17 Repeat Ex. 16.5 for
(a) north-facing,
(b) south-facing, and
(c) west-facing windows.

16.18 Use the solar-cooling-load method to determine the hourly summer design cooling load through a 10-m² double-glazed window with heat-absorbing glass as the outer glazing. The unshaded window faces south and is in a room with one exposed wall, carpet, and gypsum board partitions. The average daily outdoor temperature is 90 °F and the indoor temperature is 75 °F.

16.19 Use the cooling-load-factor method to determine the cooling load caused by the internal gains given in Prob. 15.26 at 5:00 p.m. solar time. Assume that all internal gains begin at 7:00 a.m. solar time and that they are continuous for 10 hr per day. The building is treated as a single zone with four exposed walls, carpet, and gypsum board partitions.

16.20 Consider a piece of cooking equipment in a commercial kitchen. Assume this equipment operates 12 hr per day in a type D zone. Explain why the cooling-load factors differ, depending on whether the equipment is hooded or not.

16.21 Lights are operated 12 hr per day in a type B zone. Give at least two building-design parameters that could be changed to reduce the cooling-load factor for these lights.

SYMBOLS

A	Surface area, ft² (m²).
c_{pa}	Moist air specific-heat capacity, Btu/lbm$_a$ · °F (kJ/kg$_a$ · °C).
CLF	Cooling-Load Factor.
CLTD	Cooling-Load Temperature Difference, °F (°C).
h	Heat-transfer coefficient, Btu/hr · ft² · °F (W/m² · °C).
ICL	Instantaneous Cooling Load, Btu/hr (W).
IHG	Instantaneous Heat Gain, Btu/hr · ft² (W/m²).
$\dot{m}_a$	Dry air mass flow rate, lbm$_a$/hr (kg$_a$/s).
n	Integer.
N	Number of hours in simulation.
q_i	Instantaneous heat gain through fenestration, Btu/hr · ft² (W/m²).
$q_{\tau,c}$	Instantaneous convective heat gain at time τ, Btu/hr · ft² (W/m²).
$q_{\tau,r}$	Instantaneous radiative heat gain at time τ, Btu/hr · ft² (W/m²).
$\dot{Q}$	Rate of heat transfer, Btu/hr (W).
r	Percent radiative heat gain used in tables.
r'	Actual percent radiative heat gain.
R	Thermal resistance, hr · ft² · °F/Btu (m² · °C/W).
SC	Shading Coefficient.
SHGF	Solar Heat-Gain Factor, Btu/hr ft² (W/m²).
t_a	Air dry-bulb temperature, °F (°C).
t_{ret}	Return-air temperature, °F (°C).
t_{sur}	Surface temperature, °F (°C).
t_{sup}	Supply air temperature, °F (°C).
TETD	Total Equivalent Temperature Difference, °F (°C).
t_o	Outdoor dry-bulb temperature.
t_i	Indoor dry-bulb temperature.
U	Overall heat-transfer coefficient, Btu/hr · ft² · °F (W/m² · °C).
w	Coefficients in Eqs. (16.1)–(16.2).

Greek Letters

$\Delta\tau$	Time step, usually 1 hr.
v	Coefficients in Eqs. (16.1)–(16.2).
v'	Corrected coefficients in Eqs. (16.3)–(16.4).

Subscripts

cond	Conduction.
conv	Convection.
corr	Corrected.
glass	Glazing.
i	Inside.
lat	Latent.
max	Maximum.
rad	Radiant.
sen	Sensible.
t	Total.
table	Tabulated value.
τ	Time of interest.
$\tau - n\,\Delta\tau$	n time steps (hr) prior to time of interest.

chapter 17
Energy-Estimation Methods

In Chapter 14 we determined the design, or largest, heat loss for a building based on the building construction, indoor environmental temperature, and outdoor weather conditions. In Chapter 16 we estimated the design cooling load of a building. These analyses are useful for sizing the heating and cooling equipment. The building designer often needs to estimate the building's energy requirement in addition to the design heating and cooling loads. The annual energy requirement is useful for comparison of alternate designs, for compliance with codes and standards, and as part of an economic analysis.

In this chapter we will discuss three methods of estimating energy use in buildings. The first two are considered to be steady-state methods, in which the heat gains are assumed to be average values for the time period and thermal storage effects are neglected. These methods are simple enough for hand calculations. The first is the *degree-day method*. Only a single climate-related number is required over a given time period, so this can be termed a single-measure method. The second method we will discuss is the *bin method*. This is similar to the degree-day method in that it assumes steady state, but it requires the number of hours at each interval of outdoor temperature, or bin, over the time period of interest. This method is useful when the heating-system performance is directly affected by the outdoor temperature, such as with air-to-air heat pumps. The third method is a more detailed simulation approach, in which steady-state assumptions are no longer required. This normally requires a computer because of the lengthy calculations involved. A building's energy demand can be simulated every hour of a typical year using the basic approach described in Chapter 16 for cooling-load estimation. These methods are also discussed in the *ASHRAE Handbook of Fundamentals* [1].

17.1 DEGREE-DAY METHOD

The degree-day method is the simplest one for estimating the energy required by a building over a specified time period. Often the energy use is estimated for a year or over a heating season. Monthly estimates can also be made. Assumptions inherent in this

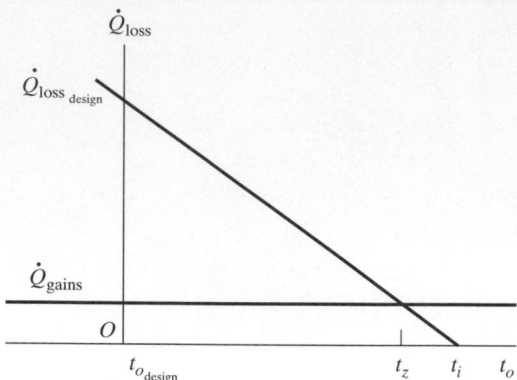

Figure 17.1 Building heat loss vs. outdoor temperature.

method are that the internal gains and other heat gains such as solar do not vary appreciably and that thermal-storage effects are negligible. These assumptions are very good for estimating heating-energy needs for small buildings in northern climates, where the solar loads are small, the thermal storage in the structure does not vary rapidly, and the indoor-to-outdoor temperature difference is large. The degree-day method has also been applied to estimate cooling-energy requirements, although the strong influence of transient solar loads and latent loads limits the usefulness of this approach. We will limit our discussion to estimates of heating-energy requirements.

The need for supplemental heating of a building is illustrated in Fig. 17.1. The horizontal axis represents the outdoor dry-bulb air temperature. Implicit in the analysis is the assumption that the outdoor dry-bulb temperature represents the environmental temperature surrounding the building. This is not strictly true, as the sky-radiant temperature may be different and ground-coupled heat transfer is not directly affected by the ambient dry-bulb temperature. The vertical axis shows the heat loss from the building. The internal heat gains are assumed to be constant and independent of the outdoor temperature. The heating provided by the internal gains is shown as a horizontal line. The building total heat loss can be expressed as

$$\dot{Q}_{loss} = (\text{UA})_{eff}(t_i - t_o) \tag{17.1}$$

where the coefficient, $(\text{UA})_{eff}$, represents the total building heat-loss coefficient. This can be determined from the design conditions

$$(\text{UA})_{eff} = \frac{\dot{Q}_{loss,\,design}}{(t_{i,\,design} - t_{o,\,design})} \tag{17.2}$$

The building total heat loss versus outdoor temperature is shown by the sloped line in Fig. 17.1. The net heating load, $\dot{Q}_{net}$, is the difference between $\dot{Q}_{loss}$ and $\dot{Q}_{gains}$.

When the indoor temperature in the building, t_i, equals the outdoor temperature, no heat loss occurs. The outdoor temperature can continue to drop until the zero-load temperature, t_z, is reached. At this point, the heat gains are equal to the building heat loss. The zero-load temperature, t_z, can be determined as

$$t_z = t_i - \frac{\dot{Q}_{gains}}{(\text{UA})_{eff}} \tag{17.3}$$

As the outdoor temperature drops below t_z, the total building heat loss is larger than the amount of heat gains, so that supplemental heating is required. This net heat loss required from a heating system can be determined by

$$\dot{Q}_{net} = (UA)_{eff}(t_z - t_o) \qquad (17.4)$$

The amount of heating energy required over a specified time period, E_p, is determined by integrating the net heat loss over the time period:

$$E_p = \int_p \dot{Q}_{net} \, d\tau$$

Substituting for the net heat loss from Eq. (17.4),

$$E_p = (UA)_{eff} \int_p \|t_z - t_o\| \, d\tau \qquad (17.5)$$

where the double vertical lines indicate that only positive values of $(t_z - t_o)$ are included in the integral. If the time period is expressed in terms of days, the value becomes the number of degree days at zero-load temperature, t_z, for the time period used:

$$(DD)_{z,p} = \int_p \|t_z - t_o\| \, d\tau \qquad (17.6)$$

Equation (17.5) can now be written in terms of the number of degree days:

$$E_p = (UA)_{eff}(DD)_{z,p} \qquad (17.7)$$

In English units, the equation is often written as

$$E_p = 24(UA)_{eff}(DD)_{z,p} \qquad (17.8)$$

where the factor 24 is used to convert the units of "hours" in the UA term (Btu/hr · °F) to "days" used in the degree-day term.

Data collected from residences in the northeastern United States in the 1930s by the American Gas Association and the National District Heating Association indicated that the zero-load temperature was about 65 °F or 18 °C. Therefore, this value of zero-load temperature has traditionally been used to compute the number of heating degree days from weather data using Eq. (17.6). Since the time of these original studies the building stock in the United States has been better insulated and weatherstripped. The effective heat-loss coefficient, $(UA)_{eff}$, is now much smaller than before. Thus the slope of the total-heat-loss line shown on Fig. 17.1 is much less steep than before. The internal heat gains are also larger than before, as many more electrical appliances and equipment are used now. Both of these factors result in a lower zero-load temperature than before. Zero-load temperatures of 55 to 60 °F (13 to 16 °C) are much more common today. For this reason, many tables of degree-day values now include data for various zero-load temperatures.

The time period of interest also varies. Many tables provide yearly values. These are useful for comparing differences in climatic factors. However, yearly degree days usually include nonzero values for the summer months, when the heating system is not usually operated, and thus overpredict the amount of supplemental heating energy needed. Monthly values are also commonly provided. This allows the user to select the months the heating system is likely to operate. One can also predict the monthly energy-usage distribution for the year, which is useful in forecasting energy use.

Table B.2 contains monthly and annual degree-day information for seven cities in the United States. The units on the degree-day data in the table are "°F·days." The monthly average outdoor-air temperature is also provided. Data are listed for five values of zero-load temperature: 50, 55, 60, 65 (the traditional value), and 70 °F. Note that the annual number of degree days increases significantly with an increase in zero-load temperature for each location. Additional degree-day data are published in reference [2]. Erbs et al. [3] presented a method to estimate the monthly and annual number of degree days based on monthly mean-ambient-temperature data. This method is useful in locations where degree-day data are unavailable.

The main use of the degree-day method is to estimate the amount of fuel that will be required to provide space heating for a building. The amount of energy required, E_p, can be used to compute the fuel requirement, F_p, by dividing it by the average heating-system efficiency and by the fuel heating value:

$$F_p = \frac{E_p}{\eta_h H} \tag{17.9}$$

The average heating-system efficiency, η_h, is defined as the ratio of the thermal output from the heating system divided by the heating value of the fuel or energy input averaged over the time period of the analysis. For fossil-fuel furnaces this efficiency is less than the steady-state value. Cycling at part load reduces the amount of heat delivered to the building, as more is lost out the flue. ASHRAE [1] gives equations to estimate η_h for various types of heating systems that are a function of oversizing of the heating-system capacity and the amount of heat lost out the ductwork. The heating-system average efficiency can be estimated from

$$\eta_h = \frac{\eta_{ss} \mathrm{CF}_{pl}}{1 + \alpha_D} \tag{17.10}$$

where η_{ss} is the steady-state or rated heating-system efficiency without cycling, CF_{pl} is a function of the type of heating system and part-load efficiency, and α_D is the fraction of heat loss from ducts. For example, the value for CF_{pl} for a natural-gas furnace with intermittent ignition is given as

$$\mathrm{CF}_{pl} = 0.7791 + 0.1983\mathrm{RLC} - 0.0711(\mathrm{RLC})^2 \tag{17.11}$$

where

$$\mathrm{RCL} = \frac{(\mathrm{UA})_{\mathrm{eff}}(t_z - t_{o,\,\mathrm{design}})(1 + \alpha_D)}{\dot{Q}_{output}} \tag{17.12}$$

and $\dot{Q}_{\mathrm{output}}$ is the rated output of the heating equipment. For fossil-fuel furnaces CF_{pl} is always less than 1.0. For electric furnaces and heaters the value is equal to 1.0.

Combining Eqs. (17.2), (17.8), and (17.9), we can obtain an expression for the amount of fuel used to provide space heating that depends on the design heat-loss and temperature conditions, the number of degree days referenced to the correct zero-load temperature, the average heating-system efficiency, and the heating value of the fuel:

$$F_p = \frac{24(\mathrm{UA})_{\mathrm{eff}}(\mathrm{DD})_{z,p}}{\eta_h H} = \frac{24\dot{Q}_{\mathrm{loss,\,design}}(\mathrm{DD})_{z,p}}{(t_{i,\,\mathrm{design}} - t_{o,\,\mathrm{design}})\eta_h H} \tag{17.13}$$

EXAMPLE 17.1

Estimate the amount of natural gas used to heat a home in Minneapolis for one year given the following information:

Winter design heat loss is 98,700 Btu/hr at $t_i = 72\ °F$ and $t_o = -16\ °F$.

Thermostat setting is maintained at 68 °F.

The building has continuous internal gains of 9000 Btu/hr.

Average furnace efficiency is 85 percent.

Heating value of natural gas is 1000 Btu/ft^3 or 100,000 Btu/ccf.

Solution: We will use Eq. (17.13) to determine the amount of fuel required. First we must determine the zero-load temperature. Using Eq. (17.2), we determine $(UA)_{eff}$,

$$UA_{eff} = \frac{\dot{Q}_{loss, design}}{t_{i, design} - t_{o, design}} = \frac{98,700\ Btu/hr}{[72 - (-16)]\ °F} = 1122\ Btu/hr \cdot °F$$

Then from Eq. (17.3) we can determine the zero-load temperature

$$t_z = 68\ °F - \frac{9000\ Btu/hr}{1122\ Btu/hr \cdot °F} = 59.979\ °F \approx 60\ °F$$

The annual number of degree days for Minneapolis at a zero-load temperature of 60 °F is obtained from Table B.2 as 6824. Now the amount of fuel required is obtained using the first part of Eq. (17.13):

$$F_{yr} = \frac{24hr/day\ (1122\ Btu/hr \cdot °F)(6824\ °F \cdot day)_{yr,\ 60F}}{0.85(100,000\ Btu/ccf)} = 2162\ ccf$$

The cost to heat a building can also be estimated by multiplying the amount of fuel required by the local fuel cost. In the previous example, if the cost of natural gas is $0.50/ccf, the annual space-heating cost estimate becomes 2162 ccf ($0.50/ccf) = $1081 for the year.

17.2 BIN METHOD

Although the degree-day method is a simple one to estimate monthly and seasonal space-heating energy use, it has some limitations:

1. A single indoor-temperature set point is assumed (no variation with time of day).
2. A constant internal heat-gain rate is assumed (no change with occupancy level or other factors).
3. No outdoor-humidity information is provided to estimate latent loads.

The bin method overcomes these limitations by dividing outdoor-temperature data into various temperature increments, or bins, and further subdividing these data into various times of the day. Traditional bin weather data are provided as the number of hours in a given time interval (month or year) that the outdoor temperature falls within a 5 °F increment, or bin. For example, the number of hours that the temperature falls between

39.5 and 44.5 °F would be included in the 40/44 °F bin with a mean value of 42 °F. The data are often further subdivided into time periods during the day, such as hours 1–8, 9–16, and 17–24. Tables B3–B9 include bin weather data in this form for seven U.S. locations [4]. Both monthly and annual totals are given. The mean coincident wet-bulb (MCWB) temperature data are also given for each month so that latent-load estimates can be made for infiltration and ventilation requirements.

Degree-day data can be determined from the bin weather data as follows. First the base, or zero-load, temperature must be selected. Then the time interval can be chosen, either monthly or annually. Calculations can be made for selected time intervals during the day (e.g., hours 1–8) or for the entire day. The following equation can then be used to compute the degree-day information from the given bin weather data:

$$(\text{DD}_{z,p}) = \frac{1}{24}\left(\sum_{\text{bins}} \|t_z - t_{\text{bin}}\| N_{\text{bin}}\right) \tag{17.14}$$

where t_{bin} is the temperature at the center of the bin, N_{bin} is the number of hours corresponding to each bin-temperature increment, and the double vertical lines indicate that only positive values of the temperature difference are included in the sum.

Two types of bin calculations will be considered. The first type assumes constant internal gains, similar to the degree day method. The second assumes that the internal gains vary during the day and that the variation is attributed to changes in building occupancy.

Constant Internal Gains

Figure 17.2 illustrates the main concepts associated with the bin calculation method. The outdoor temperature forms the horizontal axis, and the vertical coordinate is energy rate. The internal gains are assumed to be independent of outdoor temperature and are shown as a horizontal line near the bottom of the figure. The space-heating-system rated output capacity is also assumed to be independent of outdoor temperature and is shown as the horizontal line near the top of the figure. This represents heating systems such as furnaces and boilers. The rated capacity is slightly larger than the winter design load because of system oversizing. The inclined solid line shows that the building heat loss increases linearly as the outdoor temperature decreases. At the zero-load temperature and above, no supplemental heating is required, as the internal gains are sufficient to provide the necessary heat loss. However, when the outdoor temperature falls below the zero-load temperature, supplemental space heating is required. Bin calculations are required from the zero-load temperature to the minimum outdoor temperature at which data exist in the bin weather tables. The minimum bin temperature is usually below the winter design temperature, as these few hours are outside the usual $97\frac{1}{2}$ percent or 99 percent winter design limit.

The bin method also allows for variation in the heating-system efficiency with change in outdoor temperature. As shown in Fig. 17.2, the building load should always remain below the heating-system rated output capacity of a furnace or boiler. Therefore the heating system is always operating at part load. A load factor can be defined as

$$\text{LF} \equiv \frac{\dot{Q}_{\text{loss}}}{\dot{Q}_{\text{output}}} \tag{17.15}$$

where $\dot{Q}_{\text{loss}}$ is the building heat-loss rate at the bin temperature of interest and $\dot{Q}_{\text{output}}$ is the heating-system rated output capacity. The load factor is the theoretical run-time frac-

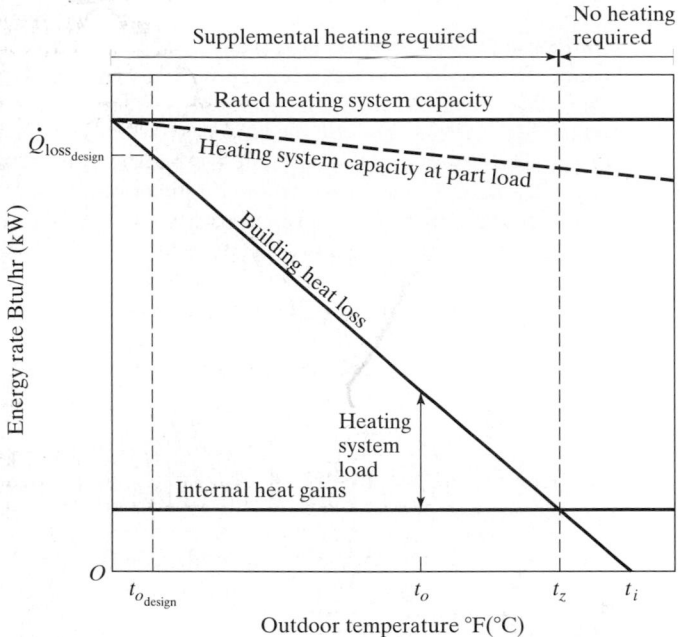

Figure 17.2 Energy rates vs. outdoor temperature for a building with a rated heating-system output independent of outdoor temperature (i.e, furnace or boiler).

tion for the heating system. Part-load conditions require cycling of equipment such as fans, burners, and pumps. Heating equipment operates less efficiently when it cycles compared to steady operation. Therefore the actual heating capacity is not constant but decreases as the cycling increases at higher outdoor temperatures. The load fraction defined by Eq. (17.15) can be modified to account for equipment-cycling losses [1, 4]:

$$\text{PLF} = 1 - C_d(1 - \text{LF}) \tag{17.16}$$

where PLF is the part-load factor and C_d is the degradation coefficient provided by the equipment manufacturer. The value for C_d usually lies between 0.15 and 0.25. If the value for the heating system under evaluation is not known, the default value of 0.25 should be used. The actual heating-system output capacity at a given bin temperature is then determined by multiplying the rated output capacity by the value for PLF obtained from Eq. (17.16). Using Eq. (17.16), the assumed actual heating-system heating capacity is shown as the inclined dashed line near the top of Fig. 17.2.

The following example illustrates the use of the bin method for a building that has a furnace for the heating system.

EXAMPLE 17.2

Estimate the annual heating-energy input requirement for the building in the previous example if the installed furnace has a rated heating capacity $\dot{Q}_{\text{capacity}} = 98{,}700$ Btu/hr and the degradation coefficient, C_d, is assumed to equal 0.25.

Solution: Recall from Ex. 17.1, $(UA)_{eff} = 1122$ Btu/hr $\cdot$ °F, $t_z = 59.979$ °F, and the rated furnace efficiency is 85 percent. It is convenient to set up the accompanying table (Table 17.1) to assist in the calculations. Note that the highest bin temperature listed in the table is 57 °F, which is the highest value given in the bin weather data that lies below the zero-load temperature.

The first column, A, lists the temperatures at the midpoint of each 5 °F bin obtained from Table B.7 below the zero-load temperature. The second column, B, lists the annual number of hours in each bin from Table B.7. One could also sum the hours in the months that the heating system would normally operate from Table B.7. The heating system would not normally be used during the summer months, although some summer hours are included in the annual totals. For ease of computation, the annual totals will be used here. Numbers in column C are the supplemental heating required from the heating system and are determined from Eq. 17.4 with the bin temperature taken as t_o. For the first bin:

$$\dot{Q}_{system} = (UA)_{eff}(t_z - t_{bin}) = 1122 \frac{Btu}{hr \cdot °F} (59.979 - 57) \text{ °F} = 3342 \text{ Btu/hr}$$

This is the difference between the heat loss from the building and the internal gains. It could also be computed from

$$\dot{Q}_{system} = \dot{Q}_{loss} - \dot{Q}_{gains} = (UA)_{eff}(t_i - t_{bin}) - \dot{Q}_{gains}$$

The heating-system capacity listed in column D is a constant for the furnace in this example. Equation (17.16) is used to compute the part-load factors in column E. For the 57 °F bin,

$$PLF = 1 - 0.25 \left(1 - \frac{3342}{98,700}\right) = 0.758$$

The part-load factor cannot exceed 1.0.

Column F lists the run time for the heating system in number of hours for each bin. The run time for the first bin becomes

$$\text{run time} = \frac{(\#hrs)(\dot{Q}_{system})}{(\dot{Q}_{capacity})PLF} = \frac{B \times C}{D \times E} = \frac{614 \text{ hr } (3342 \text{ Btu/hr})}{(98,700 \text{ Btu/hr})0.758} = 27.43 \text{ hr}$$

Note that the run time cannot exceed the hours available in the respective bin.

The last column, G, gives the required energy input into the heating system for each bin. For the first bin

$$E = \dot{Q}_{capacity}(\text{run time})/\eta_{system} = D \times F/\eta_{system}$$

$$= \frac{(98,700 \text{ Btu/hr})(27.43 \text{ hr})}{0.85} = 3.18E + 6 \text{ Btu}$$

The values in the second row of the table are computed using the same equations as outlined above for a bin temperature of 52 °F.

These equations can be programmed in a spreadsheet for rapid calculation. However, the limits in columns E and F should be included also. Table 17.2 shows the complete table of results for this example. Note that the heating-system capacity is nearly equal to the heating-system load at the lowest bin temperature of −28 °F. However, this temperature occurs for only 1 hr in a typical year. The part-load factor also becomes equal to one. Although the heating system must run almost continuously during the coldest temperatures, the energy required is largest at moderate temperatures, about 30 °F, because the number of hours there is much larger. The total amount of supplemental energy required into the furnace each year equals 2.52×10^8 Btu. Assuming 100,000 Btu/ccf of natural gas, the amount of gas required becomes

$$F_{yr} = \frac{2.52 \times 10^8 \text{ Btu}}{100,000 \text{ Btu/ccf}} = 2520 \text{ ccf}$$

TABLE 17.1 Ex. 17.2 Bin Calculation, Minneapolis Weather Data, Constant-Output Furnace

A Avg. Bin Temp., °F	B Annual Hours from Bin Data	C Heating-System Load, Btu/hr (can't be < 0)	D Heating-System Capacity, Btu/hr	E Part-Load Factor (PLF), Eq. (17.16) (can't be > 1)	F Run Time, Hr, B · C/(D · E) (can't be > col. B)	G Energy Input, Btu, F · D/(sys. eff.)
57	614	3342	98,700	0.758	27.43	3.18E + 06
52	552	8952	98,700	0.773	64.80	7.52E + 06

TABLE 17.2 Ex. 17.2 Bin Calculation, Minneapolis Weather Data, Constant-Output Furnace

$(UA)_{eff}$, Btu/hr =	1122	This is the effective (UA) value for the building
t_i, °F =	68	Indoor temperature
Internal gain, Btu/hr =	9000	This is the sum of internal gains expected
Sys. capacity, Btu/hr =	98,700	This is the constant output of the furnace at steady state
C_d =	0.25	This is the part-load degradation coefficient
Sys. efficiency =	0.85	This is the steady-state furnace efficiency
Calculated t_z, °F =	60.0	This is the calculated zero-load temperature

A	B	C	D	E	F	G
				Part Load factor	Run Time	Energy Input
		Heating-System	Heating System,	PLF	Hrs	Btu
Avg. Bin Temp.,	Annual Hours	Load, Btu/hr	Heating Capacity, Btu/hr	Eq. (17.16)	$B \cdot C/(D \cdot E)$	$F \cdot D/(\text{Sys eff.})$
°F	from Bin Data	(can't be < 0)		(can't be > 1)	(can't be > col. B)	
57	614	3342	98,700	0.758	27.41	3.18E + 06
52	552	8952	98,700	0.773	64.80	7.52E + 06
47	478	14,562	98,700	0.787	89.62	1.04E + 07
42	487	20,172	98,700	0.801	124.24	1.44E + 07
37	552	25,782	98,700	0.815	176.86	2.05E + 07
32	653	31,392	98,700	0.830	250.38	2.91E + 07
27	591	37,002	98,700	0.844	262.60	3.05E + 07
22	475	42,612	98,700	0.858	239.03	2.78E + 07
17	379	48,222	98,700	0.872	212.31	2.47E + 07
12	313	53,832	98,700	0.886	192.60	2.24E + 07
7	242	59,442	98,700	0.901	161.84	1.88E + 07
2	190	65,052	98,700	0.915	136.89	1.59E + 07
−3	131	70,662	98,700	0.929	100.96	1.17E + 07
−8	81	76,272	98,700	0.943	66.36	7.71E + 06
−13	45	81,882	98,700	0.957	38.99	4.53E + 06
−18	16	87,492	98,700	0.972	14.60	1.70E + 06
−23	6	93,102	98,700	0.986	5.74	6.67E + 05
−28	1	98,712	98,700	1.000	1.00	1.16E + 05

Annual Energy = 2.52E + 08

Note: Table was generated using a spread sheet. Use of the approximate value for t_z shown above will result in slightly different values.

The total amount of gas estimated using the bin method in Ex. 17.2 (2520 ccf) is more than the amount estimated using the degree day method in Ex. 17.1 (2162 ccf). The main difference is the use of the part-load factor in the bin calculations, which essentially derates the furnace output because of the cycling at part loads. This concept was not used in the degree-day calculation. An annual effective heating-system efficiency can be incorporated into the degree-day method to account for this.

Another important variation of the bin energy-estimation method is the use of heating equipment whose heating capacity is a strong function of the ambient temperature. An example is an air-to-air heat pump. Most heat pumps are similar to the single-stage mechanical vapor-compression refrigeration systems discussed in Chapters 3 and 4. The refrigeration coefficient of performance (C.O.P.) of these systems is shown in Fig. 3.9 as a function of the evaporating temperature with a constant condensing temperature. The C.O.P. decreases dramatically as the evaporator temperature drops. The desirable output from a heat pump is its heating capacity which is primarily the heat rejection from the condenser. The coefficient of performance of a heat pump can be described as

$$\text{C.O.P.}_{\text{h.p.}} = \frac{\dot{Q}_{\text{cond}}}{\dot{W}_{\text{net}}} \tag{17.17}$$

For theoretical cycles, the heat-pump coefficient of performance equals the refrigeration C.O.P. + 1.0. Therefore, the values for C.O.P. from Fig. 3.9 can be changed to C.O.P.$_{\text{h.p.}}$ values by adding 1.0.

The main performance parameter of interest for a heat pump in energy estimation is its heating capacity as a function of outdoor temperature. Figure 4.7 shows the refrigerating-capacity and compressor-power requirement as a function of evaporator temperature at two fixed condensing temperatures. The heating capacity as a function of evaporator temperature can be estimated from this figure by adding the refrigerating capacity and the power requirements together, as both the heat absorbed in the evaporator and most of the power added to the compressor will leave the system in the condenser.

Figure 17.3 is similar to Figure 17.2, except that the constant-rated-output heating system (i.e., furnace or boiler) has been replaced with an air-to-air heat pump. The heat-pump heating-capacity curve is shown vs. outdoor temperature, as the evaporator temperature closely corresponds to the outdoor temperature. When the outdoor temperature is greater than the zero-load temperature, no supplemental heating is required. The heat pump is needed part of the time between the zero-load temperature and the balance point where the heat-pump heating capacity is equal to the building heat loss above the internal gains. The heat pump is often selected so that the balance temperature is approximately 32 °F (0 °C). The heat-pump performance, like that of furnaces and boilers, is degraded by part-load operation. Equation (17.16) is used to estimate the effect of cycling. The dashed curve on Fig. 17.3 shows the actual heat-pump heating capacity corrected for part-load operation.

When the outdoor temperature is lower than the balance temperature, the heat-pump heating capacity is less than what is required by the building. Under these conditions, the heat pump operates continuously. In addition, an auxiliary backup heating system is also required to make up the difference between the building heating demand and the capacity of the heat pump. The backup system is often electric-resistance heating, as electric power is usually required to operate the heat pump, and other types of fuel may be unavailable. In some applications the heat pump is not operated when the outdoor temperature is below a threshold value. The reason is that the heat-pump heating capacity is very low during very cold conditions; nearly all the heating is provided by the

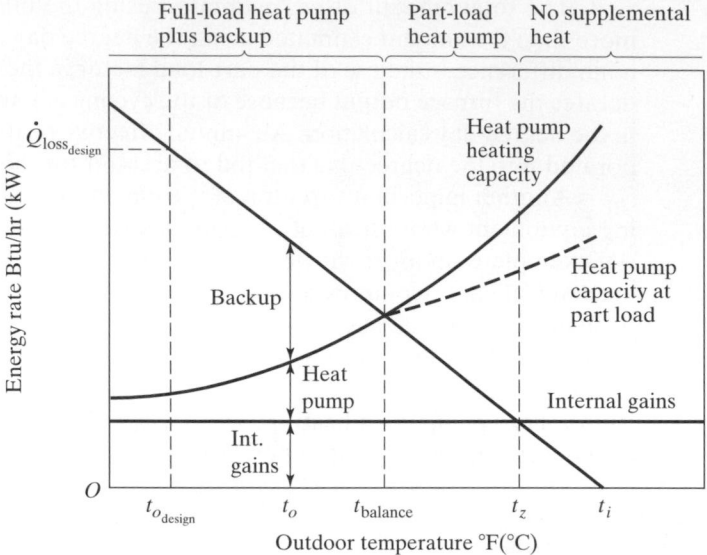

Full-load heat pump | Part-load | No supplemental
plus backup | heat pump | heat

Figure 17.3 Energy rates vs. outdoor temperature for a building with constant internal gains and an air-to-air heat pump with electric-resistance backup.

backup system. It is often uneconomical to operate the heat pump during these conditions, as the savings in backup energy do not compensate for the additional maintenance cost associated with the heat pump. This situation is not shown on Fig. 17.3 but could be included at low outdoor temperatures by setting the heat-pump heating capacity to zero and relying entirely on backup heating.

Example 17.3 illustrates the use of the bin energy-estimation method with an air-to-air heat pump and electric-resistance backup heating.

EXAMPLE 17.3

Estimate the energy required to heat the building described in Ex. 17.2 with the heating system replaced by an air-to-air heat pump with electric-resistance backup. The building is located in New York City. The heat pump has the following performance characteristics:

$$\dot{Q}_{cond}(\text{Btu/hr}) = 17{,}200 + 385t_o + 2.54t_o^2$$

$$\dot{W}_{comp}(\text{kW}) = 1.28 + 0.0093t_o - 0.000085t_o^2$$

where t_o is the outdoor temperature in °F.

Solution: As the building is the same as discussed in Ex. 17.2, many parameters remain the same, including the effective UA, the indoor temperature, the internal gains, and the zero-load temperature. The bin data are taken from Table B.8. The heat-pump heating capacity and power requirement are computed from the two equations above.

Table 17.3 shows the results of this bin calculation. Column A shows the average bin temperatures from the zero-load temperature to the minimum temperature for which annual hours are listed for New York in Table B.8. Column B lists the annual hours versus bin temperature from Table B.8. Column C is identical to column C in Table 17.2, except that the

TABLE 17.3 Ex. 17.3 Bin Calculation, New York Weather Data, Air-to-Air Heat-Pump Heating

$(UA)_{eff}$, Btu/hr =	1122	This is the effective (UA) value for the building
t_i, °F =	68	Indoor temperature
Internal gain, Btu/hr =	9000	This is the sum of internal gains expected
Heat-pump capacity =	$Q_{cond} = 17{,}200 + 385\,t_o + 2.54\,t_o^2$ Btu/hr	
$C_d =$	0.25	This is the part-load degradation coefficient
Power requirement =	$W_{comp} = 1.28 + 0.0093\,t_o - 0.000085\,t_o^2$ kW	
Calculated t_z, °F =	60.0	This is the calculated zero-load temperature

A Avg. Bin Temp, °F	B Annual Hours from Bin Data	C Heating System Load, Btu/hr (can't be < 0)	D Heat Pump Capacity, Btu/hr	E Part-Load Factor (PLF) PLF Eq. (can't be > 1)	F Run Time, Hrs, $B \cdot C/(D \cdot E)$ (can't be > col. B)	G Compressor Power kW	H Backup Energy kWh $(C - D) \cdot B/3412$ (can't be < 0)	I Energy Input kWh $(F \cdot G) + H$
57	780	3342	47,397	0.768	72	1.53	0	110
52	751	8952	44,088	0.801	190	1.53	0	292
47	771	14,562	40,906	0.839	327	1.53	0	500
42	829	20,172	37,851	0.883	500	1.52	0	761
37	842	25,782	34,922	0.935	665	1.51	0	1003
32	697	31,392	32,121	0.994	685	1.49	0	1021
27	453	37,002	29,447	1.000	453	1.47	1003	1669
22	279	42,612	26,899	1.000	279	1.44	1285	1688
17	169	48,222	24,479	1.000	169	1.41	1176	1415
12	83	53,832	22,186	1.000	83	1.38	770	884
7	35	59,442	20,019	1.000	35	1.34	404	451
2	8	65,052	17,980	1.000	8	1.30	110	121
−3	1	70,662	16,068	1.000	1	1.25	16	17

Annual Energy Input in kWh = 9932

Note: Table was generated using a spread sheet. Use of the approximate value for t_z shown above will result in slightly different values.

temperature range is smaller. Note that these values are linearly proportional to the difference between the zero-load temperature (60 °F) and the average bin temperature. The values in column D are calculated from the heat-pump-capacity equation above vs. average bin temperature. The heat pump has been sized so that its heating capacity approximately equals the building heat loss at 32 °F. For example, the heat-pump heating capacity when the average bin temperature is 57 °F becomes

$$\dot{Q}_{cond} = 17,200 + 385(57) + 2.54(57)^2 = 47,397 \text{ Btu/hr}$$

The part-load factor listed in column E is computed using Eq. (17.16) with C_d equal to the default value of 0.25 and an upper limit of 1.0. The heat-pump run time in column F is computed the same as the furnace run time in Table 17.2. In this case, however, the heat pump operates at part load for bin temperatures of 32 °F and above but operates continuously at bin temperatures below 32 °F (when the part-load factor equals 1.0). The power input is computed from the equation given above. For example, when the average bin temperature is equal to 57 °F, the power requirement is

$$\dot{W}_{comp} = 1.28 + 0.0093(57) - 0.000085(57)^2 = 1.53 \text{ kW}$$

The power requirement drops off slowly as the outdoor temperature drops. Below the balance temperature when the heat pump runs continuously, the building heat load is greater than the heat-pump heating capacity, so backup is needed. The backup energy required is given in column H. When the average bin temperature equals 27 °F, the backup energy becomes

$$E_{backup} = \frac{(37,002 - 29,447) \text{ Btu/hr} (453 \text{ hr})}{3412 \text{ Btu/hr} \cdot \text{kW}} = 1003 \text{ kWh}$$

Although the backup power requirement increases with decreasing outdoor temperature, the backup energy required decreases at very cold temperatures, because the number of hours there is small. The final column in Table 17.3 is the sum of the heat-pump electrical energy plus the backup energy for each bin. For the bin at an average outdoor temperature of 27 °F the total energy required becomes

$$E_{total} = (453 \text{ hr})(1.47 \text{ kW}) + 1003 \text{ kWh} = 1669 \text{ kWh}$$

The total annual heating-energy requirement is summed at the bottom of the table and is equal to 9932 kWh.

This example illustrates the procedure that should be used when the heating system requires backup during part of the year. Air-to-air heat pumps are usually not economical to operate in cold climates, as a considerable amount of backup energy is required. In the previous example, approximately one-half the annual heating energy was used in the backup system. In many cases the heat pump is turned off when the outdoor temperature falls below a certain value [e.g., 0 °F (−18 °C)], as the heating energy delivered by the heat pump is very low compared to the demand. The increased maintenance on the unit may more than offset the benefits of running it in cold weather. When this occurs, the backup must handle the entire heating load under the most severe heating conditions. Thus the backup system must be sized to handle the entire winter design heating load.

Variable Internal Gains (Occupied/Unoccupied)

The bin heating-energy estimation method discussed above assumes constant internal gains. This is not a good assumption in many buildings where the gains are strongly tied to occupancy. The occupants themselves, lighting, equipment, and other loads can vary significantly with time of day and day of the week. Therefore, the bin method should be modified to account for this variation.

The bin weather data given in Appendix B provide hours in 8-hour groups, hours 1–8, 9–16, and 17–24. Other bin data provide information in six 4-hour groups. These hour groups can be used to separate the hours in which the building is occupied from the unoccupied periods. The basic procedure is to determine the number of hours in each hour group that are occupied and unoccupied. Then two bin calculations are performed—one for the occupied hours, the other for the unoccupied hours. The heating-energy requirements for these two bin calculations are then summed to provide the total energy required to heat the building. The following example illustrates the methodology for using the bin method to estimate heating-energy requirements when there are two quite different internal gains based on occupancy.

EXAMPLE 17.4

Use the bin method to estimate the annual space-heating energy requirements for a building located in Denver, Colorado. The building has a winter design heat loss of 240,000 Btu/hr with a winter design temperature difference of 73 °F. The indoor temperature is to be maintained at 70 °F during the heating season when the building is occupied and at 65 °F when the building is unoccupied. The building is occupied Monday–Friday from 7 a.m. until 6 p.m. During occupancy, the internal gains are equal to 50,000 Btu/hr. During the unoccupied periods, the internal gains are 4000 Btu/hr. The building is heated using three air-to-air heat pumps with performance characteristics as described in Ex. 17.3. The backup is electric-resistance heating.

Solution: The first task is to determine the number of hours in each bin hour group that represent occupied versus unoccupied time periods. Figure 17.4 shows the occupied and unoccupied hours vs. time of day and day of the week. From this table, the number of occupied hours in hour group I (01–08) are 5 days $\times$ 1 hr/day = 5 hr. The remainder are unoccupied hours: 7 days $\times$ 8 hr/day $-$ 5 hr = 51 hr. The percentage of time in hour group I that is occu-

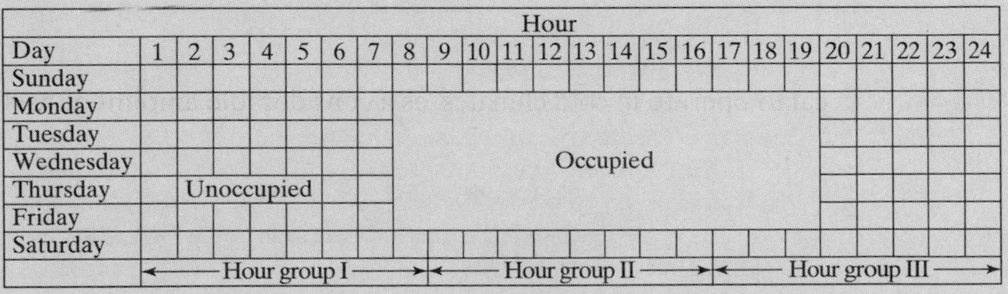

Figure 17.4 Occupied vs. unoccupied hours for Ex. 17.4.

---

pied is 5/56 = 9%. The remaining 91% of the hours in group I are unoccupied. Likewise for the other two hour groups:

$$\text{Hour group II (09–16) occupied hours} = 5 \text{ days} \times 8 \text{ hr/day} = 40 \text{ hr}$$

$$\text{Unoccupied hours} = 56 \text{ hr} - 40 \text{ hr} = 16 \text{ hr}$$

$$40/56 = 71\% \text{ occupied}, \qquad 16/56 = 29\% \text{ unoccupied}$$

$$\text{Hour group III (17–24) occupied hours} = 5 \text{ days} \times 3 \text{ hr/day} = 15 \text{ hr}$$

$$\text{Unoccupied hours} = 56 \text{ hr} - 15 \text{ hr} = 41 \text{ hr}$$

$$15/56 = 27\% \text{ occupied}, \qquad 41/56 = 73\% \text{ unoccupied}$$

The zero-load temperatures are now determined, so we know what range of bin temperatures we must include in the tables that follow. Using Eq. (17.2), we determine $(UA)_{eff}$:

$$(UA)_{eff} = \frac{\dot{Q}_{loss, design}}{(t_{i, design} - t_{o, design})} = \frac{240{,}000 \text{ Btu/hr}}{73 \text{ °F}} = 3288 \text{ Btu/hr} \cdot \text{°F}$$

Then from Eq. (17.3) we can determine the zero-load temperature for the unoccupied hours:

$$t_{z, u} = 65 \text{ °F} - (4000 \text{ Btu/hr})/(3288 \text{ Btu/hr} \cdot \text{°F}) = 63.8 \text{ °F}$$

The zero-load temperature for the occupied hours is

$$t_{z, o} = 70 \text{ °F} - (50{,}000 \text{ Btu/hr})/(3288 \text{ Btu/hr} \cdot \text{°F}) = 54.8 \text{ °F}$$

The zero-load temperature is higher during the unoccupied hours, so the bin calculations will begin with the bin at 60/64 °F that has a mean temperature of 62 °F.

Now the number of hours in each hour group that are occupied is determined as a function of bin temperature by multiplying the hour-group bin data by the percentages determined above. The results are shown in Table 17.4. The values in the first row at a mean bin temperature of 62 °F are determined as illustrated here.

The total number of hours per year in hour group I at this bin temperature (62 °F) is obtained from Table B.5, the bin weather data for Denver, Colorado, and is equal to 276 hr. The number of hours that the building is occupied is obtained by multiplying the total number of hours (276) by the percentage of time the building is occupied during this hour group from Table 17.4 (9 percent), or 276 hr × 0.09 = 25 hr. The other two hour groups are handled in a similar manner. For hour group II, the total number of hours from Table B.5 is 245, the percentage of time the building is occupied from Table 17.4 is 71 percent, so the number of occupied hours for group II becomes 245 hr × 0.71 = 174 hr. For hour group III, the total number of hours is 273, the percentage of time the building is occupied is 27 percent, so the number of hours of occupancy becomes 273 × 0.27 = 74 hours.

The total number of hours of occupancy are summed for the three hour groups: 25 + 174 + 74 = 273 hr. The total hours unoccupied are obtained by subtracting the occupied hours from the total: (276 + 245 + 273) − 273 = 322 hr. The values in the remaining rows of Table 17.4 are obtained in a similar manner. The only values that are needed in the final bin calculations are the numbers in the last two columns—the total annual hours occupied and unoccupied at each mean bin temperature.

Now that the hours have been determined for occupied and unoccupied periods, the bin calculations can be performed. The procedure for this example is identical to that used in Ex. 17.3, so the details will be omitted here. The main difference is that two bin calculations are required, one for the occupied hours and the other for the unoccupied hours. Table 17.5 shows the results of the bin calculations for the occupied hours, and Table 17.6 includes the results for the unoccupied hours. Note that the indoor temperature, the internal gains, and the zero-load temperature are different. However, the $(UA)_{eff}$, heat-pump heating capacity and power input

TABLE 17.4 Occupied-Hour Distribution for Ex. 17.4

Mean Temp.	Hour Group: % time:	I Total 100	I Occupied 9	II Total 100	II Occupied 71	III Total 100	III Occupied 27	Total Hr Occupied	Total Hr Unoccupied
						Hours			
62		276	25	245	174	273	74	273	322
57		295	27	237	168	244	66	261	337
52		264	24	244	173	231	62	259	311
47		253	23	228	162	248	67	252	296
42		291	26	193	137	268	72	236	321
37		304	27	166	118	254	69	214	325
32		299	27	158	112	247	67	206	318
27		245	22	120	85	190	51	159	258
22		187	17	73	52	134	36	105	191
17		119	11	40	28	84	23	62	120
12		72	6	24	17	41	11	35	72
7		41	4	17	12	26	7	23	42
2		25	2	10	7	19	5	14	26
−3		11	1	5	4	6	2	6	11
−8		7	1	1	1	5	1	3	7
−13		3	0	0	0	2	0	1	3
−18		1	0	1	1	1	1	1	1
−23		1	0	0	0	0	0	0	1

TABLE 17.5 Ex. 17.4 Bin Calculation, Denver Weather Data, Air-to-Air Heat-Pump Heating, Occupied

$(UA)_{eff}$, Btu/hr =	3288	This is the effective (UA) value for the building
t_i, °F =	70	Indoor temperature during occupancy
Internal gain, Btu/hr =	50,000	Sum of internal gains during occupancy
Heat-pump capacity =	$51{,}600 + 1155 t_o + 7.62 t_o^2$ Btu/hr	
C_d =	0.25	This is the part-load degradation coefficient
Power requirement =	$3.84 + 0.0279 t_o - 0.000255 t_o^2$ kW	
Calculated t_z, °F =	54.8	This is the calculated zero-load temperature

A Avg. Bin Temp., °F	B Annual Occ. Hours from Table 17.4	C Heating-System Load, Btu/hr (can't be < 0)	D Heat-Pump Capacity, Btu/hr	E Part-Load Factor (PLF), PLF Eq. (can't be > 1)	F Occ. Run Time, Hr, $B \cdot C/(D \cdot E)$ (can't be > col. B)	G Compressor Power, kW	H Backup Energy, kWh $(C - D) \cdot B/3412$ (can't be < 0)	I Energy Input, kWh, $(F \cdot G) + H$
62	273	0	152,501	0.750	0	4.59	0	0
57	261	0	142,192	0.750	0	4.60	0	0
52	259	9184	132,264	0.767	23	4.60	0	108
47	252	25,624	122,718	0.802	66	4.59	0	301
42	236	42,064	113,552	0.843	104	4.56	0	473
37	214	58,504	104,767	0.890	134	4.52	0	608
32	206	74,944	96,363	0.944	170	4.47	0	759
27	159	91,384	88,340	1.000	159	4.41	142	843
22	105	107,824	80,698	1.000	105	4.33	835	1289
17	62	124,264	73,437	1.000	62	4.24	924	1186
12	35	140,704	66,557	1.000	35	4.14	761	905
7	23	157,144	60,058	1.000	23	4.02	654	747
2	14	173,584	53,940	1.000	14	3.89	491	545
-3	6	190,024	48,204	1.000	6	3.75	249	272
-8	3	206,464	42,848	1.000	3	3.60	144	155
-13	1	222,904	37,873	1.000	1	3.43	54	58
-18	1	239,344	33,279	1.000	1	3.26	60	64
-23	0	255,784	29,066	1.000	0	3.06	0	0

Annual Energy Input in kWh = $\overline{8313}$

Note: Table was generated using a spread sheet. Use of the approximate value for t_z shown above will result in slightly different values.

TABLE 17.6 Ex. 17.4 Bin Calculation, Denver Weather Data, Air-to-Air Heat-Pump Heating, Unoccupied

$(UA)_{eff}$, Btu/hr =	3288	This is the effective (UA) value for the building
t_i, °F =	65	Indoor temperature during occupancy
Internal gain, Btu/hr =	4000	Sum of internal gains during occupancy
Heat-pump capacity =		$51{,}600 + 1155\,t_o + 7.62\,t_o^2$ Btu/hr
C_d =	0.25	This is the part-load degradation coefficient
Power requirement =		$3.84 + 0.0279\,t_o - 0.000255\,t_o^2$ kW
Calculated t_z, °F =	63.8	This is the calculated zero-load temperature

A Avg. Bin Temp., °F	B Annual Unocc. Hours from Table 17.4	C Heating-System Load, Btu/hr (can't be < 0)	D Heat-Pump Capacity, Btu/hr	E Part-Load Factor (PLF), PLF Eq. (can't be > 1)	F Unocc. Run Time, Hr, $B \cdot C/(D \cdot E)$ (can't be > col. B)	G Compressor Power, kW	H Backup Energy, kWh $(C - D) \cdot B/3412$ (can't be < 0)	I Energy Input, kWh, $(F \cdot G) + H$
62	322	5864	152,501	0.760	16	4.59	0	75
57	337	22,304	142,192	0.789	67	4.60	0	308
52	311	38,744	132,264	0.823	111	4.60	0	509
47	296	55,184	122,718	0.862	154	4.59	0	708
42	321	71,624	113,552	0.908	223	4.56	0	1018
37	325	88,064	104,767	0.960	285	4.52	0	1287
32	318	104,504	96,363	1.000	318	4.47	759	2181
27	258	120,944	88,340	1.000	258	4.41	2465	3602
22	191	137,384	80,698	1.000	191	4.33	3173	4000
17	120	153,824	73,437	1.000	120	4.24	2827	3336
12	72	170,264	66,557	1.000	72	4.14	2188	2486
7	42	186,704	60,058	1.000	42	4.02	1559	1728
2	26	203,144	53,940	1.000	26	3.89	1137	1238
−3	11	219,584	48,204	1.000	11	3.75	553	594
−8	7	236,024	42,848	1.000	7	3.60	396	422
−13	3	252,464	37,873	1.000	3	3.43	189	199
−18	1	268,904	33,279	1.000	1	3.26	69	72
−23	1	285,344	29,066	1.000	1	3.06	75	78

Annual Energy Input in kWh = 23,842

Note: Table was generated using a spread sheet. Use of the approximate value for t_z shown above will result in slightly different values.

vs. outdoor temperature, and the part-load degradation coefficient are the same. As mentioned previously, these bin calculations can be easily programmed into a spreadsheet, so that the user can readily modify any of the input values and perform comparative analyses.

The final step in this example is to sum the total energy requirements for both the occupied and unoccupied periods during the year:

$$\text{Total annual energy required} = 8313 + 23,842 = 32,155 \text{ kWh}$$

In this example, the energy required during the unoccupied hours is about three times greater than the amount needed during the occupied hours. This is primarily caused by the larger number of hours during the year that are unoccupied.

The example above illustrates a procedure that can be used for two differing periods of time, occupied and unoccupied. A similar procedure can be developed for additional periods of time when the building may be partially occupied or the internal gains are different. Examples include industrial structures, where large heat-generating processes may be used only during specified periods of the day or on certain days of the week.

17.3   MODIFIED BIN METHOD

The modified bin method was developed to provide a simplified hand-calculation method for estimating annual energy requirements including heating and cooling. This hand-calculation method has been largely replaced by computer programs that use more sophisticated methods to estimate energy usage. A brief overview is given here. References [5] and [6] provide a more detailed discussion.

The modified bin method uses heating and cooling loads divided into building zones. This allows some zones to be heated while others are cooled. Without subdivision into zones, simultaneous heating and cooling loads would tend to cancel, which is not how typical multizone HVAC systems operate. Two or more computational periods are selected, usually representing occupied and unoccupied periods. Zone loads are both time dependent and temperature dependent. For each period, the time-dependent loads are averaged and added to the temperature-dependent loads, such that the total load is characterized by a linear function of outdoor temperature for the calculation time period.

The temperature-dependent envelope load is determined by calculating the cooling load for a typical day in July and the heating load in a typical day in January. Solar gains through opaque structures are neglected in the heating mode, but solar gains through glazing are included. For the cooling load, procedures are similar to those used to determine design cooling loads given in Chapters 15 and 16. The cooling-load-temperature-difference (CLTD) approach is commonly used. Solar gains are determined using typical cloud cover rather than clear-sky conditions to represent average conditions.

The resulting linear load profiles versus outdoor temperature appear similar to those shown in Fig. 17.5. Two zero-load temperatures are shown separated by a deadband. This assumes that the indoor-temperature setpoint during space heating is less than the setpoint for cooling. Once the load profiles are determined, a bin table is set up similar to the tables used for the classical bin method, so that the annual heating and cooling energy use can be estimated. Performance data or correlations with outdoor temperature are needed for both heating and cooling equipment.

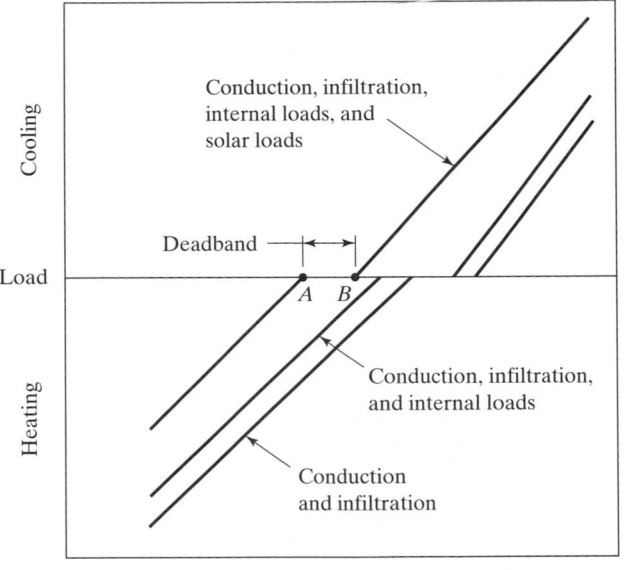

Figure 17.5 General load profile for dual-setpoint thermostats; *A* is zero-load point for heating, *B* is zero-load point for cooling. (Reprinted by permission from ASHRAE, "Simplified Energy Analysis Using the Modified Bin Method," Jan. 15, 1983.)

17.4 DETAILED ENERGY ESTIMATION METHODS

Computer technology has improved to the point where nearly everyone who wishes to simulate the thermal performance of buildings can do so with existing computer hardware and software. Simulation packages range from simple residential or small commercial codes that estimate monthly loads to simulations of large, multizone buildings with hour-by-hour calculations. Lack of time and detailed knowledge of the building on the part of the user are more of a limitation than lack of detailed modeling capability by the software. A detailed description of building energy estimation methods is beyond the scope of this chapter. Readers interested in more information on this subject are referred to the *ASHRAE Handbook of Fundamentals* [1]. A periodic listing of appropriate software is published by ASHRAE [7].

Figure 17.6 illustrates a typical flow chart for estimating the energy use and cost to operate a building. There can be feedback between some of the computations also. For example, the capacity of the equipment to deliver hot or chilled water will limit the heating/cooling that can be provided by the HVAC system. This may result in the system's being unable to maintain the indoor temperature at the setpoint. The first step in the simulation is to estimate the building heating and cooling loads, using information about the structure itself and the local weather or climate. Hourly calculations are usually used following the methods outlined in Chapters 15 and 16. These calculations require knowledge of outdoor temperature and humidity fluctuations, solar insolation, and wind for an entire year. Indoor variables include occupancy, lighting, ventilation requirements, and other internal gains, sensible and latent. Some calculations use one typical day for each month rather than an entire year of weather data. Once the heating/cooling loads are determined,

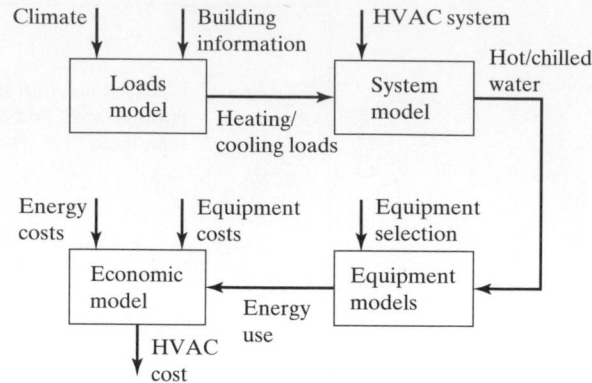

Figure 17.6 Typical flow diagram for detailed calculation of building HVAC energy use and costs.

the required system heating/cooling energy requirements are determined from a system model. For example, if a VAV system is to be used, a model to simulate the heating/cooling-coil loads must be used to estimate the hot/chilled water demands on the coils. Once the heating/cooling energy requirements are computed, the purchased-energy requirements on the HVAC equipment are determined. This includes fuel and electric energy to power fans, pumps, control motors, and auxiliary equipment. The fuel and electric-energy use and rates can then be sent to an economic model to estimate the cost of operating the building. This should include demand charges in addition to usage rates. If capital costs, depreciation, and maintenance costs are estimated, a life-cycle cost can be obtained.

An advantage of using detailed energy-estimation methods is that the user can select different heating or cooling systems, different equipment, and perhaps modify the building design to determine the effects of these variables on the total HVAC energy use and cost. The results can also be used to verify compliance with energy-efficiency standards or codes.

ENDNOTES

1. *ASHRAE Handbook of Fundamentals* (Atlanta: American Society of Heating, Refrigerating and Air Conditioning Engineers, 1993), ch. 28.

2. *Passive Solar Design Handbook*, Vol. 3: *Passive Solar Design Analysis*, Los Alamos National Laboratory, U.S. Department of Energy, DOE/CS-0127/3, July 1982.

3. D. G. Erbs, S. A. Klein, and W. A. Beckman, "Estimation of Degree Days and Ambient Temperature Bin Data From Monthly-Average Temperatures," *ASHRAE Journal*, 25:6 (1983), 60.

4. *Engineering Weather Data*, Department of the Air Force Manual AFM 88-29, July 1, 1978.

5. F. C. McQuiston and J. D. Parker, *Heating, Ventilating and Air Conditioning, Analysis and Design*, 4th ed. (New York: John Wiley & Sons, 1994).

6. *Simplified Energy Analysis Using the Modified Bin Method*, (Atlanta: American Society of Heating, Refrigerating and Air Conditioning Engineers, 1983).

7. *ASHRAE Journal's HVAC&R Software Directory*, (Atlanta: American Society of Heating, Refrigerating and Air Conditioning Engineers, 1990).

PROBLEMS

17.1 Determine the number of degree days from the hourly weather data given below for
(a) $t_z = 65\,°F$,
(b) $t_z = 60\,°F$, and
(c) $t_z = 55\,°F$.

Hour	1	2	3	4	5	6	7	8	9	10	11	12
t_o (°F)	42	40	38	36	35	35	39	41	44	47	50	54

Hour	13	14	15	16	17	18	19	20	21	22	23	24
t_o (°F)	57	59	59	58	56	53	50	46	44	41	38	36

17.2 Determine the average heating-system efficiency for an intermittent-ignition natural-gas furnace given the following information: steady-state efficiency = 0.92, rated furnace output is 1.1 times the building heat loss when $\Delta t = t_z - t_{o,\text{design}}$, and 5 percent of the heat loss is from ducts.

17.3 Repeat Ex. 17.1 if the house was located in Denver, Colorado.

17.4 Estimate the amount of natural gas required to heat a home located in New York City for one year if the value for $(UA)_{\text{eff}}$ is 1600 Btu/hr · °F, the internal gains are 10,000 Btu/hr, the thermostat is set at 70 °F when the heating system is on, and the average heating-system efficiency is 85 percent.

17.5 Determine the average heating-system efficiency for a natural-gas furnace with intermittent ignition for a home in Minneapolis, where the winter design heat loss is 105,000 Btu/hr based on $t_i = 72\,°F$ and $t_o = -12\,°F$ and the rated output of the furnace is 110,000 Btu/hr with a steady-state efficiency of 93 percent. Assume 4 percent of the heat loss is from ducts.

17.6 Compute the amount of natural gas required to heat the home described in Prob. 17.5 for one year if the internal gains are continuous and are equal to 6000 Btu/hr.

17.7 Use Eq. (17.14) and the bin weather data given in Table B.8 to compute the number of degree days in New York City in January using a zero-load temperature of 65 °F. Compute the number of degree days in each of the three daily time periods given in the bin data table and sum the results to obtain the total for the month. Compare your answer for the total number of degree days in January to the value given in Table B.2.

17.8 Repeat Prob. 17.4 using the degree-day results from Prob. 17.7 if the internal gains are 8000 Btu/hr during hours 1–8, 12,000 Btu/hr during hours 9–16,

and 10,000 Btu/hr during hours 17–24. Compare the answer to the result obtained for Prob. 17.4.

17.9 Compute the monthly average temperatures for Minneapolis, Minnesota using the bin weather data given in Table B.7, and plot the results vs. time of year.

17.10 Estimate the amount of natural gas required to heat the building described in Ex. 17.2 for the month of January in Minneapolis using the bin method.

17.11 Repeat Ex. 17.2 if the building is located in Tucson, Arizona, rather than Minneapolis.

17.12 Estimate the amount of fuel oil required to heat a home in Denver, Colorado, for one year, given the following information: design heat loss is 120,000 Btu/hr at an indoor temperature of 70 °F and an outdoor temperature of 1 °F, internal gains are 6500 Btu/hr, the interior temperature is maintained at 68 °F, and the furnace steady-state efficiency is 88 percent. Heating value of fuel oil is 140,000 Btu/lbm.

17.13 Repeat Ex. 17.3 if the building is located in Denver, Colorado, rather than in New York City.

17.14 Repeat Ex. 17.3 if the heat pump uses ground water at a constant supply temperature of 60 °F rather than outdoor air as the energy source. Redraw Fig. 17.3 for this water-source heat-pump system, showing the heat-pump heating capacity versus outdoor temperature and the balance point. Estimate the annual electrical-energy requirement using the bin method. Compare your results with those from Ex. 17.3.

17.15 Modify Ex. 17.4, assuming that the heat pump does not operate when the outdoor temperature falls below 0 °F. Determine the total annual energy requirement in this case.

17.16 Repeat the calculations for Ex. 17.4 using a different heat pump. Obtain capacity and power-input information vs. outdoor temperature from a vendor. The capacity at 32 °F should be similar to the value used in Ex. 17.4. Fit the data with linear or quadratic functions, and use these functions to estimate the heat-pump heating capacity and power input at various bin temperatures.

17.17 Repeat Ex. 17.4 with the building located in Atlanta, Georgia, rather than in Denver, Colorado.

17.18 Repeat Ex. 17.4 using bin data for your location. Additional data are available from reference [5] or can be estimated using the procedures from reference [3].

17.19 Estimate the annual heating energy required by a building located near New York City that has a natural-gas-boiler heating system. The winter design heat loss is 670,000 Btu/hr at a design temperature difference of 80 °F. The boiler has a rated steady-state heating capacity of 700,000 Btu/hr and a rated efficiency of 89 percent. The building is fully occupied from 0800 to 1700 hrs Monday through Friday with an internal gain of 120,000 Btu/hr. The building is partially occupied from 1700 to 2100 hrs Monday through Friday and from 0900 to 1700 hr Saturday with an internal gain of 50,000 Btu/hr. The remainder of the time the building is unoccupied with an internal gain of 10,000 Btu/hr.

17.20 Repeat Prob. 17.19 for your location using local bin weather data or estimated data, using the procedure in reference [3].

SYMBOLS

C_d	Degradation coefficient.
CF_{pl}	Part-load correction factor.
C.O.P.	Coefficient of performance.
DD	Degree days, °F day.
E	Energy, Btu (kJ).
E_{backup}	Energy use by backup heating system, Btu (kJ).
E_p	Energy use for time period p, Btu (kJ).
E_{total}	Total energy use, Btu (kJ).
F_p	Amount of fuel used during time period p.
F_{yr}	Amount of fuel used for one year.
H	Heating value of fuel, Btu/unit (kJ/unit).
LF	Load factor.
MCWB	Mean coincident wet-bulb temperature, °F (°C).
N_{bin}	Number of hours in each bin, hr.
$\dot{Q}_{capacity}$	Heating capacity of heating system, Btu/hr (kW).
$\dot{Q}_{cond}$	Heat-rejection rate by heat-pump condenser, Btu/hr (kW).
$\dot{Q}_{gains}$	Rate of energy addition due to internal gains, Btu/hr (kW).
$\dot{Q}_{loss}$	Heat-loss rate from the building, Btu/hr (kW).
$\dot{Q}_{net}$	Net heat loss from the building, Btu/hr (kW).
$\dot{Q}_{output}$	Rate of heat delivered by heating system at steady state, Btu/hr (kW).
$\dot{Q}_{system}$	Heating required from the heating system, Btu/hr (kW).
RLC	Relative load coefficient defined by Eq. 17.12.
t_{bin}	Mean temperature of bin, °F (°C).
t_i	Indoor temperature, °F (°C).
t_o	Outdoor temperature, °F (°C).
t_z	Zero-load temperature, °F (°C).
UA	Total heat-loss coefficient for building, Btu/hr · °F (kW/°C).
$\dot{W}_{comp}$	Power input to heat-pump compressor, kW.
$\dot{W}_{input}$	Total power required by heat pump, kW.

Greek Letters

α_D	Fraction of heat loss from ducts.
η_h	Average efficiency of heating system.

| η_{ss} | Steady-state efficiency of heating system. |
| τ | Time, hr. |

Subscripts

design	Evaluated at winter design conditions.
eff	Effective.
h.p.	Heat pump.
o	Occupied.
p	Time period.
u	Unoccupied.
z	Zero load conditions.

chapter 18

Air-Distribution Systems and Duct Design

18.1 INTRODUCTION

The object of air distribution in forced-air heating, ventilating, and air conditioning systems is to create the proper combination of temperature, air velocity, and air-contaminant concentrations in the occupied zone of the conditioned space. When the occupants are people, acceptable thermal-comfort conditions are to be maintained and the indoor air-contaminant concentrations should be controlled to assure safety and health. Similar environmental conditions can be prescribed for animals, plants, and warehouse facilities.

The basic approach used to design air-distribution systems for buildings is outlined in Fig. 18.1. The total building design heating and cooling loads (Chapters 14 and 16) and the selection of the system type (Chapter 8) are used to determine the total airflow rate supplied and returned from the building. The airflow rate supplied to each room is obtained in a similar manner. Note that the sum of the room airflow rates does not necessarily equal the total building airflow rate, as the room peak loads do not occur at the same time as the building peak. The type of diffuser and return must be selected, then their number and locations specified. Diffuser catalog information provides the pressure drop across each diffuser and return grille. Then the supply and return ductwork are laid out, connecting the supply diffusers and the return grilles to the central air-handling unit. A duct-sizing method is then used to size the ducts, round or rectangular. The transitions, fittings, and final specifications depend on the duct material chosen. From this, supports such as hangers can be chosen, and a material cost estimate can be prepared. Figure 18.1 illustrates the basic design approach for forced-air distribution systems.

Some air-distribution systems are used primarily to provide adequate ventilation air to the space. A separate heating or cooling system is used to transport the necessary energy. For these systems, the airflow rate is not determined from energy considerations but from ventilation requirements.

The remainder of this chapter will describe the process of air-distribution system design in more detail.

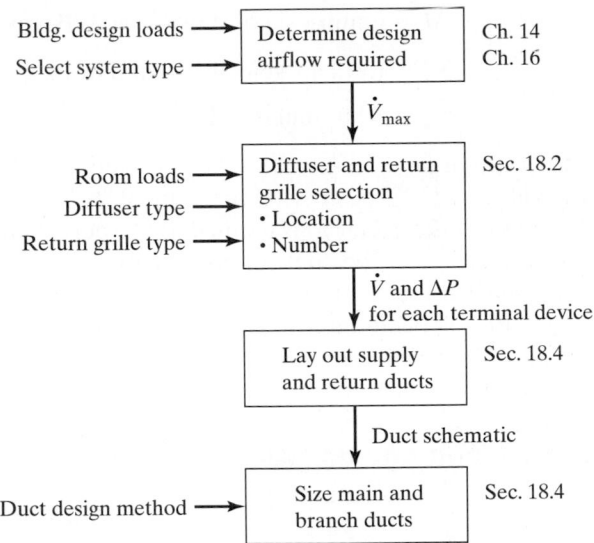

Bldg. design loads → [Determine design airflow required] Ch. 14 / Ch. 16

Select system type →

$\dot{V}_{max}$

Room loads →

Diffuser type → [Diffuser and return grille selection • Location • Number] Sec. 18.2

Return grille type →

$\dot{V}$ and ΔP for each terminal device

[Lay out supply and return ducts] Sec. 18.4

Duct schematic

Duct design method → [Size main and branch ducts] Sec. 18.4

Figure 18.1 Outline of forced-air distribution system design procedure and sections in Chapter 18 that cover the various steps.

18.2 ROOM AIR DISTRIBUTION AND DIFFUSER SELECTION

Air distribution in a room is often the primary means of distributing energy within the space. In the heating mode, warm air must be distributed to counteract heat loss by thermal radiation to cold surfaces and by convection heat loss via cold drafts. In the cooling mode, cool air needs to be distributed to absorb heat from warm building interior surfaces and from internal heat gains, such as those from occupants, lights, and equipment. The terminology used in the following discussion is that air is supplied to the room through *diffusers* and air is removed from the room through *return-air grilles*.

Air-Distribution Performance Index (ADPI)

Two factors that influence people's thermal comfort are dry-bulb temperature and air velocity, as described in Chapter 12. The higher the air dry-bulb temperature, the warmer an occupant will feel. However, the higher the air velocity, the cooler the occupant will feel. These two factors are interrelated in our approach to room air distribution. For example, when one is sitting in front of an electric fan on a hot summer day, the increased air velocity produced by the fan provides a cool sensation, even though the dry-bulb temperature of the air discharged from the fan is higher than that of the surrounding air because of the heat gained by the fan motor. An empirical relation developed by Rydberg and Norback [1] that combines the effects of dry-bulb temperature and air velocity can be written as follows:

$$\text{EDT} = (t_x - t_r) - M(\mathbf{V}_x - \mathbf{V}_r) \tag{18.1}$$

where   EDT = effective draft temperature, °F or °C

t_x = local air-stream dry-bulb temperature, °F or °C

t_r = average room dry-bulb temperature, °F or °C

M = empirical constant = 0.07 °F · min/ft or 7.0 °C · s/m

$\mathbf{V}_x$ = local air velocity, ft/min or m/s

$\mathbf{V}_r$ = 30 ft/min or 0.15 m/s

A high percentage of people engaged in sedentary activities are comfortable when the value of EDT lies between −3 and 2 °F (−1.7 and 1.1 °C). The *Air Distribution Performance Index (ADPI)* is defined as the percentage of locations in the occupied zone of a room that meet the comfort criteria above. The best value of ADPI is 100, which indicates that all the occupied locations in the room are within the comfort zone. The value of ADPI has been determined experimentally for many commercially available supply-air diffusers. However, the tabulated numbers are valid only under the conditions of test, which usually include an empty room with cold air supplied from the diffuser.

Local Mean Age and Air-Change Efficiency

Another important purpose for air distribution is to supply clean ventilation air to the occupied locations in the room and to remove airborne contaminants generated within the conditioned space. These contaminants include water vapor, particulates (dust and biological aerosol), and gases. One can envision a very good level of ventilation in the air stream discharged into the room from a diffuser. The contaminant removal is also very good adjacent to a return-air grille. However, the levels of ventilation and contaminant removal vary with location in the room and depend upon the air flow pattern. Several measures of the effectiveness of air-distribution systems have been developed. Many of these are reviewed in [2]. The discussion here will focus on the supply of clean ventilation air to a location in the room rather than contaminant removal.

Consider a point (P) in a room with one supply-air inlet and one return-air exhaust. If the flow is laminar, one can follow the streamline that passes through point P back to the inlet and determine the time required for each molecule of air to pass from the inlet to point P. However, actual room airflows are turbulent rather than laminar. Therefore, air molecules can reach point P from the inlet by following a number of different paths in the turbulent flow. The average time required for a molecule of air to pass from the inlet to point P is defined as the *local mean age of air at P* and is given the symbol $\bar{\tau}_p$. The smaller the value of $\bar{\tau}_p$, the less time is available for contaminants in the room to be mixed with the supply air. For this condition, the supply air is nearly at the same contaminant concentration as at the supply diffuser. However, large values of $\bar{\tau}_p$ indicate a significant amount of room contaminants can be mixed with the air before it reaches P. Very small values for $\bar{\tau}_p$ are found in the supply-air jet, and very large values are found in recirculation zones in the room.

Values for local mean age of air can be determined experimentally using one of three methods: (1) pulse, (2) step-up, or (3) step-down. In the *pulse method* a short pulse of a tracer gas or particle is injected into the room with the supply air. The concentration of the tracer in the room before and after the test should be equal to zero. The time-dependent concentration of the tracer is monitored at a given point P to determine the local mean age of air at that point. The average of all points in the room is defined as the room mean age of air. This can be determined by monitoring the tracer concentration at the exhaust. In the *step-up method* the procedure is similar except that the tracer is continuously added after the start of the test until the concentration in the room equals the

TABLE 18.1 Summary of Equations to Compute Local and Room Mean Age of Air

Local Mean Age $\bar{\tau}_p$	
1. Pulse method:	$\bar{\tau}_p = \dfrac{\displaystyle\int_0^\infty tC_p(t)\,dt}{\displaystyle\int_0^\infty C_p(t)\,dt}$
2. Step-up method:	$\bar{\tau}_p = \displaystyle\int_0^\infty \left(1 - \dfrac{C_p(t)}{C_s}\right) dt$
3. Step-down method:	$\bar{\tau}_p = \displaystyle\int_0^\infty \dfrac{C_p(t)}{C(0)}\,dt$

Room Mean Age $\langle \bar{\tau} \rangle$	
1. Pulse method:	$\langle \bar{\tau} \rangle = \dfrac{\dot{V}}{2V} \dfrac{\displaystyle\int_0^\infty t^2 C_e(t)\,dt}{\displaystyle\int_0^\infty C_e(t)\,dt}$
2. Step-up method:	$\langle \bar{\tau} \rangle = \dfrac{\dot{V}}{V} \displaystyle\int_0^\infty t\left(1 - \dfrac{C_e(t)}{C_s}\right) dt$
3. Step-down method:	$\langle \bar{\tau} \rangle = \dfrac{\dot{V}}{V} \displaystyle\int_0^\infty t\dfrac{C_e(t)}{C(0)}\,dt$

concentration at the inlet. In the *step-down method* the room is first filled with tracer to an initial concentration. Then a supply airflow is started that has zero concentration of tracer. Table 18.1 gives the equations that can be used to compute the local and room mean age of air for each of the three test methods.

Some additional terms commonly used to describe ventilation performance are (1) air-change rate, (2) nominal time constant, and (3) air-change efficiency. The *air-change rate* is the number of room volumes of air supplied in one hour and is defined as

$$\text{ACH} = \frac{\dot{V}}{V} \tag{18.2}$$

where $\dot{V}$ is the volumetric flow rate of air into the room, ft³/hr (m³/hr), and V is the room volume, ft³ (m³). The units on ACH are hr^{-1}. The room *nominal time constant*, τ_n, is the inverse of the air-change rate:

$$\tau_n = \frac{1}{\text{ACH}} \tag{18.3}$$

and describes the time required to change all the air in the room without mixing.

The *air-change efficiency*, ε_a, is the ratio of the nominal time constant and twice the room mean age:

$$\varepsilon_a = \frac{\tau_n}{2\langle\overline{\tau}\rangle} \times 100 \qquad (18.4)$$

and is a measure of how rapidly the air in a room is replaced.

Perfect Mixing

One limiting case of room air motion is the concept of *perfect mixing*. In this case, the entering supply air is instantaneously perfectly mixed with all the air in the room. A schematic drawing of this situation is shown in Fig. 18.2. The local mean age of air is equal for all points in the room, so the local mean age is also equal to the room mean age, which is equal to the nominal time constant for the room ($\overline{\tau}_p = \langle\overline{\tau}\rangle = \tau_n$). The results of a step-down tracer test for a room with perfect mixing are shown in Fig. 18.3. The concentration is an exponential decay with a time constant equal to the nominal time constant for the room. The ventilation efficiency for a perfectly mixed room is 50 percent.

Piston Flow

Another limiting case of room air motion is called *piston flow* or *plug flow*. In this case no mixing occurs between any of the air molecules in the room. The air molecules entering the room at a given time can be considered to be a membrane which passes through the room like the face of a piston, pushing the air ahead of it toward the exhaust. Figure 18.4 illustrates this theoretical situation. The dashed line through the room indicates the position of the "membrane" or "front" that separates the initial air in the room contain-

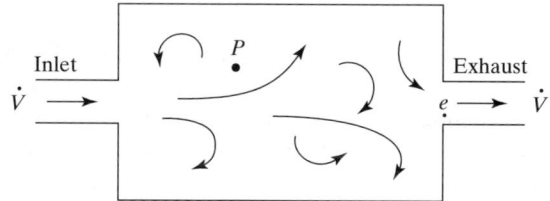

Figure 18.2 Perfect mixing of supply air in a room.

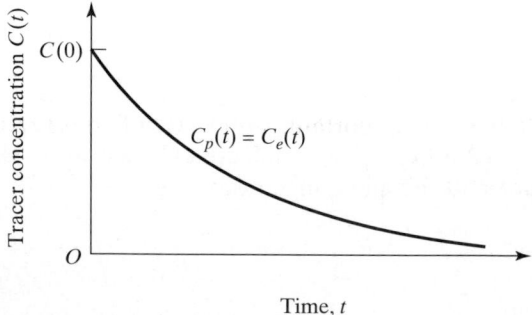

Figure 18.3 Exponential decay of tracer concentration for perfect mixing with a step-down test.

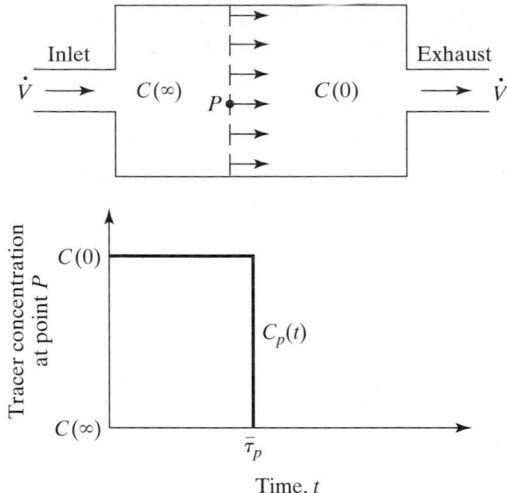

Figure 18.4 Room with piston flow and tracer concentration at point P during a step-down test.

ing a tracer from the supply air with no tracer during a step-down tracer test. The "front" is shown passing through point P. The time between the beginning of the test and the passage of the "front" through point P is the local mean age at P. The local mean age for points in the room varies linearly from zero at the plane of the inlet to the nominal time constant at the plane of the exhaust. The ventilation efficiency for a room with piston airflow can be shown to be 100 percent.

Fundamental Principles That Govern Room Airflow

Two types of force are responsible for room air motion. The first is *pumped flow*, which is the result of an external driving pressure. For rooms with open windows or doors, wind can provide the driving pressure. For rooms that are fairly well sealed from the outdoors, a mechanical means to produce airflow is necessary. This is usually a fan. Fans can be located in the room itself, such as on the ceiling, or located remotely, with ductwork connecting the fan to the room. The second force is *buoyancy*, caused primarily by temperature differences within the room. Differences in temperature create differences in air density, which in turn provides the necessary driving force. Pumped flow and buoyant flow will be considered as acting separately at first and then as acting together.

In addition to the forces that cause air motion, other forces act to retard it. The most important of these is *form drag*, created by obstructions such as furniture and partitions. Sufficient blockage can occur to prevent significant air motion in some areas of a room. Another less important force is *skin friction* or *surface drag*, caused by air viscosity.

Pumped Flow

An example of pumped flow is an isothermal free jet of air that enters a room from an opening in one of the walls. The behavior of such a jet is well known and depends on the shape of the opening—i.e., circular, rectangular, or a two-dimensional slot. The maximum velocity in the jet decreases as the jet moves away from the opening and the width of the jet increases owing to entrainment of surrounding air on all sides.

In room air distribution, designers often use the principle known as the "Coanda effect." This describes the ability of a jet to adhere to a surface. The static pressure in a jet is lower than the surrounding static pressure in the room, because the air velocity in the jet is higher than the surrounding velocity. Thus the higher room static pressure tends to push the jet toward the surface. Jets are able to follow convex curved surfaces because of this phenomenon. A good demonstration of the Coanda effect is the ability of a water jet from a faucet to change direction when the circular side of a smooth water glass is brought in contact with the jet. Not only does the jet follow the surface, it decays less rapidly than a free or unattached jet, because entrainment occurs only on the side away from the surface. This phenomenon is widely used to spread air under the ceiling in a room from supply diffusers located in the ceiling.

The dimensionless scaling parameter most useful in describing the behavior of jets is the jet Reynolds number:

$$\mathrm{Re} = \frac{\rho L \mathbf{V}}{\mu} \tag{18.5}$$

where ρ is the air density, L is a dimension of the inlet opening such as the slot width or the diameter of a round hole, $\mathbf{V}$ is the jet mean velocity at the inlet, and μ is the air dynamic viscosity. A scale model operated at the same Reynolds number as the full-scale facility should have nearly the same flow pattern. For example, a model that is half as large as a full-scale facility and that uses air as the fluid medium at the same temperature as full scale will require twice the velocity as the full-scale velocity for the two Reynolds numbers to be equal. However, the inlet-opening area is only one-fourth the size of full scale, so the actual volumetric flow rate in the model is one-half the flow rate in full scale.

Air motion in rooms is strongly affected by the location of the air inlet or supply but is not affected very much by the location of the exhaust or outlet. The reason is that the incoming air is usually in the form of a jet. The jet can maintain its direction and momentum a large distance from the inlet, especially if it is attached to a surface. The exhaust, however, acts much like a potential flow sink. There is no effect upstream from the exhaust except for the local negative pressure. There is no momentum transport as in the case of the supply jet. A good example is the airflow near a common household electric fan. The air jet discharged from the fan can be felt a large distance away, but the air flow on the inlet side can hardly be felt at all. The result is that the locations and types of supply diffusers and their respective air jets are very important in determining the motion and mixing of the air in the room, whereas the locations of the return grilles are not.

Buoyant Flow

Any change in air density will result in a buoyant force. Denser air will tend to sink downward, and less dense air will float upward. This phenomenon is called *natural* or *free convection*. Most buoyant effects in rooms are caused by temperature differences. An extreme example is a fire, in which the hot combustion gases rise rapidly upward above the combustion zone. However, much smaller temperature differences also cause air motion. Cold air near a cold window surface will tend to move downward, creating a cold draft. Warm air will rise above heat sources such as perimeter heating systems, people, and electric devices such as copy machines and computers.

The dimensionless scaling parameter often used to quantify buoyant flows is the *Grashof number*:

$$\text{Gr} = \frac{\rho^2 g \beta L^3 \Delta t}{\mu^2} \qquad (18.6)$$

where ρ is the air density, g is gravitational acceleration, β is the thermal coefficient of volumetric expansion, which for air equals the reciprocal of the absolute temperature, L is a characteristic length such as the ceiling height, Δt is the temperature difference that provides the buoyant force, and μ is the air dynamic viscosity.

EXAMPLE 18.1

Consider a half-scale model of a room in which the Grashof number is to equal that in a full-size room. The scale model uses air at the same mean temperature and pressure as the full-scale room. Compute the required temperature difference in the model compared to the temperature difference in the full-scale room.

Solution: Let subscripts M denote *model* and F denote *full scale*. Setting the two Grashof numbers equal:

$$\text{Gr}_M = \frac{\rho_M^2 g \beta_M L_M^3 \Delta t_M}{\mu_M^2} = \text{Gr}_F = \frac{\rho_F^2 g \beta_F L_F^3 \Delta t_F}{\mu_F^2}$$

As air at the same mean temperature and pressure is used, the fluid properties in the model and the full-scale application will be equal and will cancel from the equation above. Solving for the temperature-difference ratio:

$$\frac{\Delta t_M}{\Delta t_F} = \frac{L_F^3}{L_M^3} = \frac{1}{(1/2)^3} = 8$$

Therefore, the temperature difference in the half-scale model must be eight times larger than the temperature difference in the full-scale application to achieve the same Grashof number and similitude.

Mixed Flow

When both pumped flow and buoyant flow exist together, the resulting flow can be called *mixed flow*. This is often referred to in heat-transfer literature as *mixed convection*. If either the pumped flow or the buoyant flow dominates, only the dominant flow force needs to be considered and the discussion given above applies. However, in many situations these two forces are nearly the same strength, so both must be considered simultaneously. This occurs in many room air jets in which the jet temperature differs substantially from the ambient room air temperature. Warm jets tend to float upward and cool jets downward.

The dimensionless parameters governing mixed flow are the Reynolds and Grashof numbers discussed earlier. For complete similitude, a scale model must have the same Reynolds number and the same Grashof number as the full-scale application. However, many researchers have found that in the range of Reynolds and Grashof numbers frequently encountered in room air flow, which are very large owing to the large dimensions of the space, these governing parameters can be relaxed to smaller values without significantly altering the flow. The parameter that has been found to provide similitude for this situation is called the *Archimedes number*, which is defined as the ratio of the Grashof number divided by the square of the Reynolds number:

$$\mathrm{Ar} \equiv \frac{\mathrm{Gr}}{\mathrm{Re}^2} = \frac{\rho^2 g \beta L^3 \,\Delta t / \mu^2}{\rho^2 L^2 \mathbf{V}^2 / \mu^2} = \frac{g \beta L \,\Delta t}{\mathbf{V}^2} \tag{18.7}$$

Conventional Mixed-Air System Design

In conventional mixed-air systems, the objective of the design is to provide sufficient inlet-air velocity to ensure adequate mixing of the air in the room and thus provide uniform thermal conditions in as much of the occupied zone of the room as possible. For buildings occupied by people, human thermal comfort becomes the requirement on the thermal conditions.

Before describing detailed design procedures, we will review some basic terminology and concepts. Figure 18.5(a) illustrates an isothermal air jet entering a room from a side wall. There are no temperature differences between the jet and the room air, so there is no buoyant force. The jet is discharged into the room away from any of the room surfaces, so the Coanda effect is not present. The jet entrains air on all sides, which causes it to spread as indicated by the dashed lines. The entrainment also reduces the maximum velocity. If a contour of uniform velocity is plotted, say 50 ft/min (0.25 m/s), the distance between the end of this contour and the jet inlet is called the jet "throw." The throw to this velocity is commonly used for design, as velocity values below this are not felt by most people as drafts.

If the same air jet enters the room just below the ceiling, the Coanda effect is present in addition to entrainment. This is shown in Fig. 18.5(b). The lower pressure near the ceiling tends to make the jet stick to the ceiling. This eliminates entrainment on the side

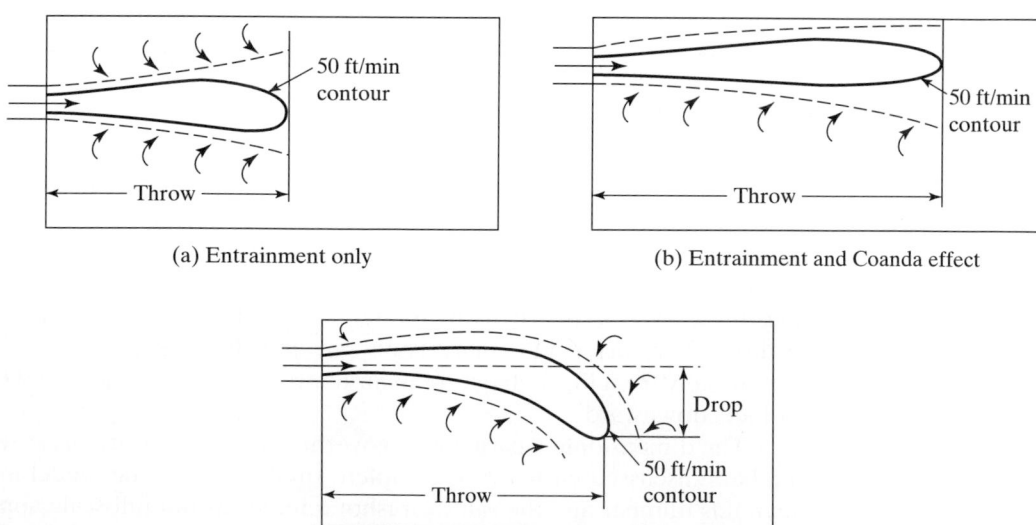

(a) Entrainment only

(b) Entrainment and Coanda effect

(c) Entrainment, Coanda effect, and buoyancy

Figure 18.5 Examples of air-jet behavior.

of the jet against the ceiling, enabling the jet to persist farther into the room. Note that the jet throw for this case is larger than in Fig. 18.5(a).

If this jet is also at a lower temperature than the room, buoyancy is included as shown in Fig. 18.5(c). Here the jet velocity near the inlet is large enough so that the Coanda effect dominates near the inlet. However, once the jet velocity decays to a low enough value, the buoyant force becomes stronger than the pressure force holding the jet to the ceiling. At this point the jet separates from the ceiling and starts to sink downward into the room. The throw is less than that of the isothermal jet near the ceiling shown in Figure 18.5(b). Another term is added to this case, and that is the jet "drop." The drop in this case is the vertical distance measured from the end of the contour to the center of the jet inlet. By knowing the throw and drop for a particular diffuser, the designer has information concerning the air-jet integrity and motion in a room. However, the throw and drop depend on the velocity or flow rate of air supplied from the diffuser, its location with respect to other room surfaces, and the temperature difference between the supply air and the room.

The next two figures illustrate in a qualitative manner the expected airflow patterns and resulting air-temperature distributions in unobstructed rooms with two different types of air-supply diffusers.

Figure 18.6 shows a room with a wall exposed to the outside and air supply upward through a diffuser located in the floor adjacent to the outside wall. The air is supplying the necessary heating or cooling energy to the room to compensate for heat loss or gain from the outside wall. The thermostat used to control the room temperature is also shown as a "T". In the upper part of the figure, the room requires heating, so the air supplied is warm. The warm air rises to and across the ceiling and down the opposite wall. The cold air near the floor is not well mixed and becomes stratified. The room air dry-bulb temperature versus height is shown to the right. The temperature at the thermostat is shown

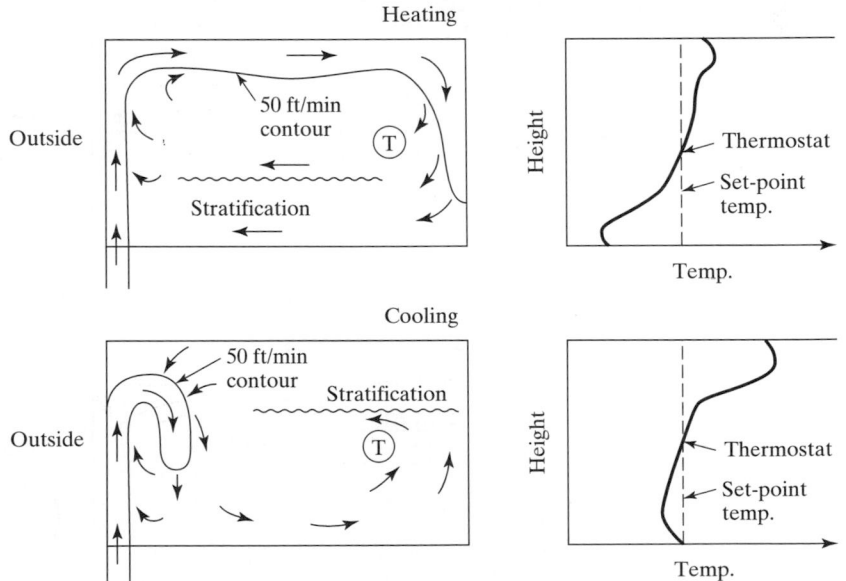

Figure 18.6 Airflow and room-temperature distributions for a floor-perimeter diffuser.

to be at the setpoint. However, the air near the floor is colder, and the warm air jet near the ceiling is warmer. This is a reasonably good method to provide space heating, provided the inlet-air velocity is sufficient to mix well with the room air and minimize stratification near the floor.

The lower portion of Fig. 18.6 shows the same configuration in the cooling mode when cold air is supplied to the room. The buoyant force opposes the upward motion of the jet along the outside wall. In this case the jet does not reach the ceiling and sinks into the lower portion of the room. This leaves a stratified region of high temperature near the ceiling. However, the ceiling is usually not occupied, so the high-temperature air near the ceiling may not be a major disadvantage to this system. The inlet jet should have sufficient velocity to mix well with the room air. If not, the jet can simply sink to the floor, resulting in a cold region near the floor and a continual temperature rise toward the ceiling.

Another common arrangement is to supply the air through ceiling diffusers. This is very common in commercial buildings, where the supply ductwork is hidden above a false ceiling and is therefore not blocked by furniture or other obstructions. The upper portion of Fig. 18.7 shows a ceiling diffuser supplying cold air during the cooling mode. The air jet is parallel to the ceiling, so that the Coanda effect is used to help spread out the jet under the ceiling. Eventually the buoyant force becomes dominant and the jet separates from the ceiling. The cool air then settles downward into the room with a low velocity. The right side of the figure shows the room air temperature distribution. The cold air in the supply jet can be seen just under the ceiling. However, the occupied zone is very nearly of uniform temperature with low velocity and should be very comfortable.

The use of ceiling difusers during the heating mode is shown in the lower portion of Fig. 18.7. In this case the warm air stays near the ceiling and does not mix with the air in

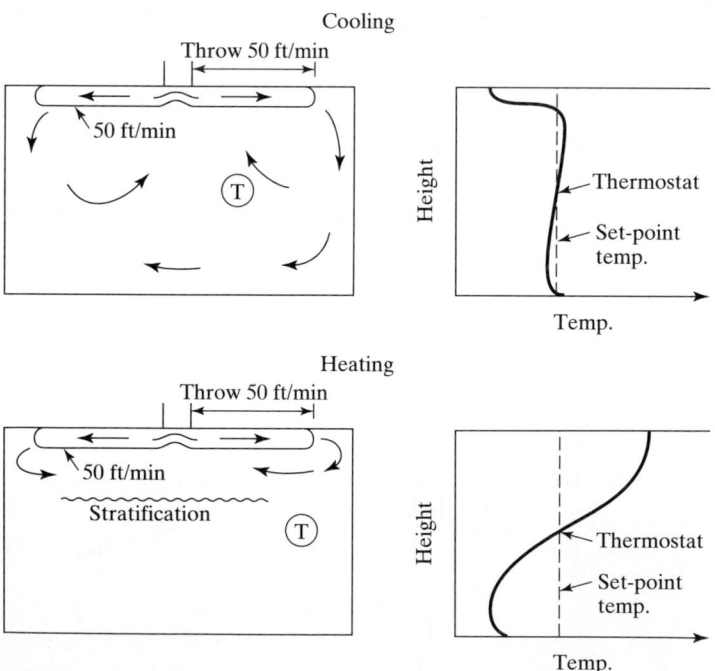

Figure 18.7 Airflow and room-temperature distributions for a ceiling diffuser.

the occupied zone. Very strong stratification develops, in which the air near the ceiling is very warm and the air near the floor is cold. This is not conducive to good human thermal comfort.

Figures 18.6 and 18.7 show that floor diffusers can result in fairly good thermal comfort during heating and also in cooling if the air-supply velocity is large enough to ensure good mixing. Ceiling air supply is excellent for cooling but not very good for heating. Common practice in commercial office buildings is to use ceiling air supply for cooling and ventilation but a separate perimeter heating system for space heating if the heating load is high.

Diffuser Selection Procedure

The procedure recommended by ASHRAE for selecting diffusers in a room is outlined below [3]:

1. Determine the room air volumetric flow rate at design conditions.

   The volumetric flow rate is determined from an energy balance on the room assuming standard air density:

   $$\dot{V} = \frac{\dot{Q}_{tot}}{\rho\,\Delta h} \tag{18.8}$$

   where $\dot{V}$ is the air volumetric flow rate supplied to the room, $\dot{Q}_{tot}$ is the total design load on the room, ρ is standard air density, and Δh is the total enthalpy difference between supply and return air. The maximum of the heating and cooling airflow rates should be used to select the diffuser if the room is both heated and cooled by the supply air. The sensible load can also be used as follows:

   $$\dot{V} = \frac{\dot{Q}_{sen}}{\rho\,\Delta t}$$

   where $\dot{Q}_{sen}$ is the sensible design load on the room and Δt is the dry bulb temperature difference between the supply and return air.

2. Select the diffuser type, and locate the diffusers at the appropriate places in the room.

   For example, if circular ceiling diffusers are to be used, each should be centered in a ceiling area that is roughly square in shape, as the air is supplied in all directions from the diffuser. If sill grilles are to be used, they should be placed near exterior walls or windows to shield the thermal load from the occupied areas of the room.

   If only one diffuser is to be used, the total air flow rate computed in part 1 should be used. If two diffusers are used, each is allocated one-half of the total supply airflow rate, and so on.

3. Based on the type of diffuser selected and their locations in the room, determine the characteristic length, L, in ft (m) for each, using Table 18.2.

4. Determine the value for $T_{50}/L\,(T_{0.25}/L)$ for maximum ADPI from Table 18.3 for each diffuser.

   The type of diffuser and the room design energy load per unit floor area, Btu/hr · ft^2 (W/m^2), are used to determine the optimum value for $T_{50}/L\,(T_{0.25}/L)$. The maximum value for the ADPI for each is also listed. The last column on the right gives the range for $T_{50}/L\,(T_{0.25}/L)$ such that the value for ADPI is greater than the value given in the column immediately to the left. Often the optimum value for

TABLE 18.2 Characteristic Length for Various Diffuser Types

Diffuser Type	Characteristic Length
High sidewall grille	Distance to opposing wall
Circular ceiling diffuser	Distance to closest wall or midway to nearest ceiling diffuser
Sill grille	Length of room in direction of airflow
Ceiling slot diffuser	Distance to nearest wall or midway to nearest diffuser
Light troffer diffuser	One-half distance to nearest diffuser plus distance from ceiling to top of occupied zone
Perforated ceiling diffusers	Distance to nearest wall or midway to nearest diffuser

TABLE 18.3 Diffuser Selection Guidelines for Mixed-Flow Rooms

Terminal Device	Room Load		T_{50}/L ($T_{0.25}/L$) for Maximum ADPI	Maximum ADPI	Range of ADPI > T_{50}/L ($T_{0.25}/L$)	
	W/m^2	$Btu/hr \cdot ft^2$				
High	250	80	1.8	68	–	–
sidewall	190	60	1.8	72	70	1.5–2.2
grilles	125	40	1.6	78	70	1.2–2.3
	65	20	1.5	85	80	1.0–1.9
Circular	250	80	0.8	76	70	0.7–1.3
ceiling	190	60	0.8	83	80	0.7–1.2
diffusers	125	40	0.8	88	80	0.5–1.5
	65	20	0.8	93	90	0.7–1.3
Sill Grille,	250	80	1.7	61	60	1.5–1.7
straight	190	60	1.7	72	70	1.4–1.7
vanes	125	40	1.3	86	80	1.2–1.8
	65	20	0.9	95	90	0.8–1.3

T_{50} = throw to 50-ft/min velocity; $T_{0.25}$ = throw to 0.25-m/s velocity.

the throw is not possible, but a throw within the range given in the last column is acceptable.

5. Calculate the throw T_{50} ($T_{0.25}$) by multiplying the value for T_{50}/L ($T_{0.25}/L$) from part 4 by the characteristic length L from part 3.

6. Select an appropriate diffuser for each location from a manufacturer's catalog, using the volumetric flow rate determined in part 1 and the throw determined in part 5.

7. Check to ensure that the diffuser selected does not exceed pressure drop limits or noise levels.

If the selection must be changed, a compromise between adequate mixing, which is determined by the throw, and the pressure drop or noise criteria must be achieved.

EXAMPLE 18.2

The room shown in Fig. 18.8 is to be heated with warm air from floor grilles. The room is in a cold climate where heating is more important than cooling. The required airflow rate is

TABLE 18.4 Typical Linear-Diffuser Catalog Data

Size	Total Pressure, in W.G.	0.009	0.02	0.036	0.057	0.08	0.109	0.143
$1\frac{1}{2}''$	Flow, ft³/min · ft	14	21	28	35	42	49	56
	Throw: Sill or floor	1	2	4	7	9	11	12
	Side wall	4	7	7	12	13	16	17
$2''$	Flow, ft³/min · ft	22	33	44	55	66	77	88
	Throw: Sill or floor	1	4	7	9	11	14	16
	Side wall	5	8	11	14	17	19	21
$2\frac{1}{2}''$	Flow, ft³/min · ft	30	44	59	74	89	104	118
	Throw: Sill or floor	1	5	9	11	14	16	19
	Side wall	6	9	13	16	18	21	24
$3''$	Flow, ft³/min · ft	38	58	77	96	115	134	154
	Throw: Sill or floor	2	7	10	13	16	19	21
	Side wall	7	10	14	17	20	24	27

Note: Tabulated throw values are given in units of ft and correspond to an active length of 4 ft. Correct other lengths as follows:
1 ft length, multiply tabulated throw by 0.7.
10 ft length, multiply tabulated throw by 1.2.
Note: Throw values assume a cooling differential of 20 °F.

600 ft³/min. The room loading is 60 Btu/ft² of floor space. Locate the supply-air grille(s), the return-air grille, and select the supply-air grille(s) from Table 18.4.

Solution: We will follow the step-by-step procedure outlined above in selecting the grilles.

1. The airflow rate has been determined to be 600 ft³/min.
2. We will assume that localized heat losses occur at the two windows shown on Fig. 18.8. Therefore, to counteract cold drafts from these windows, we will locate a supply grille below each window as shown in the figure. The warm air from the supply grilles will tend to rise toward the ceiling. To enhance room air mixing, the return grille should be located near the floor, preferably across the room from the supply. The return has been located near the floor on an interior wall as shown in the figure. This also reduces the length of the return-air duct or chase.
3. From Table 18.2, the characteristic length is the ceiling height, which is 10 ft.
4. From Table 18.3, the optimum value for T_{50}/L for a sill grille with straight vanes and 60 Btu/hr · ft² is 1.7.
5. Multiplying T_{50}/L from step 4 by L from step 3, 1.7 × 10 ft = 17 ft. The throw to 50 ft/min should be 17 ft for maximum ADPI, which in this case is 72.

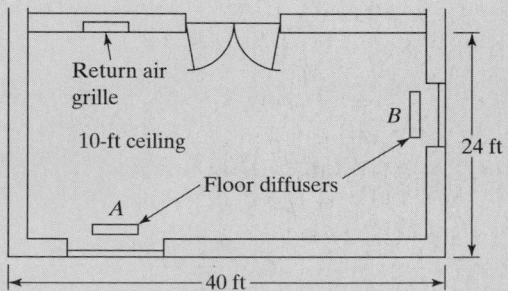

Figure 18.8 Room with floor supply grilles for Ex. 18.2.

6. We then use a catalog to determine which diffuser best meets these criteria. Using Table 18.4, we have a choice of grille widths from $1\frac{1}{2}$ to 3 in. Note that the throw values were obtained with the supply-air temperature 20 °F cooler than the room air temperature. In this example, we are using the grille for heating, so the throw should be less than the value tabulated (we are blowing warm air upward rather than cold air). Note also that the throws tabulated are for a 4-ft active length of grille. For shorter lengths entrainment on the ends of the jet is more important, so the throw should be reduced, and vice versa for longer lengths.

One choice is to use a 2-in.-wide grille. We need 300 ft³/min for each grille. If we select a throw of 11 ft, the grille will supply 66 ft³/min per foot of grille length, so we need about $4\frac{1}{2}$ ft of length. Therefore the throw will not differ very much from the tabulated value. The pressure drop will be 0.08 in. of water, which is very reasonable for a commercial building.

We could also consider other widths, for example a grille of $2\frac{1}{2}$ in. width. If we select a throw of 11 ft again, this grille will supply 74 ft³/min per foot of length, so we need about 4 ft of length. Again, we do not need to correct the throw, as this is the standard length used in the tables. The pressure drop in this case would be 0.057 in. of water.

There is no unique solution to this design problem. The designer must weigh various factors, such as throw for good mixing, pressure drop, space available, and the cost of specifying, ordering, and installing a unique diffuser at each location rather than using a few sizes for the entire building.

EXAMPLE 18.3

Consider the same room used in Ex. 18.2. However, now consider the situation where cooling is the dominant load. Specify circular ceiling diffusers to cool this room. The total design cooling load is 38,400 Btu/hr. The difference between the supply- and return-air enthalpy values is 9.5 Btu/lbm$_a$.

Solution: As with the previous example we will follow the procedure outlined above.

1. The supply airflow rate must be determined from an energy balance. Using Eq. (18.8)

$$\dot{V} = \frac{\dot{Q}_{\text{tot}}}{\rho\,\Delta h} = \frac{(38{,}400\ \text{Btu/hr})\,(\text{hr}/60\ \text{min})}{(0.0765\ \text{lbm}_a/\text{ft}^3)\,(9.5\ \text{Btu/lbm}_a)} \approx 880\ \text{ft}^3/\text{min}$$

2. The room is rectangular in plan, and its length is about twice its width. Therefore, we can divide the room into two areas that are nearly square in plan with dimensions of about 20 ft × 24 ft. We will locate a round ceiling diffuser in the center of each square, as shown in Fig. 18.9. The return grille will be placed on the interior wall. This should be placed near the floor to ensure good mixing in the room.

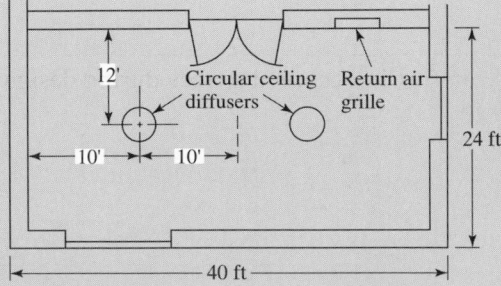

Figure 18.9 Room with circular ceiling diffusers for Ex. 18.3.

TABLE 18.5 Typical Circular Ceiling Diffuser Catalog Data

Size	Neck Velocity, ft/min	400	500	600	700	800	900	1000	1200
8″	Total press., in. W.G.	.033	.052	.075	.101	.130	.166	.205	.292
	Flow rate, ft³/min	140	175	210	245	280	315	350	420
	Throw, ft	4	4	5	6	7	8	9	11
	NC		15	21	26	31	34	37	44
10″	Total press., in. W.G.	.027	.043	.062	.084	.108	.138	.170	.243
	Flow rate, ft³/min	220	270	330	380	435	490	545	655
	Throw, ft	4	5	6	7	8	9	10	12
	NC		11	17	21	26	30	33	39
12″	Total press., in. W.G.	.026	.042	.060	.081	.105	.134	.166	.236
	Flow rate, ft³/min	315	390	470	550	630	705	785	940
	Throw, ft	5	6	7	8	10	11	12	14
	NC		11	17	22	26	30	33	39
14″	Total press., in. W.G.	.038	.061	.087	.118	.152	.194	.240	.342
	Flow rate, ft³/min	425	530	635	745	850	955	1060	1270
	Throw, ft	6	8	9	11	12	14	15	18
	NC	11	18	23	28	32	36	40	46

3. From Table 18.2 the characteristic length should be the distance to the nearest wall or midway between diffusers. The closest wall is about 10 ft, which is also equal to the distance to the midplane of the room.

4. From Table 18.3, the optimum value for T_{50}/L for circular ceiling diffusers is 0.8, regardless of the room loading.

5. Multiplying T_{50}/L from step 4 by L from step 3, 0.8×10 ft = 8 ft. The throw to 50 ft/min should be 8 ft for maximum ADPI. The room loading is $(38{,}400 \text{ Btu/hr})/(40 \text{ ft} \times 24 \text{ ft}) = 40$ Btu/hr · ft², so the maximum ADPI for this case is 88 from Table 18.3.

6. We will now use Table 18.5 to select two identical diffusers for this room. We need one-half the total flow rate through each diffuser, or 440 ft³/min per diffuser. We also desire a throw to 50 ft/min of 8 ft. From Table 18.5, the best choice is a 10-in. diffuser. The closest tabulated specifications to our needs are: neck velocity, 800 ft/min; total pressure loss, 0.108 in. of water; flow rate, 435 ft³/min; throw, 8 ft; and noise criterion (NC), 26. Larger-diameter diffusers will not have enough throw at the correct volumetric flow rate, and smaller diffusers will have throws that are too large.

The design procedure discussed above will provide adequate mixing and good thermal comfort in occupied spaces during design conditions. However, buildings normally operate at part load, in which the air quantity supplied to the space can be considerably smaller than the design value. This will reduce the throw from the diffuser, as can be seen in Tables 18.4 and 18.5, as the quantity of air is reduced. The reduced throw will result in less mixing and pehaps less efficient ventilation.

One method to overcome this part-load condition is to install active diffusers. These diffusers are not fixed in shape but change their open area in response to the supply-air pressure. When the supply airflow rate reduces, the neck area also reduces, keeping the throw approximately constant.

Another method of overcoming the lack of mixing at part load is to use a fan-powered box. This entrains return air with the supply air at low-load conditions, ensuring uniform total supply into the room from the terminal unit and retaining good mixing at part-load operation.

Displacement Ventilation System Design

Mixed-air systems are designed primarily to provide adequate thermal comfort. They are used where the air can heat or cool the space. They are not necessarily the best type of system to use for removing airborne contaminants. An air-distribution system developed specifically to address the issue of ventilation needs is the displacement ventilation system.

In many commercial buildings, cooling is needed all year or heating is provided by a second heating system, so that the air supply is strictly needed for cooling and ventilation. A ventilation system appropriate for these applications is called "displacement ventilation." In these systems, mixing is reduced as much as possible, and buoyancy becomes the main force that drives the air flow in the room. The basic concept is to provide clean, cool air near the floor at a very low velocity. Most air contaminants generated in the space also emanate from heat sources (occupants, electric equipment, etc.). Therefore, the thermal plumes rising above these heat souces will contain most of the air contaminants that must be removed from the space. The warm, contaminated air is allowed to gather under the ceiling, which is not in the occupied zone of the room. It is then removed from the room through a large, low-velocity return.

Figure 18.10 illustrates a room with displacement ventilation. The supply air enters the room through a very large opening to reduce the velocity as much as possible. The cool supply air tends to stay in the lower portion of the room because of its negative buoyancy. Air that has been heated by a heat source in the room will rise toward the ceiling, carrying contaminants from the heat source with it. If the supply airflow rate just equals the flow rate of the thermal plumes, only the heated, contaminated air will rise into the upper portion of the room. This contaminated air will mix near the ceiling because of the velocities generated by the plumes. It is then drawn from the top of the room through a large, low-velocity return.

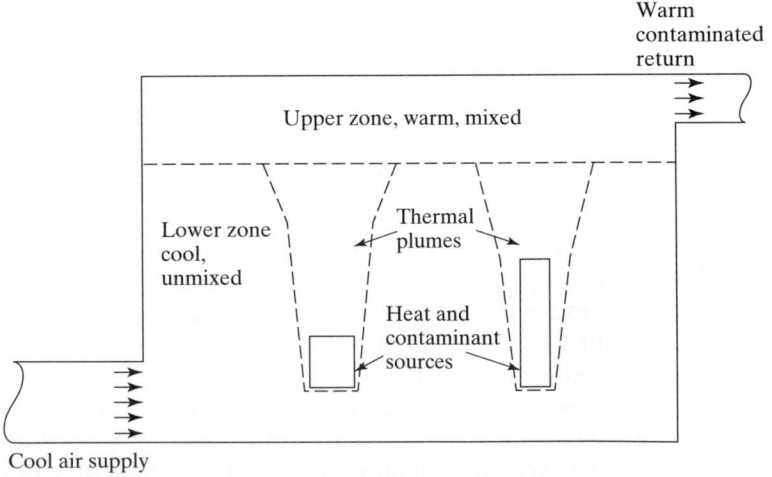

Figure 18.10 Schematic of an ideal displacement ventilation system.

In an ideal displacement ventilation system, all the occupants of the room are exposed to cool, clean ventilation air and all the contaminated air rises into the unoccupied zone of the room. However, the air velocities in the thermal plumes are very small, a few in./sec (cm/s), so that occupant motion, other motions, and other thermal plumes and boundary layers can cause considerable undesirable mixing. The supply air must not be too cold, or the occupants will complain of cold feet.

Use of Computer Codes to Simulate Room Airflow

Conventional design practice for room airflow relies on laboratory experimental data and industry empirical design guidelines. Actual rooms may differ significantly from laboratory geometries and thermal conditions. If more detailed design guidelines are required, *Computational Fluid Dynamics (CFD)* or *Finite-Element Methods (FEM)* can be used. Usually the time required and the cost associated with using these tools prevents their widespread use in conventional building design. However, some critical applications where more resources are made available have been designed using these methods.

As an example of the use of CFD methods and their accuracy, results from an experimental study of room airflow will be shown and compared with corresponding numerical simulations. The room considered here is unobstructed and rectangular in shape. A cross section of this room is shown in Fig. 18.11. The dimensions correspond to a half-scale model constructed in the laboratory. For ease of computation and for comparison of results, a two-dimensional configuration is studied. Air is supplied downward through a parallel-plate channel located above the ceiling with a velocity of 200 ft/min (1.016 m/s). The temperature of the supply air is 80.6 °F (27 °C), which matches the temperature of the room surfaces. Thus there is no thermal buoyancy. The air leaves the room through the ceiling in another parallel-plate channel.

The measured time-averaged velocity vectors in this scale-model room are shown in Fig. 18.12 [4]. These were obtained by traversing a hot-wire anemometer across the room. The supply air enters the room in a jet that can be seen below the supply-air inlet. Initially it behaves as a free jet with air entrained on both sides. The jet attaches to the right-hand wall about halfway between the wall and the ceiling. From this attachment point, the jet follows the perimeter of the room in a clockwise direction. This jet creates a large vortex

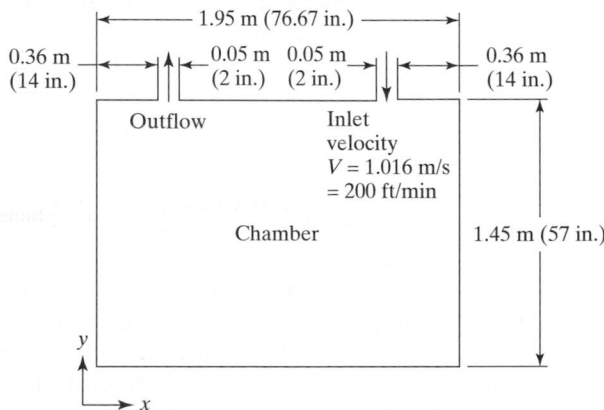

Figure 18.11 Dimensions and boundary conditions used in half-scale-model airflow measurements and simulations.

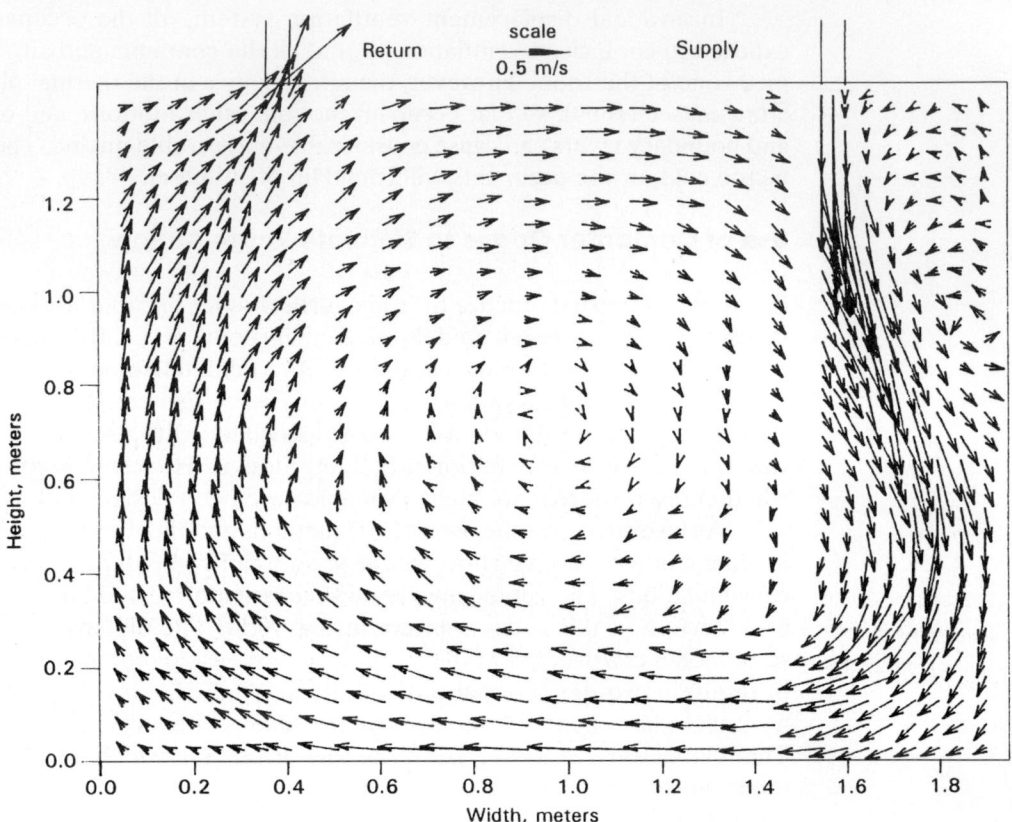

Figure 18.12 Velocity vector field obtained by experiment.

in the room with its center of rotation approximately in the center of the room. A smaller counterrotating vortex exists in the upper right-hand corner. Note that the airflow near the return is not affected very much, except immediately adjacent to the return opening.

Two simulations are shown for comparison [5, 6]. The first is a CFD simulation that uses the well-known turbulence model developed by Launder and Spalding [7]. This method uses wall functions to bridge between the wall surface and the fully turbulent region in the flow. Figure 18.13 shows the computed velocity vectors that can be compared to the experimental ones shown on Fig. 18.12. The supply-air jet behaves similarly, the large vortex in the center of the room is present, and the smaller counterrotating vortex in the upper right-hand corner can be seen. The second simulation uses a low-Reynolds-number turbulence model developed by Lam and Bremhorst [8]. The velocity-vector results from this solution are shown in Fig. 18.14. In this model, closely spaced grid points are required near the wall surfaces to simulate the complete turbulent boundary layer from the surface into the fully turbulent region. Therefore, the grid points are closer together near the wall and further apart in the interior in this solution compared to the results shown in Fig. 18.13.

Other examples of room airflow numerical simulations can be found in the published literature. Most of the applications require knowledge of airflow patterns and contaminant transport. Examples include semiconductor-manufacturing clean rooms, where particulate contamination of silicon wafer surfaces is important, and critical-care applications such as hospital operating rooms and transplant-patient recovery rooms, where fungal infection control is important.

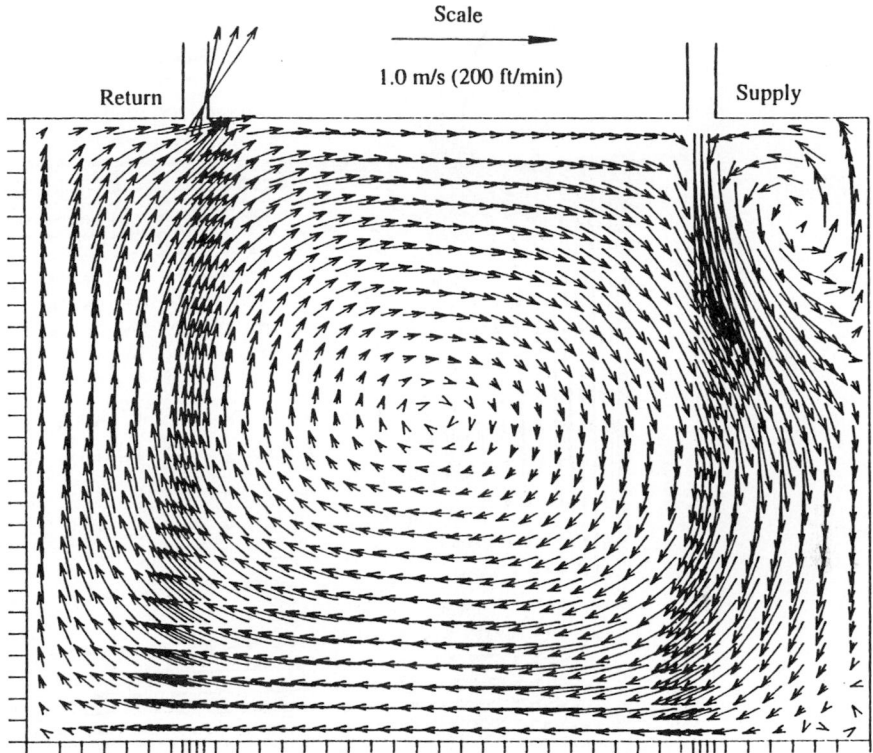

Figure 18.13 Velocity vector field simulated using high-Reynolds-number turbulence model from Launder and Spalding [5].

18.3 BASIC AIR-DISTRIBUTION SYSTEM PRINCIPLES

Duct-System Characteristic

Consider air flowing steadily through a straight, horizontal, round duct with velocity **V**. No work is transferred between the air in the duct and the exterior. If we assume that the air in the duct is incompressible, its velocity will not change. Changes in elevation are zero. Therefore, using Eq. (2.98), the pressure loss in this duct can be described by

$$\rho g \ell = \rho f \frac{L}{D} \frac{\mathbf{V}^2}{2} = (P_{s,1} - P_{s,2}) \qquad (18.9)$$

$$\underbrace{\phantom{\rho g \ell = \rho f \frac{L}{D} \frac{\mathbf{V}^2}{2}}}_{\substack{\text{total} \\ \text{pressure} \\ \text{loss}}} \qquad \underbrace{\phantom{= (P_{s,1} - P_{s,2})}}_{\substack{\text{static} \\ \text{pressure} \\ \text{loss}}}$$

Equation (18.9) shows that the total pressure drop is equal to the drop in static pressure. It also shows that it is proportional to the duct friction factor, the length of the duct in terms of number of diameters, and the square of the velocity. A common assumption used in duct flows is that the friction factor is independent of velocity. This is very nearly true for turbulent flows in the range of Reynolds numbers of most applications in buildings. Therefore, for a duct of fixed diameter and length, the total pressure drop is pro-

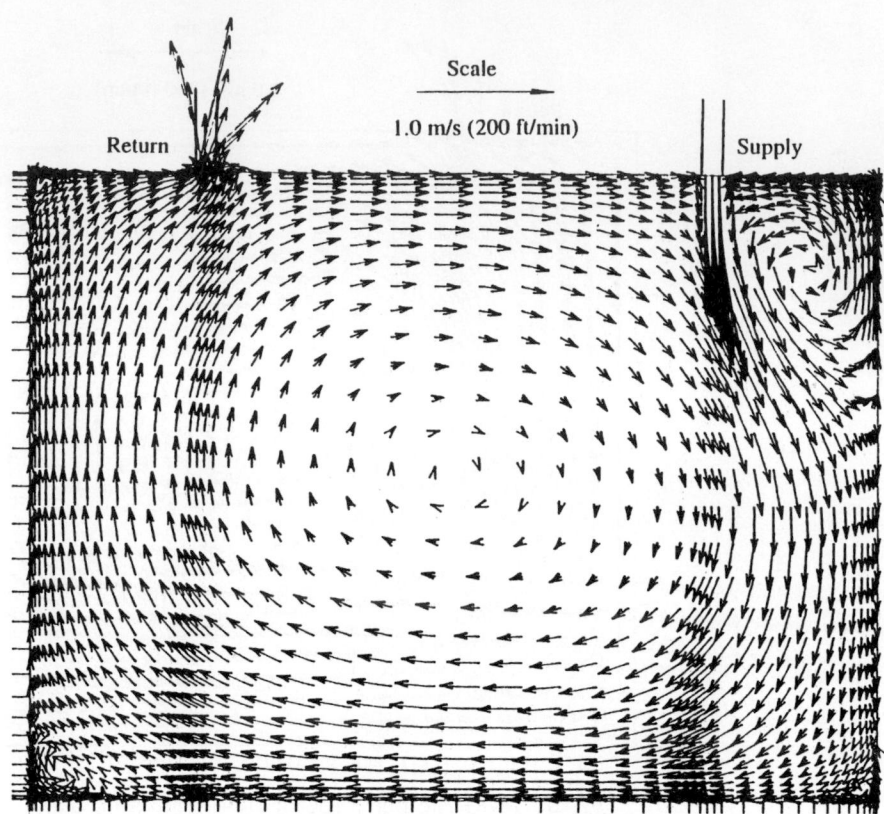

Figure 18.14 Velocity vector field simulated using Lam and Bremhorst low-Reynolds-number turbulence model [5].

portional to the square of the mean velocity. Considering Eq. (2.95), the total pressure drop is also proportional to the square of the volumetric flow rate.

A plot of total pressure loss or total head loss vs. either the mean velocity or volumetric flow rate is termed the "system characteristic." The "system" is the entire duct or piping system except for the fans or pumps. A representative system characteristic is shown in Fig. 18.15. Losses other than friction with the duct or piping wall also reduce the total pressure. Examples include bends and elbows, transitions, branches, and flow-control devices such as dampers and valves. The total pressure loss through these fittings is also essentially proportional to the square of the mean flow velocity or volumetric flow rate. Therefore, all losses in a duct or piping system, regardless of the cause, can be considered proportional to the mean velocity or volumetric flow rate squared, as indicated in Eq. (18.9).

Next we need a procedure to combine all losses in a system, regardless of their cause, into a single loss for the entire system, so that we can generate a composite system characteristic. We can achieve this by using an analogy to electric circuits, where an equivalent circuit with a given resistance is obtained from the actual, complex layout. Figure 18.16 shows how individual duct or piping-system curves can be converted into a single equivalent system. Part (a) illustrates flow in series, where the same flow rate passes through both parts of the system and the total pressure drop is the sum of the two parts.

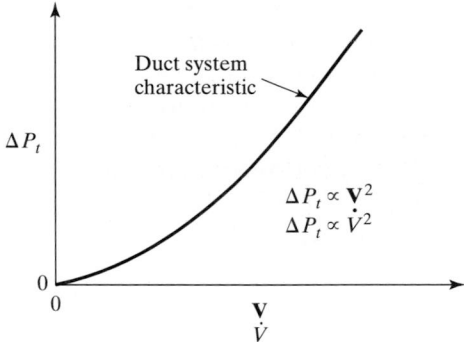

Figure 18.15 Fundamental duct-system characteristic showing relationship between total-pressure loss and duct mean velocity and volumetric flow rate.

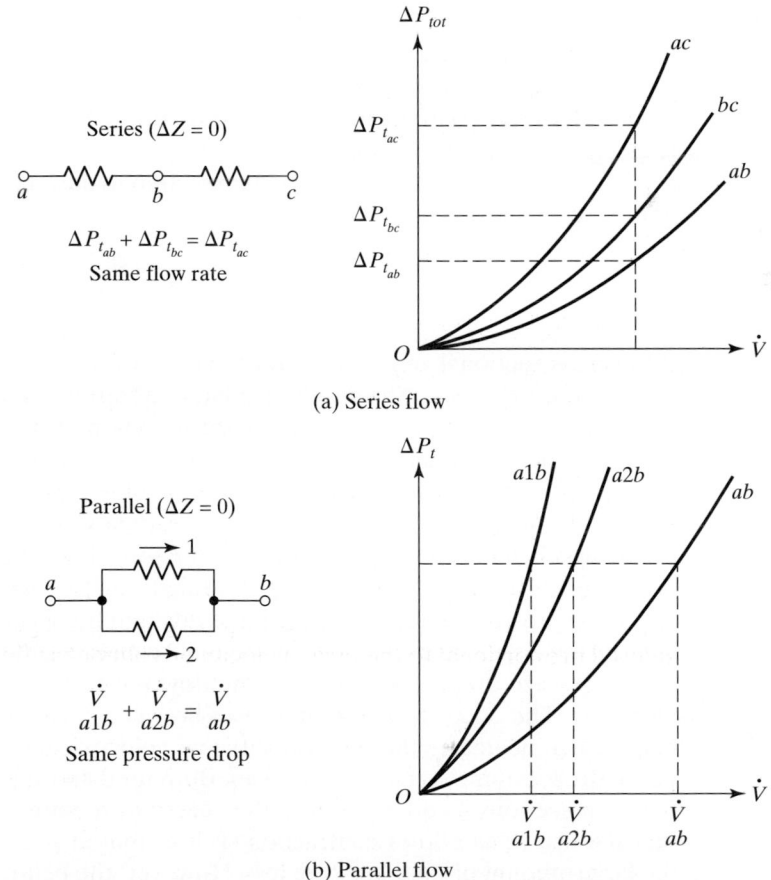

Figure 18.16 Series and parallel system characteristic curves.

In part (b), flow through parallel paths is considered. The procedure is the same as for resistive dc electric circuits if the volumetric flow is considered analogous to electric current and the pressure drop or head loss analogous to voltage drop. In actual systems, both series and parallel paths are usually present. It is often more expedient to combine the parallel paths first if possible and then the series portions. In all cases the smaller combinations must be performed before the larger ones.

Air Flow in Ducts

Consider a straight duct with air flowing at a constant rate. The relationship between pressure and air velocity at a point in the flow can be written as follows:

$$P_t \;=\; P_s \;+\; \frac{\rho \mathbf{V}^2}{2} \qquad (18.10)$$

$$\underset{\substack{\text{total} \\ \text{pressure}}}{} \qquad \underset{\substack{\text{static} \\ \text{pressure}}}{} \qquad \underset{\substack{\text{velocity} \\ \text{pressure}}}{}$$

This indicates that the difference between total and static pressure is proportional to the square of the local air velocity. Total pressure is the most important pressure in duct systems, because it is the total pressure that determines the fan pressure requirement; i.e., the rise in total pressure across the fan must equal the loss in total pressure in the entire duct system. Most duct design methods rely exclusively on total pressure. However, the static pressure is also important. The static pressure exerts force on the duct wall, causing it to bulge outward or inward. The level of the static pressure influences the strength required and therefore the duct wall thickness, weight, and material cost.

Changes in pressure from one duct section to another can also be described by modifying Eq. (18.10):

$$\Delta P_t = \Delta P_s + \frac{\rho(\mathbf{V}_1^2 - \mathbf{V}_2^2)}{2} \qquad (18.11)$$

If the cross-sectional area inside a duct is reduced and the flow rate remains constant, the velocity must increase. This results in a larger difference between total and static pressure. The total pressure always decreases in the direction of flow in a duct. However, the static pressure can either increase or decrease, depending on the magnitude of the velocity.

Figure 18.17 shows the distribution of total, static, and velocity pressure inside a short duct. A bell-mouth inlet at section 1 gradually accelerates the flow as the cross-sectional area is reduced. Thus the velocity pressure slowly increases and the static pressure drops below the total pressure. In the straight duct between locations 1 and 2, the total and static pressures decrease at the same rate and the velocity pressure remains constant. The abrupt decrease in duct size at section 2 causes a sudden drop in total pressure. The static pressure drops even more, as the flow must accelerate, and the velocity pressure increases. The duct size slowly increases between sections 3 and 4. This is a type of diffuser which decreases the flow velocity and velocity pressure. The total pressure continues to drop, but the static pressure rises. Both total and static pressure gradients are equal between sections 4 and 5, because the velocity pressure remains a constant there. Note that the use of an abrupt contraction such as that at section 2 is undesirable because of the large amount of total pressure loss. However, the bell-mouth inlet and the diffuser are gradual transitions that do not cause significant pressure drops. The designer should use

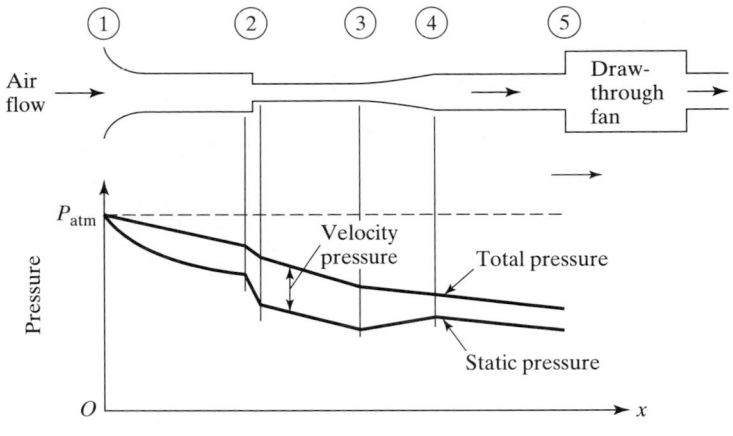

Figure 18.17 Sketch of total-pressure, static-pressure, and velocity-pressure distributions in a duct.

gradual transitions whenever possible to minimize the amount of total pressure loss in the system and therefore the fan-pressure rise and power requirement.

The velocity pressure can be calculated for air at standard conditions using the following simplified equations:

English units:

$$\text{velocity pressure} = P_v = \left(\frac{\mathbf{V}}{4005}\right)^2 \text{ (in. of water)} \tag{18.12a}$$

where $\mathbf{V}$ has units of ft/min;

SI units:

$$\text{velocity pressure} = P_v = \left(\frac{\mathbf{V}}{1.29}\right)^2 \text{ (Pa)} \tag{18.12b}$$

where $\mathbf{V}$ has units of m/s.

EXAMPLE 18.4

Determine the static pressure in a duct when the total pressure is 120 Pa and the air mean velocity is 5 m/s.

Solution: Using the approximate equation for velocity pressure above, $P_v = (5/1.29)^2 = 15.0$ Pa. From Eq. (18.11), $P_s = P_t - P_v$ or $P_s = 120 \text{ Pa} - 15 \text{ Pa} = 105 \text{ Pa}$.

Frictional Losses

Our next topic is the estimation of pressure drop in straight, round duct. The pressure loss in this case is caused by wall friction. Convectional practice has been to use data for galvanized sheet metal ducts. Figures 18.18(a) and (b) are log-log plots of total pressure loss

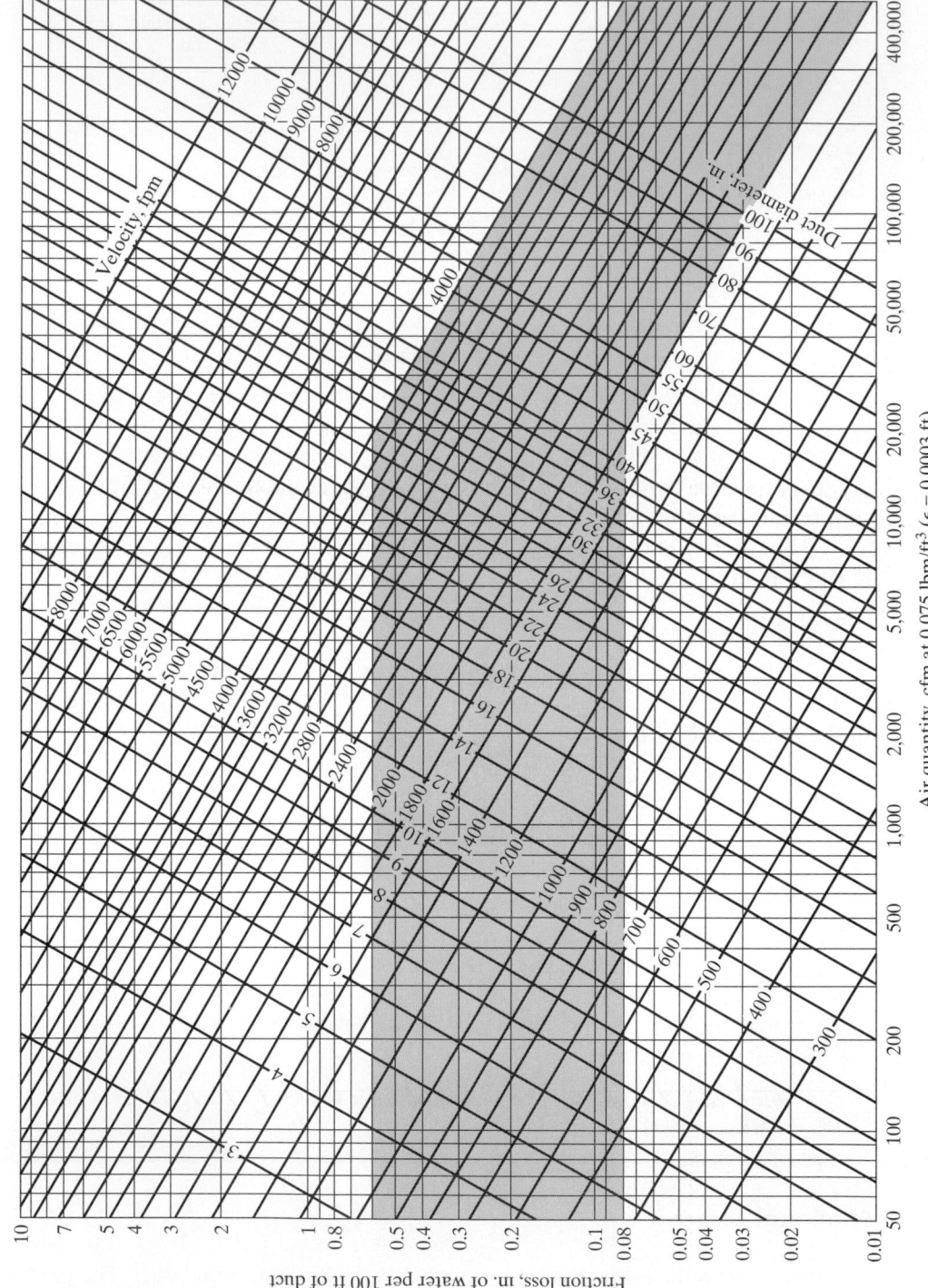

Figure 18.18a Pressure loss in typical round galvanized sheet-metal ducts. [Reprinted by permission from *ASHRAE Fundamentals 1993* (IP), p. 32.7.]

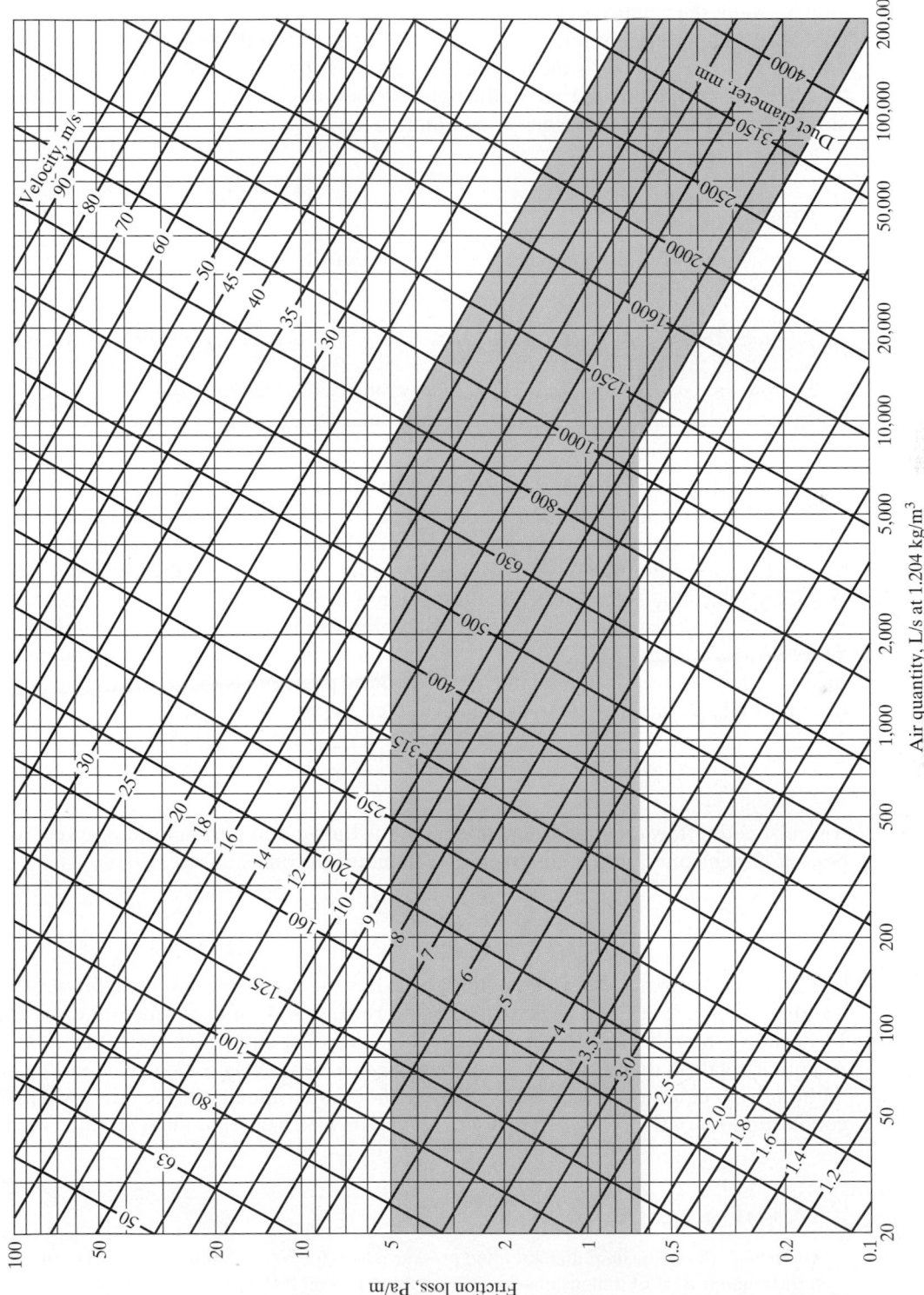

Figure 18.18b Pressure loss in typical round galvanized sheet-metal ducts. [Reprinted by permission from *ASHRAE Fundamentals 1993* (SI), p. 32.6.]

vs. air volumetric flow rate. Log-log coordinates are used because of the direct relationship between the square of the flow rate and the pressure loss. Therefore, all information should appear as straight lines on these plots. The horizontal axis is the log of the airflow rate, and the vertical axis is the log of the air-pressure drop per 100 ft (m) of straight duct. The lines that slope upward and to the right are lines of constant duct diameter. Lines that slope up and to the left are lines of constant mean velocity.

EXAMPLE 18.5

Compute the mean velocity and pressure drop for 2000 L/s of air flowing through a 400-mm diameter duct that is 5 m long.

Solution: Using Fig. 18.18(b), the intersection of the vertical line representing 2000 L/s and the inclined line representing 400-mm duct diameter occurs at a velocity of 16 m/s and at a pressure drop of 6.1 Pa/m. The total pressure drop becomes 6.1 Pa/m × 5 m = 30.5 Pa.

Figure 18.18 is very useful for round ducts of galvanized metal, but how do we determine pressure loss in ducts made from other materials or in rectangular and oval ducts? Conventional practice is to use Fig. 18.18 for nearly all duct materials and geometries but to publish correction factors. Correction for other materials and duct liners is given by ASHRAE [9]. Ducts of noncircular cross section are converted to a round duct that has the same volumetric flow rate and the same pressure loss per unit length of duct, NOT THE SAME CROSS-SECTIONAL AREA. The diameter of the corresponding round duct is called the "equivalent diameter." The basic principle is that of hydraulic diameter used in fluid mechanics. The hydraulic diameter is defined as

$$D_h \equiv \frac{4A}{p}$$

where A is the flow cross-sectional area and p is the wetted perimeter. For a rectangular duct of dimensions a and b, the hydraulic diameter becomes

$$D_h = \frac{2ab}{a + b}$$

The equivalent diameter for rectangular ducts has the same form as the hydraulic diameter given above, but the coefficients are different, based on experimental data. Figure 18.19 shows the relationship between a rectangular duct and its equivalent circular duct. A similar relation that has been developed for flat, oval ducts is also given in Fig. 18.19. Although the equations given in Fig. 18.19 are not dimensionally correct, they will provide good estimates of the equivalent diameter when inch (English) or mm (SI) units are used.

EXAMPLE 18.6

Determine the equivalent diameter and pressure drop for 5000 ft³/min of air flowing through a rectangular duct of dimensions 24 in. × 18 in. that is 30 ft long.

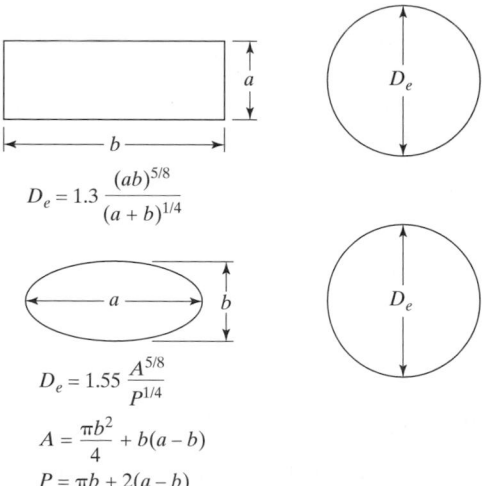

$$D_e = 1.3 \frac{(ab)^{5/8}}{(a + b)^{1/4}}$$

$$D_e = 1.55 \frac{A^{5/8}}{P^{1/4}}$$

$$A = \frac{\pi b^2}{4} + b(a - b)$$

$$P = \pi b + 2(a - b)$$

Figure 18.19 Equivalent diameter of rectangular and flat oval ducts.

Solution: First we will compute the duct equivalent diameter and then use Fig. 18.18(a) to obtain the pressure-drop information. From Fig. 18.19, the equivalent diameter of a round duct with the same flow rate and pressure drop per unit length is

$$D_e = 1.3 \frac{(24 \times 18)^{5/8}}{(24 + 18)^{1/4}} = 22.7 \text{ in.}$$

From Fig. 18.18(a), the intersection of the vertical line representing 5000 ft^3/min and the inclined line representing a duct diameter of about 22.7 in. gives a pressure drop per unit length of about 0.16 in. of water/100 ft. The total pressure loss in the 30-ft duct is 0.16 in. of water/100 ft $\times$ 30 ft = 0.05 in. of water.

Dynamic Losses

In addition to frictional losses in straight ducts, changes in flow direction, flow around obstructions, and changes in velocity also cause pressure drops. These occur in nearly all fittings and transitions. These losses are commonly referred to as "dynamic losses." As with frictional losses, the pressure drop is proportional to the square of the velocity. The losses are directly proportional to the velocity pressure, which in turn is proportional to the square of the velocity.

One method of expressing dynamic losses of fittings is to determine the length of straight duct with the same flow rate, velocity, and pressure drop as the fitting. This length is called the fitting "equivalent length." For example, if the pressure loss in a straight duct is 0.2 in. of water/100 ft (0.002 in. of water/ft) and the pressure drop in a fitting with the same volumetric flow rate and mean air velocity is 0.04 in. of water, the equivalent length of this fitting is 0.04 in. of water/(0.002 in. of water/ft) = 20 ft. Some tables list these equivalent lengths for selected duct fittings.

A more common practice, and the method adopted by ASHRAE, is to use *loss coefficients*. In this method, the pressure drop through the fitting is computed directly using the following equation:

$$\Delta P_t = C_0 P_{v,0} \tag{18.13}$$

where ΔP_t is the total pressure drop through the fitting, C_0 is the loss coefficient for the fitting, and $P_{v,0}$ is the velocity pressure at the reference location 0 for the fitting. Usually the reference location is immediately upstream of the fitting. In cases of converging sections it is the downstream location. The loss coefficient can be modified for a different reference location if necessary. If the value at reference location 0 must be changed to reference location i, use the following relation:

$$C_i = \frac{C_0}{(\mathbf{V}_i/\mathbf{V}_0)^2} \tag{18.14}$$

where $\mathbf{V}_i$ and $\mathbf{V}_0$ are the mean air velocities at locations i and 0, respectively.

Tables 18.6–18.15 list some loss coefficients for common fittings and transitions for both round and rectangular ducts. A more complete listing is provided by ASHRAE [10].

TABLE 18.6 Total Pressure-Loss Coefficients for Bell-Mouth Inlet to a Round Duct

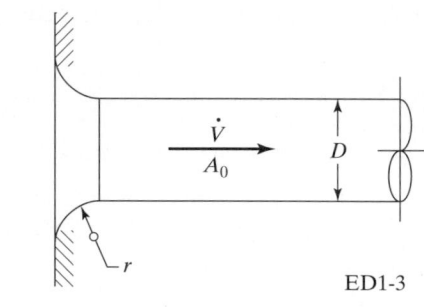

r/D	0.00	0.01	0.02	0.03	0.04	0.05	0.06	0.08	0.10	0.12	0.16	0.20	10.00
C_o	0.50	0.44	0.37	0.31	0.26	0.22	0.20	0.15	0.12	0.09	0.06	0.03	0.03

SOURCE: Reprinted by permission from *ASHRAE Duct Fitting Database, 1994.*

TABLE 18.7 Total Pressure-Loss Coefficients for a 90-deg Pleated Elbow in a Round Duct, 7 Sections, $r/D = 2.5$

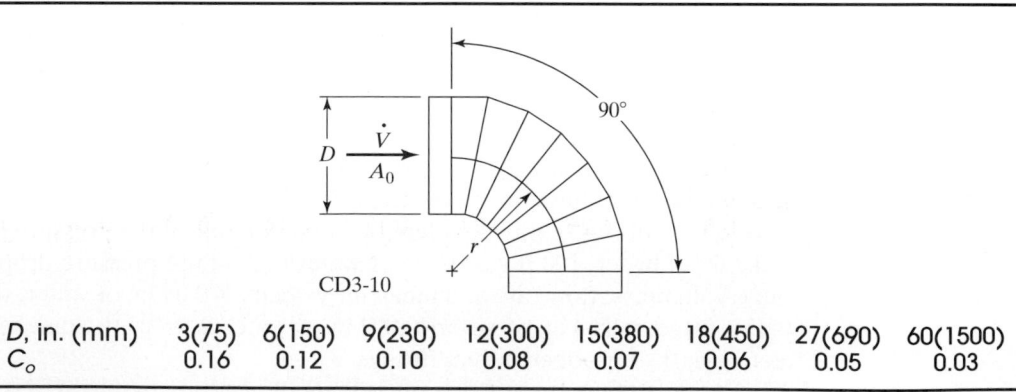

D, in. (mm)	3(75)	6(150)	9(230)	12(300)	15(380)	18(450)	27(690)	60(1500)
C_o	0.16	0.12	0.10	0.08	0.07	0.06	0.05	0.03

SOURCE: Reprinted by permission from *ASHRAE Duct Fitting Database, 1994.*

Part VI / Air- and Water-Distribution System Design

TABLE 18.8 Total Pressure-Loss Coefficients for Diverging Tee in a Round Duct with a Conical Branch Tapered into the Body

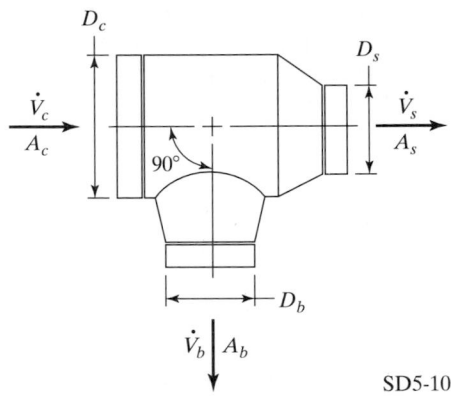

SD5-10

	C_b								
$\dot{V}_b/\dot{V}_c$ =	0.1	0.2	0.3	0.4	0.5	0.6	0.7	0.8	0.9
A_b/A_c									
0.1	0.65	0.24							
0.2	2.98	0.65	0.33	0.24	0.18				
0.3	7.36	1.56	0.65	0.39	0.29	0.24	0.20		
0.4	13.78	2.98	1.20	0.65	0.43	0.33	0.27	0.24	0.21
0.5	22.24	4.92	1.98	1.04	0.65	0.47	0.36	0.30	0.26
0.6	32.73	7.36	2.98	1.56	0.96	0.65	0.49	0.39	0.33
0.7	45.26	10.32	4.21	2.21	1.34	0.90	0.65	0.51	0.42
0.8	59.82	13.78	5.67	2.98	1.80	1.20	0.86	0.65	0.52
0.9	76.41	17.75	7.36	3.88	2.35	1.56	1.11	0.83	0.65

	C_s								
$\dot{V}_s/\dot{V}_c$ =	0.1	0.2	0.3	0.4	0.5	0.6	0.7	0.8	0.9
A_s/A_c									
0.1	0.13	0.16							
0.2	0.20	0.13	0.15	0.16	0.28				
0.3	0.90	0.13	0.13	0.14	0.15	0.16	0.20		
0.4	2.88	0.20	0.14	0.13	0.14	0.15	0.15	0.16	0.34
0.5	6.25	0.37	0.17	0.14	0.13	0.14	0.14	0.15	0.15
0.6	11.88	0.90	0.20	0.13	0.14	0.13	0.14	0.14	0.15
0.7	18.62	1.71	0.33	0.18	0.16	0.14	0.13	0.15	0.14
0.8	26.88	2.88	0.50	0.20	0.15	0.14	0.13	0.13	0.14
0.9	36.45	4.46	0.90	0.30	0.19	0.16	0.15	0.14	0.13

SOURCE: Reprinted by permission from *ASHRAE Duct Fitting Database, 1994.*

TABLE 18.9 Total Pressure-Loss Coefficients for a Conical Bell-Mouth Inlet to a Rectangular Duct

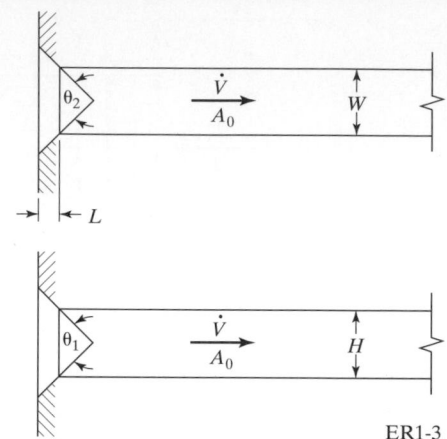

ER1-3

	C_o								
Theta =	0	10	20	30	40	60	100	140	180
L/D_h									
0.025	0.50	0.47	0.45	0.43	0.41	0.40	0.42	0.45	0.50
0.050	0.50	0.45	0.41	0.36	0.33	0.30	0.35	0.42	0.50
0.075	0.50	0.42	0.35	0.30	0.26	0.23	0.30	0.40	0.50
0.100	0.50	0.39	0.32	0.25	0.22	0.18	0.27	0.38	0.50
0.150	0.50	0.37	0.27	0.20	0.16	0.15	0.25	0.37	0.50
0.600	0.50	0.27	0.18	0.13	0.11	0.12	0.23	0.36	0.50

SOURCE: Reprinted by permission from *ASHRAE Duct Fitting Database, 1994.*

TABLE 18.10 Total Pressure-Loss Coefficients for a 90 deg Elbow in a Rectangular Duct

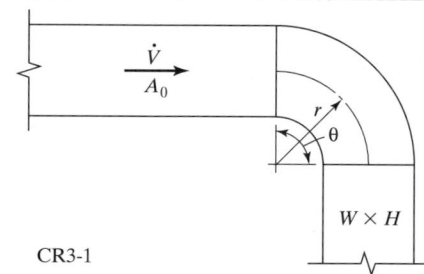

CR3-1

	C_p										
H/W =	0.25	0.50	0.75	1.00	1.50	2.00	3.00	4.00	5.00	6.00	8.00
r/W											
0.50	1.53	1.38	1.29	1.18	1.06	1.00	1.00	1.06	1.12	1.16	1.18
0.75	0.57	0.52	0.48	0.44	0.40	0.39	0.39	0.40	0.42	0.43	0.44
1.00	0.27	0.25	0.23	0.21	0.19	0.18	0.18	0.19	0.20	0.21	0.21
1.50	0.22	0.20	0.19	0.17	0.15	0.14	0.14	0.15	0.16	0.17	0.17
2.00	0.20	0.18	0.16	0.15	0.14	0.13	0.13	0.14	0.14	0.15	0.15

SOURCE: Reprinted by permission from *ASHRAE Duct Fitting Database, 1994.*

TABLE 18.11 Total Pressure-Loss Coefficients for a Mitered 90-deg Elbow in a Rectangular Duct with Single-Thickness Turning Vanes

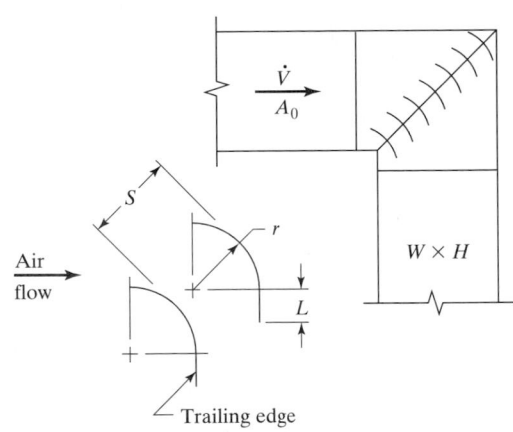

	Dimensions, in. (mm)			
Design	r	s	L	C_0
CR3-9	2.0 (50)	1.5 (40)	0.0	0.11
CR3-10	2.0 (50)	1.5 (40)	0.75 (20)	0.12
CR3-11	4.5 (110)	2.25 (60)	0.0	0.15
CR3-12	4.5 (110)	3.25 (80)	0.0	0.33

SOURCE: Reprinted by permission from *ASHRAE Duct Fitting Database, 1994.*

EXAMPLE 18.7

Use the loss coefficients given in Tables 18.6–18.15 to determine the total pressure drop from section 0 to section 4 in the duct shown in Fig. 18.20 (see pg. 620).

Solution: This is a portion of a round return duct. At section 0, 400 L/s of air enters through a bell-mouth inlet, then passes through 5 m of straight duct. Between sections 2 and 3, 600 L/s of air is added in a converging tee. The 1000 L/s of combined air passes through 3 m of straight duct to section 4. The loss in total pressure from section 0 to section 4 can be determined by adding the losses in the two fittings (ED1-3, ED5-3) and the two straight duct runs.

(a) *Fitting ED1-3.* The loss coefficient for this fitting from Table 18.6 is 0.22. The velocity pressure in the 25-cm duct can be determined using Eq. (18.12b). The velocity in the duct can be computed by dividing the volumetric flow rate by the duct cross-sectional area. However, we will use a graphical approach here and use Fig. 18.18(b). From this figure, 400 L/s flowing in a 25-cm round duct gives a velocity of about 8 m/s. Using Eq. (18.12b), the velocity pressure becomes $(8/1.29)^2 = 38.5$ Pa. The loss in total pressure is then determined by multiplying the velocity pressure by the loss coefficient, $\Delta P_{0-1} = 0.22 \times 38.5$ Pa = 8.5 Pa.

(b) *Straight run between sections 1 and 2.* The pressure drop per unit length of duct can be obtained from Fig. 18.18(b). At 400 L/s flowing through a 25-cm duct, the pressure drop per unit length becomes about 3 Pa/m. The total pressure drop between sections 1 and

TABLE 18.12 Total Pressure-Loss Coefficients for a Smooth-Radius Diverging Wye in a Rectangular Duct, $A_s + A_b \geqq A_c$, Branch 90 deg to Main

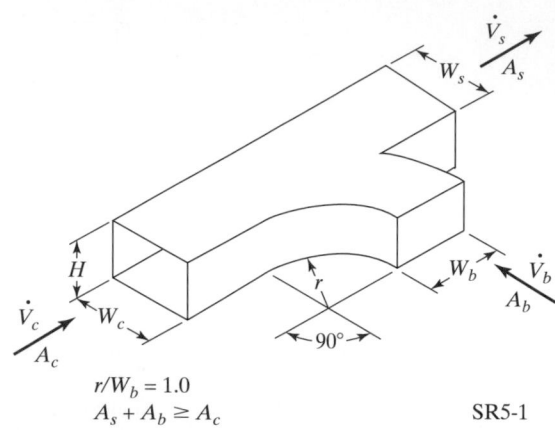

$r/W_b = 1.0$
$A_s + A_b \geqq A_c$

SR5-1

		C_b								
$\dot{V}_b/\dot{V}_c =$		0.1	0.2	0.3	0.4	0.5	0.6	0.7	0.8	0.9
A_s/A_c	A_b/A_c									
0.50	0.25	3.44	0.70	0.30	0.20	0.17	0.16	0.16	0.17	0.18
	0.50	11.00	2.37	1.06	0.64	0.52	0.47	0.47	0.47	0.48
	1.00	60.00	13.00	4.78	2.06	0.96	0.47	0.31	0.27	0.26
0.75	0.25	2.19	0.55	0.35	0.31	0.33	0.35	0.36	0.37	0.39
	0.50	13.00	2.50	0.89	0.47	0.34	0.31	0.32	0.36	0.43
	1.00	70.00	15.00	5.67	2.62	1.36	0.78	0.53	0.41	0.36
1.00	0.25	3.44	0.78	0.42	0.33	0.30	0.31	0.40	0.42	0.46
	0.50	15.50	3.00	1.11	0.62	0.48	0.42	0.40	0.42	0.46
	1.00	67.00	13.75	5.11	2.31	1.28	0.81	0.59	0.47	0.46

		C_s								
$\dot{V}_s/\dot{V}_c =$		0.1	0.2	0.3	0.4	0.5	0.6	0.7	0.8	0.9
A_s/A_c	A_b/A_c									
0.50	0.25	8.75	1.62	0.50	0.17	0.05	0.00	−0.02	−0.02	0.00
	0.50	7.50	1.12	0.25	0.06	0.05	0.09	0.14	0.19	0.22
	1.00	5.00	0.62	0.17	0.08	0.08	0.09	0.12	0.15	0.19
0.75	0.25	19.13	3.38	1.00	0.28	0.05	−0.02	−0.02	0.00	0.06
	0.50	20.81	3.23	0.75	0.14	−0.02	−0.05	−0.05	−0.02	0.03
	1.00	16.88	2.81	0.63	0.11	−0.02	−0.05	0.01	0.00	0.07
1.00	0.25	46.00	9.50	3.22	1.31	0.52	0.14	−0.02	−0.05	−0.01
	0.50	35.00	6.75	2.11	0.75	0.24	0.00	−0.10	−0.09	−0.04
	1.00	38.00	7.50	2.44	0.81	0.24	−0.03	−0.08	−0.06	−0.02

Source: Reprinted by permission from *ASHRAE Duct Fitting Database, 1994.*

Part VI / Air- and Water-Distribution System Design

TABLE 18.13 Total Pressure-Loss Coefficients for a Converging Tee in a Round Duct, $D_c \gtrsim 10$ in. (25 cm)

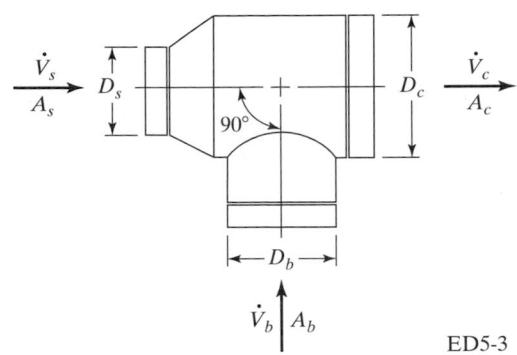

ED5-3

		C_b								
$\dot{V}_b/\dot{V}_c =$		0.1	0.2	0.3	0.4	0.5	0.6	0.7	0.8	0.9
A_s/A_c	A_b/A_c									
0.5	0.2	−7.26	−0.62	0.43	0.75	0.86	0.91	0.93	0.93	0.90
	0.3	−16.99	−2.35	−0.07	0.57	0.80	0.89	0.91	0.90	0.87
	0.4	−30.49	−4.67	−0.72	0.38	0.76	0.89	0.92	0.90	0.85
	0.5	−47.82	−7.61	−1.50	0.19	0.75	0.93	0.97	0.93	0.85
	0.6	−69.03	−11.17	−2.42	−0.03	0.76	1.01	1.05	0.98	0.88
	0.7	−94.17	−15.37	−3.49	−0.26	0.80	1.13	1.17	1.07	0.93
	0.8	−123.30	−20.22	−4.71	−0.50	0.87	1.29	1.33	1.20	1.02
	0.9	−156.48	−25.73	−6.09	−0.77	0.96	1.48	1.53	1.36	1.13
	1.0	−193.74	−31.92	−7.63	−1.07	1.06	1.71	1.77	1.56	1.28
1.0	0.2	−1.54	0.39	0.74	0.87	0.92	0.95	0.97	0.99	1.03
	0.3	−3.75	0.03	0.64	0.83	0.90	0.94	0.97	1.00	1.08
	0.4	−6.57	−0.32	0.61	0.85	0.93	0.97	0.99	1.03	1.16
	0.5	−10.05	−0.65	0.64	0.94	1.02	1.03	1.04	1.08	1.26
	0.6	−14.24	−0.98	0.74	1.10	1.16	1.15	1.13	1.16	1.40
	0.7	−19.20	−1.32	0.91	1.33	1.37	1.31	1.26	1.27	1.57
	0.8	−24.98	−1.69	1.14	1.63	1.63	1.53	1.43	1.41	1.78
	0.9	−31.62	−2.10	1.42	2.00	1.96	1.80	1.64	1.59	2.02
	1.0	−39.19	−2.55	1.76	2.43	2.35	2.12	1.90	1.81	2.30

(continued)

2 is obtained by multiplying this loss per unit length by the length of straight duct, $\Delta P_{1-2} = 3$ Pa/m $\times$ 5 m = 15 Pa.

(c) *Converging tee.* We are concerned only with the pressure loss between the straight section and the common, not the branch. Therefore, we need only a value for C_s. The ratio of the flow rate in the straight section to that in the common is (400 L/s)/(1000 L/s) = 0.4. The area ratio A_s/A_c is equal to the square of the diameter ratio = (25 cm/35 cm)^2 = 0.51. The area ratio of the branch to the common is 1.0, as they have the same diameter of 35 cm. Using these values, we can determine the loss coefficient from Table 18.13 to be 1.28. The velocity in the common duct is found by inspection using Figure 18.18(b) (1000 L/s in a 35-cm duct) to be about 11.5 m/s. The velocity pressure using Eq. (18.12b)

TABLE 18.13 (*cont.*)

$\dot{V}_s/\dot{V}_c =$		C_s								
		0.1	0.2	0.3	0.4	0.5	0.6	0.7	0.8	0.9
A_s/A_c	A_b/A_c									
0.5	0.2	126.36	16.99	5.39	2.42	1.32	0.81	0.54	0.38	0.28
	0.3	65.94	10.28	3.65	1.79	1.05	0.68	0.48	0.35	0.27
	0.4	38.84	7.27	2.87	1.51	0.93	0.63	0.45	0.34	0.27
	0.5	25.07	5.74	2.47	1.37	0.87	0.60	0.44	0.33	0.26
	0.6	17.98	4.95	2.27	1.29	0.84	0.58	0.43	0.33	0.26
	0.7	14.69	4.58	2.17	1.26	0.82	0.58	0.43	0.33	0.26
	0.8	13.78	4.48	2.15	1.25	0.82	0.57	0.43	0.33	0.26
	0.9	14.45	4.56	2.17	1.26	0.82	0.58	0.43	0.33	0.26
	1.0	16.24	4.76	2.22	1.28	0.83	0.58	0.43	0.33	0.26
1.0	0.2	−99.78	−0.17	3.15	2.40	1.58	0.98	0.56	0.25	0.02
	0.3	−75.42	2.54	3.85	2.65	1.69	1.03	0.58	0.26	0.03
	0.4	−38.31	6.66	4.92	3.04	1.86	1.11	0.62	0.28	0.03
	0.5	3.90	11.35	6.14	3.48	2.04	1.20	0.66	0.29	0.04
	0.6	48.66	16.32	7.43	3.94	2.24	1.29	0.70	0.31	0.04
	0.7	94.88	21.46	8.76	4.43	2.45	1.38	0.75	0.33	0.05
	0.8	142.01	26.70	10.12	4.92	2.66	1.48	0.79	0.35	0.06
	0.9	189.74	32.00	11.49	5.41	2.87	1.58	0.84	0.37	0.07
	1.0	237.90	37.35	12.88	5.92	3.08	1.68	0.88	0.39	0.07

SOURCE: Reprinted by permission from *ASHRAE Duct Fitting Database, 1994.*

becomes $(11.5/1.29)^2 = 79.5$ Pa. The total pressure loss is determined by multiplying the loss coefficient by the velocity pressure in the common duct, 1.28×79.5 Pa $= 102$ Pa.

(d) *Straight duct between sections 3 and 4.* The same procedure is used as between sections 1 and 2. The pressure drop per unit length of duct from Fig. 18.18(b) is about 4 Pa/m. Multiplying this by the length of duct gives the total pressure drop, 4 Pa/m $\times$ 3 m $= 12$ Pa.

Summing the individual pressure drops gives the total from section 0 to 4. $\Delta P_{t,1-4} = 8.5 + 15 + 102 + 12 = 137.5$ Pa.

The results are summarized in Table 18.16 (see pg. 620).

In some duct-sizing procedures, equivalent lengths of fittings are more useful than loss coefficients. An estimate for the equivalent length of a fitting can be determined from the loss coefficient as follows. From Eq. (2.98), the head loss for a fitting can be written as

$$\ell = f \frac{L_{eq}}{D} \frac{\mathbf{V}^2}{2g} = C_0 \frac{\mathbf{V}^2}{2g} \tag{18.15}$$

Solving for the equivalent length,

$$L_{eq} = \frac{D C_0}{f} \tag{18.16}$$

This relation is approximately valid for large-Reynolds-number flows in which the flow is highly turbulent and the friction factor is nearly independent of the Reynolds number. The friction factor must be estimated for the material used. For galvanized sheet metal, Table 18.17 (see pg. 621) gives approximate friction factors to use with Eq. (18.16).

TABLE 18.14 Total Pressure-Loss Coefficients for a Smooth-Radius Converging Wye in a Rectangular Duct, $A_s + A_b \geq A_c$, Branch 90 deg to Main

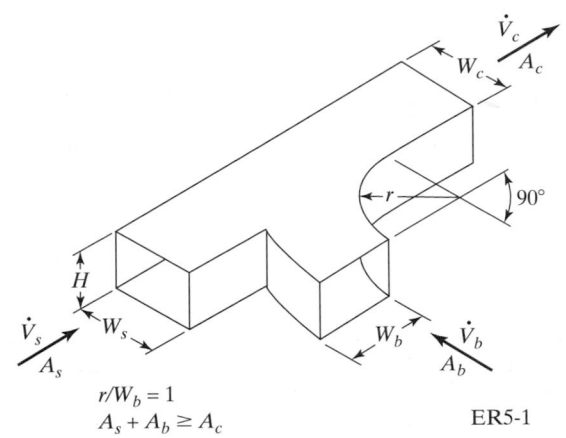

$r/W_b = 1$
$A_s + A_b \geq A_c$

ER5-1

		C_b								
$\dot{V}_b/\dot{V}_c =$		0.1	0.2	0.3	0.4	0.5	0.6	0.7	0.8	0.9
A_s/A_c	A_b/A_c									
0.50	0.25	−5.31	−0.47	0.14	0.36	0.43	0.48	0.50	0.51	0.50
	0.50	−53.75	−9.06	−2.64	−0.78	0.00	0.28	0.41	0.51	0.59
	1.00	−210.00	−35.00	−10.00	−3.12	−0.80	0.00	0.41	0.39	0.37
0.75	0.25	−7.50	−0.62	0.28	0.62	0.75	0.83	0.87	0.87	0.85
	0.50	−25.00	−3.75	−0.56	0.16	0.30	0.42	0.51	0.57	0.62
	1.00	−120.00	−20.00	−4.44	−1.25	0.00	0.44	0.49	0.50	0.47
1.00	0.25	−3.12	0.00	0.35	0.47	0.55	0.64	0.74	0.82	0.88
	0.50	−12.50	−1.25	0.00	0.39	0.45	0.49	0.51	0.59	0.62
	1.00	−60.00	−7.50	−1.11	−0.25	0.52	0.58	0.59	0.56	0.52

		C_s								
$\dot{V}_s/\dot{V}_c =$		0.1	0.2	0.3	0.4	0.5	0.6	0.7	0.8	0.9
A_s/A_c	A_b/A_c									
0.50	0.25	−35.00	−4.50	−0.42	0.62	0.90	0.92	0.87	0.77	0.62
	0.50	−32.50	−5.00	−1.11	0.00	0.35	0.45	0.46	0.43	0.35
	1.00	−2.50	0.50	0.69	0.62	0.55	0.47	0.41	0.34	0.25
0.75	0.25	−146.25	−28.12	−9.06	−3.23	−1.01	−0.16	0.23	0.26	0.21
	0.50	−32.62	−5.91	−1.44	−0.11	0.27	0.39	0.37	0.31	0.19
	1.00	−12.37	−11.25	0.31	0.63	0.61	0.55	0.44	0.32	0.17
1.00	0.25	−153.00	−29.75	−9.89	−3.44	−1.08	−0.14	0.31	0.36	0.30
	0.50	−46.00	−9.25	−3.00	−1.12	−0.32	0.00	0.20	0.25	0.21
	1.00	−12.00	0.00	1.11	1.12	0.92	0.72	0.55	0.37	0.22

SOURCE: Reprinted by permission from *ASHRAE Duct Fitting Database, 1994.*

TABLE 18.15 Total Pressure-Loss Coefficients for a Diffuser on the Discharge Side of a Centrifugal Fan

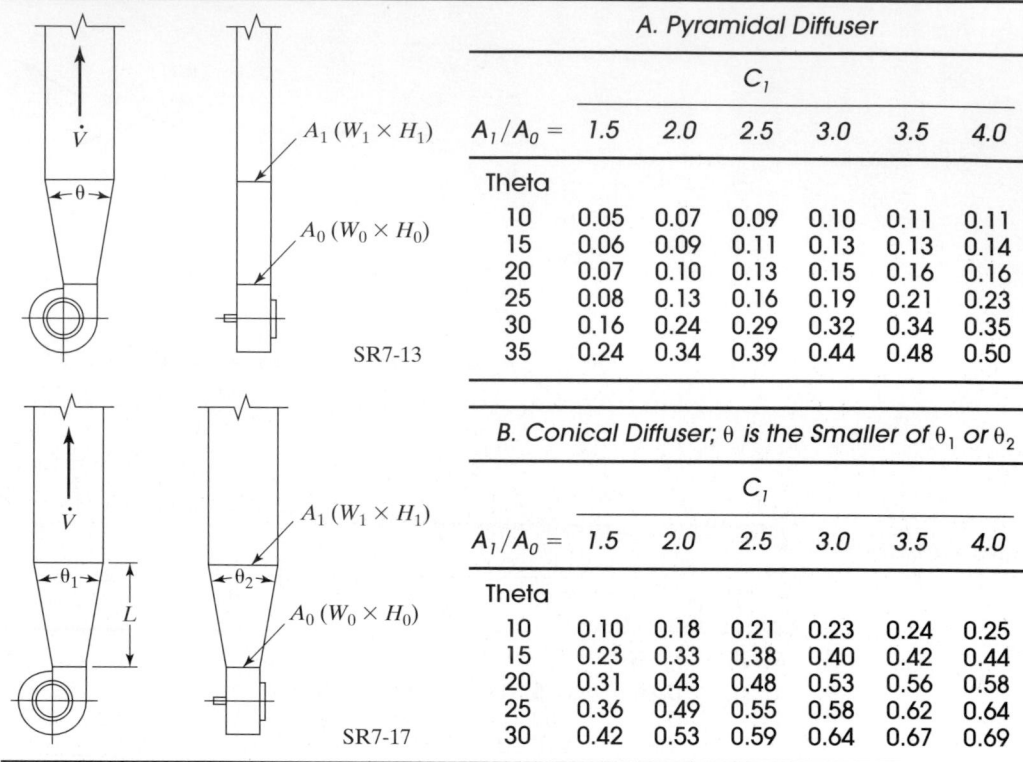

A. Pyramidal Diffuser

	C_1					
$A_1/A_0 =$	1.5	2.0	2.5	3.0	3.5	4.0
Theta						
10	0.05	0.07	0.09	0.10	0.11	0.11
15	0.06	0.09	0.11	0.13	0.13	0.14
20	0.07	0.10	0.13	0.15	0.16	0.16
25	0.08	0.13	0.16	0.19	0.21	0.23
30	0.16	0.24	0.29	0.32	0.34	0.35
35	0.24	0.34	0.39	0.44	0.48	0.50

SR7-13

B. Conical Diffuser; θ is the Smaller of θ_1 or θ_2

	C_1					
$A_1/A_0 =$	1.5	2.0	2.5	3.0	3.5	4.0
Theta						
10	0.10	0.18	0.21	0.23	0.24	0.25
15	0.23	0.33	0.38	0.40	0.42	0.44
20	0.31	0.43	0.48	0.53	0.56	0.58
25	0.36	0.49	0.55	0.58	0.62	0.64
30	0.42	0.53	0.59	0.64	0.67	0.69

SR7-17

SOURCE: Reprinted by permission from *ASHRAE Duct Fitting Database, 1994.*

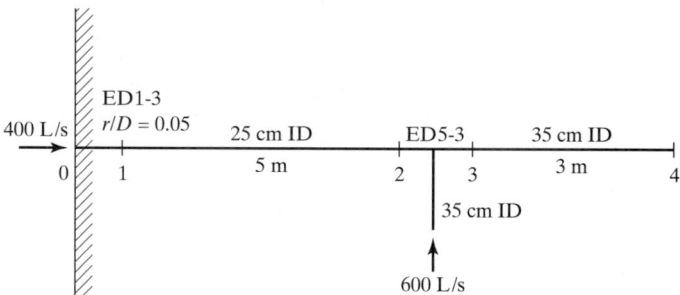

Figure 18.20 Schematic of round return duct used in Ex. 18.7.

TABLE 18.16 Summary of Results from Ex. 18.7

Section	Description	C_0	$\Delta P_t/L$, Pa/m	ΔP_t, Pa
0-1	Bell-mouth inlet	0.22	—	8.5
1-2	5-m Straight duct	—	3	15
2-3	Converging tee	1.28	—	102
3-4	3-m Straight duct	—	4	12
Total				137.5

TABLE 18.17 Friction Factors
for Galvanized Sheet Metal Ducts

Duct Diameter		Friction Factor
in.	cm.	
4	10	0.035
6	15	0.028
8	20	0.023
10	25	0.022
12	30	0.019
14	36	0.017
16	40	0.016
20	50	0.015
24	60	0.014

EXAMPLE 18.8

Repeat the pressure-loss calculations for the duct system used in Ex. 18.7, but use the equivalent-length approach.

Solution: The pressure drops for the straight sections of duct are computed using the same procedure as in the previous example, so these calculations will not be repeated here. However, the fitting losses are computed differently.

(a) *Fitting ED1-3.* The equivalent length is based on the 25-cm duct downstream of the inlet. Using Eq. (18.16), the loss coefficient determined in Ex. 18.7, and the friction factor for a 25-cm duct from Table 18.17, L_{eq} = (25 cm)(0.22/0.022) = 250 cm. The pressure drop is determined by multiplying this value by the pressure loss per cm of duct, (250 cm)(3 Pa/m)(1 m/100 cm) = 7.5 Pa.

(b) *Straight run between sections 1 and 2.* 15 Pa (same as previous example)

(c) *Converging tee.* As the loss coefficient is based on the velocity pressure of the common duct, we should use the common-duct diameter and pressure drop to determine the equivalent length. L_{eq} = (35 cm)(1.28/0.017) = 2635 cm. The total pressure drop is (2635 cm)(4 Pa/m)(1 m/100 cm) = 105 Pa.

(d) *Straight duct between sections 3 and 4.* 12 Pa (same as previous example)

The results are summarized in Table 18.18.
Comparing Tables 18.17 and 18.18 illustrates the differences between the loss-coefficient method and the equivalent-length method. Both methods give essentially the same

TABLE 18.18 Summary of Results from Ex. 18.8.

Section	Description	L_{eq}, cm	$\Delta P_t/L$, Pa/m	ΔP_t, Pa
0-1	Bell-mouth inlet	250	3	7.5
1-2	5-m Straight duct	500	3	15
2-3	Converging tee	2635	4	105
3-4	3-m Straight duct	300	4	12
Total		3685		139.5

results, although the fitting pressure losses are slightly different. For small duct diameters the designer is free to chose either method. However, for duct diameters larger than about 2 ft (60 cm) the loss-coefficient method should be used.

Fans

Given the ductwork and associated pressure drops, we need a device to overcome these pressure drops. This device is called a *fan*. There are many different types of fans, each having its benefits and drawbacks. Some are located at the discharge end of a duct and are used strictly for exhaust; others are located near the center of the ductwork and must provide air pressure to overcome losses in both supply and return duct systems.

Figure 18.21 illustrates the most common types of fans used in air-distribution systems in buildings. The first three are centrifugal fans. These all have a rotating scroll located within a volute housing. They are usually driven by an electric motor mounted nearby with the motor and fan pulleys connected by a rubber belt. The pulley sizes can be changed to change the fan rotation speed. The scroll of the backward-curved fan has blades that are angled in the direction opposite to the direction of scroll rotation. The forward-curved fan has fan blades curved in the other direction. The radial fan has blades that appear to be aligned with the scroll radius. Centrifugal fans are available in a wide variety of sizes and are used in applications ranging from residential bathroom exhaust systems to industrial power plants. The backward-curved design is the most widely used because of its high efficiency. Forward-curved designs are often used in low-pressure applications, such as residential equipment where the airflow rate is not very well known.

Axial fans have traditionally been used in systems with high flow rates. The fans operate in line and do not require a change in flow direction as the centrifugal fans do. These fans generally have a higher noise level than centrifugal fans. They are becoming more common in HVAC air-handling systems. The blades have an airfoil shape, which gives the fan a fairly high efficiency. The flow leaves the tubeaxial fan with a swirl. The velocity distribution in the downstream duct is highly nonuniform. The vaneaxial fan has stator blades in addition to the rotor to straighten out the flow leaving the fan. This reduces the amount of swirl leaving the fan and provides a higher peak pressure and higher efficiency.

The performance of fans is determined by the manufacturer by following standard rating test procedures. These performance data are often provided to the designer in tabular form, in graphical form, or sometimes in both forms. Data provided at a given fan speed and airflow rate usually include: (1) total pressure, (2) static pressure, (3) power required, (4) total efficiency, and (5) static efficiency. Each of these quantities is discussed below.

1. *Total pressure:* This is the rise in total pressure across the fan, $P_{t,1} - P_{t,2}$. Units are typically inches of water (Pa).
2. *Static pressure:* This is the rise in static pressure across the fan, $P_{s,1} - P_{s,2}$. Units are the same as for total pressure. The static pressure rise is always equal to or less than the total pressure rise.
3. *Power requirement:* This is the power necessary to drive the fan at this operating condition. Units are hp or kW. The motor rating must always be larger than the highest power required at any of the fan operating conditions.

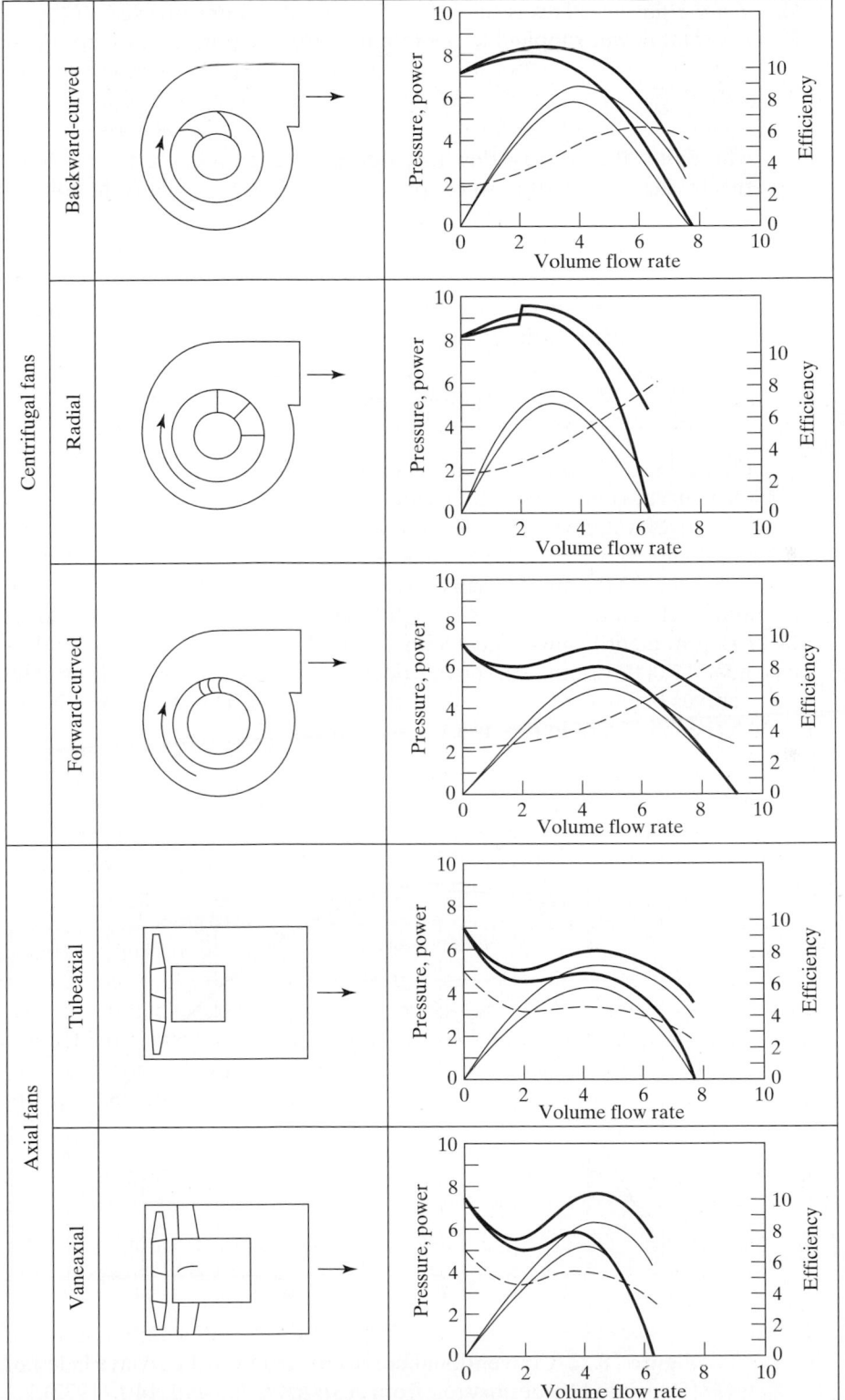

Figure 18.21 Typical centrifugal and axial fans and their performance curves. [Reprinted by permission from *ASHRAE Systems and Equipment 1996* (IP), pp. 18.2, 18.3.]

4. *Total efficiency:* This is the ratio of the total power imparted to the air divided by the total power supplied to operate the fan. The power added to the air is

$$\dot{W}_{\text{tot}} = \frac{\dot{m}(P_{t,1} - P_{t,2})}{\rho}$$

where $\dot{m}$ is the air mass flow rate through the fan and ρ is the air density. Dividing this by the total shaft power supplied to the fan, $\dot{W}_{sh}$, yields the total efficiency:

$$\eta_{\text{tot}} = \frac{\dot{W}_{\text{tot}}}{\dot{W}_{sh}} = \frac{\dot{m}(P_{t,1} - P_{t,2})}{\rho \dot{W}_{sh}} \qquad (18.17)$$

5. *Static efficiency:* This is the ratio of the power added to the air to increase the static pressure divided by the shaft power supplied to the fan:

$$\eta_{s} = \frac{\dot{W}_{s}}{\dot{W}_{sh}} = \frac{\dot{m}(P_{s,1} - P_{s,2})}{\rho \dot{W}_{sh}} \qquad (18.18)$$

Figure 18.22 illustrates typical curves by plotting the performance-test results from a backward-curved centrifugal fan. The total-pressure characteristic curve begins near a pressure of 1.8 in. of water when the flow is zero. This is the condition when the airflow leaving the fan is completely blocked. The total-pressure curve increases and reaches a peak near a flow rate of 6000 ft³/min. The curve drops and reaches zero near 17,500 ft³/min. The static-pressure characteristic curve is slightly below the total-pressure characteristic. The power requirement begins near 0.8 hp at the blocked-flow condition and increases to a peak near 12,000 ft³/min. The power decreases slightly at higher flow rates. Both efficiency curves begin at zero when the flow rate is zero. This is because the mass flow rate through the fan is zero in Eqs. (18.17) and (18.18). The efficiency curves reach maximum

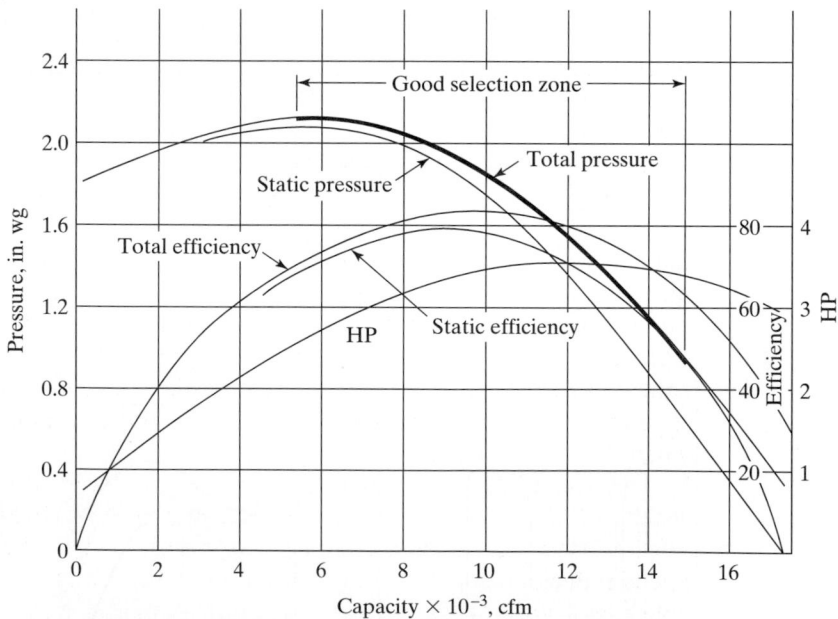

Figure 18.22 Conventional curves for backward-curved blade fan. [Reprinted by permission from *ASHRAE Journal*, 14:1 (1972).]

values near a flow rate of 10,000 ft^3/min and then decrease again toward zero. The good selection zone is indicated on this figure as a dark band on the total-pressure characteristic curve. At values of flow rate to the left of this band, the characteristic curve decreases, indicating stall or surging. To the right the efficiency becomes very low, so that a fan with a larger flow rate capacity should be used.

Figure 18.21 shows representative total and static pressure characteristics, power requirements (dashed curves), and total and static efficiency curves for the different types of fans for comparison.

Fan laws are approximate relationships that can be used to estimate the performance of a particular fan under variable operating conditions or to predict the performance of an aerodynamically similar fan to one with known performance data. Variables include: volumetric flow rate $\dot{V}$, diameter D, power $\dot{W}$, rotation speed N, total pressure P_t and gas density ρ. The three most common fan laws are given below.

1. $\dot{V}_2 = \dot{V}_1 (D_2/D_1)^3 (N_2/N_1)$
2. $P_{t,2} = P_{t,1} (D_2/D_1)^2 (N_2/N_1)^2 (\rho_2/\rho_1)$
3. $\dot{W}_2 = \dot{W}_1 (D_2/D_1)^5 (N_2/N_1)^3 (\rho_2/\rho_1)$

If the same fan is considered and the gas density does not change, some simple interpretations can be given. The first equation indicates that the volumetric flow rate is directly proportional to fan rotational speed. Although a fan is not a positive-displacement device, the first fan law gives the same result; i.e., doubling the rotation speed will double the flow rate. The second equation indicates that the pressure rise is proportional to the rotation speed squared. This is similar to the relationship between total pressure and velocity as given in Eq. (18.10). Doubling the rotation speed will double the flow rate and velocity, thereby increasing the pressure rise by a factor of four. The third equation shows that the power is proportional to the rotation speed cubed. The power is proportional to the product of the flow rate and the pressure rise. Therefore, the power should be proportional to the product of equations 1 and 2, which gives the rotation speed cubed.

These fan laws can also be applied to other similar fans for which performance data are lacking or to fans applied to gases with different density, such as hot flue gases. Other variations of these fan laws are given in ASHRAE [11].

Two methods are commonly used to control centrifugal-fan performance. One is to equip the fan with adjustable guidevanes at the inlet. When the vanes are fully open, the fan performs as if the vanes were not present. When the vanes are rotated so that they partially close the inlet, the inlet air is given a swirl before entering the rotor. This affects the attack angle of the scroll, which results in a change in the fan performance. Figure 18.23(a) illustrates the total-pressure characteristic curve for a centrifugal fan with inlet vanes. The pressure at plugged-flow conditions does not change, but the pressure rise is significantly reduced at higher flow rates. The power requirement is also slightly reduced when the vanes partially close.

Figure 18.23(b) illustrates another means of centrifugal-fan control. If a variable-speed motor or a variable-speed drive is used, the fan characteristic curve can be greatly affected. The power requirement is also reduced when the speed of the fan is reduced. Speed reduction can be accomplished using several methods, including a mechanical variable-speed drive. However, most recently installed systems use a variable-speed-drive device. This is an electrical device that rectifies line ac curent to dc, then reconverts this back into ac power but at a different frequency. The motor rotor speed is locked onto the

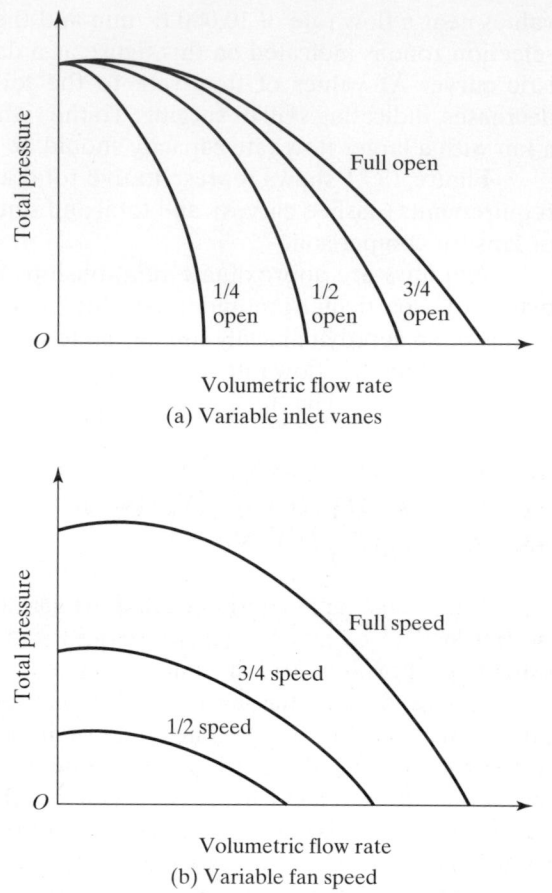

Figure 18.23 Effect of variable inlet vanes or fan speed on total-pressure characteristic curves.

frequency of the ac power supplied. By reducing the frequency supplied, the motor speed is reduced, which in turn reduces the fan speed. There are electrical losses associated with the power conversion. However, the decrease in fan power requirement at part load is significantly larger, so the total power supplied can be greatly reduced at part-load conditions.

A family of curves similar to those shown in Fig. 18.23(b) can be generated by using a vaneaxial fan with variable-pitch blades. As with the centrifugal fans, the reduced airflow and pressure rise results in a reduced power requirement.

Airflow Control

We have discussed duct systems and fans separately; now we will combine them into a single air-handling system and discuss the balance point and methods of flow control.

The *balance point* is the location where the pressure rise through the fan equals the pressure loss through the duct system and the volumetric flow rates are equal. This is the only set of operating conditions which will satisfy both the duct-system total-pressure characteristic curve and the fan total-pressure characteristic curve. Figure 18.24 illustrates the location of the balance point. It is the intersection of the fan and duct-system total-

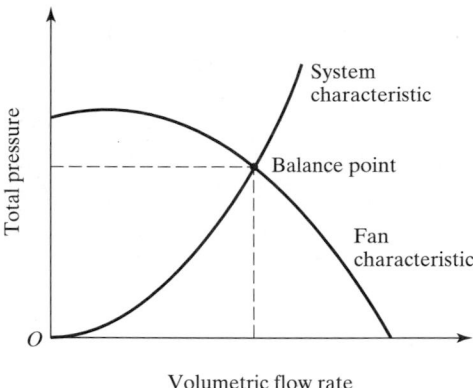

Figure 18.24 Balance point between a fan and duct system.

pressure characteristic curves. The dashed lines show the volumetric flow rate at the balance point and the total pressure across the fan, which equals the total pressure loss through the duct.

The flow rate through the duct can be changed by changing the location of the balance point. Either the duct-system characteristic or the fan characteristic can be changed. In some cases both are changed simultaneously. We will discuss duct-system changes first followed by fan change and then both.

The most widely used means of changing the duct-system characteristic is to add more airflow resistance. This is done by partially closing a set of dampers. Two basic types of dampers commonly used in air-handling systems are parallel-blade dampers and opposed-blade dampers. These are shown in Figs. 18.25 and 18.26, respectively. Also shown are the percent maximum flow versus the opening angle of the dampers. The flow is nonlinear with respect to opening angle. The nonlinearity depends on the fraction of total duct-system resistance associated with the open dampers. Dampers are usually opened or closed by the actuator of an active control system that is to maintain a set temperature somewhere in the system or some other variable.

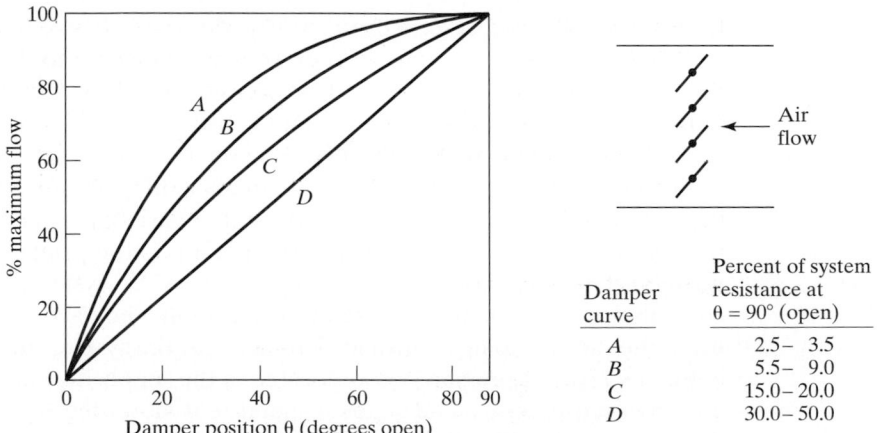

Figure 18.25 Performance of parallel-blade dampers. Used in air mixing applications where the percentage of the two air streams needs to be controlled without changing the total air volume.

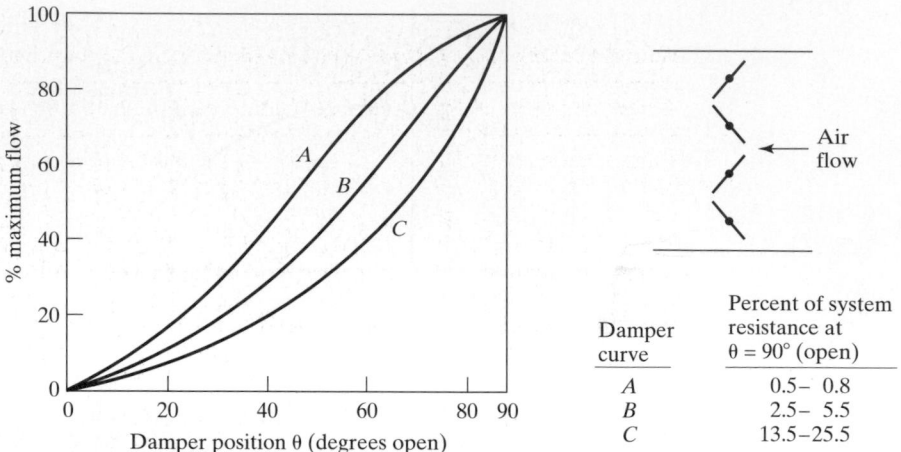

Damper curve	Percent of system resistance at $\theta = 90°$ (open)
A	0.5– 0.8
B	2.5– 5.5
C	13.5–25.5

Figure 18.26 Performance of opposed-blade dampers. Used when modulating airflow rate (variable-volume systems, bypass, ventilation air); more sensitive than parallel-blade dampers.

Figure 18.27(a) illustrates how partially closing a damper affects the balance point when the fan speed remains fixed. The fan characteristic does not change; only the duct-system characteristic changes. As the damper is partially closed, it adds more resistance to the airflow in the duct system. Thus the pressure loss in the duct increases at a given flow rate. The duct-system characteristic shifts upward, owing to this additional resistance. The balance point between the fan and the duct system also moves along the fan characteristic to a lower flow rate but a higher pressure drop. The fan power is reduced but not very much. The pressure rise across the fan increases, but the flow rate decreases. The fan power is proportional to the product of these two factors. This flow-control method is inexpensive to install and control. However, it does not significantly reduce the fan power, which is often the major operating cost.

The effect of reducing the fan speed to reduce the flow is illustrated in Fig. 18.27(b). In this case, the duct-system characteristic does not change but the fan curve changes as shown. The balance point moves along the system curve to the new location where both the airflow rate and the pressure drop are reduced. This results in a significant reduction in fan power required and the associated electrical costs.

Most building air-handling systems operate at part load much of the year except during peak-load conditions. Therefore the fan energy requirement and electrical cost for the year can be significantly reduced using a variable-speed fan rather than a damper flow-control system for many applications. This is the primary driving force behind the use of variable-air-volume (VAV) systems. In a VAV system, both damper and fan-speed controls are used. Dampers close in the air-supply boxes that supply zones of a building when the thermal load is satisfied. This changes the system curve by shifting it upward, as in Fig. 18.27(a). A pressure sensor located in the supply duct downstream of the fan senses an increased pressure and sends a signal that slows the fan. Thus the fan curve shifts downward, as in Fig. 18.27(b). The pressure in the duct where the pressure sensor is located remains constant, but the discharge pressure from the fan is reduced, as there is a smaller pressure drop between the fan and the sensor at lower airflow rates. The resulting system balance-point shift is shown in Fig. 18.28. When the system operates under

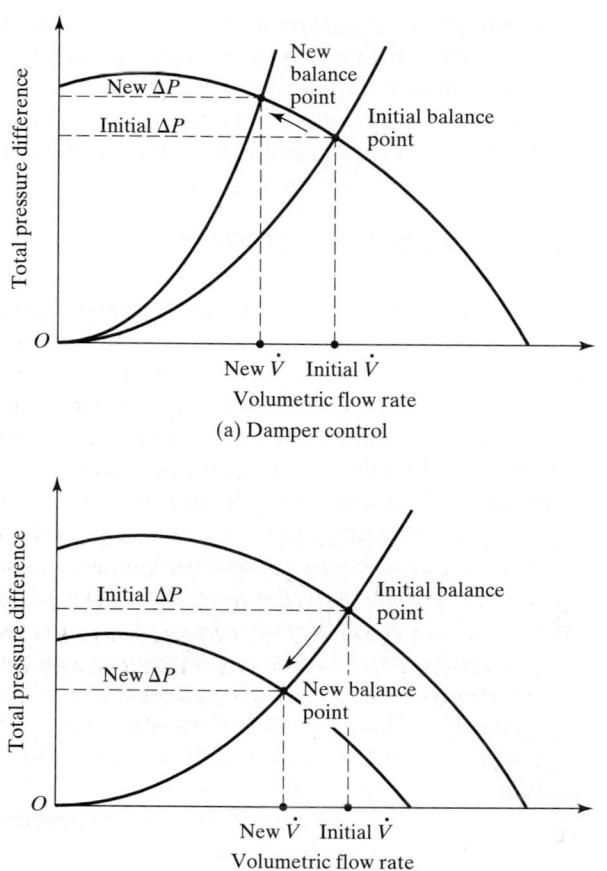

Figure 18.27 Flow control by means of (a) damper control or (b) fan-speed control.

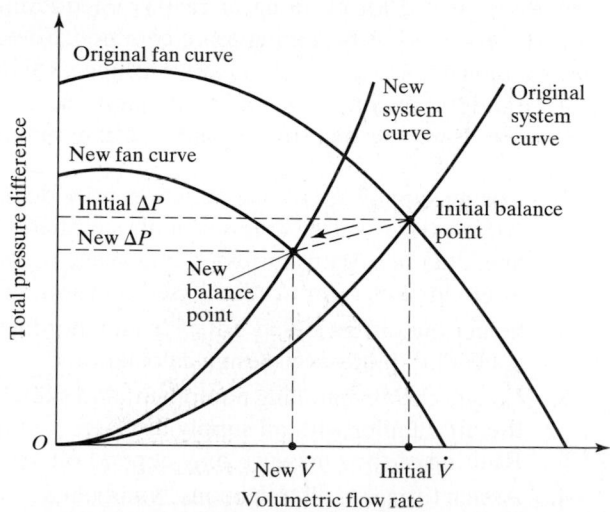

Figure 18.28 Illustration of VAV operation.

these conditions at part load, the fan power may be less than one-half the power required at full speed. Thus there is the potential for considerable energy and cost savings with this type of system compared to a constant-volume system where the fan operates at full speed. The energy control for constant-volume systems is often accomplished by reducing liquid-flow rates through heat exchangers rather than by changing the airflow rate.

18.4 DUCT LAYOUT AND SIZING PROCEDURES

Duct design principles are used to lay out and determine the size of the ductwork. This is not performed until the quantity of air required in each space has been determined from thermal load and/or ventilation requirements, the conditioned space has been subdivided into zones, and the general location of the supply and return ductwork determined. The supply diffusers need to be located and selected. The method of air return must be determined, including the use of return-air plenums if any. The outdoor ventilation air intake and exhaust locations should be known. The location of the air-handling equipment must be determined. Only after all these preliminary tasks have been accomplished is the designer ready to begin the duct design.

Several trade-offs exist. For example, a small duct with high velocity will require less material and less weight than a larger duct. Less external insulation is required. However, the pressure drop is larger, requiring more fan power, and the higher velocity generates more noise.

When packaged air-handling units are used, the fan is preselected. Thus the pressure supplied by the fan is used to determine the necessary duct size. However, in large systems where the air handler is custom built, the ducts are designed first and the total pressure required is then used to select an appropriate fan.

Additional considerations include sound transmission, access for future cleaning, air leakage through joints, and heat gain or loss. It has been standard practice to use internal lining to attenuate sound. However, these linings eventually contain dirt, become contaminated with fungi, and are very difficult to clean. The only choice in many cases is to remove the lining. It is recommended that other types of sound absorption be used that can be removed for cleaning or readily decontaminated. Providing access to clean and inspect ductwork is becoming more common. Several duct-cleaning techniques are used, including mechanical scrubbing and washing. ASHRAE [9] gives recommended practices for estimating duct leakage and heat gain/loss.

The following step-by-step duct-design approach is recommended:

1. Locate supply and return outlets to provide proper air distribution within the space. Adjust the required amount of air supplied to each space to account for duct leakage, duct heat gains or losses, and space pressurization requirements. Use the maximum airflow rates needed based on room or zone peak loads.
2. Select the diffusers and grilles from manufacturer's data. The loss of total pressure is needed in the duct-design calculations.
3. Locate the air-handling equipment and sketch ductwork to connect the discharge of the air handler with all supply diffusers and return air to the return side of the fan. Routing of the ductwork may depend on available space and aesthetics.
4. Assign fittings to all transitions. Number each section of ductwork for ready reference.
5. Size all duct sections, using a selected duct-design method.

6. Lay out the system in detail and recompute pressure losses if significant changes are made from the original sketch.

7. Analyze the system for sound levels and specify sound attenuators if necessary.

8. Ensure adequate access for future maintenance and inspection.

Several duct-design methods have been developed. Only a few of the more common ones will be discussed here. The basic principle in all methods is that the correct amount of air must be delivered to each supply diffuser at the correct pressure and the return system must handle the return airflow with the fan pressure available. In nearly all duct systems, airflow is adjusted by dampers, either manually or automatically as part of the control system. Therefore, the exact airflow rates and pressure distribution in the duct system are not critical. The dampers will perform the fine tuning of the system. This is termed "balancing" the system. This should be done when the system is first put into operation and may be required periodically afterward. However, the system should be well designed so that a minimal amount of damper control is necessary near design conditions. Thus the fan will not be oversized and will operate efficiently.

Equal-Friction Method

This method is considered to be a low-velocity duct-design method for use in residential and small to medium-sized commercial applications. A total pressure loss per unit length, $\Delta P_t/L$, is selected and used to size all ducts. ASHRAE recommends using the shaded area in Figs. 18.18(a) and (b). This is a compromise between initial cost and operating-energy cost. The horizontal portion of the shaded area below a flow rate of 15,000 ft³/min or 7000 L/s represents a value of $\Delta P/L$ of approximately 0.3 in. of water/100 ft or 2 Pa/m. The designer has some latitude in the value of $\Delta P/L$. When fan power costs are high or large duct sizes can be used, smaller values of $\Delta P/L$ can be selected.

This method will work well with little modification when symmetrical duct layouts are encountered or when the total equivalent lengths from the fan through all the terminal units are approximately equal. However, for nonsymmetrical systems, the duct sizes should be modified to balance the system. Dampers can also be used if desired. Figure 18.29 illus-

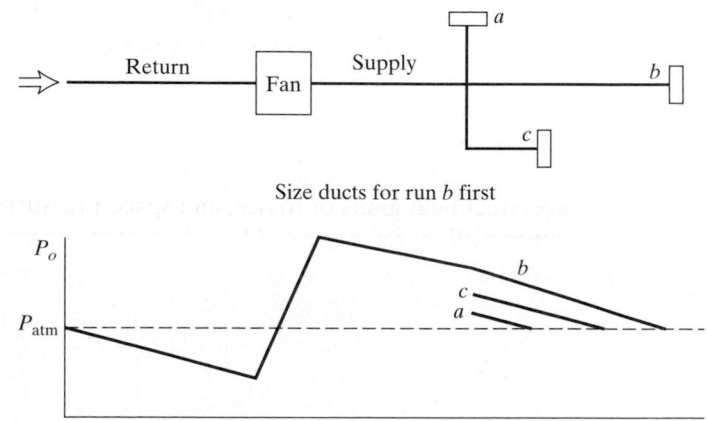

Figure 18.29 Pressure distribution in system sized with equal-friction method before balancing branches *a* and *c*.

trates a theoretical total-pressure distribution in a duct system sized using the equal-friction method. Note that the slopes of the lines connecting the three diffusers are equal. Because branches to diffusers a, b, and c are all different lengths from the common tee, and the pressure gradients in all the branches are equal, the total pressures at the junction of the branches are not equal. This cannot occur in practice; the pressure must be continuous from the fan to each diffuser. Therefore, this system will not operate with the correct flow distribution as it is designed. Additional pressure drops must be introduced into branches a and c so the total pressure drop from the junction to each diffuser is the same. This can be accomplished by reducing the size of the ducts leading to diffusers a and c. This will increase the pressure gradient in these two branches and increase the slopes of the lines shown on Fig. 18.29. Another option is to place dampers in branches a and c to provide the necessary additional pressure drop. Example 18.9 illustrates the use of the equal-friction method for a low-velocity system.

EXAMPLE 18.9

Size the ductwork shown in Fig. 18.30, using the equal-friction method. Use round sheet metal duct and select sizes to the nearest inch. Assume that all terminal units have a total pressure loss of 0.1 in. of water. Determine the fan-pressure requirement and resize the branches to balance the system to minimize the use of dampers.

Solution: We will begin by sizing the longest run of this system from the fan to terminal unit c. The section from 1 to 2 carries $200 + 300 + 500 = 1000$ ft^3/min. From the shaded region of Fig. 18.18(a) at 1000 ft^3/min, a 12-in. duct has a pressure drop of 1.9 in. of water/100 ft and a velocity of about 1300 ft/min. This is a good choice, as a 10-in. duct has a pressure loss near the maximum recommended and a 14-in. duct has a pressure loss near the lower limit. The duct section from 3 to 4 carries $300 + 500 = 800$ ft^3/min. From Fig. 18.18(a) the appropriate size could be either a 10-in. or 12-in. duct. We will select a 10-in. duct, as it is closer to the center of the recommended area. The pressure loss is 0.3 in. of water/100 ft and the velocity is about 1450 ft/min. The duct from 5 to 8 carries 500 ft^3/min of air and is selected to be a 9-in. duct with a pressure loss of 0.22 in. of water/100 ft and a velocity of about 1150 ft/min.

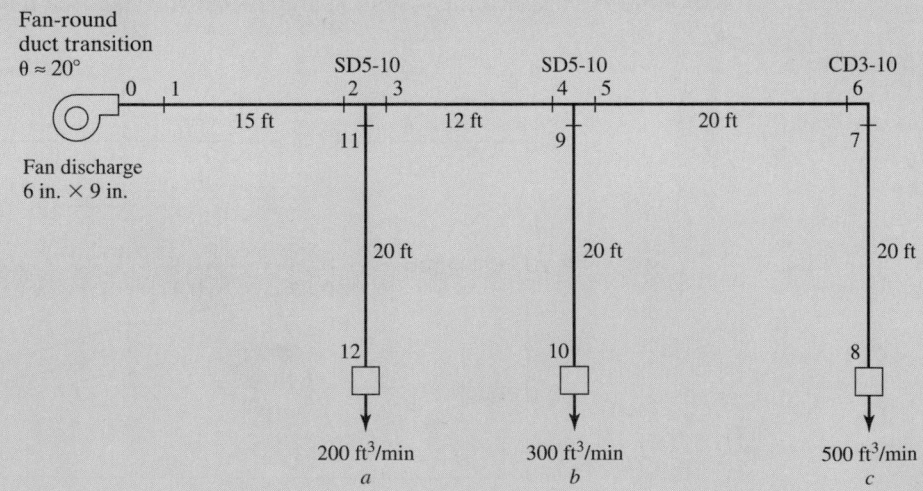

Figure 18.30 Duct-system schematic for Ex. 18.9.

Branch a is initially sized to be a 6-in. duct with a pressure loss of 0.3 in. of water/100 ft and a velocity of 1000 ft/min, and branch b is sized to be a 7-in. duct with a pressure loss of 0.3 in. of water/100 ft and a velocity of about 1130 ft/min.

Next we will determine the loss coefficients for the fittings, using the duct sizes selected above. The transition from the fan discharge to the 12-in. round duct will be assumed to be an SR7-17 fitting with $\theta = 20$ deg. Although this is not the exact fitting required, it should provide a close approximation to the actual pressure drop.

$$\frac{A_1}{A_0} = \frac{\pi D_1^2/4}{W_0 \times H_0} = \frac{\pi (12 \text{ in.})^2/4}{9 \text{ in.} \times 6 \text{ in.}} = 2.1$$

Interpolating from Table 18.15 results in the loss coefficient, $C_1 = 0.44$. The coefficient from section 2 to 3 is obtained using Table 18.8. $\dot{V}_s/\dot{V}_c = 800 \text{ ft}^3/\text{min}/1000 \text{ ft}^3/\text{min} = 0.8$, $A_s/A_c = (10 \text{ in.}/12 \text{ in.})^2 = 0.69$. Interpolating in Table 18.8 gives $C_s = 0.15$. Similarly from section 4 to 5, $\dot{V}_s/\dot{V}_c = 500 \text{ ft}^3/\text{min}/800 \text{ ft}^3/\text{min} = 0.63$, $A_s/A_c = (9 \text{ in.}/10 \text{ in.})^2 = 0.81$, $C_s = 0.14$. The loss coefficient for the 90-deg elbow between 6 and 7 is found to be $C_0 = 0.10$ from Table 18.7. Loss coefficients for the branch takeoffs are found using a similar procedure from Table 18.8. From 2 to 11, $\dot{V}_b/\dot{V}_c = 200 \text{ ft}^3/\text{min}/1000 \text{ ft}^3/\text{min} = 0.2$, $A_b/A_c = (6 \text{ in.}/12 \text{ in.})^2 = 0.25$, $C_b = 1.11$. From 4 to 9, $\dot{V}_b/\dot{V}_c = 300 \text{ ft}^3/\text{min}/800 \text{ ft}^3/\text{min} = 0.38$, $A_b/A_c = (7 \text{ in.}/10 \text{ in.})^2 = 0.49$, $C_b = 1.23$.

The following velocity pressure values are needed to compute the pressure drops in the fittings obtained from Eq. (18.12a):

$$P_{v,1} = P_{v,2} = (1300/4005)^2 = 0.11 \text{ in. of water}$$

$$P_{v,3} = P_{v,4} = (1450/4005)^2 = 0.13 \text{ in. of water}$$

$$P_{v,5} = P_{v,6} = (1150/4005)^2 = 0.08 \text{ in. of water}$$

$$P_{v,9} = (1130/4005)^2 = 0.08 \text{ in. of water}$$

$$P_{v,11} = (1000/4005)^2 = 0.06 \text{ in. of water}$$

Table 18.19 summarizes the results of the pressure-drop computations. The losses through the fittings are calculated by multiplying the velocity pressure by the loss coefficient. The losses through the straight runs are calculated by multiplying the duct length by the pressure drop per unit length.

TABLE 18.19

Duct Section	Airflow, ft^3/min	Duct Diameter, in.	Duct Length, ft	Velocity, ft/min	Velocity Pressure, in. water	Loss Coefficient	Duct Pressure Loss, in. water/100 ft	Total Pressure Loss, in. water
0-1	1000			1300	0.11	0.44		0.05
1-2	1000	12	15	1300	0.11		0.19	0.03
2-3	800				0.13	0.15		0.02
3-4	800	10	12	1450	0.13		0.3	0.04
4-5	500				0.08	0.14		0.01
5-6	500	9	20	1150	0.08		0.22	0.04
6-7	500				0.08	0.1		0.01
7-8	500	9	20	1150	0.08		0.22	0.04
0-8								0.24
4-9	300				0.08	1.23		0.10
9-10	300	7	20	1130	0.08		0.3	0.06
4-10								0.16
2-11	200				0.06	1.11		0.07
11-12	200	6	20	1000	0.06		0.3	0.06
2-12								0.13

Now we need to compare the total pressure drop from the fan to terminal units a, b, and c to determine whether we need to change any of the duct sizes used above. The total pressure losses are given below:

(a) $\Delta P_{t,a} = \Delta P_{t,0-1} + \Delta P_{t,1-2} + \Delta P_{t,2-11} + \Delta P_{t,11-12}$
$= 0.05 + 0.03 + 0.07 + 0.06 = 0.21$ in. of water

(b) $\Delta P_{t,b} = \Delta P_{t,0-1} + \Delta P_{t,1-2} + \Delta P_{t,2-3} + \Delta P_{t,3-4} + \Delta P_{t,4-9}$
$+ \Delta P_{t,9-10}$
$= 0.05 + 0.03 + 0.02 + 0.04 + 0.10 + 0.06 = 0.30$ in. of water

(c) $\Delta P_{t,c} = \Delta P_{t,0-1} + \Delta P_{t,1-2} + \Delta P_{t,2-3} + \Delta P_{t,3-4} + \Delta P_{t,4-5}$
$+ \Delta P_{t,5-6} + \Delta P_{t,6-7} + \Delta P_{t,7-8}$
$= 0.05 + 0.03 + 0.02 + 0.04 + 0.01 + 0.04 + 0.01 + 0.04 = 0.24$ in. of water

The pressure losses to terminal units a and c are nearly equal, but the loss to terminal unit b is somewhat higher. One option would be to try to reduce the loss to unit b to more nearly approximate the losses to the other two. The two losses that would be affected are the 0.10 in. of water loss through the fitting between sections 4 and 9 and the 0.06 in. of water loss through the straight duct between 9 and 10. Inspecting Table 18.8, the loss coefficient for the fitting will increase if the duct diameter to terminal b is increased in size. However, the velocity pressure will decrease. The total pressure loss should decrease when these two are multiplied together. We will try increasing the duct diameter from 7 to 8 in. from sections 9 to 10. The new values become:

$$\text{Pressure loss in duct} = 0.085 \text{ in. of water/100 ft}$$

$$\text{velocity} = 690 \text{ ft/min}$$

$$\text{velocity pressure} = (690/4005)^2 = 0.03 \text{ in. of water.}$$

$$\dot{V}_b/\dot{V}_c = (300 \text{ ft}^3/\text{min})/(800 \text{ ft}^3/\text{min}) = 0.38$$

$$A_b/A_c = (8 \text{ in.}/10 \text{ in.})^2 = 0.64$$

$$C_b = 2.15$$

The fitting pressure loss becomes 2.15(0.03 in. of water) = 0.06, and the duct loss becomes (20 ft)(0.085 in. of water/100 ft) = 0.02 in. of water. The total pressure loss from the fan to terminal unit b is now:

$$\Delta P_{t,b} = \Delta P_{t,0-1} + \Delta P_{t,1-2} + \Delta P_{t,2-3} + \Delta P_{t,3-4} + \Delta P_{t,4-9} + \Delta P_{t,9-10}$$

$$= 0.05 + 0.03 + 0.02 + 0.04 + 0.06 + 0.02$$

$$= 0.22 \text{ in. of water}$$

This value is nearly equal to the total pressure drop from the fan to the other two terminal units (0.21 and 0.24 in. of water). The system is now fairly well balanced. It may not be possible to improve the balance with round ductwork restricted to 1-in. increments in diameter.

The total pressure across the fan due to this supply-air duct system is determined using the largest total-pressure-loss value for the ductwork, which occurs to terminal unit c, 0.24 in. of water. To this we add the pressure loss through the terminal unit, 0.1 in. of water, so the total-fan-pressure requirement becomes 0.24 + 0.1 = 0.34 in. of water. Note that the total-fan-pressure rise is larger than this value, because the pressure loss through the return ductwork must be added to the value given above.

This example illustrates the basic procedure to use when sizing ductwork using the equal-friction method, including the adjusting that may be necessary to balance the system.

Static-Regain Method

In some applications, where large amounts of air must be moved or duct size is restricted, higher velocities must be used so that the duct size is not prohibitively large. These systems are commonly referred to as "high-velocity systems." Some disadvantages include a higher pressure loss per unit length than in low-velocity systems, higher noise generation, and large pressure losses at poorly designed fittings. ASHRAE recommends using the upper portion of the shaded area on Figures 18.18(a) and (b) when designing high-velocity systems. For flow rates below 15,000 ft^3/min (7000 L/s) the pressure loss per unit length is about 0.6 in. of water (5 Pa/m), and at higher flow rates the velocity should be about 4000 fpm (20 m/s).

The air velocities and noise level must be reduced before reaching the final supply terminal unit. This is often accomplished with terminal boxes. These devices control the leaving air pressure and volume and also reduce the noise passed through to the terminal units. Often short lengths of flexible duct are used to connect the terminal box to the supply diffusers.

Duct leakage could be a problem in large systems with a large amount of surface area. Stress created by the internal static pressure could also be a problem. Special types of duct have been designed to alleviate these two problems. One type commonly used is a *spiral duct*. The spiral nature of this duct provides additional structural strength. Also the fittings are designed to be very tight. This type of duct is generally available in 1-in. increments from 3 to 24 in. in diameter and in 2-in. increments between 24 and 50 in. in diameter. Metric sizes are available in 1-cm increments from 8 to 60 cm diameter and in 2-cm increments between 60 and 120 cm diameter.

Another method to minimize duct-leakage and strength issues is to design the duct system so that the static pressure throughout the entire system is nearly at atmospheric pressure. It is only the static pressure that affects the leakage and the forces on the duct wall. A high velocity duct design method to achieve this is called the "static-regain method." In this method, the air velocity is reduced through fittings in the direction of the flow, so that the decrease in velocity pressure equals the decrease in total pressure. Thus the static pressure remains unchanged. This is illustrated using Eq. (18.10):

$$P_t = P_s + \frac{\rho \mathbf{V}^2}{2} \tag{18.10}$$

or

$$P_t = P_s + P_v \tag{18.19}$$

If the total pressure is reduced the same amount as the velocity pressure, the static pressure does not change.

Figure 18.31 illustrates a duct section that is designed using the static-regain method. Applying Eq. (18.19) to sections *a* and *c*,

$$P_{t,a} = P_{s,a} + P_{v,a}$$

$$P_{t,c} = P_{s,c} + P_{v,c}$$

Rearranging these equations to solve for the respective static pressures,

$$P_{s,a} = P_{t,a} - P_{v,a}$$

$$P_{s,c} = P_{t,c} - P_{v,c}$$

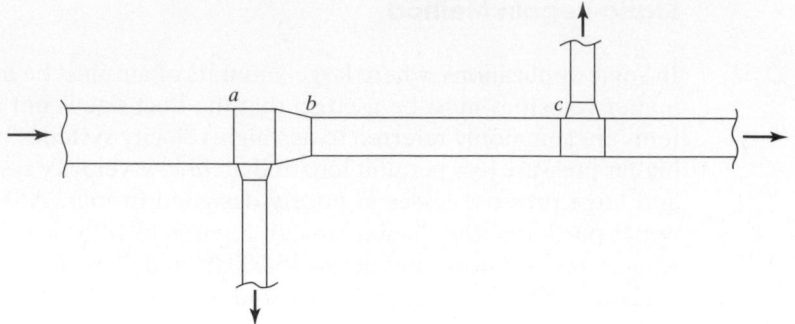

Figure 18.31 Schematic of duct section designed using the static-regain method.

Setting the difference between the two static pressures equal to zero,

$$P_{s,a} - P_{s,c} = P_{t,a} - P_{t,c} + (P_{v,c} - P_{v,a}) = 0$$

In duct design we normally work with pressure differences rather than with actual pressures. By adding and subtracting the total pressure at the intermediate section, b, we now have a simple equation that can be used to relate the change in static pressures between sections a and c with the changes in velocity pressure and the total pressure loss between these two locations:

$$P_{s,a} - P_{s,c} = (P_{t,a} - P_{t,b}) + (P_{t,b} - P_{t,c}) + (P_{v,c} - P_{v,a}) = 0 \qquad (18.20)$$

The static-regain method consists of selecting the duct size between sections b and c such that the total pressure losses and the change in velocity pressure satisfy Eq. (18.20). It is not possible to satisfy this equation exactly, as the choice of duct sizes is limited. It is not readily apparent which duct size of those available will most closely satisfy Eq. (18.20), as a change in size will affect both the fitting loss between sections a and b and the pressure drop between b and c. Therefore, the best size to use must be determined by trial.

The duct sizing is usually begun at the fan or at a main branch from a plenum. The highest possible velocity is used to size this first section. Recommended velocity limits were discussed earlier and are shown on Figs. 18.18(a) and (b). Subsequent duct sections have reduced velocity and may have increased size if the airflow rate does not change significantly in long runs.

An estimate of total fan pressure required for a duct system can be made when the static-regain method is used. As the static pressure should not change between the first fitting and the last fitting at the terminal boxes, the total pressure drop from the fan or plenum to the terminal boxes is the sum of the velocity pressure at the first fitting (which is equal to the loss of total pressure in the entire duct system after the first fitting) and the loss of total pressure from the fan or plenum to the first fitting. Adding this pressure loss to that of a terminal box will provide a good estimate of the fan pressure required for that particular duct system. Of course, the total fan pressure requirement should include both supply and return duct systems.

EXAMPLE 18.10

Size the duct between sections b and c in Fig. 18.31 using the static-regain method, given the following information. Airflow rates: $\dot{V}_a = 40{,}000$ L/s, $\dot{V}_b = \dot{V}_c = 30{,}000$ L/s, $\mathbf{V}_a = 20$ m/s. The

---

fitting between sections a and b is similar to SD5-10 shown in Table 18.8. The length of straight duct between sections b and c is 10 m.

Solution: The duct size at section a is found to be 160 cm from Fig. 18.18(a). We will assume that sizes are available in 5-cm increments. We will try a duct size of 150 cm initially. The area ratio is $A_b/A_a = (150 \text{ cm}/160 \text{ cm})^2 = 0.88$ and the flow-rate ratio is $\dot{V}_b/\dot{V}_a = (30{,}000 \text{ L/s})/(40{,}000 \text{ L/s}) = 0.75$. From Table 18.8, the loss coefficient is about $C_s = 0.14$. The velocity at b and c is about 17 m/s and the pressure drop per unit length of duct is about 1.5 Pa/m from Fig. 18.18(a). The velocity pressure at section a from Eq. 18.12(a) is 240 Pa. The velocity pressure at b and c from Eq. 18.12(a) is 174 Pa. The loss in total pressure between sections a and b is $\Delta P_{t,ab} = C_s P_{v,b} = (0.14)(174 \text{ Pa}) = 24.4 \text{ Pa}$. The loss in total pressure between sections b and c is $\Delta P_{t,bc} = (10 \text{ m})(1.5 \text{ Pa/m}) = 15 \text{ Pa}$. Substituting into Eq. 18.31,

$$P_{s,a} - P_{s,c} = 24.4 \text{ Pa} + 15 \text{ Pa} + (174 - 240)\text{Pa} = -26.6 \text{ Pa}$$

The static pressure has increased by 26.6 Pa, which indicates that the velocity has been reduced too much and a smaller-diameter duct should be used between b and c. We will next try a duct diameter of 140 cm. Repeating the calculations above:

$$A_b/A_a = (140 \text{ cm}/160 \text{ cm})^2 = 0.77, \qquad C_s = 0.13$$

$$\mathbf{V}_b = \mathbf{V}_c = 19 \text{ m/s}, \qquad P_{v,b} = 217 \text{ Pa}, \qquad \Delta P/L = 1.9 \text{ Pa/m}$$

$$\Delta P_{t,ab} = C_s P_{v,b} = (0.13)(217 \text{ Pa}) = 28.2 \text{ Pa}$$

$$\Delta P_{t,bc} = (10 \text{ m})(1.9 \text{ Pa/m}) = 19 \text{ Pa}$$

$$P_{s,a} - P_{s,c} = 28.2 \text{ Pa} + 19 \text{ Pa} + (217 - 240)\text{Pa} = 24.2 \text{ Pa}$$

This choice of duct diameter gives an increase in static pressure between sections a and c that is about the same magnitude as the loss for the first trial. Therefore, the choice of duct diameter between b and c that will provide the smallest change in static pressure is the size between these two, or 145 cm.

The previous example illustrates the procedure to be used when sizing ductwork using the static-regain method. The procedure is rather tedious to perform by hand. Several vendors have incorporated this method into computer software that makes the application much easier and reduces the likelihood of computational errors and of errors associated with estimating values such as with Fig. 18.18.

Optimization Methods

The duct-sizing methods discussed above rely on many years of practical experience and are consensus procedures that are widely used. Individual firms develop their own variations of these basic approaches to suit the local building-trades industry and to incorporate customized hardware. Because of the many variations in duct fabrication, labor costs, fitting design, fan-energy costs, and maintenance costs, no single duct-design method or sizing procedure will be best for all applications. One can consider optimizing a duct system for lowest first cost, lowest life-cycle energy costs, or some other objective function.

One such optimization method has been developed by Tsal et al. [12]. In this method, the present-worth owning and operating costs are minimized. Initial cost, first-year energy cost, and a present-worth escalation factor are used in the analysis. As with any economic calculation of this type, the results depend on the assumptions used and therefore may differ depending on who performs the analysis. As such, it is not an absolute optimum design but does consider trade-offs between initial and operating costs. References [9] and [12] provide more details concerning this particular optimization method.

ENDNOTES

1. J. Rydberg and P. Norback, "Air Distribution and Draft," ASHVE Research Report No. 1362. *ASHVE Transactions*, 55 (1949), 225.

2. AIVC Technical Note 28, "A Guide to Air Change Efficiency," Air Infiltration and Ventilation Centre, University of Warwick Science Park, February 1990.

3. *ASHRAE Handbook, Fundamentals Volume* (Atlanta: American Society of Heating, Refrigerating and Air Conditioning Engineers, 1993), ch. 31.

4. H. Liang, "Room Air Movement and Contaminant Transport," Ph.D. thesis, University of Minnesota, 1994.

5. J. Xu, "Numerical Studies of Ventilation in Rooms with Ceiling Air Supply and Return," M.S. Thesis, University of Minnesota, 1993.

6. J. Xu, H. Liang, and T. H. Kuehn, "Comparison of Numerical Predictions and Experimental Measurements of Ventilation in a Room," *Proceedings Roomvent '94*, Crakow, Poland, 1994, 213–227.

7. B. E. Lauder and D. B. Spalding, "The Numerical Computation of Turbulent Flows," *Computer Methods in Applied Mechanics and Engineering*, 3 (1974), 269–289.

8. C. K. G. Lam and K. Bremhorst, "A Modified Form of the k-e Model for Predicting Wall Turbulence," *J. Fluids Engineering, Transactions ASME*, 103 (1981), 456–460.

9. *ASHRAE Handbook, Fundamentals Volume* (Atlanta: American Society of Heating, Refrigerating and Air Conditioning Engineers, 1993), ch. 32.

10. *ASHRAE Duct Fitting Database* (Atlanta: American Society of Heating, Refrigerating and Air Conditioning Engineers, 1994).

11. *ASHRAE Handbook, Systems and Equipment* (Atlanta: American Society of Heating, Refrigerating and Air Conditioning Engineers, 1996), ch. 18.

12. R. J. Tsal, H. F. Behls, and R. Mangle, "T-Method Duct Design, Part I: Optimization Theory, Part II: Calculation Procedure and Economic Analysis," *ASHRAE Transactions*, 94, pt 2 (1988), 90–111.

PROBLEMS

18.1 Calculate the effective draft temperature (EDT) for a room where $t_r = 70$ °F, $t_x = 90$ °F, and $\mathbf{V}_x = 150$ ft/min.

18.2 Determine t_x for an effective draft temperature of 20 °C when $t_r = 25$ °C and $\mathbf{V}_x = 1$ m/s.

18.3 **(a)** Compute the local mean age and
(b) plot the contaminant concentration versus time of a perfectly mixed room when the room volume is 1600 ft³, the supply airflow rate is 320 ft³/min, and the initial concentration is 10 ppm. Assume the supply-air concentration is zero.

18.4 Plot the local mean age vs. time for a point in the center of the room for the conditions described in Prob. 18.3 if the room has ideal piston flow.

18.5 Derive the equations necessary to show that the ventilation efficiency for a perfectly mixed room is 50 percent and the value for piston flow is 100 percent.

18.6 Compute the Reynolds number of the supply-air jet shown in Fig. 18.11 and state whether this jet is laminar or turbulent. Assume that the air is at standard atmospheric pressure at a temperature of 27 °C (86 °F). Use the slot width as the length scale.

Determine the corresponding supply-air velocity in the full-scale room that is twice as large as the dimensions shown in Fig. 18.11.

18.7 Determine the air change rate for the scale model shown in Fig. 18.11 using the given air-inlet velocity. Determine the air change rate for the full-scale room that has the same inlet-jet Reynolds number.

18.8 Consider the conditions specified in Fig. 18.11. Assume the walls of the room are at 27 °C (80.6 °F) and that the supply air enters with a velocity of 1.016 m/s (200 ft/min) and is at a temperature of 0 °C (32 °F).

(a) Compute the Archimedes number for this room, using the supply-air slot width as the length scale.

(b) Determine the supply-air temperature of a full-scale room that is twice as large as the model shown in Fig. 18.11 if the Archimedes number is equal to the value from part (a) and the air change rate in the full-scale room is one-half the value in the model.

18.9 Compute the Reynolds number, Grashof number, and Archimedes number for a round air jet with an initial velocity of 0.25 m/s, an initial diameter of 0.4 m, and an intial temperature of 15 °C that enters into an ambient environment of still air at 20 °C. If this jet is discharged into the room in a horizontal direction as a free jet, sketch the direction the jet will travel.

18.10 Repeat Prob. 18.9 with the initial jet velocity at 0.1 m/s.

18.11 Compute the Reynolds number, Grashof number, and Archimedes number for a cylinder of 1 ft diam- eter. The forced flow is perpendicular to the cylinder axis with a velocity of 50 ft/min. The cylinder surface temperature is 85 °F and the room air temperature is 70 °F.

18.12 Repeat Ex. 18.3 using high sidewall diffusers and diffuser catalog information supplied by the instructor.

18.13 Repeat Ex. 18.3 using square ceiling diffusers from catalog information supplied by the instructor.

18.14 Using an available two-dimensional CFD or FEM code, simulate the airflow for the room shown in Fig. 18.11. Compare the results with the experimental data given in Fig. 18.12.

18.15 Repeat Prob. 18.14 but use a supply-air temperature 20 °C (36 °F) colder than the room walls [27 °C (80.6 °F)].

18.16 Consider the building floor plan shown in Fig. 18.32. This is a single-story commercial office building to be built in Dallas, Texas. The summer design cooling-load distribution for the various rooms is given in the figure. You may assume a sensible heat ratio of 0.8 for all rooms and assume a supply-air temperature of 60 °F and a return temperature of 75 °F. The floor is slab on grade and the walls are concrete block with face brick. The windows have tinted double glazing and the ceiling is suspended below a built-up flat roof. Interior walls do not extend to the roof except at the zone boundaries.

(a) Locate circular ceiling diffusers in the building to provide the necessary cooling. Show these on a floor plan of the structure.

(b) Select each diffuser from Table 18.5 or from catalog information supplied by the instructor.

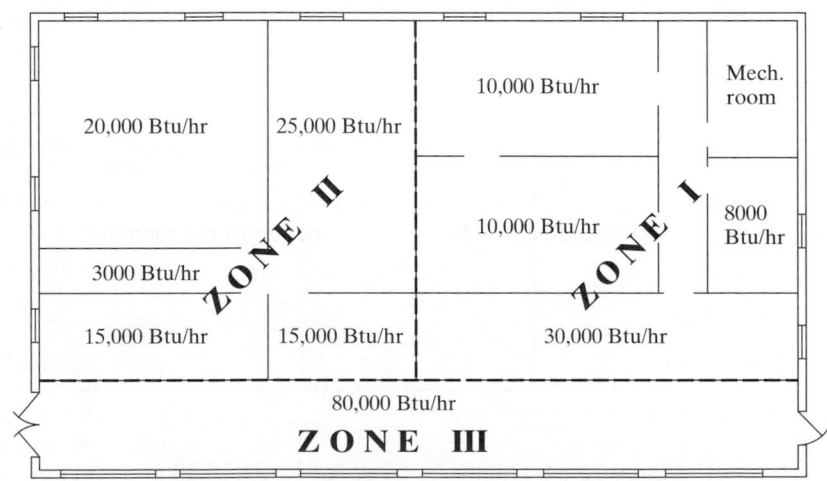

Figure 18.32 Building floor plan for Prob. 18.16.

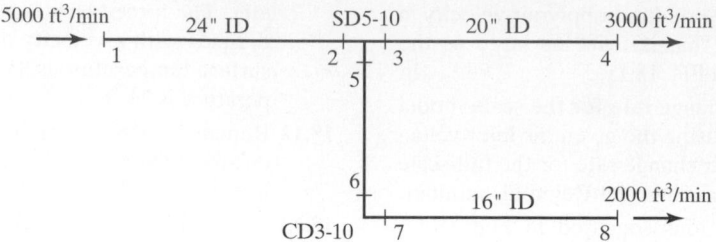

Figure 18.33 Duct schematic for Prob. 18.18.

Provide the following information for each: (i) size, (ii) neck velocity, (iii) volumetric flow rate, (iv) total pressure drop, (v) throw to 50 ft/min, and (vi) noise criteria.

18.17 Rework Ex. 18.7 with 200 L/s of air entering at section 0 and 800 L/s entering in the branch between sections 2 and 3.

18.18 Compute the total pressure at sections 4 and 8 in the duct shown in Fig. 18.33. Assume that the total pressure at section 1 is 10 in. of water.

18.19 Consider a duct of square cross section with a 90-deg elbow. Compare the loss coefficient using an elbow without turning vanes, CR3-1, vs. using an elbow with turning vanes, CR3-11. Comment on the choice of these two elbow designs.

18.20 Calculate the supply total pressure at sections 7, 11, and 15 for the rectangular duct system shown in

Fig. 18.34. Assume the fan discharge total pressure is 8 in. of water at section 1.

18.21 Compute the power required to operate a fan when the total pressure rise is 50 Pa, the volumetric flow rate is 5 m^3/sec, and the total efficiency is 75 percent.

18.22 Estimate the increase in pressure rise and power for a centrifugal fan if the pulleys are changed so that the flow rate increases by a factor of 1.4 over the previous value.

18.23 Estimate
   (a) the change in volumetric flow rate through a fan and
   (b) the change in power required from the rated values at standard air conditions if the temperature of the air entering the fan is -40 °C. Assume the pressure remains at standard sea-level pressure.

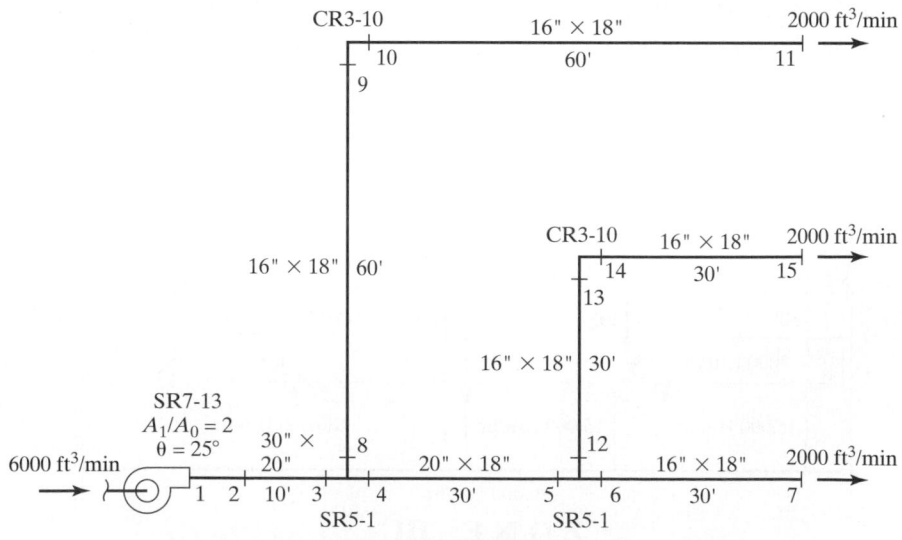

Figure 18.34 Duct schematic for Prob. 18.20.

18.24 Plot the total-pressure characteristic curve shown on Fig. 18.22 on a similar coordinate scale. Add the total-pressure characteristc curves when the fan speed is

(a) twice the original speed and

(b) one-half the original speed.

18.25 Size the supply-duct system shown in Fig. 18.35 using the equal-friction method and round duct. Specify the total pressure needed at location 1 for this system if the diffusers both have a total presure loss of 0.05 in. of water.

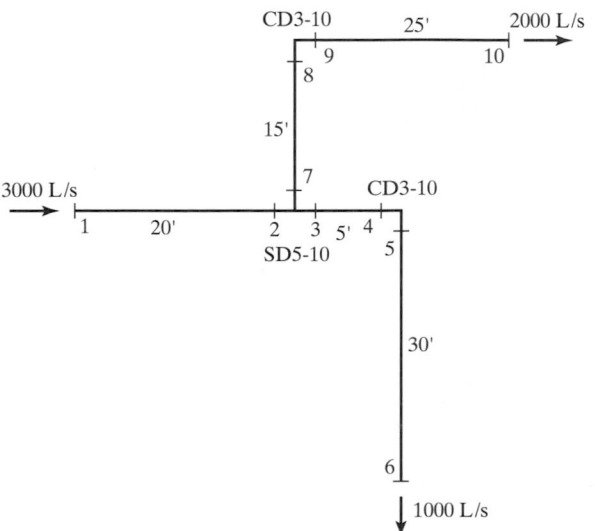

Figure 18.35 Supply-duct schematic for Prob. 18.25.

18.26 Rework Ex. 18.9 but use equivalent lengths for the fitting losses rather than the loss coefficients.

18.27 Size the ductwork in the system shown in Fig. 18.36 using the equal-friction method and rectangular sheet metal duct. The maximum dimension of the duct is limited to 18 in. Check the total-pressure drop from each inlet to the fan to ensure a balanced system, and make adjustments in duct sizes if required.

18.28 Size the ductwork shown in Fig. 18.37 using the equal-friction method and rectangular sheet metal duct. Determine the total-pressure rise across the fan needed if both return-grille pressure drops are 0.06 in. of water and all supply diffusers have a total-pressure drop of 0.04 in. of water.

18.29 Using the diffusers selected in Prob. 18.16 for the building shown in Fig. 18.32, lay out the supply and return ductwork on a floor plan of the building. Assume round duct and use fittings from Tables 18.6–18.15. Use the three zones as shown in Fig. 18.32. The three air-handling units are located in the mechanical room shown in the figure. Explain your reasoning for your duct layout.

18.30 Size the duct system you designed in Prob. 18.29 using the equal-friction method. Ensure that the pressure drops from each fan through each diffuser are nearly equal. Provide the specifications for the three fans in terms of required airflow rate and the required total-pressure rise for each.

18.31 Size the main branch of the round supply-duct system shown in Fig. 18.38 using the static-regain method. Assume the inlet is an ED1-3 entrance fit-

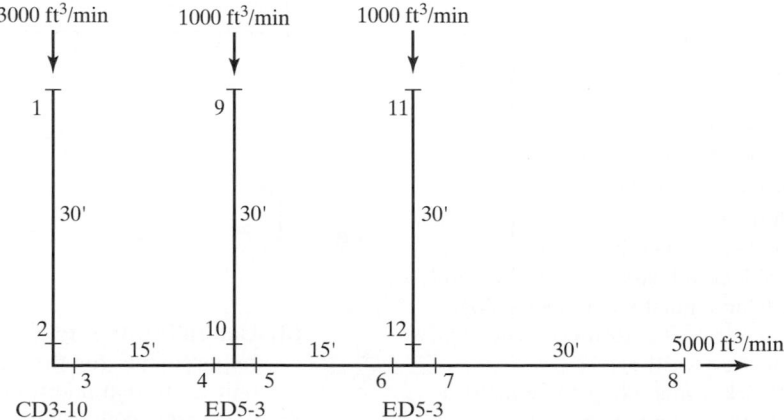

Figure 18.36 Return-duct schematic for Prob. 18.27.

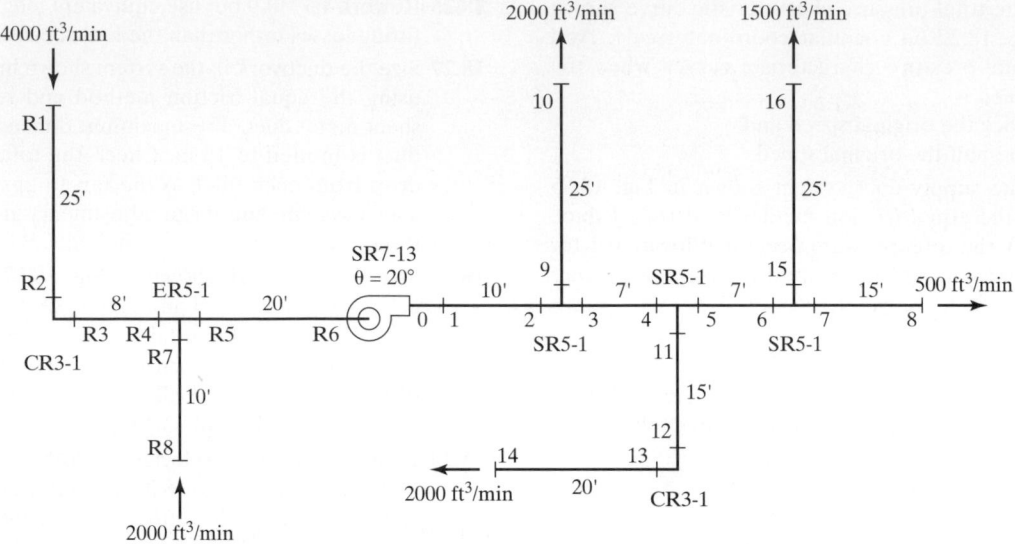

Figure 18.37 Duct-system schematic for Prob. 18.30.

ting with $r/D = 0.08$ and that each of the tees is an SD5-10 fitting. Once the duct has been sized, determine the total-pressure drop between the plenum and the terminal box.

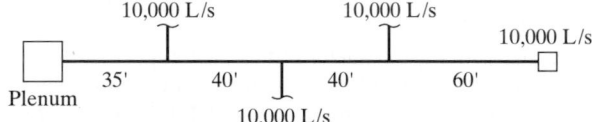

Figure 18.38 Duct-system schematic for Prob. 18.31.

18.32 Size the duct system shown in Fig. 18.39 using round duct and the static-regain method. Each terminal box requires 5000 ft³/min and has a pressure requirement of 0.1 in. of water. The entrance fitting is ED1-3 with $r/D = 0.1$, all elbows are CD3-10 fittings, and all tees are SD5-10 fittings. Determine the total pressure required in the plenum.

18.33 Obtain a set of floor plans of a building from your instructor. Use the heating/cooling loads provided.
   (a) Locate and select all supply diffusers and return grilles from a manufacturer's catalog.
   (b) Lay out all supply and return ductwork and indicate the necessary fittings.
   (c) Size the ductwork, using either the equal-friction method or the static-regain method.

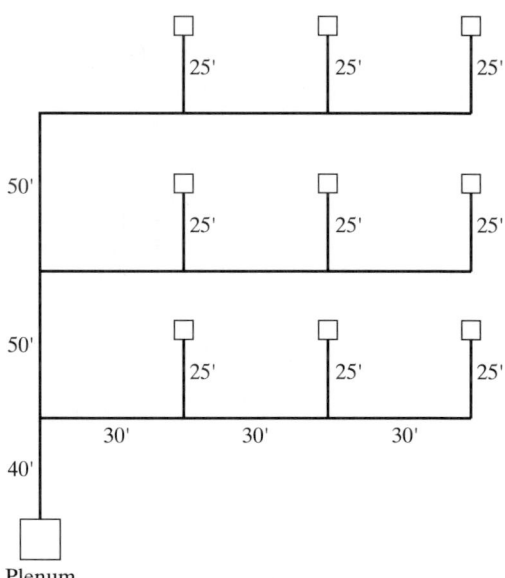

Figure 18.39 Duct-system schematic for Prob. 18.32.

   (d) Determine the total airflow and pressure requirements for the fan in the air-handling unit (note that additional pressure drops will occur from coils, filters, and dampers).

SYMBOLS

A	Flow cross-sectional area, ft^2 (m^2).
a	Major dimension of rectangular or oval duct, ft (m).
ACH	Air Changes per Hour, hr^{-1}.
ADPI	Air Diffusion Performance Index.
Ar	Archimedes number.
b	Minor dimension of rectangular or oval duct, ft (m).
C	Loss coefficient.
$C(t)$	Concentration of tracer at time t.
D	Diameter of round duct, ft (m).
D_e	Equivalent diameter of round duct, ft (m).
D_h	Hydraulic diameter, ft (m).
EDT	Effective Draft Temperature, °F (°C).
f	Friction factor.
g	Gravitational acceleration, ft/sec (m/s).
Gr	Grashof number.
H	Height of rectangular duct, ft (m).
ℓ	Head difference, ft (m).
L	Length, ft (m).
M	Empirical constant.
$\dot{m}$	Mass flow rate, lbm/sec (kg/s).
N	Rotation speed, rpm.
p	Wetted perimeter, ft (m).
P_s	Static pressure, in. water (Pa).
P_t	Total pressure, in. water (Pa).
P_v	Velocity pressure, in. water (Pa).
$\dot{Q}_{sen}$	Sensible heat load, Btu/hr (kW).
$\dot{Q}_{tot}$	Total heat load, Btu/hr (kW).
r	Radius of elbow, ft (m).
Re_L	Reynolds number based on dimension L.
T	Throw distance of jet, ft (m).
t	Time, hr.
t_r	Average room dry-bulb temperature, °F (°C).
t_x	Local air stream dry-bulb temperature, °F (°C).
V	Volume, ft^3 (m^3).
$\mathbf{V}$	Velocity, ft/min (m/s).
$\mathbf{V}_r$	Reference air velocity, ft/min (m/s).
$\mathbf{V}_x$	Local air velocity, ft/min (m/s).
$\dot{V}$	Volumetric flow rate, ft^3/min (m^3/s).
W	Rectangular duct width, ft (m).
$\dot{W}_s$	Power based on static pressure rise, hp (kW).
$\dot{W}_{sh}$	Shaft power, hp (kW).
$\dot{W}_t$	Power based on total pressure rise, hp (kW).

Greek Letters

β Coefficient of thermal volumetric expansion, $°F^{-1}$ ($°C^{-1}$).

Δh Change in enthalpy, Btu/lbm_a (kJ/kg_a).

ΔP_s Change in static pressure, in. water (Pa).

ΔP_t Change in total pressure, in. water (Pa).

Δt Change in temperature, $°F$ ($°C$).

ε_a Air-change efficiency.

η_s Fan efficiency based on static pressure rise.

η_t Fan efficiency based on total pressure rise.

μ Air dynamic viscosity, lbm/sec ft (kg/m sec).

τ_n Nominal time constant, hr.

$\bar{\tau}_p$ Local mean age of air at point p, hr.

$\langle \bar{\tau} \rangle$ Mean age of air in room, hr.

Subscripts

b	Branch
c	Common
e	Exhaust
F	Full scale
L	Characteristic length
M	Model
P	Point in room
s	Straight

chapter 19
Hydronic System Operation and Design

19.1 INTRODUCTION

One of the main methods to transport thermal energy into, out of, and within a building is with water. Water has a high specific heat and is inexpensive, nontoxic, and readily available. It is a nearly ideal energy-transport fluid for temperatures between its freezing point and boiling point. The temperature range can be extended by adding a component such as ethylene or propylene glycol. It is used to transport thermal energy from a chiller condenser to a cooling tower, from cooling and dehumidifying coils to the evaporator on a chiller, from a boiler to heating coils, and from a boiler or steam-heat exchanger to hot-water baseboard-heating elements.

The purpose of this chapter is to describe the basic principles that apply to hydronic-system operation and to discuss basic design methods. Other types of piping systems follow many of the same principles of hydronic systems. Examples include hot and cold service water, steam, and refrigerant piping. The *ASHRAE Handbook of Fundamentals* [1] discusses these other types of piping systems. We will discuss piping-system pressure drop-flow relationships, centrifugal-pump performance, system balance point and flow control, and the need for expansion tanks, strainers, and thermal expansion loops.

19.2 BASIC PRINCIPLES OF HYDRONIC SYSTEMS

We will first consider the flow-pressure drop relationship in the piping itself. We begin with Eq. (2.97):

$$\underbrace{\frac{w_{\text{actual}_{1-2}}}{g} = \Delta H_{1-2}}_{\substack{\text{total} \\ \text{head}}} = \underbrace{\frac{P_2 - P_1}{g\rho}}_{\substack{\text{pressure} \\ \text{head}}} + \underbrace{\frac{\mathbf{V}_2^2 - \mathbf{V}_1^2}{2g}}_{\substack{\text{velocity} \\ \text{head}}} + \underbrace{(z_2 - z_1)}_{\substack{\text{elevation} \\ \text{head}}} + \underbrace{\frac{w_{\text{loss}_{1-2}}}{g}}_{\substack{\text{lost} \\ \text{head}}} \qquad (2.97)$$

In hydronic systems we are almost always concerned with total pressure drop or total head loss rather than static pressure or head loss. Therefore, it is convenient to combine the terms for pressure head (static pressure) and velocity head into a total head term:

$$\frac{w_{\text{actual}_{1-2}}}{g} = \Delta H_{1-2} = \frac{P_{t,2} - P_{t,1}}{g\rho} + (z_2 - z_1) + \frac{w_{\text{loss}_{1-2}}}{g}$$

The piping alone does not add work to the fluid, so the left-hand term is zero. If we write the last term on the right-hand side as the flow head loss, ℓ_f, and rearrange this equation, we have

$$\Delta h_{1-2} = \frac{P_{t,1} - P_{t,2}}{g\rho} = (z_2 - z_1) + \ell_f \tag{19.1}$$

where Δh_{1-2} is the total head loss between locations 1 and 2 in the pipe, $z_2 - z_1$ is the change in elevation between locations 1 and 2, and ℓ_f is the flow head loss between 1 and 2. From Eq. (2.98) the flow head loss for a straight pipe is proportional to the square of the mean velocity in the pipe or the square of the volumetric flow rate:

$$\ell_f = f \frac{L}{D} \frac{\mathbf{V}^2}{2g} \propto \mathbf{V}^2 \quad \text{or} \quad \propto \dot{V}^2 \tag{19.2}$$

The same proportionality holds for head loss in fittings. Combining Eqs. (19.1) and (19.2) yields

$$\Delta h_{1-2} = (z_2 - z_1) + b_{1-2}\dot{V}^2 \tag{19.3}$$

where b_{1-2} is a constant that depends on the size and length of pipe between locations 1 and 2 and the number and type of fittings between 1 and 2.

We will consider two basic types of piping systems: open-loop systems and closed loop systems. An *open-loop system* is one in which part of the loop is in direct contact with air. Examples include pumping water from a pond into a storage tank and pumping water from the sump of a cooling tower to the top of the tower. The pressure at one point in the system must equal atmospheric pressure. In all *closed-loop systems* the fluid leaving the pump returns to the suction side of the pump without leaving the piping system. A closed-loop system can be pressurized. Examples of closed-loop systems include hot water circulating between a heating coil and a boiler and chilled water pumped between the evaporator on a chiller and a cooling coil.

For a closed-loop system there is no elevation change to consider, so the change in z shown in Eq. (19.3) is zero. This can be thought of as the elevation difference between the inlet and the discharge of the pump, which is usually negligible.

For open loop systems, however, there is a change in elevation between the inlet and the discharge of the pipe, so the change in z is not zero and is equal to the vertical elevation difference between the pipe inlet and discharge. Figure 19.1 shows a system characteristic curve for an open-loop piping system. Note that when the flow rate through the system is zero, the head is not zero but is equal to the elevation head. The curve can be represented by the quadratic equation given on the figure. System characteristic curves for closed-loop systems are the same as shown in Fig. 19.1 except that the elevation head is zero.

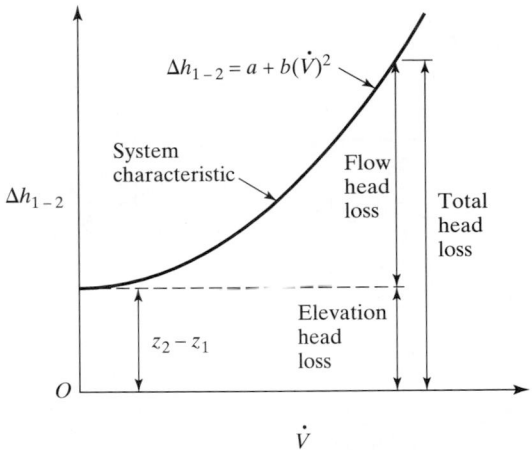

Figure 19.1 Typical characteristic curve for an open-loop hydronic system.

Head Losses in Straight Pipe

Flow head losses have been measured for common types of piping materials and sizes. Figure 19.2 shows the flow head loss vs. volumetric flow rate for water at 60 °F or 20 °C flowing in schedule-40 steel pipe. Figure 19.3 shows similar information for types K, L, and M copper tubing. Note that when the elevation head loss is zero in Eq. (19.3), the log of the head loss is linearly proportional to the log of the flow rate with a slope of 2 and an additive constant b. The head loss is given in terms of ft water/100 ft straight pipe (Pa/m straight pipe). Actual dimensions of schedule-40 steel pipe and type-L copper tubing are given in Tables 19.1 and 19.2, respectively.

EXAMPLE 19.1

Determine the flow head loss in Pa and the mean velocity for 8 L/s of water flowing through 20 m of 50-mm type-L copper tubing.

Solution: We can use Fig. 19.3 to determine the solution, if we assume that the water's properties do not differ very much from those at 20 °C used to develop the figure. The intersection of the vertical line representing 8 L/s and the line with the positive slope corresponding to 50-mm-size type-L tubing occurs near the top of the figure to the right of center. The horizontal line passing through this intersection point corresponds to about 2500 Pa/m. The total head loss then becomes:

$$\Delta h_{1-2} = 2500 \text{ Pa/m (20m)} = 50{,}000 \text{ Pa}$$

The velocity line passing through the intersection point on Fig. 19.3 is approximately 4 m/s.

Another method of solution would be to determine the velocity based on the actual pipe inner cross-sectional area from Table 19.2 and the given volumetric flow rate. Then

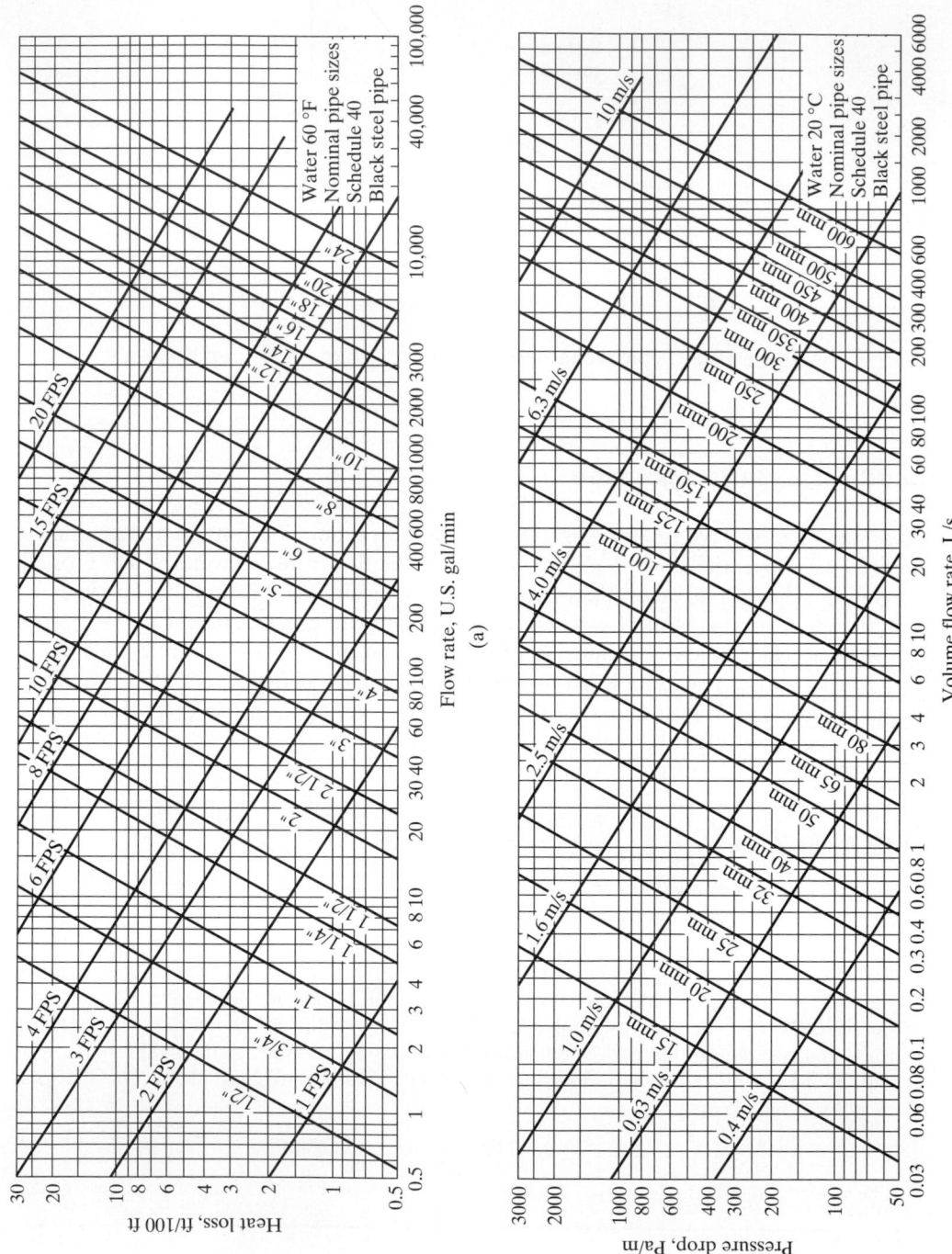

Figure 19.2 Friction loss for water in commercial steel pipe (schedule 40). [Reprinted by permission from *ASHRAE Fundamentals 1993* (IP & SI), p. 33.5.]

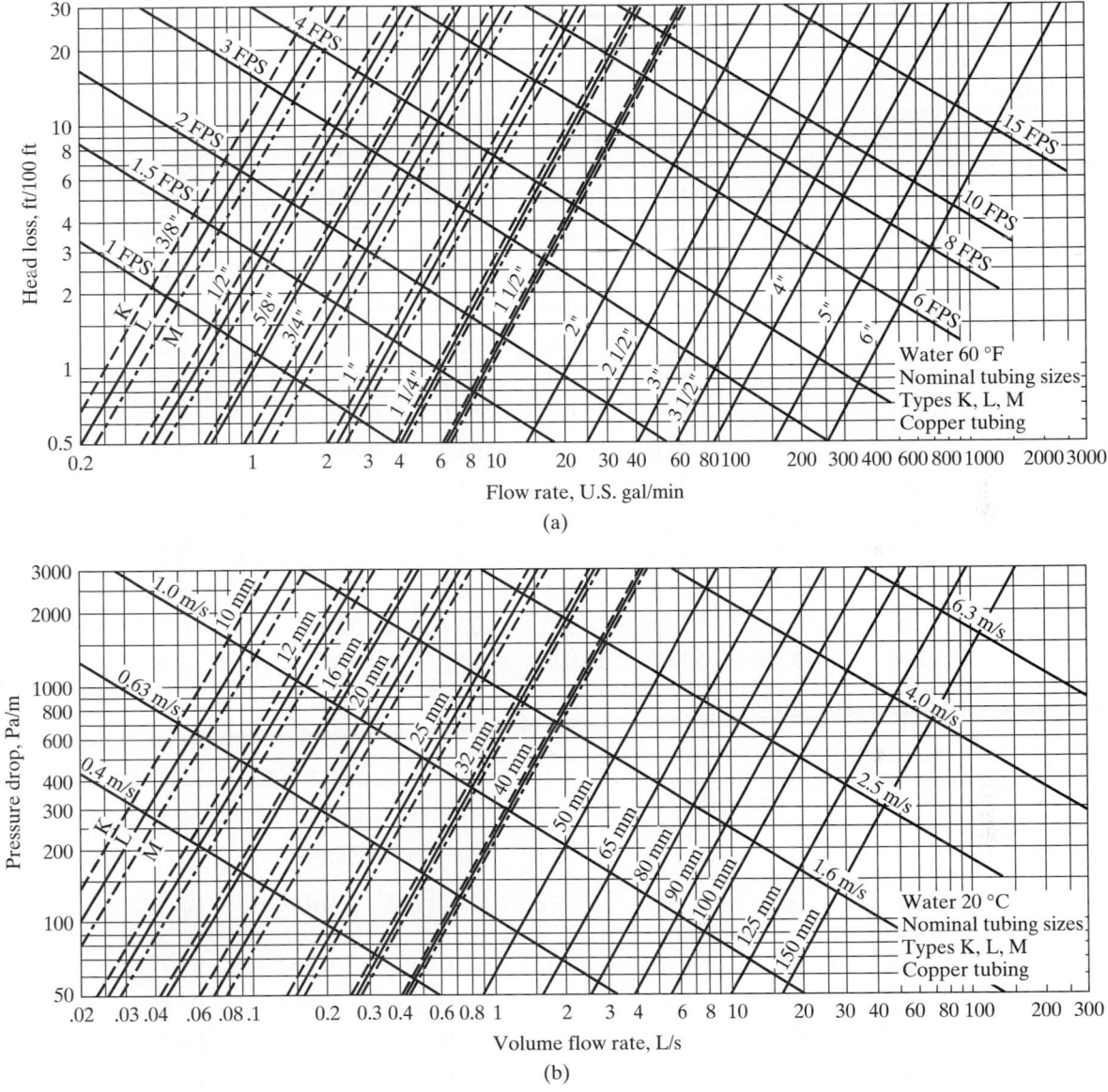

Figure 19.3 Friction loss for water in copper tubing (types K, L, M). [Reprinted by permission from *ASHRAE Fundamentals 1993* (IP & SI), p. 33.5.]

estimate the roughness of the inside pipe surface, determine the Reynolds number of the water, and estimate the friction factor. The flow head loss could then be computed from Eq. (19.2). Although the figures do not provide extremely accurate information, they can be used to quickly obtain approximate solutions that are sufficiently accurate for most design purposes.

TABLE 19.1 Schedule-40 Steel Pipe Dimensions—English and SI Units

Nominal Pipe Size, in.	Schedule Number	Diameter				Wall Thickness		Inside Cross-Sectional Area	
		O.D.		I.D.					
		in.	mm	in.	mm	in.	mm	ft²	m² × 10³
$\frac{1}{4}$	40	0.540	13.7	0.364	9.25	0.088	2.23	0.00072	0.067
$\frac{3}{8}$	40	0.675	17.1	0.493	12.5	0.091	2.31	0.00133	0.124
$\frac{1}{2}$	40	0.840	21.3	0.622	15.8	0.109	2.77	0.00211	0.196
$\frac{3}{4}$	40	1.050	26.7	0.824	20.9	0.113	2.87	0.00371	0.345
1	40	1.315	33.4	1.049	26.6	0.133	3.38	0.00600	0.557
$1\frac{1}{2}$	40	1.900	48.3	1.610	40.9	0.145	3.68	0.01414	1.314
2	40	2.375	60.3	2.067	52.5	0.154	3.91	0.02330	2.165
$2\frac{1}{2}$	40	2.875	73.0	2.469	62.7	0.203	5.16	0.03322	3.086
3	40	3.500	88.9	3.068	77.9	0.216	5.49	0.05130	4.766
4	40	4.500	114.3	4.026	102.3	0.237	6.02	0.08840	8.213
5	40	5.563	141.3	5.047	128.1	0.258	6.55	0.1390	12.91
6	40	6.625	168.3	6.065	154.1	0.280	7.11	0.2006	18.64
8	40	8.625	219.1	7.981	202.7	0.322	8.18	0.3474	32.28
10	40	10.75	273.1	10.020	254.5	0.365	9.27	0.5475	50.86

SOURCE: Adapted from *A.S.A. Standards B36.10.*

TABLE 19.2 Type-L Copper Tube Dimensions—English and SI Units

Nominal Tube Size, in.	Diameter				Wall Thickness		Inside Cross-Sectional Area	
	O.D.		I.D.					
	in.	mm	in.	mm	in.	mm	in.²	m² × 10³
$\frac{1}{4}$	0.375	9.53	0.315	8.00	0.030	0.762	0.0779	0.0503
$\frac{3}{8}$	0.500	12.7	0.43	11	0.035	0.889	0.145	0.094
$\frac{1}{2}$	0.625	15.9	0.545	13.8	0.040	1.02	0.233	0.150
$\frac{5}{8}$	0.750	19.1	0.666	16.9	0.042	1.07	0.348	0.225
$\frac{3}{4}$	0.875	22.2	0.785	19.9	0.045	1.14	0.484	0.312
1	1.125	28.58	1.025	26.04	0.050	1.27	0.825	0.532
$1\frac{1}{4}$	1.375	34.93	1.265	32.13	0.055	1.40	1.26	0.813
$1\frac{1}{2}$	1.625	41.28	1.505	38.23	0.060	1.52	1.78	1.15
2	2.125	53.98	1.985	50.42	0.070	1.78	3.10	2.00
$2\frac{1}{2}$	2.625	66.68	2.465	62.61	0.080	2.03	4.77	3.08
3	3.125	79.38	2.945	74.80	0.090	2.29	6.81	4.39
$3\frac{1}{2}$	3.625	92.08	3.425	87.00	0.100	2.54	9.21	5.94
4	4.125	104.8	3.905	99.19	0.110	2.79	12.0	7.74

SOURCE: Based on *ASTM B-88.*

Head Losses in Fittings

Three methods are commonly used to determine head loss through fittings: (a) loss coefficients (k), (b) equivalent lengths (L_{eq}), and (c) flow coefficients (C_v). We will describe each of these in detail.

The *loss-coefficient method* uses the following equation:

$$\Delta h_{1-2} = k \left(\frac{\mathbf{V}^2}{2g} \right) \qquad (19.4)$$

Values for k are given in Tables 19.3 and 19.4 for screwed and flanged welded pipe fittings, respectively. Note that the loss coefficients generally become smaller as the pipe size increases. Also note that the losses associated with globe valves are very large compared to those for other types of valves. It is important to know what type of valve is being specified when determining the head loss.

EXAMPLE 19.2

Compute the head loss in ft water through a gate valve screwed into a 2-in. pipe. The flow rate of water is 60 gal/min through the schedule-40 steel pipe.

Solution: We will use Eq. (19.4) to compute the head loss. The loss coefficient is read from Table 19.3 as $k = 0.17$. The water velocity is read from Fig. 19.2(a) as about 5.8 ft/s. Substituting into Eq. (19.4),

$$\Delta h_{1-2} = 0.17 \left[\frac{(5.8 \text{ ft/s})^2}{2(32.2 \text{ ft/s}^2)} \right] = 0.089 \text{ ft water}$$

The second method uses the concept of an *equivalent length*. In this method, the head loss through a fitting is equal to the head loss through a section of straight pipe of length L_{eq} with the same nominal pipe size as the fitting. This length of straight pipe is called the "equivalent length" of the fitting. With this method, the head loss through a fitting is obtained by knowing the equivalent length of the fitting and using Figs. 19.2 and 19.3 for the head loss per unit length of straight pipe. The loss-coefficient and equivalent-length methods are related. Looking at Eqs. (19.2) and (19.4), it can be determined that $k = f(L_{eq}/2g)$ or $L_{eq} = 2gk/f$.

EXAMPLE 19.3

Compute the head loss through a fitting whose equivalent length is given as 12. The fitting is installed into an 80-mm schedule-40 steel pipe with water flowing at 10 L/s.

Solution: We must first determine the head loss per unit length of straight pipe. From Fig. 19.2(b) we read a head loss of about 550 Pa/m. From Table 19.1 we can determine the actual inner diameter of the pipe, which is equivalent to a nominal 3-in. pipe. The I.D. is 77.9 mm. The total equivalent length of straight pipe is then 12×77.9 mm = 935 mm. The total head loss is then determined:

$$\Delta h_{1-2} = (550 \text{ Pa/m})(935 \text{ mm})/(1000 \text{ mm/m}) = 514 \text{ Pa}$$

TABLE 19.3 k-Values—Screwed Pipe Fittings

Nominal Pipe Diameter		90° Ell Reg.	90° Ell Long	45° Ell	Ret. Bend	Tee-Line	Tee-Branch	Globe Valve	Gate Valve	Angle Valve	Swing Check Valve	Bell-Mouth Inlet	Square Inlet	Proj. Inlet
in.	mm													
3/8	10	2.5	—	0.38	2.5	0.90	2.7	20	0.40	—	8.0	0.05	0.5	1.0
1/2	15	2.1	—	0.37	2.1	0.90	2.4	14	0.33	—	5.5	0.05	0.5	1.0
3/4	20	1.7	0.92	0.35	1.7	0.90	2.1	10	0.28	6.1	3.7	0.05	0.5	1.0
1	25	1.5	0.78	0.34	1.5	0.90	1.8	9	0.24	4.6	3.0	0.05	0.5	1.0
1-1/4	32	1.3	0.65	0.33	1.3	0.90	1.7	8.5	0.22	3.6	2.7	0.05	0.5	1.0
1-1/2	40	1.2	0.54	0.32	1.2	0.90	1.6	8	0.19	2.9	2.5	0.05	0.5	1.0
2	50	1.0	0.42	0.31	1.0	0.90	1.4	7	0.17	2.1	2.3	0.05	0.5	1.0
2-1/2	65	0.85	0.35	0.30	0.85	0.90	1.3	6.5	0.16	1.6	2.2	0.05	0.5	1.0
3	80	0.80	0.31	0.29	0.80	0.90	1.2	6	0.14	1.3	2.1	0.05	0.5	1.0
4	100	0.70	0.24	0.28	0.70	0.90	1.1	5.7	0.12	1.0	2.0	0.05	0.5	1.0

SOURCE: Reprinted by permission from *ASHRAE Fundamentals 1993* (IP & SI), p. 33.2.

TABLE 19.4 k-Values—Flanged Welded Pipe Fittings

Nominal Pipe Diameter		90° Ell Reg.	90° Ell Long	45° Ell Long	Ret. Bend Reg.	Ret. Bend Long	Tee-Line	Tee-Branch	Globe Valve	Gate Valve	Angle Valve	Swing Check Valve
in.	mm											
1	25	0.43	0.41	0.22	0.43	0.43	0.26	1.0	13	—	4.8	2.0
1-1/4	32	0.41	0.37	0.22	0.41	0.38	0.25	0.95	12	—	3.7	2.0
1-1/2	40	0.40	0.35	0.21	0.40	0.35	0.23	0.90	10	—	3.0	2.0
2	50	0.38	0.30	0.20	0.38	0.30	0.20	0.84	9	0.34	2.5	2.0
2-1/2	65	0.35	0.28	0.19	0.35	0.27	0.18	0.79	8	0.27	2.3	2.0
3	80	0.34	0.25	0.18	0.34	0.25	0.17	0.76	7	0.22	2.2	2.0
4	100	0.31	0.22	0.18	0.31	0.22	0.15	0.70	6.5	0.16	2.1	2.0
6	150	0.29	0.18	0.17	0.29	0.18	0.12	0.62	6	0.10	2.1	2.0
8	200	0.27	0.16	0.17	0.27	0.15	0.10	0.58	5.7	0.08	2.1	2.0
10	250	0.25	0.14	0.16	0.25	0.14	0.09	0.53	5.7	0.06	2.1	2.0
12	300	0.24	0.13	0.16	0.24	0.13	0.08	0.50	5.7	0.05	2.1	2.0

Source: Reprinted by permission from *ASHRAE Fundamentals 1993* (IP & SI), p. 33.2.

The third method of computing head loss through a fitting uses a *flow coefficient*, C_v. This coefficient is equal to the volumetric flow rate passing through the fitting required to produce a standard head loss. Therefore, this coefficient is not dimensionless but has units of gpm or L/s. The standard head loss is set at 1 lbf/in^2 (2.31 ft water) in English units and 1 kPa in SI units. The basic procedure uses the following equation:

$$\Delta h = \Delta h_{std}\left(\frac{\dot{V}}{\dot{V}_{std}}\right)^2$$

Writing this in English units,

$$\Delta h = 2.31 \text{ ft water}\left(\frac{\dot{V}}{C_v}\right)^2 \tag{19.5a}$$

where the flow rate has units of gpm and the head loss has units of ft water. In SI units this becomes

$$\Delta h = 1 \text{ kPa}\left(\frac{\dot{V}}{C_v}\right)^2 \tag{19.5b}$$

where the flow rate has units of L/s and the head loss is in kPa.

EXAMPLE 19.4

Determine the flow coefficient for the fitting in the previous example.

Solution: Substituting into Eq. (19.5b),

$$514 \text{ Pa} = 1 \text{ kPa}\left(\frac{10 \text{ L/s}}{C_v}\right)^2$$

and, solving for C_v, $C_v = 13.9$ L/s.

Centrifugal Pumps

The device usually used to circulate the fluid in hydronic systems is a *centrifugal pump*. This type of pump is a turbomachine and is not positive displacement. Therefore, although the pump may be operating at a constant rotational speed, the flow through the pump and the head produced by the pump will vary as other parts of the system change. Figure 19.4 illustrates a cross section through the impeller and volute casing of a typical centrifugal pump. Fluid enters the eye of the impeller and is given increased velocity in both the radial and angular directions because of the rotation of the impeller. The purpose of the volute is to convert as much kinetic energy as possible into increased pressure. Note that there are no valves on this type of pump. Flow can occur through the pump in either direction when the pump is not running. A check valve is often placed downstream of the pump in applications where two or more pumps may be installed to prevent backflow through a nonoperable pump. The pump is most often operated by an electric motor. The motor shaft itself can act as the pump shaft, or the pump and motor shafts can be connected with a flexible connection. Pump speed can be changed by using a two-speed motor or a motor connected to a variable-frequency power supply.

Part VI / Air- and Water-Distribution System Design

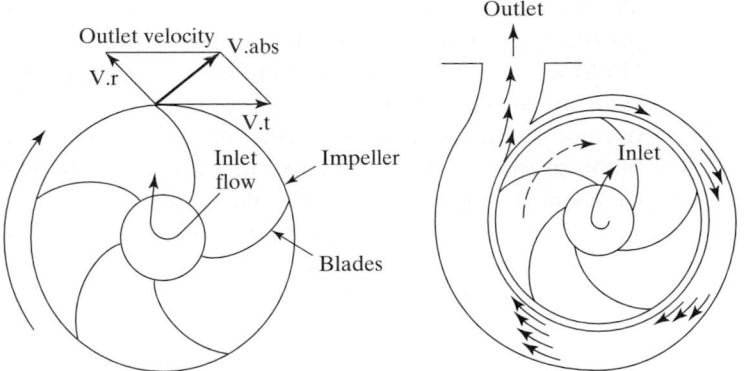

Figure 19.4 Impeller and volute interaction. [Reprinted by permission from *ASHRAE Systems and Equipment 1996* (IP & SI), p. 38.2.]

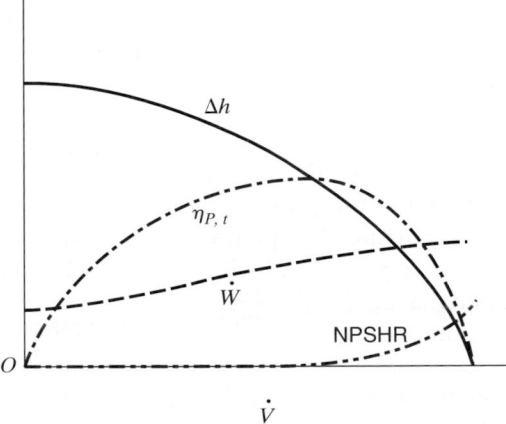

Figure 19.5 Performance curves for a typical centrifugal pump operating at constant speed.

The performance of a typical centrifugal pump operating at constant speed is shown in Fig. 19.5. When the volumetric flow rate through the pump is zero, the head across the pump is usually at its maximum. The head monatonically decreases as the flow rate increases, until it reaches zero at the maximum flow rate possible through the pump. The head-vs.-flow-rate curve is called the "pump characteristic curve" or simply the "pump characteristic." The shaft power required to operate the pump is not zero when the flow rate through the pump is zero. The power is used to stir the fluid in the pump and is all converted into viscous heating. The power increases as the flow rate though the pump increases in a nearly linear manner.

The pump efficiency is defined as the ratio of the minimum work required for the given pressure rise to the total shaft work supplied to operate the pump. The efficiency can also be written in terms of volumetric flow rate and shaft power. Two pump efficiencies are commonly used. The most common is *total efficiency*, which is based on the total pressure rise of the fluid flowing through the pump:

$$\eta_{p,t} \equiv \frac{w_{1-2,\,\text{rev}}}{w_{1-2,\,\text{actual}}} = \frac{v\,\Delta P_{t,\,1-2}}{w_{1-2,\,\text{actual}}} = \frac{\dot{V}\,\Delta P_{t,\,1-2}}{\dot{W}_{1-2,\,\text{actual}}} \qquad (19.6)$$

The second is *static efficiency*, which is defined in a similar manner except that the static pressure rise is used rather than the total pressure. By examining the last term in Eq. (19.6) we can determine limits on the pump efficiency. At zero flow rate the pump efficiency is zero, because the numerator in the equation is zero, but there is a finite amount of shaft power required to operate the pump. As the head or total pressure rise across the pump drops to zero at the maximum flow rate, the numerator is again zero, so the efficiency also drops to zero at the maximum pump flow rate. The efficiency curve plotted on Fig. 19.5 illustrates this behavior. The pump efficiency reaches a maximum value at a particular flow rate. The pump should be designed to operate near this point of maximum efficiency for minimum power consumption.

Another important parameter for pump selection is the *Net Positive Suction Head Required (NPSHR)*. The static pressure of the fluid in the impeller can decrease to very low levels as the fluid is accelerated. If the static pressure anywhere near the impeller surface becomes less than the saturation pressure of the liquid at its temperature in the pump, the liquid will begin to form vapor bubbles. The vapor bubbles are not detrimental themselves, except that they reduce the density of the fluid passing through the pump and therefore change the pump's performance. The main difficulty is that the bubbles will subsequently reach an area of higher static pressure and will collapse. Collapsing bubbles create extremely high temperatures in the center, several thousands of degrees, and strong shock waves in the liquid that can damage surrounding surfaces. Collapsing bubbles will damage the impeller of a pump by eroding the surface of the blades. Therefore, it is important to provide enough static pressure at the pump inlet to prevent cavitation from occurring. The pump manufacturer will determine the static pressure required to prevent cavitation and will often provide this information in the form of a curve, such as the one labeled "NPSHR" on Fig. 19.5. At low flow rates the requirement is modest, but it becomes much more significant at higher flow rates, where the velocities in the pump are large and the difference between static and total pressure becomes large. The *Net Positive Suction Head Available (NPSHA)* in an application must be larger than the NPSHR as provided by the pump manufacturer's data. The value for NPSHA can be determined from the following equation:

$$\text{NPSHA} \equiv \frac{P_{t,\,\text{suction}} - P_{\text{sat}}(t_{\text{suction}})}{\rho g}$$

$$= \frac{P_{\text{suction}}}{\rho g} + \frac{\mathbf{V}_{\text{suction}}^2}{2g} - \frac{P_{\text{sat}}(t_{\text{suction}})}{\rho g} \tag{19.7}$$

EXAMPLE 19.5

Determine the net positive suction head available for the following application. The pump inlet is located 3 ft below the surface of a cooling pond, the total pressure loss from the inlet in the pond to the suction side of the pump is 5 ft water, and the temperature of the water entering the pump is 85 °F.

Solution: Use Eq. (19.7) to determine the NPSHA. The total pressure at the pump suction is atmospheric pressure at the top of the cooling pond (assumed to be standard sea-level pressure) + (the increase due to the pump suction being 3 ft below the surface of the water) − (the 5-ft head loss from the inlet to the pump). The specific volume of water at 85 °F is obtained from Table A.1E.

$$P_{t, \text{suction}} = 14.696 \text{ psia} + \frac{(3 \text{ ft water})(32.2 \text{ ft/sec}^2)}{0.01609 \text{ ft}^3/\text{lbm}} \left(\frac{\text{lbf} \cdot \text{sec}^2}{32.2 \text{ ft} \cdot \text{lbm}}\right)\left(\frac{\text{ft}^2}{144 \text{ in.}^2}\right)$$

$$- \frac{(5 \text{ ft water})(32.2 \text{ ft/sec}^2)}{0.01609 \text{ ft}^3/\text{lbm}} \left(\frac{\text{lbf} \cdot \text{sec}^2}{32.2 \text{ ft} \cdot \text{lbm}}\right)\left(\frac{\text{ft}^2}{144 \text{ in.}^2}\right) = 13.83 \text{ psia}$$

The saturation pressure of water at 85 °F is obtained from Table A.1E, $P_{\text{sat}}(85) = 0.59646$ psia. The NPSHA is then found from Eq. (19.7):

$$\text{NPSHA} = \frac{(13.83 \text{ psia} - 0.59646 \text{ psia})(0.01609 \text{ ft}^3/\text{lbm})}{32.2 \text{ ft/sec}^2}\left(\frac{32.2 \text{ ft} \cdot \text{lbm}}{\text{lbf} \cdot \text{sec}^2}\right)\left(\frac{144 \text{ in.}^2}{\text{ft}^2}\right)$$

$$= 30.66 \text{ ft water}$$

Another method of plotting centrifugal-pump performance data is to plot contours of constant pump efficiency and contours of constant power input. An example of this type of plot is shown in Fig. 19.6. This plot also gives the pump characteristic curves that correspond to different-size impellers in the same pump casing. The largest size shown is a 7-in. impeller and the smallest is a 5-in. impeller. All characteristic curves are similar in shape, but they have different magnitudes. A similar family of curves can be obtained with the same impeller size by varying the impeller rotational speed. Variable-speed pumps will be discussed in a later section of this chapter. Note that only the largest impellers reach an efficiency of 68 percent, because they fill most of the volume in the casing with minimal fluid losses. The maximum efficiency occurs between about 70 and 110 gpm (1000 and 1750 L/s), regardless of the impeller size.

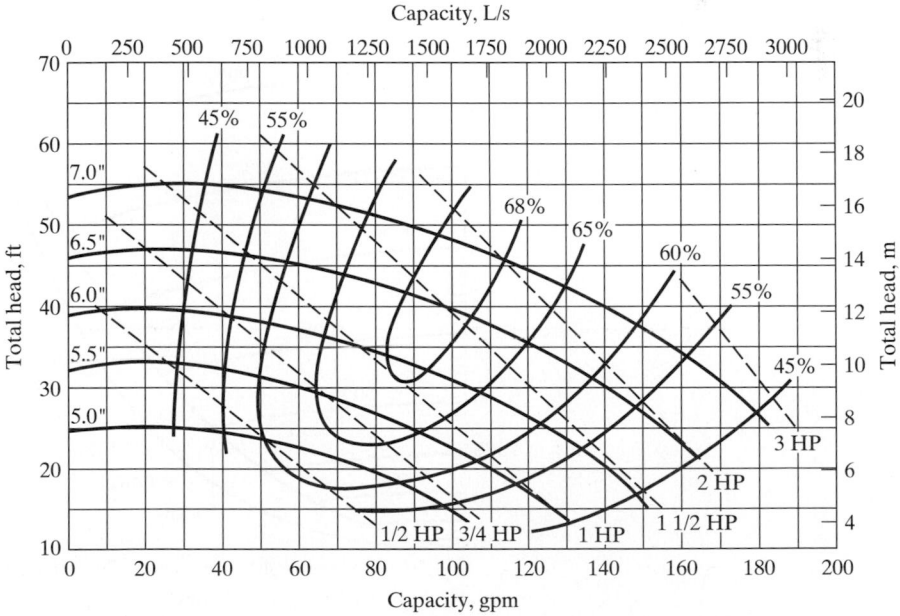

Figure 19.6 Typical centrifugal-pump performance curves. [Reprinted by permission from *ASHRAE Systems and Equipment 1996 (IP & SI)*, p. 38.4.]

Flow Control

The volumetric flow rate through a pump and piping system combination is found by the intersection of the pump and system characteristic curves. This intersection is termed the "balance point." The flow can be reduced by adding flow resistance in the system, such as by partially closing a valve. The flow can also be reduced by reducing the pump speed. Use of a variable-speed pump will save more energy than the use of a valve. However, it is more expensive to install and may not be economical for small systems. The basics of flow control for hydronic systems are similar to those of airflow control. The reader is referred to Sec. 18.3 for more detailed information on this subject.

Figure 19.7 shows the effect of a variable-speed pump on the flow, head, and power requirement of a hydronic system. Note the large reduction in power associated with a modest decrease in flow rate. At a flow rate of 50 percent of the design value, the power requirement is only about 10 percent of the design requirement. In many systems the flow is a small fraction of the design flow most of the time. The use of a variable-speed pump can greatly reduce pump energy costs compared with a constant-speed pump and valve flow control.

Another flow-control method commonly used is to place pumps in parallel. All the pumps are used to satisfy the need for a high flow rate that normally occurs during design conditions. As the flow requirement is reduced, one or more pumps can be switched off. A check valve is placed at the discharge of each pump to prevent recirculation of the fluid

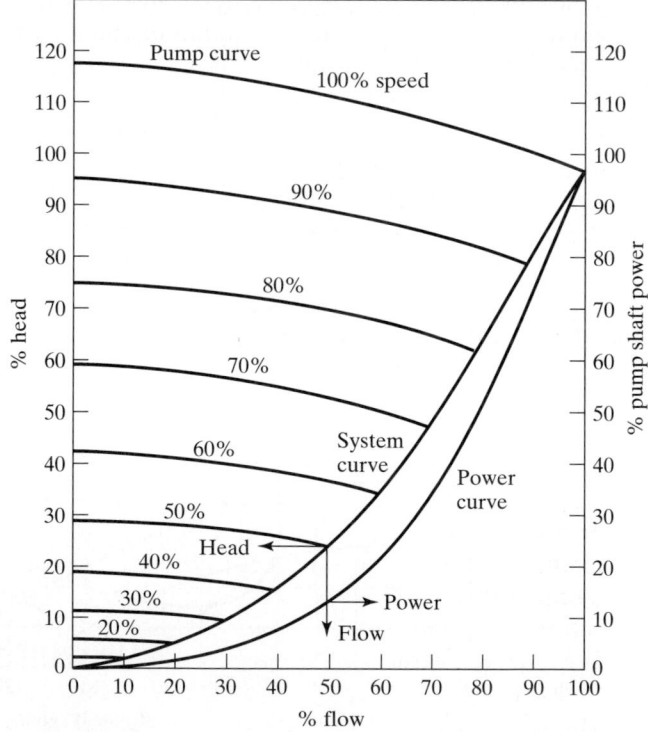

Figure 19.7 Pumping power, head, and flow vs. pump speed. [Reprinted by permission from *ASHRAE Fundamentals 1993* (IP & SI), p. 38.7.]

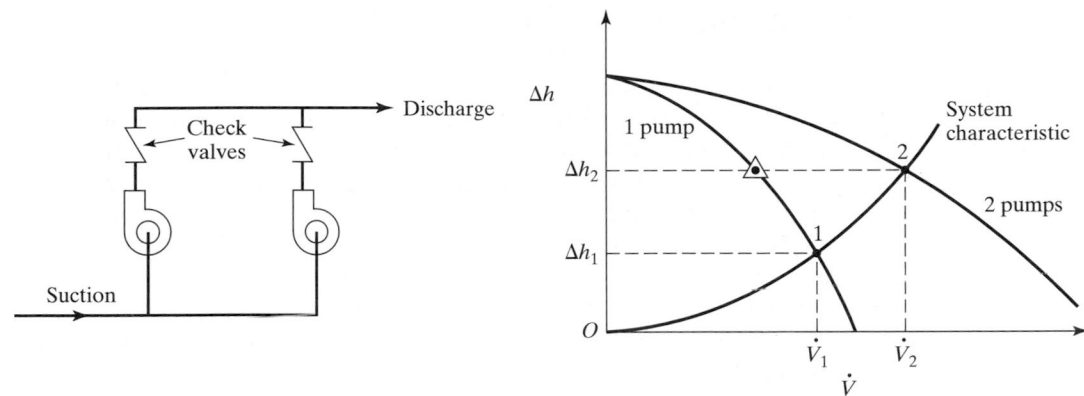

Figure 19.8 Balance points when two identical pumps are placed in parallel.

through the inoperable pumps. A sketch of this type of configuration is given in Fig. 19.8. This method of flow control is not as precise as the other methods (control-valve opening or variable-speed pump); however, it does reduce the power requirement at reduced-flow-rate conditions. Another advantage is that at low flow rate, when one or more pumps are inoperable, the unused pumps can be removed for maintenance without disrupting the flow through the system.

When two identical pumps are installed in parallel, the performance of the resulting combination is as shown on Fig. 19.8. When both pumps are operating, the combined pump/system balance point is at location 2 on the figure. The head across both pumps is Δh_2, and the total flow rate through the system is $\dot{V}_2$. However, each pump alone operates at the point indicated by the triangle at the same total head but at one-half the total flow rate. When one pump is turned off, the remaining pump-system balance point changes to location 1, where the head is now Δh_1, and the flow rate changes to $\dot{V}_1$—that is, **NOT** one-half of the previous flow rate because of the nonlinear nature of both the pump curve and the system characteristic.

19.3 PIPE LAYOUT AND SIZING PROCEDURES

Layout

As noted earlier, there are two basic forms of hydronic systems: open loop and closed loop. In an open-loop system, one portion of the system is open to the atmosphere. The pressure at the return from this opening is atmospheric pressure. The pressure at the discharge into the opening is not necessarily atmospheric pressure, as the water is often discharged through spray nozzles, which have significant pressure drop. Example applications of open-loop systems include water piping for wet cooling towers, spray humidifiers, and air washers. Water normally collects in a drain pan or sump and is recirculated to the pump. The pump must be placed below the sump elevation for adequate priming. There is usually water loss from these systems by design, so a makeup water line must be provided with a means to control the amount of water in the system.

Unlike open-loop systems, closed-loop systems are normally operated with the entire system above atmospheric pressure. This prevents air leakage into the system and

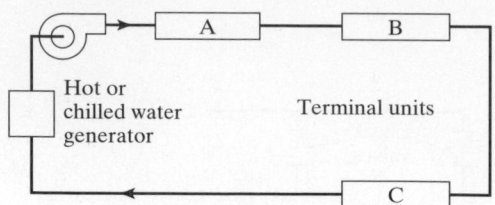

Figure 19.9 Series-loop piping system.

assists in the prevention of cavitation in the pump. Closed-loop systems require an expansion tank, because the water and pipe material have different coefficients of thermal expansion. The water makeup line is usually used only periodically to replace water that leaked from the system. Example closed-loop systems include hydronic baseboard heating systems, piping to fan coil units, and piping to heating and cooling coils in the air handler of a mechanical room. The remaining discussion in this section applies only to closed-loop systems.

There are several generic types of piping systems. Some of these are described here. Figure 19.9 illustrates a closed system in which all the terminal units are supplied hot or chilled water from the same source. The water flows through all terminal units in series. This type of system is inexpensive to install, but there is no control over any individual terminal unit. If all terminal units were installed in the same zone with a single controller, such as in a residence or small apartment, this system would be a good choice. However, if the units were located in different zones or areas with considerably different thermal loads, an alternate design with more individual control would be better.

Figures 19.10 and 19.11 show two possible alternatives that offer better control. These are called "two-pipe systems" because there are separate supply and return pipe loops. Each terminal unit could be equipped with a separate control valve to regulate the flow of water and thereby control the amount of heating or cooling provided by the unit. The two-pipe design is more complex than the series-loop design but offers much more flexibility in control. The direct-return system shown in Fig. 19.10 has a shorter length of supply and return piping connecting the pump to terminal unit A than to B or C. Therefore the water will experience less resistance passing to and from terminal unit A than B or C. If all three units have the same flow resistance (are identical units), then the flow will tend to preferentially flow through unit A rather than units B or C. This is not a problem if the flow through the units is controlled by separate thermostatic flow-control valves. The flow through unit A will be reduced by partially closing the valve, while the

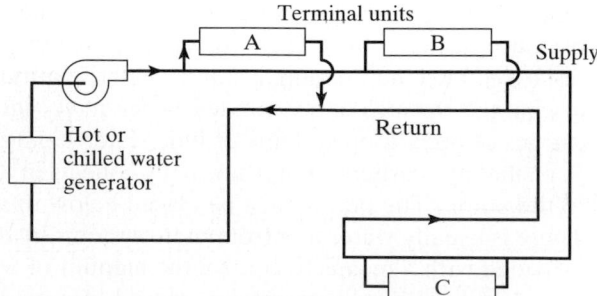

Figure 19.10 Two-pipe direct-return system.

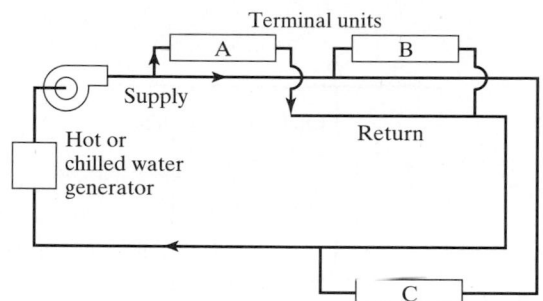

Figure 19.11 Two-pipe reverse-return system.

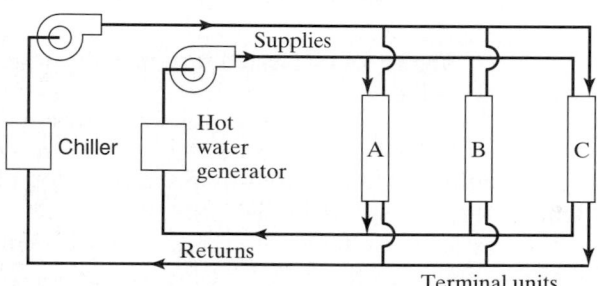

Figure 19.12 Four-pipe direct-return system.

valves for B and C will remain in a more open position with the same thermal loads. However, if all control valves are completely open, as they may be during design conditions, than the flow imbalance and the heating or cooling capacity imbalance mentioned before will occur. A design that alleviates this flow-imbalance problem is shown in Fig. 19.11. This is an example of a reverse-return system. The basic concept is that the total equivalent length from the pump discharge to the pump suction is the same through any terminal unit. Thus the total pressure loss is the same whether the flow passes through unit A, B, or C. This is a self-balancing design and does not depend on the operation of the control valves as much as the direct-return system.

Another possible design is a four-pipe system as shown on Fig. 19.12. The term "four-pipe" is derived from the need for two supply pipes and two return pipes for a total of four main supply and return pipes. In a four-pipe system, the hot- and cold-water distribution systems are completely separate. The only common feature is that both pass through the same heat exchangers, but they pass through separate tubes so the water flow streams never mix. While this system is more expensive to install than the others we have discussed, it has one feature that makes it useful in some applications. As the heating and cooling loops are separate, both can be operated simultaneously to cool one area and heat another. For example, hot water could be supplied to terminal unit A to heat a perimeter zone of a building in winter, while chilled water was supplied to units B and C to cool interior zones. A heat pump could also be installed between the two flow loops to extract heat from areas of the building that required cooling while using this energy to heat other areas.

Often a piping system in a large building will consist of combinations of the features illustrated in Figs. 19.9 to 19.12. For example, each terminal unit shown in Fig. 19.11 could be replaced by a number of terminal units in series, as shown in Fig. 19.9.

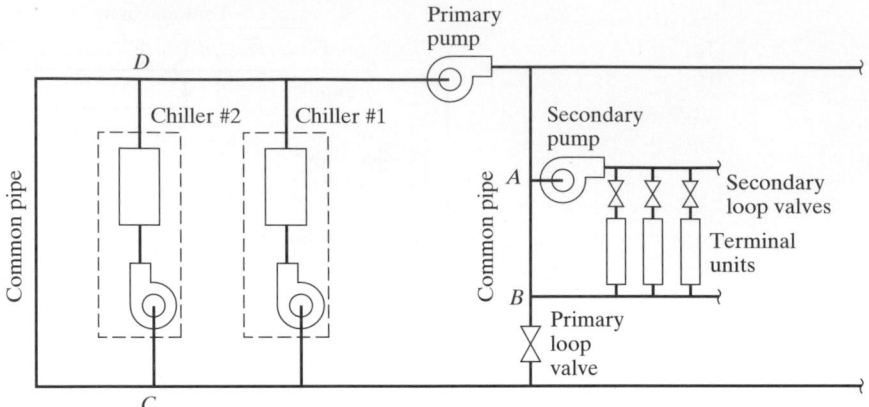

Figure 19.13 Nested system with variable flow using constant-flow chillers.

Another type of closed-loop system is found in large buildings where several systems of the type shown in Figs. 19.9, 19.10, or 19.11 are supplied with water from a single primary system. Systems with more than one pump are "nested systems" in that several small systems may lie within one larger primary system. A key to successful large system design is to ensure that each portion of the system will operate independently of the other portions. This is usually accomplished by using "common pipes" that allow water to short-circuit in a subsystem without affecting the main or primary system. These common pipes should be short and of large diameter, so that they do not add any appreciable pressure drop to the system.

Figure 19.13 shows a schematic drawing of a portion of a nested system. The primary system contains two chillers, each with their own pumps, and several secondary systems, each with a pump and several terminal units. The primary system has its own pump. We will discuss how this system responds to variations in cooling load at the terminal units. In this discussion, all pumps are assumed to be operating at constant speed with flow-control valves at each terminal unit controlling the flow through each unit. The chillers are assumed to be identical, so that each one handles one-half of the design load. When the load is at the maximum or design value, all valves will be open; each secondary pump will be supplying the design flow rate of water to the terminal units in its section. The common pipe located between points A and B will have no flow and no pressure drop. The primary pump will be supplying the maximum flow to all the secondary loops. Each chiller will be pumping its maximum flow rate, which is one-half of the total through the primary pump.

Now consider the situation when the control valves begin to close and only three-fourths of the design flow rate is needed. Each secondary pump will now experience a higher pressure drop and lower flow rate than at design conditions, owing to the partial closure of the valves on the terminal units. There should be no flow through the common pipe between points A and B if all secondary loops respond together. The primary-loop valves will begin to close as they respond to the temperature of the water in the secondary loops. The primary pump will be delivering three-fourths of the flow to the system because of the increased pressure drop from the partially closed primary-loop valves. Chiller #1 will be supplying one-half the design flow rate at full cooling capacity. Chiller

#2 will be pumping one-half the design flow rate, but only one-half of this, or one-quarter of the design flow, will be passed to the primary pump. The remaining flow will bypass the system through the common pipe located between points C and D. Chiller #2 operates at partial load because the return-water temperature is lower than the design value, owing to the short-circuiting of some of the water through the common pipe.

This response will continue until the load falls to one-half the design load. At this point, chiller #2 will be switched off. Chiller #1 continues to operate and provides one-half the design flow rate to the primary pump with no flow in the common pipe between C and D.

Further reduction in the load causes further closing of the terminal unit valves and the primary-loop valves, so that the flow in the secondary loops and in the primary loop continues to decrease. However, the flow through Chiller #1 remains at the original value, with a portion entering the primary pump and the remainder bypassing through the common pipe between points C and D.

The discussion above can be modified for the use of variable-speed pumps rather than control valves. However, the same principle of decoupling the various loops of the system should be followed. A good test of the design is that all subsystems should continue to operate at the correct flow rates if the primary pump is suddenly stopped.

Additional considerations when laying out a piping system include accounting for thermal expansion. A pipe should not be anchored at both ends without a thermal-expansion section. In general it is sufficient to anchor only one end and let the other end float in a guide support. If an expansion section is needed, guidelines are given by ASHRAE [2] for L bends, Z bends, U bends, and pipe loops.

Piping systems are usually constructed of vertical risers and horizontal runs. Often systems are better visualized in an isometric drawing that shows all risers in addition to the horizontal runs.

Sizing

Recommended pipe-sizing procedures are given by ASHRAE [1] for steam, hot and chilled water, hot and cold service water, gas, and fuel oil. Here we will discuss only hot and chilled water.

The appropriate size of pipe to use is a compromise. A pipe too small will have excessive pressure drop that results in large pumping-power and energy requirements. The velocity will also be large, which generates noise and can contribute to erosion within the pipe. A pipe too large may occupy more physical space than is available. Also the initial cost is large because of the amount of material. The weight of the pipe and contained water is also large, so that additional supports may be required. The best pipe size to use depends on the relative weighting of these various factors. The designer may wish to implement an optimization procedure for critical designs. However, this level of sophistication is not necessary in general building HVAC system designs.

ASHRAE recommends that the velocity be limited to about 4 ft/sec (1.2 m/s) for pipes 2-in. nominal size and smaller. For larger sizes, the head loss should be limited to about 4 ft water/100 ft (0.4 kPa/m). Figures 19.2 and 19.3 can be used to determine the pipe size, given the required volumetric flow rate and the design criteria above.

The general procedure to use for pipe sizing is as follows:

1. Lay out the system with all the necessary fittings and valves.

2. Size the pipe for all portions of the system based on the design or maximum flow rate anticipated through each section.

3. Determine the head loss through the fittings using one of the methods discussed in the previous section of this chapter.

4. Sum the head requirement for each pump using the longest loop or the one with the largest head loss.

5. Adjust the pipe size of branches in parallel so the head loss through each branch equals the head loss in the section used to size the pump.

The following examples illustrate the basic procedure outlined above.

EXAMPLE 19.6 Cooling-Tower Pipe Sizing/Pump Selection

Figure 19.14 shows a schematic drawing of the piping system needed to remove heat from a water-cooled refrigeration condenser and reject the heat to the atmosphere, using a wet cooling tower. The amount of heat to be rejected is 480,000 Btu/hr. The temperature rise of the water through the condenser is 10 °F. The total length of piping in the system is 60 ft. Fittings are as shown on Fig. 19.14. Select a pump size from Fig. 19.6 that will adequately handle the required flow rate and head loss. Additional information: pipe is schedule-40 steel pipe with screwed fittings; k-value of the strainer can be assumed to be twice the value of a swing check valve; the head loss through the condenser heat exchanger is 7 ft water; the difference in height between the cooling-tower spray header and the sump is 10 ft; and the piping inlet in the sump is a square inlet. Neglect the head loss through the spray nozzles in the cooling tower.

Solution: First we need to compute the amount of water flowing in this system. Then we will size the pipe. It will all be the same diameter, as the same water flow rate exists in all sections of the system. After this we will compute the head loss across the pump. Finally we will select a pump from Fig. 19.6 that will satisfy the flow and head-loss requirements.

Water flow rate: The water flow rate is determined from an energy balance across the condenser. By modifying Eq. (2.10) for steady operation, $\dot{Q} = \dot{m}c(t_2 - t_1)$, and, solving for the water mass flow rate given the heat load, the temperature difference, and the specific heat for water:

$$\dot{m} = \frac{\dot{Q}}{c(t_2 - t_1)} = \frac{480,000 \text{ Btu/hr}}{(1.0 \text{ Btu/lbm} \cdot °\text{F})(10 \text{ °F})} = 48,000 \text{ lbm/hr}$$

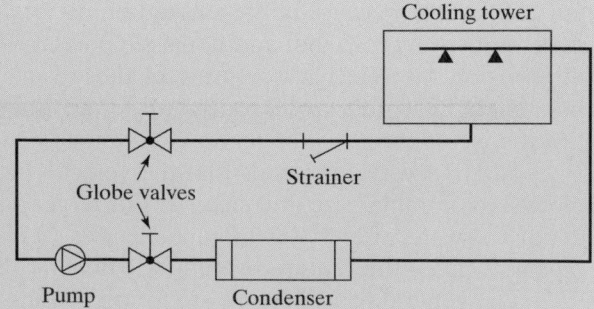

Figure 19.14 Cooling water piping system for Ex. 19.6.

Converting this to volumetric flow rate by dividing by the approximate density of water from Table A.1E near 90 °F,

$$\dot{V} = \dot{m}/\rho = (48{,}000 \text{ lbm/hr})/62.4 \text{ lbm/ft}^3 = 769 \text{ ft}^3/\text{hr}$$

Converting this to gallons per minute, gpm

$$\dot{V} = 769 \text{ ft}^3/\text{hr} \left(\frac{7.48 \text{ gal}}{\text{ft}^3}\right)\left(\frac{\text{hr}}{60 \text{ min}}\right) = 96.0 \text{ gpm}$$

Pipe size: From Fig. 19.2(a), the pipe size for 96 gpm and about 4 ft water/100 ft head loss is between $2\frac{1}{2}$ and 3 in. We could select either size. As noise is not a problem in this application and the length of pipe is relatively short (i.e., fitting losses will dominate the head loss), we will choose a $2\frac{1}{2}$-in. pipe.

Head loss across the pump: We need to determine the head loss for the straight pipe and all fittings and elbows as shown. Figure 19.2(a) is used to determine the velocity in the $2\frac{1}{2}$-in. pipe with 96-gpm flow rate, $6\frac{1}{2}$ ft/sec, and the head loss per unit length, 7 ft water/100 ft. Table 19.3 is used to obtain the *k*-values for the pipe fittings. The calculations follow the procedure demonstrated in Ex. 19.2. Table 19.5 summarizes the head-loss results for the fittings.

TABLE 19.5 Fitting Head-Loss Results for Ex. 19.6

Component or Fitting	k-Value	Head Loss, ft water
90-deg elbow	0.85	0.56
Globe valve	6.5	4.26
Strainer	4.4	2.89
Inlet	0.5	0.33

Table 19.6 summarizes the head-loss results for the complete system.

TABLE 19.6 Summary of Head Loss for Ex. 19.6

Component	Number or Length	Unit Head Loss	Total Head Loss, ft water
Straight pipe	60 ft	7 ft water/100 ft	4.2
Inlet	1	0.33 ft water	0.33
Elbows	5	0.56 ft water	2.8
Globe valves	2	4.26 ft water	8.52
Stainer	1	4.4 ft water	4.4
Condenser	1	7 ft water	7
Cooling tower	1	3 ft elevation	3
Total			30.3

Therefore, the total head requirement for the pump is 30.3 ft water.

Pump selection: The pump requirements are: 96 gpm at 30.3 ft water head. Looking at Fig. 19.6, a 6-in. pump will provide the necessary head at 96-gpm flow rate with a few extra feet of head capacity. Therefore, the best choice from Fig. 19.6 is the 6-in. pump.

The previous example shows that the designer often must make some judgment regarding the correct pipe size and pump size to specify. A variety of pumps will adequately handle the requirements; the designer can use other factors such as cost or known reliability to determine which supplier should be selected.

Most hydronic systems are not as simple as the one considered in Ex. 19.6. Most have branches with different flow rates. The system is often three dimensional, with risers and drops in addition to the layout shown on a plan view. The designer must be careful to include all elevation changes, as they will affect the pump head requirement for open systems and may affect other factors, such as net positive suction head, in closed systems.

Example 19.7 illustrates how a system with multiple branches should be handled when sizing the pipe and determining the pump requirements.

EXAMPLE 19.7   Mechanical Room Pipe Sizing/Pump Selection

Figure 19.15 shows the piping system needed to supply chilled water to two finned-tube heat exchangers. The elevation view is depicted in the figure. Size the pipe for this system, assuming schedule-40 steel pipe is used with threaded fittings. Determine the pump flow rate and head requirements. Distances are as shown in feet. Head loss through the heat exchangers is shown as provided by the manufacturer.

Solution: Although there are elevation changes in the piping, we do not have to consider them for head-loss calculations, as this is a closed-loop system. However, note that the pump is located at the bottom to minimize any cavitation and that the expansion tank is located at the top. The system has been divided into sections. Sections 1-2 and 3-4 carry the same amount of water, 70 gpm, and therefore have the same diameter. The loop from section 2 to 3 through heat exchanger (a) carries 30 gpm, and the branch through heat exchanger (b) carries 40 gpm. These two paths from section 2 to 3 may have different diameters.

We need to know which path from section 2 to section 3 has the largest head loss. This will be used to size the pump. The branch with the smaller head loss will require an increase in flow resistance to match the head loss of the other branch, so that the correct amount of

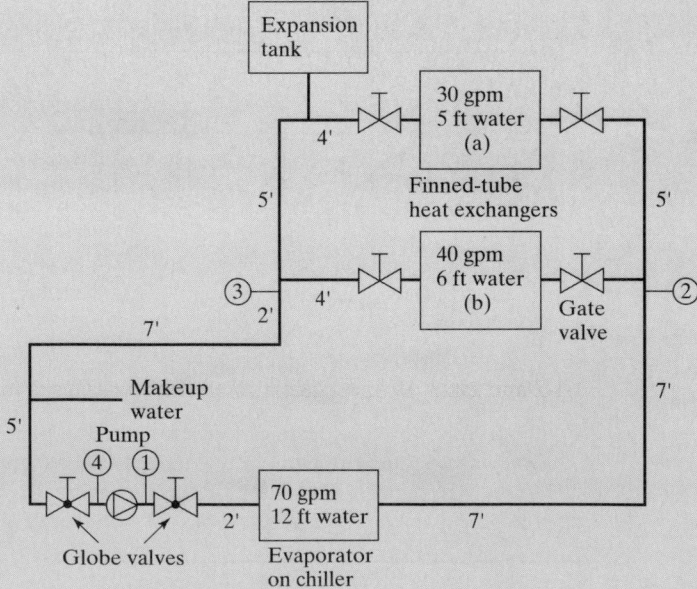

Figure 19.15 Chilled water piping schematic for Ex. 19.7.

water will flow through each branch. If these two branches are not balanced in this manner, a larger amount of water than desired will flow through the branch with the smaller resistance, leaving a smaller-than-desired flow through the other branch.

We will begin by sizing the pipe from sections 1-2 and 3-4. From Fig. 19.2(a), a $2\frac{1}{2}$-in. pipe will provide a head loss of about 3.8 ft water/100 ft with a velocity of about 4.8 ft/sec. For the 30 gpm through heat exchanger (a), we will initially select a $1\frac{1}{2}$-in. pipe with a head loss of 6.5 ft water/100 ft and a velocity of 4.8 ft/sec. For the 40 gpm through heat exchanger (b), we will select a 2-in. pipe with a head loss of 3.2 ft water/100 ft and a velocity of 3.9 ft/sec. These choices provide nearly the 4 ft water/100 ft desired. However, the size of pipe through one of the branches may need to be changed if the head loss through the two branches is not equal.

Now that the initial pipe sizes have been selected, the head loss through the various sections can be determined. Table 19.7 shows the fitting-head-loss calculations that are needed.

TABLE 19.7 Fitting Head-Loss Results for Ex. 19.7

Component	Pipe Size, in.	k-Value	Velocity, ft/sec	Head Loss, ft water
Elbow	$2\frac{1}{2}$	0.85	4.8	0.30
Elbow	$1\frac{1}{2}$	1.2	4.8	0.43
Tee line	$2\frac{1}{2}$	0.9	4.8	0.32
Tee line	$1\frac{1}{2}$	0.9	4.8	0.32
Tee branch	$1\frac{1}{2}$	1.6	4.8	0.57
Globe valve	$2\frac{1}{2}$	6.5	4.8	2.33
Gate valve	2	0.17	3.9	0.04
Gate valve	$1\frac{1}{2}$	0.19	4.8	0.07

Sections 1-2 and 3-4: $2\frac{1}{2}$-in. pipe, 4.8 ft/sec, 3.8 ft water/100 ft.

This section has a total length of $2 + 7 + 5 + 2 + 7 + 7 = 30$ ft. There are four 90-deg elbows, one tee line, two globe valves, and the chiller heat exchanger. Table 19.8 summarizes the head loss through this portion of the system.

Section 2-a-3: $1\frac{1}{2}$-in. pipe, 4.8 ft/sec, 6.5 ft water/100 ft.

This section has a total length of $5 + 4 + 5 = 14$ ft. There are two 90-deg elbows, three tee lines, two gate valves, and the heat exchanger. The head-loss results for this branch are given in Table 19.8.

Section 2-b-3: 2-in. pipe, 3.9 ft/sec, 3/2 ft water/100 ft.

This section has 4 ft of straight pipe, two tee branches, two gate valves, and the heat exchanger. Table 19.8 contains the head-loss results for this branch.

The total head requirement for the pump is the sum of the losses through sections 3-4-1-2 and 2-a-3, $19.32 + 7.87 = 27.19$ ft water. This is the loop around the system with the largest total head loss. Therefore the pump requirement becomes: 70 gpm and about 27 ft water.

Note that the total head losses through branches 2-a-3 and 2-b-3 are nearly identical for this system. The increased head loss through heat exchanger (b) compared to (a), 2 ft water, is approximately equal to the head loss through the extra elbows and the tee in the longest branch. It is unnecessary to modify this system to bring it into balance.

If the head loss through one of the branches is much smaller than the other, two options can be used to increase the resistance through this branch.

TABLE 19.8 Head-Loss Results for Ex. 19.7

Component	Number or Length	Unit Head Loss	Total Head Loss, ft water
Sections 3-4, 1-2			
Straight pipe	30 ft	3.8 ft water/100 ft	1.14
Elbows	4	0.30	1.20
Tee line	1	0.32	0.32
Globe valves	2	2.33	4.66
Chiller	1	12.00	12.00
Total 3-4, 1-2			19.32
Section 2-*a*-3			
Straight pipe	14 ft	6.5 ft water/100 ft	0.91
Elbows	2	0.43	0.86
Tee line	3	0.32	0.96
Gate valves	2	0.07	0.14
Heat exchanger *a*	1	5.00	5.00
Total 2-*a*-3			7.87
Section 2-*b*-3			
Straight pipe	4 ft	3.2 ft water/100 ft	0.13
Tee branches	2	0.57	1.14
Gate valves	2	0.04	0.08
Heat exchanger *b*	1	6.00	6.00
Total 2-*b*-3			7.35

1. A valve can be partially closed to increase the head loss in the branch, or to balance the system.

2. A smaller pipe size can be specified for the branch to increase the head loss.

Usually the first option is adopted. This may be required if the flow rates change significantly, owing to load changes, or if modifications are made to the system in the future.

19.4 ADDITIONAL COMPONENTS AND CONSIDERATIONS

An open-loop piping system has one air-water interface, at which the pressure is equal to atmospheric pressure. This serves as a reference pressure for the rest of the system; i.e., the pressure at any other point in the system can be determined by knowing the pressure difference between the point and the water free surface. Usually the pump inlet is located near and slightly below this free surface, so the rest of the system is at pressures above atmospheric.

Closed-loop systems also need a reference pressure, so that a minimum pressure can be maintained to prevent cavitation in the pump and the maximum pressure is limited so that the system does not rupture or the pressure-relief valve does not open. An expansion tank serves the purpose of providing this pressure reference. The water can be in direct contact with the atmosphere (open tank), in which case the system behaves much like an open-loop system. However, most expansion tanks are enclosed, with the water in direct contact with the air above (direct-interface tank) or in contact with a bladder that separates the air and water (diaphragm tank).

The volume of the tank can be determined as follows. First consider an open tank where the water is in direct contact with atmospheric air. The tank must be large enough to store the excess water forced out of the system due to thermal expansion or contraction of the water and piping material. Assume the water and piping will experience temperature limits t_1 and t_2. Let the difference in temperature be Δt. The change in volume of the water in the system can be calculated by

$$\Delta V_w = M_w(v_2 - v_1) \approx \frac{V_w}{v_1}(v_2 - v_1)$$

The change in internal volume of the piping system can be approximated by

$$\Delta V_s \approx 3\alpha(t_2 - t_1) = 3\alpha \, \Delta t$$

assuming the piping system is constructed of homogeneous material with linear thermal expansion coefficient α. The minimum tank volume must equal the difference between these two values. A safety factor of two is commonly recommended, so the desired tank volume becomes

$$V_t = 2\left\{ V_w\left[\left(\frac{v_2}{v_1}\right) - 1 \right] - 3\alpha \, \Delta t \right\} \tag{19.8}$$

Values for liquid-water specific volume versus temperature can be obtained from Table A.1E or Table A.1SI. Values for the linear thermal expansion coefficient, α, can be found in *Marks' Standard Handbook for Mechanical Engineers* [3]. Values commonly used are:

Steel: $\alpha = 6.5 \times 10^{-6}$ in./in. · °F $(12 \times 10^{-6}$ m/m · °C)
Copper: $\alpha = 9.5 \times 10^{-6}$ in./in. · °F $(17 \times 10^{-6}$ m/m · °C)

Most tanks today are of the bladder type. In these tanks the air does not dissolve into the water, so there are fewer problems of air coming out of solution at higher temperatures than with the direct-interface tanks. For bladder expansion tanks, the air is assumed to compress and expand isothermally in the tank, so Eq. (19.8) is modified to account for this. The equation applied to bladder expansion tanks is

$$V_t = \frac{V_w\left[\left(\frac{v_2}{v_1}\right) - 1 \right] - 3\alpha \, \Delta t}{1 - \dfrac{P_1}{P_2}} \tag{19.9}$$

EXAMPLE 19.8

Determine the size of diaphragm expansion tank for a closed piping system constructed of steel pipe with a volume of 4.3 m³. The temperature and pressure limits expected are: $t_1 = 5$ °C, $P_1 = 110$ kPa, $t_2 = 35$ °C, and $P_2 = 125$ kPa.

Solution: We will use Eq. (19.9) to determine the necessary volume of the expansion tank. The specific volume values for saturated liquid water are obtained from Table A.1SI:

$$v_1 = v_f(5 \text{ °C}) = 0.0009999 \text{ m}^3/\text{kg}$$

$$v_2 = v_f(35 \text{ °C}) = 0.001006 \text{ m}^3/\text{kg}$$

As steel pipe is used, the value for α is 12×10^{-6} m/m · °C. The temperature difference, Δt, is $35 - 5 = 30$ °C. Substituting into Eq. (19.9),

$$V_t = \frac{(4.3 \text{ m}^3)[(1.006/0.999) - 1] - 3(12 \times 10^{-6} \text{ m/m} \cdot °C)(30 \text{ °C})}{1 - 125 \text{ kPa}/110 \text{ kPa}} = 0.21 \text{ m}^3$$

ENDNOTES

1. *ASHRAE Handbook, Fundamentals Volume* (Atlanta: American Society of Heating, Refrigerating and Air Conditioning Engineers, 1997), ch. 33.
2. *ASHRAE Handbook, Systems and Equipment* (Atlanta: American Society of Heating, Refrigerating and Air Conditioning Engineers, 1996), ch. 40.
3. *Marks' Standard Handbook for Mechanical Engineers*, 8th ed. (New York: McGraw-Hill, 1978).

PROBLEMS

19.1 Determine the value for b in the equation shown in Fig. 19.1 for a straight pipe with a known friction factor.

19.2 Plot a chart similar to Fig. 19.2(a) for a pipe whose friction factor is equal to 0.03. Assume this is a constant and independent of pipe diameter. Plot lines for 1-, 2-, and 4-in. pipe inner diameters. Also plot lines for 1-, 2-, and 4-ft/s water velocities.

19.3 Determine the equivalent length for the gate valve used in Ex. 19.2.

19.4 Compute the equivalent lengths for a regular 90-deg elbow for 3-in. nominal schedule-40 steel pipe. Use velocities from 1 to 10 ft/s in increments of 1 ft/s.

19.5 Repeat Prob. 19.4 for the flow coefficients.

19.6 Determine the head loss through the branch of a tee in 1-in. schedule-40 steel pipe if the flow rate is 10 gpm.

19.7 Calculate the head loss through an angle valve in a nominal 100-mm type-L copper tube with water flowing at a rate of 20 L/s.

19.8 Develop a relationship between the loss coefficient, k, of a fitting and its flow coefficient, C_v. Give a worked example in which you convert a loss coefficient in Table 19.4 into the appropriate flow coefficient.

19.9 How high above the water can the pump in Ex. 19.5 be raised and meet the manufacturer's requirement for NPSHR = 25 ft water?

19.10 If the NPSHR for a pump is 10 m water, how large a velocity can the pump have at the inlet if the total pressure at the suction equals standard atmospheric pressure?

19.11 Compute the total efficiency for the pump shown on Fig. 19.6. Use an impeller diameter of 6.5 in., a volumetric flow rate of 100 gpm, and a head of 40 ft water. Compare your answer with the efficiency information plotted on the figure.

19.12 Consider a pump with a 7-in. impeller and the performance data given in Figs. 19.6 and 19.7. The initial flow rate is 120 gpm. Determine the head across the pump and the power requirement when the flow rate is reduced to 60 gpm when
(a) a flow-control valve is used to reduce the flow rate,
(b) the pump speed is reduced without changing the system.

19.13 Consider two identical pumps operating in parallel. Each pump has a 6.5-in. impeller and pumps 100 gpm. Using the data given on Fig. 19.6, determine
(a) the pump efficiency,
(b) the total power required, and
(c) the head across each pump.

19.14 Using the pumps and system as described in Prob. 19.13, assume that one pump is turned off without changing the system. Determine
(a) the new flow rate through the remaining operable pump,
(b) the new head across the pump, and
(c) the new power requirement.

19.15 Figure 19.16 shows an outdoor fountain to be constructed. Use Type-L copper tube and determine the water flow rate and the head across the pump when the fountain projects 10 ft above the discharge.

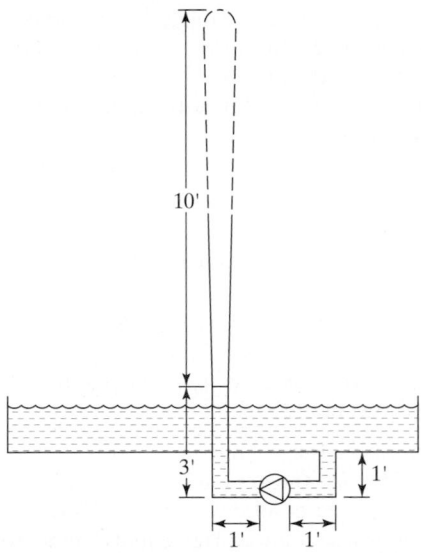

Figure 19.16 Outdoor-fountain schematic for Prob. 19.15.

19.16 A cooling-tower piping system is sketched in Fig. 19.17. Use schedule-40 steel pipe. The vertical distance between the cooling-tower sump and the spray header is 5 m. The length of pipe from (b) to (a) through the pump is 20 m and the length of pipe from (a) to (b) in each tower is 10 m.

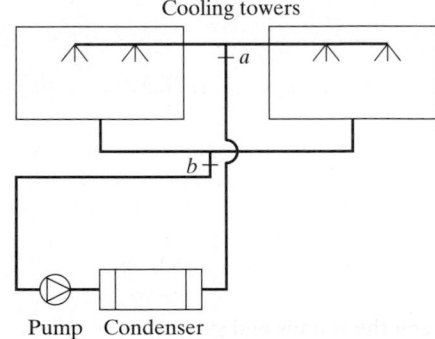

Figure 19.17 Cooling-tower piping schematic for Prob. 19.16.

(a) Size the piping in the system if 1500 L/s is required through the condenser, the condenser head loss is 3 m water at this flow rate, and this flow is equally divided between the two cooling towers.

(b) Determine the pump head requirement and select a pump from Fig. 19.6 that will satisfactorily handle this requirement. Also provide the pump power requirement and efficiency.

19.17 Consider the piping system shown in Fig. 19.17. Calculate the flow rate through the system as designed in Prob. 19.16 if only one cooling tower is installed initially. Assume the head loss through the condenser is proportional to the square of the flow rate through it.

19.18 An eight-story apartment building is to be constructed using a hot-water perimeter heating system. Figure 19.18 shows a portion of the system that uses a reverse return.

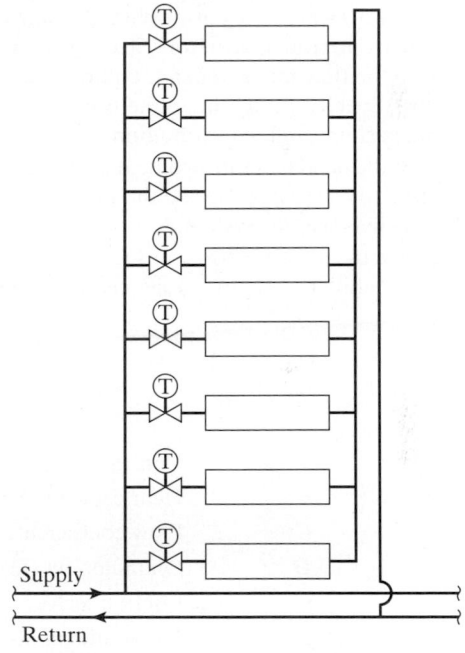

Figure 19.18 Portion of hot-water perimeter heating system for an eight-story building.

(a) Size the piping in this system, assuming each floor requires 2 gpm.

(b) Compute the head loss between the supply and return tees for your answer to part (a) if the vertical distance between floors is 10 ft, there is an additional 2 ft on the supply leg and 3 ft on the return leg to connect to the supply and return lines; each horizontal leg contains 12 ft of straight pipe, and the head loss through each heat exchanger is $2\frac{1}{2}$ ft water.

19.19 The piping system shown in Fig. 19.18 uses thermostatically operated control valves to regulate the water flow rate through each heat exchanger. Sketch a head-loss, flow-rate diagram for this system and show how the system changes as one or more control valves close.

19.20 A hot-water district heating system is to be designed to heat five buildings from a central plant. Each building will have a pump to pass hot water from the district heating system through a plate heat exchanger. A separate pump in each building will pump building hot water on the other side of the heat exchanger to be circulated to hot-water heating coils in the air-handling units. Sketch the district heating system that you would design for this application. Include all pumps, piping, and heat exchangers.

19.21 Obtain a drawing of a hot or chilled water piping system from your instructor. Size all piping and provide the flow rate and head requirements for all pumps. Specify pump manufacturer and model numbers from catalog information.

19.22 Change the head loss through heat exchanger (a) in Ex. 19.7 to 8 ft water. Resize the piping in branch 2-*b*-3 to balance this system.

19.23 Consider the system shown in Fig. 19.13. Assume the two chillers are of equal capacity and are designed to circulate 1000 gpm at a supply temperature of 45 °F and a return of 60 °F at full-load conditions. At a part-load condition during which 600 gpm are circulated through the secondary circuits,

(a) determine the flow rate in the common pipe next to the chillers,

(b) compute the inlet-water temperature to chiller #2 at point *C*, and

(c) calculate the load ratio (load/capacity) for chiller #2.

19.24 Determine the required volume of a bladder expansion tank for a piping system for a chiller that contains 1500 gallons of water. The system is at a minimum pressure of 10 psig when the water temperature is 45 °F. Maximum anticipated temperature of the water is 100 °F, and the maximum pressure should remain below 35 psig.

19.25 Determine the volume of bladder expansion tank required for the system designed in Ex. 19.7. The total volume of all three heat exchangers is 16 gal. Assume minimum and maximum temperatures and pressures are: $t_1 = 40$ °F, $P_1 = 15$ psig, $t_2 = 90$ °F, and $P_2 = 40$ psig.

SYMBOLS

b	Constant.
c	Heat capacity, Btu/lbm · °F (kJ/kg · °C).
C_v	Flow coefficient.
D	Diameter, in. (m).
f	Friction factor.
g	Gravitational acceleration, ft/sec^2 (m/sec^2).
h	Head, ft water (m water).
H	Total head, ft water (m water).
k	Loss coefficient.
ℓ_f	Flow head loss, ft water (m water).
L	Length, ft (m).
L_{eq}	Equivalent length of fitting, ft (m).
$\dot{m}$	Mass flow rate, lbm/sec (kg/sec).
M	Mass, lbm (kg).
NPSHA	Net Positive Suction Head Available, ft water (m water).
NPSHR	Net Positive Suction Head Required, ft water (m water).
P	Static pressure, psi (kPa).
P_t	Total pressure, psi (kPa).
$\dot{Q}$	Heat flow rate, Btu/hr (kW).

t	Temperature, °F (°C).
v	Specific volume, ft³/lbm (m³/kg).
v_f	Specific volume of saturated liquid, ft³/lbm (m³/kg).
$\mathbf{V}$	Velocity, ft/sec (m/sec).
V	Volume, ft³ (m³).
$\dot{V}$	Volumetric flow rate, ft³/sec (m³/sec).
w	Work, Btu (kJ).
$\dot{W}$	Power, Btu/hr (kW).
z	Elevation, ft (m).

Greek Letters

α	Thermal coefficient of linear expansion, in./in. °F (m/m °C).
Δ	Change.
$\eta_{p,t}$	Total efficiency of pump.
ρ	Density, lbm/ft³ (kg/m³).

Subscripts

rev	Reversible.
sat	Saturated.
std	Standard or reference.
suction	At pump inlet.
s	System.
t	Tank.
w	Water.

Thermodynamic and Thermophysical Property Tables

TABLE A.1E Thermodynamic Properties of Water at Saturation (English Units)

Temp. °F	Absolute Pressure $P_{w,s} * 10^5$		Specific Volume, ft³/lbm		Enthalpy, Btu/lbm		Entropy, Btu/lbm · °R	
	lbf/in^2	In. Hg	Sat. Solid v_i	Sat. Vapor $v_g * 10^{-4}$	Sat. Solid h_i	Sat. Vapor h_g	Sat. Solid s_i	Sat. Vapor s_g
-150	0.01622	0.03312	0.01725	113700	-218.72	994.76	-0.4798	3.4381
-145	0.02854	0.05828	0.01725	65670	-217.07	996.97	-0.4745	3.3829
-140	0.04935	0.1008	0.01726	38590	-215.40	999.18	-0.4692	3.3295
-135	0.08392	0.1713	0.01726	23050	-213.70	1001.39	-0.4640	3.2778
-130	0.1405	0.2868	0.01727	13980	-211.98	1003.60	-0.4587	3.2278
-125	0.2315	0.4727	0.01727	8611	-210.24	1005.81	-0.4535	3.1794
-120	0.3762	0.7680	0.01728	5379	-208.47	1008.02	-0.4482	3.1325
-115	0.6027	1.231	0.01728	3407	-206.68	1010.23	-0.4430	3.0870
-110	0.9529	1.946	0.01729	2186	-204.87	1012.44	-0.4378	3.0429
-105	1.487	3.037	0.01729	1420	-203.03	1014.65	-0.4326	3.0001
-100	2.293	4.683	0.01730	934.2	-201.18	1016.86	-0.4274	2.9585
-95	3.495	7.136	0.01730	621.5	-199.29	1019.07	-0.4222	2.9182
-90	5.266	10.75	0.01731	418.1	-197.38	1021.28	-0.4170	2.8790
-85	7.850	16.03	0.01731	284.3	-195.45	1023.49	-0.4118	2.8410
-80	11.58	23.64	0.01732	195.3	-193.50	1025.70	-0.4066	2.8040
-75	16.91	34.53	0.01733	135.5	-191.52	1027.91	-0.4014	2.7680
-70	24.46	49.94	0.01733	94.91	-189.52	1030.12	-0.3963	2.7331
-65	35.05	71.56	0.01734	67.08	-187.49	1032.33	-0.3911	2.6991
-60	49.77	101.6	0.01734	47.83	-185.44	1034.54	-0.3859	2.6660
-55	70.08	143.1	0.01735	34.40	-183.37	1036.76	-0.3808	2.6338
-50	97.85	199.8	0.01736	24.94	-181.27	1038.97	-0.3756	2.6024
-48	111.6	227.8	0.01736	21.98	-180.42	1039.85	-0.3736	2.5901
-46	127.1	259.5	0.01736	19.39	-179.57	1040.74	-0.3715	2.5779
-44	144.5	295.1	0.01736	17.13	-178.72	1041.62	-0.3694	2.5658
-42	164.2	335.3	0.01737	15.15	-177.86	1042.50	-0.3674	2.5539
-40	186.3	380.4	0.01737	13.42	-177.00	1043.39	-0.3653	2.5421
-38	211.1	431.1	0.01737	11.90	-176.13	1044.27	-0.3633	2.5304
-36	239.0	488.0	0.01738	10.56	-175.26	1045.16	-0.3612	2.5188
-34	270.2	551.7	0.01738	9.383	-174.39	1046.04	-0.3592	2.5074
-32	305.2	623.1	0.01738	8.348	-173.51	1046.93	-0.3571	2.4961
-30	344.3	702.9	0.01738	7.434	-172.63	1047.81	-0.3550	2.4848
-28	387.9	792.0	0.01739	6.628	-171.74	1048.70	-0.3530	2.4737
-26	436.6	891.5	0.01739	5.916	-170.86	1049.58	-0.3509	2.4627
-24	490.9	1002	0.01739	5.286	-169.96	1050.46	-0.3489	2.4519
-22	551.4	1126	0.01739	4.728	-169.07	1051.35	-0.3468	2.4411
-20	618.6	1263	0.01740	4.233	-168.16	1052.23	-0.3448	2.4304
-19	655.0	1337	0.01740	4.007	-167.71	1052.67	-0.3437	2.4251
-18	693.3	1416	0.01740	3.794	-167.26	1053.12	-0.3427	2.4199
-17	733.7	1498	0.01740	3.594	-166.81	1053.56	-0.3417	2.4146
-16	776.3	1585	0.01740	3.404	-166.35	1054.00	-0.3407	2.4094
-15	821.1	1676	0.01740	3.226	-165.90	1054.44	-0.3396	2.4042
-14	868.3	1773	0.01741	3.057	-165.44	1054.88	-0.3386	2.3990
-13	917.9	1874	0.01741	2.898	-164.98	1055.33	-0.3376	2.3939
-12	970.2	1981	0.01741	2.748	-164.52	1055.77	-0.3366	2.3888
-11	1025	2093	0.01741	2.607	-164.06	1056.21	-0.3355	2.3837
-10	1083	2211	0.01741	2.473	-163.60	1056.65	-0.3345	2.3786
-9	1144	2335	0.01741	2.347	-163.14	1057.09	-0.3335	2.3736
-8	1208	2466	0.01741	2.227	-162.68	1057.53	-0.3325	2.3686
-7	1275	2603	0.01742	2.115	-162.21	1057.98	-0.3314	2.3636
-6	1346	2747	0.01742	2.008	-161.75	1058.42	-0.3304	2.3586
-5	1420	2899	0.01742	1.907	-161.28	1058.86	-0.3294	2.3537
-4	1498	3058	0.01742	1.812	-160.82	1059.30	-0.3284	2.3488
-3	1579	3225	0.01742	1.722	-160.35	1059.74	-0.3273	2.3439
-2	1665	3400	0.01742	1.637	-159.88	1060.18	-0.3263	2.3390
-1	1756	3585	0.01742	1.556	-159.41	1060.62	-0.3253	2.3341
0	1850	3778	0.01743	1.480	-158.94	1061.07	-0.3243	2.3293
1	1950	3980	0.01743	1.407	-158.47	1061.51	-0.3232	2.3245
2	2054	4193	0.01743	1.339	-157.99	1061.95	-0.3222	2.3197
3	2163	4416	0.01743	1.274	-157.52	1062.39	-0.3212	2.3150
4	2277	4650	0.01743	1.213	-157.05	1062.83	-0.3202	2.3103
5	2397	4895	0.01743	1.154	-156.57	1063.27	-0.3191	2.3055

(continued)

TABLE A.1E (cont.)

Temp. °F	Absolute Pressure $P_{w,s} * 10^5$		Specific Volume, ft³/lbm		Enthalpy, Btu/lbm		Entropy, Btu/lbm · °R	
	lbf/in²	In. Hg	Sat. Solid v_i	Sat. Vapor $v_g * 10^{-4}$	Sat. Solid h_i	Sat. Vapor h_g	Sat. Solid s_i	Sat. Vapor s_g
6	2523	5152	0.01743	1.099	-156.09	1063.71	-0.3181	2.3009
7	2655	5421	0.01744	1.047	-155.62	1064.15	-0.3171	2.2962
8	2793	5703	0.01744	0.9971	-155.14	1064.59	-0.3161	2.2915
9	2938	5999	0.01744	0.9500	-154.66	1065.03	-0.3150	2.2869
10	3089	6308	0.01744	0.9054	-154.18	1065.47	-0.3140	2.2823
11	3248	6632	0.01744	0.8630	-153.70	1065.92	-0.3130	2.2777
12	3414	6971	0.01744	0.8228	-153.21	1066.36	-0.3120	2.2732
13	3588	7325	0.01745	0.7846	-152.73	1066.80	-0.3109	2.2686
14	3770	7697	0.01745	0.7483	-152.24	1067.24	-0.3099	2.2641
15	3960	8085	0.01745	0.7139	-151.76	1067.68	-0.3089	2.2596
16	4159	8491	0.01745	0.6812	-151.27	1068.12	-0.3079	2.2552
17	4367	8916	0.01745	0.6501	-150.78	1068.56	-0.3068	2.2507
18	4584	9360	0.01745	0.6205	-150.30	1069.00	-0.3058	2.2463
19	4811	9824	0.01745	0.5924	-149.81	1069.44	-0.3048	2.2419
20	5049	10310	0.01746	0.5657	-149.32	1069.88	-0.3038	2.2375
21	5297	10820	0.01746	0.5404	-148.82	1070.32	-0.3027	2.2331
22	5556	11340	0.01746	0.5162	-148.33	1070.76	-0.3017	2.2287
23	5827	11900	0.01746	0.4932	-147.84	1071.20	-0.3007	2.2244
24	6110	12470	0.01746	0.4714	-147.34	1071.64	-0.2997	2.2201
25	6405	13080	0.01746	0.4506	-146.85	1072.07	-0.2987	2.2158
26	6713	13710	0.01747	0.4308	-146.35	1072.51	-0.2976	2.2115
27	7035	14360	0.01747	0.4119	-145.85	1072.95	-0.2966	2.2073
28	7371	15050	0.01747	0.3940	-145.35	1073.39	-0.2956	2.2031
29	7721	15760	0.01747	0.3769	-144.85	1073.83	-0.2946	2.1988
30	8086	16510	0.01747	0.3606	-144.35	1074.27	-0.2935	2.1946
31	8467	17290	0.01747	0.3450	-143.85	1074.71	-0.2925	2.1905
32	8864	18100	0.01747	0.3302	-143.35	1075.15	-0.2915	2.1863

(continued)

TABLE A.1E (cont.)

Temp. °F	Absolute Pressure $P_{w,s}$		Specific Volume, ft³/lbm		Enthalpy, Btu/lbm		Entropy, Btu/lbm · °R	
	lbf/in²	In. Hg	Sat. Liquid v_f	Sat. Vapor v_g	Sat. Liquid h_f	Sat. Vapor h_g	Sat. Liquid s_f	Sat. Vapor s_g
32	0.088602	0.18091	0.01601	3304.2	0.0	1075.19	0.0	2.1864
33	0.092244	0.18834	0.01601	3180.2	0.96	1075.63	0.00194	2.1828
34	0.096017	0.19604	0.01601	3061.4	1.93	1076.07	0.00392	2.1793
35	0.099926	0.20403	0.01602	2947.6	2.91	1076.51	0.00589	2.1758
36	0.10398	0.21230	0.01602	2838.4	3.89	1076.95	0.00786	2.1723
37	0.10817	0.22086	0.01602	2733.8	4.86	1077.39	0.00984	2.1689
38	0.11251	0.22973	0.01602	2633.5	5.84	1077.83	0.01181	2.1654
39	0.11701	0.23891	0.01602	2537.4	6.82	1078.27	0.01377	2.1620
40	0.12167	0.24842	0.01602	2445.1	7.81	1078.71	0.01574	2.1585
41	0.12649	0.25826	0.01602	2356.6	8.79	1079.15	0.0177	2.1551
42	0.13147	0.26844	0.01602	2271.7	9.77	1079.59	0.01967	2.1518
43	0.13663	0.27898	0.01602	2190.3	10.76	1080.03	0.02163	2.1484
44	0.14197	0.28988	0.01602	2112.1	11.74	1080.47	0.02359	2.1451
45	0.14750	0.30115	0.01602	2037.0	12.73	1080.91	0.02554	2.1417
46	0.15321	0.31281	0.01602	1964.9	13.72	1081.35	0.02750	2.1384
47	0.15911	0.32487	0.01602	1895.7	14.71	1081.79	0.02945	2.1351
48	0.16521	0.33733	0.01602	1829.2	15.70	1082.22	0.03140	2.1318
49	0.17152	0.35021	0.01602	1765.4	16.69	1082.66	0.03335	2.1286
50	0.17805	0.36353	0.01603	1704.0	17.68	1083.10	0.03530	2.1253
51	0.18479	0.37729	0.01603	1645.1	18.67	1083.54	0.03724	2.1221
52	0.19175	0.39151	0.01603	1588.4	19.66	1083.98	0.03919	2.1189
53	0.19894	0.40619	0.01603	1533.9	20.66	1084.42	0.04113	2.1157
54	0.20637	0.42136	0.01603	1481.6	21.65	1084.86	0.04306	2.1125
55	0.21404	0.43703	0.01603	1431.2	22.65	1085.29	0.04500	2.1093
56	0.22197	0.45320	0.01603	1382.8	23.64	1085.73	0.04693	2.1062
57	0.23014	0.46990	0.01603	1336.2	24.64	1086.17	0.04886	2.1030
58	0.23859	0.48714	0.01603	1291.4	25.64	1086.61	0.05079	2.0999
59	0.24730	0.50493	0.01604	1248.3	26.64	1087.05	0.05272	2.0968
60	0.25629	0.52329	0.01604	1206.8	27.63	1087.48	0.05464	2.0937
61	0.26557	0.54223	0.01604	1166.8	28.63	1087.92	0.05656	2.0906
62	0.27514	0.56177	0.01604	1128.4	29.63	1088.36	0.05848	2.0876
63	0.28501	0.58192	0.01604	1091.3	30.63	1088.80	0.06039	2.0845
64	0.29519	0.60271	0.01604	1055.7	31.63	1089.23	0.06230	2.0815
65	0.30569	0.62415	0.01605	1021.4	32.63	1089.67	0.06421	2.0785
66	0.31651	0.64625	0.01605	988.28	33.64	1090.11	0.06612	2.0755
67	0.32767	0.66903	0.01605	956.42	34.64	1090.54	0.06802	2.0725
68	0.33917	0.69251	0.01605	925.71	35.64	1090.98	0.06993	2.0695
69	0.35102	0.71671	0.01605	896.13	36.64	1091.41	0.07182	2.0666
70	0.36324	0.74165	0.01605	867.61	37.65	1091.85	0.07372	2.0636
71	0.37582	0.76734	0.01606	840.12	38.65	1092.29	0.07561	2.0607
72	0.38878	0.79381	0.01606	813.61	39.66	1092.72	0.07750	2.0578
73	0.40214	0.82107	0.01606	788.05	40.66	1093.16	0.07939	2.0549
74	0.41589	0.84915	0.01606	763.40	41.67	1093.59	0.08128	2.0520
75	0.43005	0.87806	0.01607	739.62	42.67	1094.03	0.08316	2.0491
76	0.44462	0.90782	0.01607	716.69	43.68	1094.46	0.08504	2.0463
77	0.45963	0.93846	0.01607	694.56	44.68	1094.90	0.08691	2.0434
78	0.47508	0.97000	0.01607	673.20	45.69	1095.33	0.08879	2.0406
79	0.49098	1.0025	0.01607	652.59	46.70	1095.76	0.09065	2.0378
80	0.50734	1.0359	0.01608	632.70	47.70	1096.20	0.09252	2.0350
81	0.52417	1.0702	0.01608	613.49	48.71	1096.63	0.09439	2.0322
82	0.54149	1.1056	0.01608	595.00	49.63	1097.07	0.09609	2.0294
83	0.55930	1.1420	0.01608	577.08	50.65	1097.50	0.09798	2.0267
84	0.57762	1.1794	0.01609	559.78	51.68	1097.93	0.09987	2.0239
85	0.59646	1.2178	0.01609	543.06	52.71	1098.36	0.10176	2.0212
86	0.61583	1.2574	0.01609	526.91	53.74	1098.80	0.10364	2.0185
87	0.63575	1.2981	0.01609	511.31	54.76	1099.23	0.10551	2.0157
88	0.65623	1.3399	0.01610	496.23	55.79	1099.66	0.10739	2.0130
89	0.67727	1.3828	0.01610	481.67	56.82	1100.09	0.10926	2.0104
90	0.69890	1.4270	0.01610	467.58	57.84	1100.53	0.11112	2.0077

(continued)

Temp. °F	Absolute Pressure $P_{w,s}$		Specific Volume, ft³/lbm		Enthalpy, Btu/lbm		Entropy, Btu/lbm · °R	
	lbf/in²	In. Hg	Sat. Liquid v_f	Sat. Vapor v_g	Sat. Liquid h_f	Sat. Vapor h_g	Sat. Liquid s_f	Sat. Vapor s_g
91	0.72112	1.4724	0.01611	453.97	58.87	1100.96	0.11299	2.0050
92	0.74396	1.5190	0.01611	440.81	59.89	1101.39	0.11485	2.0024
93	0.76742	1.5669	0.01611	428.13	60.81	1101.82	0.11651	1.9998
94	0.79151	1.6161	0.01611	415.82	61.82	1102.25	0.11833	1.9971
95	0.81627	1.6666	0.01612	403.92	62.82	1102.68	0.12015	1.9945
96	0.84168	1.7185	0.01612	392.41	63.83	1103.11	0.12197	1.9919
97	0.86779	1.7718	0.01612	381.28	64.84	1103.54	0.12378	1.9893
98	0.89458	1.8265	0.01613	370.50	65.85	1103.97	0.12560	1.9868
99	0.92210	1.8827	0.01613	360.07	66.86	1104.40	0.12740	1.9842
100	0.95034	1.9404	0.01613	349.98	67.87	1104.83	0.12921	1.9817
101	0.97932	1.9996	0.01614	340.21	68.88	1105.26	0.13101	1.9791
102	1.0091	2.0603	0.01614	330.76	69.89	1105.69	0.13281	1.9766
103	1.0396	2.1226	0.01614	321.60	70.90	1106.11	0.13460	1.9741
104	1.0709	2.1865	0.01615	312.73	71.91	1106.54	0.13639	1.9716
105	1.1030	2.2521	0.01615	304.15	72.92	1106.97	0.13818	1.9691
106	1.1360	2.3194	0.01615	295.83	73.93	1107.40	0.13997	1.9666
107	1.1698	2.3884	0.01616	287.77	74.94	1107.82	0.14175	1.9641
108	1.2045	2.4592	0.01616	279.97	75.94	1108.25	0.14353	1.9617
109	1.2400	2.5318	0.01616	272.40	76.95	1108.68	0.14530	1.9592
110	1.2765	2.6062	0.01617	265.07	77.96	1109.10	0.14707	1.9568
111	1.3138	2.6825	0.01617	257.97	78.97	1109.53	0.14884	1.9544
112	1.3521	2.7607	0.01617	251.09	79.98	1109.96	0.15061	1.9520
113	1.3914	2.8409	0.01618	244.41	80.99	1110.38	0.15237	1.9496
114	1.4316	2.9230	0.01618	237.95	82.00	1110.81	0.15413	1.9472
115	1.4729	3.0072	0.01619	231.67	83.01	1111.23	0.15589	1.9448
116	1.5151	3.0935	0.01619	225.59	84.02	1111.65	0.15764	1.9424
117	1.5584	3.1819	0.01619	219.69	85.02	1112.08	0.15939	1.9401
118	1.6027	3.2724	0.01620	213.97	86.03	1112.50	0.16114	1.9377
119	1.6482	3.3652	0.01620	208.42	87.04	1112.93	0.16288	1.9354
120	1.6947	3.4602	0.01620	203.03	88.05	1113.35	0.16462	1.9330
121	1.7423	3.5574	0.01621	197.80	89.06	1113.77	0.16636	1.9307
122	1.7911	3.6571	0.01621	192.73	90.06	1114.19	0.16809	1.9284
123	1.8411	3.7591	0.01622	187.81	91.07	1114.62	0.16982	1.9261
124	1.8922	3.8635	0.01622	183.04	92.08	1115.04	0.17155	1.9238
125	1.9446	3.9704	0.01623	178.40	93.09	1115.46	0.17327	1.9216
126	1.9982	4.0798	0.01623	173.90	94.10	1115.88	0.17500	1.9193
127	2.0530	4.1918	0.01623	169.53	95.10	1116.30	0.17671	1.9170
128	2.1092	4.3064	0.01624	165.28	96.11	1116.72	0.17843	1.9148
129	2.1666	4.4237	0.01624	161.16	97.12	1117.14	0.18014	1.9126
130	2.2254	4.5437	0.01625	157.16	98.12	1117.56	0.18185	1.9103
131	2.2855	4.6665	0.01625	153.27	99.13	1117.98	0.18355	1.9081
132	2.3470	4.7921	0.01626	149.50	100.14	1118.40	0.18526	1.9059
133	2.4100	4.9206	0.01626	145.83	101.14	1118.81	0.18696	1.9037
134	2.4743	5.0520	0.01626	142.26	102.15	1119.23	0.18865	1.9015
135	2.5401	5.1864	0.01627	138.80	103.16	1119.65	0.19035	1.8994
136	2.6075	5.3238	0.01627	135.43	104.16	1120.07	0.19204	1.8972
137	2.6763	5.4644	0.01628	132.16	105.17	1120.48	0.19372	1.8950
138	2.7466	5.6080	0.01628	128.97	106.18	1120.90	0.19541	1.8929
139	2.8186	5.7549	0.01629	125.88	107.18	1121.32	0.19709	1.8907
140	2.8921	5.9050	0.01629	122.87	108.19	1121.73	0.19876	1.8886
141	2.9673	6.0585	0.01630	119.95	109.19	1122.15	0.20044	1.8865
142	3.0441	6.2153	0.01630	117.11	110.20	1122.56	0.20211	1.8844
143	3.1226	6.3756	0.01631	114.34	111.20	1122.97	0.20378	1.8823
144	3.2028	6.5394	0.01631	111.65	112.21	1123.39	0.20545	1.8802
145	3.2848	6.7068	0.01632	109.04	113.21	1123.80	0.20711	1.8781
146	3.3685	6.8777	0.01632	106.49	114.22	1124.21	0.20877	1.8760
147	3.4540	7.0524	0.01632	104.02	115.22	1124.63	0.21043	1.8739
148	3.5414	7.2308	0.01633	101.61	116.23	1125.04	0.21208	1.8719
149	3.6307	7.4130	0.01633	99.261	117.23	1125.45	0.21373	1.8698
150	3.7218	7.5991	0.01634	96.979	118.24	1125.86	0.21538	1.8678

(continued)

Temp. °F	Absolute Pressure $P_{w,s}$		Specific Volume, ft³/lbm		Enthalpy, Btu/lbm		Entropy, Btu/lbm · °R	
	lbf/in²	In. Hg	Sat. Liquid v_f	Sat. Vapor v_g	Sat. Liquid h_f	Sat. Vapor h_g	Sat. Liquid s_f	Sat. Vapor s_g
151	3.8149	7.7892	0.01634	94.758	119.24	1126.27	0.21702	1.8658
152	3.9100	7.9832	0.01635	92.596	120.25	1126.68	0.21866	1.8637
153	4.0070	8.1814	0.01636	90.491	121.25	1127.09	0.22030	1.8617
154	4.1061	8.3837	0.01636	88.442	122.25	1127.50	0.22194	1.8597
155	4.2072	8.5902	0.01637	86.447	123.26	1127.91	0.22357	1.8577
156	4.3105	8.8010	0.01637	84.504	124.26	1128.32	0.22520	1.8557
157	4.4159	9.0162	0.01638	82.612	125.26	1128.73	0.22683	1.8537
158	4.5234	9.2358	0.01638	80.769	126.27	1129.13	0.22846	1.8518
159	4.6332	9.4599	0.01639	78.974	127.27	1129.54	0.23008	1.8498
160	4.7452	9.6886	0.01639	77.225	128.27	1129.95	0.23170	1.8478
161	4.8595	9.9219	0.01640	75.522	129.28	1130.35	0.23331	1.8459
162	4.9761	10.160	0.01640	73.862	130.28	1130.76	0.23493	1.8440
163	5.0950	10.403	0.01641	72.245	131.28	1131.16	0.23654	1.8420
164	5.2164	10.651	0.01641	70.669	132.28	1131.57	0.23815	1.8401
165	5.3401	10.903	0.01642	69.133	133.29	1131.97	0.23975	1.8382
166	5.4664	11.161	0.01642	67.636	134.29	1132.37	0.24135	1.8363
167	5.5951	11.424	0.01643	66.177	135.29	1132.78	0.24295	1.8344
168	5.7264	11.692	0.01644	64.755	136.29	1133.18	0.24455	1.8325
169	5.8603	11.965	0.01644	63.368	137.30	1133.58	0.24614	1.8306
170	5.9968	12.244	0.01645	62.016	138.30	1133.98	0.24774	1.8287
171	6.1360	12.528	0.01645	60.697	139.30	1134.38	0.24932	1.8268
172	6.2778	12.818	0.01646	59.412	140.30	1134.78	0.25091	1.8250
173	6.4225	13.113	0.01646	58.158	141.30	1135.18	0.25249	1.8231
174	6.5699	13.414	0.01647	56.935	142.30	1135.58	0.25408	1.8213
175	6.7201	13.721	0.01648	55.742	143.31	1135.98	0.25565	1.8194
176	6.8732	14.034	0.01648	54.578	144.31	1136.38	0.25723	1.8176
177	7.0293	14.352	0.01649	53.443	145.31	1136.77	0.25880	1.8158
178	7.1883	14.677	0.01649	52.335	146.31	1137.17	0.26037	1.8140
179	7.3503	15.008	0.01650	51.255	147.31	1137.57	0.26194	1.8121
180	7.5153	15.345	0.01651	50.200	148.31	1137.96	0.26351	1.8103
181	7.6835	15.688	0.01651	49.171	149.31	1138.36	0.26507	1.8085
182	7.8548	16.038	0.01652	48.166	150.32	1138.75	0.26663	1.8067
183	8.0293	16.394	0.01652	47.186	151.32	1139.15	0.26819	1.8050
184	8.2070	16.757	0.01653	46.228	152.32	1139.54	0.26974	1.8032
185	8.3880	17.126	0.01654	45.294	153.32	1139.93	0.27130	1.8014
186	8.5723	17.503	0.01654	44.382	154.32	1140.32	0.27285	1.7997
187	8.7600	17.886	0.01655	43.491	155.32	1140.72	0.27440	1.7979
188	8.9512	18.276	0.01655	42.621	156.32	1141.11	0.27594	1.7962
189	9.1458	18.674	0.01656	41.772	157.32	1141.50	0.27748	1.7944
190	9.3439	19.078	0.01657	40.943	158.32	1141.89	0.27903	1.7927
191	9.5456	19.490	0.01657	40.133	159.33	1142.28	0.28056	1.7909
192	9.7509	19.909	0.01658	39.341	160.33	1142.67	0.28210	1.7892
193	9.9599	20.336	0.01659	38.568	161.33	1143.05	0.28363	1.7875
194	10.173	20.770	0.01659	37.813	162.33	1143.44	0.28516	1.7858
195	10.389	21.212	0.01660	37.075	163.33	1143.83	0.28669	1.7841
196	10.609	21.662	0.01661	36.355	164.33	1144.21	0.28822	1.7824
197	10.834	22.120	0.01661	35.650	165.33	1144.60	0.28974	1.7807
198	11.062	22.586	0.01662	34.962	166.33	1144.98	0.29127	1.7790
199	11.294	23.060	0.01663	34.289	167.33	1145.37	0.29279	1.7774
200	11.530	23.542	0.01663	33.632	168.34	1145.75	0.29430	1.7757
201	11.770	24.032	0.01664	32.989	169.34	1146.13	0.29582	1.7740
202	12.015	24.531	0.01665	32.361	170.34	1146.52	0.29733	1.7724
203	12.263	25.039	0.01665	31.747	171.34	1146.90	0.29884	1.7707
204	12.516	25.555	0.01666	31.147	172.34	1147.28	0.30035	1.7691
205	12.773	26.080	0.01667	30.559	173.34	1147.66	0.30186	1.7674
206	13.035	26.614	0.01667	29.986	174.34	1148.04	0.30336	1.7658
207	13.301	27.157	0.01668	29.424	175.35	1148.42	0.30487	1.7642
208	13.571	27.710	0.01669	28.875	176.35	1148.80	0.30637	1.7626
209	13.846	28.271	0.01669	28.339	177.35	1149.17	0.30786	1.7610
210	14.126	28.842	0.01670	27.814	178.35	1149.55	0.30936	1.7593

(*continued*)

TABLE A.1E (cont.)

Temp. °F	Absolute Pressure $P_{w,s}$ lbf/in²	Absolute Pressure $P_{w,s}$ In. Hg	Specific Volume, ft³/lbm Sat. Liquid v_f	Specific Volume, ft³/lbm Sat. Vapor v_g	Enthalpy, Btu/lbm Sat. Liquid h_f	Enthalpy, Btu/lbm Sat. Vapor h_g	Entropy, Btu/lbm·°R Sat. Liquid s_f	Entropy, Btu/lbm·°R Sat. Vapor s_g
211	14.410	29.422	0.01671	27.300	179.35	1149.93	0.31085	1.7577
212	14.699	30.013	0.01671	26.798	180.36	1150.30	0.31234	1.7561
215	15.595	31.842	0.01674	25.355	183.36	1151.42	0.31681	1.7514
220	17.189	35.095	0.01677	23.150	188.38	1153.28	0.32420	1.7436
225	18.914	38.619	0.01681	21.168	193.40	1155.12	0.33155	1.7359
230	20.78	42.429	0.01685	19.385	198.42	1156.93	0.33885	1.7284
235	22.795	46.542	0.01688	17.777	203.45	1158.73	0.34610	1.7210
240	24.967	50.978	0.01692	16.326	208.48	1160.51	0.35331	1.7137
245	27.307	55.754	0.01696	15.012	213.61	1162.26	0.36059	1.7066
250	29.822	60.890	0.01700	13.824	218.61	1164.00	0.36766	1.6996
255	32.524	66.406	0.01704	12.747	223.62	1165.71	0.37468	1.6926
260	35.422	72.323	0.01709	11.768	228.64	1167.40	0.38167	1.6859
265	38.526	78.661	0.01713	10.878	233.66	1169.07	0.38862	1.6792
270	41.848	85.444	0.01717	10.066	238.84	1170.71	0.39571	1.6726
275	45.398	92.692	0.01722	9.3257	243.93	1172.32	0.40264	1.6661
280	49.188	100.43	0.01726	8.6499	249.03	1173.91	0.40954	1.6597
285	53.229	108.68	0.01731	8.0319	254.14	1175.47	0.41641	1.6534
290	57.534	117.47	0.01736	7.4661	259.26	1177.00	0.42325	1.6472
295	62.115	126.82	0.01740	6.9474	264.39	1178.51	0.43005	1.6411
300	66.984	136.77	0.01745	6.4713	269.53	1179.98	0.43682	1.6351
305	72.154	147.32	0.01750	6.0338	274.69	1181.43	0.44356	1.6291
310	77.639	158.52	0.01755	5.6313	279.86	1182.84	0.45027	1.6233
315	83.453	170.39	0.01760	5.2605	285.04	1184.22	0.45696	1.6175
320	89.608	182.96	0.01766	4.9187	290.23	1185.57	0.46361	1.6118
325	96.119	196.25	0.01771	4.6032	295.43	1186.89	0.47024	1.6061
330	103.00	210.31	0.01776	4.3116	300.65	1188.17	0.47683	1.6005
335	110.27	225.14	0.01782	4.0419	305.88	1189.41	0.48341	1.5950
340	117.94	240.80	0.01787	3.7922	311.13	1190.61	0.48995	1.5895
345	126.02	257.31	0.01793	3.5607	316.39	1191.78	0.49647	1.5842
350	134.54	274.69	0.01799	3.3459	321.66	1192.91	0.50297	1.5788
355	143.50	292.99	0.01805	3.1465	326.94	1193.99	0.50943	1.5735
360	152.93	312.24	0.01811	2.9611	332.24	1195.04	0.51588	1.5683
365	162.83	332.47	0.01817	2.7887	337.55	1196.04	0.52230	1.5631
370	173.24	353.71	0.01823	2.6281	342.88	1197.00	0.52869	1.5580
375	184.16	376.00	0.01830	2.4784	348.22	1197.91	0.53506	1.5529
380	195.60	399.38	0.01836	2.3387	353.57	1198.78	0.54141	1.5478
385	207.60	423.88	0.01843	2.2084	358.94	1199.60	0.54774	1.5428
390	220.17	449.54	0.01850	2.0866	364.33	1200.37	0.55404	1.5378
395	233.32	476.39	0.01856	1.9726	369.73	1201.09	0.56033	1.5329
400	247.08	504.47	0.01863	1.8660	375.14	1201.76	0.56659	1.5280
405	261.45	533.83	0.01871	1.7662	380.57	1202.38	0.57283	1.5231
410	276.47	564.50	0.01878	1.6726	386.02	1202.94	0.57905	1.5182
415	292.16	596.51	0.01885	1.5847	391.49	1203.45	0.58525	1.5134
420	308.52	629.92	0.01893	1.5023	396.97	1203.90	0.59144	1.5086
425	325.58	664.76	0.01901	1.4249	402.47	1204.30	0.59760	1.5038
430	343.36	701.07	0.01909	1.3521	407.99	1204.63	0.60375	1.4990
435	361.89	738.89	0.01917	1.2836	413.53	1204.91	0.60989	1.4943
440	381.18	778.27	0.01925	1.2191	419.09	1205.12	0.61601	1.4895
445	401.25	819.25	0.01934	1.1583	424.67	1205.27	0.62211	1.4848
450	422.12	861.87	0.01943	1.1011	430.28	1205.35	0.62821	1.4801
455	443.82	906.18	0.01952	1.0471	435.90	1205.37	0.63429	1.4754
460	466.37	952.22	0.01961	0.99607	441.55	1205.32	0.64036	1.4707
465	489.79	1000.0	0.01970	0.94792	447.23	1205.20	0.64642	1.4660
470	514.10	1049.7	0.01980	0.90240	452.94	1205.01	0.65248	1.4613
475	539.33	1101.2	0.01990	0.85936	458.67	1204.75	0.65853	1.4566
480	565.49	1154.6	0.02000	0.81864	464.43	1204.41	0.66457	1.4519
485	592.62	1210.0	0.02010	0.78007	470.23	1203.99	0.67062	1.4472
490	620.73	1267.4	0.02021	0.74354	476.06	1203.49	0.67666	1.4425
495	649.85	1326.9	0.02032	0.70891	481.92	1202.92	0.68270	1.4378
500	680.01	1388.4	0.02043	0.67606	487.82	1202.26	0.68874	1.4331

TABLE A.1SI Thermodynamic Properties of Water at Saturation (SI Units)

Temp. °C	Absolute Pressure, Pa $P_{w,s}$	Specific Volume, m^3/kg		Enthalpy, kJ/kg		Entropy, kJ/kg · K	
		Sat. Solid v_i	Sat. Vapor v_g	Sat. Solid h_i	Sat. Vapor h_g	Sat. Solid s_i	Sat. Vapor s_g
-100	0.001405	0.001077	56870000	-507.22	2315.87	-2.0003	14.3039
-98	0.002103	0.001077	38430000	-504.44	2319.57	-1.9844	14.1389
-96	0.003121	0.001077	26200000	-501.64	2323.27	-1.9685	13.9779
-94	0.004590	0.001077	18010000	-498.81	2326.97	-1.9526	13.8206
-92	0.006694	0.001078	12490000	-495.95	2330.67	-1.9368	13.6670
-90	0.009684	0.001078	8729000	-493.07	2334.37	-1.9209	13.5169
-88	0.01390	0.001078	6148000	-490.16	2338.07	-1.9051	13.3702
-86	0.01980	0.001078	4363000	-487.22	2341.77	-1.8893	13.2268
-84	0.02799	0.001079	3119000	-484.25	2345.47	-1.8735	13.0866
-82	0.03930	0.001079	2245000	-481.25	2349.17	-1.8578	12.9496
-80	0.05478	0.001079	1627000	-478.22	2352.87	-1.8420	12.8155
-78	0.07586	0.001079	1187000	-475.17	2356.57	-1.8263	12.6843
-76	0.1044	0.001080	871800	-472.09	2360.27	-1.8106	12.5560
-74	0.1427	0.001080	644200	-468.98	2363.97	-1.7949	12.4304
-72	0.1938	0.001080	478900	-465.84	2367.68	-1.7792	12.3074
-70	0.2618	0.001080	358100	-462.67	2371.38	-1.7635	12.1870
-68	0.3515	0.001081	269300	-459.47	2375.08	-1.7479	12.0691
-66	0.4693	0.001081	203700	-456.25	2378.78	-1.7322	11.9537
-64	0.6232	0.001081	154900	-452.99	2382.48	-1.7166	11.8406
-62	0.8232	0.001081	118400	-449.71	2386.19	-1.7010	11.7298
-60	1.082	0.001082	90940	-446.40	2389.89	-1.6854	11.6212
-58	1.414	0.001082	70210	-443.06	2393.59	-1.6698	11.5148
-56	1.840	0.001082	54470	-439.69	2397.30	-1.6542	11.4105
-54	2.382	0.001082	42460	-436.29	2401.00	-1.6386	11.3083
-52	3.070	0.001083	33240	-432.87	2404.70	-1.6230	11.2080
-50	3.939	0.001083	26150	-429.41	2408.41	-1.6075	11.1097
-49	4.454	0.001083	23220	-427.67	2410.26	-1.5997	11.0612
-48	5.031	0.001083	20650	-425.93	2412.11	-1.5919	11.0132
-47	5.677	0.001083	18380	-424.17	2413.96	-1.5842	10.9657
-46	6.399	0.001083	16380	-422.41	2415.81	-1.5764	10.9186
-45	7.206	0.001084	14610	-420.65	2417.67	-1.5686	10.8720
-44	8.105	0.001084	13050	-418.87	2419.52	-1.5609	10.8258
-43	9.108	0.001084	11660	-417.09	2421.37	-1.5531	10.7800
-42	10.22	0.001084	10430	-415.30	2423.22	-1.5453	10.7347
-41	11.47	0.001084	9345	-413.50	2425.07	-1.5376	10.6898
-40	12.85	0.001084	8377	-411.70	2426.92	-1.5298	10.6453
-39	14.38	0.001085	7516	-409.88	2428.78	-1.5221	10.6012
-38	16.08	0.001085	6750	-408.07	2430.63	-1.5143	10.5576
-37	17.96	0.001085	6068	-406.24	2432.48	-1.5066	10.5143
-36	20.04	0.001085	5460	-404.40	2434.33	-1.4988	10.4714
-35	22.35	0.001085	4917	-402.56	2436.18	-1.4911	10.4290
-34	24.90	0.001085	4432	-400.72	2438.03	-1.4833	10.3869
-33	27.71	0.001085	3999	-398.86	2439.88	-1.4756	10.3452
-32	30.82	0.001086	3611	-397.00	2441.73	-1.4678	10.3038
-31	34.24	0.001086	3263	-395.12	2443.59	-1.4601	10.2629
-30	38.02	0.001086	2952	-393.25	2445.44	-1.4524	10.2223
-29	42.17	0.001086	2672	-391.36	2447.29	-1.4446	10.1821
-28	46.73	0.001086	2421	-389.47	2449.14	-1.4369	10.1422
-27	51.74	0.001086	2195	-387.57	2450.99	-1.4291	10.1027
-26	57.25	0.001087	1992	-385.66	2452.84	-1.4214	10.0635
-25	63.29	0.001087	1809	-383.74	2454.69	-1.4137	10.0247
-24	69.91	0.001087	1645	-381.82	2456.54	-1.4059	9.9862
-23	77.16	0.001087	1496	-379.89	2458.39	-1.3982	9.9481
-22	85.10	0.001087	1362	-377.95	2460.23	-1.3905	9.9103
-21	93.78	0.001087	1241	-376.01	2462.08	-1.3828	9.8728
-20	103.3	0.001087	1131	-374.06	2463.93	-1.3750	9.8357
-19	113.6	0.001088	1032	-372.10	2465.78	-1.3673	9.7989
-18	124.9	0.001088	942.5	-370.13	2467.63	-1.3596	9.7624
-17	137.2	0.001088	861.2	-368.15	2469.47	-1.3518	9.7262
-16	150.7	0.001088	787.5	-366.17	2471.32	-1.3441	9.6903

(continued)

TABLE A.1SI *(cont.)*

Temp. °C	Absolute Pressure, Pa $P_{w,s}$	Specific Volume, m^3/kg		Enthalpy, kJ/kg		Entropy, kJ/kg · K	
		Sat. Solid v_i	Sat. Vapor v_g	Sat. Solid h_i	Sat. Vapor h_g	Sat. Solid s_i	Sat. Vapor s_g
-15	165.3	0.001088	720.6	-364.18	2473.17	-1.3364	9.6547
-14	181.2	0.001088	659.9	-362.18	2475.01	-1.3287	9.6194
-13	198.5	0.001089	604.7	-360.18	2476.86	-1.3210	9.5845
-12	217.3	0.001089	554.5	-358.17	2478.70	-1.3132	9.5498
-11	237.7	0.001089	508.8	-356.15	2480.55	-1.3055	9.5154
-10	259.9	0.001089	467.2	-354.12	2482.39	-1.2978	9.4813
-9	283.9	0.001089	429.2	-352.08	2484.24	-1.2901	9.4475
-8	310.0	0.001090	394.6	-350.04	2486.08	-1.2824	9.4139
-7	338.2	0.001090	363.1	-347.99	2487.92	-1.2746	9.3807
-6	368.7	0.001090	334.3	-345.93	2489.76	-1.2669	9.3477
-5	401.8	0.001090	307.9	-343.87	2491.60	-1.2592	9.3150
-4	437.5	0.001090	283.8	-341.80	2493.44	-1.2515	9.2825
-3	476.1	0.001090	261.8	-339.72	2495.28	-1.2438	9.2504
-2	517.7	0.001091	241.6	-337.63	2497.12	-1.2361	9.2185
-1	562.7	0.001091	223.1	-335.53	2498.95	-1.2284	9.1868
0	611.2	0.001091	206.2	-333.43	2500.79	-1.2206	9.1554

(continued)

TABLE A.1SI *(cont.)*

Temp. °C	Absolute Pressure, MPa $P_{w,s}$	Specific Volume, m^3/kg		Enthalpy, kJ/kg		Entropy, kJ/kg · K	
		Sat. Liquid v_f	Sat. Vapor v_g	Sat. Liquid h_f	Sat. Vapor h_g	Sat. Liquid s_f	Sat. Vapor s_g
0	0.0006109	0.0009997	206.3	0.0	2500.89	0.0	9.1557
1	0.0006567	0.0009998	192.6	4.04	2502.73	0.0147	9.1291
2	0.0007056	0.0009998	179.9	8.13	2504.57	0.0296	9.1027
3	0.0007577	0.0009998	168.1	12.23	2506.42	0.0445	9.0765
4	0.0008131	0.0009999	157.2	16.33	2508.26	0.0593	9.0506
5	0.0008721	0.0009999	147.1	20.44	2510.10	0.0741	9.0249
6	0.0009349	0.001000	137.7	24.56	2511.94	0.0889	8.9995
7	0.001002	0.001000	129.0	28.69	2513.78	0.1037	8.9743
8	0.001072	0.001000	120.9	32.83	2515.62	0.1184	8.9493
9	0.001148	0.001000	113.4	36.97	2517.46	0.1331	8.9245
10	0.001228	0.001000	106.4	41.12	2519.30	0.1478	8.9000
11	0.001312	0.001001	99.86	45.28	2521.13	0.1625	8.8757
12	0.001402	0.001001	93.78	49.44	2522.97	0.1771	8.8516
13	0.001497	0.001001	88.12	53.61	2524.80	0.1917	8.8277
14	0.001598	0.001001	82.85	57.78	2526.64	0.2062	8.8040
15	0.001705	0.001001	77.93	61.95	2528.47	0.2207	8.7806
16	0.001818	0.001001	73.33	66.13	2530.30	0.2352	8.7573
17	0.001938	0.001001	69.04	70.32	2532.13	0.2497	8.7343
18	0.002064	0.001002	65.04	74.51	2533.96	0.2641	8.7115
19	0.002197	0.001002	61.29	78.71	2535.79	0.2785	8.6888
20	0.002339	0.001002	57.79	82.90	2537.61	0.2928	8.6664
21	0.002487	0.001002	54.51	87.10	2539.44	0.3071	8.6442
22	0.002644	0.001002	51.45	91.31	2541.26	0.3214	8.6221
23	0.002810	0.001003	48.57	95.51	2543.09	0.3356	8.6003
24	0.002985	0.001003	45.88	99.72	2544.91	0.3498	8.5786
25	0.003169	0.001003	43.36	103.93	2546.73	0.3640	8.5571
26	0.003363	0.001003	40.99	108.15	2548.55	0.3781	8.5358
27	0.003567	0.001004	38.77	112.36	2550.36	0.3921	8.5147
28	0.003782	0.001004	36.69	116.39	2552.18	0.4056	8.4939
29	0.004008	0.001004	34.73	120.69	2553.99	0.4198	8.4731
30	0.004246	0.001005	32.89	124.99	2555.80	0.4340	8.4525
31	0.004496	0.001005	31.16	129.29	2557.61	0.4481	8.4321
32	0.004759	0.001005	29.54	133.58	2559.42	0.4622	8.4119
33	0.005034	0.001006	28.01	137.88	2561.23	0.4763	8.3918
34	0.005324	0.001006	26.57	141.91	2563.03	0.4894	8.3720
35	0.005628	0.001006	25.22	146.13	2564.83	0.5032	8.3523
36	0.005947	0.001007	23.94	150.36	2566.64	0.5168	8.3327
37	0.006281	0.001007	22.74	154.58	2568.43	0.5305	8.3133
38	0.006632	0.001007	21.60	158.81	2570.23	0.5441	8.2941
39	0.006999	0.001008	20.53	163.03	2572.02	0.5576	8.2751
40	0.007384	0.001008	19.52	167.26	2573.82	0.5712	8.2562
41	0.007786	0.001008	18.57	171.48	2575.61	0.5846	8.2374
42	0.008208	0.001009	17.67	175.71	2577.40	0.5981	8.2188
43	0.008649	0.001009	16.82	179.93	2579.18	0.6114	8.2004
44	0.009111	0.001010	16.02	184.16	2580.97	0.6248	8.1821
45	0.009593	0.001010	15.26	188.38	2582.75	0.6381	8.1640
46	0.01010	0.001010	14.54	192.60	2584.52	0.6513	8.1460
47	0.01062	0.001011	13.86	196.83	2586.30	0.6645	8.1281
48	0.01117	0.001011	13.22	201.05	2588.07	0.6777	8.1104
49	0.01175	0.001012	12.61	205.27	2589.85	0.6908	8.0929
50	0.01235	0.001012	12.03	209.49	2591.61	0.7039	8.0755
51	0.01298	0.001013	11.49	213.71	2593.38	0.7169	8.0582
52	0.01363	0.001013	10.97	217.93	2595.14	0.7299	8.0411
53	0.01431	0.001014	10.48	222.15	2596.90	0.7429	8.0241
54	0.01502	0.001014	10.01	226.36	2598.66	0.7558	8.0072
55	0.01576	0.001014	9.569	230.58	2600.42	0.7687	7.9905
56	0.01653	0.001015	9.149	234.80	2602.17	0.7815	7.9739
57	0.01733	0.001015	8.751	239.01	2603.92	0.7943	7.9574
58	0.01817	0.001016	8.372	243.22	2605.66	0.8070	7.9410
59	0.01904	0.001017	8.013	247.43	2607.41	0.8197	7.9248

(continued)

Appendix A / Thermodynamic and Thermophysical Property Tables

TABLE A.1SI (*cont.*)

Temp. °C	Absolute Pressure, MPa $P_{w,s}$	Specific Volume, m^3/kg		Enthalpy, kJ/kg		Entropy, kJ/kg · K	
		Sat. Liquid v_f	Sat. Vapor v_g	Sat. Liquid h_f	Sat. Vapor h_g	Sat. Liquid s_f	Sat. Vapor s_g
60	0.01994	0.001017	7.671	251.64	2609.15	0.8323	7.9087
61	0.02088	0.001018	7.346	255.85	2610.88	0.8450	7.8928
62	0.02186	0.001018	7.037	260.06	2612.62	0.8575	7.8769
63	0.02288	0.001019	6.743	264.27	2614.35	0.8701	7.8612
64	0.02393	0.001019	6.463	268.48	2616.07	0.8826	7.8456
65	0.02503	0.001020	6.197	272.68	2617.80	0.8950	7.8301
66	0.02617	0.001020	5.943	276.89	2619.52	0.9074	7.8148
67	0.02736	0.001021	5.701	281.09	2621.23	0.9198	7.7995
68	0.02859	0.001021	5.471	285.29	2622.95	0.9321	7.7844
69	0.02986	0.001022	5.252	289.50	2624.66	0.9444	7.7694
70	0.03119	0.001023	5.042	293.70	2626.36	0.9567	7.7545
71	0.03256	0.001023	4.843	297.90	2628.06	0.9689	7.7397
72	0.03399	0.001024	4.652	302.10	2629.76	0.9811	7.7250
73	0.03546	0.001024	4.470	306.29	2631.46	0.9932	7.7104
74	0.03699	0.001025	4.297	310.49	2633.15	1.0053	7.6960
75	0.03858	0.001026	4.131	314.69	2634.84	1.0174	7.6816
76	0.04022	0.001026	3.973	318.88	2636.52	1.0294	7.6674
77	0.04192	0.001027	3.822	323.08	2638.2	1.0414	7.6532
78	0.04368	0.001028	3.677	327.27	2639.87	1.0534	7.6392
79	0.04550	0.001028	3.539	331.47	2641.55	1.0653	7.6252
80	0.04739	0.001029	3.407	335.66	2643.21	1.0772	7.6114
81	0.04934	0.001030	3.281	339.85	2644.88	1.0890	7.5976
82	0.05136	0.001030	3.160	344.04	2646.53	1.1008	7.5840
83	0.05345	0.001031	3.044	348.24	2648.19	1.1126	7.5704
84	0.05560	0.001032	2.934	352.43	2649.84	1.1244	7.5570
85	0.05783	0.001032	2.828	356.62	2651.48	1.1361	7.5436
86	0.06014	0.001033	2.726	360.81	2653.12	1.1478	7.5304
87	0.06252	0.001034	2.629	365.00	2654.76	1.1594	7.5172
88	0.06498	0.001034	2.536	369.20	2656.39	1.1710	7.5041
89	0.06752	0.001035	2.446	373.39	2658.02	1.1826	7.4911
90	0.07014	0.001036	2.361	377.58	2659.64	1.1942	7.4782
91	0.07284	0.001037	2.278	381.77	2661.26	1.2057	7.4654
92	0.07564	0.001037	2.200	385.96	2662.88	1.2172	7.4527
93	0.07852	0.001038	2.124	390.15	2664.48	1.2286	7.4401
94	0.08149	0.001039	2.052	394.34	2666.09	1.2400	7.4276
95	0.08455	0.001040	1.982	398.54	2667.69	1.2514	7.4151
96	0.08771	0.001040	1.915	402.73	2669.28	1.2628	7.4027
97	0.09097	0.001041	1.851	406.92	2670.87	1.2741	7.3904
98	0.09433	0.001042	1.789	411.12	2672.45	1.2854	7.3782
99	0.09778	0.001043	1.730	415.31	2674.03	1.2967	7.3661
100	0.1013	0.001043	1.673	419.51	2675.60	1.3080	7.3541
102	0.1088	0.001045	1.566	427.90	2678.73	1.3304	7.3302
104	0.1167	0.001047	1.466	436.30	2681.84	1.3527	7.3067
106	0.1251	0.001048	1.374	444.70	2684.93	1.3749	7.2834
108	0.1339	0.001050	1.289	453.11	2687.99	1.3970	7.2605
110	0.1433	0.001052	1.210	461.52	2691.03	1.4190	7.2379
112	0.1532	0.001053	1.137	469.94	2694.04	1.4409	7.2155
114	0.1636	0.001055	1.069	478.37	2697.03	1.4626	7.1934
116	0.1746	0.001057	1.006	486.80	2700.00	1.4843	7.1716
118	0.1863	0.001059	0.9466	495.46	2702.94	1.5065	7.1500
120	0.1985	0.001060	0.8918	503.83	2705.85	1.5278	7.1288
122	0.2114	0.001062	0.8408	512.21	2708.74	1.5490	7.1078
124	0.2250	0.001064	0.7932	520.61	2711.60	1.5702	7.0870
126	0.2393	0.001066	0.7488	529.01	2714.44	1.5913	7.0665
128	0.2543	0.001068	0.7073	537.42	2717.24	1.6123	7.0462
130	0.2701	0.001070	0.6685	546.08	2720.01	1.6337	7.0261
132	0.2866	0.001072	0.6323	554.59	2722.76	1.6547	7.0063
134	0.3040	0.001074	0.5983	563.11	2725.47	1.6757	6.9867
136	0.3222	0.001076	0.5666	571.64	2728.15	1.6965	6.9673
138	0.3413	0.001078	0.5368	580.18	2730.80	1.7173	6.9481

(*continued*)

Appendix A / Thermodynamic and Thermophysical Property Tables

TABLE A.1SI (cont.)

Temp. °C	Absolute Pressure, MPa $P_{w,s}$	Specific Volume, m³/kg		Enthalpy, kJ/kg		Entropy, kJ/kg · K	
		Sat. Liquid v_f	Sat. Vapor v_g	Sat. Liquid h_f	Sat. Vapor h_g	Sat. Liquid s_f	Sat. Vapor s_g
140	0.3613	0.001080	0.5089	588.74	2733.42	1.7380	6.9291
142	0.3822	0.001082	0.4827	597.31	2736.01	1.7587	6.9103
144	0.4041	0.001084	0.4581	605.90	2738.56	1.7792	6.8917
146	0.4270	0.001086	0.4350	614.49	2741.07	1.7997	6.8733
148	0.4509	0.001089	0.4132	623.10	2743.55	1.8202	6.8551
150	0.4758	0.001091	0.3928	631.73	2746.00	1.8405	6.8370
152	0.5019	0.001093	0.3735	640.37	2748.40	1.8608	6.8192
154	0.5291	0.001095	0.3554	649.02	2750.77	1.8811	6.8015
156	0.5575	0.001098	0.3384	657.69	2753.10	1.9012	6.7840
158	0.5870	0.001100	0.3223	666.38	2755.39	1.9214	6.7666
160	0.6178	0.001102	0.3071	675.07	2757.64	1.9414	6.7494
162	0.6499	0.001105	0.2927	683.79	2759.85	1.9614	6.7323
164	0.6833	0.001107	0.2792	692.52	2762.02	1.9813	6.7154
166	0.7180	0.001109	0.2664	701.26	2764.14	2.0012	6.6986
168	0.7541	0.001112	0.2543	710.02	2766.22	2.0210	6.6820
170	0.7917	0.001114	0.2428	718.80	2768.25	2.0408	6.6655
172	0.8307	0.001117	0.2320	727.59	2770.24	2.0605	6.6491
174	0.8712	0.001120	0.2217	736.40	2772.18	2.0801	6.6329
176	0.9132	0.001122	0.2120	745.23	2774.08	2.0997	6.6168
178	0.9569	0.001125	0.2028	754.07	2775.92	2.1192	6.6008
180	1.002	0.001127	0.1940	762.93	2777.72	2.1387	6.5849
182	1.049	0.001130	0.1858	771.80	2779.47	2.1581	6.5691
184	1.098	0.001133	0.1779	780.69	2781.16	2.1775	6.5535
186	1.148	0.001136	0.1704	789.60	2782.80	2.1968	6.5379
188	1.200	0.001139	0.1633	798.53	2784.39	2.2161	6.5224
190	1.254	0.001141	0.1565	807.47	2785.92	2.2353	6.5071
192	1.310	0.001144	0.1501	816.43	2787.40	2.2545	6.4918
194	1.368	0.001147	0.1440	825.40	2788.82	2.2736	6.4766
196	1.428	0.001150	0.1382	834.40	2790.19	2.2927	6.4615
198	1.490	0.001153	0.1326	843.41	2791.49	2.3117	6.4464
200	1.554	0.001156	0.1274	852.44	2792.74	2.3306	6.4315
202	1.620	0.001159	0.1223	861.49	2793.93	2.3496	6.4166
204	1.688	0.001163	0.1175	870.56	2795.05	2.3685	6.4017
206	1.759	0.001166	0.1130	879.65	2796.11	2.3873	6.3870
208	1.831	0.001169	0.1086	888.76	2797.11	2.4061	6.3723
210	1.906	0.001172	0.1044	897.88	2798.04	2.4248	6.3577
212	1.984	0.001176	0.1004	907.04	2798.90	2.4435	6.3431
214	2.063	0.001179	0.09663	916.20	2799.70	2.4622	6.3286
216	2.146	0.001183	0.09299	925.39	2800.43	2.4808	6.3141
218	2.230	0.001186	0.08952	934.61	2801.09	2.4994	6.2996
220	2.318	0.001190	0.08619	943.84	2801.68	2.5180	6.2853
222	2.408	0.001193	0.08301	953.10	2802.19	2.5365	6.2709
224	2.500	0.001197	0.07996	962.38	2802.63	2.5550	6.2566
226	2.596	0.001201	0.07705	971.70	2803.00	2.5735	6.2423
228	2.694	0.001205	0.07426	981.03	2803.29	2.5919	6.2280
230	2.795	0.001208	0.07158	990.39	2803.51	2.6103	6.2138
232	2.899	0.001212	0.06902	999.78	2803.64	2.6286	6.1996
234	3.006	0.001216	0.06656	1009.19	2803.70	2.6470	6.1854
236	3.115	0.001220	0.06420	1018.64	2803.67	2.6653	6.1712
238	3.228	0.001225	0.06194	1028.11	2803.56	2.6836	6.1571
240	3.344	0.001229	0.05976	1037.61	2803.37	2.7019	6.1429
242	3.463	0.001233	0.05768	1047.16	2803.09	2.7202	6.1288
244	3.586	0.001237	0.05568	1056.73	2802.72	2.7384	6.1146
246	3.711	0.001242	0.05375	1066.33	2802.27	2.7566	6.1005
248	3.840	0.001246	0.05190	1075.98	2801.72	2.7749	6.0863
250	3.973	0.001251	0.05013	1085.66	2801.08	2.7931	6.0721

TABLE A.2E Thermodynamic Properties of Ammonia at Saturation (English Units)

Temp. °F	Pressure		Liquid, Density lbm/ft³ 1/v_f	Vapor, Sp. Vol. ft³/lbm v_g	Enthalpy, Datum −40 °F, Btu/lbm		Entropy, Datum −40 °F, Btu/lbm · °R	
	psia	psig			Liquid h_f	Vapor h_g	Liquid s_f	Vapor s_g
-100.	1.24	*27.5	45.51	182.342	-63.0	572.2	-0.1618	1.6039
-95.	1.52	*26.9	45.32	149.990	-57.7	574.4	-0.1473	1.5858
-90.	1.86	*26.2	45.12	124.140	-52.4	576.6	-0.1330	1.5683
-85.	2.27	*25.4	44.92	103.350	-47.2	578.8	-0.1188	1.5515
-80.	2.74	*24.4	44.72	86.526	-41.9	580.9	-0.1048	1.5352
-75.	3.29	*23.3	44.52	72.838	-36.7	583.0	-0.0912	1.5195
-70.	3.94	*22.0	44.31	61.624	-31.5	585.1	-0.0777	1.5042
-65.	4.69	*20.4	44.11	52.393	-26.2	587.1	-0.0643	1.4895
-60.	5.55	*18.7	43.91	44.753	-21.0	589.2	-0.0511	1.4752
-58.	5.93	*17.9	43.82	42.072	-18.9	590.0	-0.0459	1.4696
-56.	6.33	*17.1	43.74	39.579	-16.8	590.8	-0.0407	1.4641
-54.	6.75	*16.2	43.66	37.260	-14.7	591.6	-0.0355	1.4586
-52.	7.20	*15.3	43.58	35.100	-12.6	592.4	-0.0304	1.4532
-50.	7.66	*14.4	43.49	33.087	-10.5	593.1	-0.0253	1.4479
-48.	8.16	*13.3	43.41	31.211	-8.4	593.9	-0.0202	1.4426
-46.	8.68	*12.3	43.33	29.460	-6.3	594.7	-0.0151	1.4374
-44.	9.23	*11.2	43.24	27.824	-4.2	595.4	-0.0100	1.4323
-42.	9.80	*10.0	43.16	26.296	-2.1	596.2	-0.0050	1.4272
-40.	10.40	*8.8	43.07	24.867	0.0	597.0	0.0000	1.4222
-38.	11.04	*7.5	42.99	23.529	2.1	597.7	0.0050	1.4172
-36.	11.70	*6.1	42.90	22.277	4.2	598.5	0.0100	1.4123
-34.	12.40	*4.7	42.82	21.104	6.3	599.2	0.0149	1.4075
-32.	13.13	*3.2	42.73	20.003	8.4	599.9	0.0198	1.4027
-30.	13.90	*1.6	42.65	18.971	10.5	600.7	0.0247	1.3979
-29.	14.29	*0.8	42.61	18.479	11.6	601.0	0.0272	1.3956
-28.	14.70	0.0	42.56	18.002	12.6	601.4	0.0296	1.3932
-27.	15.11	0.4	42.52	17.540	13.7	601.7	0.0321	1.3909
-26.	15.54	0.8	42.48	17.092	14.7	602.1	0.0345	1.3886
-25.	15.97	1.3	42.43	16.657	15.8	602.4	0.0369	1.3863
-24.	16.42	1.7	42.39	16.236	16.9	602.8	0.0393	1.3840
-23.	16.87	2.2	42.35	15.828	17.9	603.2	0.0418	1.3817
-22.	17.33	2.6	42.30	15.431	19.0	603.5	0.0442	1.3795
-21.	17.81	3.1	42.26	15.047	20.0	603.9	0.0466	1.3772
-20.	18.29	3.6	42.22	14.674	21.1	604.2	0.0490	1.3750
-19.	18.78	4.1	42.17	14.312	22.1	604.5	0.0514	1.3728
-18.	19.29	4.6	42.13	13.961	23.2	604.9	0.0538	1.3706
-17.	19.81	5.1	42.09	13.620	24.3	605.2	0.0562	1.3684
-16.	20.33	5.6	42.04	13.289	25.3	605.6	0.0586	1.3662
-15.	20.87	6.2	42.00	12.968	26.4	605.9	0.0610	1.3640
-14.	21.42	6.7	41.96	12.656	27.5	606.3	0.0634	1.3618
-13.	21.98	7.3	41.91	12.353	28.5	606.6	0.0657	1.3597
-12.	22.56	7.9	41.87	12.059	29.6	606.9	0.0681	1.3575
-11.	23.14	8.4	41.82	11.773	30.7	607.3	0.0705	1.3554
-10.	23.74	9.0	41.78	11.495	31.7	607.6	0.0729	1.3533
-9.	24.35	9.7	41.74	11.225	32.8	607.9	0.0752	1.3511
-8.	24.97	10.3	41.69	10.963	33.9	608.3	0.0776	1.3490
-7.	25.60	10.9	41.65	10.708	34.9	608.6	0.0800	1.3470
-6.	26.25	11.6	41.60	10.460	36.0	608.9	0.0823	1.3449
-5.	26.91	12.2	41.56	10.219	37.1	609.2	0.0847	1.3428
-4.	27.59	12.9	41.51	9.985	38.1	609.5	0.0870	1.3408
-3.	28.27	13.6	41.47	9.757	39.2	609.9	0.0894	1.3387
-2.	28.97	14.3	41.43	9.535	40.3	610.2	0.0917	1.3367
-1.	29.69	15.0	41.38	9.320	41.4	610.5	0.0940	1.3346
0.	30.42	15.7	41.34	9.110	42.4	610.8	0.0964	1.3326
1.	31.16	16.5	41.29	8.906	43.5	611.1	0.0987	1.3306
2.	31.92	17.2	41.25	8.707	44.6	611.4	0.1010	1.3286
3.	32.69	18.0	41.20	8.514	45.7	611.8	0.1034	1.3266
4.	33.47	18.8	41.16	8.326	46.8	612.1	0.1057	1.3246
5.	34.27	19.6	41.11	8.143	47.8	612.4	0.1080	1.3227
6.	35.09	20.4	41.07	7.965	48.9	612.7	0.1103	1.3207
7.	35.92	21.2	41.02	7.791	50.0	613.0	0.1126	1.3188
8.	36.77	22.1	40.98	7.622	51.1	613.3	0.1149	1.3168
9.	37.63	22.9	40.93	7.458	52.2	613.6	0.1173	1.3149
10.	38.51	23.8	40.89	7.297	53.3	613.9	0.1196	1.3130

* Inches of mercury below one standard atmosphere (29.92 in)

(continued)

Appendix A / Thermodynamic and Thermophysical Property Tables

TABLE A.2E (cont.)

Temp. °F	Pressure		Liquid, Density lbm/ft³ $1/v_f$	Vapor, Sp. Vol. ft³/lbm v_g	Enthalpy, Datum −40 °F, Btu/lbm		Entropy, Datum −40 °F, Btu/lbm · °R	
	psia	psig			Liquid h_f	Vapor h_g	Liquid s_f	Vapor s_g
11.	39.41	24.7	40.84	7.141	54.3	614.2	0.1219	1.3111
12.	40.32	25.6	40.79	6.989	55.4	614.5	0.1242	1.3092
13.	41.25	26.5	40.75	6.841	56.5	614.8	0.1265	1.3073
14.	42.19	27.5	40.70	6.696	57.6	615.0	0.1288	1.3054
15.	43.15	28.5	40.66	6.556	58.7	615.3	0.1311	1.3035
16.	44.13	29.4	40.61	6.418	59.8	615.6	0.1333	1.3016
17.	45.13	30.4	40.56	6.285	60.9	615.9	0.1356	1.2998
18.	46.14	31.4	40.52	6.154	62.0	616.2	0.1379	1.2979
19.	47.17	32.5	40.47	6.027	63.1	616.5	0.1402	1.2961
20.	48.22	33.5	40.43	5.903	64.2	616.8	0.1425	1.2942
21.	49.29	34.6	40.38	5.782	65.3	617.0	0.1447	1.2924
22.	50.38	35.7	40.33	5.664	66.4	617.3	0.1470	1.2906
23.	51.48	36.8	40.29	5.549	67.5	617.6	0.1493	1.2888
24.	52.61	37.9	40.24	5.437	68.6	617.8	0.1515	1.2870
25.	53.75	39.1	40.19	5.328	69.7	618.1	0.1538	1.2852
26.	54.91	40.2	40.15	5.221	70.8	618.4	0.1561	1.2834
27.	56.10	41.4	40.10	5.117	71.9	618.7	0.1583	1.2816
28.	57.30	42.6	40.05	5.015	73.0	618.9	0.1606	1.2798
29.	58.52	43.8	40.01	4.916	74.1	619.2	0.1628	1.2781
30.	59.76	45.1	39.96	4.819	75.2	619.4	0.1651	1.2763
31.	61.03	46.3	39.91	4.724	76.3	619.7	0.1673	1.2746
32.	62.31	47.6	39.86	4.632	77.4	619.9	0.1696	1.2728
33.	63.62	48.9	39.82	4.541	78.5	620.2	0.1718	1.2711
34.	64.94	50.2	39.77	4.453	79.6	620.5	0.1741	1.2694
35.	66.29	51.6	39.72	4.367	80.8	620.7	0.1763	1.2676
36.	67.66	53.0	39.67	4.283	81.9	621.0	0.1785	1.2659
37.	69.05	54.4	39.63	4.201	83.0	621.2	0.1808	1.2642
38.	70.46	55.8	39.58	4.121	84.1	621.4	0.1830	1.2625
39.	71.90	57.2	39.53	4.043	85.2	621.7	0.1852	1.2608
40.	73.36	58.7	39.48	3.966	86.3	621.9	0.1875	1.2591
41.	74.84	60.1	39.44	3.891	87.4	622.2	0.1897	1.2575
42.	76.34	61.6	39.39	3.818	88.6	622.4	0.1919	1.2558
43.	77.87	63.2	39.34	3.747	89.7	622.6	0.1941	1.2541
44.	79.42	64.7	39.29	3.677	90.8	622.9	0.1963	1.2525
45.	81.00	66.3	39.24	3.609	91.9	623.1	0.1985	1.2508
46.	82.60	67.9	39.19	3.542	93.1	623.3	0.2007	1.2492
47.	84.22	69.5	39.14	3.477	94.2	623.5	0.2029	1.2475
48.	85.87	71.2	39.10	3.413	95.3	623.8	0.2052	1.2459
49.	87.54	72.8	39.05	3.351	96.4	624.0	0.2074	1.2443
50.	89.24	74.5	39.00	3.290	97.6	624.2	0.2096	1.2426
51.	90.96	76.3	38.95	3.230	98.7	624.4	0.2117	1.2410
52.	92.71	78.0	38.90	3.171	99.8	624.6	0.2139	1.2394
53.	94.49	79.8	38.85	3.114	100.9	624.8	0.2161	1.2378
54.	96.29	81.6	38.80	3.058	102.1	625.0	0.2183	1.2362
55.	98.12	83.4	38.75	3.004	103.2	625.2	0.2205	1.2346
56.	99.97	85.3	38.70	2.950	104.3	625.4	0.2227	1.2330
57.	101.85	87.2	38.65	2.898	105.5	625.6	0.2249	1.2314
58.	103.76	89.1	38.60	2.847	106.6	625.8	0.2270	1.2299
59.	105.69	91.0	38.55	2.796	107.8	626.0	0.2292	1.2283
60.	107.66	93.0	38.50	2.747	108.9	626.2	0.2314	1.2267
61.	109.65	95.0	38.45	2.699	110.0	626.4	0.2336	1.2251
62.	111.67	97.0	38.40	2.652	111.2	626.6	0.2357	1.2236
63.	113.71	99.0	38.35	2.606	112.3	626.8	0.2379	1.2220
64.	115.79	101.1	38.30	2.561	113.4	627.0	0.2400	1.2205
65.	117.90	103.2	38.25	2.517	114.6	627.2	0.2422	1.2189
66.	120.03	105.3	38.20	2.474	115.7	627.3	0.2444	1.2174
67.	122.19	107.5	38.15	2.431	116.8	627.5	0.2464	1.2159
68.	124.39	109.7	38.10	2.390	118.0	627.7	0.2486	1.2144
69.	126.61	111.9	38.05	2.349	119.1	627.9	0.2507	1.2128
70.	128.86	114.2	38.00	2.309	120.3	628.0	0.2529	1.2113

(continued)

TABLE A.2E (*cont.*)

Temp. °F	Pressure		Liquid, Density lbm/ft³ $1/v_f$	Vapor, Sp. Vol. ft³/lbm v_g	Enthalpy, Datum −40 °F, Btu/lbm		Entropy, Datum −40 °F, Btu/lbm · °R	
	psia	psig			Liquid h_f	Vapor h_g	Liquid s_f	Vapor s_g
71.	131.15	116.5	37.94	2.270	121.4	628.2	0.2550	1.2098
72.	133.46	118.8	37.89	2.232	122.5	628.4	0.2571	1.2083
73.	135.81	121.1	37.84	2.195	123.7	628.5	0.2593	1.2068
74.	138.19	123.5	37.79	2.158	124.8	628.7	0.2614	1.2053
75.	140.60	125.9	37.74	2.122	126.0	628.8	0.2635	1.2038
76.	143.04	128.3	37.69	2.087	127.1	629.0	0.2656	1.2023
77.	145.51	130.8	37.63	2.052	128.3	629.1	0.2678	1.2008
78.	148.02	133.3	37.58	2.018	129.4	629.3	0.2699	1.1993
79.	150.56	135.9	37.53	1.985	130.6	629.4	0.2720	1.1979
80.	153.13	138.4	37.48	1.952	131.7	629.6	0.2741	1.1964
81.	155.74	141.0	37.42	1.920	132.9	629.7	0.2762	1.1949
82.	158.37	143.7	37.37	1.889	134.1	629.8	0.2783	1.1934
83.	161.05	146.4	37.32	1.858	135.2	630.0	0.2804	1.1920
84.	163.75	149.1	37.27	1.828	136.4	630.1	0.2826	1.1905
85.	166.50	151.8	37.21	1.799	137.5	630.2	0.2847	1.1891
86.	169.27	154.6	37.16	1.770	138.7	630.3	0.2868	1.1876
87.	172.08	157.4	37.11	1.741	139.8	630.5	0.2889	1.1861
88.	174.93	160.2	37.05	1.713	141.0	630.6	0.2909	1.1847
89.	177.81	163.1	37.00	1.686	142.2	630.7	0.2930	1.1833
90.	180.73	166.0	36.95	1.659	143.3	630.8	0.2951	1.1818
91.	183.68	169.0	36.89	1.633	144.5	630.9	0.2972	1.1804
92.	186.67	172.0	36.84	1.607	145.7	631.0	0.2993	1.1789
93.	189.70	175.0	36.78	1.582	146.8	631.1	0.3014	1.1775
94.	192.77	178.1	36.73	1.557	148.0	631.2	0.3035	1.1761
95.	195.87	181.2	36.67	1.532	149.2	631.3	0.3055	1.1746
96.	199.01	184.3	36.62	1.508	150.3	631.4	0.3076	1.1732
97.	202.19	187.5	36.57	1.485	151.5	631.5	0.3097	1.1718
98.	205.40	190.7	36.51	1.462	152.7	631.6	0.3118	1.1704
99.	208.66	194.0	36.46	1.439	153.8	631.7	0.3138	1.1690
100.	211.95	197.3	36.40	1.417	155.0	631.7	0.3159	1.1675
101.	215.28	200.6	36.34	1.395	156.2	631.8	0.3180	1.1661
102.	218.66	204.0	36.29	1.373	157.4	631.9	0.3200	1.1647
103.	222.07	207.4	36.23	1.352	158.5	632.0	0.3221	1.1633
104.	225.52	210.8	36.18	1.332	159.7	632.0	0.3241	1.1619
105.	229.01	214.3	36.12	1.311	160.9	632.1	0.3262	1.1605
106.	232.55	217.9	36.06	1.291	162.1	632.1	0.3283	1.1591
107.	236.12	221.4	36.01	1.272	163.2	632.2	0.3303	1.1577
108.	239.74	225.0	35.95	1.252	164.4	632.2	0.3324	1.1563
109.	243.40	228.7	35.90	1.233	165.6	632.3	0.3344	1.1549
110.	247.10	232.4	35.84	1.215	166.8	632.3	0.3365	1.1535
111.	250.84	236.1	35.78	1.197	168.0	632.4	0.3385	1.1521
112.	254.62	239.9	35.72	1.179	169.2	632.4	0.3406	1.1507
113.	258.45	243.8	35.67	1.161	170.4	632.4	0.3426	1.1493
114.	262.32	247.6	35.61	1.144	171.6	632.5	0.3446	1.1480
115.	266.23	251.5	35.55	1.127	172.7	632.5	0.3467	1.1466
116.	270.19	255.5	35.49	1.110	173.9	632.5	0.3487	1.1452
117.	274.19	259.5	35.43	1.093	175.1	632.5	0.3508	1.1438
118.	278.24	263.5	35.38	1.077	176.3	632.6	0.3528	1.1424
119.	282.33	267.6	35.32	1.061	177.5	632.6	0.3548	1.1410
120.	286.47	271.8	35.26	1.046	178.7	632.6	0.3569	1.1396
122.	294.88	280.2	35.14	1.015	181.1	632.6	0.3609	1.1369
124.	303.47	288.8	35.02	0.986	183.5	632.6	0.3650	1.1341
126.	312.25	297.6	34.90	0.957	186.0	632.5	0.3691	1.1314
128.	321.23	306.5	34.78	0.929	188.4	632.5	0.3731	1.1286
130.	330.39	315.7	34.66	0.903	190.8	632.4	0.3771	1.1259
132.	339.75	325.1	34.54	0.877	193.3	632.3	0.3812	1.1231
134.	349.31	334.6	34.41	0.852	195.7	632.2	0.3852	1.1204
136.	359.06	344.4	34.29	0.828	198.2	632.1	0.3893	1.1176
138.	369.02	354.3	34.17	0.804	200.6	632.0	0.3933	1.1149
140.	379.19	364.5	34.04	0.782	203.1	631.8	0.3973	1.1121

Appendix A / Thermodynamic and Thermophysical Property Tables

TABLE A.2SI Thermodynamic Properties of Ammonia at Saturation (SI Units)

Temp. °C	Pressure MPa	Liquid Density kg/m³ $1/v_f$	Vapor Sp. Vol. m³/kg v_g	Enthalpy, Datum −40 °C, kJ/kg Sat. Liquid h_f	Enthalpy, Datum −40 °C, kJ/kg Sat. Vapor h_g	Entropy, Datum −40 °C, kJ/(kg · K) Sat. Liquid s_f	Entropy, Datum −40 °C, kJ/(kg · K) Sat. Vapor s_g
-70	0.0109	725.2	9.01161	-131.80	1337.11	-0.6048	6.6259
-68	0.0126	723.0	7.86626	-122.97	1340.76	-0.5615	6.5734
-66	0.0146	720.7	6.88750	-114.14	1344.38	-0.5187	6.5222
-64	0.0167	718.4	6.04836	-105.32	1347.98	-0.4764	6.4722
-62	0.0192	716.1	5.32667	-96.49	1351.55	-0.4344	6.4235
-60	0.0219	713.7	4.70458	-87.82	1355.08	-0.3935	6.3759
-58	0.0249	711.4	4.16592	-79.03	1358.59	-0.3525	6.3295
-56	0.0283	709.1	3.69850	-70.24	1362.06	-0.3118	6.2841
-54	0.0321	706.7	3.29177	-61.46	1365.50	-0.2716	6.2398
-52	0.0363	704.4	2.93690	-52.68	1368.91	-0.2317	6.1964
-50	0.0409	702.0	2.62645	-43.90	1372.28	-0.1923	6.1540
-49	0.0433	700.8	2.48589	-39.51	1373.95	-0.1727	6.1332
-48	0.0459	699.6	2.35417	-35.13	1375.61	-0.1531	6.1126
-47	0.0487	698.4	2.23066	-30.74	1377.26	-0.1337	6.0922
-46	0.0515	697.2	2.11478	-26.35	1378.90	-0.1144	6.0721
-45	0.0545	696.0	2.00599	-21.96	1380.53	-0.0951	6.0521
-44	0.0577	694.8	1.90379	-17.57	1382.15	-0.0759	6.0324
-43	0.0609	693.6	1.80774	-13.18	1383.77	-0.0568	6.0129
-42	0.0644	692.4	1.71741	-8.79	1385.37	-0.0378	5.9936
-41	0.068	691.2	1.63241	-4.40	1386.96	-0.0189	5.9745
-40	0.0717	690.0	1.55238	0.00	1388.54	0.0000	5.9556
-39	0.0757	688.8	1.47700	4.40	1390.11	0.0188	5.9369
-38	0.0798	687.5	1.40594	8.80	1391.67	0.0375	5.9183
-37	0.0840	686.3	1.33894	13.20	1393.22	0.0562	5.9000
-36	0.0885	685.1	1.27572	17.61	1394.76	0.0748	5.8819
-35	0.0932	683.9	1.21604	22.02	1396.29	0.0933	5.8639
-34	0.0980	682.6	1.15968	26.43	1397.80	0.1118	5.8461
-33	0.1031	681.4	1.10641	30.85	1399.31	0.1302	5.8285
-32	0.1083	680.1	1.05606	35.27	1400.80	0.1485	5.8111
-31	0.1138	678.9	1.00843	39.70	1402.28	0.1668	5.7938
-30	0.1195	677.7	0.96335	44.13	1403.76	0.1850	5.7768
-29	0.1254	676.4	0.92068	48.56	1405.22	0.2032	5.7598
-28	0.1316	675.1	0.88025	53.00	1406.66	0.2213	5.7431
-27	0.1380	673.9	0.84194	57.45	1408.10	0.2393	5.7265
-26	0.1447	672.6	0.80562	61.90	1409.52	0.2574	5.7100
-25	0.1516	671.4	0.77117	66.35	1410.94	0.2753	5.6937
-24	0.1587	670.1	0.73847	70.82	1412.34	0.2932	5.6776
-23	0.1662	668.8	0.70744	75.29	1413.72	0.3111	5.6616
-22	0.1739	667.6	0.67796	79.76	1415.10	0.3289	5.6458
-21	0.1819	666.3	0.64994	84.24	1416.46	0.3466	5.6301
-20	0.1902	665.0	0.62332	88.73	1417.81	0.3644	5.6145
-19	0.1988	663.7	0.59799	93.23	1419.15	0.3820	5.5991
-18	0.2077	662.4	0.57390	97.73	1420.48	0.3997	5.5839
-17	0.2169	661.1	0.55097	102.24	1421.79	0.4172	5.5687
-16	0.2264	659.8	0.52914	106.76	1423.09	0.4348	5.5537
-15	0.2363	658.5	0.50834	111.28	1424.37	0.4523	5.5388
-14	0.2465	657.2	0.48853	115.81	1425.65	0.4697	5.5241
-13	0.2571	655.9	0.46964	120.35	1426.91	0.4872	5.5095
-12	0.2680	654.6	0.45162	124.90	1428.15	0.5045	5.4950
-11	0.2792	653.3	0.43444	129.45	1429.38	0.5219	5.4806
-10	0.2909	652.0	0.41804	134.02	1430.60	0.5392	5.4664
-9	0.3029	650.7	0.40238	138.59	1431.81	0.5565	5.4522
-8	0.3153	649.3	0.38742	143.16	1433.00	0.5737	5.4382
-7	0.3281	648.0	0.37314	147.75	1434.17	0.5909	5.4243
-6	0.3413	646.7	0.35948	152.34	1435.34	0.6080	5.4105
-5	0.3550	645.3	0.34643	156.94	1436.49	0.6251	5.3969
-4	0.3690	644.0	0.33395	161.55	1437.62	0.6422	5.3833
-3	0.3835	642.6	0.32200	166.17	1438.74	0.6593	5.3698
-2	0.3984	641.3	0.31057	170.80	1439.84	0.6763	5.3565
-1	0.4138	639.9	0.29963	175.43	1440.93	0.6932	5.3432

(continued)

TABLE A.2SI (*cont.*)

Temp. °C	Pressure MPa	Liquid Density kg/m³ $1/v_f$	Vapor Sp. Vol. m³/kg v_g	Enthalpy, Datum −40°C, kJ/kg		Entropy, Datum −40°C, kJ/(kg · K)	
				Sat. Liquid h_f	Sat. Vapor h_g	Sat. Liquid s_f	Sat. Vapor s_g
0	0.4296	638.6	0.28915	180.07	1442.00	0.7102	5.3301
1	0.4459	637.2	0.27911	184.72	1443.06	0.7271	5.3170
2	0.4627	635.8	0.26948	189.38	1444.10	0.7439	5.3041
3	0.4800	634.5	0.26026	194.05	1445.13	0.7607	5.2912
4	0.4977	633.1	0.25142	198.72	1446.14	0.7775	5.2784
5	0.5160	631.7	0.24293	203.40	1447.14	0.7943	5.2657
6	0.5348	630.3	0.23479	208.09	1448.11	0.8110	5.2531
7	0.5541	628.9	0.22698	212.78	1449.08	0.8277	5.2406
8	0.5739	627.5	0.21948	217.49	1450.02	0.8443	5.2282
9	0.5943	626.1	0.21228	222.20	1450.95	0.8609	5.2159
10	0.6153	624.7	0.20536	226.92	1451.86	0.8775	5.2036
11	0.6368	623.3	0.19871	231.65	1452.76	0.8941	5.1915
12	0.6589	621.8	0.19232	236.38	1453.63	0.9106	5.1794
13	0.6816	620.4	0.18617	241.12	1454.49	0.9270	5.1674
14	0.7049	619.0	0.18026	245.87	1455.33	0.9435	5.1554
15	0.7287	617.5	0.17458	250.63	1456.16	0.9599	5.1435
16	0.7532	616.1	0.16910	255.39	1456.96	0.9762	5.1317
17	0.7784	614.7	0.16384	260.16	1457.75	0.9925	5.1200
18	0.8041	613.2	0.15877	264.94	1458.51	1.0088	5.1084
19	0.8305	611.7	0.15388	269.72	1459.26	1.0251	5.0968
20	0.8576	610.3	0.1492	274.38	1460.00	1.0409	5.0853
21	0.8854	608.8	0.14467	279.17	1460.71	1.0570	5.0738
22	0.9138	607.3	0.14030	283.97	1461.40	1.0732	5.0624
23	0.9429	605.8	0.13609	288.77	1462.07	1.0892	5.0511
24	0.9727	604.3	0.13203	293.58	1462.72	1.1053	5.0398
25	1.0033	602.8	0.12811	298.39	1463.35	1.1213	5.0286
26	1.0345	601.3	0.12433	303.22	1463.95	1.1373	5.0174
27	1.0665	599.8	0.12068	308.05	1464.54	1.1532	5.0063
28	1.0993	598.3	0.11716	312.88	1465.11	1.1691	4.9952
29	1.1328	596.8	0.11377	317.73	1465.65	1.1850	4.9842
30	1.1671	595.2	0.11048	322.58	1466.17	1.2008	4.9732
31	1.2022	593.7	0.10731	327.43	1466.67	1.2166	4.9623
32	1.2380	592.2	0.10425	332.30	1467.14	1.2324	4.9514
33	1.2747	590.6	0.10129	337.17	1467.59	1.2481	4.9405
34	1.3122	589.0	0.09843	342.05	1468.02	1.2638	4.9297
35	1.3505	587.5	0.09566	346.93	1468.43	1.2795	4.9189
36	1.3896	585.9	0.09299	351.83	1468.81	1.2951	4.9082
37	1.4296	584.3	0.09040	356.73	1469.16	1.3108	4.8975
38	1.4705	582.7	0.08790	361.64	1469.49	1.3263	4.8868
39	1.5123	581.1	0.08548	366.56	1469.80	1.3419	4.8762
40	1.5549	579.5	0.08313	371.49	1470.08	1.3574	4.8656
41	1.5985	577.9	0.08086	376.42	1470.33	1.3729	4.8550
42	1.6429	576.3	0.07866	381.37	1470.56	1.3884	4.8445
43	1.6883	574.6	0.07654	386.32	1470.76	1.4038	4.8340
44	1.7347	573.0	0.07447	391.29	1470.93	1.4193	4.8235
45	1.7819	571.3	0.07248	396.26	1471.08	1.4347	4.8130
46	1.8302	569.7	0.07054	401.25	1471.20	1.4500	4.8025
47	1.8794	568.0	0.06866	406.25	1471.28	1.4654	4.7921
48	1.9296	566.3	0.06684	411.25	1471.34	1.4807	4.7817
49	1.9809	564.6	0.06508	416.27	1471.37	1.4961	4.7712
50	2.0331	562.9	0.06337	421.31	1471.37	1.5114	4.7608
51	2.0864	561.2	0.06171	426.35	1471.34	1.5267	4.7504
52	2.1407	559.5	0.06009	431.51	1471.26	1.5422	4.7400
53	2.1961	557.7	0.05853	436.53	1471.18	1.5573	4.7297
54	2.2525	556.0	0.05702	441.58	1471.05	1.5725	4.7193
55	2.3101	554.2	0.05555	446.66	1470.90	1.5877	4.7090
56	2.3687	552.5	0.05412	451.78	1470.70	1.6030	4.6986
57	2.4284	550.7	0.05273	456.94	1470.47	1.6183	4.6882
58	2.4893	548.9	0.05138	462.13	1470.21	1.6336	4.6778
59	2.5513	547.1	0.05007	467.21	1469.92	1.6486	4.6675
60	2.6144	545.3	0.04880	472.39	1469.59	1.6639	4.6571

Appendix A / Thermodynamic and Thermophysical Property Tables

TABLE A.3E Thermodynamic Properties of R-22 at Saturation (English Units)

Temp. °F	Pressure psia	Pressure psig	Liquid, Density lbm/ft³ 1/v_f	Vapor, Sp. Vol. ft³/lbm v_g	Enthalpy, Datum −40°F, Btu/lbm Liquid h_f	Enthalpy, Datum −40°F, Btu/lbm Vapor h_g	Entropy, Datum −40°F, Btu/lbm·°R Liquid s_f	Entropy, Datum −40°F, Btu/lbm·°R Vapor s_g
-100.	2.40	*25.1	93.77	18.444	-14.6	93.4	-0.0373	0.2627
-95.	2.87	*24.1	93.31	15.588	-13.4	94.0	-0.0341	0.2602
-90.	3.42	*23.0	92.85	13.243	-12.2	94.5	-0.0309	0.2578
-85.	4.05	*21.7	92.38	11.307	-11.0	95.1	-0.0277	0.2556
-80.	4.78	*20.2	91.91	9.700	-9.8	95.7	-0.0246	0.2534
-75.	5.61	*18.6	91.43	8.360	-8.6	96.3	-0.0214	0.2513
-70.	6.55	*16.6	90.95	7.236	-7.4	96.9	-0.0183	0.2493
-65.	7.61	*14.5	90.47	6.289	-6.2	97.4	-0.0152	0.2474
-60.	8.81	*12.0	89.99	5.487	-5.0	98.0	-0.0121	0.2455
-58.	9.33	*10.9	89.79	5.201	-4.5	98.2	-0.0109	0.2448
-56.	9.88	*9.8	89.60	4.934	-4.0	98.5	-0.0097	0.2441
-54.	10.45	*8.7	89.40	4.682	-3.5	98.7	-0.0085	0.2434
-52.	11.04	*7.5	89.20	4.446	-3.0	98.9	-0.0073	0.2427
-50.	11.67	*6.2	89.01	4.224	-2.5	99.1	-0.0060	0.2421
-48.	12.32	*4.9	88.81	4.016	-2.0	99.4	-0.0048	0.2414
-46.	13.00	*3.5	88.61	3.820	-1.5	99.6	-0.0036	0.2407
-44.	13.71	*2.0	88.41	3.635	-1.0	99.8	-0.0024	0.2401
-42.	14.44	*0.5	88.21	3.461	-0.5	100.0	-0.0012	0.2395
-40.	15.21	0.5	88.01	3.297	0.0	100.3	0.0000	0.2388
-38.	16.02	1.3	87.81	3.143	0.5	100.5	0.0012	0.2382
-36.	16.85	2.2	87.60	2.997	1.0	100.7	0.0024	0.2376
-34.	17.72	3.0	87.40	2.859	1.5	100.9	0.0036	0.2370
-32.	18.62	3.9	87.20	2.729	2.0	101.1	0.0048	0.2365
-30.	19.56	4.9	86.99	2.606	2.5	101.3	0.0060	0.2359
-29.	20.05	5.4	86.89	2.547	2.8	101.5	0.0066	0.2356
-28.	20.54	5.8	86.79	2.490	3.1	101.6	0.0072	0.2353
-27.	21.04	6.3	86.68	2.434	3.3	101.7	0.0078	0.2350
-26.	21.55	6.9	86.58	2.380	3.6	101.8	0.0083	0.2348
-25.	22.08	7.4	86.48	2.327	3.8	101.9	0.0089	0.2345
-24.	22.61	7.9	86.37	2.276	4.1	102.0	0.0095	0.2342
-23.	23.15	8.5	86.27	2.226	4.4	102.1	0.0101	0.2339
-22.	23.70	9.0	86.17	2.177	4.6	102.2	0.0107	0.2337
-21.	24.26	9.6	86.06	2.130	4.9	102.3	0.0113	0.2334
-20.	24.83	10.1	85.96	2.083	5.1	102.4	0.0119	0.2331
-19.	25.42	10.7	85.85	2.038	5.4	102.5	0.0125	0.2329
-18.	26.01	11.3	85.75	1.995	5.7	102.6	0.0131	0.2326
-17.	26.61	11.9	85.64	1.952	5.9	102.7	0.0137	0.2323
-16.	27.23	12.5	85.54	1.911	6.2	102.8	0.0142	0.2321
-15.	27.85	13.2	85.43	1.870	6.4	102.9	0.0148	0.2318
-14.	28.49	13.8	85.33	1.831	6.7	103.0	0.0154	0.2316
-13.	29.14	14.4	85.22	1.792	7.0	103.1	0.0160	0.2313
-12.	29.80	15.1	85.12	1.755	7.2	103.3	0.0166	0.2310
-11.	30.47	15.8	85.01	1.719	7.5	103.4	0.0172	0.2308
-10.	31.15	16.5	84.90	1.683	7.8	103.5	0.0178	0.2305
-9.	31.84	17.1	84.80	1.648	8.0	103.6	0.0183	0.2303
-8.	32.55	17.9	84.69	1.615	8.3	103.7	0.0189	0.2301
-7.	33.27	18.6	84.58	1.582	8.5	103.8	0.0195	0.2298
-6.	34.00	19.3	84.48	1.550	8.8	103.9	0.0201	0.2296
-5.	34.74	20.0	84.37	1.518	9.1	104.0	0.0207	0.2293
-4.	35.49	20.8	84.26	1.488	9.3	104.1	0.0212	0.2291
-3.	36.26	21.6	84.15	1.458	9.6	104.2	0.0218	0.2288
-2.	37.04	22.3	84.04	1.429	9.9	104.3	0.0224	0.2286
-1.	37.84	23.1	83.94	1.400	10.1	104.4	0.0230	0.2284
0.	38.64	23.9	83.83	1.373	10.4	104.5	0.0236	0.2281
1.	39.46	24.8	83.72	1.346	10.7	104.6	0.0241	0.2279
2.	40.29	25.6	83.61	1.319	10.9	104.7	0.0247	0.2277
3.	41.14	26.4	83.50	1.294	11.2	104.8	0.0253	0.2274
4.	42.00	27.3	83.39	1.268	11.5	104.9	0.0259	0.2272
5.	42.87	28.2	83.28	1.244	11.8	105.0	0.0265	0.2270
6.	43.76	29.1	83.17	1.220	12.0	105.1	0.0270	0.2268
7.	44.66	30.0	83.06	1.196	12.3	105.2	0.0276	0.2265
8.	45.57	30.9	82.95	1.174	12.6	105.2	0.0282	0.2263
9.	46.50	31.8	82.84	1.151	12.8	105.3	0.0288	0.2261
10.	47.45	32.8	82.73	1.129	13.1	105.4	0.0293	0.2259

* Inches of mercury below one standard atmosphere (29.92 in).

(continued)

Temp. °F	Pressure		Liquid, Density lbm/ft³ 1/v_f	Vapor, Sp. Vol. ft³/lbm v_g	Enthalpy, Datum −40 °F, Btu/lbm		Entropy, Datum −40 °F, Btu/lbm · °R	
	psia	psig			Liquid h_f	Vapor h_g	Liquid s_f	Vapor s_g
11.	48.40	33.7	82.62	1.108	13.4	105.5	0.0299	0.2257
12.	49.38	34.7	82.50	1.087	13.7	105.6	0.0305	0.2254
13.	50.37	35.7	82.39	1.067	13.9	105.7	0.0310	0.2252
14.	51.37	36.7	82.28	1.047	14.2	105.8	0.0316	0.2250
15.	52.39	37.7	82.17	1.027	14.5	105.9	0.0322	0.2248
16.	53.42	38.7	82.05	1.008	14.7	106.0	0.0328	0.2246
17.	54.47	39.8	81.94	0.990	15.0	106.1	0.0333	0.2244
18.	55.53	40.8	81.83	0.972	15.3	106.2	0.0339	0.2242
19.	56.61	41.9	81.71	0.954	15.6	106.3	0.0345	0.2240
20.	57.71	43.0	81.60	0.937	15.8	106.4	0.0350	0.2238
21.	58.82	44.1	81.49	0.920	16.1	106.5	0.0356	0.2236
22.	59.95	45.3	81.37	0.903	16.4	106.6	0.0362	0.2233
23.	61.09	46.4	81.26	0.887	16.7	106.7	0.0367	0.2231
24.	62.25	47.6	81.14	0.871	16.9	106.7	0.0373	0.2229
25.	63.43	48.7	81.03	0.855	17.2	106.8	0.0379	0.2227
26.	64.62	49.9	80.91	0.840	17.5	106.9	0.0384	0.2225
27.	65.83	51.1	80.79	0.825	17.8	107.0	0.0390	0.2223
28.	67.06	52.4	80.68	0.810	18.1	107.1	0.0396	0.2221
29.	68.31	53.6	80.56	0.796	18.3	107.2	0.0401	0.2219
30.	69.57	54.9	80.44	0.782	18.6	107.3	0.0407	0.2217
31.	70.85	56.2	80.33	0.769	18.9	107.4	0.0413	0.2216
32.	72.14	57.4	80.21	0.755	19.2	107.5	0.0418	0.2214
33.	73.46	58.8	80.09	0.742	19.4	107.5	0.0424	0.2212
34.	74.79	60.1	79.97	0.729	19.7	107.6	0.0429	0.2210
35.	76.14	61.4	79.86	0.717	20.0	107.7	0.0435	0.2208
36.	77.51	62.8	79.74	0.704	20.3	107.8	0.0441	0.2206
37.	78.90	64.2	79.62	0.692	20.6	107.9	0.0446	0.2204
38.	80.31	65.6	79.50	0.681	20.9	108.0	0.0452	0.2202
39.	81.73	67.0	79.38	0.669	21.1	108.1	0.0457	0.2200
40.	83.18	68.5	79.26	0.658	21.4	108.1	0.0463	0.2198
41.	84.64	69.9	79.14	0.647	21.7	108.2	0.0469	0.2196
42.	86.13	71.4	79.02	0.636	22.0	108.3	0.0474	0.2195
43.	87.63	72.9	78.90	0.625	22.3	108.4	0.0480	0.2193
44.	89.15	74.5	78.77	0.615	22.6	108.5	0.0485	0.2191
45.	90.69	76.0	78.65	0.604	22.8	108.6	0.0491	0.2189
46.	92.25	77.6	78.53	0.594	23.1	108.6	0.0497	0.2187
47.	93.83	79.1	78.41	0.585	23.4	108.7	0.0502	0.2185
48.	95.43	80.7	78.28	0.575	23.7	108.8	0.0508	0.2184
49.	97.05	82.4	78.16	0.565	24.0	108.9	0.0513	0.2182
50.	98.70	84.0	78.04	0.556	24.3	109.0	0.0519	0.2180
51.	100.36	85.7	77.91	0.547	24.6	109.0	0.0524	0.2178
52.	102.04	87.3	77.79	0.538	24.9	109.1	0.0530	0.2176
53.	103.75	89.1	77.66	0.529	25.1	109.2	0.0536	0.2175
54.	105.47	90.8	77.54	0.521	25.4	109.3	0.0541	0.2173
55.	107.22	92.5	77.41	0.513	25.7	109.3	0.0547	0.2171
56.	108.99	94.3	77.28	0.504	26.0	109.4	0.0552	0.2169
57.	110.78	96.1	77.16	0.496	26.3	109.5	0.0558	0.2168
58.	112.59	97.9	77.03	0.488	26.6	109.6	0.0563	0.2166
59.	114.42	99.7	76.90	0.480	26.9	109.6	0.0569	0.2164
60.	116.28	101.6	76.78	0.473	27.2	109.7	0.0574	0.2162
61.	118.16	103.5	76.65	0.465	27.5	109.8	0.0580	0.2161
62.	120.06	105.4	76.52	0.458	27.8	109.9	0.0585	0.2159
63.	121.98	107.3	76.39	0.451	28.0	109.9	0.0591	0.2157
64.	123.92	109.2	76.26	0.444	28.3	110.0	0.0596	0.2155
65.	125.89	111.2	76.13	0.437	28.6	110.1	0.0602	0.2154
66.	127.88	113.2	76.00	0.430	28.9	110.1	0.0607	0.2152
67.	129.90	115.2	75.87	0.423	29.2	110.2	0.0613	0.2150
68.	131.94	117.2	75.74	0.417	29.5	110.3	0.0619	0.2149
69.	134.00	119.3	75.60	0.410	29.8	110.3	0.0624	0.2147
70.	136.08	121.4	75.47	0.404	30.1	110.4	0.0630	0.2145

(continued)

TABLE A.3E (*cont.*)

Temp. °F	Pressure		Liquid, Density lbm/ft³ $1/v_f$	Vapor, Sp. Vol. ft³/lbm v_g	Enthalpy, Datum −40 °F, Btu/lbm		Entropy, Datum −40 °F, Btu/lbm · °R	
	psia	psig			Liquid h_f	Vapor h_g	Liquid s_f	Vapor s_g
71.	138.19	123.5	75.34	0.398	30.4	110.5	0.0635	0.2144
72.	140.33	125.6	75.21	0.391	30.7	110.5	0.0641	0.2142
73.	142.48	127.8	75.07	0.385	31.0	110.6	0.0646	0.2140
74.	144.67	130.0	74.94	0.380	31.3	110.7	0.0652	0.2139
75.	146.87	132.2	74.80	0.374	31.6	110.7	0.0657	0.2137
76.	149.11	134.4	74.67	0.368	31.9	110.8	0.0663	0.2135
77.	151.36	136.7	74.53	0.363	32.2	110.9	0.0668	0.2134
78.	153.64	138.9	74.39	0.357	32.5	110.9	0.0674	0.2132
79.	155.95	141.3	74.26	0.352	32.8	111.0	0.0679	0.2130
80.	158.28	143.6	74.12	0.346	33.1	111.1	0.0684	0.2129
81.	160.64	145.9	73.98	0.341	33.4	111.1	0.0690	0.2127
82.	163.02	148.3	73.84	0.336	33.7	111.2	0.0695	0.2125
83.	165.43	150.7	73.70	0.331	34.0	111.2	0.0701	0.2124
84.	167.87	153.2	73.56	0.326	34.3	111.3	0.0707	0.2122
85.	170.33	155.6	73.42	0.321	34.6	111.3	0.0712	0.2120
86.	172.82	158.1	73.28	0.316	34.9	111.4	0.0717	0.2119
87.	175.34	160.6	73.14	0.312	35.2	111.5	0.0723	0.2117
88.	177.88	163.2	73.00	0.307	35.5	111.5	0.0728	0.2115
89.	180.45	165.8	72.86	0.302	35.8	111.6	0.0734	0.2114
90.	183.05	168.4	72.71	0.298	36.2	111.6	0.0739	0.2112
91.	185.67	171.0	72.57	0.294	36.5	111.7	0.0745	0.2110
92.	188.32	173.6	72.42	0.289	36.8	111.7	0.0750	0.2109
93.	191.00	176.3	72.28	0.285	37.1	111.8	0.0756	0.2107
94.	193.71	179.0	72.13	0.281	37.4	111.8	0.0761	0.2105
95.	196.44	181.7	71.98	0.277	37.7	111.9	0.0767	0.2104
96.	199.21	184.5	71.84	0.273	38.0	111.9	0.0772	0.2102
97.	202.00	187.3	71.69	0.269	38.3	112.0	0.0778	0.2100
98.	204.82	190.1	71.54	0.265	38.6	112.0	0.0783	0.2099
99.	207.67	193.0	71.39	0.261	39.0	112.1	0.0789	0.2097
100.	210.55	195.9	71.24	0.257	39.3	112.1	0.0794	0.2095
101.	213.46	198.8	71.09	0.253	39.6	112.2	0.0800	0.2094
102.	216.40	201.7	70.94	0.250	39.9	112.2	0.0805	0.2092
103.	219.36	204.7	70.78	0.246	40.2	112.2	0.0811	0.2090
104.	222.36	207.7	70.63	0.243	40.5	112.3	0.0816	0.2089
105.	225.39	210.7	70.48	0.239	40.8	112.3	0.0821	0.2087
106.	228.45	213.8	70.32	0.236	41.2	112.4	0.0827	0.2085
107.	231.53	216.8	70.17	0.232	41.5	112.4	0.0832	0.2084
108.	234.65	220.0	70.01	0.229	41.8	112.4	0.0838	0.2082
109.	237.80	223.1	69.85	0.226	42.1	112.5	0.0843	0.2080
110.	240.98	226.3	69.69	0.222	42.4	112.5	0.0849	0.2078
111.	244.20	229.5	69.53	0.219	42.8	112.5	0.0854	0.2077
112.	247.44	232.7	69.37	0.216	43.1	112.6	0.0860	0.2075
113.	250.71	236.0	69.21	0.213	43.4	112.6	0.0865	0.2073
114.	254.02	239.3	69.05	0.210	43.7	112.6	0.0871	0.2072
115.	257.36	242.7	68.89	0.207	44.1	112.7	0.0876	0.2070
116.	260.73	246.0	68.72	0.204	44.4	112.7	0.0882	0.2068
117.	264.13	249.4	68.56	0.201	44.7	112.7	0.0887	0.2066
118.	267.57	252.9	68.39	0.198	45.0	112.7	0.0893	0.2065
119.	271.04	256.3	68.23	0.195	45.4	112.8	0.0899	0.2063
120.	274.54	259.8	68.06	0.192	45.7	112.8	0.0904	0.2061
122.	281.64	266.9	67.72	0.187	46.4	112.8	0.0915	0.2057
124.	288.88	274.2	67.38	0.182	47.0	112.9	0.0926	0.2054
126.	296.26	281.6	67.03	0.177	47.7	112.9	0.0937	0.2050
128.	303.77	289.1	66.67	0.172	48.4	112.9	0.0949	0.2047
130.	311.42	296.7	66.32	0.167	49.1	112.9	0.0960	0.2043
132.	319.22	304.5	65.95	0.162	49.7	113.0	0.0971	0.2039
134.	327.16	312.5	65.59	0.157	50.4	113.0	0.0982	0.2035
136.	335.25	320.6	65.21	0.153	51.1	113.0	0.0993	0.2031
138.	343.48	328.8	64.83	0.148	51.8	112.9	0.1005	0.2027
140.	351.87	337.2	64.45	0.144	52.5	112.9	0.1016	0.2023

TABLE A.3SI Thermodynamic Properties of R-22 at Saturation (SI Units)

Temp. °C	Pressure MPa	Liquid Density kg/m³ $1/v_f$	Vapor Sp. Vol. m³/kg v_g	Enthalpy, Datum −40 °C, kJ/kg		Entropy, Datum −40 °C, kJ/kg · K	
				Sat. Liquid h_f	Sat. Vapor h_g	Sat. Liquid s_f	Sat. Vapor s_g
-70	0.0205	1493.2	0.94151	-30.61	218.82	-0.1401	1.0877
-68	0.0233	1487.8	0.83743	-28.63	219.80	-0.1305	1.0805
-66	0.0263	1482.5	0.74678	-26.65	220.78	-0.1209	1.0736
-64	0.0297	1477.5	0.66763	-24.66	221.76	-0.1113	1.0669
-62	0.0334	1471.6	0.59831	-22.66	222.73	-0.1018	1.0603
-60	0.0375	1466.1	0.53745	-20.65	223.70	-0.0924	1.0540
-58	0.0419	1460.6	0.48387	-18.63	224.67	-0.0829	1.0479
-56	0.0468	1455.1	0.43659	-16.60	225.64	-0.0736	1.0420
-54	0.0522	1449.6	0.39476	-14.56	226.60	-0.0642	1.0362
-52	0.0580	1444.0	0.35768	-12.51	227.56	-0.0550	1.0306
-50	0.0644	1438.4	0.32472	-10.46	228.51	-0.0457	1.0252
-49	0.0677	1435.5	0.30962	-9.42	228.98	-0.0411	1.0225
-48	0.0712	1432.7	0.29536	-8.39	229.46	-0.0365	1.0199
-47	0.0749	1429.9	0.28189	-7.35	229.93	-0.0319	1.0173
-46	0.0787	1427.0	0.26915	-6.30	230.40	-0.0273	1.0148
-45	0.0827	1424.2	0.25711	-5.26	230.87	-0.0227	1.0122
-44	0.0868	1421.3	0.24571	-4.21	231.34	-0.0182	1.0098
-43	0.0911	1418.4	0.23492	-3.16	231.81	-0.0136	1.0073
-42	0.0955	1415.6	0.22470	-2.11	232.27	-0.0091	1.0049
-41	0.1001	1412.7	0.21502	-1.06	232.74	-0.0045	1.0025
-40	0.1049	1409.8	0.20584	0.00	233.20	0.0000	1.0002
-39	0.1099	1406.9	0.19713	1.06	233.66	0.0045	0.9979
-38	0.1150	1403.9	0.18886	2.12	234.12	0.0090	0.9956
-37	0.1204	1401.0	0.18101	3.19	234.58	0.0135	0.9934
-36	0.1259	1398.1	0.17355	4.26	235.03	0.0180	0.9912
-35	0.1316	1395.1	0.16647	5.33	235.49	0.0225	0.9890
-34	0.1375	1392.2	0.15973	6.40	235.94	0.0270	0.9868
-33	0.1437	1389.2	0.15333	7.48	236.39	0.0315	0.9847
-32	0.1500	1386.2	0.14723	8.56	236.84	0.0360	0.9826
-31	0.1566	1383.3	0.14143	9.64	237.28	0.0404	0.9805
-30	0.1634	1380.3	0.13590	10.73	237.73	0.0449	0.9785
-29	0.1704	1377.3	0.13063	11.81	238.17	0.0493	0.9764
-28	0.1777	1374.2	0.12561	12.90	238.61	0.0537	0.9744
-27	0.1852	1371.2	0.12082	14.00	239.05	0.0582	0.9725
-26	0.1929	1368.2	0.11626	15.09	239.49	0.0626	0.9705
-25	0.2009	1365.1	0.11190	16.19	239.92	0.0670	0.9686
-24	0.2091	1362.1	0.10774	17.30	240.35	0.0714	0.9667
-23	0.2176	1359.0	0.10377	18.40	240.78	0.0758	0.9648
-22	0.2264	1355.9	0.09997	19.51	241.21	0.0802	0.9630
-21	0.2354	1352.8	0.09634	20.62	241.63	0.0846	0.9611
-20	0.2447	1349.7	0.09288	21.73	242.06	0.0890	0.9593
-19	0.2543	1346.6	0.08956	22.85	242.48	0.0933	0.9575
-18	0.2642	1343.5	0.08638	23.97	242.89	0.0977	0.9558
-17	0.2744	1340.3	0.08335	25.09	243.31	0.1021	0.9540
-16	0.2848	1337.2	0.08044	26.21	243.72	0.1064	0.9523
-15	0.2956	1334.0	0.07765	27.34	244.13	0.1108	0.9506
-14	0.3067	1330.8	0.07498	28.47	244.54	0.1151	0.9489
-13	0.3181	1327.6	0.07242	29.60	244.95	0.1194	0.9472
-12	0.3298	1324.4	0.06997	30.74	245.35	0.1238	0.9455
-11	0.3418	1321.2	0.06762	31.88	245.75	0.1281	0.9439
-10	0.3542	1318.0	0.06536	33.02	246.15	0.1324	0.9423
-9	0.3669	1314.7	0.06319	34.17	246.54	0.1367	0.9407
-8	0.3799	1311.5	0.06111	35.32	246.93	0.1410	0.9391
-7	0.3933	1308.2	0.05912	36.47	247.32	0.1453	0.9375
-6	0.4071	1304.9	0.05720	37.62	247.70	0.1496	0.9360
-5	0.4212	1301.6	0.05535	38.78	248.09	0.1539	0.9344
-4	0.4357	1298.3	0.05358	39.94	248.47	0.1581	0.9329
-3	0.4505	1294.9	0.05188	41.10	248.84	0.1624	0.9314
-2	0.4658	1291.6	0.05024	42.27	249.21	0.1667	0.9299
-1	0.4814	1288.2	0.04867	43.41	249.58	0.1709	0.9284

(continued)

TABLE A.3SI *(cont.)*

Temp. °C	Pressure MPa	Liquid Density kg/m^3 $1/v_f$	Vapor Sp. Vol. m^3/kg v_g	Enthalpy, Datum −40 °C, kJ/kg		Entropy, Datum −40 °C, kJ/kg · K	
				Sat. Liquid h_f	Sat. Vapor h_g	Sat. Liquid s_f	Sat. Vapor s_g
0	0.4974	1284.8	0.04715	44.59	249.95	0.1751	0.9270
1	0.5138	1281.4	0.04569	45.76	250.31	0.1794	0.9255
2	0.5307	1278.0	0.04428	46.94	250.67	0.1836	0.9241
3	0.5479	1274.6	0.04293	48.12	251.03	0.1878	0.9226
4	0.5655	1271.1	0.04163	49.30	251.38	0.1920	0.9212
5	0.5836	1267.7	0.04037	50.48	251.73	0.1963	0.9198
6	0.6021	1264.2	0.03916	51.67	252.08	0.2005	0.9184
7	0.6210	1260.7	0.03799	52.87	252.42	0.2047	0.9170
8	0.6404	1257.1	0.03686	54.06	252.76	0.2089	0.9156
9	0.6602	1253.6	0.03577	55.26	253.09	0.2131	0.9143
10	0.6805	1250.0	0.03472	56.46	253.43	0.2173	0.9129
11	0.7012	1246.4	0.03371	57.67	253.75	0.2215	0.9116
12	0.7224	1242.8	0.03273	58.88	254.08	0.2257	0.9102
13	0.7441	1239.2	0.03179	60.09	254.39	0.2298	0.9089
14	0.7663	1235.5	0.03088	61.30	254.71	0.2340	0.9076
15	0.7889	1231.9	0.03000	62.52	255.02	0.2382	0.9062
16	0.8121	1228.2	0.02914	63.74	255.33	0.2424	0.9049
17	0.8357	1224.5	0.02832	64.97	255.63	0.2465	0.9036
18	0.8598	1220.7	0.02753	66.20	255.93	0.2507	0.9023
19	0.8845	1217.0	0.02676	67.43	256.22	0.2549	0.9011
20	0.9097	1213.2	0.02601	68.67	256.51	0.2590	0.8998
21	0.9354	1209.4	0.02529	69.91	256.79	0.2632	0.8985
22	0.9616	1205.5	0.02459	71.15	257.07	0.2673	0.8972
23	0.9884	1201.7	0.02392	72.40	257.35	0.2715	0.8960
24	1.0157	1197.8	0.02326	73.65	257.62	0.2756	0.8947
25	1.0436	1193.9	0.02263	74.91	257.88	0.2797	0.8934
26	1.0720	1189.9	0.02202	76.17	258.14	0.2839	0.8922
27	1.1011	1186.0	0.02142	77.43	258.39	0.2880	0.8909
28	1.1306	1182.0	0.02085	78.70	258.64	0.2921	0.8897
29	1.1608	1177.9	0.02029	79.99	258.88	0.2963	0.8884
30	1.1916	1173.9	0.01975	81.26	259.12	0.3004	0.8872
31	1.2229	1169.8	0.01922	82.53	259.35	0.3045	0.8859
32	1.2549	1165.7	0.01871	83.81	259.58	0.3087	0.8847
33	1.2874	1161.5	0.01822	85.11	259.80	0.3128	0.8834
34	1.3206	1157.3	0.01774	86.41	260.01	0.3169	0.8822
35	1.3544	1153.1	0.01727	87.72	260.22	0.3211	0.8809
36	1.3889	1148.8	0.01682	89.00	260.42	0.3252	0.8797
37	1.4240	1144.5	0.01638	90.31	260.61	0.3293	0.8784
38	1.4597	1140.2	0.01596	91.63	260.80	0.3334	0.8771
39	1.4961	1135.8	0.01554	92.94	260.98	0.3376	0.8759
40	1.5331	1131.4	0.01514	94.27	261.15	0.3417	0.8746
41	1.5709	1126.9	0.01475	95.60	261.32	0.3458	0.8734
42	1.6093	1122.4	0.01437	96.94	261.48	0.3500	0.8721
43	1.6483	1117.9	0.01400	98.28	261.63	0.3541	0.8708
44	1.6881	1113.3	0.01364	99.63	261.77	0.3583	0.8695
45	1.7286	1108.7	0.01329	100.98	261.90	0.3624	0.8682
46	1.7698	1104.0	0.01295	102.34	262.02	0.3666	0.8669
47	1.8117	1099.3	0.01262	103.71	262.14	0.3707	0.8656
48	1.8544	1094.5	0.01229	105.08	262.25	0.3749	0.8643
49	1.8977	1089.6	0.01198	106.46	262.34	0.3791	0.8629
50	1.9419	1084.8	0.01167	107.85	262.43	0.3832	0.8616
51	1.9867	1079.8	0.01137	109.24	262.51	0.3874	0.8602
52	2.0324	1074.8	0.01108	110.65	262.58	0.3916	0.8589
53	2.0788	1069.7	0.01080	112.06	262.63	0.3958	0.8575
54	2.1260	1064.6	0.01052	113.48	262.68	0.4000	0.8561
55	2.1740	1059.4	0.01025	114.91	262.71	0.4042	0.8546
56	2.2227	1054.1	0.00999	116.34	262.73	0.4085	0.8532
57	2.2723	1048.8	0.00974	117.79	262.74	0.4127	0.8517
58	2.3227	1043.4	0.00949	119.24	262.73	0.4170	0.8503
59	2.3740	1037.9	0.00924	120.71	262.71	0.4212	0.8488
60	2.4260	1032.3	0.00900	122.18	262.68	0.4255	0.8472

TABLE A.4E Thermodynamic Properties of Moist Air, Standard Atmospheric Pressure, 14.696 psi (29.921 in. Hg)

Temp., °F t	Humidity Ratio lbm_w/lbm_a W_s	Specific Volume, ft^3/lbm dry air			Specific Enthalpy, Btu/lbm dry air			Specific Entropy, Btu/lbm dry air · °F		
		v_a	v_{as}	v_s	h_a	h_{as}	h_s	s_a	s_{as}	s_s
−80	0.0000049	9.553	0.000	9.553	−19.221	0.005	−19.215	−0.04594	0.00001	−0.04592
−75	0.0000072	9.680	0.000	9.680	−18.019	0.007	−18.011	−0.04279	0.00002	−0.04277
−70	0.0000104	9.807	0.000	9.807	−16.806	0.011	−16.817	−0.03969	0.00003	−0.03966
−65	0.0000149	9.933	0.000	9.934	−15.616	0.015	−15.600	−0.03663	0.00004	−0.03659
−60	0.0000212	10.060	0.000	10.060	−14.414	0.022	−14.392	−0.03360	0.00006	−0.03354
−55	0.0000298	10.187	0.000	10.187	−13.213	0.031	−13.182	−0.03061	0.00008	−0.03053
−50	0.0000416	10.313	0.001	10.314	−12.011	0.043	−11.968	−0.02766	0.00011	−0.02755
−45	0.0000577	10.440	0.001	10.441	−10.810	0.060	−10.750	−0.02475	0.00016	−0.02459
−40	0.0000793	10.567	0.001	10.568	−9.609	0.083	−9.526	−0.02187	0.00021	−0.02166
−35	0.0001081	10.693	0.002	10.695	−8.407	0.113	−8.294	−0.01902	0.00028	−0.01874
−30	0.0001465	10.820	0.003	10.822	−7.206	0.154	−7.053	−0.01621	0.00038	−0.01583
−25	0.0001970	10.947	0.003	10.950	−6.005	0.207	−5.798	−0.01343	0.00051	−0.01293
−20	0.0002632	11.073	0.005	11.078	−4.804	0.277	−4.527	−0.01069	0.00067	−0.01002
−15	0.0003493	11.200	0.006	11.206	−3.603	0.368	−3.235	−0.00797	0.00088	−0.00709
−10	0.0004608	11.326	0.008	11.335	−2.402	0.487	−1.915	−0.00528	0.00115	−0.00414
−5	0.0006041	11.453	0.011	11.464	−1.201	0.640	−0.561	−0.00263	0.00149	−0.00114
0	0.0007875	11.579	0.015	11.594	0.0	0.835	0.835	0.00000	0.00192	0.00192
5	0.0010207	11.706	0.019	11.725	1.201	1.085	2.286	0.00260	0.00247	0.00506
10	0.0013158	11.832	0.025	11.857	2.402	1.402	3.804	0.00517	0.00315	0.00832
15	0.0016874	11.959	0.032	11.991	3.603	1.801	5.404	0.00771	0.00400	0.01171
20	0.0021531	12.085	0.042	12.127	4.804	2.303	7.107	0.01023	0.00505	0.01528
25	0.0027339	12.212	0.054	12.265	6.005	2.930	8.935	0.01272	0.00636	0.01908
30	0.0034552	12.338	0.068	12.406	7.206	3.711	10.917	0.01519	0.00796	0.02315
35	0.004277	12.464	0.085	12.550	8.408	4.603	13.010	0.01763	0.00977	0.02740
40	0.005216	12.591	0.105	12.696	9.609	5.624	15.233	0.02004	0.01183	0.03187
45	0.006334	12.717	0.129	12.846	10.810	6.843	17.653	0.02244	0.01426	0.03669
50	0.007661	12.844	0.158	13.001	12.012	8.295	20.306	0.02480	0.01712	0.04192
55	0.009233	12.970	0.192	13.162	13.213	10.016	23.229	0.02715	0.02048	0.04763
60	0.011087	13.096	0.233	13.329	14.415	12.052	26.467	0.02947	0.02442	0.05389
65	0.013270	13.223	0.281	13.504	15.616	14.454	30.071	0.03178	0.02902	0.06080
70	0.015832	13.349	0.339	13.688	16.818	17.279	34.097	0.03406	0.03438	0.06844
75	0.018833	13.476	0.407	13.882	18.020	20.595	38.615	0.03631	0.04062	0.07694
80	0.022340	13.602	0.487	14.089	19.222	24.479	43.701	0.03855	0.04787	0.08642
85	0.026433	13.728	0.581	14.310	20.424	29.021	49.445	0.04077	0.05626	0.09703
90	0.031203	13.855	0.692	14.547	21.626	34.325	55.951	0.04297	0.06598	0.10895
95	0.036757	13.981	0.823	14.804	22.828	40.515	63.343	0.04514	0.07722	0.12237
100	0.043219	14.107	0.976	15.084	24.031	47.730	71.761	0.04730	0.09022	0.13752
105	0.050737	14.234	1.156	15.390	25.233	56.142	81.375	0.04944	0.10525	0.15469
110	0.059486	14.360	1.367	15.727	26.436	65.950	92.386	0.05156	0.12262	0.17418

115	0.069676	14.486	1.615	16.101	27.639	77.396	105.035	0.05366	0.14274	0.19640
120	0.081560	14.613	1.906	16.519	28.842	90.770	119.612	0.05575	0.16605	0.22180
125	0.095456	14.739	2.250	16.989	30.045	106.437	136.482	0.05781	0.19314	0.25096
130	0.111738	14.865	2.655	17.520	31.249	124.828	156.076	0.05986	0.22470	0.28457
135	0.130895	14.992	3.136	18.127	32.452	146.504	178.957	0.06190	0.26161	0.32351
140	0.153538	15.118	3.708	18.825	33.656	172.168	205.824	0.06391	0.30498	0.36890
145	0.180467	15.244	4.392	19.637	34.860	202.740	237.600	0.06591	0.35626	0.42218
150	0.212730	15.370	5.218	20.589	36.064	239.426	275.490	0.06790	0.41735	0.48524
155	0.251738	15.497	6.223	21.720	37.269	283.849	321.118	0.06986	0.49077	0.56064
160	0.29945	15.623	7.459	23.082	38.474	338.263	376.737	0.07181	0.58007	0.65188
165	0.35865	15.749	9.001	24.750	39.679	405.865	445.544	0.07375	0.69022	0.76397
170	0.43343	15.875	10.959	26.834	40.884	491.372	532.256	0.07567	0.82858	0.90425
175	0.53019	16.002	13.504	29.505	42.089	602.139	644.229	0.07758	1.00657	1.08416
180	0.65911	16.128	16.909	33.037	43.295	749.871	793.166	0.07947	1.24236	1.32183
185	0.83817	16.254	21.656	37.910	44.501	955.261	999.763	0.08135	1.56797	1.64932
190	1.10154	16.381	28.661	45.042	45.707	1257.614	1303.321	0.08321	2.04412	2.12733
195	1.52396	16.507	39.928	56.435	46.914	1742.879	1789.793	0.08506	2.80332	2.88838
200	2.30454	16.633	60.793	77.426	48.121	2640.084	2688.205	0.08690	4.19787	4.28477

SOURCE: Reprinted by permission from *ASHRAE Fundamentals 1996 (IP)*, p. 6.2–5.

TABLE A.4SI Thermodynamic Properties of Moist Air, Standard Atmospheric Pressure, 101.325 kPa

Temp., °C	Humidity Ratio, kg_w/kg_a W_s	Specific Volume, m^3/kg dry air v_a	v_{as}	v_s	Specific Enthalpy, kJ/kg dry air h_a	h_{as}	h_s	Specific Entropy, kJ/(kg dry air) (K) s_a	s_{as}	s_s	Temp., °C
−60	0.0000067	0.6027	0.0000	0.6027	−60.351	0.017	−60.334	−0.2495	0.0001	−0.2494	−60
−58	0.0000087	0.6084	0.0000	0.6084	−58.338	0.021	−58.317	−0.2401	0.0001	−0.2400	−58
−56	0.0000114	0.6141	0.0000	0.6141	−56.326	0.028	−56.298	−0.2308	0.0001	−0.2306	−56
−54	0.0000147	0.6198	0.0000	0.6198	−54.313	0.036	−54.278	−0.2215	0.0002	−0.2214	−54
−52	0.0000190	0.6255	0.0000	0.6255	−52.301	0.046	−52.255	−0.2124	0.0002	−0.2122	−52
−50	0.0000243	0.6312	0.0000	0.6312	−50.289	0.059	−50.230	−0.2033	0.0003	−0.2031	−50
−48	0.0000311	0.6369	0.0000	0.6369	−48.277	0.075	−48.202	−0.1944	0.0004	−0.1940	−48
−46	0.0000395	0.6426	0.0000	0.6426	−46.265	0.095	−46.170	−0.1855	0.0004	−0.1850	−46
−44	0.0000500	0.6483	0.0001	0.6483	−44.253	0.121	−44.132	−0.1767	0.0006	−0.1761	−44
−42	0.0000631	0.6540	0.0001	0.6540	−42.241	0.153	−42.088	−0.1679	0.0007	−0.1672	−42
−40	0.0000793	0.6597	0.0001	0.6597	−40.229	0.192	−40.037	−0.1592	0.0009	−0.1584	−40
−38	0.0000992	0.6653	0.0001	0.6654	−38.218	0.241	−37.976	−0.1507	0.0011	−0.1496	−38
−36	0.0001237	0.6710	0.0001	0.6712	−36.206	0.302	−35.905	−0.1421	0.0014	−0.1408	−36
−34	0.0001536	0.6767	0.0002	0.6769	−34.195	0.375	−33.820	−0.1337	0.0017	−0.1320	−34
−32	0.0001902	0.6824	0.0002	0.6826	−32.183	0.464	−31.718	−0.1253	0.0020	−0.1233	−32
−30	0.0002346	0.6881	0.0003	0.6884	−30.171	0.574	−29.597	−0.1170	0.0025	−0.1145	−30
−28	0.0002883	0.6938	0.0003	0.6941	−28.160	0.707	−27.454	−0.1088	0.0031	−0.1057	−28
−26	0.0003533	0.6995	0.0004	0.6999	−26.149	0.867	−25.282	−0.1006	0.0037	−0.0969	−26
−24	0.0004314	0.7052	0.0005	0.7057	−24.137	1.059	−23.078	−0.0925	0.0045	−0.0880	−24
−22	0.0005251	0.7109	0.0006	0.7115	−22.126	1.292	−20.834	−0.0845	0.0054	−0.0790	−22
−20	0.0006373	0.7165	0.0007	0.7173	−20.115	1.570	−18.545	−0.0765	0.0066	−0.0699	−20
−18	0.0007711	0.7222	0.0009	0.7231	−18.103	1.902	−16.201	−0.0686	0.0079	−0.0607	−18
−16	0.0009303	0.7279	0.0011	0.7290	−16.092	2.299	−13.793	−0.0607	0.0094	−0.0513	−16
−14	0.0011191	0.7336	0.0013	0.7349	−14.080	2.769	−11.311	−0.0529	0.0113	−0.0416	−14
−12	0.0013425	0.7393	0.0016	0.7409	−12.069	3.327	−8.742	−0.0452	0.0134	−0.0318	−12
−10	0.0016062	0.7450	0.0019	0.7469	−10.057	3.986	−6.072	−0.0375	0.0160	−0.0215	−10
−8	0.0019166	0.7507	0.0023	0.7530	−8.046	4.764	−3.283	−0.0299	0.0189	−0.0110	−8
−6	0.0022811	0.7563	0.0028	0.7591	−6.035	5.677	−0.357	−0.0223	0.0224	0.0000	−6
−4	0.0027081	0.7620	0.0033	0.7653	−4.023	6.751	2.728	−0.0148	0.0264	0.0115	−4
−2	0.0032074	0.7677	0.0039	0.7717	−2.011	8.007	5.995	−0.0074	0.0310	0.0236	−2
0	0.0037895	0.7734	0.0047	0.7781	−0.000	9.473	9.473	0.0000	0.0364	0.0364	0
0 *	0.003789	0.7734	0.0047	0.7781	−0.000	9.473	9.473	0.0000	0.0364	0.0364	0
2	0.004381	0.7791	0.0055	0.7845	2.012	10.970	12.982	0.0073	0.0419	0.0492	2
4	0.005054	0.7848	0.0064	0.7911	4.024	12.672	16.696	0.0146	0.0480	0.0627	4
6	0.005818	0.7904	0.0074	0.7978	6.036	14.608	20.644	0.0219	0.0550	0.0769	6
8	0.006683	0.7961	0.0085	0.8046	8.047	16.805	24.852	0.0290	0.0628	0.0919	8
10	0.007661	0.8018	0.0098	0.8116	10.059	19.293	29.352	0.0362	0.0717	0.1078	10
12	0.008766	0.8075	0.0113	0.8188	12.071	22.108	34.179	0.0433	0.0816	0.1248	12
14	0.010012	0.8132	0.0131	0.8262	14.084	25.286	39.370	0.0503	0.0927	0.1430	14

16	0.011413	0.8188	0.0150	0.8338	16.096	28.867	44.963	0.0573	0.1051	0.1624	16
18	0.012989	0.8245	0.0172	0.8417	18.108	32.900	51.008	0.0642	0.1190	0.1832	18
20	0.014758	0.8302	0.0196	0.8498	20.121	37.434	57.555	0.0711	0.1346	0.2057	20
22	0.016741	0.8359	0.0224	0.8583	22.133	42.527	64.660	0.0779	0.1519	0.2298	22
24	0.018963	0.8416	0.0256	0.8671	24.146	48.239	72.385	0.0847	0.1712	0.2559	24
26	0.021448	0.8472	0.0291	0.8764	26.159	54.638	80.798	0.0915	0.1927	0.2842	26
28	0.024226	0.8529	0.0331	0.8860	28.172	61.804	89.976	0.0982	0.2166	0.3148	28
30	0.027329	0.8586	0.0376	0.8962	30.185	69.820	100.006	0.1048	0.2432	0.3481	30
32	0.030793	0.8643	0.0426	0.9069	32.198	78.780	110.979	0.1115	0.2728	0.3842	32
34	0.034660	0.8700	0.0483	0.9183	34.212	88.799	123.011	0.1180	0.3056	0.4236	34
36	0.038971	0.8756	0.0546	0.9303	36.226	99.983	136.209	0.1246	0.3420	0.4666	36
38	0.043778	0.8813	0.0618	0.9431	38.239	112.474	150.713	0.1311	0.3824	0.5135	38
40	0.049141	0.8870	0.0698	0.9568	40.253	126.430	166.683	0.1375	0.4273	0.5649	40
42	0.055119	0.8927	0.0788	0.9714	42.268	142.007	184.275	0.1439	0.4771	0.6211	42
44	0.061791	0.8983	0.0888	0.9872	44.282	159.417	203.699	0.1503	0.5325	0.6828	44
46	0.069239	0.9040	0.1002	1.0042	46.296	178.882	225.179	0.1566	0.5940	0.7507	46
48	0.077556	0.9097	0.1129	1.0226	48.311	200.644	248.955	0.1629	0.6624	0.8253	48
50	0.086858	0.9154	0.1272	1.0425	50.326	225.019	275.345	0.1692	0.7385	0.9077	50
52	0.097272	0.9211	0.1433	1.0643	52.341	252.340	304.682	0.1754	0.8234	0.9988	52
54	0.108954	0.9267	0.1614	1.0882	54.357	283.031	337.388	0.1816	0.9182	1.0998	54
56	0.122077	0.9324	0.1819	1.1143	56.373	317.549	373.922	0.1877	1.0243	1.2120	56
58	0.136851	0.9381	0.2051	1.1432	58.389	356.461	414.850	0.1938	1.1432	1.3370	58
60	0.15354	0.9438	0.2315	1.1752	60.405	400.458	460.863	0.1999	1.2769	1.4768	60
62	0.17244	0.9494	0.2614	1.2109	62.421	450.377	512.798	0.2059	1.4278	1.6337	62
64	0.19393	0.9551	0.2957	1.2508	64.438	507.177	571.615	0.2119	1.5985	1.8105	64
66	0.21848	0.9608	0.3350	1.2958	66.455	572.116	638.571	0.2179	1.7927	2.0106	66
68	0.24664	0.9665	0.3803	1.3467	68.472	646.724	715.196	0.2238	2.0147	2.2385	68
70	0.27916	0.9721	0.4328	1.4049	70.489	732.959	803.448	0.2297	2.2699	2.4996	70
72	0.31698	0.9778	0.4941	1.4719	72.507	833.335	905.842	0.2356	2.5655	2.8010	72
74	0.36130	0.9835	0.5662	1.5497	74.525	951.077	1025.603	0.2414	2.9104	3.1518	74
76	0.41377	0.9892	0.6519	1.6411	76.543	1090.628	1167.172	0.2472	3.3171	3.5644	76
78	0.47663	0.9948	0.7550	1.7498	78.562	1257.921	1336.483	0.2530	3.8023	4.0553	78
80	0.55295	1.0005	0.8805	1.8810	80.581	1461.200	1541.781	0.2587	4.3890	4.6477	80
82	0.64724	1.0062	1.0360	2.0422	82.600	1712.547	1795.148	0.2644	5.1108	5.3753	82
84	0.76624	1.0119	1.2328	2.2446	84.620	2029.983	2114.603	0.2701	6.0181	6.2882	84
86	0.92062	1.0175	1.4887	2.5062	86.640	2442.036	2528.677	0.2757	7.1901	7.4658	86
88	1.12800	1.0232	1.8333	2.8565	88.661	2995.890	3084.551	0.2813	8.7580	9.0393	88
90	1.42031	1.0289	2.3199	3.3488	90.681	3776.918	3867.599	0.2869	10.9586	11.2455	90

* Extrapolated to represent metastable equilibrium with undercooled liquid.

Source: Reprinted by permission from *ASHRAE Fundamentals 1996* (SI), p. 6.2-3.

TABLE A.5E Thermophysical Properties of Air at Atmospheric Pressure

T, °F	Density, lbm/ft³	c_p, Btu/lbm · °R	Viscosity, μ lbm/ft · hr	Thermal Cond., k Btu/hr · ft · °F	Pr
−40	0.09468	0.2402	0.03655	0.01207	0.728
−30	0.09245	0.2402	0.03726	0.01233	0.726
−20	0.09036	0.2401	0.03797	0.01260	0.724
−10	0.08829	0.2401	0.03866	0.01286	0.722
0	0.08640	0.2401	0.03935	0.01312	0.720
10	0.08453	0.2402	0.04004	0.01338	0.719
20	0.08277	0.2402	0.04071	0.01364	0.717
30	0.08107	0.2402	0.04138	0.01390	0.715
40	0.07944	0.2402	0.04204	0.01415	0.714
50	0.07788	0.2402	0.04270	0.01441	0.712
60	0.07637	0.2403	0.04335	0.01466	0.711
70	0.07493	0.2403	0.04400	0.01491	0.709
80	0.07352	0.2404	0.04464	0.01516	0.708
90	0.07220	0.2404	0.04527	0.01541	0.706
100	0.07089	0.2405	0.04590	0.01566	0.705
110	0.06965	0.2405	0.04652	0.01590	0.704
120	0.06844	0.2406	0.04714	0.01615	0.702
130	0.06728	0.2407	0.04775	0.01639	0.701
140	0.06616	0.2408	0.04835	0.01663	0.700
150	0.06506	0.2409	0.04896	0.01687	0.699
160	0.06402	0.2410	0.04955	0.01711	0.698
170	0.06300	0.2411	0.05014	0.01735	0.697
180	0.06202	0.2412	0.05073	0.01759	0.696
190	0.06106	0.2413	0.05131	0.01782	0.695
200	0.06013	0.2414	0.05189	0.01806	0.694
210	0.05923	0.2415	0.05246	0.01829	0.693
220	0.05836	0.2416	0.05303	0.01852	0.692
230	0.05751	0.2418	0.05360	0.01875	0.691
240	0.05669	0.2419	0.05416	0.01898	0.690
250	0.05589	0.2421	0.05471	0.01921	0.689
260	0.05510	0.2422	0.05527	0.01944	0.689
270	0.05436	0.2424	0.05581	0.01967	0.688
280	0.05362	0.2426	0.05636	0.01989	0.687

SOURCE: *Tables of Thermal Properties of Gases*, NBS Circular 564, 1955.

TABLE A.5SI Thermophysical Properties of Air at Atmospheric Pressure

T, °C	Density, kg/m^3	c_p, $kJ/kg \cdot K$	Viscosity, $\mu \times 10^5$ $kg/m \cdot s$	Thermal Cond., $k \times 10^3$ $W/m \cdot K$	Pr
−40	1.517	1.006	1.511	20.88	0.728
−35	1.485	1.005	1.537	21.30	0.725
−30	1.454	1.005	1.564	21.71	0.724
−25	1.424	1.005	1.590	22.12	0.722
−20	1.396	1.005	1.615	22.53	0.721
−15	1.369	1.005	1.641	22.93	0.719
−10	1.343	1.005	1.666	23.34	0.718
−5	1.317	1.006	1.691	23.74	0.717
0	1.293	1.006	1.716	24.14	0.715
5	1.270	1.006	1.741	24.53	0.714
10	1.248	1.006	1.765	24.93	0.712
15	1.226	1.006	1.789	25.32	0.711
20	1.205	1.006	1.813	25.72	0.709
25	1.184	1.006	1.837	26.11	0.708
30	1.165	1.006	1.861	26.50	0.707
35	1.146	1.007	1.884	26.88	0.706
40	1.128	1.007	1.908	27.27	0.705
45	1.110	1.007	1.931	27.65	0.703
50	1.093	1.007	1.954	28.03	0.702
55	1.076	1.008	1.976	28.41	0.701
60	1.060	1.008	1.999	28.79	0.700
65	1.044	1.008	2.021	29.16	0.699
70	1.029	1.009	2.043	29.53	0.698
75	1.014	1.009	2.066	29.91	0.697
80	1.000	1.009	2.087	30.27	0.696
85	0.986	1.010	2.109	30.64	0.695
90	0.972	1.010	2.131	31.01	0.694
95	0.959	1.011	2.152	31.37	0.694
100	0.946	1.011	2.173	31.74	0.692
105	0.933	1.012	2.195	32.10	0.692
110	0.921	1.012	2.216	32.46	0.691
115	0.909	1.013	2.236	32.81	0.690
120	0.898	1.013	2.257	33.17	0.689

SOURCE: *Tables of Thermal Properties of Gases*, NBS Circular 564, 1955.

TABLE A.6E Thermophysical Properties of Water

Temp., °F	Saturated Liquid				Saturated Vapor			
	Specific Heat, Btu/lbm·°F	Dynamic Viscosity, lbm/ft·hr	Thermal Conductivity, Btu/hr·ft·°F	Pr	Specific Heat, Btu/lbm·°F	Dynamic Viscosity, lbm/hr·ft	Thermal Conductivity, Btu/hr·ft·°F	Pr
32.02[a]	1.0100	4.336	0.3241	13.51	0.4461	0.0223	0.00986	1.009
40	1.0037	3.740	0.3290	11.41	0.4467	0.0226	0.01000	1.010
60	0.9993	2.713	0.3411	7.95	0.4486	0.0232	0.01037	1.004
80	0.9992	2.075	0.3524	5.88	0.4511	0.0240	0.01078	1.004
100	0.9990	1.648	0.3625	4.54	0.4542	0.0248	0.01123	1.003
120	0.9988	1.348	0.3710	3.63	0.4580	0.0256	0.01171	1.001
140	0.9990	1.129	0.3781	2.98	0.4627	0.0264	0.01224	0.998
160	1.0002	0.963	0.3836	2.51	0.4683	0.0273	0.01281	0.998
180	1.0023	0.834	0.3879	2.15	0.4750	0.0282	0.01342	0.998
200	1.0052	0.733	0.3910	1.88	0.4829	0.0291	0.01408	0.998
212.00[b]	1.0072	0.682	0.3924	1.75	0.4882	0.0297	0.01450	1.000
220	1.0087	0.651	0.3931	1.67	0.4921	0.0300	0.01479	0.998
240	1.0129	0.585	0.3944	1.50	0.5029	0.0310	0.01555	1.003
260	1.0177	0.530	0.3950	1.37	0.5155	0.0319	0.01636	1.005
280	1.0231	0.484	0.3949	1.25	0.5301	0.0328	0.01723	1.009
300	1.0292	0.445	0.3942	1.16	0.5472	0.0338	0.01816	1.018

[a] Triple point.
[b] Boiling point.
Source: Reprinted by permission from *ASHRAE Fundamentals 1993* (IP), p. 17.47.

TABLE A.6SI Thermophysical Properties of Water

	Saturated Liquid				Saturated Vapor			
Temp., °C	Specific Heat, kJ/kg·°C	Dynamic Viscosity, kg/m·s ×10⁶	Thermal Conductivity, W/m·°C	Pr	Specific Heat, kJ/kg·°C	Dynamic Viscosity, kg/m·s ×10⁶	Thermal Conductivity, W/m·°C	Pr
0.01ᵃ	4.229	1792.4	0.5610	13.51	1.868	9.22	0.01707	1.009
10	4.188	1306.6	0.5800	9.43	1.874	9.46	0.01762	1.006
20	4.183	1002.1	0.5984	7.00	1.882	9.73	0.01823	1.004
30	4.813	797.7	0.6154	6.24	1.892	10.01	0.01888	1.003
40	4.182	653.2	0.6305	4.33	1.905	10.31	0.01960	1.002
50	4.182	547.0	0.6435	3.55	1.919	10.62	0.02036	1.001
60	4.183	466.5	0.6543	2.98	1.937	10.93	0.02118	1.000
70	4.187	404.0	0.6631	2.55	1.958	11.26	0.02207	0.999
80	4.194	354.5	0.6700	2.22	1.983	11.59	0.02301	0.999
90	4.204	314.5	0.6753	1.96	2.011	11.93	0.02402	0.999
100ᵇ	4.217	281.8	0.6791	1.75	2.044	12.27	0.02509	1.000
110	4.232	254.8	0.6817	1.58	2.082	12.61	0.02624	1.001
120	4.249	232.1	0.6832	1.44	2.600	12.96	0.02746	1.227
130	4.268	213.0	0.6837	1.33	2.176	13.30	0.02876	1.006
140	4.288	196.6	0.6833	1.23	2.233	13.65	0.03013	1.012
150	4.312	182.5	0.6821	1.15	2.299	13.99	0.03159	1.018

ᵃ Triple point.
ᵇ Boiling point.

Source: Reprinted by permission from *ASHRAE Fundamentals 1993 (SI)*, p. 17.47.

TABLE A.7E Thermophysical Properties of Gases at 32 °F and Atmospheric Pressure

Gas	Gas Constant, R, Btu/lbm · °R	Density, ρ, lbm/ft³	Specific Heat, c_p, Btu/lbm · °R	Viscosity, μ, lbm/ft · hr	Thermal Conductivity, k, Btu/hr · ft · °F	Prandtl Number, Pr
Argon	0.04972	0.1114	0.1246	0.0514	0.00944	0.678
Carbon dioxide	0.04512	0.1234	0.1975	0.0331	0.00841	0.782
Carbon monoxide	0.07090	0.0781	0.2489	0.0401	0.01342	0.744
Hydrogen (H_2)	0.9851	0.00561	3.392	0.0204	0.0972	0.710
Nitrogen (N_2)	0.07088	0.0781	0.2487	0.0402	0.0140	0.721
Oxygen (O_2)	0.06206	0.0892	0.2189	0.0464	0.01419	0.718

Source: *Tables of Thermal Properties of Gases*, NBS Circular 564, 1955.

TABLE A.7SI Thermophysical Properties of Gases at 0 °C and Atmospheric Pressure

Gas	Gas Constant, R, kJ/kg · K	Density, ρ, kg/m³	Specific Heat, c_p, kJ/kg · K	Viscosity, $\mu \times 10^5$, kg/m · s	Thermal Conductivity, $k \times 10^3$, W/m · K	Prandtl Number, Pr
Argon	0.2082	1.784	0.522	2.125	16.34	0.678
Carbon dioxide	0.1889	1.977	0.827	1.370	14.55	0.782
Carbon monoxide	0.2968	1.250	1.042	1.657	23.22	0.744
Hydrogen (H_2)	4.124	0.0899	14.20	0.841	168.2	0.710
Nitrogen (N_2)	0.2968	1.250	1.041	1.663	24.1	0.721
Oxygen (O_2)	0.2598	1.429	0.917	1.919	24.55	0.718

Source: *Tables of Thermal Properties of Gases*, NBS Circular 564, 1955.

TABLE A.8E Thermophysical Properties of R-22

Temp., °F	Saturated Liquid				Saturated Vapor			
	Specific Heat, Btu/lbm·°F	Dynamic Viscosity, lbm/ft·hr	Thermal Conductivity, Btu/hr·ft·°F	Pr	Specific Heat, Btu/lbm·°F	Dynamic Viscosity, lbm/hr·ft	Thermal Conductivity, Btu/hr·ft·°F	Pr
0	0.2697	0.615	0.0600	2.76	0.1611	0.0268	0.00486	0.888
20	0.2756	0.546	0.0572	2.63	0.1709	0.0279	0.00526	0.906
40	0.2829	0.484	0.0545	2.51	0.1825	0.0290	0.00565	0.937
60	0.2916	0.429	0.0518	2.41	0.1964	0.0301	0.00604	0.979
80	0.3024	0.380	0.0492	2.34	0.2135		0.00642	
100	0.3162	0.338	0.0466	2.29	0.2356		0.00680	
120	0.3353		0.0441		0.2660		0.00719	

TABLE A.8SI Thermophysical Properties of R-22

Temp., °C	Saturated Liquid				Saturated Vapor			
	Specific Heat, kJ/kg·°C	Dynamic Viscosity, $kg/m \cdot s \times 10^6$	Thermal Conductivity, W/m·°C	Pr	Specific Heat, kJ/kg·°C	Dynamic Viscosity, $kg/m \cdot s \times 10^6$	Thermal Conductivity, W/m·°C	Pr
−20	1.125	260.1	0.1048	2.79	0.667	11.4	0.00827	0.901
−10	1.146	234.1	0.1004	2.67	0.703	11.8	0.00889	0.924
0	1.171	210.1	0.0962	2.56	0.744	12.2	0.00950	0.956
10	1.202	188.5	0.0920	2.46	0.792		0.01011	
20	1.238	169.1	0.0878	2.38	0.849		0.01071	
30	1.282	151.7	0.0838	2.32	0.919		0.01130	
40	1.338	136.3	0.0798	2.29	1.009		0.01190	

SOURCE: Reprinted by permission from *ASHRAE Fundamentals 1993 (SI)*, p. 17.13.

TABLE A.9E Thermophysical Properties of Ammonia

Temp., °F	Saturated Liquid				Saturated Vapor			
	Specific Heat, Btu/lbm·°F	Dynamic Viscosity, lbm/ft·hr	Thermal Conductivity, Btu/hr·ft·°F	Pr	Specific Heat, Btu/lbm·°F	Dynamic Viscosity, lbm/hr·ft	Thermal Conductivity, Btu/hr·ft·°F	Pr
0	1.0786	0.521	0.3216	1.75	0.5752	0.0207	0.01112	1.071
20	1.0937	0.458	0.3086	1.62	0.6113	0.0215	0.01202	1.093
40	1.1103	0.405	0.2956	1.52	0.6533	0.0223	0.01305	1.116
60	1.1290	0.359	0.2824	1.44	0.7018	0.0231	0.01419	1.142
80	1.1504	0.320	0.2690	1.37	0.7584	0.0240	0.01541	1.181
100	1.1757	0.285	0.2555	1.31	0.8253	0.0249	0.01669	1.231
120	1.2072	0.255	0.2417	1.27	0.9067	0.0258	0.01809	1.293

SOURCE: Reprinted by permission from *ASHRAE Fundamentals 1993* (IP), p. 17.45.

TABLE A.9SI Thermophysical Properties of Ammonia

Temp., °C	Saturated Liquid				Saturated Vapor			
	Specific Heat, kJ/kg·°C	Dynamic Viscosity, kg/m·s × 10^6	Thermal Conductivity, W/m·°C	Pr	Specific Heat, kJ/kg·°C	Dynamic Viscosity, kg/m·s × 10^6	Thermal Conductivity, W/m·°C	Pr
−20	4.501	221.3	0.5607	1.78	2.379	8.49	0.01896	1.065
−10	4.556	196.8	0.5405	1.66	2.510	8.79	0.02029	1.087
0	4.617	175.8	0.5202	1.56	2.660	9.09	0.02676	0.904
10	4.683	157.6	0.4998	1.48	2.831	9.40	0.02355	1.130
20	4.758	141.6	0.4792	1.41	3.027	9.71	0.02538	1.158
30	4.843	127.6	0.4583	1.35	3.252	10.02	0.02730	1.194
40	4.943	115.2	0.4371	1.30	3.516	10.35	0.02934	1.240

SOURCE: Reprinted by permission from *ASHRAE Fundamentals 1993* (SI), p. 17.45.

appendix B
Weather Data

TABLE B.1 Winter and Summer Design Conditions for Selected Locations

U.S. State and Station	Latitude, deg	Longitude, deg	Elevation, ft	Winter, °F Design Dry-Bulb 99%	Winter, °F Design Dry-Bulb 97.5%	Summer, °F Design Dry-Bulb/Mean Coincident Wet-Bulb 1%	Summer, °F Design Dry-Bulb/Mean Coincident Wet-Bulb 2.5%	Mean Daily Range	Design Wet-Bulb 1%	Design Wet-Bulb 2.5%
Alabama, Birmingham AP	33.6 N	86.8 W	620	17	21	96/74	94/75	21	78	77
Alaska, Anchorage AP	61.2 N	150.0 W	114	-23	-18	71/59	68/58	15	60	59
Arizona, Tucson AP	32.1 N	110.9 W	2558	28	32	104/66	102/66	26	72	71
Arkansas, Little Rock AP	34.7 N	92.2 W	257	15	20	99/76	96/77	22	80	79
California, Los Angeles AP	33.9 N	118.4 W	97	41	43	83/68	80/68	15	70	69
Colorado, Denver AP	39.8 N	104.9 W	5283	-5	1	93/59	91/59	28	64	63
Connecticut, Bridgeport AP	41.2 N	73.2 W	25	6	9	86/73	84/71	18	75	74
Delaware, Wilmington AP	39.7 N	75.6 W	74	10	14	92/74	89/74	20	77	76
Dist. of Columbia, Washington Nat. AP	38.9 N	77.0 W	14	14	17	93/75	91/74	18	78	77
Florida, Miami AP	25.8 N	80.3 W	7	44	47	91/77	90/77	15	79	79
Georgia, Atlanta AP	33.7 N	84.4 W	1010	17	22	94/74	92/74	19	77	76
Hawaii, Honolulu AP	21.3 N	157.9 W	13	62	63	87/73	86/73	12	76	75
Idaho, Boise AP	43.6 N	116.2 W	2838	3	10	96/65	94/64	31	68	66
Illinois, Chicago, O'Hare AP	42.0 N	87.9 W	658	-8	-4	91/74	89/74	20	77	76
Indiana, Indianapolis AP	39.7 N	86.3 W	792	-2	2	92/74	90/74	22	78	76
Iowa, Des Moines AP	41.5 N	93.7 W	938	-10	-5	94/75	91/74	23	78	77
Kansas, Wichita AP	37.7 N	97.4 W	1321	3	7	101/72	98/73	23	77	76
Kentucky, Louisville AP	38.2 N	85.7 W	477	5	10	95/74	93/74	23	79	77
Louisiana, Shreveport AP	32.5 N	93.8 W	254	20	25	99/97	96/76	20	79	79
Maine, Caribou AP	46.9 N	68.0 W	624	-18	-13	84/69	81/67	21	71	69
Maryland, Baltimore AP	39.2 N	76.7 W	148	10	13	94/75	91/75	21	78	77
Massachusetts, Boston AP	42.4 N	71.0 W	15	6	9	91/73	88/71	16	75	74
Michigan, Lansing AP	42.8 N	84.6 W	873	-3	1	90/73	87/72	24	75	74
Minnesota, Minneapolis/St. Paul AP	44.9 N	93.2 W	834	-16	-12	92/75	89/73	22	77	75
Mississippi, Jackson AP	32.3 N	90.1 W	310	21	25	97/76	95/76	21	79	78
Missouri, St. Louis AP	38.8 N	90.4 W	535	2	6	97/75	94/75	21	78	77
Montana, Billings AP	45.8 N	108.5 W	3567	-15	-10	94/64	91/64	31	67	66
Nebraska, Lincoln Co	40.9 N	96.8 W	1180	-5	-2	99/75	95/74	24	78	77
Nevada, Las Vegas AP	36.1 N	115.2 W	2178	25	28	108/66	106/65	30	71	70
New Hampshire, Concord AP	43.2 N	71.5 W	342	-8	-3	90/72	87/70	26	74	73
New Jersey, Atlantic City Co	39.4 N	74.4 W	11	10	13	92/74	89/74	18	78	77
New Mexico, Albuquerque AP	35.1 N	106.6 W	5313	12	16	96/61	94/61	27	66	65
New York, NYC-La Guardia AP	40.8 N	73.9 W	11	11	15	92/74	89/73	16	76	75
North Carolina, Charlotte AP	35.2 N	80.9 W	736	18	22	95/74	93/74	20	77	76
North Dakota, Bismark AP	46.8 N	100.8 W	1647	-23	-19	95/68	91/68	27	73	71
Ohio, Columbus AP	40.0 N	82.9 W	812	0	5	92/73	90/73	24	77	75
Oklahoma, Stillwater	36.2 N	97.1 W	984	8	13	100/74	96/74	24	77	76
Oregon, Portland AP	45.6 N	122.6 W	21	17	23	89/68	85/67	23	69	67
Pennsylvania, Pittsburg AP	40.5 N	80.2 W	1137	1	5	89/72	86/71	22	74	73
Rhode Island, Providence AP	41.7 N	71.4 W	51	5	9	89/73	86/72	19	75	74
South Carolina, Charleston AFB	32.9 N	80.0 W	45	24	27	93/78	91/78	18	81	80

Station	Latitude	Longitude	Elevation							
South Dakota, Rapid City AP	44.1 N	103.1 W	3162	−11	−7	95/66	92/65	28	71	69
Tennessee, Memphis AP	35.1 N	90.0 W	258	13	18	98/77	95/76	21	80	79
Texas, Dallas AP	32.9 N	96.9 W	481	18	22	102/75	100/75	20	78	78
Utah, Salt Lake City AP	40.8 N	112.0 W	4220	3	8	97/62	95/62	32	66	65
Vermont, Burlington AP	44.5 N	73.2 W	332	−12	−7	88/72	85/70	23	74	72
Virginia, Norfolk AP	36.9 N	76.2 W	22	20	22	93/77	91/76	18	79	78
Washington, Seattle-Tacoma AP	47.5 N	122.3 W	400	21	26	84/65	80/64	22	66	64
West Virginia, Charleston AP	38.4 N	81.6 W	939	7	11	92/74	90/73	20	76	75
Wisconsin, Madison AP	43.1 N	89.3 W	858	−11	−7	91/74	88/73	22	77	75
Wyoming, Casper AP	42.9 N	106.5 W	5338	−11	−5	92/58	90/57	31	63	61

Canadian Province and Station

Station	Latitude	Longitude	Elevation							
Alberta, Edmonton AP	53.6 N	113.5 W	2219	−29	−25	85/66	82/65	23	68	66
Manitoba, Winnipeg AP	49.9 N	97.2 W	786	−30	−27	89/73	86/71	22	75	73
Ontario, Ottawa AP	45.3 N	75.7 W	413	−17	−13	90/72	87/71	21	75	73
Quebec, Quebec AP	46.8 N	71.4 W	245	−19	−14	87/72	84/70	20	74	72

Country and Station

Station	Latitude	Longitude	Elevation							
Australia, Sydney	33.9 S	151.2 E	138	40	42	89	84	13	74	73
Brazil, Rio De Janeiro	22.9 S	43.2 E	201	58	60	94	92	11	80	79
France, Paris	48.8 N	2.5 E	164	22	25	89	86	21	70	68
India, Bombay	18.9 N	72.8 E	37	65	67	96	94	13	82	81
Japan, Tokyo	35.7 N	139.8 E	56	1	5	86	83	20	76	74
Mexico, Mexico City	19.4 N	99.2 W	7575	37	39	83	81	25	61	60
Russia, Moscow	55.8 N	37.7 E	505	−11	−6	84	81	21	69	67
South Africa, Cape Town	33.9 S	18.5 E	55	40	42	93	90	20	72	71
Sweden, Stockholm	59.4 N	18.1 E	146	5	8	78	74	15	64	62
Tunisia, Tunis	36.8 N	10.2 E	217	39	41	102	99	22	77	76
Ukraine, Kiev	50.5 N	30.5 E	600	−5	1	87	84	22	69	68

SOURCE: Reprinted by permission from *ASHRAE Fundamentals 1993* (IP & SI), p. 24.4.

To convert elevation to meters: m = 0.3048 ft.

To convert temperatures to °C: °C = (°F − 32)/1.8.

To convert mean daily range (i.e., temperature difference) to °C: (temp. diff. °C) = (temp. diff. °F)/1.8.

TABLE B.2 Average Monthly Degree Days for Selected U.S. Cities

	Jan.	Feb.	Mar.	Apr.	May	Jun.	Jul.	Aug.	Sep.	Oct.	Nov.	Dec.	Yr.
Tucson, AZ													
Avg. Temp. (°F)	51	54	58	66	74	82	86	84	80	70	59	52	68
DD 50F	80	41	15	2	0	0	0	0	0	0	12	64	214
DD 55F	166	106	56	7	0	0	0	0	0	2	44	144	525
DD 60F	292	201	133	26	0	0	0	0	0	8	114	262	1036
DD 65F	442	333	243	81	0	0	0	0	0	29	221	403	1752
DD 70F	593	463	388	169	44	4	1	3	7	91	350	559	2673
San Francisco, CA													
Avg. Temp. (°F)	48	51	53	55	58	62	63	63	64	61	55	50	57
DD 50F	82	32	17	5	1	0	0	0	0	0	5	58	202
DD 55F	210	117	88	47	15	3	2	1	1	4	47	170	705
DD 60F	363	247	219	148	82	30	21	17	10	39	148	320	1643
DD 65F	518	386	372	291	210	120	93	84	66	137	291	474	3042
DD 70F	673	526	527	441	363	253	234	219	181	280	441	629	4768
Denver, CO													
Avg. Temp. (°F)	30	33	37	48	57	66	73	72	63	52	39	33	50
DD 50F	623	482	406	130	18	1	0	0	3	63	324	540	2592
DD 55F	778	622	559	240	63	5	0	0	14	143	469	695	3588
DD 60F	933	762	713	379	143	23	0	0	51	261	618	849	4733
DD 65F	1088	902	868	525	253	80	0	0	120	408	768	1004	6016
DD 70F	1243	1042	1023	675	406	158	50	69	232	559	918	1159	7535
Atlanta, GA													
Avg. Temp. (°F)	42	45	51	61	69	76	78	78	72	62	51	44	61
DD 50F	246	161	67	3	0	0	0	0	0	2	61	217	758
DD 55F	393	284	153	16	1	0	0	0	0	11	142	360	1362
DD 60F	546	421	283	65	7	0	0	0	2	49	266	512	2150
DD 65F	701	560	443	144	27	0	0	0	8	137	408	667	3095
DD 70F	856	700	586	274	98	19	9	11	49	246	558	822	4228
Minneapolis, MN													
Avg. Temp. (°F)	12	17	28	45	57	67	72	70	60	50	32	19	44
DD 50F	1172	938	673	178	18	1	0	0	8	93	529	973	4584
DD 55F	1327	1078	828	305	63	4	1	2	30	186	678	1128	5631
DD 60F	1482	1218	983	449	143	18	5	8	90	318	828	1283	6824
DD 65F	1637	1358	1138	597	271	65	11	21	173	472	978	1438	8159
DD 70F	1792	1498	1293	747	403	142	66	90	308	620	1128	1593	9680
New York, NY													
Avg. Temp. (°F)	32	33	41	52	62	72	77	75	68	59	47	36	55
DD 50F	552	465	282	50	2	0	0	0	0	7	122	451	1931
DD 55F	707	605	432	127	11	0	0	0	1	33	238	605	2759
DD 60F	862	745	586	246	49	0	0	0	7	103	380	760	3737
DD 65F	1017	885	741	387	137	0	0	0	29	209	528	915	4848
DD 70F	1172	1025	896	537	248	56	14	23	104	353	678	1070	6177
Houston, TX													
Avg. Temp. (°F)	52	55	61	69	76	81	83	83	79	71	61	55	69
DD 50F	71	31	9	1	0	0	0	0	0	1	8	41	161
DD 55F	150	87	31	3	0	0	0	0	0	2	28	108	409
DD 60F	263	168	89	12	0	0	0	0	0	8	82	201	825
DD 65F	416	294	189	23	0	0	0	0	0	24	155	333	1434
DD 70F	556	414	298	107	31	8	4	4	13	88	281	480	2285

TABLE B.3 Davis-Monthan AFB/Tucson Arizona—Lat. 32° 11′ N, Long. 110° 54′ W, Elev. 2654 ft: Mean Frequency of Occurrence of Dry-Bulb Temperature (°F) with Mean Coincident Wet-Bulb Temperature (°F) for Each Dry-Bulb Temperature Range

Temperature Range	MAY					JUNE					JULY					AUGUST				
	Obsn. Hour Gp. 01 to 08	09 to 16	17 to 24	Total Obsn.	MCWB	01 to 08	09 to 16	17 to 24	Total Obsn.	MCWB	01 to 08	09 to 16	17 to 24	Total Obsn.	MCWB	01 to 08	09 to 16	17 to 24	Total Obsn.	MCWB
110/114									0	68										
105/109							5	1	6	67		3	1	4	68				0	68
100/104		3	1	4	64	0	28	13	41	66		28	14	42	68		9	3	12	69
95/99		15	7	22	62		50	32	82	64		61	34	95	68		37	20	57	69
90/94		43	21	64	59	1	63	47	111	62	2	71	50	123	68	0	75	39	114	69
85/89	1	56	34	91	57	12	47	49	108	60	20	51	55	126	67	5	67	51	123	69
80/84	5	53	42	100	55	39	30	42	111	59	72	23	49	144	67	39	42	62	143	68
75/79	15	38	46	99	54	50	12	30	92	57	100	8	31	139	66	110	14	49	173	67
70/74	36	22	43	101	51	57	3	18	78	54	51	3	13	67	66	84	3	20	107	66
65/69	57	11	30	98	49	48	0	6	54	50	3	0	1	4	63	10	0	3	13	63
60/64	67	4	15	86	46	26		1	27	47								0	0	61
55/59	41	2	6	49	44	7			7	44										
50/54	19	1	2	22	41	0			0	41										
45/49	6	0	1	7	38															
40/44	2	0		2	34															
35/39	0			0	28															

(continued)

TABLE B.3 (cont.)

Temperature Range	SEPTEMBER Obsn. Hour Gp. 01 to 08	09 to 16	17 to 24	Total Obsn.	MCWB	OCTOBER Obsn. Hour Gp. 01 to 08	09 to 16	17 to 24	Total Obsn.	MCWB	NOVEMBER Obsn. Hour Gp. 01 to 08	09 to 16	17 to 24	Total Obsn.	MCWB	DECEMBER Obsn. Hour Gp. 01 to 08	09 to 16	17 to 24	Total Obsn.	MCWB
110/114																				
105/109																				
100/104		3	0	3	69		2	0	2	61										
95/99		34	10	44	66		18	4	22	61		1		1	55		0		0	63
90/94		54	26	80	65		41	14	55	59		16	2	18	55		1		1	53
85/89	1	64	44	109	65	0	56	23	79	58		34	8	42	53		8	1	9	52
80/84	20	47	54	121	64	3	52	40	95	57	0	45	17	62	52		24	4	28	50
75/79	57	23	52	132	63	21	37	50	108	55	3	47	31	81	50	0	41	12	53	49
70/74	80	11	37	128	61	53	21	49	123	53	18	42	49	109	48	3	41	26	70	47
65/69	55	3	13	71	57	70	11	37	118	50	51	27	54	132	46	16	46	46	108	46
60/64	21	0	3	24	52	54	6	19	79	47	65	18	42	125	43	42	39	59	140	43
55/59	5		1	6	48	29	2	8	39	44	58	7	23	88	40	60	28	50	138	40
50/54	1			1	43	12	1	3	16	40	28	3	10	41	37	68	12	32	112	37
45/49						4	1	1	6	38	13	1	3	17	34	38	6	14	58	33
40/44						0		0	0	30	4	0	1	5	31	16	1	3	20	29
35/39						0			0	26	0	0		0	26	4	0	1	5	25
30/34																1			1	21
25/29																				
20/24																				

Temperature Range	JANUARY					FEBRUARY					MARCH					APRIL					ANNUAL TOTAL				
	Obsn. Hour Gp. 01 to 08	09 to 16	17 to 24	Total Obsn.	MCWB	Obsn. Hour Gp. 01 to 08	09 to 16	17 to 24	Total Obsn.	MCWB	Obsn. Hour Gp. 01 to 08	09 to 16	17 to 24	Total Obsn.	MCWB	Obsn. Hour Gp. 01 to 08	09 to 16	17 to 24	Total Obsn.	MCWB	Obsn. Hour Gp. 01 to 08	09 to 16	17 to 24	Total Obsn.	MCWB
110/114																							0	0	68
105/109																						8	2	10	68
100/104																	0		0	62		71	31	102	67
95/99																	1		1	61	0	200	103	303	66
90/94																	9	3	12	58	3	333	190	526	65
85/89							1		1	59		0	0	0	58		19	8	27	56	39	350	256	645	63
80/84		1		1	55		3	1	4	56		3	1	4	56		42	21	63	54	175	330	301	806	62
75/79		5	1	6	52		12	3	15	53		16	5	21	54		50	33	83	52	335	284	309	928	60
70/74		19	5	24	50		24	11	35	51		28	15	43	52	6	46	42	94	50	336	280	285	901	56
65/69		34	12	46	48	1	38	21	60	49	1	43	25	69	50	20	33	46	99	48	252	270	259	781	51
60/64	1	51	28	80	47	4	43	37	84	47	2	42	35	79	48	44	21	40	105	46	265	256	281	802	48
55/59	12	49	48	109	45	21	43	44	108	45	11	43	45	99	46	61	11	28	100	44	300	217	292	809	45
50/54	37	41	55	133	43	40	32	46	118	42	32	33	46	111	44	58	6	12	76	41	357	161	262	780	42
45/49	53	26	47	126	39	54	17	32	103	39	66	22	38	126	42	35	1	5	41	39	341	92	185	618	40
40/44	70	14	30	114	36	48	8	18	74	36	63	12	24	99	39	14		1	15	36	279	42	102	423	36
35/39	43	7	15	65	32	36	3	8	47	32	45	4	10	59	36	3			3	34	153	18	44	215	32
30/34	20	2	6	28	28	17	1	2	20	28	20	1	4	25	32	0			0	33	64	4	13	81	29
25/29	9	1	2	12	23	3			3	24	7		1	8	28						17	1	3	21	24
20/24	3			3	18	0			0	22	1			1	22						4	0	0	4	19
15/19	1			1	16						0			0	17						1			1	16

713

TABLE B.4 Moffett Field NAS California—Lat. 37° 25' N, Long. 122° 03' W, Elev. 34 ft: Mean Frequency of Occurrence of Dry-Bulb Temperature (°F) with Mean Coincident Wet-Bulb Temperature (°F) for Each Dry-Bulb Temperature Range

Temperature Range	MAY 01 to 08	MAY 09 to 16	MAY 17 to 24	MAY Total Obsn.	MAY MCWB	JUNE 01 to 08	JUNE 09 to 16	JUNE 17 to 24	JUNE Total Obsn.	JUNE MCWB	JULY 01 to 08	JULY 09 to 16	JULY 17 to 24	JULY Total Obsn.	JULY MCWB	AUGUST 01 to 08	AUGUST 09 to 16	AUGUST 17 to 24	AUGUST Total Obsn.	AUGUST MCWB
105/109																				
100/104							0		0	70		0	0	0	70		0		0	69
95/99		0	0	0	64		0		0	71		1		1	68		0	0	0	66
90/94		1	0	1	64		2	0	2	69		3	1	4	67		2	0	2	68
85/89		4	1	5	64		4	1	5	67		10	2	12	66		14	1	15	66
80/84							9	3	12	66		25	7	32	65	0	32	9	41	65
75/79		11	3	14	63	0	22	8	30	65	3	66	26	95	63	3	79	35	117	63
70/74	1	22	9	32	61	3	52	23	78	62	15	102	72	189	61	19	89	79	187	61
65/69	3	64	26	93	58	14	86	56	156	59	98	39	106	243	58	114	30	99	243	58
60/64	21	98	65	184	55	64	55	82	201	57	116	2	33	151	55	98	2	22	122	55
55/59	78	42	91	211	52	104	10	60	174	54	16		1	17	51	14		1	15	51
50/54	104	5	49	158	49	52	0	8	60	50						0			0	47
45/49	38	0	5	43	45	3		0	3	45										
40/44	3			3	41															
35/39																				

Tempera-ture Range	SEPTEMBER Obsn. Hour Gp. 01 to 08	09 to 16	17 to 24	Total Obsn.	MCWB	OCTOBER Obsn. Hour Gp. 01 to 08	09 to 16	17 to 24	Total Obsn.	MCWB	NOVEMBER Obsn. Hour Gp. 01 to 08	09 to 16	17 to 24	Total Obsn.	MCWB	DECEMBER Obsn. Hour Gp. 01 to 08	09 to 16	17 to 24	Total Obsn.	MCWB
105/109		0	0	0	71															
100/104		0	0	0	71															
95/99		0	0	0	67															
90/94		3	0	3	66		0		0	66										
85/89		8	1	9	65		3	0	3	65		0	0	0	65					
80/84		16	4	20	64		11	2	13	63		1	0	1	63					
75/79	0	32	12	44	63	0	22	6	28	61		1	0	1	59		0		0	60
70/74	2	66	32	100	62	0	42	16	58	60		11	2	13	58	0	2	0	2	58
65/69	18	77	68	163	60	4	77	42	123	58	0	35	8	43	57	2	7	3	12	57
60/64	86	34	84	204	58	42	67	87	196	56	6	71	42	119	55	5	30	12	47	55
55/59	97	4	36	137	54	102	23	71	196	53	49	77	84	210	52	19	60	40	119	52
50/54	34	0	3	37	50	74	2	21	97	49	79	33	68	180	49	38	75	73	186	48
45/49	2	0	0	2	46	22	0	2	24	44	58	9	30	97	44	62	50	75	187	44
40/44						4	0	0	4	39	37	1	5	43	40	76	20	39	135	40
35/39						0			0	32	9	0	0	9	35	40	4	6	50	36
30/34											1	0	0	1	31	6	0	0	6	32

(continued)

TABLE B.4 (cont.)

Temperature Range	JANUARY					FEBRUARY					MARCH					APRIL					ANNUAL TOTAL				
	Obsn. Hour Gp.			Total Obsn.	M C W B	Obsn. Hour Gp.			Total Obsn.	M C W B	Obsn. Hour Gp.			Total Obsn.	M C W B	Obsn. Hour Gp.			Total Obsn.	M C W B	Obsn. Hour Gp.			Total Obsn.	M C W B
	01 to 08	09 to 16	17 to 24			01 to 08	09 to 16	17 to 24			01 to 08	09 to 16	17 to 24			01 to 08	09 to 16	17 to 24			01 to 08	09 to 16	17 to 24		
105/109																							0	0	71
100/104																							0	0	70
95/99																							0	0	69
90/94																						0	6	6	67
85/89																0	0	0	0	65		21	3	24	66
80/84												1	0	1	62		4	1	5	64		70	14	84	65
75/79		1	0	1	54		0		0	61		3	1	4	60		10	3	13	61	0	158	49	207	63
70/74		5	1	6	57		3	1	4	58		9	3	12	57		17	6	23	59	12	370	153	535	62
65/69	4	27	11	42	55	3	16	4	20	57		21	9	30	55	1	37	16	54	57	76	616	384	1076	59
60/64	19	64	42	125	52	29	44	23	70	55	3	58	25	86	53	8	71	40	119	54	454	624	676	1754	56
55/59	38	75	77	190	48	55	77	65	171	52	21	91	69	181	51	39	67	72	178	51	771	519	685	1975	53
50/54	57	50	69	176	44	75	60	79	194	48	82	54	94	230	48	82	29	70	181	48	668	333	544	1545	48
45/49	74	21	36	131	40	48	21	44	140	44	91	10	42	143	44	85	4	30	119	44	493	144	297	934	44
40/44	42	5	11	58	35	12	3	8	59	40	43	1	5	49	40	25	0	2	27	41	310	46	95	451	40
35/39	14	1	1	16	31	1	0	0	12	36	7		0	7	36	2			2	37	112	9	17	138	35
30/34	1			1	26		0	0	1	30	0			0	32						22	1	1	24	31
25/29																					1			1	26

716

TABLE B.5 Buckley ANGB/Denver Colorado—Lat. 39° 42′ N, Long. 104° 45′ W, Elev. 5663 ft: Mean Frequency of Occurrence of Dry-Bulb Temperature (°F) with Mean Coincident Wet-Bulb Temperature (°F) for Each Dry-Bulb Temperature Range

Column groups: for each month — Obsn. Hour Gp. (01 to 08, 09 to 16, 17 to 24), Total Obsn., MCWB (Mean Coincident Wet-Bulb).

Temperature Range	MAY 01–08	MAY 09–16	MAY 17–24	MAY Total Obsn.	MAY MCWB	JUNE 01–08	JUNE 09–16	JUNE 17–24	JUNE Total Obsn.	JUNE MCWB	JULY 01–08	JULY 09–16	JULY 17–24	JULY Total Obsn.	JULY MCWB	AUG 01–08	AUG 09–16	AUG 17–24	AUG Total Obsn.	AUG MCWB
100/104																	0	0	0	64
95/99																	3	0	3	63
90/94							2	0	2	60		2	0	2	61		18	2	20	60
85/89		4	0	4	55		12	3	15	58		31	5	36	60		41	9	50	60
80/84		20	4	24	53	0	23	6	29	58	0	62	14	76	60	2	62	19	83	59
75/79	0	32	9	41	52	1	32	12	45	57	3	60	25	88	61	6	51	30	87	59
70/74	2	41	18	61	51	3	44	21	68	56	10	44	37	91	60	18	37	48	103	58
65/69	5	38	25	68	50	9	42	30	81	55	26	29	56	111	59	48	20	59	127	56
60/64	17	39	39	95	48	19	38	37	94	54	62	12	60	134	57	82	11	54	147	55
55/59	34	23	43	100	46	45	23	48	116	52	90	5	37	132	56	63	5	20	88	52
50/54	50	22	39	111	44	63	11	29	115	51	45	3	13	61	53	25	1	5	31	48
45/49	56	14	34	104	42	58	9	10	96	48	11	0	2	13	50	5	0	1	6	43
40/44	46	11	23	80	38	33	5	10	48	44	1	1	1	3	45	0			0	43
35/39	24	3	9	36	35	9	1	2	12	41	1			1	43					
30/34	10	1	5	16	30	1	0	1	2	37										
25/29	3	0	0	3	26															

(continued)

TABLE B.5 (cont.)

Each month group lists the "Obsn. Hour Gp." columns (01 to 08, 09 to 16, 17 to 24), "Total Obsn.", and "MCWB".

Temperature Range	Sep 01–08	Sep 09–16	Sep 17–24	Sep Total Obsn.	Sep MCWB	Oct 01–08	Oct 09–16	Oct 17–24	Oct Total Obsn.	Oct MCWB	Nov 01–08	Nov 09–16	Nov 17–24	Nov Total Obsn.	Nov MCWB	Dec 01–08	Dec 09–16	Dec 17–24	Dec Total Obsn.	Dec MCWB
90/94				0	60															
85/89		13	2	15	56		0		0	54										
80/84		38	6	44	56		7		7	52										
75/79	0	48	13	61	55	0	23	2	25	51		0		0	48					
70/74	3	43	23	69	54	1	32	4	36	49		3		3	47		0		0	51
65/69	10	32	36	78	52	6	35	10	46	48		12		12	46	1	2	0	3	45
60/64	33	22	51	106	51	17	39	20	65	46	0	28	1	29	44	0	10	1	11	43
55/59	62	16	45	123	49	33	30	35	82	44	1	37	6	44	42	1	20	2	23	40
50/54	54	10	27	91	46	48	27	42	102	41	6	32	19	57	39	2	27	5	34	38
45/49	38	7	17	62	42	55	18	47	113	39	16	33	31	80	37	6	28	11	45	35
40/44	21	4	8	33	39	41	12	39	106	36	42	26	49	117	34	16	34	25	75	32
35/39	11	4	6	21	35	27	10	20	71	33	48	21	45	114	31	29	31	35	95	29
30/34	5	4	4	13	31	12	10	17	54	30	47	22	33	102	28	43	33	46	122	27
25/29	2		1	3	27	6	3	7	22	25	41	19	30	90	25	42	25	41	108	23
20/24	0			0	23	1	2	4	12	21	21	6	18	45	20	42	15	32	89	20
15/19						1	1	1	3	17	13	1	8	22	16	30	10	25	65	15
10/14							0	1	2	12	4	0	1	5	11	19	7	11	37	11
5/9									1	6	1			1	8	10	4	7	21	7
0/4																5	1	4	10	2
-5/-1																2	0	1	3	-3
-10/-6																1	0	1	2	-8
-15/-11																0			0	-14

718

Temperature Range	JAN 01 to 08	JAN 09 to 16	JAN 17 to 24	JAN Total Obsn.	JAN MCWB	FEB 01 to 08	FEB 09 to 16	FEB 17 to 24	FEB Total Obsn.	FEB MCWB	MAR 01 to 08	MAR 09 to 16	MAR 17 to 24	MAR Total Obsn.	MAR MCWB	APR 01 to 08	APR 09 to 16	APR 17 to 24	APR Total Obsn.	APR MCWB	ANN 01 to 08	ANN 09 to 16	ANN 17 to 24	ANN Total Obsn.	ANN MCWB
100/104																								0	64
95/99																						7	0	7	61
90/94																						61	10	71	60
85/89																	0		0	50	0	143	31	174	59
80/84												0		0	53		8	2	10	49	6	219	66	291	58
75/79												1	0	1	51		18	5	23	47	19	251	114	384	56
70/74							1	0	1	48		5	1	6	47	0	27	8	35	46	58	251	185	494	55
65/69		1		1	46		4	0	4	46		13	3	16	44	2	32	14	48	44	146	234	238	618	53
60/64		7	1	8	43		7	1	8	43	1	22	6	29	42	7	39	26	72	42	276	245	273	794	51
55/59	1	16	1	18	41	0	12	3	15	41	1	25	9	35	41	18	37	36	91	40	295	237	244	776	47
50/54	2	27	5	34	38	1	26	8	35	38	4	26	14	44	38	32	31	43	106	38	264	244	231	739	43
45/49	5	31	13	49	36	5	34	16	55	36	8	26	24	58	36	48	17	40	105	35	253	228	248	729	39
40/44	16	30	26	72	33	12	30	24	66	33	25	28	32	85	33	52	15	31	98	33	291	193	268	752	35
35/39	32	30	40	102	30	29	30	35	94	30	37	22	32	91	31	49	10	23	82	30	304	166	254	724	31
30/34	38	30	35	103	27	36	23	42	101	27	44	25	42	111	28	19	4	9	32	25	299	158	247	704	28
25/29	41	24	35	100	24	43	20	34	97	24	42	25	33	100	24	8	2	2	12	21	245	120	190	555	24
20/24	33	16	27	76	19	41	18	26	85	20	36	14	25	75	20	3	1	1	5	17	187	73	134	394	20
15/19	27	11	22	60	15	24	7	16	47	15	21	9	11	41	16	1			1	13	119	40	84	243	15
10/14	19	6	13	38	10	15	6	8	29	11	13	5	7	25	11						72	24	41	137	11
5/9	13	7	9	29	6	8	3	4	15	6	8	3	6	17	7						41	17	26	84	6
0/4	8	6	8	22	1	7	2	5	14	-1	5	1	2	8	1						25	10	19	54	2
-5/-1	6	4	4	14	-3	2	1	0	3	-4	2	0	0	2	-3						11	5	6	22	-3
-10/-6	5	1	4	10	-8	1	0	0	1	-9	0	0	0	0	-6						7	1	5	13	-8
-15/-11	2	0	2	4	-13	1			1	-13											3	1	1	5	-13
-20/-16	1	1	1	3	-18																1	1	1	3	-18
-25/-21	1	0	0	1	-24																1			1	-24
-30/-26				0	-28																			0	-28

TABLE B.6 Atlanta/Hartsfield IAP Georgia—Lat. 33° 39′ N, Long. 84° 26′ W, Elev. 1010 ft: Mean Frequency of Occurrence of Dry-Bulb Temperature (°F) with Mean Coincident Wet-Bulb Temperature (°F) for Each Dry-Bulb Temperature Range

Temperature Range	MAY Obsn. Hour Gp. 01 to 08	09 to 16	17 to 24	Total Obsn.	MCWB	JUNE Obsn. Hour Gp. 01 to 08	09 to 16	17 to 24	Total Obsn.	MCWB	JULY Obsn. Hour Gp. 01 to 08	09 to 16	17 to 24	Total Obsn.	MCWB	AUGUST Obsn. Hour Gp. 01 to 08	09 to 16	17 to 24	Total Obsn.	MCWB
100/104																	0		0	76
95/99							0		0	77		1	0	1	72		6	1	7	74
90/94		4	1	5	70		4	1	5	74		5	1	6	74		30	9	39	74
85/89		33	13	46	69		29	10	39	74		30	10	40	74		72	33	105	73
80/84	0	53	28	81	67	0	52	25	77	72	0	61	30	91	74	3	76	58	137	72
75/79	3	57	46	106	66	5	65	44	114	71	5	81	57	143	73	47	42	76	165	71
70/74	32	47	62	141	64	31	50	62	143	69	46	51	74	171	72	133	19	59	211	69
65/69	94	30	52	176	62	93	28	64	185	68	147	18	69	234	70	54	3	11	68	65
60/64	59	15	26	100	58	82	11	28	121	64	46	2	7	55	66	10	0	1	11	60
55/59	30	7	15	52	53	23	1	4	28	59	5	0	1	6	60	0	0		0	56
50/54	19	1	5	25	48	5	1	1	7	55										
45/49	8	0	1	9	44	1	0	0	1	49										
40/44	3	0	0	3	40	0	0	0	0	44										

Temperature Range	SEPTEMBER Obsn. Hour Gp. 01 to 08	09 to 16	17 to 24	Total Obsn.	M C W B	OCTOBER Obsn. Hour Gp. 01 to 08	09 to 16	17 to 24	Total Obsn.	M C W B	NOVEMBER Obsn. Hour Gp. 01 to 08	09 to 16	17 to 24	Total Obsn.	M C W B	DECEMBER Obsn. Hour Gp. 01 to 08	09 to 16	17 to 24	Total Obsn.	M C W B
95/99		2	0	2	72		0	0	0	74										
90/94		9	2	11	71		1	0	1	73										
85/89		31	11	42	71		3	1	4	70										
80/84	0	60	31	91	70		16	4	20	68	0	1	0	1	67					
75/79	8	57	53	118	68	1	42	14	57	64	0	5	1	6	64		3	0	3	60
70/74	72	43	71	186	67	8	52	34	94	63	0	20	5	25	60	1	13	5	19	60
65/69	82	21	40	143	63	28	53	54	135	60	5	32	17	54	59	9	22	16	47	57
60/64	49	14	23	86	59	51	39	59	149	57	18	39	35	92	56	13	26	22	61	52
55/59	22	2	6	30	54	60	23	40	123	52	28	40	43	111	51	17	36	33	86	47
50/54	6	1	1	8	49	51	12	24	87	48	34	38	41	113	47	29	44	41	114	42
45/49	1		0	1	45	28	4	11	43	43	45	30	42	117	42	39	44	53	136	38
40/44	0			0	42	14	2	4	20	39	44	20	30	94	38	55	29	37	121	34
35/39						7	0	1	8	33	35	9	16	60	33	43	19	25	87	29
30/34						2	0	0	2	29	20	4	7	31	29	23	7	11	41	24
25/29											8	1	3	12	25	13	2	3	18	20
20/24											3	0	0	3	21	3	1	1	5	15
15/19											1	0	1	2	14	2	1	0	3	11
10/14											0	0	0	0	10	0	0	0	0	5
5/9											0	0	0	0	6	0	0	0	0	
0/4											0			0	3	1			1	1

(continued)

TABLE B.6 (cont.)

Temperature Range	JANUARY 01 to 08	09 to 16	17 to 24	Total Obsn.	MCWB	FEBRUARY 01 to 08	09 to 16	17 to 24	Total Obsn.	MCWB	MARCH 01 to 08	09 to 16	17 to 24	Total Obsn.	MCWB	APRIL 01 to 08	09 to 16	17 to 24	Total Obsn.	MCWB	ANNUAL TOTAL 01 to 08	09 to 16	17 to 24	Total Obsn.	MCWB
100/104																						1	0	1	73
95/99																						17	3	20	74
90/94																	2	0	2	66	0	103	32	135	74
85/89																	17	7	24	64	0	254	113	367	72
80/84												1	0	1	64		38	20	58	62	13	370	229	612	70
75/79	0	0		0	64		2	0	2	63		9	4	13	62	2	41	36	79	61	136	353	350	839	69
70/74		4	1	5	63		8	3	11	60		18	9	27	59	22	43	48	113	59	487	301	413	1201	67
65/69	2	11	5	18	60	2	16	10	28	57	5	27	24	56	57	55	40	48	143	56	423	262	301	986	62
60/64	9	19	19	47	57	7	25	20	52	55	16	34	34	84	54	54	29	38	121	52	311	248	286	845	57
55/59	17	29	23	69	53	17	32	33	82	51	30	43	44	117	50	42	17	22	81	46	276	234	263	773	52
50/54	21	34	34	89	47	26	34	37	97	47	38	47	37	122	47	31	9	14	54	42	255	213	241	709	47
45/49	25	42	39	106	43	32	37	37	106	42	44	41	30	115	42	22	3	6	31	38	243	200	222	665	42
40/44	36	38	43	117	38	43	30	35	108	38	48	36	15	99	38	10	0	1	11	34	249	158	201	608	38
35/39	46	37	40	123	34	41	21	25	87	34	34	21	6	61	33	2	0	0	2	29	228	108	135	471	34
30/34	45	18	28	91	29	29	10	14	53	29	25	9	3	37	29						166	57	80	303	29
25/25	28	9	9	46	24	15	5	6	26	24	5	3	1	9	25						79	23	32	134	24
20/24	9	4	4	17	20	7	2	2	11	19	1	1	0	2	20						33	8	10	51	20
15/19	5	1	3	9	15	3	1	2	6	15	1	0	0	1	16						13	3	7	23	15
10/14	3	0	1	4	11	2	0	0	2	11	0	0	0	0	11						7	2	0	9	11
5/9	0	0	0	0	6	1	0	0	1	6											1	0	0	1	6
0/4	0	0	0	0	0																1	0	0	1	1
-5/-1	0	0	0	0	-3																0	0	0	0	-3

TABLE B.7 Minneapolis-St. Paul IAP Minnesota—Lat. 44° 53′ N, Long. 93° 13′ W, Elev. 834 ft: Mean Frequency of Occurrence of Dry-Bulb Temperature (°F) with Mean Coincident Wet-Bulb Temperature (°F) for Each Dry-Bulb Temperature Range

Temperature Range	MAY					JUNE					JULY					AUGUST				
	Obsn. Hour Gp.			Total Obsn.	MCWB	Obsn. Hour Gp.			Total Obsn.	MCWB	Obsn. Hour Gp.			Total Obsn.	MCWB	Obsn. Hour Gp.			Total Obsn.	MCWB
	01 to 08	09 to 16	17 to 24			01 to 08	09 to 16	17 to 24			01 to 08	09 to 16	17 to 24			01 to 08	09 to 16	17 to 24		
100/104												0		0	80					
95/99		1	0	1	67		2	1	3	74		3	1	4	77		1	0	1	75
90/94		5	2	7	66		9	4	13	72		11	5	16	75		12	3	15	74
85/89		14	6	20	64	0	17	10	27	71	0	37	15	52	71	0	27	11	38	72
80/84	1	25	14	40	62	2	37	20	59	68	4	58	36	98	69	3	46	27	76	69
75/79	5	35	24	64	59	10	48	34	92	65	16	60	52	128	67	12	59	43	114	67
70/74	16	39	34	89	57	28	49	47	124	63	53	48	63	164	65	41	48	61	150	65
65/69	35	42	46	123	54	50	36	50	136	60	78	23	48	149	63	69	33	57	159	62
60/64	48	34	47	129	51	62	25	41	128	57	61	8	23	92	59	67	16	29	112	59
55/59	51	24	36	111	47	43	10	22	75	53	29	1	5	35	55	37	5	13	55	54
50/54	45	18	22	85	43	29	5	10	44	49	7	0	1	8	51	16	0	3	19	50
45/49	29	8	12	49	39	13	0	2	15	45	0	0	0	0	47	4	0	0	4	46
40/44	13	2	3	18	34	2	0	0	2	42						0	0	0	0	41
35/39	5	0	2	7	30															
30/34	1	0	0	1	27															
25/29																				

(continued)

TABLE B.7 (cont.)

Temperature Range	SEP 01 to 08	SEP 09 to 16	SEP 17 to 24	SEP Total Obsn.	SEP MCWB	OCT 01 to 08	OCT 09 to 16	OCT 17 to 24	OCT Total Obsn.	OCT MCWB	NOV 01 to 08	NOV 09 to 16	NOV 17 to 24	NOV Total Obsn.	NOV MCWB	DEC 01 to 08	DEC 09 to 16	DEC 17 to 24	DEC Total Obsn.	DEC MCWB
95/99		0		0	76															
90/94		4	1	5	75															
85/89		7	2	9	73		1	0	1	65										
80/84	1	13	7	21	69		7	0	7	63										
75/79	6	26	13	45	65	0	11	3	14	62										
70/74	8	40	24	72	62	1	17	7	25	59		0		0	55					
65/69	16	41	35	92	59	3	28	16	47	56	0	2		2	54					
60/64	34	41	47	122	56	12	34	28	74	54	0	6	2	8	53		0		0	53
55/59	51	36	47	134	53	27	37	36	100	51	2	12	7	21	50	0	0	1	1	50
50/54	56	23	35	114	49	37	37	45	119	47	7	18	11	36	47	2	1	0	3	49
45/49	35	8	22	65	44	50	33	40	123	43	11	28	19	58	42	1	2	2	5	44
40/44	25	2	7	34	40	43	26	36	105	39	23	32	32	87	38	3	11	4	18	38
35/39	6	0	1	7	36	38	11	22	71	34	39	37	42	118	34	11	20	18	49	34
30/34	1			1	32	23	6	11	40	30	48	36	42	126	30	24	33	34	91	30
25/29	0			0	28	9	1	3	13	25	39	25	30	94	25	38	43	39	120	26
20/24						4	0	1	5	21	29	18	22	69	21	36	36	34	106	21
15/19						0		0	0	18	15	14	17	46	16	31	27	30	88	16
10/14											13	6	11	30	11	28	21	25	74	11
5/9											7	3	2	12	7	22	21	21	64	6
0/4											4	2	2	8	1	19	17	18	54	1
-5/-1											2	1	1	4	-3	14	8	11	33	-3
-10/-6											0	0		0	-8	11	4	6	21	-8
-15/-11											0			0	-13	5	1	2	8	-12
-20/-16																2	0	0	2	-18

Temperature Range	JANUARY					FEBRUARY					MARCH					APRIL					ANNUAL TOTAL				
	Obsn. Hour Gp.			Total Obsn.	M C W B	Obsn. Hour Gp.			Total Obsn.	M C W B	Obsn. Hour Gp.			Total Obsn.	M C W B	Obsn. Hour Gp.			Total Obsn.	M C W B	Obsn. Hour Gp.			Total Obsn.	M C W B
	01 to 08	09 to 16	17 to 24			01 to 08	09 to 16	17 to 24			01 to 08	09 to 16	17 to 24			01 to 08	09 to 16	17 to 24			01 to 08	09 to 16	17 to 24		
100/104																								0	80
95/99																					0	6	2	8	76
90/94																			0	63	0	37	13	50	74
85/89																0	2	0	2	61	0	96	40	136	71
80/84																0	3	1	4	61	10	178	97	285	68
75/79														0	54	0	6	3	9	59	45	235	162	442	66
70/74														0	58	0	9	5	14	57	136	246	231	613	63
65/69											0	1	0	1	54	1	15	11	27	54	233	218	251	702	60
60/64											0	1	1	2	52	5	21	17	43	51	276	194	234	704	56
55/59						0	1	0	1	47	0	4	2	6	48	12	22	23	57	48	249	162	203	614	52
50/54				0	38	1	0	0	1	43	1	10	5	16	44	20	33	28	81	45	226	152	174	552	47
45/49	0	3	1	4	36	1	3	0	4	40	1	13	9	23	40	28	35	33	96	41	188	140	150	478	43
40/44	4	12	9	25	33	1	10	4	15	37	10	24	20	54	37	39	39	41	119	38	175	155	157	487	38
35/39	14	20	18	52	30	8	20	16	44	34	24	41	37	102	34	50	29	39	118	34	193	172	187	552	34
30/34	22	38	33	93	25	23	36	37	96	30	40	54	58	152	29	45	18	25	88	30	223	203	227	653	30
25/29	31	33	35	99	21	30	32	32	94	25	52	38	43	133	25	26	6	11	43	25	217	183	191	591	25
20/24	31	27	28	86	16	25	32	30	87	21	41	25	29	95	20	10	1	3	14	20	176	145	154	475	21
15/19	29	29	30	88	11	30	31	33	94	16	27	16	20	63	16	2	0	0	2	16	136	115	128	379	16
10/14	29	25	27	81	6	33	22	25	80	11	18	10	12	40	11	1	0	0	1	10	122	88	103	313	11
5/9	24	24	23	71	1	21	16	19	56	6	15	7	7	29	6	0	0	0	0	8	94	72	76	242	6
0/4	21	16	20	57	-3	19	10	12	41	1	10	3	3	16	1						76	56	58	190	1
-5/-1	17	11	13	41	-8	16	7	7	30	-4	5	0	2	7	-3						58	32	41	131	-3
-10/-6	15	7	7	29	-13	10	3	4	17	-8	2	0	0	2	-8						40	18	23	81	-8
-15/-11	6	2	3	11	-18	4	1	2	7	-13	1	0	0	1	-13						25	9	11	45	-13
-20/-16	4	0	2	6	-23	2	0	1	3	-17				0	-18						10	2	4	16	-18
-25/-21	1	0	0	1	-27	0	0	0	0	-23				0	-24						4	0	2	6	-23
-30/-26						0	0	0	0	-26				0	-27						1	0	0	1	-27

TABLE B.8 Suffolk Co. Westhampton Bch. New York—Lat. 40° 51′ N, Long. 72° 38′ W, Elev. 67 ft: Mean Frequency of Occurrence of Dry-Bulb Temperature (°F) with Mean Coincident Wet-Bulb Temperature (°F) for Each Dry-Bulb Temperature Range

Temperature Range	MAY					JUNE					JULY					AUGUST				
	Obsn. Hour Gp. 01–08	09–16	17–24	Total Obsn.	MCWB	Obsn. Hour Gp. 01–08	09–16	17–24	Total Obsn.	MCWB	Obsn. Hour Gp. 01–08	09–16	17–24	Total Obsn.	MCWB	Obsn. Hour Gp. 01–08	09–16	17–24	Total Obsn.	MCWB
95/99												1		1	71					
90/94		1	0	1	67	0	3	0	3	71		4	0	4	73	0	1	0	1	77
85/89		4	0	4	66		9	1	10	70	1	13	1	15	74		7	0	7	74
80/84						2	16	3	21	68	3	39	7	49	71	1	41	3	45	71
75/79	0	7	2	9	63	5	41	11	57	67	14	96	36	146	69	13	86	35	134	70
70/74	2	19	4	25	60	18	64	33	115	64	77	71	99	247	67	73	80	95	248	64
65/69	9	45	14	68	58	54	60	62	176	62	86	21	73	180	64	81	27	72	180	59
60/64	24	64	39	127	55	77	37	75	189	59	46	4	25	75	59	44	5	31	80	54
55/59	54	60	69	183	52	53	11	43	107	55	16	0	6	22	55	23	0	9	32	50
50/54	66	34	68	168	48	23	0	9	32	50	5		1	6	50	10		2	12	45
45/49	54	12	39	105	44	7		1	8	45	1			1	45	3		1	4	41
40/44	28	3	12	43	40	1		0	1	42						0			0	41
35/39	9	0	2	11	35	1		0	1	36						0			0	38
30/34	2			2	31	0			0	32										
25/29	0			0	25															

726

Temperature Range	SEPTEMBER Obsn. Hour Gp. 01 to 08	09 to 16	17 to 24	Total Obsn.	MCWB	OCTOBER Obsn. Hour Gp. 01 to 08	09 to 16	17 to 24	Total Obsn.	MCWB	NOVEMBER Obsn. Hour Gp. 01 to 08	09 to 16	17 to 24	Total Obsn.	MCWB	DECEMBER Obsn. Hour Gp. 01 to 08	09 to 16	17 to 24	Total Obsn.	MCWB
90/94		0		0	71															
85/89		2	0	2	72															
80/84		11	1	12	71		1		1	68										
75/79	3	37	9	49	69		5	0	5	67	0	1		1	61					
70/74	25	68	36	129	66	2	18	3	23	64		2	0	2	62					
65/69	53	63	61	177	62	10	44	16	70	62	5	15	8	28	58		0		0	49
60/64	55	40	60	155	58	32	70	46	148	57	22	45	23	90	54	1	4	1	6	53
55/59	47	18	40	105	53	48	55	58	161	52	31	54	41	126	48	11	21	9	41	49
50/54	34	2	22	58	49	46	35	49	130	48	39	56	44	139	43	15	29	22	66	44
45/49	16	0	8	24	44	43	15	36	94	43	42	39	49	130	38	24	46	33	103	39
40/44	5		2	7	40	35	4	24	63	39	43	21	38	102	33	41	51	43	135	34
35/39	2		1	3	36	23	1	11	35	35	34	6	24	64	29	43	44	51	138	29
30/34						6	0	4	10	30	17	1	10	28	25	42	28	38	108	25
25/29						1		1	2	26	6	1	3	10	21	30	16	26	72	20
20/24						0		0	0	21	1		0	1	16	24	6	16	46	15
15/19						0		0	0	17	0			0	14	11	2	6	19	11
10/14																5	1	2	8	6
5/9																1	0	0	1	2
0/4																				

(continued)

TABLE B.8 (cont.)

Tempera-ture Range	JANUARY Obsn. Hour Gp. 01 to 08	09 to 16	17 to 24	Total Obsn.	MCWB	FEBRUARY Obsn. Hour Gp. 01 to 08	09 to 16	17 to 24	Total Obsn.	MCWB	MARCH Obsn. Hour Gp. 01 to 08	09 to 16	17 to 24	Total Obsn.	MCWB	APRIL Obsn. Hour Gp. 01 to 08	09 to 16	17 to 24	Total Obsn.	MCWB	ANNUAL TOTAL Obsn. Hour Gp. 01 to 08	09 to 16	17 to 24	Total Obsn.	MCWB
95/99																						1	0	1	71
90/94																						8	0	8	73
85/89																					1	32	2	35	72
80/84																0	1	0	1	64	6	113	14	133	70
75/79																0	2	0	2	61	35	274	93	402	69
70/74																0	3	0	3	58	197	324	270	791	66
65/69	0	0		0	55						0	1	0	1	53	0	10	1	11	55	293	273	299	865	62
60/64	0	0	0	0	52						0	1	0	1	49	2	18	4	24	52	285	254	288	827	58
55/59	0	1	1	2	52		1		1	51	0	6	1	7	47	8	40	16	64	50	272	241	267	780	53
50/54	3	3	2	8	48	0	6	1	7	47	2	23	7	32	45	31	59	41	131	47	262	237	252	751	48
45/49	6	17	7	30	44	4	18	8	30	43	11	47	24	82	42	55	65	68	188	43	254	259	258	771	43
40/44	18	36	23	77	39	18	42	23	83	39	35	65	53	153	38	66	34	69	169	39	272	269	288	829	39
35/39	35	57	44	136	34	37	53	48	138	34	67	58	81	206	34	40	8	27	75	34	298	249	295	842	34
30/34	45	50	54	149	29	49	44	58	151	29	63	29	51	143	29	28	2	10	40	30	270	175	252	697	29
25/29	42	37	47	126	24	39	29	38	106	24	41	12	21	74	25	7	0	2	9	25	189	107	157	453	25
20/24	39	27	33	99	20	30	18	20	68	20	18	3	7	28	20	2	0	0	2	21	125	65	89	279	20
15/19	28	13	23	64	15	21	9	15	45	15	8	2	2	12	15	1			1	16	83	30	56	169	15
10/14	18	4	10	32	11	16	4	9	29	11	2	0	1	3	11						47	10	26	83	11
5/9	10	2	4	16	6	7	1	2	10	6	1		0	1	7						23	4	8	35	6
0/4	3	0	1	4	2	3	0	0	3	2	0			0	2						7	0	1	8	2
−5/−1	0			0	−2	1	0	0	1	−3	0			0	−1						1		0	1	−2
−10/−6						0			0	−9											0			0	−9
−15/−11	0			0	−12																0			0	−12

728

TABLE B.9 Ellington AFB/Houston Texas—Lat. 29° 37' N, Long. 95° 10' W, Elev. 40 ft: Mean Frequency of Occurrence of Dry-Bulb Temperature (°F) with Mean Coincident Wet-Bulb Temperature (°F) for Each Dry-Bulb Temperature Range

Temperature Range	MAY Obsn. Hour Gp. 01 to 08	MAY Obsn. Hour Gp. 09 to 16	MAY Obsn. Hour Gp. 17 to 24	MAY Total Obsn.	MAY MCWB	JUNE Obsn. Hour Gp. 01 to 08	JUNE Obsn. Hour Gp. 09 to 16	JUNE Obsn. Hour Gp. 17 to 24	JUNE Total Obsn.	JUNE MCWB	JULY Obsn. Hour Gp. 01 to 08	JULY Obsn. Hour Gp. 09 to 16	JULY Obsn. Hour Gp. 17 to 24	JULY Total Obsn.	JULY MCWB	AUGUST Obsn. Hour Gp. 01 to 08	AUGUST Obsn. Hour Gp. 09 to 16	AUGUST Obsn. Hour Gp. 17 to 24	AUGUST Total Obsn.	AUGUST MCWB
100/104												0		0	80		1	0	1	77
95/99		0		0	78		4	0	4	78		16	2	18	79		15	2	17	78
90/94		4	1	5	76		56	7	63	77		87	19	106	78		81	16	97	78
85/89		52	6	58	74	4	103	36	143	76	5	93	44	142	77	3	90	38	131	77
80/84	4	103	32	139	72	28	48	64	140	75	39	34	80	153	76	33	41	81	155	76
75/79	50	57	82	189	71	100	19	92	211	73	148	13	89	250	74	152	17	95	264	74
70/74	95	21	79	195	69	83	9	34	126	70	51	4	13	68	70	55	3	14	72	71
65/69	55	9	33	97	64	21	1	5	27	65	4	0	1	5	65	5	1	1	7	65
60/64	27	2	13	42	59	3	0	1	4	59	1			1	59	1		0	1	59
55/59	15	1	2	18	54	1			1	54						0			0	55
50/54	3	0		3	50															
45/49	0			0	48															

(continued)

TABLE B.9 (cont.)

Temperature Range	September 01 to 08	September 09 to 16	September 17 to 24	September Total Obsn.	September MCWB	October 01 to 08	October 09 to 16	October 17 to 24	October Total Obsn.	October MCWB	November 01 to 08	November 09 to 16	November 17 to 24	November Total Obsn.	November MCWB	December 01 to 08	December 09 to 16	December 17 to 24	December Total Obsn.	December MCWB
100/104	0	0	0	0	77				0	78										
95/99		2	0	2	77		0	2	2	77										
90/94		29	2	31	77		34	2	36	74										
85/89		89	20	109	76	1	65	13	79	72		2		2	71					
80/84	14	67	54	135	75	17	65	43	125	69		20	1	21	71		1		1	69
75/79	92	35	94	221	73	47	39	68	154	67	4	48	10	62	69	6	19	1	20	68
70/74	79	14	50	143	70	58	25	51	134	63	19	45	32	96	66	21	33	14	53	66
65/69	35	4	16	55	64	48	12	36	96	58	38	40	47	125	63	31	38	28	87	63
60/64	14	1	3	18	59	36	5	22	63	53	37	30	42	109	58	33	41	40	112	58
55/59	5	0	1	6	56	26	2	10	38	49	32	24	35	91	53	33	38	43	114	53
50/54	0		0	0	49	11	0	2	13	45	36	18	35	89	48	35	32	42	109	48
45/49	0			0	45	2		0	2	40	35	8	22	65	44	42	25	41	108	44
40/44						1			1	37	22	4	10	36	39	36	13	23	72	39
35/39											10	1	4	15	34	26	6	12	44	35
30/34											5	0	0	5	30	13	2	4	19	30
25/29											0			0	26	4	0	1	5	25
20/24																1	0		1	21

Temperature Range	Jan 01 to 08	Jan 09 to 16	Jan 17 to 24	Jan Total Obsn.	Jan MCWB	Feb 01 to 08	Feb 09 to 16	Feb 17 to 24	Feb Total Obsn.	Feb MCWB	Mar 01 to 08	Mar 09 to 16	Mar 17 to 24	Mar Total Obsn.	Mar MCWB	Apr 01 to 08	Apr 09 to 16	Apr 17 to 24	Apr Total Obsn.	Apr MCWB	Ann 01 to 08	Ann 09 to 16	Ann 17 to 24	Ann Total Obsn.	Ann MCWB
100/104																						1	0	1	77
95/99																						37	4	41	78
90/94														0	70				0	74		259	45	304	78
85/89									0	66		7	1	8	67		4	0	4	73	12	467	146	625	76
80/84		2	0	2	70		3	0	3	69		30	6	36	66		50	6	56	71	119	441	332	892	74
75/79		12	1	13	68		10	1	11	67	8	54	25	87	65	10	82	36	128	70	573	407	550	1530	72
70/74	2	28	10	40	65	2	26	10	38	65	34	52	50	136	62	72	53	82	207	68	519	329	431	1279	68
65/69	21	28	26	75	63	19	34	24	77	62	44	42	55	141	57	53	31	55	139	63	364	263	337	964	63
60/64	24	32	28	84	58	19	35	30	84	57	45	27	44	116	52	41	14	35	90	58	290	209	283	782	58
55/59	26	34	36	96	53	27	38	40	105	53	41	17	32	90	47	35	5	18	58	53	255	172	241	668	53
50/54	29	36	43	108	48	41	34	44	119	48	34	11	23	68	43	17	1	6	24	48	228	140	212	580	48
45/49	33	28	37	98	43	39	21	37	97	43	24	5	9	38	39	10		1	11	44	204	93	163	460	43
40/44	42	24	32	98	39	37	14	22	73	39	13	1	3	17	34	2			2	38	165	60	96	321	39
35/39	33	14	23	70	34	22	6	11	39	35	4	0	0	4	30	0			0	37	105	28	53	186	34
30/34	24	6	9	39	29	14	2	4	20	30	0			0	26						60	10	17	87	30
25/29	10	3	3	16	25	4	0	0	4	26											18	3	4	25	25
20/24	3	1	2	6	20																4	1	2	7	20
15/19	2	0	0	2	15																2	0	0	2	15

appendix C
Refrigerant, Cryogenic, and Psychrometric Charts

Please see the foldout section following the Index for the charts listed below.

REFRIGERANT AND CRYOGENIC CHARTS

Chart	Substance	Coordinates	Units
C-1E	NH_3	P-h	English
C-1SI	NH_3	P-h	SI
C-2E	R22	P-h	English
C-2SI	R22	P-h	SI
C-3E	NH_3-H_2O	h-x	English
C-3SI	NH_3-H_2O	h-x	SI
C-4E	$LiBr$-H_2O	h-x	English
C-4SI	$LiBr$-H_2O	h-x	SI
C-5E	Air	T-s	English
C-5SI	Air	T-s	SI
C-6E	N_2-O_2	h-x	English

PSYCHROMETRIC CHARTS

Chart	Elevation	Range	Units
C-7E	Sea Level	low	English
C-7SI	Sea Level	low	SI
C-8E	Sea Level	normal	English
C-8SI	Sea Level	normal	SI
C-9E	Sea Level	high	English
C-9SI	Sea Level	high	SI
C-10E	5000 ft	normal	English
C-10SI	1500 m	normal	SI

Index

Vapor, water (*See* Water vapor)
Vapor compression cycles, 49–69
Vapor retarders, 462
Vapor transport through building
    materials and structures, 459–64
Variable air volume (VAV), 228–33,
    628–30
Ventilation, 2, 369–72, 584–603
    air change efficiency, 588
    efficiency, 372
    local mean age, 586–87
    nominal time constant, 369, 587
Vesley, D., 368
Virial equation of state, 16
Visible radiation, 32, 396
Volume lines on psychrometric
    chart, 195

Walker, G., 152
Walls:
    heat loss through, 421–39
    instantaneous cooling loads, 534–44
    moisture transport through, 459–64
    periodic heat transfer
        through, 481–503
Waste heat, utilization of, 148

Water:
    properties of, 16–19, 702–3 (tables)
    as a refrigerant, 78
    thermodynamic properties at
        saturation, 675–85 (tables)
Water vapor:
    low pressure, 19–20
Webb, R. L., 280
Wet-bulb depression, 254–57 (diagrams)
Wet-bulb temperature:
    correlation of two types, 252–57
    thermodynamic, 181
Wet-bulb thermometer, 246–51
Wexler, A., 17, 183, 258
Whetstone, J. R., 259
Whitacker, S., 26
Wile, D. D., 252
Wind chill temperature, 349–51 (table)
Windows:
    heat transmission, 431–35
        center of glazing, 432
        edge of glazing, 432
        frame, 432
        total, 432
    infiltration, 443–54
    shading from solar radiation, 292–96
        (diagrams)

solar heat gain factor (SHGF), 513–22
Winter design heat loss, 415–65
    above grade structures, 416–31
    basements, 435–42
    design temperatures, 456–57, 708–9
        (table)
    doors, 435
    infiltration, 443–54
    slab on grade floors, 442–43
    unheated spaces, 455–56
    windows, 431–35
Woods, J., 2
Work, 12
Workers, effect of thermal environment
    upon, 351–53
Wright, L. T. Jr., 486

Xenon:
    pressure and temperature data, 152
    separation from air, 170–71
Xu, J., 602

Z-transform, 490
Zarr, R. R., 431
Zenith angle, solar, 388 (diagram)
Zenner, G. H., 170
Zero load temperature, 560